Fluid Mechanics

Eighth Edition in SI Units

Frank M. White
University of Rhode Island

Adapted by

Professor Rhim Yoon Chul
School of Mechanical Engineering, Yonsei University

McGraw Hill Education (India) Private Limited
CHENNAI

McGraw Hill Education Offices
Chennai New York St Louis San Francisco Auckland Bogotá Caracas
Kuala Lumpur Lisbon London Madrid Mexico City Milan Montreal
San Juan Santiago Singapore Sydney Tokyo Toronto

McGraw Hill Education (India) Private Limited

Published by McGraw Hill Education (India) Private Limited,
444/1, Sri Ekambara Naicker Industrial Estate, Alapakkam, Porur, Chennai 600116, Tamil Nadu, India.

Fluid Mechanics, 8e in SI Units (SIE)

Sixth reprint 2018
RCBRRRDREDDYB

ISBN (13 digits): 978-93-85965-49-4
ISBN (10 digits): 93-85965-49-2

07 08 09 10 11 APO 22 21 20 19 18

Director—Product (Higher Education and Professional): *Vibha Mahajan*
Head—Production (Higher Education and Professional): *Satinder S Baveja*
AGM—Product Management (Higher Education and Professional): *Shalini Jha*
General Manager—Production: *Rajender P Ghansela*
Manager—Production: *Reji Kumar*

Typeset at Text-O-Graphics, B-1/56, Arawali Apartments, Sector 34, Noida, and
Printed and Bound in India at A.P Offset Pvt. Ltd., Zulfe Bengal, Dilshad Garden, Delhi 110095.

Visit us at: www.mheducation.co.in ; (Toll free in India): **1800 103 5875**
Write to us at: info.india@mheducation.com

CIN: U22200TN1970PTC111531

About the Author

Frank M. White is Professor Emeritus of Mechanical and Ocean Engineering at the University of Rhode Island. He studied at Georgia Tech and M.I.T. In 1966 he helped found, at URI, the first department of ocean engineering in the country. Known primarily as a teacher and writer, he has received eight teaching awards and has written four textbooks on fluid mechanics and heat transfer.

From 1979 to 1990, he was editor-in-chief of the *ASME Journal of Fluids Engineering* and then served from 1991 to 1997 as chairman of the ASME Board of Editors and of the Publications Committee. He is a Fellow of ASME and in 1991 received the ASME Fluids Engineering Award. He lives with his wife, Jeanne, in Narragansett, Rhode Island.

To Jeanne

Contents

Preface

General Approach

The eighth edition of *Fluid Mechanics* sees some additions and deletions but no philosophical change. The basic outline of eleven chapters, plus appendices, remains the same. The triad of integral, differential, and experimental approaches is retained. Many problem exercises, and some fully worked examples, have been changed. The informal, student-oriented style is retained. A number of new photographs and figures have been added. Many new references have been added, for a total of 445. The writer is a firm believer in "further reading," especially in the postgraduate years.

Learning Tools

The total number of problem exercises continues to increase, from 1089 in the first edition, to 1683 in this eighth edition. There are approximately 20 new problems in each chapter. Most of these are basic end-of-chapter problems, classified according to topic. There are also Word Problems, multiple-choice Fundamentals of Engineering Problems, Comprehensive Problems, and Design Projects. The appendix lists approximately 700 Answers to Selected Problems.

The example problems are structured in the text to follow the sequence of recommended steps outlined in Section 1.7.

Most of the problems in this text can be solved with a hand calculator. Some can even be simply explained in words. A few problems, especially in Chapters 6, 9, and 10, involve solving complicated algebraic expressions, laborious for a hand calculator. Check to see if your institution has a license for equation-solving software. Here the writer solves complicated example problems by using the iterative power of Microsoft Office Excel, as illustrated, for example, in Example 6.5. For further use in your work, Excel also contains several hundred special mathematical functions for engineering and statistics. Another benefit: Excel is free.

Content Changes

There are some revisions in each chapter.

Chapter 1 has been substantially revised. The pre-reviewers felt, correctly, that it was too long, too detailed, and at too high a level for an introduction. Former Section 1.2, History of Fluid Mechanics, has been shortened and moved to the end of the chapter. Former Section 1.3, Problem-Solving Techniques, has been moved to appear just before Example 1.7, where these techniques are first used. Eulerian and Lagrangian descriptions have been moved to Chapter 4. A temperature-entropy chart for steam has been added, to illustrate when steam can and cannot be approximated as an ideal gas. Former Section 1.11, Flow Patterns, has been cut sharply and mostly moved to **Chapter 4**. Former Section 1.13, Uncertainty in Experimental Data, has been moved to a new Appendix E. No one teaches "uncertainty" in introductory fluid mechanics, but the writer feels it is extremely important in all engineering fields involving experimental or numerical data.

Chapter 2 adds a brief discussion of the fact that pressure is a thermodynamic property, not a *force*, has no direction, and is not a vector. The arrow, on a surface force caused by pressure, causes confusion for beginning students. The subsection of Section 2.8 entitled Stability Related to Waterline Area has been shortened to omit the complicated derivations. The final metacenter formula is retained; the writer does not think it is sufficient just to show a sketch of a floating body falling over. This book should have reference value.

Chapter 3 was substantially revised in the last edition, especially by moving Bernoulli's equation to follow the linear momentum section. This time the only changes are improvements in the example problems.

Chapter 4 now discusses the Eulerian and Lagrangian systems, moved from Chapter 1. The no-slip and no-temperature-jump boundary conditions are added, with problem assignments.

Chapter 5 explains a bit more about drag force before assigning dimensional analysis problems. It retains Ipsen's method as an interesting alternative which, of course, may be skipped by pi theorem adherents.

Chapter 6 downplays the Moody chart a bit, suggesting that students use either iteration or Excel. For rough walls, the chart is awkward to read, although it gives an approximation for use in iteration. The author's fancy rearrangement of pi groups to solve type 2, flow rate, and type 3, pipe diameter problems is removed from the main text and assigned as problems. For noncircular ducts, the hydraulic *radius* is omitted and moved to **Chapter 10**. There is a new Example 6.11, which solves for pipe diameter and determines if Schedule 40 pipe is strong enough. A general discussion of pipe strength is added. There is a new subsection on *laminar-flow* minor losses, appropriate for micro- and nano-tube flows.

Chapter 7 has more treatment of vehicle drag and rolling resistance, and a rolling resistance coefficient is defined. There is additional discussion of the Kline-Fogelman airfoil, extremely popular now for model aircraft.

Chapter 8 has backed off from extensive discussion of CFD methods, as proposed by the pre-reviewers. Only a few CFD examples are now given. The inviscid duct-expansion example and the implicit boundary layer method are now omitted, but the explicit method is retained. For airfoil theory, the writer considers thin-airfoil vortex-sheet theory to be obsolete and has deleted it.

Chapter 9 now has a better discussion of the normal shock wave. New supersonic wave photographs are added. The "new trend in aeronautics" is the Air Force X-35 Joint Strike Fighter.

Chapter 10 improves the definition of normal depth of a channel. There is a new subsection on the water-channel compressible flow analogy, and problems are assigned to find the oblique wave angle for supercritical water flow past a wedge.

Chapter 11 greatly expands the discussion of wind turbines, with examples and problems taken from the author's own experience.

Appendices B and D are unchanged. **Appendix A** adds a list of liquid kinematic viscosities to Table A.4. A few more conversion factors are added to **Appendix C**. There is a new **Appendix E**, Estimating Uncertainty in Experimental Data, which was moved from its inappropriate position in **Chapter 1**. The writer believes that "uncertainty" is vital to reporting measurements and always insisted upon it when he was an engineering journal editor.

Supplements

The following supplements are related to users of this **SI edition**.

Solutions Manual

The **Solutions Manual** that accompanies this book offers typeset, one-per-page solutions with detail explanations, to end-of-chapter problems.

Powerpoint Slides

PowerPoint presentation slides for all chapters in the text are available for use in lectures.

Text Website

Web support is provided for the text specific website at
http://www.mhhe.com/white/fm8/sie

Visit the website for general text information, errata, and author information. The instructor side of the site includes the solutions manual and the PowerPoint slides.

More from McGraw-Hill

Adaptive Online Learning Tools

LEARNSMART®

McGraw-Hill LearnSmart® is available as a standalone product or an integrated feature of McGraw-Hill Connect Engineering. It is an adaptive learning system designed to help students learn faster, study more efficiently, and retain more knowledge for greater success. LearnSmart assesses a student's knowledge of course content through a series of adaptive questions. It pinpoints concepts the student does not understand and maps out a personalized study plan for success. This innovative study tool also has features that allow instructors to see exactly what students have accomplished and a built-in assessment tool for graded assignments. Visit the following site for a demonstration: www.LearnSmartAdvantage.com

SMARTBOOK®

Powered by the intelligent and adaptive LearnSmart engine, **SmartBook™** is the first and only continuously adaptive reading experience available today. Distinguishing what students know from what they don't, and honing in on concepts they are most likely to forget, SmartBook personalizes the reading experience for each student. Reading is no longer a passive and linear experience but an engaging and dynamic one, where students are more likely to master and retain important concepts, coming to class better prepared. SmartBook includes powerful reports that identify specific topics and learning objectives students need to study. www.LearnSmartAdvantage.com

Acknowledgments

As usual, so many people have helped me that I may fail to list them all. Material help, in the form of photos, articles, and problems, came from Scott Larwood of the University of the Pacific; Sukanta Dash of the Indian Institute of Technology at Kharagpur; Mark Coffey of the Colorado School of Mines; Mac Stevens of Oregon State University; Stephen Carrington of Malvern Instruments; Carla Cioffi of NASA; Lisa Lee and Robert Pacquette of the Rhode Island Department of Environmental Management; Vanessa Blakeley and Samuel Schweighart of Terrafugia Inc.; Beric Skews of the University of the Witwatersrand, South Africa; Kelly Irene Knorr and John Merrill of the School of Oceanography at the University of Rhode Island; Adam Rein of Altaeros Energies Inc.; Dasari Abhinav of Anna University, India; Kris Allen of Transcanada Corporation; Bruce Findlayson of the University of Washington; Wendy Koch of USA Today; Liz Boardman of the South County Independent; Beth Darchi and Colin McAteer of the American Society of Mechanical Engineers; Catherine Hines of the William Beebe Web Site; Laura Garrison of York College of Pennsylvania.

The following pre-reviewers gave many excellent suggestions for improving the manuscript: Steve Baker, Naval Postgraduate School; Suresh Aggarwal, University of Illinois at Chicago; Edgar Caraballo, Miami University; Chang-Hwan Choi, Stevens Institute of Technology; Drazen Fabris, Santa Clara University; James Liburdy, Oregon State University; Daniel Maynes, Brigham Young University; Santosh Sahu, Indian Institute of Technology Indore; Brian Savilonis, Worcester Polytechnic Institute; Eric Savory, University of Western Ontario; Rick Sellens, Queen's University; Gordon Stubley, University of Waterloo.

Many others have supported me, throughout my revision efforts, with comments and suggestions: Gordon Holloway of the University of New Brunswick; David Taggart, Donna Meyer, Arun Shukla, and Richard Lessmann of the University of Rhode Island; Debendra K. Das of the University of Alaska–Fairbanks; Elizabeth Kenyon of Mathworks; Deborah V. Pence of Oregon State University; Sheldon Green of the University of British Columbia; Elena Bingham of the DuPont Corporation; Jane Bates of Broad Rock School; Kim Mather of West Kingston School; Nancy Dreier of Curtiss Corner School; Richard Kline, co-inventor of the Kline-Fogelman airfoil.

The McGraw-Hill staff was, as usual, very helpful. Thanks are due to Bill Stenquist, Katherine Neubauer, Lorraine Buczek, Samantha Donisi-Hamm, Raghu Srinivasan, Tammy Juran, Thomas Scaife, and Lisa Bruflodt.

Finally, I am thankful for the continuing support of my family, especially Jeanne, who remains in my heart, and my sister Sally White GNSH, my dog Jack, and my cats Cole and Kerry.

Publisher's Note

The publishers would like to acknowledge with appreciation the suggestions and praise received from reviewers of the Special Indian Edition (SIE).

Ashwinkumar S. Dhoble
Visvesvaraya National Institute of Technology, Nagpur

Husain S.Maghrabi
M.H.Saboo Siddik College of Engineering, Mumbai

S. Jayaraj
National Institute of Technology, Calicut

Manish Gopichand Walecha
P. R. Pote(Patil) College of Engineering and Management, Amravati, Maharashtra

Manabendra Pathak
Indian Institute of Technology, Patna

Fluid Mechanics

Eighth Edition in SI Units

Falls on the Nesowadnehunk Stream in Baxter State Park, Maine, which is the northern terminus of the Appalachian Trail. Such flows, open to the atmosphere, are driven simply by gravity and do not depend much upon fluid properties such as density and viscosity. They are discussed later in Chap. 10. To the writer, one of the joys of fluid mechanics is that visualization of a fluid-flow process is simple and beautiful [*Photo Credit: Design Pics/Natural Selection Robert Cable*].

Chapter 1
Introduction

1.1 Preliminary Remarks

Fluid mechanics is the study of fluids either in motion (fluid *dynamics*) or at rest (fluid *statics*). Both gases and liquids are classified as fluids, and the number of fluid engineering applications is enormous: breathing, blood flow, swimming, pumps, fans, turbines, airplanes, ships, rivers, windmills, pipes, missiles, icebergs, engines, filters, jets, and sprinklers, to name a few. When you think about it, almost everything on this planet either is a fluid or moves within or near a fluid.

The essence of the subject of fluid flow is a judicious compromise between theory and experiment. Since fluid flow is a branch of mechanics, it satisfies a set of well-documented basic laws, and thus a great deal of theoretical treatment is available. However, the theory is often frustrating because it applies mainly to idealized situations, which may be invalid in practical problems. The two chief obstacles to a workable theory are geometry and viscosity. The basic equations of fluid motion (Chap. 4) are too difficult to enable the analyst to attack arbitrary geometric configurations. Thus, most textbooks concentrate on flat plates, circular pipes, and other easy geometries. It is possible to apply numerical computer techniques to complex geometries, and specialized textbooks are now available to explain the new *computational fluid dynamics* (CFD) approximations and methods [1–4].[1] This book will present many theoretical results while keeping their limitations in mind.

The second obstacle to a workable theory is the action of viscosity, which can be neglected only in certain idealized flows (Chap. 8). First, viscosity increases the difficulty of the basic equations, although the boundary-layer approximation found by Ludwig Prandtl in 1904 (Chap. 7) has greatly simplified viscous-flow analyses. Second, viscosity has a destabilizing effect on all fluids, giving rise, at frustratingly small velocities, to a disorderly, random phenomenon called *turbulence*. The theory of turbulent flow is crude and heavily backed up by experiment (Chap. 6), yet it can be quite serviceable as an engineering estimate. This textbook only introduces the standard experimental correlations for turbulent time-mean flow. Meanwhile, there are advanced texts on both time-mean *turbulence and turbulence modeling* [5, 6] and on the newer, computer-intensive *direct numerical simulation* (DNS) of fluctuating turbulence [7, 8].

Thus, there is theory available for fluid flow problems, but in all cases it should be backed up by experiment. Often the experimental data provide the main source of information about specific flows, such as the drag and lift of immersed bodies (Chap. 7). Fortunately, fluid mechanics is a highly visual subject, with good

[1]Numbered references appear at the end of each chapter.

instrumentation [9–11], and the use of dimensional analysis and modeling concepts (Chap. 5) is widespread. Thus experimentation provides a natural and easy complement to the theory. You should keep in mind that theory and experiment should go hand in hand in all studies of fluid mechanics.

1.2 The Concept of a Fluid

From the point of view of fluid mechanics, all matter consists of only two states, fluid and solid. The difference between the two is perfectly obvious to the layperson, and it is an interesting exercise to ask a layperson to put this difference into words. The technical distinction lies with the reaction of the two to an applied shear or tangential stress. *A solid can resist a shear stress by a static deflection; a fluid cannot.* Any shear stress applied to a fluid, no matter how small, will result in motion of that fluid. The fluid moves and deforms continuously as long as the shear stress is applied. As a corollary, we can say that a fluid at rest must be in a state of zero shear stress, a state often called the hydrostatic stress condition in structural analysis. In this condition, Mohr's circle for stress reduces to a point, and there is no shear stress on any plane cut through the element under stress.

Given this definition of a fluid, every layperson also knows that there are two classes of fluids, *liquids* and *gases.* Again the distinction is a technical one concerning the effect of cohesive forces. A liquid, being composed of relatively close-packed molecules with strong cohesive forces, tends to retain its volume and will form a free surface in a gravitational field if unconfined from above. Free-surface flows are dominated by gravitational effects and are studied in Chaps. 5 and 10. Since gas molecules are widely spaced with negligible cohesive forces, a gas is free to expand until it encounters confining walls. A gas has no definite volume, and when left to itself without confinement, a gas forms an atmosphere that is essentially hydrostatic. The hydrostatic behavior of liquids and gases is taken up in Chap. 2. Gases cannot form a free surface, and thus gas flows are rarely concerned with gravitational effects other than buoyancy.

Figure 1.1 illustrates a solid block resting on a rigid plane and stressed by its own weight. The solid sags into a static deflection, shown as a highly exaggerated dashed line, resisting shear without flow. A free-body diagram of element A on the side of the block shows that there is shear in the block along a plane cut at an angle θ through A. Since the block sides are unsupported, element A has zero stress on the left and right sides and compression stress $\sigma = -p$ on the top and bottom. Mohr's circle does not reduce to a point, and there is nonzero shear stress in the block.

By contrast, the liquid and gas at rest in Fig. 1.1 require the supporting walls in order to eliminate shear stress. The walls exert a compression stress of $-p$ and reduce Mohr's circle to a point with zero shear everywhere—that is, the hydrostatic condition. The liquid retains its volume and forms a free surface in the container. If the walls are removed, shear develops in the liquid and a big splash results. If the container is tilted, shear again develops, waves form, and the free surface seeks a horizontal configuration, pouring out over the lip if necessary. Meanwhile, the gas is unrestrained and expands out of the container, filling all available space. Element A in the gas is also hydrostatic and exerts a compression stress $-p$ on the walls.

In the previous discussion, clear decisions could be made about solids, liquids, and gases. Most engineering fluid mechanics problems deal with these clear cases—that is, the common liquids, such as water, oil, mercury, gasoline, and alcohol, and the common gases, such as air, helium, hydrogen, and steam, in their common temperature and pressure ranges. There are many borderline cases, however, of which you should be aware. Some apparently "solid" substances such as asphalt and lead resist shear stress for short periods but actually deform slowly and exhibit definite fluid behavior over long periods. Other substances, notably colloid and slurry mixtures, resist small

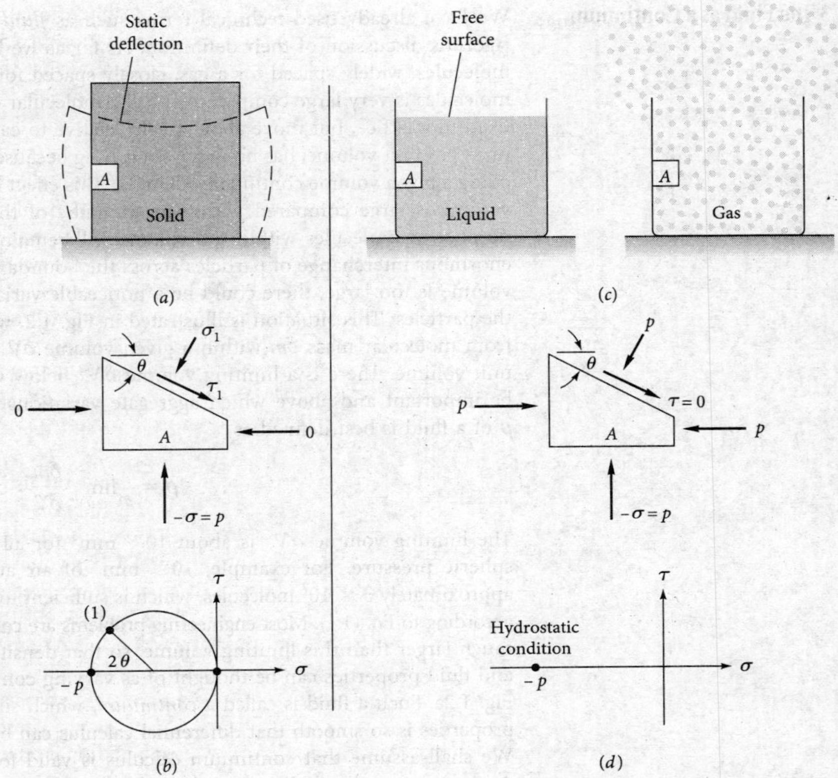

Fig. 1.1 A solid at rest can resist shear. (*a*) Static deflection of the solid; (*b*) equilibrium and Mohr's circle for solid element *A*. A fluid cannot resist shear. (*c*) Containing walls are needed; (*d*) equilibrium and Mohr's circle for fluid element *A*.

shear stresses but "yield" at large stress and begin to flow as fluids do. Specialized textbooks are devoted to this study of more general deformation and flow, a field called *rheology* [16]. Also, liquids and gases can coexist in two-phase mixtures, such as steam–water mixtures or water with entrapped air bubbles. Specialized textbooks present the analysis of such *multiphase flows* [17]. Finally, in some situations the distinction between a liquid and a gas blurs. This is the case at temperatures and pressures above the so-called *critical point* of a substance, where only a single phase exists, primarily resembling a gas. As pressure increases far above the critical point, the gaslike substance becomes so dense that there is some resemblance to a liquid, and the usual thermodynamic approximations like the perfect-gas law become inaccurate. The critical temperature and pressure of water are $T_c = 647$ K and $p_c = 219$ atm (atmosphere)[2] so that typical problems involving water and steam are below the critical point. Air, being a mixture of gases, has no distinct critical point, but its principal component, nitrogen, has $T_c = 126$ K and $p_c = 34$ atm. Thus, typical problems involving air are in the range of high temperature and low pressure, where air is distinctly and definitely a gas. This text will be concerned solely with clearly identifiable liquids and gases, and the borderline cases just discussed will be beyond our scope.

[2]One atmosphere equals 2116 lbf/ft^2 = 101,300 Pa.

1.3 The Fluid as a Continuum

We have already used technical terms such as *fluid pressure* and *density* without a rigorous discussion of their definition. As far as we know, fluids are aggregations of molecules, widely spaced for a gas, closely spaced for a liquid. The distance between molecules is very large compared with the molecular diameter. The molecules are not fixed in a lattice, but move about freely relative to each other. Thus, fluid density, or mass per unit volume, has no precise meaning because the number of molecules occupying a given volume continually changes. This effect becomes unimportant if the unit volume is large compared with, say, the cube of the molecular spacing, when the number of molecules within the volume will remain nearly constant in spite of the enormous interchange of particles across the boundaries. If, however, the chosen unit volume is too large, there could be a noticeable variation in the bulk aggregation of the particles. This situation is illustrated in Fig. 1.2, where the "density" as calculated from molecular mass δm within a given volume $\delta \mathcal{V}$ is plotted versus the size of the unit volume. There is a limiting volume $\delta \mathcal{V}^*$ below which molecular variations may be important and above which aggregate variations may be important. The *density* ρ of a fluid is best defined as

$$\rho = \lim_{\delta \mathcal{V} \to \delta \mathcal{V}^*} \frac{\delta m}{\delta \mathcal{V}} \tag{1.1}$$

The limiting volume $\delta \mathcal{V}^*$ is about 10^{-9} mm^3 for all liquids and for gases at atmospheric pressure. For example, 10^{-9} mm^3 of air at standard conditions contains approximately 3×10^7 molecules, which is sufficient to define a nearly constant density according to Eq. (1.1). Most engineering problems are concerned with physical dimensions much larger than this limiting volume, so that density is essentially a point function and fluid properties can be thought of as varying continually in space, as sketched in Fig. 1.2a. Such a fluid is called a *continuum,* which simply means that its variation in properties is so smooth that differential calculus can be used to analyze the substance. We shall assume that continuum calculus is valid for all the analyses in this book. Again, there are borderline cases for gases at such low pressures that molecular spacing and mean free path[3] are comparable to, or larger than, the physical size of the system. This requires that the continuum approximation be dropped in favor of a molecular theory of rarefied gas flow [18]. In principle, all fluid mechanics problems can be attacked from the molecular viewpoint, but no such attempt will be made here. Note that the use of continuum calculus does not preclude the possibility of discontinuous jumps in fluid properties across a free surface or fluid interface or across a shock wave in a compressible fluid (Chap. 9). Our calculus in analyzing fluid flow must be flexible enough to handle discontinuous boundary conditions.

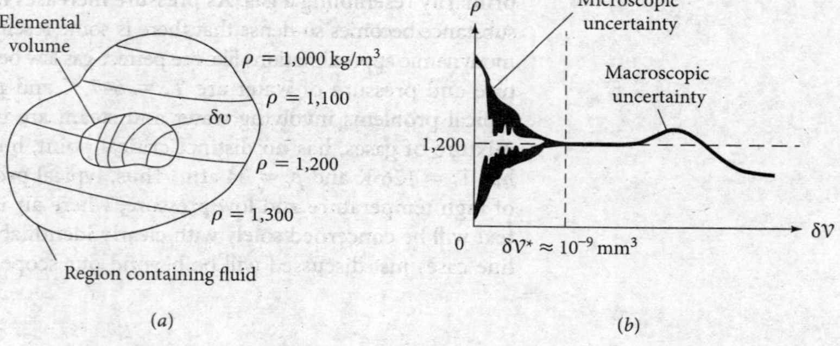

Fig. 1.2 The limit definition of continuum fluid density: (*a*) an elemental volume in a fluid region of variable continuum density; (*b*) calculated density versus size of the elemental volume.

[3] The mean distance traveled by molecules between collisions (see Prob. P1.5).

1.4 Dimensions and Units

A *dimension* is the measure by which a physical variable is expressed quantitatively. A *unit* is a particular way of attaching a number to the quantitative dimension. Thus, length is a dimension associated with such variables as distance, displacement, width, deflection, and height, while centimeters and inches are both numerical units for expressing length. Dimension is a powerful concept about which a splendid tool called *dimensional analysis* has been developed (Chap. 5), while units are the numerical quantity that the customer wants as the final answer.

In 1872, an international meeting in France proposed a treaty called the Metric Convention, which was signed in 1875 by 17 countries including the United States. It was an improvement over British systems because its use of base 10 is the foundation of our number system, learned from childhood by all. Problems still remained because even the metric countries differed in their use of kiloponds instead of dynes or newtons, kilograms instead of grams, or calories instead of joules. To standardize the metric system, a General Conference of Weights and Measures, attended in 1960 by 40 countries, proposed the *International System of Units* (SI). We are now undergoing a painful period of transition to SI, an adjustment that may take many more years to complete. The professional societies have led the way. Since July 1, 1974, SI units have been required by all papers published by the American Society of Mechanical Engineers, and there is a textbook explaining the SI [19]. This metricated text is wholly in SI units. In this chapter however, SI units will be used with British gravitational units.

Primary Dimensions

In fluid mechanics, there are only four *primary dimensions* from which all other dimensions can be derived: mass, length, time, and temperature.[4] These dimensions and their units in both systems are given in Table 1.1. Note that the Kelvin unit uses no degree symbol. The braces around a symbol like $\{M\}$ mean "the dimension" of mass. All other variables in fluid mechanics can be expressed in terms of $\{M\}$, $\{L\}$, $\{T\}$, and $\{\Theta\}$. For example, acceleration has the dimensions $\{LT^{-2}\}$. The most crucial of these secondary dimensions is force, which is directly related to mass, length, and time by Newton's second law. Force equals the time rate of change of momentum or, for constant mass,

$$\mathbf{F} = m\mathbf{a} \tag{1.2}$$

From this we see that, dimensionally, $\{F\} = \{MLT^{-2}\}$.

The International System (SI)

The use of a constant of proportionality in Newton's law, Eq. (1.2), is avoided by defining the force unit exactly in terms of the other basic units. In the SI system, the basic units are newtons $\{F\}$, kilograms $\{M\}$, meters $\{L\}$, and seconds $\{T\}$. We define

$$1 \text{ newton of force} = 1 \text{ N} = 1 \text{ kg} \cdot 1 \text{ m/s}^2$$

Table 1.1 Primary Dimensions in SI and BG Systems

Primary dimension	SI unit	BG unit	Conversion factor
Mass $\{M\}$	Kilogram (kg)	Slug	1 slug = 14.5939 kg
Length $\{L\}$	Meter (m)	Foot (ft)	1 ft = 0.3048 m
Time $\{T\}$	Second (s)	Second (s)	1 s = 1 s
Temperature $\{\Theta\}$	Kelvin (K)	Rankine (°R)	1 K = 1.8°R

[4]If electromagnetic effects are important, a fifth primary dimension must be included, electric current $\{I\}$, whose SI unit is the ampere (A).

The newton is a relatively small force, about the weight of an apple (0.225 lbf). In addition, the basic unit of temperature $\{\Theta\}$ in the SI system is the degree Kelvin, K. Use of these SI units (N, kg, m, s, K) will require no conversion factors in our equations.

The British Gravitational (BG) System

In the BG system also, a constant of proportionality in Eq. (1.2) is avoided by defining the force unit exactly in terms of the other basic units. In the BG system, the basic units are pound-force $\{F\}$, slugs $\{M\}$, feet $\{L\}$, and seconds $\{T\}$. We define

$$1 \text{ pound of force} = 1 \text{ lbf} \equiv 1 \text{ slug} \cdot 1 \text{ ft/s}^2$$

One lbf $\approx$ 4.4482 N (Approximately this is the weight of four apples.) We will use the abbreviation *lbf* for pound-force and *lbm* for pound-mass. The slug is a rather hefty mass, equal to 32.174 lbm. The basic unit of temperature $\{\Theta\}$ in the BG system is the degree Rankine, °R. Recall that a temperature difference 1 K = 1.8°R. Use of these BG units (lbf, slug, ft, s, °R) will require no conversion factors in our equations.

Other Unit Systems

There are other unit systems still in use. At least one needs no proportionality constant: the CGS system (dyne, gram, cm, s, K). However, CGS units are too small for most applications (1 dyne $= 10^{-5}$ N) and will not be used here.

In the USA, some still use the English Engineering system (lbf, lbm, ft, s, °R), where the basic mass unit is the *pound of mass*. Newton's law (1.2) must be rewritten:

$$\mathbf{F} = \frac{m\mathbf{a}}{g_c}, \text{ where } g_c = 32.174 \frac{\text{ft} \cdot \text{lbm}}{\text{lbf} \cdot \text{s}^2} \tag{1.3}$$

The constant of proportionality, g_c, has both dimensions and a numerical value not equal to 1.0. The present text uses only the SI and BG systems and will not solve problems or examples in the English Engineering system. Because Americans still use them, a few problems in the text will be stated in truly awkward units: acres, gallons, ounces, or miles. Your assignment will be to convert these and solve in the SI or BG systems.

The Principle of Dimensional Homogeneity

In engineering and science, *all* equations must be *dimensionally homogeneous*, that is, each additive term in an equation must have the same dimensions. For example, take Bernoulli's incompressible equation, to be studied and used throughout this text:

$$p + \frac{1}{2}\rho V^2 + \rho g Z = \text{constant}$$

Each and every term in this equation *must* have dimensions of pressure $\{ML^{-1}T^{-2}\}$. We will examine the dimensional homogeneity of this equation in detail in Example 1.3.

A list of some important secondary variables in fluid mechanics, with dimensions derived as combinations of the four primary dimensions, is given in Table 1.2. A more complete list of conversion factors is given in App. C.

EXAMPLE 1.1

A body weighs 1,000 lbf when exposed to a standard earth gravity $g = 32.174$ ft/s². (*a*) What is its mass in kg? (*b*) What will the weight of this body be in N if it is exposed to the moon's standard acceleration $g_{moon} = 1.62$ m/s²? (*c*) How fast will the body accelerate if a net force of 400 lbf is applied to it on the moon or on the earth?

Table 1.2 Secondary Dimensions in Fluid Mechanics

Secondary dimension	SI unit	BG unit	Conversion factor
Area $\{L^2\}$	m^2	ft^2	$1\ m^2 = 10.764\ ft^2$
Volume $\{L^3\}$	m^3	ft^3	$1\ m^3 = 35.315\ ft^3$
Velocity $\{LT^{-1}\}$	m/s	ft/s	1 ft/s = 0.3048 m/s
Acceleration $\{LT^{-2}\}$	m/s^2	ft/s^2	$1\ ft/s^2 = 0.3048\ m/s^2$
Pressure or stress $\{ML^{-1}T^{-2}\}$	$Pa = N/m^2$	lbf/ft^2	$1\ lbf/ft^2 = 47.88\ Pa$
Angular velocity $\{T^{-1}\}$	s^{-1}	s^{-1}	$1\ s^{-1} = 1\ s^{-1}$
Energy, heat, work $\{ML^2T^{-2}\}$	J = N · m	ft · lbf	1 ft · lbf = 1.3558 J
Power $\{ML^2T^{-3}\}$	W = J/s	ft · lbf/s	1 ft · lbf/s = 1.3558 W
Density $\{ML^{-3}\}$	kg/m^3	$slugs/ft^3$	$1\ slug/ft^3 = 515.4\ kg/m^3$
Viscosity $\{ML^{-1}T^{-1}\}$	kg/(m · s)	slugs/(ft · s)	1 slug/(ft · s) = 47.88 kg/(m · s)
Specific heat $\{L^2T^{-2}\Theta^{-1}\}$	$m^2/(s^2 \cdot K)$	$ft^2/(s^2 \cdot °R)$	$1\ m^2/(s^2 \cdot K) = 5.980\ ft^2/(s^2 \cdot °R)$

Solution

We need to find the (*a*) mass; (*b*) weight on the moon; and (*c*) acceleration of this body. This is a fairly simple example of conversion factors for differing unit systems. No property data is needed. The example is too low-level for a sketch.

Part (a) Newton's law (1.2) holds with known weight and gravitational acceleration. Solve for *m*:

$$F = W = 1{,}000 \text{ lbf} = mg = (m)(32.174 \text{ ft/s}^2), \text{ or } m = \frac{1{,}000 \text{ lbf}}{32.174 \text{ ft/s}^2} = 31.08 \text{ slugs}$$

Convert this to kilograms:

$$m = 31.08 \text{ slugs} = (31.08 \text{ slugs})(14.5939 \text{ kg/slug}) = 454 \text{ kg} \qquad \textit{Ans. (a)}$$

Part (b) The mass of the body remains 454 kg regardless of its location. Equation (1.2) applies with a new gravitational acceleration and hence a new weight:

$$F = W_{moon} = mg_{moon} = (454 \text{ kg})(1.62 \text{ m/s}^2) = 735 \text{ N} = 165 \text{ lbf} \qquad \textit{Ans. (b)}$$

Part (c) This part does not involve weight or gravity or location. It is simply an application of Newton's law with a known mass and known force:

$$F = 400 \text{ lbf} = ma = (31.08 \text{ slugs})\ a$$

Solve for

$$a = \frac{400 \text{ lbf}}{31.08 \text{ slugs}} = 12.87 \frac{\text{ft}}{\text{s}^2}\left(0.3048 \frac{\text{m}}{\text{ft}}\right) = 3.92 \frac{\text{m}}{\text{s}^2} \qquad \textit{Ans. (c)}$$

Comment (c): This acceleration would be the same on the earth or moon or anywhere.

Many data in the literature are reported in inconvenient or arcane units suitable only to some industry or specialty or country. The engineer should convert these data to the SI or BG system before using them. This requires the systematic application of conversion factors, as in the following example.

EXAMPLE 1.2

Industries involved in viscosity measurement [27, 29] continue to use the CGS system of units, since centimeters and grams yield convenient numbers for many fluids. The absolute viscosity (μ) unit is the *poise*, named after J. L. M. Poiseuille, a French physician who in 1840 performed pioneering experiments on water flow in pipes; 1 poise = 1 g/(cm-s). The kinematic viscosity (ν) unit is the *stokes,* named after G. G. Stokes, a British physicist who in 1845 helped develop the basic partial differential equations of fluid momentum; 1 stokes = 1 cm²/s. Water at 20°C has $\mu \approx 0.01$ poise and also $\nu \approx 0.01$ stokes. Express these results in (*a*) SI and (*b*) BG units.

Solution

Part (a)

- *Approach:* Systematically change grams to kg or slugs and change centimeters to meters or feet.
- *Property values:* Given $\mu = 0.01$ g/(cm-s) and $\nu = 0.01$ cm²/s.
- *Solution steps:* (*a*) For conversion to SI units,

$$\mu = 0.01 \frac{\text{g}}{\text{cm} \cdot \text{s}} = 0.01 \frac{\text{g}(1 \text{ kg}/1{,}000 \text{ g})}{\text{cm}(0.01 \text{ m/cm})\text{s}} = 0.001 \frac{\text{kg}}{\text{m} \cdot \text{s}}$$

$$\nu = 0.01 \frac{\text{cm}^2}{\text{s}} = 0.01 \frac{\text{cm}^2(0.01 \text{ m/cm})^2}{\text{s}} = 0.000001 \frac{\text{m}^2}{\text{s}} \qquad \textit{Ans. (a)}$$

Part (b)

- For conversion to BG units

$$\mu = 0.01 \frac{\text{g}}{\text{cm} \cdot \text{s}} = 0.01 \frac{\text{g}(1 \text{ kg}/1{,}000 \text{ g})(1 \text{ slug}/14.5939 \text{ kg})}{(0.01 \text{ m/cm})(1 \text{ ft}/0.3048 \text{ m})\text{s}} = 0.0000209 \frac{\text{slug}}{\text{ft} \cdot \text{s}}$$

$$\nu = 0.01 \frac{\text{cm}^2}{\text{s}} = 0.01 \frac{\text{cm}^2(0.01 \text{ m/cm})^2(1 \text{ ft}/0.3048 \text{ m})^2}{\text{s}} = 0.0000108 \frac{\text{ft}^2}{\text{s}} \qquad \textit{Ans. (b)}$$

- *Comments:* This was a laborious conversion that could have been shortened by using the direct viscosity conversion factors in App. C or the inside front cover. For example, $\mu_{BG} = \mu_{SI}/47.88$.

We repeat our advice: Faced with data in unusual units, convert them immediately to either SI or BG units because (1) it is more professional and (2) theoretical equations in fluid mechanics are *dimensionally consistent* and require no further conversion factors when these two fundamental unit systems are used, as the following example shows.

EXAMPLE 1.3

A useful theoretical equation for computing the relation between pressure, velocity, and altitude in a steady flow of a nearly inviscid, nearly incompressible fluid with negligible heat transfer and shaft work[5] is the *Bernoulli relation,* named after Daniel Bernoulli, who published a hydrodynamics textbook in 1738:

$$p_0 = p + \tfrac{1}{2}\rho V^2 + \rho g Z \qquad (1)$$

where p_0 = stagnation pressure
p = pressure in moving fluid
V = velocity
ρ = density
Z = altitude
g = gravitational acceleration

(*a*) Show that Eq. (1) satisfies the principle of dimensional homogeneity, which states that all additive terms in a physical equation must have the same dimensions. (*b*) Show that consistent units result without additional conversion factors in SI units. (*c*) Repeat (*b*) for BG units.

[5]That's an awful lot of assumptions, which need further study in Chap. 3.

Solution

Part (a) We can express Eq. (1) dimensionally, using braces, by entering the dimensions of each term from Table 1.2:

$$\{ML^{-1}T^{-2}\} = \{ML^{-1}T^{-2}\} + \{ML^{-3}\}\{L^2T^{-2}\} + \{ML^{-3}\}\{LT^{-2}\}\{L\}$$

$$= \{ML^{-1}T^{-2}\} \text{ for all terms} \qquad \textit{Ans. (a)}$$

Part (b) Enter the SI units for each quantity from Table 1.2:

$$\{\text{N/m}^2\} = \{\text{N/m}^2\} + \{\text{kg/m}^3\}\{\text{m}^2/\text{s}^2\} + \{\text{kg/m}^3\}\{\text{m/s}^2\}\{\text{m}\}$$

$$= \{\text{N/m}^2\} + \{\text{kg/(m} \cdot \text{s}^2)\}$$

The right-hand side looks bad until we remember from Eq. (1.3) that 1 kg = 1 N · s^2/m.

$$\{\text{kg/(m} \cdot \text{s}^2)\} = \frac{\{\text{N} \cdot \text{s}^2/\text{m}\}}{\{\text{m} \cdot \text{s}^2\}} = \{\text{N/m}^2\} \qquad \textit{Ans. (b)}$$

Thus, all terms in Bernoulli's equation will have units of pascals, or newtons per square meter, when SI units are used. No conversion factors are needed, which is true of all theoretical equations in fluid mechanics.

Part (c) Introducing BG units for each term, we have

$$\{\text{lbf/ft}^2\} = \{\text{lbf/ft}^2\} + \{\text{slugs/ft}^3\}\{\text{ft}^2/\text{s}^2\} + \{\text{slugs/ft}^3\}\{\text{ft/s}^2\}\{\text{ft}\}$$

$$= \{\text{lbf/ft}^2\} + \{\text{slugs/(ft} \cdot \text{s}^2)\}$$

But, from Eq. (1.3), 1 slug = 1 lbf · s^2/ft, so that

$$\{\text{slugs/(ft} \cdot \text{s}^2)\} = \frac{\{\text{lbf} \cdot \text{s}^2/\text{ft}\}}{\{\text{ft} \cdot \text{s}^2\}} = \{\text{lbf/ft}^2\} \qquad \textit{Ans. (c)}$$

All terms have the unit of pounds-force per square foot. No conversion factors are needed in the BG system either.

There is still a tendency in English-speaking countries to use pound-force per square inch as a pressure unit because the numbers are more manageable. For example, standard atmospheric pressure is 14.7 lbf/in^2 = 2,116 lbf/ft^2 = 101,300 Pa. The pascal is a small unit because the newton is less than $\frac{1}{4}$ lbf and a square meter is a very large area.

Consistent Units

Note that not only must all (fluid) mechanics equations be dimensionally homogeneous, one must also use *consistent units;* that is, each additive term must have the same units. There is no trouble doing this with the SI and BG systems, as in Example 1.3, but woe unto those who try to mix colloquial English units. For example, in Chap. 9, we often use the assumption of steady adiabatic compressible gas flow:

$$h + \tfrac{1}{2}V^2 = \text{constant}$$

where h is the fluid enthalpy and $V^2/2$ is its kinetic energy per unit mass. Colloquial thermodynamic tables might list h in units of British thermal units per pound mass (Btu/lb), whereas V is likely used in ft/s. It is completely erroneous to add Btu/lb to ft^2/s^2. The proper unit for h in this case is ft · lbf/slug, which is identical to ft^2/s^2. The conversion factor is 1 Btu/lb ≈ 25,040 ft^2/s^2 = 25,040 ft · lbf/slug.

Homogeneous versus Dimensionally Inconsistent Equations

All theoretical equations in mechanics (and in other physical sciences) are *dimensionally homogeneous;* that is, each additive term in the equation has the same dimensions. However, the reader should be warned that many empirical formulas in the engineering literature, arising primarily from correlations of data, are dimensionally inconsistent. Their units cannot be reconciled simply, and some terms may contain hidden variables. An example is the formula that pipe valve manufacturers cite for liquid volume flow rate Q (m^3/s) through a partially open valve:

$$Q = C_V\left(\frac{\Delta p}{\text{SG}}\right)^{1/2}$$

where Δp is the pressure drop across the valve and SG is the specific gravity of the liquid (the ratio of its density to that of water). The quantity C_V is the *valve flow coefficient,* which manufacturers tabulate in their valve brochures. Since SG is dimensionless {1}, we see that this formula is totally inconsistent, with one side being a flow rate $\{L^3/T\}$ and the other being the square root of a pressure drop $\{M^{1/2}/L^{1/2}T\}$. It follows that C_V must have dimensions, and rather odd ones at that: $\{L^{7/2}/M^{1/2}\}$. Nor is the resolution of this discrepancy clear, although one hint is that the values of C_V in the literature increase nearly as the square of the size of the valve. The presentation of experimental data in homogeneous form is the subject of *dimensional analysis* (Chap. 5). There we shall learn that a homogeneous form for the valve flow relation is

$$Q = C_d A_{\text{opening}}\left(\frac{\Delta p}{\rho}\right)^{1/2}$$

where ρ is the liquid density and A the area of the valve opening. The *discharge coefficient* C_d is dimensionless and changes only moderately with valve size. Please believe—until we establish the fact in Chap. 5—that this latter is a *much* better formulation of the data.

Meanwhile, we conclude that dimensionally inconsistent equations, though they occur in engineering practice, are misleading and vague and even dangerous, in the sense that they are often misused outside their range of applicability.

Table 1.3 Convenient Prefixes for Engineering Units

Multiplicative factor	Prefix	Symbol
10^{12}	tera	T
10^9	giga	G
10^6	mega	M
10^3	kilo	k
10^2	hecto	h
10	deka	da
10^{-1}	deci	d
10^{-2}	centi	c
10^{-3}	milli	m
10^{-6}	micro	μ
10^{-9}	nano	n
10^{-12}	pico	p
10^{-15}	femto	f
10^{-18}	atto	a

Convenient Prefixes in Powers of 10

Engineering results often are too small or too large for the common units, with too many zeros one way or the other. For example, to write $p = 114{,}000{,}000$ Pa is long and awkward. Using the prefix "M" to mean 10^6, we convert this to a concise $p = 114$ MPa (megapascals). Similarly, $t = 0.000000003$ s is a proofreader's nightmare compared to the equivalent $t = 3$ ns (nanoseconds). Such prefixes are common and convenient, in both the SI and BG systems. A complete list is given in Table 1.3.

EXAMPLE 1.4

In 1890 Robert Manning, an Irish engineer, proposed the following empirical formula for the average velocity V in uniform flow due to gravity down an open channel (BG units):

$$V = \frac{1.49}{n}R^{2/3}S^{1/2} \tag{1}$$

where R = hydraulic radius of channel (Chaps. 6 and 10)
S = channel slope (tangent of angle that bottom makes with horizontal)
n = Manning's roughness factor (Chap. 10)

and n is a constant for a given surface condition for the walls and bottom of the channel. (*a*) Is Manning's formula dimensionally consistent? (*b*) Equation (1) is commonly taken to be valid in BG units with n taken as dimensionless. Rewrite it in SI form.

Solution

- *Assumption:* The channel slope S is the tangent of an angle and is thus a dimensionless ratio with the dimensional notation {1}—that is, not containing M, L, or T.
- *Approach (a):* Rewrite the dimensions of each term in Manning's equation, using brackets {}:

$$\{V\} = \left\{\frac{1.49}{n}\right\}\{R^{2/3}\}\{S^{1/2}\} \quad \text{or} \quad \left\{\frac{L}{T}\right\} = \left\{\frac{1.49}{n}\right\}\{L^{2/3}\}\{1\}$$

This formula is incompatible unless $\{1.49/n\} = \{L^{1/3}/T\}$. If n is dimensionless (and it is never listed with units in textbooks), the number 1.49 must carry the dimensions of $\{L^{1/3}/T\}$. *Ans. (a)*

- *Comment (a):* Formulas whose numerical coefficients have units can be disastrous for engineers working in a different system or another fluid. Manning's formula, though popular, is inconsistent both dimensionally and physically and is valid only for water flow with certain wall roughnesses. The effects of water viscosity and density are hidden in the numerical value 1.49.
- *Approach (b):* Part (*a*) showed that 1.49 has dimensions. If the formula is valid in BG units, then it must equal 1.49 $\text{ft}^{1/3}/\text{s}$. By using the SI conversion for length, we obtain

$$(1.49\ \text{ft}^{1/3}/\text{s})(0.3048\ \text{m/ft})^{1/3} = 1.00\ \text{m}^{1/3}/\text{s}$$

Therefore, Manning's inconsistent formula changes form when converted to the SI system:

$$\text{SI units: } V = \frac{1.0}{n} R^{2/3} S^{1/2} \qquad \textit{Ans. (b)}$$

with R in meters and V in meters per second.

- *Comment (b):* Actually, we misled you: This is the way Manning, a metric user, first proposed the formula. It was later converted to BG units. Such dimensionally inconsistent formulas are dangerous and should either be reanalyzed or treated as having very limited application.

1.5 Properties of the Velocity Field

In a given flow situation, the determination, by experiment or theory, of the properties of the fluid as a function of position and time is considered to be the *solution* to the problem. In almost all cases, the emphasis is on the space–time distribution of the fluid properties. One rarely keeps track of the actual fate of the specific fluid particles. This treatment of properties as continuum-field functions distinguishes fluid mechanics from solid mechanics, where we are more likely to be interested in the trajectories of individual particles or systems.

The Velocity Field

Foremost among the properties of a flow is the velocity field $\mathbf{V}(x, y, z, t)$. In fact, determining the velocity is often tantamount to solving a flow problem, since other properties follow directly from the velocity field. Chapter 2 is devoted to the calculation of the pressure field once the velocity field is known. Books on heat transfer (for example, Ref. 20) are largely devoted to finding the temperature field from known velocity fields.

In general, velocity is a vector function of position and time, and thus has three components u, v, and w, each a scalar field in itself:

$$\mathbf{V}(x, y, z, t) = \mathbf{i}u(x, y, z, t) + \mathbf{j}v(x, y, z, t) + \mathbf{k}w(x, y, z, t) \tag{1.4}$$

The use of u, v, and w instead of the more logical component notation V_x, V_y, and V_z is the result of an almost unbreakable custom in fluid mechanics. Much of this textbook, especially Chaps. 4, 7, 8, and 9, is concerned with finding the distribution of the velocity vector **V** for a variety of practical flows.

The Acceleration Field

The acceleration vector, $\mathbf{a} = d\mathbf{V}/dt$, occurs in Newton's law for a fluid and thus is very important. In order to follow a particle in the Eulerian frame of reference, the final result for acceleration is nonlinear and quite complicated. Here we only give the formula:

$$\mathbf{a} = \frac{d\mathbf{V}}{dt} = \frac{\partial \mathbf{V}}{\partial t} + u\frac{\partial \mathbf{V}}{\partial x} + v\frac{\partial \mathbf{V}}{\partial y} + w\frac{\partial \mathbf{V}}{\partial z} \tag{1.5}$$

where (u, v, w) are the velocity components from Eq. (1.4). We shall study this formula in detail in Chap. 4. The last three terms in Eq. (1.5) are nonlinear products and greatly complicate the analysis of general fluid motions, especially viscous flows.

1.6 Thermodynamic Properties of a Fluid

While the velocity field **V** is the most important fluid property, it interacts closely with the thermodynamic properties of the fluid. We have already introduced into the discussion the three most common such properties:

1. Pressure p
2. Density ρ
3. Temperature T

These three are constant companions of the velocity vector in flow analyses. Four other intensive thermodynamic properties become important when work, heat, and energy balances are treated (Chaps. 3 and 4):

4. Internal energy $\hat{u}$
5. Enthalpy $h = \hat{u} + p/\rho$
6. Entropy s
7. Specific heats c_p and c_v

In addition, friction and heat conduction effects are governed by the two so-called *transport properties:*

8. Coefficient of viscosity μ
9. Thermal conductivity k

All nine of these quantities are true thermodynamic properties that are determined by the thermodynamic condition or *state* of the fluid. For example, for a single-phase substance such as water or oxygen, two basic properties such as pressure and temperature are sufficient to fix the value of all the others:

$$\rho = \rho(p, T) \qquad h = h(p, T) \qquad \mu = \mu(p, T)$$

and so on for every quantity in the list. Note that the specific volume, so important in thermodynamic analyses, is omitted here in favor of its inverse, the density ρ.

Recall that thermodynamic properties describe the state of a *system*—that is, a collection of matter of fixed identity that interacts with its surroundings. In most cases here, the system will be a small fluid element, and all properties will be assumed to be continuum properties of the flow field: $\rho = \rho(x, y, z, t)$, and so on.

Recall also that thermodynamics is normally concerned with *static* systems, whereas fluids are usually in variable motion with constantly changing properties. Do the properties retain their meaning in a fluid flow that is technically not in equilibrium? The answer is yes, from a statistical argument. In gases at normal pressure (and even more so for liquids), an enormous number of molecular collisions occur over a very short distance of the order of 1 μm, so that a fluid subjected to sudden changes rapidly adjusts itself toward equilibrium. We therefore assume that all the thermodynamic properties just listed exist as point functions in a flowing fluid and follow all the laws and state relations of ordinary equilibrium thermodynamics. There are, of course, important nonequilibrium effects such as chemical and nuclear reactions in flowing fluids, which are not treated in this text.

Pressure

Pressure is the (compression) stress at a point in a static fluid (Fig. 1.3). Next to velocity, the pressure p is the most dynamic variable in fluid mechanics. Differences or *gradients* in pressure often drive a fluid flow, especially in ducts. In low-speed flows, the actual magnitude of the pressure is often not important, unless it drops so low as to cause vapor bubbles to form in a liquid. For convenience, we set many such problem assignments at the level of 1 atm = 2,116 lbf/ft^2 = 101,300 Pa. High-speed (compressible) gas flows (Chap. 9), however, are indeed sensitive to the magnitude of pressure.

Temperature

Temperature T is related to the internal energy level of a fluid. It may vary considerably during high-speed flow of a gas (Chap. 9). Although engineers often use Celsius or Fahrenheit scales for convenience, many applications in this text require *absolute* (Kelvin or Rankine) temperature scales:

$$^\circ R = {}^\circ F + 459.69$$

$$K = {}^\circ C + 273.16$$

If temperature differences are strong, *heat transfer* may be important [20], but our concern here is mainly with dynamic effects.

Density

The density of a fluid, denoted by ρ (lowercase Greek rho), is its mass per unit volume. Density is highly variable in gases and increases nearly proportionally to the pressure level. Density in liquids is nearly constant; the density of water (about 1000 kg/m^3) increases only 1 percent if the pressure is increased by a factor of 220. Thus, most liquid flows are treated analytically as nearly "incompressible."

In general, liquids are about three orders of magnitude more dense than gases at atmospheric pressure. The heaviest common liquid is mercury, and the lightest gas is hydrogen. Compare their densities at 20°C and 1 atm:

$$\text{Mercury: } \rho = 13{,}580 \text{ kg/m}^3 \qquad \text{Hydrogen: } \rho = 0.0838 \text{ kg/m}^3$$

They differ by a factor of 162,000! Thus, the physical parameters in various liquid and gas flows might vary considerably. The differences are often resolved by the use of *dimensional analysis* (Chap. 5). Other fluid densities are listed in Tables A.3 and A.4 (in App. A), and in Ref. 21.

Specific Weight

The *specific weight* of a fluid, denoted by γ (lowercase Greek gamma), is its weight per unit volume. Just as a mass has a weight $W = mg$, density and specific weight are simply related by gravity:

$$\gamma = \rho g \tag{1.6}$$

The units of γ are weight per unit volume, in lbf/ft³ or N/m³. In standard earth gravity, $g = 32.174$ ft/s² $= 9.807$ m/s². Thus, for example, the specific weights of air and water at 20°C and 1 atm are approximately

$$\gamma_{air} = (1.205\ \text{kg/m}^3)(9.807\ \text{m/s}^2) = 11.8\ \text{N/m}^3 = 0.0752\ \text{lbf/ft}^3$$

$$\gamma_{water} = (998\ \text{kg/m}^3)(9.807\ \text{m/s}^2) = 9{,}790\ \text{N/m}^3 = 62.4\ \text{lbf/ft}^3$$

Specific weight is very useful in the hydrostatic pressure applications of Chap. 2. Specific weights of other fluids are given in Tables A.3 and A.4.

Specific Gravity

Specific gravity, denoted by SG, is the ratio of a fluid density to a standard reference fluid, usually water at 4°C (for liquids) and air (for gases):

$$\text{SG}_{gas} = \frac{\rho_{gas}}{\rho_{air}} = \frac{\rho_{gas}}{1.205\ \text{kg/m}^3} \tag{1.7}$$

$$\text{SG}_{liquid} = \frac{\rho_{liquid}}{\rho_{water}} = \frac{\rho_{liquid}}{1{,}000\ \text{kg/m}^3}$$

For example, the specific gravity of mercury (Hg) is $\text{SG}_{Hg} = 13{,}580/1{,}000 \approx 13.6$. Engineers find these dimensionless ratios easier to remember than the actual numerical values of density of a variety of fluids.

Potential and Kinetic Energies

In thermostatics, the only energy in a substance is that stored in a system by molecular activity and molecular bonding forces. This is commonly denoted as *internal energy* $\hat{u}$. A commonly accepted adjustment to this static situation for fluid flow is to add two more energy terms that arise from newtonian mechanics: potential energy and kinetic energy.

The potential energy equals the work required to move the system of mass m from the origin to a position vector $\mathbf{r} = \mathbf{i}x + \mathbf{j}y + \mathbf{k}z$ against a gravity field $\mathbf{g}$. Its value is $-m\mathbf{g} \cdot \mathbf{r}$, or $-\mathbf{g} \cdot \mathbf{r}$ per unit mass. The kinetic energy equals the work required to change the speed of the mass from zero to velocity V. Its value is $\frac{1}{2}mV^2$ or $\frac{1}{2}V^2$ per unit mass. Then by common convention, the total stored energy e per unit mass in fluid mechanics is the sum of three terms:

$$e = \hat{u} + \tfrac{1}{2}V^2 + (-\mathbf{g} \cdot \mathbf{r}) \tag{1.8}$$

Also, throughout this book we shall define z as upward, so that $\mathbf{g} = -g\mathbf{k}$ and $\mathbf{g} \cdot \mathbf{r} = -gz$. Then Eq. (1.8) becomes

$$e = \hat{u} + \tfrac{1}{2}V^2 + gz \tag{1.9}$$

The molecular internal energy $\hat{u}$ is a function of T and p for the single-phase pure substance, whereas the potential and kinetic energies are kinematic quantities.

State Relations for Gases

Thermodynamic properties are found both theoretically and experimentally to be related to each other by state relations that differ for each substance. As mentioned, we shall confine ourselves here to single-phase pure substances, such as water in its liquid phase. The second most common fluid, air, is a mixture of gases, but since the mixture ratios remain nearly constant between 160 and 2,200 K, in this temperature range air can be considered to be a pure substance.

All gases at high temperatures and low pressures (relative to their critical point) are in good agreement with the *perfect-gas law*

$$p = \rho R T \quad R = c_p - c_v = \text{gas constant} \tag{1.10}$$

where the specific heats c_p and c_v are defined in Eqs. (1.14) and (1.15).

Since Eq. (1.10) is dimensionally consistent, R has the same dimensions as specific heat, $\{L^2T^{-2}\Theta^{-1}\}$, or velocity squared per temperature unit (kelvin or degree Rankine). Each gas has its own constant R, equal to a universal constant Λ divided by the molecular weight

$$R_{\text{gas}} = \frac{\Lambda}{M_{\text{gas}}} \tag{1.11}$$

where Λ = 8314 J/(kg-mol · K) = 49,700 ft-lbf/(slugmol · °R). Most applications in this book are for air, whose molecular weight is M = 28.97/mol:

$$R_{\text{air}} = \frac{8{,}314\ \text{J}/(\text{kg-mol}\cdot\text{K})}{28.97/\text{mol}} = 287\,\frac{\text{N}\cdot\text{m}}{\text{kg}\cdot\text{K}} = 287\,\frac{\text{m}^2}{\text{s}^2\cdot\text{K}} = 1{,}716\,\frac{\text{ft}^2}{\text{s}^2\cdot{}^\circ\text{R}} \tag{1.12}$$

Standard atmospheric pressure is 101,300 Pa = 101,300 kg/(m · s²), and standard temperature is 20°C = 293K. Thus, standard air density is

$$\rho_{\text{air}} = \frac{101{,}300\ \text{kg}/(\text{m}\cdot\text{s}^2)}{[287\ \text{m}^2/(\text{s}^2\cdot\text{K})](293\text{K})} = 1.205\ \text{kg/m}^3 = 0.00237\ \text{slug/ft}^3 \tag{1.13}$$

This is a nominal value suitable for problems. For other gases, see Table A.4.

Most of the common gases—oxygen, nitrogen, hydrogen, helium, argon—are nearly ideal. This is not so true for steam, whose simplified temperature-entropy chart is shown in Fig. 1.3. Unless you are sure that the steam temperature is "high" and the pressure "low," it is best to use the Steam Tables to make accurate calculations.

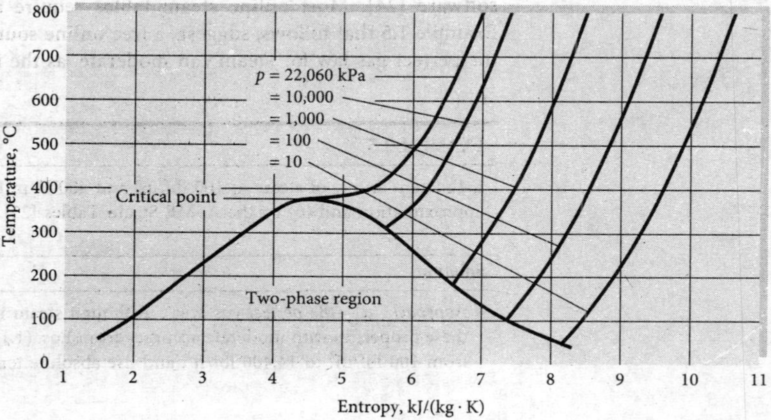

Fig. 1.3 Temperature-entropy chart for steam. The critical point is p_c = 22,060 kPa, T_c = 374°C, S_c = 4.41 kJ/(kg · K). Except near the critical point, the smooth isobars tempt one to assume, often incorrectly, that the ideal-gas law is valid for steam. It is *not*, except at low pressure and high temperature: the upper right of the graph.

One proves in thermodynamics that Eq. (1.10) requires that the internal molecular energy $\hat{u}$ of a perfect gas vary only with temperature: $\hat{u} = \hat{u}(T)$. Therefore, the specific heat c_v also varies only with temperature:

$$c_v = \left(\frac{\partial \hat{u}}{\partial T}\right)_\rho = \frac{d\hat{u}}{dT} = c_v(T)$$

or

$$d\hat{u} = c_v(T)dT \tag{1.14}$$

In like manner h and c_p of a perfect gas also vary only with temperature:

$$h = \hat{u} + \frac{p}{\rho} = \hat{u} + RT = h(T)$$

$$c_p = \left(\frac{\partial h}{\partial T}\right)_p = \frac{dh}{dT} = c_p(T) \tag{1.15}$$

$$dh = c_p(T)dT$$

The ratio of specific heats of a perfect gas is an important dimensionless parameter in compressible flow analysis (Chap. 9).

$$k = \frac{c_p}{c_v} = k(T) \geq 1 \tag{1.16}$$

As a first approximation in airflow analysis we commonly take c_p, c_v, and k to be constant:

$$k_{\text{air}} \approx 1.4$$

$$c_v = \frac{R}{k-1} \approx 718\ \text{m}^2/(\text{s}^2 \cdot \text{K}) = 4{,}293\ \text{ft}^2/(\text{s}^2 \cdot {}^\circ\text{R}) \tag{1.17}$$

$$c_p = \frac{kR}{k-1} \approx 1{,}005\ \text{m}^2/(\text{s}^2 \cdot \text{K}) = 6{,}010\ \text{ft}^2/(\text{s}^2 \cdot {}^\circ\text{R})$$

Actually, for all gases, c_p and c_v increase gradually with temperature, and k decreases gradually. Experimental values of the specific-heat ratio for eight common gases are shown in Fig. 1.4. Nominal values are in Table A.4.

Many flow problems involve steam. Typical steam operating conditions are often close to the critical point, so that the perfect-gas approximation is inaccurate. Then we must turn to the steam tables, either in tabular or CD-ROM form [23] or as online software [24]. Most online steam tables require a license fee, but the writer, in Example 1.5 that follows, suggests a free online source. Sometimes the error of using the perfect-gas law for steam can moderate, as the following example shows.

EXAMPLE 1.5

Estimate ρ and c_p of steam at 100 lbf/in^2 and 400°F, in English units, (*a*) by the perfect-gas approximation and (*b*) by the ASME Steam Tables [23].

Solution

- *Approach (a)—the perfect-gas law:* Although steam is not an ideal gas, we can estimate these properties with moderate accuracy from Eqs. (1.10) and (1.17). First convert pressure from 100 lbf/in^2 to 14,400 lbf/ft^2, and use absolute temperature, (400°F + 460) = 860°R.

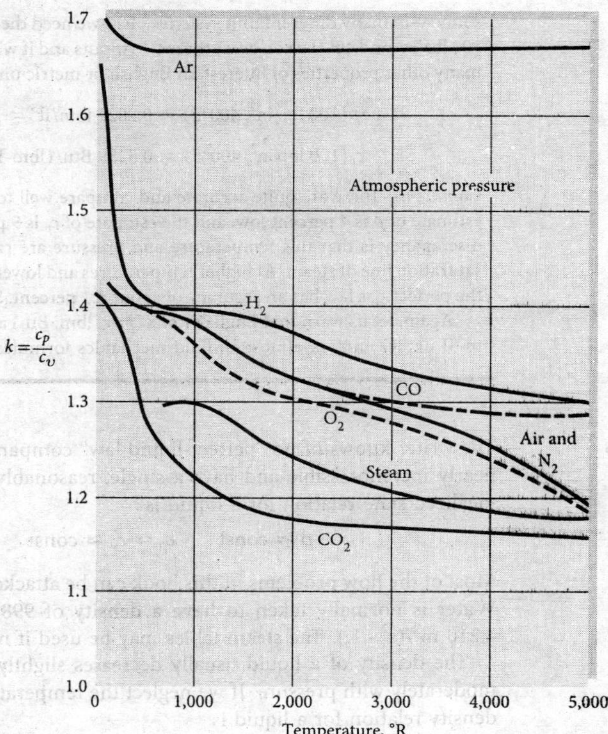

Fig. 1.4 Specific-heat ratio of eight common gases as a function of temperature. *(Data from Ref. 22.)*

Then, we need the gas constant for steam in English units. From Table A.4, the molecular weight of H_2O is 18.02, whence

$$R_{steam} = \frac{\Lambda_{English}}{M_{H_2O}} = \frac{49{,}700 \text{ ft} \cdot \text{lbf/(slugmol °R)}}{18.02/\text{mol}} = 2{,}758 \frac{\text{ft} \cdot \text{lbf}}{\text{slug °R}}$$

Then the density estimate follows from the perfect-gas law, Eq. (1.10):

$$\rho \approx \frac{p}{RT} = \frac{14{,}400 \text{ lbf/ft}^2}{[2{,}758 \text{ ft} \cdot \text{lbf/(slug} \cdot \text{°R)}](860 \text{ °R})} \approx 0.00607 \frac{\text{slug}}{\text{ft}^3} \qquad \textit{Ans. (a)}$$

At 860°R, from Fig. 1.5, $k_{steam} = c_p/c_v \approx 1.30$. Then, from Eq. (1.17),

$$c_p \approx \frac{kR}{k-1} = \frac{(1.3)(2{,}758 \text{ ft} \cdot \text{lbf/(slug °R)})}{(1.3-1)} \approx 12{,}000 \frac{\text{ft} \cdot \text{lbf}}{\text{slug °R}} \qquad \textit{Ans. (a)}$$

- *Approach (b)—tables or software:* One can either read the ASME Steam Tables [23] or use online software. Online software, such as [24], calculates the properties of steam without reading a table. Most of these require a license fee, which your institution may or may not possess. For work at home, the writer has found success with this free commercial online site:

www.spiraxsarco.com/esc/SH_Properties.aspx

The software calculates superheated steam properties, as required in this example. The Spirax Sarco Company makes many types of steam equipment: boilers, condensers, valves, pumps, regulators. This site provides many steam properties—density, specific heat, enthalpy, speed of sound—in many different unit systems. Here we need the density and specific heat of steam at 100 lbf/in² and 400°F. You enter these two inputs and it will calculate not only ρ and c_p but also many other properties of interest, in English or metric units. The software results are:

$$\rho(100\ \text{lbf/in}^2, 400°\text{F}) = 0.2027\ \text{lbm/ft}^3 = 3.247\ \text{kg/m}^3 \qquad \textit{Ans. (b)}$$

$$c_p(100\ \text{lbf/in}^2, 400°\text{F}) = 0.5289\ \text{Btu/(lbm-F)} = 2215\ \text{J/(kg-K)} \qquad \textit{Ans. (b)}$$

Comments: These are quite accurate and compare well to other steam tables. The perfect gas estimate of ρ is 4 percent low, and the estimate of c_p is 9 percent low. The chief reason for the discrepancy is that this temperature and pressure are rather close to the critical point and saturation line of steam. At higher temperatures and lower pressures, say, 800°F and 50 lbf/in², the perfect-gas law has an accuracy of about ±1 percent. See Fig. 1.3.

Again, let us warn that English units (psia, lbm, Btu) are awkward, and must be converted to SI or BG units in almost all fluid mechanics formulas.

State Relations for Liquids

The writer knows of no "perfect-liquid law" comparable to that for gases. Liquids are nearly incompressible and have a single, reasonably constant specific heat. Thus, an idealized state relation for a liquid is

$$\rho \approx \text{const} \qquad c_p \approx c_v \approx \text{const} \qquad dh \approx c_p\, dT \tag{1.18}$$

Most of the flow problems in this book can be attacked with these simple assumptions. Water is normally taken to have a density of 998 kg/m³ and a specific heat c_p = 4,210 m²/(s² · K). The steam tables may be used if more accuracy is required.

The density of a liquid usually decreases slightly with temperature and increases moderately with pressure. If we neglect the temperature effect, an empirical pressure–density relation for a liquid is

$$\frac{p}{p_a} \approx (B+1)\left(\frac{\rho}{\rho_a}\right)^n - B \tag{1.19}$$

where B and n are dimensionless parameters that vary slightly with temperature and p_a and ρ_a are standard atmospheric values. Water can be fitted approximately to the values $B \approx 3{,}000$ and $n \approx 7$.

Seawater is a variable mixture of water and salt and thus requires three thermodynamic properties to define its state. These are normally taken as pressure, temperature, and the *salinity* $\hat{S}$, defined as the weight of the dissolved salt divided by the weight of the mixture. The average salinity of seawater is 0.035, usually written as 35 parts per 1,000, or 35 ‰. The average density of seawater is 2.00 slugs/ft³ ≈ 1,030 kg/m³. Strictly speaking, seawater has three specific heats, all approximately equal to the value for pure water of 25,200 ft²/(s² · °R) = 4,210 m²/(s² · K).

EXAMPLE 1.6

The pressure at the deepest part of the ocean is approximately 1,100 atm. Estimate the density of seawater in slug/ft³ at this pressure.

Solution

Equation (1.19) holds for either water or seawater. The ratio p/p_a is given as 1,100:

$$1{,}100 \approx (3{,}001)\left(\frac{\rho}{\rho_a}\right)^7 - 3{,}000$$

or
$$\frac{\rho}{\rho_a} = (4{,}100/3{,}001)^{1/7} = 1.046$$

Assuming an average surface seawater density $\rho_a = 2.00$ slugs/ft^3, we compute

$$\rho \approx 1.046(2.00) = 2.09 \text{ slugs/ft}^3 \qquad \textit{Ans.}$$

Even at these immense pressures, the density increase is less than 5 percent, which justifies the treatment of a liquid flow as essentially incompressible.

1.7 Viscosity and Other Secondary Properties

The quantities such as pressure, temperature, and density discussed in the previous section are *primary* thermodynamic variables characteristic of any system. Certain secondary variables also characterize specific fluid mechanical behavior. The most important of these is viscosity, which relates the local stresses in a moving fluid to the strain rate of the fluid element.

Viscosity

Viscosity is a quantitative measure of a fluid's resistance to flow. More specifically, it determines the fluid strain rate that is generated by a given applied shear stress. We can easily move through air, which has very low viscosity. Movement is more difficult in water, which has 50 times higher viscosity. Still more resistance is found in SAE 30 oil, which is 300 times more viscous than water. Try to slide your hand through glycerin, which is five times more viscous than SAE 30 oil, or blackstrap molasses, another factor of five higher than glycerin. Fluids may have a vast range of viscosities.

Consider a fluid element sheared in one plane by a single shear stress τ, as in Fig. 1.5*a*. The shear strain angle $\delta\theta$ will continuously grow with time as long as the stress τ is maintained, the upper surface moving at speed δu larger than the lower. Such common fluids as water, oil, and air show a linear relation between applied shear and resulting strain rate:

$$\tau \propto \frac{\delta\theta}{\delta t} \tag{1.20}$$

From the geometry of Fig. 1.5*a*, we see that

$$\tan \delta\theta = \frac{\delta u\, \delta t}{\delta y} \tag{1.21}$$

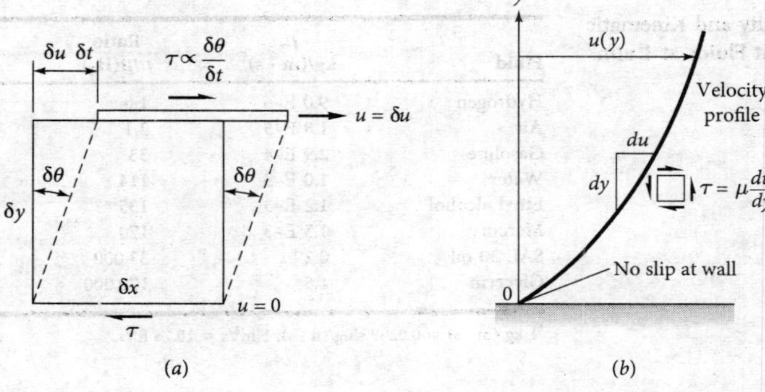

Fig. 1.5 Shear stress causes continuous shear deformation in a fluid: (*a*) a fluid element straining at a rate $\delta\theta/\delta t$; (*b*) newtonian shear distribution in a shear layer near a wall.

In the limit of infinitesimal changes, this becomes a relation between shear strain rate and velocity gradient:

$$\frac{d\theta}{dt} = \frac{du}{dy} \tag{1.22}$$

From Eq. (1.20), then, the applied shear is also proportional to the velocity gradient for the common linear fluids. The constant of proportionality is the viscosity coefficient μ:

$$\tau = \mu\frac{d\theta}{dt} = \mu\frac{du}{dy} \tag{1.23}$$

Equation (1.23) is dimensionally consistent; therefore μ has dimensions of stress–time: $\{FT/L^2\}$ or $\{M/(LT)\}$. The BG unit is slugs per foot-second, and the SI unit is kilograms per meter-second. The linear fluids that follow Eq. (1.23) are called *newtonian fluids*, after Sir Isaac Newton, who first postulated this resistance law in 1687.

We do not really care about the strain angle $\theta(t)$ in fluid mechanics, concentrating instead on the velocity distribution $u(y)$, as in Fig. 1.5*b*. We shall use Eq. (1.23) in Chap. 4 to derive a differential equation for finding the velocity distribution $u(y)$—and, more generally, $\mathbf{V}(x, y, z, t)$—in a viscous fluid. Figure 1.5*b* illustrates a shear layer, or *boundary layer*, near a solid wall. The shear stress is proportional to the slope of the velocity profile and is greatest at the wall. Further, at the wall, the velocity u is zero relative to the wall: This is called the *no-slip condition* and is characteristic of all viscous fluid flows.

The viscosity of newtonian fluids is a true thermodynamic property and varies with temperature and pressure. At a given state (p, T) there is a vast range of values among the common fluids. Table 1.4 lists the viscosity of eight fluids at standard pressure and temperature. There is a variation of six orders of magnitude from hydrogen up to glycerin. Thus, there will be wide differences between fluids subjected to the same applied stresses.

Generally speaking, the viscosity of a fluid increases only weakly with pressure. For example, increasing p from 1 to 50 atm will increase μ of air only 10 percent. Temperature, however, has a strong effect, with μ increasing with T for gases and decreasing for liquids. Figure A.1 (in App. A) shows this temperature variation for various common fluids. It is customary in most engineering work to neglect the pressure variation.

The variation $\mu(p, T)$ for a typical fluid is nicely shown by Fig. 1.6, from Ref. 25, which normalizes the data with the *critical-point state* (μ_c, p_c, T_c). This behavior, called the *principle of corresponding states*, is characteristic of all fluids, but the actual

Table 1.4 Viscosity and Kinematic Viscosity of Eight Fluids at 1 atm and 20°C

Fluid	μ, kg/(m · s)†	Ratio $\mu/\mu(H_2)$	ρ, kg/m³	ν m²/s†	Ratio ν/ν(Hg)
Hydrogen	9.0 E–6	1.0	0.084	1.05 E–4	910
Air	1.8 E–5	2.1	1.20	1.50 E–5	130
Gasoline	2.9 E–4	33	680	4.22 E–7	3.7
Water	1.0 E–3	114	998	1.01 E–6	8.7
Ethyl alcohol	1.2 E–3	135	789	1.52 E–6	13
Mercury	1.5 E–3	170	13,550	1.16 E–7	1.0
SAE 30 oil	0.29	33,000	891	3.25 E–4	2,850
Glycerin	1.5	170,000	1,260	1.18 E–3	10,300

†1 kg/(m · s) = 0.0209 slug/(ft · s); 1 m²/s = 10.76 ft²/s.

numerical values are uncertain to ±20 percent for any given fluid. For example, values of $\mu(T)$ for air at 1 atm, from Table A.2, fall about 8 percent low compared to the "low-density limit" in Fig. 1.6.

Note in Fig. 1.6 that changes with temperature occur very rapidly near the critical point. In general, critical-point measurements are extremely difficult and uncertain.

The Reynolds Number

The primary parameter correlating the viscous behavior of all newtonian fluids is the dimensionless *Reynolds number:*

$$\mathrm{Re} = \frac{\rho V L}{\mu} = \frac{VL}{\nu} \tag{1.24}$$

where V and L are characteristic velocity and length scales of the flow. The second form of Re illustrates that the ratio of μ to ρ has its own name, the *kinematic viscosity:*

$$\nu = \frac{\mu}{\rho} \tag{1.25}$$

It is called kinematic because the mass units cancel, leaving only the dimensions $\{L^2/T\}$.

Generally, the first thing a fluids engineer should do is estimate the Reynolds number range of the flow under study. Very low Re indicates viscous *creeping* motion, where inertia effects are negligible. Moderate Re implies a smoothly varying *laminar* flow. High Re probably spells *turbulent* flow, which is slowly varying in the time-mean but has superimposed strong random high-frequency fluctuations. Explicit numerical

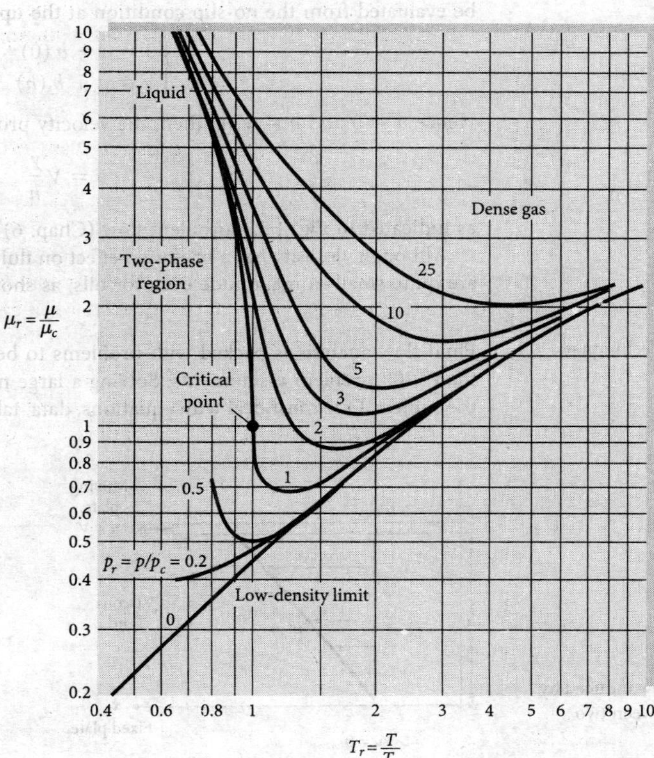

Fig. 1.6 Fluid viscosity nondimensionalized by critical-point properties. This generalized chart is characteristic of all fluids but is accurate only to ±20 percent. *(From Ref. 25.)*

values for low, moderate, and high Reynolds numbers cannot be stated here. They depend on flow geometry and will be discussed in Chaps. 5 through 7.

Table 1.4 also lists values of ν for the same eight fluids. The pecking order changes considerably, and mercury, the heaviest, has the smallest viscosity relative to its own weight. All gases have high ν relative to thin liquids such as gasoline, water, and alcohol. Oil and glycerin still have the highest ν, but the ratio is smaller. For given values of V and L in a flow, these fluids exhibit a spread of four orders of magnitude in the Reynolds number.

Flow between Plates

A classic problem is the flow induced between a fixed lower plate and an upper plate moving steadily at velocity V, as shown in Fig. 1.7. The clearance between plates is h, and the fluid is newtonian and does not slip at either plate. If the plates are large, this steady shearing motion will set up a velocity distribution $u(y)$, as shown, with $v = w = 0$. The fluid acceleration is zero everywhere.

With zero acceleration and assuming no pressure variation in the flow direction, you should show that a force balance on a small fluid element leads to the result that the shear stress is constant throughout the fluid. Then Eq. (1.23) becomes

$$\frac{du}{dy} = \frac{\tau}{\mu} = \text{const}$$

which we can integrate to obtain

$$u = a + by$$

The velocity distribution is linear, as shown in Fig. 1.7, and the constants a and b can be evaluated from the no-slip condition at the upper and lower walls:

$$u = \begin{cases} 0 = a + b\,(0) & \text{at } y = 0 \\ V = a + b\,(h) & \text{at } y = h \end{cases}$$

Hence $a = 0$ and $b = V/h$. Then, the velocity profile between the plates is given by

$$u = V\frac{y}{h} \tag{1.26}$$

as indicated in Fig. 1.7. Turbulent flow (Chap. 6) does not have this shape.

Although viscosity has a profound effect on fluid motion, the actual viscous stresses are quite small in magnitude even for oils, as shown in the next example.

Problem-Solving Techniques

Fluid flow analysis is packed with problems to be solved. The present text has more than 1700 problem assignments. Solving a large number of these is a key to learning the subject. One must deal with equations, data, tables, assumptions, unit systems, and

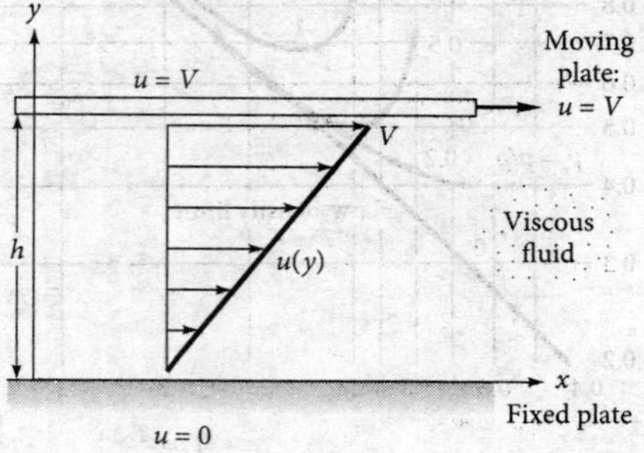

Fig. 1.7 Viscous flow induced by relative motion between two parallel plates.

solution schemes. The degree of difficulty will vary, and we urge you to sample the whole spectrum of assignments, with or without the answers in the Appendix. Here are the recommended steps for problem solution:

1. Read the problem and restate it with your summary of the results desired.
2. From tables or charts, gather the needed property data: density, viscosity, etc.
3. Make sure you understand what is *asked.* Students are apt to answer the wrong question—for example, pressure instead of pressure gradient, lift force instead of drag force, or mass flow instead of volume flow. Read the problem carefully.
4. Make a detailed, *labeled* sketch of the system or control volume needed.
5. Think carefully and list your *assumptions.* You must decide if the flow is steady or unsteady, compressible or incompressible, viscous or inviscid, and whether a control volume or partial differential equations are needed.
6. Find an algebraic solution if possible. Then, if a numerical value is needed, use either the SI or BG unit systems reviewed in Sec. 1.4.
7. Report your solution, *labeled,* with the proper units and the proper number of significant figures (usually two or three) that the data uncertainty allows.

We shall follow these steps, where appropriate, in our example problems.

EXAMPLE 1.7

Suppose that the fluid being sheared in Fig. 1.7 is SAE 30 oil at 20°C, compute the shear stress in the oil if $V = 3$ m/s and $h = 2$ cm.

Solution

- *System sketch:* This is shown earlier in Fig. 1.7.
- *Assumptions:* Linear velocity profile, laminar newtonian fluid, no slip at either plate surface.
- *Approach:* The analysis of Fig. 1.7 leads to Eq. (1.26) for laminar flow.
- *Property values:* From Table 1.4 for SAE 30 oil, the oil viscosity $\mu = 0.29$ kg/(m-s).
- *Solution steps:* In Eq. (1.26), the only unknown is the fluid shear stress:

$$\tau = \mu\frac{V}{h} = \left(0.29\frac{\text{kg}}{\text{m}\cdot\text{s}}\right)\frac{(3\text{ m/s})}{(0.02\text{ m})} = 43.5\frac{\text{kg}\cdot\text{m/s}^2}{\text{m}^2} = 43.5\frac{\text{N}}{\text{m}^2} \approx 44\text{Pa} \qquad \textit{Ans.}$$

- *Comments:* Note the unit identities, 1 kg-m/s^2 ≡ 1 N and 1 N/m^2 ≡ 1 Pa. Although oil is very viscous, this shear stress is modest, about 2,400 times less than atmospheric pressure. Viscous stresses in gases and thin (watery) liquids are even smaller.

Variation of Viscosity with Temperature

Temperature has a strong effect and pressure a moderate effect on viscosity. The viscosity of gases and most liquids increases slowly with pressure. Water is anomalous in showing a very slight decrease below 30°C. Since the change in viscosity is only a few percent up to 100 atm, we shall neglect pressure effects in this book.

Gas viscosity increases with temperature. Two common approximations are the power law and the Sutherland law:

$$\frac{\mu}{\mu_0} \approx \begin{cases} \left(\dfrac{T}{T_0}\right)^n & \text{power law} \\ \dfrac{(T/T_0)^{3/2}(T_0 + S)}{T + S} & \text{Sutherland law} \end{cases} \qquad (1.27)$$

where μ_0 is a known viscosity at a known absolute temperature T_0 (usually 273 K). The constants n and S are fit to the data, and both formulas are adequate over a wide range of temperatures. For air, $n \approx 0.7$ and $S \approx 110$ K $= 199°$R. Other values are given in Ref. 26.

Liquid viscosity decreases with temperature and is roughly exponential, $\mu \approx ae^{-bT}$; but a better fit is the empirical result that ln μ is quadratic in $1/T$, where T is absolute temperature:

$$\ln \frac{\mu}{\mu_0} \approx a + b\left(\frac{T_0}{T}\right) + c\left(\frac{T_0}{T}\right)^2 \tag{1.28}$$

For water, with $T_0 = 273.16$ K, $\mu_0 = 0.001792$ kg/(m · s), suggested values are $a = -1.94$, $b = -4.80$, and $c = 6.74$, with accuracy about ± 1 percent. The viscosity of water is tabulated in Table A.1. For further viscosity data, see Refs. 21, 28, and 29.

Nonnewtonian Fluids

Fluids that do not follow the linear law of Eq. (1.23) are called *nonnewtonian* and are treated in books on *rheology* [16]. Figure 1.8*a* compares some examples to a newtonian fluid. For the nonlinear curves, the slope at any point is called the *apparent viscosity*.

Dilatant. This fluid is *shear-thickening*, increasing its resistance with increasing strain rate. Examples are suspensions of corn starch or sand in water. The classic case is *quicksand*, which stiffens up if one thrashes about.

Pseudoplastic. A *shear-thinning* fluid is less resistant at higher strain rates. A very strong thinning is called *plastic*. Some of the many examples are polymer solutions, colloidal suspensions, paper pulp in water, latex paint, blood plasma, syrup, and molasses. The classic case is *paint*, which is thick when poured but thin when brushed at a high strain rate.

Bingham plastic. The limiting case of a plastic substance is one that requires a finite yield stress before it begins to flow. Figure 1.8*a* shows yielding followed by linear

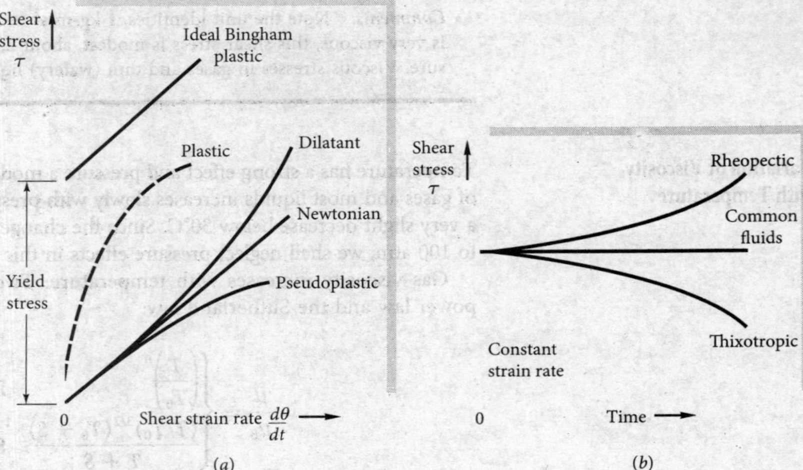

Fig. 1.8 Rheological behavior of various viscous materials: (*a*) stress versus strain rate; (*b*) effect of time on applied stress.

behavior, but nonlinear flow can also occur. Some examples are clay suspensions, drilling mud, toothpaste, mayonnaise, chocolate, and mustard. The classic case is *catsup,* which will not come out of the bottle until you stress it by shaking.

A further complication of nonnewtonian behavior is the transient effect shown in Fig. 1.8*b*. Some fluids require a gradually increasing shear stress to maintain a constant strain rate and are called *rheopectic.* The opposite case of a fluid that thins out with time and requires decreasing stress is termed *thixotropic.* We neglect nonnewtonian effects in this book; see Ref. 16 for further study.

Rheometers

There are many commercial devices for measuring the shear stress versus strain rate behavior of both newtonian and nonnewtonian fluids. They are generically called *rheometers* and have various designs: parallel disks, cone-plate, rotating coaxial cylinders, torsion, extensional, and capillary tubes. Reference 29 gives a good discussion. A popular device is the parallel-disk rheometer, shown in Fig. 1.9. A thin layer of fluid is placed between the disks, one of which rotates. The resisting torque on the rotating disk is proportional to the viscosity of the fluid. A simplified theory for this device is given in Example 1.10.

Surface Tension

Fig. 1.9 A rotating parallel-disk rheometer (*Image of Kinexus rheometer, used with kind permission of Malvern Instruments*).

A liquid, being unable to expand freely, will form an *interface* with a second liquid or gas. The physical chemistry of such interfacial surfaces is quite complex, and whole textbooks are devoted to this specialty [30]. Molecules deep within the liquid repel each other because of their close packing. Molecules at the surface are less dense and attract each other. Since half of their neighbors are missing, the mechanical effect is that the surface is in tension. We can account adequately for surface effects in fluid mechanics with the concept of surface tension.

If a cut of length dL is made in an interfacial surface, equal and opposite forces of magnitude $\Upsilon\ dL$ are exposed normal to the cut and parallel to the surface, where Υ is called the *coefficient of surface tension.* The dimensions of Υ are $\{F/L\}$, with SI units of newtons per meter and BG units of pounds-force per foot. An alternate concept is to open up the cut to an area dA; this requires work to be done of amount $\Upsilon\ dA$. Thus the coefficient Υ can also be regarded as the surface energy per unit area of the interface, in $N \cdot m/m^2$ or $ft \cdot lbf/ft^2$.

The two most common interfaces are water–air and mercury–air. For a clean surface at 20°C = 68°F, the measured surface tension is

$$\gamma = \begin{cases} 0.073 \text{ N/m} = 0.0050 \text{ lbf/ft} & \text{air–water} \\ 0.48 \text{ N/m} = 0.033 \text{ lbf/ft} & \text{air–mercury} \end{cases} \tag{1.29}$$

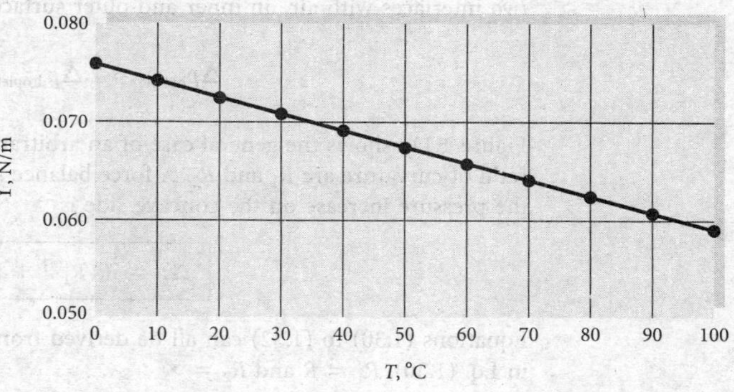

Fig. 1.10 Surface tension of a clean air–water interface. Data from Table A.5.

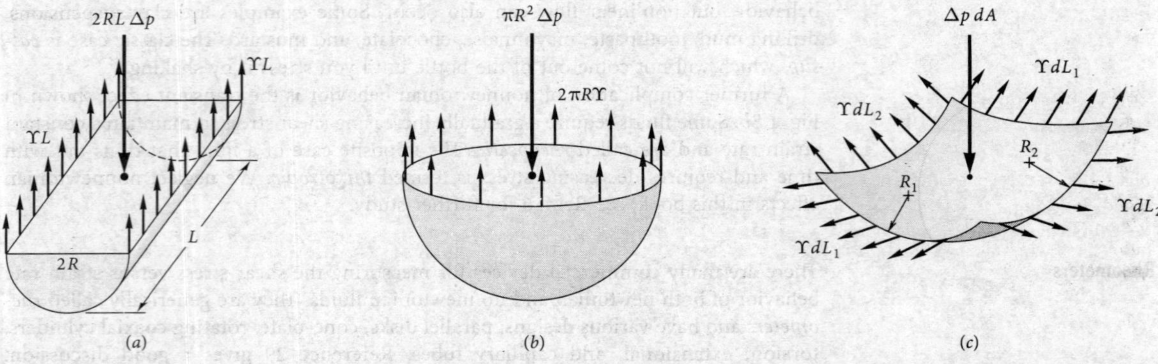

Fig. 1.11 Pressure change across a curved interface due to surface tension: (*a*) interior of a liquid cylinder; (*b*) interior of a spherical droplet; (*c*) general curved interface.

These are design values and can change considerably if the surface contains contaminants like detergents or slicks. Generally, Υ decreases with liquid temperature and is zero at the critical point. Values of Υ for water are given in Fig. 1.10 and Table A.5.

If the interface is curved, a mechanical balance shows that there is a pressure difference across the interface, the pressure being higher on the concave side, as illustrated in Fig. 1.11. In Fig. 1.11*a*, the pressure increase in the interior of a liquid cylinder is balanced by two surface-tension forces:

$$2RL\,\Delta p = 2\Upsilon L$$

or

$$\Delta p = \frac{\Upsilon}{R} \tag{1.30}$$

We are not considering the weight of the liquid in this calculation. In Fig. 1.11*b*, the pressure increase in the interior of a spherical droplet balances a ring of surface-tension force:

$$\pi R^2\,\Delta p = 2\pi R\Upsilon$$

or

$$\Delta p = \frac{2\Upsilon}{R} \tag{1.31}$$

We can use this result to predict the pressure increase inside a soap bubble, which has *two* interfaces with air, an inner and outer surface of nearly the same radius R:

$$\Delta p_{\text{bubble}} \approx 2\,\Delta p_{\text{droplet}} = \frac{4\Upsilon}{R} \tag{1.32}$$

Figure 1.11*c* shows the general case of an arbitrarily curved interface whose principal radii of curvature are R_1 and R_2. A force balance normal to the surface will show that the pressure increase on the concave side is

$$\boxed{\Delta p = \Upsilon(R_1^{-1} + R_2^{-1})} \tag{1.33}$$

Equations (1.30) to (1.32) can all be derived from this general relation; for example, in Eq. (1.30), $R_1 = R$ and $R_2 = \infty$.

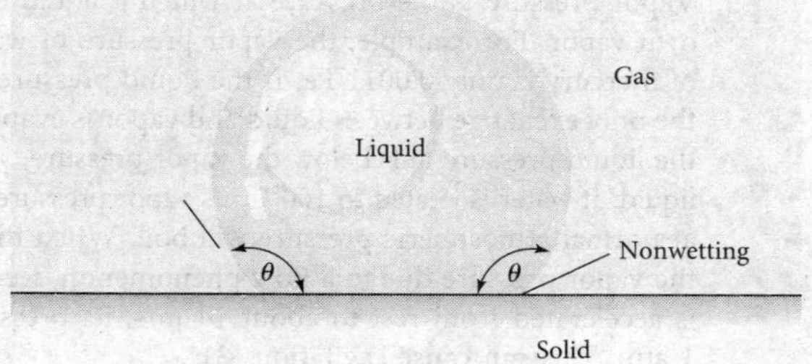

Fig. 1.12 Contact-angle effects at liquid–gas–solid interface. If $\theta < 90°$, the liquid "wets" the solid; if $\theta > 90°$, the liquid is nonwetting.

A second important surface effect is the *contact angle* θ, which appears when a liquid interface intersects with a solid surface, as in Fig. 1.12. The force balance would then involve both Υ and θ. If the contact angle is less than 90°, the liquid is said to *wet* the solid; if $\theta > 90°$, the liquid is termed *nonwetting*. For example, water wets soap but does not wet wax. Water is extremely wetting to a clean glass surface, with $\theta \approx 0°$. Like Υ, the contact angle θ is sensitive to the actual physicochemical conditions of the solid–liquid interface. For a clean mercury–air–glass interface, $\theta = 130°$.

Example 1.8 illustrates how surface tension causes a fluid interface to rise or fall in a capillary tube.

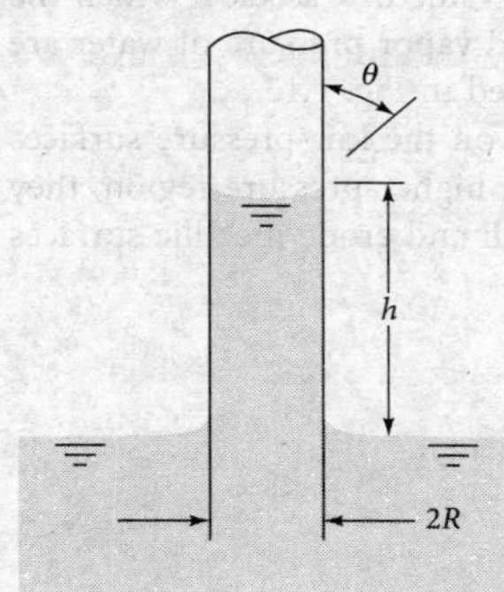

E1.8

EXAMPLE 1.8

Derive an expression for the change in height h in a circular tube of a liquid with surface tension Υ and contact angle θ, as in Fig. E1.8.

Solution

The vertical component of the ring surface-tension force at the interface in the tube must balance the weight of the column of fluid of height h:

$$2\pi R\Upsilon \cos\theta = \gamma\pi R^2 h$$

Solving for h, we have the desired result:

$$h = \frac{2\Upsilon\cos\theta}{\gamma R} \qquad \textit{Ans.}$$

Thus, the capillary height increases inversely with tube radius R and is positive if $\theta < 90°$ (wetting liquid) and negative (capillary depression) if $\theta > 90°$.

Suppose that $R = 1$ mm. Then, the capillary rise for a water–air–glass interface, $\theta \approx 0°$, $\Upsilon = 0.073$ N/m, and $\rho = 1{,}000$ kg/m^3 is

$$h = \frac{2(0.073\ \text{N/m})(\cos 0°)}{(1{,}000\ \text{kg/m}^3)(9.81\ \text{m/s}^2)(0.001\ \text{m})} = 0.015\ (\text{N}\cdot\text{s}^2)/\text{kg} = 0.015\ \text{m} = 1.5\ \text{cm}$$

For a mercury–air–glass interface, with $\theta = 130°$, $\Upsilon = 0.48$ N/m, and $\rho = 13{,}600$ kg/m^3, the capillary rise is

$$h = \frac{2(0.48)(\cos 130°)}{13{,}600(9.81)(0.001)} = -0.0046\ \text{m} = -0.46\ \text{cm}$$

When a small-diameter tube is used to make pressure measurements (Chap. 2), these capillary effects must be corrected for.

Vapor Pressure

Vapor pressure is the pressure at which a liquid boils and is in equilibrium with its own vapor. For example, the vapor pressure of water at 20°C is 2,340 Pa, while that of mercury is only 0.0011 Pa. If the liquid pressure is greater than the vapor pressure, the only exchange between liquid and vapor is evaporation at the interface. If, however, the liquid pressure falls below the vapor pressure, vapor bubbles begin to appear in the liquid. If water is heated to 100°C, its vapor pressure rises to 101,300 Pa, and thus water at normal atmospheric pressure will boil. When the liquid pressure is dropped below the vapor pressure due to a flow phenomenon, we call the process *cavitation*. If water is accelerated from rest to about 14 m/s, its pressure drops by about 100,000 Pa, or 1 atm. This can cause cavitation [31].

The dimensionless parameter describing flow-induced boiling is the *cavitation number*

$$\mathrm{Ca} = \frac{p_a - p_v}{\frac{1}{2}\rho V^2} \tag{1.34}$$

where p_a = ambient pressure
p_v = vapor pressure
V = characteristic flow velocity
ρ = fluid density

Depending on the geometry, a given flow has a critical value of Ca below which the flow will begin to cavitate. Values of surface tension and vapor pressure of water are given in Table A.5. The vapor pressure of water is plotted in Fig. 1.13.

Figure 1.14*a* shows cavitation bubbles being formed on the low-pressure surfaces of a marine propeller. When these bubbles move into a higher-pressure region, they collapse implosively. Cavitation collapse can rapidly spall and erode metallic surfaces and eventually destroy them, as shown in Fig. 1.14*b*.

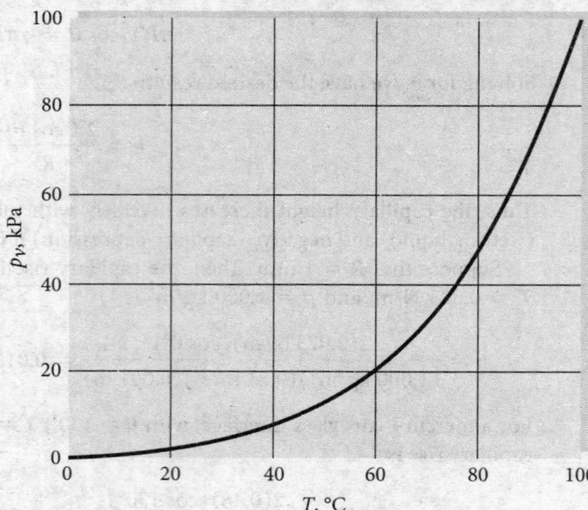

Fig. 1.13 Vapor pressure of water. Data from Table A.5.

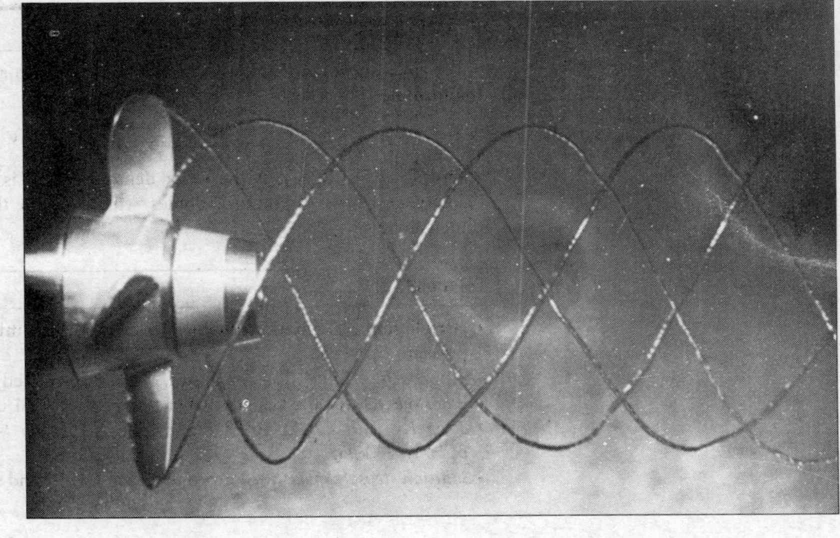

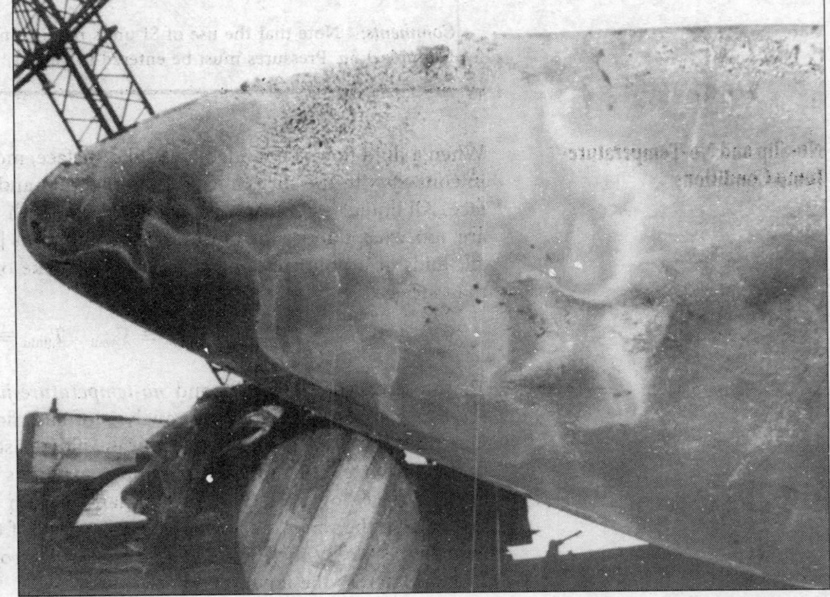

Fig. 1.14 Two aspects of cavitation bubble formation in liquid flows: (*a*) Beauty: spiral bubble sheets form from the surface of a marine propeller *(courtesy of the Garfield Thomas Water Tunnel, Pennsylvania State University)*; (*b*) ugliness: collapsing bubbles erode a propeller surface *(courtesy of Thomas T. Huang, David Taylor Research Center)*.

EXAMPLE 1.9

A certain torpedo, moving in fresh water at 10°C, has a minimum-pressure point given by the formula

$$p_{min} = p_0 - 0.35\,\rho V^2 \quad (1)$$

where p_0 = 115 kPa, ρ is the water density, and V is the torpedo velocity. Estimate the velocity at which cavitation bubbles will form on the torpedo. The constant 0.35 is dimensionless.

Solution

- *Assumption:* Cavitation bubbles form when the minimum pressure equals the vapor pressure p_v.
- *Approach:* Solve Eq. (1) above, which is related to the Bernoulli equation from Example 1.3, for the velocity when $p_{min} = p_v$. Use SI units (m, N, kg, s).
- *Property values:* At 10°C, read Table A.1 for ρ = 1,000 kg/m^3 and Table A.5 for p_v = 1.227 kPa.
- *Solution steps:* Insert the known data into Eq. (1) and solve for the velocity, using SI units:

$$p_{min} = p_v = 1{,}227 \text{ Pa} = 115{,}000 \text{ Pa} - 0.35\left(1{,}000\,\frac{\text{kg}}{\text{m}^3}\right)V^2, \text{ with } V \text{ in m/s}$$

$$\text{Solve } V^2 = \frac{(115{,}000 - 1227)}{0.35(1{,}000)} = 325\,\frac{\text{m}^2}{\text{s}^2} \text{ or } V = \sqrt{325} \approx 18.0\text{m/s} \qquad \textit{Ans.}$$

- *Comments:* Note that the use of SI units requires no conversion factors, as discussed in Example 1.3*b*. Pressures must be entered in pascals, not kilopascals.

No-Slip and No-Temperature-Jump Conditions

When a fluid flow is bounded by a solid surface, molecular interactions cause the fluid in contact with the surface to seek momentum and energy equilibrium with that surface. All liquids essentially are in equilibrium with the surfaces they contact. All gases are, too, except under the most rarefied conditions [18]. Excluding rarefied gases, then, all fluids at a point of contact with a solid take on the velocity and temperature of that surface:

$$V_{fluid} \equiv V_{wall} \qquad T_{fluid} \equiv T_{wall} \quad (1.35)$$

These are called the *no-slip* and *no-temperature-jump conditions,* respectively. They serve as *boundary conditions* for analysis of fluid flow past a solid surface. Figure 1.15 illustrates the no-slip condition for water flow past the top and bottom surfaces of a fixed thin plate. The flow past the upper surface is disorderly, or turbulent, while the lower surface flow is smooth, or laminar.[6] In both cases there is clearly no slip at the wall, where the water takes on the zero velocity of the fixed plate. The velocity profile is made visible by the discharge of a line of hydrogen bubbles from the wire shown stretched across the flow.

To decrease the mathematical difficulty, the no-slip condition is partially relaxed in the analysis of inviscid flow (Chap. 8). The flow is allowed to "slip" past the surface, but not to permeate through the surface

$$V_{normal}(\text{fluid}) \equiv V_{normal}(\text{solid}) \quad (1.36)$$

[6]Laminar and turbulent flows are studied in Chaps. 6 and 7.

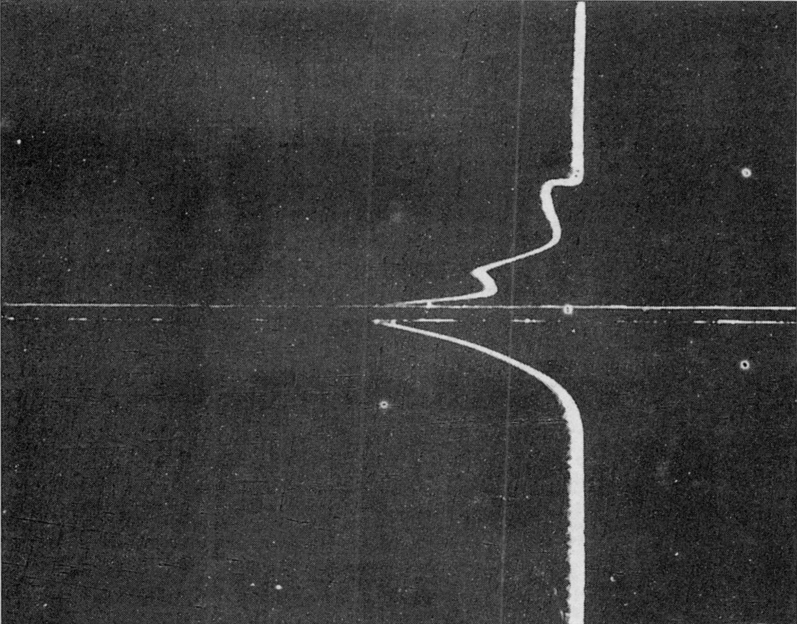

Fig. 1.15 The no-slip condition in water flow past a thin fixed plate. The upper flow is turbulent; the lower flow is laminar. The velocity profile is made visible by a line of hydrogen bubbles discharged from the wire across the flow. *(National Committee for Fluid Mechanics Films, Education Development Center, Inc., © 1972.)*

while the tangential velocity V_t is allowed to be independent of the wall. The analysis is much simpler, but the flow patterns are highly idealized.

For high-viscosity newtonian fluids, the linear velocity assumption and the no-slip conditions can yield some sophisticated approximate analyses for two- and three-dimensional viscous flows. The following example, for a type of rotating-disk viscometer, will illustrate.

EXAMPLE 1.10

An oil film of viscosity μ and thickness $h \ll R$ lies between a solid wall and a circular disk, as in Fig. E1.10. The disk is rotated steadily at angular velocity Ω. Noting that both velocity and shear stress vary with radius r, derive a formula for the torque M required to rotate the disk. Neglect air drag.

Solution

- *System sketch:* Figure E1.10 shows a side view (a) and a top view (b) of the system.
- *Assumptions:* Linear velocity profile, laminar flow, no-slip, local shear stress given by Eq. (1.23).
- *Approach:* Estimate the shear stress on a circular strip of width dr and area $dA = 2\pi r\, dr$ in Fig. E1.10b, then find the moment dM about the origin caused by this shear stress. Integrate over the entire disk to find the total moment M.
- *Property values:* Constant oil viscosity μ. In this steady flow, oil density is not relevant.
- *Solution steps:* At radius r, the velocity in the oil is tangential, varying from zero at the fixed wall (no-slip) to $u = \Omega r$ at the disk surface (also no-slip). The shear stress at this position is thus

$$\tau = \mu \frac{du}{dy} \approx \mu \frac{\Omega r}{h}$$

E1.10

This shear stress is everywhere perpendicular to the radius from the origin (see Fig. E1.10*b*). Then the total moment about the disk origin, caused by shearing this circular strip, can be found and integrated:

$$dM = (\tau)(dA)r = \left(\frac{\mu \Omega r}{h}\right)(2\pi r\, dr)r, \ M = \int dM = \frac{2\pi\mu\Omega}{h}\int_0^R r^3 dr = \frac{\pi\mu\Omega R^4}{2h} \quad \textit{Ans.}$$

- *Comments:* This is a simplified engineering analysis, which neglects possible edge effects, air drag on the top of the disk, and the turbulence that might ensue if the disk rotates too fast.

Slip Flow in Gases

The "free slip" boundary condition, Eq. (1.36), is an unrealistic mathematical artifice to enable inviscid-flow solutions. However, actual, realistic wall slip occurs in rarefied gases, where there are too few molecules to establish momentum equilibrium with the wall. In 1879, the physicist James Clerk Maxwell used the kinetic theory of gases to predict a *slip velocity* at the wall:

$$\delta u_{\text{wall}} \approx \ell \frac{\partial u}{\partial y}\Big|_{\text{wall}} \tag{1.37}$$

where ℓ is the mean free path of the gas, and u and x are along the wall. If ℓ is very small compared to the lateral scale L of the flow, the *Knudsen number*, $\text{Kn} = \ell/L$, is small, and the slip velocity is near zero. We will assign a few slip problems, but the details of rarefied gas flow are left for further reading in Refs. 18 and 52.

Speed of Sound

In gas flow, one must be aware of *compressibility* effects (significant density changes caused by the flow). We shall see in Sec. 4.2 and in Chap. 9 that compressibility becomes important when the flow velocity reaches a significant fraction of the speed of sound of the fluid. The *speed of sound a* of a fluid is the rate of propagation of small-disturbance pressure pulses ("sound waves") through the fluid. In Chap. 9, we shall show, from momentum and thermodynamic arguments, that the speed of sound is defined by a pressure-density derivative proportional to the *isentropic bulk modulus*:

$$a^2 = \frac{\beta}{\rho} = \left(\frac{\partial p}{\partial \rho}\right)_s = k\left(\frac{\partial p}{\partial \rho}\right)_T, \ k = \frac{c_p}{c_v}$$

where β = isentropic bulk modulus $= \rho\left(\dfrac{\partial p}{\partial \rho}\right)_s$.

This is true for either a liquid or a gas, but it is for *gases* that the problem of compressibility occurs. For an ideal gas, Eq. (1.10), we obtain the simple formula

$$a_{\text{ideal gas}} = (kRT)^{1/2} \tag{1.38}$$

where R is the gas constant, Eq. (1.11), and T the absolute temperature. For example, for air at 20°C, $a = \{(1.40)[287\ \text{m}^2/(\text{s}^2 \cdot \text{K})](293\ \text{K})\}^{1/2} \approx 343$ m/s (1126 ft/s = 768 mi/h). If, in this case, the air velocity reaches a significant fraction of a, say, 100 m/s, then we must account for compressibility effects (Chap. 9). Another way to state this is to account for compressibility when the *Mach number* Ma = V/a of the flow reaches about 0.3.

The speed of sound of water is tabulated in Table A.5. For near perfect gases, like air, the speed of sound is simply calculated by Eq. (1.38). Many liquids have their bulk modulus listed in Table A.3. Note, however, as discussed in Ref. 51, even a very small amount of dissolved gas in a liquid can reduce the mixture speed of sound by up to 80 percent.

EXAMPLE 1.11

A commercial airplane flies at 870 km/h at a standard altitude of 9,000 m. What is its Mach number?

Solution

- *Approach:* Find the "standard" speed of sound; divide it into the velocity, using proper units.
- *Property values:* From Table A.6, at 9,000 m, $a \approx 304$ m/s. Check this against the standard temperature, estimated from the table to be 230 K. From Eq. (1.38) for air,

$$a = [kR_{\text{air}}T]^{1/2} = [1.4(287)(230)]^{1/2} \approx 304\ \text{m/s}.$$

- *Solution steps:* Convert the airplane velocity to m/s:

$$V = 870\ \text{km/h} \approx 242\ \text{m/s}.$$

Then, the Mach number is given by

$$\text{Ma} = V/a = (242\ \text{m/s})/(304\ \text{m/s}) = 0.80 \qquad \textit{Ans.}$$

- *Comments:* This value, Ma = 0.80, is typical of present-day commercial airliners.

1.8 Basic Flow Analysis Techniques

There are three basic ways to attack a fluid flow problem. They are equally important for a student learning the subject, and this book tries to give adequate coverage to each method:

1. Control-volume, or *integral* analysis (Chap. 3).
2. Infinitesimal system, or *differential* analysis (Chap. 4).
3. Experimental study, or *dimensional* analysis (Chap. 5).

In all cases, the flow must satisfy the three basic laws of mechanics plus a thermodynamic state relation and associated boundary conditions:

1. Conservation of mass (continuity).
2. Linear momentum (Newton's second law).

3. First law of thermodynamics (conservation of energy).
4. A state relation like $\rho = \rho(p, T)$.
5. Appropriate boundary conditions at solid surfaces, interfaces, inlets, and exits.

In integral and differential analyses, these five relations are modeled mathematically and solved by computational methods. In an experimental study, the fluid itself performs this task without the use of any mathematics. In other words, these laws are believed to be fundamental to physics, and no fluid flow is known to violate them.

1.9 Flow Patterns: Streamlines, Streaklines, and Pathlines

Fluid mechanics is a highly visual subject. The patterns of flow can be visualized in a dozen different ways, and you can view these sketches or photographs and learn a great deal qualitatively and often quantitatively about the flow.

Four basic types of line patterns are used to visualize flows:

1. A *streamline* is a line everywhere tangent to the velocity vector at a given instant.
2. A *pathline* is the actual path traversed by a given fluid particle.
3. A *streakline* is the locus of particles that have earlier passed through a prescribed point. [33]
4. A *timeline* is a set of fluid particles that form a line at a given instant.

The streamline is convenient to calculate mathematically, while the other three are easier to generate experimentally. Note that a streamline and a timeline are instantaneous lines, while the pathline and the streakline are generated by the passage of time. The velocity profile shown in Fig. 1.15 is really a timeline generated earlier by a single discharge of bubbles from the wire. A pathline can be found by a time exposure of a single marked particle moving through the flow. Streamlines are difficult to generate experimentally in unsteady flow unless one marks a great many particles and notes their direction of motion during a very short time interval [32]. In steady flow, where velocity varies only with position, the situation simplifies greatly:

Streamlines, pathlines, and streaklines are identical in steady flow.

In fluid mechanics, the most common mathematical result for visualization purposes is the streamline pattern. Figure 1.16*a* shows a typical set of streamlines, and Fig. 1.16*b* shows a closed pattern called a *streamtube*. By definition the fluid within a streamtube is confined there because it cannot cross the streamlines; thus the streamtube walls need not be solid but may be fluid surfaces.

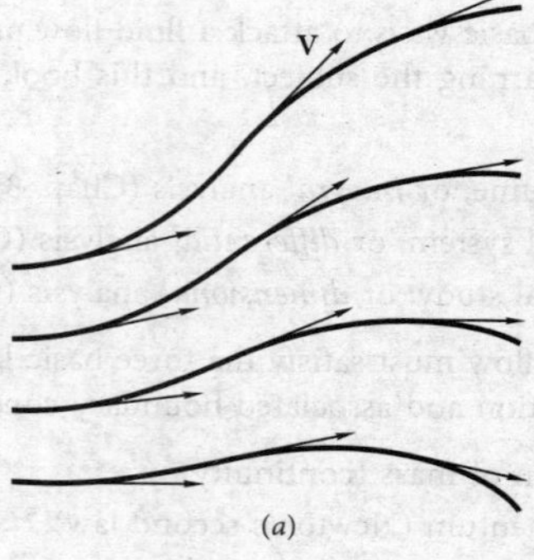

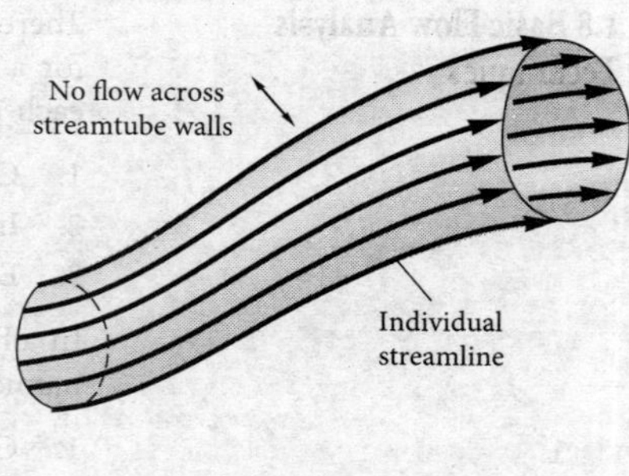

Fig. 1.16 The most common method of flow-pattern presentation: (*a*) Streamlines are everywhere tangent to the local velocity vector; (*b*) a streamtube is formed by a closed collection of streamlines.

Figure 1.17 shows an arbitrary velocity vector. If the elemental arc length dr of a streamline is to be parallel to $\mathbf{V}$, their respective components must be in proportion:

Streamline:
$$\frac{dx}{u} = \frac{dy}{v} = \frac{dz}{w} = \frac{dr}{V} \tag{1.39}$$

If the velocities (u, v, w) are known functions of position and time, Eq. (1.39) can be integrated to find the streamline passing through the initial point (x_0, y_0, z_0, t_0). The method is straightforward for steady flows, but may be laborious for unsteady flow.

Flow Visualization

Clever experimentation can produce revealing images of a fluid flow pattern, as shown earlier in Figs. 1.14*a* and 1.15. For example, streaklines are produced by the continuous release of marked particles (dye, smoke, or bubbles) from a given point. If the flow is steady, the streaklines will be identical to the streamlines and pathlines of the flow.

Some methods of flow visualization include the following [34–36]:

1. Dye, smoke, or bubble discharges.
2. Surface powder or flakes on liquid flows.
3. Floating or neutral-density particles.
4. Optical techniques that detect density changes in gas flows: shadowgraph, schlieren, and interferometer.
5. Tufts of yarn attached to boundary surfaces.
6. Evaporative coatings on boundary surfaces.
7. Luminescent fluids, additives, or bioluminescence.
8. Particle image velocimetry (PIV).

Figures 1.14*a* and 1.15 were both visualized by bubble releases. Another example is the use of particles in Fig. 1.18 to visualize a flow negotiating a 180° turn in a serpentine channel [42].

Figure 1.18*a* is at a low, laminar Reynolds number of 1,000. The flow is steady, and the particles form streaklines showing that the flow cannot make the sharp turn without separating away from the bottom wall.

Figure 1.18*b* is at a higher, turbulent Reynolds number of 30,000. The flow is unsteady, and the streaklines would be chaotic and smeared, unsuitable for visualization. The

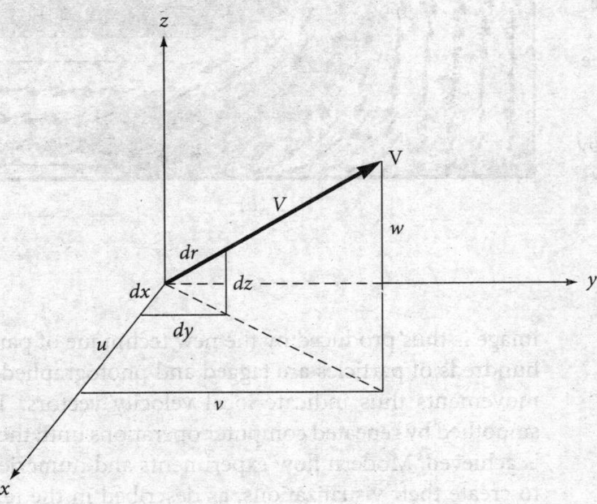

Fig. 1.17 Geometric relations for defining a streamline.

(a)

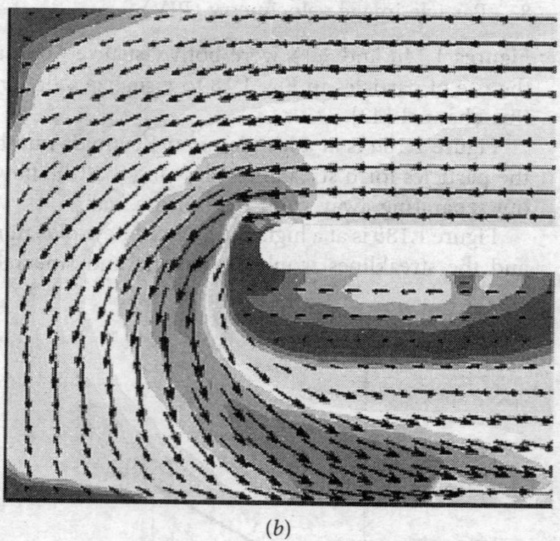

(b)

Fig. 1.18 Two visualizations of flow making a 180° turn in a serpentine channel: (*a*) particle streaklines at a Reynolds number of 1,000; (*b*) time-mean particle image velocimetry (PIV) at a turbulent Reynolds number of 30,000. *(From Ref. 42, by permission of the American Society of Mechanical Engineers.)*

image is thus produced by the new technique of particle image velocimetry [37]. In PIV, hundreds of particles are tagged and photographed at two closely spaced times. Particle movements thus indicate local velocity vectors. These hundreds of vectors are then smoothed by repeated computer operations until the time-mean flow pattern in Fig. 1.18*b* is achieved. Modern flow experiments and numerical models use computers extensively to create their visualizations, as described in the text by Yang [38].

Mathematical details of streamline/streakline/pathline analysis are given in Ref. 33. References 39–41 are beautiful albums of flow photographs. References 34–36 are monographs on flow visualization techniques.

Fluid mechanics is a marvelous subject for visualization, not just for still (steady) patterns, but also for moving (unsteady) motion studies. An outstanding list of available flow movies and videotapes is given by Carr and Young [43].

1.10 The Fundamentals of Engineering (FE) Examination

The road toward a professional engineer's license has a first stop, the Fundamentals of Engineering Examination, known as the FE exam. It was formerly known as the Engineer-in-Training (E-I-T) Examination. This eight-hour national test will probably soon be required of all engineering graduates, not just for licensure, but as a student assessment tool. The 120-problem, four-hour morning session covers many general studies:

Mathematics—15%	Ethics and business practices—7%	Material properties—7%
Engineering probability and statistics—7%	Engineering economics—8%	**Fluid mechanics—7%**
Chemistry—9%	Engineering mechanics—10%	Electricity and magnetism—9%
Computers—7%	Strength of materials—7%	Thermodynamics—7%

For the 60-problem, four-hour afternoon session you may choose one of seven modules: chemical, civil, electrical, environmental, industrial, mechanical, and other/general engineering. Note that fluid mechanics is an integral topic of the examination. Therefore, for practice, this text includes a number of end-of-chapter FE problems where appropriate.

The format for the FE exam questions is multiple-choice, usually with five selections, chosen carefully to tempt you with plausible answers if you used incorrect units, forgot to double or halve something, are missing a factor of π, or the like. In some cases, the selections are unintentionally ambiguous, such as the following example from a previous exam:

Transition from laminar to turbulent flow occurs at a Reynolds number of
(A) 900 (B) 1,200 (C) 1,500 (D) 2,100 (E) 3,000

The "correct" answer was graded as (D), Re = 2,100. Clearly the examiner was thinking, but forgot to specify, Re_d for *flow in a smooth circular pipe*, since (see Chaps. 6 and 7) transition is highly dependent on geometry, surface roughness, and the length scale used in the definition of Re. The moral is not to get peevish about the exam, but simply to go with the flow (pun intended) and decide which answer best fits an undergraduate training situation. Every effort has been made to keep the FE exam questions in this text unambiguous.

1.11 The History of Fluid Mechanics

Many distinguished workers have contributed to the development of fluid mechanics. If you are a student, however, this is probably not the time to be studying history. Later, during your career, you will enjoy reading about the history of, not just fluid mechanics, but all of science. Here are some names that will be mentioned as we encounter their contributions in the rest of this book.

Name	Important Contribution
Archimedes (285–212 BC)	Established laws of buoyancy and floating bodies.
Leonardo da Vinci (1452–1519)	Formulated the first equation of continuity.
Isaac Newton (1642–1727)	Postulated the law of linear viscous stresses.
Leonhard Euler (1707–1783)	Developed Bernoulli's equation by solving the basic equations.
L. M. H. Navier (1785–1836)	Formulated the basic differential equations of viscous flow.
Jean Louis Poiseuille (1799–1869)	Performed first experiments on laminar flow in tubes.
Osborne Reynolds (1842–1912)	Explained the phenomenon of transition to turbulence.
Ludwig Prandtl (1875–1953)	Formulated boundary layer theory, predicting flow separation.
Theodore von Kármán (1881–1963)	Major advances in aerodynamics and turbulence theory.

References [12] through [15] provide a comprehensive treatment of the history of fluid mechanics.

Summary

This chapter has discussed the behavior of a fluid—which, unlike a solid, must move if subjected to a shear stress—and the important fluid properties. The writer believes the most important property to be the velocity vector field $\mathbf{V}(x, y, z, t)$. Following closely are the pressure p, density ρ, and temperature T. Many secondary properties enter into various flow problems: viscosity μ, thermal conductivity k, specific weight γ, surface tension Υ, speed of sound a, and vapor pressure p_v. You must learn to locate and use all these properties to become proficient in fluid mechanics.

There was a brief discussion of the five different kinds of mathematical relations we will use to solve flow problems—mass conservation, linear momentum, first law of thermodynamics, equations of state, and appropriate boundary conditions at walls and other boundaries.

Flow patterns are also discussed briefly. The most popular, and useful, scheme is to plot the field of streamlines, that is, lines everywhere parallel to the local velocity vector.

Since the earth is 75 percent covered with water and 100 percent covered with air, the scope of fluid mechanics is vast and touches nearly every human endeavor. The sciences of meteorology, physical oceanography, and hydrology are concerned with naturally occurring fluid flows, as are medical studies of breathing and blood circulation. All transportation problems involve fluid motion, with well-developed specialties in aerodynamics of aircraft and rockets and in naval hydrodynamics of ships and submarines. Almost all our electric energy is developed either from water flow, air flow through wind turbines, or from steam flow through turbine generators. All combustion problems involve fluid motion as do the more classic problems of irrigation, flood control, water supply, sewage disposal, projectile motion, and oil and gas pipelines. The aim of this book is to present enough fundamental concepts and practical applications in fluid mechanics to prepare you to move smoothly into any of these specialized fields of the science of flow—and then be prepared to move out again as new technologies develop.

Problems

Most of the problems herein are fairly straightforward. More difficult or open-ended assignments are labeled with an asterisk as in Prob. 1.18. Problems labeled with a computer icon may require the use of a computer. The standard end-of-chapter problems 1.1 to 1.86 (categorized in the problem list below) are followed by fundamentals of engineering (FE) exam problems FE1.1 to FE1.10 and comprehensive problems C1.1 to C1.12.

Problem Distribution

Section	Topic	Problems
1.1, 1.4, 1.5	Fluid continuum concept	1.1–1.4
1.6	Dimensions and units	1.5–1.23
1.8	Thermodynamic properties	1.24–1.37
1.9	Viscosity, no-slip condition	1.38–1.61
1.9	Surface tension	1.62–1.71
1.9	Vapor pressure; cavitation	1.72–1.74
1.9	Speed of sound, Mach number	1.75–1.80
1.11	Streamlines	1.81–1.83
1.2	History of fluid mechanics	1.84–1.85*a–n*
1.13	Experimental uncertainty	1.86–1.90

The concept of a fluid

P1.1 A gas at 20°C may be considered *rarefied*, deviating from the continuum concept, when it contains less than 10^{12} molecules per cubic millimeter. If Avogadro's number is 6.023 E23 molecules per mole, what absolute pressure (in Pa) for air does this represent?

P1.2 Table A.6 lists the density of the standard atmosphere as a function of altitude. Use these values to estimate, crudely—say, within a factor of 2—the number of molecules of air in the entire atmosphere of the earth.

P1.3 For the triangular element in Fig. P1.3, show that a *tilted* free liquid surface, in contact with an atmosphere at pressure p_a, must undergo shear stress and hence begin to flow. *Hint:* Account for the weight of the fluid and show that a no-shear condition will cause horizontal forces to be out of balance.

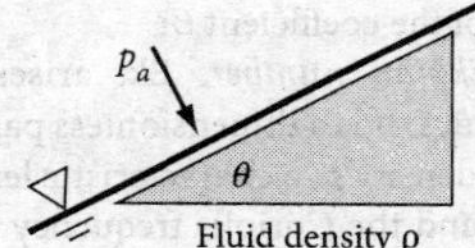

P1.3

P1.4 Sand, and other granular materials, appear to *flow;* that is, you can pour them from a container or a hopper. There are whole textbooks on the "transport" of granular materials [54]. Therefore, is sand a *fluid*? Explain.

Dimensions and units

P1.5 The *mean free path* of a gas, *l*, is defined as the average distance traveled by molecules between collisions. A proposed formula for estimating *l* of an ideal gas is

$$l = 1.26 \frac{\mu}{\rho\sqrt{RT}}$$

What are the dimensions of the constant 1.26? Use the formula to estimate the mean free path of air at 20°C and 7 kPa. Would you consider air rarefied at this condition?

P1.6 Henri Darcy, a French engineer, proposed that the pressure drop Δp for flow at velocity *V* through a tube of length *L* could be correlated in the form

$$\frac{\Delta p}{\rho} = \alpha L V^2$$

If Darcy's formulation is consistent, what are the dimensions of the coefficient α?

P1.7 Convert the following inappropriate quantities into SI units: (*a*) 2.283 E7 U.S. gallons per day; (*b*) 4.5 furlongs per minute (racehorse speed); and (*c*) 72,800 avoirdupois ounces per acre.

P1.8 Suppose we know little about the strength of materials but are told that the bending stress σ in a beam is *proportional* to the beam half-thickness *y* and also depends on the bending moment *M* and the beam area moment of inertia *I*. We also learn that, for the particular case $M = 2{,}900$ in · lbf, $y = 1.5$ in, and $I = 0.4$ in^4, the predicted stress is 75 MPa. Using this information and dimensional reasoning only, find, to three significant figures, the only possible dimensionally homogeneous formula $\sigma = y\,f(M, I)$.

P1.9 A hemispherical container, 26 inches in diameter, is filled with a liquid at 20°C and weighed. The liquid weight is found to be 1,617 ounces. (*a*) What is the density of the fluid, in kg/m^3? (*b*) What fluid might this be? Assume standard gravity, $g = 9.807$ m/s^2.

P1.10 The Stokes-Oseen formula [33] for drag force *F* on a sphere of diameter *D* in a fluid stream of low velocity *V*, density ρ, and viscosity μ is

$$F = 3\pi\mu DV + \frac{9\pi}{16}\rho V^2 D^2$$

Is this formula dimensionally homogeneous?

P1.11 In English Engineering units, the specific heat c_p of air at room temperature is approximately 0.24 Btu/(lbm-°F). When working with kinetic energy relations, it is more appropriate to express c_p as a velocity-squared per absolute degree. Give the numerical value, in this form, of c_p for air in (*a*) SI units, and (*b*) BG units.

P1.12 For low-speed (laminar) steady flow through a circular pipe, as shown in Fig. P1.12, the velocity *u* varies with radius and takes the form

$$u = B\frac{\Delta p}{\mu}(r_0^2 - r^2)$$

where μ is the fluid viscosity and Δp is the pressure drop from entrance to exit. What are the dimensions of the constant *B*?

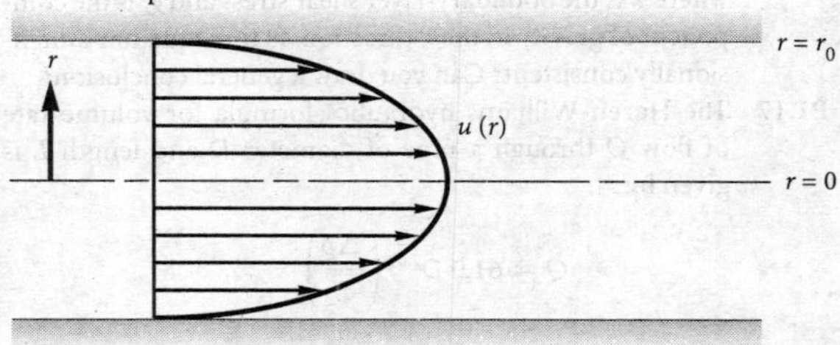

P1.12

P1.13 The efficiency η of a pump is defined as the (dimensionless) ratio of the power developed by the flow to the power required to drive the pump:

$$\eta = \frac{Q\Delta p}{\text{input power}}$$

where Q is the volume rate of flow and Δp is the pressure rise produced by the pump. Suppose that a certain pump develops a pressure rise of 35 lbf/in^2 when its flow rate is 40 L/s. If the input power is 16 hp, what is the efficiency?

***P1.14** Figure P1.14 shows the flow of water over a dam. The volume flow Q is known to depend only on crest width B, acceleration of gravity g, and upstream water height H above the dam crest. It is further known that Q is proportional to B. What is the form of the only possible dimensionally homogeneous relation for this flow rate?

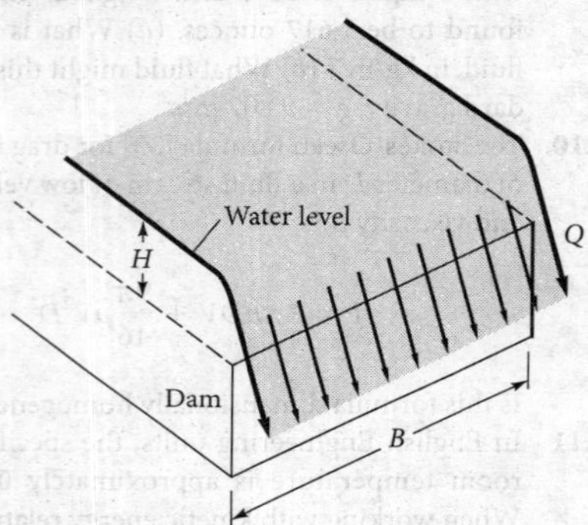

P1.14

P1.15 The height H that fluid rises in a liquid barometer tube depends upon the liquid density ρ, the barometric pressure p, and the acceleration of gravity g. (*a*) Arrange these four variables into a single dimensionless group. (*b*) Can you deduce (or guess) the numerical value of your group?

P1.16 Algebraic equations such as Bernoulli's relation, Eq. (1) of Example 1.3, are dimensionally consistent, but what about differential equations? Consider, for example, the boundary-layer x-momentum equation, first derived by Ludwig Prandtl in 1904:

$$\rho u \frac{\partial u}{\partial x} + \rho v \frac{\partial u}{\partial y} = -\frac{\partial p}{\partial x} + \rho g_x + \frac{\partial \tau}{\partial y}$$

where τ is the boundary-layer shear stress and g_x is the component of gravity in the x direction. Is this equation dimensionally consistent? Can you draw a general conclusion?

P1.17 The Hazen-Williams hydraulics formula for volume rate of flow Q through a pipe of diameter D and length L is given by

$$Q \approx 61.9\, D^{2.63} \left(\frac{\Delta p}{L}\right)^{0.54}$$

where Δp is the pressure drop required to drive the flow. What are the dimensions of the constant 61.9? Can this formula be used with confidence for various liquids and gases?

***P1.18** For small particles at low velocities, the first term in the Stokes-Oseen drag law, Prob. 1.10, is dominant; hence, $F \approx KV$, where K is a constant. Suppose a particle of mass m is constrained to move horizontally from the initial position $x = 0$ with initial velocity V_0. Show (*a*) that its velocity will decrease exponentially with time and (*b*) that it will stop after traveling a distance $x = mV_0/K$.

P1.19 In his study of the circular hydraulic jump formed by a faucet flowing into a sink, Watson [53] proposed a parameter combining volume flow rate Q, density ρ, and viscosity μ of the fluid, and depth h of the water in the sink. He claims that his grouping is dimensionless, with Q in the numerator. Can you verify this?

P1.20 Books on porous media and atomization claim that the viscosity μ and surface tension Υ of a fluid can be combined with a characteristic velocity U to form an important dimensionless parameter. (*a*) Verify that this is so. (*b*) Evaluate this parameter for water at 20°C and a velocity of 3.5 cm/s. *Note:* You get extra credit if you know the name of this parameter.

P1.21 Aeronautical engineers measure the pitching moment M_0 of a wing and then write it in the following form for use in other cases:

$$M_0 = \beta V^2 AC\rho$$

where V is the wing velocity, A the wing area, C the wing chord length, and ρ the air density. What are the dimensions of the coefficient β?

P1.22 The *Ekman number,* Ek, arises in geophysical fluid dynamics. It is a dimensionless parameter combining seawater density ρ, a characteristic length L, seawater viscosity μ, and the Coriolis frequency $\Omega \sin\varphi$, where Ω is the rotation rate of the earth and φ is the latitude angle. Determine the correct form of Ek if the viscosity is in the numerator.

P1.23 During World War II, Sir Geoffrey Taylor, a British fluid dynamicist, used dimensional analysis to estimate the energy released by an atomic bomb explosion. He assumed that the energy released E, was a function of blast wave radius R, air density ρ, and time t. Arrange these variables into a single dimensionless group, which we may term the *blast wave number*.

Thermodynamic properties

P1.24 Air, assumed to be an ideal gas with $k = 1.40$, flows isentropically through a nozzle. At section 1, conditions are sea level standard (see Table A.6). At section 2, the temperature is −50°C. Estimate (*a*) the pressure, and (*b*) the density of the air at section 2.

P1.25 On a summer day in Narragansett, Rhode Island, the air temperature is 74°F and the barometric pressure is 14.5 lbf/in^2. Estimate the air density in kg/m^3.

P1.26 When we in the United States say a car's tire is filled "to 32 lb," we mean that its internal pressure is 32 lbf/in^2 above the ambient atmosphere. If the tire is at sea level, has a volume of 3.0 ft^3, and is at 75°F, estimate the total weight of air, in lbf, inside the tire.

P1.27 For steam at a pressure of 45 atm, some values of temperature and specific volume are as follows, from Ref. 23:

T, °F	500	600	700	800	900
v, ft^3/lbm	0.7014	0.8464	0.9653	1.074	1.177

Find an average value of the predicted gas constant R in m^2/(s$^2 \cdot$ K). Does this data reasonably approximate an ideal gas? If not, explain.

P1.28 Wet atmospheric air at 100 percent relative humidity contains saturated water vapor and, by Dalton's law of partial pressures,

$$p_{\text{atm}} = p_{\text{dry air}} + p_{\text{water vapor}}$$

Suppose this wet atmosphere is at 40°C and 1 atm. Calculate the density of this 100 percent humid air, and compare it with the density of dry air at the same conditions.

P1.29 A compressed-air tank holds 5 ft^3 of air at 120 lbf/in^2 "gage," that is, above atmospheric pressure. Estimate the energy, in ft-lbf, required to compress this air from the atmosphere, assuming an ideal isothermal process.

P1.30 Repeat Prob. 1.29 if the tank is filled with compressed *water* instead of air. Why is the result thousands of times less than the result of 215,000 ft · lbf in Prob. 1.29?

P1.31 One cubic foot of argon gas at 10°C and 1 atm is compressed isentropically to a pressure of 600 kPa. (*a*) What will be its new pressure and temperature? (*b*) If it is allowed to cool at this new volume back to 10°C, what will be the final pressure?

P1.32 A blimp is approximated by a prolate spheroid 90 m long and 30 m in diameter. Estimate the weight of 20°C gas within the blimp for (*a*) helium at 1.1 atm and (*b*) air at 1.0 atm. What might the *difference* between these two values represent (see Chap. 2)?

P1.33 A tank contains 9 kg of CO_2 at 20°C and 2.0 MPa. Estimate the volume of the tank, in m^3.

P1.34 Consider steam at the following state near the saturation line: $(p_1, T_1) = (1.31 \text{ MPa}, 290°\text{C})$. Calculate and compare, for an ideal gas (Table A.4) and the steam tables (*a*) the density ρ_1 and (*b*) the density ρ_2 if the steam expands isentropically to a new pressure of 414 kPa. Discuss your results.

P1.35 In Table A.4, most common gases (air, nitrogen, oxygen, hydrogen) have a specific heat ratio $k \approx 1.40$. Why do argon and helium have such high values? Why does NH_3 have such a low value? What is the lowest k for any gas that you know of?

P1.36 Experimental data [55] for the density of n-pentane liquid for high pressures, at 50°C, are listed as follows:

Pressure, kPa	100	10,230	20,700	34,310
Density, kg/m^3	586.3	604.1	617.8	632.8

(*a*) Fit this data to reasonably accurate values of B and n from Eq. (1.19). (*b*) Evaluate ρ at 30 MPa.

P1.37 A near-ideal gas has a molecular weight of 44 and a specific heat $c_v = 610$ J/(kg · K). What are (*a*) its specific heat ratio, k, and (*b*) its speed of sound at 100°C?

Viscosity, no-slip condition

P1.38 In Fig. 1.7, if the fluid is glycerin at 20°C and the width between plates is 6 mm, what shear stress (in Pa) is required to move the upper plate at 5.5 m/s? What is the Reynolds number if L is taken to be the distance between plates?

P1.39 Knowing μ for air at 20°C from Table 1.4, estimate its viscosity at 500°C by (*a*) the power law and (*b*) the Sutherland law. Also make an estimate from (*c*) Fig. 1.6. Compare with the accepted value of $\mu \approx 3.58$ E-5 kg/m · s.

P1.40 Glycerin at 20°C fills the space between a hollow sleeve of diameter 12 cm and a fixed coaxial solid rod of diameter 11.8 cm. The outer sleeve is rotated at 120 rev/min. Assuming no temperature change, estimate the torque required, in N · m per meter of rod length, to hold the inner rod fixed.

P1.41 An aluminum cylinder weighing 30 N, 6 cm in diameter and 40 cm long, is falling concentrically through a long vertical sleeve of diameter 6.04 cm. The clearance is filled with SAE 50 oil at 20°C. Estimate the *terminal* (zero acceleration) fall velocity. Neglect air drag and assume a linear velocity distribution in the oil. *Hint:* You are given diameters, not radii.

P1.42 Helium at 20°C has a viscosity of 1.97 E-5 kg/(m · s). Use the data of Table A.4 to estimate the temperature, in °C, at which helium's viscosity will double.

P1.43 For the flow of gas between two parallel plates of Fig. 1.7, reanalyze for the case of *slip flow* at both walls. Use the simple slip condition, $\delta u_{\text{wall}} = \ell\,(du/dy)_{\text{wall}}$, where ℓ is the mean free path of the fluid. Sketch the expected velocity profile and find an expression for the shear stress at each wall.

P1.44 One type of viscometer is simply a long capillary tube. A commercial device is shown in Prob. C1.10. One measures the volume flow rate Q and the pressure drop Δp and, of course, the radius and length of the tube. The theoretical formula, which will be discussed in Chap. 6, is $\Delta p \approx 8\mu QL/(\pi R^4)$. For a capillary of diameter 4 mm and length 10 inches, the test fluid flows at 0.9 m^3/h when the pressure drop is 58 lbf/in^2. Find the predicted viscosity in kg/m · s.

P1.45 A block of weight W slides down an inclined plane while lubricated by a thin film of oil, as in Fig. P1.45. The film contact area is A and its thickness is h. Assuming a linear velocity distribution in the film, derive an expression for the "terminal" (zero-acceleration) velocity V of the block. Find the terminal velocity of the block if the block mass is 6 kg, $A = 35$ cm^2, $\theta = 15°$, and the film is 1-mm-thick SAE 30 oil at 20°C.

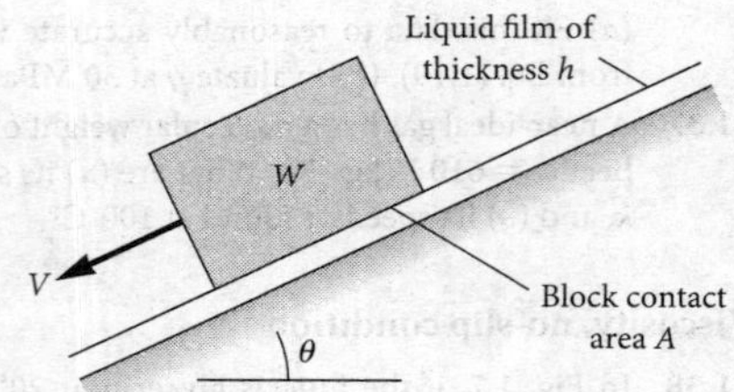

P1.45

P1.46 A simple and popular model for two nonnewtonian fluids in Fig. 1.8*a* is the *power-law*:

$$\tau \approx C\left(\frac{du}{dy}\right)^n$$

where C and n are constants fit to the fluid [16]. From Fig. 1.8*a*, deduce the values of the exponent n for which the fluid is (*a*) newtonian, (*b*) dilatant, and (*c*) pseudoplastic. Consider the specific model constant $C = 0.4\ \text{N}\cdot\text{s}^n/\text{m}^2$, with the fluid being sheared between two parallel plates as in Fig. 1.7. If the shear stress in the fluid is 1,200 Pa, find the velocity V of the upper plate for the cases (*d*) $n = 1.0$, (*e*) $n = 1.2$, and (*f*) $n = 0.8$.

P1.47 Data for the apparent viscosity of average human blood, at normal body temperature of 37°C, varies with shear strain rate, as shown in the following table.

Strain rate, s^{-1}	1	10	100	1,000
Apparent viscosity, kg/(m · s)	0.011	0.009	0.006	0.004

(*a*) Is blood a nonnewtonian fluid? (*b*) If so, what type of fluid is it? (*c*) How do these viscosities compare with plain water at 37°C?

P1.48 A thin plate is separated from two fixed plates by very viscous liquids μ_1 and μ_2, respectively, as in Fig. P1.48. The plate spacings h_1 and h_2 are unequal, as shown. The contact area is A between the center plate and each fluid. (*a*) Assuming a linear velocity distribution in each fluid, derive the force F required to pull the plate at velocity V. (*b*) Is there a necessary *relation* between the two viscosities, μ_1 and μ_2?

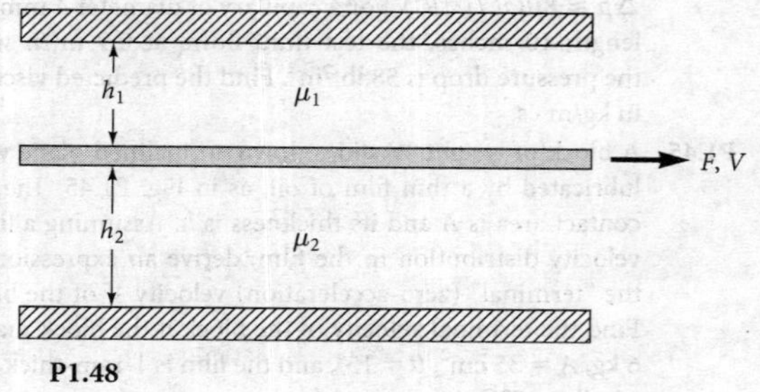

P1.48

P1.49 An amazing number of commercial and laboratory devices have been developed to measure fluid viscosity, as described in Refs. 29 and 49. Consider a concentric shaft, fixed axially and rotated inside the sleeve. Let the inner and outer cylinders have radii r_i and r_o, respectively, with total sleeve length L. Let the rotational rate be Ω (rad/s) and the applied torque be M. Using these parameters, derive a theoretical relation for the viscosity μ of the fluid between the cylinders.

P1.50 A simple viscometer measures the time t for a solid sphere to fall a distance L through a test fluid of density ρ. The fluid viscosity μ is then given by

$$\mu \approx \frac{W_{\text{net}}t}{3\pi DL} \quad \text{if} \quad t \geq \frac{2\rho DL}{\mu}$$

where D is the sphere diameter and W_{net} is the sphere net weight in the fluid. (*a*) Prove that both of these formulas are dimensionally homogeneous. (*b*) Suppose that a 2.5 mm diameter aluminum sphere (density 2,700 kg/m³) falls in an oil of density 875 kg/m³. If the time to fall 50 cm is 32 s, estimate the oil viscosity and verify that the inequality is valid.

P1.51 An approximation for the boundary-layer shape in Figs. 1.5*b* and P1.51 is the formula

$$u(y) \approx U\sin\left(\frac{\pi y}{2\delta}\right), \quad 0 \leq y \leq \delta$$

where U is the stream velocity far from the wall and δ is the boundary layer thickness, as in Fig. P1.51. If the fluid is helium at 20°C and 1 atm, and if $U = 10.8$ m/s and $\delta = 3$ cm, use the formula to (*a*) estimate the wall shear stress τ_w in Pa, and (*b*) find the position in the boundary layer where τ is one-half of τ_w.

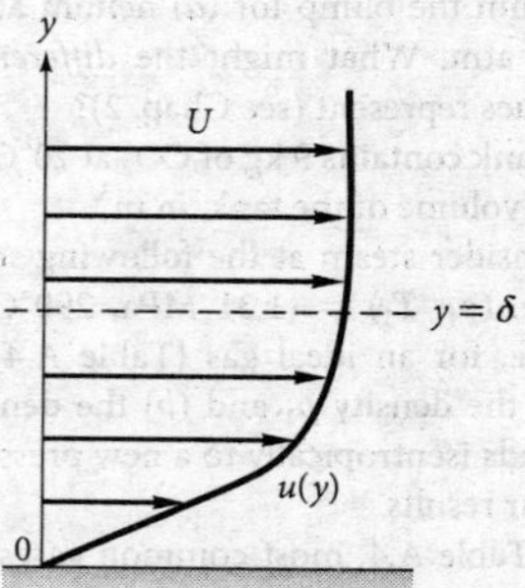

P1.51

P1.52 The belt in Fig. P1.52 moves at a steady velocity V and skims the top of a tank of oil of viscosity μ, as shown. Assuming a linear velocity profile in the oil, develop a simple formula for the required belt-drive power P as a function of (h, L, V, b, μ). What belt-drive power P, in watts, is required if the belt moves at 2.5 m/s over SAE 30W oil at 20°C, with $L = 2$ m, $b = 60$ cm, and $h = 3$ cm?

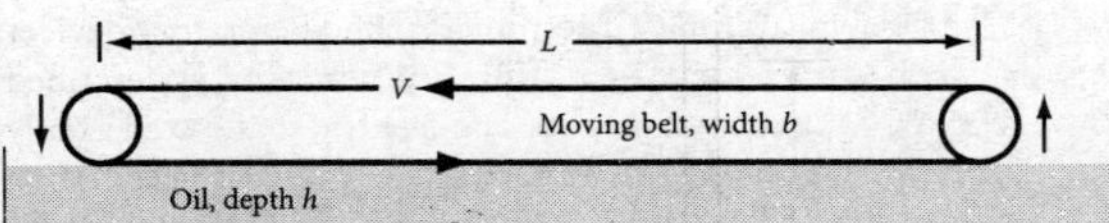

P1.52

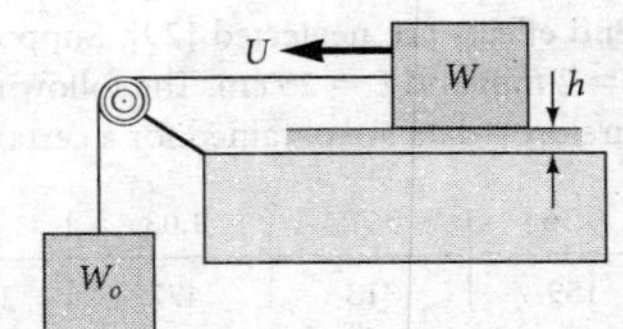

P1.55

*P1.53 A solid cone of angle 2θ, base r_0, and density ρ_c is rotating with initial angular velocity ω_0 inside a conical seat, as shown in Fig. P1.53. The clearance h is filled with oil of viscosity μ. Neglecting air drag, derive an analytical expression for the cone's angular velocity $\omega(t)$ if there is no applied torque.

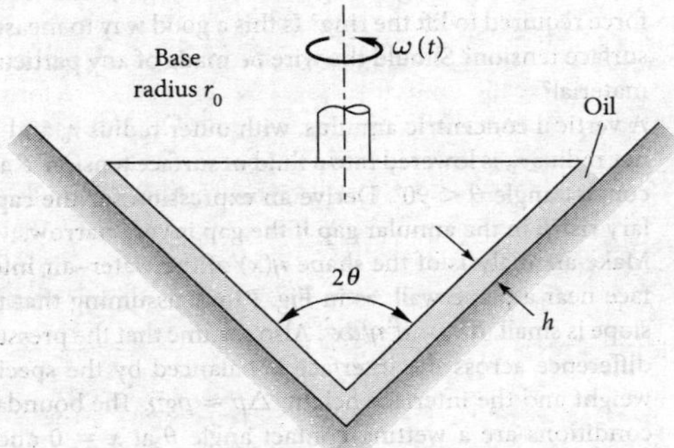

P1.53

*P1.54 A disk of radius R rotates at an angular velocity Ω inside a disk-shaped container filled with oil of viscosity μ, as shown in Fig. P1.54. Assuming a linear velocity profile and neglecting shear stress on the outer disk edges, derive a formula for the viscous torque on the disk.

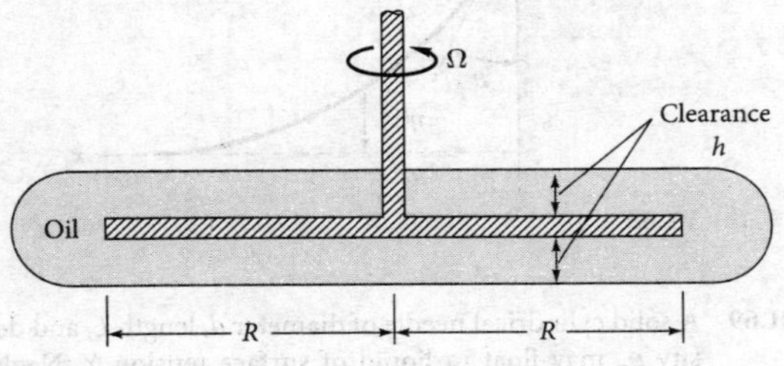

P1.54

P1.55 A block of weight W is being pulled over a table by another weight W_o, as shown in Fig. P1.55. Find an algebraic formula for the steady velocity U of the block if it slides on an oil film of thickness h and viscosity μ. The block bottom area A is in contact with the oil. Neglect the cord weight and the pulley friction. Assume a linear velocity profile in the oil film.

*P1.56 The device in Fig. P1.56 is called a *cone-plate viscometer* [29]. The angle of the cone is very small, so that $\sin\theta \approx \theta$, and the gap is filled with the test liquid. The torque M to rotate the cone at a rate Ω is measured. Assuming a linear velocity profile in the fluid film, derive an expression for fluid viscosity μ as a function of (M, R, Ω, θ).

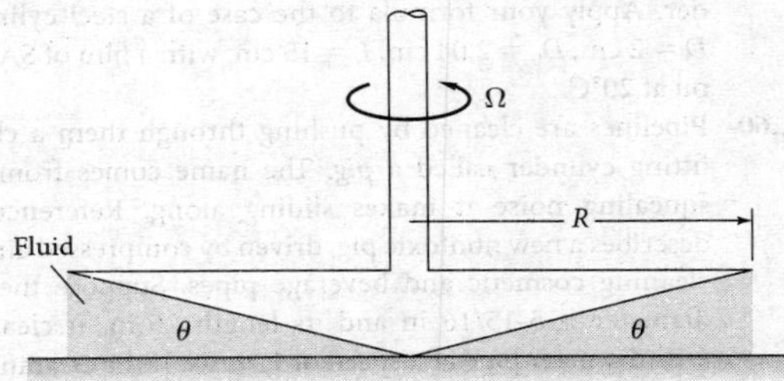

P1.56

P1.57 Extend the steady flow between a fixed lower plate and a moving upper plate, from Fig. 1.7, to the case of two immiscible liquids between the plates, as in Fig. P1.57.

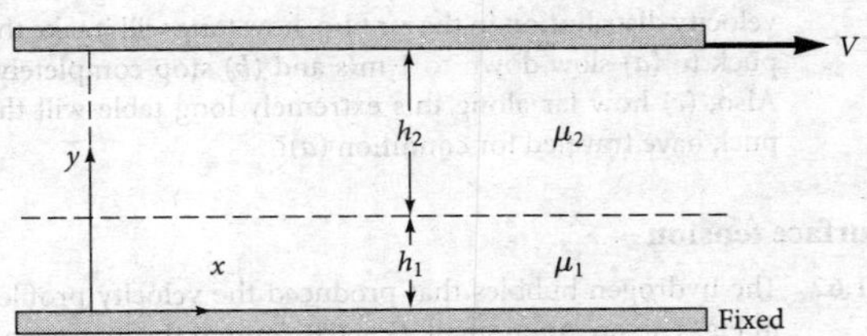

P1.57

(*a*) Sketch the expected no-slip velocity distribution $u(y)$ between the plates. (*b*) Find an analytic expression for the velocity U at the interface between the two liquid layers. (*c*) What is the result of (*b*) if the viscosities and layer thicknesses are equal?

*P1.58 The laminar pipe flow example of Prob. 1.12 can be used to design a *capillary viscometer* [29]. If Q is the volume flow rate, L is the pipe length, and Δp is the pressure drop from entrance to exit, the theory of Chap. 6 yields a formula for viscosity:

$$\mu = \frac{\pi r_0^4 \Delta p}{8LQ}$$

Pipe end effects are neglected [29]. Suppose our capillary has $r_0 = 2$ mm and $L = 25$ cm. The following flow rate and pressure drop data are obtained for a certain fluid:

Q, m³/h	0.36	0.72	1.08	1.44	1.80
Δp, kPa	159	318	477	1,274	1,851

What is the viscosity of the fluid? *Note:* Only the first three points give the proper viscosity. What is peculiar about the last two points, which were measured accurately?

P1.59 A solid cylinder of diameter D, length L, and density ρ_s falls due to gravity inside a tube of diameter D_0. The clearance, $D_0 - D << D$, is filled with fluid of density ρ and viscosity μ. Neglect the air above and below the cylinder. Derive a formula for the terminal fall velocity of the cylinder. Apply your formula to the case of a steel cylinder, $D = 2$ cm, $D_0 = 2.04$ cm, $L = 15$ cm, with a film of SAE 30 oil at 20°C.

P1.60 Pipelines are cleaned by pushing through them a close-fitting cylinder called a *pig*. The name comes from the squealing noise it makes sliding along. Reference 50 describes a new nontoxic pig, driven by compressed air, for cleaning cosmetic and beverage pipes. Suppose the pig diameter is 5-15/16 in and its length 26 in. It cleans a 6-in-diameter pipe at a speed of 1.2 m/s. If the clearance is filled with glycerin at 20°C, what pressure difference, in pascals, is needed to drive the pig? Assume a linear velocity profile in the oil and neglect air drag.

***P1.61** An air-hockey puck has a mass of 50 g and is 9 cm in diameter. When placed on the air table, a 20°C air film, of 0.12-mm thickness, forms under the puck. The puck is struck with an initial velocity of 10 m/s. Assuming a linear velocity distribution in the air film, how long will it take the puck to (*a*) slow down to 1 m/s and (*b*) stop completely? Also, (*c*) how far along this extremely long table will the puck have traveled for condition (*a*)?

Surface tension

P1.62 The hydrogen bubbles that produced the velocity profiles in Fig. 1.15 are quite small, $D \approx 0.01$ mm. If the hydrogen–water interface is comparable to air–water and the water temperature is 30°C, estimate the excess pressure within the bubble.

P1.63 Derive Eq. (1.33) by making a force balance on the fluid interface in Fig. 1.11*c*.

P1.64 Pressure in a water container can be measured by an open vertical tube—see Fig. P2.11 for a sketch. If the expected water rise is about 20 cm, what tube diameter is needed to ensure that the error due to capillarity will be less than 3 percent?

P1.65 The system in Fig. P1.65 is used to calculate the pressure p_1 in the tank by measuring the 15-cm height of liquid in the 1-mm-diameter tube. The fluid is at 60°C. Calculate the true fluid height in the tube and the percentage error due to capillarity if the fluid is (*a*) water or (*b*) mercury.

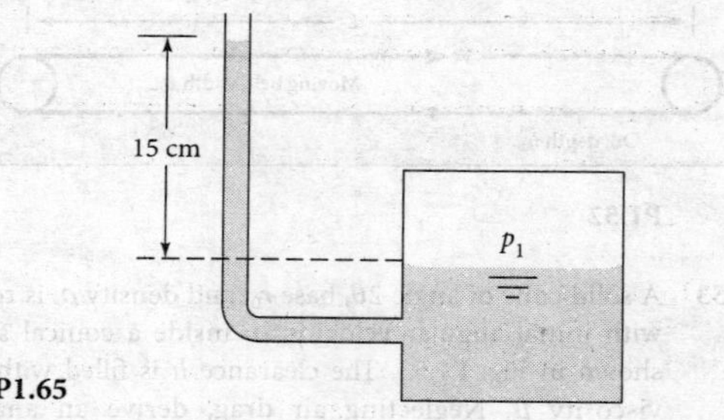

P1.65

P1.66 A thin wire ring, 3 cm in diameter, is lifted from a water surface at 20°C. Neglecting the wire weight, what is the force required to lift the ring? Is this a good way to measure surface tension? Should the wire be made of any particular material?

P1.67 A vertical concentric annulus, with outer radius r_o and inner radius r_i, is lowered into a fluid of surface tension Υ and contact angle $\theta < 90°$. Derive an expression for the capillary rise h in the annular gap if the gap is very narrow.

***P1.68** Make an analysis of the shape $\eta(x)$ of the water–air interface near a plane wall, as in Fig. P1.68, assuming that the slope is small, $R^{-1} \approx d^2\eta/dx^2$. Also assume that the pressure difference across the interface is balanced by the specific weight and the interface height, $\Delta p \approx \rho g \eta$. The boundary conditions are a wetting contact angle θ at $x = 0$ and a horizontal surface $\eta = 0$ as $x \to \infty$. What is the maximum height h at the wall?

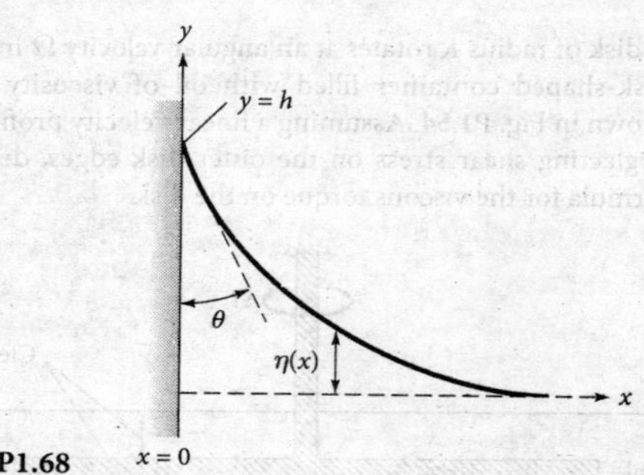

P1.68

P1.69 A solid cylindrical needle of diameter d, length L, and density ρ_n may float in liquid of surface tension Y. Neglect buoyancy and assume a contact angle of 0°. Derive a formula for the maximum diameter d_{max} able to float in the liquid. Calculate d_{max} for a steel needle (SG = 7.84) in water at 20°C.

P1.70 Derive an expression for the capillary height change h for a fluid of surface tension Υ and contact angle θ between two vertical parallel plates a distance W apart, as in Fig. P1.70. What will h be for water at 20°C if $W = 0.5$ mm?

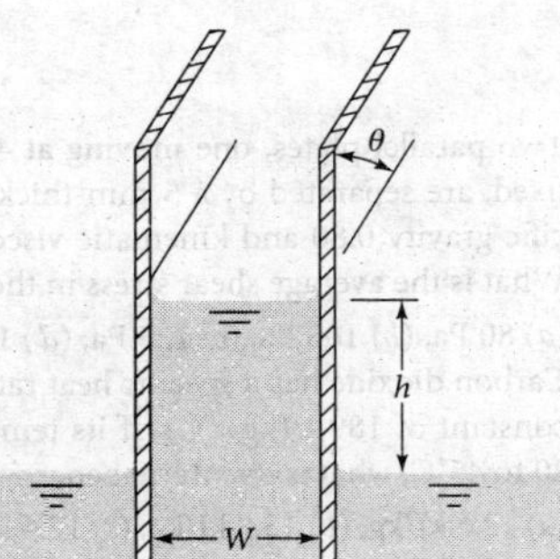

P1.70

***P1.71** A soap bubble of diameter D_1 coalesces with another bubble of diameter D_2 to form a single bubble D_3 with the same amount of air. Assuming an isothermal process, derive an expression for finding D_3 as a function of D_1, D_2, p_{atm}, and Υ.

Vapor pressure

P1.72 Early mountaineers boiled water to estimate their altitude. If they reach the top and find that water boils at 84°C, approximately how high is the mountain?

P1.73 A small submersible moves at velocity V, in fresh water at 20°C, at a 2-m depth, where ambient pressure is 131 kPa. Its critical cavitation number is known to be $C_a = 0.25$. At what velocity will cavitation bubbles begin to form on the body? Will the body cavitate if $V = 30$ m/s and the water is cold (5°C)?

P1.74 Oil, with a vapor pressure of 20 kPa, is delivered through a pipeline by equally spaced pumps, each of which increases the oil pressure by 1.3 MPa. Friction losses in the pipe are 150 Pa per meter of pipe. What is the maximum possible pump spacing to avoid cavitation of the oil?

Speed of sound, Mach number

P1.75 An airplane flies at 555 mi/h. At what altitude in the standard atmosphere will the airplane's Mach number be exactly 0.8?

P1.76 Derive a formula for the bulk modulus of an ideal gas, with constant specific heats, and calculate it for steam at 300°C and 200 kPa. Compare your result to the steam tables.

P1.77 Assume that the n-pentane data of Prob. P1.36 represents isentropic conditions. Estimate the value of the speed of sound at a pressure of 30 MPa. [*Hint:* The data approximately fit Eq. (1.19) with $B = 260$ and $n = 11$.]

P1.78 Sir Isaac Newton measured the speed of sound by timing the difference between seeing a cannon's puff of smoke and hearing its boom. If the cannon is on a mountain 5.2 mi away, estimate the air temperature in degrees Celsius if the time difference is (*a*) 24.2 s and (*b*) 25.1 s.

P1.79 From Table A.3, the density of glycerin at standard conditions is about 1,260 kg/m^3. At a very high pressure of 8,000 lb/in^2, its density increases to approximately 1,275 kg/m^3. Use this data to estimate the speed of sound of glycerin, in ft/s.

P1.80 In Problem P1.24, for the given data, the air velocity at section 2 is 1,180 ft/s. What is the Mach number at that section?

Streamlines

P1.81 Use Eq. (1.39) to find and sketch the streamlines of the following flow field:

$$u = Kx;\ v = -Ky;\ w = 0, \text{ where } K \text{ is a constant.}$$

P1.82 A velocity field is given by $u = V\cos\theta$, $v = V\sin\theta$, and $w = 0$, where V and θ are constants. Derive a formula for the streamlines of this flow.

***P1.83** Use Eq. (1.39) to find and sketch the streamlines of the following flow field:

$$u = K(x^2 - y^2);\ v = -2Kxy;\ w = 0, \text{ where } K \text{ is a constant.}$$

Hint: This is a first-order *exact* differential equation.

History of fluid mechanics

P1.84 In the early 1900s, the British chemist Sir Cyril Hinshelwood quipped that fluid dynamics study was divided into "workers who observed things they could not explain and workers who explained things they could not observe." To what historic situation was he referring?

P1.85 Do some reading and report to the class on the life and achievements, especially vis-à-vis fluid mechanics, of

(*a*) Evangelista Torricelli (1608–1647)
(*b*) Henri de Pitot (1695–1771)
(*c*) Antoine Chézy (1718–1798)
(*d*) Gotthilf Heinrich Ludwig Hagen (1797–1884)
(*e*) Julius Weisbach (1806–1871)
(*f*) George Gabriel Stokes (1819–1903)
(*g*) Moritz Weber (1871–1951)
(*h*) Theodor von Kármán (1881–1963)
(*i*) Paul Richard Heinrich Blasius (1883–1970)
(*j*) Ludwig Prandtl (1875–1953)
(*k*) Osborne Reynolds (1842–1912)
(*l*) John William Strutt, Lord Rayleigh (1842–1919)
(*m*) Daniel Bernoulli (1700–1782)
(*n*) Leonhard Euler (1707–1783)

Experimental uncertainty

P1.86 A right circular cylinder volume υ is to be calculated from the measured base radius R and height H. If the uncertainty in R is 2 percent and the uncertainty in H is 3 percent, estimate the overall uncertainty in the calculated volume. *Hint:* Read Appendix E.

Fundamentals of Engineering Exam Problems

FE1.1 The absolute viscosity μ of a fluid is primarily a function of

(*a*) Density, (*b*) Temperature, (*c*) Pressure, (*d*) Velocity, (*e*) Surface tension

FE1.2 Carbon dioxide, at 20°C and 1 atm, is compressed isentropically to 4 atm. Assume CO_2 is an ideal gas. The final temperature would be

(*a*) 130°C, (*b*) 162°C, (*c*) 171°C, (*d*) 237°C, (*e*) 313°C

FE1.3 Helium has a molecular weight of 4.003. What is the weight of 2 m^3 of helium at 1 atm and 20°C?

(*a*) 3.3 N, (*b*) 6.5 N, (*c*) 11.8 N, (*d*) 23.5 N, (*e*) 94.2 N

FE1.4 An oil has a kinematic viscosity of 1.25 E-4 m^2/s and a specific gravity of 0.80. What is its dynamic (absolute) viscosity in kg/(m · s)?

(*a*) 0.08, (*b*) 0.10, (*c*) 0.125, (*d*) 1.0, (*e*) 1.25

FE1.5 Consider a soap bubble of diameter 3 mm. If the surface tension coefficient is 0.072 N/m and external pressure is 0 Pa gage, what is the bubble's internal gage pressure?

(*a*) −24 Pa, (*b*) +48 Pa, (*c*) +96 Pa, (*d*) +192 Pa, (*e*) −192 Pa

FE1.6 The only possible dimensionless group that combines velocity V, body size L, fluid density ρ, and surface tension coefficient σ is

(*a*) $L\rho\sigma/V$, (*b*) $\rho VL^2/\sigma$, (*c*) $\rho\sigma V^2/L$, (*d*) $\sigma LV^2/\rho$, (*e*) $\rho LV^2/\sigma$

FE1.7 Two parallel plates, one moving at 4 m/s and the other fixed, are separated by a 5-mm-thick layer of oil of specific gravity 0.80 and kinematic viscosity 1.25 E-4 m^2/s. What is the average shear stress in the oil?

(*a*) 80 Pa, (*b*) 100 Pa, (*c*) 125 Pa, (*d*) 160 Pa, (*e*) 200 Pa

FE1.8 Carbon dioxide has a specific heat ratio of 1.30 and a gas constant of 189 J/(kg · °C). If its temperature rises from 20 to 45°C, what is its internal energy rise?

(*a*) 12.6 kJ/kg, (*b*) 15.8 kJ/kg, (*c*) 17.6 kJ/kg, (*d*) 20.5 kJ/kg, (*e*) 25.1 kJ/kg

FE1.9 A certain water flow at 20°C has a critical cavitation number, where bubbles form, $Ca \approx 0.25$, where $Ca = 2(p_a - p_{vap})/\rho V^2$. If p_a = 1 atm and the vapor pressure is 0.34 pounds per square inch absolute (psia), for what water velocity will bubbles form?

(*a*) 12 mi/h, (*b*) 28 mi/h, (*c*) 36 mi/h, (*d*) 55 mi/h, (*e*) 63 mi/h

FE1.10 Example 1.10 gave an analysis that predicted that the viscous moment on a rotating disk $M = \pi\mu\Omega R^4/(2h)$. If the uncertainty of each of the four variables (μ, Ω, R, h) is 1.0 percent, what is the estimated overall uncertainty of the moment M?

(*a*) 4.0 percent (*b*) 4.4 percent (*c*) 5.0 percent
(*d*) 6.0 percent (*e*) 7.0 percent

Comprehensive Problems

C1.1 Sometimes we can develop equations and solve practical problems by knowing nothing more than the dimensions of the key parameters in the problem. For example, consider the heat loss through a window in a building. Window efficiency is rated in terms of "R value," which has units of (ft^2 · h · °F)/Btu. A certain manufacturer advertises a double-pane window with an R value of 2.5. The same company produces a triple-pane window with an R value of 3.4. In either case the window dimensions are 3 ft by 5 ft. On a given winter day, the temperature difference between the inside and outside of the building is 45°F.

(*a*) Develop an equation for the amount of heat lost in a given time period Δt, through a window of area A, with a given R value, and temperature difference ΔT. How much heat (in Btu) is lost through the double-pane window in one 24-h period?

(*b*) How much heat (in Btu) is lost through the triple-pane window in one 24-h period?

(*c*) Suppose the building is heated with propane gas, which costs \$3.25 per gallon. The propane burner is 80 percent efficient. Propane has approximately 90,000 Btu of available energy per gallon. In that same 24-h period, how much money would a homeowner save per window by installing triple-pane rather than double-pane windows?

(*d*) Finally, suppose the homeowner buys 20 such triple-pane windows for the house. A typical winter has the equivalent of about 120 heating days at a temperature difference of 45°F. Each triple-pane window costs \$85 more than the double-pane window. Ignoring interest and inflation, how many years will it take the homeowner to make up the additional cost of the triple-pane windows from heating bill savings?

C1.2 When a person ice skates, the surface of the ice actually melts beneath the blades, so that he or she skates on a thin sheet of water between the blade and the ice.

(*a*) Find an expression for total friction force on the bottom of the blade as a function of skater velocity V, blade length L, water thickness (between the blade and the ice) h, water viscosity μ, and blade width W.

(*b*) Suppose an ice skater of total mass m is skating along at a constant speed of V_0 when she suddenly stands stiff with her skates pointed directly forward, allowing herself to coast to a stop. Neglecting friction due to air resistance, how far will she travel before she comes to a stop? (Remember, she is coasting on *two* skate blades.)

Give your answer for the total distance traveled, x, as a function of V_0, m, L, h, μ, and W.

(c) Find x for the case where $V_0 = 4.0$ m/s, $m = 100$ kg, $L = 30$ cm, $W = 5.0$ mm, and $h = 0.10$ mm. Do you think our assumption of negligible air resistance is a good one?

C1.3 Two thin flat plates, tilted at an angle α, are placed in a tank of liquid of known surface tension Υ and contact angle θ, as shown in Fig. C1.3. At the free surface of the liquid in the tank, the two plates are a distance L apart and have width b into the page. The liquid rises a distance h between the plates, as shown.

(a) What is the total upward (z-directed) force, due to surface tension, acting on the liquid column between the plates?

(b) If the liquid density is ρ, find an expression for surface tension Υ in terms of the other variables.

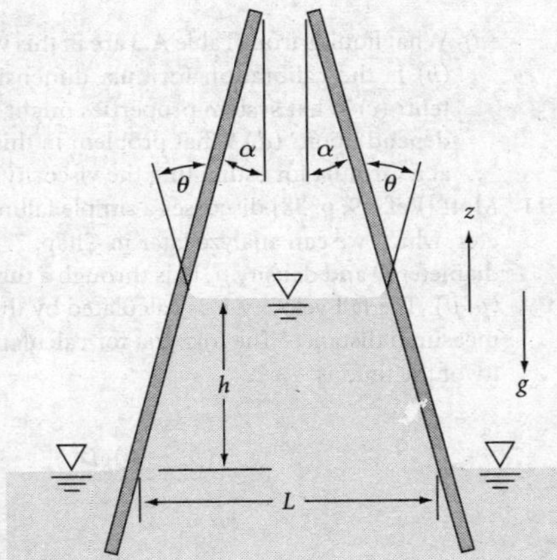

C1.3

C1.4 Oil of viscosity μ and density ρ drains steadily down the side of a tall, wide vertical plate, as shown in Fig. C1.4. In the region shown, *fully developed* conditions exist; that is, the velocity profile shape and the film thickness Δ are independent of distance z along the plate. The vertical velocity w becomes a function only of x, and the shear resistance from the atmosphere is negligible.

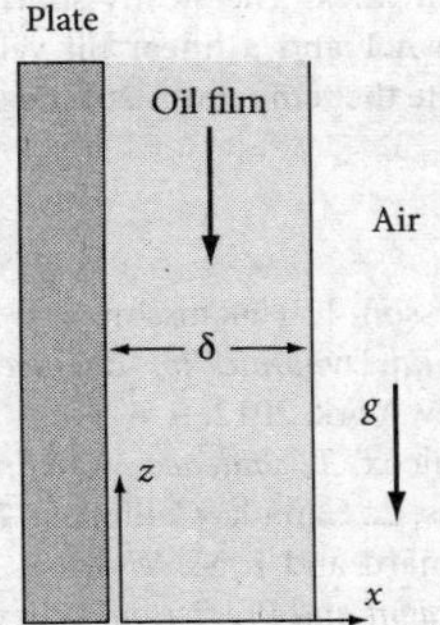

C1.4

(a) Sketch the approximate shape of the velocity profile $w(x)$, considering the boundary conditions at the wall and at the film surface.

(b) Suppose film thickness δ, and the slope of the velocity profile at the wall, $(dw/dx)_{wall}$, are measured by a laser Doppler anemometer (to be discussed in Chap. 6). Find an expression for the viscosity of the oil as a function of ρ, δ, $(dw/dx)_{wall}$, and the gravitational acceleration g. Note that, for the coordinate system given, both w and $(dw/dx)_{wall}$ are negative.

C1.5 Viscosity can be measured by flow through a thin-bore or *capillary* tube if the flow rate is low. For length L, (small) diameter $D \ll L$, pressure drop Δp, and (low) volume flow rate Q, the formula for viscosity is $\mu = D^4\Delta p/(CLQ)$, where C is a constant.

(a) Verify that C is dimensionless. The following data are for water flowing through a 2-mm-diameter tube which is 1 meter long. The pressure drop is held constant at $\Delta p = 5$ kPa.

T, °C	10.0	40.0	70.0
Q, L/min	0.091	0.179	0.292

(b) Using proper SI units, determine an average value of C by accounting for the variation with temperature of the viscosity of water.

C1.6 The *rotating-cylinder viscometer* in Fig. C1.6 shears the fluid in a narrow clearance Δr, as shown. Assuming a linear velocity distribution in the gaps, if the driving torque M is measured, find an expression for μ by (a) neglecting and (b) including the bottom friction.

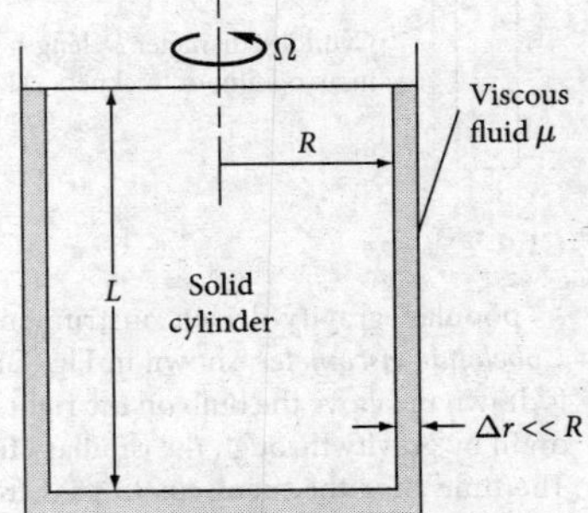

C1.6

C1.7 Make an analytical study of the transient behavior of the sliding block in Prob. 1.45. (a) Solve for $V(t)$ if the block starts from rest, $V = 0$ at $t = 0$. (b) Calculate the time t_1 when the block has reached 98 percent of its terminal velocity.

C1.8 A mechanical device that uses the rotating cylinder of Fig. C1.6 is the *Stormer viscometer* [29]. Instead of being driven at constant Ω, a cord is wrapped around the shaft and attached to a falling weight W. The time t to turn the shaft a given number of revolutions (usually five) is measured and correlated with viscosity. The formula is

$$t \approx \frac{A\mu}{W - B}$$

where A and B are constants that are determined by calibrating the device with a known fluid. Here are calibration data for a Stormer viscometer tested in glycerol, using a weight of 50 N:

μ, kg/(m-s)	0.23	0.34	0.57	0.84	1.15
t, sec	15	23	38	56	77

(*a*) Find reasonable values of A and B to fit this calibration data. *Hint:* The data are not very sensitive to the value of B.

(*b*) A more viscous fluid is tested with a 100 N weight and the measured time is 44 s. Estimate the viscosity of this fluid.

C1.9 The lever in Fig. C1.9 has a weight W at one end and is tied to a cylinder at the left end. The cylinder has negligible weight and buoyancy and slides upward through a film of heavy oil of viscosity μ. (*a*) If there is no acceleration (uniform lever rotation), derive a formula for the rate of fall V_2 of the weight. Neglect the lever weight. Assume a linear velocity profile in the oil film. (*b*) Estimate the fall velocity of the weight if $W = 20$ N, $L_1 = 75$ cm, $L_2 = 50$ cm, $D = 10$ cm, $L = 22$ cm, $\Delta R = 1$ mm, and the oil is glycerin at 20°C.

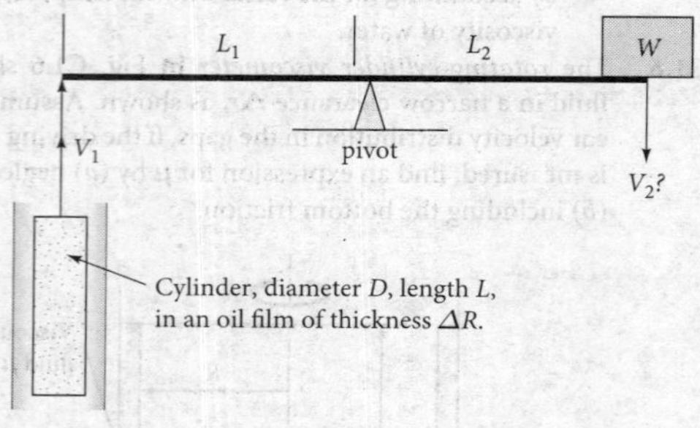

C1.9

C1.10 A popular gravity-driven instrument is the *Cannon-Ubbelohde viscometer*, shown in Fig. C1.10. The test liquid is drawn up above the bulb on the right side and allowed to drain by gravity through the capillary tube below the bulb. The time t for the meniscus to pass from upper to lower timing marks is recorded. The kinematic viscosity is computed by the simple formula:

$$\nu = Ct$$

where C is a calibration constant. For ν in the range of 100–500 mm²/s, the recommended constant is $C = 0.50$ mm²/s², with an accuracy less than 0.5 percent.

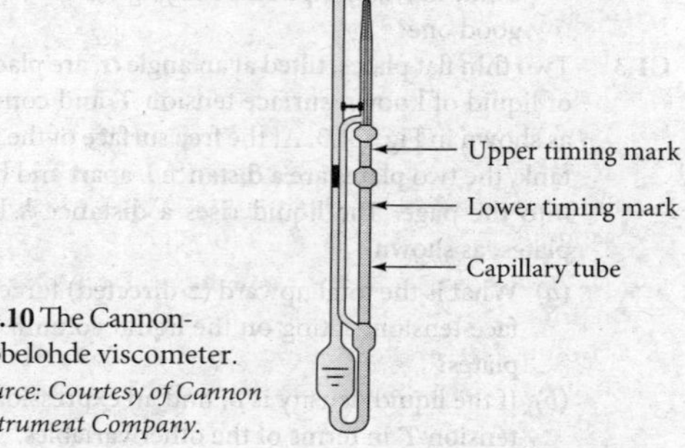

C1.10 The Cannon-Ubbelohde viscometer.
Source: Courtesy of Cannon Instrument Company.

(*a*) What liquids from Table A.3 are in this viscosity range? (*b*) Is the calibration formula dimensionally consistent? (*c*) What system properties might the constant C depend upon? (*d*) What problem in this chapter hints at a formula for estimating the viscosity?

C1.11 Mott [Ref. 49, p. 38] discusses a simple falling-ball viscometer, which we can analyze later in Chap. 7. A small ball of diameter D and density ρ_b falls through a tube of test liquid (ρ, μ). The fall velocity V is calculated by the time to fall a measured distance. The formula for calculating the viscosity of the fluid is

$$\mu = \frac{(\rho_b - \rho)gD^2}{18\ V}$$

This result is limited by the requirement that the Reynolds number $(\rho VD/\mu)$ be less than 1.0. Suppose a steel ball (SG = 7.87) of diameter 2.2 mm falls in SAE 25W oil (SG = 0.88) at 20°C. The measured fall velocity is 8.4 cm/s. (*a*) What is the viscosity of the oil, in kg/m-s? (*b*) Is the Reynolds number small enough for a valid estimate?

C1.12 A solid aluminum disk (SG = 2.7) is 2 in in diameter and 3/16 in thick. It slides steadily down a 14° incline that is coated with a castor oil (SG = 0.96) film one hundredth of an inch thick. The steady slide velocity is 2 cm/s. Using Figure A.1 and a linear oil velocity profile assumption, estimate the temperature of the castor oil.

References

1. T. J. Chung, *Computational Fluid Dynamics*, 2d ed., Cambridge University Press, New York, 2010.
2. J. D. Anderson, *Computational Fluid Dynamics: An Introduction*, 3d ed., Springer, New York, 2010.
3. H. Lomax, T. H. Pulliam, and D. W. Zingg, *Fundamentals of Computational Fluid Dynamics*, Springer, New York, 2011.
4. B. Andersson, L. Håkansson, and M. Mortensen, *Computational Fluid Dynamics for Engineers*, Cambridge University Press, New York, 2012.
5. D. C. Wilcox, *Turbulence Modeling for CFD*, 3d ed., DCW Industries, La Cañada, California, 2006.
6. P. S. Bernard and J. M. Wallace, *Turbulent Flow: Analysis, Measurement and Prediction*, Wiley, New York, 2002.

7. X. Jiang and C. H. Lai, *Numerical Techniques for Direct and Large-Eddy Simulations*, CRC Press, Boca Raton, Florida, 2009.
8. B. Geurts, *Elements of Direct and Large Eddy Simulation*, R. T. Edwards Inc., Flourtown, PA, 2003.
9. S. Tavoularis, *Measurement in Fluid Mechanics*, Cambridge University Press, New York, 2005.
10. R. C. Baker, *Introductory Guide to Flow Measurement*, Wiley, New York, 2002.
11. R. W. Miller, *Flow Measurement Engineering Handbook*, 3d ed., McGraw-Hill, New York, 1996.
12. H. Rouse and S. Ince, *History of Hydraulics*, Iowa Institute of Hydraulic Research, Univ. of Iowa, Iowa City, IA, 1957; reprinted by Dover, New York, 1963.
13. H. Rouse, *Hydraulics in the United States 1776–1976*, Iowa Institute of Hydraulic Research, Univ. of Iowa, Iowa City, IA, 1976.
14. G. A. Tokaty, *A History and Philosophy of Fluid Mechanics*, Dover Publications, New York, 1994.
15. Cambridge University Press, "Ludwig Prandtl—Father of Modern Fluid Mechanics," URL <www.fluidmech.net/msc/prandtl.htm>.
16. R. I. Tanner, *Engineering Rheology*, 2d ed., Oxford University Press, New York, 2000.
17. C. E. Brennen, *Fundamentals of Multiphase Flow*, Cambridge University Press, New York, 2009.
18. C. Shen, *Rarefied Gas Dynamics*, Springer, New York, 2010.
19. F. Carderelli and M. J. Shields, *Scientific Unit Conversion: A Practical Guide to Metrification*, 2d ed., Springer-Verlag, New York, 1999.
20. J. P. Holman, *Heat Transfer*, 10th ed., McGraw-Hill, New York, 2009.
21. B. E. Poling, J. M. Prausnitz, and J. P. O'Connell, *The Properties of Gases and Liquids*, 5th ed., McGraw-Hill, New York, 2000.
22. J. Hilsenrath et al., "Tables of Thermodynamic and Transport Properties," *U.S. Nat. Bur. Standards Circular 564*, 1955; reprinted by Pergamon, New York, 1960.
23. W. T. Parry, *ASME International Steam Tables for Industrial Use*, 2d ed., ASME Press, New York, 2009.
24. Steam Tables URL: http://www.steamtablesonline.com/
25. O. A. Hougen and K. M. Watson, *Chemical Process Principles Charts*, Wiley, New York, 1960.
26. F. M. White, *Viscous Fluid Flow*, 3d ed., McGraw-Hill, New York, 2005.
27. M. Bourne, *Food Texture and Viscosity: Concept and Measurement*, 2d ed., Academic Press, Salt Lake City, Utah, 2002.
28. *SAE Fuels and Lubricants Standards Manual*, Society of Automotive Engineers, Warrendale, PA, 2001.
29. C. L. Yaws, *Handbook of Viscosity*, 3 vols., Elsevier Science and Technology, New York, 1994.
30. A. W. Adamson and A. P. Gast, *Physical Chemistry of Surfaces*, Wiley, New York, 1999.
31. C. E. Brennen, *Fundamentals of Multiphase Flow*, Cambridge University Press, New York, 2009.
32. National Committee for Fluid Mechanics Films, *Illustrated Experiments in Fluid Mechanics*, M.I.T. Press, Cambridge, MA, 1972.
33. I. G. Currie, *Fundamental Mechanics of Fluids*, 3d ed., Marcel Dekker, New York, 2003.
34. W.-J. Yang (ed.), *Handbook of Flow Visualization*, 2d ed., Taylor and Francis, New York, 2001.
35. F.T. Nieuwstadt (ed.), *Flow Visualization and Image Analysis*, Springer, New York, 2007.
36. A. J. Smits and T. T. Lim, *Flow Visualization: Techniques and Examples*, 2d ed., Imperial College Press, London, 2011.
37. R. J. Adrian and J. Westerweel, *Particle Image Velocimetry*, Cambridge University Press, New York, 2010.
38. Wen-Jai Yang, *Computer-Assisted Flow Visualization*, Begell House, New York, 1994.
39. M. van Dyke, *An Album of Fluid Motion*, Parabolic Press, Stanford, CA, 1982.
40. Y. Nakayama and Y. Tanida (eds.), *Visualized Flow*, vol. 1, Elsevier, New York, 1993; vols. 2 and 3, CRC Press, Boca Raton, FL, 1996.
41. M. Samimy, K. S. Breuer, L. G. Leal, and P. H. Steen, *A Gallery of Fluid Motion*, Cambridge University Press, New York, 2003.
42. S. Y. Son et al., "Coolant Flow Field Measurements in a Two-Pass Channel Using Particle Image Velocimetry," 1999 Heat Transfer Gallery, *Journal of Heat Transfer*, vol. 121, August, 1999.
43. B. Carr and V. E. Young, "Videotapes and Movies on Fluid Dynamics and Fluid Machines," in *Handbook of Fluid Dynamics and Fluid Machinery*, vol. II, J. A. Schetz and A. E. Fuhs (eds.), Wiley, New York, 1996, pp. 1171–1189.
44. Online Steam Tables URL: http://www.spiraxsarco.com/esc/SH_Properties.aspx.
45. H. W. Coleman and W. G. Steele, *Experimentation and Uncertainty Analysis for Engineers*, 3d ed., Wiley, New York, 2009.
46. I. Hughes and T. Hase, *Measurements and Their Uncertainties*, Oxford University Press, New York, 2010.
47. A. Thom, "The Flow Past Circular Cylinders at Low Speeds," *Proc. Royal Society*, A141, London, 1933, pp. 651–666.
48. S. J. Kline and F. A. McClintock, "Describing Uncertainties in Single-Sample Experiments," *Mechanical Engineering*, January, 1953, pp. 3–9.
49. R. L. Mott, *Applied Fluid Mechanics*, Pearson Prentice-Hall, Upper Saddle River, NJ, 2006.
50. "Putting Porky to Work," Technology Focus, *Mechanical Engineering*, August 2002, p. 24.
51. R. M. Olson and S. J. Wright, *Essentials of Engineering Fluid Mechanics*, 5th ed., HarperCollins, New York, 1990.
52. B. Kirby, *Micro- and Nanoscale Fluid Mechanics*, Cambridge University Press, New York, 2010.
53. E. J. Watson, "The Spread of a Liquid Jet over a Horizontal Plane," *J. Fluid Mechanics*, vol. 20, 1964, pp. 481–499.
54. J. Dean and A. Reisinger, *Sands, Powders, and Grains: An Introduction to the Physics of Granular Materials*, Springer-Verlag, New York, 1999.
55. E. Kiran and Y. L. Sen, "High Pressure Density and Viscosity of n-alkanes," *Int. J. Thermophysics*, vol. 13, no. 3, 1992, pp. 411–442.

The bathyscaphe *Trieste* was built by explorers August and Jacques Picard in 1953. They made many deep-sea dives over the next five years. In 1958 the U.S. Navy purchased it and, in 1960, descended to the deepest part of the ocean, the Marianas Trench, near Guam. The recorded depth was 10,900 m, or nearly seven miles. The photo shows the vessel being lifted from the water. The pressure sphere is on the bottom, and the boat-shaped structure above is filled with gasoline, which provides buoyancy even at those depths. [*Image courtesy of the U.S. Navy.*]

Chapter 2 Pressure Distribution in a Fluid

Motivation. Many fluid problems do not involve motion. They concern the pressure distribution in a static fluid and its effect on solid surfaces and on floating and submerged bodies.

When the fluid velocity is zero, denoted as the *hydrostatic condition*, the pressure variation is due only to the weight of the fluid. Assuming a known fluid in a given gravity field, the pressure may easily be calculated by integration. Important applications in this chapter are (1) pressure distribution in the atmosphere and the oceans, (2) the design of manometer, mechanical, and electronic pressure instruments, (3) forces on submerged flat and curved surfaces, (4) buoyancy on a submerged body, and (5) the behavior of floating bodies. The last two result in Archimedes' principles.

If the fluid is moving in *rigid-body motion*, such as a tank of liquid that has been spinning for a long time, the pressure also can be easily calculated because the fluid is free of shear stress. We apply this idea here to simple rigid-body accelerations in Sec. 2.9. Pressure measurement instruments are discussed in Sec. 2.10. As a matter of fact, pressure also can be analyzed in arbitrary (nonrigid-body) motions $\mathbf{V}(x, y, z, t)$, but we defer that subject to Chap. 4.

2.1 Pressure and Pressure Gradient

In Fig. 1.1, we saw that a fluid at rest cannot support shear stress, and thus Mohr's circle reduces to a point. In other words, the normal stress on any plane through a fluid element at rest is a point property called the *fluid pressure p*, taken positive for compression by common convention. This is such an important concept that we shall review it with another approach.

First, let us emphasize that pressure is a thermodynamic property of the fluid, like temperature or density. It is not a *force*. Pressure has no direction and is not a vector. The concept of force only arises when considering a surface immersed in fluid pressure. The pressure creates a force due to fluid molecules bombarding the surface, and it is normal to that surface.

Figure 2.1 shows a small wedge of fluid at rest of size Δx by Δz by Δs and depth b into the paper. There is no shear by definition, but we postulate that the pressures p_x, p_z, and p_n may be different on each face. The weight of the element also may be

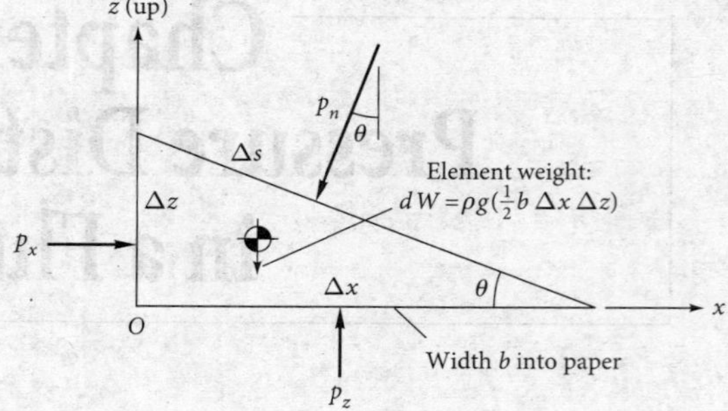

Fig. 2.1 Equilibrium of a small wedge of fluid at rest.

important. The element is assumed to be small, so the pressure is constant on each face. Summation of forces must equal zero (no acceleration) in both the x and z directions.

$$\sum F_x = 0 = p_x b\,\Delta z - p_n b\,\Delta s \sin\theta$$
$$\sum F_z = 0 = p_z b\,\Delta x - p_n b\,\Delta s \cos\theta - \tfrac{1}{2}\rho g b\,\Delta x\,\Delta z \tag{2.1}$$

But the geometry of the wedge is such that

$$\Delta s \sin\theta = \Delta z \qquad \Delta s \cos\theta = \Delta x \tag{2.2}$$

Substitution into Eq. (2.1) and rearrangement give

$$p_x = p_n \qquad p_z = p_n + \tfrac{1}{2}\rho g\,\Delta z \tag{2.3}$$

These relations illustrate two important principles of the hydrostatic, or shear-free, condition: (1) There is no pressure change in the horizontal direction, and (2) there is a vertical change in pressure proportional to the density, gravity, and depth change. We shall exploit these results to the fullest, starting in Sec. 2.3.

In the limit as the fluid wedge shrinks to a "point," $\Delta z \to 0$ and Eq. (2.3) become

$$p_x = p_z = p_n = p \tag{2.4}$$

Since θ is arbitrary, we conclude that the pressure p in a static fluid is a point property, independent of orientation.

Pressure Force on a Fluid Element

Pressure (or any other stress, for that matter) causes a net force on a fluid element. To see this, consider the pressure acting on the two x faces in Fig. 2.2. Let the pressure vary arbitrarily

$$p = p(x, y, z, t)$$

The net force in the x direction on the element in Fig. 2.2 is given by

$$dF_x = p\,dy\,dz - \left(p + \frac{\partial p}{\partial x}dx\right)dy\,dz = -\frac{\partial p}{\partial x}dx\,dy\,dz$$

In like manner the net force dF_y involves $-\partial p/\partial y$, and the net force dF_z concerns $-\partial p/\partial z$. The total net-force vector on the element due to pressure is

$$d\mathbf{F}_{\text{press}} = \left(-\mathbf{i}\frac{\partial p}{\partial x} - \mathbf{j}\frac{\partial p}{\partial y} - \mathbf{k}\frac{\partial p}{\partial z}\right)dx\,dy\,dz \tag{2.5}$$

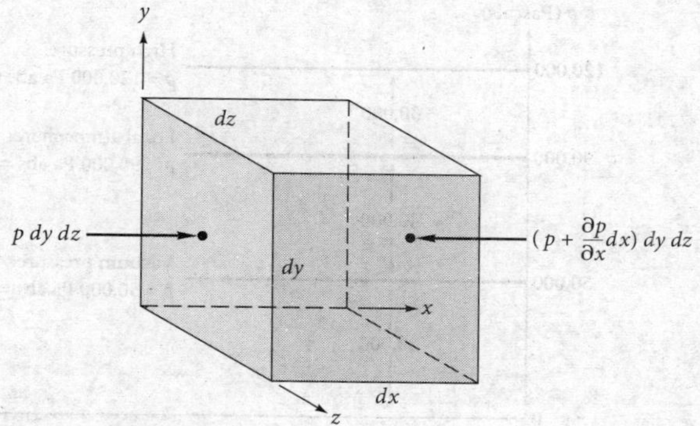

Fig. 2.2 Net x force on an element due to pressure variation.

We recognize the term in parentheses as the negative vector gradient of p. Denoting $\mathbf{f}$ as the net force per unit element volume, we rewrite Eq. (2.5) as

$$\mathbf{f}_{\text{press}} = -\nabla p \tag{2.6}$$

where $\nabla = \text{gradient operator} = \mathbf{i}\dfrac{\partial}{\partial x} + \mathbf{j}\dfrac{\partial}{\partial y} + \mathbf{k}\dfrac{\partial}{\partial z}$

Thus, it is not the pressure but the pressure *gradient* causing a net force that must be balanced by gravity or acceleration or some other effect in the fluid.

2.2 Equilibrium of a Fluid Element

The pressure gradient is a *surface* force that acts on the sides of the element. There may also be a *body* force, due to electromagnetic or gravitational potentials, acting on the entire mass of the element. Here we consider only the gravity force, or weight of the element:

$$d\mathbf{F}_{\text{grav}} = \rho\mathbf{g}\,dx\,dy\,dz$$

or

$$\mathbf{f}_{\text{grav}} = \rho\mathbf{g} \tag{2.7}$$

In addition to gravity, a fluid in motion will have *surface* forces due to viscous stresses. By Newton's law, Eq. (1.2), the sum of these per-unit-volume forces equals the mass per unit volume (density) times the acceleration $\mathbf{a}$ of the fluid element:

$$\sum \mathbf{f} = \mathbf{f}_{\text{press}} + \mathbf{f}_{\text{grav}} + \mathbf{f}_{\text{visc}} = -\nabla p + \rho\mathbf{g} + \mathbf{f}_{\text{visc}} = \rho\mathbf{a} \tag{2.8}$$

This general equation will be studied in detail in Chap. 4. Note that Eq. (2.8) is a *vector* relation, and the acceleration may not be in the same vector direction as the velocity. For our present topic, *hydrostatics*, the viscous stresses and the acceleration are zero.

Gage Pressure and Vacuum Pressure: Relative Terms

Before embarking on examples, we should note that engineers are apt to specify pressures as (1) the *absolute* or total magnitude or (2) the value *relative* to the local ambient atmosphere. The second case occurs because many pressure instruments are of *differential* type and record, not an absolute magnitude, but the difference between the fluid pressure and the atmosphere. The measured pressure may be either higher or lower than the local atmosphere, and each case is given a name:

1. $p > p_a$ *Gage* pressure: $p(\text{gage}) = p - p_a$
2. $p < p_a$ *Vacuum* pressure: $p(\text{vacuum}) = p_a - p$

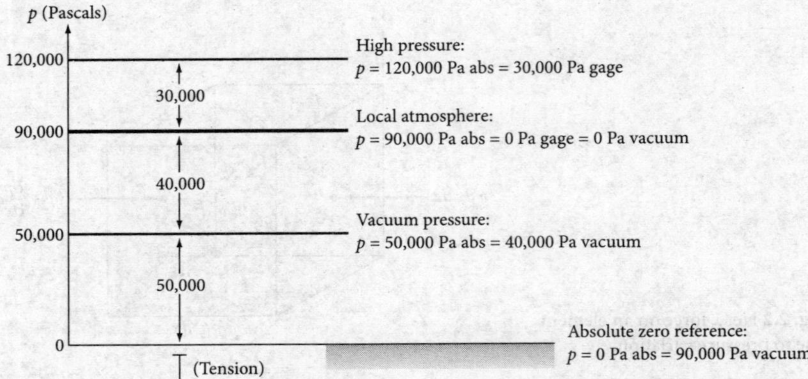

Fig. 2.3 Illustration of absolute, gage, and vacuum pressure readings.

This is a convenient shorthand, and one later adds (or subtracts) atmospheric pressure to determine the absolute fluid pressure.

A typical situation is shown in Fig. 2.3. The local atmosphere is at, say, 90,000 Pa, which might reflect a storm condition in a sea-level location or normal conditions at an altitude of 1,000 m. Thus, on this day, $p_a = 90{,}000$ Pa absolute $= 0$ Pa gage $= 0$ Pa vacuum. Suppose gage 1 in a laboratory reads $p_1 = 120{,}000$ Pa absolute. This value may be reported as a *gage* pressure, $p_1 = 120{,}000 - 90{,}000 = 30{,}000$ Pa *gage*. (One must also record the atmospheric pressure in the laboratory, since p_a changes gradually.) Suppose gage 2 reads $p_2 = 50{,}000$ Pa absolute. Locally, this is a *vacuum* pressure and might be reported as $p_2 = 90{,}000 - 50{,}000 = 40{,}000$ Pa *vacuum*. Occasionally, in the problems section, we will specify gage or vacuum pressure to keep you alert to this common engineering practice. If a pressure is listed without the modifier gage or vacuum, we assume it is absolute pressure.

2.3 Hydrostatic Pressure Distributions

If the fluid is at rest or at constant velocity, $\mathbf{a} = 0$ and $\mathbf{f}_{visc} = 0$. Equation (2.8) for the pressure distribution reduces to

$$\nabla p = \rho \mathbf{g} \tag{2.9}$$

This is a *hydrostatic* distribution and is correct for all fluids at rest, regardless of their viscosity, because the viscous term vanishes identically.

Recall from vector analysis that the vector ∇p expresses the magnitude and direction of the maximum spatial rate of increase of the scalar property p. As a result, ∇p is perpendicular everywhere to surfaces of constant p. Thus, Eq. (2.9) states that a fluid in hydrostatic equilibrium will align its constant-pressure surfaces everywhere normal to the local-gravity vector. The maximum pressure increase will be in the direction of gravity—that is, "down." If the fluid is a liquid, its free surface, being at atmospheric pressure, will be normal to local gravity, or "horizontal." You probably knew all this before, but Eq. (2.9) is the proof of it.

In our customary coordinate system, z is "up." Thus, the local-gravity vector for small-scale problems is

$$\mathbf{g} = -g\mathbf{k} \tag{2.10}$$

where g is the magnitude of local gravity, for example, 9.807 m/s^2. For these coordinates Eq. (2.9) has the components

$$\frac{\partial p}{\partial x} = 0 \qquad \frac{\partial p}{\partial y} = 0 \qquad \frac{\partial p}{\partial z} = -\rho g = -\gamma \tag{2.11}$$

the first two of which tell us that p is independent of x and y. Hence, $\partial p/\partial z$ can be replaced by the total derivative dp/dz, and the hydrostatic condition reduces to

$$\frac{dp}{dz} = -\gamma$$

or

$$p_2 - p_1 = -\int_1^2 \gamma\, dz \tag{2.12}$$

Equation (2.12) is the solution to the hydrostatic problem. The integration requires an assumption about the density and gravity distribution. Gases and liquids are usually treated differently.

We state the following conclusions about a hydrostatic condition:

> Pressure in a continuously distributed uniform static fluid varies only with vertical distance and is independent of the shape of the container. The pressure is the same at all points on a given horizontal plane in the fluid. The pressure increases with depth in the fluid.

An illustration of this is shown in Fig. 2.4. The free surface of the container is atmospheric and forms a horizontal plane. Points a, b, c, and d are at equal depth in a horizontal plane and are interconnected by the same fluid, water; therefore, all points have the same pressure. The same is true of points A, B, and C on the bottom, which all have the same higher pressure than at a, b, c, and d. However, point D, although at the same depth as A, B, and C, has a different pressure because it lies beneath a different fluid, mercury.

Effect of Variable Gravity

For a spherical planet of uniform density, the acceleration of gravity varies inversely as the square of the radius from its center

$$g = g_0\left(\frac{r_0}{r}\right)^2 \tag{2.13}$$

where r_0 is the planet radius and g_0 is the surface value of g. For earth, $r_0 \approx 3{,}960$ statute mi $\approx 6{,}400$ km. In typical engineering problems the deviation from r_0 extends from the deepest ocean, about 11 km, to the atmospheric height of supersonic transport operation, about 20 km. This gives a maximum variation in g of $(6{,}400/6{,}420)^2$, or 0.6 percent. We therefore neglect the variation of g in most problems.

Hydrostatic Pressure in Liquids

Liquids are so nearly incompressible that we can neglect their density variation in hydrostatics. In Example 1.6, we saw that water density increases only 4.6 percent at

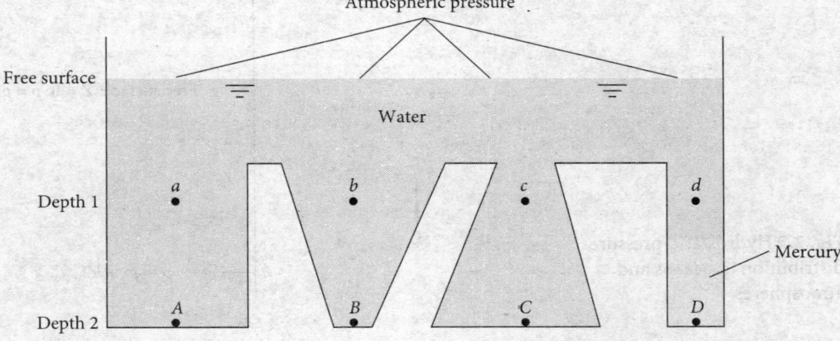

Fig. 2.4 Hydrostatic-pressure distribution. Points a, b, c, and d are at equal depths in water and therefore have identical pressures. Points A, B, and C are also at equal depths in water and have identical pressures higher than a, b, c, and d. Point D has a different pressure from A, B, and C because it is not connected to them by a water path.

the deepest part of the ocean. Its effect on hydrostatics would be about half of this, or 2.3 percent. Thus, we assume constant density in liquid hydrostatic calculations, for which Eq. (2.12) integrates to

$$\text{Liquids:} \qquad p_2 - p_1 = -\gamma(z_2 - z_1) \tag{2.14}$$

$$\text{or} \qquad z_1 - z_2 = \frac{p_2}{\gamma} - \frac{p_1}{\gamma}$$

We use the first form in most problems. The quantity γ is called the *specific weight* of the fluid, with dimensions of weight per unit volume; some values are tabulated in Table 2.1. The quantity p/γ is a length called the *pressure head* of the fluid.

For lakes and oceans, the coordinate system is usually chosen as in Fig. 2.5, with $z = 0$ at the free surface, where p equals the surface atmospheric pressure p_a. When we introduce the reference value $(p_1, z_1) = (p_a, 0)$, Eq. (2.14) becomes, for p at any (negative) depth z,

$$\text{Lakes and oceans:} \qquad p = p_a - \gamma z \tag{2.15}$$

where γ is the average specific weight of the lake or ocean. As we shall see, Eq. (2.15) holds in the atmosphere also with an accuracy of 2 percent for heights z up to 1000 m.

Table 2.1 Specific Weight of Some Common Fluids

Fluid	Specific weight γ at 68°F = 20°C	
	lbf/ft³	N/m³
Air (at 1 atm)	0.0752	11.8
Ethyl alcohol	49.2	7,733
SAE 30 oil	55.5	8,720
Water	62.4	9,790
Seawater	64.0	10,050
Glycerin	78.7	12,360
Carbon tetrachloride	99.1	15,570
Mercury	846	133,100

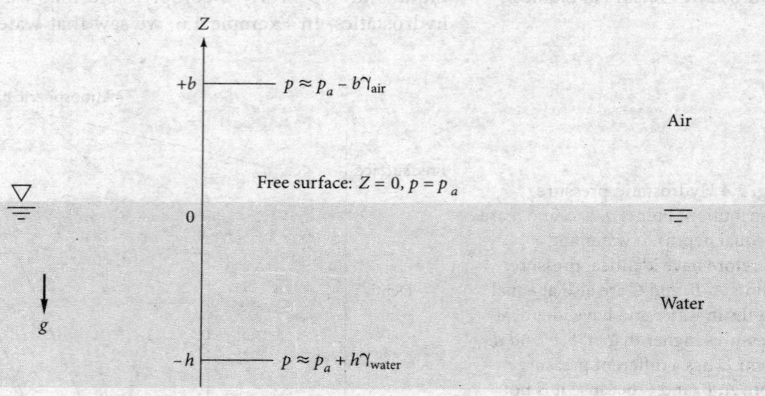

Fig. 2.5 Hydrostatic-pressure distribution in oceans and atmospheres.

EXAMPLE 2.1

Newfound Lake, a freshwater lake near Bristol, New Hampshire, has a maximum depth of 60 m, and the mean atmospheric pressure is 91 kPa. Estimate the absolute pressure in kPa at this maximum depth.

Solution

- *System sketch:* Imagine that Fig. 2.5 is Newfound Lake, with $h = 60$ m and $z = 0$ at the surface.
- *Property values:* From Table 2.1, $\gamma_{water} = 9790 \text{ N/m}^3$. We are given that $p_{atmos} = 91$ kPa.
- *Solution steps:* Apply Eq. (2.15) to the deepest point. Use SI units, pascals, not kilopascals:

$$p_{max} = p_a - \gamma z = 91{,}000 \text{ Pa} - \left(9790 \frac{\text{N}}{\text{m}^3}\right)(-60 \text{ m}) = 678{,}400 \text{ Pa} \approx 678 \text{ kPa} \qquad \textit{Ans.}$$

- *Comments:* Kilopascals are awkward. Use pascals in the formula, then convert the answer.

The Mercury Barometer

The simplest practical application of the hydrostatic formula (2.14) is the barometer (Fig. 2.6), which measures atmospheric pressure. A tube is filled with mercury and inverted while submerged in a reservoir. This causes a near vacuum in the closed upper end because mercury has an extremely small vapor pressure at room temperatures (0.16 Pa at 20°C). Since atmospheric pressure forces a mercury column to rise a distance h into the tube, the upper mercury surface is at zero pressure.

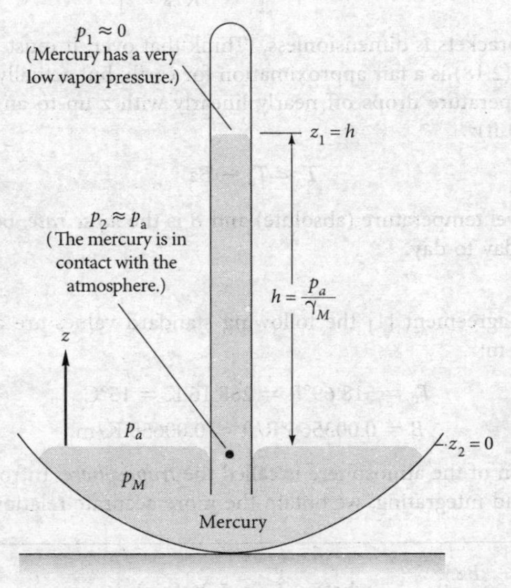

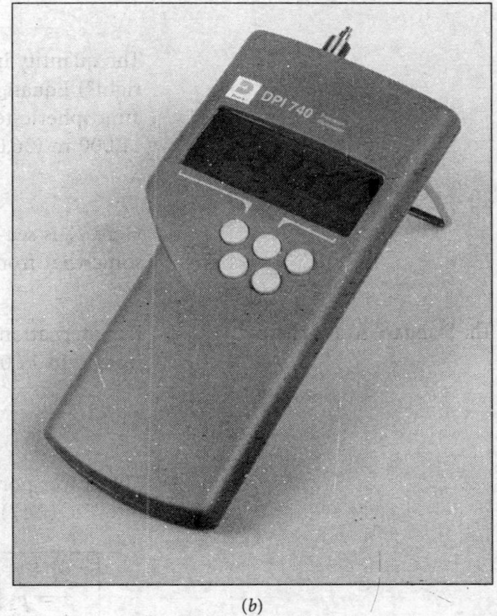

Fig. 2.6 A barometer measures local absolute atmospheric pressure: (*a*) the height of a mercury column is proportional to p_{atm}; (*b*) a modern portable barometer, with digital readout, uses the resonating silicon element of Fig. 2.28*c*. *(Courtesy of Paul Lupke, Druck, Inc.)*

From Fig. 2.6, Eq. (2.14) applies with $p_1 = 0$ at $z_1 = h$ and $p_2 = p_a$ at $z_2 = 0$:

$$p_a - 0 = -\gamma_M (0 - h)$$

or

$$h = \frac{p_a}{\gamma_M} \tag{2.16}$$

At sea-level standard, with $p_a = 101{,}350$ Pa and $\gamma_M = 133{,}100\ \text{N/m}^3$ from Table 2.1, the barometric height is $h = 101{,}350/133{,}100 = 0.761$ m or 761 mm. In the United States, the weather service reports this as an atmospheric "pressure" of 761 mmHg (millimeters of mercury). Mercury is used because it is the heaviest common liquid. A water barometer would be 10.363 meters high.

Hydrostatic Pressure in Gases

Gases are compressible, with density nearly proportional to pressure. Thus, density must be considered as a variable in Eq. (2.12) if the integration carries over large pressure changes. It is sufficiently accurate to introduce the perfect-gas law $p = \rho RT$ in Eq. (2.12):

$$\frac{dp}{dz} = -\rho g = -\frac{p}{RT} g$$

Separate the variables and integrate between points 1 and 2:

$$\int_1^2 \frac{dp}{p} = \ln \frac{p_2}{p_1} = -\frac{g}{R} \int_1^2 \frac{dz}{T} \tag{2.17}$$

The integral over z requires an assumption about the temperature variation $T(z)$. One common approximation is the *isothermal atmosphere,* where $T = T_0$:

$$p_2 = p_1 \exp\left[-\frac{g(z_2 - z_1)}{RT_0}\right] \tag{2.18}$$

The quantity in brackets is dimensionless. (Think that over; it must be dimensionless, right?) Equation (2.18) is a fair approximation for earth, but actually the earth's mean atmospheric temperature drops off nearly linearly with z up to an altitude of about 11,000 m (36,000 ft):

$$T \approx T_0 - Bz \tag{2.19}$$

Here T_0 is sea-level temperature (absolute) and B is the *lapse rate,* both of which vary somewhat from day to day.

The Standard Atmosphere

By international agreement [1] the following standard values are assumed to apply from 0 to 11,000 m:

$$T_0 = 518.69°\text{R} = 288.16\ \text{K} = 15°\text{C}$$

$$B = 0.003566°\text{R/ft} = 0.00650\ \text{K/m}$$

This lower portion of the atmosphere is called the *troposphere.* Introducing Eq. (2.19) into Eq. (2.17) and integrating, we obtain the more accurate relation

$$p = p_a\left(1 - \frac{Bz}{T_0}\right)^{g/(RB)} \quad \text{where} \quad \frac{g}{RB} = 5.26\ \text{(air)}$$

$$\rho = \rho_o\left(1 - \frac{Bz}{T_o}\right)^{\frac{g}{RB}-1} \quad \text{where } \rho_o = 1.2255\frac{kg}{m^3},\ p_o = 101{,}350\ p_a \tag{2.20}$$

in the troposphere, with $z = 0$ at sea level. The exponent $g/(RB)$ is dimensionless (again it must be) and has the standard value of 5.26 for air, with $R = 287\ \text{m}^2/(\text{s}^2 \cdot \text{K})$.

The U.S. standard atmosphere [1] is sketched in Fig. 2.7. The pressure is seen to be nearly zero at $z = 30$ km. For tabulated properties see Table A.6.

EXAMPLE 2.2

If sea-level pressure is 101,350 Pa, compute the standard pressure at an altitude of 5,000 m, using (*a*) the exact formula and (*b*) an isothermal assumption at a standard sea-level temperature of 15°C. Is the isothermal approximation adequate?

Solution

Part (a) Use absolute temperature in the exact formula, Eq. (2.20):

$$p = p_a\left[1 - \frac{(0.00650\ \text{K/m})(5{,}000\ \text{m})}{288.16\ \text{K}}\right]^{5.26} = (101{,}350\ \text{Pa})(0.8872)^{5.26}$$

$$= 101{,}350(0.5328) = 54{,}000\ \text{Pa} \qquad \textit{Ans. (a)}$$

This is the standard-pressure result given at $z = 5{,}000$ m in Table A.6.

Part (b) If the atmosphere were isothermal at 288.16 K, Eq. (2.18) would apply:

$$p \approx p_a \exp\left(-\frac{gz}{RT}\right) = (101{,}350\ \text{Pa})\exp\left\{-\frac{(9.807\ \text{m/s}^2)(5{,}000\ \text{m})}{[287\ \text{m}^2/(\text{s}^2 \cdot \text{K})](288.16\ \text{K})}\right\}$$

$$= (101{,}350\ \text{Pa})\exp(-0.5929) \approx 56{,}000\ \text{Pa} \qquad \textit{Ans. (b)}$$

This is 4 percent higher than the exact result. The isothermal formula is inaccurate in the troposphere.

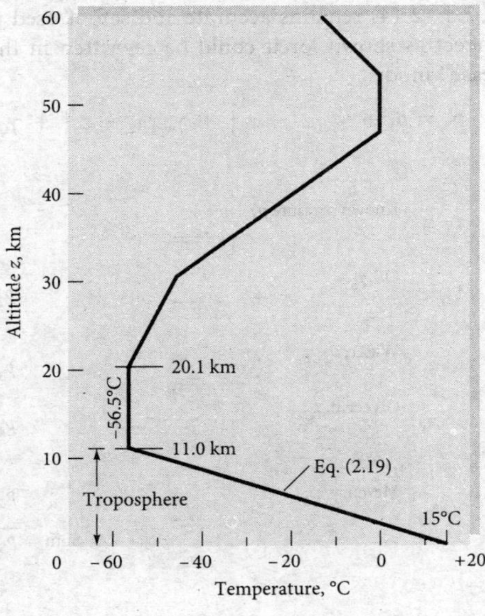

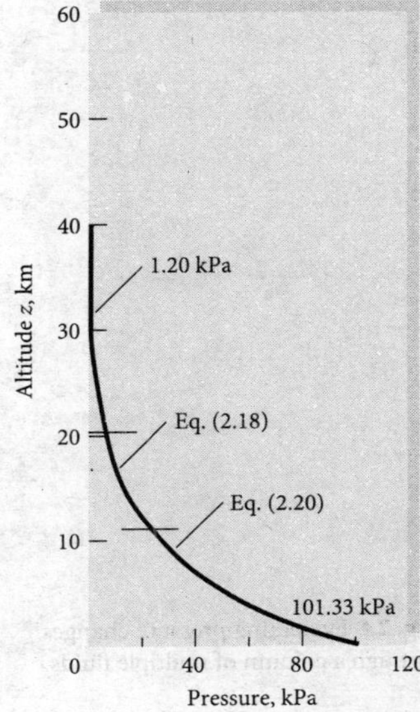

Fig. 2.7 Temperature and pressure distribution in the U.S. standard atmosphere.
Source: U.S. Standard Atmosphere, 1976, Government Printing Office, Washington DC, 1976.

Is the Linear Formula Adequate for Gases?

The linear approximation from Eq. (2.14), $\delta p \approx -\rho g\, \delta z$, is satisfactory for liquids, which are nearly incompressible. For gases, it is inaccurate unless δz is rather small. Problem P2.26 asks you to show, by binomial expansion of Eq. (2.20), that the error in using constant gas density to estimate δp from Eq. (2.14) is small if

$$\delta z \ll \frac{2T_0}{(n-1)B} \tag{2.21}$$

where T_0 is the local absolute temperature, B is the lapse rate from Eq. (2.19), and $n = g/(RB)$ is the exponent in Eq. (2.20). The error is less than 1 percent if $\delta z < 200$ m.

2.4 Application to Manometry

From the hydrostatic formula (2.14), a change in elevation $z_2 - z_1$ of a liquid is equivalent to a change in pressure $(p_2 - p_1)/\gamma$. Thus, a static column of one or more liquids or gases can be used to measure pressure differences between two points. Such a device is called a *manometer*. If multiple fluids are used, we must change the density in the formula as we move from one fluid to another. Figure 2.8 illustrates the use of the formula with a column of multiple fluids. The pressure change through each fluid is calculated separately. If we wish to know the total change $p_5 - p_1$, we add the successive changes $p_2 - p_1$, $p_3 - p_2$, $p_4 - p_3$, and $p_5 - p_4$. The intermediate values of p cancel, and we have, for the example of Fig. 2.8,

$$p_5 - p_1 = -\gamma_0(z_2 - z_1) - \gamma_w(z_3 - z_2) - \gamma_G(z_4 - z_3) - \gamma_M(z_5 - z_4) \tag{2.22}$$

No additional simplification is possible on the right-hand side because of the different densities. Notice that we have placed the fluids in order from the lightest on top to the heaviest at bottom. This is the only stable configuration. If we attempt to layer them in any other manner, the fluids will overturn and seek the stable arrangement.

Pressure Increases Downward

The basic hydrostatic relation, Eq. (2.14), is mathematically correct but vexing to engineers because it combines two negative signs to have the pressure increase downward. When calculating hydrostatic pressure changes, engineers work instinctively by simply having the pressure increase downward and decrease upward. If point 2 is a distance h below point 1 in a uniform liquid, then $p_2 = p_1 + \rho g h$. In the meantime, Eq. (2.14) remains accurate and safe if used properly. For example, Eq. (2.22) is correct as shown, or it could be rewritten in the following "multiple downward increase" mode:

$$p_5 = p_1 + \gamma_0\,|z_1 - z_2| + \gamma_w\,|z_2 - z_3| + \gamma_G\,|z_3 - z_4| + \gamma_M\,|z_4 - z_5|$$

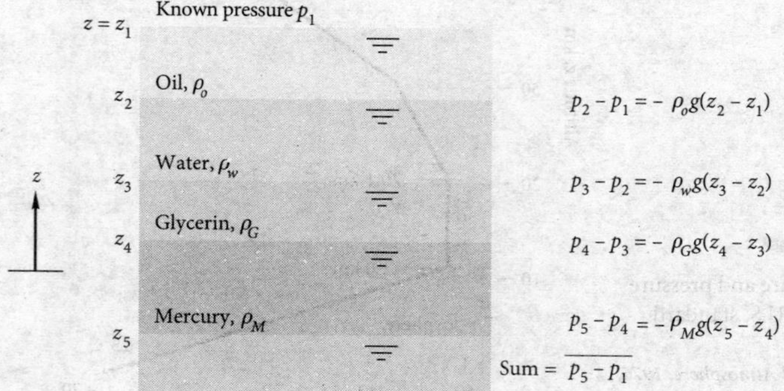

Fig. 2.8 Evaluating pressure changes through a column of multiple fluids.

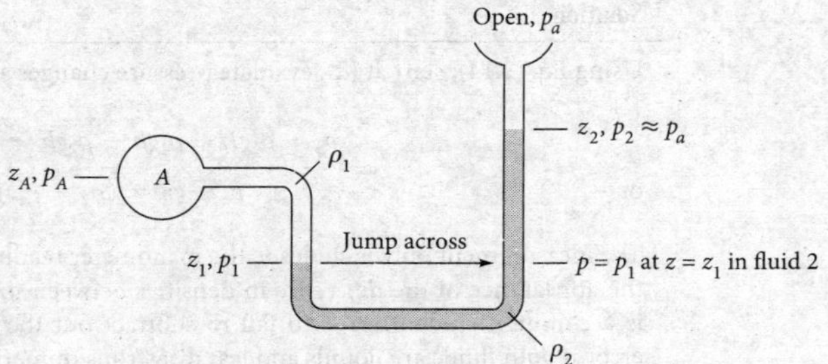

Fig. 2.9 Simple open manometer for measuring p_A relative to atmospheric pressure.

That is, keep adding on pressure increments as you move down through the layered fluid. A different application is a manometer, which involves both "up" and "down" calculations.

Application: A Simple Manometer

Figure 2.9 shows a simple U-tube open manometer that measures the *gage* pressure p_A relative to the atmosphere, p_a. The chamber fluid ρ_1 is separated from the atmosphere by a second, heavier fluid ρ_2, perhaps because fluid A is corrosive, or more likely because a heavier fluid ρ_2 will keep z_2 small and the open tube can be shorter.

We first apply the hydrostatic formula (2.14) from A down to z_1. Note that we can then go down to the bottom of the U-tube and back up on the right side to z_1, and the pressure will be the same, $p = p_1$. Thus, we can "jump across" and then up to level z_2:

$$p_A + \gamma_1 |z_A - z_1| - \gamma_2 |z_1 - z_2| = p_2 \approx p_{\text{atm}} \tag{2.23}$$

Another physical reason that we can "jump across" at section 1 is that a continuous length of the same fluid connects these two equal elevations. The hydrostatic relation (2.14) requires this equality as a form of Pascal's law:

> Any two points at the same elevation in a continuous mass of the same static fluid will be at the same pressure.

This idea of jumping across to equal pressures facilitates multiple-fluid problems. It will be inaccurate, however, if there are bubbles in the fluid.

EXAMPLE 2.3

The classic use of a manometer is when two U-tube legs are of equal length, as in Fig. E2.3, and the measurement involves a pressure difference across two horizontal points. The typical application is to measure pressure change across a flow device, as shown. Derive a formula for the pressure difference $p_a - p_b$ in terms of the system parameters in Fig. E2.3.

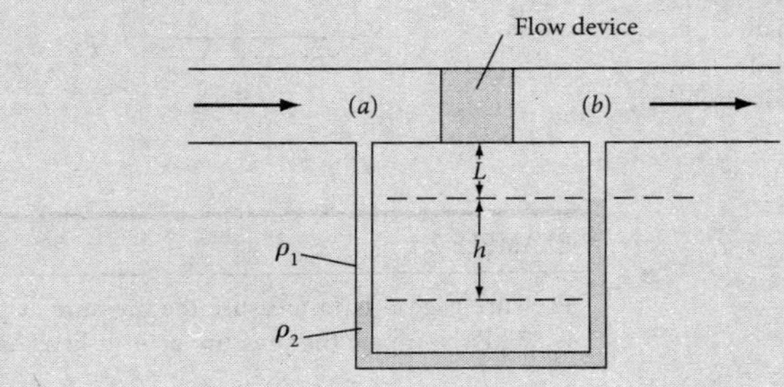

E2.3

Solution

Using Eq. (2.14), start at (*a*), evaluate pressure changes around the U-tube, and end up at (*b*):

$$p_a + \rho_1 gL + \rho_1 gh - \rho_2 gh - \rho_1 gL = p_b$$

or

$$p_a - p_b = (\rho_2 - \rho_1)gh \qquad \textit{Ans.}$$

The measurement only includes *h*, the manometer reading. Terms involving *L* drop out. Note the appearance of the *difference* in densities between manometer fluid and working fluid. It is a common student error to fail to subtract out the working fluid density ρ_1—a serious error if both fluids are liquids and less disastrous numerically if fluid 1 is a gas. Academically, of course, such an error is always considered serious by fluid mechanics instructors.

Although Example 2.3, because of its popularity in engineering experiments, is sometimes considered to be the "manometer formula," it is best *not* to memorize it but rather to adapt Eq. (2.14) to each new multiple-fluid hydrostatics problem. For example, Fig. 2.10 illustrates a multiple-fluid manometer problem for finding the difference in pressure between two chambers *A* and *B*. We repeatedly apply Eq. (2.14), jumping across at equal pressures when we come to a continuous mass of the same fluid. Thus, in Fig. 2.10, we compute four pressure differences while making three jumps:

$$\begin{aligned} p_A - p_B &= (p_A - p_1) + (p_1 - p_2) + (p_2 - p_3) + (p_3 - p_B) \\ &= -\gamma_1(z_A - z_1) - \gamma_2(z_1 - z_2) - \gamma_3(z_2 - z_3) - \gamma_4(z_3 - z_B) \end{aligned} \tag{2.24}$$

The intermediate pressures $p_{1,2,3}$ cancel. It looks complicated, but really it is merely *sequential*. One starts at *A*, goes down to 1, jumps across, goes up to 2, jumps across, goes down to 3, jumps across, and finally goes up to *B*.

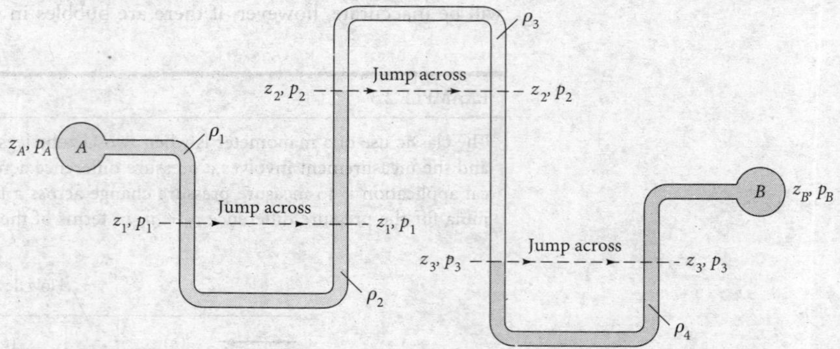

Fig. 2.10 A complicated multiple-fluid manometer to relate p_A to p_B. This system is not especially practical but makes a good homework or examination problem.

EXAMPLE 2.4

Pressure gage *B* is to measure the pressure at point *A* in a water flow. If the pressure at *B* is 87 kPa, estimate the pressure at *A* in kPa. Assume all fluids are at 20°C. See Fig. E2.4.

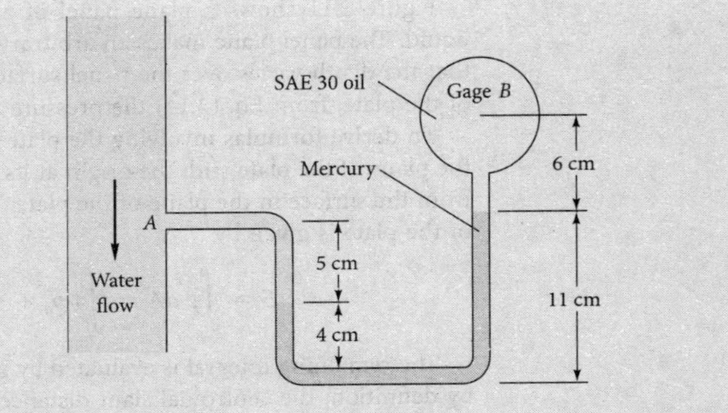

E2.4

Solution

- *System sketch:* The system is shown in Fig. E2.4.
- *Assumptions:* Hydrostatic fluids, no mixing, vertical "up" in Fig. E2.4.
- *Approach:* Sequential use of Eq. (2.14) to go from A to B.
- *Property values:* From Table 2.1 or Table A.3:

$$\gamma_{water} = 9{,}790\ \text{N/m}^3; \quad \gamma_{mercury} = 133{,}100\ \text{N/m}^3; \quad \gamma_{oil} = 8{,}720\ \text{N/m}^3$$

- *Solution steps:* Proceed from A to B, "down" then "up," jumping across at the left mercury meniscus:

$$p_A + \rho_w \mid \Delta z \mid_w - \gamma_m \mid \Delta z_m \mid - \gamma_o \mid \Delta z \mid_o = p_B$$

or $p_A + (9{,}790\ \text{N/m}^3)(0.05\ \text{m}) - (133{,}100\ \text{N/m}^3)(0.07\ \text{m}) - (8{,}720\ \text{N/m}^3)(0.06\ \text{m}) = 87{,}000$

or $p_A + 490 - 9{,}317 - 523 = 87{,}000$ Solve for $p_A = 96{,}350\ \text{N/m}^2 \approx 96.4\ \text{kPa}$ *Ans.*

- *Comments:* Note that we abbreviated the units N/m^2 to pascals, or Pa. The intermediate five-figure result, $p_A = 96{,}350$ Pa, is unrealistic, since the data are known to only about three significant figures.

In making these manometer calculations we have neglected the capillary height changes due to surface tension, which were discussed in Example 1.8. These effects cancel if there is a fluid interface, or *meniscus,* between similar fluids on both sides of the U-tube. Otherwise, as in the right-hand U-tube of Fig. 2.10, a capillary correction can be made or the effect can be made negligible by using large-bore ($\geq$ 1 cm) tubes.

2.5 Hydrostatic Forces on Plane Surfaces

The design of containment structures requires computation of the hydrostatic forces on various solid surfaces adjacent to the fluid. These forces relate to the weight of fluid bearing on the surface. For example, a container with a flat, horizontal bottom of area A_b and water depth H will experience a downward bottom force $F_b = \gamma H A_b$. If the surface is not horizontal, additional computations are needed to find the horizontal components of the hydrostatic force.

If we neglect density changes in the fluid, Eq. (2.14) applies and the pressure on any submerged surface varies linearly with depth. For a plane surface, the linear stress distribution is exactly analogous to combined bending and compression of a beam in strength-of-materials theory. The hydrostatic problem thus reduces to simple formulas involving the centroid and moments of inertia of the plate cross-sectional area.

Figure 2.11 shows a plane panel of arbitrary shape completely submerged in a liquid. The panel plane makes an arbitrary angle θ with the horizontal free surface, so that the depth varies over the panel surface. If h is the depth to any element area dA of the plate, from Eq. (2.14) the pressure there is $p = p_a + \gamma h$.

To derive formulas involving the plate shape, establish an xy coordinate system in the plane of the plate with the origin at its centroid, plus a dummy coordinate ξ down from the surface in the plane of the plate. Then the total hydrostatic force on one side of the plate is given by

$$F = \int p\, dA = \int (p_a + \gamma h)\, dA = p_a A + \gamma \int h\, dA \tag{2.25}$$

The remaining integral is evaluated by noticing from Fig. 2.11 that $h = \xi \sin\theta$ and, by definition, the centroidal slant distance from the surface to the plate is

$$\xi_{CG} = \frac{1}{A}\int \xi\, dA$$

Therefore, since θ is constant along the plate, Eq. (2.25) becomes

$$F = p_a A + \gamma \sin\theta \int \xi\, dA = p_a A + \gamma \sin\theta\, \xi_{CG} A$$

Finally, unravel this by noticing that $\xi_{CG} \sin\theta = h_{CG}$, the depth straight down from the surface to the plate centroid. Thus

$$\boxed{F = p_a A + \gamma h_{CG} A = (p_a + \gamma h_{CG}) A = p_{CG} A} \tag{2.26}$$

The force on one side of any plane submerged surface in a uniform fluid equals the pressure at the plate centroid times the plate area, independent of the shape of the plate or the angle θ at which it is slanted.

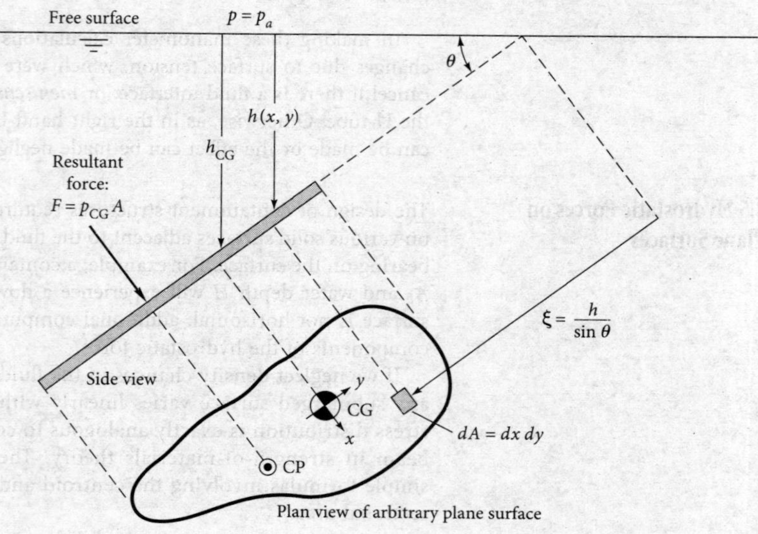

Fig. 2.11 Hydrostatic force and center of pressure on an arbitrary plane surface of area A inclined at an angle θ below the free surface.

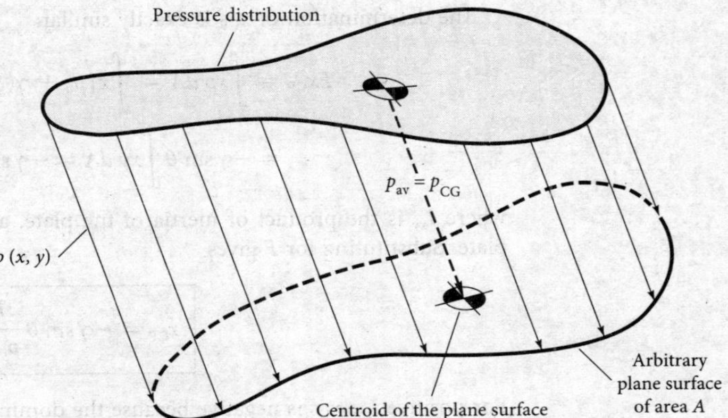

Fig. 2.12 The hydrostatic pressure force on a plane surface is equal, regardless of its shape, to the resultant of the three-dimensional linear pressure distribution on that surface $F = p_{CG}A$.

Equation (2.26) can be visualized physically in Fig. 2.12 as the resultant of a linear stress distribution over the plate area. This simulates combined compression and bending of a beam of the same cross section. It follows that the "bending" portion of the stress causes no force if its "neutral axis" passes through the plate centroid of area. Thus, the remaining "compression" part must equal the centroid stress times the plate area. This is the result of Eq. (2.26).

However, to balance the bending-moment portion of the stress, the resultant force F acts not through the centroid but below it toward the high-pressure side. Its line of action passes through the *center of pressure* CP of the plate, as sketched in Fig. 2.11. To find the coordinates (x_{CP}, y_{CP}), we sum moments of the elemental force $p\,dA$ about the centroid and equate to the moment of the resultant F. To compute y_{CP}, we equate

$$Fy_{CP} = \int yp\,dA = \int y(p_a + \gamma\xi \sin\theta)\,dA = \gamma \sin\theta \int y\xi\,dA$$

The term $\int p_a y\,dA$ vanishes by definition of centroidal axes. Introducing $\xi = \xi_{CG} - y$, we obtain

$$Fy_{CP} = \gamma \sin\theta \left(\xi_{CG} \int y\,dA - \int y^2\,dA \right) = -\gamma \sin\theta\, I_{xx}$$

where again $\int y\,dA = 0$ and I_{xx} is the area moment of inertia of the plate area about its centroidal x axis, computed in the plane of the plate. Substituting for F gives the result

$$y_{CP} = -\gamma \sin\theta \frac{I_{xx}}{p_{CG}A} \tag{2.27}$$

The negative sign in Eq. (2.27) shows that y_{CP} is below the centroid at a deeper level and, unlike F, depends on angle θ. If we move the plate deeper, y_{CP} approaches the centroid because every term in Eq. (2.27) remains constant except p_{CG}, which increases.

The determination of x_{CP} is exactly similar:

$$Fx_{CP} = \int xp\, dA = \int x[p_a + \gamma(\xi_{CG} - y)\sin\theta]\, dA$$

$$= -\gamma \sin\theta \int xy\, dA = -\gamma \sin\theta\, I_{xy}$$

where I_{xy} is the product of inertia of the plate, again computed in the plane of the plate. Substituting for F gives

$$x_{CP} = -\gamma \sin\theta \frac{I_{xy}}{p_{CG}A} \tag{2.28}$$

For positive I_{xy}, x_{CP} is negative because the dominant pressure force acts in the third, or lower left, quadrant of the panel. If $I_{xy} = 0$, usually implying symmetry, $x_{CP} = 0$ and the center of pressure lies directly below the centroid on the y axis.

Gage Pressure Formulas

In most cases the ambient pressure p_a is neglected because it acts on both sides of the plate; for example, the other side of the plate is inside a ship or on the dry side of a gate or dam. In this case $p_{CG} = \gamma h_{CG}$, and the center of pressure becomes independent of specific weight:

$$F = \gamma h_{CG} A \qquad y_{CP} = -\frac{I_{xx}\sin\theta}{h_{CG}A} \qquad x_{CP} = -\frac{I_{xy}\sin\theta}{h_{CG}A} \tag{2.29}$$

Figure 2.13 gives the area and moments of inertia of several common cross sections for use with these formulas. Note that θ is the angle between the plate and the horizon.

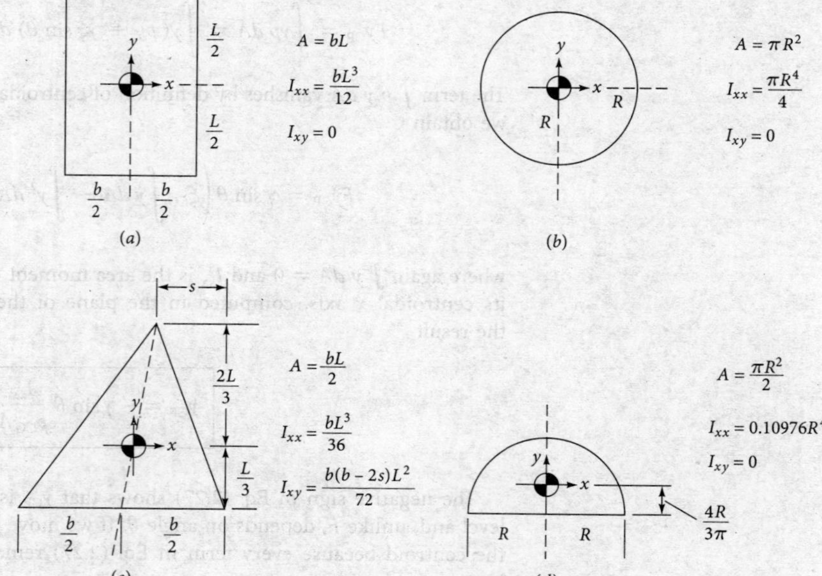

Fig. 2.13 Centroidal moments of inertia for various cross sections: (*a*) rectangle, (*b*) circle, (*c*) triangle, and (*d*) semicircle.

EXAMPLE 2.5

The gate in Fig. E2.5*a* is 1.5 m wide, is hinged at point *B*, and rests against a smooth wall at point *A*. Compute (*a*) the force on the gate due to seawater pressure, (*b*) the horizontal force *P* exerted by the wall at point *A*, and (*c*) the reactions at the hinge *B*.

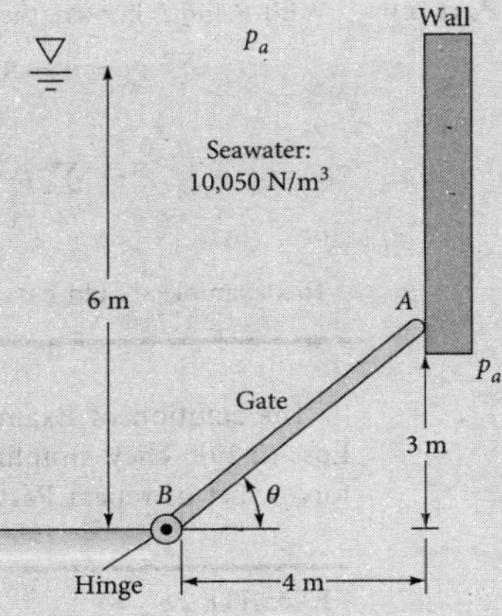

E2.5*a*

Solution

Part (a) By geometry the gate is 5 m long from *A* to *B*, and its centroid is halfway between, or at elevation 1.5 m above point *B*. The depth h_{CG} is thus 6 m − 1.5 m = 4.5 m. The gate area is 1.5 m (5 m) = 7.5 m². Neglect p_a as acting on both sides of the gate. From Eq. (2.26) the hydrostatic force on the gate is

$$F = p_{CG}A = \gamma h_{CG}A = (10{,}050\ \text{N/m}^3)(4.5\ \text{m})(7.5\ \text{m}^2) = 339{,}200\ \text{N} \qquad \textit{Ans. (a)}$$

Part (b) First we must find the center of pressure of *F*. A free-body diagram of the gate is shown in Fig. E2.5*b*. The gate is a rectangle, hence

$$I_{xy} = 0 \text{ and } I_{xx} = \frac{bL^3}{12} = \frac{(1.5\ \text{m})(5\ \text{m})^3}{12} = 15.63\ \text{m}^4$$

The distance *l* from the CG to the CP is given by Eqs. (2.29) since p_a is neglected.

$$l = -y_{CP} = +\frac{I_{xx}\sin\theta}{h_{CG}A} = \frac{(15.63\ \text{m}^4)(3\ \text{m}/5\ \text{m})}{(4.5\ m)(7.5\ m^2)} = 0.278\ \text{m}$$

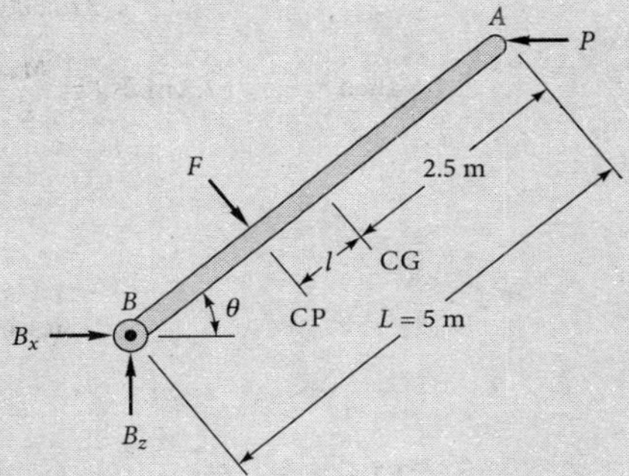

E2.5*b*

The distance from point B to force F is thus 5 m $- l -$ 2.5 m = 2.222 m. Summing the moments counterclockwise about B gives

$$PL \sin \theta - F(5 - l) = P(3\text{ m}) - (339{,}200\text{ N})(2.222\text{ m}) = 0$$

or
$$P = 251{,}200\text{ N} \qquad \textit{Ans. (b)}$$

Part (c) With F and P known, the reactions B_x and B_z are found by summing forces on the gate:

$$\sum F_x = 0 = B_x + F \sin \theta - P = B_x + (339{,}200\text{ N})\ (0.6) - 251{,}200\text{ N}$$

or
$$B_x = 47{,}700\text{ N}$$

$$\sum F_z = 0 = B_z - F \cos \theta = B_z - (339{,}200\text{ N})\ (0.8)$$

or
$$B_z = 271{,}400\text{ N} \qquad \textit{Ans. (c)}$$

This example should have reviewed your knowledge of statics.

The solution of Example 2.5 was achieved with the moment of inertia formulas, Eqs. (2.29). They simplify the calculations, but one loses a physical feeling for the forces. Let us repeat Parts (*a*) and (*b*) of Example 2.5 using a more visual approach.

EXAMPLE 2.6

Repeat Example 2.5 to sketch the pressure distribution on plate AB, and break this distribution into rectangular and triangular parts to solve for (*a*) the force on the plate and (*b*) the center of pressure.

Solution

Part (a) Point A is 3 m deep, hence $p_A = \gamma h_A = (10{,}050\text{ N/m}^3)(3\text{ m}) = 30{,}150\text{ N/m}^2$. Similarly, Point B is 6 m deep, hence $p_B = \gamma h_B = (10{,}050\text{ N/m}^3)(6\text{ m}) = 60{,}300\text{ N/m}^2$. This defines the linear pressure distribution in Fig. E2.6. The rectangle is 30,150 N/m^2 by (5 m) (1.5 m) into the paper. The triangle is (60,300 N/m^2 − 30,150 N/m^2) = 30,150 N/m^2 × (5 m) (1.5 m). The centroid of the rectangle is 2.5 m down the plate from A. The centroid of the triangle is 3.33 m down from A. The total force is the rectangle force plus the triangle force:

$$F = (30{,}150\text{ N/m}^2)\ (5\text{ m})(1.5\text{ m}) + (30{,}150/2\text{ N/m}^2)(5\text{ m})(1.5\text{ m})$$
$$= 226{,}130\text{ N} + 113{,}060\text{ N} = 339{,}200\text{ N} \qquad \textit{Ans. (a)}$$

Part (b) The moments of these forces about point A are

$$\sum M_A = (226{,}130\text{ N})(2.5\text{ m}) + (113{,}060\text{ N})(3.33\text{ m}) = 565{,}300\text{Nm}$$
$$+ 376{,}500\text{ Nm} = 941{,}800\text{ Nm}$$

Then
$$2.5\text{ m} + l = \frac{M_A}{F} = \frac{941{,}800\text{ Nm}}{339{,}200\text{ N}} = 2.777\text{ m} \text{ hence } l = 0.277\text{ m} \qquad \textit{Ans. (b)}$$

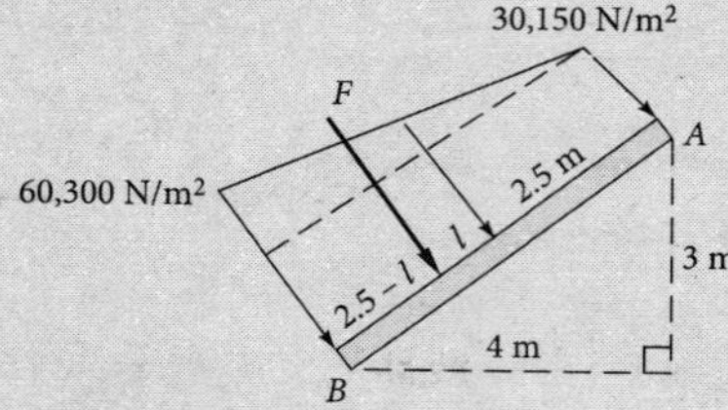

E2.6

Comment: We obtain the same force and center of pressure as in Example 2.5 but with more understanding. However, this approach is awkward and laborious if the plate is not a rectangle. It would be difficult to solve Example 2.7 with the pressure distribution alone because the plate is a triangle. Thus, moments of inertia can be a useful simplification.

EXAMPLE 2.7

A tank of oil has a right-triangular panel near the bottom, as in Fig. E2.7. Omitting p_a, find the (*a*) hydrostatic force and (*b*) CP on the panel.

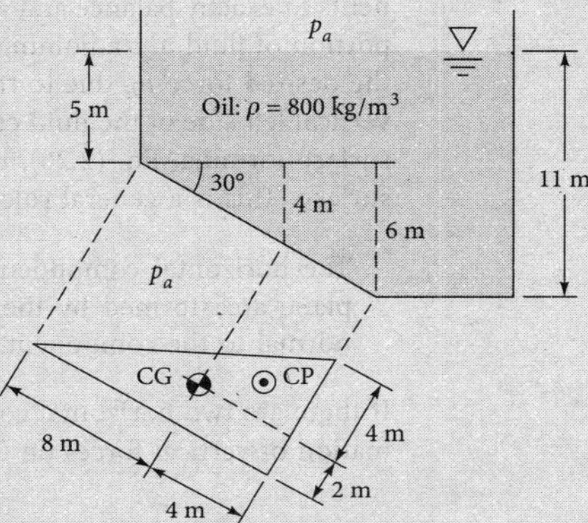

E2.7

Solution

Part (a) The triangle has properties given in Fig. 2.13*c*. The centroid is one-third up (4 m) and one-third over (2 m) from the lower left corner, as shown. The area is

$$\tfrac{1}{2}(6\text{ m})(12\text{ m}) = 36\text{ m}^2$$

The moments of inertia are

$$I_{xx} = \frac{bL^3}{36} = \frac{(6\text{ m})(12\text{ m})^3}{36} = 288\text{ m}^4$$

and
$$I_{xy} = \frac{b(b-2s)L^2}{72} = \frac{(6\text{ m})[6\text{ m} - 2(6\text{ m})](12\text{ m})^2}{72} = -72\text{ m}^4$$

The depth to the centroid is $h_{CG} = 5 + 4 = 9$ m; thus, the hydrostatic force from Eq. (2.26) is

$$F = \rho g h_{CG} A = (800\text{ kg/m}^3)(9.807\text{ m/s}^2)(9\text{ m})(36\text{ m}^2)$$

$$= 2.54 \times 10^6\ (\text{kg}\cdot\text{m})/\text{s}^2 = 2.54 \times 10^6\text{ N} = 2.54\text{ MN} \qquad \textit{Ans. (a)}$$

Part (b) The CP position is given by Eqs. (2.29):

$$y_{CP} = -\frac{I_{xx}\sin\theta}{h_{CG}A} = -\frac{(288\text{ m}^4)(\sin 30°)}{(9\text{ m})(36\text{ m}^2)} = -0.444\text{ m}$$

$$x_{CP} = -\frac{I_{xy}\sin\theta}{h_{CG}A} = -\frac{(-72\text{ m}^4)(\sin 30°)}{(9\text{ m})(36\text{ m}^2)} = +0.111\text{ m} \qquad \textit{Ans. (b)}$$

The resultant force $F = 2.54$ MN acts through this point, which is down and to the right of the centroid, as shown in Fig. E2.7.

2.6 Hydrostatic Forces on Curved Surfaces

The resultant pressure force on a curved surface is most easily computed by separating it into horizontal and vertical components. Consider the arbitrary curved surface sketched in Fig. 2.14*a*. The incremental pressure forces, being normal to the local area element, vary in direction along the surface, and thus cannot be added numerically. We could sum the separate three components of these elemental pressure forces, but it turns out that we need not perform a laborious three-way integration.

Figure 2.14*b* shows a free-body diagram of the column of fluid contained in the vertical projection above the curved surface. The desired forces F_H and F_V are exerted by the surface on the fluid column. Other forces are shown due to fluid weight and horizontal pressure on the vertical sides of this column. The column of fluid must be in static equilibrium. On the upper part of the column *bcde*, the horizontal components F_1 exactly balance and are not relevant to the discussion. On the lower, irregular portion of fluid *abc* adjoining the surface, summation of horizontal forces shows that the desired force F_H due to the curved surface is exactly equal to the force F_H on the vertical left side of the fluid column. This left-side force can be computed by the plane surface formula, Eq. (2.26), based on a vertical projection of the area of the curved surface. This is a general rule and simplifies the analysis:

> The horizontal component of force on a curved surface equals the force on the plane area formed by the projection of the curved surface onto a vertical plane normal to the component.

If there are two horizontal components, both can be computed by this scheme. Summation of vertical forces on the fluid free body then shows that

$$F_V = W_1 + W_2 + W_{\text{air}} \tag{2.30}$$

We can state this in words as our second general rule:

> The vertical component of pressure force on a curved surface equals in magnitude and direction the weight of the entire column of fluid, both liquid and atmosphere, above the curved surface.

Thus, the calculation of F_V involves little more than finding centers of mass of a column of fluid—perhaps a little integration if the lower portion *abc* in Fig. 2.14*b* has a particularly vexing shape.

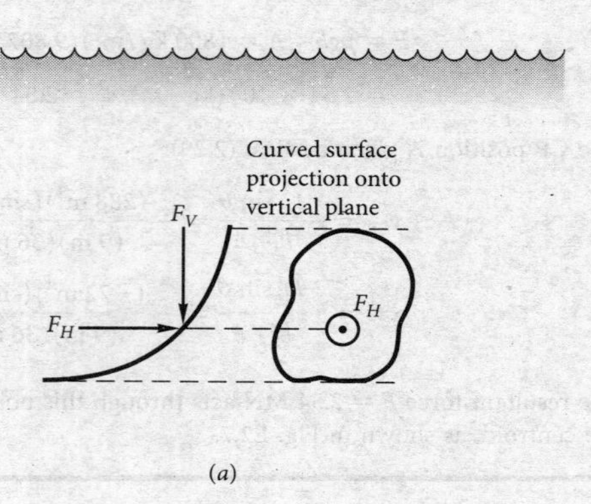

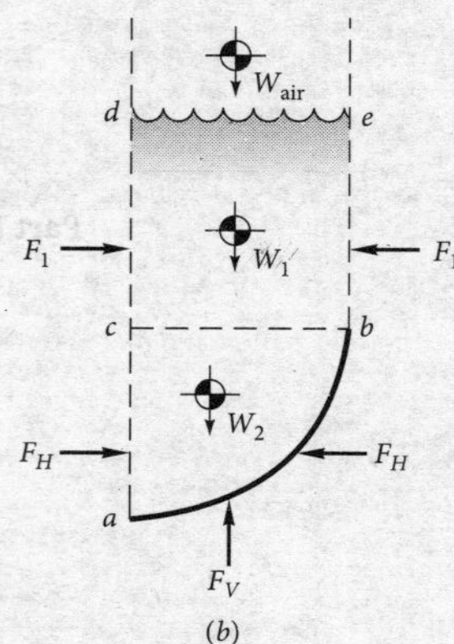

Fig. 2.14 Computation of hydrostatic force on a curved surface: (*a*) submerged curved surface; (*b*) free-body diagram of fluid above the curved surface.

EXAMPLE 2.8

A dam has a parabolic shape $z/z_0 = (x/x_0)^2$ as shown in Fig. E2.8*a*, with $x_0 = 3$ m and $z_0 =$ 7 m. The fluid is water, $\gamma = 9{,}810$ N/m^3, and atmospheric pressure may be omitted. Compute the forces F_H and F_V on the dam and their line of action. The width of the dam is 15 m.

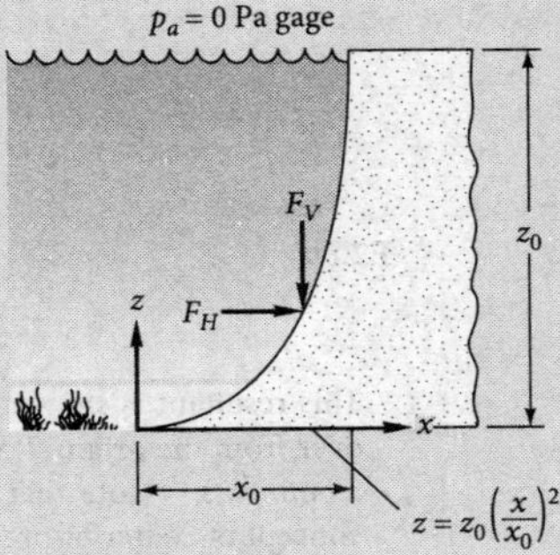

E2.8*a*

Solution

- *System sketch:* Figure E2.8*b* shows the various dimensions. The dam width is $b = 15$ m.
- *Approach:* Calculate F_H and its line of action from Eqs. (2.26) and (2.29). Calculate F_V and its line of action by finding the weight of fluid above the parabola and the centroid of this weight.
- *Solution steps for the horizontal component:* The vertical projection of the parabola lies along the z axis in Fig. E2.8*b* and is a rectangle 7 m high and 15 m wide. Its centroid is halfway down, or $h_{CG} = (7\text{ m})/2 = 3.5$ m. Its area is $A_{proj} = (7\text{ m})(15\text{ m}) = 105\text{ m}^2$. Then, from Eq. (2.26),

$$F_H = \gamma h_{CG} A_{proj} = (9{,}810\text{ N/m}^3)(3.5\text{ m})(105\text{ m}^2) = 3{,}605{,}000\text{ N} \approx 3{,}605\text{ kN}$$

The line of action of F_H is below the centroid of A_{proj}, as given by Eq. (2.29):

$$y_{CP,\,proj} = -\frac{I_{xx}\sin\theta}{h_{CG}A_{proj}} = -\frac{(1/12)(15\text{ m})(7\text{ m})^3\sin 90°}{(3.5\text{ m})(105\text{ m}^2)} = -1.167\text{ m}$$

Thus, F_H is 3.5 m + 1.167 m = 4.667 m, or two-thirds of the way down from the surface (2.333 m up from the bottom).

- *Comments:* Note that you calculate F_H and its line of action from the *vertical projection* of the parabola, not from the parabola itself. Since this projection is *vertical,* its angle $\theta = 90°$.
- *Solution steps for the vertical component:* The vertical force F_V equals the weight of water above the parabola. Alas, a parabolic section is not in Fig. 2.13, so we had to look it up in another book. The area and centroid are shown in Fig. E2.8*b*. The weight of this parabolic amount of water is

$$F_V = \gamma A_{section} b = (9{,}810\text{ N/m}^3)\left[\frac{2}{3}(7\text{ m})(3\text{ m})\right](15\text{ m}) = 2{,}060{,}000\text{ N} = 2{,}060\text{ kN}$$

This force acts downward, through the centroid of the parabolic section, or at a distance $3x_0/8 = 1.125$ m over from the origin, as shown in Figs. E2.8*b,c*. The resultant hydrostatic force on the dam is

$$F = (F_H^2 + F_V^2)^{1/2} = [(3{,}605\text{ kN})^2 + (2{,}060\text{ kN})^2]^{1/2} = 4{,}152\text{ kN at } 29° \qquad \textit{Ans.}$$

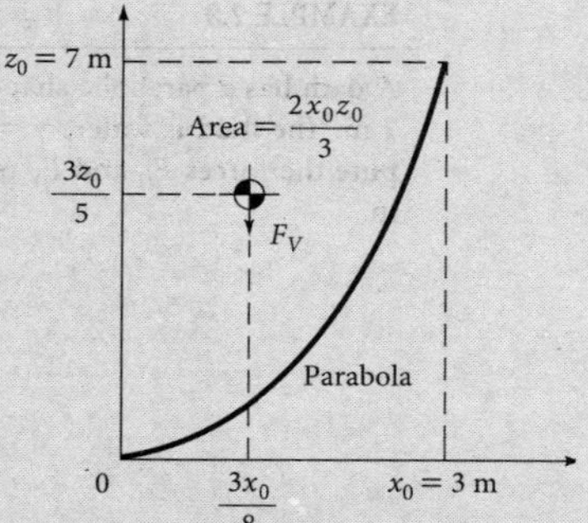

E2.8*b*

This resultant is shown in Fig. E2.8*c* and passes through a point 2.333 m up and 1.125 m over from the origin. It strikes the dam at a point 1.623 m over and 2.049 m up, as shown.

- *Comments:* Note that entirely different formulas are used to calculate F_H and F_V. The concept of center of pressure CP is, in the writer's opinion, stretched too far when applied to curved surfaces.

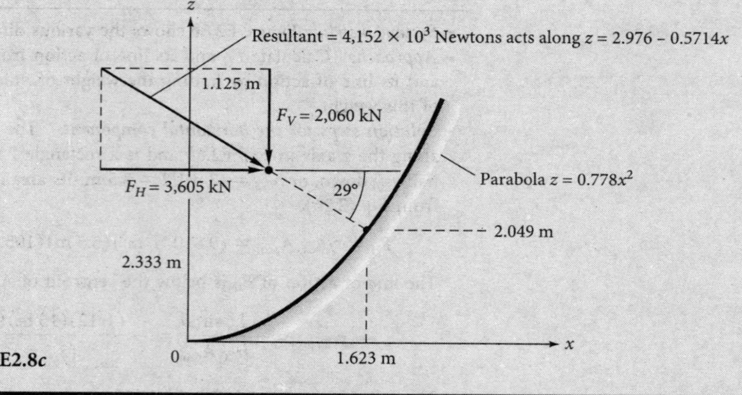

E2.8*c*

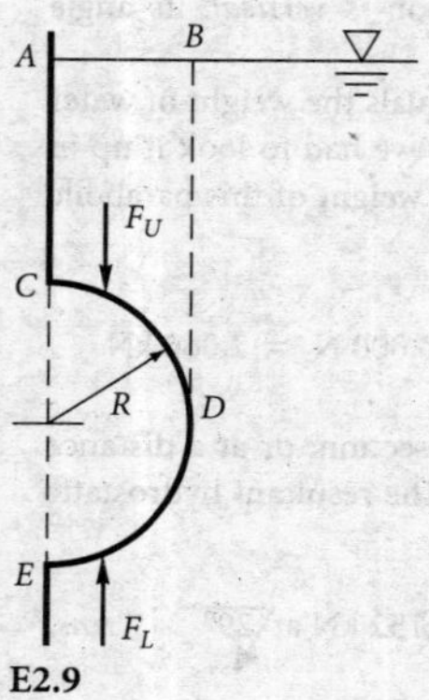

E2.9

EXAMPLE 2.9

Find an algebraic formula for the net vertical force *F* on the submerged semicircular projecting structure *CDE* in Fig. E2.9. The structure has uniform width *b* into the paper. The liquid has specific weight γ.

Solution

The net force is the difference between the upward force F_L on the lower surface *DE* and the downward force F_U on the upper surface *CD*, as shown in Fig. E2.9. The force F_U equals γ times the volume *ABDC* above surface *CD*. The force F_L equals γ times the volume *ABDEC* above surface *DE*. The latter is clearly larger. The difference is γ times the volume of the structure itself. Thus, the net upward fluid force on the semicylinder is

$$F = \gamma_{\text{fluid}}\,(\text{volume } CDE) = \gamma_{\text{fluid}}\,\frac{\pi}{2}R^2 b \qquad \textit{Ans.}$$

This is the principle upon which the laws of buoyancy, Sec. 2.8, are founded. Note that the result is independent of the depth of the structure and depends upon the specific weight of the *fluid*, not the material within the structure.

2.7 Hydrostatic Forces in Layered Fluids

The formulas for plane and curved surfaces in Secs. 2.5 and 2.6 are valid only for a fluid of uniform density. If the fluid is layered with different densities, as in Fig. 2.15, a single formula cannot solve the problem because the slope of the linear pressure distribution changes between layers. However, the formulas apply separately to each layer, and thus the appropriate remedy is to compute and sum the separate layer forces and moments.

Consider the slanted plane surface immersed in a two-layer fluid in Fig. 2.15. The slope of the pressure distribution becomes steeper as we move down into the denser second layer. The total force on the plate does *not* equal the pressure at the centroid times the plate area, but the plate portion in each layer does satisfy the formula, so that we can sum forces to find the total:

$$F = \sum F_i = \sum p_{CG_i} A_i \tag{2.31}$$

Similarly, the centroid of the plate portion in each layer can be used to locate the center of pressure on that portion:

$$y_{CP_i} = -\frac{\rho_i g \sin\theta_i\, I_{xx_i}}{p_{CG_i} A_i} \qquad x_{CP_i} = -\frac{\rho_i g \sin\theta_i\, I_{xy_i}}{p_{CG_i} A_i} \tag{2.32}$$

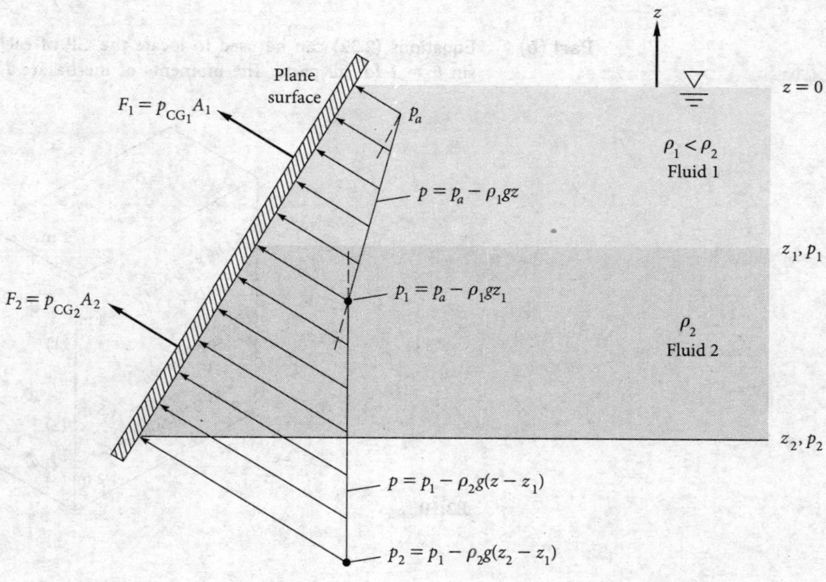

Fig. 2.15 Hydrostatic forces on a surface immersed in a layered fluid must be summed in separate pieces.

These formulas locate the center of pressure of that particular F_i with respect to the centroid of that particular portion of plate in the layer, not with respect to the centroid of the entire plate. The center of pressure of the total force $F = \sum F_i$ can then be found by summing moments about some convenient point such as the surface. The following example will illustrate this.

EXAMPLE 2.10

A tank 6 m deep and 2 m wide is layered with 2.4 m of oil, 1.8 m of water, and 1.2 m of mercury. Compute (*a*) the total hydrostatic force and (*b*) the resultant center of pressure of the fluid on the right-hand side of the tank.

Solution

Part (a) Divide the end panel into three parts as sketched in Fig. E2.10, and find the hydrostatic pressure at the centroid of each part, using the relation (2.26) in steps as in Fig. E2.10:

$$p_{CG_1} = (8{,}650\ \text{N/m}^3)(1.2\ \text{m}) = 10{,}380\ \text{Pa}$$

$$p_{CG_2} = (8{,}650\ \text{N/m}^3)(2.4\ \text{m}) + (9{,}810\ \text{N/m}^3)(0.9\ \text{m}) = 29{,}590\ \text{Pa}$$

$$p_{CG_3} = (8{,}650\ \text{N/m}^3)(2.4\ \text{m}) + (9{,}810\ \text{N/m}^3)(1.8\ \text{m}) + (133{,}000\ \text{N/m}^3)(0.6\ \text{m}) = 118{,}200\ \text{Pa}$$

These pressures are then multiplied by the respective panel areas to find the force on each portion:

$$F_1 = p_{CG_1}A_1 = (10{,}380\ \text{Pa})(2.4\ \text{m})(2\ \text{m}) = 49{,}800\ \text{N}$$

$$F_2 = p_{CG_2}A_2 = (29{,}590\ \text{Pa})(1.8\ \text{m})(2\ \text{m}) = 106{,}500\ \text{N}$$

$$F_3 = p_{CG_3}A_3 = (118{,}200\ \text{Pa})(1.2\ \text{m})(2\ \text{m}) = \underline{283{,}700\ \text{N}}$$

$$F = \sum F_i = 440{,}000\ \text{N} \qquad \textit{Ans. (a)}$$

Part (b) Equations (2.32) can be used to locate the CP of each force F_i, noting that $\theta = 90°$ and $\sin\theta = 1$ for all parts. The moments of inertia are $I_{xx_1} = (2\ \text{m})(2.4\ \text{m})^3/12 = 2.304\ \text{m}^4$,

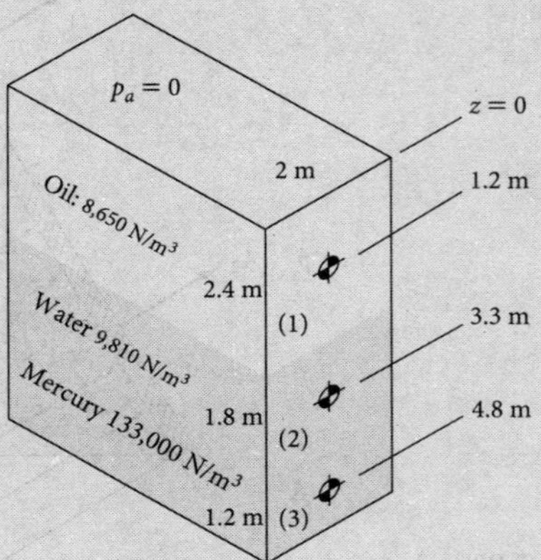

E2.10

$I_{xx_2} = (2\text{ m})(1.8\text{ m})^3/12 = 0.972\text{ m}^4$, and $I_{xx_3} = (2\text{ m})(1.2\text{ m})^3/12 = 0.288\text{ m}^4$. The centers of pressure are thus at

$$y_{CP_1} = -\frac{\rho_1 g I_{xx_1}}{F_1} = -\frac{(8{,}650\text{ N/m}^3)(2.304\text{ m}^4)}{49{,}800\text{ N}} = -0.446\text{ m}$$

$$y_{CP_2} = -\frac{(9{,}810\text{ N/m}^3)(0.972\text{ m}^4)}{106{,}500\text{ N}} = -0.0895\text{ m}$$

$$y_{CP_3} = -\frac{(133{,}000\text{ N/m}^3)(0.288\text{ m}^4)}{283{,}700\text{ N}} = -0.135\text{ m}$$

This locates $z_{CP_1} = -1.2\text{ m} - 0.446\text{ m} = -1.646\text{ m}$, $z_{CP_2} = 3.3\text{ m} - 0.0895\text{ m} = -3.390\text{ m}$, and $z_{CP_3} = -4.8\text{ m} - 0.135\text{ m} = -4.935\text{ m}$. Summing moments about the surface then gives

$$\sum F_i z_{CP_i} = F z_{CP}$$

or $(49{,}800\text{ N})(-1.646\text{ m}) + (106{,}500\text{ N})(-3.390\text{ m}) + (283{,}700\text{ N})(-4.935\text{ m}) = 440{,}000\text{ N}z_{CP}$

or $$z_{CP} = -\frac{1{,}843{,}000\text{ Nm}}{440{,}000\text{ N}} = -4.189\text{ m}$$ *Ans. (b)*

The center of pressure of the total resultant force on the right side of the tank lies 4.189 m below the surface.

2.8 Buoyancy and Stability

The same principles used to compute hydrostatic forces on surfaces can be applied to the net pressure force on a completely submerged or floating body. The results are the two laws of buoyancy discovered by Archimedes in the third century B.C.:

1. A body immersed in a fluid experiences a vertical buoyant force equal to the weight of the fluid it displaces.
2. A floating body displaces its own weight in the fluid in which it floats.

Archimedes (287–212 B.C.) was born and lived in the Greek city-state of Syracuse, on what is now the island of Sicily. He was a brilliant mathematician and engineer, two millennia ahead of his time. He calculated an accurate value for pi and approximated areas and volumes of various bodies by summing elemental shapes. In other words, he invented the integral calculus. He developed levers, pulleys, catapults, and a screw pump. Archimedes was the first to write large numbers as powers of 10, avoiding Roman numerals. And he deduced the principles of buoyancy, which we study here, when he realized how light he was when sitting in a bathtub.

Archimedes' two laws are easily derived by referring to Fig. 2.16. In Fig. 2.16*a*, the body lies between an upper curved surface 1 and a lower curved surface 2. From Eq. (2.30) for vertical force, the body experiences a net upward force

$$\begin{aligned} F_B &= F_V(2) - F_V(1) \\ &= (\text{fluid weight above 2}) - (\text{fluid weight above 1}) \\ &= \text{weight of fluid equivalent to body volume} \end{aligned} \tag{2.33}$$

Alternatively, from Fig. 2.16*b*, we can sum the vertical forces on elemental vertical slices through the immersed body:

$$F_B = \int_{\text{body}} (p_2 - p_1)\, dA_H = -\gamma \int (z_2 - z_1)\, dA_H = (\gamma)(\text{body volume}) \tag{2.34}$$

These are identical results and equivalent to Archimedes' law 1.

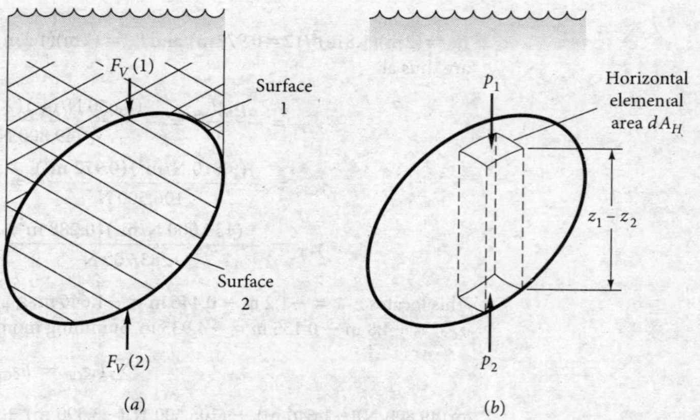

Fig. 2.16 Two different approaches to the buoyant force on an arbitrary immersed body: (*a*) forces on upper and lower curved surfaces; (*b*) summation of elemental vertical-pressure forces.

Equation (2.34) assumes that the fluid has uniform specific weight. The line of action of the buoyant force passes through the center of volume of the displaced body; that is, its center of mass computed as if it had uniform density. This point through which F_B acts is called the *center of buoyancy*, commonly labeled B or CB on a drawing. Of course, the point B may or may not correspond to the actual center of mass of the body's own material, which may have variable density.

Equation (2.34) can be generalized to a layered fluid (LF) by summing the weights of each layer of density ρ_i displaced by the immersed body:

$$(F_B)_{LF} = \sum \rho_i g(\text{displaced volume})_i \tag{2.35}$$

Each displaced layer would have its own center of volume, and one would have to sum moments of the incremental buoyant forces to find the center of buoyancy of the immersed body.

Since liquids are relatively heavy, we are conscious of their buoyant forces, but gases also exert buoyancy on any body immersed in them. For example, human beings have an average specific weight of about 9,400 N/m^3. We may record the weight of a person as 800 N and thus estimate the person's total volume as 0.085 m^3. However, in so doing we are neglecting the buoyant force of the air surrounding the person. At standard conditions, the specific weight of air is 11.47 N/m^3; hence the buoyant force is approximately 0.975 N. If measured in a vacuum, the person would weigh about 0.975 N more. For balloons and blimps the buoyant force of air, instead of being negligible, is the controlling factor in the design. Also, many flow phenomena, such as natural convection of heat and vertical mixing in the ocean, are strongly dependent on seemingly small buoyant forces.

Floating bodies are a special case; only a portion of the body is submerged, with the remainder poking up out of the free surface. This is illustrated in Fig. 2.17, where the shaded portion is the displaced volume. Equation (2.34) is modified to apply to this smaller volume:

$$F_B = (\gamma)(\text{displaced volume}) = \text{floating-body weight} \tag{2.36}$$

Not only does the buoyant force equal the body weight, but also they are *collinear* since there can be no net moments for static equilibrium. Equation (2.36) is the mathematical equivalent of Archimedes' law 2, previously stated.

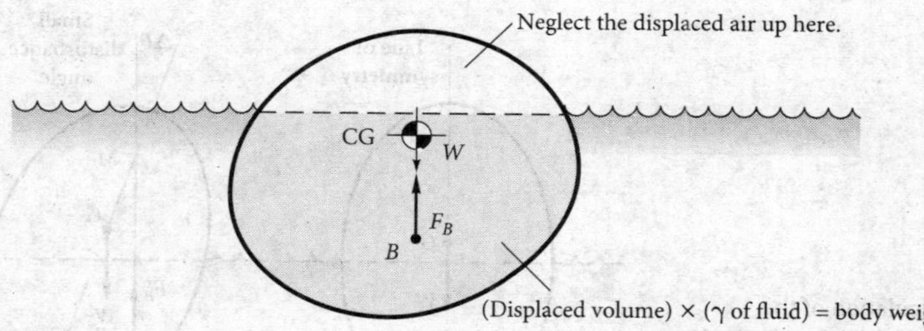

Fig. 2.17 Static equilibrium of a floating body.

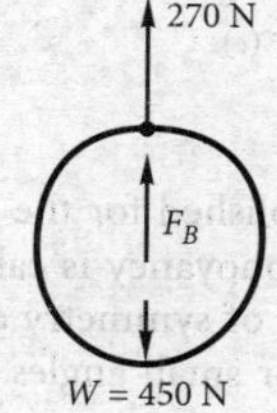

E2.11

EXAMPLE 2.11

A block of concrete weighs 450 N in air and "weighs" only 270 N when immersed in fresh water 9,810 N/m³. What is the average specific weight of the block?

Solution

A free-body diagram of the submerged block (see Fig. E2.11) shows a balance between the apparent weight, the buoyant force, and the actual weight:

$$\sum F_z = 0 = 270 \text{ N} + F_B - 450 \text{ N}$$

or

$$F_B = 180 \text{ N} = (9{,}810 \text{ N/m}^3)(\text{block volume, m}^3)$$

Solving gives the volume of the block as (180 N)/(9,810 N/m³) = 0.0183 m³. Therefore, the specific weight of the block is

$$\gamma_{\text{block}} = \frac{450 \text{ N}}{0.0183 \text{ m}^3} = 24{,}600 \text{ N/m}^3 \qquad \textit{Ans.}$$

Occasionally, a body will have exactly the right weight and volume for its ratio to equal the specific weight of the fluid. If so, the body will be *neutrally buoyant* and will remain at rest at any point where it is immersed in the fluid. Small, neutrally buoyant particles are sometimes used in flow visualization, and a neutrally buoyant body called a *Swallow float* [2] is used to track oceanographic currents. A submarine can achieve positive, neutral, or negative buoyancy by pumping water into or out of its ballast tanks.

Stability

A floating body as in Fig. 2.17 may not approve of the position in which it is floating. If so, it will overturn at the first opportunity and is said to be statically *unstable*, like a pencil balanced on its point. The least disturbance will cause it to seek another equilibrium position that is stable. Engineers must design to avoid floating instability. The only way to tell for sure whether a floating position is stable is to "disturb" the body a slight amount mathematically and see whether it develops a restoring moment that will return it to its original position. If so, it is stable; if not, unstable. Such calculations for arbitrary floating bodies have been honed to a fine art by naval architects [3], but we can at least outline the basic principle of the static stability calculation. Figure 2.18 illustrates the computation for the usual case of a symmetric floating body. The steps are as follows:

1. The basic floating position is calculated from Eq. (2.36). The body's center of mass G and center of buoyancy B are computed.

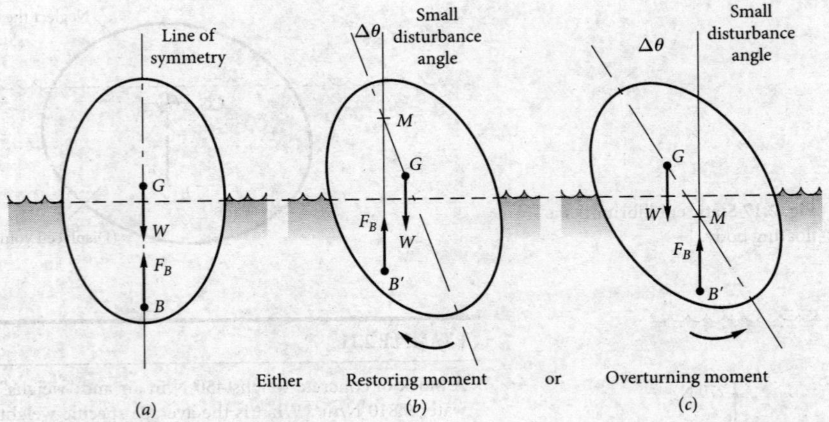

Fig. 2.18 Calculation of the metacenter M of the floating body shown in (*a*). Tilt the body a small angle $\Delta\theta$. Either (*b*) B' moves far out (point M above G denotes stability); or (*c*) B' moves slightly (point M below G denotes instability).

2. The body is tilted a small angle $\Delta\theta$, and a new waterline is established for the body to float at this angle. The new position B' of the center of buoyancy is calculated. A vertical line drawn upward from B' intersects the line of symmetry at a point M, called the *metacenter*, which is independent of $\Delta\theta$ for small angles.
3. If point M is above G (that is, if the *metacentric height* $\overline{MG}$ is positive), a restoring moment is present and the original position is stable. If M is below G (negative $\overline{MG}$), the body is unstable and will overturn if disturbed. Stability increases with increasing $\overline{MG}$.

Thus, the metacentric height is a property of the cross section for the given weight, and its value gives an indication of the stability of the body. For a body of varying cross section and draft, such as a ship, the computation of the metacenter can be very involved.

Stability Related to Waterline Area[1]

Naval architects [3] have developed the general stability concepts from Fig. 2.18 into a simple computation involving the area moment of inertia of the *waterline area* (as seen from above) about the axis of tilt. The derivation—see [3] for details—assumes that the body has a smooth shape variation (no discontinuities) near the waterline. Recall that M is the metacenter, B is the center of buoyancy, and G is the center of gravity. The final elegant formula relates the distances between these points:

$$\overline{MG} = \frac{I_O}{\upsilon_{sub}} - \overline{GB} \tag{2.37}$$

Where I_O is the area moment of inertia of the waterline area about the tilt axis O and υ_{sub} is the volume of the submerged portion of the floating body. It is desirable, of course, that $\overline{MG}$ be positive for the body to be stable.

The engineer locates G and B from the basic shape and design of the floating body and then calculates I_O and υ_{sub} to determine if $\overline{MG}$ is positive.

Engineering design counts upon effective operation of the results. A stability analysis is useless if the floating body runs aground on rocks, as in Fig. 2.19.

[1]This section may be omitted without loss of continuity.

Fig. 2.19 The Italian liner *Costa Concordia* aground on January 14, 2012. Stability analysis may fail when operator mistakes occur (*Associated Press photo/Gregorio Borgia*).

EXAMPLE 2.12

A barge has a uniform rectangular cross section of width $2L$ and vertical draft of height H, as in Fig. E2.12. Determine (*a*) the metacentric height for a small tilt angle and (*b*) the range of ratio L/H for which the barge is statically stable if G is exactly at the waterline as shown.

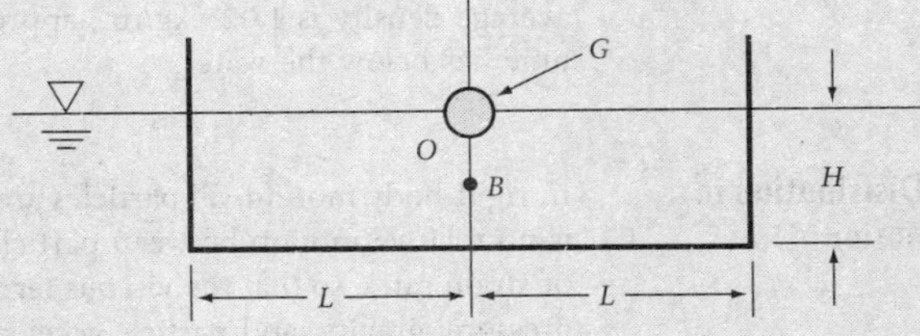

E2.12

Solution

If the barge has length b into the paper, the waterline area, relative to tilt axis O, has a base b and a height $2L$; therefore, $I_O = b(2L)^3/12$. Meanwhile, $\upsilon_{sub} = 2LbH$. Equation (2.37) predicts

$$\overline{MG} = \frac{I_o}{\upsilon_{sub}} - \overline{GB} = \frac{8bL^3/12}{2LbH} - \frac{H}{2} = \frac{L^2}{3H} - \frac{H}{2} \qquad \textit{Ans. (a)}$$

The barge can thus be stable only if

$$L^2 > 3H^2/2 \text{ or } 2L > 2.45H \qquad \textit{Ans. (b)}$$

The wider the barge relative to its draft, the more stable it is. Lowering G would help also.

Even an expert will have difficulty determining the floating stability of a buoyant body of irregular shape. Such bodies may have two or more stable positions. For example, a ship may float the way we like it, so that we can sit on the deck, or it may float upside down (capsized). An interesting mathematical approach to floating

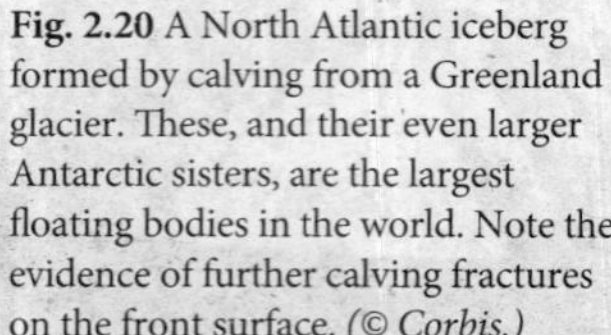

Fig. 2.20 A North Atlantic iceberg formed by calving from a Greenland glacier. These, and their even larger Antarctic sisters, are the largest floating bodies in the world. Note the evidence of further calving fractures on the front surface. *(© Corbis.)*

stability is given in Ref. 11. The author of this reference points out that even simple shapes, such as a cube of uniform density, may have a great many stable floating orientations, not necessarily symmetric. Homogeneous circular cylinders can float with the axis of symmetry tilted from the vertical.

Floating instability occurs in nature. Fish generally swim with their planes of symmetry vertical. After death, this position is unstable and they float with their flat sides up. Giant icebergs may overturn after becoming unstable when their shapes change due to underwater melting. Iceberg overturning is a dramatic, rarely seen event.

Figure 2.20 shows a typical North Atlantic iceberg formed by calving from a Greenland glacier that protruded into the ocean. The exposed surface is rough, indicating that it has undergone further calving. Icebergs are frozen fresh, bubbly, glacial water of average density 900 kg/m^3. Thus, when an iceberg is floating in seawater, whose average density is 1,025 kg/m^3, approximately 900/1,025, or seven-eighths, of its volume lies below the water.

2.9 Pressure Distribution in Rigid-Body Motion

In rigid-body motion, all particles are in combined translation and rotation, and there is no relative motion between particles. With no relative motion, there are no strains or strain rates, so that the viscous term in Eq. (2.8) vanishes, leaving a balance between pressure, gravity, and particle acceleration:

$$\nabla p = \rho(\mathbf{g} - \mathbf{a}) \tag{2.38}$$

The pressure gradient acts in the direction $\mathbf{g} - \mathbf{a}$, and lines of constant pressure (including the free surface, if any) are perpendicular to this direction. The general case of combined translation and rotation of a rigid body is discussed in Chap. 3, Fig. 3.11.

Fluids can rarely move in rigid-body motion unless restrained by confining walls for a long time. For example, suppose a tank of water is in a car that starts a constant acceleration. The water in the tank would begin to slosh about, and that sloshing would damp out very slowly until finally the particles of water would be in approximately rigid-body acceleration. This would take so long that the car would have reached hypersonic speeds. Nevertheless, we can at least discuss the pressure distribution in a tank of rigidly accelerating water.

Uniform Linear Acceleration

In the case of uniform rigid-body acceleration, Eq. (2.38) applies, **a** having the same magnitude and direction for all particles. With reference to Fig. 2.21, the parallelogram sum of **g** and $-\mathbf{a}$ gives the direction of the pressure gradient or greatest rate of increase

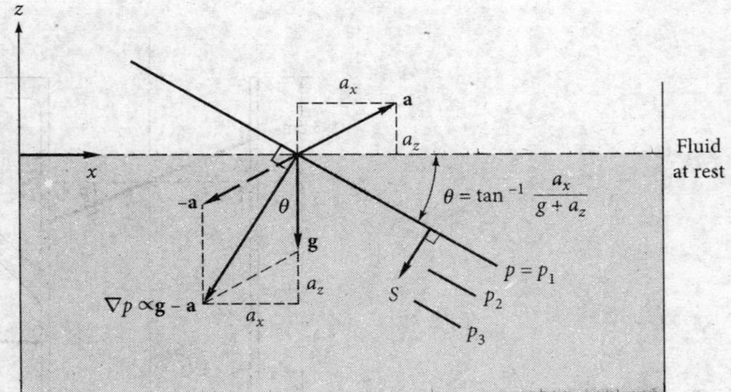

Fig. 2.21 Tilting of constant-pressure surfaces in a tank of liquid in rigid-body acceleration.

of p. The surfaces of constant pressure must be perpendicular to this, and are thus tilted at a downward angle θ such that

$$\theta = \tan^{-1}\frac{a_x}{g + a_z} \tag{2.39}$$

One of these tilted lines is the free surface, which is found by the requirement that the fluid retain its volume unless it spills out. The rate of increase of pressure in the direction $\mathbf{g} - \mathbf{a}$ is greater than in ordinary hydrostatics and is given by

$$\frac{dp}{ds} = \rho G \quad \text{where } G = [a_x^2 + (g + a_z)^2]^{1/2} \tag{2.40}$$

These results are independent of the size or shape of the container as long as the fluid is continuously connected throughout the container.

EXAMPLE 2.13

A drag racer rests her coffee mug on a horizontal tray while she accelerates at 7 m/s^2. The mug is 10 cm deep and 6 cm in diameter and contains coffee 7 cm deep at rest. (*a*) Assuming rigid-body acceleration of the coffee, determine whether it will spill out of the mug. (*b*) Calculate the gage pressure in the corner at point *A* if the density of coffee is 1,010 kg/m^3.

Solution

- *System sketch:* Figure E2.13 shows the coffee tilted during the acceleration.
- *Assumptions:* Rigid-body horizontal acceleration, $a_x = 7$ m/s^2. Symmetric coffee cup.
- *Property values:* Density of coffee given as 1,010 kg/m^3.
- *Approach (a):* Determine the angle of tilt from the known acceleration, then find the height rise.
- *Solution steps:* From Eq. (2.39), the angle of tilt is given by

$$\theta = \tan^{-1}\frac{a_x}{g} = \tan^{-1}\frac{7.0\ \text{m/s}^2}{9.81\ \text{m/s}^2} = 35.5°$$

If the mug is symmetric, the tilted surface will pass through the center point of the rest position, as shown in Fig. E2.13. Then the rear side of the coffee free surface will rise an amount Δz given by

$$\Delta z = (3\ \text{cm})(\tan 35.5°) = 2.14\ \text{cm} < 3\ \text{cm} \quad \text{therefore no spilling} \qquad \textit{Ans. (a)}$$

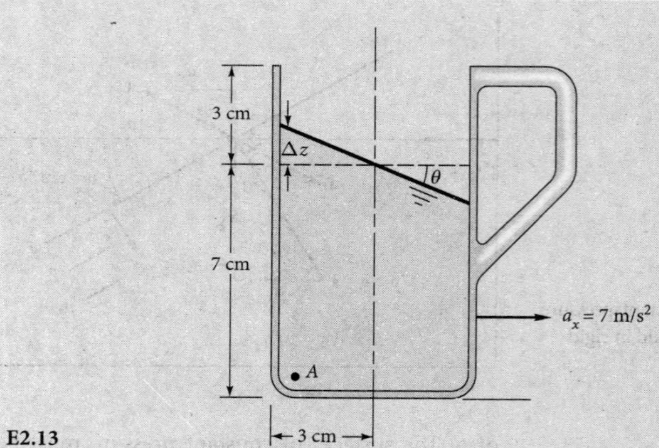

E2.13

- *Comment (a):* This solution neglects sloshing, which might occur if the start-up is uneven.
- *Approach (b):* The pressure at A can be computed from Eq. (2.40), using the perpendicular distance Δs from the surface to A. When at rest, $p_A = \rho g h_{\text{rest}} = (1{,}010\ \text{kg/m}^3)(9.81\ \text{m/s}^2)(0.07\ \text{m}) = 694\ \text{Pa}$.

$$p_A = \rho G\,\Delta s = \left(1{,}010\frac{\text{kg}}{\text{m}^3}\right)\left[\sqrt{(9.81)^2 + (7.0)^2}\right][(0.07 + 0.0214)\cos 35.5°] \approx 906\ \text{Pa} \quad \textit{Ans. (b)}$$

- *Comment (b):* The acceleration has increased the pressure at A by 31 percent. Think about this alternative: why does it work? Since $a_z = 0$, we may proceed vertically down the left side to compute

$$p_A = \rho g(z_{\text{surf}} - z_A) = (1{,}010\ \text{kg/m}^3)(9.81\ \text{m/s}^2)(0.0214\ \text{m} + 0.07\ \text{m}) = 906\ \text{Pa}$$

Rigid-Body Rotation

As a second special case, consider rotation of the fluid about the z axis without any translation, as sketched in Fig. 2.22. We assume that the container has been rotating long enough at constant Ω for the fluid to have attained rigid-body rotation. The fluid

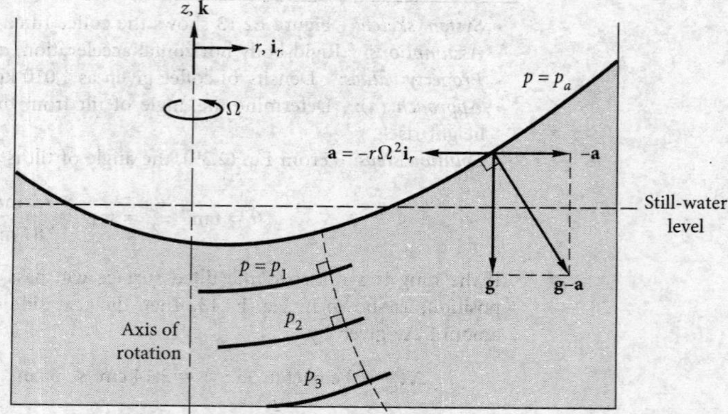

Fig. 2.22 Development of paraboloid constant-pressure surfaces in a fluid in rigid-body rotation. The dashed line along the direction of maximum pressure increase is an exponential curve.

acceleration will then be a centripetal term. In the coordinates of Fig. 2.22, the angular-velocity and position vectors are given by

$$\mathbf{\Omega} = \mathbf{k}\Omega \quad \mathbf{r}_0 = \mathbf{i}_r r \tag{2.41}$$

Then the acceleration is given by

$$\mathbf{\Omega} \times (\mathbf{\Omega} \times \mathbf{r}_0) = -r\Omega^2\mathbf{i}_r \tag{2.42}$$

as marked in the figure, and Eq. (2.38) for the force balance becomes

$$\nabla p = \mathbf{i}_r \frac{\partial p}{\partial r} + \mathbf{k}\frac{\partial p}{\partial z} = \rho(\mathbf{g} - \mathbf{a}) = \rho(-g\mathbf{k} + r\Omega^2\mathbf{i}_r)$$

Equating like components, we find the pressure field by solving two first-order partial differential equations:

$$\frac{\partial p}{\partial r} = \rho r\Omega^2 \quad \frac{\partial p}{\partial z} = -\gamma \tag{2.43}$$

The right-hand sides of (2.43) are known functions of r and z. One can proceed as follows: Integrate the first equation "partially," holding z constant, with respect to r. The result is

$$p = \tfrac{1}{2}\rho r^2\Omega^2 + f(z) \tag{2.44}$$

where the "constant" of integration is actually a function $f(z)$.[2] Now differentiate this with respect to z and compare with the second relation of (2.43):

$$\frac{\partial p}{\partial z} = 0 + f'(z) = -\gamma$$

or

$$f(z) = -\gamma z + C$$

where C is a constant. Thus, Eq. (2.44) now becomes

$$p = \text{const} - \gamma z + \tfrac{1}{2}\rho r^2\Omega^2 \tag{2.45}$$

This is the pressure distribution in the fluid. The value of C is found by specifying the pressure at one point. If $p = p_0$ at $(r, z) = (0, 0)$, then $C = p_0$. The final desired distribution is

$$\boxed{p = p_0 - \gamma z + \tfrac{1}{2}\rho r^2\Omega^2} \tag{2.46}$$

The pressure is linear in z and parabolic in r. If we wish to plot a constant-pressure surface, say, $p = p_1$, Eq. (2.45) becomes

$$z = \frac{p_0 - p_1}{\gamma} + \frac{r^2\Omega^2}{2g} = a + br^2 \tag{2.47}$$

[2]This is because $f(z)$ vanishes when differentiated with respect to r. If you don't see this, you should review your calculus.

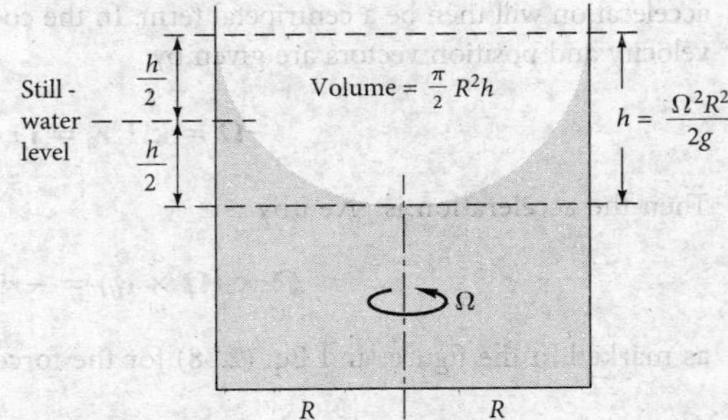

Fig. 2.23 Determining the free-surface position for rotation of a cylinder of fluid about its central axis.

Thus, the surfaces are paraboloids of revolution, concave upward, with their minimum points on the axis of rotation. Some examples are sketched in Fig. 2.22.

As in the previous example of linear acceleration, the position of the free surface is found by conserving the volume of fluid. For a noncircular container with the axis of rotation off-center, as in Fig. 2.22, a lot of laborious mensuration is required, and a single problem will take you all weekend. However, the calculation is easy for a cylinder rotating about its central axis, as in Fig. 2.23. Since the volume of a paraboloid is one-half the base area times its height, the still-water level is exactly halfway between the high and low points of the free surface. The center of the fluid drops an amount $h/2 = \Omega^2 R^2/(4g)$, and the edges rise an equal amount.

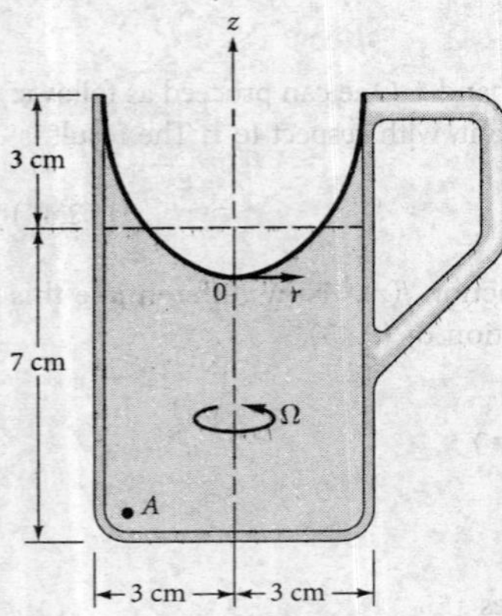

E2.14

EXAMPLE 2.14

The coffee cup in Example 2.13 is removed from the drag racer, placed on a turntable, and rotated about its central axis until a rigid-body mode occurs. Find (*a*) the angular velocity that will cause the coffee to just reach the lip of the cup and (*b*) the gage pressure at point *A* for this condition.

Solution

Part (a) The cup contains 7 cm of coffee. The remaining distance of 3 cm up to the lip must equal the distance $h/2$ in Fig. 2.23. Thus

$$\frac{h}{2} = 0.03\text{ m} = \frac{\Omega^2 R^2}{4g} = \frac{\Omega^2 (0.03\text{ m})^2}{4(9.81\text{ m/s}^2)}$$

Solving, we obtain

$$\Omega^2 = 1{,}308 \quad \text{or} \quad \Omega = 36.2\text{ rad/s} = 345\text{ r/min} \qquad \textit{Ans. (a)}$$

Part (b) To compute the pressure, it is convenient to put the origin of coordinates r and z at the bottom of the free-surface depression, as shown in Fig. E2.14. The gage pressure here is $p_0 = 0$, and point A is at $(r, z) = (3\text{ cm}, -4\text{ cm})$. Equation (2.46) can then be evaluated:

$$\begin{aligned} p_A &= 0 - (1{,}010\text{ kg/m}^3)(9.81\text{ m/s}^2)(-0.04\text{ m}) \\ &\quad + \tfrac{1}{2}(1{,}010\text{ kg/m}^3)(0.03\text{ m})^2(1{,}308\text{ rad}^2/\text{s}^2) \\ &= 396\text{ N/m}^2 + 594\text{ N/m}^2 = 990\text{ Pa} \end{aligned} \qquad \textit{Ans. (b)}$$

This is about 43 percent greater than the still-water pressure $p_A = 694$ Pa.

Here, as in the linear acceleration case, it should be emphasized that the paraboloid pressure distribution (2.46) sets up in *any* fluid under rigid-body rotation, regardless of the shape or size of the container. The container may even be closed and filled with fluid. It is only necessary that the fluid be continuously interconnected throughout the container. The following example will illustrate a peculiar case in which one can visualize an imaginary free surface extending outside the walls of the container.

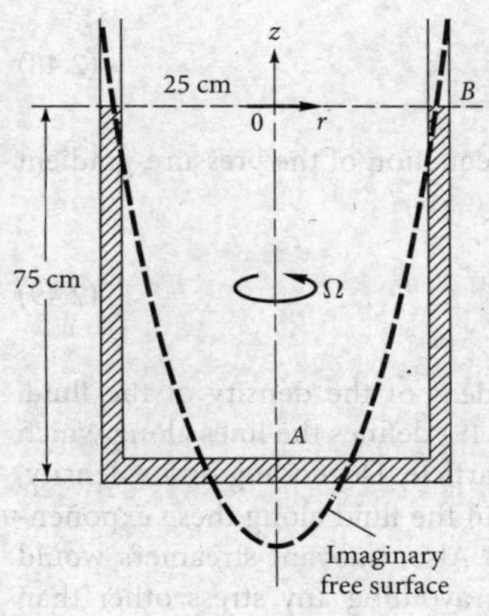

E2.15

EXAMPLE 2.15

A U-tube with a radius of 25 cm and containing mercury to a height of 75 cm is rotated about its center at 180 r/min until a rigid-body mode is achieved. The diameter of the tubing is negligible. Atmospheric pressure is 101.3 kPa. Find the pressure at point A in the rotating condition. See Fig. E2.15.

Solution

Convert the angular velocity to radians per second:

$$\Omega = (180 \text{ r/min})\frac{2\pi \text{ rad/r}}{60 \text{ s/min}} = 18.85 \text{ rad/s}$$

From Table 2.1 we find for mercury that $\gamma = 133{,}000 \text{ N/m}^3$ and hence $\rho = (133{,}000 \text{ N/m}^3)/(9.81 \text{ m/s}^2) = 13{,}560 \text{ kg/m}^3$. At this high rotation rate, the free surface will slant upward at a fierce angle [about 84°; check this from Eq. (2.47)], but the tubing is so thin that the free surface will remain at approximately the same 0.75 m height, point B. Placing our origin of coordinates at this height, we can calculate the constant C in Eq. (2.45) from the condition $p_B = 101.3$ kPa at $(r, z) = (0.25 \text{ m}, 0)$:

$$p_B = 101{,}300 \text{ Pa} = C - 0 + \tfrac{1}{2}(13{,}560 \text{ kg/m}^3)(0.25 \text{ m})^2(18.85 \text{ rad/s})^2$$

or $$C = 101{,}300 \text{ Pa} - 150{,}600 \text{ Pa} = -49{,}300 \text{ Pa}$$

We then obtain p_A by evaluating Eq. (2.46) at $(r, z) = (0, -0.75 \text{ m})$:

$$p_A = -49{,}300 \text{ Pa} - (133{,}000 \text{ N/m}^3)(7.5 \text{ m}) = -49{,}300 \text{ Pa} + 99{,}800 \text{ Pa} = 50{,}500 \text{ Pa} \quad \textit{Ans.}$$

This is less than atmospheric pressure, and we can see why if we follow the free-surface paraboloid down from point B along the dashed line in the figure. It will cross the horizontal portion of the U-tube (where p will be atmospheric) and fall *below* point A. From Fig. 2.23 the actual drop from point B will be

$$h = \frac{\Omega^2 R^2}{2g} = \frac{(18.85 \text{ rad/s})^2(0.25 \text{ m})^2}{2(9.81 \text{ m/s}^2)} = 1.132 \text{ m}$$

Thus, p_A is about 382 mmHg below atmospheric pressure, or about (0.382 m) $(133{,}000 \text{ N/m}^3) = 50{,}800$ Pa below $p_a = 101{,}300$ Pa, which checks with the answer above. When the tube is at rest,

$$p_A = 101{,}300 \text{ Pa} - (133{,}000 \text{ N/m}^3)(-0.75 \text{ m}) = 201{,}000 \text{ Pa}$$

Hence rotation has reduced the pressure at point A by 77 percent. Further rotation can reduce p_A to near-zero pressure, and cavitation can occur.

An interesting by-product of this analysis for rigid-body rotation is that the lines everywhere parallel to the pressure gradient form a family of curved surfaces, as sketched in Fig. 2.22. They are everywhere orthogonal to the constant-pressure surfaces, and hence their slope is the negative inverse of the slope computed from Eq. (2.47):

$$\left.\frac{dz}{dr}\right|_{\mathrm{GL}} = -\frac{1}{(dz/dr)_{p=\mathrm{const}}} = -\frac{1}{r\Omega^2/g}$$

where GL stands for gradient line

or
$$\frac{dz}{dr} = -\frac{g}{r\Omega^2} \tag{2.48}$$

Separating the variables and integrating, we find the equation of the pressure-gradient surfaces:

$$r = C_1 \exp\left(-\frac{\Omega^2 z}{g}\right) \tag{2.49}$$

Notice that this result and Eq. (2.47) are independent of the density of the fluid. In the absence of friction and Coriolis effects, Eq. (2.49) defines the lines along which the apparent net gravitational field would act on a particle. Depending on its density, a small particle or bubble would tend to rise or fall in the fluid along these exponential lines, as demonstrated experimentally in Ref. 5. Also, buoyant streamers would align themselves with these exponential lines, thus avoiding any stress other than pure tension. Figure 2.24 shows the configuration of such streamers before and during rotation.

2.10 Pressure Measurement

Pressure is a derived property. It is the force per unit area as related to fluid molecular bombardment of a surface. Thus, most pressure instruments only *infer* the pressure by calibration with a primary device such as a deadweight piston tester. There are many such instruments, for both a static fluid and a moving stream. The instrumentation texts in Refs. 7 to 10, 12, 13, and 16–17 list over 20 designs for pressure measurement instruments. These instruments may be grouped into four categories:

1. *Gravity-based:* barometer, manometer, deadweight piston.
2. *Elastic deformation:* bourdon tube (metal and quartz), diaphragm, bellows, strain-gage, optical beam displacement.
3. *Gas behavior:* gas compression (McLeod gage), thermal conductance (Pirani gage), molecular impact (Knudsen gage), ionization, thermal conductivity, air piston.
4. *Electric output:* resistance (Bridgman wire gage), diffused strain gage, capacitative, piezoelectric, potentiometric, magnetic inductance, magnetic reluctance, linear variable differential transformer (LVDT), resonant frequency.
5. *Luminescent coatings* for surface pressures [15].

The gas-behavior gages are mostly special-purpose instruments used for certain scientific experiments. The deadweight tester is the instrument used most often for calibrations; for example, it is used by the U.S. National Institute for Standards and Technology (NIST). The barometer is described in Fig. 2.6.

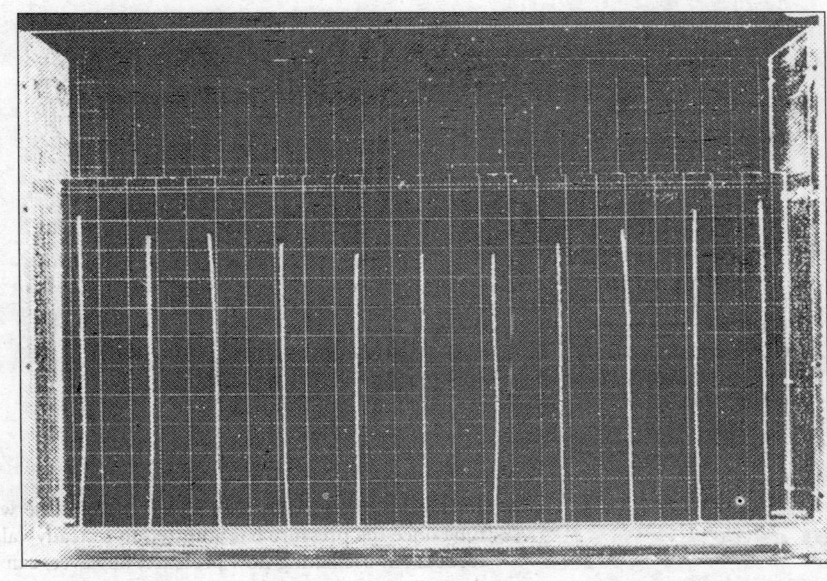

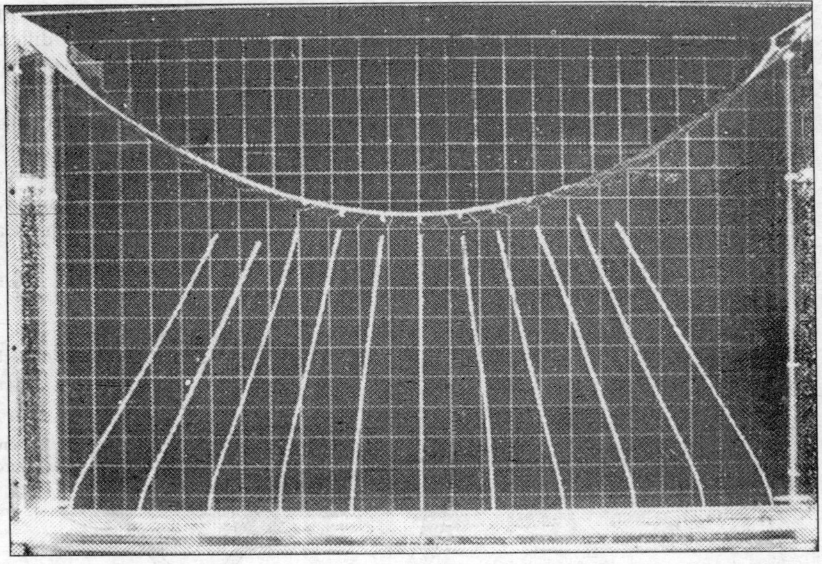

Fig. 2.24 Experimental demonstration with buoyant streamers of the fluid force field in rigid-body rotation: (*top*) fluid at rest (streamers hang vertically upward); (*bottom*) rigid-body rotation (streamers are aligned with the direction of maximum pressure gradient). (*© The American Association of Physics Teachers. Reprinted with permission from "The Apparent Field of Gravity in a Rotating Fluid System" by R. Ian Fletcher. American Journal of Physics vol. 40, pp. 959–965, July 1972.*)

The manometer, analyzed in Sec. 2.4, is a simple and inexpensive hydrostatic-principle device with no moving parts except the liquid column itself. Manometer measurements must not disturb the flow. The best way to do this is to take the measurement through a *static hole* in the wall of the flow, as illustrated in Fig. 2.25*a*. The hole should be normal to the wall, and burrs should be avoided. If the hole is

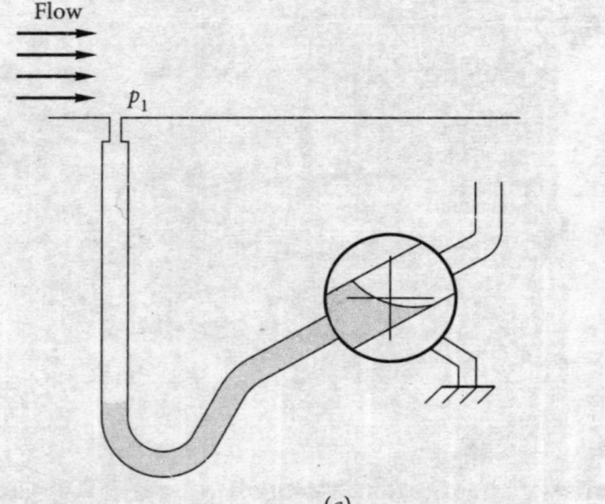

Fig. 2.25 Two types of accurate manometers for precise measurements: (*a*) tilted tube with eyepiece; (*b*) a capacitive-type digital manometer of rated accuracy ±0.1 percent. *(Courtesy of Dwyer Instruments, Inc.)*

small enough (typically 1-mm diameter), there will be no flow into the measuring tube once the pressure has adjusted to a steady value. Thus, the flow is almost undisturbed. An oscillating flow pressure, however, can cause a large error due to possible dynamic response of the tubing. Other devices of smaller dimensions are used for dynamic-pressure measurements. The manometer in Fig. 2.25*a* measures the gage pressure p_1. The instrument in Fig. 2.25*b* is a digital *differential* manometer, which can measure the difference between two different points in the flow, with stated accuracy of 0.1 percent of full scale. The world of instrumentation is moving quickly toward digital readings.

In category 2, elastic-deformation instruments, a popular, inexpensive, and reliable device is the *bourdon tube*, sketched in Fig. 2.26. When pressurized internally, a curved tube with flattened cross section will deflect outward. The deflection can be measured by a linkage attached to a calibrated dial pointer, as shown. Or the deflection can be used to drive electric-output sensors, such as a variable transformer. Similarly, a

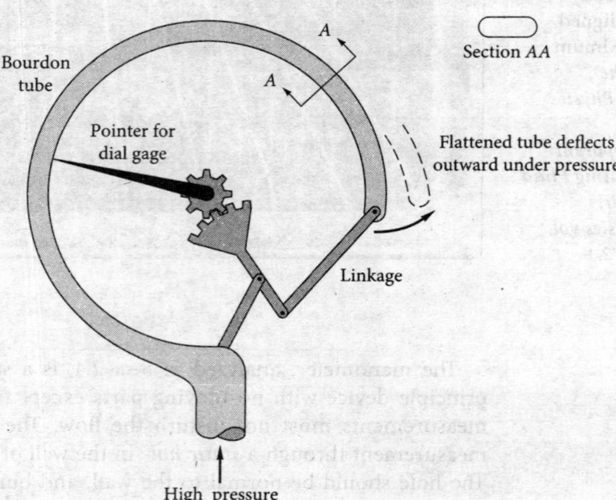

Fig. 2.26 Schematic of a bourdon-tube device for mechanical measurement of high pressures.

membrane or *diaphragm* will deflect under pressure and can either be sensed directly or used to drive another sensor.

An interesting variation of Fig. 2.26 is the *fused-quartz, force-balanced bourdon tube,* shown in Fig. 2.27, whose spiral-tube deflection is sensed optically and returned to a zero reference state by a magnetic element whose output is proportional to the fluid pressure. The fused-quartz, force-balanced bourdon tube is reported to be one of the most accurate pressure sensors ever devised, with uncertainty on the order of ±0.003 percent.

The quartz gages, both the bourdon type and the resonant type, are expensive but extremely accurate, stable, and reliable [14]. They are often used for deep-ocean pressure measurements, which detect long waves and tsunami activity over extensive time periods.

The last category, *electric-output sensors,* is extremely important in engineering because the data can be stored on computers and freely manipulated, plotted, and analyzed. Three examples are shown in Fig. 2.28, the first being the *capacitive* sensor in Fig. 2.28*a*. The differential pressure deflects the silicon diaphragm and changes the capacitance of the liquid in the cavity. Note that the cavity has spherical end caps to prevent overpressure damage. In the second type, Fig. 2.28*b*, strain gages and other sensors are chemically diffused or etched onto a chip, which is stressed by the applied pressure. Finally, in Fig. 2.28*c*, a micromachined silicon sensor is arranged to deform under pressure such that its natural vibration frequency is proportional to the pressure. An oscillator excites the element's resonant frequency and converts it into appropriate pressure units.

Another kind of dynamic electric-output sensor is the *piezoelectric transducer,* shown in Fig. 2.29. The sensing elements are thin layers of quartz, which generate an electric charge when subjected to stress. The design in Fig. 2.29 is flush-mounted on a solid surface and can sense rapidly varying pressures, such as blast waves. Other designs are of the cavity type. This type of sensor primarily detects transient pressures, not steady stress, but if highly insulated can also be used for short-term static events. Note also that it measures *gage* pressure—that is, it detects only a change from ambient conditions.

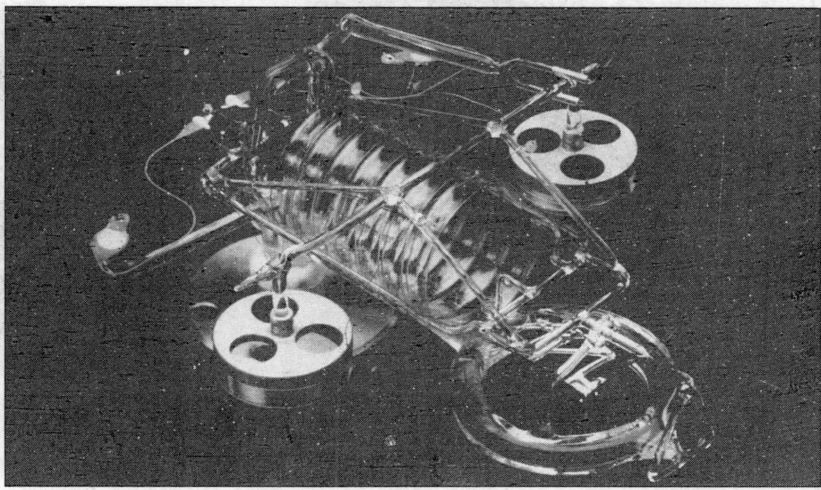

Fig. 2.27 The fused-quartz, force-balanced bourdon tube is the most accurate pressure sensor used in commercial applications today. *(Courtesy of Ruska Instrument Corporation, Houston, TX.)*

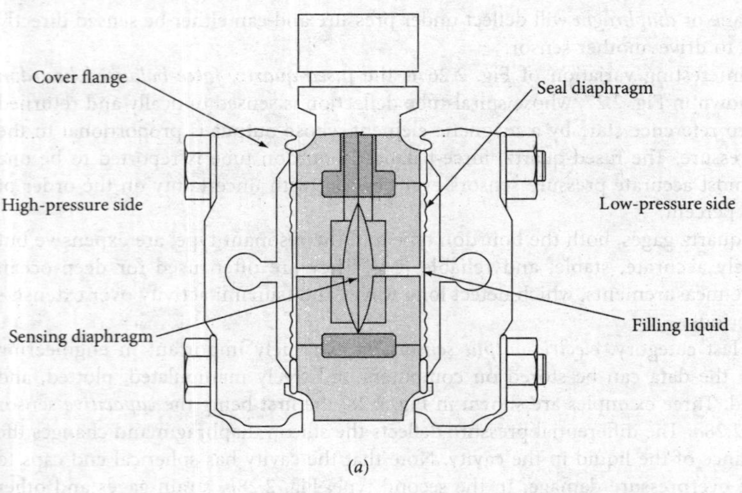

(a)

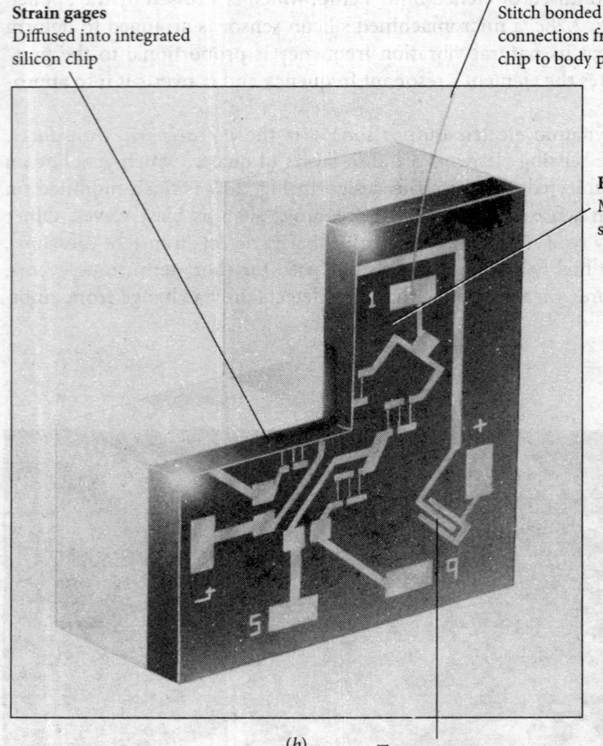

(b)

(c)

Fig. 2.28 Pressure sensors with electric output: (*a*) a silicon diaphragm whose deflection changes the cavity capacitance (*b*) a silicon strain gage that is stressed by applied pressure; (*c*) a micromachined silicon element that resonates at a frequency proportional to applied pressure.

Source: (a) Courtesy of Yokogawa Corporation of America. (b) and (c) are courtesy of Druck, Inc., Fairfield, CT.

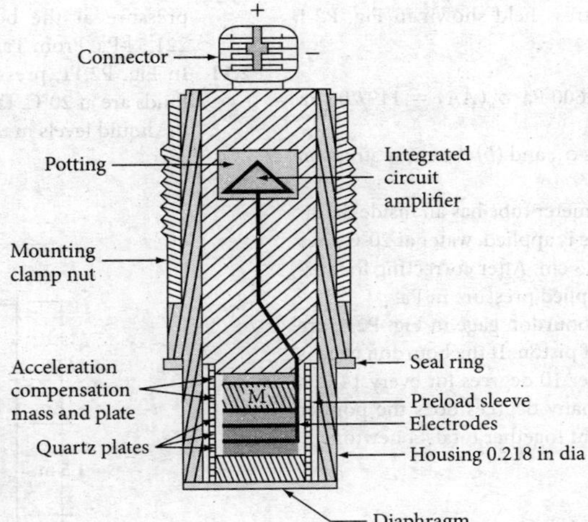

Fig. 2.29 A piezoelectric transducer measures rapidly changing pressures. *Source: Courtesy of PCB Piezorronics, Inc, Depew, New York.*

Summary

This chapter has been devoted entirely to the computation of pressure distributions and the resulting forces and moments in a static fluid or a fluid with a known velocity field. All hydrostatic (Secs. 2.3 to 2.8) and rigid-body (Sec. 2.9) problems are solved in this manner and are classic cases that every student should understand. In arbitrary viscous flows, both pressure and velocity are unknowns and are solved together as a system of equations in the chapters that follow.

Problems

Most of the problems herein are fairly straightforward. More difficult or open-ended assignments are indicated with an asterisk, as in Prob. 2.9. Problems labeled with a computer icon may require the use of a computer. The standard end-of-chapter problems 2.1 to 2.161 (categorized in the problem distribution) are followed by word problems W2.1 to W2.9, fundamentals of engineering exam problems FE2.1 to FE2.10, comprehensive problems C2.1 to C2.9, and design projects D2.1 to D2.3.

Problem Distribution

Section	Topic	Problems
2.1, 2.2	Stresses; pressure gradient; gage pressure	2.1–2.6
2.3	Hydrostatic pressure; barometers	2.7–2.23
2.3	The atmosphere	2.24–2.29
2.4	Manometers; multiple fluids	2.30–2.47
2.5	Forces on plane surfaces	2.48–2.80
2.6	Forces on curved surfaces	2.81–2.100
2.7	Forces in layered fluids	2.101–2.102
2.8	Buoyancy; Archimedes' principles	2.103–2.126
2.8	Stability of floating bodies	2.127–2.136
2.9	Uniform acceleration	2.137–2.151
2.9	Rigid-body rotation	2.152–2.159
2.10	Pressure measurements	2.160–2.161

Stresses; pressure gradient; gage pressure

P2.1 For the two-dimensional stress field shown in Fig. P2.1 it is found that

$$\sigma_{xx} = 143{,}600 \text{ Pa} \quad \sigma_{yy} = 95{,}800 \text{ Pa} \quad \sigma_{xy} = 23{,}900 \text{ Pa}$$

Find the shear and normal stresses (in Pa) acting on plane *AA* cutting through the element at a 30° angle as shown.

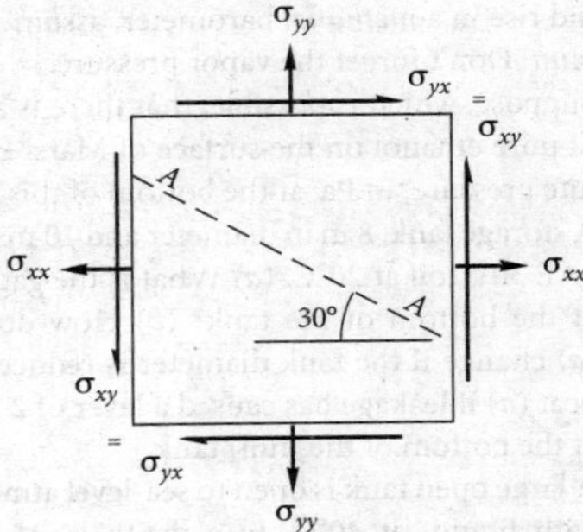

P2.1

P2.2 For the two-dimensional stress field shown in Fig. P2.1 suppose that

$$\sigma_{xx} = 95{,}800 \text{ Pa} \quad \sigma_{yy} = 143{,}600 \text{ Pa} \quad \sigma_n(AA) = 119{,}700 \text{ Pa}$$

Compute (*a*) the shear stress σ_{xy} and (*b*) the shear stress on plane *AA*.

P2.3 A vertical, clean, glass piezometer tube has an inside diameter of 1 mm. When pressure is applied, water at 20°C rises into the tube to a height of 25 cm. After correcting for surface tension, estimate the applied pressure in Pa.

P2.4 Pressure gages, such as the bourdon gage in Fig. P2.4, are calibrated with a deadweight piston. If the bourdon gage is designed to rotate the pointer 10 degrees for every 14 kPa of internal pressure, how many degrees does the pointer rotate if the piston and weight together total 44 newtons?

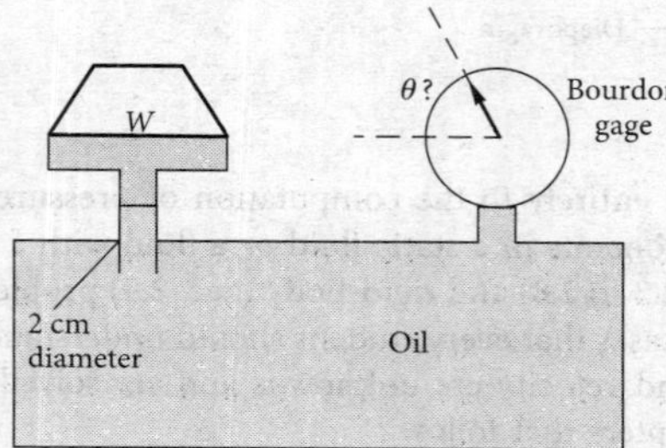

P2.4

P2.5 Quito, Ecuador, has an average altitude of 2,850 m. On a standard day, pressure gage A in a laboratory experiment reads 63 kPa and gage B reads 105 kPa. Express these readings in gage pressure or vacuum pressure, whichever is appropriate.

P2.6 Any pressure reading can be expressed as a length or *head*, $h = p/\rho g$. What is standard sea-level pressure expressed in (*a*) ft of glycerin, (*b*) inHg, (*c*) m of water, and (*d*) mm of ethanol? Assume all fluids are at 20°C.

Hydrostatic pressure; barometers

P2.7 La Paz, Bolivia, is at an altitude of approximately 3,700 m. Assume a standard atmosphere. How high would the liquid rise in a *methanol* barometer, assumed at 20°C? *Hint:* Don't forget the vapor pressure.

P2.8 Suppose, which is possible, that there is a 800 m deep lake of pure ethanol on the surface of Mars. Estimate the absolute pressure, in Pa, at the bottom of this speculative lake.

P2.9 A storage tank, 8 m in diameter and 10 m high, is filled with SAE 30W oil at 20°C. (*a*) What is the gage pressure, in Pa, at the bottom of the tank? (*b*) How does your result in (*a*) change if the tank diameter is reduced to 5 m? (*c*) Repeat (*a*) if leakage has caused a layer of 2 m of water to rest at the bottom of the (full) tank.

P2.10 A large open tank is open to sea-level atmosphere and filled with liquid, at 20°C, to a depth of 15 m. The absolute pressure at the bottom of the tank is approximately 221.5 kPa. From Table A.3, what might this liquid be?

P2.11 In Fig. P2.11, pressure gage *A* reads 1.5 kPa (gage). The fluids are at 20°C. Determine the elevations *z*, in meters, of the liquid levels in the open piezometer tubes *B* and *C*.

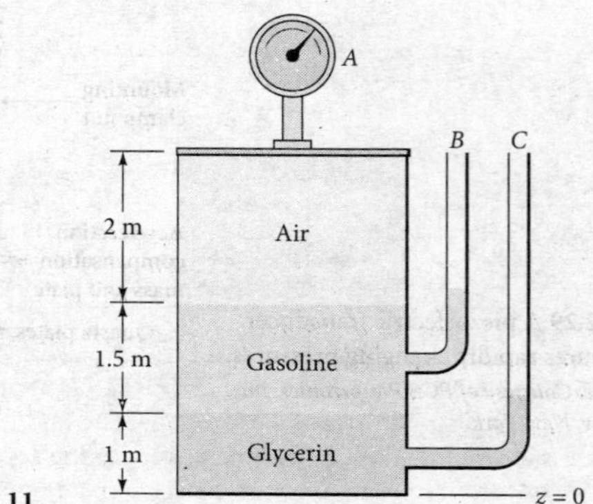

P2.11

P2.12 In Fig. P2.12 the tank contains water and immiscible oil at 20°C. What is *h* in cm if the density of the oil is 898 kg/m³?

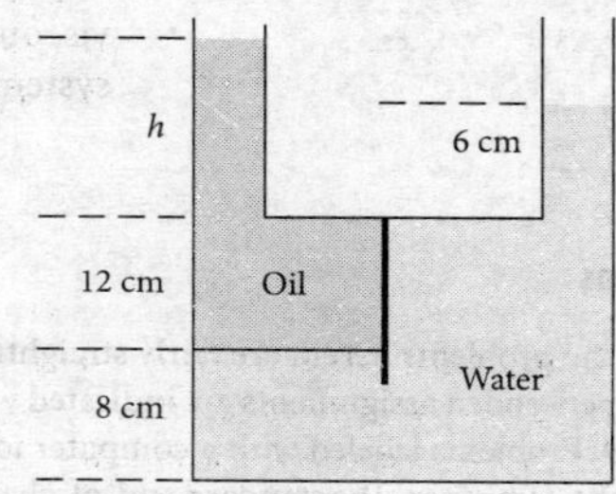

P2.12

P2.13 In Fig. P2.13 the 20°C water and gasoline surfaces are open to the atmosphere and at the same elevation. What is the height *h* of the third liquid in the right leg?

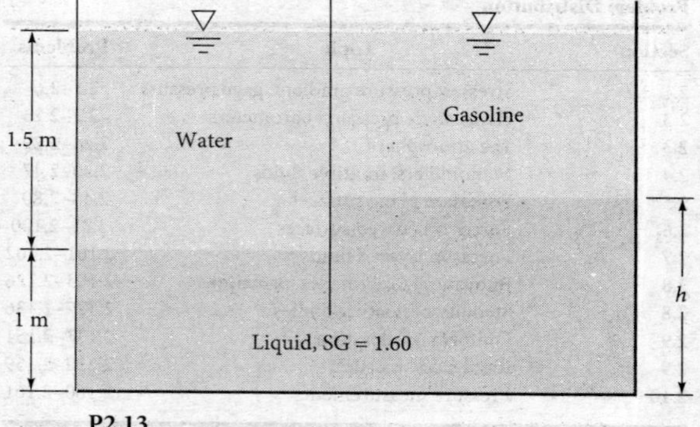

P2.13

P2.14 For the three-liquid system shown, compute h_1 and h_2. Neglect the air density.

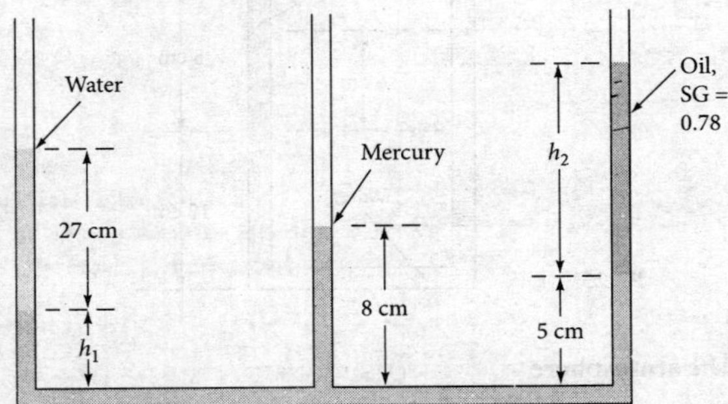

P2.14

P2.15 The air–oil–water system in Fig. P2.15 is at 20°C. Knowing that gage *A* reads 100 kPa absolute and gage *B* reads 8.5 kPa less than gage *C*, compute (*a*) the specific weight of the oil in N/m³ and (*b*) the actual reading of gage *C* in kPa absolute.

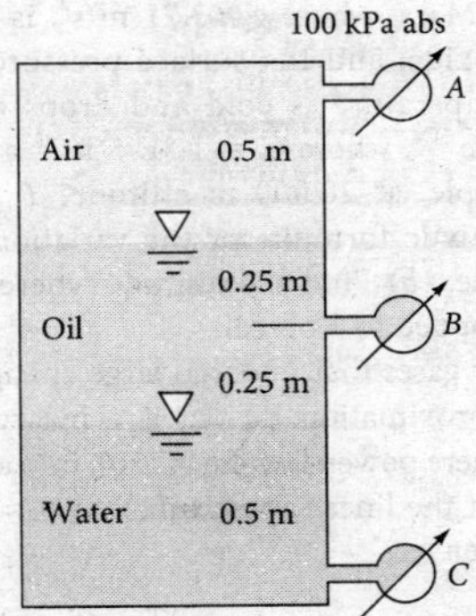

P2.15

P2.16 If the absolute pressure at the interface between water and mercury in Fig. P2.16 is 93 kPa, what, in Pa, is (*a*) the pressure at the surface and (*b*) the pressure at the bottom of the container?

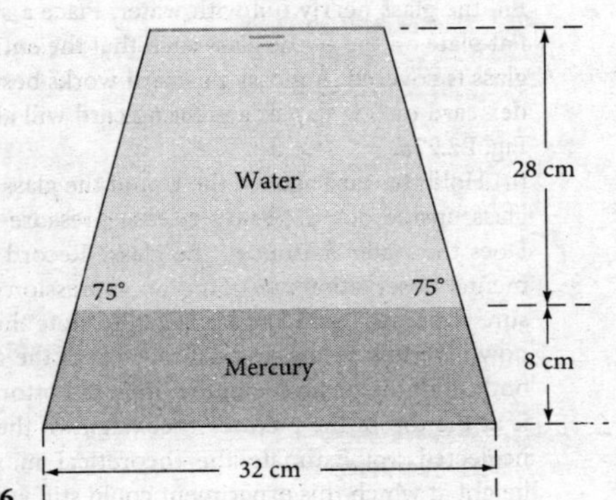

P2.16

P2.17 The system in Fig. P2.17 is at 20°C. Determine the height *h* of the water in the left side.

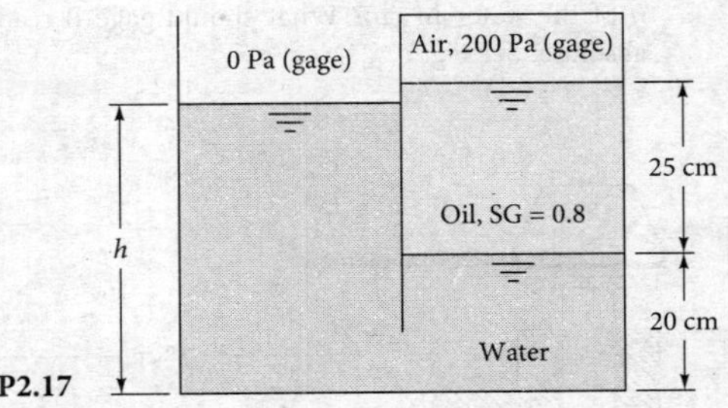

P2.17

P2.18 The system in Fig. P2.18 is at 20°C. If atmospheric pressure is 101.33 kPa and the pressure at the bottom of the tank is 242 kPa, what is the specific gravity of fluid *X*?

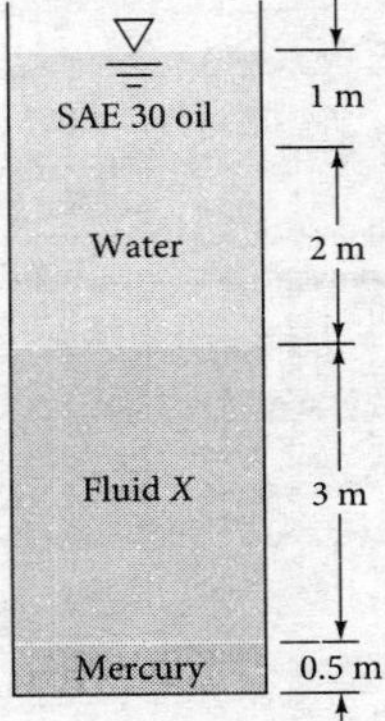

P2.18

P2.19 The U-tube in Fig. P2.19 has a 1-cm ID and contains mercury as shown. If 20 cm³ of water is poured into the right-hand leg, what will the free-surface height in each leg be after the sloshing has died down?

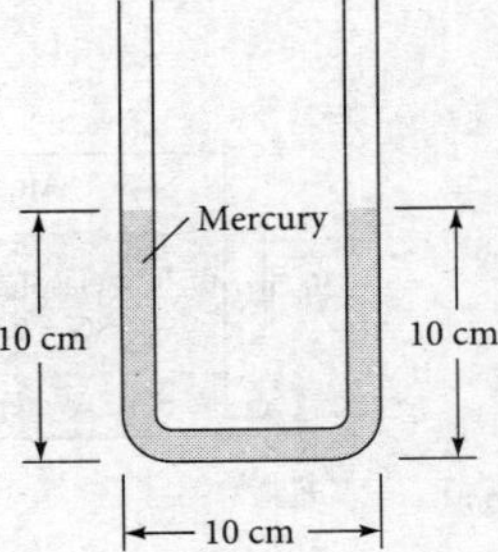

P2.19

P2.20 The hydraulic jack in Fig. P2.20 is filled with oil at 8,800 N/m³. Neglecting the weight of the two pistons, what force F on the handle is required to support the 8,900 N weight for this design?

P2.21 At 20°C gage A reads 350 kPa absolute. What is the height h of the water in cm? What should gage B read in kPa absolute? See Fig. P2.21.

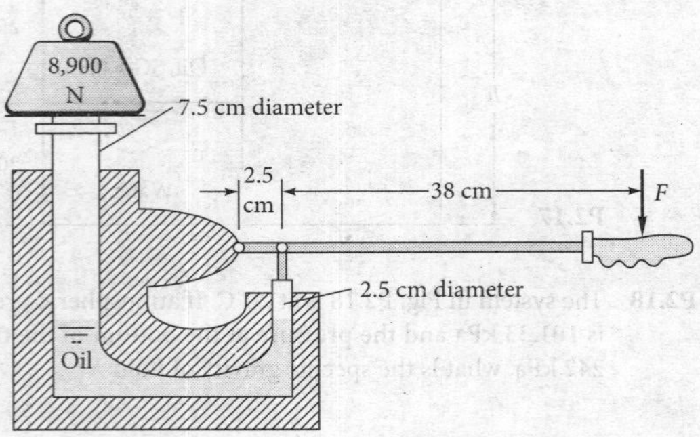

P2.20

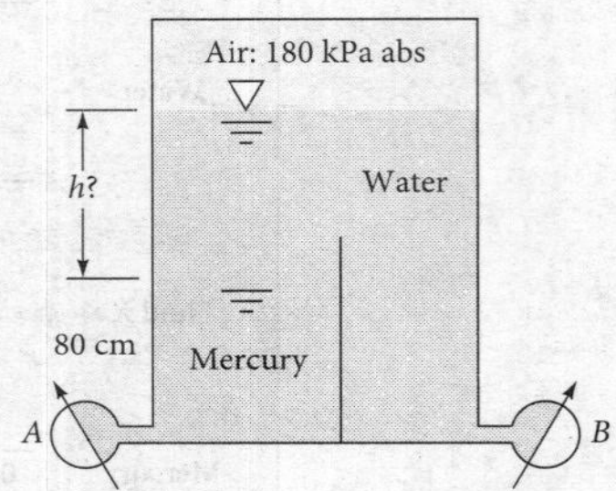

P2.21

P2.22 The fuel gage for a gasoline tank in a car reads proportional to the bottom gage pressure as in Fig. P2.22. If the tank is 30 cm deep and accidentally contains 2 cm of water plus gasoline, how many centimeters of air remain at the top when the gage erroneously reads "full"?

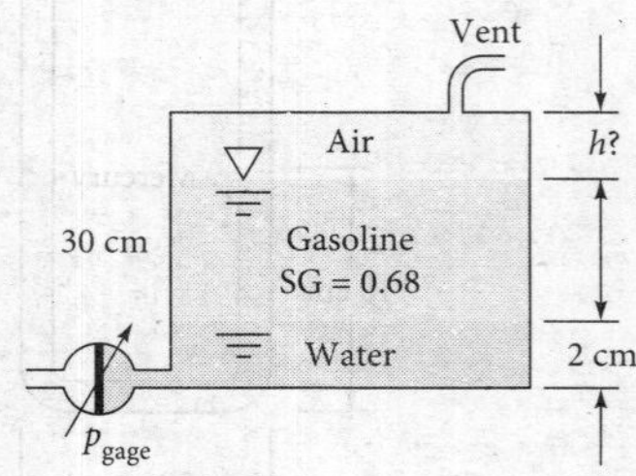

P2.22

P2.23 In Fig. P2.23 both fluids are at 20°C. If surface tension effects are negligible, what is the density of the oil, in kg/m^3?

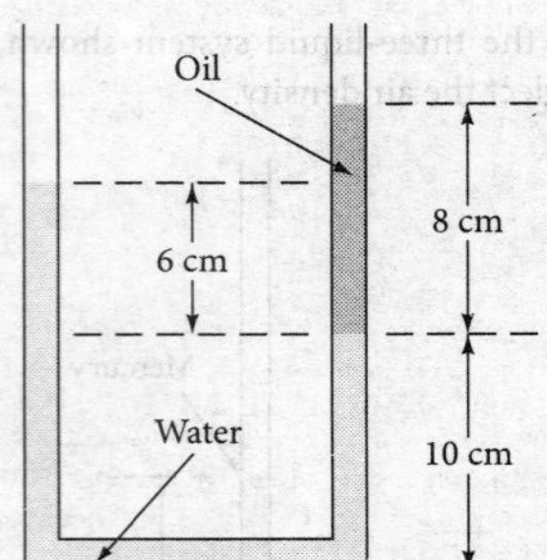

P2.23

The atmosphere

P2.24 In Prob. 1.2 we made a crude integration of the density distribution $\rho(z)$ in Table A.6 and estimated the mass of the earth's atmosphere to be $m \approx 6$ E18 kg. Can this result be used to estimate sea-level pressure on the earth? Conversely, can the actual sea-level pressure of 101.35 kPa be used to make a more accurate estimate of the atmospheric mass?

***P2.25** As measured by NASA's Viking landers, the atmosphere of Mars, where $g \approx 3.71$ m/s^2, is almost entirely carbon dioxide, and the surface pressure averages 700 Pa. The temperature is cold and drops off exponentially: $T \approx T_o e^{-Cz}$, where C = 1.3E-5 m^{-1} and T_o = 250 K. For example, at 20,000 m altitude, $T \approx$ 193 K. (*a*) Find an analytic formula for the variation of pressure with altitude. (*b*) Find the altitude where pressure on Mars has dropped to 1 pascal.

P2.26 For gases that undergo large changes in height, the linear approximation, Eq. (2.14), is inaccurate. Expand the troposphere power-law, Eq. (2.20), into a power series, and show that the linear approximation $p \approx p_a - \rho_a gz$ is adequate when

$$\delta z \ll \frac{2T_0}{(n-1)B} \quad \text{where } n = \frac{g}{RB}$$

P2.27 Conduct an experiment to illustrate atmospheric pressure. *Note:* Do this over a sink or you may get wet! Find a drinking glass with a very smooth, uniform rim at the top. Fill the glass nearly full with water. Place a smooth, light, flat plate on top of the glass such that the entire rim of the glass is covered. A glossy postcard works best. A small index card or one flap of a greeting card will also work. See Fig. P2.27*a*.

(*a*) Hold the card against the rim of the glass and turn the glass upside down. Slowly release pressure on the card. Does the water fall out of the glass? Record your experimental observations. (*b*) Find an expression for the pressure at points 1 and 2 in Fig. P2.27*b*. Note that the glass is now inverted, so the original top rim of the glass is at the bottom of the picture, and the original bottom of the glass is at the top of the picture. The weight of the card can be neglected. (*c*) Estimate the theoretical maximum glass height at which this experiment could still work, such that the water would not fall out of the glass.

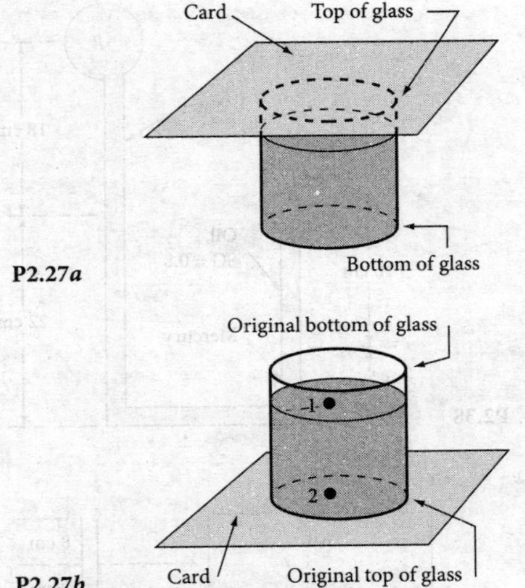

P2.27a

P2.27b

P2.28 A correlation of computational fluid dynamics results indicates that, all other things being equal, the distance traveled by a well-hit baseball varies inversely as the 0.36 power of the air density. If a home-run ball hit in Citi Field in New York travels 120 m, estimate the distance it would travel in (*a*) Quito, Ecuador, and (*b*) Colorado Springs, CO.

P2.29 Follow up on Prob. P2.8 by estimating the altitude on Mars where the pressure has dropped to 20 percent of its surface value. Assume an *isothermal* atmosphere, not the exponential variation of P2.25.

Manometers; multiple fluids

P2.30 For the traditional equal-level manometer measurement in Fig. E2.3, water at 20°C flows through the plug device from *a* to *b*. The manometer fluid is mercury. If $L = 12$ cm and $h = 24$ cm, (*a*) what is the pressure drop through the device? (*b*) If the water flows through the pipe at a velocity $V = 5.5$ m/s, what is the *dimensionless loss coefficient* of the device, defined by $K = \Delta p/(\rho V^2)$? We will study loss coefficients in Chap. 6.

P2.31 In Fig. P2.31 all fluids are at 20°C. Determine the pressure difference (Pa) between points *A* and *B*.

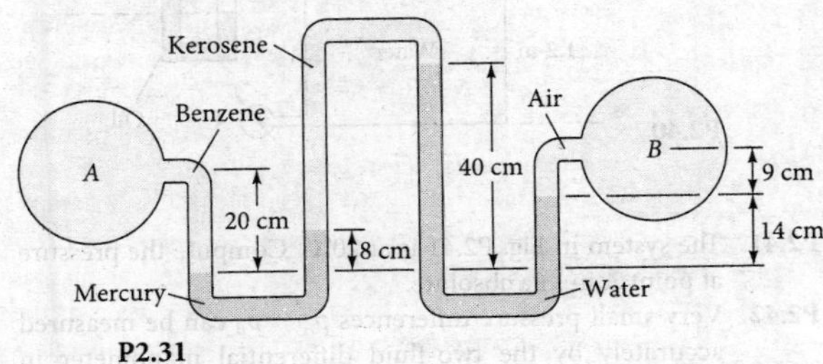

P2.31

P2.32 For the inverted manometer of Fig. P2.32, all fluids are at 20°C. If $p_B - p_A = 97$ kPa, what must the height *H* be in cm?

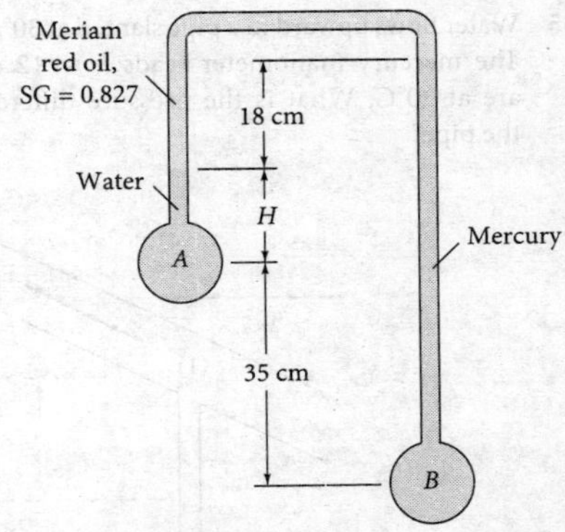

P2.32

P2.33 In Fig. P2.33 the pressure at point *A* is 170 kPa. All fluids are at 20°C. What is the air pressure in the closed chamber *B*, in Pa?

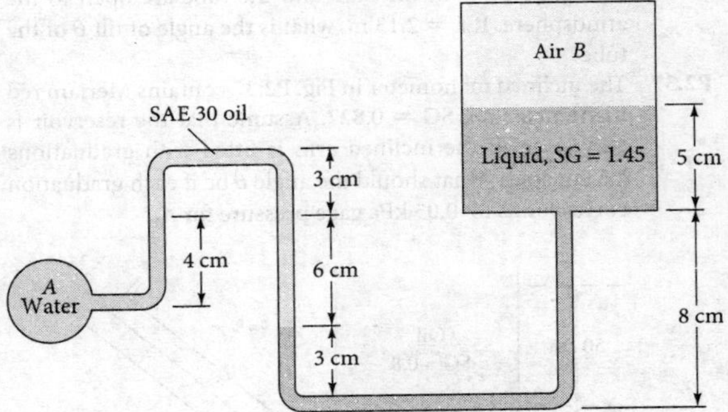

P2.33

***P2.34** Sometimes manometer dimensions have a significant effect. In Fig. P2.34 containers (*a*) and (*b*) are cylindrical and conditions are such that $p_a = p_b$. Derive a formula for the pressure difference $p_a - p_b$ when the oil–water interface on the right rises a distance $\Delta h < h$, for (*a*) $d \ll D$ and (*b*) $d = 0.15D$. What is the percentage change in the value of Δp?

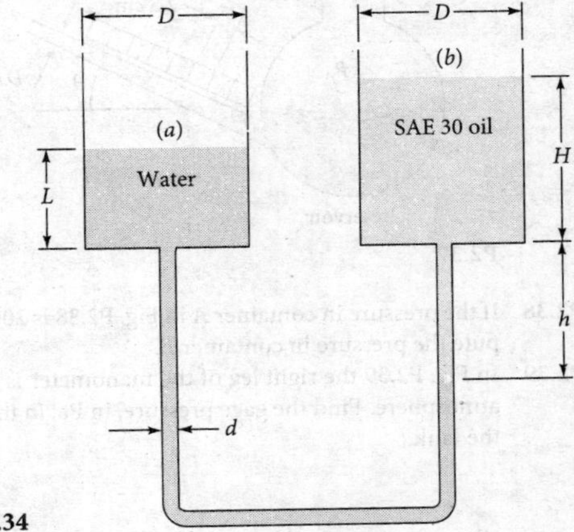

P2.34

P2.35 Water flows upward in a pipe slanted at 30°, as in Fig. P2.35. The mercury manometer reads $h = 12$ cm. Both fluids are at 20°C. What is the pressure difference $p_1 - p_2$ in the pipe?

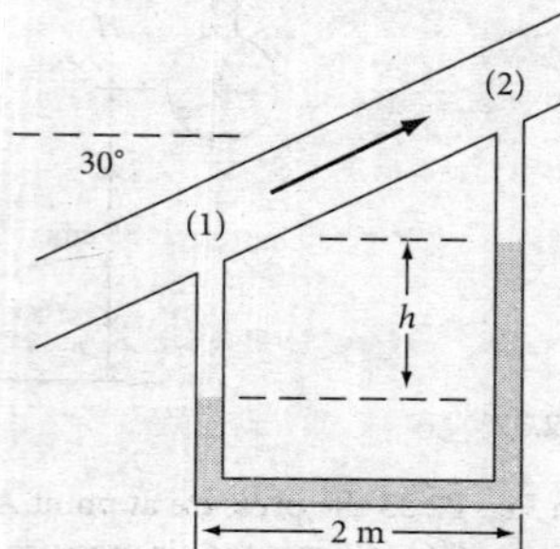

P2.35

P2.36 In Fig. P2.36 both the tank and the tube are open to the atmosphere. If $L = 2.13$ m, what is the angle of tilt θ of the tube?

P2.37 The inclined manometer in Fig. P2.37 contains Meriam red manometer oil, SG = 0.827. Assume that the reservoir is very large. If the inclined arm is fitted with graduations 2.5 cm apart, what should the angle θ be if each graduation corresponds to 0.05 kPa gage pressure for p_A?

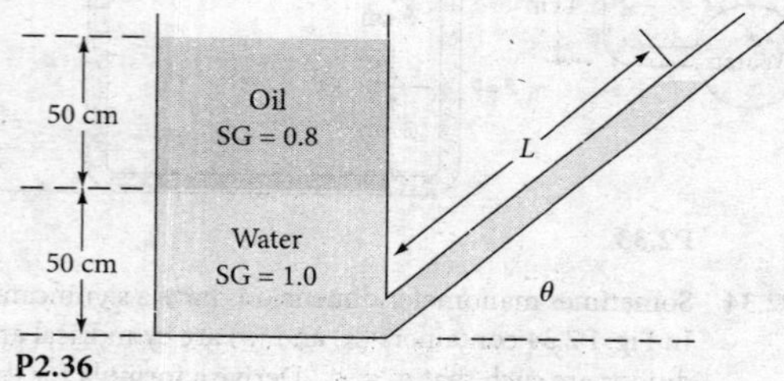

P2.36

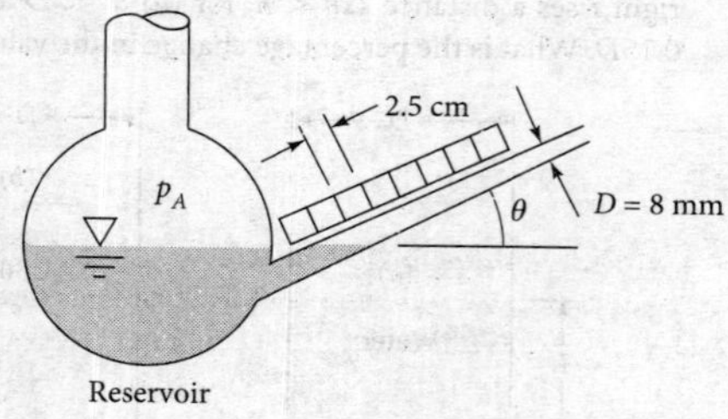

P2.37

P2.38 If the pressure in container A in Fig. P2.38 is 200 kPa, compute the pressure in container B.

P2.39 In Fig. P2.39 the right leg of the manometer is open to the atmosphere. Find the gage pressure, in Pa, in the air gap in the tank.

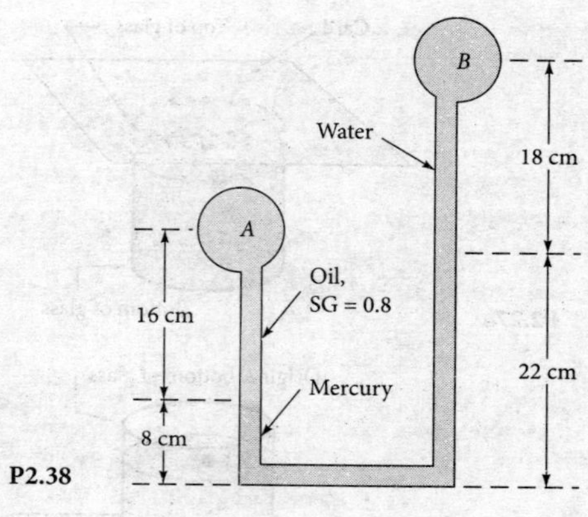

P2.38

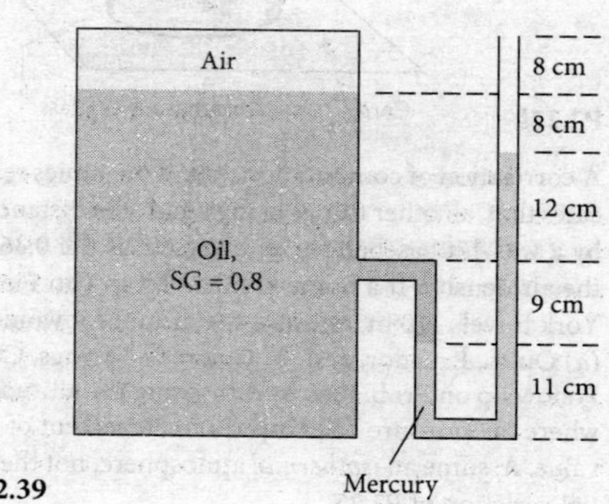

P2.39

P2.40 In Fig. P2.40, if pressure gage A reads 140 kPa absolute, find the pressure in the closed air space B. The manometer fluid is Meriam red oil, SG = 0.827.

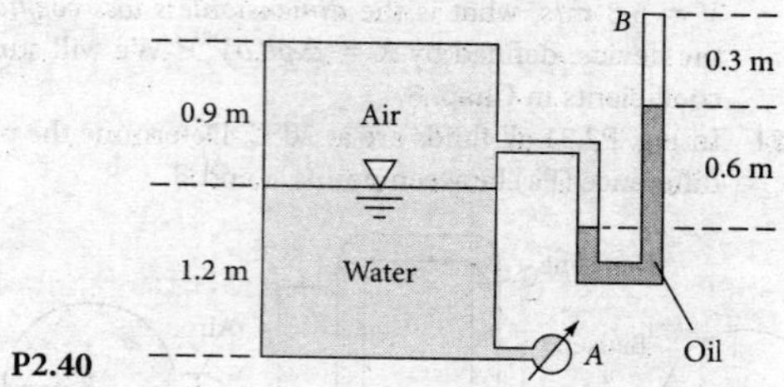

P2.40

P2.41 The system in Fig. P2.41 is at 20°C. Compute the pressure at point A in kPa absolute.

P2.42 Very small pressure differences $p_A - p_B$ can be measured accurately by the two-fluid differential manometer in Fig. P2.42. Density ρ_2 is only slightly larger than that of the upper fluid ρ_1. Derive an expression for the proportionality between h and $p_A - p_B$ if the reservoirs are very large.

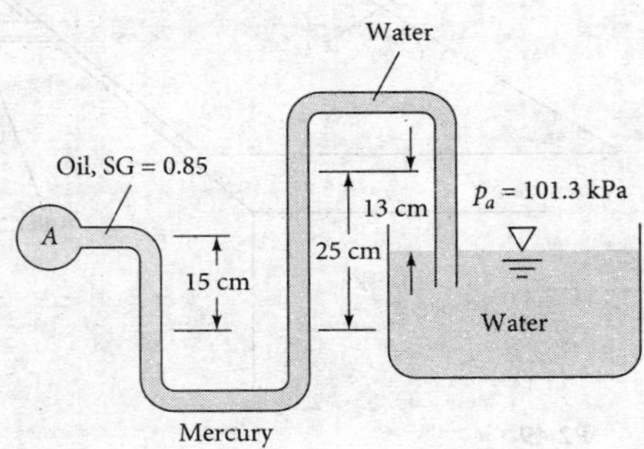

P2.41

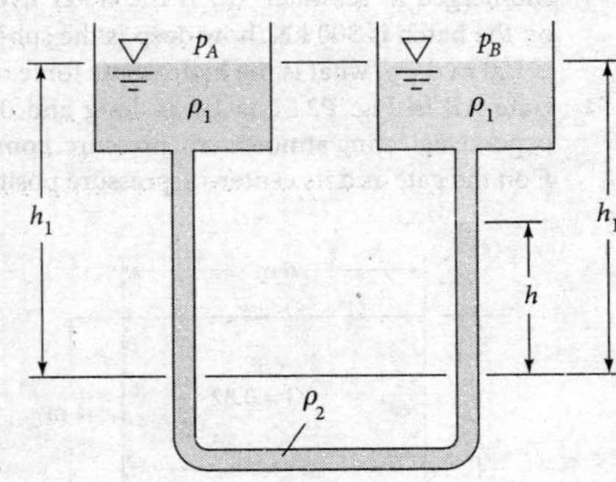

P2.42

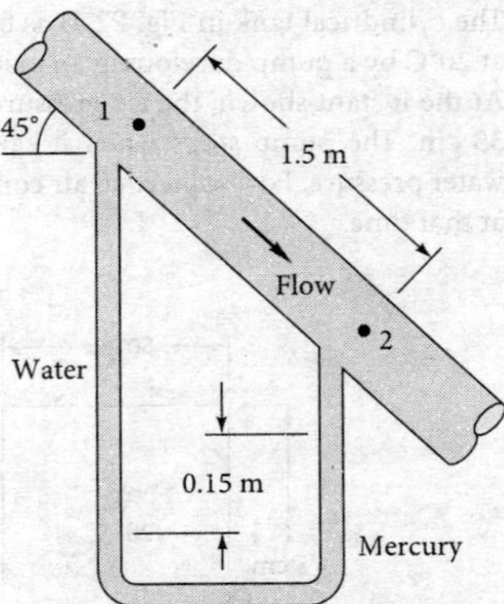

P2.44

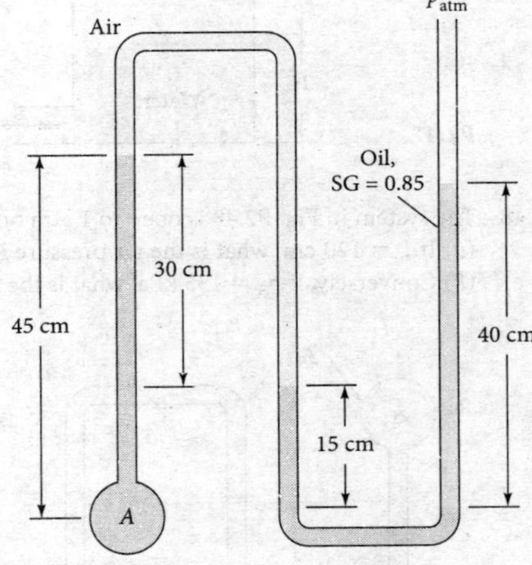

P2.45

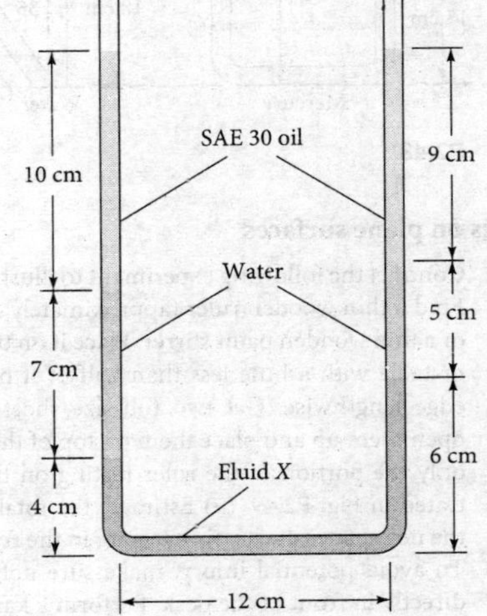

P2.46

P2.43 The traditional method of measuring blood pressure uses a *sphygmomanometer*, first recording the highest (*systolic*) and then the lowest (*diastolic*) pressure from which flowing "Korotkoff" sounds can be heard. Patients with dangerous hypertension can exhibit systolic pressures as high as 34,500 Pa. Normal levels, however, are 18,600 Pa and 11,700 Pa respectively, for systolic and diastolic pressures. The manometer uses mercury and air as fluids.
(*a*) How high in cm should the manometer tube be?
(*b*) Express normal systolic and diastolic blood pressure in millimeters of mercury.

P2.44 Water flows downward in a pipe at 45°, as shown in Fig. P2.44. The pressure drop $p_1 - p_2$ is partly due to gravity and partly due to friction. The mercury manometer reads a 0.15 m height difference. What is the total pressure drop $p_1 - p_2$ in Pa? What is the pressure drop due to friction only between 1 and 2 in Pa? Does the manometer reading correspond only to friction drop? Why?

P2.45 In Fig. P2.45, determine the gage pressure at point A in Pa. Is it higher or lower than atmospheric?

P2.46 In Fig. P2.46 both ends of the manometer are open to the atmosphere. Estimate the specific gravity of fluid X.

P2.47 The cylindrical tank in Fig. P2.47 is being filled with water at 20°C by a pump developing an exit pressure of 175 kPa. At the instant shown, the air pressure is 110 kPa and H = 35 cm. The pump stops when it can no longer raise the water pressure. For isothermal air compression, estimate H at that time.

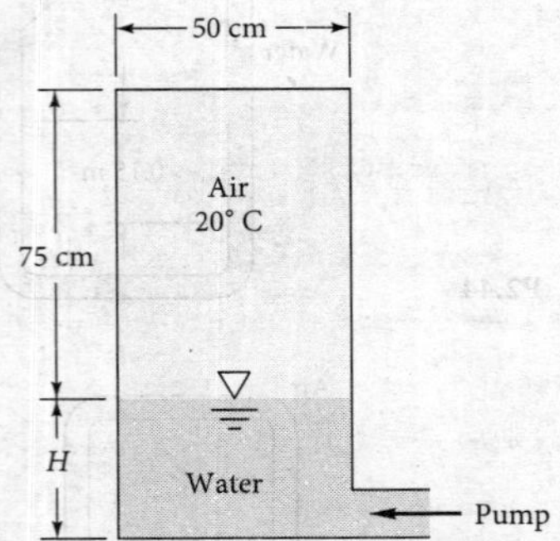

P2.47

P2.48 The system in Fig. P2.48 is open to 1 atm on the right side. (*a*) If L = 120 cm, what is the air pressure in container A? (*b*) Conversely, if p_A = 135 kPa, what is the length L?

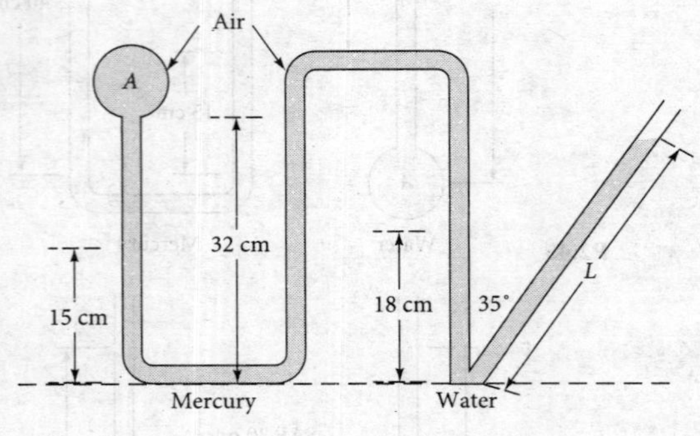

P2.48

Forces on plane surfaces

P2.49 Conduct the following experiment to illustrate air pressure. Find a thin wooden ruler (approximately 30 cm in length) or a thin wooden paint stirrer. Place it on the edge of a desk or table with a little less than half of it hanging over the edge lengthwise. Get two full-size sheets of newspaper; open them up and place them on top of the ruler, covering only the portion of the ruler resting on the desk as illustrated in Fig. P2.49. (*a*) Estimate the total force on top of the newspaper due to air pressure in the room. (*b*) *Careful!* To avoid potential injury, make sure nobody is standing directly in front of the desk. Perform a karate chop on the portion of the ruler sticking out over the edge of the desk. Record your results. (*c*) Explain your results.

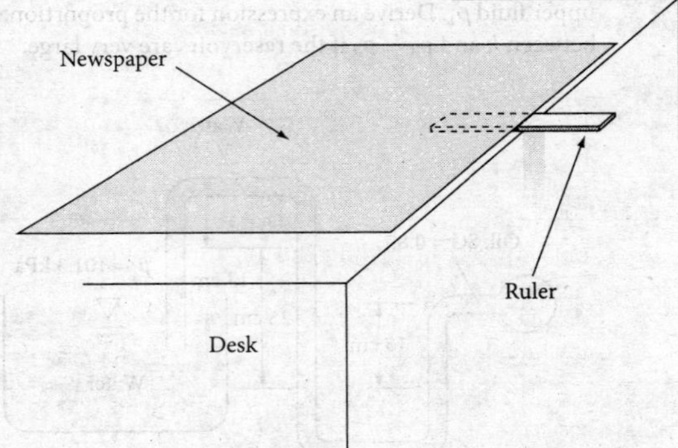

P2.49

P2.50 A small submarine, with a hatch door 0.8 m in diameter, is submerged in seawater. (*a*) If the water hydrostatic force on the hatch is 300 kN, how deep is the sub? (*b*) If the sub is 100 m deep, what is the hydrostatic force on the hatch?

P2.51 Gate AB in Fig. P2.51 is 1.2 m long and 0.8 m into the paper. Neglecting atmospheric pressure, compute the force F on the gate and its center-of-pressure position X.

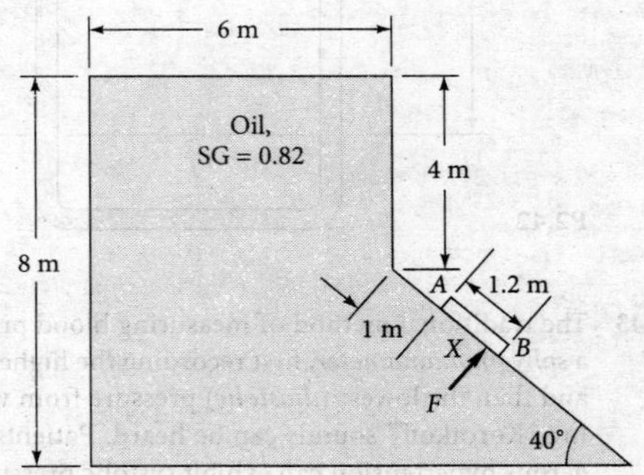

P2.51

P2.52 Example 2.5 calculated the force on plate AB and its line of action, using the moment-of-inertia approach. Some teachers say it is more instructive to calculate these by *direct integration* of the pressure forces. Using Figs. P2.52 and E2.5*a*, (*a*) find an expression for the pressure variation $p(\xi)$ along the plate; (*b*) integrate this expression to find the total force F; (*c*) integrate the moments about point A to find the position of the center of pressure.

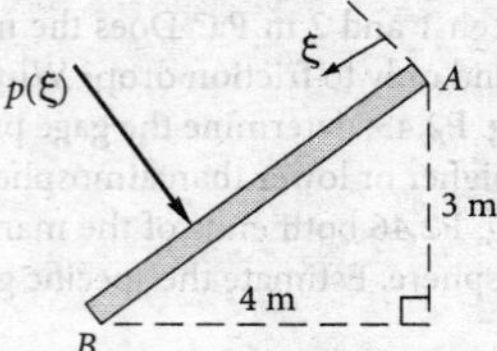

P2.52

P2.53 The Hoover Dam, in Arizona, encloses Lake Mead, which contains 38 trillion liters of water. The dam is 366 m wide and the lake is 152 m deep. (*a*) Estimate the hydrostatic force on the dam, in MN. (*b*) Explain how you might analyze the stress in the dam due to this hydrostatic force.

P2.54 In Fig. P2.54, the hydrostatic force F is the same on the bottom of all three containers, even though the weights of liquid above are quite different. The three bottom shapes and the fluids are the same. This is called the *hydrostatic paradox*. Explain why it is true and sketch a free body of each of the liquid columns.

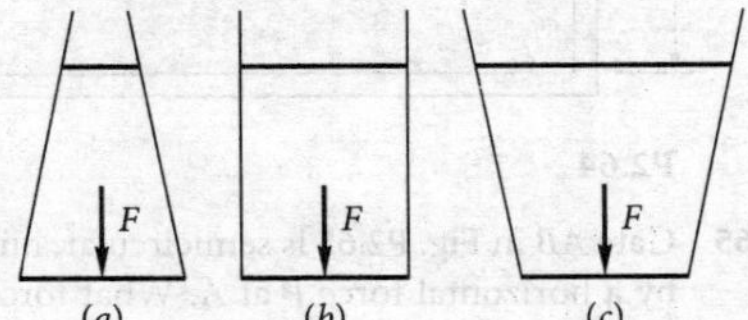

P2.54

P2.55 Gate AB in Fig. P2.55 is 1.5 m wide into the paper, hinged at A, and restrained by a stop at B. The water is at 20°C. Compute (*a*) the force on stop B and (*b*) the reactions at A if the water depth $h = 3$ m.

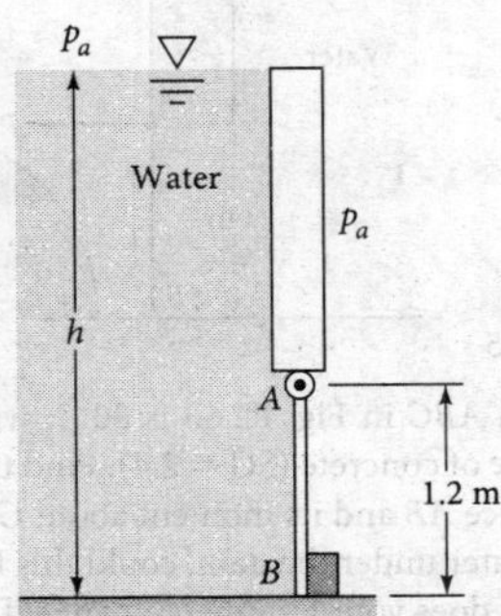

P2.55

P2.56 In Fig. P2.55, gate AB is 1.5 m wide into the paper, and stop B will break if the water force on it equals 40,900 N. For what water depth h is this condition reached?

P2.57 The square vertical panel $ABCD$ in Fig. P2.57 is submerged in water at 20°C. Side AB is at least 1.7 m below the surface. Determine the difference between the hydrostatic forces on subpanels ABD and BCD.

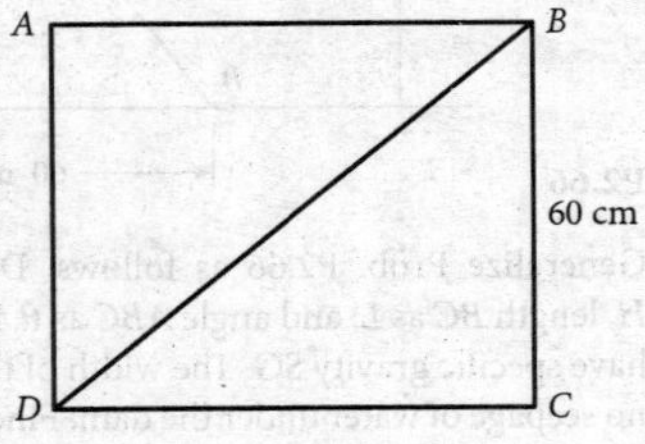

P2.57

P2.58 In Fig. P2.58, the cover gate AB closes a circular opening 80 cm in diameter. The gate is held closed by a 200-kg mass as shown. Assume standard gravity at 20°C. At what water level h will the gate be dislodged? Neglect the weight of the gate.

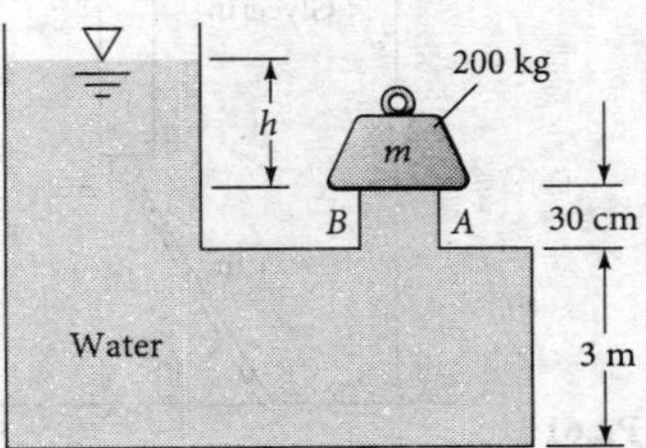

P2.58

***P2.59** Gate AB has length L and width b into the paper, is hinged at B, and has negligible weight. The liquid level h remains at the top of the gate for any angle θ. Find an analytic expression for the force P, perpendicular to AB, required to keep the gate in equilibrium in Fig. P2.59.

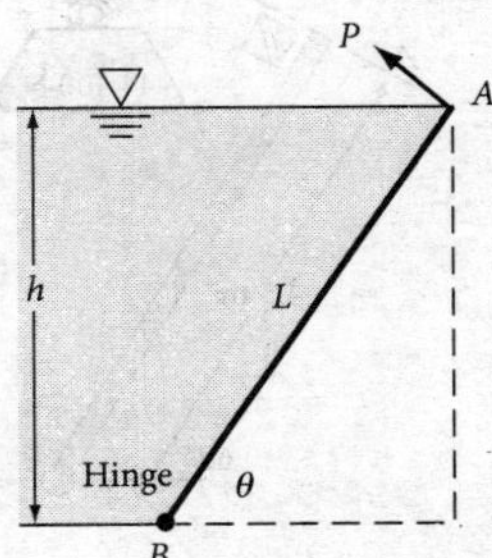

P2.59

P2.60 In Fig. P2.60, vertical, unsymmetrical trapezoidal panel $ABCD$ is submerged in fresh water with side AB 4 m below the surface. Since trapezoid formulas are complicated, (*a*) estimate, reasonably, the water force on the panel, in N, neglecting atmospheric pressure. For extra credit, (*b*) look up the formula and compute the exact force on the panel.

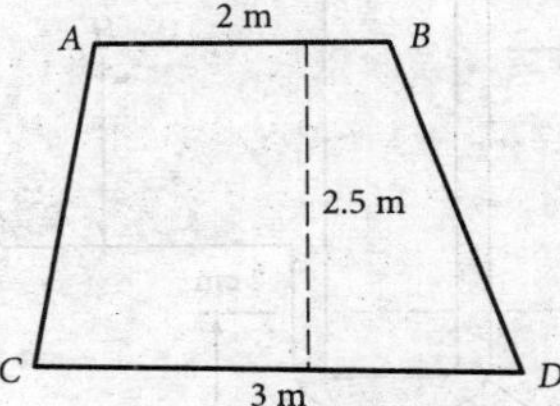

P2.60

***P2.61** Gate AB in Fig. P2.61 is a homogeneous mass of 180 kg, 1.2 m wide into the paper, hinged at A, and resting on a smooth bottom at B. All fluids are at 20°C. For what water depth h will the force at point B be zero?

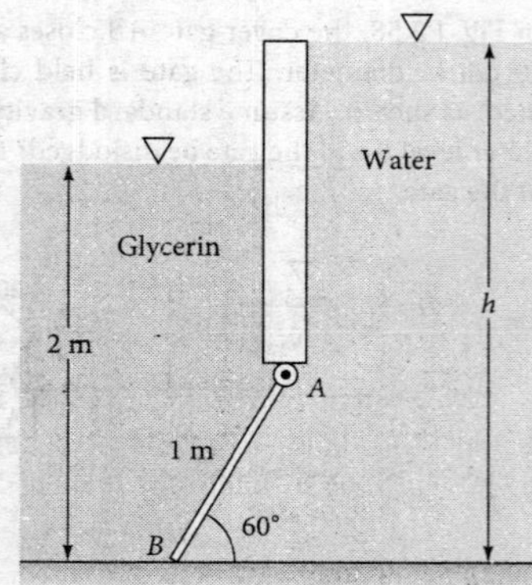

P2.61

P2.62 Gate AB in Fig. P2.62 is 4.5 m long and 2.5 m wide into the paper and is hinged at B with a stop at A. The water is at 20°C. The gate is 0.025 m-thick steel, SG = 7.85. Compute the water level h for which the gate will start to fall.

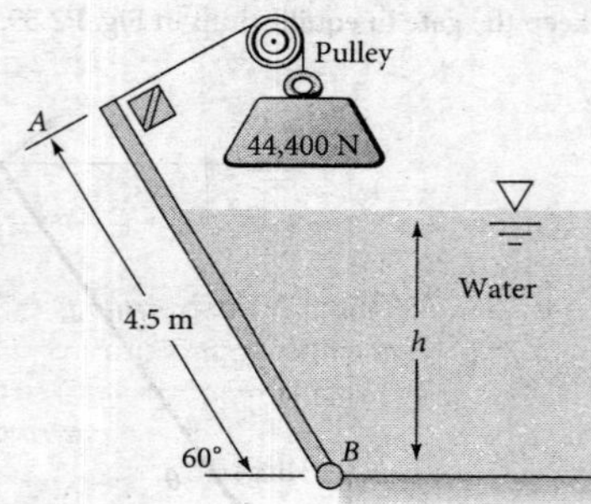

P2.62

P2.63 The tank in Fig. P2.63 has a 4-cm-diameter plug at the bottom on the right. All fluids are at 20°C. The plug will pop out if the hydrostatic force on it is 25 N. For this condition, what will be the reading h on the mercury manometer on the left side?

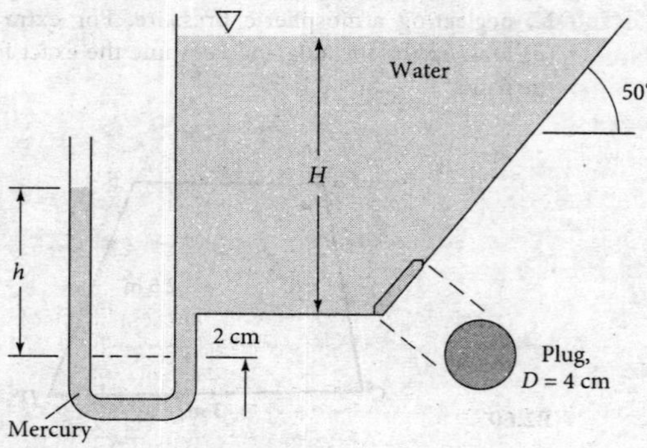

P2.63

***P2.64** Gate ABC in Fig. P2.64 has a fixed hinge line at B and is 2 m wide into the paper. The gate will open at A to release water if the water depth is high enough. Compute the depth h for which the gate will begin to open.

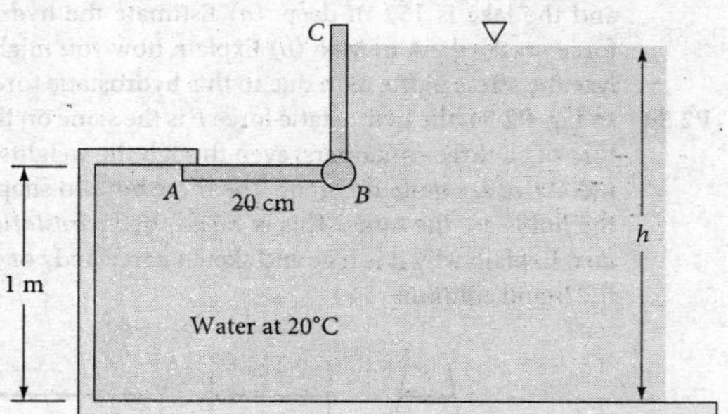

P2.64

***P2.65** Gate AB in Fig. P2.65 is semicircular, hinged at B, and held by a horizontal force P at A. What force P is required for equilibrium?

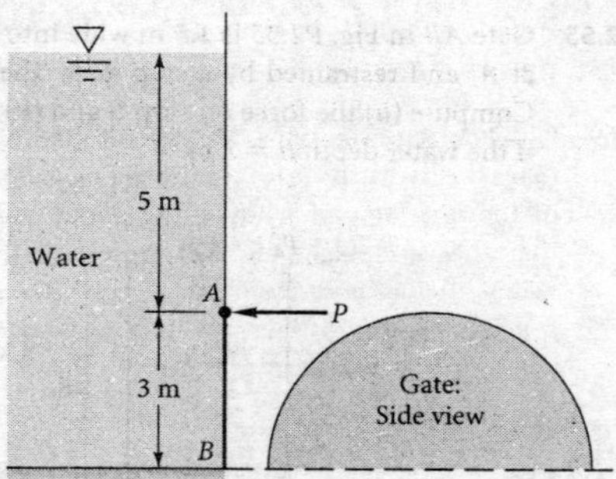

P2.65

P2.66 Dam ABC in Fig. P2.66 is 30 m wide into the paper and made of concrete (SG = 2.4). Find the hydrostatic force on surface AB and its moment about C. Assuming no seepage of water under the dam, could this force tip the dam over? How does your argument change if there is seepage under the dam?

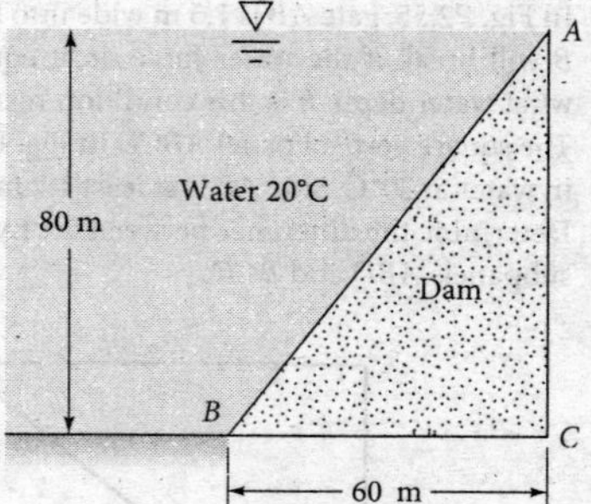

P2.66

***P2.67** Generalize Prob. P2.66 as follows. Denote length AB as H, length BC as L, and angle ABC as θ. Let the dam material have specific gravity SG. The width of the dam is b. Assume no seepage of water under the dam. Find an analytic relation between SG and the critical angle θ_c for which the dam will just tip over to the right. Use your relation to compute θ_c for the special case SG = 2.4 (concrete).

P2.68 Isosceles triangle gate *AB* in Fig. P2.68 is hinged at *A* and weighs 1500 N. What horizontal force *P* is required at point *B* for equilibrium?

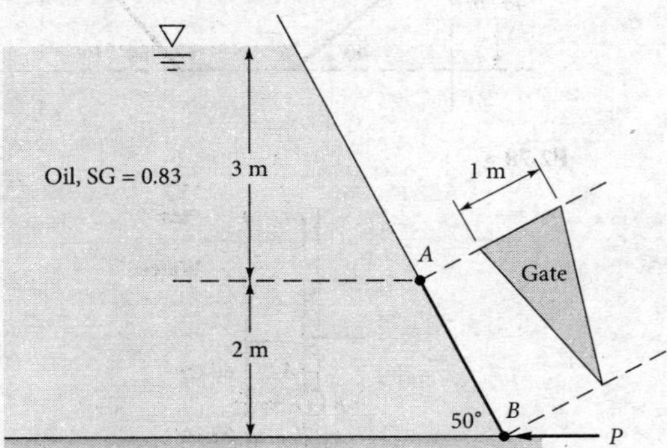

P2.68

P2.69 Consider the slanted plate *AB* of length *L* in Fig. P2.69. (*a*) Is the hydrostatic force *F* on the plate equal to the weight of the *missing water* above the plate? If not, correct this hypothesis. Neglect the atmosphere. (*b*) Can a "missing water" theory be generalized to *curved* surfaces of this type?

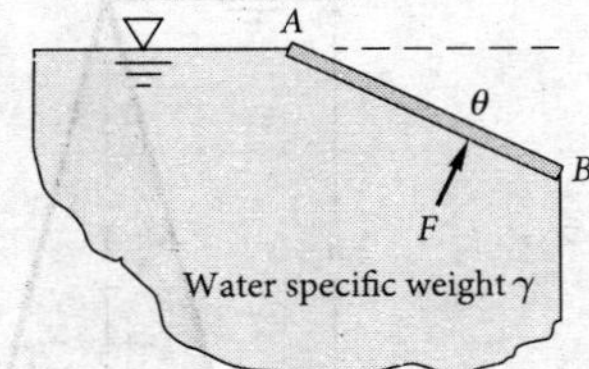

P2.69

P2.70 The swing-check valve in Fig. P2.70 covers a 22.86-cm diameter opening in the slanted wall. The hinge is 15 cm from the centerline, as shown. The valve will open when the hinge moment is 50 N · m. Find the value of *h* for the water to cause this condition.

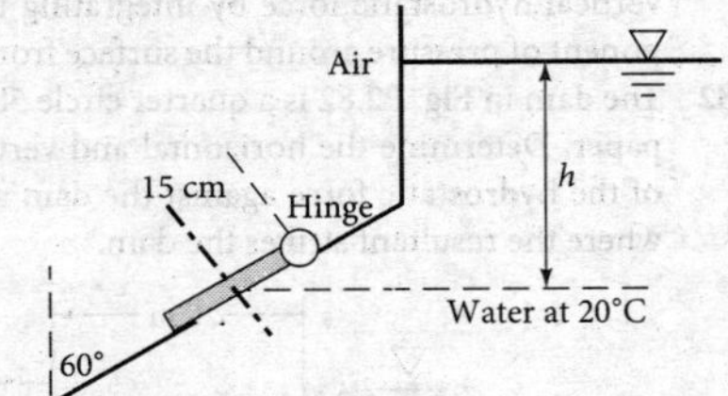

P2.70

***P2.71** In Fig. P2.71 gate *AB* is 3 m wide into the paper and is connected by a rod and pulley to a concrete sphere (SG = 2.40). What diameter of the sphere is just sufficient to keep the gate closed?

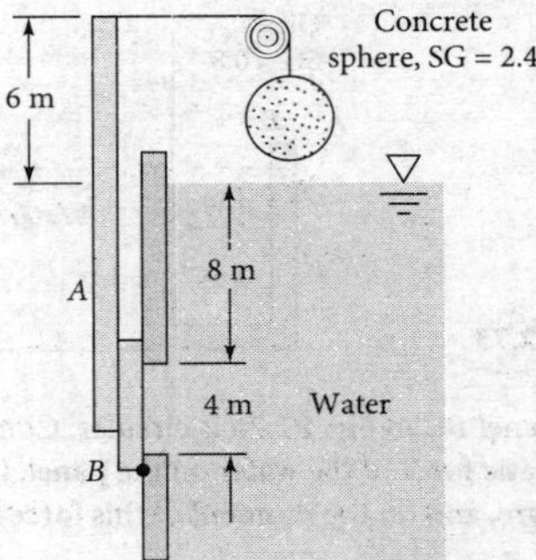

P2.71

P2.72 In Fig. P2.72, gate *AB* is circular. Find the moment of the hydrostatic force on this gate about axis *A*.

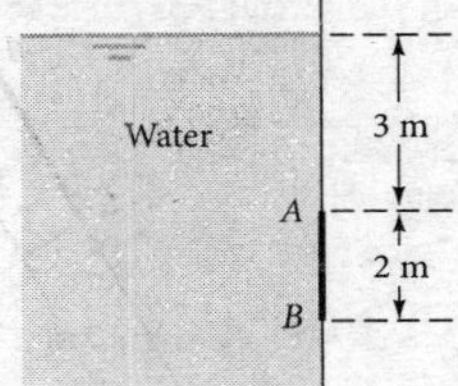

P2.72

P2.73 Gate *AB* is 1.5 m wide into the paper and opens to let fresh water out when the ocean tide is dropping. The hinge at *A* is 0.6 m above the freshwater level. At what ocean level *h* will the gate first open? Neglect the gate weight.

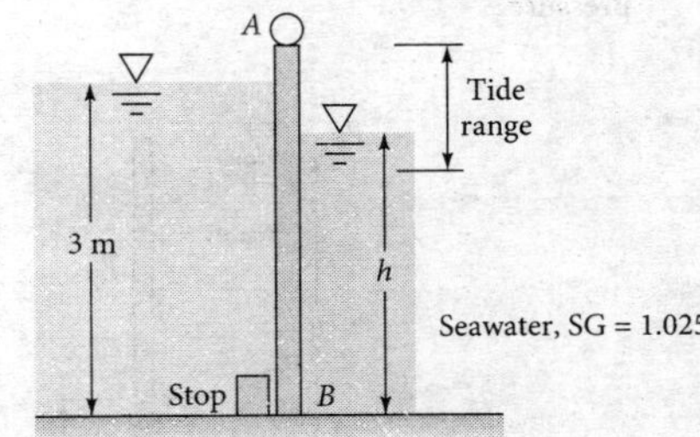

P2.73

P2.74 Find the height *H* in Fig. P2.74 for which the hydrostatic force on the rectangular panel is the same as the force on the semicircular panel below.

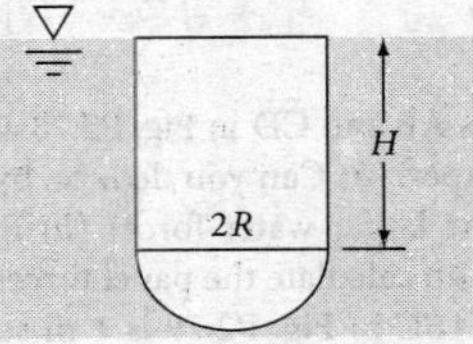

P2.74

P2.75 The cap at point B on the 5-cm-diameter tube in Fig. P2.75 will be dislodged when the hydrostatic force on its base reaches 100 N. For what water depth *h* does this occur?

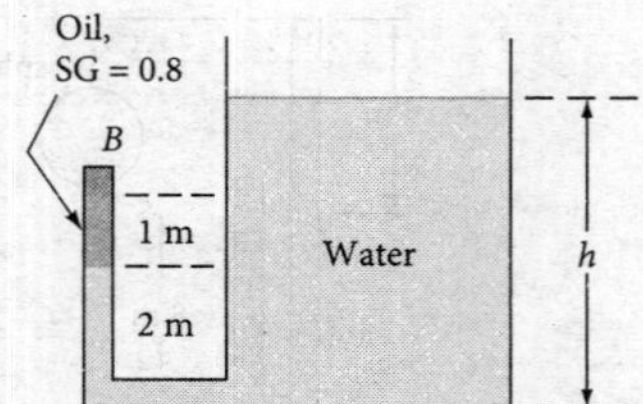

P2.75

P2.76 Panel *BC* in Fig. P2.76 is circular. Compute (*a*) the hydrostatic force of the water on the panel, (*b*) its center of pressure, and (*c*) the moment of this force about point *B*.

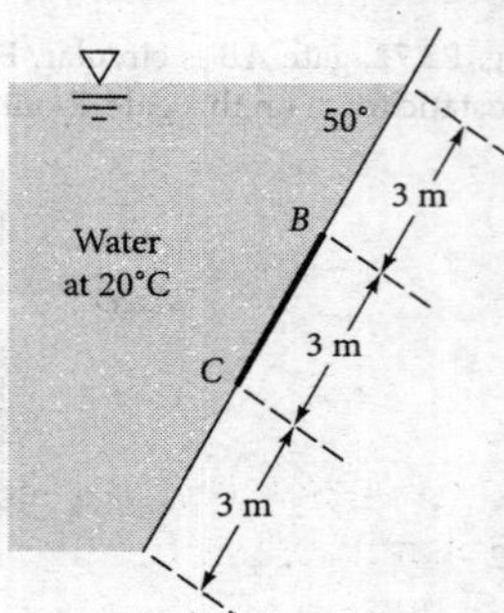

P2.76

P2.77 The circular gate *ABC* in Fig. P2.77 has a 1-m radius and is hinged at *B*. Compute the force *P* just sufficient to keep the gate from opening when $h = 8$ m. Neglect atmospheric pressure.

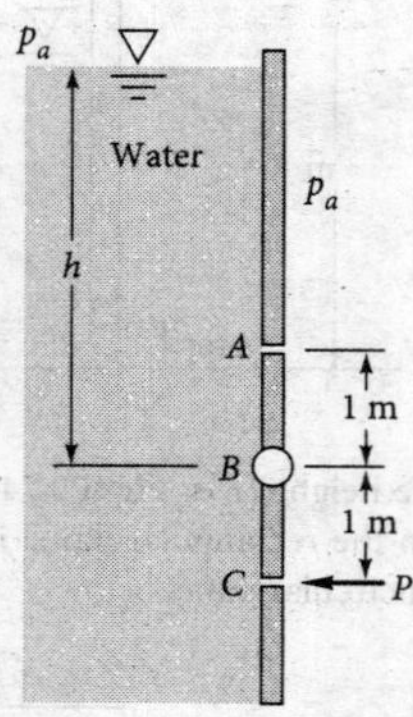

P2.77

P2.78 Panels AB and CD in Fig. P2.78 are each 120 cm wide into the paper. (*a*) Can you deduce, by inspection, which panel has the larger water force? (*b*) Even if your deduction is brilliant, calculate the panel forces anyway.

P2.79 Gate *ABC* in Fig. P2.79 is 1 m square and is hinged at *B*. It will open automatically when the water level *h* becomes high enough. Determine the lowest height for which the gate will open. Neglect atmospheric pressure. Is this result independent of the liquid density?

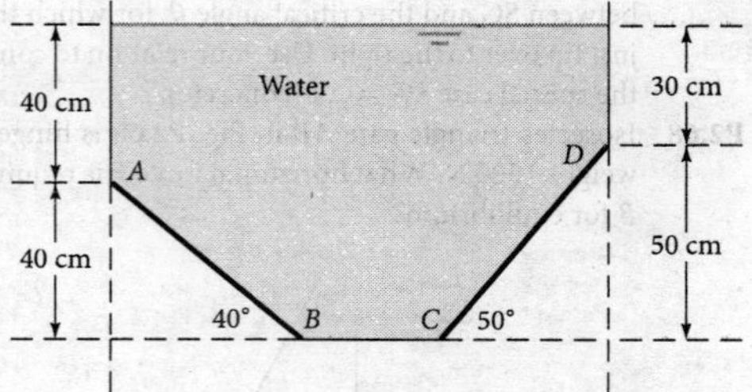

P2.78

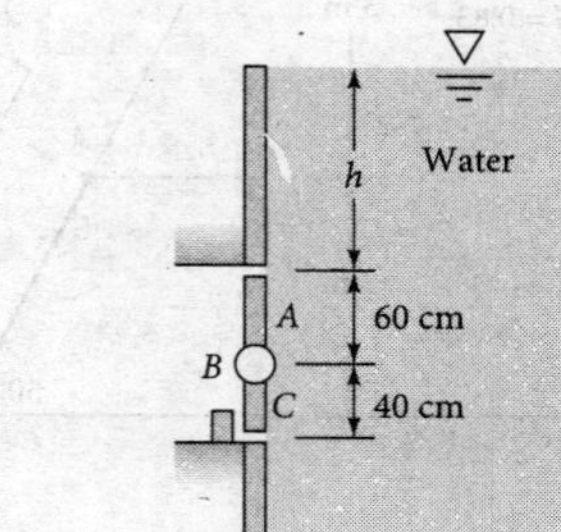

P2.79

***P2.80** A concrete dam (SG = 2.5) is made in the shape of an isosceles triangle, as in Fig. P2.80. Analyze this geometry to find the range of angles θ for which the hydrostatic force will tend to tip the dam over at point *B*. The width into the paper is *b*.

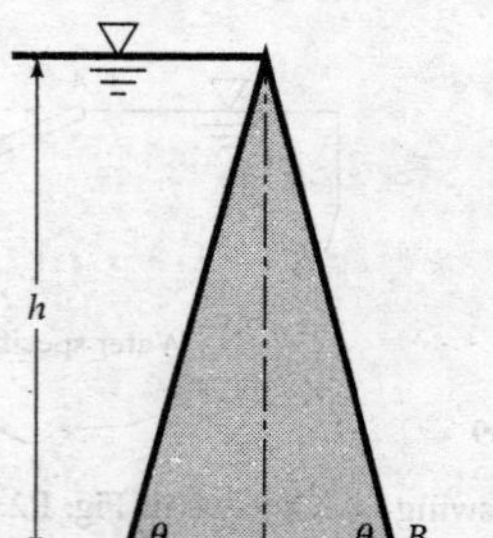

P2.80

Forces on curved surfaces

P2.81 For the semicircular cylinder *CDE* in Example 2.9, find the vertical hydrostatic force by integrating the vertical component of pressure around the surface from $\theta = 0$ to $\theta = \pi$.

***P2.82** The dam in Fig. P2.82 is a quarter circle 50 m wide into the paper. Determine the horizontal and vertical components of the hydrostatic force against the dam and the point CP where the resultant strikes the dam.

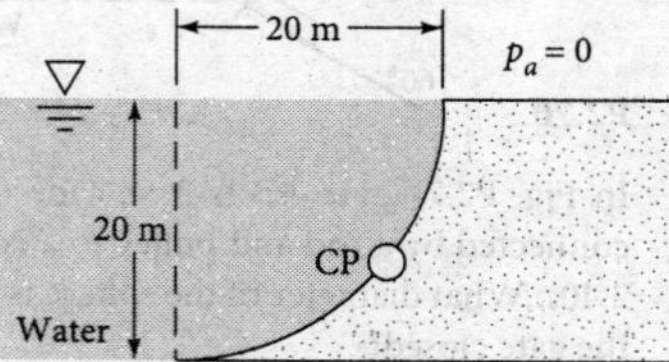

P2.82

***P2.83** Gate *AB* in Fig. P2.83 is a quarter circle 3 m wide into the paper and hinged at *B*. Find the force *F* just sufficient to keep the gate from opening. The gate is uniform and weighs 13,300 N.

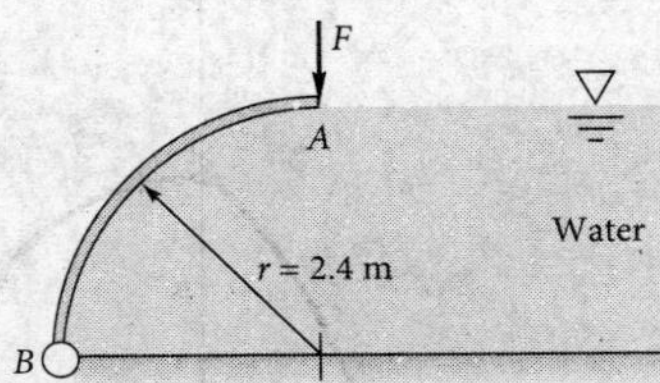

P2.83

P2.84 Panel AB in Fig. P2.84 is a parabola with its maximum at point *A*. It is 150 cm wide into the paper. Neglect atmospheric pressure. Find (*a*) the vertical and (*b*) the horizontal water forces on the panel.

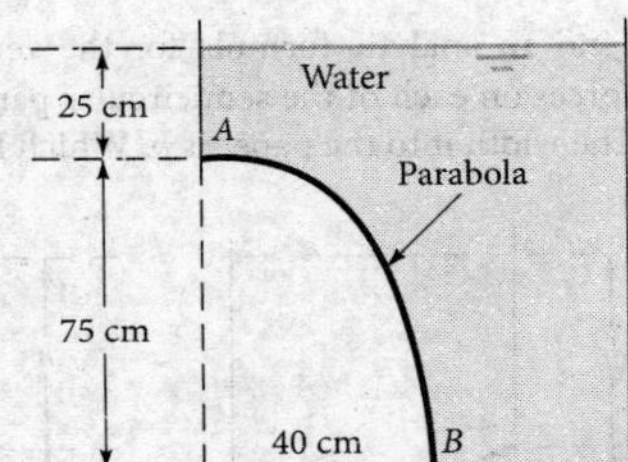

P2.84

P2.85 Compute the horizontal and vertical components of the hydrostatic force on the quarter-circle panel at the bottom of the water tank in Fig. P2.85.

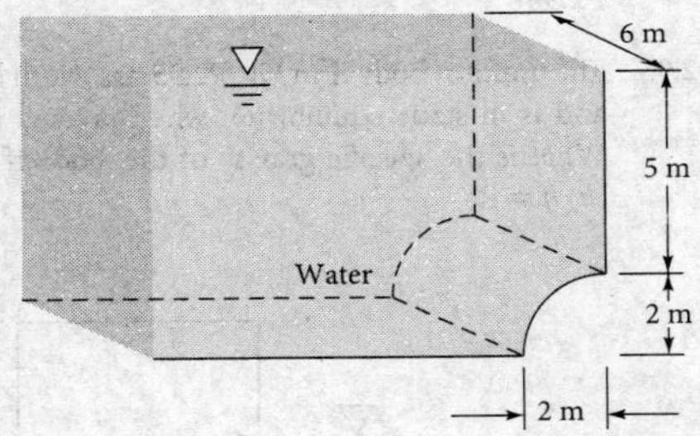

P2.85

P2.86 The quarter circle gate *BC* in Fig. P2.86 is hinged at *C*. Find the horizontal force *P* required to hold the gate stationary. Neglect the weight of the gate.

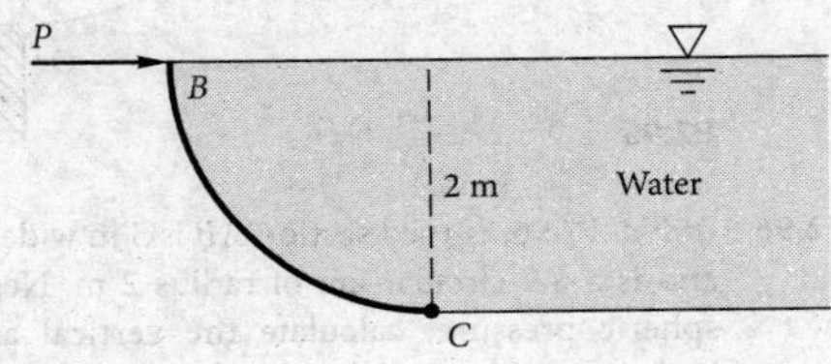

P2.86

P2.87 The bottle of champagne (SG = 0.96) in Fig. P2.87 is under pressure, as shown by the mercury-manometer reading. Compute the net force on the 5 cm radius hemispherical end cap at the bottom of the bottle.

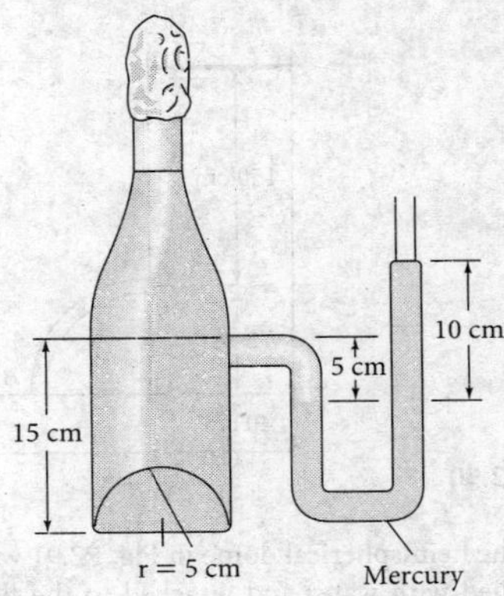

P2.87

***P2.88** Gate *ABC* is a circular arc, sometimes called a *Tainter gate*, which can be raised and lowered by pivoting about point *O*. See Fig. P2.88. For the position shown, determine (*a*) the hydrostatic force of the water on the gate and (*b*) its line of action. Does the force pass through point *O*?

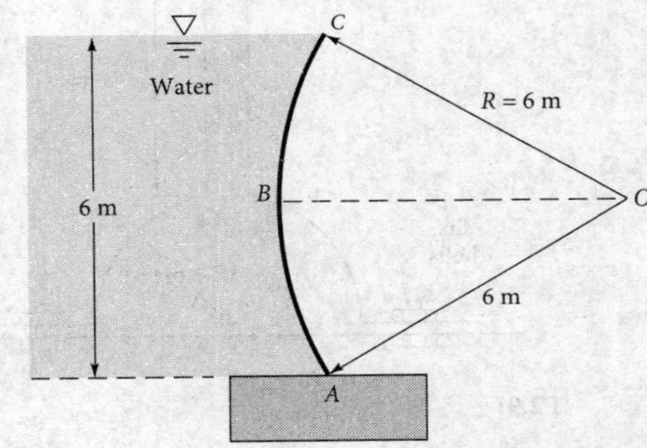

P2.88

P2.89 The tank in Fig. P2.89 contains benzene and is pressurized to 200 kPa (gage) in the air gap. Determine the vertical hydrostatic force on circular-arc section *AB* and its line of action.

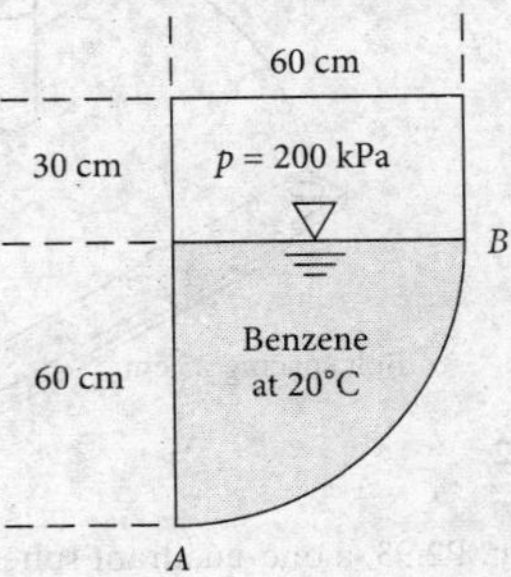

P2.89

P2.90 The tank in Fig. P2.90 is 120 cm long into the paper. Determine the horizontal and vertical hydrostatic forces on the quarter-circle panel AB. The fluid is water at 20°C. Neglect atmospheric pressure.

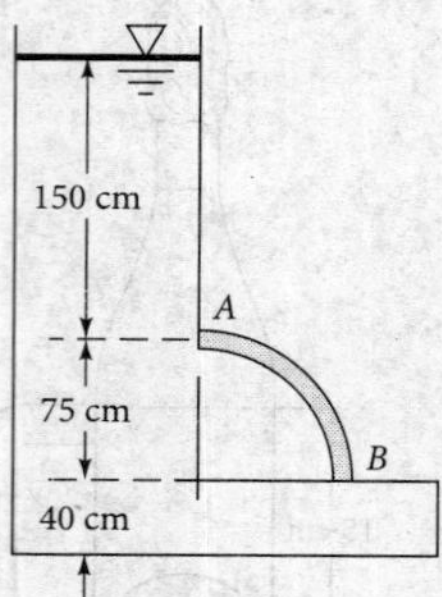

P2.90

P2.91 The hemispherical dome in Fig. P2.91 weighs 30 kN and is filled with water and attached to the floor by six equally spaced bolts. What is the force in each bolt required to hold down the dome?

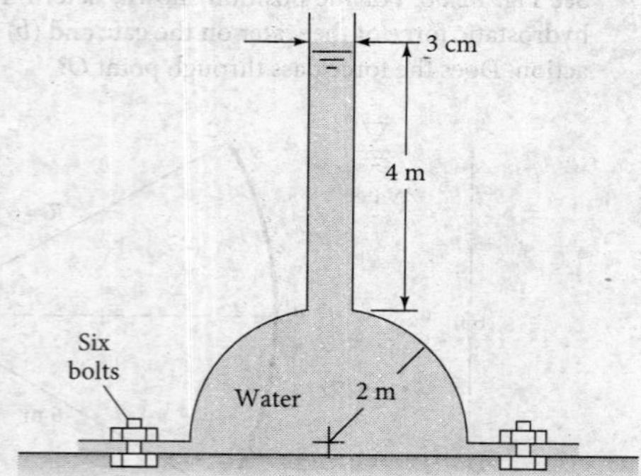

P2.91

P2.92 A 4-m-diameter water tank consists of two half cylinders, each weighing 4.5 kN/m, bolted together as shown in Fig. P2.92. If the support of the end caps is neglected, determine the force induced in each bolt.

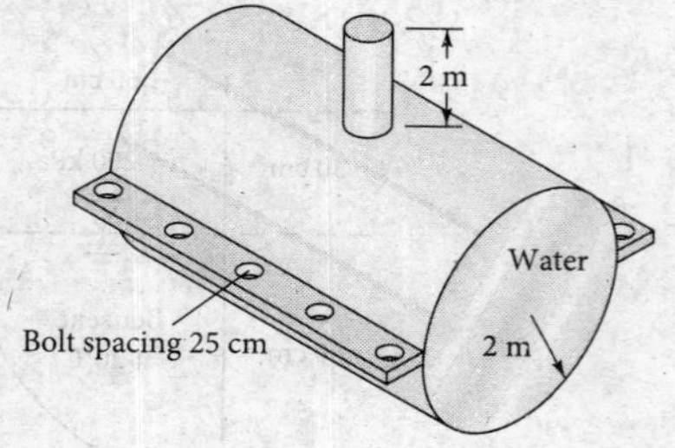

P2.92

***P2.93** In Fig. P2.93, a one-quadrant spherical shell of radius R is submerged in liquid of specific weight γ and depth $h > R$. Find an analytic expression for the resultant hydrostatic force, and its line of action, on the shell surface.

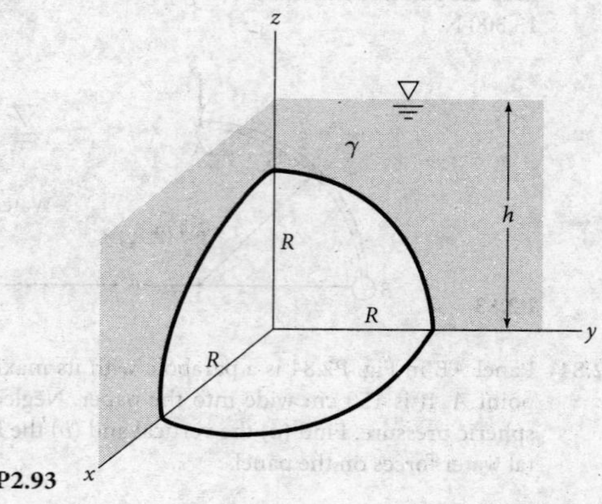

P2.93

P2.94 Find an analytic formula for the vertical and horizontal forces on each of the semicircular panels AB in Fig. P2.94. The width into the paper is b. Which force is larger? Why?

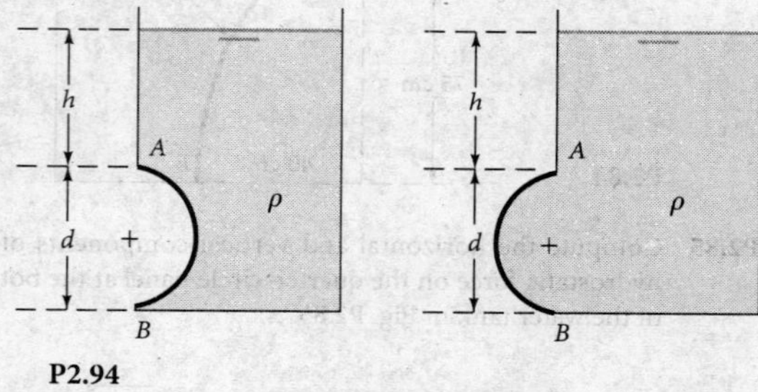

P2.94

***P2.95** The uniform body A in Fig. P2.95 has width b into the paper and is in static equilibrium when pivoted about hinge O. What is the specific gravity of this body if (*a*) $h = 0$ and (*b*) $h = R$?

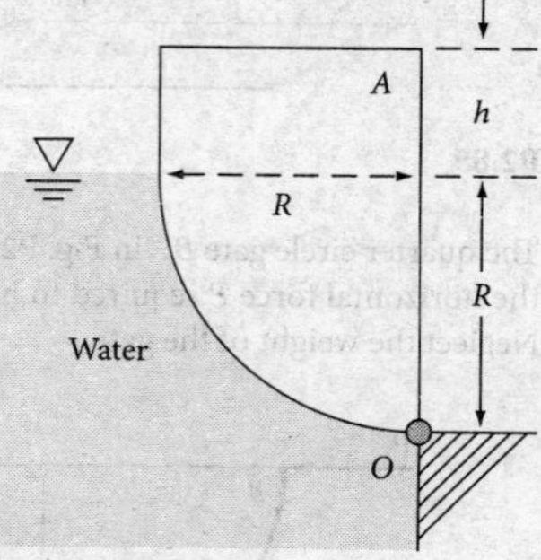

P2.95

P2.96 In Fig. P2.96, curved section AB is 5 m wide into the paper and is a 60° circular arc of radius 2 m. Neglecting atmospheric pressure, calculate the vertical and horizontal hydrostatic forces on arc AB.

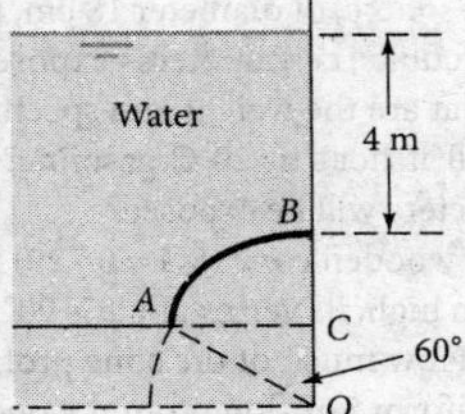

P2.96

P2.97 The contractor ran out of gunite mixture and finished the deep corner of a 5-m-wide swimming pool with a quarter-circle piece of PVC pipe, labeled *AB* in Fig. P2.97. Compute the horizontal and vertical water forces on the curved panel *AB*.

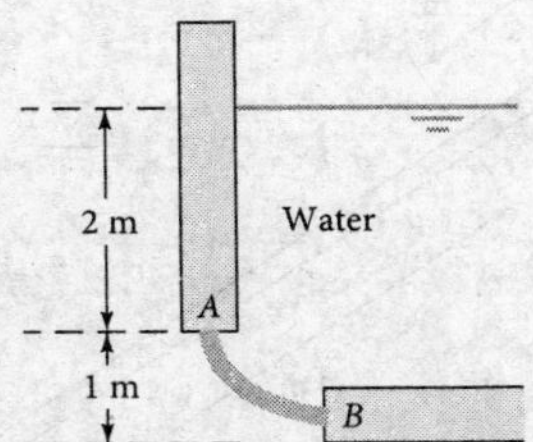

P2.97

P2.98 The curved surface in Fig. P2.98 consists of two quarter-spheres and a half cylinder. A side view and front view are shown. Calculate the horizontal and vertical forces on the surface.

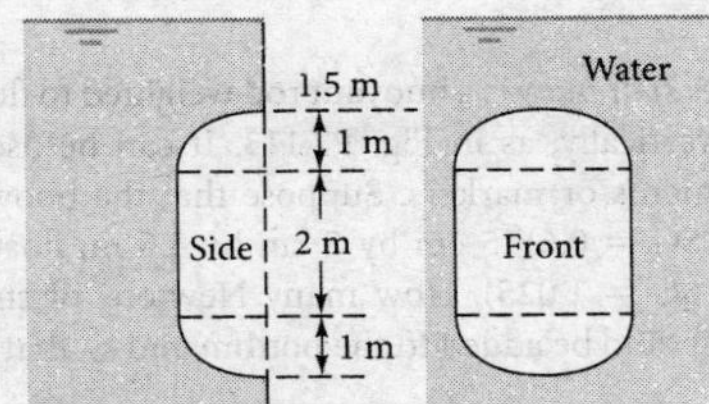

P2.98

P2.99 The mega-magnum cylinder in Fig. P2.99 has a hemispherical bottom and is pressurized with air to 75 kPa (gage) at the top. Determine (*a*) the horizontal and (*b*) the vertical hydrostatic forces on the hemisphere, in N.

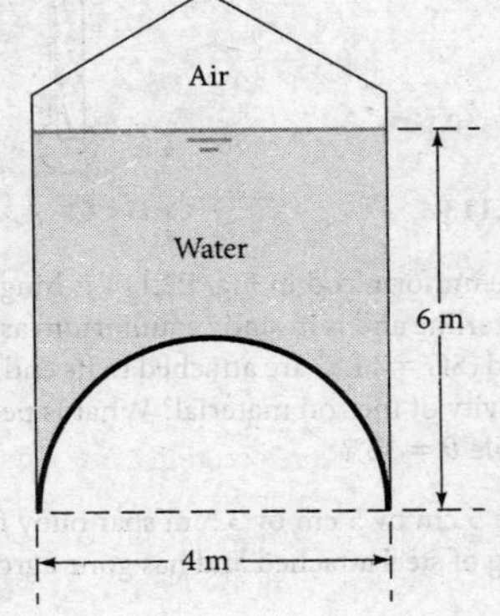

P2.99

P2.100 Pressurized water fills the tank in Fig. P2.100. Compute the net hydrostatic force on the conical surface *ABC*.

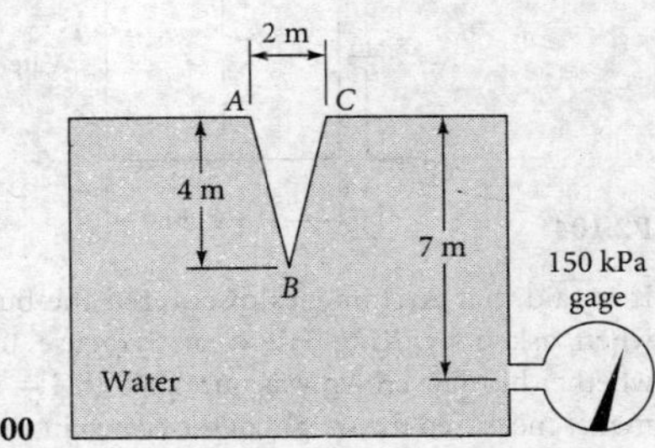

P2.100

Forces on layered surfaces

P2.101 The closed layered box in Fig. P2.101 has square horizontal cross sections everywhere. All fluids are at 20°C. Estimate the gage pressure of the air if (*a*) the hydrostatic force on panel *AB* is 48 kN or (*b*) the hydrostatic force on the bottom panel *BC* is 97 kN.

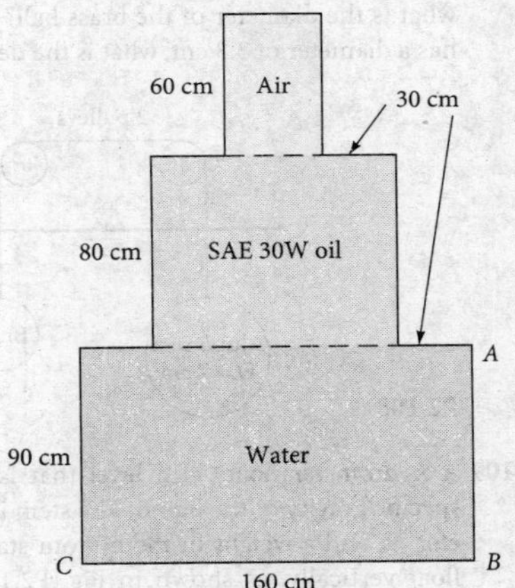

P2.101

P2.102 A cubical tank is 3 m × 3 m × 3 m and is layered with 1 meter of fluid of specific gravity 1.0, 1 meter of fluid with SG = 0.9, and 1 meter of fluid with SG = 0.8. Neglect atmospheric pressure. Find (*a*) the hydrostatic force on the bottom and (*b*) the force on a side panel.

Buoyancy; Archimedes' principles

P2.103 A solid block, of specific gravity 0.9, floats such that 75 percent of its volume is in water and 25 percent of its volume is in fluid *X*, which is layered above the water. What is the specific gravity of fluid *X*?

P2.104 The can in Fig. P2.104 floats in the position shown. What is its weight in N?

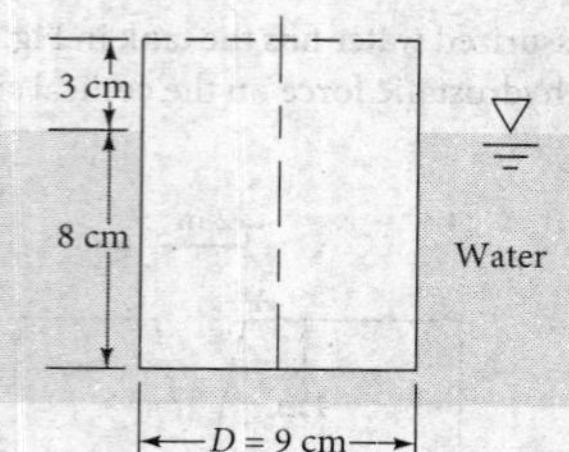

P2.104

P2.105 It is said that Archimedes discovered the buoyancy laws when asked by King Hiero of Syracuse to determine whether his new crown was pure gold (SG = 19.3). Archimedes measured the weight of the crown in air to be 11.8 N and its weight in water to be 10.9 N. Was it pure gold?

P2.106 A spherical helium balloon has a total mass of 3 kg. It settles in a calm standard atmosphere at an altitude of 5500 m. Estimate the diameter of the balloon.

P2.107 Repeat Prob. 2.62, assuming that the 44,400 N weight is aluminum (SG = 2.71) and is hanging submerged in the water.

P2.108 A 7-cm-diameter solid aluminum ball (SG = 2.7) and a solid brass ball (SG = 8.5) balance nicely when submerged in a liquid, as in Fig. P2.108. (*a*) If the fluid is water at 20°C, what is the diameter of the brass ball? (*b*) If the brass ball has a diameter of 3.8 cm, what is the density of the fluid?

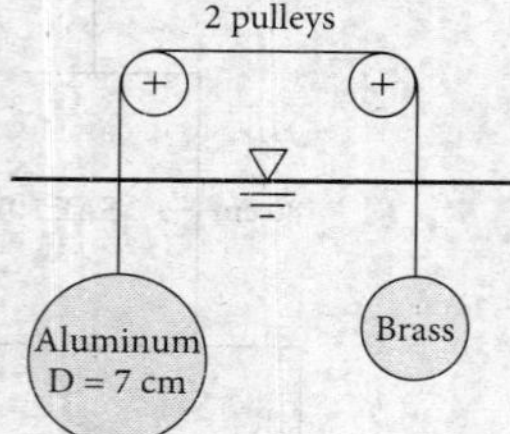

P2.108

P2.109 A *hydrometer* floats at a level that is a measure of the specific gravity of the liquid. The stem is of constant diameter D, and a weight in the bottom stabilizes the body to float vertically, as shown in Fig. P2.109. If the position $h = 0$ is pure water (SG = 1.0), derive a formula for h as a function of total weight W, D, SG, and the specific weight γ_0 of water.

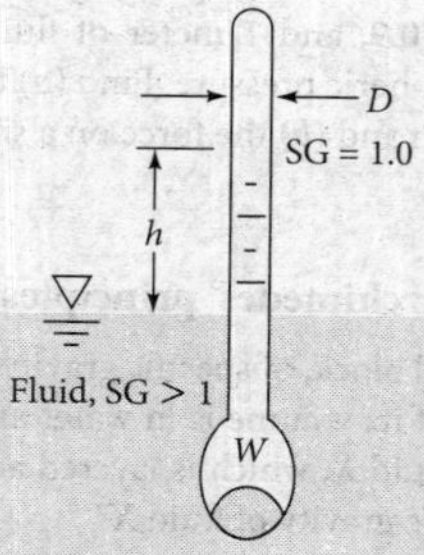

P2.109

P2.110 A solid sphere, of diameter 18 cm, floats in 20°C water with 1,527 cubic centimeters exposed above the surface. (*a*) What are the weight and specific gravity of this sphere? (*b*) Will it float in 20°C gasoline? If so, how many cubic centimeters will be exposed?

P2.111 A solid wooden cone (SG = 0.729) floats in water. The cone is 30 cm high, its vertex angle is 90°, and it floats with vertex down. How much of the cone protrudes above the water?

P2.112 The uniform 5-m-long round wooden rod in Fig. P2.112 is tied to the bottom by a string. Determine (*a*) the tension in the string and (*b*) the specific gravity of the wood. Is it possible for the given information to determine the inclination angle θ? Explain.

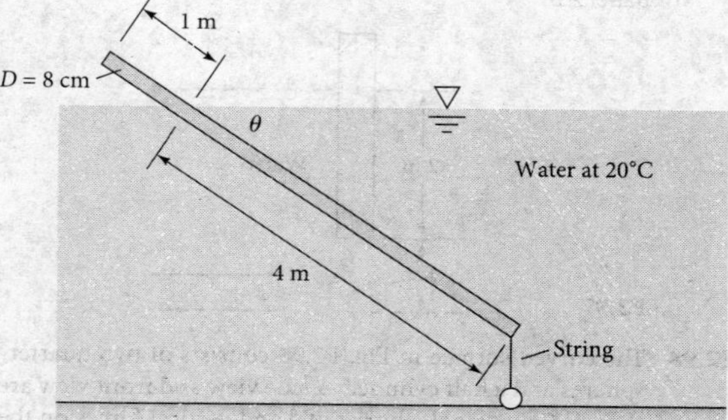

P2.112

P2.113 A *spar buoy* is a buoyant rod weighted to float and protrude vertically, as in Fig. P2.113. It can be used for measurements or markers. Suppose that the buoy is maple wood (SG = 0.6), 5 cm by 5 cm by 3.5 m, floating in seawater (SG = 1.025). How many Newtons of steel (SG = 7.85) should be added to the bottom end so that $h = 0.5$ m?

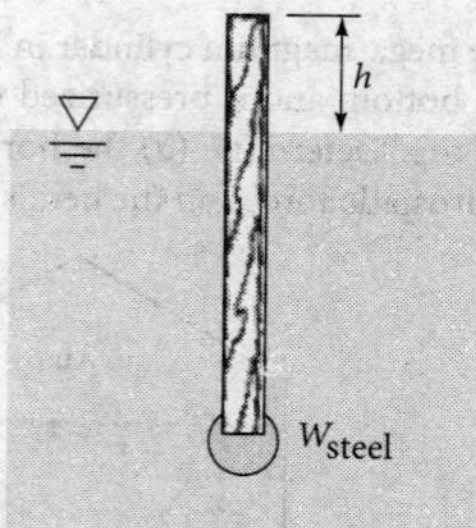

P2.113

P2.114 The uniform rod in Fig. P2.114 is hinged at point B on the waterline and is in static equilibrium as shown when 2 kg of lead (SG = 11.4) are attached to its end. What is the specific gravity of the rod material? What is peculiar about the rest angle $\theta = 30°$?

P2.115 The 5 cm by 5 cm by 3.5 m spar buoy from Fig. P2.113 has 2 kg of steel attached and has gone aground on a rock, as in

Fig. P2.115. Compute the angle θ at which the buoy will lean, assuming that the rock exerts no moments on the spar.

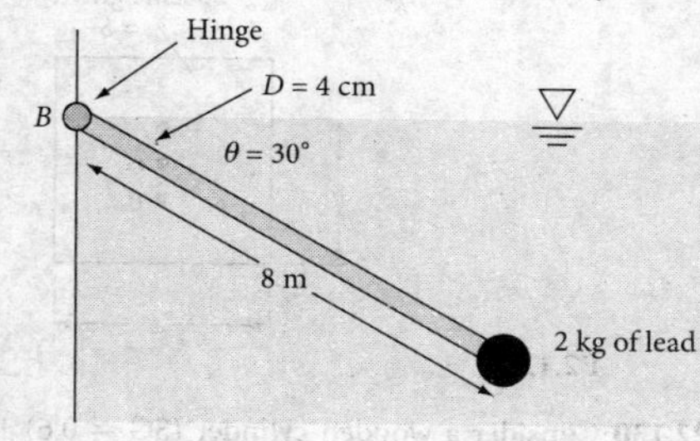

P2.114

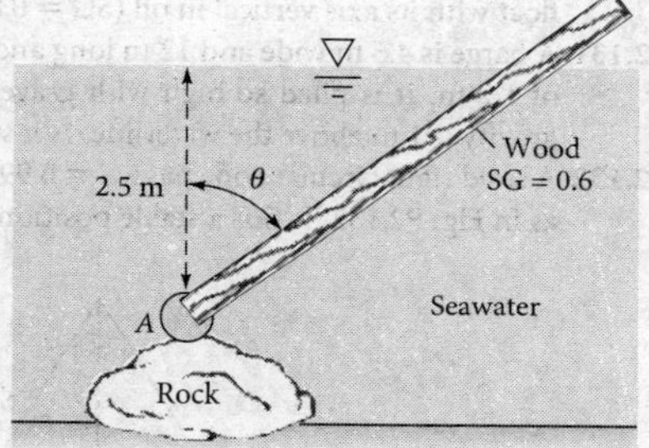

P2.115

P2.116 The bathysphere of the chapter-opener photo is steel, SG $\approx$ 7.85, with inside diameter 1.4 m and wall thickness 4 cm. Will the empty sphere float in seawater?

P2.117 The solid sphere in Fig. P2.117 is iron (SG $\approx$ 7.9). The tension in the cable is 3,000 N. Estimate the diameter of the sphere, in cm.

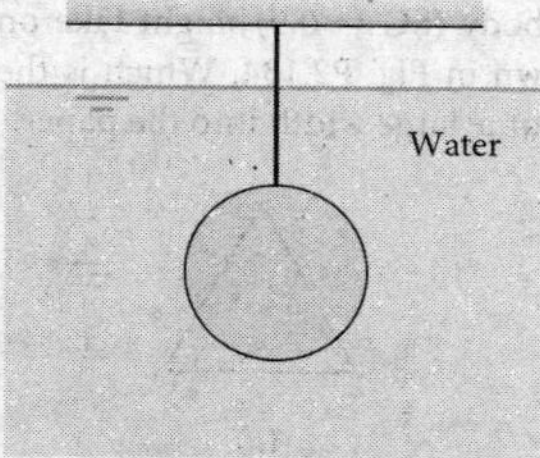

P2.117

P2.118 An intrepid treasure-salvage group has discovered a steel box, containing gold doubloons and other valuables, resting in 25 m of seawater. They estimate the weight of the box and treasure (in air) at 30,000 N. Their plan is to attach the box to a sturdy balloon, inflated with air to 3 atm pressure. The empty balloon weighs 1,000 N. The box is 0.6 m wide, 1.5 m long, and 0.5 m high. What is the proper diameter of the balloon to ensure an upward lift force on the box that is 20 percent more than required?

P2.119 When a 20 N weight is placed on the end of the uniform floating wooden beam in Fig. P2.119, the beam tilts at an angle θ with its upper right corner at the surface, as shown. Determine (*a*) the angle θ and (*b*) the specific gravity of the wood. *Hint:* Both the vertical forces and the moments about the beam centroid must be balanced.

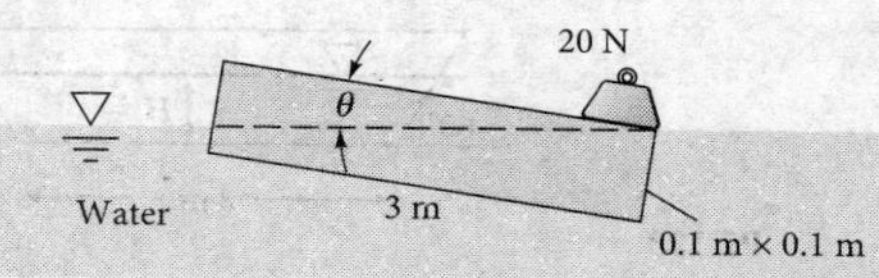

P2.119

P2.120 A uniform wooden beam (SG = 0.65) is 10 cm by 10 cm by 3 m and is hinged at *A*, as in Fig. P2.120. At what angle θ will the beam float in the 20°C water?

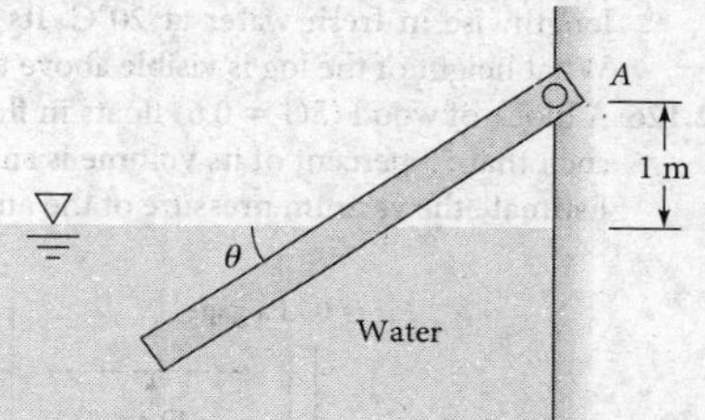

P2.120

P2.121 The uniform beam in Fig. P2.121, of size *L* by *h* by *b* and with specific weight γ_b, floats exactly on its diagonal when a heavy uniform sphere is tied to the left corner, as shown.

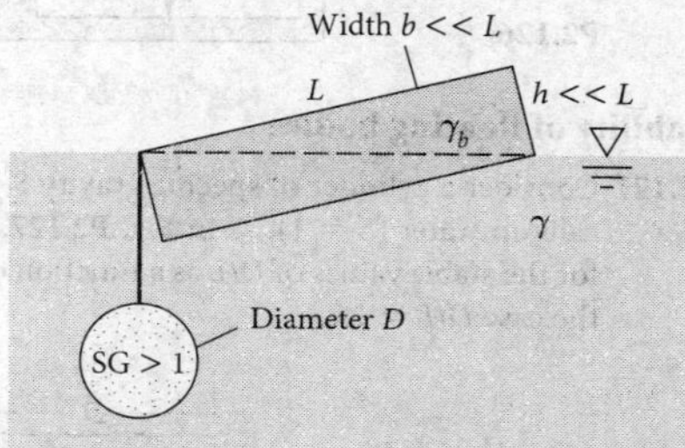

P2.121

Show that this can happen only (*a*) when $\gamma_b = \gamma/3$ and (*b*) when the sphere has size

$$D = \left[\frac{Lhb}{\pi(\text{SG} - 1)}\right]^{1/3}$$

P2.122 A uniform block of steel (SG = 7.85) will "float" at a mercury–water interface as in Fig. P2.122. What is the ratio of the distances *a* and *b* for this condition?

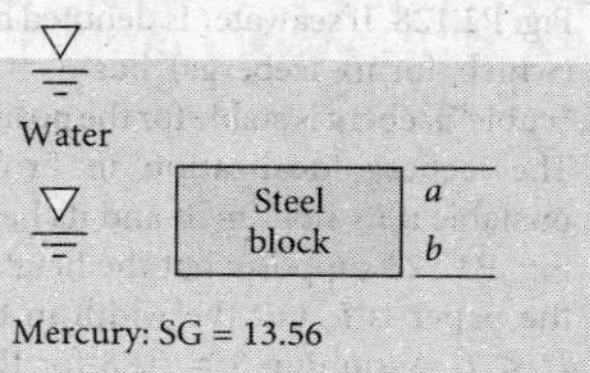

P2.122

P2.123 A barge has the trapezoidal shape shown in Fig. P2.123 and is 22 m long into the paper. If the total weight of barge and cargo is 3,500 kN, what is the draft H of the barge when floating in seawater?

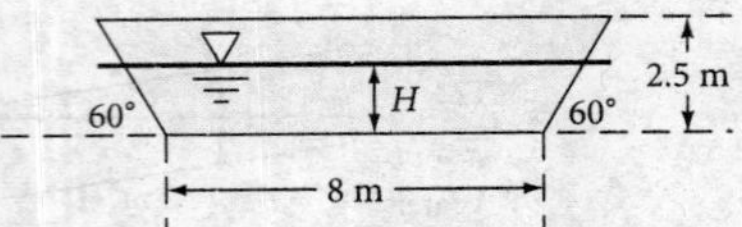

P2.123

P2.124 A balloon weighing 16 N is 2 m in diameter. It is filled with hydrogen at 124 kPa absolute and 290 K and is released. At what altitude in the U.S. standard atmosphere will this balloon be neutrally buoyant?

P2.125 A uniform cylindrical white oak log, $\rho = 710$ kg/m³, floats lengthwise in fresh water at 20°C. Its diameter is 0.6 m. What height of the log is visible above the surface?

P2.126 A block of wood (SG = 0.6) floats in fluid X in Fig. P2.126 such that 75 percent of its volume is submerged in fluid X. Estimate the vacuum pressure of the air in the tank.

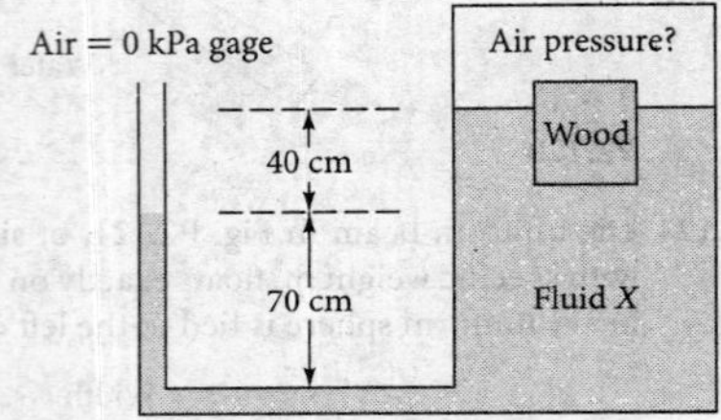

P2.126

Stability of floating bodies

***P2.127** Consider a cylinder of specific gravity $S < 1$ floating vertically in water ($S = 1$), as in Fig. P2.127. Derive a formula for the stable values of D/L as a function of S and apply it to the case $D/L = 1.2$.

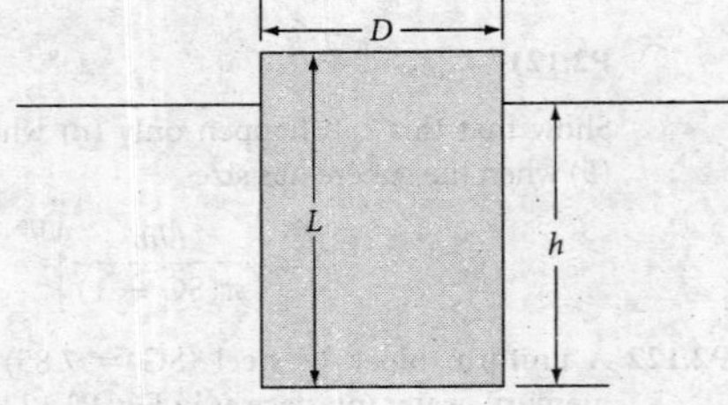

P2.127

P2.128 An iceberg can be idealized as a cube of side length L, as in Fig. P2.128. If seawater is denoted by $S = 1.0$, then glacier ice (which forms icebergs) has $S = 0.88$. Determine if this "cubic" iceberg is stable for the position shown in Fig. P2.128.

P2.129 The iceberg idealization in Prob. P2.128 may become unstable if its sides melt and its height exceeds its width. In Fig. P2.128 suppose that the height is L and the depth into the paper is L, but the width in the plane of the paper is $H < L$. Assuming $S = 0.88$ for the iceberg, find the ratio H/L for which it becomes neutrally stable (about to overturn).

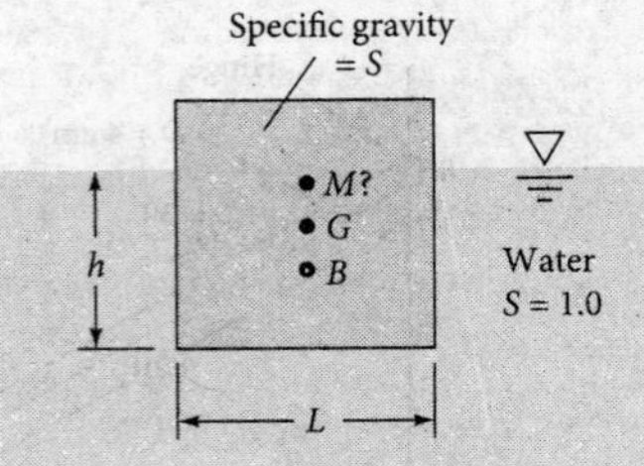

P2.128

P2.130 Consider a wooden cylinder (SG = 0.6) 1 m in diameter and 0.8 m long. Would this cylinder be stable if placed to float with its axis vertical in oil (SG = 0.8)?

P2.131 A barge is 4.5 m wide and 12 m long and floats with a draft of 1.2 m. It is piled so high with gravel that its center of gravity is 1 m above the waterline. Is it stable?

P2.132 A solid right circular cone has SG = 0.99 and floats ertically as in Fig. P2.132. Is this a stable position for the cone?

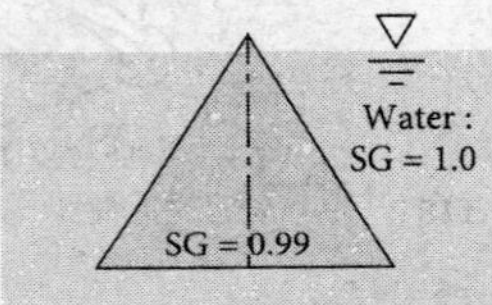

P2.132

P2.133 Consider a uniform right circular cone of specific gravity $S < 1$, floating with its vertex down in water ($S = 1$). The base radius is R and the cone height is H. Calculate and plot the stability MG of this cone, in dimensionless form, versus H/R for a range of $S < 1$.

P2.134 When floating in water (SG = 1.0), an equilateral triangular body (SG = 0.9) might take one of the two positions shown in Fig. P2.134. Which is the more stable position? Assume large width into the paper.

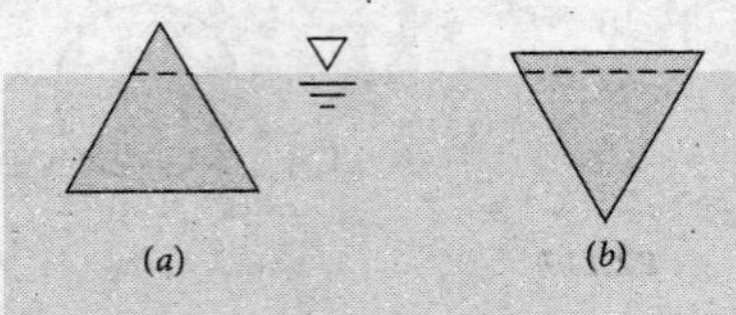

P2.134

P2.135 Consider a homogeneous right circular cylinder of length L, radius R, and specific gravity SG, floating in water (SG = 1). Show that the body will be stable with its axis vertical if

$$\frac{R}{L} > [2SG(1 - SG)]^{1/2}$$

P2.136 Consider a homogeneous right circular cylinder of length L, radius R, and specific gravity SG = 0.5, floating in water (SG = 1). Show that the body will be stable with its axis horizontal if $L/R > 2.0$.

Uniform acceleration

P2.137 A tank of water 4 m deep receives a constant upward acceleration a_z. Determine (*a*) the gage pressure at the tank bottom if $a_z = 5\ m^2/s$ and (*b*) the value of a_z that causes the gage pressure at the tank bottom to be 1 atm.

P2.138 A 350 ml glass, of 75 mm diameter, partly full of water, is attached to the edge of an 2.5 m-diameter merry-go-round, which is rotated at 12 r/min. How full can the glass be before water spills? *Hint:* Assume that the glass is much smaller than the radius of the merry-go-round.

P2.139 The tank of liquid in Fig. P2.139 accelerates to the right with the fluid in rigid-body motion. (*a*) Compute a_x in m/s^2. (*b*) Why doesn't the solution to part (*a*) depend on the density of the fluid? (*c*) Determine the gage pressure at point *A* if the fluid is glycerin at 20°C.

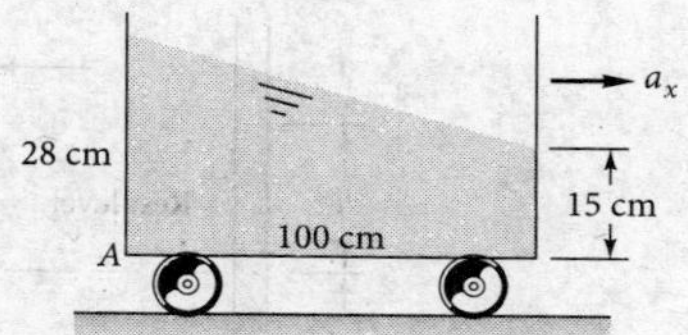

P2.139

P2.140 The U-tube in Fig. P2.140 is moving to the right with variable velocity. The water level in the left tube is 6 cm, and the level in the right tube is 16 cm. Determine the acceleration and its direction.

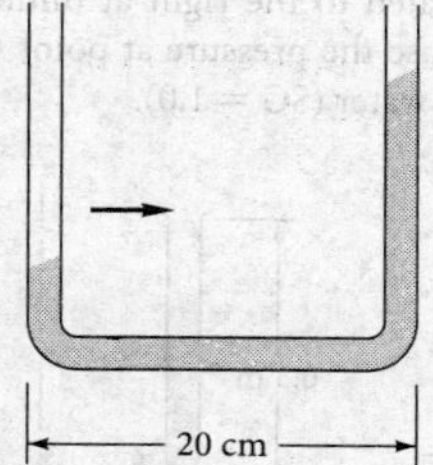

P2.140

P2.141 The same tank from Prob. P2.139 is now moving with constant acceleration up a 30° inclined plane, as in Fig. P2.141. Assuming rigid-body motion, compute (*a*) the value of the acceleration *a*, (*b*) whether the acceleration is up or down, and (*c*) the gage pressure at point *A* if the fluid is mercury at 20°C.

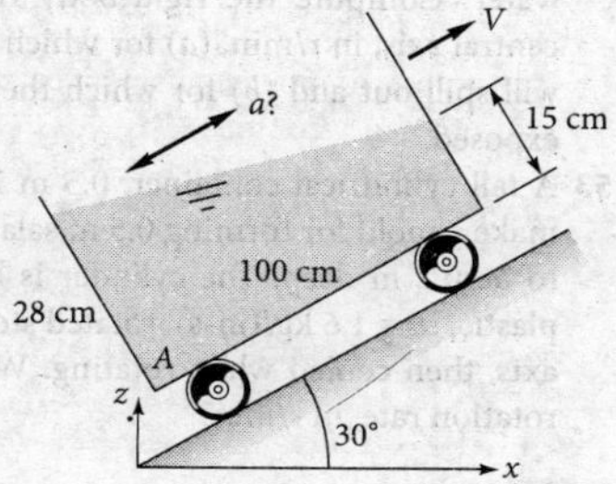

P2.141

P2.142 The tank of water in Fig. P2.142 is 12 cm wide into the paper. If the tank is accelerated to the right in rigid-body motion at 6.0 m/s^2, compute (*a*) the water depth on side *AB* and (*b*) the water-pressure force on panel *AB*. Assume no spilling.

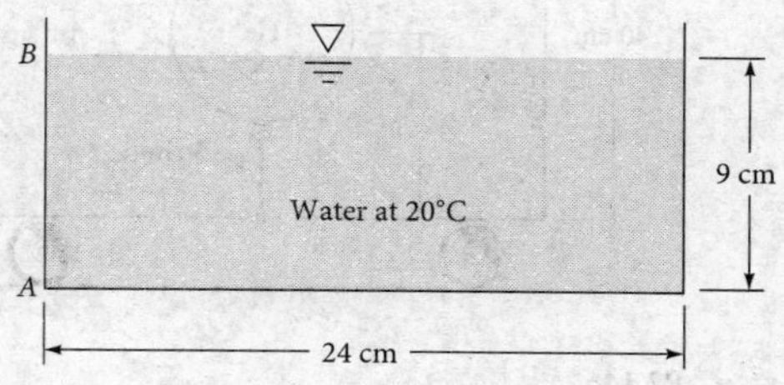

P2.142

P2.143 The tank of water in Fig. P2.143 is full and open to the atmosphere at point *A*. For what acceleration a_x in m/s^2 will the pressure at point *B* be (*a*) atmospheric and (*b*) zero absolute?

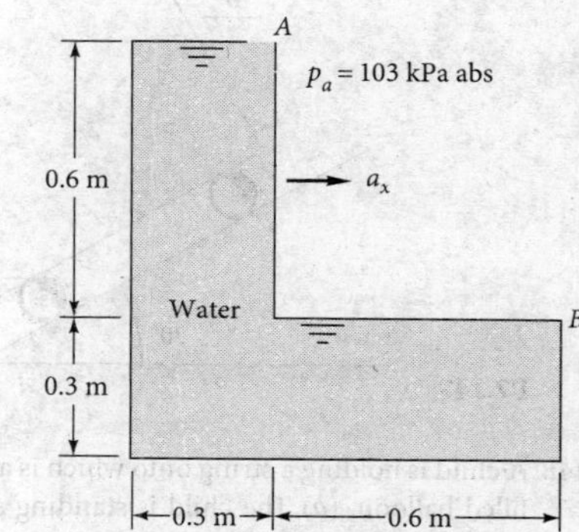

P2.143

P2.144 Consider a hollow cube of side length 22 cm, filled completely with water at 20°C. The top surface of the cube is horizontal. One top corner, point *A*, is open through a small hole to a pressure of 1 atm. Diagonally opposite to point *A* is top corner *B*. Determine and discuss the various rigid-body accelerations for which the water at point *B* begins to cavitate, for (*a*) horizontal motion and (*b*) vertical motion.

P2.145 A fish tank 0.35 m deep by 0.4 m by 0.7 m is to be carried in a car that may experience accelerations as high as 6 m/s^2. What is the maximum water depth that will avoid spilling in rigid-body motion? What is the proper alignment of the tank with respect to the car motion?

P2.146 The tank in Fig. P2.146 is filled with water and has a vent hole at point *A*. The tank is 1 m wide into the paper. Inside the tank, a 10-cm balloon, filled with helium at 130 kPa, is tethered centrally by a string. If the tank accelerates to the right at 5 m/s^2 in rigid-body motion, at what angle will the balloon lean? Will it lean to the right or to the left?

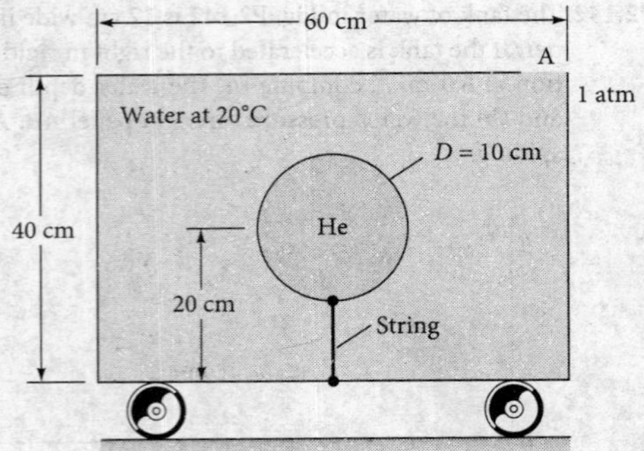

P2.146

P2.147 The tank of water in Fig. P2.147 accelerates uniformly by freely rolling down a 30° incline. If the wheels are frictionless, what is the angle θ? Can you explain this interesting result?

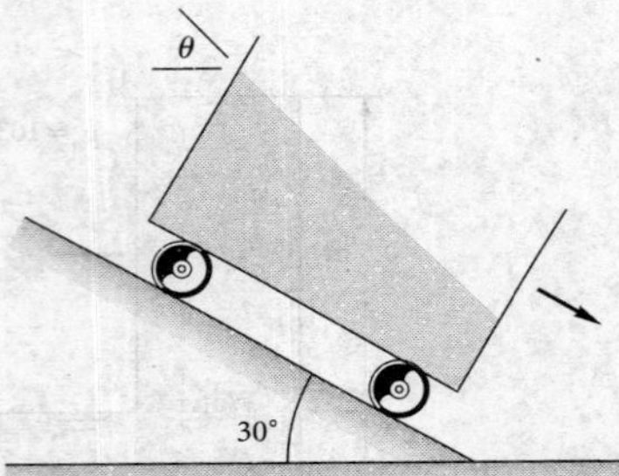

P2.147

P2.148 A child is holding a string onto which is attached a helium-filled balloon. (*a*) The child is standing still and suddenly accelerates forward. In a frame of reference moving with the child, which way will the balloon tilt, forward or backward? Explain. (*b*) The child is now sitting in a car that is stopped at a red light. The helium-filled balloon is not in contact with any part of the car (seats, ceiling, etc.) but is held in place by the string, which is in turn held by the child. All the windows in the car are closed. When the traffic light turns green, the car accelerates forward. In a frame of reference moving with the car and child, which way will the balloon tilt, forward or backward? Explain. (*c*) Purchase or borrow a helium-filled balloon. Conduct a scientific experiment to see if your predictions in parts (*a*) and (*b*) above are correct. If not, explain.

P2.149 The 1.8 m-radius waterwheel in Fig. P2.149 is being used to lift water with its 0.3 m-diameter half-cylinder blades. If the wheel rotates at 10 r/min and rigid-body motion is assumed, what is the water surface angle θ at position A?

P2.150 A cheap accelerometer, probably worth the price, can be made from a U-tube as in Fig. P2.150. If L = 18 cm and D = 5 mm, what will h be if a_x = 6 m/s^2? Can the scale markings on the tube be linear multiples of a_x?

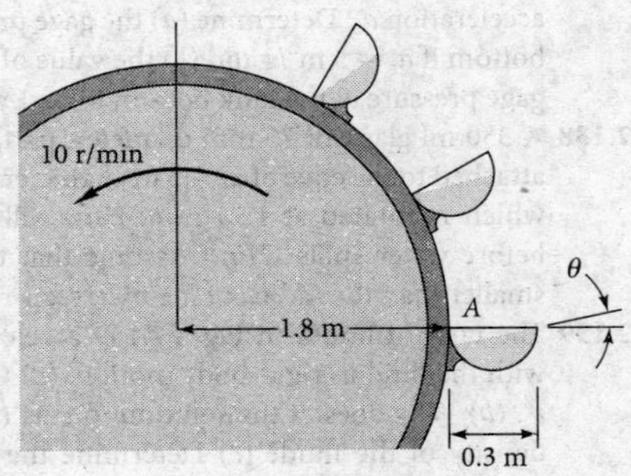

P2.149

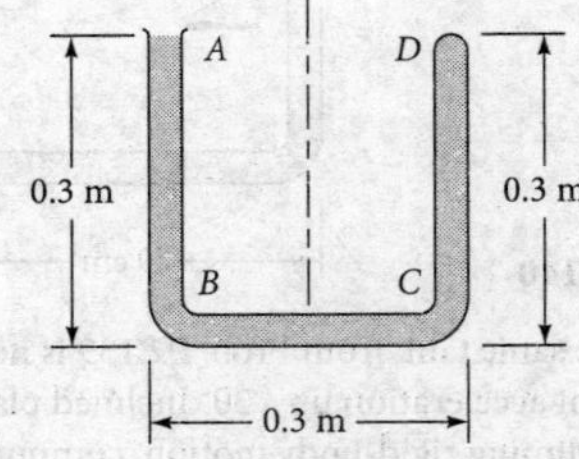

P2.150

P2.151 The U-tube in Fig. P2.151 is open at A and closed at D. If accelerated to the right at uniform a_x, what acceleration will cause the pressure at point C to be atmospheric? The fluid is water (SG = 1.0).

P2.151

Rigid-body rotation

P2.152 A 16-cm-diameter open cylinder 27 cm high is full of water. Compute the rigid-body rotation rate about its central axis, in r/min, (*a*) for which one-third of the water will spill out and (*b*) for which the bottom will be barely exposed.

P2.153 A tall cylindrical container, 0.5 m in diameter, is used to make a mold for forming 0.5 m salad bowls. The bowls are to be 0.2 m deep. The cylinder is half-filled with molten plastic, μ = 1.6 kg/(m-s), rotated steadily about the central axis, then cooled while rotating. What is the appropriate rotation rate, in r/min?

P2.154 A very tall 10-cm-diameter vase contains 1,178 cm^3 of water. When spun steadily to achieve rigid-body rotation, a 4-cm-diameter dry spot appears at the bottom of the vase. What is the rotation rate, in r/min, for this condition?

P2.155 For what uniform rotation rate in r/min about axis *C* will the U-tube in Fig. P2.155 take the configuration shown? The fluid is mercury at 20°C.

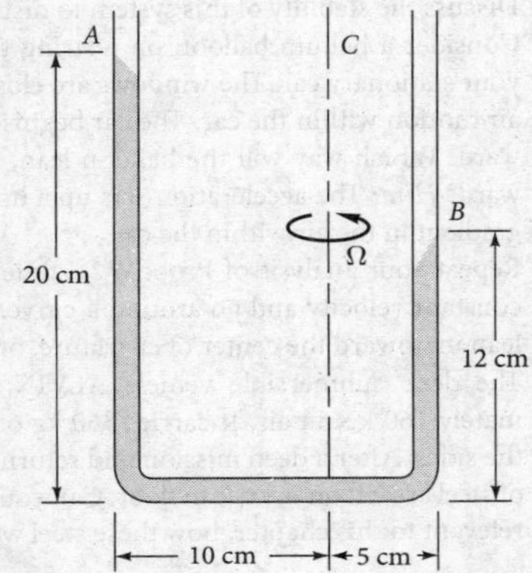

P2.155

P2.156 Suppose that the U-tube of Fig. P2.151 is rotated about axis *DC*. If the fluid is water at 50°C and atmospheric pressure is 101,300 Pa absolute, at what rotation rate will the fluid within the tube begin to vaporize? At what point will this occur?

P2.157 The 45° V-tube in Fig. P2.157 contains water and is open at *A* and closed at *C*. What uniform rotation rate in r/min about axis *AB* will cause the pressure to be equal at points *B* and *C*? For this condition, at what point in leg *BC* will the pressure be a minimum?

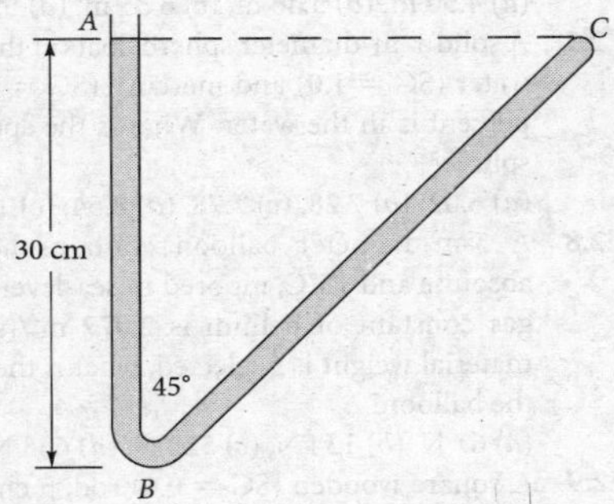

P2.157

***P2.158** It is desired to make a 3-m-diameter parabolic telescope mirror by rotating molten glass in rigid-body motion until the desired shape is achieved and then cooling the glass to a solid. The focus of the mirror is to be 4 m from the mirror, measured along the centerline. What is the proper mirror rotation rate, in r/min, for this task?

P2.159 The three-legged manometer in Fig. P2.159 is filled with water to a depth of 20 cm. All tubes are long and have equal small diameters. If the system spins at angular velocity Ω about the central tube, (*a*) derive a formula to find the change of height in the tubes; (*b*) find the height in cm in each tube if Ω = 120 r/min. *Hint:* The central tube must supply water to *both* the outer legs.

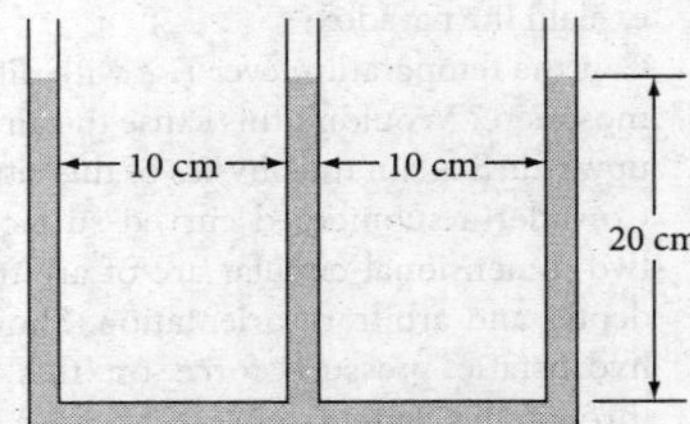

P2.159

Pressure measurements

P2.160 Figure P2.160 shows a gage for very low pressures, invented in 1874 by Herbert McLeod. (*a*) Can you deduce, from the figure, how it works? (*b*) If not, read about it and explain it to the class.

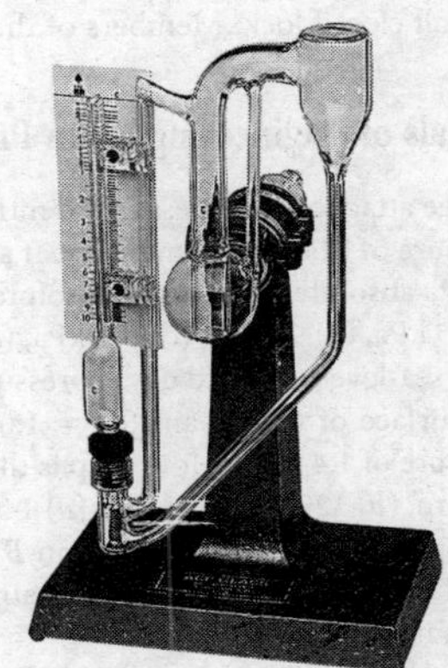

P2.160

P2.161 Figure P2.161 shows a sketch of a commercial pressure gage. (*a*) Can you deduce, from the figure, how it works?

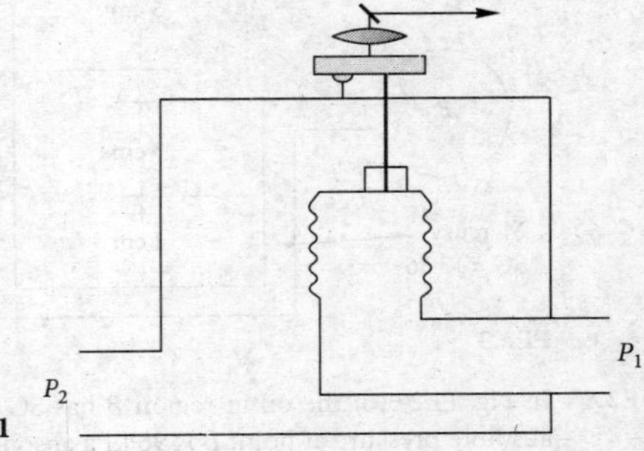

P2.161

Word Problems

W2.1 Consider a hollow cone with a vent hole in the vertex at the top, along with a hollow cylinder, open at the top, with the same base area as the cone. Fill both with water to the top. The *hydrostatic paradox* is that both containers have the same force on the bottom due to the water pressure, although the cone contains 67 percent less water. Can you explain the paradox?

W2.2 Can the temperature ever *rise* with altitude in the real atmosphere? Wouldn't this cause the air pressure to *increase* upward? Explain the physics of this situation.

W2.3 Consider a submerged curved surface that consists of a two-dimensional circular arc of arbitrary angle, arbitrary depth, and arbitrary orientation. Show that the resultant hydrostatic pressure force on this surface must pass through the center of curvature of the arc.

W2.4 Fill a glass approximately 80 percent with water, and add a large ice cube. Mark the water level. The ice cube, having SG $\approx$ 0.9, sticks up out of the water. Let the ice cube melt with negligible evaporation from the water surface. Will the water level be higher than, lower than, or the same as before?

W2.5 A ship, carrying a load of steel, is trapped while floating in a small closed lock. Members of the crew want to get out, but they can't quite reach the top wall of the lock. A crew member suggests throwing the steel overboard in the lock, claiming the ship will then rise and they can climb out. Will this plan work?

W2.6 Consider a balloon of mass m floating neutrally in the atmosphere, carrying a person/basket of mass $M > m$. Discuss the stability of this system to disturbances.

W2.7 Consider a helium balloon on a string tied to the seat of your stationary car. The windows are closed, so there is no air motion within the car. The car begins to accelerate forward. Which way will the balloon lean, forward or backward? *Hint:* The acceleration sets up a horizontal pressure gradient in the air within the car.

W2.8 Repeat your analysis of Prob. W2.7 to let the car move at constant velocity and go around a curve. Will the balloon lean in, toward the center of curvature, or out?

W2.9 The deep submersible vehicle ALVIN weighs approximately 160 kN in air. It carries 360 kg of steel weights on the sides. After a deep mission and return, two 180 kg piles of steel are left on the ocean floor. Can you explain, in terms relevant to this chapter, how these steel weights are used?

Fundamentals of Engineering Exam Problems

FE2.1 A gage attached to a pressurized nitrogen tank reads a gage pressure of 711 mm of mercury. If atmospheric pressure is 99 kPa absolute, what is the absolute pressure in the tank?
(*a*) 95 kPa, (*b*) 99 kPa, (*c*) 101 kPa, (*d*) 194 kPa, (*e*) 203 kPa

FE2.2 On a sea-level standard day, a pressure gage, moored below the surface of the ocean (SG = 1.025), reads an absolute pressure of 1.4 MPa. How deep is the instrument?
(*a*) 4 m, (*b*) 129 m, (*c*) 133 m, (*d*) 140 m, (*e*) 2,080 m

FE2.3 In Fig. FE2.3, if the oil in region *B* has SG = 0.8 and the absolute pressure at point *A* is 1 atm, what is the absolute pressure at point *B*?
(*a*) 5.6 kPa, (*b*) 10.9 kPa, (*c*) 107 kPa, (*d*) 112 kPa, (*e*) 157 kPa

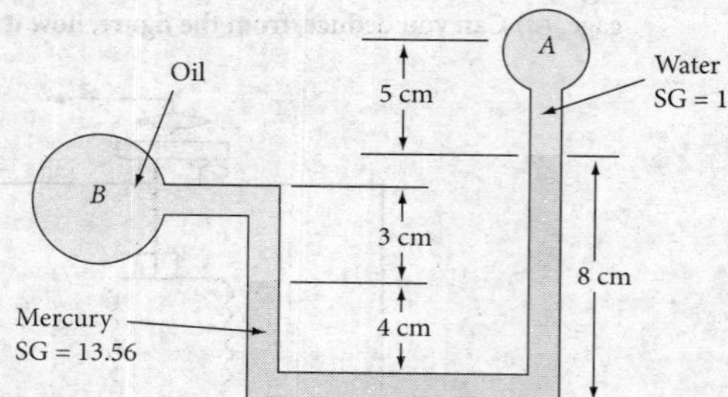

FE2.3

FE2.4 In Fig. FE2.3, if the oil in region *B* has SG = 0.8 and the absolute pressure at point *B* is 96 kPa absolute, what is the absolute pressure at point *A*?
(*a*) 11 kPa, (*b*) 41 kPa, (*c*) 86 kPa, (*d*) 91 kPa, (*e*) 101 kPa

FE2.5 A tank of water (SG = 1.0) has a gate in its vertical wall 5 m high and 3 m wide. The top edge of the gate is 2 m below the surface. What is the hydrostatic force on the gate?
(*a*) 147 kN, (*b*) 367 kN, (*c*) 490 kN, (*d*) 661 kN, (*e*) 1,028 kN

FE2.6 In Prob. FE2.5, how far below the surface is the center of pressure of the hydrostatic force?
(*a*) 4.50 m, (*b*) 5.46 m, (*c*) 6.35 m, (*d*) 5.33 m, (*e*) 4.96 m

FE2.7 A solid 1-m-diameter sphere floats at the interface between water (SG = 1.0) and mercury (SG = 13.56) such that 40 percent is in the water. What is the specific gravity of the sphere?
(*a*) 6.02, (*b*) 7.28, (*c*) 7.78, (*d*) 8.54, (*e*) 12.56

FE2.8 A 5-m-diameter balloon contains helium at 125 kPa absolute and 15°C, moored in sea-level standard air. If the gas constant of helium is 2,077 $m^2/(s^2 \cdot K)$ and balloon material weight is neglected, what is the net lifting force of the balloon?
(*a*) 67 N, (*b*) 134 N, (*c*) 522 N, (*d*) 653 N, (*e*) 787 N

FE2.9 A square wooden (SG = 0.6) rod, 5 cm by 5 cm by 10 m long, floats vertically in water at 20°C when 6 kg of steel (SG = 7.84) are attached to one end. How high above the water surface does the wooden end of the rod protrude?
(*a*) 0.6 m, (*b*) 1.6 m, (*c*) 1.9 m, (*d*) 2.4 m, (*e*) 4.0 m

FE2.10 A floating body will be stable when its
(*a*) center of gravity is above its center of buoyancy, (*b*) center of buoyancy is below the waterline, (*c*) center of buoyancy is above its metacenter, (*d*) metacenter is above its center of buoyancy, (*e*) metacenter is above its center of gravity.

Comprehensive Problems

C2.1 Some manometers are constructed as in Fig. C2.1, where one side is a large reservoir (diameter D) and the other side is a small tube of diameter d, open to the atmosphere. In such a case, the height of manometer liquid on the reservoir side does not change appreciably. This has the advantage that only one height needs to be measured rather than two. The manometer liquid has density ρ_m while the air has density ρ_a. Ignore the effects of surface tension. When there is no pressure difference across the manometer, the elevations on both sides are the same, as indicated by the dashed line. Height h is measured from the zero pressure level as shown. (*a*) When a high pressure is applied to the left side, the manometer liquid in the large reservoir goes down, while that in the tube at the right goes up to conserve mass. Write an exact expression for p_{1gage}, taking into account the movement of the surface of the reservoir. Your equation should give p_{1gage} as a function of h, ρ_m, and the physical parameters in the problem, h, d, D, and gravity constant g. (*b*) Write an approximate expression for p_{1gage}, neglecting the change in elevation of the surface of the reservoir liquid. (*c*) Suppose $h = 0.26$ m in a certain application. If $p_a = 101{,}000$ Pa and the manometer liquid has a density of 820 kg/m^3, estimate the ratio D/d required to keep the error of the approximation of part (*b*) within 1 percent of the exact measurement of part (*a*). Repeat for an error within 0.1 percent.

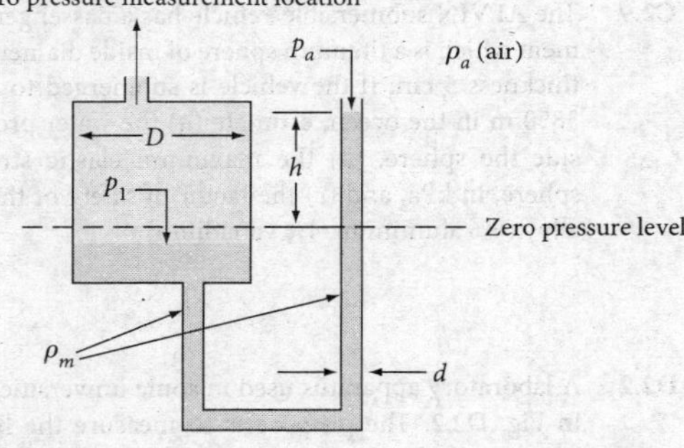

C2.1

C2.2 A prankster has added oil, of specific gravity SG_0, to the left leg of the manometer in Fig. C2.2. Nevertheless, the U-tube is still useful as a pressure-measuring device. It is attached to a pressurized tank as shown in the figure. (*a*) Find an expression for h as a function of H and other parameters in the problem. (*b*) Find the special case of your result in (*a*) when $p_{tank} = p_a$. (*c*) Suppose $H = 5.0$ cm, p_a is 101.2 kPa, p_{tank} is 1.82 kPa higher than p_a, and $SG_0 = 0.85$. Calculate h in cm, ignoring surface tension effects and neglecting air density effects.

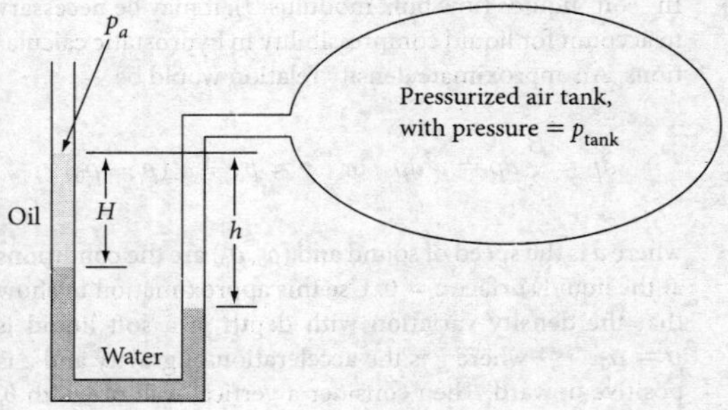

C2.2

C2.3 Professor F. Dynamics, riding the merry-go-round with his son, has brought along his U-tube manometer. (You never know when a manometer might come in handy.) As shown in Fig. C2.3, the merry-go-round spins at constant angular velocity and the manometer legs are 7 cm apart. The manometer center is 5.8 m from the axis of rotation. Determine the height difference h in two ways: (*a*) approximately, by assuming rigid-body translation with a equal to the average manometer acceleration; and (*b*) exactly, using rigid-body rotation theory. How good is the approximation?

C2.4 A student sneaks a glass of cola onto a roller coaster ride. The glass is cylindrical, twice as tall as it is wide, and filled to the brim. He wants to know what percent of the cola he should drink before the ride begins, so that none of it spills during the big drop, in which the roller coaster achieves 0.55-g acceleration at a 45° angle below the horizontal. Make the calculation for him, neglecting sloshing and assuming that the glass is vertical at all times.

C2.5 *Dry adiabatic lapse rate* (DALR) is defined as the negative value of atmospheric temperature gradient, dT/dz, when

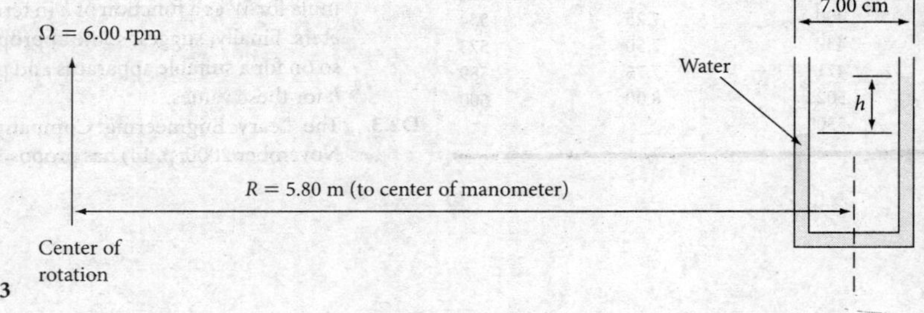

C2.3

temperature and pressure vary in an isentropic fashion. Assuming air is an ideal gas, DALR $= -dT/dz$ when $T = T_0(p/p_0)^a$, where exponent $a = (k-1)/k$, $k = c_p/c_v$ is the ratio of specific heats, and T_0 and p_0 are the temperature and pressure at sea level, respectively. (*a*) Assuming that hydrostatic conditions exist in the atmosphere, show that the dry adiabatic lapse rate is constant and is given by DALR $= g(k-1)/(kR)$, where R is the ideal gas constant for air. (*b*) Calculate the numerical value of DALR for air in units of °C/km.

C2.6 In "soft" liquids (low bulk modulus β), it may be necessary to account for liquid compressibility in hydrostatic calculations. An approximate density relation would be

$$dp \approx \frac{\beta}{\rho}\,d\rho = a^2 d\rho \quad \text{or} \quad p \approx p_0 + a^2(\rho - \rho_0)$$

where a is the speed of sound and (p_0, ρ_0) are the conditions at the liquid surface $z = 0$. Use this approximation to show that the density variation with depth in a soft liquid is $\rho = \rho_0 e^{-gz/a^2}$ where g is the acceleration of gravity and z is positive upward. Then consider a vertical wall of width b, extending from the surface ($z = 0$) down to depth $z = -h$. Find an analytic expression for the hydrostatic force F on this wall, and compare it with the incompressible result $F = \rho_0 g h^2 b/2$. Would the center of pressure be below the incompressible position $z = -2h/3$?

C2.7 Venice, Italy, is slowly sinking, so now, especially in winter, piazas and walkways are flooded during storms. The proposed solution is the floating levee of Fig. C2.7. When filled with air, it rises to block off the sea. The levee is 30 m high, 5 m wide, and 20 m deep. Assume a uniform density of 300 kg/m³ when floating. For the 1-m sea–lagoon difference shown, estimate the angle at which the levee floats.

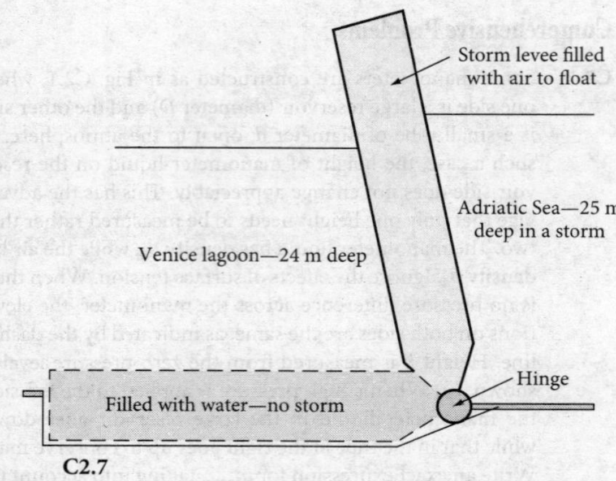

C2.7

C2.8 In the U.S. Standard Atmosphere, the lapse rate B may vary from day to day. It is not a fundamental quantity like, say, Planck's constant. Suppose that, on a certain day in Rhode Island, with $T_o = 288$ K, the following pressures are measured by weather balloons:

Altitude *z*, km	0	2	5	8
Pressure *p*, kPa	100	78	53	34

Estimate the best-fit value of B for this data. Explain any difficulties. [*Hint:* EES is recommended.]

C2.9 The ALVIN submersible vehicle has a passenger compartment which is a titanium sphere of inside diameter 2 m and thickness 5 cm. If the vehicle is submerged to a depth of 3850 m in the ocean, estimate (*a*) the water pressure outside the sphere, (*b*) the maximum elastic stress in the sphere, in kPa, and (*c*) the factor of safety of the titanium alloy (6% aluminum, 4% vanadium).

Design Projects

D2.1 It is desired to have a bottom-moored, floating system that creates a nonlinear force in the mooring line as the water level rises. The design force F need only be accurate in the range of seawater depths h between 6 and 8 m, as shown in the accompanying table. Design a buoyant system that will provide this force distribution. The system should be practical (of inexpensive materials and simple construction).

h, m	*F*, N	*h*, m	*F*, N
6.00	400	7.25	554
6.25	437	7.50	573
6.50	471	7.75	589
6.75	502	8.00	600
7.00	530		

D2.2 A laboratory apparatus used in some universities is shown in Fig. D2.2. The purpose is to measure the hydrostatic force on the flat face of the circular-arc block and compare it with the theoretical value for given depth h. The counterweight is arranged so that the pivot arm is horizontal when the block is not submerged, whence the weight W can be correlated with the hydrostatic force when the submerged arm is again brought to horizontal. First show that the apparatus concept is valid in principle; then derive a formula for W as a function of h in terms of the system parameters. Finally, suggest some appropriate values of Y, L, and so on for a suitable apparatus and plot theoretical W versus h for these values.

D2.3 The Leary Engineering Company (see *Popular Science*, November 2000, p. 14) has proposed a ship hull with hinges

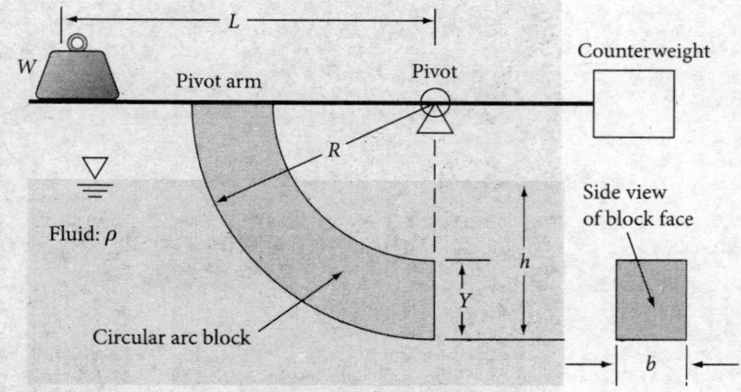

D2.2

that allow it to open into a flatter shape when entering shallow water. A simplified version is shown in Fig. D2.3. In deep water, the hull cross section would be triangular, with large draft. In shallow water, the hinges would open to an angle as high as $\theta = 45°$. The dashed line indicates that the bow and stern would be closed. Make a parametric study of this configuration for various θ, assuming a reasonable weight and center of gravity location. Show how the draft, the metacentric height, and the ship's stability vary as the hinges are opened. Comment on the effectiveness of this concept.

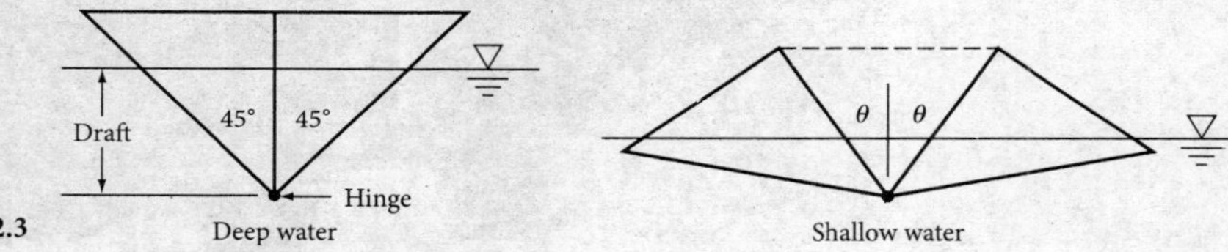

D2.3

References

1. *U.S. Standard Atmosphere,* 1976, Government Printing Office, Washington, DC, 1976.
2. J. A. Knauss, *Introduction to Physical Oceanography*, 2d ed., Waveland Press, Long Grove, IL, 2005.
3. E. C. Tupper, *Introduction to Naval Architecture,* 4th ed., Elsevier, New York, 2004.
4. D. T. Greenwood, *Advanced Dynamics*, Cambridge University Press, New York, 2006.
5. R. I. Fletcher, "The Apparent Field of Gravity in a Rotating Fluid System," *Am. J. Phys.*, vol. 40, July 1972, pp. 959–965.
6. National Committee for Fluid Mechanics Films, *Illustrated Experiments in Fluid Mechanics,* M.I.T. Press, Cambridge, MA, 1972.
7. J. P. Holman, *Experimental Methods for Engineers,* 8th ed., McGraw-Hill, New York, 2011.
8. R. C. Baker, *Flow Measurement Handbook*, Cambridge University Press, New York, 2005.
9. T. G. Beckwith, R. G. Marangoni, and J. H. Lienhard V, *Mechanical Measurements*, 6th ed., Prentice-Hall, Upper Saddle River, NJ, 2006.
10. J. W. Dally, W. F. Riley, and K. G. McConnell, *Instrumentation for Engineering Measurements,* 2d ed., Wiley, New York, 1993.
11. E. N. Gilbert, "How Things Float," *Am. Math. Monthly,* vol. 98, no. 3, 1991, pp. 201–216.
12. R. J. Figliola and D. E. Beasley, *Theory and Design for Mechanical Measurements,* 4th ed., Wiley, New York, 2005.
13. R. W. Miller, *Flow Measurement Engineering Handbook,* 3d ed., McGraw-Hill, New York, 1996.
14. L. D. Clayton, E. P. EerNisse, R. W. Ward, and R. B. Wiggins, "Miniature Crystalline Quartz Electromechanical Structures," *Sensors and Actuators,* vol. 20, Nov. 15, 1989, pp. 171–177.
15. A. Kitai (ed.), *Luminescent Materials and Applications*, John Wiley, New York, 2008.
16. B. G. Liptak (ed.), *Instrument Engineer's Handbook: Process Measurement and Analysis,* 4th ed., vol. 1, CRC Press, Boca Raton, FL, 2003.
17. A. von Beckerath, *WIKA Handbook—Pressure and Temperature Measurement*, WIKA Instrument Corp., Lawrenceville, GA, 2008.

On July 16, 1969, a massive Saturn V rocket lifted *Apollo 11* from the NASA Kennedy Space Center, carrying astronauts Neil Armstrong, Michael Collins, and Edwin Aldrin to the first landing on the moon, four days later. The photo is filled with fluid momentum. In this chapter we learn how to analyze both the thrust of the rocket and the force of the exit jet on the solid surface. [*Photo credit: NASA*]

Chapter 3
Integral Relations for a Control Volume

Motivation. In analyzing fluid motion, we might take one of two paths: (1) seeking to describe the detailed flow pattern at every point (x, y, z) in the field or (2) working with a finite region, making a balance of flow in versus flow out, and determining gross flow effects such as the force or torque on a body or the total energy exchange. The second is the "control volume" method and is the subject of this chapter. The first is the "differential" approach and is developed in Chap. 4.

We first develop the concept of the control volume, in nearly the same manner as one does in a thermodynamics course, and we find the rate of change of an arbitrary gross fluid property, a result called the *Reynolds transport theorem.* We then apply this theorem, in sequence, to mass, linear momentum, angular momentum, and energy, thus deriving the four basic control volume relations of fluid mechanics. There are many applications, of course. The chapter includes a special case of frictionless, shaft-work-free momentum and energy: the *Bernoulli equation.* The Bernoulli equation is a wonderful, historic relation, but it is extremely restrictive and should always be viewed with skepticism and care in applying it to a real (viscous) fluid motion.

3.1 Basic Physical Laws of Fluid Mechanics

It is time now to really get serious about flow problems. The fluid statics applications of Chap. 2 were more like fun than work, at least in this writer's opinion. Statics problems basically require only the density of the fluid and knowledge of the position of the free surface, but most flow problems require the analysis of an arbitrary state of variable fluid motion defined by the geometry, the boundary conditions, and the laws of mechanics. This chapter and the next two outline the three basic approaches to the analysis of arbitrary flow problems:

1. Control volume, or large-scale, analysis (Chap. 3).
2. Differential, or small-scale, analysis (Chap. 4).
3. Experimental, or dimensional, analysis (Chap. 5).

The three approaches are roughly equal in importance. Control volume analysis, the present topic, is accurate for any flow distribution but is often based on average or "one-dimensional" property values at the boundaries. It always gives useful "engineering" estimates. In principle, the differential equation approach of Chap. 4 can be applied to any problem. Only a few problems, such as straight pipe flow, yield to exact analytical solutions. But the differential equations can be modeled numerically, and the flourishing field of computational fluid dynamics (CFD)[8] can now be used to give good estimates for almost any geometry. Finally, the dimensional analysis of Chap. 5 applies to any problem, whether analytical, numerical, or experimental. It is particularly useful to reduce the cost of experimentation. Differential analysis of hydrodynamics began with Euler and d'Alembert in the late eighteenth century. Lord Rayleigh and E. Buckingham pioneered dimensional analysis at the end of the nineteenth century. The control volume was described in words, on an ad hoc one-case basis, by Daniel Bernoulli in 1753. Ludwig Prandtl, the celebrated founder of modern fluid mechanics, developed the control volume as a systematic tool in the early 1900s. The writer's teachers at M.I.T. introduced control volume analysis into American textbooks, for thermodynamics by Keenan in 1941 [10], and for fluids by Hunsaker and Rightmire in 1947 [11]. For a complete history of the control volume, see Vincenti [9].

Systems versus Control Volumes

All the laws of mechanics are written for a *system,* which is defined as an arbitrary quantity of mass of fixed identity. Everything external to this system is denoted by the term *surroundings,* and the system is separated from its surroundings by its *boundaries.* The laws of mechanics then state what happens when there is an interaction between the system and its surroundings.

First, the system is a fixed quantity of mass, denoted by m. Thus the mass of the system is conserved and does not change.[1] This is a law of mechanics and has a very simple mathematical form, called *conservation of mass:*

$$m_{\text{syst}} = \text{const}$$

or

$$\frac{dm}{dt} = 0 \tag{3.1}$$

This is so obvious in solid mechanics problems that we often forget about it. In fluid mechanics, we must pay a lot of attention to mass conservation, and it takes a little analysis to make it hold.

Second, if the surroundings exert a net force **F** on the system, Newton's second law states that the mass in the system will begin to accelerate:[2]

$$\mathbf{F} = m\mathbf{a} = m\frac{d\mathbf{V}}{dt} = \frac{d}{dt}(m\mathbf{V}) \tag{3.2}$$

In Eq. (2.8) we saw this relation applied to a differential element of viscous incompressible fluid. In fluid mechanics, Newton's second law is called the linear momentum relation. Note that it is a vector law that implies the three scalar equations $F_x = ma_x$, $F_y = ma_y$, and $F_z = ma_z$.

Third, if the surroundings exert a net moment **M** about the center of mass of the system, there will be a rotation effect

$$\mathbf{M} = \frac{d\mathbf{H}}{dt} \tag{3.3}$$

where $\mathbf{H} = \sum(\mathbf{r} \times \mathbf{V})\delta m$ is the angular momentum of the system about its center of mass. Here we call Eq. (3.3) the angular momentum relation. Note that it is also a vector equation implying three scalar equations such as $M_x = dH_x/dt$.

[1] We are neglecting nuclear reactions, where mass can be changed to energy.
[2] We are neglecting relativistic effects, where Newton's law must be modified.

For an arbitrary mass and arbitrary moment, **H** is quite complicated and contains nine terms (see, for example, Ref. 1). In elementary dynamics we commonly treat only a rigid body rotating about a fixed x axis, for which Eq. (3.3) reduces to

$$M_x = I_x \frac{d}{dt}(\omega_x) \tag{3.4}$$

where ω_x is the angular velocity of the body and I_x is its mass moment of inertia about the x axis. Unfortunately, fluid systems are not rigid and rarely reduce to such a simple relation, as we shall see in Sec. 3.6.

Fourth, if heat δQ is added to the system or work δW is done by the system, the system energy dE must change according to the energy relation, or first law of thermodynamics:

$$\delta Q - \delta W = dE$$

or

$$\dot{Q} - \dot{W} = \frac{dE}{dt} \tag{3.5}$$

Like mass conservation, Eq. (3.1), this is a scalar relation having only a single component.

Finally, the second law of thermodynamics relates entropy change dS to heat added dQ and absolute temperature T:

$$dS \geq \frac{\delta Q}{T} \tag{3.6}$$

This is valid for a system and can be written in control volume form, but there are almost no practical applications in fluid mechanics except to analyze flow loss details (see Sec. 9.5).

All these laws involve thermodynamic properties, and thus we must supplement them with state relations $p = p(\rho, T)$ and $e = e(\rho, T)$ for the particular fluid being studied, as in Sec. 1.8. Although thermodynamics is not the main topic of this book, it is very important to the general study of fluid mechanics. Thermodynamics is crucial to compressible flow, Chap. 9. The student should review the first law and the state relations, as discussed in Refs. 6 and 7.

The purpose of this chapter is to put our four basic laws into the control volume form suitable for arbitrary regions in a flow:

1. Conservation of mass (Sec. 3.3).
2. The linear momentum relation (Sec. 3.4).
3. The angular momentum relation (Sec. 3.6).
4. The energy equation (Sec. 3.7).

Wherever necessary to complete the analysis we also introduce a state relation such as the perfect-gas law.

Equations (3.1) to (3.6) apply to either fluid or solid systems. They are ideal for solid mechanics, where we follow the same system forever because it represents the product we are designing and building. For example, we follow a beam as it deflects under load. We follow a piston as it oscillates. We follow a rocket system all the way to Mars.

But, fluid systems do not demand this concentrated attention. It is rare that we wish to follow the ultimate path of a specific particle of fluid. Instead, it is likely that the fluid forms the environment whose effect on our product we wish to know. For the three examples just cited, we wish to know the wind loads on the beam, the fluid pressures on the piston, and the drag and lift loads on the rocket. This requires that the basic laws be rewritten to apply to a specific *region* in the neighborhood of our product. In other words, where the fluid particles in the wind go after they leave the beam is of little interest to a beam designer. The user's point of view underlies the need for the control volume analysis of this chapter.

In analyzing a control volume, we convert the system laws to apply to a specific region, which the system may occupy for only an instant. The system passes on, and other systems come along, but no matter. The basic laws are reformulated to apply to this local region called a control volume. All we need to know is the flow field in this region, and often simple assumptions will be accurate enough (such as uniform inlet and/or outlet flows). The flow conditions away from the control volume are then irrelevant. The technique for making such localized analyses is the subject of this chapter.

Volume and Mass Rate of Flow

All the analyses in this chapter involve evaluation of the volume flow Q or mass flow $\dot{m}$ passing through a surface (imaginary) defined in the flow.

Suppose that the surface S in Fig. 3.1a is a sort of (imaginary) wire mesh through which the fluid passes without resistance. How much volume of fluid passes through S in unit time? If, typically, $\mathbf{V}$ varies with position, we must integrate over the elemental surface dA in Fig. 3.1a. Also, typically $\mathbf{V}$ may pass through dA at an angle θ off the normal. Let $\mathbf{n}$ be defined as the unit vector normal to dA. Then the amount of fluid swept through dA in time dt is the volume of the slanted parallelepiped in Fig. 3.1b:

$$d\mathcal{V} = V\,dt\,dA\cos\theta = (\mathbf{V}\cdot\mathbf{n})\,dA\,dt$$

The integral of $d\mathcal{V}/dt$ is the total volume rate of flow Q through the surface S:

$$Q = \int_s (\mathbf{V}\cdot\mathbf{n})\,dA = \int_s V_n\,dA \tag{3.7}$$

We could replace $\mathbf{V}\cdot\mathbf{n}$ by its equivalent, V_n, the component of $\mathbf{V}$ normal to dA, but the use of the dot product allows Q to have a sign to distinguish between inflow and outflow. By convention throughout this book we consider $\mathbf{n}$ to be the *outward* normal unit vector. Therefore $\mathbf{V}\cdot\mathbf{n}$ denotes outflow if it is positive and inflow if negative. This will be an extremely useful housekeeping device when we are computing volume and mass flow in the basic control volume relations.

Volume flow can be multiplied by density to obtain the mass flow $\dot{m}$. If density varies over the surface, it must be part of the surface integral:

$$\dot{m} = \int_s \rho(\mathbf{V}\cdot\mathbf{n})\,dA = \int_s \rho V_n\,dA$$

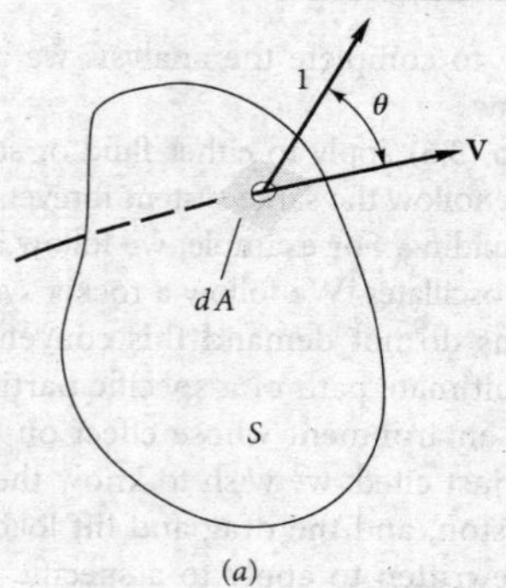

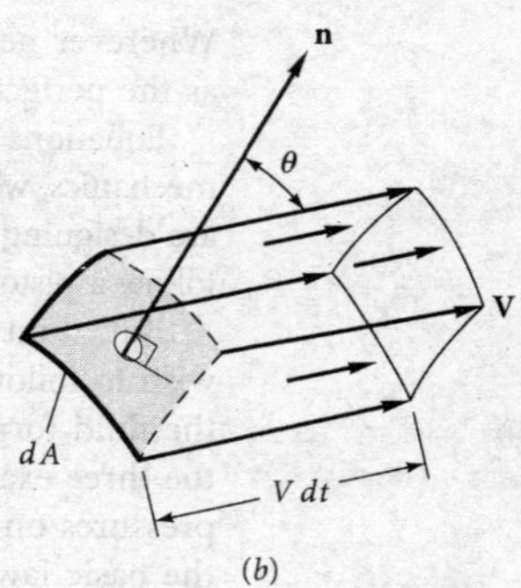

Fig. 3.1 Volume rate of flow through an arbitrary surface: (a) an elemental area dA on the surface; (b) the incremental volume swept through dA equals $V\,dt\,dA\cos\theta$.

If density and velocity are constant over the surface S, a simple expression results:

One-dimensional approximation: $\dot{m} = \rho Q = \rho AV$

Typical units for Q are m^3/s and for $\dot{m}$ kg/s.

3.2 The Reynolds Transport Theorem

To convert a system analysis to a control volume analysis, we must convert our mathematics to apply to a specific region rather than to individual masses. This conversion, called the *Reynolds transport theorem*, can be applied to all the basic laws. Examining the basic laws (3.1) to (3.3) and (3.5), we see that they are all concerned with the time derivative of fluid properties m, $\mathbf{V}$, $\mathbf{H}$, and E. Therefore what we need is to relate the time derivative of a system property to the rate of change of that property within a certain region.

The desired conversion formula differs slightly according to whether the control volume is fixed, moving, or deformable. Figure 3.2 illustrates these three cases. The fixed control volume in Fig. 3.2*a* encloses a stationary region of interest to a nozzle designer. The control surface is an abstract concept and does not hinder the flow in any way. It slices through the jet leaving the nozzle, encloses the surrounding atmosphere, and slices through the flange bolts and the fluid within the nozzle. This particular control volume exposes the stresses in the flange bolts, which contribute to applied forces in the momentum analysis. In this sense, the control volume resembles the *free-body* concept, which is applied to systems in solid mechanics analyses.

Figure 3.2*b* illustrates a moving control volume. Here the ship is of interest, not the ocean, so that the control surface chases the ship at ship speed V. The control volume is of fixed volume, but the relative motion between water and ship must be considered. If V is constant, this relative motion is a steady flow pattern, which simplifies the analysis.[3] If V is variable, the relative motion is unsteady, so that the computed results are time-variable and certain terms enter the momentum analysis to reflect the noninertial (accelerating) frame of reference.

Figure 3.2*c* shows a deforming control volume. Varying relative motion at the boundaries becomes a factor, and the rate of change of shape of the control volume enters the analysis. We begin by deriving the fixed control volume case, and we consider the other cases as advanced topics. An interesting history of control volume analysis is given by Vincenti [9].

Arbitrary Fixed Control Volume

Figure 3.3 shows a fixed control volume with an arbitrary flow pattern passing through. There are variable slivers of inflow and outflow of fluid all about the control surface. In general, each differential area dA of surface will have a different velocity $\mathbf{V}$ making

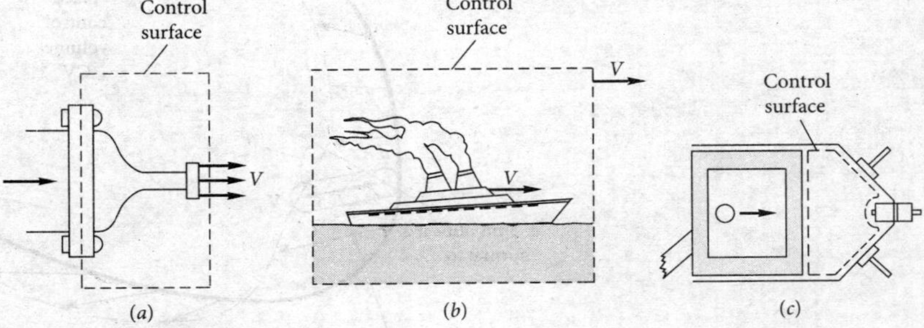

Fig. 3.2 Fixed, moving, and deformable control volumes: (*a*) fixed control volume for nozzle stress analysis; (*b*) control volume moving at ship speed for drag force analysis; (*c*) control volume deforming within cylinder for transient pressure variation analysis.

[3] A *wind tunnel* uses a fixed model to simulate flow over a body moving through a fluid. A *tow tank* uses a moving model to simulate the same situation.

a different angle θ with the local normal to dA. Some elemental areas will have inflow volume $(VA \cos\theta)_{in}\, dt$, and others will have outflow volume $(VA \cos\theta)_{out}\, dt$, as seen in Fig. 3.3. Some surfaces might correspond to streamlines ($\theta = 90°$) or solid walls ($\mathbf{V} = 0$) with neither inflow nor outflow.

Let B be any property of the fluid (energy, momentum, enthalpy, etc.) and let $\beta = dB/dm$ be the *intensive* value, or the amount of B per unit mass in any small element of the fluid. The total amount of B in the control volume (the solid curve in Fig. 3.3) is thus

$$B_{CV} = \int_{CV} \beta\, dm = \int_{CV} \beta\rho\, d\mathcal{V} \quad \beta = \frac{dB}{dm} \tag{3.8}$$

Examining Fig. 3.3, we see three sources of changes in B relating to the control volume:

$$\text{A change within the control volume} \quad \frac{d}{dt}\left(\int_{CV} \beta\rho\, d\mathcal{V}\right)$$

$$\text{Outflow of } \beta \text{ from the control volume} \quad \int_{CS} \beta\rho V \cos\theta\, dA_{out} \tag{3.9}$$

$$\text{Inflow of } \beta \text{ to the control volume} \quad \int_{CS} \beta\rho V \cos\theta\, dA_{in}$$

The notations CV and CS refer to the control volume and control surface, respectively. Note, in Fig. 3.3, that the *system* has moved a bit. In the limit as $dt \to 0$, the instantaneous change of B in the system is the sum of the change within, plus the outflow, minus the inflow:

$$\frac{d}{dt}(B_{syst}) = \frac{d}{dt}\left(\int_{CV} \beta\rho\, d\mathcal{V}\right) + \int_{CS} \beta\rho V \cos\theta\, dA_{out} - \int_{CS} \beta\rho V \cos\theta\, dA_{in} \tag{3.10}$$

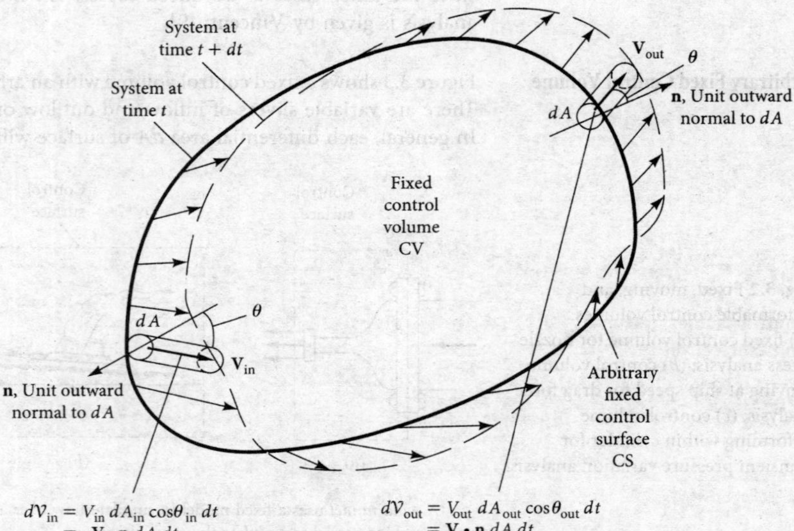

Fig. 3.3 An arbitrary control volume with an arbitrary flow pattern.

This is the *Reynolds transport theorem* for an arbitrary fixed control volume. By letting the property B be mass, momentum, angular momentum, or energy, we can rewrite all the basic laws in control volume form. Note that all three of the integrals are concerned with the intensive property β. Since the control volume is fixed in space, the elemental volumes $d\mathcal{V}$ do not vary with time, so that the time derivative of the volume integral vanishes unless either β or ρ varies with time (unsteady flow).

Equation (3.10) expresses the basic formula that a system derivative equals the rate of change of B within the control volume plus the flow of B out of the control surface minus the flow of B into the control surface. The quantity B (or β) may be any vector or scalar property of the fluid. Two alternate forms are possible for the flow terms. First, we may notice that $V \cos \theta$ is the component of V normal to the area element of the control surface. Thus, we can write

$$\text{Flow terms} = \int_{CS} \beta \rho V_n \, dA_{\text{out}} - \int_{CS} \beta \rho V_n \, dA_{\text{in}} = \int_{CS} \beta \, d\dot{m}_{\text{out}} - \int_{CS} \beta \, d\dot{m}_{\text{in}} \tag{3.10a}$$

where $d\dot{m} = \rho V_n \, dA$ is the differential mass flow through the surface. Form (3.10a) helps us visualize what is being calculated.

A second, alternative form offers elegance and compactness as advantages. If $\mathbf{n}$ is defined as the *outward* normal unit vector everywhere on the control surface, then $\mathbf{V} \cdot \mathbf{n} = V_n$ for outflow and $\mathbf{V} \cdot \mathbf{n} = -V_n$ for inflow. Therefore, the flow terms can be represented by a single integral involving $\mathbf{V} \cdot \mathbf{n}$ that accounts for both positive outflow and negative inflow:

$$\text{Flow terms} = \int_{CS} \beta \rho (\mathbf{V} \cdot \mathbf{n}) \, dA \tag{3.11}$$

The compact form of the Reynolds transport theorem is thus

$$\frac{d}{dt}(B_{\text{syst}}) = \frac{d}{dt}\left(\int_{CV} \beta \rho \, d\mathcal{V} \right) + \int_{CS} \beta \rho (\mathbf{V} \cdot \mathbf{n}) \, dA \tag{3.12}$$

This is beautiful, but only occasionally useful, when the coordinate system is ideally suited to the control volume selected. Otherwise, the computations are easier when the flow of B out is added and the flow of B in is subtracted, according to Eqs. (3.10) or (3.11).

The time derivative term can be written in the equivalent form

$$\frac{d}{dt}\left(\int_{CV} \beta \rho \, d\mathcal{V} \right) = \int_{CV} \frac{\partial}{\partial t} (\beta \rho) \, d\mathcal{V} \tag{3.13}$$

for the fixed control volume, since the volume elements do not vary.

Control Volume Moving at Constant Velocity

If the control volume is moving uniformly at velocity $\mathbf{V}_s$, as in Fig. 3.2b, an observer fixed to the control volume will see a relative velocity $\mathbf{V}_r$ of fluid crossing the control surface, defined by

$$\mathbf{V}_r = \mathbf{V} - \mathbf{V}_s \tag{3.14}$$

where $\mathbf{V}$ is the fluid velocity relative to the same coordinate system in which the control volume motion $\mathbf{V}_s$ is observed. Note that Eq. (3.14) is a vector subtraction. The flow terms will be proportional to $\mathbf{V}_r$, but the volume integral of Eq. (3.12) is unchanged because the control volume moves as a fixed shape without deforming. The Reynolds transport theorem for this case of a uniformly moving control volume is

$$\frac{d}{dt}(B_{syst}) = \frac{d}{dt}\left(\int_{CV} \beta\rho\, d\mathcal{V}\right) + \int_{CS} \beta\rho(\mathbf{V}_r \cdot \mathbf{n})\, dA \tag{3.15}$$

which, reduces to Eq. (3.12) if $\mathbf{V}_s \equiv 0$.

Control Volume of Constant Shape but Variable Velocity[4]

If the control volume moves with a velocity $\mathbf{V}_s(t)$ that retains its shape, then the volume elements do not change with time, but the boundary relative velocity $\mathbf{V}_r = \mathbf{V}(\mathbf{r}, t) - \mathbf{V}_s(t)$ becomes a somewhat more complicated function. Equation (3.15) is unchanged in form, but the area integral may be more laborious to evaluate.

Arbitrarily Moving and Deformable Control Volume[5]

The most general situation is when the control volume is both moving and deforming arbitrarily, as illustrated in Fig. 3.4. The flow of volume across the control surface is again proportional to the relative normal velocity component $\mathbf{V}_r \cdot \mathbf{n}$, as in Eq. (3.15). However, since the control surface has a deformation, its velocity $\mathbf{V}_s = \mathbf{V}_s(\mathbf{r}, t)$, so that the relative velocity $\mathbf{V}_r = \mathbf{V}(\mathbf{r}, t) - \mathbf{V}_s(\mathbf{r}, t)$ is or can be a complicated function, even though the flow integral is the same as in Eq. (3.15). Meanwhile, the volume integral in Eq. (3.15) must allow the volume elements to distort with time. Thus, the time derivative must be applied *after* integration. For the deforming control volume, then, the transport theorem takes the form

$$\frac{d}{dt}(B_{syst}) = \frac{d}{dt}\left(\int_{CV} \beta\rho\, d\mathcal{V}\right) + \int_{CS} \beta\rho(\mathbf{V}_r \cdot \mathbf{n})\, dA \tag{3.16}$$

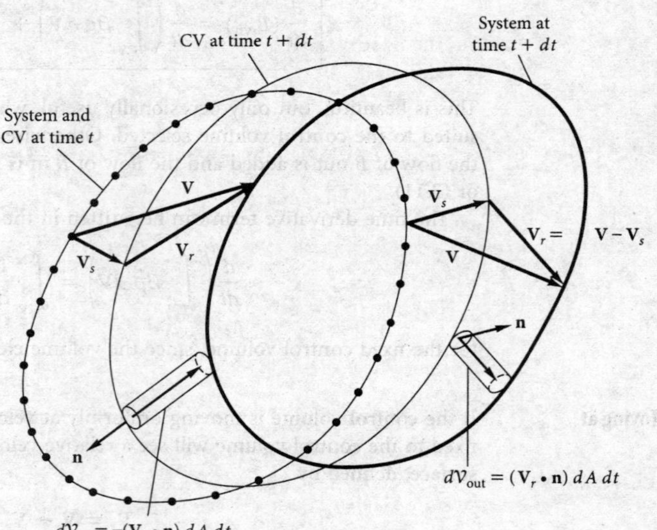

Fig. 3.4 Relative velocity effects between a system and a control volume when both move and deform. The system boundaries move at velocity $\mathbf{V}$, and the control surface moves at velocity $\mathbf{V}_s$.

[4]This section may be omitted without loss of continuity.
[5]This section may be omitted without loss of continuity.

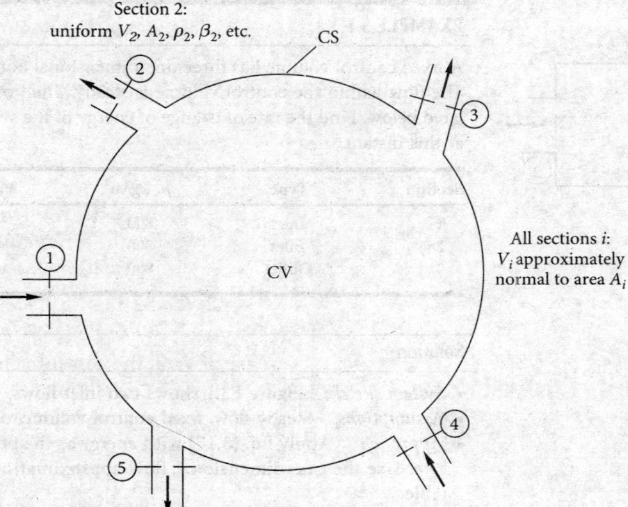

Fig. 3.5 A control volume with simplified one-dimensional inlets and exits.

This is the most general case, which we can compare with the equivalent form for a fixed control volume:

$$\frac{d}{dt}(B_{\text{syst}}) = \int_{\text{CV}} \frac{\partial}{\partial t}(\beta\rho)\, d\mathcal{V} + \int_{\text{CS}} \beta\rho(\mathbf{V}\cdot\mathbf{n})\, dA \tag{3.17}$$

The moving and deforming control volume, Eq. (3.16), contains only two complications: (1) The time derivative of the first integral on the right must be taken outside, and (2) the second integral involves the *relative* velocity $\mathbf{V}_r$ between the fluid system and the control surface. These differences and mathematical subtleties are best shown by examples.

One-Dimensional Flux Term Approximations

In many situations, the flow crosses the boundaries of the control surface only at simplified inlets and exits that are approximately *one-dimensional;* that is, flow properties are nearly uniform over the cross section. For a fixed control volume, the surface integral in Eq. (3.12) reduces to a sum of positive (outlet) and negative (inlet) product terms for each cross section:

$$\frac{d}{dt}(B_{\text{syst}}) = \frac{d}{dt}\left(\int_{\text{CV}} \beta\, dm\right) + \sum_{\text{outlets}} \beta_i \dot{m}_i \mid_{\text{out}} - \sum_{\text{inlets}} \beta_i \dot{m}_i \mid_{\text{in}} \quad \text{where } \dot{m}_i = \rho_i A_i V_i \tag{3.18}$$

To the writer, this is an attractive way to set up a control volume analysis without using the dot product notation. An example of multiple one-dimensional flows is shown in Fig. 3.5. There are inlet flows at sections 1 and 4 and outflows at sections 2, 3, and 5. Equation (3.18) becomes

$$\frac{d}{dt}(B_{\text{syst}}) = \frac{d}{dt}\left(\int_{\text{CV}} \beta\, dm\right) + \beta_2(\rho AV)_2 + \beta_3(\rho AV)_3 + \beta_5(\rho AV)_5 - \beta_1(\rho AV)_1 - \beta_4(\rho AV)_4 \tag{3.19}$$

with no contribution from any other portion of the control surface because there is no flow across the boundary.

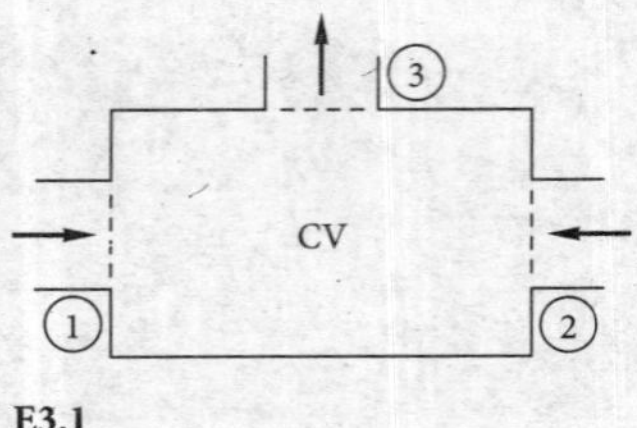

E3.1

EXAMPLE 3.1

A fixed control volume has three one-dimensional boundary sections, as shown in Fig. E3.1. The flow within the control volume is steady. The flow properties at each section are tabulated below. Find the rate of change of energy of the system that occupies the control volume at this instant.

Section	Type	ρ, kg/m^3	V, m/s	A, m^2	e, J/kg
1	Inlet	800	5.0	2.0	300
2	Inlet	800	8.0	3.0	100
3	Outlet	800	17.0	2.0	150

Solution

- *System sketch:* Figure E3.1 shows two inlet flows, 1 and 2, and a single outlet flow, 3.
- *Assumptions:* Steady flow, fixed control volume, one-dimensional inlet and exit flows.
- *Approach:* Apply Eq. (3.17) with *energy* as the property, where $B = E$ and $\beta = dE/dm = e$. Use the one-dimensional flow approximation and then insert the data from the table.
- *Solution steps:* Outlet 3 contributes a positive term, and inlets 1 and 2 are negative. The appropriate form of Eq. (3.12) is

$$\left(\frac{dE}{dt}\right)_{\text{syst}} = \frac{d}{dt}\left(\int_{CV} e\rho\, d\upsilon\right) + e_3\dot{m}_3 - e_1\dot{m}_1 - e_2\dot{m}_2$$

Since the flow is steady, the time-derivative volume integral term is zero. Introducing $(\rho AV)_i$ as the mass flow grouping, we obtain

$$\left(\frac{dE}{dt}\right)_{\text{syst}} = -e_1\rho_1A_1V_1 - e_2\rho_2A_2V_2 + e_3\rho_3A_3V_3$$

Introducing the numerical values from the table, we have

$$\left(\frac{dE}{dt}\right)_{\text{syst}} = -(300\text{ J/kg})(800\text{ kg/m}^3)(2\text{ m}^2)(5\text{ m/s}) - 100(800)(3)(8) + 150(800)(2)(17)$$

$$= (-2{,}400{,}000 - 1{,}920{,}000 + 4{,}080{,}000)\text{ J/s}$$

$$= -240{,}000\text{ J/s} = -0.24\text{ MJ/s} \qquad \textit{Ans.}$$

Thus, the system is losing energy at the rate of 0.24 MJ/s = 0.24 MW. Since, we have accounted for all fluid energy crossing the boundary, we conclude from the first law that there must be heat loss through the control surface, or the system must be doing work on the environment through some device not shown. Notice that the use of SI units leads to a consistent result in joules per second without any conversion factors. We promised in Chap. 1 that this would be the case.

- *Comments:* This problem involves energy, but suppose we check the balance of mass also. Then B = mass m, and $\beta = dm/dm$ = unity. Again, the volume integral vanishes for steady flow, and Eq. (3.17) reduces to

$$\left(\frac{dm}{dt}\right)_{\text{syst}} = \int_{CS} \rho(\mathbf{V}\cdot\mathbf{n})\,dA = -\rho_1A_1V_1 - \rho_2A_2V_2 + \rho_3A_3V_3$$

$$= -(800\text{ kg/m}^3)(2\text{ m}^2)(5\text{ m/s}) - 800(3)(8) + 800(17)(2)$$

$$= (-8{,}000 - 19{,}200 + 27{,}200)\text{ kg/s} = 0\text{ kg/s}$$

Thus, the system mass does not change, which correctly expresses the law of conservation of system mass, Eq. (3.1).

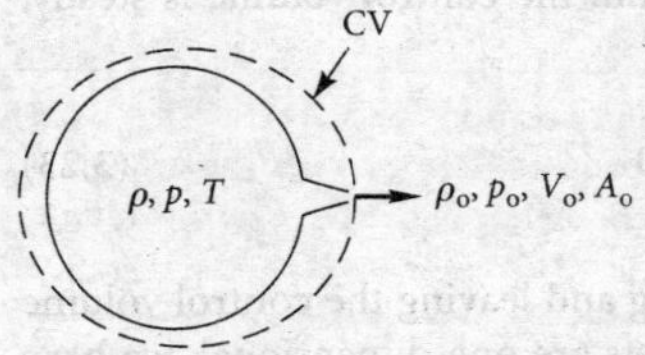

E3.2

EXAMPLE 3.2

Compressed air in a rigid tank of volume $\mathcal{V}$ exhausts through a small nozzle as in Fig. E3.2. Air properties change through the nozzle, and the flow exits at ρ_o, V_o, A_o. Find an expression for the rate of change of tank density.

Solution

- *System sketch:* Fig. E3.2 shows one exit, no inlets. The constant exit area is A_o.
- *Control volume:* As shown, we choose a CV that encircles the entire tank and nozzle.
- *Assumptions:* Unsteady flow (the tank mass decreases), one-dimensional exit flow.
- *Approach:* Apply Eq. (3.16) for mass, $B = m$ and $\beta = dm/dm =$ unity.
- *Solution steps:* Write out the Reynolds transport relation (3.16) for this problem:

$$\left(\frac{dm}{dt}\right)_{syst} = 0 = \frac{d}{dt}\left(\int_{CV} \rho\, d\mathcal{V}\right) + \int_{CS} \rho(\mathbf{V}\cdot\mathbf{n})\, dA = \mathcal{V}\frac{d\rho}{dt} + \rho_o V_o A_o$$

Solve for the rate of change of tank density:

$$\frac{d\rho}{dt} = -\frac{\rho_o V_o A_o}{\mathcal{V}} \qquad \textit{Ans.}$$

- *Comments:* This is a first-order ordinary differential equation for the tank density. If we account for changes in ρ_o and V_o from the compressible-flow theories of Chap. 9, we can readily solve this equation for the tank density $\rho(t)$.

For advanced study, many more details of the analysis of deformable control volumes can be found in Hansen [4] and Potter et al. [5].

3.3 Conservation of Mass

The Reynolds transport theorem, Eq. (3.16) or (3.17), establishes a relation between system rates of change and control volume surface and volume integrals. But, system derivatives are related to the basic laws of mechanics, Eqs. (3.1) to (3.5). Eliminating system derivatives between the two gives the control volume, or *integral*, forms of the laws of mechanics of fluids. The dummy variable B becomes, respectively, mass, linear momentum, angular momentum, and energy.

For conservation of mass, as discussed in Examples 3.1 and 3.2, $B = m$ and $\beta = dm/dm = 1$. Equation (3.1) becomes

$$\left(\frac{dm}{dt}\right)_{syst} = 0 = \frac{d}{dt}\left(\int_{CV} \rho\, d\mathcal{V}\right) + \int_{CS} \rho(\mathbf{V}_r\cdot\mathbf{n})\, dA \qquad (3.20)$$

This is the integral mass conservation law for a deformable control volume. For a fixed control volume, we have

$$\int_{CV} \frac{\partial \rho}{\partial t}\, d\mathcal{V} + \int_{CS} \rho(\mathbf{V}\cdot\mathbf{n})\, dA = 0 \qquad (3.21)$$

If the control volume has only a number of one-dimensional inlets and outlets, we can write

$$\int_{CV} \frac{\partial \rho}{\partial t}\, d\mathcal{V} + \sum_i (\rho_i A_i V_i)_{out} - \sum_i (\rho_i A_i V_i)_{in} = 0 \qquad (3.22)$$

Other special cases occur. Suppose that the flow within the control volume is steady; then $\partial\rho/\partial t \equiv 0$, and Eq. (3.21) reduces to

$$\int_{CS} \rho(\mathbf{V} \cdot \mathbf{n})\, dA = 0 \tag{3.23}$$

This states that in steady flow the mass flows entering and leaving the control volume must balance exactly.[6] If, further, the inlets and outlets are one-dimensional, we have for steady flow

$$\sum_i (\rho_i A_i V_i)_{in} = \sum_i (\rho_i A_i V_i)_{out} \tag{3.24}$$

This simple approximation is widely used in engineering analyses. For example, referring to Fig. 3.5, we see that if the flow in that control volume is steady, the three outlet mass flows balance the two inlets:

$$\text{Outflow} = \text{Inflow}$$

$$\rho_2 A_2 V_2 + \rho_3 A_3 V_3 + \rho_5 A_5 V_5 = \rho_1 A_1 V_1 + \rho_4 A_4 V_4 \tag{3.25}$$

The quantity ρAV is called the *mass flow* $\dot{m}$ passing through the one-dimensional cross section and has consistent units of kilograms per second. Equation (3.25) can be rewritten in the short form

$$\dot{m}_2 + \dot{m}_3 + \dot{m}_5 = \dot{m}_1 + \dot{m}_4 \tag{3.26}$$

and, in general, the steady-flow–mass-conservation relation (3.23) can be written as

$$\sum_i (\dot{m}_i)_{out} = \sum_i (\dot{m}_i)_{in} \tag{3.27}$$

If the inlets and outlets are not one-dimensional, one has to compute $\dot{m}$ by integration over the section

$$\dot{m}_{cs} = \int_{cs} \rho(\mathbf{V} \cdot \mathbf{n})\, dA \tag{3.28}$$

where "cs" stands for cross section. An illustration of this is given in Example 3.4.

Incompressible Flow

Still, further simplification is possible if the fluid is incompressible, which we may define as having density variations that are negligible in the mass conservation requirement.[7] As we saw in Chap. 1, all liquids are nearly incompressible, and gas flows can *behave* as if they were incompressible, particularly if the gas velocity is less than about 30 percent of the speed of sound of the gas.

[6]Throughout this section, we are neglecting *sources* or *sinks* of mass that might be embedded in the control volume. Equations (3.20) and (3.21) can readily be modified to add source and sink terms, but this is rarely necessary.

[7]Be warned that there is subjectivity in specifying incompressibility. Oceanographers consider a 0.1 percent density variation very significant, while aerodynamicists may neglect density variations in highly compressible, even hypersonic, gas flows. Your task is to justify the incompressible approximation when you make it.

Again, consider the fixed control volume. For nearly incompressible flow, the term $\partial\rho/\partial t$ is small, so the time-derivative volume integral in Eq. (3.21) can be neglected. The constant density can then be removed from the surface integral for a nice simplification:

$$\frac{d}{dt}\left(\int_{CV}\frac{\partial\rho}{\partial t}\,d\upsilon\right) + \int_{CS}\rho(\mathbf{V}\cdot\mathbf{n})\,dA = 0 = \int_{CS}\rho(\mathbf{V}\cdot\mathbf{n})\,dA = \rho\int_{CS}(\mathbf{V}\cdot\mathbf{n})\,dA$$

or

$$\int_{CS}(\mathbf{V}\cdot\mathbf{n})\,dA = 0 \tag{3.29}$$

If the inlets and outlets are one-dimensional, we have

$$\sum_i (V_i A_i)_{\text{out}} = \sum_i (V_i A_i)_{\text{in}} \tag{3.30}$$

or

$$\Sigma Q_{\text{out}} = \Sigma Q_{\text{in}}$$

where $Q_i = V_i A_i$ is called the *volume flow* passing through the given cross section.

Again, if consistent units are used, $Q = VA$ will have units of cubic meters per second (SI) or cubic feet per second (BG). If the cross section is not one-dimensional, we have to integrate

$$Q_{CS} = \int_{CS}(\mathbf{V}\cdot\mathbf{n})\,dA \tag{3.31}$$

Equation (3.31) allows us to define an *average velocity* V_{av} that, when multiplied by the section area, gives the correct volume flow:

$$V_{\text{av}} = \frac{Q}{A} = \frac{1}{A}\int(\mathbf{V}\cdot\mathbf{n})\,dA \tag{3.32}$$

This could be called the *volume-average velocity*. If the density varies across the section, we can define an average density in the same manner:

$$\rho_{\text{av}} = \frac{1}{A}\int\rho\,dA \tag{3.33}$$

But, the mass flow would contain the product of density and velocity, and the average product $(\rho V)_{\text{av}}$ would in general have a different value from the product of the averages:

$$(\rho V)_{\text{av}} = \frac{1}{A}\int\rho(\mathbf{V}\cdot\mathbf{n})\,dA \approx \rho_{\text{av}}V_{\text{av}} \tag{3.34}$$

We illustrate average velocity in Example 3.4. We can often neglect the difference or, if necessary, use a correction factor between mass average and volume average.

EXAMPLE 3.3

Write the conservation-of-mass relation for steady flow through a streamtube (flow everywhere parallel to the walls) with a single one-dimensional inlet 1 and exit 2 (Fig. E3.3).

E3.3

Solution

For steady flow, Eq. (3.24) applies with the single inlet and exit:

$$\dot{m} = \rho_1 A_1 V_1 = \rho_2 A_2 V_2 = \text{const}$$

Thus, in a streamtube in steady flow, the mass flow is constant across every section of the tube. If the density is constant, then

$$Q = A_1V_1 = A_2V_2 = \text{const} \quad \text{or} \quad V_2 = \frac{A_1}{A_2}V_1$$

The volume flow is constant in the tube in steady incompressible flow, and the velocity increases as the section area decreases. This relation was derived by Leonardo da Vinci in 1500.

EXAMPLE 3.4

For steady viscous flow through a circular tube (Fig. E3.4), the axial velocity profile is given approximately by

$$u = U_0\left(1 - \frac{r}{R}\right)^m$$

so that u varies from zero at the wall ($r = R$), or no slip, up to a maximum $u = U_0$ at the centerline $r = 0$. For highly viscous (laminar) flow $m \approx \frac{1}{2}$, while for less viscous (turbulent) flow $m \approx \frac{1}{7}$. Compute the average velocity if the density is constant.

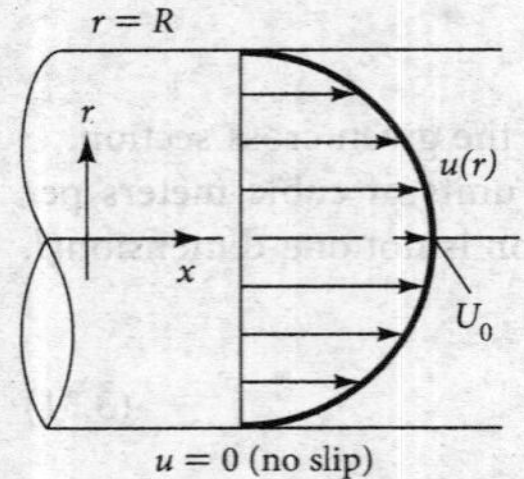

E3.4

Solution

The average velocity is defined by Eq. (3.32). Here, $\mathbf{V} = \mathbf{i}u$ and $\mathbf{n} = \mathbf{i}$, and thus $\mathbf{V} \cdot \mathbf{n} = u$. Since the flow is symmetric, the differential area can be taken as a circular strip $dA = 2\,\pi r\,dr$. Equation (3.32) becomes

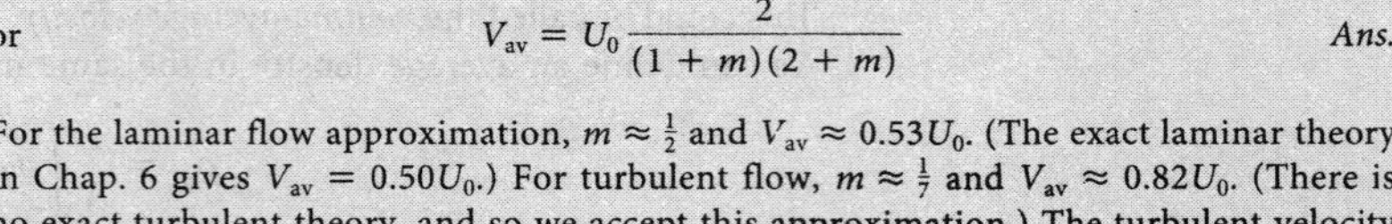

$$V_{av} = \frac{1}{A}\int u\,dA = \frac{1}{\pi R^2}\int_0^R U_0\left(1 - \frac{r}{R}\right)^m 2\pi r\,dr$$

or

$$V_{av} = U_0\frac{2}{(1+m)(2+m)} \qquad \textit{Ans.}$$

For the laminar flow approximation, $m \approx \frac{1}{2}$ and $V_{av} \approx 0.53U_0$. (The exact laminar theory in Chap. 6 gives $V_{av} = 0.50U_0$.) For turbulent flow, $m \approx \frac{1}{7}$ and $V_{av} \approx 0.82U_0$. (There is no exact turbulent theory, and so we accept this approximation.) The turbulent velocity profile is more uniform across the section, and thus the average velocity is only slightly less than maximum.

EXAMPLE 3.5

The tank in Fig. E3.5 is being filled with water by two one-dimensional inlets. Air is trapped at the top of the tank. The water height is h. (*a*) Find an expression for the change in water height dh/dt. (*b*) Compute dh/dt if $D_1 = 2.5$ cm, $D_2 = 8$ cm, $V_1 = 1$ m/s, $V_2 = 0.5$ m/s, and $A_t = 0.2$ m^2, assuming water at 20°C.

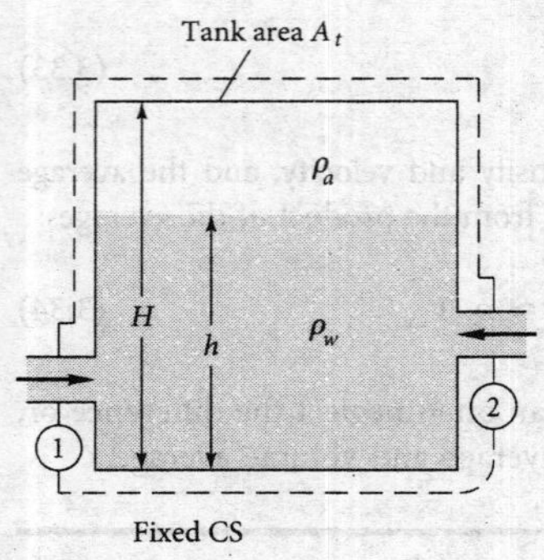

E3.5

Solution

Part (a)

A suggested control volume encircles the tank and cuts through the two inlets. The flow within is unsteady, and Eq. (3.22) applies with no outlets and two inlets:

$$\frac{d}{dt}\left(\int_{CV} \rho\,d\mathcal{V}\right) - \rho_1A_1V_1 - \rho_2A_2V_2 = 0 \qquad (1)$$

Now, if A_t is the tank cross-sectional area, the unsteady term can be evaluated as follows:

$$\frac{d}{dt}\left(\int_{CV} \rho \, d\mathcal{V}\right) = \frac{d}{dt}(\rho_w A_t h) + \frac{d}{dt}[\rho_a A_t(H - h)] = \rho_w A_t \frac{dh}{dt} \qquad (2)$$

The ρ_a term vanishes because it is the rate of change of air mass and is zero because the air is trapped at the top. Substituting (2) into (1), we find the change of water height

$$\frac{dh}{dt} = \frac{\rho_1 A_1 V_1 + \rho_2 A_2 V_2}{\rho_w A_t} \qquad \textit{Ans. (a)}$$

For water, $\rho_1 = \rho_2 = \rho_w$, and this result reduces to

$$\frac{dh}{dt} = \frac{A_1 V_1 + A_2 V_2}{A_t} = \frac{Q_1 + Q_2}{A_t} \qquad (3)$$

Part (b) The two inlet volume flows are

$$Q_1 = A_1 V_1 = \tfrac{1}{4}\pi(0.025 \text{ m})^2(1 \text{ m/s}) = 0.000491 \text{m}^3\text{/s}$$

$$Q_2 = A_2 V_2 = \tfrac{1}{4}\pi(0.08 \text{ m})^2(0.5 \text{ m/s}) = 0.00251 \text{ m}^3\text{/s}$$

Then, from Eq. (3),

$$\frac{dh}{dt} = \frac{(0.000491 + 0.00251) \text{ m}^3\text{/s}}{0.2 \text{ m}^2} = 0.0150 \text{ m/s} \qquad \textit{Ans. (b)}$$

Suggestion: Repeat this problem with the top of the tank open.

The control volume mass relations, Eq. (3.20) or (3.21), are fundamental to all fluid flow analyses. They involve only velocity and density. Vector directions are of no consequence except to determine the normal velocity at the surface and hence whether the flow is *in* or *out*. Although your specific analysis may concern forces or moments or energy, you must always make sure that mass is balanced as part of the analysis; otherwise the results will be unrealistic and probably incorrect. We shall see in the examples that follow how mass conservation is constantly checked in performing an analysis of other fluid properties.

3.4 The Linear Momentum Equation

In Newton's second law, Eq. (3.2), the property being differentiated is the linear momentum $m\mathbf{V}$. Therefore, our dummy variable is $\mathbf{B} = m\mathbf{V}$ and $\beta = d\mathbf{B}/dm = \mathbf{V}$, and application of the Reynolds transport theorem gives the linear momentum relation for a deformable control volume:

$$\frac{d}{dt}(m\mathbf{V})_{\text{syst}} = \sum \mathbf{F} = \frac{d}{dt}\left(\int_{CV} \mathbf{V}\rho \, d\mathcal{V}\right) + \int_{CS} \mathbf{V}\rho(\mathbf{V}_r \cdot \mathbf{n}) \, dA \qquad (3.35)$$

The following points concerning this relation should be strongly emphasized:

1. The term $\mathbf{V}$ is the fluid velocity relative to an *inertial* (nonaccelerating) coordinate system; otherwise Newton's second law must be modified to include noninertial relative acceleration terms (see the end of this section).

2. The term $\sum \mathbf{F}$ is the *vector* sum of all forces acting on the system material considered as a free body; that is, it includes surface forces on all fluids and solids cut by the control surface plus all body forces (gravity and electromagnetic) acting on the masses within the control volume.
3. The entire equation is a vector relation; both the integrals are vectors due to the term $\mathbf{V}$ in the integrands. The equation thus has three components. If we want only, say, the x component, the equation reduces to

$$\sum F_x = \frac{d}{dt}\left(\int_{CV} u\rho \, d\mathcal{V}\right) + \int_{CS} u\rho(\mathbf{V}_r \cdot \mathbf{n}) \, dA \tag{3.36}$$

and similarly, $\sum F_y$ and $\sum F_z$ would involve v and w, respectively. Failure to account for the vector nature of the linear momentum relation (3.35) is probably the greatest source of student error in control volume analyses.

For a fixed control volume, the relative velocity $\mathbf{V}_r \equiv \mathbf{V}$, and Eq. (3.35) becomes

$$\sum \mathbf{F} = \frac{d}{dt}\left(\int_{CV} \mathbf{V}\rho \, d\mathcal{V}\right) + \int_{CS} \mathbf{V}\rho(\mathbf{V} \cdot \mathbf{n}) \, dA \tag{3.37}$$

Again, we stress that this is a vector relation and that $\mathbf{V}$ must be an inertial-frame velocity. Most of the momentum analyses in this text are concerned with Eq. (3.37).

One-Dimensional Momentum Flux

By analogy with the term *mass flow* used in Eq. (3.28), the surface integral in Eq. (3.37) is called the *momentum flow term.* If we denote momentum by $\mathbf{M}$, then

$$\dot{\mathbf{M}}_{CS} = \int_{sec} \mathbf{V}\rho(\mathbf{V} \cdot \mathbf{n}) \, dA \tag{3.38}$$

Because of the dot product, the result will be negative for inlet momentum flow and positive for outlet flow. If the cross section is one-dimensional, $\mathbf{V}$ and ρ are uniform over the area and the integrated result is

$$\dot{\mathbf{M}}_{seci} = \mathbf{V}_i(\rho_i V_{ni} A_i) = \dot{m}_i \mathbf{V}_i \tag{3.39}$$

for outlet flow and $-\dot{m}_i \mathbf{V}_i$ for inlet flow. Thus, if the control volume has only one-dimensional inlets and outlets, Eq. (3.37) reduces to

$$\sum \mathbf{F} = \frac{d}{dt}\left(\int_{CV} \mathbf{V}\rho \, d\mathcal{V}\right) + \sum (\dot{m}_i \mathbf{V}_i)_{out} - \sum (\dot{m}_i \mathbf{V}_i)_{in} \tag{3.40}$$

This is a commonly used approximation in engineering analyses. It is crucial to realize that we are dealing with vector sums. Equation (3.40) states that the net vector force on a fixed control volume equals the rate of change of vector momentum within the control volume plus the vector sum of outlet momentum flows minus the vector sum of inlet flows.

Net Pressure Force on a Closed Control Surface

Generally speaking, the surface forces on a control volume are due to (1) forces exposed by cutting through solid bodies that protrude through the surface and (2) forces due to pressure and viscous stresses of the surrounding fluid. The computation of pressure force is relatively simple, as shown in Fig. 3.6. Recall from Chap. 2

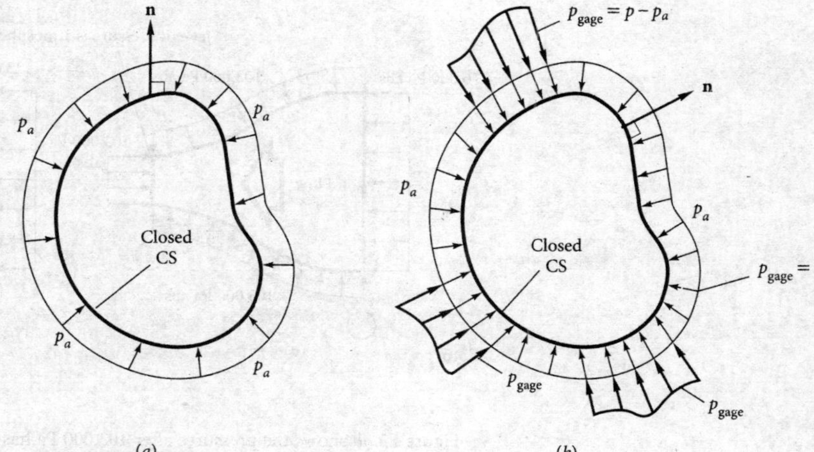

Fig. 3.6 Pressure force computation by subtracting a uniform distribution: (*a*) uniform pressure, $\mathbf{F} = -p_a \int \mathbf{n}\, dA \equiv 0$; (*b*) nonuniform pressure, $\mathbf{F} = -\int (p - p_a)\mathbf{n}\, dA$.

that the external pressure force on a surface is normal to the surface and *inward.* Since the unit vector **n** is defined as *outward,* one way to write the pressure force is

$$\mathbf{F}_{\text{press}} = \int_{CS} p(-\mathbf{n})\, dA \tag{3.41}$$

Now if the pressure has a uniform value p_a all around the surface, as in Fig. 3.7*a*, the net pressure force is zero:

$$\mathbf{F}_{\text{UP}} = \int p_a(-\mathbf{n})dA = -p_a \int \mathbf{n}\, dA \equiv 0 \tag{3.42}$$

where the subscript UP stands for uniform pressure. This result is *independent of the shape of the surface*[8] as long as the surface is closed and all our control volumes are closed. Thus, a seemingly complicated pressure force problem can be simplified by subtracting any convenient uniform pressure p_a and working only with the pieces of gage pressure that remain, as illustrated in Fig. 3.6*b*. So Eq. (3.41) is entirely equivalent to

$$\mathrm{F}_{\text{press}} = \int_{CS} (p - p_a)(-\mathbf{n})dA = \int_{CS} p_{\text{gage}}(-\mathbf{n})\, dA$$

This trick can mean quite a savings in computation.

EXAMPLE 3.6

A control volume of a nozzle section has surface pressures of 276,000 Pa absolute at section 1 and atmospheric pressure of 103,000 Pa absolute at section 2 and on the external rounded part of the nozzle, as in Fig. E3.6*a*. Compute the net pressure force if $D_1 = 8$ cm and $D_2 = 2.5$ cm

Solution

- *System sketch:* The control volume is the *outside* of the nozzle, plus the cut sections (1) and (2). There would also be *stresses* in the cut nozzle wall at section 1, which we are neglecting here. The pressures acting on the control volume are shown in Fig. E3.6*a*.

[8]Can you prove this? It is a consequence of Gauss's theorem from vector analysis.

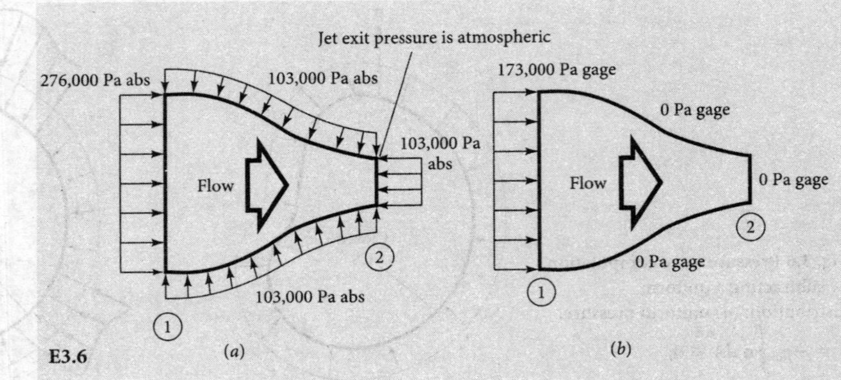

E3.6

Figure E3.6*b* shows the pressures after 103,000 Pa has been subtracted from all sides. Here we compute the net pressure force only.

- *Assumptions:* Known pressures, as shown, on all surfaces of the control volume.
- *Approach:* Since three surfaces have $p = 103{,}000$ Pa, subtract this amount everywhere so that these three sides reduce to zero "gage pressure" for convenience. This is allowable because of Eq. (3.42).
- *Solution steps:* For the modified pressure distribution, Fig. E3.6*b*, only section 1 is needed:

$$\mathbf{F}_{\text{press}} = p_{\text{gage},1}(-\mathbf{n})_1 A_1 = (173{,}000 \text{ Pa})\left[-(-\mathbf{i})\right]\left[\frac{\pi}{4}(0.08 \text{ m})^2\right] = 870\,\mathbf{i}\text{ N} \qquad \textit{Ans.}$$

- *Comments:* This "uniform subtraction" artifice, which is entirely legal, has greatly simplified the calculation of pressure force. *Note:* We were a bit too informal when multiplying pressure in N/m^2 times area in square meters. We achieved force in Newtons correctly. *Further note:* In addition to $\mathbf{F}_{\text{press}}$, there are other forces involved in this flow, due to tension stresses in the cut nozzle wall and the fluid weight inside the control volume.

Pressure Condition at a Jet Exit

Figure E3.6 illustrates a pressure boundary condition commonly used for jet exit flow problems. When a fluid flow leaves a confined internal duct and exits into an ambient "atmosphere," its free surface is exposed to that atmosphere. Therefore the jet itself will essentially be at atmospheric pressure also. This condition was used at section 2 in Fig. E3.6.

Only two effects could maintain a pressure difference between the atmosphere and a free exit jet. The first is surface tension, Eq. (1.31), which is usually negligible. The second effect is a *supersonic* jet, which can separate itself from an atmosphere with expansion or compression waves (Chap. 9). For the majority of applications, therefore, we shall set the pressure in an exit jet as atmospheric.

EXAMPLE 3.7

A fixed control volume of a streamtube in steady flow has a uniform inlet flow (ρ_1, A_1, V_1) and a uniform exit flow (ρ_2, A_2, V_2), as shown in Fig. 3.7. Find an expression for the net force on the control volume.

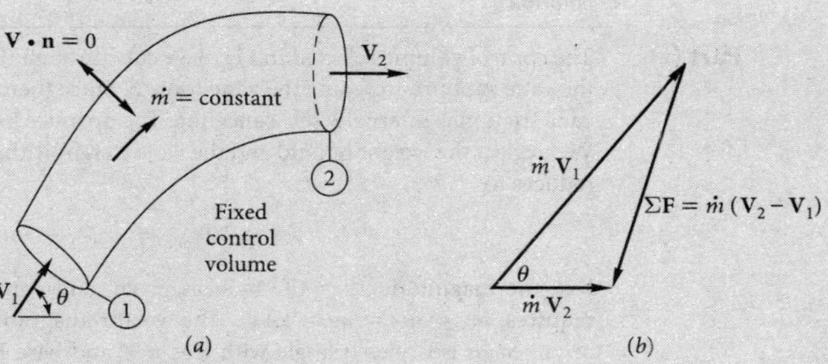

Fig. 3.7 Net force on a one-dimensional streamtube in steady flow: (*a*) streamtube in steady flow; (*b*) vector diagram for computing net force.

Solution

Equation (3.40) applies with one inlet and exit:

$$\sum \mathbf{F} = \dot{m}_2\mathbf{V}_2 - \dot{m}_1\mathbf{V}_1 = (\rho_2 A_2 V_2)\mathbf{V}_2 - (\rho_1 A_1 V_1)\mathbf{V}_1$$

The volume integral term vanishes for steady flow, but from conservation of mass in Example 3.3 we saw that

$$\dot{m}_1 = \dot{m}_2 = \dot{m} = \text{const}$$

Therefore, a simple form for the desired result is

$$\sum \mathbf{F} = \dot{m}(\mathbf{V}_2 - \mathbf{V}_1) \qquad \textit{Ans.}$$

This is a *vector* relation and is sketched in Fig. 3.7*b*. The term $\sum \mathbf{F}$ represents the net force acting on the control volume due to all causes; it is needed to balance the change in momentum of the fluid as it turns and decelerates while passing through the control volume.

EXAMPLE 3.8

As shown in Fig. 3.8*a*, a fixed vane turns a water jet of area A through an angle θ without changing its velocity magnitude. The flow is steady, pressure is p_a everywhere, and friction on the vane is negligible. (*a*) Find the components F_x and F_y of the applied vane force. (*b*) Find expressions for the force magnitude F and the angle ϕ between F and the horizontal; plot them versus θ.

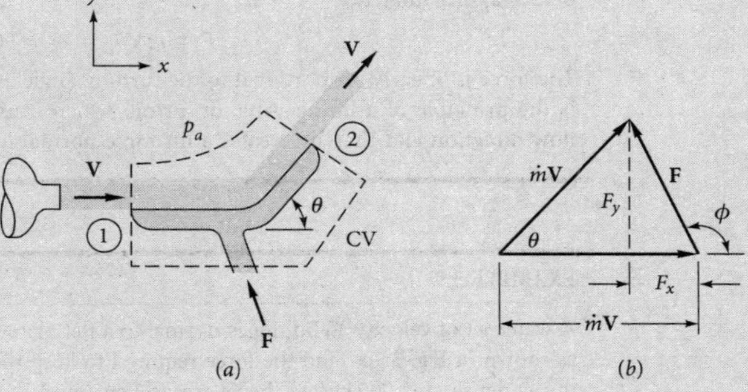

Fig. 3.8 Net applied force on a fixed jet-turning vane: (*a*) geometry of the vane turning the water jet; (*b*) vector diagram for the net force.

Solution

Part (a) The control volume selected in Fig. 3.8*a* cuts through the inlet and exit of the jet and through the vane support, exposing the vane force **F**. Since there is no cut along the vane–jet interface, vane friction is internally self-canceling. The pressure force is zero in the uniform atmosphere. We neglect the weight of fluid and the vane weight within the control volume. Then, Eq. (3.40) reduces to

$$\mathbf{F}_{\text{vane}} = \dot{m}_2 \mathbf{V}_2 - \dot{m}_1 \mathbf{V}_1$$

But, the magnitude $V_1 = V_2 = V$ as given, and conservation of mass for the streamtube requires $\dot{m}_1 = \dot{m}_2 = \dot{m} = \rho A V$. The vector diagram for force and momentum change becomes an isosceles triangle with legs $\dot{m}\mathbf{V}$ and base **F**, as in Fig. 3.8*b*. We can readily find the force components from this diagram:

$$F_x = \dot{m}V(\cos\theta - 1) \quad F_y = \dot{m}V\sin\theta \qquad \textit{Ans. (a)}$$

where $\dot{m}V = \rho A V^2$ for this case. This is the desired result.

Part (b) The force magnitude is obtained from part (*a*):

$$F = (F_x^2 + F_y^2)^{1/2} = \dot{m}V[\sin^2\theta + (\cos\theta - 1)^2]^{1/2} = 2\dot{m}V\sin\frac{\theta}{2} \qquad \textit{Ans. (b)}$$

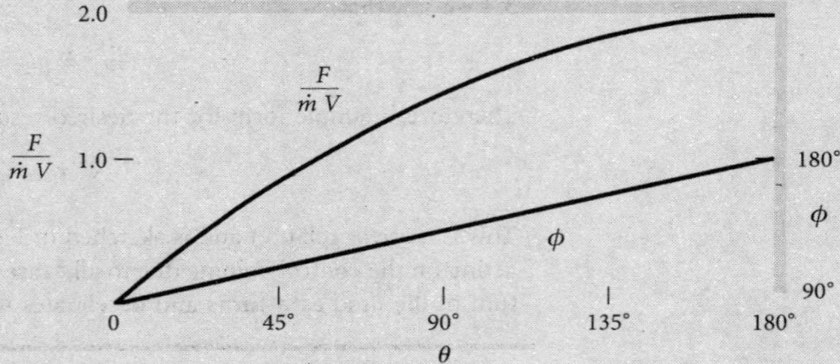

E3.8

From the geometry of Fig. 3.8*b* we obtain

$$\phi = 180° - \tan^{-1}\frac{F_y}{F_x} = 90° + \frac{\theta}{2} \qquad \textit{Ans. (b)}$$

These can be plotted versus θ as shown in Fig. E3.8. Two special cases are of interest. First, the maximum force occurs at $\theta = 180°$—that is, when the jet is turned around and thrown back in the opposite direction with its momentum completely reversed. This force is $2\dot{m}V$ and acts to the *left*; that is, $\phi = 180°$. Second, at very small turning angles ($\theta < 10°$) we obtain approximately

$$F \approx \dot{m}V\theta \qquad \phi \approx 90°$$

The force is linearly proportional to the turning angle and acts nearly normal to the jet. This is the principle of a lifting vane, or airfoil, which causes a slight change in the oncoming flow direction and thereby creates a lift force normal to the basic flow.

EXAMPLE 3.9

A water jet of velocity V_j impinges normal to a flat plate that moves to the right at velocity V_c, as shown in Fig. 3.9*a*. Find the force required to keep the plate moving at constant velocity if the jet density is 1,000 kg/m^3, the jet area is 3 cm^2, and V_j and V_c are 20 and 15 m/s, respectively.

Neglect the weight of the jet and plate, and assume steady flow with respect to the moving plate with the jet splitting into an equal upward and downward half-jet.

Solution

The suggested control volume in Fig. 3.9*a* cuts through the plate support to expose the desired forces R_x and R_y. This control volume moves at speed V_c and thus is fixed relative to the plate, as in Fig. 3.9*b*. We must satisfy both mass and momentum conservation for the assumed steady flow pattern in Fig. 3.9*b*. There are two outlets and one inlet, and Eq. (3.30) applies for mass conservation:

$$\dot{m}_{\text{out}} = \dot{m}_{\text{in}}$$

or
$$\rho_1 A_1 V_1 + \rho_2 A_2 V_2 = \rho_j A_j (V_j - V_c) \tag{1}$$

We assume that the water is incompressible $\rho_1 = \rho_2 = \rho_j$, and we are given that $A_1 = A_2 = \frac{1}{2}A_j$. Therefore Eq. (1) reduces to

$$V_1 + V_2 = 2(V_j - V_c) \tag{2}$$

Strictly speaking, this is all that mass conservation tells us. However, from the symmetry of the jet deflection and the neglect of gravity on the fluid trajectory, we conclude that the two velocities V_1 and V_2 must be equal, and hence Eq. (2) becomes

$$V_1 = V_2 = V_j - V_c \tag{3}$$

This equality can also be predicted by Bernoulli's equation in Sec. 3.5. For the given numerical values, we have

$$V_1 = V_2 = 20 - 15 = 5 \text{ m/s}$$

Now, we can compute R_x and R_y from the two components of momentum conservation. Equation (3.40) applies with the unsteady term zero:

$$\sum F_x = R_x = \dot{m}_1 u_1 + \dot{m}_2 u_2 - \dot{m}_j u_j \tag{4}$$

where from the mass analysis, $\dot{m}_1 = \dot{m}_2 = \frac{1}{2}\dot{m}_j = \frac{1}{2}\rho_j A_j (V_j - V_c)$. Now, check the flow directions at each section: $u_1 = u_2 = 0$, and $u_j = V_j - V_c = 5$ m/s. Thus, Eq. (4) becomes

$$R_x = -\dot{m}_j u_j = -[\rho_j A_j (V_j - V_c)](V_j - V_c) \tag{5}$$

For the given numerical values we have

$$R_x = -(1{,}000 \text{ kg/m}^3)(0.0003 \text{ m}^2)(5 \text{ m/s})^2 = -7.5 \text{ (kg} \cdot \text{m)/s}^2 = -7.5 \text{ N} \qquad \textit{Ans.}$$

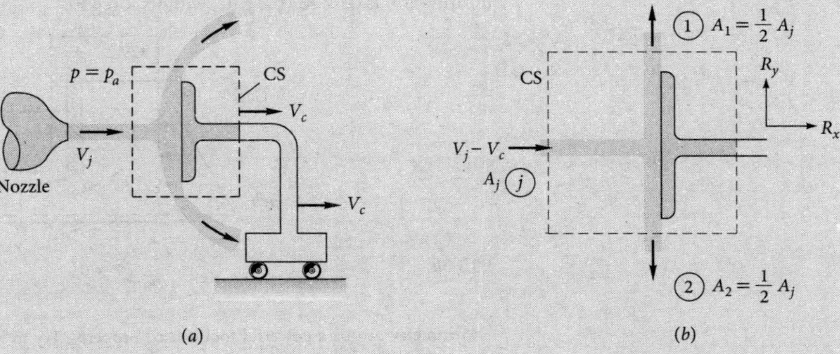

Fig. 3.9 Force on a plate moving at constant velocity: (*a*) jet striking a moving plate normally; (*b*) control volume fixed relative to the plate.

This acts to the *left*; that is, it requires a restraining force to keep the plate from accelerating to the right due to the continuous impact of the jet. The vertical force is

$$F_y = R_y = \dot{m}_1 \upsilon_1 + \dot{m}_2 \upsilon_2 - \dot{m}_j \upsilon_j$$

Check directions again: $\upsilon_1 = V_1$, $\upsilon_2 = -V_2$, $\upsilon_j = 0$. Thus

$$R_y = \dot{m}_1(V_1) + \dot{m}_2(-V_2) = \tfrac{1}{2}\dot{m}_j(V_1 - V_2) \qquad (6)$$

But, since we found earlier that $V_1 = V_2$, this means that $R_y = 0$, as we could expect from the symmetry of the jet deflection.[9] Two other results are of interest. First, the relative velocity at section 1 was found to be 5 m/s up, from Eq. (3). If we convert this to absolute motion by adding on the control-volume speed $V_c = 15$ m/s to the right, we find that the absolute velocity $\mathbf{V}_1 = 15\mathbf{i} + 5\mathbf{j}$ m/s, or 15.8 m/s at an angle of 18.4° upward, as indicated in Fig. 3.9*a*. Thus, the absolute jet speed changes after hitting the plate. Second, the computed force R_x does not change if we assume the jet deflects in all radial directions along the plate surface rather than just up and down. Since the plate is normal to the x axis, there would still be zero outlet x-momentum flow when Eq. (4) was rewritten for a radial deflection condition.

EXAMPLE 3.10

The sluice gate in Fig. E3.10*a* controls flow in open channels. At sections 1 and 2, the flow is uniform and the pressure is hydrostatic. Neglecting bottom friction and atmospheric pressure, derive a formula for the horizontal force F required to hold the gate. Express your final formula in terms of the inlet velocity V_1, eliminating V_2.

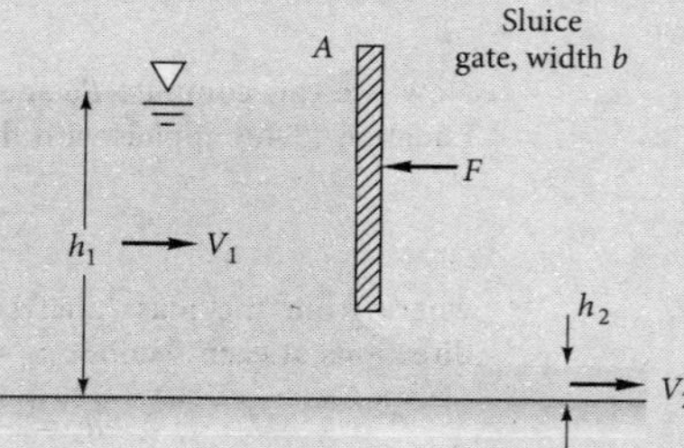

E3.10*a*

Solution

Choose a control volume, Fig. E3.10b, that cuts through known regions (section 1 and section 2 just above the bottom, and the atmosphere) and that cuts along regions where unknown information is desired (the gate, with its force F).

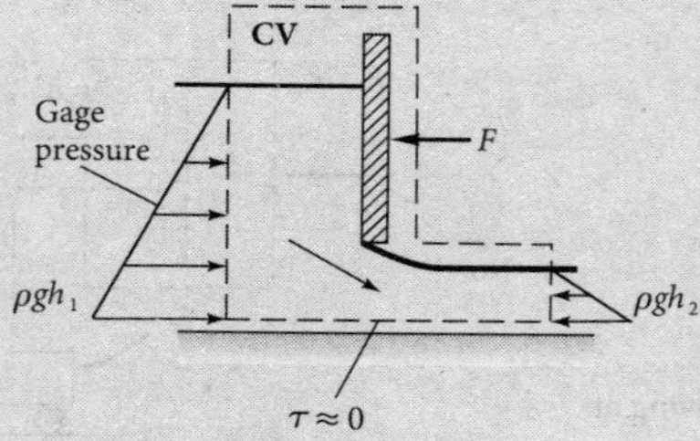

E3.10*b*

[9]Symmetry can be a powerful tool if used properly. Try to learn more about the uses and misuses of symmetry conditions.

Assume steady incompressible flow with no variation across the width b. The inlet and outlet mass flows balance:

$$\dot{m} = \rho V_1 h_1 b = \rho V_2 h_2 b \text{ or } V_2 = V_1(h_1/h_2)$$

We may use gage pressures for convenience because a uniform atmospheric pressure causes no force, as shown earlier in Fig. 3.6. With x positive to the right, equate the net horizontal force to the x-directed momentum change:

$$\sum F_x = -F_{\text{gate}} + \frac{\rho}{2} g h_1(h_1 b) - \frac{\rho}{2} g h_2(h_2 b) = \dot{m}\,(V_2 - V_1)$$
$$\dot{m} = \rho h_1 b V_1$$

Solve for F_{gate}, and eliminate V_2 using the mass flow relation. The desired result is:

$$F_{\text{gate}} = \frac{\rho}{2} g b h_1^2 \left[1 - \left(\frac{h_2}{h_1}\right)^2\right] - \rho h_1 b V_1^2 \left(\frac{h_1}{h_2} - 1\right) \qquad \textit{Ans.}$$

This is a powerful result from a relatively simple analysis. Later, in Sec. 10.4, we will be able to calculate the actual flow rate from the water depths and the gate opening height.

EXAMPLE 3.11

Example 3.9 treated a plate at normal incidence to an oncoming flow. In Fig. 3.10 the plate is parallel to the flow. The stream is not a jet but a broad river, or *free stream*, of uniform velocity $\mathbf{V} = U_0\mathbf{i}$. The pressure is assumed uniform, and so it has no net force on the plate. The plate does not block the flow as in Fig. 3.9, so the only effect is due to boundary shear, which was neglected in the previous example. The no-slip condition at the wall brings the fluid there to a halt, and these slowly moving particles retard their neighbors above, so that at the end of the plate there is a significant retarded shear layer, or *boundary layer*, of thickness $y = \delta$. The viscous stresses along the wall can sum to a finite drag force on the plate. These effects are illustrated in Fig. 3.10. The problem is to make an integral analysis and find the drag force D in terms of the flow properties ρ, U_0, and δ and the plate dimensions L and b.[10]

Solution

Like most practical cases, this problem requires a combined mass and momentum balance. A proper selection of control volume is essential, and we select the four-sided region from

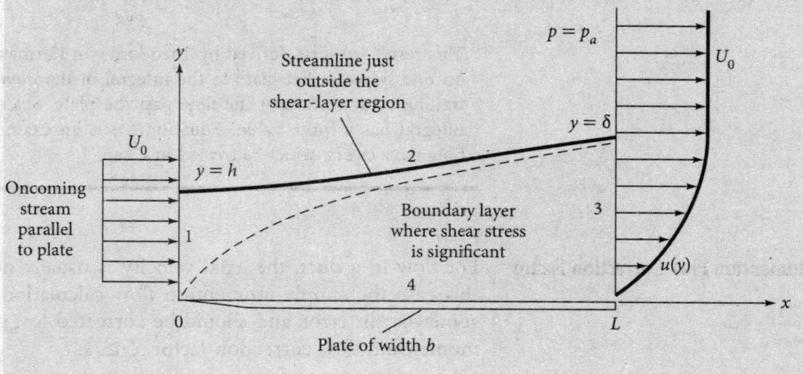

Fig. 3.10 Control volume analysis of drag force on a flat plate due to boundary shear. The control volume is bounded by sections 1, 2, 3, and 4.

[10]The general analysis of such wall shear problems, called ***boundary-layer theory,*** is treated in Sec. 7.3.

0 to h to δ to L and back to the origin 0, as shown in Fig. 3.10. Had we chosen to cut across horizontally from left to right along the height $y = h$, we would have cut through the shear layer and exposed unknown shear stresses. Instead we follow the streamline passing through $(x, y) = (0, h)$, which is outside the shear layer and also has no mass flow across it. The four control volume sides are thus

1. From (0, 0) to (0, h): a one-dimensional inlet, $\mathbf{V} \cdot \mathbf{n} = -U_0$.
2. From (0, h) to (L, δ): a streamline, no shear, $\mathbf{V} \cdot \mathbf{n} = 0$.
3. From (L, δ) to (L, 0): a two-dimensional outlet, $\mathbf{V} \cdot \mathbf{n} = +u(y)$.
4. From (L, 0) to (0, 0): a streamline just above the plate surface, $\mathbf{V} \cdot \mathbf{n} = 0$, shear forces summing to the drag force $-D\mathbf{i}$ acting from the plate onto the retarded fluid.

The pressure is uniform, and so there is no net pressure force. Since the flow is assumed incompressible and steady, Eq. (3.37) applies with no unsteady term and flows only across sections 1 and 3:

$$\sum F_x = -D = \rho \int_1 u(0, y)(\mathbf{V} \cdot \mathbf{n})\, dA \; + \; \rho \int_3 u(L, y)\, (\mathbf{V} \cdot \mathbf{n})\, dA$$

$$= \rho \int_0^h U_0(-U_0) b\, dy + \rho \int_0^\delta u(L, y)\,[+u(L, y)]\, b\, dy$$

Evaluating the first integral and rearranging give

$$D = \rho U_0^2 bh - \rho b \int_0^\delta u^2 dy \;|_{x=L} \tag{1}$$

This could be considered the answer to the problem, but it is not useful because the height h is not known with respect to the shear layer thickness δ. This is found by applying mass conservation, since the control volume forms a streamtube:

$$\rho \int_{CS} (\mathbf{V} \cdot \mathbf{n})\, dA = 0 = \rho \int_0^h (-U_0) b dy + \rho \int_0^\delta ub\, dy \;|_{x=L}$$

or

$$U_0 h = \int_0^\delta u\, dy \;|_{x=L} \tag{2}$$

after canceling b and ρ and evaluating the first integral. Introduce this value of h into Eq. (1) for a much cleaner result:

$$D = \rho b \int_0^\delta u(U_0 - u)\, dy \;|_{x=L} \qquad \textit{Ans. } (3)$$

This result was first derived by Theodore von Kármán in 1921.[11] It relates the friction drag on one side of a flat plate to the integral of the *momentum deficit* $\rho u(U_0 - u)$ across the trailing cross section of the flow past the plate. Since $U_0 - u$ vanishes as y increases, the integral has a finite value. Equation (3) is an example of *momentum integral theory* for boundary layers, which is treated in Chap. 7.

Momentum Flux Correction Factor

For flow in a duct, the axial velocity is usually nonuniform, as in Example 3.4. For this case the simple momentum flow calculation $\int u\rho(\mathbf{V} \cdot \mathbf{n})\, dA = \dot{m}V = \rho A V^2$ is somewhat in error and should be corrected to $\zeta \rho A V^2$, where ζ is the dimensionless momentum flow correction factor, $\zeta \geq 1$.

[11]The autobiography of this great twentieth-century engineer and teacher [2] is recommended for its historical and scientific insight.

The factor ζ accounts for the variation of u^2 across the duct section. That is, we compute the exact flow and set it equal to a flow based on average velocity in the duct:

$$\rho \int u^2 dA = \zeta \dot{m}\, V_{av} = \zeta \rho A V_{av}^2$$

or

$$\zeta = \frac{1}{A}\int \left(\frac{u}{V_{av}}\right)^2 dA \tag{3.43a}$$

Values of ζ can be computed based on typical duct velocity profiles similar to those in Example 3.4. The results are as follows:

Laminar flow: $$u = U_0\left(1 - \frac{r^2}{R^2}\right) \qquad \zeta = \frac{4}{3} \tag{3.43b}$$

Turbulent flow: $$u \approx U_0\left(1 - \frac{r}{R}\right)^m \qquad \frac{1}{9} \le m \le \frac{1}{5}$$

$$\zeta = \frac{(1+m)^2(2+m)^2}{2(1+2m)(2+2m)} \tag{3.43c}$$

The turbulent correction factors have the following range of values:

Turbulent flow:					
m	$\frac{1}{5}$	$\frac{1}{6}$	$\frac{1}{7}$	$\frac{1}{8}$	$\frac{1}{9}$
ζ	1.037	1.027	1.020	1.016	1.013

These are so close to unity that they are normally neglected. The laminar correction is often important.

To illustrate a typical use of these correction factors, the solution to Example 3.8 for nonuniform velocities at sections 1 and 2 would be modified as

$$\sum \mathbf{F} = \dot{m}(\zeta_2 \mathbf{V}_2 - \zeta_1 \mathbf{V}_1) \tag{3.43d}$$

Note that the basic parameters and vector character of the result are not changed at all by this correction.

Linear Momentum Tips

The previous examples make it clear that the vector momentum equation is more difficult to handle than the scalar mass and energy equations. Here are some momentum tips to remember:

- The momentum relation is a *vector* equation. The forces and the momentum terms are directional and can have three components. A *sketch* of these vectors will be indispensable for the analysis.
- The momentum flow terms, such as $\int \mathbf{V}(\rho \mathbf{V} \cdot \mathbf{n})dA$, link *two* different sign conventions, so special care is needed. First, the vector coefficient $\mathbf{V}$ will have a sign depending on its direction. Second, the mass flow term $(\rho \mathbf{V} \cdot \mathbf{n})$ will have a sign (+ , −) depending on whether it is (out, in). For example, in Fig. 3.8, the x-components of $\mathbf{V}_2$ and $\mathbf{V}_1$, u_2 and u_1, are both positive; that is, they both act to the right. Meanwhile, the mass flow at (2) is positive (out) and at (1) is negative (in).
- The *one-dimensional approximation*, Eq. (3.40), is glorious, because nonuniform velocity distributions require laborious integration, as in Eq. (3.11). Thus the momentum flow correction factors ζ are very useful in avoiding this integration, especially for pipe flow.

- The applied forces $\sum \mathbf{F}$ act on *all the material in the control volume*—that is, the surfaces (pressure and shear stresses), the solid supports that are cut through, and the weight of the interior masses. Stresses on non-control-surface parts of the interior are self-canceling and should be ignored.
- If the fluid exits subsonically to an atmosphere, the fluid pressure there is *atmospheric.*
- Where possible, choose inlet and outlet surfaces *normal to the flow,* so that pressure is the dominant force and the normal velocity equals the actual velocity.

Clearly, with that many helpful tips, substantial practice is needed to achieve momentum skills.

Noninertial Reference Frame[12]

All previous derivations and examples in this section have assumed that the coordinate system is inertial—that is, at rest or moving at constant velocity. In this case, the rate of change of velocity equals the absolute acceleration of the system, and Newton's law applies directly in the form of Eqs. (3.2) and (3.35).

In many cases, it is convenient to use a *noninertial,* or accelerating, coordinate system. An example would be coordinates fixed to a rocket during takeoff. A second example is any flow on the earth's surface, which is accelerating relative to the fixed stars because of the rotation of the earth. Atmospheric and oceanographic flows experience the so-called *Coriolis acceleration,* outlined next. It is typically less than $10^{-5}g$, where g is the acceleration of gravity, but its accumulated effect over distances of many kilometers can be dominant in geophysical flows. By contrast, the Coriolis acceleration is negligible in small-scale problems like pipe or airfoil flows.

Suppose that the fluid flow has velocity $\mathbf{V}$ relative to a noninertial xyz coordinate system, as shown in Fig. 3.11. Then $d\mathbf{V}/dt$ will represent a noninertial acceleration that must be added vectorially to a relative acceleration $\mathbf{a}_{\text{rel}}$ to give the absolute acceleration $\mathbf{a}_i$ relative to some inertial coordinate system XYZ, as in Fig. 3.11. Thus

$$\mathbf{a}_i = \frac{d\mathbf{V}}{dt} + \mathbf{a}_{\text{rel}} \tag{3.44}$$

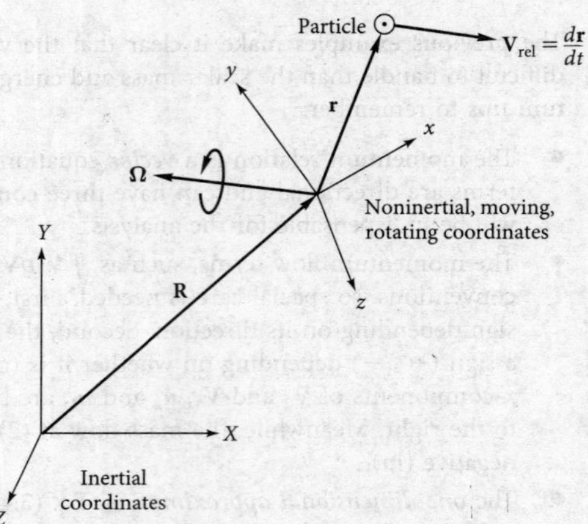

Fig. 3.11 Geometry of fixed versus accelerating coordinates.

[12]This section may be omitted without loss of continuity.

Since Newton's law applies to the absolute acceleration,

$$\sum \mathbf{F} = m\mathbf{a}_i = m\left(\frac{d\mathbf{V}}{dt} + \mathbf{a}_{\text{rel}}\right)$$

or

$$\sum \mathbf{F} - m\mathbf{a}_{\text{rel}} = m\frac{d\mathbf{V}}{dt} \tag{3.45}$$

Thus, Newton's law in noninertial coordinates xyz is analogous to adding more "force" terms $-m\mathbf{a}_{\text{rel}}$ to account for noninertial effects. In the most general case, sketched in Fig. 3.11, the term $\mathbf{a}_{\text{rel}}$ contains four parts, three of which account for the angular velocity $\Omega(t)$ of the inertial coordinates. By inspection of Fig. 3.11, the absolute displacement of a particle is

$$\mathbf{S}_i = \mathbf{r} + \mathbf{R} \tag{3.46}$$

Differentiation gives the absolute velocity

$$\mathbf{V}_i = \mathbf{V} + \frac{d\mathbf{R}}{dt} + \boldsymbol{\Omega} \times \mathbf{r} \tag{3.47}$$

A second differentiation gives the absolute acceleration:

$$\mathbf{a}_i = \frac{d\mathbf{V}}{dt} + \frac{d^2\mathbf{R}}{dt^2} + \frac{d\boldsymbol{\Omega}}{dt} \times \mathbf{r} + 2\boldsymbol{\Omega} \times \mathbf{V} + \boldsymbol{\Omega} \times (\boldsymbol{\Omega} \times \mathbf{r}) \tag{3.48}$$

By comparison with Eq. (3.44), we see that the last four terms on the right represent the additional relative acceleration:

1. $d^2\mathbf{R}/dt^2$ is the acceleration of the noninertial origin of coordinates xyz.
2. $(d\boldsymbol{\Omega}/dt) \times \mathbf{r}$ is the angular acceleration effect.
3. $2\boldsymbol{\Omega} \times \mathbf{V}$ is the Coriolis acceleration.
4. $\boldsymbol{\Omega} \times (\boldsymbol{\Omega} \times \mathbf{r})$ is the centripetal acceleration, directed from the particle normal to the axis of rotation with magnitude $\Omega^2 L$, where L is the normal distance to the axis.[13]

Equation (3.45) differs from Eq. (3.2) only in the added inertial forces on the left-hand side. Thus, the control volume formulation of linear momentum in noninertial coordinates merely adds inertial terms by integrating the added relative acceleration over each differential mass in the control volume:

$$\sum \mathbf{F} - \int_{\text{CV}} \mathbf{a}_{\text{rel}}\, dm = \frac{d}{dt}\left(\int_{\text{CV}} \mathbf{V}\rho\, d\mathcal{V}\right) + \int_{\text{CS}} \mathbf{V}\rho(\mathbf{V}_r \cdot \mathbf{n})\, dA \tag{3.49}$$

where

$$\mathbf{a}_{\text{rel}} = \frac{d^2\mathbf{R}}{dt^2} + \frac{d\boldsymbol{\Omega}}{dt} \times \mathbf{r} + 2\boldsymbol{\Omega} \times \mathbf{V} + \boldsymbol{\Omega} \times (\boldsymbol{\Omega} \times \mathbf{r})$$

This is the noninertial analog of the inertial form given in Eq. (3.35). To analyze such problems, one must know the displacement $\mathbf{R}$ and angular velocity $\boldsymbol{\Omega}$ of the noninertial coordinates.

[13]A complete discussion of these noninertial coordinate terms is given, for example, in Ref. 4, pp. 49–51.

If the control volume is fixed in a moving frame, Eq. (3.49) reduces to

$$\sum \mathbf{F} - \int_{CV} \mathbf{a}_{rel}\, dm = \frac{d}{dt}\left(\int_{CV} \mathbf{V}\rho\, d\mathcal{V}\right) + \int_{CS} \mathbf{V}\rho(\mathbf{V}\cdot\mathbf{n})\, dA \tag{3.50}$$

In other words, the right-hand side reduces to that of Eq. (3.37).

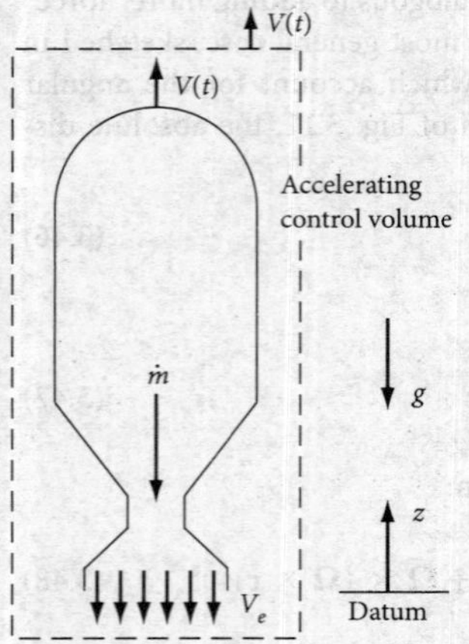

E3.12

EXAMPLE 3.12

A classic example of an accelerating control volume is a rocket moving straight up, as in Fig. E3.12. Let the initial mass be M_0, and assume a steady exhaust mass flow $\dot{m}$ and exhaust velocity V_e relative to the rocket, as shown. If the flow pattern within the rocket motor is steady and air drag is neglected, derive the differential equation of vertical rocket motion $V(t)$ and integrate using the initial condition $V = 0$ at $t = 0$.

Solution

The appropriate control volume in Fig. E3.12 encloses the rocket, cuts through the exit jet, and accelerates upward at rocket speed $V(t)$. The z-momentum Eq. (3.49) becomes

$$\sum F_z - \int a_{rel}\, dm = \frac{d}{dt}\left(\int_{CV} w\, d\dot{m}\right) + (\dot{m}w)_e$$

or

$$-mg - m\frac{dV}{dt} = 0 + \dot{m}(-V_e) \quad \text{with } m = m(t) = M_0 - \dot{m}t$$

The term $a_{rel} = dV/dt$ of the rocket. The control volume integral vanishes because of the steady rocket flow conditions. Separate the variables and integrate, assuming $V = 0$ at $t = 0$:

$$\int_0^V dV = \dot{m}\, V_e \int_0^t \frac{dt}{M_0 - \dot{m}t} - g\int_0^t dt \text{ or } V(t) = -V_e \ln\left(1 - \frac{\dot{m}t}{M_0}\right) - gt \qquad \textit{Ans.}$$

This is a classic approximate formula in rocket dynamics. The first term is positive and, if the fuel mass burned is a large fraction of initial mass, the final rocket velocity can exceed V_e.

3.5 Frictionless Flow: The Bernoulli Equation

A classic linear momentum analysis is a relation between pressure, velocity, and elevation in a frictionless flow, now called the *Bernoulli equation*. It was stated (vaguely) in words in 1738 in a textbook by Daniel Bernoulli. A complete derivation of the equation was given in 1755 by Leonhard Euler. The Bernoulli equation is very famous and very widely used, but one should be wary of its restrictions—all fluids are viscous and thus all flows have friction to some extent. To use the Bernoulli equation correctly, one must confine it to regions of the flow that are nearly frictionless. This section (and, in more detail, Chap. 8) will address the proper use of the Bernoulli relation.

Consider Fig. 3.12, which is an elemental fixed streamtube control volume of variable area $A(s)$ and length ds, where s is the streamline direction. The properties (ρ, V, p) may vary with s and time but are assumed to be uniform over the cross section A. The streamtube orientation θ is arbitrary, with an elevation change $dz = ds \sin\theta$. Friction on the streamtube walls is shown and then neglected—a very restrictive assumption. Note that the limit of a vanishingly small area means that the streamtube is equivalent to a *streamline* of the flow. Bernoulli's equation is valid for both and is usually stated as holding "along a streamline" in frictionless flow.

Conservation of mass [Eq. (3.20)] for this elemental control volume yields

$$\frac{d}{dt}\left(\int_{CV} \rho\, d\mathcal{V}\right) + \dot{m}_{out} - \dot{m}_{in} = 0 \approx \frac{\partial \rho}{\partial t}\, d\mathcal{V} + d\dot{m}$$

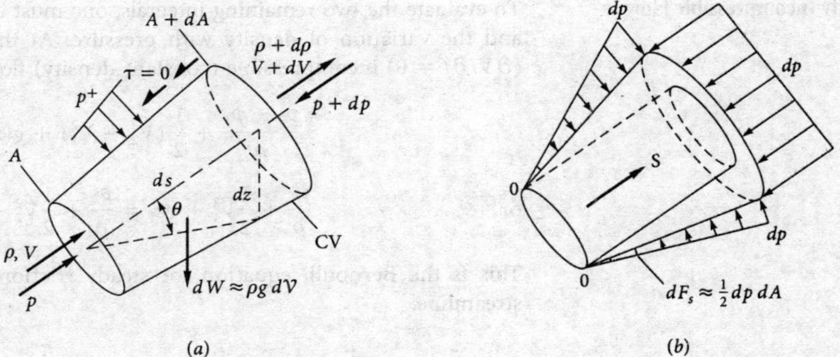

Fig. 3.12 The Bernoulli equation for frictionless flow along a streamline: (*a*) forces and flows; (*b*) net pressure force after uniform subtraction of *p*.

where $\dot{m} = \rho A V$ and $d\mathcal{V} \approx A\, ds$. Then our desired form of mass conservation is

$$d\dot{m} = d(\rho A V) = -\frac{\partial \rho}{\partial t} A\, ds \tag{3.51}$$

This relation does not require an assumption of frictionless flow.

Now, write the linear momentum relation [Eq. (3.37)] in the streamwise direction:

$$\sum dF_s = \frac{d}{dt}\left(\int_{CV} V\rho\, d\mathcal{V}\right) + (\dot{m}V)_{out} - (\dot{m}V)_{in} \approx \frac{\partial}{\partial t}(\rho V)\, A\, ds + d(\dot{m}V)$$

where $V_s = V$ itself because s is the streamline direction. If we neglect the shear force on the walls (frictionless flow), the forces are due to pressure and gravity. The streamwise gravity force is due to the weight component of the fluid within the control volume:

$$dF_{s,\text{grav}} = -dW \sin\theta = -\gamma A\, ds \sin\theta = -\gamma A\, dz$$

The pressure force is more easily visualized, in Fig. 3.12*b*, by first subtracting a uniform value p from all surfaces, remembering from Fig. 3.6 that the net force is not changed. The pressure along the slanted side of the streamtube has a streamwise component that acts not on A itself but on the outer ring of area increase dA. The net pressure force is thus

$$dF_{s,\text{press}} = \tfrac{1}{2}\, dp\, dA - dp(A + dA) \approx -A\, dp$$

to first order. Substitute these two force terms into the linear momentum relation:

$$\sum dF_s = -\gamma A\, dz - A\, dp = \frac{\partial}{\partial t}(\rho V)\, A\, ds + d(\dot{m}V)$$

$$= \frac{\partial \rho}{\partial t} VA\, ds + \frac{\partial V}{\partial t}\rho A\, ds + \dot{m}\, dV + V\, d\dot{m}$$

The first and last terms on the right cancel by virtue of the continuity relation [Eq. (3.51)]. Divide what remains by ρA and rearrange into the final desired relation:

$$\frac{\partial V}{\partial t} ds + \frac{dp}{\rho} + V\, dV + g\, dz = 0 \tag{3.52}$$

This is Bernoulli's equation for *unsteady frictionless flow along a streamline.* It is in differential form and can be integrated between any two points 1 and 2 on the streamline:

$$\int_1^2 \frac{\partial V}{\partial t} ds + \int_1^2 \frac{dp}{\rho} + \frac{1}{2}(V_2^2 - V_1^2) + g(z_2 - z_1) = 0 \tag{3.53}$$

Steady Incompressible Flow

To evaluate the two remaining integrals, one must estimate the unsteady effect $\partial V/\partial t$ and the variation of density with pressure. At this time, we consider only steady ($\partial V/\partial t = 0$) incompressible (constant-density) flow, for which Eq. (3.53) becomes

$$\frac{p_2 - p_1}{\rho} + \frac{1}{2}(V_2^2 - V_1^2) + g(z_2 - z_1) = 0$$

or

$$\frac{p_1}{\rho} + \frac{1}{2}V_1^2 + gz_1 = \frac{p_2}{\rho} + \frac{1}{2}V_2^2 + gz_2 = \text{const} \qquad (3.54)$$

This is the Bernoulli equation for steady frictionless incompressible flow along a streamline.

Bernoulli Interpreted as an Energy Relation

The Bernoulli relation, Eq. (3.54), is a classic *momentum* result, Newton's law for a frictionless, incompressible fluid. It may also be interpreted, however, as an idealized *energy* relation. The changes from 1 to 2 in Eq. (3.54) represent reversible pressure work, kinetic energy change, and potential energy change. The fact that the total remains the same means that there is no energy exchange due to viscous dissipation, heat transfer, or shaft work. Section 3.7 will add these effects by making a control volume analysis of the first law of thermodynamics.

Restrictions on the Bernoulli Equation

The Bernoulli equation is a momentum-based force relation and was derived using the following restrictive assumptions:

1. *Steady flow:* a common situation, application to most flows in this text.
2. *Incompressible flow:* appropriate if the flow Mach number is less than 0.3. This restriction is removed in Chap. 9 by allowing for compressibility.
3. *Frictionless flow:* restrictive—solid walls and mixing introduce friction effects.
4. *Flow along a single streamline:* different streamlines may have different "Bernoulli constants" $w_o = p/\rho + V^2/2 + gz$, but this is rare. In most cases, as we shall prove in Chap. 4, a frictionless flow region is *irrotational*; that is, curl(**V**) = 0. For irrotational flow, the Bernoulli constant is the same everywhere.

The Bernoulli derivation does not account for possible energy exchange due to heat or work. These thermodynamic effects are accounted for in the steady flow energy equation. We are thus warned that the Bernoulli equation may be modified by such an energy exchange.

Figure 3.13 illustrates some practical limitations on the use of Bernoulli's equation (3.54). For the wind tunnel model test of Fig. 3.13*a*, the Bernoulli equation is valid in the core flow of the tunnel but not in the tunnel wall boundary layers, the model surface boundary layers, or the wake of the model, all of which are regions with high friction.

In the propeller flow of Fig. 3.13*b*, Bernoulli's equation is valid both upstream and downstream, but with a different constant $w_0 = p/\rho + V^2/2 + gz$, caused by the work addition of the propeller. The Bernoulli relation (3.54) is not valid near the propeller blades or in the helical vortices (not shown, see Fig. 1.14) shed downstream of the blade edges. Also, the Bernoulli constants are higher in the flowing "slipstream" than in the ambient atmosphere because of the slipstream kinetic energy.

For the chimney flow of Fig. 3.13*c*, Eq. (3.54) is valid before and after the fire, but with a change in Bernoulli constant that is caused by heat addition. The Bernoulli equation is not valid within the fire itself or in the chimney wall boundary layers.

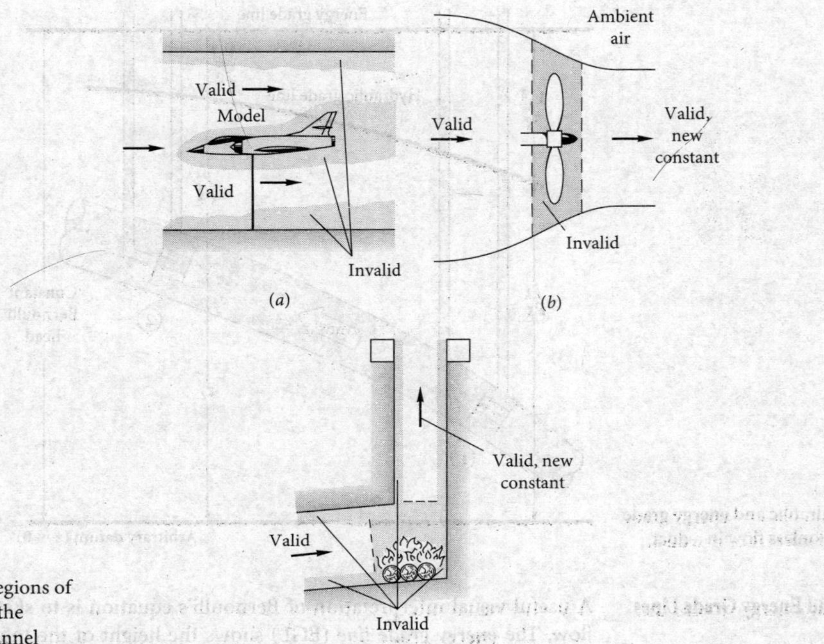

Fig. 3.13 Illustration of regions of validity and invalidity of the Bernoulli equation: (*a*) tunnel model, (*b*) propeller, (*c*) chimney.

Jet Exit Pressure Equals Atmospheric Pressure

When a subsonic jet of liquid or gas exits from a duct into the free atmosphere, it immediately takes on the pressure of that atmosphere. This is a very important boundary condition in solving Bernoulli problems, since the pressure at that point is known. The interior of the free jet will also be atmospheric, except for small effects due to surface tension and streamline curvature.

Stagnation, Static, and Dynamic Pressures

In many incompressible-flow Bernoulli analyses, elevation changes are negligible. Thus, Eq. (3.54) reduces to a balance between pressure and kinetic energy. We can write this as

$$p_1 + \frac{1}{2}\rho V_1^2 = p_2 + \frac{1}{2}\rho V_2^2 = p_o = \text{constant}$$

The quantity p_o is the pressure at any point in the frictionless flow where the velocity is zero. It is called the *stagnation pressure* and is the highest pressure possible in the flow, if elevation changes are neglected. The place where zero-velocity occurs is called a *stagnation point*. For example, on a moving aircraft, the front nose and the wing leading edges are points of highest pressure. The pressures p_1 and p_2 are called *static* pressures, in the moving fluid. The grouping $(1/2)\rho V^2$ has dimensions of pressure and is called the *dynamic* pressure. A popular device called a *Pitot-static tube* (Fig. 6.30) measures $(p_o - p)$ and then calculates V from the dynamic pressure.

Note, however, that one particular zero-velocity condition, no-slip flow along a fixed wall, does *not* result in stagnation pressure. The no-slip condition is a *frictional* effect, and the Bernoulli equation does not apply.

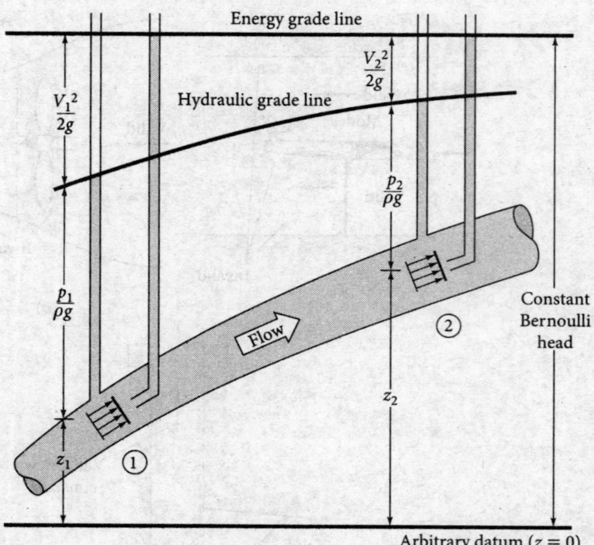

Fig. 3.14 Hydraulic and energy grade lines for frictionless flow in a duct.

Hydraulic and Energy Grade Lines

A useful visual interpretation of Bernoulli's equation is to sketch two grade lines of a flow. The *energy grade line* (EGL) shows the height of the total Bernoulli constant $h_0 = z + p/\gamma + V^2/(2g)$. In frictionless flow with no work or heat transfer [Eq. (3.54)] the EGL has constant height. The *hydraulic grade line* (HGL) shows the height corresponding to elevation and pressure head $z + p/\gamma$—that is, the EGL minus the velocity head $V^2/(2g)$. The HGL is the height to which liquid would rise in a piezometer tube (see Prob. 2.11) attached to the flow. In an open-channel flow, the HGL is identical to the free surface of the water.

Figure 3.14 illustrates the EGL and HGL for frictionless flow at sections 1 and 2 of a duct. The piezometer tubes measure the static pressure head $z + p/\gamma$ and thus outline the HGL. The pitot stagnation-velocity tubes measure the total head $z + p/\gamma + V^2/(2g)$, which corresponds to the EGL. In this particular case the EGL is constant, and the HGL rises due to a drop in velocity.

In more general flow conditions, the EGL will drop slowly due to friction losses and will drop sharply due to a substantial loss (a valve or obstruction) or due to work extraction (to a turbine). The EGL can rise only if there is work addition (as from a pump or propeller). The HGL generally follows the behavior of the EGL with respect to losses or work transfer, and it rises and/or falls if the velocity decreases and/or increases.

As mentioned before, no conversion factors are needed in computations with the Bernoulli equation if consistent SI or BG units are used, as the following examples will show.

In all Bernoulli-type problems in this text, we consistently take point 1 upstream and point 2 downstream.

EXAMPLE 3.13

Find a relation between nozzle discharge velocity V_2 and tank free surface height h as in Fig. E3.13. Assume steady frictionless flow.

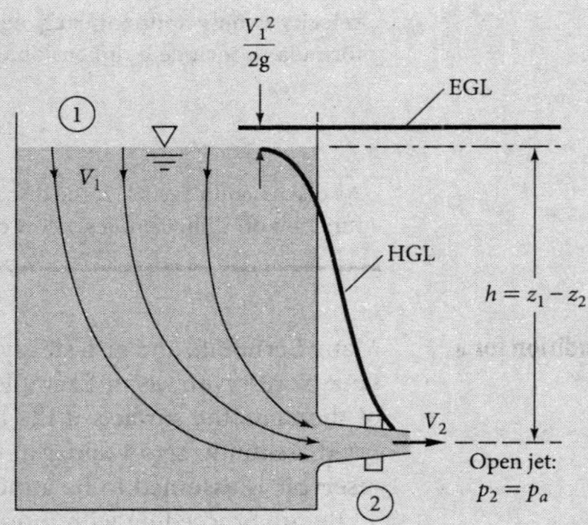

E3.13

Solution

As mentioned, we always choose point 1 upstream and point 2 downstream. Try to choose points 1 and 2 where maximum information is known or desired. Here, we select point 1 as the tank free surface, where elevation and pressure are known, and point 2 as the nozzle exit, where again pressure and elevation are known. The two unknowns are V_1 and V_2.

Mass conservation is usually a vital part of Bernoulli analyses. If A_1 is the tank cross section and A_2 the nozzle area, this is approximately a one-dimensional flow with constant density, Eq. (3.30):

$$A_1V_1 = A_2V_2 \tag{1}$$

Bernoulli's equation (3.54) gives

$$\frac{p_1}{\rho} + \tfrac{1}{2}V_1^2 + gz_1 = \frac{p_2}{\rho} + \tfrac{1}{2}V_2^2 + gz_2$$

But, since sections 1 and 2 are both exposed to atmospheric pressure $p_1 = p_2 = p_a$, the pressure terms cancel, leaving

$$V_2^2 - V_1^2 = 2g(z_1 - z_2) = 2gh \tag{2}$$

Eliminating V_1 between Eqs. (1) and (2), we obtain the desired result:

$$V_2^2 = \frac{2gh}{1 - A_2^2/A_1^2} \qquad \textit{Ans. } (3)$$

Generally, the nozzle area A_2 is very much smaller than the tank area A_1, so that the ratio A_2^2/A_1^2 is doubly negligible, and an accurate approximation for the outlet velocity is

$$V_2 \approx (2gh)^{1/2} \qquad \textit{Ans. } (4)$$

This formula, discovered by Evangelista Torricelli in 1644, states that the discharge velocity equals the speed that a frictionless particle would attain if it fell freely from point 1 to point 2. In other words, the potential energy of the surface fluid is entirely converted to kinetic energy of efflux, which is consistent with the neglect of friction and the fact that no net pressure work is done. Note that Eq. (4) is independent of the fluid density, a characteristic of gravity-driven flows.

Except for the wall boundary layers, the streamlines from 1 to 2 all behave in the same way, and we can assume that the Bernoulli constant h_0 is the same for all the core flow. However, the outlet flow is likely to be nonuniform, not one-dimensional, so that the average velocity is only approximately equal to Torricelli's result. The engineer will then adjust the formula to include a dimensionless *discharge coefficient* c_d:

$$(V_2)_{av} = \frac{Q}{A_2} = c_d(2gh)^{1/2} \tag{5}$$

As discussed in Sec. 6.12, the discharge coefficient of a nozzle varies from about 0.6 to 1.0 as a function of (dimensionless) flow conditions and nozzle shape.

Surface Velocity Condition for a Large Tank

Many Bernoulli, and also steady flow energy, problems involve liquid flow from a large tank or reservoir, as in Example 3.13. If the outflow is small compared to the volume of the tank, the surface of the tank hardly moves. Therefore these problems are analyzed assuming *zero velocity* at the tank surface. The *pressure* at the top of the tank or reservoir is assumed to be atmospheric.

Before proceeding with more examples, we should note carefully that a solution by Bernoulli's equation (3.54) does *not* require a second control volume analysis, only a selection of two points 1 and 2 along a given streamline. The control volume was used to derive the differential relation (3.52), but the integrated form (3.54) is valid all along the streamline for frictionless flow with no heat transfer or shaft work, and a control volume is not necessary.

A classic Bernoulli application is the familiar process of siphoning a fluid from one container to another. No pump is involved; a hydrostatic pressure difference provides the motive force. We analyze this in the following example.

EXAMPLE 3.14

Consider the water siphon shown in Fig. E3.14. Assuming that Bernoulli's equation is valid, (*a*) find an expression for the velocity V_2 exiting the siphon tube. (*b*) If the tube is 1 cm in diameter and $z_1 = 60$ cm, $z_2 = -25$ cm, $z_3 = 90$ cm, and $z_4 = 35$ cm, estimate the flow rate in cm^3/s.

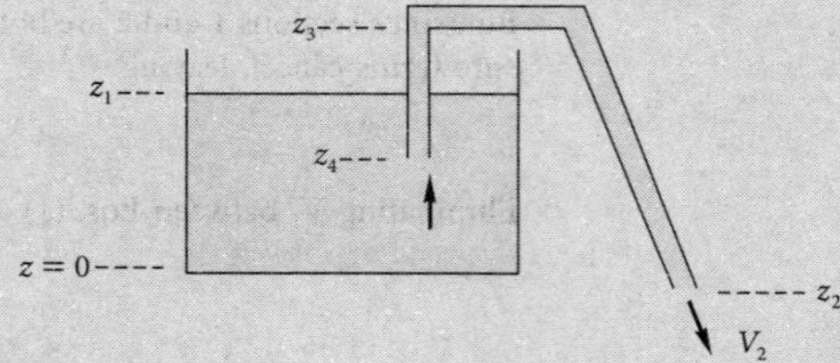

E3.14

Solution

- *Assumptions:* Frictionless, steady, incompressible flow. Write Bernoulli's equation starting from where information is known (the surface, z_1) and proceeding to where information is desired (the tube exit, z_2).

$$\frac{p_1}{\rho} + \cancel{\frac{V_1^2}{2}} + gz_1 = \frac{p_2}{\rho} + \frac{V_2^2}{2} + gz_2$$

Note that the velocity is approximately zero at z_1, and a streamline goes from z_1 to z_2. Note further that p_1 and p_2 are both atmospheric, $p = p_{atm}$, and therefore cancel. (*a*) Solve for the exit velocity from the tube:

$$V_2 = \sqrt{2g(z_1 - z_2)} \qquad \textit{Ans. (a)}$$

The velocity exiting the siphon increases as the tube exit is lowered below the tank surface. There is no siphon effect if the exit is at or above the tank surface. Note that z_3 and z_4 do not directly enter the analysis. However, z_3 should not be too high because the pressure there will be lower than atmospheric, and the liquid might vaporize. (*b*) For the given numerical information, we need only z_1 and z_2 and calculate, in SI units,

$$V_2 = \sqrt{2(9.81\ \text{m/s}^2)[0.6\ \text{m} - (-0.25)\ \text{m}]} = 4.08\ \text{m/s}$$

$$Q = V_2 A_2 = (4.08\ \text{m/s})(\pi/4)(0.01\ \text{m})^2 = 321\ \text{E–6}\ \text{m}^3/\text{s} = 321\ \text{cm}^3/\text{s} \quad \textit{Ans. (b)}$$

- *Comments:* Note that this result is independent of the density of the fluid. As an exercise, you may check that, for water (998 kg/m^3), p_3 is 11,300 Pa *below* atmospheric pressure.

  In Chap. 6 we will modify this example to include friction effects.

EXAMPLE 3.15

A constriction in a pipe will cause the velocity to rise and the pressure to fall at section 2 in the throat. The pressure difference is a measure of the flow rate through the pipe. The smoothly necked-down system shown in Fig. E3.15 is called a *venturi tube*. Find an expression for the mass flow in the tube as a function of the pressure change.

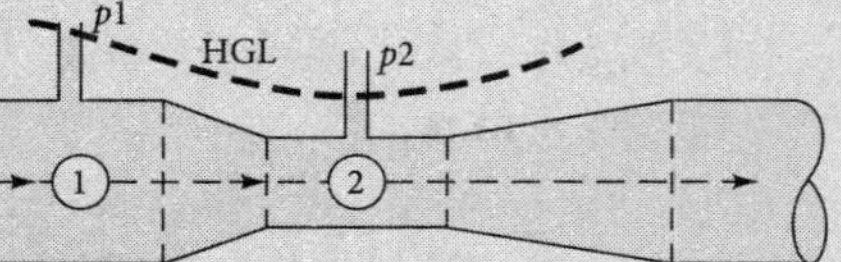

E3.15

Solution

Bernoulli's equation is assumed to hold along the center streamline:

$$\frac{p_1}{\rho} + \tfrac{1}{2}V_1^2 + gz_1 = \frac{p_2}{\rho} + \tfrac{1}{2}V_2^2 + gz_2$$

If the tube is horizontal, $z_1 = z_2$ and we can solve for V_2:

$$V_2^2 - V_1^2 = \frac{2\,\Delta p}{\rho} \qquad \Delta p = p_1 - p_2 \tag{1}$$

We relate the velocities from the incompressible continuity relation:

$$A_1 V_1 = A_2 V_2$$

or

$$V_1 = \beta^2 V_2 \qquad \beta = \frac{D_2}{D_1} \tag{2}$$

Combining (1) and (2), we obtain a formula for the velocity in the throat:

$$V_2 = \left[\frac{2\,\Delta p}{\rho(1-\beta^4)}\right]^{1/2} \tag{3}$$

The mass flow is given by

$$\dot{m} = \rho A_2 V_2 = A_2\left(\frac{2\rho\,\Delta p}{1-\beta^4}\right)^{1/2} \tag{4}$$

This is the ideal frictionless mass flow. In practice, we measure $\dot{m}_{\text{actual}} = c_d \dot{m}_{\text{ideal}}$ and correlate the dimensionless discharge coefficient c_d.

EXAMPLE 3.16

A 10-cm fire hose with a 3-cm nozzle discharges 1.5 m^3/min to the atmosphere. Assuming frictionless flow, find the force F_B exerted by the flange bolts to hold the nozzle on the hose.

Solution

We use Bernoulli's equation and continuity to find the pressure p_1 upstream of the nozzle, and then we use a control volume momentum analysis to compute the bolt force, as in Fig. E3.16.

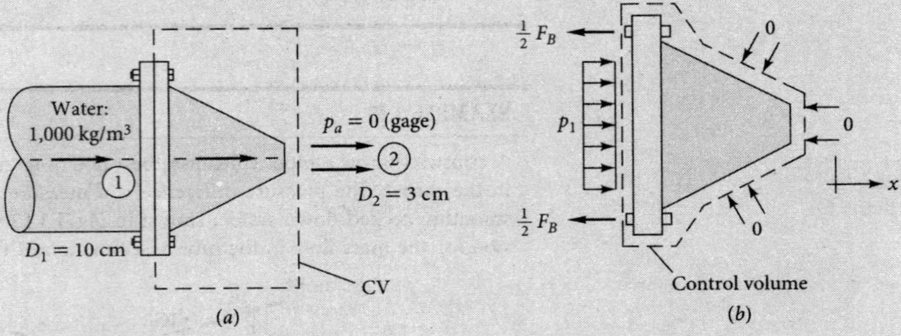

E3.16

The flow from 1 to 2 is a constriction exactly similar in effect to the venturi in Example 3.15, for which Eq. (1) gave

$$p_1 = p_2 + \tfrac{1}{2}\rho(V_2^2 - V_1^2) \tag{1}$$

The velocities are found from the known flow rate Q = 1.5 m^3/min or 0.025 m^3/s:

$$V_2 = \frac{Q}{A_2} = \frac{0.025 \text{ m}^3/\text{s}}{(\pi/4)(0.03 \text{ m})^2} = 35.4 \text{ m/s}$$

$$V_1 = \frac{Q}{A_1} = \frac{0.025 \text{ m}^3/\text{s}}{(\pi/4)(0.1 \text{ m})^2} = 3.2 \text{ m/s}$$

We are given $p_2 = p_a = 0$ gage pressure. Then, Eq. (1) becomes

$$p_1 = \tfrac{1}{2}(1{,}000 \text{ kg/m}^3)[(35.4^2 - 3.2^2)\text{m}^2/\text{s}^2]$$

$$= 620{,}000 \text{ kg/(m} \cdot \text{s}^2) = 620{,}000 \text{ Pa gage}$$

The control volume force balance is shown in Fig. E3.16*b*:

$$\sum F_x = -F_B + p_1 A_1$$

and the zero gage pressure on all other surfaces contributes no force. The x-momentum flow is $+\dot{m}V_2$ at the outlet and $-\dot{m}V_1$ at the inlet. The steady flow momentum relation (3.40) thus gives

$$-F_B + p_1 A_1 = \dot{m}(V_2 - V_1)$$

or

$$F_B = p_1 A_1 - \dot{m}(V_2 - V_1) \tag{2}$$

Substituting the given numerical values, we find

$$\dot{m} = \rho Q = (1{,}000\ \text{kg/m}^3)(0.025\ \text{m}^3/\text{s}) = 25\ \text{kg/s}$$

$$A_1 = \frac{\pi}{4} D_1^2 = \frac{\pi}{4}(0.1\ \text{m})^2 = 0.00785\ \text{m}^2$$

$$F_B = (620{,}000\ \text{N/m}^2)(0.00785\ \text{m}^2) - (25\ \text{kg/s})[(35.4 - 3.2)\text{m/s}]$$
$$= 4{,}872\ \text{N} - 805\ (\text{kg} \cdot \text{m})/\text{s}^2 = 4{,}067\ \text{N} \qquad \textit{Ans.}$$

Notice from these examples that the solution of a typical problem involving Bernoulli's equation almost always leads to a consideration of the continuity equation as an equal partner in the analysis. The only exception is when the complete velocity distribution is already known from a previous or given analysis, but that means the continuity relation has already been used to obtain the given information. The point is that the continuity relation is always an important element in a flow analysis.

3.6 The Angular Momentum Theorem[14]

A control volume analysis can be applied to the angular momentum relation, Eq. (3.3), by letting our dummy variable **B** be the angular-momentum vector **H**. However, since the system considered here is typically a group of nonrigid fluid particles of variable velocity, the concept of mass moment of inertia is of no help, and we have to calculate the instantaneous angular momentum by integration over the elemental masses *dm*. If *O* is the point about which moments are desired, the angular momentum about *O* is given by

$$\mathbf{H}_o = \int_{\text{syst}} (\mathbf{r} \times \mathbf{V})\, dm \tag{3.55}$$

where **r** is the position vector from 0 to the elemental mass *dm* and **V** is the velocity of that element. The amount of angular momentum per unit mass is thus seen to be

$$\beta = \frac{d\mathbf{H}_o}{dm} = \mathbf{r} \times \mathbf{V}$$

The Reynolds transport theorem (3.16) then tells us that

$$\left.\frac{d\mathbf{H}_o}{dt}\right|_{\text{syst}} = \frac{d}{dt}\left[\int_{\text{CV}} (\mathbf{r} \times \mathbf{V})\rho\, d\mathcal{V}\right] + \int_{\text{CS}} (\mathbf{r} \times \mathbf{V})\rho(\mathbf{V}_r \cdot \mathbf{n})\, dA \tag{3.56}$$

for the most general case of a deformable control volume. But, from the angular momentum theorem (3.3), this must equal the sum of all the moments about point *O* applied to the control volume

$$\frac{d\mathbf{H}_o}{dt} = \sum \mathbf{M}_o = \sum (\mathbf{r} \times \mathbf{F})_o$$

Note that the total moment equals the summation of moments of all applied forces about point *O*. Recall, however, that this law, like Newton's law (3.2), assumes that the particle velocity **V** is relative to an *inertial* coordinate system. If not, the moments about point *O* of the relative acceleration terms $\mathbf{a}_{\text{rel}}$ in Eq. (3.49) must also be included:

$$\sum \mathbf{M}_o = \sum (\mathbf{r} \times \mathbf{F})_o - \int_{\text{CV}} (\mathbf{r} \times \mathbf{a}_{\text{rel}})\, dm \tag{3.57}$$

[14]This section may be omitted without loss of continuity.

where the four terms constituting $\mathbf{a}_{rel}$ are given in Eq. (3.49). Thus, the most general case of the angular momentum theorem is for a deformable control volume associated with a noninertial coordinate system. We combine Eqs. (3.56) and (3.57) to obtain

$$\sum (\mathbf{r} \times \mathbf{F})_o - \int_{CV} (\mathbf{r} \times \mathbf{a}_{rel})\, dm = \frac{d}{dt}\left[\int_{CV} (\mathbf{r} \times \mathbf{V})\rho\, d\mathcal{V}\right] + \int_{CS} (\mathbf{r} \times \mathbf{V})\rho(\mathbf{V}_r \cdot \mathbf{n})\, dA \tag{3.58}$$

For a nondeformable inertial control volume, this reduces to

$$\sum \mathbf{M}_0 = \frac{\partial}{\partial t}\left[\int_{CV} (\mathbf{r} \times \mathbf{V})\rho\, d\mathcal{V}\right] + \int_{CS} (\mathbf{r} \times \mathbf{V})\rho(\mathbf{V} \cdot \mathbf{n})\, dA \tag{3.59}$$

Further, if there are only one-dimensional inlets and exits, the angular momentum flow terms evaluated on the control surface become

$$\int_{CS} (\mathbf{r} \times \mathbf{V})\rho(\mathbf{V} \cdot \mathbf{n})dA = \sum (\mathbf{r} \times \mathbf{V})_{out}\dot{m}_{out} - \sum (\mathbf{r} \times \mathbf{V})_{in}\dot{m}_{in} \tag{3.60}$$

Although at this stage the angular momentum theorem can be considered a supplementary topic, it has direct application to many important fluid flow problems involving torques or moments. A particularly important case is the analysis of rotating fluid flow devices, usually called *turbomachines* (Chap. 11).

EXAMPLE 3.17

As shown in Fig. E3.17*a*, a pipe bend is supported at point *A* and connected to a flow system by flexible couplings at sections 1 and 2. The fluid is incompressible, and ambient pressure p_a is zero. (*a*) Find an expression for the torque *T* that must be resisted by the support at *A*, in terms of the flow properties at sections 1 and 2 and the distances h_1 and h_2. (*b*) Compute this torque if $D_1 = D_2 = 8$ cm, $p_1 = 690{,}000$ Pa gage, $p_2 = 550{,}000$ Pa gage, $V_1 = 12$m/s, $h_1 = 5$ cm, $h_2 = 25$ cm, and $\rho = 1{,}000$ kg/m^3.

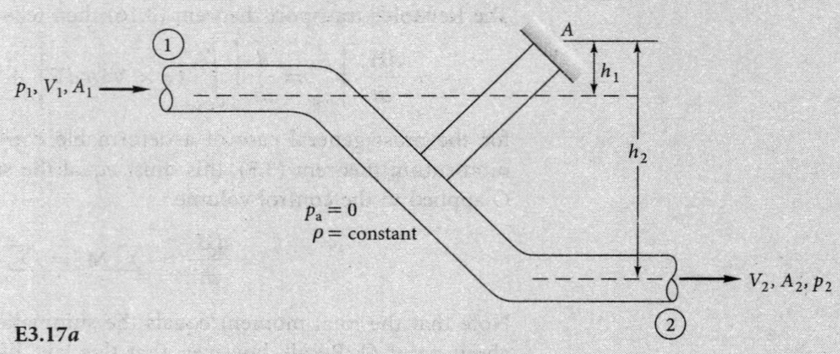

E3.17*a*

Solution

Part (a) The control volume chosen in Fig. E3.17*b* cuts through sections 1 and 2 and through the support at *A*, where the torque T_A is desired. The flexible couplings description specifies that there is no torque at either section 1 or 2, and so the cuts there expose no moments. For the angular momentum terms $\mathbf{r} \times \mathbf{V}$, $\mathbf{r}$ should be taken from point *A* to sections 1 and 2. Note that the gage pressure forces p_1A_1 and p_2A_2 both have moments about A. Equation (3.59) with one-dimensional flow terms becomes

$$\sum \mathbf{M}_A = \mathbf{T}_A + \mathbf{r}_1 \times (-p_1A_1\mathbf{n}_1) + \mathbf{r}_2 \times (-p_2A_2\mathbf{n}_2)$$
$$= (\mathbf{r}_2 \times \mathbf{V}_2)(+\dot{m}_{out}) + (\mathbf{r}_1 \times \mathbf{V}_1)(-\dot{m}_{in}) \quad (1)$$

Figure E3.17*c* shows that all the cross products are associated with either $r_1 \sin\theta_1 = h_1$ or $r_2 \sin\theta_2 = h_2$, the perpendicular distances from point A to the pipe axes at 1 and 2. Remember that $\dot{m}_{in} = \dot{m}_{out}$ from the steady flow continuity relation. In terms of counterclockwise moments, Eq. (1) then becomes

$$T_A + p_1A_1h_1 - p_2A_2h_2 = \dot{m}(h_2V_2 - h_1V_1) \quad (2)$$

Rewriting this, we find the desired torque to be

$$T_A = h_2(p_2A_2 + \dot{m}V_2) - h_1(p_1A_1 + \dot{m}V_1) \qquad \textit{Ans. (a)} \ (3)$$

counterclockwise. The quantities p_1 and p_2 are gage pressures. Note that this result is independent of the shape of the pipe bend and varies only with the properties at sections 1 and 2 and the distances h_1 and h_2.[15]

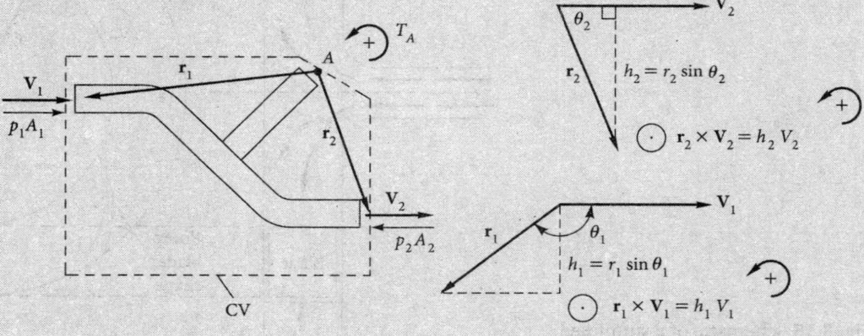

E3.17*b* E3.17*c*

Part (b)

$$D_1 = D_2 = 0.08 \text{ m} \quad p_1 = 690{,}000 \text{ Pa} \quad p_2 = 550{,}000 \text{ Pa}$$
$$h_1 = 0.05 \text{ m} \quad h_2 = 0.25 \text{ m} \quad \rho = 1{,}000 \text{ kg/m}^3$$

The inlet and exit areas are the same, $A_1 = A_2 = (\pi/4)(0.08 \text{ m})^2 = 0.005 \text{ m}^2$. Since the density is constant, we conclude from mass conservation, $\rho A_1V_1 = \rho A_2V_2$, that $V_1 = V_2 = 12$ m/s. The mass flow is

$$\dot{m} = \rho A_1V_1 = (1{,}000 \text{ kg/m}^3)(0.005 \text{ m}^2)(12 \text{ m/s}) = 60 \text{ kg/s}$$

- *Evaluation of the torque:* The data can now be substituted into Eq. (3):

$$T_A = (0.25 \text{ m})[(550{,}000 \text{ Pa})(0.005 \text{ m}^2) + (60 \text{ kg/s})(12 \text{ m/s})]$$
$$- (0.05 \text{ m})[(690{,}000 \text{ Pa})(0.005 \text{ m}^2) + (60 \text{ kg/s})(12 \text{ m/s})]$$
$$= 867.5 \text{ Nm} - 208.5 \text{ Nm} = 659 \text{ Nm counterclockwise} \qquad \textit{Ans. (b)}$$

- *Comments:* The use of standard SI units is crucial when combining dissimilar terms, such as pressure times area and mass flow times velocity, into proper additive units for a numerical solution.

[15]Indirectly, the pipe bend shape probably affects the pressure change from p_1 to p_2.

EXAMPLE 3.18

Figure 3.15 shows a schematic of a centrifugal pump. The fluid enters axially and passes through the pump blades, which rotate at angular velocity ω; the velocity of the fluid is changed from V_1 to V_2 and its pressure from p_1 to p_2. (*a*) Find an expression for the torque T_o that must be applied to these blades to maintain this flow. (*b*) The power supplied to the pump would be $P = \omega T_o$. To illustrate numerically, suppose $r_1 = 0.2$ m, $r_2 = 0.5$ m, and $b = 0.15$ m. Let the pump rotate at 600 r/min and deliver water at 2.5 m^3/s with a density of 1,000 kg/m^3. Compute the torque and power supplied.

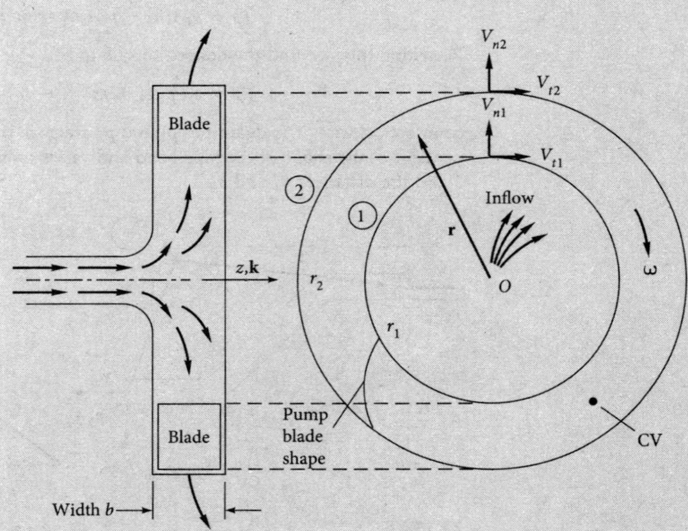

Fig. 3.15 Schematic of a simplified centrifugal pump.

Solution

Part (a)

The control volume is chosen to be the annular region between sections 1 and 2 where the flow passes through the pump blades (see Fig. 3.15). The flow is steady and assumed incompressible. The contribution of pressure to the torque about axis O is zero since the pressure forces at 1 and 2 act radially through O. Equation (3.59) becomes

$$\sum \mathbf{M}_o = \mathbf{T}_o = (\mathbf{r}_2 \times \mathbf{V}_2)\dot{m}_{\text{out}} - (\mathbf{r}_1 \times V_1)\dot{m}_{\text{in}} \quad (1)$$

where steady flow continuity tells us that

$$\dot{m}_{\text{in}} = \rho V_{n1} 2\pi r_1 b = \dot{m}_{\text{out}} = \rho V_{n2} 2\pi r_2 b = \rho Q$$

The cross product $\mathbf{r} \times \mathbf{V}$ is found to be clockwise about O at both sections:

$$\mathbf{r}_2 \times \mathbf{V}_2 = r_2 V_{t2} \sin 90° \, \mathbf{k} = r_2 V_{t2} \mathbf{k} \quad \text{clockwise}$$
$$\mathbf{r}_1 \times \mathbf{V}_1 = r_1 V_{t1} \mathbf{k} \quad \text{clockwise}$$

Equation (1) thus becomes the desired formula for torque:

$$T_o = \rho Q (r_2 V_{t2} - r_1 V_{t1})\mathbf{k} \quad \text{clockwise}$$ *Ans.* (*a*) (2*a*)

This relation is called *Euler's turbine formula*. In an idealized pump, the inlet and outlet tangential velocities would match the blade rotational speeds $V_{t1} = \omega r_1$ and $V_{t2} = \omega r_2$. Then the formula for torque supplied becomes

$$T_o = \rho Q \omega (r_2^2 - r_1^2) \quad \text{clockwise} \tag{2b}$$

Part (b) Convert ω to $600(2\pi/60) = 62.8$ rad/s. The normal velocities are not needed here, but follow from the flow rate

$$V_{n1} = \frac{Q}{2\pi r_1 b} = \frac{2.5 \text{ m}^3/\text{s}}{2\pi(0.2 \text{ m})(0.15 \text{ m})} = 13.3 \text{ m/s}$$

$$V_{n2} = \frac{Q}{2\pi r_2 b} = \frac{2.5}{2\pi(0.5)(0.15)} = 5.3 \text{ m/s}$$

For the idealized inlet and outlet, tangential velocity equals tip speed:

$$V_{t1} = \omega r_1 = (62.8 \text{ rad/s})(0.2 \text{ m}) = 12.6 \text{ m/s}$$
$$V_{t2} = \omega r_2 = 62.8(0.5) = 31.4 \text{ m/s}$$

Equation (2*a*) predicts the required torque to be

$$T_o = (1{,}000 \text{ kg/m}^3)(2.5 \text{ m}^3/\text{s})[(0.5 \text{ m})(31.4 \text{ m/s}) - (0.2 \text{ m})(12.6 \text{ m/s})]$$
$$= 33{,}000 \text{ (kg} \cdot \text{m}^2)/\text{s}^2 = 33{,}000 \text{ N} \cdot \text{m} \qquad \textit{Ans.}$$

The power required is

$$P = \omega T_o = (62.8 \text{ rad/s})(33{,}000 \text{ N} \cdot \text{m}) = 2{,}070{,}000 \text{ (N} \cdot \text{m)/s}$$
$$= 2.07 \text{ MW } (2{,}780 \text{ hp}) \qquad \textit{Ans.}$$

In actual practice, the tangential velocities are considerably less than the impeller-tip speeds, and the design power requirements for this pump may be only 1 MW or less.

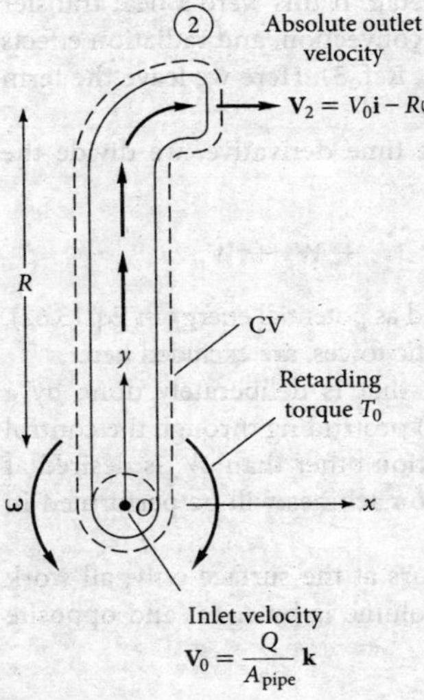

Fig. 3.16 View from above of a single arm of a rotating lawn sprinkler.

EXAMPLE 3.19

Figure 3.16 shows a lawn sprinkler arm viewed from above. The arm rotates about *O* at constant angular velocity ω. The volume flow entering the arm at *O* is *Q*, and the fluid is incompressible. There is a retarding torque at *O*, due to bearing friction, of amount $-T_o\mathbf{k}$. Find an expression for the rotation ω in terms of the arm and flow properties.

Solution

The entering velocity is $V_0\mathbf{k}$, where $V_0 = Q/A_{\text{pipe}}$. Equation (3.59) applies to the control volume sketched in Fig. 3.16 only if **V** is the absolute velocity relative to an inertial frame. Thus, the exit velocity at section 2 is

$$\mathbf{V}_2 = V_0\mathbf{i} - R\omega\mathbf{i}$$

Equation (3.59) then predicts that, for steady flow,

$$\sum \mathbf{M}_o = -T_o\mathbf{k} = (\mathbf{r}_2 \times \mathbf{V}_2)\dot{m}_{\text{out}} - (\mathbf{r}_1 \times \mathbf{V}_1)\dot{m}_{\text{in}} \tag{1}$$

where, from continuity, $\dot{m}_{\text{out}} = \dot{m}_{\text{in}} = \rho Q$. The cross products with reference to point *O* are

$$\mathbf{r}_2 \times \mathbf{V}_2 = R\mathbf{j} \times (V_0 - R\omega)\mathbf{i} = (R^2\omega - RV_0)\mathbf{k}$$
$$\mathbf{r}_1 \times \mathbf{V}_1 = 0\mathbf{j} \times V_0\mathbf{k} = 0$$

Equation (1) thus becomes

$$-T_o\mathbf{k} = \rho Q(R^2\omega - RV_0)\mathbf{k}$$

$$\omega = \frac{V_o}{R} - \frac{T_o}{\rho Q R^2} \qquad \textit{Ans.}$$

The result may surprise you: Even if the retarding torque T_o is negligible, the arm rotational speed is limited to the value V_0/R imposed by the outlet speed and the arm length.

3.7 The Energy Equation[16]

As our fourth and final basic law, we apply the Reynolds transport theorem (3.12) to the first law of thermodynamics, Eq. (3.5). The dummy variable B becomes energy E, and the energy per unit mass is $\beta = dE/dm = e$. Equation (3.5) can then be written for a fixed control volume as follows:[17]

$$\frac{dQ}{dt} - \frac{dW}{dt} = \frac{dE}{dt} = \frac{d}{dt}\left(\int_{CV} e\rho \, d\mathcal{V}\right) + \int_{CS} e\rho(\mathbf{V} \cdot \mathbf{n})\, dA \tag{3.61}$$

Recall that positive Q denotes heat added to the system and positive W denotes work done by the system.

The system energy per unit mass e may be of several types:

$$e = e_{\text{internal}} + e_{\text{kinetic}} + e_{\text{potential}} + e_{\text{other}}$$

where e_{other} could encompass chemical reactions, nuclear reactions, and electrostatic or magnetic field effects. We neglect e_{other} here and consider only the first three terms as discussed in Eq. (1.9), with z defined as "up":

$$e = \hat{u} + \tfrac{1}{2}V^2 + gz \tag{3.62}$$

The heat and work terms could be examined in detail. If this were a heat transfer book, dQ/dt would be broken down into conduction, convection, and radiation effects and whole chapters written on each (see, for example, Ref. 3). Here we leave the term untouched and consider it only occasionally.

Using for convenience the overdot to denote the time derivative, we divide the work term into three parts:

$$\dot{W} = \dot{W}_{\text{shaft}} + \dot{W}_{\text{press}} + \dot{W}_{\text{viscous stresses}} = \dot{W}_s + \dot{W}_p + \dot{W}_v$$

The work of gravitational forces has already been included as potential energy in Eq. (3.62). Other types of work, such as those due to electromagnetic forces, are excluded here.

The shaft work isolates the portion of the work that is deliberately done by a machine (pump impeller, fan blade, piston, or the like) protruding through the control surface into the control volume. No further specification other than $\dot{W}_s$ is desired at this point, but calculations of the work done by turbomachines will be performed in Chap. 11.

The rate of work $\dot{W}_p$ done by pressure forces occurs at the surface only; all work on internal portions of the material in the control volume is by equal and opposite

[16]This section should be read for information and enrichment even if you lack formal background in thermodynamics.

[17]The energy equation for a deformable control volume is rather complicated and is not discussed here. See Refs. 4 and 5 for further details.

forces and is self-canceling. The pressure work equals the pressure force on a small surface element dA times the normal velocity component into the control volume:

$$d\dot{W}_p = -(p\, dA)V_{n,\text{in}} = -p(-\mathbf{V} \cdot \mathbf{n})\, dA$$

The total pressure work is the integral over the control surface:

$$\dot{W}_p = \int_{\text{CS}} p(\mathbf{V} \cdot \mathbf{n})\, dA \tag{3.63}$$

A cautionary remark: If part of the control surface is the surface of a machine part, we prefer to delegate that portion of the pressure to the *shaft work* term $\dot{W}_s$, not to $\dot{W}_p$, which is primarily meant to isolate the fluid flow pressure work terms.

Finally, the shear work due to viscous stresses occurs at the control surface and consists of the product of each viscous stress (one normal and two tangential) and the respective velocity component:

$$d\dot{W}_v = -\boldsymbol{\tau} \cdot \mathbf{V}\, dA$$

or

$$\dot{W}_v = -\int_{\text{CS}} \boldsymbol{\tau} \cdot \mathbf{V}\, dA \tag{3.64}$$

where $\boldsymbol{\tau}$ is the stress vector on the elemental surface dA. This term may vanish or be negligible according to the particular type of surface at that part of the control volume:

Solid surface. For all parts of the control surface that are solid confining walls, $\mathbf{V} = 0$ from the viscous no-slip condition; hence $\dot{W}_v =$ zero identically.

Surface of a machine. Here, the viscous work is contributed by the machine, and so we absorb this work in the term $\dot{W}_s$.

An inlet or outlet. At an inlet or outlet, the flow is approximately normal to the element dA; hence the only viscous work term comes from the normal stress $\tau_{nn}V_n\, dA$. Since viscous normal stresses are extremely small in all but rare cases, such as the interior of a shock wave, it is customary to neglect viscous work at inlets and outlets of the control volume.

Streamline surface. If the control surface is a streamline, such as the upper curve in the boundary layer analysis of Fig. 3.11, the viscous work term must be evaluated and retained if shear stresses are significant along this line. In the particular case of Fig. 3.11, the streamline is outside the boundary layer, and viscous work is negligible.

The net result of this discussion is that the rate-of-work term in Eq. (3.61) consists essentially of

$$\dot{W} = \dot{W}_s + \int_{\text{CS}} p(\mathbf{V} \cdot \mathbf{n})\, dA - \int_{\text{CS}} (\boldsymbol{\tau} \cdot \mathbf{V})_{ss}\, dA \tag{3.65}$$

where the subscript SS stands for stream surface. When we introduce (3.65) and (3.62) into (3.61), we find that the pressure work term can be combined with the energy flow term since both involve surface integrals of $\mathbf{V} \cdot \mathbf{n}$. The control volume energy equation thus becomes

$$\dot{Q} - \dot{W}_s - \dot{W}_v = \frac{\partial}{\partial t}\left(\int_{\text{CV}} e\rho\, d\mathcal{V}\right) + \int_{\text{CS}} \left(e + \frac{p}{\rho}\right)\rho(\mathbf{V} \cdot \mathbf{n})\, dA \tag{3.66}$$

Using e from (3.62), we see that the enthalpy $\hat{h} = \hat{u} + p/\rho$ occurs in the control surface integral. The final general form for the energy equation for a fixed control volume becomes

$$\dot{Q} - \dot{W}_s - \dot{W}_v = \frac{\partial}{\partial t}\left[\int_{CV}\left(\hat{u} + \tfrac{1}{2}V^2 + gz\right)\rho\, d\mathcal{V}\right] + \int_{CS}\left(\hat{h} + \tfrac{1}{2}V^2 + gz\right)\rho(\mathbf{V}\cdot\mathbf{n})\, dA \tag{3.67}$$

As mentioned, the shear work term $\dot{W}_v$ is rarely important.

One-Dimensional Energy-Flux Terms

If the control volume has a series of one-dimensional inlets and outlets, as in Fig. 3.5, the surface integral in (3.67) reduces to a summation of outlet flows minus inlet flows:

$$\int_{CS}(\hat{h} + \tfrac{1}{2}V^2 + gz)\rho(\mathbf{V}\cdot\mathbf{n})\, dA$$

$$= \sum(\hat{h} + \tfrac{1}{2}V^2 + gz)_{out}\dot{m}_{out} - \sum(\hat{h} + \tfrac{1}{2}V^2 + gz)_{in}\dot{m}_{in} \tag{3.68}$$

where the values of $\hat{h}$, $\tfrac{1}{2}V^2$, and gz are taken to be averages over each cross section.

EXAMPLE 3.20

A steady flow machine (Fig. E3.20) takes in air at section 1 and discharges it at sections 2 and 3. The properties at each section are as follows:

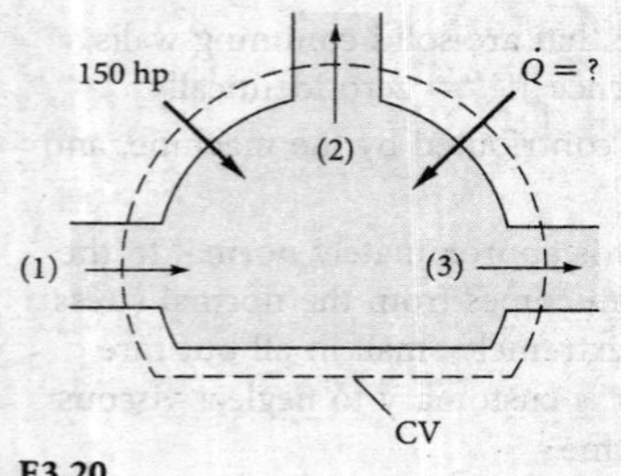

E3.20

Section	A, cm²	Q, m³/s	T, Kelvin	p, Pa abs	z, m
1	370	2.8	294	138,000	0.3
2	900	1.1	311	207,000	1.2
3	230	1.4	366	?	0.5

Work is provided to the machine at the rate of 150 hp. Find the pressure p_3 in pa absolute and the heat transfer $\dot{Q}$ in Joule/s. Assume that air is a perfect gas with $R = 287$ joule/kg · K and $c_p = 1{,}003.7$ J/(kg · K).

Solution

- *System sketch:* Figure E3.20 shows inlet 1 (negative flow) and outlets 2 and 3 (positive flows).
- *Assumptions:* Steady flow, one-dimensional inlets and outlets, ideal gas, negligible shear work. The flow is *not* incompressible. Note that $Q_1 \neq Q_2 + Q_3$ because the densities are different.
- *Approach:* Evaluate the velocities and densities and enthalpies and substitute into Eq. (3.67). With Q_i given, we evaluate $V_i = Q_i/A_i$:

$$V_1 = \frac{Q_1}{A_1} = \frac{2.8\ \text{m}^3/\text{s}}{0.037\ \text{m}^2} = 75.7\ \text{m/s} \quad V_2 = \frac{1.1\ \text{m}^3/\text{s}}{0.09\ \text{m}^2} = 12.2\ \text{m/s} \quad V_3 = \frac{1.4\ \text{m}^3/\text{s}}{0.023\ \text{m}^2} = 60.9\ \text{m/s}$$

The densities at sections 1 and 2 follow from the ideal-gas law:

$$\rho_1 = \frac{p_1}{RT_1} = \frac{138{,}000\ \text{Pa}}{287\ \text{J/(kg}\cdot\text{K)}\ (294\ \text{K})} = 1.635\ \text{kg/m}^3$$

$$\rho_2 = \frac{207{,}000\ \text{Pa}}{287\ \text{J/(kg}\cdot\text{K)}\ (311\ \text{K})} = 2.319\ \text{kg/m}^3$$

However, p_3 is unknown, so how do we find ρ_3? Use the steady flow continuity relation:

$$\dot{m}_1 = \dot{m}_2 + \dot{m}_3 \quad \text{or} \quad \rho_1 Q_1 = \rho_2 Q_2 + \rho_3 Q_3 \tag{1}$$

$$(1.635\ \text{kg/m}^3)(2.8\ \text{m}^3/\text{s}) = 2.319\ \text{kg/m}^3(1.1\ \text{m}^3/\text{s}) + \rho_3(1.4\ \text{m}^3/\text{s}) \quad \text{solve for } \rho_3 = 1.448\ \text{kg/m}^3$$

Knowing ρ_3 enables us to find p_3 from the ideal-gas law:

$$p_3 = \rho_3 R T_3 = (1.448\ \text{kg/m}^3)(287\ \text{J/(kg}\cdot\text{K)})(366\ \text{K}) = 152{,}000\ \text{Pa} \qquad \textit{Ans.}$$

- *Final solution steps:* For an ideal gas, simply approximate enthalpies as $h_i = c_p T_i$. The shaft work is *negative* (into the control volume) and viscous work is neglected for this solid-wall machine:

$$\dot{W}_v \approx 0 \qquad \dot{W}_s = (-150\ \text{hp})(746\ \text{Nm/hp}) = -112{,}000\ \text{Nm/s} \ \ (\text{work } on \text{ the system})$$

For steady flow, the volume integral in Eq. (3.67) vanishes, and the energy equation becomes

$$\dot{Q} - \dot{W}_s = -\dot{m}_1(c_p T_1 + \tfrac{1}{2}V_1^2 + g z_1) + \dot{m}_2(c_p T_2 + \tfrac{1}{2}V_2^2 + g z_2) + \dot{m}_3(c_p T_3 + \tfrac{1}{2}V_3^2 + g z_3) \tag{2}$$

From our continuity calculations in Eq. (1) above, the mass flows are

$$\dot{m}_1 = \rho_1 Q_1 = (1.635\ \text{kg/m}^3)(2.8\ \text{m}^3/\text{s}) = 4.58\ \text{kg/s} \qquad \dot{m}_2 = \rho_2 Q_2 = 2.55\ \text{kg/s}$$

$$\dot{m}_3 = \rho_3 Q_3 = 2.03\ \text{kg/s}$$

It is instructive to separate the flow terms in the energy equation (2) for examination:

$$\begin{aligned}\text{Enthalpy flow} &= c_p(-\dot{m}_1 T_1 + \dot{m}_2 T_2 + \dot{m}_3 T_3)\\ &= (1{,}003.7\ \text{Nm/(kg}\cdot\text{K)})[(-4.58\ \text{kg/s})(294\ \text{K}) + (2.55\ \text{kg/s})(311\ \text{K})\\ &\quad + (2.03\ \text{kg/s})(366\ \text{K})]\\ &= -1{,}352{,}000 + 796{,}000 + 746{,}000 \approx 190{,}000\ \text{Nm/s}\end{aligned}$$

$$\begin{aligned}\text{Kinetic energy flow} &= \tfrac{1}{2}(-\dot{m}_1 V_1^2 + \dot{m}_2 V_2^2 + \dot{m}_3 V_3^2)\\ &= \tfrac{1}{2}[(-4.58\ \text{kg/s})(75.7\ \text{m/s})^2 + (2.55\ \text{kg/s})(12.2\ \text{m/s})^2\\ &\quad + (2.03\ \text{kg/s})(60.9\ \text{m/s})^2]\\ &= -13{,}100\ \text{Nm/s} + 190\ \text{Nm/s} + 3{,}760\ \text{Nm/s} \approx -9{,}150\ \text{Nm/s}\end{aligned}$$

$$\begin{aligned}\text{Potential energy flow} &= g(-\dot{m}_1 z_1 + \dot{m}_2 z_2 + \dot{m}_3 z_3)\\ &= (9.81\ \text{m/s})^2[(4.58\ \text{kg/s})(0.3\ \text{meters}) + (2.55\ \text{kg/s})(1.2\ \text{meters})\\ &\quad + (2.03\ \text{kg/s})(0.5\ \text{m})]\\ &= -13.5\ \text{Nm/s} + 30.0\ \text{Nm/s} + 9.96\ \text{Nm/s} \approx 26.5\ \text{Nm/s}\end{aligned}$$

Equation (2) may now be evaluated for the heat transfer:

$$\dot{Q} - (-112{,}000)\ \text{Nm/s} = 190{,}000\ \text{Nm/s} - 9{,}150\ \text{Nm/s} + 26.5\ \text{Nm/s}$$

or

$$\dot{Q} \approx 68{,}900\ \text{J/s} \qquad \textit{Ans.}$$

- *Comments:* The heat transfer is positive, which means *into* the control volume. It is typical of gas flows that potential energy flow is negligible, enthalpy flow is dominant, and kinetic energy flow is small unless the velocities are very high (that is, high subsonic or supersonic).

The Steady Flow Energy Equation

For steady flow with one inlet and one outlet, both assumed one-dimensional, Eq. (3.67) reduces to a celebrated relation used in many engineering analyses. Let section 1 be the inlet and section 2 the outlet. Then

$$\dot{Q} - \dot{W}_s - \dot{W}_v = -\dot{m}_1(\hat{h}_1 + \tfrac{1}{2}V_1^2 + g z_1) + \dot{m}_2(\hat{h}_2 + \tfrac{1}{2}V_2^2 + g z_2) \tag{3.69}$$

But, from continuity, $\dot{m}_1 = \dot{m}_2 = \dot{m}$, we can rearrange (3.69) as follows:

$$\hat{h}_1 + \tfrac{1}{2}V_1^2 + gz_1 = (\hat{h}_2 + \tfrac{1}{2}V_2^2 + gz_2) - q + w_s + w_v \tag{3.70}$$

where $q = \dot{Q}/\dot{m} = dQ/dm$, the heat transferred to the fluid per unit mass. Similarly, $w_s = \dot{W}_s/\dot{m} = dW_s/dm$ and $w_v = \dot{W}_v/\dot{m} = dW_v/dm$. Equation (3.70) is a general form of the *steady flow energy equation*, which states that the upstream *stagnation enthalpy* $H_1 = (h + \tfrac{1}{2}V^2 + gz)_1$ differs from the downstream value H_2 only if there is heat transfer, shaft work, or viscous work as the fluid passes between sections 1 and 2. Recall that q is positive if heat is added to the control volume and that w_s and w_v are positive if work is done by the fluid on the surroundings.

Each term in Eq. (3.70) has the dimensions of energy per unit mass, or velocity squared, which is a form commonly used by mechanical engineers. If we divide through by g, each term becomes a length, or head, which is a form preferred by civil engineers. The traditional symbol for head is h, which we do not wish to confuse with enthalpy. Therefore, we use internal energy in rewriting the head form of the energy relation:

$$\frac{p_1}{\gamma} + \frac{\hat{u}_1}{g} + \frac{V_1^2}{2g} + z_1 = \frac{p_2}{\gamma} + \frac{\hat{u}_2}{g} + \frac{V_2^2}{2g} + z_2 - h_q + h_s + h_v \tag{3.71}$$

where $h_q = q/g$, $h_s = w_s/g$, and $h_v = w_v/g$ are the head forms of the heat added, shaft work done, and viscous work done, respectively. The term p/γ is called *pressure head*, and the term $V^2/2g$ is denoted as *velocity head*.

Friction and Shaft Work in Low-Speed Flow

A common application of the steady flow energy equation is for low-speed (incompressible) flow through a pipe or duct. A pump or turbine may be included in the pipe system. The pipe and machine walls are solid, so the viscous work is zero. Equation (3.71) may be written as

$$\left(\frac{p_1}{\gamma} + \frac{V_1^2}{2g} + z_1\right) = \left(\frac{p_2}{\gamma} + \frac{V_2^2}{2g} + z_2\right) + \frac{\hat{u}_2 - \hat{u}_1 - q}{g} \tag{3.72}$$

Every term in this equation is a length, or *head*. The terms in parentheses are the upstream (1) and downstream (2) values of the useful or *available head* or *total head* of the flow, denoted by h_0. The last term on the right is the difference $(h_{01} - h_{02})$, which can include pump head input, turbine head extraction, and the friction head loss h_f, always *positive*. Thus, in incompressible flow with one inlet and one outlet, we may write

$$\left(\frac{p}{\gamma} + \frac{V^2}{2g} + z\right)_{\text{in}} = \left(\frac{p}{\gamma} + \frac{V^2}{2g} + z\right)_{\text{out}} + h_{\text{friction}} - h_{\text{pump}} + h_{\text{turbine}} \tag{3.73}$$

Most of our internal flow problems will be solved with the aid of Eq. (3.73). The h terms are all positive; that is, friction loss is always positive in real (viscous) flows, a pump adds energy (increases the left-hand side), and a turbine extracts energy from the flow. If h_p and/or h_t are included, the pump and/or turbine must lie *between* points 1 and 2. In Chaps. 5 and 6 we shall develop methods of correlating h_f losses with flow parameters in pipes, valves, fittings, and other internal flow devices.

EXAMPLE 3.21

Gasoline at 20°C is pumped through a smooth 12-cm-diameter pipe 10 km long, at a flow rate of 75 m^3/h. The inlet is fed by a pump at an absolute pressure of 24 atm. The exit is at standard atmospheric pressure and is 150 m higher. Estimate the frictional head loss h_f, and compare it to the velocity head of the flow $V^2/(2g)$. (These numbers are quite realistic for liquid flow through long pipelines.)

Solution

- *Property values:* From Table A.3 for gasoline at 20°C, $\rho = 680\ \text{kg/m}^3$, or $\gamma = (680)(9.81) = 6{,}670\ \text{N/m}^3$.
- *Assumptions:* Steady flow. No shaft work, thus $h_p = h_t = 0$. If $z_1 = 0$, then $z_2 = 150$ m.
- *Approach:* Find the velocity and the velocity head. These are needed for comparison. Then evaluate the friction loss from Eq. (3.73).
- *Solution steps:* Since the pipe diameter is constant, the average velocity is the same everywhere:

$$V_{in} = V_{out} = \frac{Q}{A} = \frac{Q}{(\pi/4)D^2} = \frac{(75\ \text{m}^3/\text{h})/(3{,}600\ \text{s/h})}{(\pi/4)(0.12\ \text{m})^2} \approx 1.84\ \frac{\text{m}}{\text{s}}$$

$$\text{Velocity head} = \frac{V^2}{2g} = \frac{(1.84\ \text{m/s})^2}{2(9.81\ \text{m/s}^2)} \approx 0.173\ \text{m}$$

Substitute into Eq. (3.73) and solve for the friction head loss. Use pascals for the pressures and note that the velocity heads cancel because of the constant-area pipe.

$$\frac{p_{in}}{\gamma} + \frac{V_{in}^2}{2g} + z_{in} = \frac{p_{out}}{\gamma} + \frac{V_{out}^2}{2g} + z_{out} + h_f$$

$$\frac{(24)(101{,}350\ \text{N/m}^2)}{6{,}670\ \text{N/m}^3} + 0.173\ \text{m} + 0\ \text{m} = \frac{101{,}350\ \text{N/m}^2}{6{,}670\ \text{N/m}^3} + 0.173\ \text{m} + 150\ \text{m} + h_f$$

or $$h_f = 364.7 - 15.2 - 150 \approx 199\ \text{m} \qquad \textit{Ans.}$$

The friction head is larger than the elevation change Δz, and the pump must drive the flow against both changes, hence the high inlet pressure. The ratio of friction to velocity head is

$$\frac{h_f}{V^2/(2g)} \approx \frac{199\ \text{m}}{0.173\ \text{m}} \approx 1{,}150 \qquad \textit{Ans.}$$

- *Comments:* This high ratio is typical of long pipelines. (Note that we did not make direct use of the 10,000-m pipe length, whose effect is hidden within h_f.) In Chap. 6, we can state this problem in a more direct fashion: Given the flow rate, fluid, and pipe size, what inlet pressure is needed? Our correlations for h_f will lead to the estimate $p_{inlet} \approx 24$ atm, as stated here.

EXAMPLE 3.22

Air [$R = 287$ J/(kg · K), $c_p = 1{,}003.7$ Nm/(kg · K)] flows steadily, as shown in Fig. E3.22, through a turbine that produces 700 hp. For the inlet and exit conditions shown, estimate (*a*) the exit velocity V_2 and (*b*) the heat transferred Q in J/s.

Solution

Part (a) The inlet and exit densities can be computed from the perfect-gas law:

$$\rho_1 = \frac{p_1}{RT_1} = \frac{1{,}034{,}000\ \text{Pa}}{287\ \text{J/(kg}\cdot\text{K)}(422\ \text{K})} = 8.455\ \text{kg/m}^3$$

$$\rho_2 = \frac{p_2}{RT_2} = \frac{(276{,}000\ \text{Pa})}{287\ \text{J/(kg}\cdot\text{K)}(275\ \text{K})} = 3.497\ \text{kg/m}^3$$

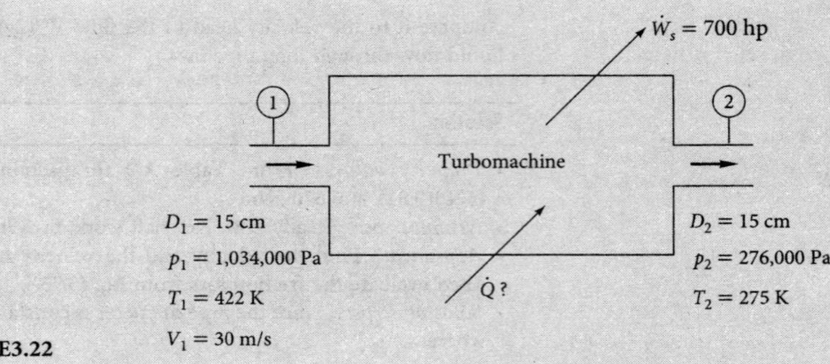

E3.22

The mass flow is determined by the inlet conditions

$$\dot{m} = \rho_1 A_1 V_1 = (8.455 \text{ kg/m}^3)\frac{\pi}{4}(0.15 \text{ m})^2(30 \text{ m/s}) = 4.482 \text{ kg/s}$$

Knowing mass flow, we compute the exit velocity

$$\dot{m} = 4.482 \text{ kg/s} = \rho_2 A_2 V_2 = (3.497 \text{ kg/m}^3)\frac{\pi}{4}(0.15 \text{ m})^2 V_2$$

or $$V_2 = 72.5 \text{ m/s}$$ *Ans. (a)*

Part (b) The steady flow energy equation (3.69) applies with $\dot{W}_v = 0$, $z_1 = z_2$, and $\hat{h} = c_p T$:

$$\dot{Q} - \dot{W}_s = \dot{m}(c_p T_2 + \tfrac{1}{2}V_2^2 - c_p T_1 - \tfrac{1}{2}V_1^2)$$

Convert the turbine work to Joule per second with the conversion factor 1 hp = 746 J/s. The turbine work $\dot{W}_s$ is positive

$$\dot{Q} - (700 \text{ hp})(746) = 4.482 \text{ kg/s}[1{,}003.7 \text{ Nm/(kg}\cdot\text{K)}(275 \text{ K}) + \tfrac{1}{2}(72.5 \text{ m/s})^2 - 1{,}003.7 \text{ Nm/(kg}\cdot\text{K)}(422 \text{ K}) - \tfrac{1}{2}(30 \text{ m/s})^2]$$

$$= -651{,}500 \text{ Nm/s}$$

or $$\dot{Q} = -129{,}500 \text{ Nm/s}$$ *Ans. (b)*

The negative sign indicates that this heat transfer is a *loss* from the control volume.

Kinetic Energy Correction Factor

Often, the flow entering or leaving a port is not strictly one-dimensional. In particular, the velocity may vary over the cross section, as in Fig. E3.4. In this case, the kinetic energy term in Eq. (3.68) for a given port should be modified by a dimensionless correction factor α so that the integral can be proportional to the square of the average velocity through the port:

$$\int_{\text{port}} (\tfrac{1}{2}V^2)\rho(\mathbf{V}\cdot\mathbf{n})\,dA \equiv \alpha(\tfrac{1}{2}V_{\text{av}}^2)\dot{m}$$

where $$V_{\text{av}} = \frac{1}{A}\int u\,dA \qquad \text{for incompressible flow}$$

If the density is also variable, the integration is very cumbersome; we shall not treat this complication. By letting u be the velocity normal to the port, the first equation above becomes, for incompressible flow,

$$\tfrac{1}{2}\rho \int u^3 dA = \tfrac{1}{2}\rho\alpha V_{av}^3 A$$

or

$$\alpha = \frac{1}{A}\int\left(\frac{u}{V_{av}}\right)^3 dA \tag{3.74}$$

The term α is the kinetic energy correction factor, having a value of about 2.0 for fully developed laminar pipe flow and from 1.04 to 1.11 for turbulent pipe flow. The complete incompressible steady flow energy equation (3.73), including pumps, turbines, and losses, would generalize to

$$\left(\frac{p}{\rho g} + \frac{\alpha}{2g}V^2 + z\right)_{in} = \left(\frac{p}{\rho g} + \frac{\alpha}{2g}V^2 + z\right)_{out} + h_{turbine} - h_{pump} + h_{friction} \tag{3.75}$$

where the head terms on the right (h_t, h_p, h_f) are all numerically positive. All additive terms in Eq. (3.75) have dimensions of length $\{L\}$. In problems involving turbulent pipe flow, it is common to assume that $\alpha \approx 1.0$. To compute numerical values, we can use these approximations to be discussed in Chap. 6:

Laminar flow: $$u = U_0\left[1 - \left(\frac{r}{R}\right)^2\right]$$

from which $$V_{av} = 0.5U_0$$

and $$\alpha = 2.0 \tag{3.76}$$

Turbulent flow: $$u \approx U_0\left(1 - \frac{r}{R}\right)^m \qquad m \approx \frac{1}{7}$$

from which, in Example 3.4,

$$V_{av} = \frac{2U_0}{(1+m)(2+m)}$$

Substituting into Eq. (3.74) gives

$$\alpha = \frac{(1+m)^3(2+m)^3}{4(1+3m)(2+3m)} \tag{3.77}$$

and numerical values are as follows:

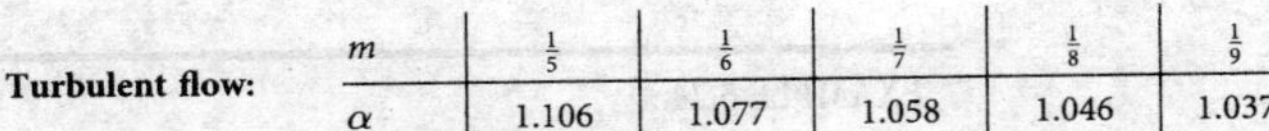

Turbulent flow:						
	m	$\frac{1}{5}$	$\frac{1}{6}$	$\frac{1}{7}$	$\frac{1}{8}$	$\frac{1}{9}$
	α	1.106	1.077	1.058	1.046	1.037

These values are only slightly different from unity and are often neglected in elementary turbulent flow analyses. However, α should never be neglected in laminar flow.

EXAMPLE 3.23

A hydroelectric power plant (Fig. E3.23) takes in 30 m^3/s of water through its turbine and discharges it to the atmosphere at $V_2 = 2$ m/s. The head loss in the turbine and penstock system is $h_f = 20$ m. Assuming turbulent flow, $\alpha \approx 1.06$, estimate the power in MW extracted by the turbine.

Solution

We neglect viscous work and heat transfer and take section 1 at the reservoir surface (Fig. E3.23), where $V_1 \approx 0$, $p_1 = p_{atm}$, and $z_1 = 100$ m. Section 2 is at the turbine outlet.

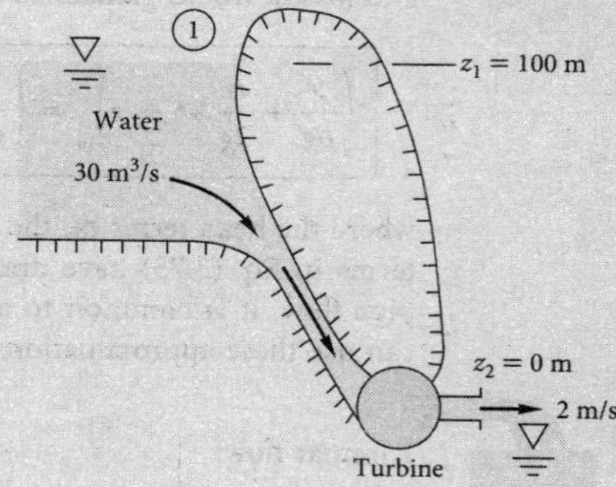

E3.23

The steady flow energy equation (3.75) becomes, in head form,

$$\frac{p_1}{\gamma} + \frac{\alpha_1 V_1^2}{2g} + z_1 = \frac{p_2}{\gamma} + \frac{\alpha_2 V_2^2}{2g} + z_2 + h_t + h_f$$

$$\frac{p_a}{\gamma} + \frac{1.06(0)^2}{2(9.81)} + 100 \text{ m} = \frac{p_a}{\gamma} + \frac{1.06(2.0 \text{ m/s})^2}{2(9.81 \text{ m/s}^2)} + 0 \text{ m} + h_t + 20 \text{ m}$$

The pressure terms cancel, and we may solve for the turbine head (which is positive):

$$h_t = 100 - 20 - 0.2 \approx 79.8 \text{ m}$$

The turbine extracts about 79.8 percent of the 100-m head available from the dam. The total power extracted may be evaluated from the water mass flow:

$$P = \dot{m}w_s = (\rho Q)(gh_t) = (998 \text{ kg/m}^3)(30 \text{ m}^3/\text{s})(9.81 \text{ m/s}^2)(79.8 \text{ m})$$

$$= 23.4 \text{ E6 kg} \cdot \text{m}^2/\text{s}^3 = 23.4 \text{ E6 N} \cdot \text{m/s} = 23.4 \text{ MW} \qquad \textit{Ans.}$$

The turbine drives an electric generator that probably has losses of about 15 percent, so the net power generated by this hydroelectric plant is about 20 MW.

EXAMPLE 3.24

The pump in Fig. E3.24 delivers water (1,000 kg/m^3) at 0.04 m^3/s to a machine at section 2, which is 6 m higher than the reservoir surface. The losses between 1 and 2 are given by $h_f = KV_2^2/(2g)$, where $K \approx 7.5$ is a dimensionless loss coefficient (see Sec. 6.7). Take $\alpha \approx 1.07$. Find the horsepower required for the pump if it is 80 percent efficient.

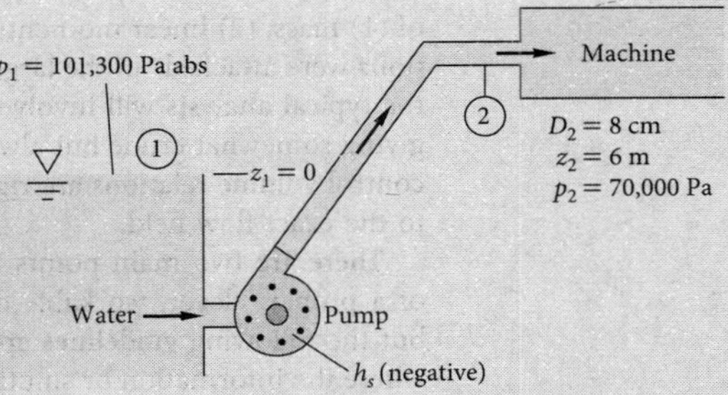

E3.24

Solution

- *System sketch:* Figure E3.24 shows the proper selection for sections 1 and 2.
- *Assumptions:* Steady flow, negligible viscous work, large reservoir ($V_1 \approx 0$).
- *Approach:* First find the velocity V_2 at the exit, then apply the steady flow energy equation.

Find V_2 from the known flow rate and the pipe diameter:

$$V_2 = \frac{Q}{A_2} = \frac{0.04\ \text{m}^3/\text{s}}{(\pi/4)(0.08\ \text{m})^2} = 7.96\ \text{m/s}$$

The steady flow energy equation (3.75), with a pump (no turbine) plus $z_1 \approx 0$ and $V_1 \approx 0$, becomes

$$\frac{p_1}{\gamma} + \frac{\alpha_1 V_1^2}{2g} + z_1 = \frac{p_2}{\gamma} + \frac{\alpha_2 V_2^2}{2g} + z_2 - h_p + h_f,\ h_f = K\frac{V_2^2}{2g}$$

or

$$h_p = \frac{p_2 - p_1}{\gamma} + z_2 + (\alpha_2 + K)\frac{V_2^2}{2g}$$

- *Comment:* The pump must balance four different effects: the pressure change, the elevation change, the exit jet kinetic energy, and the friction losses.
- *Final solution:* For the given data, we can evaluate the required pump head:

$$h_p = \frac{(70{,}000 - 101{,}300)\ \text{Pa}}{(9{,}810\ \text{N/m}^3)} + 6\ \text{m} + (1.07 + 7.5)\frac{(7.96\ \text{m/s})^2}{2(9.81\ \text{m/s}^2)} = -3.19 + 6$$

$$+27.7 = 30.5\ \text{m}$$

With the pump head known, the delivered pump power is computed similar to the turbine in Example 3.23:

$$P_{\text{pump}} = \dot{m}w_s = \gamma Q h_p = (9{,}810\ \text{N/m}^3)(0.04\ \text{m}^3/\text{s})(30.5\ \text{m})$$

$$= 11{,}970\ \text{Nm/s} = \frac{11{,}970\ \text{Nm/s}}{746\ (\text{Nm/s})/\text{hp}} = 16\ \text{hp}$$

If the pump is 80 percent efficient, then we divide by the efficiency to find the input power required:

$$P_{\text{input}} = \frac{P_{\text{pump}}}{\text{efficiency}} = \frac{16\ \text{hp}}{0.80} = 20\ \text{hp} \qquad \textit{Ans.}$$

- *Comment:* The inclusion of the kinetic energy correction factor α in this case made a difference of about 1 percent in the result. The friction loss, not the exit jet, was the dominant parameter.

Summary

This chapter has analyzed the four basic equations of fluid mechanics: conservation of (1) mass, (2) linear momentum, (3) angular momentum, and (4) energy. The equations were attacked "in the large"—that is, applied to whole regions of a flow. As such, the typical analysis will involve an approximation of the flow field within the region, giving somewhat crude but always instructive quantitative results. However, the basic control volume relations are rigorous and correct and will give exact results if applied to the exact flow field.

There are two main points to a control volume analysis. The first is the selection of a proper, clever, workable control volume. There is no substitute for experience, but the following guidelines apply. The control volume should cut through the place where the information or solution is desired. It should cut through places where maximum information is already known. If the momentum equation is to be used, it should *not* cut through solid walls unless absolutely necessary, since this will expose possible unknown stresses and forces and moments that make the solution for the desired force difficult or impossible. Finally, every attempt should be made to place the control volume in a frame of reference where the flow is steady or quasi-steady, since the steady formulation is much simpler to evaluate.

The second main point to a control volume analysis is the reduction of the analysis to a case that applies to the problem at hand. The 24 examples in this chapter give only an introduction to the search for appropriate simplifying assumptions. You will need to solve 24 or 124 more examples to become truly experienced in simplifying the problem just enough and no more. In the meantime, it would be wise for the beginner to adopt a very general form of the control volume conservation laws and then make a series of simplifications to achieve the final analysis. Starting with the general form, one can ask a series of questions:

1. Is the control volume nondeforming or nonaccelerating?
2. Is the flow field steady? Can we change to a steady flow frame?
3. Can friction be neglected?
4. Is the fluid incompressible? If not, is the perfect-gas law applicable?
5. Are gravity or other body forces negligible?
6. Is there heat transfer, shaft work, or viscous work?
7. Are the inlet and outlet flows approximately one-dimensional?
8. Is atmospheric pressure important to the analysis? Is the pressure hydrostatic on any portions of the control surface?
9. Are there reservoir conditions that change so slowly that the velocity and time rates of change can be neglected?

In this way, by approving or rejecting a list of basic simplifications like these, one can avoid pulling Bernoulli's equation off the shelf when it does not apply.

Problems

Most of the problems herein are fairly straightforward. More difficult or open-ended assignments are labeled with an asterisk. Problems labeled with a computer icon may require the use of a computer. The standard end-of-chapter problems P3.1 to P3.185 (categorized in the problem list here) are followed by word problems W3.1 to W3.7, fundamentals of engineering (FE) exam problems FE3.1 to FE3.10, comprehensive problems C3.1 to C3.5, and design project D3.1.

Problem Distribution

Section	Topic	Problems
3.1	Basic physical laws; volume flow	P3.1–P3.5
3.2	The Reynolds transport theorem	P3.6–P3.9
3.3	Conservation of mass	P3.10–P3.38
3.4	The linear momentum equation	P3.39–P3.109
3.5	The Bernoulli equation	P3.110–P3.148
3.6	The angular momentum theorem	P3.149–P3.164
3.7	The energy equation	P3.165–P3.185

Basic physical laws; volume flow

P3.1 Discuss Newton's second law (the linear momentum relation) in these three forms:

$$\sum \mathbf{F} = m\mathbf{a} \qquad \sum \mathbf{F} = \frac{d}{dt}(m\mathbf{V})$$

$$\sum \mathbf{F} = \frac{d}{dt}\left(\int_{\text{system}} \mathbf{V}\rho\, d\mathcal{V}\right)$$

Are they all equally valid? Are they equivalent? Are some forms better for fluid mechanics as opposed to solid mechanics?

P3.2 Consider the angular momentum relation in the form

$$\sum \mathbf{M}_O = \frac{d}{dt}\left[\int_{\text{system}} (\mathbf{r} \times \mathbf{V})\rho\, d\mathcal{V}\right]$$

What does $\mathbf{r}$ mean in this relation? Is this relation valid in both solid and fluid mechanics? Is it related to the *linear* momentum equation (Prob. 3.1)? In what manner?

P3.3 For steady low-Reynolds-number (laminar) flow through a long tube (see Prob. 1.12), the axial velocity distribution is given by $u = C(R^2 - r^2)$, where R is the tube radius and $r \leq R$. Integrate $u(r)$ to find the total volume flow Q through the tube.

P3.4 Water at 20°C flows through a long elliptical duct 30 cm wide and 22 cm high. What average velocity, in m/s, would cause the weight flow to be 2,225 N/s?

P3.5 Water at 20°C flows through a 13 cm-diameter smooth pipe at a high Reynolds number, for which the velocity profile is approximated by $u \approx U_o(y/R)^{1/8}$, where U_o is the centerline velocity, R is the pipe radius, and y is the distance measured from the wall toward the centerline. If the centerline velocity is 8 m/s, estimate the volume flow rate in m^3/min.

The Reynolds transport theorem

P3.6 Water fills a cylindrical tank to depth h. The tank has diameter D. The water flows out at average velocity V_o from a hole in the bottom of area A_o. Use the Reynolds transport theorem to find an expression for the instantaneous depth change dh/dt.

P3.7 A spherical tank, of diameter 35 cm, is leaking air through a 5-mm-diameter hole in its side. The air exits the hole at 360 m/s and a density of 2.5 kg/m^3. Assuming uniform mixing, (*a*) find a formula for the rate of change of average density in the tank and (*b*) calculate a numerical value for $(d\rho/dt)$ in the tank for the given data.

P3.8 Three pipes steadily deliver water at 20°C to a large exit pipe in Fig. P3.8. The velocity $V_2 = 5$ m/s, and the exit flow rate $Q_4 = 120$ m^3/h. Find (*a*) V_1, (*b*) V_3, and (*c*) V_4 if it is known that increasing Q_3 by 20 percent would increase Q_4 by 10 percent.

P3.9 A laboratory test tank contains seawater of salinity S and density ρ. Water enters the tank at conditions (S_1, ρ_1, A_1, V_1) and is assumed to mix immediately in the tank. Tank water leaves through an outlet A_2 at velocity V_2. If salt is a "conservative" property (neither created nor destroyed), use the Reynolds transport theorem to find an expression for the rate of change of salt mass M_{salt} within the tank.

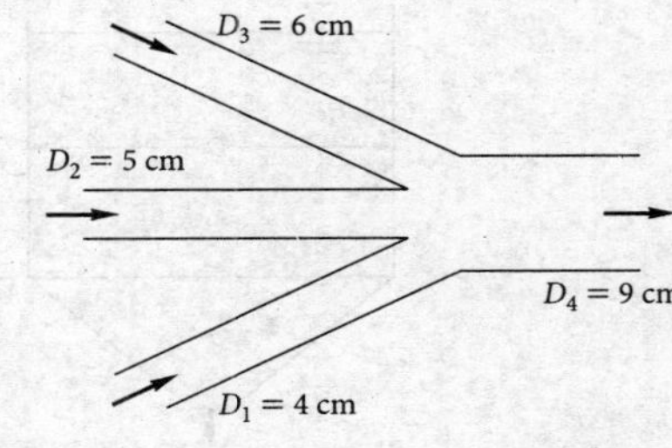

P3.8

Conservation of mass

P3.10 Water flowing through an 8-cm-diameter pipe enters a porous section, as in Fig. P3.10, which allows a uniform radial velocity v_w through the wall surfaces for a distance of 1.2 m. If the entrance average velocity V_1 is 12 m/s, find the exit velocity V_2 if (*a*) $v_w = 15$ cm/s out of the pipe walls or (*b*) $v_w = 10$ cm/s into the pipe. (*c*) What value of v_w will make $V_2 = 9$ m/s?

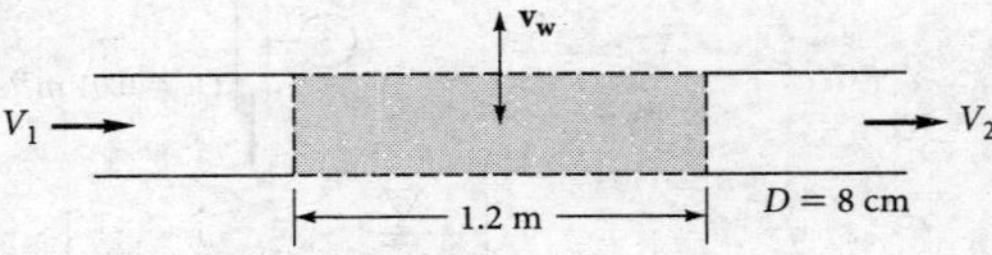

P3.10

P3.11 Water flows from a faucet into a sink at 11L/min. The stopper is closed, and the sink has two rectangular overflow drains, each 9 mm by 32 mm. If the sink water level remains constant, estimate the average overflow velocity, in m/s.

P3.12 The pipe flow in Fig. P3.12 fills a cylindrical surge tank as shown. At time $t = 0$, the water depth in the tank is 30 cm. Estimate the time required to fill the remainder of the tank.

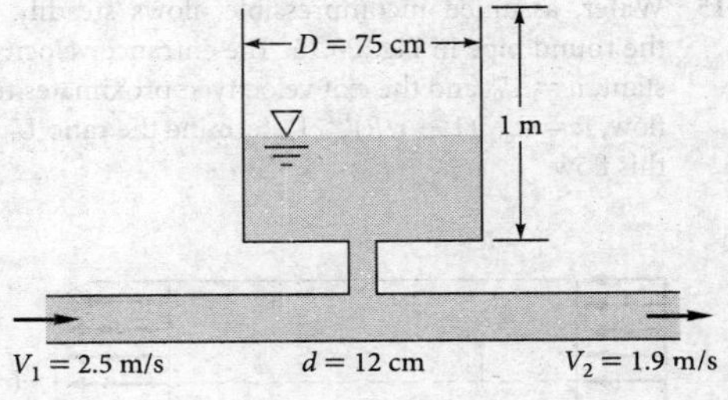

P3.12

P3.13 The cylindrical container in Fig. P3.13 is 20 cm in diameter and has a conical contraction at the bottom with an exit hole 3 cm in diameter. The tank contains fresh water at standard sea-level conditions. If the water surface is falling at the nearly steady rate $dh/dt \approx -0.072$ m/s, estimate the average velocity V out of the bottom exit.

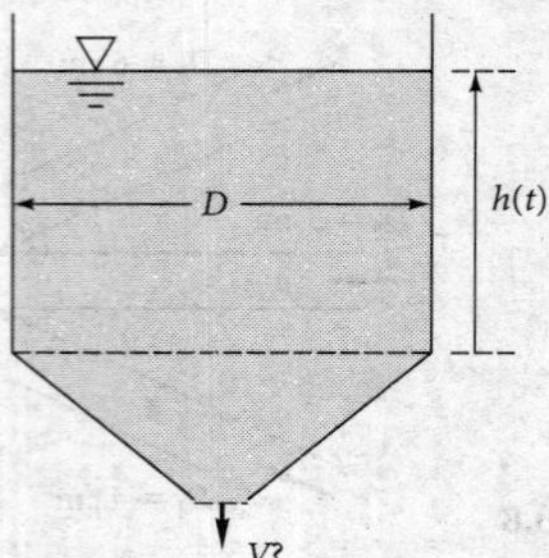

P3.13

P3.14 The open tank in Fig. P3.14 contains water at 20°C and is being filled through section 1. Assume incompressible flow. First derive an analytic expression for the water-level change dh/dt in terms of arbitrary volume flows (Q_1, Q_2, Q_3) and tank diameter d. Then, if the water level h is constant, determine the exit velocity V_2 for the given data $V_1 = 3$ m/s and $Q_3 = 0.01$ m^3/s.

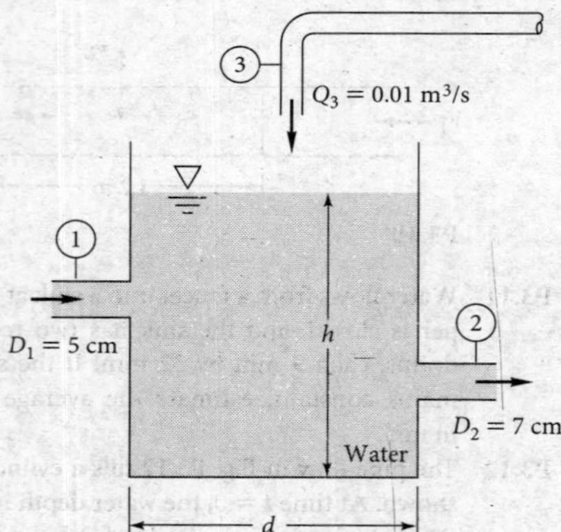

P3.14

P3.15 Water, assumed incompressible, flows steadily through the round pipe in Fig. P3.15. The entrance velocity is constant, $u = U_0$, and the exit velocity approximates turbulent flow, $u = u_{max}(1 - r/R)^{1/7}$. Determine the ratio U_0/u_{max} for this flow.

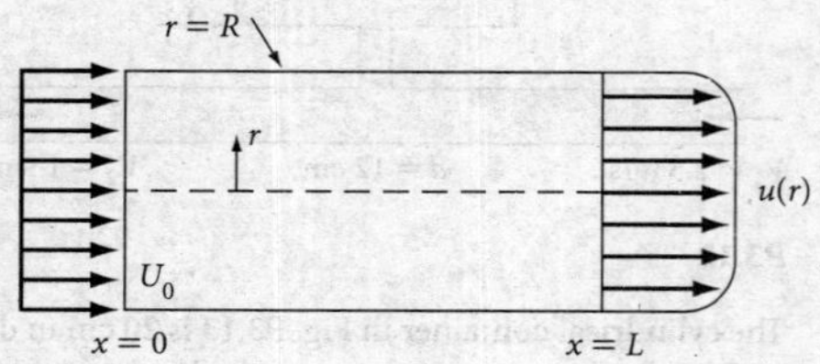

P3.15

P3.16 An incompressible fluid flows past an impermeable flat plate, as in Fig. P3.16, with a uniform inlet profile $u = U_0$ and a cubic polynomial exit profile

$$u \approx U_0\left(\frac{3\eta - \eta^3}{2}\right) \quad \text{where } \eta = \frac{y}{\delta}$$

Compute the volume flow Q across the top surface of the control volume.

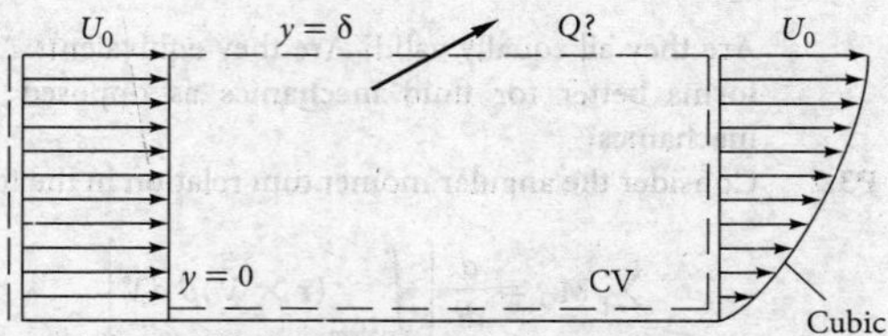

P3.16

P3.17 Incompressible steady flow in the inlet between parallel plates in Fig. P3.17 is uniform, $u = U_0 = 8$ cm/s, while downstream the flow develops into the parabolic laminar profile $u = az(z_0 - z)$, where a is a constant. If $z_0 = 4$ cm and the fluid is SAE 30 oil at 20°C, what is the value of u_{max} in cm/s?

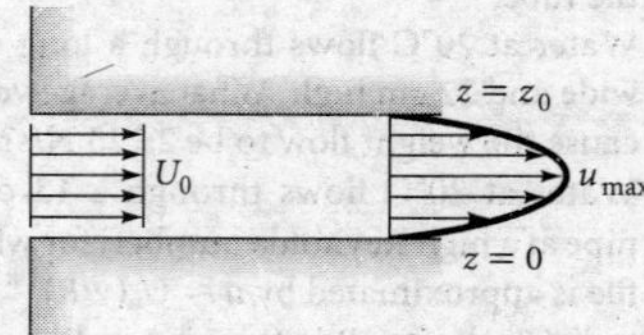

P3.17

P3.18 Gasoline enters section 1 in Fig. P3.18 at 0.5 m^3/s. It leaves section 2 at an average velocity of 12 m/s. What is the average velocity at section 3? Is it in or out?

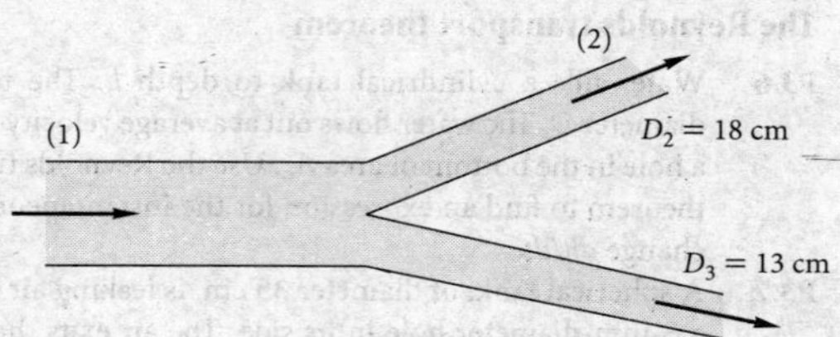

Fig. P3.18

P3.19 Water from a storm drain flows over an outfall onto a porous bed that absorbs the water at a uniform vertical velocity of 8 mm/s, as shown in Fig. P3.19. The system is 5 m deep into the paper. Find the length L of the bed that will completely absorb the storm water.

P3.20 Oil (SG = 0.89) enters at section 1 in Fig. P3.20 at a weight flow of 250 N/h to lubricate a thrust bearing. The steady oil flow exits radially through the narrow clearance between thrust plates. Compute (*a*) the outlet volume flow in mL/s and (*b*) the average outlet velocity in cm/s.

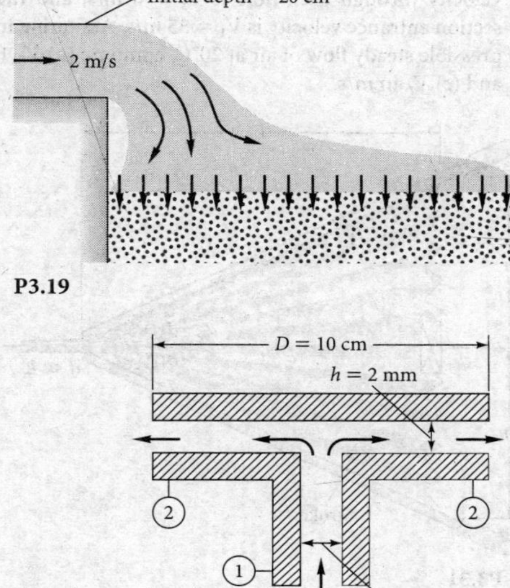

P3.19

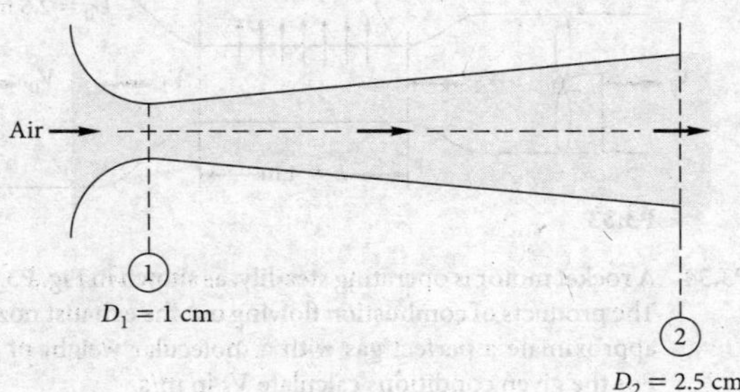

P3.20

P3.21 For the two-port tank of Fig. E3.5, assume $D_1 = 4$ cm, $V_1 = 18$ m/s, $D_2 = 7$ cm, and $V_2 = 8$ m/s. If the tank surface is rising at 17 mm/s, estimate the tank diameter.

P3.22 The converging–diverging nozzle shown in Fig. P3.22 expands and accelerates dry air to supersonic speeds at the exit, where $p_2 = 8$ kPa and $T_2 = 240$ K. At the throat, $p_1 = 284$ kPa, $T_1 = 665$ K, and $V_1 = 517$ m/s. For steady compressible flow of an ideal gas, estimate (*a*) the mass flow in kg/h, (*b*) the velocity V_2, and (*c*) the Mach number Ma_2.

Air
1
$D_1 = 1$ cm
2
$D_2 = 2.5$ cm

P3.22

P3.23 The hypodermic needle in Fig. P3.23 contains a liquid serum (SG = 1.05). If the serum is to be injected steadily at 6 cm³/s, how fast in in/s should the plunger be advanced (*a*) if leakage in the plunger clearance is neglected and (*b*) if leakage is 10 percent of the needle flow?

***P3.24** Water enters the bottom of the cone in Fig. P3.24 at a uniformly increasing average velocity $V = Kt$. If d is very small, derive an analytic formula for the water surface rise $h(t)$ for the condition $h = 0$ at $t = 0$. Assume incompressible flow.

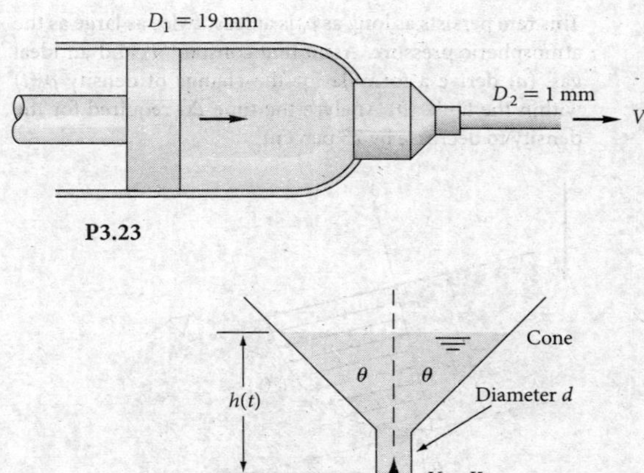

P3.23

Cone
$h(t)$
θ θ
Diameter d
$V = Kt$

P3.24

P3.25 As will be discussed in Chaps. 7 and 8, the flow of a stream U_0 past a blunt flat plate creates a broad low-velocity *wake* behind the plate. A simple model is given in Fig. P3.25, with only half of the flow shown due to symmetry. The velocity profile behind the plate is idealized as "dead air" (near-zero velocity) behind the plate, plus a higher velocity, decaying vertically above the wake according to the variation $u \approx U_0 + \Delta U\, e^{-z/L}$, where L is the plate height and $z = 0$ is the top of the wake. Find ΔU as a function of stream speed U_0.

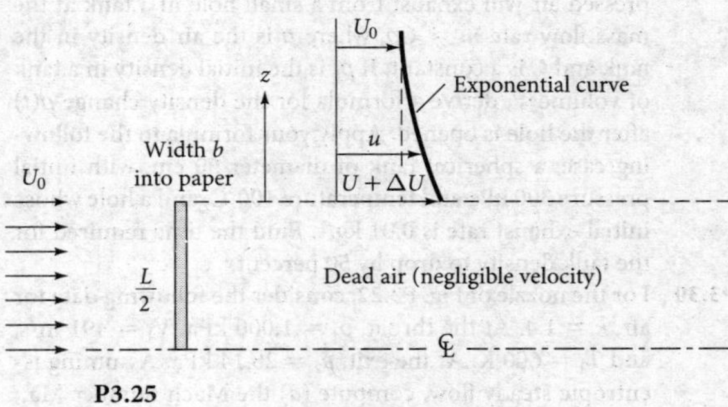

P3.25

P3.26 A thin layer of liquid, draining from an inclined plane, as in Fig. P3.26, will have a laminar velocity profile $u \approx U_0(2y/h - y^2/h^2)$, where U_0 is the surface velocity. If the plane has width b into the paper, determine the volume rate of flow in the film. Suppose that $h = 13$ mm and the flow rate per meter of channel width is 5 L/min. Estimate U_0 in mm/s.

P3.27 Consider a highly pressurized air tank at conditions (p_0, ρ_0, T_0) and volume υ_0. In Chap. 9 we will learn that, if the tank is allowed to exhaust to the atmosphere through a well-designed converging nozzle of exit area A, the outgoing mass flow rate will be

$$\dot{m} = \frac{\alpha\, p_0 A}{\sqrt{RT_0}} \quad \text{where } \alpha \approx 0.685 \text{ for air}$$

This rate persists as long as p_0 is at least twice as large as the atmospheric pressure. Assuming constant T_0 and an ideal gas, (*a*) derive a formula for the change of density $\rho_0(t)$ within the tank. (*b*) Analyze the time Δt required for the density to decrease by 25 percent.

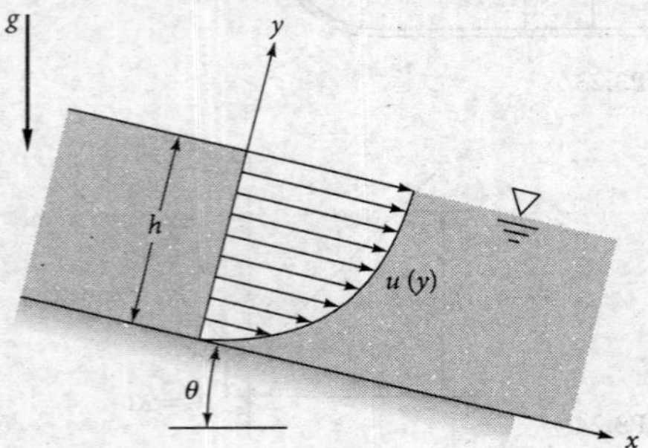

P3.26

P3.28 Air, assumed to be a perfect gas from Table A.4, flows through a long, 2-cm-diameter insulated tube. At section 1, the pressure is 1.1 MPa and the temperature is 345 K. At section 2, 67 meters further downstream, the density is 1.34 kg/m^3, the temperature 298 K, and the Mach number is 0.90. For one-dimensional flow, calculate (*a*) the mass flow; (*b*) p_2; (c) V_2; and (*d*) the change in entropy between 1 and 2. (*e*) How do you explain the entropy change?

P3.29 In elementary compressible flow theory (Chap. 9), compressed air will exhaust from a small hole in a tank at the mass flow rate $\dot{m} \approx C\rho$, where ρ is the air density in the tank and C is a constant. If ρ_0 is the initial density in a tank of volume $\mathcal{V}$, derive a formula for the density change $\rho(t)$ after the hole is opened. Apply your formula to the following case: a spherical tank of diameter 50 cm, with initial pressure 300 kPa and temperature 100°C, and a hole whose initial exhaust rate is 0.01 kg/s. Find the time required for the tank density to drop by 50 percent.

P3.30 For the nozzle of Fig. P3.22, consider the following data for air, $k = 1.4$. At the throat, $p_1 = 1{,}000$ kPa, $V_1 = 491$ m/s, and $T_1 = 600$ K. At the exit, $p_2 = 28.14$ kPa. Assuming isentropic steady flow, compute (*a*) the Mach number Ma$_1$; (*b*) T_2; (*c*) the mass flow; and (*d*) V_2.

P3.31 A bellows may be modeled as a deforming wedge-shaped volume as in Fig. P3.31. The check valve on the left (pleated) end is closed during the stroke. If b is the bellows width into the paper, derive an expression for outlet mass flow $\dot{m}_0$ as a function of stroke $\theta(t)$.

P3.32 Water at 20°C flows steadily through the piping junction in Fig. P3.32, entering section 1 at 75 L/min. The average velocity at section 2 is 2.5 m/s. A portion of the flow is diverted through the showerhead, which contains 100 holes of 1-mm diameter. Assuming uniform shower flow, estimate the exit velocity from the showerhead jets.

P3.33 In some wind tunnels the test section is perforated to suck out fluid and provide a thin viscous boundary layer. The test section wall in Fig. P3.33 contains 1,200 holes of 5-mm diameter each per square meter of wall area. The suction velocity through each hole is $V_s = 8$ m/s, and the test-section entrance velocity is $V_1 = 35$ m/s. Assuming incompressible steady flow of air at 20°C, compute (*a*) V_0, (*b*) V_2, and (*c*) V_f, in m/s.

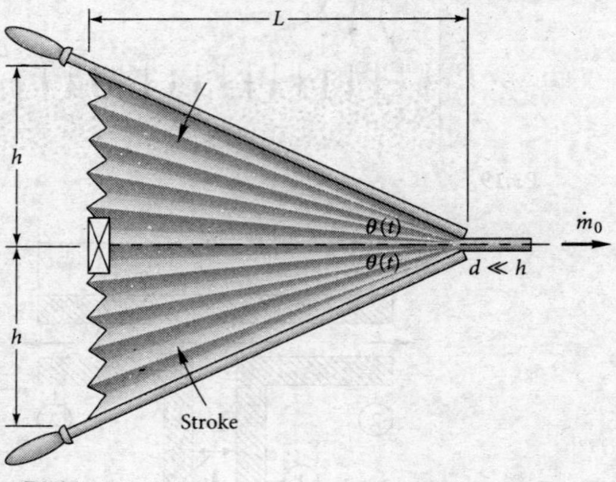

P3.31

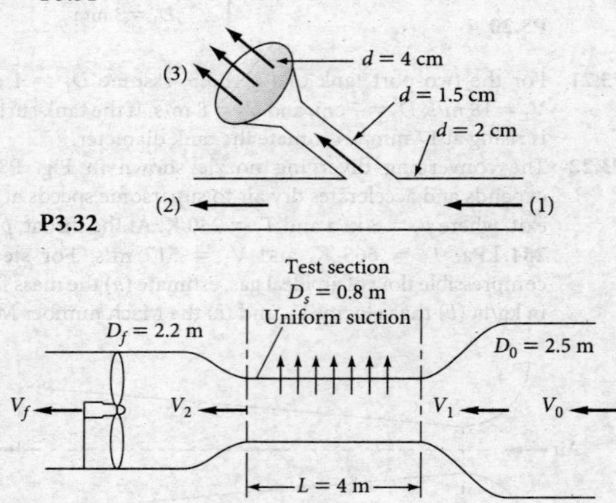

P3.32

P3.33

P3.34 A rocket motor is operating steadily, as shown in Fig. P3.34. The products of combustion flowing out the exhaust nozzle approximate a perfect gas with a molecular weight of 28. For the given conditions calculate V_2 in m/s.

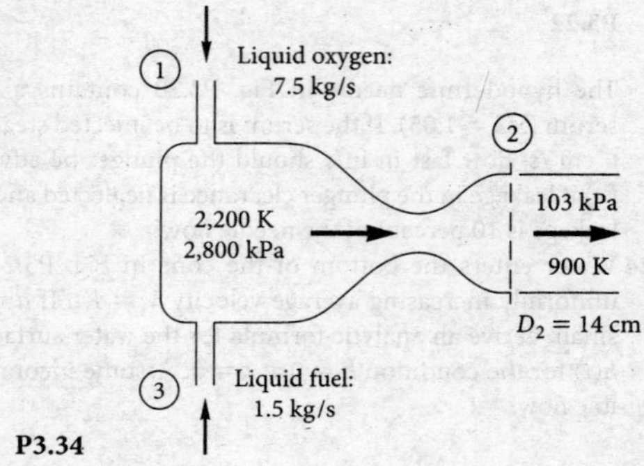

P3.34

P3.35 In contrast to the liquid rocket in Fig. P3.34, the solid-propellant rocket in Fig. P3.35 is self-contained and has no entrance ducts. Using a control volume analysis for the conditions shown in Fig. P3.35, compute the rate of mass loss of the propellant, assuming that the exit gas has a molecular weight of 28.

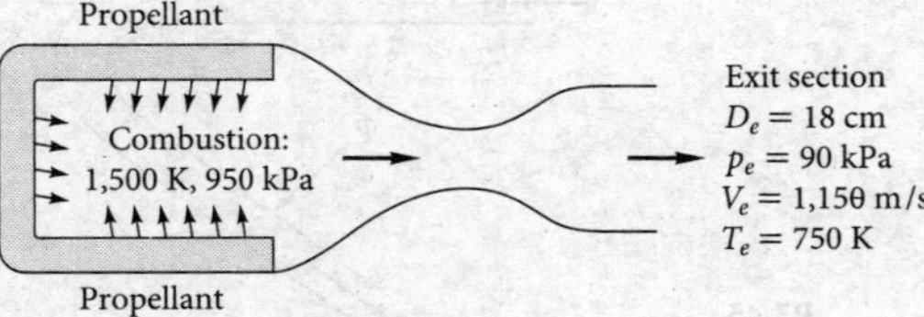

P3.35

P3.36 The jet pump in Fig. P3.36 injects water at $U_1 = 40$ m/s through a 7.5 cm pipe and entrains a secondary flow of water $U_2 = 3$ m/s in the annular region around the small pipe. The two flows become fully mixed downstream, where U_3 is approximately constant. For steady incompressible flow, compute U_3 in m/s.

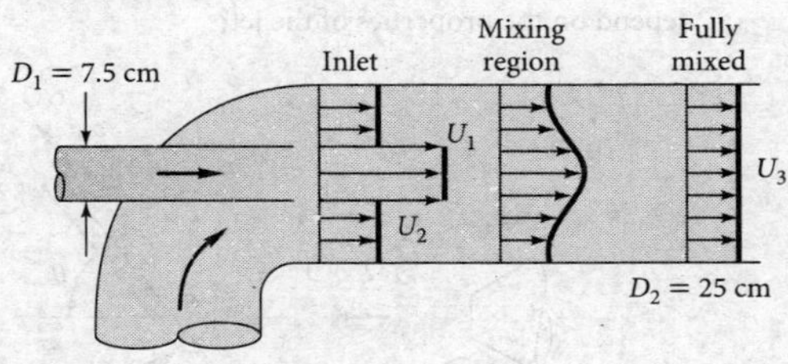

P3.36

P3.37 If the rectangular tank full of water in Fig. P3.37 has its right-hand wall lowered by an amount δ, as shown, water will flow out as it would over a weir or dam. In Prob. P1.14 we deduced that the outflow Q would be given by

$$Q = Cbg^{1/2}\delta^{3/2}$$

where b is the tank width into the paper, g is the acceleration of gravity, and C is a dimensionless constant. Assume that the water surface is horizontal, not slightly curved as in the figure. Let the initial excess water level be δ_o. Derive a formula for the time required to reduce the excess water level to (*a*) $\delta_o/10$ and (*b*) zero.

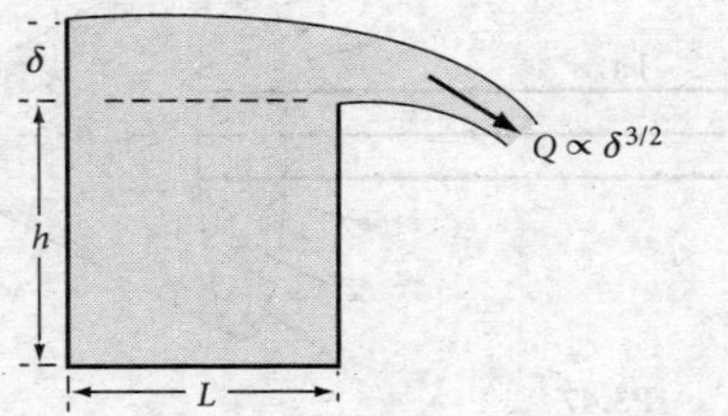

P3.37

P3.38 An incompressible fluid in Fig. P3.38 is being squeezed outward between two large circular disks by the uniform downward motion V_0 of the upper disk. Assuming one-dimensional radial outflow, use the control volume shown to derive an expression for $V(r)$.

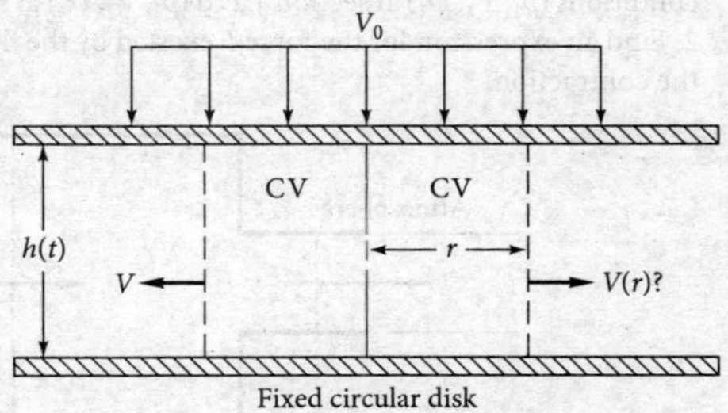

P3.38

The linear momentum equation

P3.39 A wedge splits a sheet of 20°C water, as shown in Fig. P3.39. Both wedge and sheet are very long into the paper. If the force required to hold the wedge stationary is $F = 124$ N per meter of depth into the paper, what is the angle θ of the wedge?

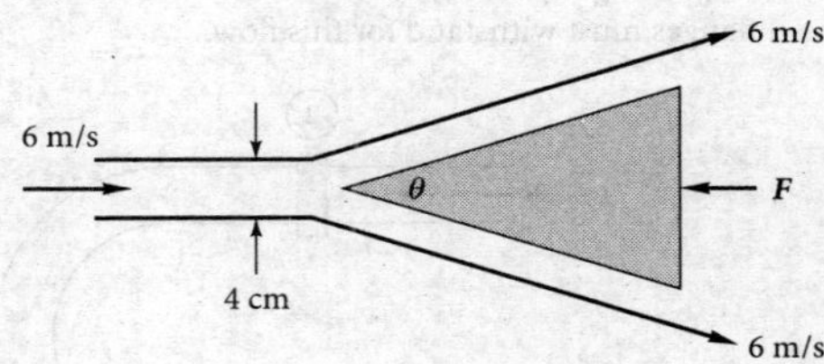

P3.39

P3.40 The water jet in Fig. P3.40 strikes normal to a fixed plate. Neglect gravity and friction, and compute the force F in newtons required to hold the plate fixed.

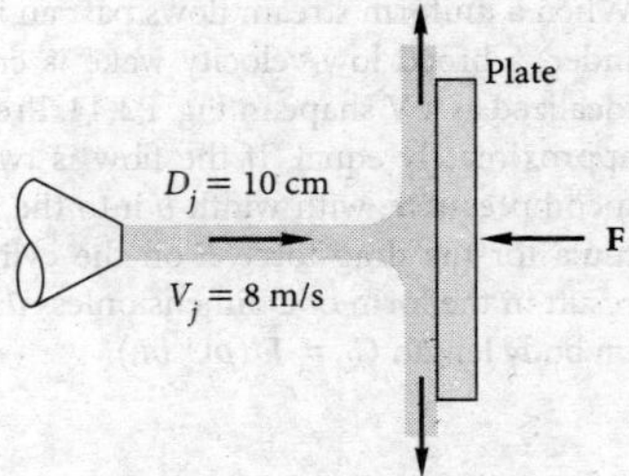

P3.40

P3.41 In Fig. P3.41 the vane turns the water jet completely around. Find an expression for the maximum jet velocity V_0 if the maximum possible support force is F_0.

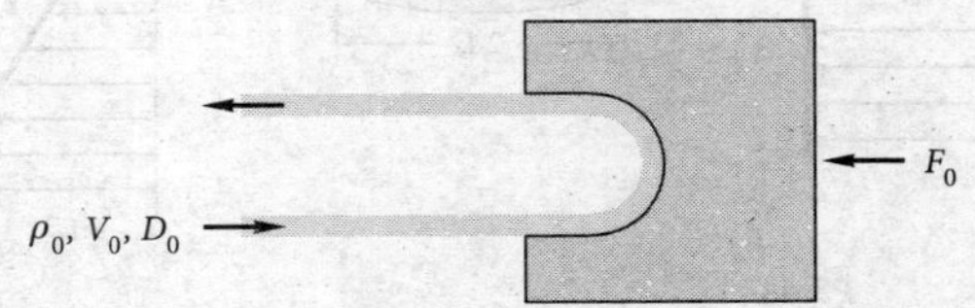

P3.41

P3.42 A liquid of density ρ flows through the sudden contraction in Fig. P3.42 and exits to the atmosphere. Assume uniform conditions (p_1, V_1, D_1) at section 1 and (p_2, V_2, D_2) at section 2. Find an expression for the force F exerted by the fluid on the contraction.

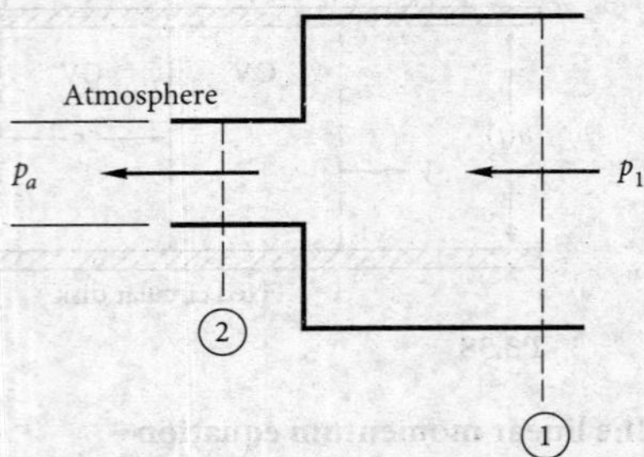

P3.42

P3.43 Water at 20°C flows through a 5-cm-diameter pipe that has a 180° vertical bend, as in Fig. P3.43. The total length of pipe between flanges 1 and 2 is 75 cm. When the weight flow rate is 230 N/s, p_1 = 165 kPa and p_2 = 134 kPa. Neglecting pipe weight, determine the total force that the flanges must withstand for this flow.

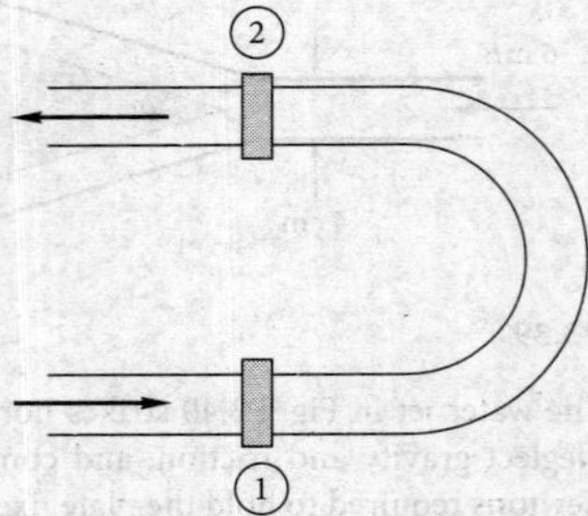

P3.43

***P3.44** When a uniform stream flows past an immersed thick cylinder, a broad low-velocity wake is created downstream, idealized as a V shape in Fig. P3.44. Pressures p_1 and p_2 are approximately equal. If the flow is two-dimensional and incompressible, with width b into the paper, derive a formula for the drag force F on the cylinder. Rewrite your result in the form of a dimensionless *drag coefficient* based on body length $C_D = F/(\rho U^2 bL)$.

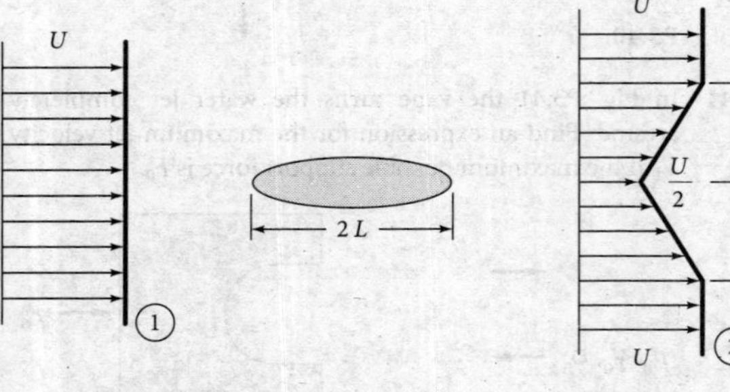

P3.44

P3.45 Water enters and leaves the 6-cm-diameter pipe bend in Fig. P3.45 at an average velocity of 8.5 m/s. The horizontal force to support the bend against momentum change is 300 N. Find (*a*) the angle ϕ; and (*b*) the vertical force on the bend.

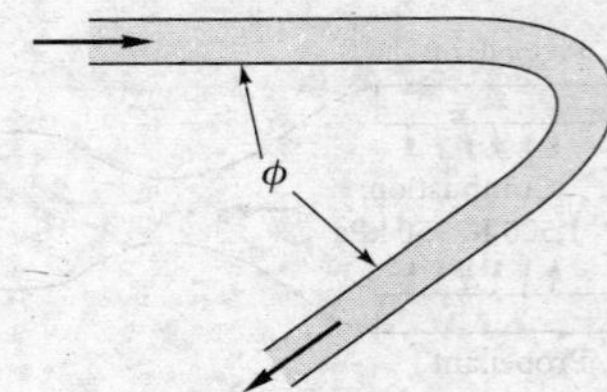

P3.45

P3.46 When a jet strikes an inclined fixed plate, as in Fig. P3.46, it breaks into two jets at 2 and 3 of equal velocity $V = V_{jet}$ but unequal flows αQ at 2 and $(1 - \alpha)Q$ at section 3, α being a fraction. The reason is that for frictionless flow the fluid can exert no tangential force F_t on the plate. The condition F_t = 0 enables us to solve for α. Perform this analysis, and find α as a function of the plate angle θ. Why doesn't the answer depend on the properties of the jet?

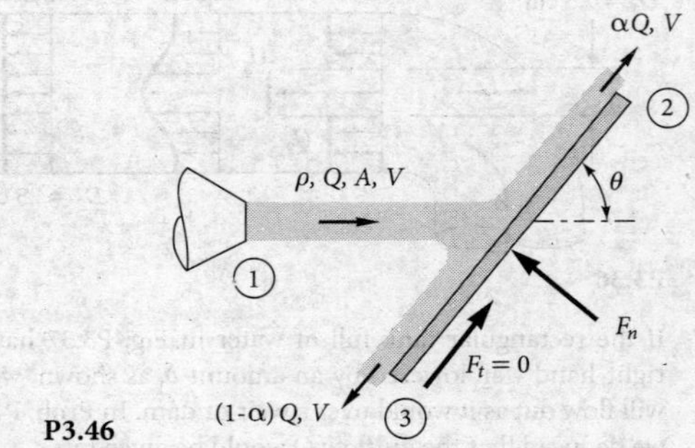

P3.46

P3.47 A liquid jet of velocity V_j and diameter D_j strikes a fixed hollow cone, as in Fig. P3.47, and deflects back as a conical sheet at the same velocity. Find the cone angle θ for which the restraining force $F = \frac{3}{2}\rho A_j V_j^2$.

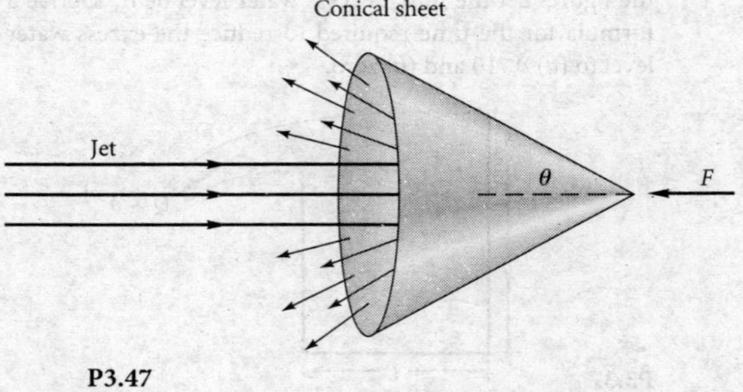

P3.47

P3.48 The small boat in Fig. P3.48 is driven at a steady speed V_0 by a jet of compressed air issuing from a 3-cm-diameter hole at $V_e = 343$ m/s. Jet exit conditions are $p_e = 1$ atm and $T_e = 30°C$. Air drag is negligible, and the hull drag is kV_0^2, where $k \approx 19\ N \cdot s^2/m^2$. Estimate the boat speed V_0 in m/s.

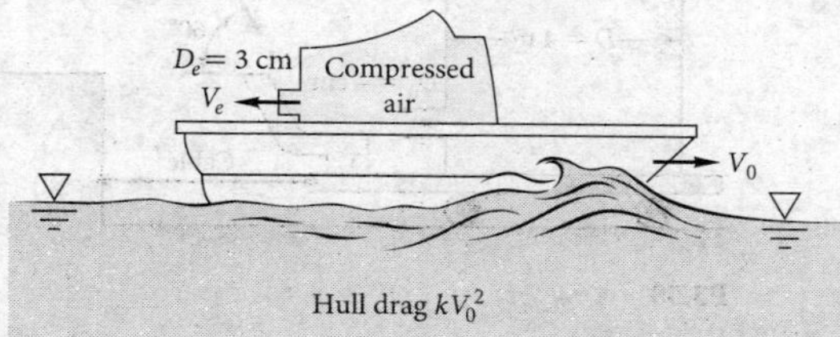

P3.48

P3.49 The horizontal nozzle in Fig. P3.49 has $D_1 = 30$ cm and $D_2 = 15$ cm, with inlet pressure $p_1 = 260$ kPa absolute and $V_2 = 17$ m/s. For water at 20°C, compute the horizontal force provided by the flange bolts to hold the nozzle fixed.

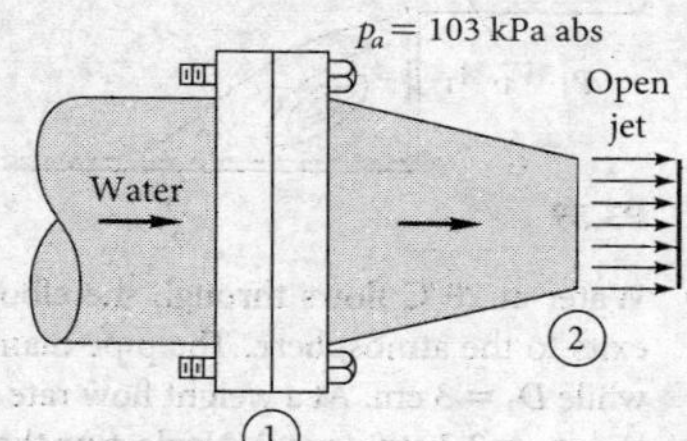

P3.49

P3.50 The jet engine on a test stand in Fig. P3.50 admits air at 20°C and 1 atm at section 1, where $A_1 = 0.5\ m^2$ and $V_1 = 250$ m/s. The fuel-to-air ratio is 1:30. The air leaves section 2 at atmospheric pressure and higher temperature, where $V_2 = 900$ m/s and $A_2 = 0.4\ m^2$. Compute the horizontal test stand reaction R_x needed to hold this engine fixed.

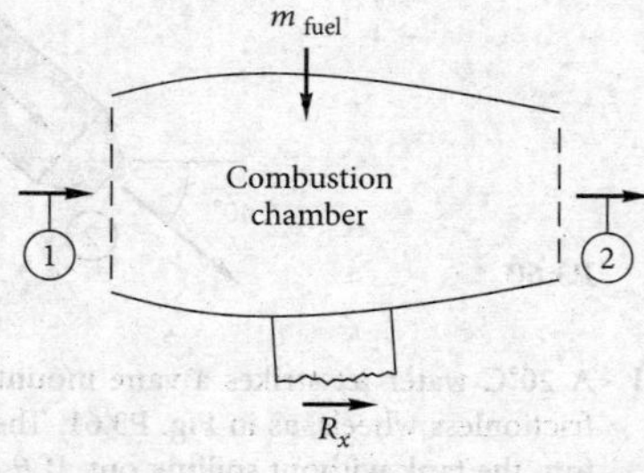

P3.50

P3.51 A liquid jet of velocity V_j and area A_j strikes a single 180° bucket on a turbine wheel rotating at angular velocity Ω, as in Fig. P3.51. Derive an expression for the power P delivered to this wheel at this instant as a function of the system parameters. At what angular velocity is the maximum power delivered? How would your analysis differ if there were many, many buckets on the wheel, so that the jet was continually striking at least one bucket?

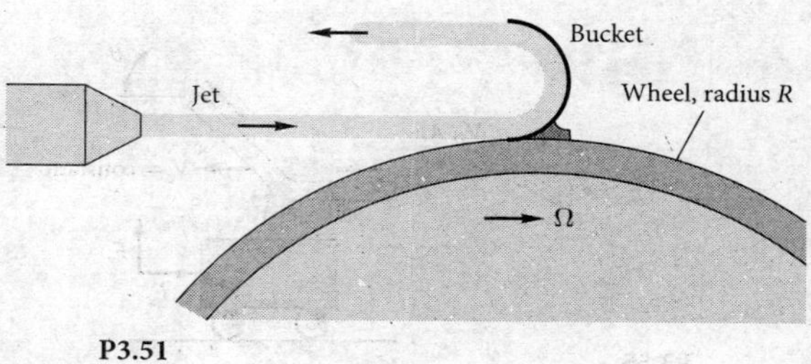

P3.51

P3.52 A large commercial power washer delivers 80 L/min of water through a nozzle of exit diameter of 0.8 cm. Estimate the force of the water jet on a wall normal to the jet.

P3.53 Consider incompressible flow in the entrance of a circular tube, as in Fig. P3.53. The inlet flow is uniform, $u_1 = U_0$. The flow at section 2 is developed pipe flow. Find the wall drag force F as a function of (p_1, p_2, ρ, U_0, R) if the flow at section 2 is

(*a*) Laminar: $u_2 = u_{max}\left(1 - \frac{r^2}{R^2}\right)$

(*b*) Turbulent: $u_2 \approx u_{max}\left(1 - \frac{r}{R}\right)^{1/7}$

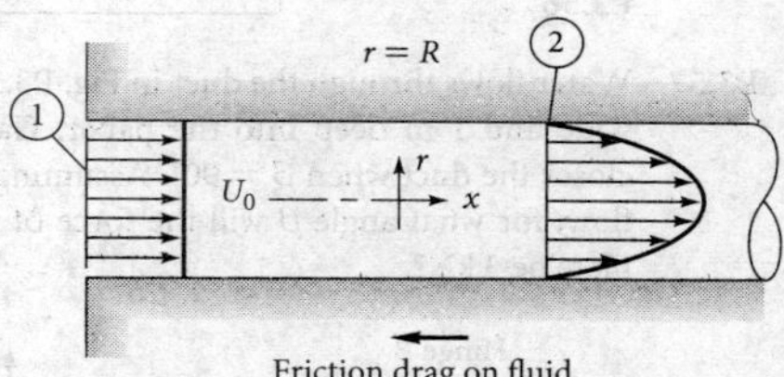

P3.53

P3.54 For the pipe-flow-reducing section of Fig. P3.54, $D_1 = 8$ cm, $D_2 = 5$ cm, and $p_2 = 1$ atm. All fluids are at 20°C. If $V_1 = 5$ m/s and the manometer reading is $h = 58$ cm, estimate the total force resisted by the flange bolts.

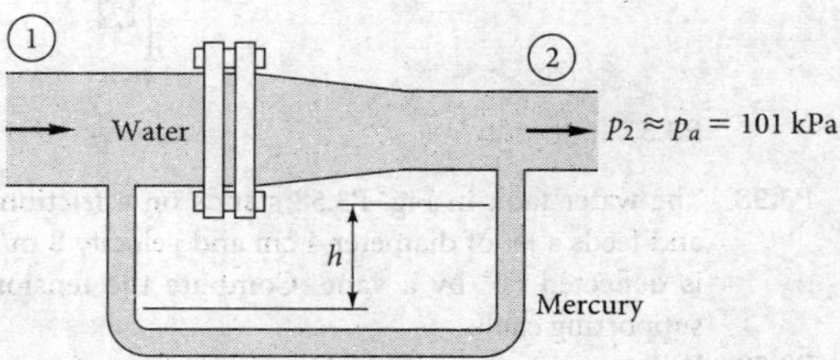

P3.54

P3.55 In Fig. P3.55 the jet strikes a vane that moves to the right at constant velocity V_c on a frictionless cart. Compute (*a*) the force F_x required to restrain the cart and (*b*) the power P delivered to the cart. Also find the cart velocity for which (*c*) the force F_x is a maximum and (*d*) the power P is a maximum.

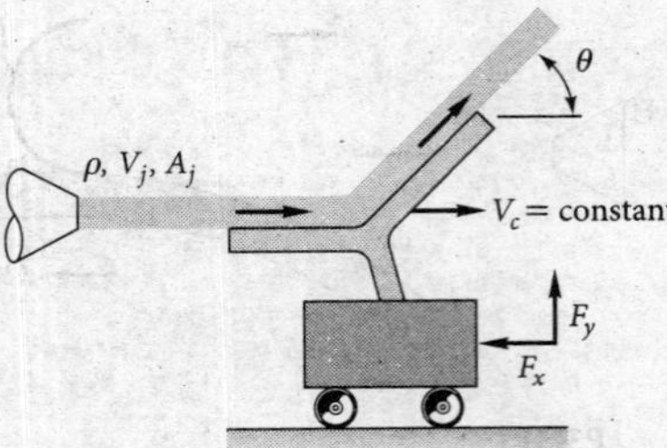

P3.55

P3.56 Water at 20°C flows steadily through the box in Fig. P3.56, entering station (1) at 2 m/s. Calculate the (*a*) horizontal and (*b*) vertical forces required to hold the box stationary against the flow momentum.

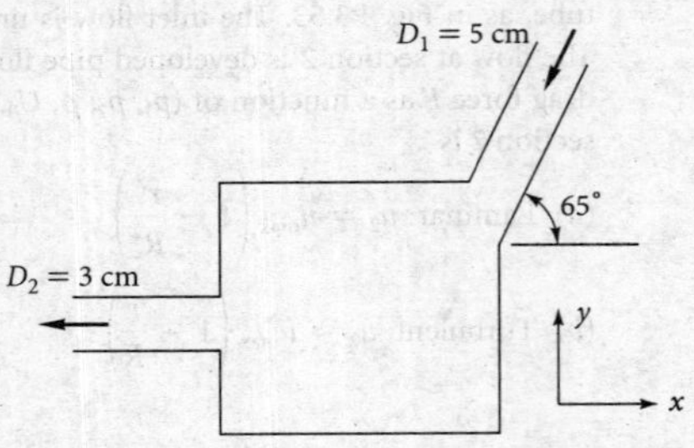

P3.56

P3.57 Water flows through the duct in Fig. P3.57, which is 50 cm wide and 1 m deep into the paper. Gate *BC* completely closes the duct when $\beta = 90°$. Assuming one-dimensional flow, for what angle β will the force of the exit jet on the plate be 3 kN?

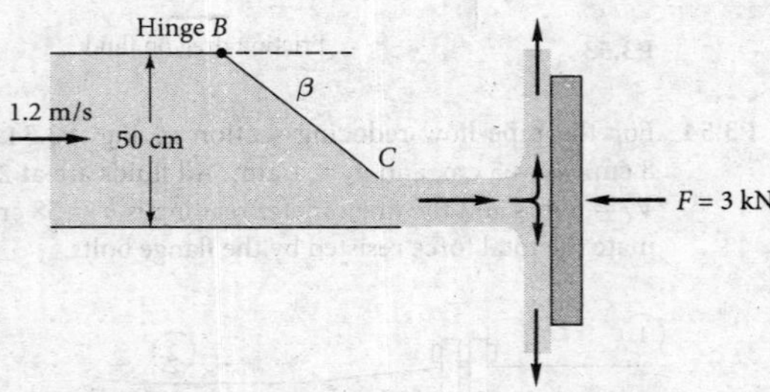

P3.57

P3.58 The water tank in Fig. P3.58 stands on a frictionless cart and feeds a jet of diameter 4 cm and velocity 8 m/s, which is deflected 60° by a vane. Compute the tension in the supporting cable.

P3.59 When a pipe flow suddenly expands from A_1 to A_2, as in Fig. P3.59, low-speed, low-friction eddies appear in the corners and the flow gradually expands to A_2 downstream. Using the suggested control volume for incompressible steady flow and assuming that $p \approx p_1$ on the corner annular ring as shown, show that the downstream pressure is given by

$$p_2 = p_1 + \rho V_1^2 \frac{A_1}{A_2}\left(1 - \frac{A_1}{A_2}\right)$$

Neglect wall friction.

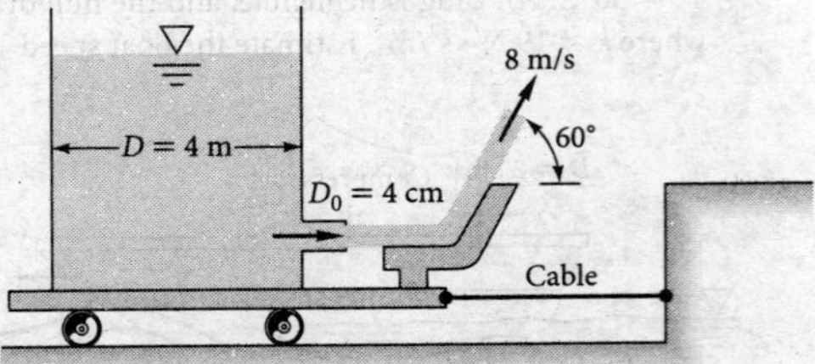

P3.58

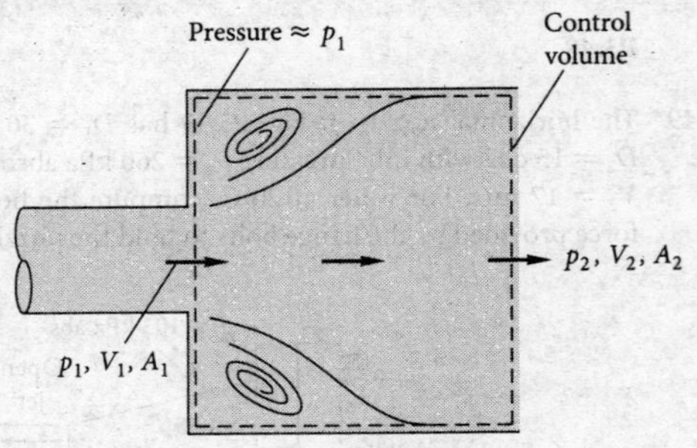

P3.59

P3.60 Water at 20°C flows through the elbow in Fig. P3.60 and exits to the atmosphere. The pipe diameter is $D_1 = 10$ cm, while $D_2 = 3$ cm. At a weight flow rate of 150 N/s, the pressure $p_1 = 2.3$ atm (gage). Neglecting the weight of water and elbow, estimate the force on the flange bolts at section 1.

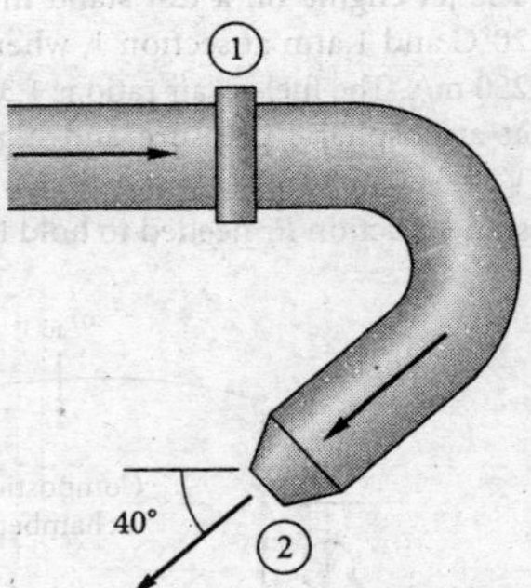

P3.60

P3.61 A 20°C water jet strikes a vane mounted on a tank with frictionless wheels, as in Fig. P3.61. The jet turns and falls into the tank without spilling out. If $\theta = 30°$, evaluate the horizontal force *F* required to hold the tank stationary.

P3.62 Water at 20°C exits to the standard sea-level atmosphere through the split nozzle in Fig. P3.62. Duct areas are $A_1 = 0.02$ m^2 and $A_2 = A_3 = 0.008$ m^2. If $p_1 = 135$ kPa (absolute) and the flow rate is $Q_2 = Q_3 = 275$ m^3/h, compute the force on the flange bolts at section 1.

P3.63 Water flows steadily through the box in Fig. P3.63. Average velocity at all ports is 7 m/s. The vertical momentum force on the box is 36 N. What is the inlet mass flow?

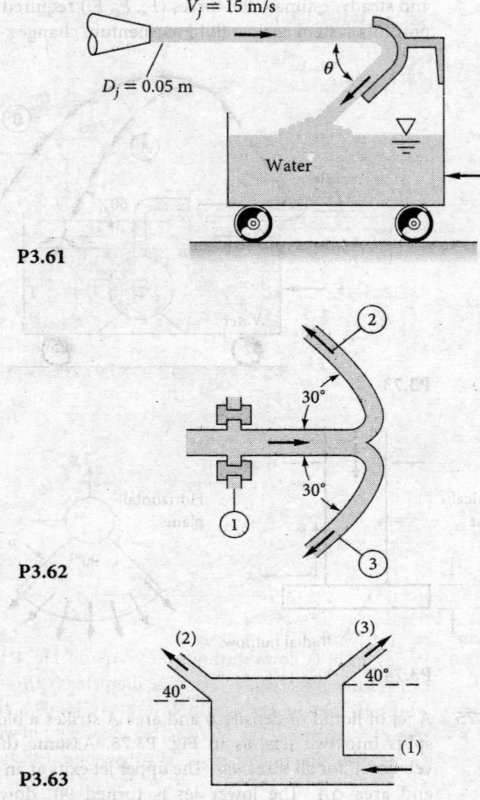

P3.61

P3.62

P3.63

P3.64 The 6-cm-diameter 20°C water jet in Fig. P3.64 strikes a plate containing a hole of 4-cm diameter. Part of the jet passes through the hole, and part is deflected. Determine the horizontal force required to hold the plate.

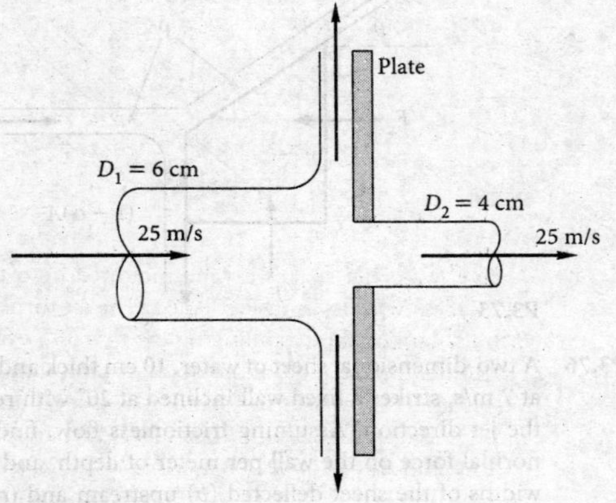

P3.64

P3.65 The box in Fig. P3.65 has three 13 mm holes on the right side. The volume flows of 20°C water shown are steady, but the details of the interior are not known. Compute the force, if any, that this water flow causes on the box.

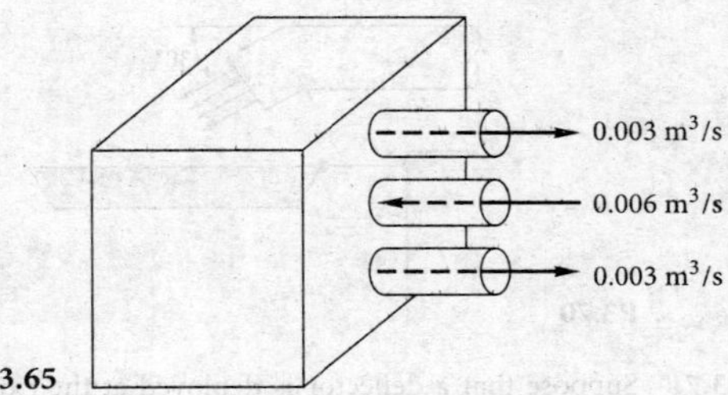

P3.65

P3.66 The tank in Fig. P3.66 weighs 500 N empty and contains 600 L of water at 20°C. Pipes 1 and 2 have equal diameters of 6 cm and equal steady volume flows of 300 m^3/h. What should the scale reading W be in N?

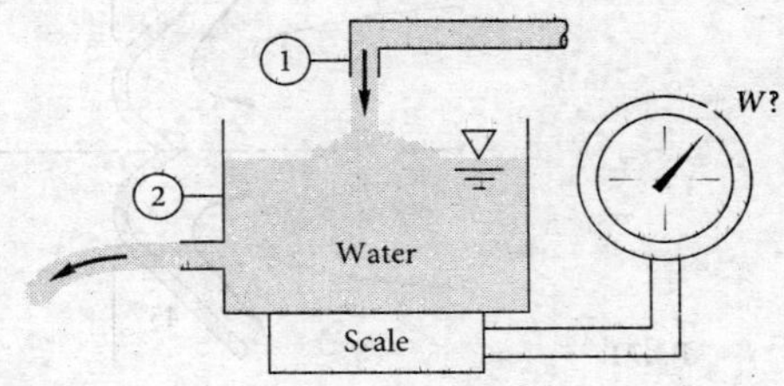

P3.66

P3.67 For the boundary layer of Fig. 3.10, for air, $\rho = 1.2$ kg/m^3, let $h = 7$ cm, $U_o = 12$ m/s, $b = 2$ m, and $L = 1$ m. Let the velocity at the exit, $x = L$, approximate a turbulent flow: $u/U_o \approx (y/\delta)^{1/7}$. Calculate (*a*) δ; and (*b*) the friction drag D.

P3.68 The rocket in Fig. P3.68 has a supersonic exhaust, and the exit pressure p_e is not necessarily equal to p_a. Show that the force F required to hold this rocket on the test stand is $F = \rho_e A_e V_e^2 + A_e(p_e - p_a)$. Is this force F what we term the *thrust* of the rocket?

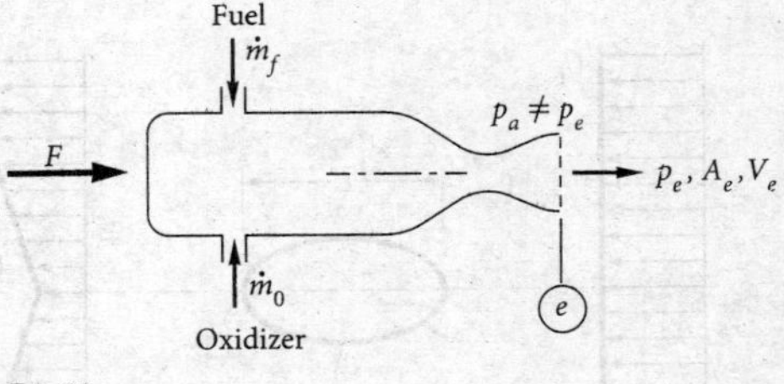

P3.68

P3.69 A uniform rectangular plate, 40 cm long and 30 cm deep into the paper, hangs in air from a hinge at its top (the 30-cm side). It is struck in its center by a horizontal 3-cm-diameter jet of water moving at 8 m/s. If the gate has a mass of 16 kg, estimate the angle at which the plate will hang from the vertical.

P3.70 The dredger in Fig. P3.70 is loading sand (SG = 2.6) onto a barge. The sand leaves the dredger pipe at 1 m/s with a weight flow of 4 kN/s. Estimate the tension on the mooring line caused by this loading process.

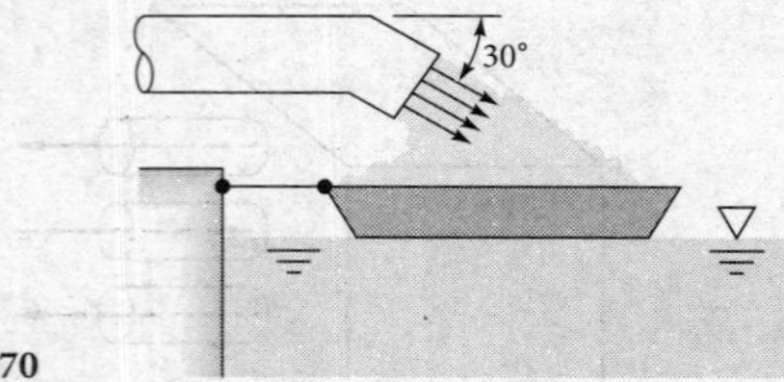

P3.70

P3.71 Suppose that a deflector is deployed at the exit of the jet engine of Prob. P3.50, as shown in Fig. P3.71. What will the reaction R_x on the test stand be now? Is this reaction sufficient to serve as a braking force during airplane landing?

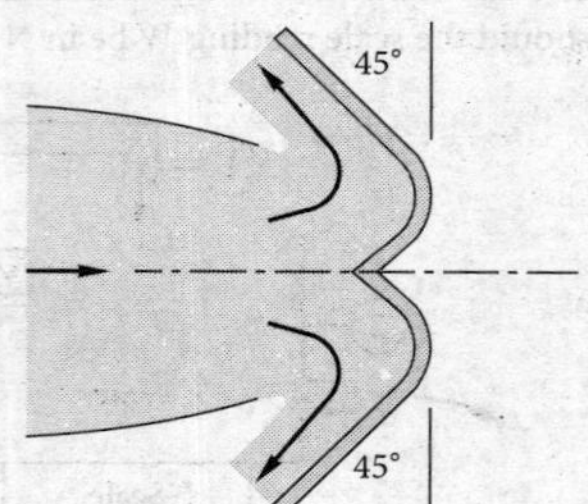

P3.71

***P3.72** When immersed in a uniform stream, a thick elliptical cylinder creates a broad downstream wake, as idealized in Fig. P3.72. The pressure at the upstream and downstream sections are approximately equal, and the fluid is water at 20°C. If $U_0 = 4$ m/s and $L = 80$ cm, estimate the drag force on the cylinder per unit width into the paper. Also compute the dimensionless drag coefficient $C_D = 2F/(\rho U_0^2 bL)$.

P3.73 A pump in a tank of water at 20°C directs a jet at 14 m/s and 0.8 m^3/min against a vane, as shown in Fig. P3.73. Compute the force F to hold the cart stationary if the jet follows (*a*) path *A* or (*b*) path *B*. The tank holds 2 m^3 of water at this instant.

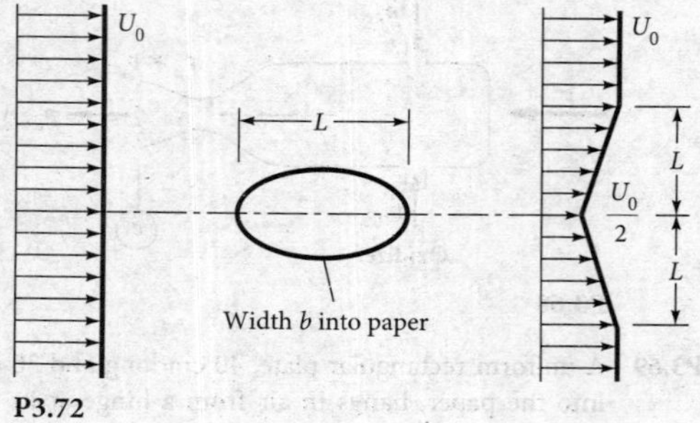

P3.72

P3.74 Water at 20°C flows down through a vertical, 6-cm-diameter tube at 1.2 m^3/min, as in Fig. P3.74. The flow then turns horizontally and exits through a 90° radial duct segment 1 cm thick, as shown. If the radial outflow is uniform and steady, estimate the forces (F_x, F_y, F_z) required to support this system against fluid momentum changes.

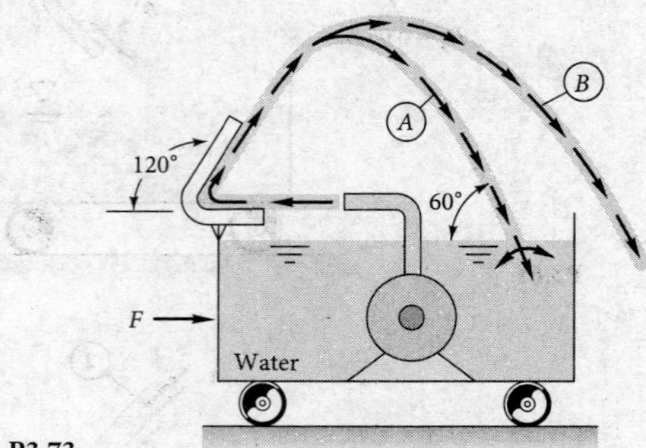

P3.73

P3.74

***P3.75** A jet of liquid of density ρ and area A strikes a block and splits into two jets, as in Fig. P3.75. Assume the same velocity V for all three jets. The upper jet exits at an angle θ and area αA. The lower jet is turned 90° downward. Neglecting fluid weight, (*a*) derive a formula for the forces (F_x, F_y) required to support the block against fluid momentum changes. (*b*) Show that $F_y = 0$ only if $\alpha \geq 0.5$. (*c*) Find the values of α and θ for which both F_x and F_y are zero.

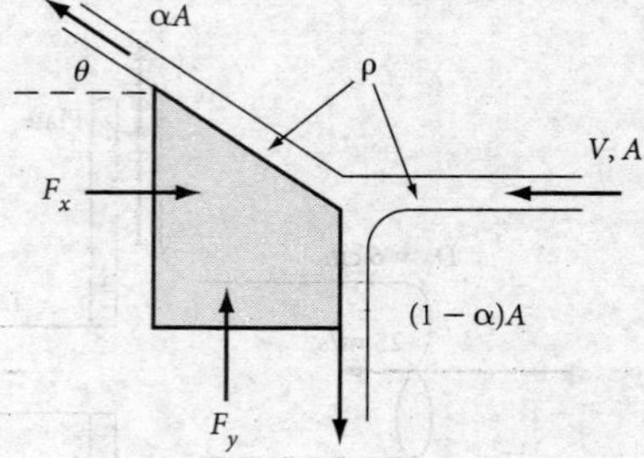

P3.75

P3.76 A two-dimensional sheet of water, 10 cm thick and moving at 7 m/s, strikes a fixed wall inclined at 20° with respect to the jet direction. Assuming frictionless flow, find (*a*) the normal force on the wall per meter of depth, and find the widths of the sheet deflected (*b*) upstream and (*c*) downstream along the wall.

P3.77 Water at 20°C flows steadily through a reducing pipe bend, as in Fig. P3.77. Known conditions are $p_1 = 350$ kPa,

$D_1 = 25$ cm, $V_1 = 2.2$ m/s, $p_2 = 120$ kPa, and $D_2 = 8$ cm. Neglecting bend and water weight, estimate the total force that must be resisted by the flange bolts.

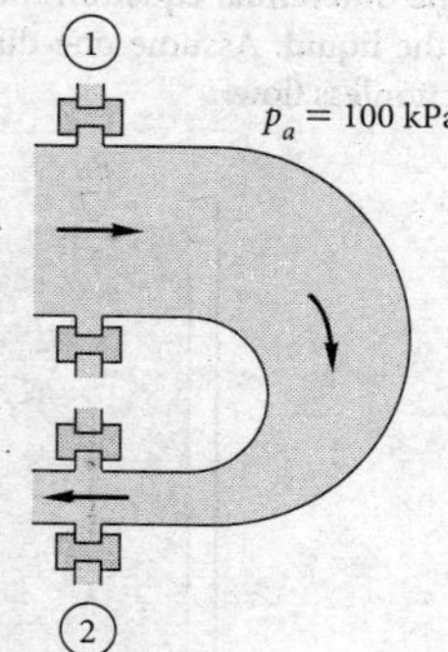

P3.77

P3.78 A fluid jet of diameter D_1 enters a cascade of moving blades at absolute velocity V_1 and angle β_1, and it leaves at absolute velocity V_2 and angle β_2, as in Fig. P3.78. The blades move at velocity u. Derive a formula for the power P delivered to the blades as a function of these parameters.

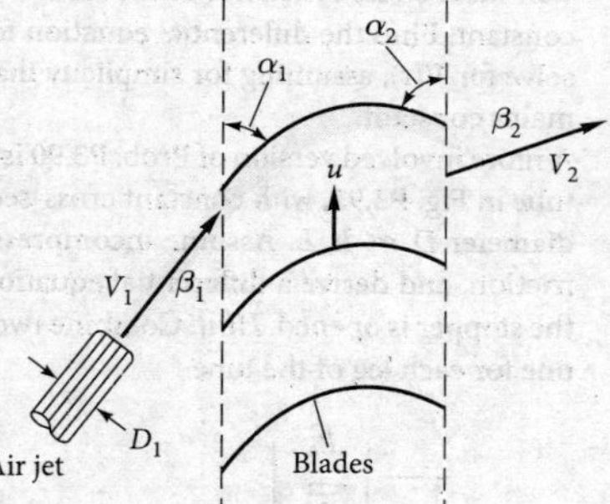

P3.78

P3.79 The Saturn V rocket in the chapter opener photo was powered by five F-1 engines, each of which burned 1,800 kg of liquid oxygen and 800 kg of kerosene per second. The exit velocity of burned gases was approximately 2,600 m/s. In the spirit of Prob. P3.34, neglecting external pressure forces, estimate the total thrust of the rocket, in N.

P3.80 A river of width b and depth h_1 passes over a submerged obstacle, or "drowned weir," in Fig. P3.80, emerging at a new flow condition (V_2, h_2). Neglect atmospheric pressure, and assume that the water pressure is hydrostatic at both sections 1 and 2. Derive an expression for the force exerted by the river on the obstacle in terms of V_1, h_1, h_2, b, ρ, and g. Neglect water friction on the river bottom.

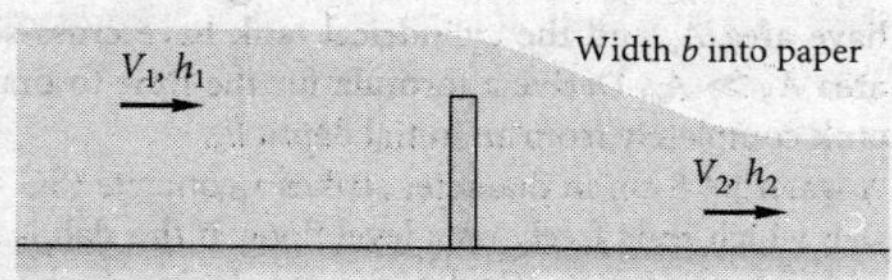

P3.80

P3.81 Torricelli's idealization of efflux from a hole in the side of a tank is $V = \sqrt{2\,gh}$, as shown in Fig. P3.81. The cylindrical tank weighs 150 N when empty and contains water at 20°C. The tank bottom is on very smooth ice (static friction coefficient $\zeta \approx 0.01$). The hole diameter is 9 cm. For what water depth h will the tank just begin to move to the right?

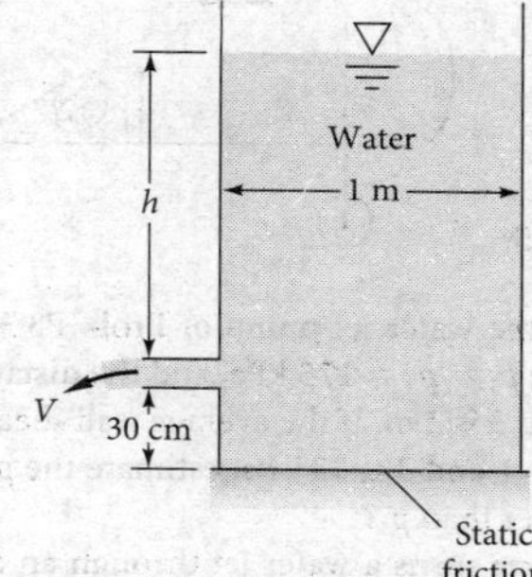

P3.81

***P3.82** The model car in Fig. P3.82 weighs 17 N and is to be accelerated from rest by a 1-cm-diameter water jet moving at 75 m/s. Neglecting air drag and wheel friction, estimate the velocity of the car after it has moved forward 1 m.

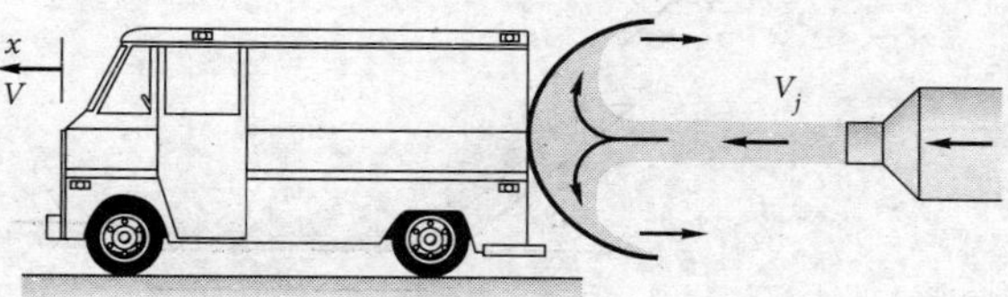

P3.82

P3.83 Gasoline at 20°C is flowing at $V_1 = 12$ m/s in a 5-cm-diameter pipe when it encounters a 1-m length of uniform radial wall suction. At the end of this suction region, the average fluid velocity has dropped to $V_2 = 10$ m/s. If $p_1 = 120$ kPa, estimate p_2 if the wall friction losses are neglected.

P3.84 Air at 20°C and 1 atm flows in a 25-cm-diameter duct at 15 m/s, as in Fig. P3.84. The exit is choked by a 90° cone, as shown. Estimate the force of the airflow on the cone.

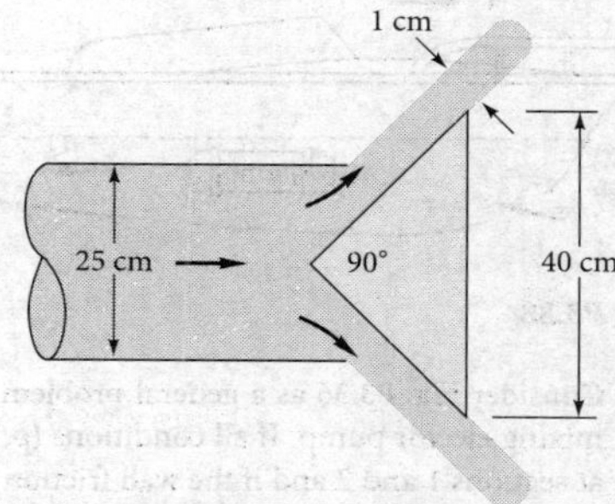

P3.84

P3.85 The thin-plate orifice in Fig. P3.85 causes a large pressure drop. For 20°C water flow at 2 m³/min, with pipe

$D = 10$ cm and orifice $d = 6$ cm, $p_1 - p_2 \approx 145$ kPa. If the wall friction is negligible, estimate the force of the water on the orifice plate.

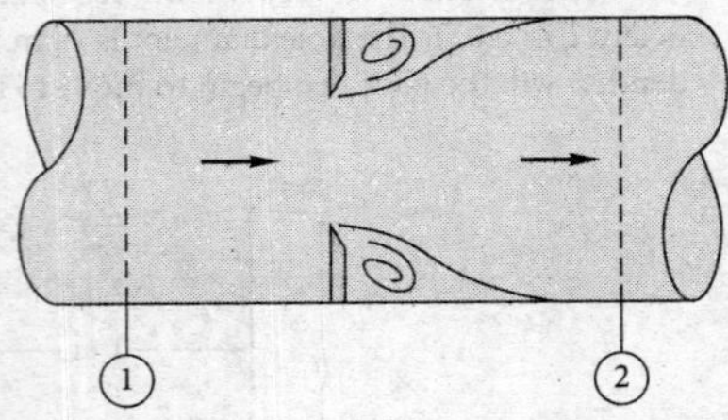

P3.85

P3.86 For the water jet pump of Prob. P3.36, add the following data: $p_1 = p_2 = 175$ kPa, and the distance between sections 1 and 3 is 2 m. If the average wall shear stress between sections 1 and 3 is 335 Pa, estimate the pressure p_3. Why is it higher than p_1?

P3.87 A vane turns a water jet through an angle α, as shown in Fig. P3.87. Neglect friction on the vane walls. (*a*) What is the angle α for the support force to be in pure compression? (*b*) Calculate this compression force if the water velocity is 7 m/s and the jet cross section is 0.003 m^2.

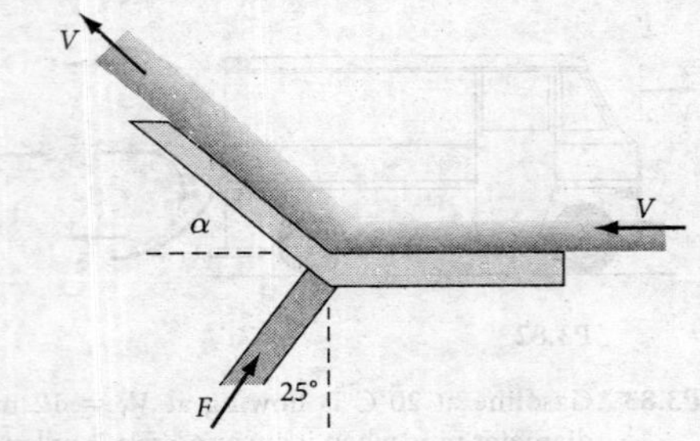

P3.87

P3.88 The boat in Fig. P3.88 is jet-propelled by a pump that develops a volume flow rate Q and ejects water out the stern at velocity V_j. If the boat drag force is $F = kV^2$, where k is a constant, develop a formula for the steady forward speed V of the boat.

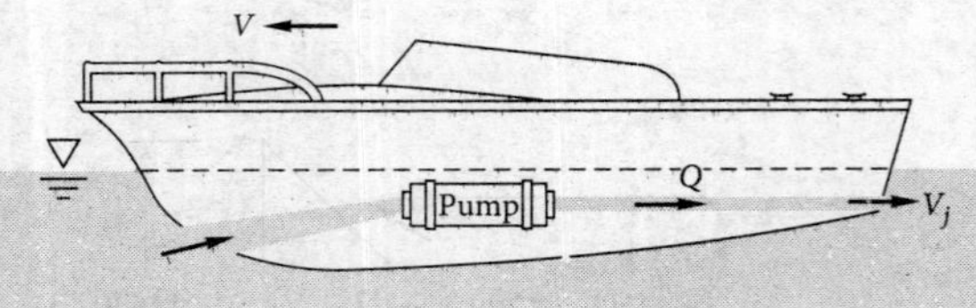

P3.88

P3.89 Consider Fig. P3.36 as a general problem for analysis of a mixing ejector pump. If all conditions (p, ρ, V) are known at sections 1 and 2 and if the wall friction is negligible, derive formulas for estimating (*a*) V_3 and (*b*) p_3.

P3.90 As shown in Fig. P3.90, a liquid column of height h is confined in a vertical tube of cross-sectional area A by a stopper. At $t = 0$ the stopper is suddenly removed, exposing the bottom of the liquid to atmospheric pressure. Using a control volume analysis of mass and vertical momentum, derive the differential equation for the downward motion $V(t)$ of the liquid. Assume one-dimensional, incompressible, frictionless flow.

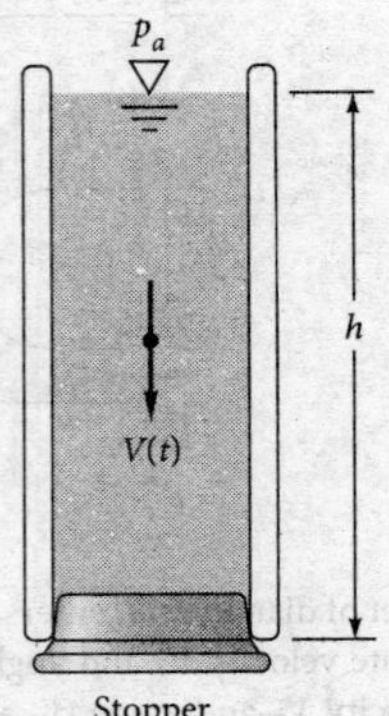

P3.90

P3.91 Extend Prob. P3.90 to include a linear (laminar) average wall shear stress resistance of the form $\tau \approx cV$, where c is a constant. Find the differential equation for dV/dt and then solve for $V(t)$, assuming for simplicity that the wall area remains constant.

***P3.92** A more involved version of Prob. P3.90 is the elbow-shaped tube in Fig. P3.92, with constant cross-sectional area A and diameter $D \ll h$, L. Assume incompressible flow, neglect friction, and derive a differential equation for dV/dt when the stopper is opened. *Hint:* Combine two control volumes, one for each leg of the tube.

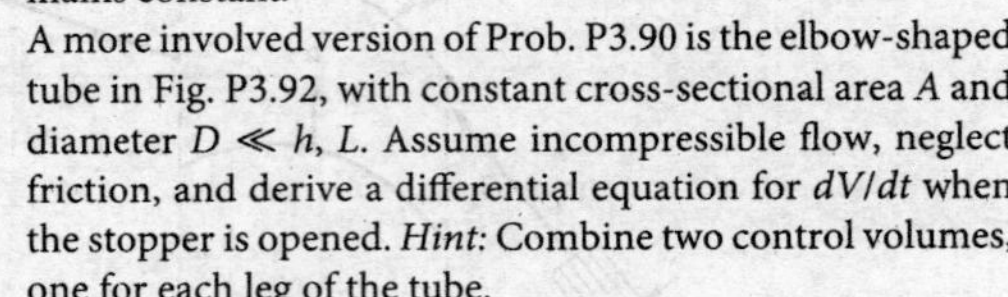

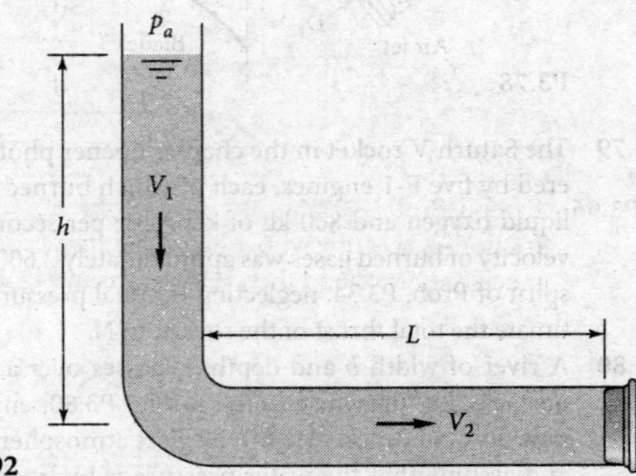

P3.92

P3.93 According to Torricelli's theorem, the velocity of a fluid draining from a hole in a tank is $V \approx (2gh)^{1/2}$, where h is the depth of water above the hole, as in Fig. P3.93. Let the hole have area A_o and the cylindrical tank have cross-section area $A_b \gg A_o$. Derive a formula for the time to drain the tank completely from an initial depth h_o.

P3.94 A water jet 8 cm in diameter strikes a concrete (SG = 2.3) slab which rests freely on a level floor. If the slab is 30 cm wide into the paper, calculate the jet velocity which will just begin to tip the slab over.

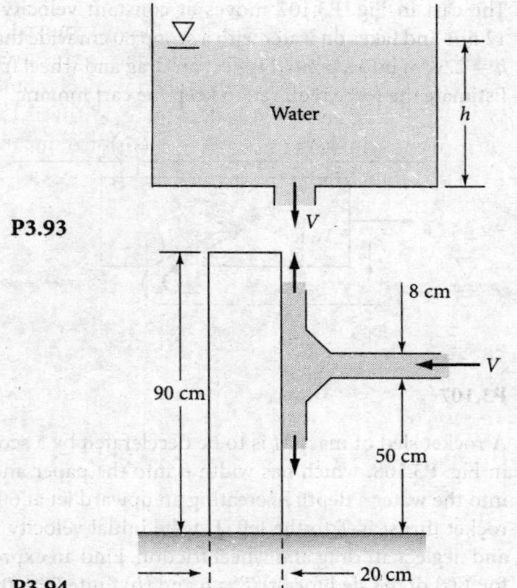

P3.93

P3.94

P3.95 A tall water tank discharges through a well-rounded orifice, as in Fig. P3.95. Use the Torricelli formula of Prob. P3.81 to estimate the exit velocity. (*a*) If, at this instant, the force F required to hold the plate is 40 N, what is the depth h? (*b*) If the tank surface is dropping at the rate of 2.5 cm/s, what is the tank diameter D?

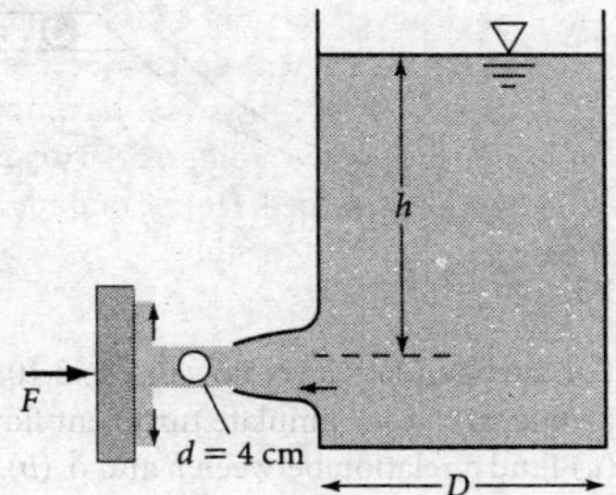

P3.95

P3.96 Extend Prob. P3.90 to the case of the liquid motion in a frictionless U-tube whose liquid column is displaced a distance Z upward and then released, as in Fig. P3.96.

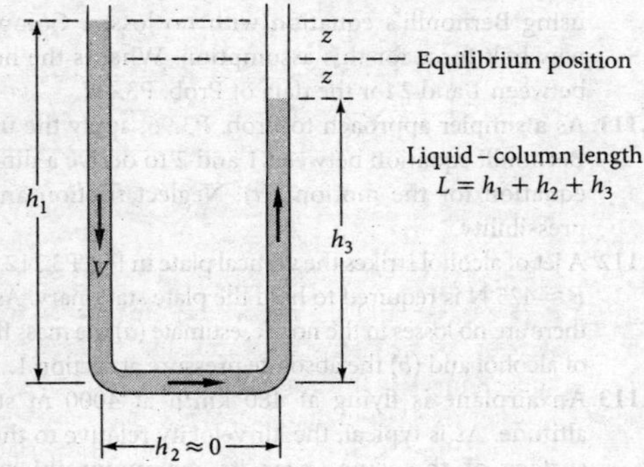

P3.96

Neglect the short horizontal leg, and combine control volume analyses for the left and right legs to derive a single differential equation for $V(t)$ of the liquid column.

***P3.97** Extend Prob. P3.96 to include a linear (laminar) average wall shear stress resistance of the form $\tau \approx 8\mu V/D$, where μ is the fluid viscosity. Find the differential equation for dV/dt and then solve for $V(t)$, assuming an initial displacement $z = z_o$, $V = 0$ at $t = 0$. The result should be a damped oscillation tending toward $z = 0$.

***P3.98** As an extension of Example 3.9, let the plate and its cart (see Fig. 3.9*a*) be unrestrained horizontally, with frictionless wheels. Derive (*a*) the equation of motion for cart velocity $V_c(t)$ and (*b*) a formula for the time required for the cart to accelerate from rest to 90 percent of the jet velocity (assuming the jet continues to strike the plate horizontally). (*c*) Compute numerical values for part (*b*) using the conditions of Example 3.9 and a cart mass of 2 kg.

P3.99 Let the rocket of Fig. E3.12 start at $z = 0$, with constant exit velocity and exit mass flow, and rise vertically with zero drag. (*a*) Show that, as long as fuel burning continues, the vertical height $S(t)$ reached is given by

$$S = \frac{V_e M_o}{\dot{m}}\left[\zeta \ln\zeta - \zeta + 1\right], \text{ where } \zeta = 1 - \frac{\dot{m}t}{M_o}$$

(*b*) Apply this to the case $V_e = 1{,}500$ m/s and $M_o = 1{,}000$ kg to find the height reached after a burn of 30 seconds, when the final rocket mass is 400 kg.

P3.100 Suppose that the solid-propellant rocket of Prob. P3.35 is built into a missile of diameter 70 cm and length 4 m. The system weighs 1,800 N, which includes 700 N of propellant. Neglect air drag. If the missile is fired vertically from rest at sea level, estimate (*a*) its velocity and height at fuel burnout and (*b*) the maximum height it will attain.

P3.101 Water at 20°C flows steadily through the tank in Fig. P3.101. Known conditions are $D_1 = 8$ cm, $V_1 = 6$ m/s, and $D_2 = 4$ cm. A rightward force $F = 70$ N is required to keep the tank fixed. (*a*) What is the velocity leaving section 2? (*b*) If the tank cross section is 1.2 m², how fast is the water surface $h(t)$ rising or falling?

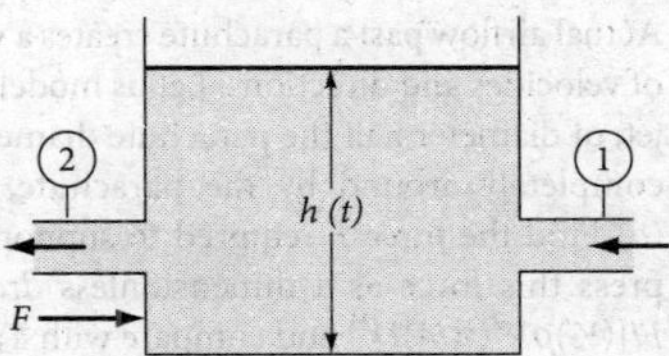

P3.101

P3.102 As can often be seen in a kitchen sink when the faucet is running, a high-speed channel flow (V_1, h_1) may "jump" to a low-speed, low-energy condition (V_2, h_2) as in Fig. P3.102. The pressure at sections 1 and 2 is approximately hydrostatic, and wall friction is negligible. Use the continuity and momentum relations to find h_2 and V_2 in terms of (h_1, V_1).

***P3.103** Suppose that the solid-propellant rocket of Prob. P3.35 is mounted on a 1,000-kg car to propel it up a long slope of 15°. The rocket motor weighs 900 N, which includes 500 N of propellant. If the car starts from rest when the rocket is fired, and if air drag and wheel friction are neglected, estimate the maximum distance that the car will travel up the hill.

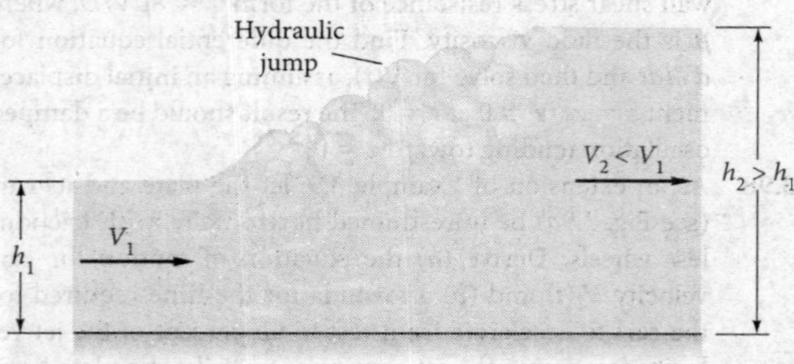

P3.102

P3.104 A rocket is attached to a rigid horizontal rod hinged at the origin as in Fig. P3.104. Its initial mass is M_0, and its exit properties are $\dot{m}$ and V_e relative to the rocket. Set up the differential equation for rocket motion, and solve for the angular velocity $\omega(t)$ of the rod. Neglect gravity, air drag, and the rod mass.

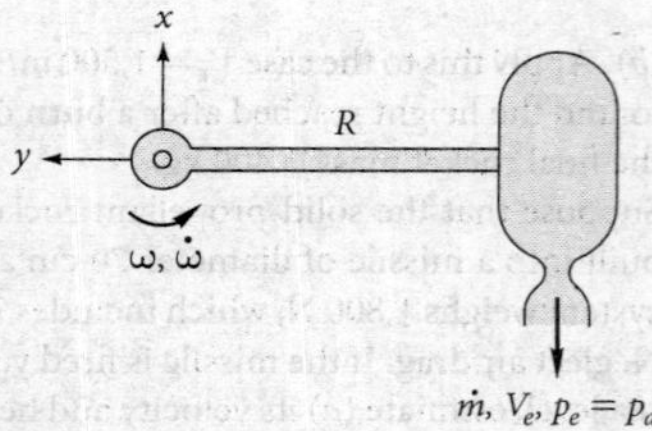

P3.104

P3.105 Extend Prob. P3.104 to the case where the rocket has a linear air drag force $F = cV$, where c is a constant. Assuming no burnout, solve for $\omega(t)$ and find the *terminal* angular velocity—that is, the final motion when the angular acceleration is zero. Apply to the case $M_0 = 6$ kg, $R = 3$ m, $\dot{m} = 0.05$ kg/s, $V_e = 1{,}100$ m/s, and $c = 0.075$ N · s/m to find the angular velocity after 12 s of burning.

P3.106 Actual airflow past a parachute creates a variable distribution of velocities and directions. Let us model this as a circular air jet, of diameter half the parachute diameter, which is turned completely around by the parachute, as in Fig. P3.106. (*a*) Find the force F required to support the chute. (*b*) Express this force as a dimensionless *drag coefficient*, $C_D = F/[(½)\rho V^2(\pi/4)D^2]$ and compare with Table 7.3.

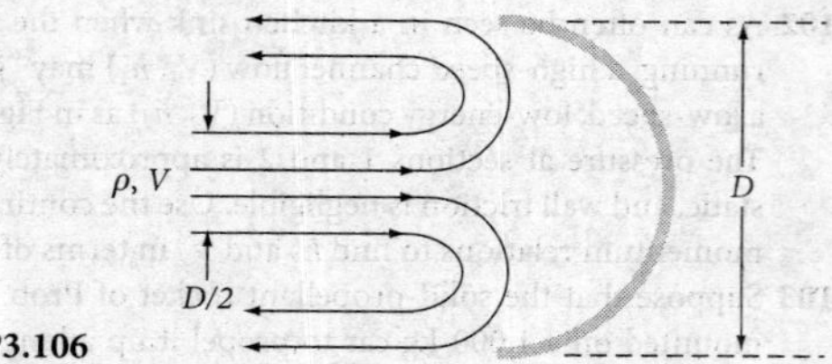

P3.106

P3.107 The cart in Fig. P3.107 moves at constant velocity $V_0 = 12$ m/s and takes on water with a scoop 80 cm wide that dips $h = 2.5$ cm into a pond. Neglect air drag and wheel friction. Estimate the force required to keep the cart moving.

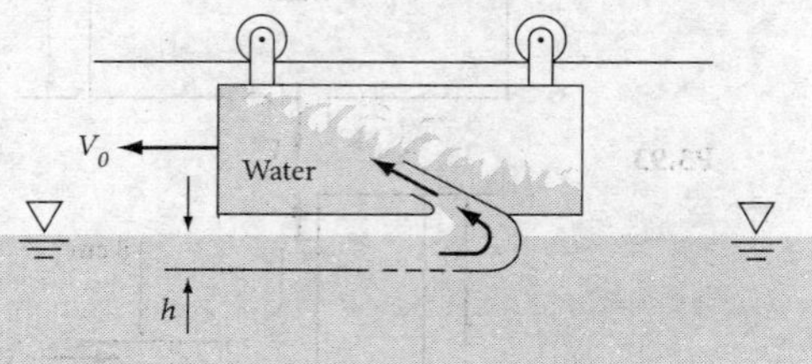

P3.107

***P3.108** A rocket sled of mass M is to be decelerated by a scoop, as in Fig. P3.108, which has width b into the paper and dips into the water a depth h, creating an upward jet at 60°. The rocket thrust is T to the left. Let the initial velocity be V_0, and neglect air drag and wheel friction. Find an expression for $V(t)$ of the sled for (*a*) $T = 0$ and (*b*) finite $T \neq 0$.

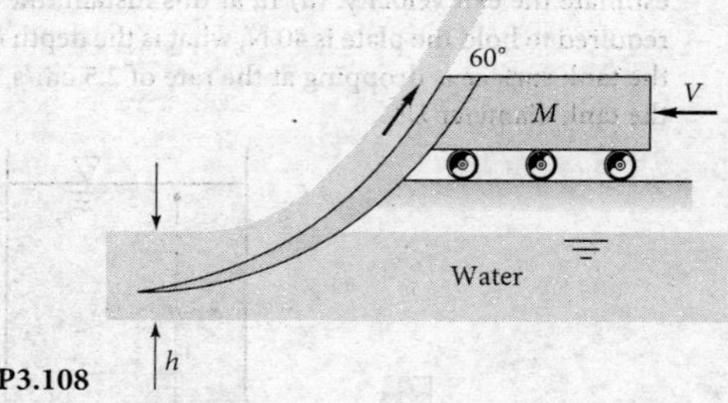

P3.108

P3.109 For the boundary layer flow in Fig. 3.10, let the exit velocity profile, at $x = L$, simulate turbulent flow, $u \approx U_0(y/\delta)^{1/7}$. (*a*) Find a relation between h and δ. (*b*) Find an expression for the drag force F on the plate between 0 and L.

The Bernoulli Equation

P3.110 Repeat Prob. P3.49 by assuming that p_1 is unknown and using Bernoulli's equation with no losses. Compute the new bolt force for this assumption. What is the head loss between 1 and 2 for the data of Prob. P3.49?

P3.111 As a simpler approach to Prob. P3.96, apply the unsteady Bernoulli equation between 1 and 2 to derive a differential equation for the motion $z(t)$. Neglect friction and compressibility.

P3.112 A jet of alcohol strikes the vertical plate in Fig. P3.112. A force $F \approx 425$ N is required to hold the plate stationary. Assuming there are no losses in the nozzle, estimate (*a*) the mass flow rate of alcohol and (*b*) the absolute pressure at section 1.

P3.113 An airplane is flying at 480 km/h at 4000 m standard altitude. As is typical, the air velocity relative to the upper surface of the wing, near its maximum thickness, is

26 percent higher than the plane's velocity. Using Bernoulli's equation, calculate the absolute pressure at this point on the wing. Neglect elevation changes and compressibility.

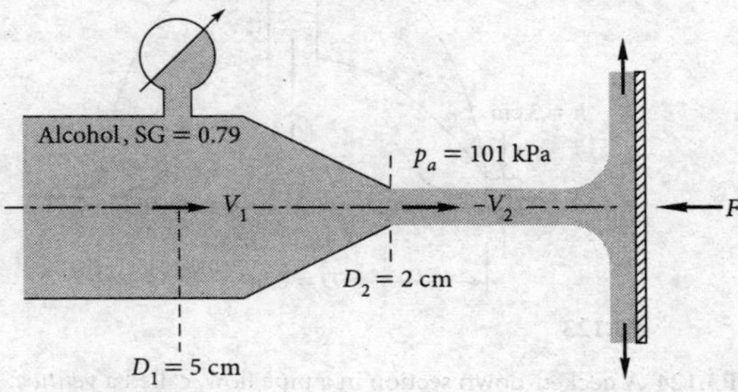

P3.112

P3.114 Water flows through a circular nozzle, exits into the air as a jet, and strikes a plate, as shown in Fig. P3.114. The force required to hold the plate steady is 70 N. Assuming steady, frictionless, one-dimensional flow, estimate (*a*) the velocities at sections (1) and (2) and (*b*) the mercury manometer reading *h*.

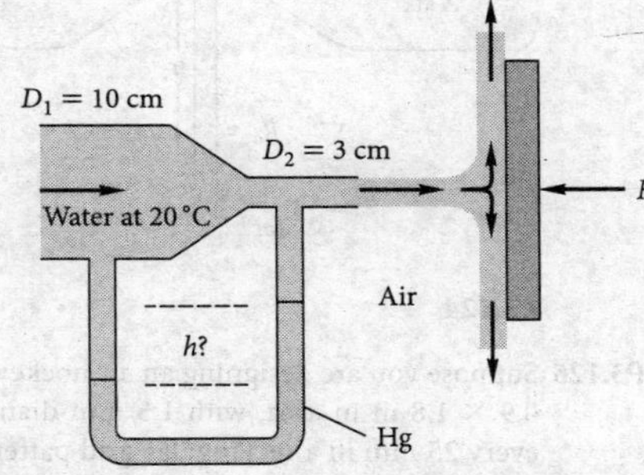

P3.114

P3.115 A free liquid jet, as in Fig. P3.115, has constant ambient pressure and small losses; hence from Bernoulli's equation $z + V^2/(2g)$ is constant along the jet. For the fire nozzle in the figure, what are (*a*) the minimum and (*b*) the maximum values of θ for which the water jet will clear the corner of the building? For which case will the jet velocity be higher when it strikes the roof of the building?

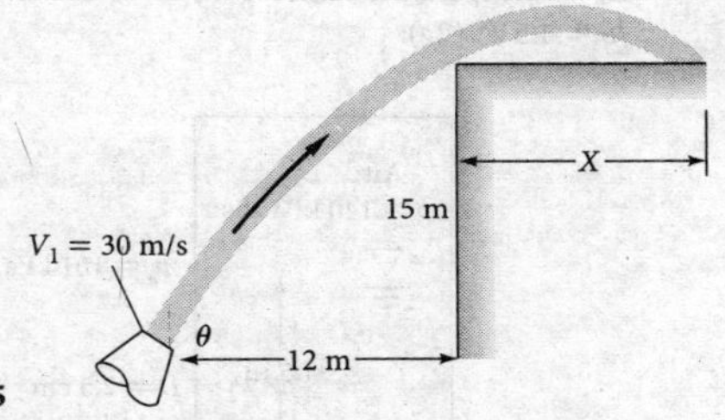

P3.115

P3.116 For the container of Fig. P3.116 use Bernoulli's equation to derive a formula for the distance *X* where the free jet leaving horizontally will strike the floor, as a function of *h* and *H*. For what ratio *h*/*H* will *X* be maximum? Sketch the three trajectories for *h*/*H* = 0.25, 0.5, and 0.75.

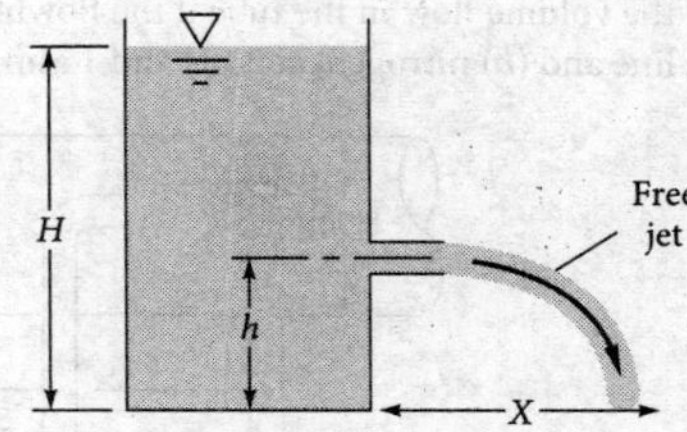

P3.116

P3.117 Water at 20°C, in the pressurized tank of Fig. P3.117, flows out and creates a vertical jet as shown. Assuming steady frictionless flow, determine the height *H* to which the jet rises.

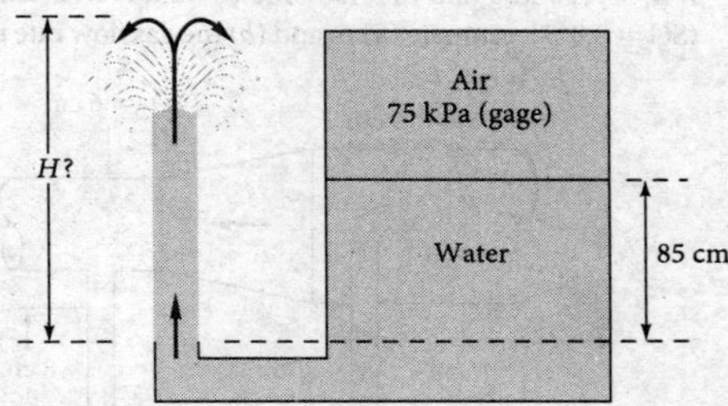

P3.117

P3.118 Bernoulli's 1738 treatise *Hydrodynamica* contains many excellent sketches of flow patterns related to his frictionless relation. One, however, redrawn here as Fig. P3.118, seems physically misleading. Can you explain what might be wrong with the figure?

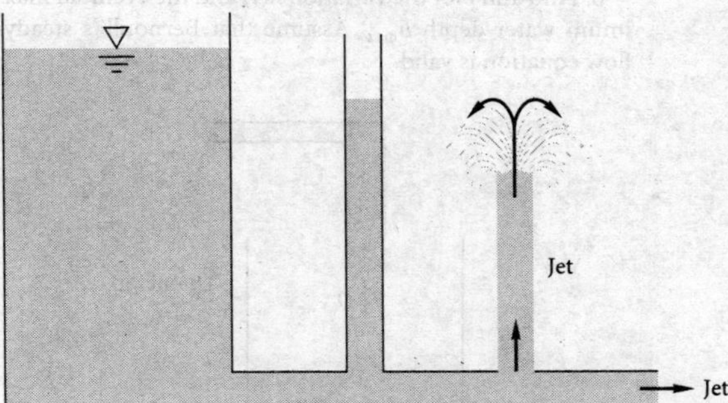

P3.118

P3.119 A long fixed tube with a rounded nose, aligned with an oncoming flow, can be used to measure velocity. Measurements are made of the pressure at (1) the front nose and (2) a hole in the side of the tube further along, where the pressure nearly equals stream pressure.
(*a*) Make a sketch of this device and show how the velocity is calculated. (*b*) For a particular sea-level airflow, the

difference between nose pressure and side pressure is 10 kPa. What is the air velocity, in m/s?

P3.120 The manometer fluid in Fig. P3.120 is mercury. Estimate the volume flow in the tube if the flowing fluid is (*a*) gasoline and (*b*) nitrogen, at 20°C and 1 atm.

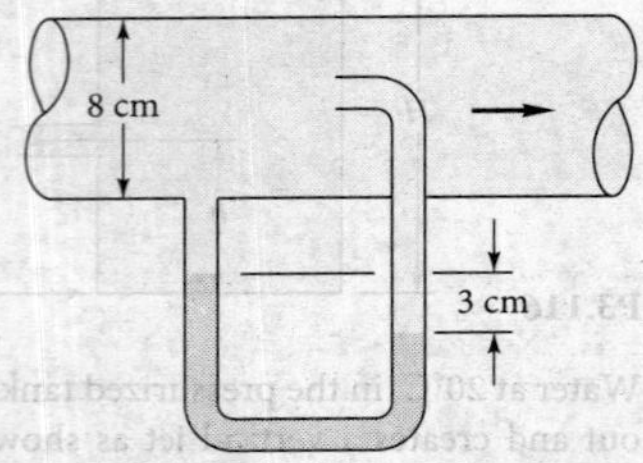

P3.120

P3.121 In Fig. P3.121 the flowing fluid is CO_2 at 20°C. Neglect losses. If p_1 = 170 kPa and the manometer fluid is Meriam red oil (SG = 0.827), estimate (*a*) p_2 and (*b*) the gas flow rate in m³/h.

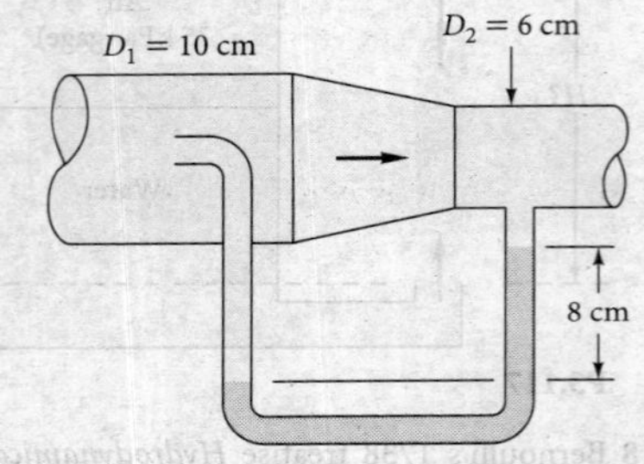

P3.121

P3.122 The cylindrical water tank in Fig. P3.122 is being filled at a volume flow Q_1 = 4 L/min, while the water also drains from a bottom hole of diameter d = 6 mm. At time t = 0, h = 0. Find and plot the variation $h(t)$ and the eventual maximum water depth h_{max}. Assume that Bernoulli's steady-flow equation is valid.

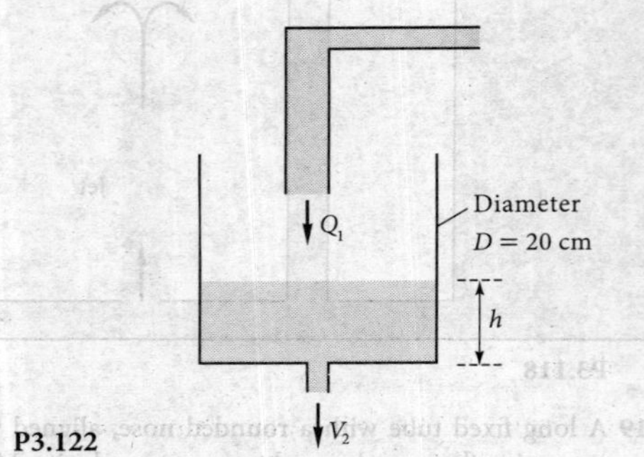

P3.122

P3.123 The air-cushion vehicle in Fig. P3.123 brings in sea-level standard air through a fan and discharges it at high velocity through an annular skirt of 3-cm clearance. If the vehicle weighs 50 kN, estimate (*a*) the required airflow rate and (*b*) the fan power in kW.

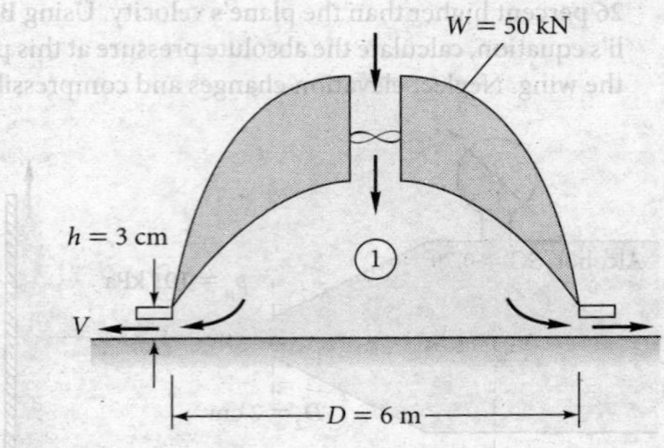

P3.123

P3.124 A necked-down section in a pipe flow, called a *venturi*, develops a low throat pressure that can aspirate fluid upward from a reservoir, as in Fig. P3.124. Using Bernoulli's equation with no losses, derive an expression for the velocity V_1 that is just sufficient to bring reservoir fluid into the throat.

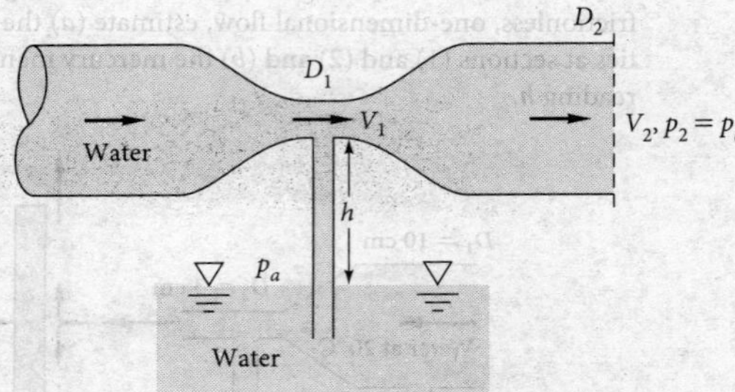

P3.124

P3.125 Suppose you are designing an air hockey table. The table is 0.9 × 1.8 m in area, with 1.5 mm-diameter holes spaced every 25 mm in a rectangular grid pattern (2,592 holes total). The required jet speed from each hole is estimated to be 15 m/s. Your job is to select an appropriate blower that will meet the requirements. Estimate the volumetric flow rate (in m³/min) and pressure rise (in N/m²) required of the blower. *Hint:* Assume that the air is stagnant in the large volume of the manifold under the table surface, and neglect any frictional losses.

P3.126 The liquid in Fig. P3.126 is kerosene at 20°C. Estimate the flow rate from the tank for (*a*) no losses and (*b*) pipe losses $h_f \approx 4.5V^2/(2g)$.

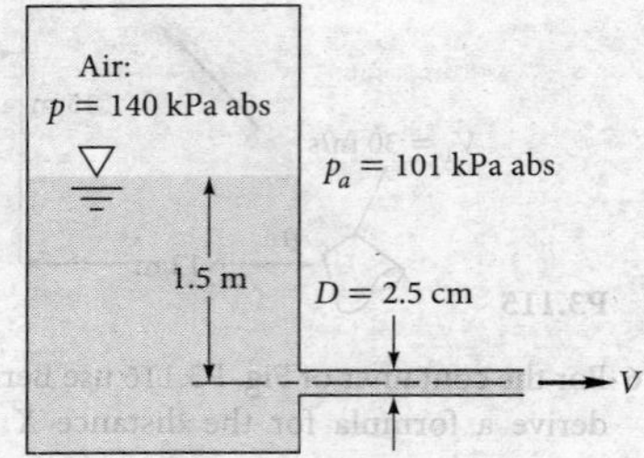

P3.126

P3.127 In Fig. P3.127 the open jet of water at 20°C exits a nozzle into sea-level air and strikes a stagnation tube as shown.

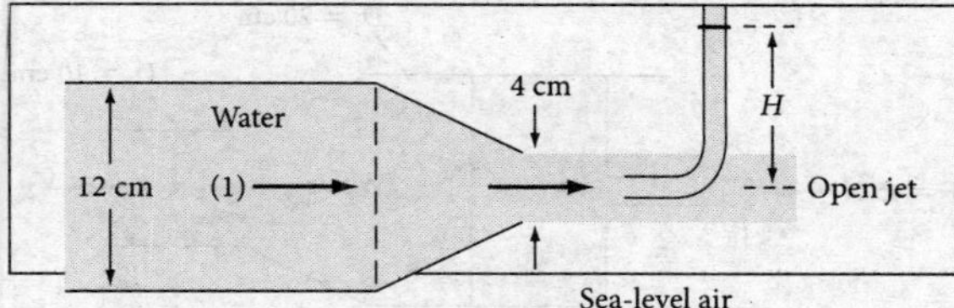

P3.127

If the pressure at the centerline at section 1 is 110 kPa, and losses are neglected, estimate (*a*) the mass flow in kg/s and (*b*) the height *H* of the fluid in the stagnation tube.

P3.128 A *venturi meter*, shown in Fig. P3.128, is a carefully designed constriction whose pressure difference is a measure of the flow rate in a pipe. Using Bernoulli's equation for steady incompressible flow with no losses, show that the flow rate *Q* is related to the manometer reading *h* by

$$Q = \frac{A_2}{\sqrt{1 - (D_2/D_1)^4}}\sqrt{\frac{2gh(\rho_M - \rho)}{\rho}}$$

where ρ_M is the density of the manometer fluid.

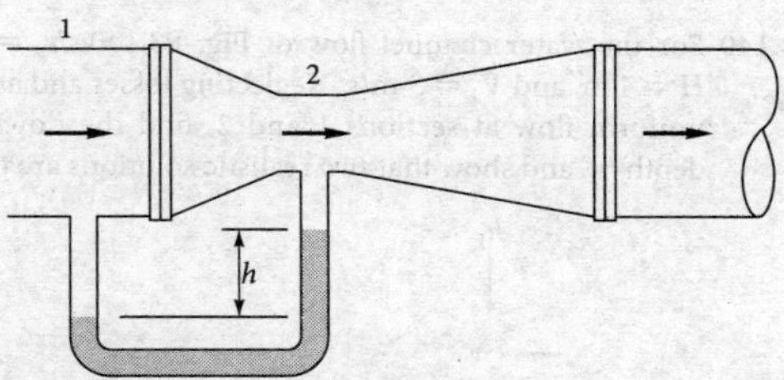

P3.128

P3.129 A water stream flows past a small circular cylinder at 7 m/s, approaching the cylinder at 150 kPa. Measurements at low (laminar flow) Reynolds numbers indicate a maximum surface velocity 60 percent higher than the stream velocity at point *B* on the cylinder. Estimate the pressure at *B*.

P3.130 In Fig. P3.130 the fluid is gasoline at 20°C at a weight flow of 120 N/s. Assuming no losses, estimate the gage pressure at section 1.

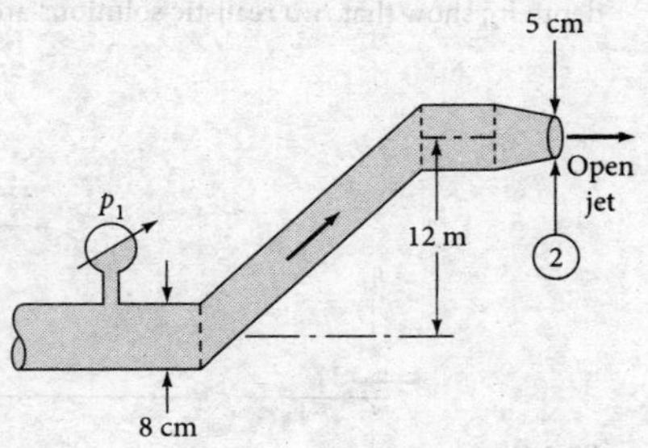

P3.130

P3.131 In Fig. P3.131 both fluids are at 20°C. If V_1 = 0.5 m/s and losses are neglected, what should the manometer reading *h* m be?

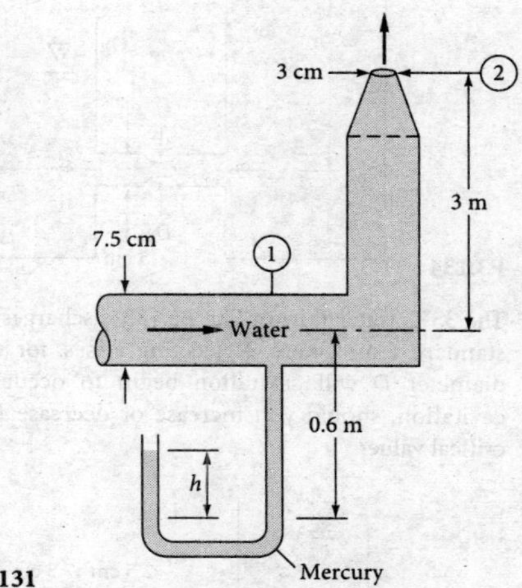

P3.131

P3.132 Extend the siphon analysis of Example 3.14 to account for friction in the tube, as follows. Let the friction head loss in the tube be correlated as $5.4(V_{tube})^2/(2g)$, which approximates turbulent flow in a 2-m-long tube. Calculate the exit velocity in m/s and the volume flow rate in cm³/s, and compare to Example 3.14.

P3.133 If losses are neglected in Fig. P3.133, for what water level *h* will the flow begin to form vapor cavities at the throat of the nozzle?

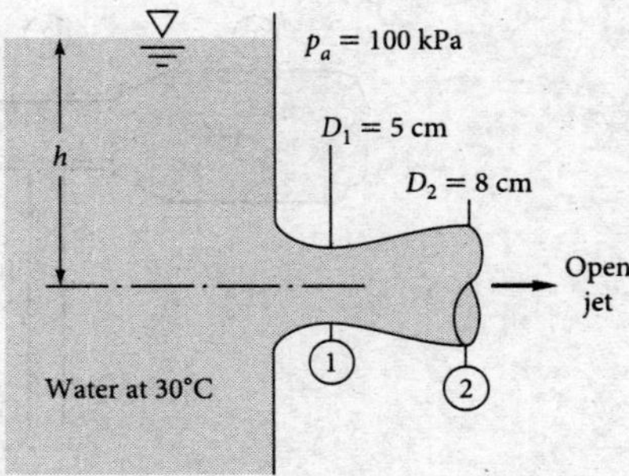

P3.133

***P3.134** For the 40°C water flow in Fig. P3.134, estimate the volume flow through the pipe, assuming no losses; then explain what is wrong with this seemingly innocent question. If the actual flow rate is *Q* = 40 m³/h, compute (*a*) the head loss in m and (*b*) the constriction diameter *D* that causes cavitation, assuming that the throat divides the head loss equally and that changing the constriction causes no additional losses.

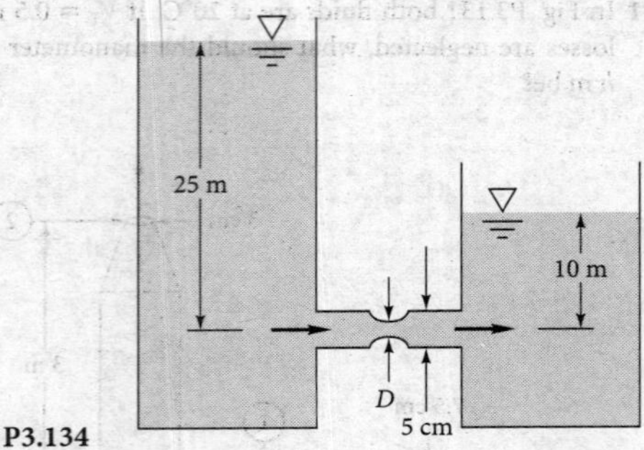

P3.134

P3.135 The 35°C water flow of Fig. P3.135 discharges to sea-level standard atmosphere. Neglecting losses, for what nozzle diameter D will cavitation begin to occur? To avoid cavitation, should you increase or decrease D from this critical value?

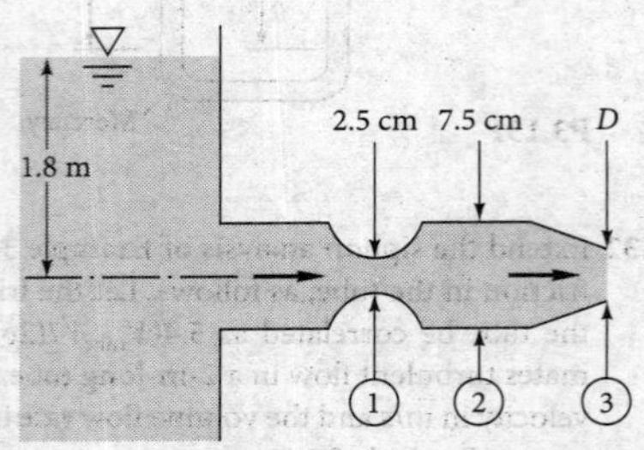

P3.135

P3.136 Air, assumed frictionless, flows through a tube, exiting to sea-level atmosphere. Diameters at 1 and 3 are 5 cm, while $D_2 = 3$ cm. What mass flow of air is required to suck water up 10 cm into section 2 of Fig. P3.136?

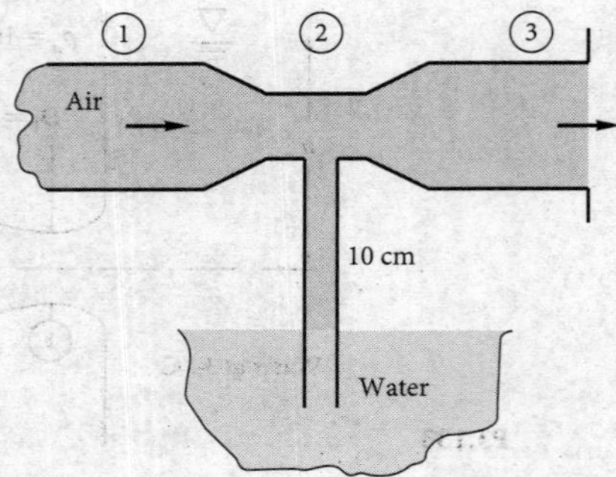

P3.136

P3.137 In Fig. P3.137 the piston drives water at 20°C. Neglecting losses, estimate the exit velocity V_2 m/s. If D_2 is further constricted, what is the limiting possible value of V_2?

P3.138 For the sluice gate flow of Example 3.10, use Bernoulli's equation, along the surface, to estimate the flow rate Q as a function of the two water depths. Assume constant width b.

P3.139 In the spillway flow of Fig. P3.139, the flow is assumed uniform and hydrostatic at sections 1 and 2. If losses are neglected, compute (*a*) V_2 and (*b*) the force per unit width of the water on the spillway.

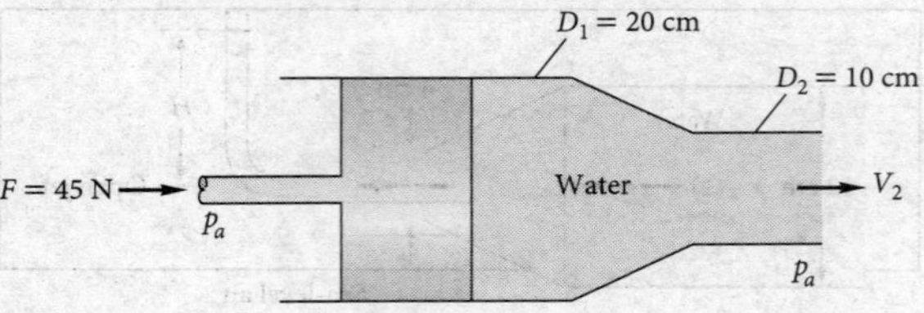

P3.137

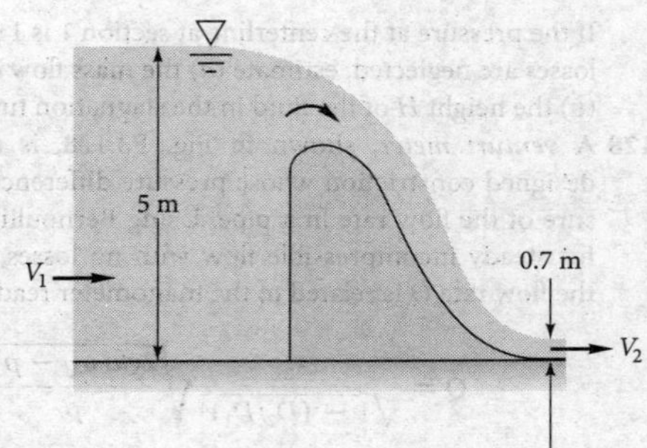

P3.139

P3.140 For the water channel flow of Fig. P3.140, $h_1 = 1.5$ m, $H = 4$ m, and $V_1 = 3$ m/s. Neglecting losses and assuming uniform flow at sections 1 and 2, find the downstream depth h_2, and show that *two* realistic solutions are possible.

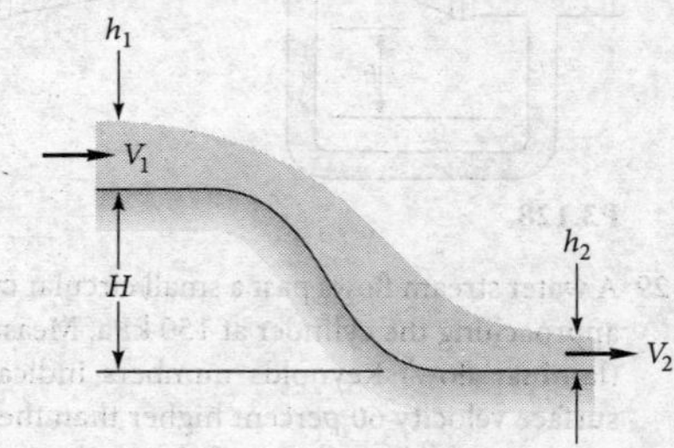

P3.140

P3.141 For the water channel flow of Fig. P3.141, $h_1 = 14$ cm, $H = 0.7$ m, and $V_1 = 5$ m/s. Neglecting losses and assuming uniform flow at sections 1 and 2, find the downstream depth h_2; show that *two* realistic solutions are possible.

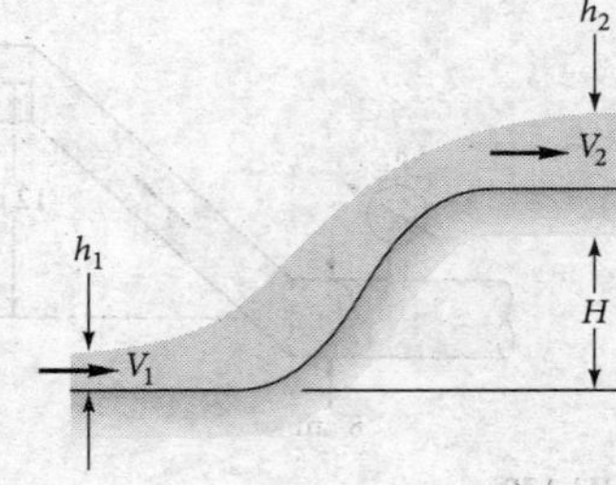

P3.141

***P3.142** A cylindrical tank of diameter D contains liquid to an initial height h_0. At time $t = 0$ a small stopper of diameter d is removed from the bottom. Using Bernoulli's equation with no losses, derive (*a*) a differential equation for the free-surface height $h(t)$ during draining and (*b*) an expression for the time t_0 to drain the entire tank.

***P3.143** The large tank of incompressible liquid in Fig. P3.143 is at rest when, at $t = 0$, the valve is opened to the atmosphere. Assuming $h \approx$ constant (negligible velocities and accelerations in the tank), use the unsteady frictionless Bernoulli equation to derive and solve a differential equation for $V(t)$ in the pipe.

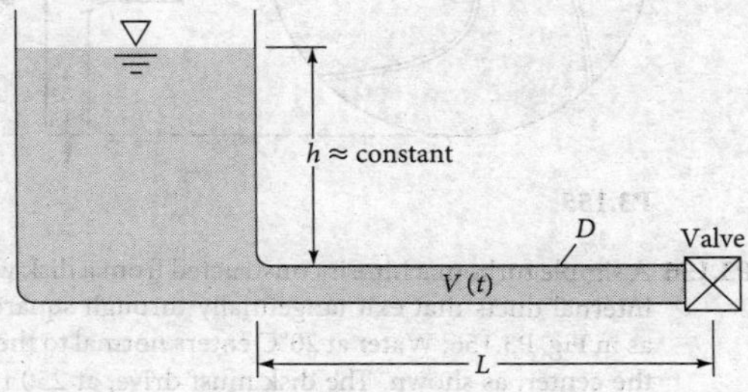

P3.143

P3.144 A fire hose, with a 5 cm-diameter nozzle, delivers a water jet straight up against a ceiling 2.4 m higher. The force on the ceiling, due to momentum change, is 111 N. Use Bernoulli's equation to estimate the hose flow rate, in m^3/s. [*Hint:* The water jet area expands upward.]

P3.145 The incompressible flow form of Bernoulli's relation, Eq. (3.54), is accurate only for Mach numbers less than about 0.3. At higher speeds, variable density must be accounted for. The most common assumption for compressible fluids is *isentropic flow of an ideal gas*, or $p = C\rho^k$, where $k = c_p/c_v$. Substitute this relation into Eq. (3.52), integrate, and eliminate the constant C. Compare your compressible result with Eq. (3.54) and comment.

P3.146 The pump in Fig. P3.146 draws gasoline at 20°C from a reservoir. Pumps are in big trouble if the liquid vaporizes (cavitates) before it enters the pump. (*a*) Neglecting losses and assuming a flow rate of 4 L/s, find the limitations on (x, y, z) for avoiding cavitation. (*b*) If pipe friction losses are included, what additional limitations might be important?

P3.147 The very large water tank in Fig. P3.147 is discharging through a 10 cm-diameter pipe. The pump is running, with a performance curve $h_p \approx 12 - 150\,Q^2$, with h_p in m and Q in m^3/s. Estimate the discharge flow rate in m^3/s if the pipe friction loss is $1.5(V^2/2g)$.

P3.148 By neglecting friction, (*a*) use the Bernoulli equation between surfaces 1 and 2 to estimate the volume flow through the orifice, whose diameter is 3 cm. (*b*) Why is the result to part (*a*) absurd? (*c*) Suggest a way to resolve this paradox and find the true flow rate.

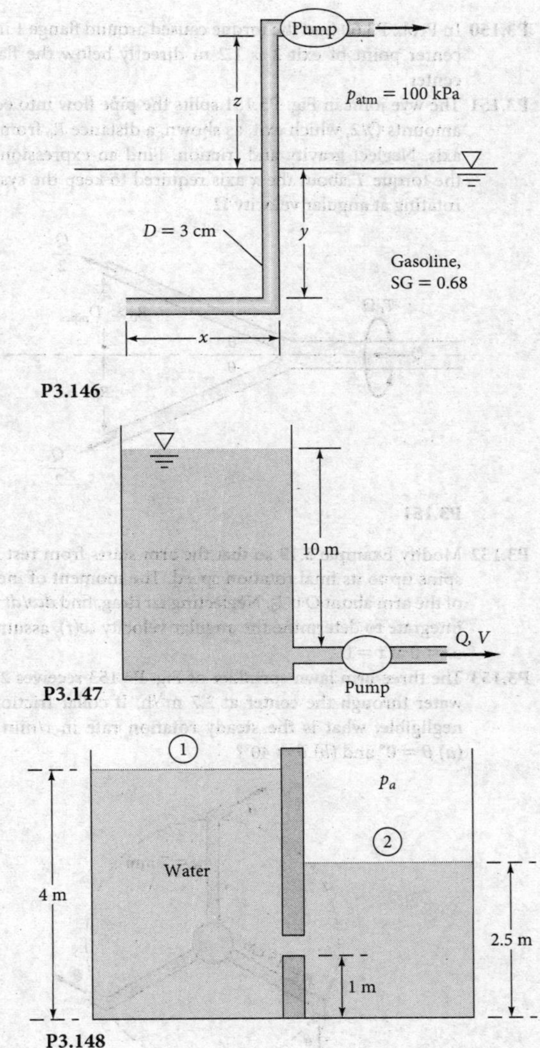

P3.146

P3.147

P3.148

The angular momentum theorem

P3.149 The horizontal lawn sprinkler in Fig. P3.149 has a water flow rate of 15 L/min introduced vertically through the center. Estimate (*a*) the retarding torque required to keep the arms from rotating and (*b*) the rotation rate (r/min) if there is no retarding torque.

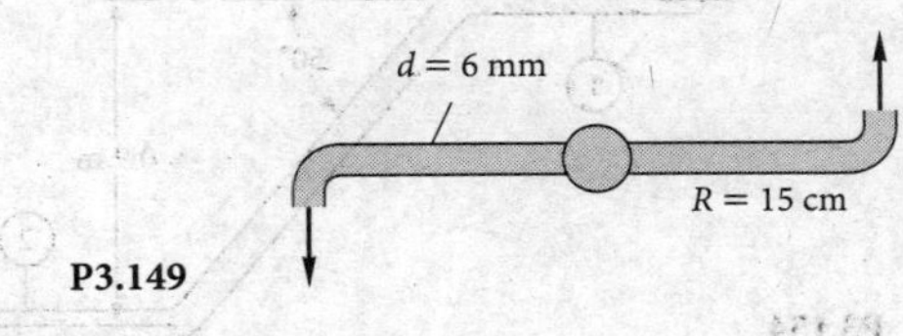

P3.149

P3.150 In Prob. P3.60 find the torque caused around flange 1 if the center point of exit 2 is 1.2 m directly below the flange center.

P3.151 The wye joint in Fig. P3.151 splits the pipe flow into equal amounts $Q/2$, which exit, as shown, a distance R_0 from the axis. Neglect gravity and friction. Find an expression for the torque T about the x axis required to keep the system rotating at angular velocity Ω.

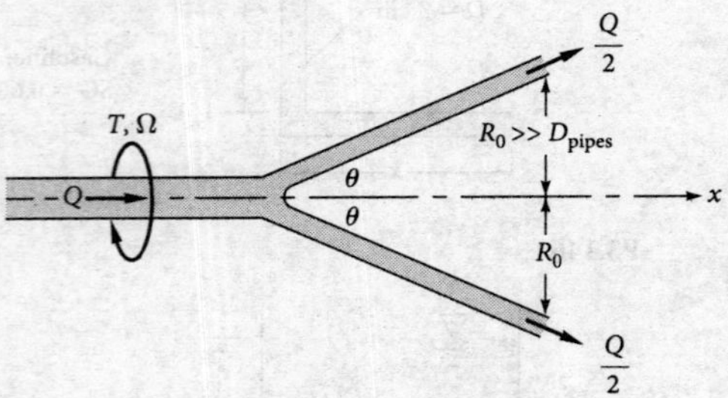

P3.151

P3.152 Modify Example 3.19 so that the arm starts from rest and spins up to its final rotation speed. The moment of inertia of the arm about O is I_0. Neglecting air drag, find $d\omega/dt$ and integrate to determine the angular velocity $\omega(t)$, assuming $\omega = 0$ at $t = 0$.

P3.153 The three-arm lawn sprinkler of Fig. P3.153 receives 20°C water through the center at 2.7 m^3/h. If collar friction is negligible, what is the steady rotation rate in r/min for (*a*) $\theta = 0°$ and (*b*) $\theta = 40°$?

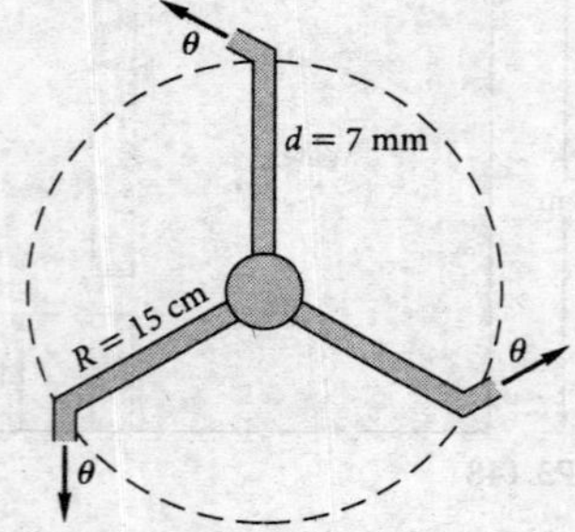

P3.153

P3.154 Water at 20°C flows at 120 L/min through the 2-cm-diameter double pipe bend of Fig. P3.154. The pressures are $p_1 = 210$ kPa and $p_2 = 170$ kPa. Compute the torque T at point B necessary to keep the pipe from rotating.

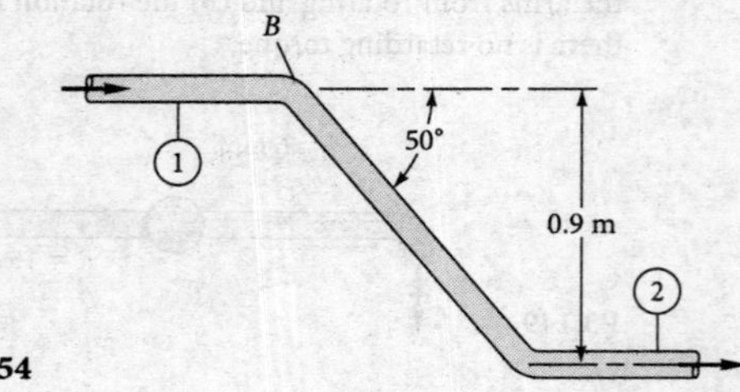

P3.154

P3.155 The centrifugal pump of Fig. P3.155 has a flow rate Q and exits the impeller at an angle θ_2 relative to the blades, as shown. The fluid enters axially at section 1. Assuming incompressible flow at shaft angular velocity ω, derive a formula for the power P required to drive the impeller.

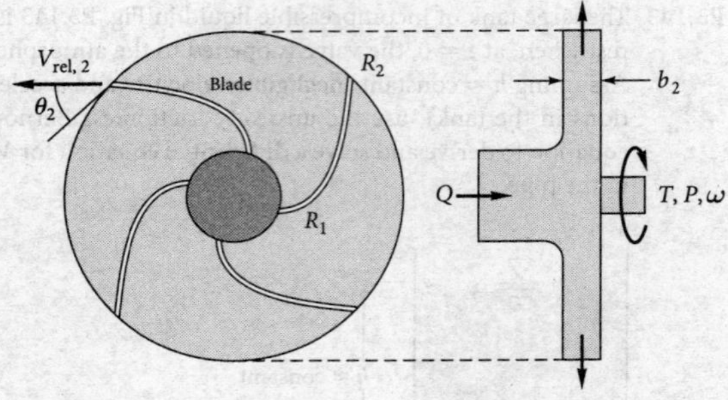

P3.155

P3.156 A simple turbomachine is constructed from a disk with two internal ducts that exit tangentially through square holes, as in Fig. P3.156. Water at 20°C enters normal to the disk at the center, as shown. The disk must drive, at 250 r/min, a small device whose retarding torque is 1.5 N · m. What is the proper mass flow of water, in kg/s?

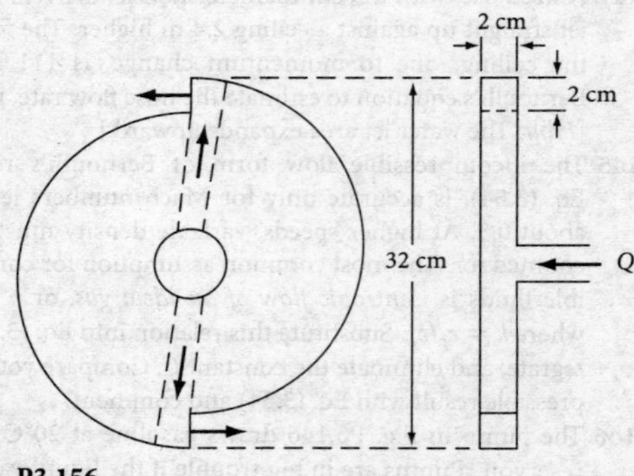

P3.156

P3.157 Reverse the flow in Fig. P3.155, so that the system operates as a radial-inflow *turbine*. Assuming that the outflow into section 1 has no tangential velocity, derive an expression for the power P extracted by the turbine.

P3.158 Revisit the turbine cascade system of Prob. P3.78, and derive a formula for the power P delivered, using the *angular* momentum theorem of Eq. (3.59).

P3.159 A centrifugal pump impeller delivers 15 m^3/min of water at 20°C with a shaft rotation rate of 1,750 r/min. Neglect losses. If $r_1 = 0.15$ m, $r_2 = 0.4$ m, $b_1 = b_2 = 4$ cm, $V_{t1} = 3$ m/s, and $V_{t2} = 34$ m/s, compute the absolute velocities (*a*) V_1 and (*b*) V_2 and (*c*) the horsepower required. (*d*) Compare with the ideal horsepower required.

P3.160 The pipe bend of Fig. P3.160 has $D_1 = 27$ cm and $D_2 = 13$ cm. When water at 20°C flows through the pipe at

15 m³/min, p_1 = 194 kPa (gage). Compute the torque required at point *B* to hold the bend stationary.

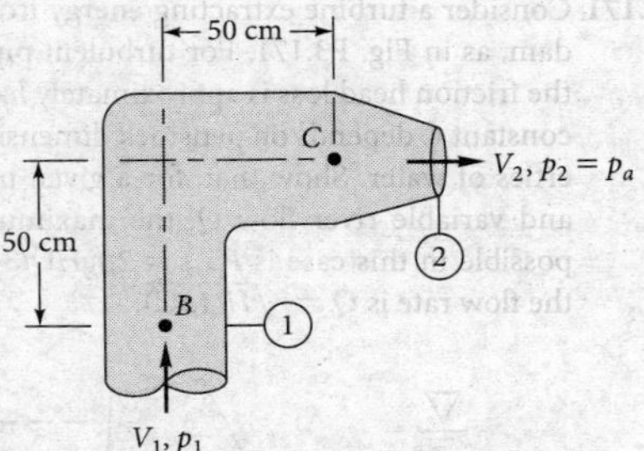

P3.160

***P3.161** Extend Prob. P3.46 to the problem of computing the center of pressure *L* of the normal face F_n, as in Fig. P3.161. (At the center of pressure, no moments are required to hold the plate at rest.) Neglect friction. Express your result in terms of the sheet thickness h_1 and the angle θ between the plate and the oncoming jet 1.

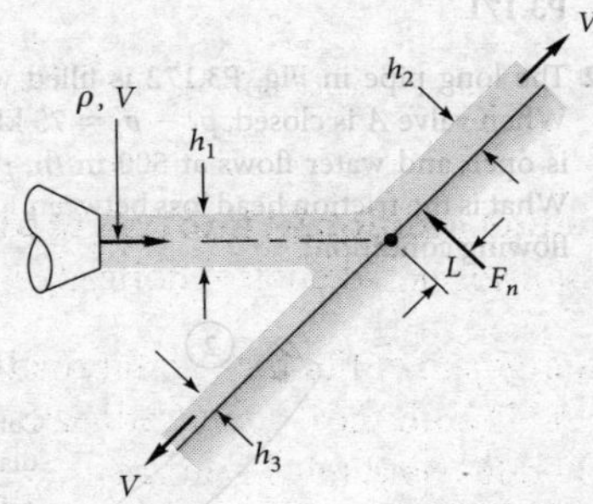

P3.161

P3.162 The waterwheel in Fig. P3.162 is being driven at 200 r/min by a 45 m/s jet of water at 20°C. The jet diameter is 6 cm. Assuming no losses, what is the horsepower developed by the wheel? For what speed Ω r/min will the horsepower developed be a maximum? Assume that there are many buckets on the waterwheel.

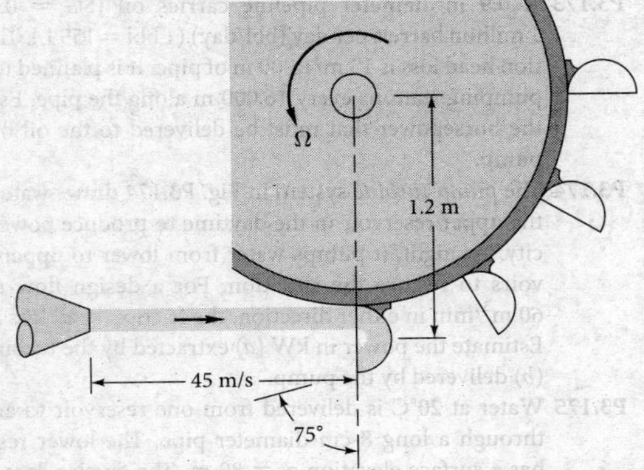

P3.162

P3.163 A rotating dishwasher arm delivers at 60°C to six nozzles, as in Fig. P3.163. The total flow rate is 0.2 L/s. Each nozzle has a diameter of 5 mm. If the nozzle flows are equal and friction is neglected, estimate the steady rotation rate of the arm, in r/min.

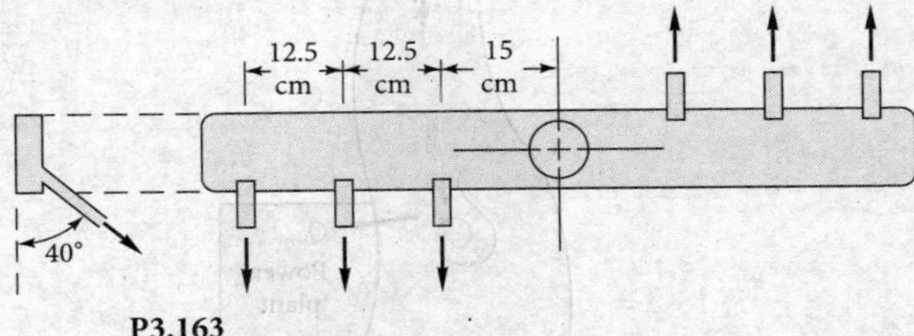

P3.163

***P3.164** A liquid of density ρ flows through a 90° bend as shown in Fig. P3.164 and issues vertically from a uniformly porous section of length *L*. Neglecting pipe and liquid weight, derive an expression for the torque *M* at point 0 required to hold the pipe stationary.

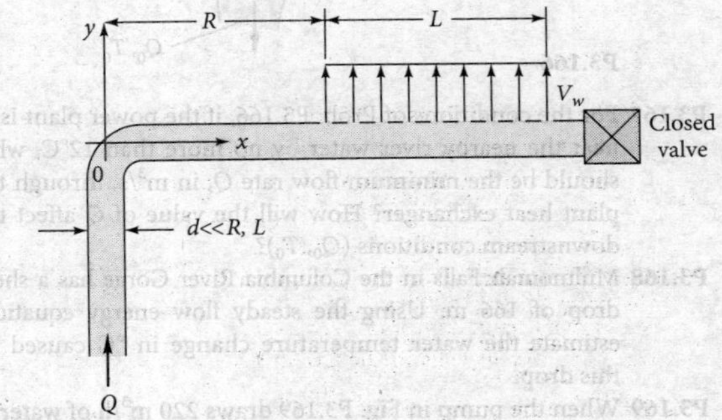

P3.164

The energy equation

P3.165 There is a steady isothermal flow of water at 20°C through the device in Fig. P3.165. Heat-transfer, gravity, and temperature effects are negligible. Known data are D_1 = 9 cm, Q_1 = 220 m³/h, p_1 = 150 kPa, D_2 = 7 cm, Q_2 = 100 m³/h, p_2 = 225 kPa, D_3 = 4 cm, and p_3 = 265 kPa. Compute the rate of shaft work done for this device and its direction.

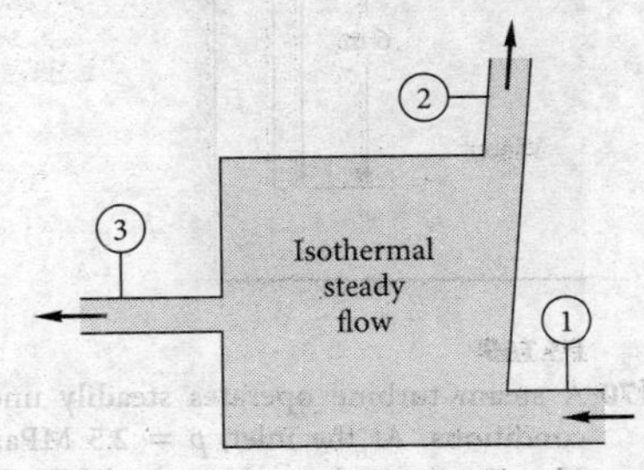

P3.165

P3.166 A power plant on a river, as in Fig. P3.166, must eliminate 55 MW of waste heat to the river. The river conditions upstream are Q_i = 2.5 m³/s and T_i = 18°C. The river is 45 m wide and 2.7 m deep. If heat losses to the atmosphere and ground are negligible, estimate the downstream river conditions (Q_0, T_0).

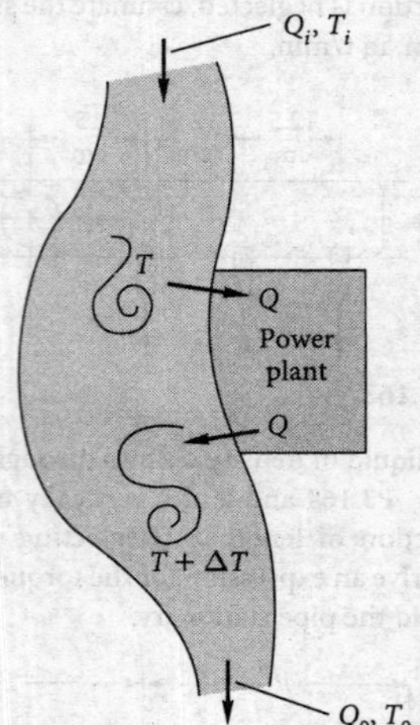

P3.166

P3.167 For the conditions of Prob. P3.166, if the power plant is to heat the nearby river water by no more than 12°C, what should be the minimum flow rate Q, in m³/s, through the plant heat exchanger? How will the value of Q affect the downstream conditions (Q_0, T_0)?

P3.168 Multnomah Falls in the Columbia River Gorge has a sheer drop of 166 m. Using the steady flow energy equation, estimate the water temperature change in °C caused by this drop.

P3.169 When the pump in Fig. P3.169 draws 220 m³/h of water at 20°C from the reservoir, the total friction head loss is 5 m. The flow discharges through a nozzle to the atmosphere. Estimate the pump power in kW delivered to the water.

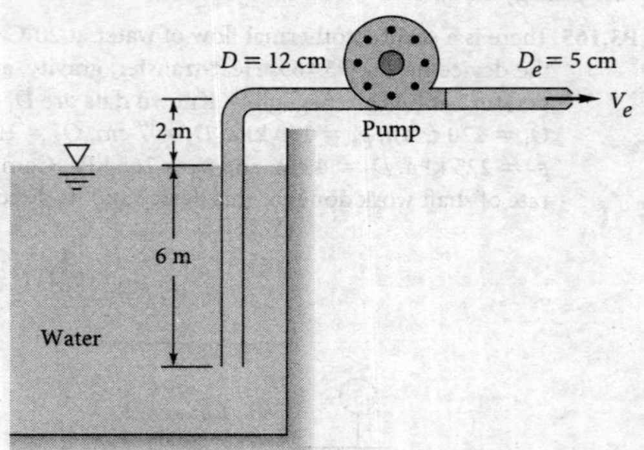

P3.169

P3.170 A steam turbine operates steadily under the following conditions. At the inlet, p = 2.5 MPa, T = 450°C, and V = 40 m/s. At the outlet, p = 22 kPa, T = 70°C, and V = 225 m/s. (*a*) If we neglect elevation changes and heat transfer, how much work is delivered to the turbine blades, in kJ/kg? (*b*) If the mass flow is 10 kg/s, how much total power is delivered? (*c*) Is the steam wet as it leaves the exit?

P3.171 Consider a turbine extracting energy from a penstock in a dam, as in Fig. P3.171. For turbulent pipe flow (Chap. 6), the friction head loss is approximately $h_f = CQ^2$, where the constant C depends on penstock dimensions and the properties of water. Show that, for a given penstock geometry and variable river flow Q, the maximum turbine power possible in this case is $P_{max} = 2\rho gHQ/3$ and occurs when the flow rate is $Q = \sqrt{H/(3C)}$.

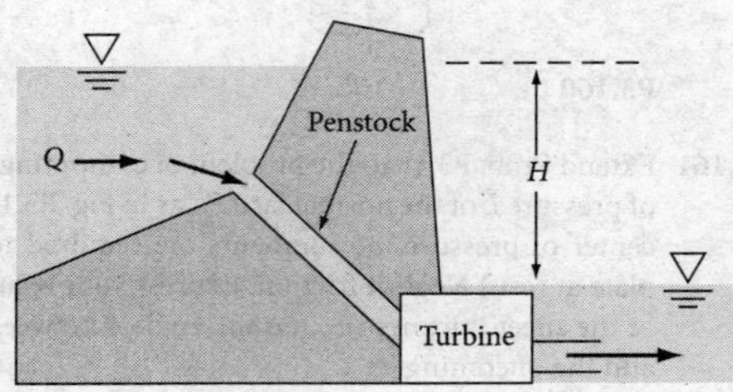

P3.171

P3.172 The long pipe in Fig. P3.172 is filled with water at 20°C. When valve A is closed, $p_1 - p_2$ = 75 kPa. When the valve is open and water flows at 500 m³/h, $p_1 - p_2$ = 160 kPa. What is the friction head loss between 1 and 2, in m, for the flowing condition?

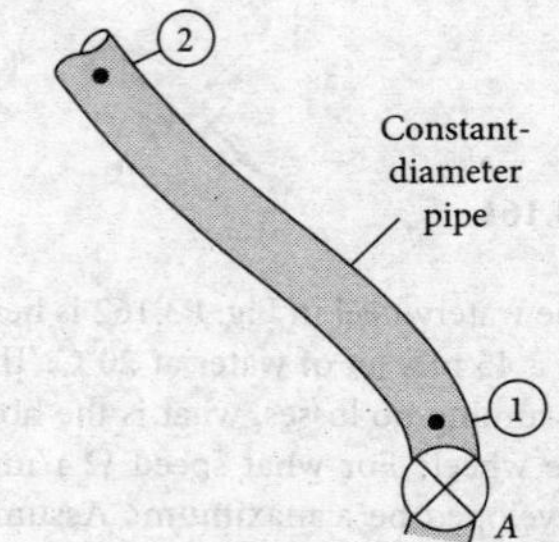

P3.172

P3.173 A 0.9 m-diameter pipeline carries oil (SG = 0.89) at 1 million barrels per day (bbl/day) (1 bbl = 159 L). The friction head loss is 13 m/1,000 m of pipe. It is planned to place pumping stations every 16,000 m along the pipe. Estimate the horsepower that must be delivered to the oil by each pump.

P3.174 The *pump-turbine* system in Fig. P3.174 draws water from the upper reservoir in the daytime to produce power for a city. At night, it pumps water from lower to upper reservoirs to restore the situation. For a design flow rate of 60 m³/min in either direction, the friction head loss is 5 m. Estimate the power in kW (*a*) extracted by the turbine and (*b*) delivered by the pump.

P3.175 Water at 20°C is delivered from one reservoir to another through a long 8-cm-diameter pipe. The lower reservoir has a surface elevation z_2 = 80 m. The friction loss in the pipe is correlated by the formula $h_{loss} \approx 17.5(V^2/2g)$, where V is the average velocity in the pipe. If the steady flow rate through the pipe is 0.0315 m^3/s, estimate the surface elevation of the higher reservoir.

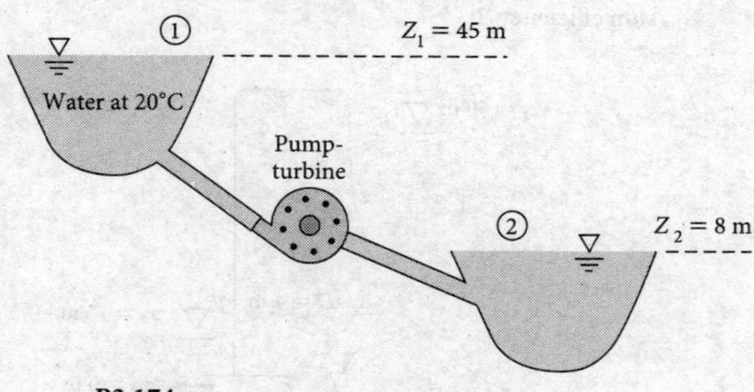

P3.174

P3.176 A fireboat draws seawater (SG = 1.025) from a submerged pipe and discharges it through a nozzle, as in Fig. P3.176. The total head loss is 2 m. If the pump efficiency is 75 percent, what horsepower motor is required to drive it?

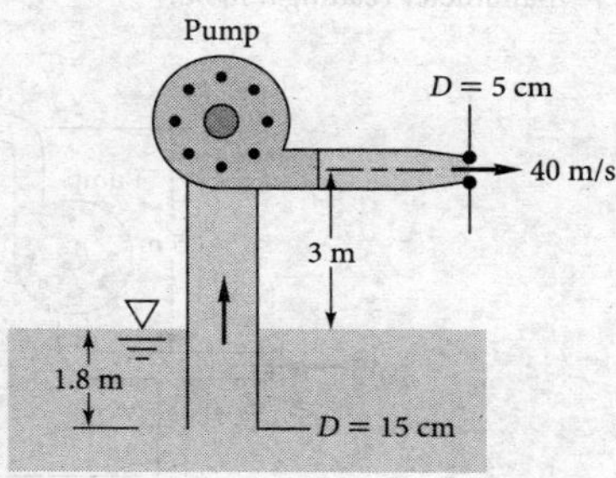

P3.176

P3.177 A device for measuring liquid viscosity is shown in Fig. P3.177. With the parameters (ρ, L, H, d) known, the flow rate Q is measured and the viscosity calculated, assuming a laminar-flow pipe loss from Chap. 6, $h_f = (32\mu LV)/(\rho g d^2)$. Heat transfer and all other losses are negligible. (*a*) Derive a formula for the viscosity μ of the fluid. (*b*) Calculate μ for the case $d = 2$ mm, $\rho = 800$ kg/m^3, $L = 95$ cm, $H = 30$ cm, and $Q = 760$ cm^3/h. (*c*) What is your guess of the fluid in part (*b*)? (*d*) Verify that the Reynolds number Re_d is less than 2,000 (laminar pipe flow).

P3.178 The horizontal pump in Fig. P3.178 discharges 20°C water at 57 m^3/h. Neglecting losses, what power in kW is delivered to the water by the pump?

P3.179 Steam enters a horizontal turbine at 2,400 kPa absolute, 580°C, and 4 m/s and is discharged at 35 m/s and 25°C saturated conditions. The mass flow is 1.5 kg/s, and the heat losses are 16 kJ/kg of steam. If head losses are negligible, how much horsepower does the turbine develop?

P3.180 Water at 20°C is pumped at 6 m^3/min from the lower to the upper reservoir, as in Fig. P3.180. Pipe friction losses are approximated by $h_f \approx 27V^2/(2g)$, where V is the average velocity in the pipe. If the pump is 75 percent efficient, what horsepower is needed to drive it?

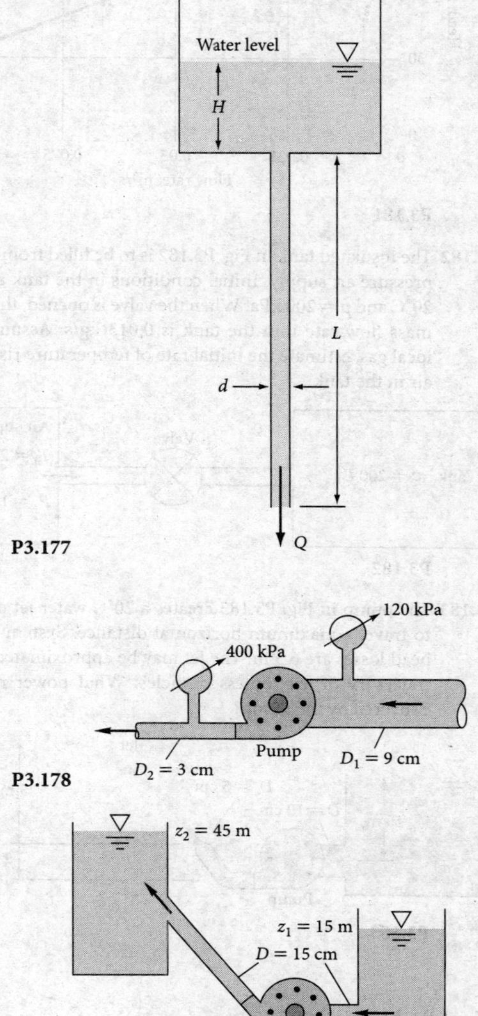

P3.177

P3.178

P3.180

P3.181 A typical pump has a head that, for a given shaft rotation rate, varies with the flow rate, resulting in a *pump performance curve* as in Fig. P3.181. Suppose that this pump is 75 percent efficient and is used for the system in Prob. 3.180. Estimate (*a*) the flow rate, in m^3/s, and (*b*) the horsepower needed to drive the pump.

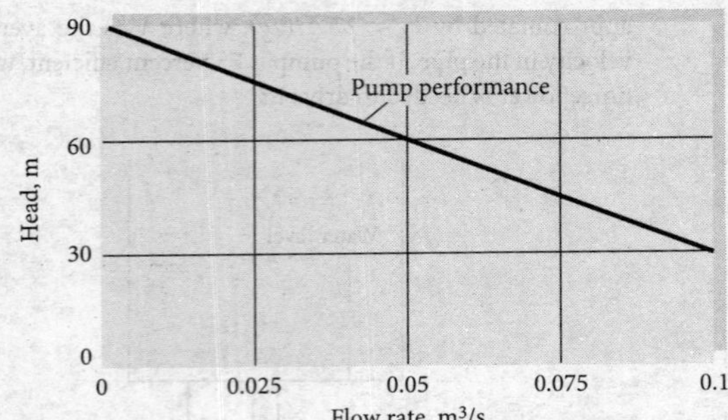

P3.181

P3.182 The insulated tank in Fig. P3.182 is to be filled from a high-pressure air supply. Initial conditions in the tank are $T = 20°C$ and $p = 200$ kPa. When the valve is opened, the initial mass flow rate into the tank is 0.013 kg/s. Assuming an ideal gas, estimate the initial rate of temperature rise of the air in the tank.

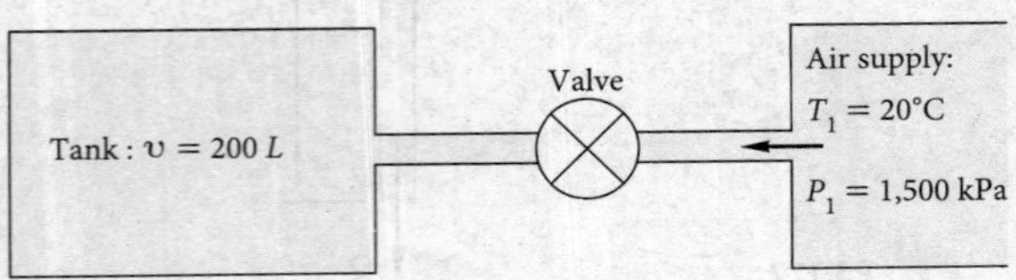

P3.182

P3.183 The pump in Fig. P3.183 creates a 20°C water jet oriented to travel a maximum horizontal distance. System friction head losses are 6.5 m. The jet may be approximated by the trajectory of frictionless particles. What power must be delivered by the pump?

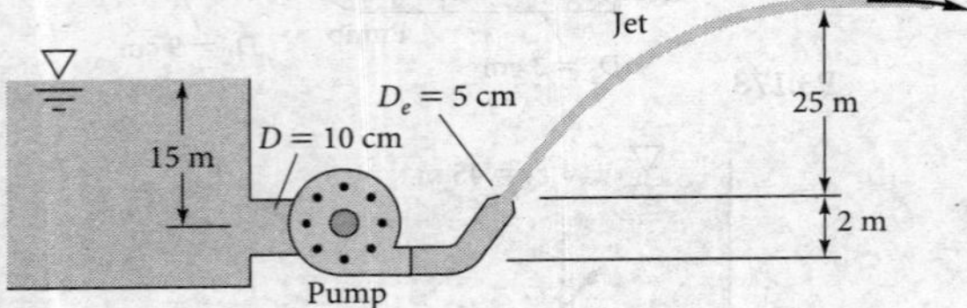

P3.183

P3.184 The large turbine in Fig. P3.184 diverts the river flow under a dam as shown. System friction losses are $h_f = 3.5V^2/(2g)$, where V is the average velocity in the supply pipe. For what river flow rate in m^3/s will the power extracted be 25 MW? Which of the *two* possible solutions has a better "conversion efficiency"?

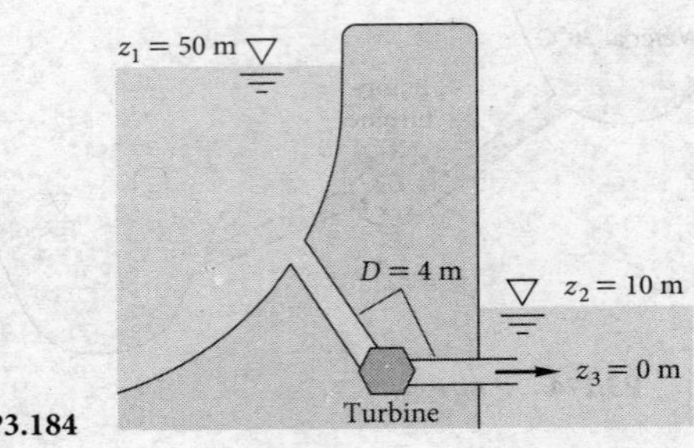

P3.184

P3.185 Kerosine at 20°C flows through the pump in Fig. P3.185 at 0.065 m^3/s. Head losses between 1 and 2 are 2.5 m, and the pump delivers 8 hp to the flow. What should the mercury manometer reading h m be?

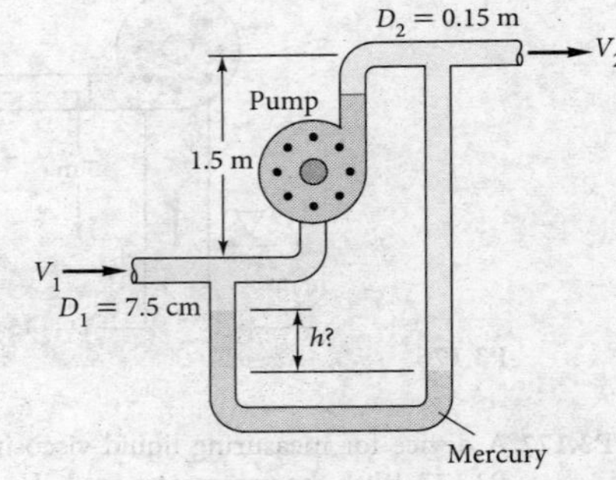

P3.185

Word Problems

W3.1 Derive a control volume form of the *second* law of thermodynamics. Suggest some practical uses for your relation in analyzing real fluid flows.

W3.2 Suppose that it is desired to estimate volume flow Q in a pipe by measuring the axial velocity $u(r)$ at specific points. For cost reasons only *three* measuring points are to be used. What are the best radii selections for these three points?

W3.3 Consider water flowing by gravity through a short pipe connecting two reservoirs whose surface levels differ by an amount Δz. Why does the incompressible frictionless Bernoulli equation lead to an absurdity when the flow rate through the pipe is computed? Does the paradox have something to do with the length of the short pipe? Does the paradox disappear if we round the entrance and exit edges of the pipe?

W3.4 Use the steady flow energy equation to analyze flow through a water faucet whose supply pressure is p_0. What physical mechanism causes the flow to vary continuously from zero to maximum as we open the faucet valve?

W3.5 Consider a long sewer pipe, half full of water, sloping downward at angle θ. Antoine Chézy in 1768 determined that the average velocity of such an open channel flow should be $V \approx C\sqrt{R \tan \theta}$, where R is the pipe radius and C is a constant. How does this famous formula relate to the steady flow energy equation applied to a length L of the channel?

W3.6 Put a table tennis ball in a funnel, and attach the small end of the funnel to an air supply. You probably won't be able to blow the ball either up or down out of the funnel. Explain why.

W3.7 How does a *siphon* work? Are there any limitations (such as how high or how low can you siphon water away from a tank)? Also, how far—could you use a flexible tube to siphon water from a tank to a point 30 m away?

Fundamentals of Engineering Exam Problems

FE3.1 In Fig. FE3.1 water exits from a nozzle into atmospheric pressure of 101 kPa. If the flow rate is 0.01 m^3/s, what is the average velocity at section 1?
(*a*) 2.6 m/s, (*b*) 0.81 m/s, (*c*) 93 m/s, (*d*) 23 m/s, (*e*) 1.62 m/s

FE3.2 In Fig. FE3.1 water exits from a nozzle into atmospheric pressure of 101 kPa. If the flow rate is 0.01 m^3/s and friction is neglected, what is the gage pressure at section 1?
(*a*) 1.4 kPa, (*b*) 32 kPa, (*c*) 43 kPa, (*d*) 29 kPa, (*e*) 123 kPa

FE3.3 In Fig. FE3.1 water exits from a nozzle into atmospheric pressure of 101 kPa. If the exit velocity is $V_2 = 8$ m/s and friction is neglected, what is the axial flange force required to keep the nozzle attached to pipe 1?
(*a*) 11 N, (*b*) 56 N, (*c*) 83 N, (*d*) 123 N, (*e*) 110 N

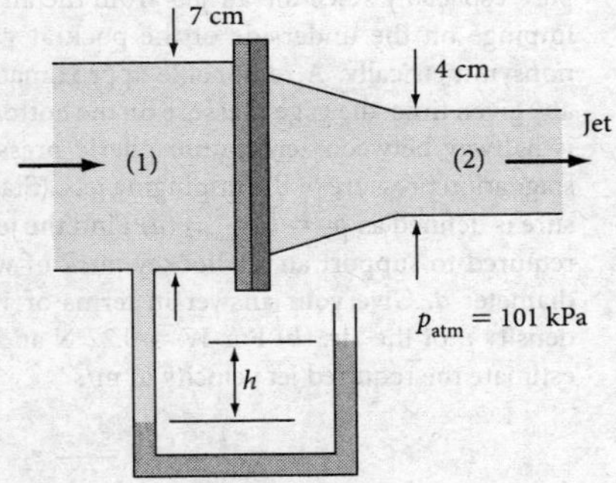

FE3.1

FE3.4 In Fig. FE3.1 water exits from a nozzle into atmospheric pressure of 101 kPa. If the manometer fluid has a specific gravity of 1.6 and $h = 66$ cm, with friction neglected, what is the average velocity at section 2?
(*a*) 4.55 m/s, (*b*) 2.4 m/s, (*c*) 2.95 m/s, (*d*) 5.55 m/s, (*e*) 3.4 m/s

FE3.5 A jet of water 3 cm in diameter strikes normal to a plate as in Fig. FE3.5. If the force required to hold the plate is 23 N, what is the jet velocity?
(*a*) 2.85 m/s, (*b*) 5.7 m/s, (*c*) 8.1 m/s, (*d*) 4.0 m/s, (*e*) 23 m/s

FE3.6 A fireboat pump delivers water to a vertical nozzle with a 3:1 diameter ratio, as in Fig. FE3.6. If friction is neglected and the flow rate is 0.03 m^3/s, how high will the outlet water jet rise?
(*a*) 2.0 m, (*b*) 9.8 m, (*c*) 32 m, (*d*) 64 m, (*e*) 98 m

FE3.7 A fireboat pump delivers water to a vertical nozzle with a 3:1 diameter ratio, as in Fig. FE3.6. If friction is neglected and the pump increases the pressure at section 1 to 51 kPa (gage), what will be the resulting flow rate?
(*a*) 0.71 m^3/min, (*b*) 0.75 m^3/min, (*c*) 0.81 m^3/min, (*d*) 1.35 m^3/min, (*e*) 0.53 m^3/min

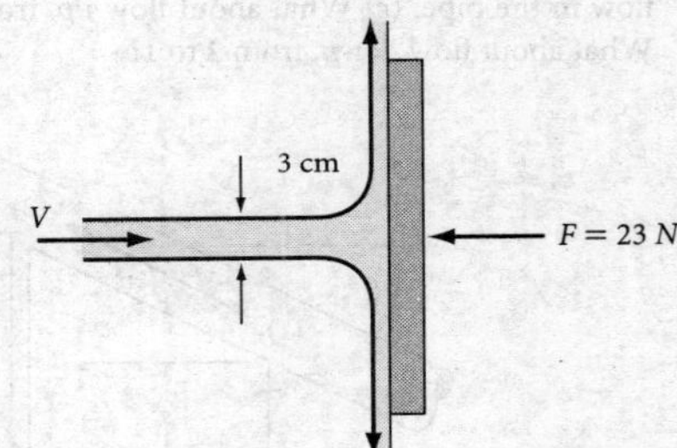

FE3.5

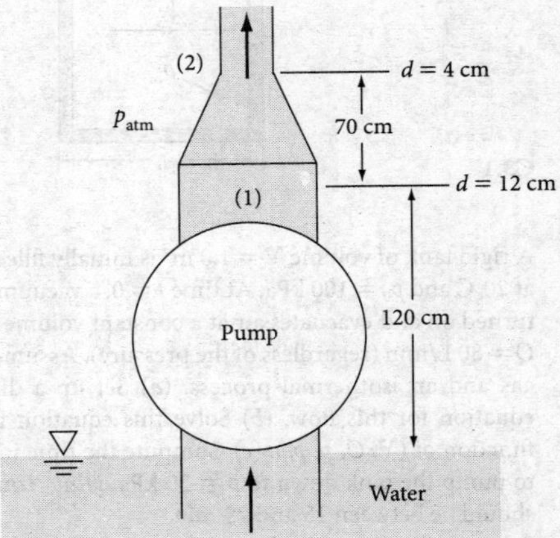

FE3.6

FE3.8 A fireboat pump delivers water to a vertical nozzle with a 3:1 diameter ratio, as in Fig. FE3.6. If duct and nozzle friction are neglected and the pump provides 3.75 m of head to the flow, what will be the outlet flow rate?
(*a*) 0.32 m^3/min, (*b*) 0.45 m^3/min, (*c*) 0.58 m^3/min, (*d*) 0.82m^3/min, (*e*) 1.07 m^3/min

FE3.9 Water flowing in a smooth 6-cm-diameter pipe enters a venturi contraction with a throat diameter of 3 cm.

Upstream pressure is 120 kPa. If cavitation occurs in the throat at a flow rate of 0.01 m^3/s, what is the estimated fluid vapor pressure, assuming ideal frictionless flow?
(*a*) 6 kPa, (*b*) 12 kPa, (*c*) 24 kPa, (*d*) 31 kPa, (*e*) 52 kPa

FE3.10 Water flowing in a smooth 6-cm-diameter pipe enters a venturi contraction with a throat diameter of 4 cm. Upstream pressure is 120 kPa. If the pressure in the throat is 50 kPa, what is the flow rate, assuming ideal frictionless flow?
(*a*) 0.03 m^3/min, (*b*) 0.89 m^3/min, (*c*) 0.99 m^3/min, (*d*) 2.82 m^3/min, (*e*) 3.98 m^3/min

Comprehensive Problems

C3.1 In a certain industrial process, oil of density ρ flows through the inclined pipe in Fig. C3.1. A *U*-tube manometer, with fluid density ρ_m, measures the pressure difference between points 1 and 2, as shown. The pipe flow is steady, so that the fluids in the manometer are stationary. (*a*) Find an analytic expression for $p_1 - p_2$ in terms of the system parameters. (*b*) Discuss the conditions on *h* necessary for there to be no flow in the pipe. (*c*) What about flow *up*, from 1 to 2? (*d*) What about flow *down*, from 2 to 1?

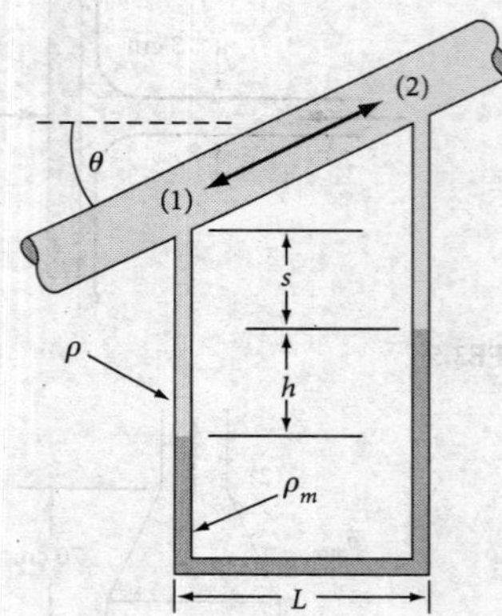

C3.1

C3.2 A rigid tank of volume $\mathcal{V} = 1.0$ m^3 is initially filled with air at 20°C and $p_0 = 100$ kPa. At time $t = 0$, a vacuum pump is turned on and evacuates air at a constant volume flow rate $Q = 80$ L/min (regardless of the pressure). Assume an ideal gas and an isothermal process. (*a*) Set up a differential equation for this flow. (*b*) Solve this equation for *t* as a function of ($\mathcal{V}$, *Q*, *p*, p_0). (*c*) Compute the time in minutes to pump the tank down to $p = 20$ kPa. *Hint:* Your answer should lie between 15 and 25 min.

C3.3 Suppose the same steady water jet as in Prob. P3.40 (jet velocity 8 m/s and jet diameter 10 cm) impinges instead on a cup cavity as shown in Fig. C3.3. The water is turned 180° and exits, due to friction, at lower velocity, $V_e = 4$ m/s. (Looking from the left, the exit jet is a circular annulus of outer radius *R* and thickness *h*, flowing toward the viewer.) The cup has a radius of curvature of 25 cm. Find (*a*) the thickness *h* of the exit jet and (*b*) the force *F* required to hold the cupped object in place. (*c*) Compare part (*b*) to Prob. 3.40, where $F \approx 500$ N, and give a physical explanation as to why *F* has changed.

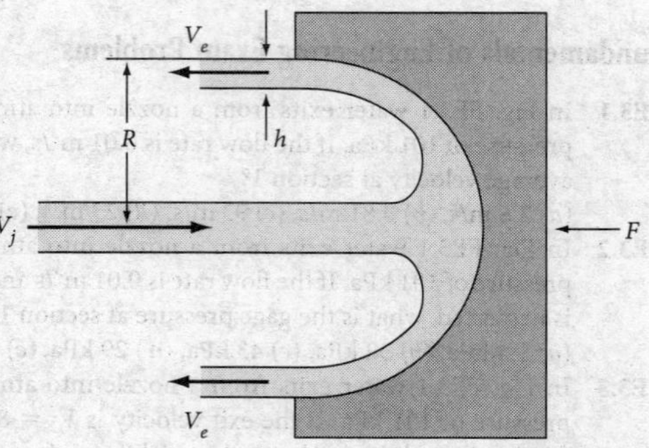

C3.3

C3.4 The airflow underneath an air hockey puck is very complex, especially since the air jets from the air hockey table impinge on the underside of the puck at various points nonsymmetrically. A reasonable approximation is that at any given time, the gage pressure on the bottom of the puck is halfway between zero (atmospheric pressure) and the stagnation pressure of the impinging jets. (Stagnation pressure is defined as $p_0 = \frac{1}{2}\rho V_{jet}^2$.) (*a*) Find the jet velocity V_{jet} required to support an air hockey puck of weight *W* and diameter *d*. Give your answer in terms of *W*, *d*, and the density ρ of the air. (*b*) For $W = 0.22$ N and $d = 0.06$ m, estimate the required jet velocity in m/s.

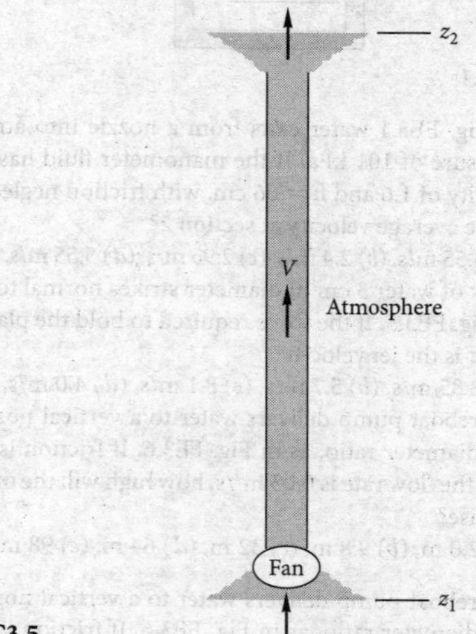

C3.5

C3.5 Neglecting friction sometimes leads to odd results. You are asked to analyze and discuss the following example in Fig. C3.5. A fan blows air through a duct from section 1 to section 2, as shown. Assume constant air density ρ. Neglecting frictional losses, find a relation between the required fan head h_p and the flow rate and the elevation change. Then explain what may be an unexpected result.

Design Project

D3.1 Let us generalize Probs. P3.180 and P3.181, in which a pump performance curve was used to determine the flow rate between reservoirs. The particular pump in Fig. P3.181 is one of a family of pumps of similar shape, whose dimensionless performance is as follows:

Head:

$$\phi \approx 6.13 - 176\zeta \qquad \phi = \frac{gh_p}{n^2 D_p^2} \text{ and } \zeta = \frac{Q}{nD_p^3}$$

Efficiency:

$$\eta \approx 70\zeta - 91{,}500\zeta^3 \qquad \eta = \frac{\text{power to water}}{\text{power input}}$$

where h_p is the pump head (m), n is the shaft rotation rate (r/s), and D_p is the impeller diameter (m). The range of validity is $0 < \zeta < 0.027$. The pump of Fig. P3.181 had D_p = 60 cm in diameter and rotated at n = 20 r/s (1,200 r/min). The solution to Prob. P3.181, namely, $Q \approx 0.067$ m^3/s and $h_p \approx 50$ m, corresponds to $\phi \approx 3.40$, $\zeta \approx 0.0155$, $\eta \approx 0.74$ (or 74 percent), and power to the water $= \rho g Q h_p \approx$ 32,800 Nm/s (44 hp). Please check these numerical values before beginning this project.

Now revisit Prob. P3.181 and select a *low-cost* pump that rotates at a rate no slower than 600 r/min and delivers no less than 0.03 m^3/s of water. Assume that the cost of the pump is linearly proportional to the power input required. Comment on any limitations to your results.

References

1. D. T. Greenwood, *Advanced Dynamics*, Cambridge University Press, New York, 2006.
2. T. von Kármán, *The Wind and Beyond*, Little, Brown, Boston, 1967.
3. J. P. Holman, *Heat Transfer*, 10th ed., McGraw-Hill, New York, 2009.
4. A. G. Hansen, *Fluid Mechanics*, Wiley, New York, 1967.
5. M. C. Potter, D. C. Wiggert, and M. Hondzo, *Mechanics of Fluids*, Brooks/Cole, Chicago, 2001.
6. S. Klein and G. Nellis, *Thermodynamics*, Cambridge University Press, New York, 2011.
7. Y. A. Cengel and M. A. Boles, *Thermodynamics: An Engineering Approach*, 7th ed., McGraw-Hill, New York, 2010.
8. J. F. Wendt, *Computational Fluid Dynamics: An Introduction*, Springer, 3d ed., New York, 2009.
9. W. G. Vincenti, "Control Volume Analysis: A Difference in Thinking between Engineering and Physics," *Technology and Culture*, vol. 23, no. 2, 1982, pp. 145–174.
10. J. Keenan, *Thermodynamics*, Wiley, New York, 1941.
11. J. Hunsaker and B. Rightmire, *Engineering Applications of Fluid Mechanics*, McGraw-Hill, New York, 1947.

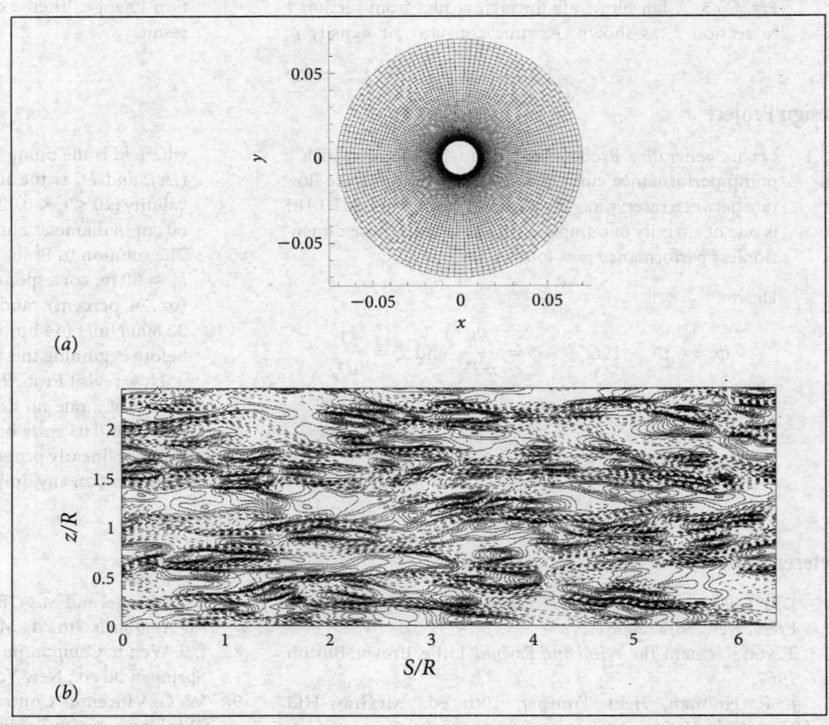

The differential equations to be studied in this chapter can be modeled numerically by computational fluid dynamics (CFD). This study, from Ref. 21, models turbulent flow near a rotating cylinder at a Reynolds number $\mathrm{Re}_D \approx 8{,}960$. The grid (*a*) contains 3.1 million nodes, very finely spaced near the cylinder. The results (*b*) show turbulent velocity fluctuations, obtained by direct numerical simulation (DNS), at $y^+ = 10$ away from the cylinder surface.

Source: From ASME J. Fluids Engineering, J-Y. Hwang, K-S. Yang, and K. Bremhorst, "Direct Numerical Simulation of Turbulent Flow Around a Rotating Circular Cylinder," Vol. 129, Jan. 2007, pp 40–47, by permission of the American Society of Mechanical Engineers.

Chapter 4
Differential Relations for Fluid Flow

Motivation. In analyzing fluid motion, we might take one of two paths: (1) seeking an estimate of gross effects (mass flow, induced force, energy change) over a *finite* region or control volume or (2) seeking the point-by-point details of a flow pattern by analyzing an *infinitesimal* region of the flow. The former or gross-average viewpoint was the subject of Chap. 3.

This chapter treats the second in our trio of techniques for analyzing fluid motion: small-scale, or *differential*, analysis. That is, we apply our four basic conservation laws to an infinitesimally small control volume or, alternately, to an infinitesimal fluid system. In either case the results yield the basic *differential equations* of fluid motion. Appropriate *boundary conditions* are also developed.

In their most basic form, these differential equations of motion are quite difficult to solve, and very little is known about their general mathematical properties. However, certain things can be done that have great educational value. First, as shown in Chap. 5, the equations (even if unsolved) reveal the basic dimensionless parameters that govern fluid motion. Second, as shown in Chap. 6, a great number of useful solutions can be found if one makes two simplifying assumptions: (1) steady flow and (2) incompressible flow. A third and rather drastic simplification, frictionless flow, makes our old friend the Bernoulli equation valid and yields a wide variety of idealized, or *perfect-fluid,* possible solutions. These idealized flows are treated in Chap. 8, and we must be careful to ascertain whether such solutions are in fact realistic when compared with actual fluid motion. Finally, even the difficult general differential equations now yield to the approximating technique known as computational fluid dynamics (CFD) whereby the derivatives are simulated by algebraic relations between a finite number of grid points in the flow field, which are then solved on a computer. Reference 1 is an example of a textbook devoted entirely to numerical analysis of fluid motion.

4.1 The Acceleration Field of a Fluid

In Sec. 1.7 we established the cartesian vector form of a velocity field that varies in space and time:

$$\mathbf{V}(\mathbf{r}, t) = \mathbf{i}u(x, y, z, t) + \mathbf{j}\upsilon(x, y, z, t) + \mathbf{k}w(x, y, z, t) \tag{1.4}$$

This is the most important variable in fluid mechanics: Knowledge of the velocity vector field is nearly equivalent to *solving* a fluid flow problem. Our coordinates are fixed in space, and we observe the fluid as it passes by—as if we had scribed a set of coordinate lines on a glass window in a wind tunnel. This is the *Eulerian* frame of reference, as opposed to the Lagrangian frame, which follows the moving position of individual particles.

The Eulerian system can be visualized as a window through which we watch a flow. The coordinates (x, y, z) are fixed, and the flow passes by. A fixed instrument placed in the flow takes an Eulerian measurement. In contrast, Lagrangian coordinates follow the moving particles and are common in solid mechanics. Almost all articles and books about fluid mechanics use the Eulerian system. Writers often use *traffic* as an example. A traffic engineer will remain fixed and will measure the flow of cars going by—a Eulerian viewpoint. Conversely, the police will follow specific cars as a function of time—a Lagrangian viewpoint.

To write Newton's second law for an infinitesimal fluid system, we need to calculate the acceleration vector field **a** of the flow. Thus, we compute the total time derivative of the velocity vector:

$$\mathbf{a} = \frac{d\mathbf{V}}{dt} = \mathbf{i}\frac{du}{dt} + \mathbf{j}\frac{dv}{dt} + \mathbf{k}\frac{dw}{dt}$$

Since each scalar component (u, v, w) is a function of the four variables (x, y, z, t), we use the chain rule to obtain each scalar time derivative. For example,

$$\frac{du(x, y, z, t)}{dt} = \frac{\partial u}{\partial t} + \frac{\partial u}{\partial x}\frac{dx}{dt} + \frac{\partial u}{\partial y}\frac{dy}{dt} + \frac{\partial u}{\partial z}\frac{dz}{dt}$$

But, by definition, dx/dt is the local velocity component u, and $dy/dt = v$, and $dz/dt = w$. The total time derivative of u may thus be written as follows, with exactly similar expressions for the time derivatives of v and w:

$$\begin{aligned} a_x &= \frac{du}{dt} = \frac{\partial u}{\partial t} + u\frac{\partial u}{\partial x} + v\frac{\partial u}{\partial y} + w\frac{\partial u}{\partial z} = \frac{\partial u}{\partial t} + (\mathbf{V}\cdot\nabla)u \\ a_y &= \frac{dv}{dt} = \frac{\partial v}{\partial t} + u\frac{\partial v}{\partial x} + v\frac{\partial v}{\partial y} + w\frac{\partial v}{\partial z} = \frac{\partial v}{\partial t} + (\mathbf{V}\cdot\nabla)v \\ a_z &= \frac{dw}{dt} = \frac{\partial w}{\partial t} + u\frac{\partial w}{\partial x} + v\frac{\partial w}{\partial y} + w\frac{\partial w}{\partial z} = \frac{\partial w}{\partial t} + (\mathbf{V}\cdot\nabla)w \end{aligned} \tag{4.1}$$

Summing these into a vector, we obtain the total acceleration:

$$\mathbf{a} = \frac{d\mathbf{V}}{dt} = \underbrace{\frac{\partial \mathbf{V}}{\partial t}}_{\text{Local}} + \underbrace{\left(u\frac{\partial \mathbf{V}}{\partial x} + v\frac{\partial \mathbf{V}}{\partial y} + w\frac{\partial \mathbf{V}}{\partial z}\right)}_{\text{Convective}} = \frac{\partial \mathbf{V}}{\partial t} + (\mathbf{V}\cdot\nabla)\mathbf{V} \tag{4.2}$$

The term $\partial\mathbf{V}/\partial t$ is called the *local acceleration,* which vanishes if the flow is steady—that is, independent of time. The three terms in parentheses are called the *convective acceleration,* which arises when the particle moves through regions of spatially varying velocity, as in a nozzle or diffuser. Flows that are nominally "steady" may have large accelerations due to the convective terms.

Note our use of the compact dot product involving **V** and the gradient operator ∇:

$$u\frac{\partial}{\partial x} + v\frac{\partial}{\partial y} + w\frac{\partial}{\partial z} = \mathbf{V}\cdot\nabla \quad \text{where} \quad \nabla = \mathbf{i}\frac{\partial}{\partial x} + \mathbf{j}\frac{\partial}{\partial y} + \mathbf{k}\frac{\partial}{\partial z}$$

The total time derivative—sometimes called the *substantial* or *material* derivative—concept may be applied to any variable, such as the pressure:

$$\frac{dp}{dt} = \frac{\partial p}{\partial t} + u\frac{\partial p}{\partial x} + v\frac{\partial p}{\partial y} + w\frac{\partial p}{\partial z} = \frac{\partial p}{\partial t} + (\mathbf{V} \cdot \nabla)p \quad (4.3)$$

Wherever convective effects occur in the basic laws involving mass, momentum, or energy, the basic differential equations become nonlinear and are usually more complicated than flows that do not involve convective changes.

We emphasize that this total time derivative follows a particle of fixed identity, making it convenient for expressing laws of particle mechanics in the eulerian fluid field description. The operator d/dt is sometimes assigned a special symbol such as D/Dt as a further reminder that it contains four terms and follows a fixed particle.

As another reminder of the special nature of d/dt, some writers give it the name *substantial* or *material derivative.*

EXAMPLE 4.1

Given the Eulerian velocity vector field

$$\mathbf{V} = 3t\mathbf{i} + xz\mathbf{j} + ty^2\mathbf{k}$$

find the total acceleration of a particle.

Solution

- *Assumptions:* Given three known unsteady velocity components, $u = 3t$, $v = xz$, and $w = ty^2$.
- *Approach:* Carry out all the required derivatives with respect to (x, y, z, t), substitute into the total acceleration vector, Eq. (4.2), and collect terms.
- *Solution step 1:* First work out the local acceleration $\partial\mathbf{V}/\partial t$:

$$\frac{\partial \mathbf{V}}{\partial t} = \mathbf{i}\frac{\partial u}{\partial t} + \mathbf{j}\frac{\partial v}{\partial t} + \mathbf{k}\frac{\partial w}{\partial t} = \mathbf{i}\frac{\partial}{\partial t}(3t) + \mathbf{j}\frac{\partial}{\partial t}(xz) + \mathbf{k}\frac{\partial}{\partial t}(ty^2) = 3\mathbf{i} + 0\mathbf{j} + y^2\mathbf{k}$$

- *Solution step 2:* In a similar manner, the convective acceleration terms, from Eq. (4.2), are

$$u\frac{\partial \mathbf{V}}{\partial x} = (3t)\frac{\partial}{\partial x}(3t\mathbf{i} + xz\mathbf{j} + ty^2\mathbf{k}) = (3t)(0\mathbf{i} + z\mathbf{j} + 0\mathbf{k}) = 3tz\,\mathbf{j}$$

$$v\frac{\partial \mathbf{V}}{\partial y} = (xz)\frac{\partial}{\partial y}(3t\mathbf{i} + xz\mathbf{j} + ty^2\mathbf{k}) = (xz)(0\mathbf{i} + 0\mathbf{j} + 2ty\mathbf{k}) = 2txyz\,\mathbf{k}$$

$$w\frac{\partial \mathbf{V}}{\partial z} = (ty^2)\frac{\partial}{\partial z}(3t\mathbf{i} + xz\mathbf{j} + ty^2\mathbf{k}) = (ty^2)(0\mathbf{i} + x\mathbf{j} + 0\mathbf{k}) = txy^2\,\mathbf{j}$$

- *Solution step 3:* Combine all four terms above into the single "total" or "substantial" derivative:

$$\frac{d\mathbf{V}}{dt} = \frac{\partial \mathbf{V}}{\partial t} + u\frac{\partial \mathbf{V}}{\partial x} + v\frac{\partial \mathbf{V}}{\partial y} + w\frac{\partial \mathbf{V}}{\partial z} = (3\mathbf{i} + y^2\mathbf{k}) + 3tz\mathbf{j} + 2txyz\mathbf{k} + txy^2\mathbf{j}$$

$$= 3\mathbf{i} + (3tz + txy^2)\mathbf{j} + (y^2 + 2txyz)\mathbf{k} \quad \textit{Ans.}$$

- *Comments:* Assuming that **V** is valid everywhere as given, this total acceleration vector $d\mathbf{V}/dt$ applies to all positions and times within the flow field.

4.2 The Differential Equation of Mass Conservation

Conservation of mass, often called the *continuity* relation, states that the fluid mass cannot change. We apply this concept to a very small region. All the basic differential equations can be derived by considering either an elemental control volume or an elemental system. We choose an infinitesimal fixed control volume (dx, dy, dz), as in Fig. 4.1, and use our basic control volume relations from Chap. 3. The flow through each side of the element is approximately one-dimensional, and so the appropriate mass conservation relation to use here is

$$\int_{CV} \frac{\partial \rho}{\partial t}\, d\mathcal{V} + \sum_i (\rho_i A_i V_i)_{out} - \sum_i (\rho_i A_i V_i)_{in} = 0 \tag{3.22}$$

The element is so small that the volume integral simply reduces to a differential term:

$$\int_{CV} \frac{\partial \rho}{\partial t}\, d\mathcal{V} \approx \frac{\partial \rho}{\partial t}\, dx\, dy\, dz$$

The mass flow terms occur on all six faces, three inlets and three outlets. We make use of the field or continuum concept from Chap. 1, where all fluid properties are considered to be uniformly varying functions of time and position, such as $\rho = \rho(x, y, z, t)$. Thus, if T is the temperature on the left face of the element in Fig. 4.1, the right face will have a slightly different temperature $T + (\partial T/\partial x)\, dx$. For mass conservation, if ρu is known on the left face, the value of this product on the right face is $\rho u + (\partial \rho u/\partial x)\, dx$.

Figure 4.1 shows only the mass flows on the x or left and right faces. The flows on the y (bottom and top) and the z (back and front) faces have been omitted to avoid cluttering up the drawing. We can list all these six flows as follows:

Face	Inlet mass flow	Outlet mass flow
x	$\rho u\, dy\, dz$	$\left[\rho u + \frac{\partial}{\partial x}(\rho u)\, dx\right] dy\, dz$
y	$\rho v\, dx\, dz$	$\left[\rho v + \frac{\partial}{\partial y}(\rho v)\, dy\right] dx\, dz$
z	$\rho w\, dx\, dy$	$\left[\rho w + \frac{\partial}{\partial z}(\rho w)\, dz\right] dx\, dy$

Introduce these terms into Eq. (3.22) and we have

$$\frac{\partial \rho}{\partial t}\, dx\, dy\, dz + \frac{\partial}{\partial x}(\rho u)\, dx\, dy\, dz + \frac{\partial}{\partial y}(\rho v)\, dx\, dy\, dz + \frac{\partial}{\partial z}(\rho w)\, dx\, dy\, dz = 0$$

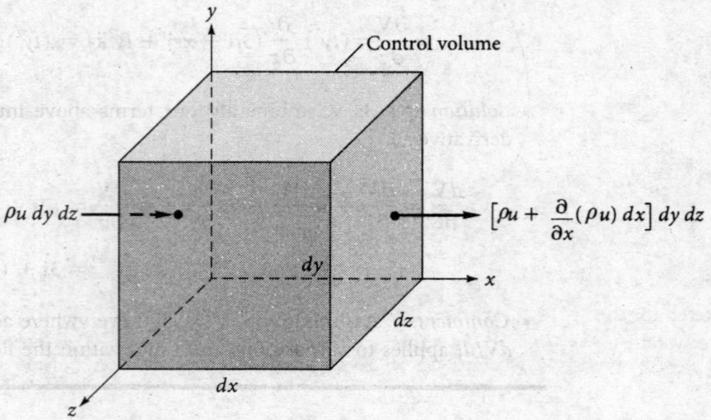

Fig. 4.1 Elemental cartesian fixed control volume showing the inlet and outlet mass flows on the x faces.

The element volume cancels out of all terms, leaving a partial differential equation involving the derivatives of density and velocity:

$$\frac{\partial \rho}{\partial t} + \frac{\partial}{\partial x}(\rho u) + \frac{\partial}{\partial y}(\rho \upsilon) + \frac{\partial}{\partial z}(\rho w) = 0 \tag{4.4}$$

This is the desired result: conservation of mass for an infinitesimal control volume. It is often called the *equation of continuity* because it requires no assumptions except that the density and velocity are continuum functions. That is, the flow may be either steady or unsteady, viscous or frictionless, compressible or incompressible.[1] However, the equation does not allow for any source or sink singularities within the element.

The vector gradient operator

$$\nabla = \mathbf{i}\frac{\partial}{\partial x} + \mathbf{j}\frac{\partial}{\partial y} + \mathbf{k}\frac{\partial}{\partial z}$$

enables us to rewrite the equation of continuity in a compact form, not that it helps much in finding a solution. The last three terms of Eq. (4.4) are equivalent to the divergence of the vector $\rho\mathbf{V}$

$$\frac{\partial}{\partial x}(\rho u) + \frac{\partial}{\partial y}(\rho \upsilon) + \frac{\partial}{\partial z}(\rho w) \equiv \nabla \cdot (\rho \mathbf{V}) \tag{4.5}$$

so the compact form of the continuity relation is

$$\frac{\partial \rho}{\partial t} + \nabla \cdot (\rho \mathbf{V}) = 0 \tag{4.6}$$

In this vector form the equation is still quite general and can readily be converted to other coordinate systems.

Cylindrical Polar Coordinates

The most common alternative to the cartesian system is the *cylindrical polar* coordinate system, sketched in Fig. 4.2. An arbitrary point P is defined by a distance z along the axis, a radial distance r from the axis, and a rotation angle θ about the axis. The

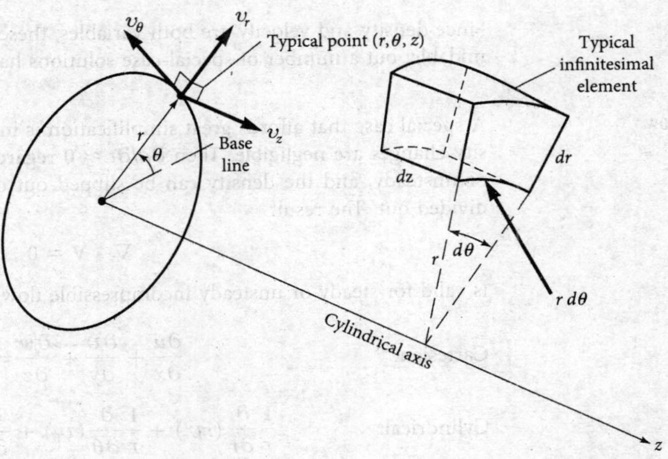

Fig. 4.2 Definition sketch for the cylindrical coordinate system.

[1]One case where Eq. (4.4) might need special care is *two-phase flow,* where the density is discontinuous between the phases. For further details on this case, see Ref. 2, for example.

three independent orthogonal velocity components are an axial velocity υ_z, a radial velocity υ_r, and a circumferential velocity υ_θ, which is positive counterclockwise—that is, in the direction of increasing θ. In general, all components, as well as pressure and density and other fluid properties, are continuous functions of r, θ, z, and t.

The divergence of any vector function $\mathbf{A}(r, \theta, z, t)$ is found by making the transformation of coordinates

$$r = (x^2 + y^2)^{1/2} \qquad \theta = \tan^{-1}\frac{y}{x} \qquad z = z \tag{4.7}$$

and the result is given here without proof[2]

$$\nabla \cdot \mathbf{A} = \frac{1}{r}\frac{\partial}{\partial r}(rA_r) + \frac{1}{r}\frac{\partial}{\partial \theta}(A_\theta) + \frac{\partial}{\partial z}(A_z) \tag{4.8}$$

The general continuity equation (4.6) in cylindrical polar coordinates is thus

$$\frac{\partial \rho}{\partial t} + \frac{1}{r}\frac{\partial}{\partial r}(r\rho\upsilon_r) + \frac{1}{r}\frac{\partial}{\partial \theta}(\rho\upsilon_\theta) + \frac{\partial}{\partial z}(\rho\upsilon_z) = 0 \tag{4.9}$$

There are other orthogonal curvilinear coordinate systems, notably *spherical polar coordinates*, which occasionally merit use in a fluid mechanics problem. We shall not treat these systems here except in Prob. P4.12.

There are also other ways to derive the basic continuity equation (4.6) that are interesting and instructive. One example is the use of the divergence theorem. Ask your instructor about these alternative approaches.

Steady Compressible Flow

If the flow is steady, $\partial/\partial t \equiv 0$ and all properties are functions of position only. Equation (4.6) reduces to

Cartesian: $$\frac{\partial}{\partial x}(\rho u) + \frac{\partial}{\partial y}(\rho \upsilon) + \frac{\partial}{\partial z}(\rho w) = 0$$

Cylindrical: $$\frac{1}{r}\frac{\partial}{\partial r}(r\rho\upsilon_r) + \frac{1}{r}\frac{\partial}{\partial \theta}(\rho\upsilon_\theta) + \frac{\partial}{\partial z}(\rho\upsilon_z) = 0 \tag{4.10}$$

Since density and velocity are both variables, these are still nonlinear and rather formidable, but a number of special-case solutions have been found.

Incompressible Flow

A special case that affords great simplification is incompressible flow, where the density changes are negligible. Then $\partial\rho/\partial t \approx 0$ regardless of whether the flow is steady or unsteady, and the density can be slipped out of the divergence in Eq. (4.6) and divided out. The result

$$\nabla \cdot \mathbf{V} = 0 \tag{4.11}$$

is valid for steady or unsteady incompressible flow. The two coordinate forms are

Cartesian: $$\frac{\partial u}{\partial x} + \frac{\partial \upsilon}{\partial y} + \frac{\partial w}{\partial z} = 0 \tag{4.12a}$$

Cylindrical: $$\frac{1}{r}\frac{\partial}{\partial r}(r\upsilon_r) + \frac{1}{r}\frac{\partial}{\partial \theta}(\upsilon_\theta) + \frac{\partial}{\partial z}(\upsilon_z) = 0 \tag{4.12b}$$

[2]See, for example, Ref. 3.

These are *linear* differential equations, and a wide variety of solutions are known, as discussed in Chaps. 6 to 8. Since no author or instructor can resist a wide variety of solutions, it follows that a great deal of time is spent studying incompressible flows. Fortunately, this is exactly what should be done, because most practical engineering flows are approximately incompressible, the chief exception being the high-speed gas flows treated in Chap. 9.

When is a given flow approximately incompressible? We can derive a nice criterion by using some density approximations. In essence, we wish to slip the density out of the divergence in Eq. (4.6) and approximate a typical term such as

$$\frac{\partial}{\partial x}(\rho u) \approx \rho \frac{\partial u}{\partial x} \tag{4.13}$$

This is equivalent to the strong inequality

$$\left| u \frac{\partial \rho}{\partial x} \right| \ll \left| \rho \frac{\partial u}{\partial x} \right|$$

or

$$\left| \frac{\delta \rho}{\rho} \right| \ll \left| \frac{\delta V}{V} \right| \tag{4.14}$$

As shown in Eq. (1.38), the pressure change is approximately proportional to the density change and the square of the speed of sound a of the fluid:

$$\delta p \approx a^2 \, \delta \rho \tag{4.15}$$

Meanwhile, if elevation changes are negligible, the pressure is related to the velocity change by Bernoulli's equation (3.52):

$$\delta p \approx -\rho V \, \delta V \tag{4.16}$$

Combining Eqs. (4.14) to (4.16), we obtain an explicit criterion for incompressible flow:

$$\frac{V^2}{a^2} = \text{Ma}^2 \ll 1 \tag{4.17}$$

where $\text{Ma} = V/a$ is the dimensionless *Mach number* of the flow. How small is small? The commonly accepted limit is

$$\text{Ma} \le 0.3 \tag{4.18}$$

For air at standard conditions, a flow can thus be considered incompressible if the velocity is less than about 100 m/s. This encompasses a wide variety of airflows: automobile and train motions, light aircraft, landing and takeoff of high-speed aircraft, most pipe flows, and turbomachinery at moderate rotational speeds. Further, it is clear that almost all liquid flows are incompressible, since flow velocities are small and the speed of sound is very large.[3]

Before attempting to analyze the continuity equation, we shall proceed with the derivation of the momentum and energy equations, so that we can analyze them as a group. A very clever device called the *stream function* can often make short work of the continuity equation, but we shall save it until Sec. 4.7.

One further remark is appropriate: The continuity equation is always important and must always be satisfied for a rational analysis of a flow pattern. Any newly discovered momentum or energy "solution" will ultimately fail when subjected to critical analysis if it does not also satisfy the continuity equation.

[3]An exception occurs in geophysical flows, where a density change is imposed thermally or mechanically rather than by the flow conditions themselves. An example is fresh water layered upon saltwater or warm air layered upon cold air in the atmosphere. We say that the fluid is *stratified,* and we must account for vertical density changes in Eq. (4.6) even if the velocities are small.

EXAMPLE 4.2

Under what conditions does the velocity field

$$\mathbf{V} = (a_1x + b_1y + c_1z)\mathbf{i} + (a_2x + b_2y + c_2z)\mathbf{j} + (a_3x + b_3y + c_3z)\mathbf{k}$$

where a_1, b_1, etc. = const, represent an incompressible flow that conserves mass?

Solution

Recalling that $\mathbf{V} = u\mathbf{i} + v\mathbf{j} + w\mathbf{k}$, we see that $u = (a_1x + b_1y + c_1z)$, etc. Substituting into Eq. (4.12*a*) for incompressible continuity, we obtain

$$\frac{\partial}{\partial x}(a_1x + b_1y + c_1z) + \frac{\partial}{\partial y}(a_2x + b_2y + c_2z) + \frac{\partial}{\partial z}(a_3x + b_3y + c_3z) = 0$$

or $$a_1 + b_2 + c_3 = 0 \qquad \textit{Ans.}$$

At least two of constants a_1, b_2, and c_3 must have opposite signs. Continuity imposes no restrictions whatever on constants b_1, c_1, a_2, c_2, a_3, and b_3, which do not contribute to a volume increase or decrease of a differential element.

EXAMPLE 4.3

An incompressible velocity field is given by

$$u = a(x^2 - y^2) \qquad v \text{ unknown} \qquad w = b$$

where a and b are constants. What must the form of the velocity component v be?

Solution

Again Eq. (4.12*a*) applies:

$$\frac{\partial}{\partial x}(ax^2 - ay^2) + \frac{\partial v}{\partial y} + \frac{\partial b}{\partial z} = 0$$

or $$\frac{\partial v}{\partial y} = -2ax \qquad (1)$$

This is easily integrated partially with respect to y:

$$v(x, y, z, t) = -2axy + f(x, z, t) \qquad \textit{Ans.}$$

This is the only possible form for v that satisfies the incompressible continuity equation. The function of integration f is entirely arbitrary since it vanishes when v is differentiated with respect to y.[4]

EXAMPLE 4.4

A centrifugal impeller of 40-cm diameter is used to pump hydrogen at 15°C and 1-atm pressure. Estimate the maximum allowable impeller rotational speed to avoid compressibility effects at the blade tips.

[4]This is a very realistic flow that simulates the turning of an inviscid fluid through a 60° angle; see Examples 4.7 and 4.9.

Solution

- *Assumptions:* The maximum fluid velocity is approximately equal to the impeller tip speed:

$$V_{max} \approx \Omega r_{max} \qquad \text{where } r_{max} = D/2 = 0.20 \text{ m}$$

- *Approach:* Find the speed of sound of hydrogen and make sure that V_{max} is much less.
- *Property values:* From Table A.4 for hydrogen, $R = 4{,}124 \text{ m}^2/(\text{s}^2 - \text{K})$ and $k = 1.41$. From Eq. (1.39) at 15°C = 288K, compute the speed of sound:

$$a_{H_2} = \sqrt{kRT} = \sqrt{1.41\,[4{,}124 \text{ m}^2/(\text{s}^2 - \text{K})]\,(288 \text{ K})} \approx 1{,}294 \text{ m/s}$$

- *Final solution step:* Use our rule of thumb, Eq. (4.18), to estimate the maximum impeller speed:

$$V = \Omega r_{max} \le 0.3a \quad \text{or} \quad \Omega(0.2 \text{ m}) \le 0.3(1{,}294 \text{ m/s})$$

$$\text{Solve for} \quad \Omega \le 1{,}940 \frac{\text{rad}}{\text{s}} \approx 18{,}500 \frac{\text{rev}}{\text{min}} \qquad \textit{Ans.}$$

- *Comments:* This is a high rate because the speed of sound of hydrogen, a light gas, is nearly four times greater than that of air. An impeller moving at this speed in air would create tip shock waves.

4.3 The Differential Equation of Linear Momentum

This section uses an elemental volume to derive Newton's law for a moving fluid. An alternate approach, which the reader might pursue, would be a force balance on an elemental moving particle. Having done it once in Sec. 4.2 for mass conservation, we can move along a little faster this time. We use the same elemental control volume as in Fig. 4.1, for which the appropriate form of the linear momentum relation is

$$\sum \mathbf{F} = \frac{\partial}{\partial t}\left(\int_{CV} \mathbf{V}\rho\, d\mathcal{V}\right) + \sum (\dot{m}_i \mathbf{V}_i)_{out} - \sum (\dot{m}_i \mathbf{V}_i)_{in} \tag{3.40}$$

Again, the element is so small that the volume integral simply reduces to a derivative term:

$$\frac{\partial}{\partial t}(\mathbf{V}\rho\, d\mathcal{V}) \approx \frac{\partial}{\partial t}(\rho \mathbf{V})\, dx\, dy\, dz \tag{4.19}$$

The momentum fluxes occur on all six faces, three inlets and three outlets. Referring again to Fig. 4.1, we can form a table of momentum fluxes by exact analogy with the discussion that led up to the equation for net mass flux:

Faces	Inlet momentum flux	Outlet momentum flux
x	$\rho u \mathbf{V}\, dy\, dz$	$\left[\rho u \mathbf{V} + \frac{\partial}{\partial x}(\rho u \mathbf{V})\, dx\right] dy\, dz$
y	$\rho v \mathbf{V}\, dx\, dz$	$\left[\rho v \mathbf{V} + \frac{\partial}{\partial y}(\rho v \mathbf{V})\, dy\right] dx\, dz$
z	$\rho w \mathbf{V}\, dx\, dy$	$\left[\rho w \mathbf{V} + \frac{\partial}{\partial z}(\rho w \mathbf{V})\, dz\right] dx\, dy$

Introduce these terms and Eq. (4.19) into Eq. (3.40), and get this intermediate result:

$$\sum \mathbf{F} = dx\, dy\, dz\left[\frac{\partial}{\partial t}(\rho \mathbf{V}) + \frac{\partial}{\partial x}(\rho u \mathbf{V}) + \frac{\partial}{\partial y}(\rho v \mathbf{V}) + \frac{\partial}{\partial z}(\rho w \mathbf{V})\right] \tag{4.20}$$

Note that this is a vector relation. A simplification occurs if we split up the term in brackets as follows:

$$\frac{\partial}{\partial t}(\rho\mathbf{V}) + \frac{\partial}{\partial x}(\rho u\mathbf{V}) + \frac{\partial}{\partial y}(\rho v\mathbf{V}) + \frac{\partial}{\partial z}(\rho w\mathbf{V})$$
$$= \mathbf{V}\left[\frac{\partial\rho}{\partial t} + \nabla\cdot(\rho\mathbf{V})\right] + \rho\left(\frac{\partial\mathbf{V}}{\partial t} + u\frac{\partial\mathbf{V}}{\partial x} + v\frac{\partial\mathbf{V}}{\partial y} + w\frac{\partial\mathbf{V}}{\partial z}\right) \tag{4.21}$$

The term in brackets on the right-hand side is seen to be the equation of continuity, Eq. (4.6), which vanishes identically. The long term in parentheses on the right-hand side is seen from Eq. (4.2) to be the total acceleration of a particle that instantaneously occupies the control volume:

$$\frac{\partial\mathbf{V}}{\partial t} + u\frac{\partial\mathbf{V}}{\partial x} + v\frac{\partial\mathbf{V}}{\partial y} + w\frac{\partial\mathbf{V}}{\partial z} = \frac{d\mathbf{V}}{dt} \tag{4.2}$$

Thus, we have now reduced Eq. (4.20) to

$$\sum\mathbf{F} = \rho\frac{d\mathbf{V}}{dt}\,dx\,dy\,dz \tag{4.22}$$

It might be good for you to stop and rest now and think about what we have just done. What is the relation between Eqs. (4.22) and (3.40) for an infinitesimal control volume? Could we have *begun* the analysis at Eq. (4.22)?

Equation (4.22) points out that the net force on the control volume must be of differential size and proportional to the element volume. These forces are of two types, *body* forces and *surface* forces. Body forces are due to external fields (gravity, magnetism, electric potential) that act on the entire mass within the element. The only body force we shall consider in this book is gravity. The gravity force on the differential mass $\rho\,dx\,dy\,dz$ within the control volume is

$$d\mathbf{F}_{\text{grav}} = \rho\mathbf{g}\,dx\,dy\,dz \tag{4.23}$$

where $\mathbf{g}$ may in general have an arbitrary orientation with respect to the coordinate system. In many applications, such as Bernoulli's equation, we take z "up," and $\mathbf{g} = -g\mathbf{k}$.

The surface forces are due to the stresses on the sides of the control surface. These stresses are the sum of hydrostatic pressure plus viscous stresses τ_{ij} that arise from motion with velocity gradients:

$$\sigma_{ij} = \begin{vmatrix} -p+\tau_{xx} & \tau_{yx} & \tau_{zx} \\ \tau_{xy} & -p+\tau_{yy} & \tau_{zy} \\ \tau_{xz} & \tau_{yz} & -p+\tau_{zz} \end{vmatrix} \tag{4.24}$$

The subscript notation for stresses is given in Fig. 4.3. Unlike velocity $\mathbf{V}$, which is a three-component *vector,* stresses σ_{ij} and τ_{ij} and strain rates ε_{ij} are nine-component *tensors* and require two subscripts to define each component. For further study of *tensor analysis,* see Refs. 6, 11, or 13.

It is not these stresses, but their *gradients,* or differences, that cause a net force on the differential control surface. This is seen by referring to Fig. 4.4, which shows only the x-directed stresses to avoid cluttering up the drawing. For example, the leftward force $\sigma_{xx}\,dy\,dz$ on the left face is balanced by the rightward force $\sigma_{xx}\,dy\,dz$ on the right face, leaving only the net rightward force $(\partial\sigma_{xx}/\partial x)\,dx\,dy\,dz$ on the right face.

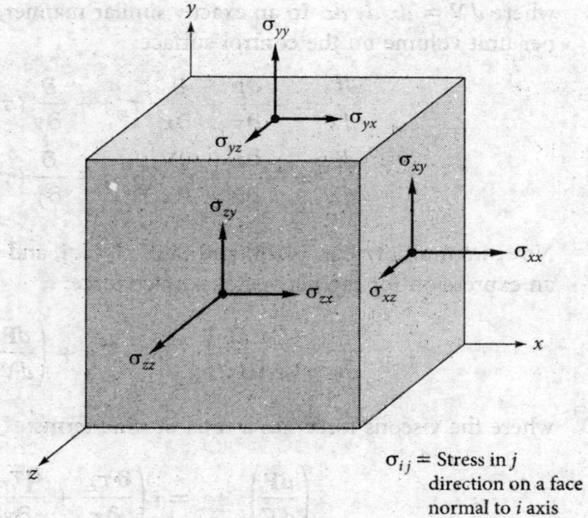

Fig. 4.3 Notation for stresses.

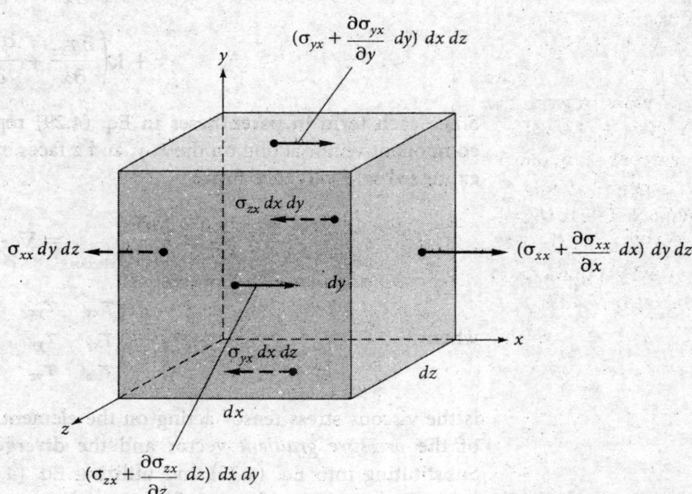

Fig. 4.4 Elemental cartesian fixed control volume showing the surface forces in the x direction only.

The same thing happens on the other four faces, so the net surface force in the x direction is given by

$$dF_{x,\text{surf}} = \left[\frac{\partial}{\partial x}(\sigma_{xx}) + \frac{\partial}{\partial y}(\sigma_{yx}) + \frac{\partial}{\partial z}(\sigma_{zx})\right] dx\,dy\,dz \tag{4.25}$$

We see that this force is proportional to the element volume. Notice that the stress terms are taken from the *top row* of the array in Eq. (4.24). Splitting this row into pressure plus viscous stresses, we can rewrite Eq. (4.25) as

$$\frac{dF_x}{d\mathcal{V}} = -\frac{\partial p}{\partial x} + \frac{\partial}{\partial x}(\tau_{xx}) + \frac{\partial}{\partial y}(\tau_{yx}) + \frac{\partial}{\partial z}(\tau_{zx}) \tag{4.26}$$

where $d\mathcal{V} = dx\, dy\, dz$. In an exactly similar manner, we can derive the y and z forces per unit volume on the control surface:

$$\frac{dF_y}{d\mathcal{V}} = -\frac{\partial p}{\partial y} + \frac{\partial}{\partial x}(\tau_{xy}) + \frac{\partial}{\partial y}(\tau_{yy}) + \frac{\partial}{\partial z}(\tau_{zy})$$

$$\frac{dF_z}{d\mathcal{V}} = -\frac{\partial p}{\partial z} + \frac{\partial}{\partial x}(\tau_{xz}) + \frac{\partial}{\partial y}(\tau_{yz}) + \frac{\partial}{\partial z}(\tau_{zz}) \qquad (4.27)$$

Now, we multiply Eqs. (4.26) and (4.27) by **i**, **j**, and **k**, respectively, and add to obtain an expression for the net vector surface force:

$$\left(\frac{d\mathbf{F}}{d\mathcal{V}}\right)_{\text{surf}} = -\nabla p + \left(\frac{d\mathbf{F}}{d\mathcal{V}}\right)_{\text{viscous}} \qquad (4.28)$$

where the viscous force has a total of nine terms:

$$\left(\frac{d\mathbf{F}}{d\mathcal{V}}\right)_{\text{viscous}} = \mathbf{i}\left(\frac{\partial \tau_{xx}}{\partial x} + \frac{\partial \tau_{yx}}{\partial y} + \frac{\partial \tau_{zx}}{\partial z}\right) + \mathbf{j}\left(\frac{\partial \tau_{xy}}{\partial x} + \frac{\partial \tau_{yy}}{\partial y} + \frac{\partial \tau_{zy}}{\partial z}\right) + \mathbf{k}\left(\frac{\partial \tau_{xz}}{\partial x} + \frac{\partial \tau_{yz}}{\partial y} + \frac{\partial \tau_{zz}}{\partial z}\right) \qquad (4.29)$$

Since each term in parentheses in Eq. (4.29) represents the divergence of a stress component vector acting on the x, y, and z faces, respectively, Eq. (4.29) is sometimes expressed in divergence form:

$$\left(\frac{d\mathbf{F}}{d\mathcal{V}}\right)_{\text{viscous}} = \nabla \cdot \tau_{ij} \qquad (4.30)$$

where

$$\tau_{ij} = \begin{bmatrix} \tau_{xx} & \tau_{yx} & \tau_{zx} \\ \tau_{xy} & \tau_{yy} & \tau_{zy} \\ \tau_{xz} & \tau_{yz} & \tau_{zz} \end{bmatrix} \qquad (4.31)$$

is the viscous stress tensor acting on the element. The surface force is thus the sum of the *pressure gradient* vector and the divergence of the viscous stress tensor. Substituting into Eq. (4.22) and utilizing Eq. (4.23), we have the basic differential momentum equation for an infinitesimal element:

$$\rho\mathbf{g} - \nabla p + \nabla \cdot \tau_{ij} = \rho \frac{d\mathbf{V}}{dt} \qquad (4.32)$$

where

$$\frac{d\mathbf{V}}{dt} = \frac{\partial \mathbf{V}}{\partial t} + u\frac{\partial \mathbf{V}}{\partial x} + \upsilon\frac{\partial \mathbf{V}}{\partial y} + w\frac{\partial \mathbf{V}}{\partial z} \qquad (4.33)$$

We can also express Eq. (4.32) in words:

Gravity force per unit volume + pressure force per unit volume
+ viscous force per unit volume = density × acceleration (4.34)

Equation (4.32) is so brief and compact that its inherent complexity is almost invisible. It is a *vector* equation, each of whose component equations contains nine terms. Let us therefore write out the component equations in full to illustrate the mathematical difficulties inherent in the momentum equation:

$$\rho g_x - \frac{\partial p}{\partial x} + \frac{\partial \tau_{xx}}{\partial x} + \frac{\partial \tau_{yx}}{\partial y} + \frac{\partial \tau_{zx}}{\partial z} = \rho\left(\frac{\partial u}{\partial t} + u\frac{\partial u}{\partial x} + \upsilon\frac{\partial u}{\partial y} + w\frac{\partial u}{\partial z}\right)$$
$$\rho g_y - \frac{\partial p}{\partial y} + \frac{\partial \tau_{xy}}{\partial x} + \frac{\partial \tau_{yy}}{\partial y} + \frac{\partial \tau_{zy}}{\partial z} = \rho\left(\frac{\partial \upsilon}{\partial t} + u\frac{\partial \upsilon}{\partial x} + \upsilon\frac{\partial \upsilon}{\partial y} + w\frac{\partial \upsilon}{\partial z}\right) \quad (4.35)$$
$$\rho g_z - \frac{\partial p}{\partial z} + \frac{\partial \tau_{xz}}{\partial x} + \frac{\partial \tau_{yz}}{\partial y} + \frac{\partial \tau_{zz}}{\partial z} = \rho\left(\frac{\partial w}{\partial t} + u\frac{\partial w}{\partial x} + \upsilon\frac{\partial w}{\partial y} + w\frac{\partial w}{\partial z}\right)$$

This is the differential momentum equation in its full glory, and it is valid for any fluid in any general motion, particular fluids being characterized by particular viscous stress terms. Note that the last three "convective" terms on the right-hand side of each component equation in (4.35) are nonlinear, which complicates the general mathematical analysis.

Inviscid Flow: Euler's Equation

Equation (4.35) is not ready to use until we write the viscous stresses in terms of velocity components. The simplest assumption is frictionless flow $\tau_{ij} = 0$, for which Eq. (4.32) reduces to

$$\rho \mathbf{g} - \nabla p = \rho \frac{d\mathbf{V}}{dt} \quad (4.36)$$

This is *Euler's equation* for inviscid flow. We show in Sec. 4.9 that Euler's equation can be integrated along a streamline to yield the frictionless Bernoulli equation, (3.52) or (3.54). The complete analysis of inviscid flow fields, using continuity and the Bernoulli relation, is given in Chap. 8.

Newtonian Fluid: Navier-Stokes Equations

For a newtonian fluid, as discussed in Sec. 1.9, the viscous stresses are proportional to the element strain rates and the coefficient of viscosity. For incompressible flow, the generalization of Eq. (1.23) to three-dimensional viscous flow is[5]

$$\tau_{xx} = 2\mu\frac{\partial u}{\partial x} \qquad \tau_{yy} = 2\mu\frac{\partial \upsilon}{\partial y} \qquad \tau_{zz} = 2\mu\frac{\partial w}{\partial z}$$
$$\tau_{xy} = \tau_{yx} = \mu\left(\frac{\partial u}{\partial y} + \frac{\partial \upsilon}{\partial x}\right) \qquad \tau_{xz} = \tau_{zx} = \mu\left(\frac{\partial w}{\partial x} + \frac{\partial u}{\partial z}\right) \quad (4.37)$$
$$\tau_{yz} = \tau_{zy} = \mu\left(\frac{\partial \upsilon}{\partial z} + \frac{\partial w}{\partial y}\right)$$

where μ is the viscosity coefficient. Substitution into Eq. (4.35) gives the differential momentum equation for a newtonian fluid with constant density and viscosity:

$$\rho g_x - \frac{\partial p}{\partial x} + \mu\left(\frac{\partial^2 u}{\partial x^2} + \frac{\partial^2 u}{\partial y^2} + \frac{\partial^2 u}{\partial z^2}\right) = \rho\frac{du}{dt}$$
$$\rho g_y - \frac{\partial p}{\partial y} + \mu\left(\frac{\partial^2 \upsilon}{\partial x^2} + \frac{\partial^2 \upsilon}{\partial y^2} + \frac{\partial^2 \upsilon}{\partial z^2}\right) = \rho\frac{d\upsilon}{dt} \quad (4.38)$$
$$\rho g_z - \frac{\partial p}{\partial z} + \mu\left(\frac{\partial^2 w}{\partial x^2} + \frac{\partial^2 w}{\partial y^2} + \frac{\partial^2 w}{\partial z^2}\right) = \rho\frac{dw}{dt}$$

[5]When compressibility is significant, additional small terms arise containing the element volume expansion rate and a *second* coefficient of viscosity; see Refs. 4 and 5 for details.

These are the incompressible flow *Navier-Stokes equations,* named after C. L. M. H. Navier (1785–1836) and Sir George G. Stokes (1819–1903), who are credited with their derivation. They are second-order nonlinear partial differential equations and are quite formidable, but solutions have been found to a variety of interesting viscous flow problems, some of which are discussed in Sec. 4.11 and in Chap. 6 (see also Refs. 4 and 5). For compressible flow, see Eq. (2.29) of Ref. 5.

Equations (4.38) have four unknowns: p, u, v, and w. They should be combined with the incompressible continuity relation [Eqs. (4.12)] to form four equations in these four unknowns. We shall discuss this again in Sec. 4.6, which presents the appropriate boundary conditions for these equations.

Even though the Navier-Stokes equations have only a limited number of known analytical solutions, they are amenable to fine-gridded computer modeling [1]. The field of CFD is maturing fast, with many commercial software tools available. It is possible now to achieve approximate, but realistic, CFD results for a wide variety of complex two- and three-dimensional viscous flows.

EXAMPLE 4.5

Take the velocity field of Example 4.3, with $b = 0$ for algebraic convenience

$$u = a(x^2 - y^2) \qquad v = -2axy \qquad w = 0$$

and determine under what conditions it is a solution to the Navier-Stokes momentum equations (4.38). Assuming that these conditions are met, determine the resulting pressure distribution when z is "up" ($g_x = 0$, $g_y = 0$, $g_z = -g$).

Solution

- *Assumptions:* Constant density and viscosity, steady flow (u and v independent of time).
- *Approach:* Substitute the known (u, v, w) into Eqs. (4.38) and solve for the pressure gradients. If a unique pressure function $p(x, y, z)$ can then be found, the given solution is exact.
- *Solution step 1:* Substitute (u, v, w) into Eqs. (4.38) in sequence:

$$\rho(0) - \frac{\partial p}{\partial x} + \mu(2a - 2a + 0) = \rho\left(u\frac{\partial u}{\partial x} + v\frac{\partial u}{\partial y}\right) = 2a^2\rho\,(x^3 + xy^2)$$

$$\rho(0) - \frac{\partial p}{\partial y} + \mu(0 + 0 + 0) = \rho\left(u\frac{\partial v}{\partial x} + v\frac{\partial v}{\partial y}\right) = 2a^2\rho(x^2y + y^3)$$

$$\rho(-g) - \frac{\partial p}{\partial z} + \mu(0 + 0 + 0) = \rho\left(u\frac{\partial w}{\partial x} + v\frac{\partial w}{\partial y}\right) = 0$$

Rearrange and solve for the three pressure gradients:

$$\frac{\partial p}{\partial x} = -2a^2\rho(x^3 + xy^2) \qquad \frac{\partial p}{\partial y} = -2a^2\rho(x^2y + y^3) \qquad \frac{\partial p}{\partial z} = -\rho g \tag{1}$$

- *Comment 1:* The vertical pressure gradient is *hydrostatic.* [Could you have predicted this by noting in Eqs. (4.38) that $w = 0$?] However, the pressure is velocity-dependent in the xy plane.
- *Solution step 2:* To determine if the x and y gradients of pressure in Eq. (1) are compatible, evaluate the mixed derivative $(\partial^2 p/\partial x\,\partial y)$; that is, cross-differentiate these two equations:

$$\frac{\partial}{\partial y}\left(\frac{\partial p}{\partial x}\right) = \frac{\partial}{\partial y}\left[-2a^2\rho(x^3 + xy^2)\right] = -4a^2\rho xy$$

$$\frac{\partial}{\partial x}\left(\frac{\partial p}{\partial y}\right) = \frac{\partial}{\partial x}\left[-2a^2\rho(x^2y + y^3)\right] = -4a^2\rho xy$$

- *Comment 2:* Since these are equal, the given velocity distribution is indeed an *exact* solution of the Navier-Stokes equations.
- *Solution step 3:* To find the pressure, integrate Eqs. (1), collect, and compare. Start with $\partial p/\partial x$. The procedure requires care! Integrate *partially* with respect to x, holding y and z constant:

$$p = \int \frac{\partial p}{\partial x} dx \,|_{y,z} = \int -2a^2\rho(x^3 + xy^2)\, dx \,|_{y,z} = -2a^2\rho\left(\frac{x^4}{4} + \frac{x^2y^2}{2}\right) + f_1(y, z) \qquad (2)$$

Note that the "constant" of integration f_1 is a *function* of the variables that were not integrated. Now differentiate Eq. (2) with respect to y and compare with $\partial p/\partial y$ from Eq. (1):

$$\frac{\partial p}{\partial y}\,|_{(2)} = -2a^2\rho\, x^2y + \frac{\partial f_1}{\partial y} = \frac{\partial p}{\partial y}\,|_{(1)} = -2a^2\rho(x^2y + y^3)$$

$$\text{Compare: } \frac{\partial f_1}{\partial y} = -2a^2\rho\, y^3 \quad \text{or} \quad f_1 = \int \frac{\partial f_1}{\partial y} dy \,|_z = -2a^2\rho \frac{y^4}{4} + f_2(z)$$

$$\text{Collect terms: So far} \quad p = -2a^2\rho\left(\frac{x^4}{4} + \frac{x^2y^2}{2} + \frac{y^4}{4}\right) + f_2(z) \qquad (3)$$

This time the "constant" of integration f_2 is a function of z only (the variable not integrated). Now differentiate Eq. (3) with respect to z and compare with $\partial p/\partial z$ from Eq. (1):

$$\frac{\partial p}{\partial z}\,|_{(3)} = \frac{df_2}{dz} = \frac{\partial p}{\partial z}\,|_{(1)} = -\rho g \quad \text{or} \quad f_2 = -\rho g z + C \qquad (4)$$

where C is a constant. This completes our three integrations. Combine Eqs. (3) and (4) to obtain the full expression for the pressure distribution in this flow:

$$p(x, y, z) = -\rho g z - \tfrac{1}{2} a^2\rho(x^4 + y^4 + 2x^2y^2) + C \qquad \textit{Ans. } (5)$$

This is the desired solution. Do you recognize it? Not unless you go back to the beginning and square the velocity components:

$$u^2 + \upsilon^2 + w^2 = V^2 = a^2(x^4 + y^4 + 2x^2y^2) \qquad (6)$$

Comparing with Eq. (5), we can rewrite the pressure distribution as

$$p + \tfrac{1}{2}\rho V^2 + \rho g z = C \qquad (7)$$

- *Comment:* This is Bernoulli's equation (3.54). That is no accident, because the velocity distribution given in this problem is one of a family of flows that are solutions to the Navier-Stokes equations and that satisfy Bernoulli's incompressible equation everywhere in the flow field. They are called *irrotational flows,* for which curl $\mathbf{V} = \nabla \times \mathbf{V} \equiv 0$. This subject is discussed again in Sec. 4.9.

4.4 The Differential Equation of Angular Momentum

Having now been through the same approach for both mass and linear momentum, we can go rapidly through a derivation of the differential angular momentum relation. The appropriate form of the integral angular momentum equation for a fixed control volume is

$$\sum \mathbf{M}_o = \frac{\partial}{\partial t}\left[\int_{CV} (\mathbf{r} \times \mathbf{V})\rho\, d\mathcal{V}\right] + \int_{CS} (\mathbf{r} \times \mathbf{V})\rho(\mathbf{V} \cdot \mathbf{n})\, dA \qquad (3.59)$$

We shall confine ourselves to an axis through O that is parallel to the z axis and passes through the centroid of the elemental control volume. This is shown in Fig. 4.5. Let θ be the angle of rotation about O of the fluid within the control volume. The only

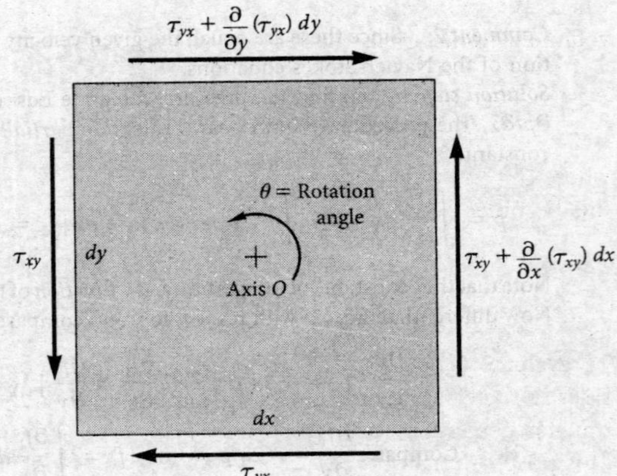

Fig. 4.5 Elemental cartesian fixed control volume showing shear stresses that may cause a net angular acceleration about axis *O*.

stresses that have moments about *O* are the shear stresses τ_{xy} and τ_{yx}. We can evaluate the moments about *O* and the angular momentum terms about *O*. A lot of algebra is involved, and we give here only the result:

$$\left[\tau_{xy} - \tau_{yx} + \frac{1}{2}\frac{\partial}{\partial x}(\tau_{xy})\, dx - \frac{1}{2}\frac{\partial}{\partial y}(\tau_{yx})\, dy\right] dx\, dy\, dz = \frac{1}{12}\rho(dx\, dy\, dz)(dx^2 + dy^2)\frac{d^2\theta}{dt^2} \qquad (4.39)$$

Assuming that the angular acceleration $d^2\theta/dt^2$ is not infinite, we can neglect all higher-order differential terms, which leaves a finite and interesting result:

$$\tau_{xy} \approx \tau_{yx} \qquad (4.40)$$

Had we summed moments about axes parallel to *y* or *x*, we would have obtained exactly analogous results:

$$\tau_{xz} \approx \tau_{zx} \qquad \tau_{yz} \approx \tau_{zy} \qquad (4.41)$$

There is *no* differential angular momentum equation. Application of the integral theorem to a differential element gives the result, well known to students of stress analysis or strength of materials, that the shear stresses are symmetric: $\tau_{ij} = \tau_{ji}$. This is the only result of this section.[6] There is no differential equation to remember, which leaves room in your brain for the next topic, the differential energy equation.

4.5 The Differential Equation of Energy[7]

We are now so used to this type of derivation that we can race through the energy equation at a bewildering pace. The appropriate integral relation for the fixed control volume of Fig. 4.1 is

$$\dot{Q} - \dot{W}_s - \dot{W}_v = \frac{\partial}{\partial t}\left(\int_{CV} e\rho\, d\mathcal{V}\right) + \int_{CS}\left(e + \frac{p}{\rho}\right)\rho(\mathbf{V}\cdot\mathbf{n})\, dA \qquad (3.66)$$

[6]We are neglecting the possibility of a finite *couple* being applied to the element by some powerful external force field. See, for example, Ref. 6.

[7]This section may be omitted without loss of continuity.

where $\dot{W}_s = 0$ because there can be no infinitesimal shaft protruding into the control volume. By analogy with Eq. (4.20), the right-hand side becomes, for this tiny element,

$$\dot{Q} - \dot{W}_v = \left[\frac{\partial}{\partial t}(\rho e) + \frac{\partial}{\partial x}(\rho u \zeta) + \frac{\partial}{\partial y}(\rho \upsilon \zeta) + \frac{\partial}{\partial z}(\rho w \zeta)\right] dx\,dy\,dz$$

where $\zeta = e + p/\rho$. When we use the continuity equation by analogy with Eq. (4.21), this becomes

$$\dot{Q} - \dot{W}_v = \left(\rho \frac{de}{dt} + \mathbf{V} \cdot \nabla p + p \nabla \cdot \mathbf{V}\right) dx\,dy\,dz \tag{4.42}$$

Thermal Conductivity; Fourier's Law

To evaluate $\dot{Q}$, we neglect radiation and consider only heat conduction through the sides of the element. Experiments for both fluids and solids show that the vector heat transfer per unit area, $\mathbf{q}$, is proportional to the vector gradient of temperature, ∇T. This proportionality is called *Fourier's law of conduction,* which is analogous to Newton's viscosity law:

$$\mathbf{q} = -k \nabla T$$

$$\text{or:} \quad q_x = -k\frac{\partial T}{\partial x}, \quad q_y = -k\frac{\partial T}{\partial y}, \quad q_z = -k\frac{\partial T}{\partial z} \tag{4.43}$$

where k is called the *thermal conductivity*, a fluid property that varies with temperature and pressure in much the same way as viscosity. The minus sign satisfies the convention that heat flux is positive in the direction of decreasing temperature. Fourier's law is dimensionally consistent, and k has SI units of joules per (sec-meter-kelvin) and can be correlated with T in much the same way as Eqs. (1.27) and (1.28) for gases and liquids, respectively.

Figure 4.6 shows the heat flow passing through the x faces, the y and z heat flows being omitted for clarity. We can list these six heat flux terms:

Faces	Inlet heat flux	Outlet heat flux
x	$q_x\,dy\,dz$	$\left[q_x + \frac{\partial}{\partial x}(q_x)\,dx\right] dy\,dz$
y	$q_y\,dx\,dz$	$\left[q_y + \frac{\partial}{\partial y}(q_y)\,dy\right] dx\,dz$
z	$q_z\,dx\,dy$	$\left[q_z + \frac{\partial}{\partial z}(q_z)\,dz\right] dx\,dy$

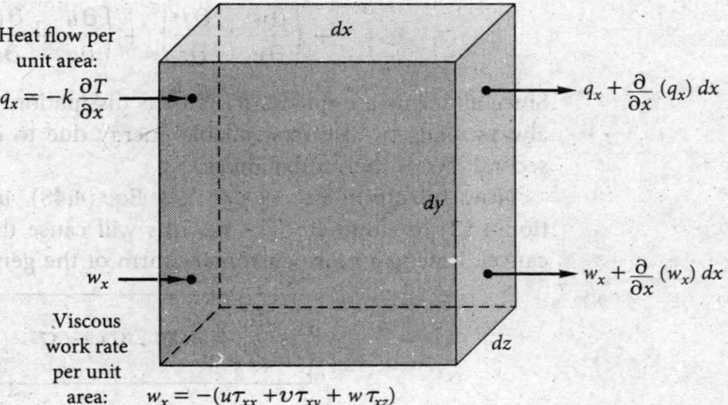

Fig. 4.6 Elemental cartesian control volume showing heat flow and viscous work rate terms in the x direction.

By adding the inlet terms and subtracting the outlet terms, we obtain the net heat added to the element:

$$\dot{Q} = -\left[\frac{\partial}{\partial x}(q_x) + \frac{\partial}{\partial y}(q_y) + \frac{\partial}{\partial z}(q_z)\right] dx\,dy\,dz = -\nabla \cdot \mathbf{q}\,dx\,dy\,dz \quad (4.44)$$

As expected, the heat flux is proportional to the element volume. Introducing Fourier's law from Eq. (4.43), we have

$$\dot{Q} = \nabla \cdot (k\nabla T)\,dx\,dy\,dz \quad (4.45)$$

The rate of work done by viscous stresses equals the product of the stress component, its corresponding velocity component, and the area of the element face. Figure 4.6 shows the work rate on the left x face is

$$\dot{W}_{v,\mathrm{LF}} = w_x\,dy\,dz \quad \text{where } w_x = -(u\tau_{xx} + v\tau_{xy} + w\tau_{xz}) \quad (4.46)$$

(where the subscript LF stands for left face) and a slightly different work on the right face due to the gradient in w_x. These work fluxes could be tabulated in exactly the same manner as the heat fluxes in the previous table, with w_x replacing q_x, and so on. After outlet terms are subtracted from inlet terms, the net viscous work rate becomes

$$\dot{W}_v = -\left[\frac{\partial}{\partial x}(u\tau_{xx} + v\tau_{xy} + w\tau_{xz}) + \frac{\partial}{\partial y}(u\tau_{yx} + v\tau_{yy} + w\tau_{yz}) + \frac{\partial}{\partial z}(u\tau_{zx} + v\tau_{zy} + w\tau_{zz})\right] dx\,dy\,dz$$

$$= -\nabla \cdot (\mathbf{V} \cdot \boldsymbol{\tau}_{ij})\,dx\,dy\,dz \quad (4.47)$$

We now substitute Eqs. (4.45) and (4.47) into Eq. (4.43) to obtain one form of the differential energy equation:

$$\rho\frac{de}{dt} + \mathbf{V} \cdot \nabla p + p\nabla \cdot \mathbf{V} = \nabla \cdot (k\,\nabla T) + \nabla \cdot (\mathbf{V} \cdot \boldsymbol{\tau}_{ij}) \quad (4.48)$$

where $e = \hat{u} + \frac{1}{2}V^2 + gz$

A more useful form is obtained if we split up the viscous work term:

$$\nabla \cdot (\mathbf{V} \cdot \boldsymbol{\tau}_{ij}) \equiv \mathbf{V} \cdot (\nabla \cdot \boldsymbol{\tau}_{ij}) + \Phi \quad (4.49)$$

where Φ is short for the *viscous-dissipation function*.[8] For a newtonian incompressible viscous fluid, this function has the form

$$\phi = \mu\left[2\left(\frac{\partial u}{\partial x}\right)^2 + 2\left(\frac{\partial v}{\partial y}\right)^2 + 2\left(\frac{\partial w}{\partial z}\right)^2 + \left(\frac{\partial v}{\partial x} + \frac{\partial u}{\partial y}\right)^2 + \left(\frac{\partial w}{\partial y} + \frac{\partial v}{\partial z}\right)^2 + \left(\frac{\partial u}{\partial z} + \frac{\partial w}{\partial x}\right)^2\right] \quad (4.50)$$

Since all terms are quadratic, viscous dissipation is always positive, so a viscous flow always tends to lose its available energy due to dissipation, in accordance with the second law of thermodynamics.

Now substitute Eq. (4.49) into Eq. (4.48), using the linear momentum equation(4.32) to eliminate $\nabla \cdot \boldsymbol{\tau}_{ij}$. This will cause the kinetic and potential energies to cancel, leaving a more customary form of the general differential energy equation:

$$\boxed{\rho\frac{d\hat{u}}{dt} + p(\nabla \cdot \mathbf{V}) = \nabla \cdot (k\,\nabla T) + \Phi} \quad (4.51)$$

[8]For further details, see, Ref. 5, p. 72.

This equation is valid for a newtonian fluid under very general conditions of unsteady, compressible, viscous, heat-conducting flow, except that it neglects radiation heat transfer and internal *sources* of heat that might occur during a chemical or nuclear reaction.

Equation (4.51) is too difficult to analyze except on a digital computer [1]. It is customary to make the following approximations:

$$d\hat{u} \approx c_v\, dT \qquad c_v, \mu, k, \rho \approx \text{const} \tag{4.52}$$

Equation (4.51) then takes the simpler form, for $\nabla \cdot \mathbf{V} = 0$,

$$\rho c_v \frac{dT}{dt} = k\nabla^2 T + \Phi \tag{4.53}$$

which involves temperature T as the sole primary variable plus velocity as a secondary variable through the total time-derivative operator:

$$\frac{dT}{dt} = \frac{\partial T}{\partial t} + u\frac{\partial T}{\partial x} + v\frac{\partial T}{\partial y} + w\frac{\partial T}{\partial z} \tag{4.54}$$

A great many interesting solutions to Eq. (4.53) are known for various flow conditions, and extended treatments are given in advanced books on viscous flow [4, 5] and books on heat transfer [7, 8].

One well-known special case of Eq. (4.53) occurs when the fluid is at rest or has negligible velocity, where the dissipation Φ and convective terms become negligible:

$$\rho c_p \frac{\partial T}{\partial t} = k\nabla^2 T \tag{4.55}$$

The change from c_v to c_p is correct and justified by the fact that, when pressure terms are neglected from a gas flow energy equation [4, 5], what remains is approximately an enthalpy change, not an internal energy change. This is called the *heat conduction equation* in applied mathematics and is valid for solids and fluids at rest. The solution to Eq. (4.55) for various conditions is a large part of courses and books on heat transfer.

This completes the derivation of the basic differential equations of fluid motion.

4.6 Boundary Conditions for the Basic Equations

There are three basic differential equations of fluid motion, just derived. Let us summarize those here:

Continuity:
$$\frac{\partial \rho}{\partial t} + \nabla \cdot (\rho \mathbf{V}) = 0 \tag{4.56}$$

Momentum:
$$\rho \frac{d\mathbf{V}}{dt} = \rho \mathbf{g} - \nabla p + \nabla \cdot \tau_{ij} \tag{4.57}$$

Energy:
$$\rho \frac{d\hat{u}}{dt} + p(\nabla \cdot \mathbf{V}) = \nabla \cdot (k\,\nabla T) + \Phi \tag{4.58}$$

where Φ is given by Eq. (4.50). In general, the density is variable, so these three equations contain five unknowns, ρ, $\mathbf{V}$, p, $\hat{u}$, and T. Therefore, we need two additional relations to complete the system of equations. These are provided by data or algebraic expressions for the state relations of the thermodynamic properties:

$$\rho = \rho(p, T) \qquad \hat{u} = \hat{u}(p, T) \tag{4.59}$$

For example, for a perfect gas with constant specific heats, we complete the system with

$$\rho = \frac{p}{RT} \qquad \hat{u} = \int c_v \, dT \approx c_v T + \text{const} \tag{4.60}$$

It is shown in advanced books [4, 5] that this system of equations (4.56) to (4.59) is well posed and can be solved analytically or numerically, subject to the proper boundary conditions.

What are the proper boundary conditions? First, if the flow is unsteady, there must be an *initial condition* or initial spatial distribution known for each variable:

$$\text{At } t = 0: \qquad \rho, \mathbf{V}, p, \hat{u}, T = \text{known } f(x, y, z) \tag{4.61}$$

Thereafter, for all times t to be analyzed, we must know something about the variables at each *boundary* enclosing the flow.

Figure 4.7 illustrates the three most common types of boundaries encountered in fluid flow analysis: a solid wall, an inlet or outlet, and a liquid–gas interface.

First, for a solid, impermeable wall, there can be slip and temperature jump in a viscous heat-conducting fluid:

$$\text{No-slip:} \qquad \mathbf{V}_{\text{fluid}} = \mathbf{V}_{\text{wall}} \qquad T_{\text{fluid}} = T_{\text{wall}}$$

$$\text{Rarefied gas: } u_{\text{fluid}} - u_{\text{wall}} \approx \ell \frac{\partial u}{\partial n}\Big|_{\text{wall}} \quad T_{\text{fluid}} - T_{\text{wall}} \approx \left(\frac{2\zeta}{\zeta + 1}\right) \frac{k}{\mu c_p} \ell \frac{\partial T}{\partial n}\Big|_{\text{wall}} \tag{4.62}$$

where, for the rarefied gas, n is normal to the wall, u is parallel to the wall, ℓ is the mean free path of the gas [see Eq. (1.37)], and ζ denotes, just this one time, the specific heat ratio. The above so-called *temperature-jump relation* for gases is given here only for completeness and will not be studied (see page 48 of Ref. 5). A few velocity-jump assignments will be given.

Second, at any inlet or outlet section of the flow, the complete distribution of velocity, pressure, and temperature must be known for all times:

$$\text{Inlet or outlet:} \qquad \text{Known } \mathbf{V}, p, T \tag{4.63}$$

These inlet and outlet sections can be and often are at $\pm\,\infty$, simulating a body immersed in an infinite expanse of fluid.

Finally, the most complex conditions occur at a liquid–gas interface, or free surface, as sketched in Fig. 4.7. Let us denote the interface by

$$\text{Interface:} \qquad z = \eta(x, y, t) \tag{4.64}$$

Then there must be equality of vertical velocity across the interface, so that no holes appear between liquid and gas:

$$w_{\text{liq}} = w_{\text{gas}} = \frac{d\eta}{dt} = \frac{\partial \eta}{\partial t} + u \frac{\partial \eta}{\partial x} + v \frac{\partial \eta}{\partial y} \tag{4.65}$$

This is called the *kinematic boundary condition*.

There must be mechanical equilibrium across the interface. The viscous shear stresses must balance:

$$(\tau_{zy})_{\text{liq}} = (\tau_{zy})_{\text{gas}} \qquad (\tau_{zx})_{\text{liq}} = (\tau_{zx})_{\text{gas}} \tag{4.66}$$

Neglecting the viscous normal stresses, the pressures must balance at the interface except for surface tension effects:

$$p_{\text{liq}} = p_{\text{gas}} - \Upsilon(R_x^{-1} + R_y^{-1}) \tag{4.67}$$

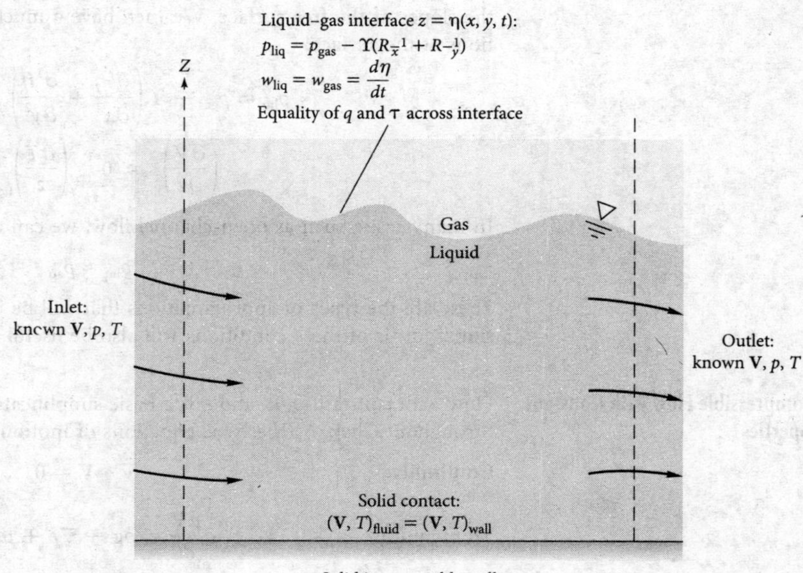

Fig. 4.7 Typical boundary conditions in a viscous heat-conducting fluid flow analysis.

which is equivalent to Eq. (1.33). The radii of curvature can be written in terms of the free surface position η:

$$R_x^{-1} + R_y^{-1} = \frac{\partial}{\partial x}\left[\frac{\partial\eta/\partial x}{\sqrt{1 + (\partial\eta/\partial x)^2 + (\partial\eta/\partial y)^2}}\right] + \frac{\partial}{\partial y}\left[\frac{\partial\eta/\partial y}{\sqrt{1 + (\partial\eta/\partial x)^2 + (\partial\eta/\partial y)^2}}\right] \quad (4.68)$$

Finally, the heat transfer must be the same on both sides of the interface, since no heat can be stored in the infinitesimally thin interface:

$$(q_z)_{liq} = (q_z)_{gas} \quad (4.69)$$

Neglecting radiation, this is equivalent to

$$\left(k\frac{\partial T}{\partial z}\right)_{liq} = \left(k\frac{\partial T}{\partial z}\right)_{gas} \quad (4.70)$$

This is as much detail as we wish to give at this level of exposition. Further and even more complicated details on fluid flow boundary conditions are given in Refs. 5 and 9.

Simplified Free Surface Conditions

In the introductory analyses given in this book, such as open-channel flows in Chap. 10, we shall back away from the exact conditions (4.65) to (4.69) and assume that the upper fluid is an "atmosphere" that merely exerts pressure on the lower fluid, with shear and heat conduction negligible. We also neglect nonlinear terms involving the slopes of the free surface. We then have a much simpler and linear set of conditions at the surface:

$$p_{liq} \approx p_{gas} - \Upsilon\left(\frac{\partial^2\eta}{\partial x^2} + \frac{\partial^2\eta}{\partial y^2}\right) \qquad w_{liq} \approx \frac{\partial \eta}{\partial t}$$

$$\left(\frac{\partial V}{\partial z}\right)_{liq} \approx 0 \qquad \left(\frac{\partial T}{z}\right)_{liq} \approx 0 \tag{4.71}$$

In many cases, such as open-channel flow, we can also neglect surface tension, so

$$p_{liq} \approx p_{atm} \tag{4.72}$$

These are the types of approximations that will be used in Chap. 10. The nondimensional forms of these conditions will also be useful in Chap. 5.

Incompressible Flow with Constant Properties

Flow with constant ρ, μ, and k is a basic simplification that will be used, for example, throughout Chap. 6. The basic equations of motion (4.56) to (4.58) reduce to

Continuity:
$$\nabla \cdot \mathbf{V} = 0 \tag{4.73}$$

Momentum:
$$\rho\frac{d\mathbf{V}}{dt} = \rho\mathbf{g} - \nabla p + \mu\nabla^2\mathbf{V} \tag{4.74}$$

Energy:
$$\rho c_p \frac{dT}{dt} = k\nabla^2 T + \Phi \tag{4.75}$$

Since ρ is constant, there are only three unknowns: p, $\mathbf{V}$, and T. The system is closed.[9] Not only that, the system splits apart: Continuity and momentum are independent of T. Thus we can solve Eqs. (4.73) and (4.74) entirely separately for the pressure and velocity, using such boundary conditions as

Solid surface:
$$\mathbf{V} = \mathbf{V}_{wall} \tag{4.76}$$

Inlet or outlet:
$$\text{Known } \mathbf{V}, p \tag{4.77}$$

Free surface:
$$p \approx p_a \qquad w \approx \frac{\partial \eta}{\partial t} \tag{4.78}$$

Later, usually in another course,[10] we can solve for the temperature distribution from Eq. (4.75), which depends on velocity $\mathbf{V}$ through the dissipation Φ and the total time-derivative operator d/dt.

Inviscid Flow Approximations

Chapter 8 assumes inviscid flow throughout, for which the viscosity $\mu = 0$. The momentum equation (4.74) reduces to

$$\rho\frac{d\mathbf{V}}{dt} = \rho\mathbf{g} - \nabla p \tag{4.79}$$

This is *Euler's equation;* it can be integrated along a streamline to obtain Bernoulli's equation (see Sec. 4.9). By neglecting viscosity we have lost the second-order derivative of $\mathbf{V}$ in Eq. (4.74); therefore, we must relax one boundary condition on velocity. The only mathematically sensible condition to drop is the no-slip condition at the wall. We

[9] For this system, what are the thermodynamic equivalents to Eq. (4.59)?

[10] Since temperature is entirely *uncoupled* by this assumption, we may never get around to solving for it here and may ask you to wait until you take a course on heat transfer.

let the flow slip parallel to the wall but do not allow it to flow into the wall. The proper inviscid condition is that the normal velocities must match at any solid surface:

Inviscid flow: $$(V_n)_{\text{fluid}} = (V_n)_{\text{wall}} \tag{4.80}$$

In most cases, the wall is fixed; therefore, the proper inviscid flow condition is

$$V_n = 0 \tag{4.81}$$

There is *no* condition whatever on the tangential velocity component at the wall in inviscid flow. The tangential velocity will be part of the solution to an inviscid flow analysis (see Chap. 8).

EXAMPLE 4.6

For steady incompressible laminar flow through a long tube, the velocity distribution is given by

$$v_z = U\left(1 - \frac{r^2}{R^2}\right) \qquad v_r = v_\theta = 0$$

where U is the maximum, or centerline, velocity and R is the tube radius. If the wall temperature is constant at T_w and the temperature $T = T(r)$ only, find $T(r)$ for this flow.

Solution

With $T = T(r)$, Eq. (4.75) reduces for steady flow to

$$\rho c_p v_r \frac{dT}{dr} = \frac{k}{r}\frac{d}{dr}\left(r\frac{dT}{dr}\right) + \mu\left(\frac{dv_z}{dr}\right)^2 \tag{1}$$

But, since $v_r = 0$ for this flow, the convective term on the left vanishes. Introduce v_z into Eq. (1) to obtain

$$\frac{k}{r}\frac{d}{dr}\left(r\frac{dT}{dr}\right) = -\mu\left(\frac{dv_z}{dr}\right)^2 = -\frac{4U^2\mu r^2}{R^4} \tag{2}$$

Multiply through by r/k and integrate once:

$$r\frac{dT}{dr} = -\frac{\mu U^2 r^4}{kR^4} + C_1 \tag{3}$$

Divide through by r and integrate once again:

$$T = -\frac{\mu U^2 r^4}{4kR^4} + C_1 \ln r + C_2 \tag{4}$$

Now, we are in position to apply our boundary conditions to evaluate C_1 and C_2.

First, since the logarithm of zero is $-\infty$, the temperature at $r = 0$ will be infinite unless

$$C_1 = 0 \tag{5}$$

Thus, we eliminate the possibility of a logarithmic singularity. The same thing will happen if we apply the *symmetry* condition $dT/dr = 0$ at $r = 0$ to Eq. (3). The constant C_2 is then found by the wall-temperature condition at $r = R$:

$$T = T_w = -\frac{\mu U^2}{4k} + C_2$$

or

$$C_2 = T_w + \frac{\mu U^2}{4k} \tag{6}$$

The correct solution is thus

$$T(r) = T_w + \frac{\mu U^2}{4k}\left(1 - \frac{r^4}{R^4}\right) \qquad \textit{Ans. (7)}$$

which is a fourth-order parabolic distribution with a maximum value $T_0 = T_w + \mu U^2/(4k)$ at the centerline.

4.7 The Stream Function

We have seen in Sec. 4.6 that even if the temperature is uncoupled from our system of equations of motion, we must solve the continuity and momentum equations simultaneously for pressure and velocity. The *stream function* ψ is a clever device that allows us to satisfy the continuity equation and then solve the momentum equation directly for the single variable ψ. Lines of constant ψ are streamlines of the flow.

The stream function idea works only if the continuity equation (4.56) can be reduced to *two* terms. In general, we have *four* terms:

Cartesian:
$$\frac{\partial \rho}{\partial t} + \frac{\partial}{\partial x}(\rho u) + \frac{\partial}{\partial y}(\rho v) + \frac{\partial}{\partial z}(\rho w) = 0 \qquad (4.82a)$$

Cylindrical:
$$\frac{\partial \rho}{\partial t} + \frac{1}{r}\frac{\partial}{\partial r}(r\rho v_r) + \frac{1}{r}\frac{\partial}{\partial \theta}(\rho v_\theta) + \frac{\partial}{\partial z}(\rho v_z) = 0 \qquad (4.82b)$$

First, let us eliminate unsteady flow, which is a peculiar and unrealistic application of the stream function idea. Reduce either of Eqs. (4.82) to any *two* terms. The most common application is incompressible flow in the *xy* plane:

$$\frac{\partial u}{\partial x} + \frac{\partial v}{\partial y} = 0 \qquad (4.83)$$

This equation is satisfied *identically* if a function $\psi(x, y)$ is defined such that Eq. (4.83) becomes

$$\frac{\partial}{\partial x}\left(\frac{\partial \psi}{\partial y}\right) + \frac{\partial}{\partial y}\left(-\frac{\partial \psi}{\partial x}\right) \equiv 0 \qquad (4.84)$$

Comparison of (4.83) and (4.84) shows that this new function ψ must be defined such that

$$u = \frac{\partial \psi}{\partial y} \qquad v = -\frac{\partial \psi}{\partial x} \qquad (4.85)$$

or

$$\mathbf{V} = \mathbf{i}\frac{\partial \psi}{\partial y} - \mathbf{j}\frac{\partial \psi}{\partial x}$$

Is this legitimate? Yes, it is just a mathematical trick of replacing two variables (u and v) by a single higher-order function ψ. The vorticity[11] or curl **V** is an interesting function:

$$\text{curl } \mathbf{V} = -\mathbf{k}\nabla^2\psi \qquad \text{where} \qquad \nabla^2\psi = \frac{\partial^2 \psi}{\partial x^2} + \frac{\partial^2 \psi}{\partial y^2} \qquad (4.86)$$

[11]See Section 4.8.

Thus, if we take the curl of the momentum equation (4.74) and utilize Eq. (4.86), we obtain a single equation for ψ for incompressible flow:

$$\frac{\partial \psi}{\partial y}\frac{\partial}{\partial x}(\nabla^2\psi) - \frac{\partial \psi}{\partial x}\frac{\partial}{\partial y}(\nabla^2\psi) = \nu\nabla^2(\nabla^2\psi) \tag{4.87}$$

where $\nu = \mu/\rho$ is the kinematic viscosity. This is partly a victory and partly a defeat: Eq. (4.87) is scalar and has only one variable, ψ, but it now contains *fourth*-order derivatives and probably will require computer analysis. There will be four boundary conditions required on ψ. For example, for the flow of a uniform stream in the x direction past a solid body, the four conditions would be

$$\text{At infinity:} \quad \frac{\partial \psi}{\partial y} = U_\infty \quad \frac{\partial \psi}{\partial x} = 0$$

$$\text{At the body:} \quad \frac{\partial \psi}{\partial y} = \frac{\partial \psi}{\partial x} = 0 \tag{4.88}$$

Many examples of numerical solution of Eqs. (4.87) and (4.88) are given in **Ref. 1**.

One important application is inviscid, incompressible, *irrotational* flow[12] in **the** xy plane, where curl $\mathbf{V} \equiv 0$. Equations (4.86) and (4.87) reduce to

$$\nabla^2\psi = \frac{\partial^2 \psi}{\partial x^2} + \frac{\partial^2 \psi}{\partial y^2} = 0 \tag{4.89}$$

This is the second-order *Laplace equation* (Chap. 8), for which many solutions and analytical techniques are known. Also, boundary conditions like Eq. (4.88) reduce to

$$\text{At infinity:} \quad \psi = U_\infty y + \text{const} \tag{4.90}$$

$$\text{At the body:} \quad \psi = \text{const}$$

It is well within our capability to find some useful solutions to Eqs. (4.89) and (4.90), which we shall do in Chap. 8.

Geometric Interpretation of ψ

The fancy mathematics above would serve alone to make the stream function immortal and always useful to engineers. Even better, though, ψ has a beautiful geometric interpretation: Lines of constant ψ are *streamlines* of the flow. This can be shown as follows: From Eq. (1.41) the definition of a streamline in two-dimensional flow is

$$\frac{dx}{u} = \frac{dy}{v}$$

or

$$u\,dy - v\,dx = 0 \quad \text{streamline} \tag{4.91}$$

Introducing the stream function from Eq. (4.85), we have

$$\frac{\partial \psi}{\partial x}dx + \frac{\partial \psi}{\partial y}dy = 0 = d\psi \tag{4.92}$$

Thus, the change in ψ is zero along a streamline, or

$$\psi = \text{const along a streamline} \tag{4.93}$$

Having found a given solution $\psi(x, y)$, we can plot lines of constant ψ to give the streamlines of the flow.

[12]See Section 4.8.

There is also a physical interpretation that relates ψ to volume flow. From Fig. 4.8, we can compute the volume flow dQ through an element ds of control surface of unit depth:

$$dQ = (\mathbf{V} \cdot \mathbf{n})\, dA = \left(\mathbf{i}\frac{\partial \psi}{\partial y} - \mathbf{j}\frac{\partial \psi}{\partial x}\right) \cdot \left(\mathbf{i}\frac{dy}{ds} - \mathbf{j}\frac{dx}{ds}\right) ds(1)$$

$$= \frac{\partial \psi}{\partial x} dx + \frac{\partial \psi}{\partial y} dy = d\psi \tag{4.94}$$

Thus, the change in ψ across the element is numerically equal to the volume flow through the element. The volume flow between any two streamlines in the flow field is equal to the change in stream function between those streamlines:

$$Q_{1\to 2} = \int_1^2 (\mathbf{V} \cdot \mathbf{n})\, dA = \int_1^2 d\psi = \psi_2 - \psi_1 \tag{4.95}$$

Further, the direction of the flow can be ascertained by noting whether ψ increases or decreases. As sketched in Fig. 4.9, the flow is to the right if ψ_U is greater than ψ_L, where the subscripts stand for upper and lower, as before; otherwise the flow is to the left.

Both the stream function and the velocity potential were invented by the French mathematician Joseph Louis Lagrange and published in his treatise on fluid mechanics in 1781.

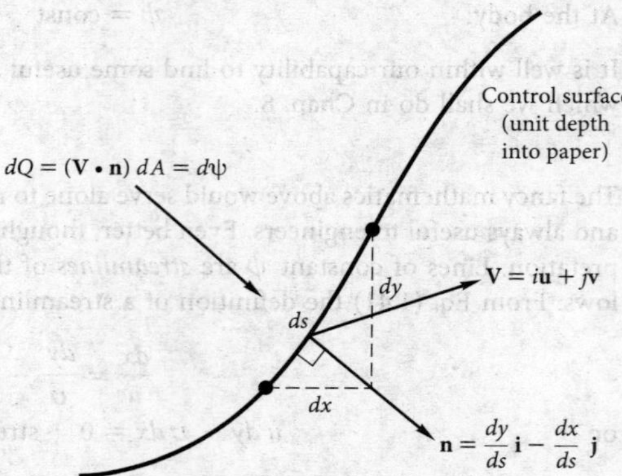

Fig. 4.8 Geometric interpretation of stream function: volume flow through a differential portion of a control surface.

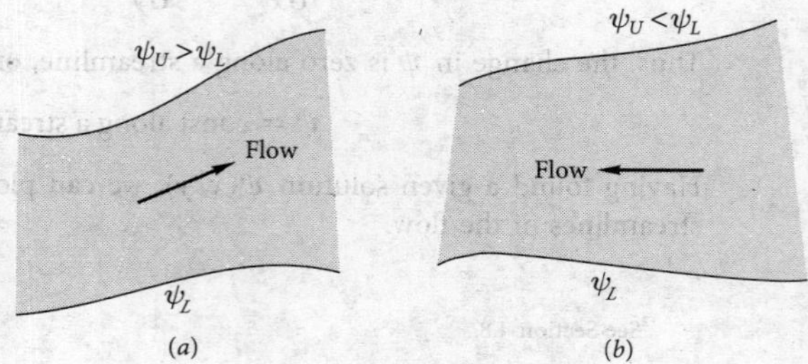

Fig. 4.9 Sign convention for flow in terms of change in stream function: (*a*) flow to the right if ψ_U is greater; (*b*) flow to the left if ψ_L is greater.

EXAMPLE 4.7

If a stream function exists for the velocity field of Example 4.5

$$u = a(x^2 - y^2) \quad v = -2axy \quad w = 0$$

find it, plot it, and interpret it.

Solution

- *Assumptions:* Incompressible, two-dimensional flow.
- *Approach:* Use the definition of stream function derivatives, Eqs. (4.85), to find $\psi(x, y)$.
- *Solution step 1:* Note that this velocity distribution was also examined in Example 4.3. It satisfies continuity, Eq. (4.83), but let's check that; otherwise ψ will not exist:

$$\frac{\partial u}{\partial x} + \frac{\partial v}{\partial y} = \frac{\partial}{\partial x}[a(x^2 - y^2)] + \frac{\partial}{\partial y}(-2axy) = 2ax + (-2ax) \equiv 0 \quad \text{checks}$$

Thus, we are certain that a stream function exists.

- *Solution step 2:* To find ψ, write out Eqs. (4.85) and integrate:

$$u = \frac{\partial \psi}{\partial y} = ax^2 - ay^2 \tag{1}$$

$$v = -\frac{\partial \psi}{\partial x} = -2axy \tag{2}$$

and work from either one toward the other. Integrate (1) partially

$$\psi = ax^2y - \frac{ay^3}{3} + f(x) \tag{3}$$

Differentiate (3) with respect to x and compare with (2)

$$\frac{\partial \psi}{\partial x} = 2axy + f'(x) = 2axy \tag{4}$$

Therefore, $f'(x) = 0$, or $f =$ constant. The complete stream function is thus found:

$$\psi = a\left(x^2y - \frac{y^3}{3}\right) + C \qquad \textit{Ans.} \ (5)$$

To plot this, set $C = 0$ for convenience and plot the function

$$3x^2y - y^3 = \frac{3\psi}{a} \tag{6}$$

for constant values of ψ. The result is shown in Fig. E4.7*a* to be six 60° wedges of circulating motion, each with identical flow patterns except for the arrows. Once the streamlines are

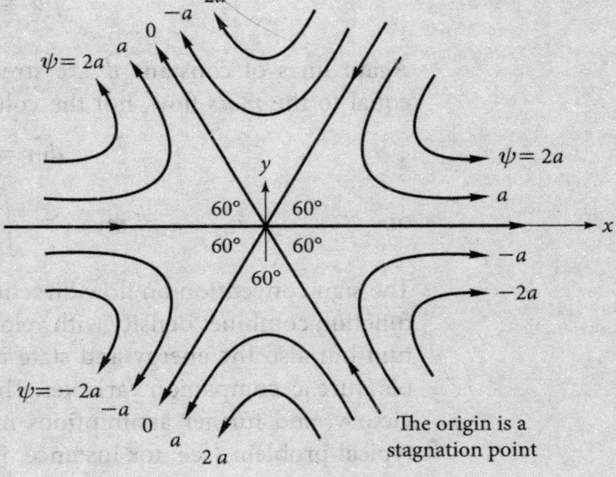

E4.7*a*

labeled, the flow directions follow from the sign convention of Fig. 4.9. How can the flow be interpreted? Since there is slip along all streamlines, no streamline can truly represent a solid surface in a viscous flow. However, the flow could represent the impingement of three incoming streams at 60, 180, and 300°. This would be a rather unrealistic yet exact solution to the Navier-Stokes equations, as we showed in Example 4.5.

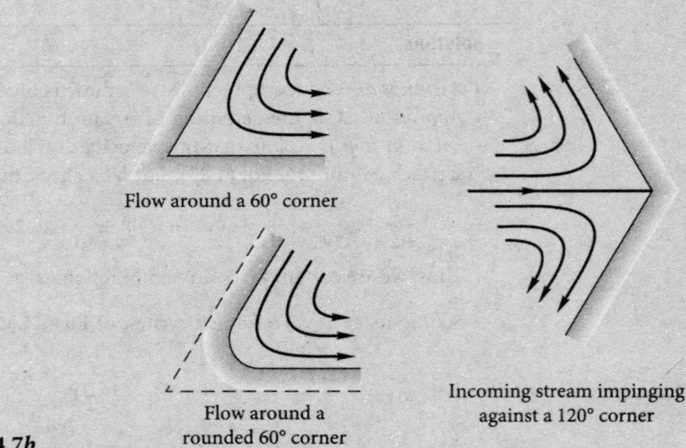

E4.7b

By allowing the flow to slip as a frictionless approximation, we could let any given streamline be a body shape. Some examples are shown in Fig. E4.7*b*.

A stream function also exists in a variety of other physical situations where only two coordinates are needed to define the flow. Three examples are illustrated here.

Steady Plane Compressible Flow

Suppose now that the density is variable but that $w = 0$, so that the flow is in the xy plane. Then the equation of continuity becomes

$$\frac{\partial}{\partial x}(\rho u) + \frac{\partial}{\partial y}(\rho v) = 0 \tag{4.96}$$

We see that this is in exactly the same form as Eq. (4.84). Therefore, a compressible flow stream function can be defined such that

$$\rho u = \frac{\partial \psi}{\partial y} \qquad \rho v = -\frac{\partial \psi}{\partial x} \tag{4.97}$$

Again lines of constant ψ are streamlines of the flow, but the change in ψ is now equal to the *mass* flow, not the volume flow:

$$d\dot{m} = \rho(\mathbf{V} \cdot \mathbf{n})\, dA = d\psi$$

or

$$\dot{m}_{1\to 2} = \int_1^2 \rho(\mathbf{V} \cdot \mathbf{n})\, dA = \psi_2 - \psi_1 \tag{4.98}$$

The sign convention on flow direction is the same as in Fig. 4.9. This particular stream function combines density with velocity and must be substituted into not only momentum but also the energy and state relations (4.58) and (4.59) with pressure and temperature as companion variables. Thus, the compressible stream function is not a great victory, and further assumptions must be made to effect an analytical solution to a typical problem (see, for instance, Ref. 5, Chap. 7).

Incompressible Plane Flow in Polar Coordinates

Suppose that the important coordinates are r and θ, with $\upsilon_z = 0$, and that the density is constant. Then, Eq. (4.82*b*) reduces to

$$\frac{1}{r}\frac{\partial}{\partial r}(r\upsilon_r) + \frac{1}{r}\frac{\partial}{\partial \theta}(\upsilon_\theta) = 0 \tag{4.99}$$

After multiplying through by r, we see that this is the analogous form of Eq. (4.84):

$$\frac{\partial}{\partial r}\left(\frac{\partial \psi}{\partial \theta}\right) + \frac{\partial}{\partial \theta}\left(-\frac{\partial \psi}{\partial r}\right) = 0 \tag{4.100}$$

By comparison of (4.99) and (4.100), we deduce the form of the incompressible polar coordinate stream function:

$$\upsilon_r = \frac{1}{r}\frac{\partial \psi}{\partial \theta} \qquad \upsilon_\theta = -\frac{\partial \psi}{\partial r} \tag{4.101}$$

Once again lines of constant ψ are streamlines, and the change in ψ is the *volume flow* $Q_{1\to 2} = \psi_2 - \psi_1$. The sign convention is the same as in Fig. 4.9. This type of stream function is very useful in analyzing flows with cylinders, vortices, sources, and sinks (Chap. 8).

Incompressible Axisymmetric Flow

As a final example, suppose that the flow is three-dimensional (υ_r, υ_z) but with no circumferential variations, $\upsilon_\theta = \partial/\partial\theta = 0$ (see Fig. 4.2 for definition of coordinates). Such a flow is termed *axisymmetric*, and the flow pattern is the same when viewed on any meridional plane through the axis of revolution z. For incompressible flow, Eq. (4.82*b*) becomes

$$\frac{1}{r}\frac{\partial}{\partial r}(r\upsilon_r) + \frac{\partial}{\partial z}(\upsilon_z) = 0 \tag{4.102}$$

This doesn't seem to work: Can't we get rid of the one r outside? But when we realize that r and z are independent coordinates, Eq. (4.102) can be rewritten as

$$\frac{\partial}{\partial r}(r\upsilon_r) + \frac{\partial}{\partial z}(r\upsilon_z) = 0 \tag{4.103}$$

By analogy with Eq. (4.84), this has the form

$$\frac{\partial}{\partial r}\left(-\frac{\partial \psi}{\partial z}\right) + \frac{\partial}{\partial z}\left(\frac{\partial \psi}{\partial r}\right) = 0 \tag{4.104}$$

By comparing (4.103) and (4.104), we deduce the form of an incompressible axisymmetric stream function $\psi(r, z)$

$$\upsilon_r = -\frac{1}{r}\frac{\partial \psi}{\partial z} \qquad \upsilon_z = \frac{1}{r}\frac{\partial \psi}{\partial r} \tag{4.105}$$

Here again lines of constant ψ are streamlines, but there is a factor (2π) in the volume flow: $Q_{1\to 2} = 2\pi(\psi_2 - \psi_1)$. The sign convention on flow is the same as in Fig. 4.9.

EXAMPLE 4.8

Investigate the stream function in polar coordinates

$$\psi = U \sin\theta\left(r - \frac{R^2}{r}\right) \tag{1}$$

where U and R are constants, a velocity and a length, respectively. Plot the streamlines. What does the flow represent? Is it a realistic solution to the basic equations?

Solution

The streamlines are lines of constant ψ, which has units of square meters per second. Note that $\psi/(UR)$ is dimensionless. Rewrite Eq. (1) in dimensionless form

$$\frac{\psi}{UR} = \sin\theta\left(\eta - \frac{1}{\eta}\right) \qquad \eta = \frac{r}{R} \tag{2}$$

Of particular interest is the special line $\psi = 0$. From Eq. (1) or (2) this occurs when (*a*) $\theta = 0$ or 180° and (*b*) $r = R$. Case (*a*) is the x axis, and case (*b*) is a circle of radius R, both of which are plotted in Fig. E4.8.

For any other nonzero value of ψ it is easiest to pick a value of r and solve for θ:

$$\sin\theta = \frac{\psi/(UR)}{r/R - R/r} \tag{3}$$

In general, there will be two solutions for θ because of the symmetry about the y axis. For example, take $\psi/(UR) = +1.0$:

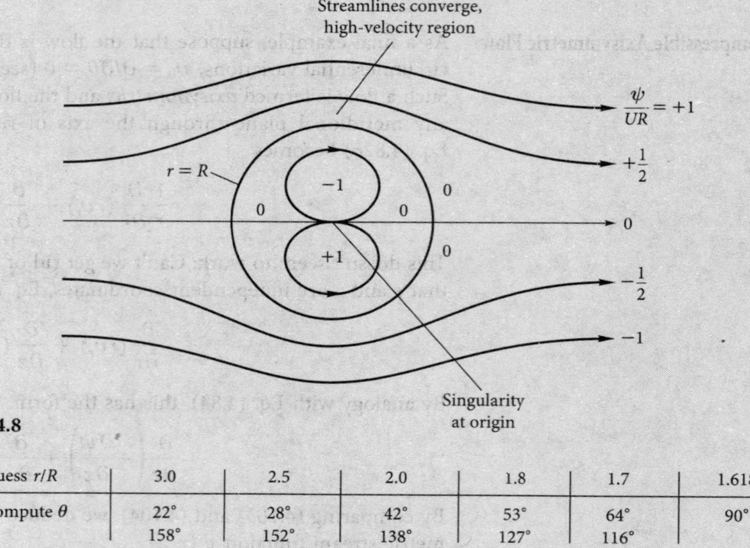

E4.8

Guess r/R	3.0	2.5	2.0	1.8	1.7	1.618
Compute θ	22° 158°	28° 152°	42° 138°	53° 127°	64° 116°	90°

This line is plotted in Fig. E4.8 and passes over the circle $r = R$. Be careful, though, because there is a second curve for $\psi/(UR) = +1.0$ for small $r < R$ below the x axis:

Guess r/R	0.618	0.6	0.5	0.4	0.3	0.2	0.1
Compute θ	−90°	−70° −110°	−42° −138°	−28° −152°	−19° −161°	−12° −168°	−6° −174°

This second curve plots as a closed curve inside the circle $r = R$. There is a singularity of infinite velocity and indeterminate flow direction at the origin. Figure E4.8 shows the full pattern.

The given stream function, Eq. (1), is an exact and classic solution to the momentum equation (4.38) for frictionless flow. Outside the circle $r = R$ it represents two-dimensional inviscid flow of a uniform stream past a circular cylinder (Sec. 8.4). Inside the circle it represents a rather unrealistic trapped circulating motion of what is called a *line doublet*.

4.8 Vorticity and Irrotationality

The assumption of zero fluid angular velocity, or irrotationality, is a very useful simplification. Here, we show that angular velocity is associated with the curl of the local velocity vector.

The differential relations for deformation of a fluid element can be derived by examining Fig. 4.10. Two fluid lines *AB* and *BC*, initially perpendicular at time *t*, move and deform so that at $t + dt$ they have slightly different lengths $A'B'$ and $B'C'$ and are slightly off the perpendicular by angles $d\alpha$ and $d\beta$. Such deformation occurs kinematically because *A*, *B*, and *C* have slightly different velocities when the velocity field **V** has spatial gradients. All these differential changes in the motion of *A*, *B*, and *C* are noted in Fig. 4.10.

We define the angular velocity ω_z about the *z* axis as the average rate of counterclockwise turning of the two lines:

$$\omega_z = \frac{1}{2}\left(\frac{d\alpha}{dt} - \frac{d\beta}{dt}\right) \tag{4.106}$$

But, from Fig. 4.10, $d\alpha$ and $d\beta$ are each directly related to velocity derivatives in the limit of small *dt*:

$$\begin{aligned} d\alpha &= \lim_{dt\to 0}\left[\tan^{-1}\frac{(\partial v/\partial x)\,dx\,dt}{dx + (\partial u/\partial x)\,dx\,dt}\right] = \frac{\partial v}{\partial x}dt \\ d\beta &= \lim_{dt\to 0}\left[\tan^{-1}\frac{(\partial u/\partial y)\,dy\,dt}{dy + (\partial v/\partial y)\,dy\,dt}\right] = \frac{\partial u}{\partial y}dt \end{aligned} \tag{4.107}$$

Combining Eqs. (4.106) and (4.107) gives the desired result:

$$\omega_z = \frac{1}{2}\left(\frac{\partial v}{\partial x} - \frac{\partial u}{\partial y}\right) \tag{4.108}$$

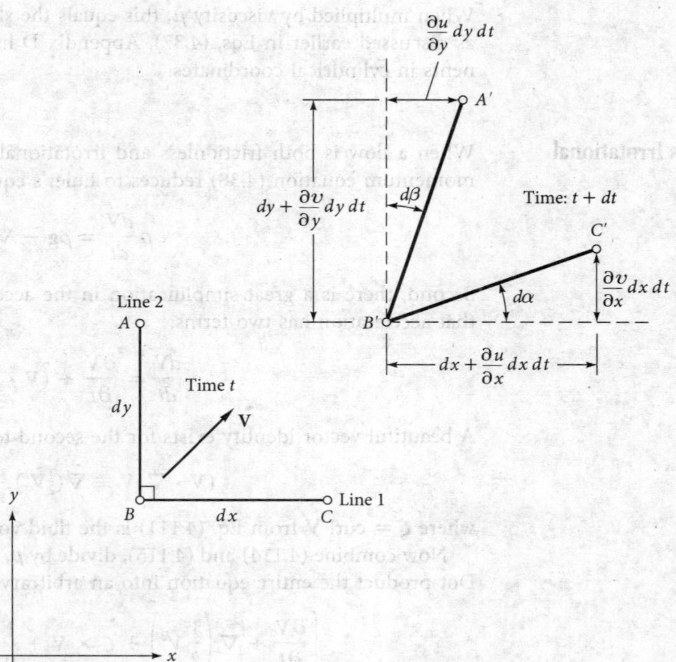

Fig. 4.10 Angular velocity and strain rate of two fluid lines deforming in the *xy* plane.

In exactly similar manner, we determine the other two rates:

$$\omega_x = \frac{1}{2}\left(\frac{\partial w}{\partial y} - \frac{\partial v}{\partial z}\right) \qquad \omega_y = \frac{1}{2}\left(\frac{\partial u}{\partial z} - \frac{\partial w}{\partial x}\right) \tag{4.109}$$

The vector $\boldsymbol{\omega} = \mathbf{i}\omega_x + \mathbf{j}\omega_y + \mathbf{k}\omega_z$ is thus one-half the curl of the velocity vector

$$\boldsymbol{\omega} = \frac{1}{2}(\text{curl } \mathbf{V}) = \frac{1}{2}\begin{vmatrix} \mathbf{i} & \mathbf{j} & \mathbf{k} \\ \frac{\partial}{\partial x} & \frac{\partial}{\partial y} & \frac{\partial}{\partial z} \\ u & v & w \end{vmatrix} \tag{4.110}$$

Since the factor of $\frac{1}{2}$ is annoying, many workers prefer to use a vector twice as large, called the *vorticity*:

$$\boldsymbol{\zeta} = 2\boldsymbol{\omega} = \text{curl } \mathbf{V} \tag{4.111}$$

Many flows have negligible or zero vorticity and are called *irrotational*:

$$\text{curl } \mathbf{V} \equiv 0 \tag{4.112}$$

The next section expands on this idea. Such flows can be incompressible or compressible, steady or unsteady.

We may also note that Fig. 4.10 demonstrates the *shear strain rate* of the element, which is defined as the rate of closure of the initially perpendicular lines:

$$\dot{\varepsilon}_{xy} = \frac{d\alpha}{dt} + \frac{d\beta}{dt} = \frac{\partial v}{\partial x} + \frac{\partial u}{\partial y} \tag{4.113}$$

When multiplied by viscosity μ, this equals the shear stress τ_{xy} in a newtonian fluid, as discussed earlier in Eqs. (4.37). Appendix D lists strain rate and vorticity components in cylindrical coordinates.

4.9 Frictionless Irrotational Flows

When a flow is both frictionless and irrotational, pleasant things happen. First, the momentum equation (4.38) reduces to Euler's equation:

$$\rho \frac{d\mathbf{V}}{dt} = \rho\mathbf{g} - \nabla p \tag{4.114}$$

Second, there is a great simplification in the acceleration term. Recall from Sec. 4.1 that acceleration has two terms:

$$\frac{d\mathbf{V}}{dt} = \frac{\partial \mathbf{V}}{\partial t} + (\mathbf{V} \cdot \nabla)\mathbf{V} \tag{4.2}$$

A beautiful vector identity exists for the second term [11]:

$$(\mathbf{V} \cdot \nabla)\mathbf{V} \equiv \nabla(\tfrac{1}{2}V^2) + \boldsymbol{\zeta} \times \mathbf{V} \tag{4.115}$$

where $\boldsymbol{\zeta} = \text{curl } \mathbf{V}$ from Eq. (4.111) is the fluid vorticity.

Now combine (4.114) and (4.115), divide by ρ, and rearrange on the left-hand side. Dot-product the entire equation into an arbitrary vector displacement $d\mathbf{r}$:

$$\left[\frac{\partial \mathbf{V}}{\partial t} + \nabla\left(\frac{1}{2}V^2\right) + \boldsymbol{\zeta} \times \mathbf{V} + \frac{1}{\rho}\nabla p - \mathbf{g}\right] \cdot d\mathbf{r} = 0 \tag{4.116}$$

Nothing works right unless we can get rid of the third term. We want

$$(\boldsymbol{\zeta} \times \mathbf{V}) \cdot (d\mathbf{r}) \equiv 0 \tag{4.117}$$

This will be true under various conditions:

1. **V** is zero; trivial, no flow (hydrostatics).
2. $\boldsymbol{\zeta}$ is zero; irrotational flow.
3. $d\mathbf{r}$ is perpendicular to $\boldsymbol{\zeta} \times \mathbf{V}$; this is rather specialized and rare.
4. $d\mathbf{r}$ is parallel to **V**; we integrate *along a streamline* (see Sec. 3.5).

Condition 4 is the common assumption. If we integrate along a streamline in frictionless compressible flow and take, for convenience, $\mathbf{g} = -g\mathbf{k}$, Eq. (4.116) reduces to

$$\frac{\partial \mathbf{V}}{\partial t} \cdot d\mathbf{r} + d\left(\frac{1}{2}V^2\right) + \frac{dp}{\rho} + g\,dz = 0 \tag{4.118}$$

Except for the first term, these are exact differentials. Integrate between any two points 1 and 2 along the streamline:

$$\int_1^2 \frac{\partial V}{\partial t}\,ds + \int_1^2 \frac{dp}{\rho} + \frac{1}{2}(V_2^2 - V_1^2) + g(z_2 - z_1) = 0 \tag{4.119}$$

where ds is the arc length along the streamline. Equation (4.119) is Bernoulli's equation for frictionless unsteady flow along a streamline and is identical to Eq. (3.53). For incompressible steady flow, it reduces to

$$\frac{p}{\rho} + \frac{1}{2}V^2 + gz = \text{constant along streamline} \tag{4.120}$$

The constant may vary from streamline to streamline unless the flow is also irrotational (assumption 2). For irrotational flow $\zeta = 0$, the offending term Eq. (4.117) vanishes regardless of the direction of $d\mathbf{r}$, and Eq. (4.120) then holds all over the flow field with the same constant.

Velocity Potential

Irrotationality gives rise to a scalar function ϕ similar and complementary to the stream function ψ. From a theorem in vector analysis [11], a vector with zero curl must be the gradient of a scalar function

$$\text{If} \quad \nabla \times \mathbf{V} \equiv 0 \quad \text{then} \quad \mathbf{V} = \nabla\phi \tag{4.121}$$

where $\phi = \phi(x, y, z, t)$ is called the *velocity potential function*. Knowledge of ϕ thus immediately gives the velocity components

$$u = \frac{\partial \phi}{\partial x} \qquad v = \frac{\partial \phi}{\partial y} \qquad w = \frac{\partial \phi}{\partial z} \tag{4.122}$$

Lines of constant ϕ are called the *potential lines* of the flow.

Note that ϕ, unlike the stream function, is fully three-dimensional and is not limited to two coordinates. It reduces a velocity problem with three unknowns u, v, and w to a single unknown potential ϕ; many examples are given in Chap. 8. The velocity potential also simplifies the unsteady Bernoulli equation (4.118) because if ϕ exists, we obtain

$$\frac{\partial \mathbf{V}}{\partial t} \cdot d\mathbf{r} = \frac{\partial}{\partial t}(\nabla\phi) \cdot d\mathbf{r} = d\left(\frac{\partial \phi}{\partial t}\right) \tag{4.123}$$

along any arbitrary direction. Equation (4.118) then becomes a relation between ϕ and p:

$$\frac{\partial \phi}{\partial t} + \int \frac{dp}{\rho} + \frac{1}{2}|\nabla\phi|^2 + gz = \text{const} \tag{4.124}$$

This is the unsteady irrotational Bernoulli equation. It is very important in the analysis of accelerating flow fields (see Refs. 10 and 15), but the only application in this text will be in Sec. 9.3 for steady flow.

Orthogonality of Streamlines and Potential Lines

If a flow is both irrotational and described by only two coordinates, ψ and ϕ both exist, and the streamlines and potential lines are everywhere mutually perpendicular except at a stagnation point. For example, for incompressible flow in the xy plane, we would have

$$u = \frac{\partial \psi}{\partial y} = \frac{\partial \phi}{\partial x} \tag{4.125}$$

$$v = -\frac{\partial \psi}{\partial x} = \frac{\partial \phi}{\partial y} \tag{4.126}$$

Can you tell by inspection not only that these relations imply orthogonality but also that ϕ and ψ satisfy Laplace's equation?[13] A line of constant ϕ would be such that the change in ϕ is zero:

$$d\phi = \frac{\partial \phi}{\partial x}dx + \frac{\partial \phi}{\partial y}dy = 0 = u\,dx + v\,dy \tag{4.127}$$

Solving, we have

$$\left(\frac{dy}{dx}\right)_{\phi=\text{const}} = -\frac{u}{v} = -\frac{1}{(dy/dx)_{\psi=\text{const}}} \tag{4.128}$$

Equation (4.128) is the mathematical condition that lines of constant ϕ and ψ be mutually orthogonal. It may not be true at a stagnation point, where both u and v are zero, so their ratio in Eq. (4.128) is indeterminate.

Generation of Rotationality[14]

This is the second time we have discussed Bernoulli's equation under different circumstances (the first was in Sec. 3.5). Such reinforcement is useful, since this is probably the most widely used equation in fluid mechanics. It requires frictionless flow with no shaft work or heat transfer between sections 1 and 2. The flow may or may not be irrotational, the former being an easier condition, allowing a universal Bernoulli constant.

The only remaining question is this: *When* is a flow irrotational? In other words, when does a flow have negligible angular velocity? The exact analysis of fluid rotationality under arbitrary conditions is a topic for advanced study (for example, Ref. 10, Sec. 8.5; Ref. 9, Sec. 5.2; and Ref. 5, Sec. 2.10). We shall simply state those results here without proof.

[13]Equations (4.125) and (4.126) are called the *Cauchy-Riemann equations* and are studied in complex variable theory.

[14]This section may be omitted without loss of continuity.

A fluid flow that is initially irrotational may become rotational if

1. There are significant viscous forces induced by jets, wakes, or solid boundaries. In this case Bernoulli's equation will not be valid in such viscous regions.
2. There are entropy gradients caused by curved shock waves (see Fig. 4.11*b*).
3. There are density gradients caused by *stratification* (uneven heating) rather than by pressure gradients.
4. There are significant *noninertial* effects such as the earth's rotation (the Coriolis acceleration).

In cases 2 to 4, Bernoulli's equation still holds along a streamline if friction is negligible. We shall not study cases 3 and 4 in this book. Case 2 will be treated briefly in Chap. 9 on gas dynamics. Primarily we are concerned with case 1, where rotation is induced by viscous stresses. This occurs near solid surfaces, where the no-slip condition creates a boundary layer through which the stream velocity drops to zero, and in jets and wakes, where streams of different velocities meet in a region of high shear.

Internal flows, such as pipes and ducts, are mostly viscous, and the wall layers grow to meet in the core of the duct. Bernoulli's equation does not hold in such flows unless it is modified for viscous losses.

External flows, such as a body immersed in a stream, are partly viscous and partly inviscid, the two regions being patched together at the edge of the shear layer or boundary layer. Two examples are shown in Fig. 4.11. Figure 4.11*a* shows a low-speed

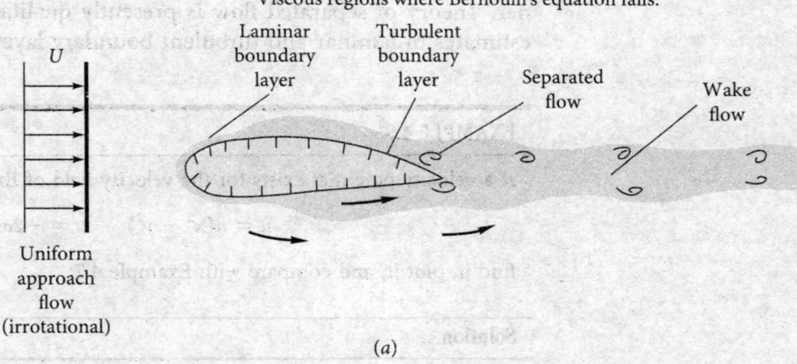

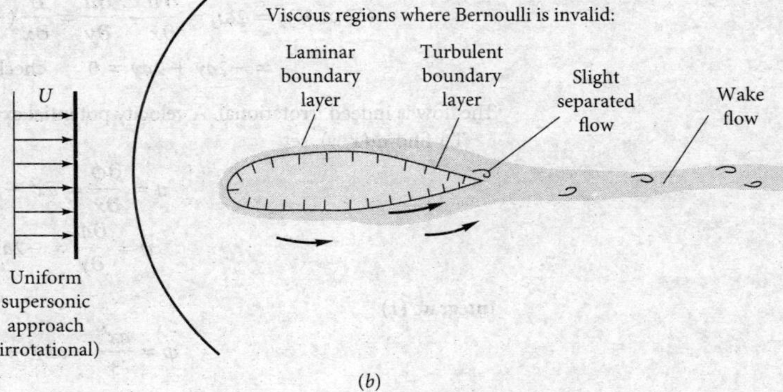

Fig. 4.11 Typical flow patterns illustrating viscous regions patched onto nearly frictionless regions: (*a*) low subsonic flow past a body ($U \ll a$); frictionless, irrotational potential flow outside the boundary layer (Bernoulli and Laplace equations valid); (*b*) supersonic flow past a body ($U > a$); frictionless, rotational flow outside the boundary layer (Bernoulli equation valid, potential flow invalid).

subsonic flow past a body. The approach stream is irrotational; that is, the curl of a constant is zero, but viscous stresses create a rotational shear layer beside and downstream of the body. Generally speaking (see Chap. 7), the shear layer is laminar, or smooth, near the front of the body and turbulent, or disorderly, toward the rear. A separated, or deadwater, region usually occurs near the trailing edge, followed by an unsteady turbulent wake extending far downstream. Some sort of laminar or turbulent viscous theory must be applied to these viscous regions; they are then patched onto the outer flow, which is frictionless and irrotational. If the stream Mach number is less than about 0.3, we can combine Eq. (4.122) with the incompressible continuity equation (4.73):

$$\nabla \cdot \mathbf{V} = \nabla \cdot (\nabla \phi) = 0$$

or

$$\nabla^2 \phi = 0 = \frac{\partial^2 \phi}{\partial x^2} + \frac{\partial^2 \phi}{\partial y^2} + \frac{\partial^2 \phi}{\partial z^2} \tag{4.129}$$

This is Laplace's equation in three dimensions, there being no restraint on the number of coordinates in potential flow. A great deal of Chap. 8 will be concerned with solving Eq. (4.129) for practical engineering problems; it holds in the entire region of Fig. 4.11*a* outside the shear layer.

Figure 4.11*b* shows a supersonic flow past a round-nosed body. A curved shock wave generally forms in front, and the flow downstream is *rotational* due to entropy gradients (case 2). We can use Euler's equation (4.114) in this frictionless region but not potential theory. The shear layers have the same general character as in Fig. 4.11*a* except that the separation zone is slight or often absent and the wake is usually thinner. Theory of separated flow is presently qualitative, but we can make quantitative estimates of laminar and turbulent boundary layers and wakes.

EXAMPLE 4.9

If a velocity potential exists for the velocity field of Example 4.5

$$u = a(x^2 - y^2) \qquad v = -2axy \qquad w = 0$$

find it, plot it, and compare with Example 4.7.

Solution

Since $w = 0$, the curl of **V** has only one z component, and we must show that it is zero:

$$(\nabla \times \mathbf{V})_z = 2\omega_z = \frac{\partial v}{\partial x} - \frac{\partial u}{\partial y} = \frac{\partial}{\partial x}(-2axy) - \frac{\partial}{\partial y}(ax^2 - ay^2)$$

$$= -2ay + 2ay = 0 \qquad \text{checks} \qquad \textit{Ans.}$$

The flow is indeed irrotational. A velocity potential exists.

To find $\phi(x, y)$, set

$$u = \frac{\partial \phi}{\partial x} = ax^2 - ay^2 \tag{1}$$

$$v = \frac{\partial \phi}{\partial y} = -2axy \tag{2}$$

Integrate (1)

$$\phi = \frac{ax^3}{3} - axy^2 + f(y) \tag{3}$$

Differentiate (3) and compare with (2)

$$\frac{\partial \phi}{\partial y} = -2axy + f'(y) = -2axy \tag{4}$$

Therefore $f' = 0$, or $f =$ constant. The velocity potential is

$$\phi = \frac{ax^3}{3} - axy^2 + C \qquad \textit{Ans.}$$

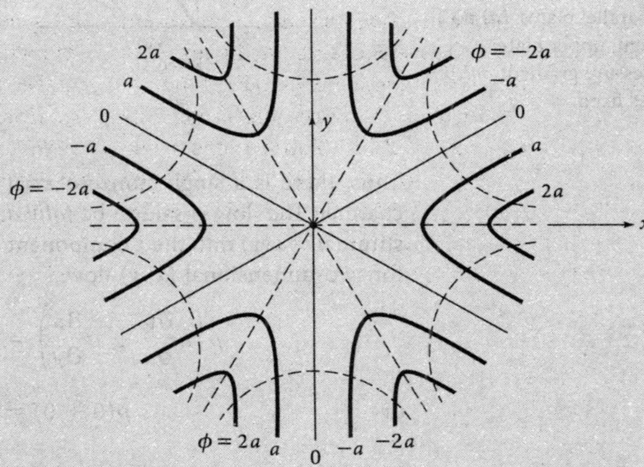

E4.9

Letting $C = 0$, we can plot the ϕ lines in the same fashion as in Example 4.7. The result is shown in Fig. E4.9 (no arrows on ϕ). For this particular problem, the ϕ lines form the same pattern as the ψ lines of Example 4.7 (which are shown here as dashed lines) but are displaced 30°. The ϕ and ψ lines are everywhere perpendicular except at the origin, a stagnation point, where they are 30° apart. We expected trouble at the stagnation point, and there is no general rule for determining the behavior of the lines at that point.

4.10 Some Illustrative Incompressible Viscous Flows

Inviscid flows do *not* satisfy the no-slip condition. They "slip" at the wall, but do not flow through the wall. To look at fully viscous no-slip conditions, we must attack the complete Navier-Stokes equation (4.74), and the result is usually not at all irrotational, nor does a velocity potential exist. We look here at three cases: (1) flow between parallel plates due to a moving upper wall, (2) flow between parallel plates due to pressure gradient, and (3) flow between concentric cylinders when the inner one rotates. Other cases will be given as problem assignments or considered in Chap. 6. Extensive solutions for viscous flows are discussed in Refs. 4 and 5. All flows in this section are viscous and rotational.

Couette Flow between a Fixed and a Moving Plate

Consider two-dimensional incompressible plane ($\partial/\partial z = 0$) viscous flow between parallel plates a distance $2h$ apart, as shown in Fig. 4.12. We assume that the plates are very wide and very long, so the flow is essentially axial, $u \neq 0$ but $v = w = 0$. The present case is Fig. 4.12*a*, where the upper plate moves at velocity V but there is no pressure gradient. Neglect gravity effects. We learn from the continuity equation (4.73) that

$$\frac{\partial u}{\partial x} + \frac{\partial v}{\partial y} + \frac{\partial w}{\partial z} = 0 = \frac{\partial u}{\partial x} + 0 + 0 \qquad \text{or} \qquad u = u(y) \text{ only}$$

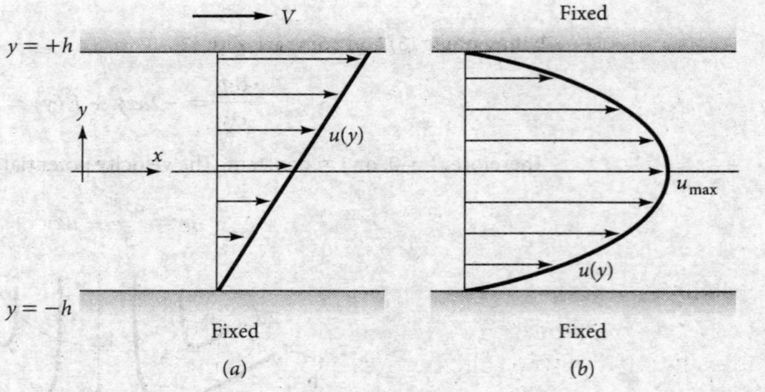

Fig. 4.12 Incompressible viscous flow between parallel plates: (*a*) no pressure gradient, upper plate moving; (*b*) pressure gradient $\partial p/\partial x$ with both plates fixed.

Thus, there is a single nonzero axial velocity component that varies only across the channel. The flow is said to be *fully developed* (far downstream of the entrance). Substitute $u = u(y)$ into the x component of the Navier-Stokes momentum equation (4.74) for two-dimensional (x, y) flow:

$$\rho\left(u\frac{\partial u}{\partial x} + \upsilon\frac{\partial u}{\partial y}\right) = -\frac{\partial p}{\partial x} + \rho g_x + \mu\left(\frac{\partial^2 u}{\partial x^2} + \frac{\partial^2 u}{\partial y^2}\right)$$

or

$$\rho(0 + 0) = 0 + 0 + \mu\left(0 + \frac{d^2 u}{dy^2}\right) \tag{4.130}$$

Most of the terms drop out, and the momentum equation reduces to simply

$$\frac{d^2 u}{dy^2} = 0 \quad \text{or} \quad u = C_1 y + C_2$$

The two constants are found by applying the no-slip condition at the upper and lower plates:

At $y = +h$: $\quad u = V = C_1 h + C_2$

At $y = -h$: $\quad u = 0 = C_1(-h) + C_2$

or $\quad C_1 = \dfrac{V}{2h} \quad$ and $\quad C_2 = \dfrac{V}{2}$

Therefore, the solution for this case (*a*), flow between plates with a moving upper wall, is

$$u = \frac{V}{2h}y + \frac{V}{2} \qquad -h \le y \le +h \tag{4.131}$$

This is *Couette flow* due to a moving wall: a linear velocity profile with no slip at each wall, as anticipated and sketched in Fig. 4.12*a*. Note that the origin has been placed in the center of the channel for convenience in case (*b*) which follows.

What we have just presented is a rigorous derivation of the more informally discussed flow of Fig. 1.7 (where y and h were defined differently).

Flow Due to Pressure Gradient between Two Fixed Plates

Case (*b*) is sketched in Fig. 4.12*b*. Both plates are fixed ($V = 0$), but the pressure varies in the x direction. If $\upsilon = w = 0$, the continuity equation leads to the same conclusion as case (*a*)—namely, that $u = u(y)$ only. The x-momentum equation (4.130) changes only because the pressure is variable:

$$\mu\frac{d^2 u}{dy^2} = \frac{\partial p}{\partial x} \tag{4.132}$$

Also, since $v = w = 0$ and gravity is neglected, the y- and z-momentum equations lead to

$$\frac{\partial p}{\partial y} = 0 \quad \text{and} \quad \frac{\partial p}{\partial z} = 0 \quad \text{or} \quad p = p(x) \text{ only}$$

Thus, the pressure gradient in Eq. (4.132) is the total and only gradient:

$$\mu \frac{d^2 u}{dy^2} = \frac{dp}{dx} = \text{const} < 0 \tag{4.133}$$

Why did we add the fact that dp/dx is *constant*? Recall a useful conclusion from the theory of separation of variables: If two quantities are equal and one varies only with y and the other varies only with x, then they must both equal the same constant. Otherwise they would not be independent of each other.

Why did we state that the constant is *negative*? Physically, the pressure must decrease in the flow direction in order to drive the flow against resisting wall shear stress. Thus the velocity profile $u(y)$ must have negative curvature everywhere, as anticipated and sketched in Fig. 4.12*b*.

The solution to Eq. (4.133) is accomplished by double integration:

$$u = \frac{1}{\mu}\frac{dp}{dx}\frac{y^2}{2} + C_1 y + C_2$$

The constants are found from the no-slip condition at each wall:

$$\text{At } y = \pm h: \quad u = 0 \quad \text{or} \quad C_1 = 0 \quad \text{and} \quad C_2 = -\frac{dp}{dx}\frac{h^2}{2\mu}$$

Thus, the solution to case (*b*), flow in a channel due to pressure gradient, is

$$u = -\frac{dp}{dx}\frac{h^2}{2\mu}\left(1 - \frac{y^2}{h^2}\right) \tag{4.134}$$

The flow forms a *Poiseuille* parabola of constant negative curvature. The maximum velocity occurs at the centerline $y = 0$:

$$u_{\max} = -\frac{dp}{dx}\frac{h^2}{2\mu} \tag{4.135}$$

Other (laminar) flow parameters are computed in the following example.

EXAMPLE 4.10

For case (*b*) in Fig. 4.12*b*, flow between parallel plates due to the pressure gradient, compute (*a*) the wall shear stress, (*b*) the stream function, (*c*) the vorticity, (*d*) the velocity potential, and (*e*) the average velocity.

Solution

All parameters can be computed from the basic solution, Eq. (4.134), by mathematical manipulation.

Part (a) The wall shear follows from the definition of a newtonian fluid, Eq. (4.37):

$$\tau_w = \tau_{xy\,\text{wall}} = \mu\left(\frac{\partial u}{\partial y} + \frac{\partial v}{\partial x}\right)\Bigg|_{y=\pm h} = \mu\frac{\partial}{\partial y}\left[\left(-\frac{dp}{dx}\right)\left(\frac{h^2}{2\mu}\right)\left(1 - \frac{y^2}{h^2}\right)\right]\Bigg|_{y=\pm h}$$

$$= \pm\frac{dp}{dx}h = \mp\frac{2\mu u_{\max}}{h} \qquad \textit{Ans. (a)}$$

The wall shear has the same magnitude at each wall, but by our sign convention of Fig. 4.3, the upper wall has negative shear stress.

Part (b) Since the flow is plane, steady, and incompressible, a stream function exists:

$$u = \frac{\partial \psi}{\partial y} = u_{max}\left(1 - \frac{y^2}{h^2}\right) \qquad v = -\frac{\partial \psi}{\partial x} = 0$$

Integrating and setting $\psi = 0$ at the centerline for convenience, we obtain

$$\psi = u_{max}\left(y - \frac{y^3}{3h^2}\right) \qquad \textit{Ans. (b)}$$

At the walls, $y = \pm h$ and $\psi = \pm 2u_{max}h/3$, respectively.

Part (c) In plane flow, there is only a single nonzero vorticity component:

$$\zeta_z = (\text{curl } \mathbf{V})_z = \frac{\partial v}{\partial x} - \frac{\partial u}{\partial y} = \frac{2u_{max}}{h^2}y \qquad \textit{Ans. (c)}$$

The vorticity is highest at the wall and is positive (counterclockwise) in the upper half and negative (clockwise) in the lower half of the fluid. Viscous flows are typically full of vorticity and are not at all irrotational.

Part (d) From part (*c*), the vorticity is finite. Therefore the flow is not irrotational, and the velocity potential *does not exist.* *Ans. (d)*

Part (e) The average velocity is defined as $V_{av} = Q/A$, where $Q = \int u\, dA$ over the cross section. For our particular distribution $u(y)$ from Eq. (4.134), we obtain

$$V_{av} = \frac{1}{A}\int u\, dA = \frac{1}{b(2h)}\int_{-h}^{+h} u_{max}\left(1 - \frac{y^2}{h^2}\right) b\, dy = \frac{2}{3}u_{max} \qquad \textit{Ans. (e)}$$

In plane Poiseuille flow between parallel plates, the average velocity is two-thirds of the maximum (or centerline) value. This result could also have been obtained from the stream function derived in part (*b*). From Eq. (4.95),

$$Q_{channel} = \psi_{upper} - \psi_{lower} = \frac{2u_{max}h}{3} - \left(-\frac{2u_{max}h}{3}\right) = \frac{4}{3}u_{max}h \text{ per unit width}$$

whence $V_{av} = Q/A_{b=1} = (4u_{max}h/3)/(2h) = 2u_{max}/3$, the same result.

This example illustrates a statement made earlier: Knowledge of the velocity vector **V** [as in Eq. (4.134)] is essentially the *solution* to a fluid mechanics problem, since all other flow properties can then be calculated.

Fully Developed Laminar Pipe Flow

Perhaps the most useful exact solution of the Navier-Stokes equation is for incompressible flow in a straight circular pipe of radius R, first studied experimentally by G. Hagen in 1839 and J. L. Poiseuille in 1840. By *fully developed* we mean that the region studied is far enough from the entrance that the flow is purely axial, $v_z \neq 0$, while v_r and v_θ are zero. We neglect gravity and also assume axial symmetry—that is, $\partial/\partial\theta = 0$. The equation of continuity in cylindrical coordinates, Eq. (4.12*b*), reduces to

$$\frac{\partial}{\partial z}(v_z) = 0 \qquad \text{or} \qquad v_z = v_z(r) \qquad \text{only}$$

The flow proceeds straight down the pipe without radial motion. The r-momentum equation in cylindrical coordinates, Eq. (D.5), simplifies to $\partial p/\partial r = 0$, or

$p = p(z)$ only. The z-momentum equation in cylindrical coordinates, Eq. (D.7), reduces to

$$\rho v_z \frac{\partial v_z}{\partial z} = -\frac{dp}{dz} + \mu \nabla^2 v_z = -\frac{dp}{dz} + \frac{\mu}{r}\frac{d}{dr}\left(r\frac{dv_z}{dr}\right)$$

The convective acceleration term on the left vanishes because of the previously given continuity equation. Thus the momentum equation may be rearranged as follows:

$$\frac{\mu}{r}\frac{d}{dr}\left(r\frac{dv_z}{dr}\right) = \frac{dp}{dz} = \text{const} < 0 \tag{4.136}$$

This is exactly the situation that occurred for flow between flat plates in Eq. (4.132). Again the "separation" constant is negative, and pipe flow will look much like the plate flow in Fig. 4.12*b*.

Equation (4.136) is linear and may be integrated twice, with the result

$$v_z = \frac{dp}{dz}\frac{r^2}{4\mu} + C_1 \ln(r) + C_2$$

where C_1 and C_2 are constants. The boundary conditions are no slip at the wall and finite velocity at the centerline:

$$\text{No slip at } r = R\text{:}\ v_z = 0 = \frac{dp}{dz}\frac{R^2}{4\mu} + C_1 \ln(R) + C_2$$

$$\text{Finite velocity at } r = 0\text{:}\ v_z = \text{finite} = 0 + C_1 \ln(0) + C_2$$

To avoid a logarithmic singularity, the centerline condition requires that $C_1 = 0$. Then, from no slip, $C_2 = (-dp/dz)(R^2/4\mu)$. The final, and famous, solution for fully developed *Hagen-Poiseuille flow* is

$$v_z = \left(-\frac{dp}{dz}\right)\frac{1}{4\mu}(R^2 - r^2) \tag{4.137}$$

The velocity profile is a paraboloid with a maximum at the centerline. Just as in Example 4.10, knowledge of the velocity distribution enables other parameters to be calculated:

$$V_{\max} = v_z(r = 0) = \left(-\frac{dp}{dz}\right)\frac{R^2}{4\mu}$$

$$V_{\text{avg}} = \frac{1}{A}\int v_z\, dA = \frac{1}{\pi R^2}\int_0^R V_{\max}\left(1 - \frac{r^2}{R^2}\right) 2\pi r\, dr = \frac{V_{\max}}{2} = \left(-\frac{dp}{dz}\right)\frac{R^2}{8\mu}$$

$$Q = \int v_z\, dA = \int_0^R V_{\max}\left(1 - \frac{r^2}{R^2}\right) 2\pi r\, dr = \pi R^2 V_{\text{avg}} = \frac{\pi R^4}{8\mu}\left(-\frac{dp}{dz}\right) = \frac{\pi R^4 \Delta p}{8\mu L}$$

$$\tau_{\text{wall}} = \mu \left|\frac{\partial v_z}{\partial r}\right|_{r=R} = \frac{4\mu V_{\text{avg}}}{R} = \frac{R}{2}\left(-\frac{dp}{dz}\right) = \frac{R}{2}\frac{\Delta p}{L} \tag{4.138}$$

Note that we have substituted the equality $(-dp/dz) = \Delta p/L$, where Δp is the pressure drop along the entire length L of the pipe.

These formulas are valid as long as the flow is *laminar*—that is, when the dimensionless Reynolds number of the flow, $\text{Re}_D = \rho V_{\text{avg}}(2R)/\mu$, is less than about 2100. Note also that the formulas do not depend on density, the reason being that the convective acceleration of this flow is zero.

EXAMPLE 4.11

SAE 10W oil at 20°C flows at 1.1 m^3/h through a horizontal pipe with $d = 2$ cm and $L = 12$ m. Find (*a*) the average velocity, (*b*) the Reynolds number, (*c*) the pressure drop, and (*d*) the power required.

Solution

- *Assumptions:* Laminar, steady, Hagen-Poiseuille pipe flow.
- *Approach:* The formulas of Eqs. (4.138) are appropriate for this problem. Note that $R = 0.01$ m.
- *Property values:* From Table A.3 for SAE 10W oil, $\rho = 870\ \text{kg/m}^3$ and $\mu = 0.104$ kg/(m-s).
- *Solution steps:* The average velocity follows easily from the flow rate and the pipe area:

$$V_{\text{avg}} = \frac{Q}{\pi R^2} = \frac{(1.1/3{,}600)\ \text{m}^3/\text{s}}{\pi(0.01\ \text{m})^2} = 0.973\ \frac{\text{m}}{\text{s}} \qquad \textit{Ans. (a)}$$

We had to convert Q to m^3/s. The (diameter) Reynolds number follows from the average velocity:

$$\text{Re}_d = \frac{\rho V_{\text{avg}} d}{\mu} = \frac{(870\ \text{kg/m}^3)(0.973\ \text{m/s})(0.02\ \text{m})}{0.104\ \text{kg/(m-s)}} = 163 \qquad \textit{Ans. (b)}$$

This is less than the "transition" value of 2100; so the flow is indeed *laminar*, and the formulas are valid. The pressure drop is computed from the third of Eqs. (4.138):

$$Q = \frac{1.1}{3{,}600}\frac{\text{m}^3}{\text{s}} = \frac{\pi R^4 \Delta p}{8\mu L} = \frac{\pi(0.01\ \text{m})^4 \Delta p}{8(0.104\ \text{kg/(m-s)})(12\ \text{m})} \quad \text{solve for } \Delta p = 97{,}100\ \text{Pa} \qquad \textit{Ans. (c)}$$

When using SI units, the answer returns in pascals; no conversion factors are needed. Finally, the power required is the product of flow rate and pressure drop:

$$\text{Power} = Q\Delta p = \left(\frac{1.1}{3{,}600}\ \text{m}^3/\text{s}\right)(97{,}100\ \text{N/m}^2) = 29.7\ \frac{\text{N-m}}{\text{s}} = 29.7\ \text{W} \qquad \textit{Ans. (d)}$$

- *Comments:* Pipe flow problems are straightforward algebraic exercises if the data are compatible. Note again that SI units can be used in the formulas without conversion factors.

Flow between Long Concentric Cylinders

Consider a fluid of constant (ρ, μ) between two concentric cylinders, as in Fig. 4.13. There is no axial motion or end effect $\upsilon_z = \partial/\partial z = 0$. Let the inner cylinder rotate at angular velocity Ω_i. Let the outer cylinder be fixed. There is circular symmetry, so the velocity does not vary with θ and varies only with r.

The continuity equation for this problem is Eq. (4.12*b*) with $\upsilon_z = 0$:

$$\frac{1}{r}\frac{\partial}{\partial r}(r\upsilon_r) + \frac{1}{r}\frac{\partial \upsilon_\theta}{\partial \theta} = 0 = \frac{1}{r}\frac{d}{dr}(r\upsilon_r) \quad \text{or} \quad r\upsilon_r = \text{const}$$

Note that υ_θ does not vary with θ. Since $\upsilon_r = 0$ at both the inner and outer cylinders, it follows that $\upsilon_r = 0$ everywhere and the motion can only be purely circumferential, $\upsilon_\theta = \upsilon_\theta(r)$. The θ-momentum equation (D.6) becomes

$$\rho(\mathbf{V}\cdot\nabla)\upsilon_\theta + \frac{\rho \upsilon_r \upsilon_\theta}{r} = -\frac{1}{r}\frac{\partial p}{\partial \theta} + \rho g_\theta + \mu\left(\nabla^2 \upsilon_\theta - \frac{\upsilon_\theta}{r^2}\right)$$

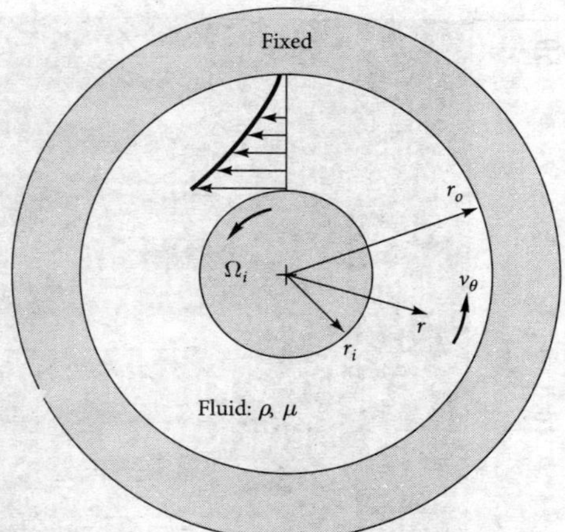

Fig. 4.13 Coordinate system for incompressible viscous flow between a fixed outer cylinder and a steadily rotating inner cylinder.

For the conditions of the present problem, all terms are zero except the last. Therefore, the basic differential equation for flow between rotating cylinders is

$$\nabla^2 v_\theta = \frac{1}{r}\frac{d}{dr}\left(r\frac{dv_\theta}{dr}\right) = \frac{v_\theta}{r^2} \tag{4.139}$$

This is a linear second-order ordinary differential equation with the solution

$$v_\theta = C_1 r + \frac{C_2}{r}$$

The constants are found by the no-slip condition at the inner and outer cylinders:

Outer, at $r = r_o$: $\qquad v_\theta = 0 = C_1 r_o + \frac{C_2}{r_o}$

Inner, at $r = r_i$: $\qquad v_\theta = \Omega_i r_i = C_1 r_i + \frac{C_2}{r_i}$

The final solution for the velocity distribution is

Rotating inner cylinder: $\qquad v_\theta = \Omega_i r_i \dfrac{r_o/r - r/r_o}{r_o/r_i - r_i/r_o} \qquad (4.140)$

The velocity profile closely resembles the sketch in Fig. 4.13. Variations of this case, such as a rotating outer cylinder, are given in the problem assignments.

Instability of Rotating Inner[15] Cylinder Flow

The classic *Couette flow* solution[16] of Eq. (4.140) describes a physically satisfying concave, two-dimensional, laminar flow velocity profile as in Fig. 4.13. The solution is mathematically exact for an incompressible fluid. However, it becomes unstable at a relatively low rate of rotation of the inner cylinder, as shown in 1923 in a classic paper by G. I. Taylor [17]. At a critical value of what is now called the dimensionless *Taylor number,* denoted Ta,

$$\text{Ta}_{\text{crit}} = \frac{r_i(r_o - r_i)^3\Omega_i^2}{\nu^2} \approx 1{,}700 \tag{4.141}$$

[15]This section may be omitted without loss of continuity.

[16]Named after M. Couette, whose pioneering paper in 1890 established rotating cylinders as a method, still used today, for measuring the viscosity of fluids.

(a)

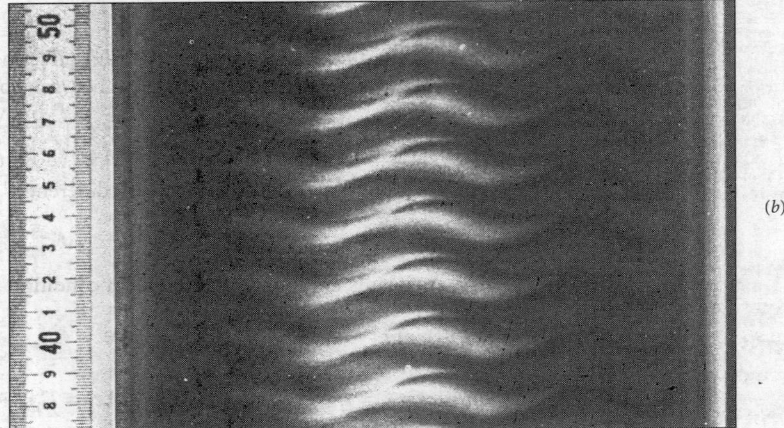

(b)

Fig. 4.14 Experimental verification of the instability of flow between a fixed outer and a rotating inner cylinder. (*a*) Toroidal Taylor vortices exist at 1.16 times the critical speed; (*b*) at 8.5 times the critical speed, the vortices are doubly periodic. (*Courtesy of Cambridge University Press—E.L. Koschmieder, "Turbulent Taylor Vortex Flow," Journal of Fluid Mechanics, vol. 93. pt. 3, 1979, pp. 515–527.*) This instability does not occur if only the outer cylinder rotates.

the plane flow of Fig. 4.13 vanishes and is replaced by a laminar *three-dimensional* flow pattern consisting of rows of nearly square alternating toroidal vortices. An experimental demonstration of toroidal "Taylor vortices" is shown in Fig. 4.14*a*, measured at Ta $\approx$ 1.16 Ta_{crit} by Koschmieder [18]. At higher Taylor numbers, the vortices also develop a circumferential periodicity, but are still laminar, as illustrated in Fig. 4.14*b*. At still higher Ta, turbulence ensues. This interesting instability reminds us that the Navier-Stokes equations, being nonlinear, do admit to multiple (nonunique) laminar solutions in addition to the usual instabilities associated with turbulence and chaotic dynamic systems.

Summary

This chapter complements Chap. 3 by using an infinitesimal control volume to derive the basic partial differential equations of mass, momentum, and energy for a fluid. These equations, together with thermodynamic state relations for the fluid and appropriate boundary conditions, in principle can be solved for the complete flow field in any given fluid mechanics problem. Except for Chap. 9, in most of the problems to be studied here an incompressible fluid with constant viscosity is assumed.

In addition to deriving the basic equations of mass, momentum, and energy, this chapter introduced some supplementary ideas—the stream function, vorticity, irrotationality, and the velocity potential—which will be useful in coming chapters, especially Chap. 8. Temperature and density variations will be neglected except in Chap. 9, where compressibility is studied.

This chapter ended by discussing a few classic solutions for laminar viscous flows (Couette flow due to moving walls, Poiseuille duct flow due to pressure gradient, and flow between rotating cylinders). Whole books [4, 5, 9–11, 15] discuss classic approaches to fluid mechanics, and other texts [6, 12–14] extend these studies to the realm of continuum mechanics. This does not mean that all problems can be solved analytically. The new field of computational fluid dynamics [1] shows great promise of achieving approximate solutions to a wide variety of flow problems. In addition, when the geometry and boundary conditions are truly complex, experimentation (Chap. 5) is a preferred alternative.

Problems

Most of the problems herein are fairly straightforward. More difficult or open-ended assignments are labeled with an asterisk. Problems labeled with a computer icon may require the use of a computer. The standard end-of-chapter problems P4.1 to P4.99 (categorized in the problem list here) are followed by word problems W4.1 to W4.10, fundamentals of engineering exam problems FE4.1 to FE4.6, and comprehensive problems C4.1 and C4.2.

Problem Distribution

Section	Topic	Problems
4.1	The acceleration of a fluid	P4.1–P4.8
4.2	The continuity equation	P4.9–P4.25
4.3	Linear momentum: Navier-Stokes	P4.26–P4.38
4.4	Angular momentum: couple stresses	P4.39
4.5	The differential energy equation	P4.40–P4.41
4.6	Boundary conditions	P4.42–P4.46
4.7	Stream function	P4.47–P4.55
4.8 and 4.9	Velocity potential, vorticity	P4.56–P4.67
4.7 and 4.9	Stream function and velocity potential	P4.68–P4.78
4.10	Incompressible viscous flows	P4.79–P4.96
4.10	Slip flows	P4.97–P4.99

The acceleration of a fluid

P4.1 An idealized velocity field is given by the formula

$$\mathbf{V} = 4tx\mathbf{i} - 2t^2y\mathbf{j} + 4xz\mathbf{k}$$

Is this flow field steady or unsteady? Is it two- or three-dimensional? At the point $(x, y, z) = (-1, 1, 0)$, compute (*a*) the acceleration vector and (*b*) any unit vector normal to the acceleration.

P4.2 Flow through the converging nozzle in Fig. P4.2 can be approximated by the one-dimensional velocity distribution

$$u \approx V_0\left(1 + \frac{2x}{L}\right) \quad v \approx 0 \quad w \approx 0$$

(*a*) Find a general expression for the fluid acceleration in the nozzle. (*b*) For the specific case $V_0 = 3$ m/s and $L = 0.15$ m, compute the acceleration, in *g*'s, at the entrance and at the exit.

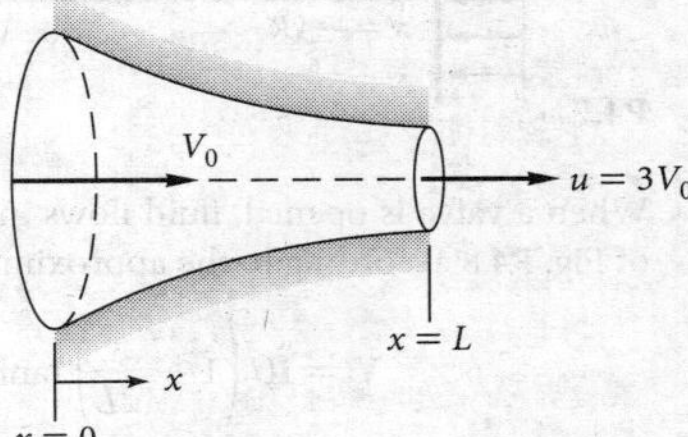

P4.2

P4.3 A two-dimensional velocity field is given by

$$\mathbf{V} = (x^2 - y^2 + x)\mathbf{i} - (2xy + y)\mathbf{j}$$

in arbitrary units. At $(x, y) = (1, 2)$, compute (*a*) the accelerations a_x and a_y, (*b*) the velocity component in the direction $\theta = 40°$, (*c*) the direction of maximum velocity, and (*d*) the direction of maximum acceleration.

P4.4 A simple flow model for a two-dimensional converging nozzle is the distribution

$$u = U_0\left(1 + \frac{x}{L}\right) \quad v = -U_0\frac{y}{L} \quad w = 0$$

(*a*) Sketch a few streamlines in the region $0 < x/L < 1$ and $0 < y/L < 1$, using the method of Sec. 1.11. (*b*) Find expressions for the horizontal and vertical accelerations. (*c*) Where is the largest resultant acceleration and its numerical value?

P4.5 The velocity field near a stagnation point may be written in the form

$$u = \frac{U_0 x}{L} \quad v = -\frac{U_0 y}{L} \quad U_0 \text{ and } L \text{ are constants}$$

(*a*) Show that the acceleration vector is purely radial. (*b*) For the particular case $L = 1.5$ m, if the acceleration at $(x, y) = (1$ m, 1 m$)$ is 25 m/s^2, what is the value of U_0?

P4.6 In deriving the continuity equation, we assumed, for simplicity, that the mass flow per unit area on the left face was just ρu. In fact, ρu varies also with y and z, and thus it must be different on the four corners of the left face. Account for these variations, average the four corners, and determine how this might change the inlet mass flow from $\rho u\, dy\, dz$.

P4.7 Consider a sphere of radius R immersed in a uniform stream U_0, as shown in Fig. P4.7. According to the theory of Chap. 8, the fluid velocity along streamline AB is given by

$$\mathbf{V} = u\mathbf{i} = U_0\left(1 + \frac{R^3}{x^3}\right)\mathbf{i}$$

Find (*a*) the position of maximum fluid acceleration along AB and (*b*) the time required for a fluid particle to travel from A to B.

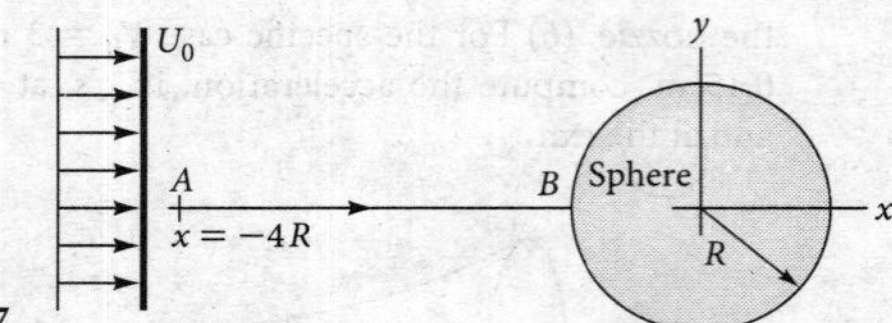

P4.7

P4.8 When a valve is opened, fluid flows in the expansion duct of Fig. P4.8 according to the approximation

$$\mathbf{V} = \mathbf{i}U\left(1 - \frac{x}{2L}\right)\tanh\frac{Ut}{L}$$

Find (*a*) the fluid acceleration at $(x, t) = (L, L/U)$ and (*b*) the time for which the fluid acceleration at $x = L$ is zero. Why does the fluid acceleration become negative after condition (*b*)?

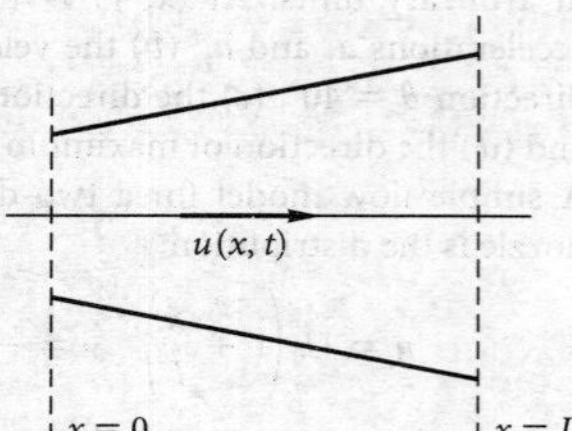

P4.8

The continuity equation

P4.9 An idealized incompressible flow has the proposed three-dimensional velocity distribution

$$\mathbf{V} = 4xy^2\mathbf{i} + f(y)\mathbf{j} - zy^2\mathbf{k}$$

Find the appropriate form of the function $f(y)$ that satisfies the continuity relation.

P4.10 A two-dimensional, incompressible flow has the velocity components $u = 4y$ and $v = 2x$. (*a*) Find the acceleration components. (*b*) Is the vector acceleration radial? (*c*) Sketch a few streamlines in the first quadrant and determine if any are straight lines.

P4.11 Derive Eq. (4.12*b*) for cylindrical coordinates by considering the flux of an incompressible fluid in and out of the elemental control volume in Fig. 4.2.

P4.12 Spherical polar coordinates (r, θ, ϕ) are defined in Fig. P4.12. The cartesian transformations are

$$x = r\sin\theta\cos\phi$$
$$y = r\sin\theta\sin\phi$$
$$z = r\cos\theta$$

Do not show that the cartesian incompressible continuity relation [Eq. (4.12*a*)] can be transformed to the spherical polar form

$$\frac{1}{r^2}\frac{\partial}{\partial r}(r^2 \upsilon_r) + \frac{1}{r\sin\theta}\frac{\partial}{\partial\theta}(\upsilon_\theta\sin\theta) + \frac{1}{r\sin\theta}\frac{\partial}{\partial\phi}(\upsilon_\phi) = 0$$

What is the most general form of υ_r when the flow is purely radial—that is, υ_θ and υ_ϕ are zero?

P4.13 For an incompressible plane flow in polar coordinates, we are given

$$\upsilon_r = r^3\cos\theta + r^2\sin\theta$$

Find the appropriate form of circumferential velocity for which continuity is satisfied.

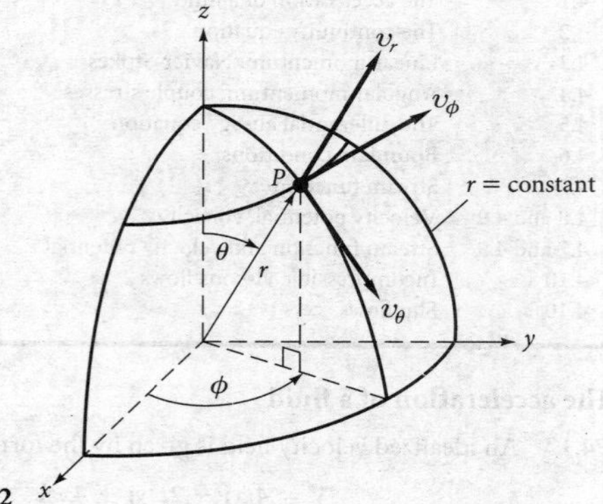

P4.12

P4.14 For incompressible polar coordinate flow, what is the most general form of a purely circulatory motion, $\upsilon_\theta = \upsilon_\theta(r, \theta, t)$ and $\upsilon_r = 0$, that satisfies continuity?

P4.15 What is the most general form of a purely radial polar coordinate incompressible flow pattern, $\upsilon_r = \upsilon_r(r, \theta, t)$ and $\upsilon_\theta = 0$, that satisfies continuity?

P4.16 Consider the plane polar coordinate velocity distribution

$$v_r = \frac{C}{r} \qquad v_\theta = \frac{K}{r} \qquad v_z = 0$$

where C and K are constants. (*a*) Determine if the equation of continuity is satisfied. (*b*) By sketching some velocity vector directions, plot a single streamline for $C = K$. What might this flow field simulate?

P4.17 An excellent approximation for the two-dimensional incompressible laminar boundary layer on the flat surface in Fig. P4.17 is

$$u \approx U\left(2\frac{y}{\delta} - 2\frac{y^3}{\delta^3} + \frac{y^4}{\delta^4}\right) \quad \text{for } y \le \delta$$

$$\text{where } \delta = Cx^{1/2},\ C = \text{const}$$

(*a*) Assuming a no-slip condition at the wall, find an expression for the velocity component $v(x, y)$ for $y \le \delta$. (*b*) Then find the maximum value of v at the station $x = 1$ m, for the particular case of airflow, when $U = 3$ m/s and $\delta = 1.1$ cm.

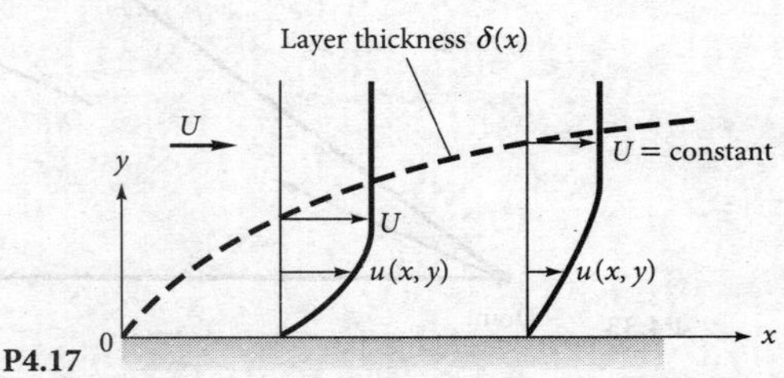

P4.17

P4.18 A piston compresses gas in a cylinder by moving at constant speed V, as in Fig. P4.18. Let the gas density and length at $t = 0$ be ρ_0 and L_0, respectively. Let the gas velocity vary linearly from $u = V$ at the piston face to $u = 0$ at $x = L$. If the gas density varies only with time, find an expression for $\rho(t)$.

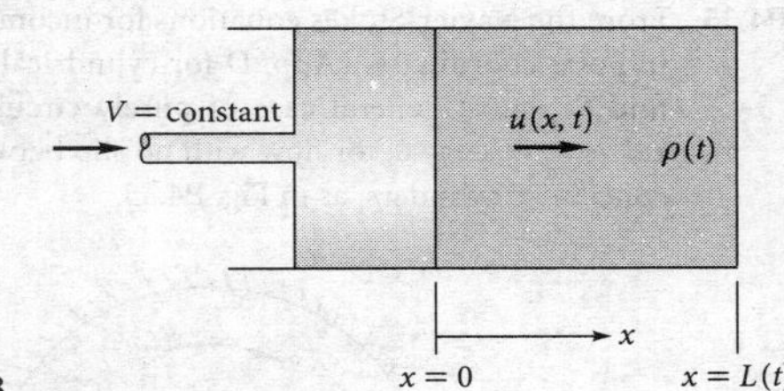

P4.18

P4.19 A proposed incompressible plane flow in polar coordinates is given by

$$v_r = 2r\cos(2\theta); \quad v_\theta = -2r\sin(2\theta)$$

(*a*) Determine if this flow satisfies the equation of continuity. (*b*) If so, sketch a possible streamline in the first quadrant by finding the velocity vectors at $(r, \theta) = (1.25, 20°)$, $(1.0, 45°)$, and $(1.25, 70°)$. (*c*) Speculate on what this flow might represent.

P4.20 A two-dimensional incompressible velocity field has $u = K(1 - e^{-ay})$, for $x \le L$ and $0 \le y \le \infty$. What is the most general form of $v(x, y)$ for which continuity is satisfied and $v = v_0$ at $y = 0$? What are the proper dimensions for constants K and a?

P4.21 Air flows under steady, approximately one-dimensional conditions through the conical nozzle in Fig. P4.21. If the speed of sound is approximately 340 m/s, what is the minimum nozzle-diameter ratio D_e/D_0 for which we can safely neglect compressibility effects if $V_0 =$ (*a*) 10 m/s and (*b*) 30 m/s?

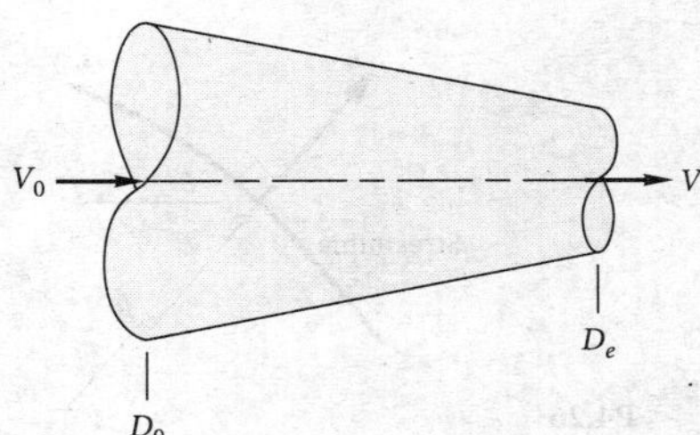

P4.21

P4.22 In an *axisymmetric* flow, nothing varies with θ, and the only nonzero velocities are v_r and v_z (see Fig. 4.2). If the flow is steady and incompressible and $v_z = Bz$, where B is constant, find the most general form of v_r which satisfies continuity.

P4.23 A tank volume $\mathcal{V}$ contains gas at conditions (ρ_0, p_0, T_0). At time $t = 0$ it is punctured by a small hole of area A. According to the theory of Chap. 9, the mass flow out of such a hole is approximately proportional to A and to the tank pressure. If the tank temperature is assumed constant and the gas is ideal, find an expression for the variation of density within the tank.

P4.24 For laminar flow between parallel plates (see Fig. 4.12*b*), the flow is two-dimensional $(v \ne 0)$ if the walls are porous. A special case solution is $u = (A - Bx)(h^2 - y^2)$, where A and B are constants. (*a*) Find a general formula for velocity v if $v = 0$ at $y = 0$. (*b*) What is the value of the constant B if $v = v_w$ at $y = +h$?

P4.25 An incompressible flow in polar coordinates is given by

$$v_r = K\cos\theta\left(1 - \frac{b}{r^2}\right)$$

$$v_\theta = -K\sin\theta\left(1 + \frac{b}{r^2}\right)$$

Does this field satisfy continuity? For consistency, what should the dimensions of constants K and b be? Sketch the surface where $v_r = 0$ and interpret.

Linear momentum: Navier-Stokes

***P4.26** Curvilinear, or streamline, coordinates are defined in Fig. P4.26, where n is normal to the streamline in the plane of the radius of curvature R. Euler's frictionless momentum equation (4.36) in streamline coordinates becomes

$$\frac{\partial V}{\partial t} + V\frac{\partial V}{\partial s} = -\frac{1}{\rho}\frac{\partial p}{\partial s} + g_s \qquad (1)$$

$$-V\frac{\partial \theta}{\partial t} - \frac{V^2}{R} = -\frac{1}{\rho}\frac{\partial p}{\partial n} + g_n \qquad (2)$$

Show that the integral of Eq. (1) with respect to s is none other than our old friend Bernoulli's equation (3.54).

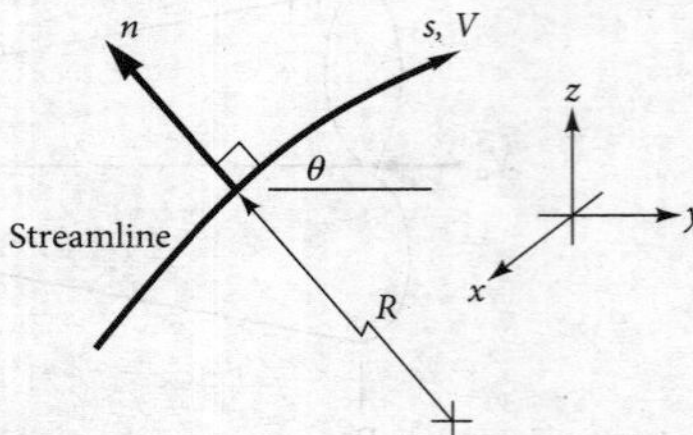

P4.26

P4.27 A frictionless, incompressible steady flow field is given by

$$\mathbf{V} = 2xy\mathbf{i} - y^2\mathbf{j}$$

in arbitrary units. Let the density be ρ_0 = constant and neglect gravity. Find an expression for the pressure gradient in the x direction.

P4.28 For the velocity distribution of Prob. 4.10, (*a*) check continuity. (*b*) Are the Navier-Stokes equations valid? (*c*) If so, determine $p(x, y)$ if the pressure at the origin is p_0.

P4.29 Consider a steady, two-dimensional, incompressible flow of a newtonian fluid in which the velocity field is known: $u = -2xy$, $v = y^2 - x^2$, $w = 0$. (*a*) Does this flow satisfy conservation of mass? (*b*) Find the pressure field, $p(x, y)$ if the pressure at the point ($x = 0, y = 0$) is equal to p_a.

P4.30 For the velocity distribution of Prob. P4.4, determine if (*a*) the equation of continuity and (*b*) the Navier-Stokes equation are satisfied. (*c*) If the latter is true, find the pressure distribution $p(x, y)$ when the pressure at the origin equals p_o.

P4.31 According to potential theory (Chap. 8) for the flow approaching a rounded two-dimensional body, as in Fig. P4.31, the velocity approaching the stagnation point is given by $u = U(1 - a^2/x^2)$, where a is the nose radius and U is the velocity far upstream. Compute the value and position of the maximum viscous normal stress along this streamline.

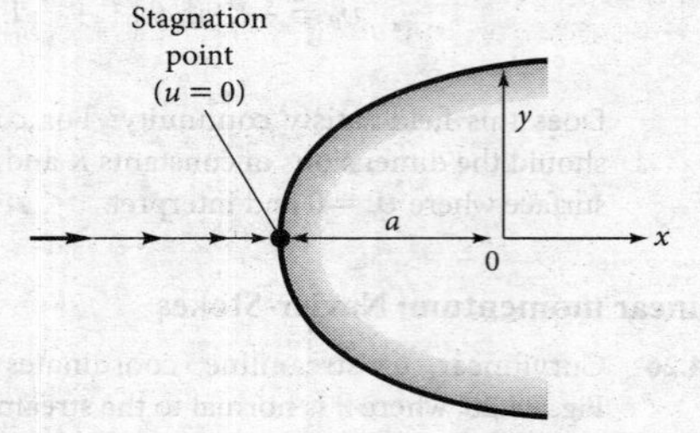

P4.31

Is this also the position of maximum fluid deceleration? Evaluate the maximum viscous normal stress if the fluid is SAE 30 oil at 20°C, with U = 2 m/s and a = 6 cm.

P4.32 The answer to Prob. P4.14 is v_θ = f(r) only. Do not reveal this to your friends if they are still working on Prob. P4.14. Show that this flow field is an exact solution to the Navier-Stokes equations (4.38) for only two special cases of the function f(r). Neglect gravity. Interpret these two cases physically.

P4.33 Consider incompressible flow at a volume rate Q toward a drain at the vertex of a 45° wedge of width b, as in Fig. P4.33. Neglect gravity and friction and assume purely radial inflow. (*a*) Find an expression for $v_r(r)$. (*b*) Show that the viscous term in the r-momentum equation is zero. (*c*) Find the pressure distribution $p(r)$ if $p = p_o$ at $r = R$.

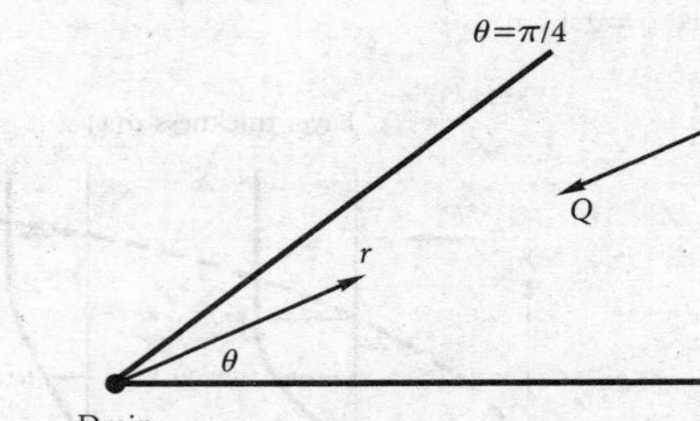

P4.33

P4.34 A proposed three-dimensional incompressible flow field has the following vector form:

$$\mathbf{V} = Kx\mathbf{i} + Ky\mathbf{j} - 2Kz\mathbf{k}$$

(*a*) Determine if this field is a valid solution to continuity and Navier-Stokes. (*b*) If $\mathbf{g} = -g\mathbf{k}$, find the pressure field $p(x, y, z)$. (*c*) Is the flow irrotational?

P4.35 From the Navier-Stokes equations for incompressible flow in polar coordinates (App. D for cylindrical coordinates), find the most general case of purely circulating motion $v_\theta(r)$, $v_r = v_z = 0$, for flow with no slip between two fixed concentric cylinders, as in Fig. P4.35.

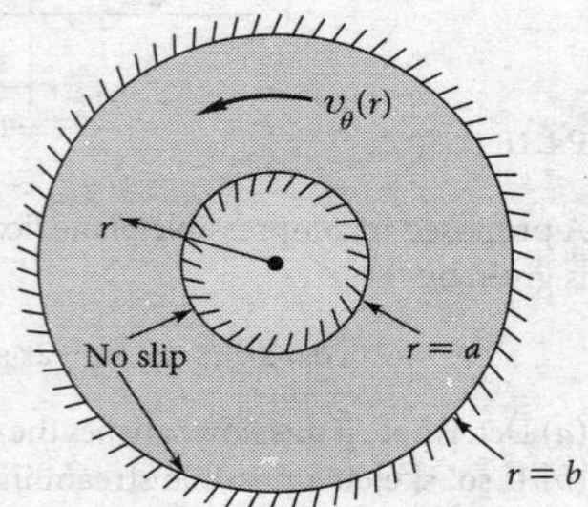

P4.35

P4.36 A constant-thickness film of viscous liquid flows in laminar motion down a plate inclined at angle θ, as in Fig. P4.36. The velocity profile is

$$u = Cy(2h - y) \quad v = w = 0$$

Find the constant C in terms of the specific weight and viscosity and the angle θ. Find the volume flux Q per unit width in terms of these parameters.

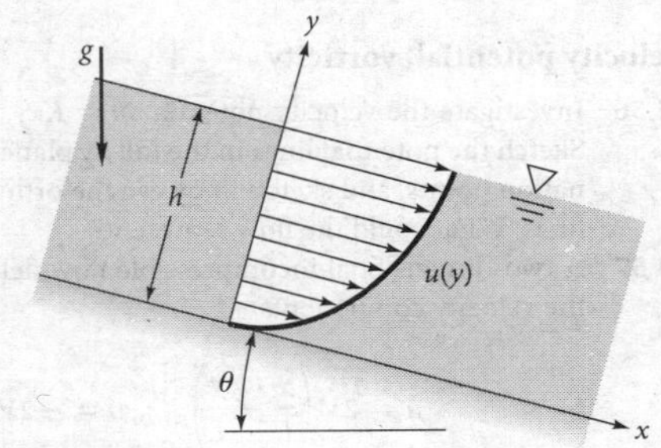

P4.36

***P4.37** A viscous liquid of constant ρ and μ falls due to gravity between two plates a distance $2h$ apart, as in Fig. P4.37. The flow is fully developed, with a single velocity component $w = w(x)$. There are no applied pressure gradients, only gravity. Solve the Navier-Stokes equation for the velocity profile between the plates.

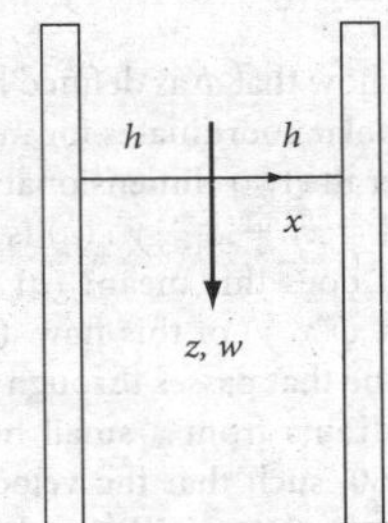

P4.37

P4.38 Show that the incompressible flow distribution, in cylindrical coordinates,

$$v_r = 0 \qquad v_\theta = Cr^n \qquad v_z = 0$$

where C is a constant, (*a*) satisfies the Navier-Stokes equation for only two values of n. Neglect gravity. (*b*) Knowing that $p = p(r)$ only, find the pressure distribution for each case, assuming that the pressure at $r = R$ is p_0. What might these two cases represent?

Angular momentum: couple stresses

P4.39 Reconsider the angular momentum balance of Fig. 4.5 by adding a concentrated *body couple* C_z about the z axis [6]. Determine a relation between the body couple and shear stress for equilibrium. What are the proper dimensions for C_z? (Body couples are important in continuous media with microstructure, such as granular materials.)

The differential energy equation

P4.40 For pressure-driven laminar flow between parallel plates (see Fig. 4.12*b*), the velocity components are $u = U(1 - y^2/h^2)$, $v = 0$, and $w = 0$, where U is the centerline velocity. In the spirit of Ex. 4.6, find the temperature distribution $T(y)$ for a constant wall temperature T_w.

P4.41 As mentioned in Sec. 4.10, the velocity profile for laminar flow between two plates, as in Fig. P4.41, is

$$u = \frac{4u_{max}y(h - y)}{h^2} \qquad v = w = 0$$

If the wall temperature is T_w at both walls, use the incompressible flow energy equation (4.75) to solve for the temperature distribution $T(y)$ between the walls for steady flow.

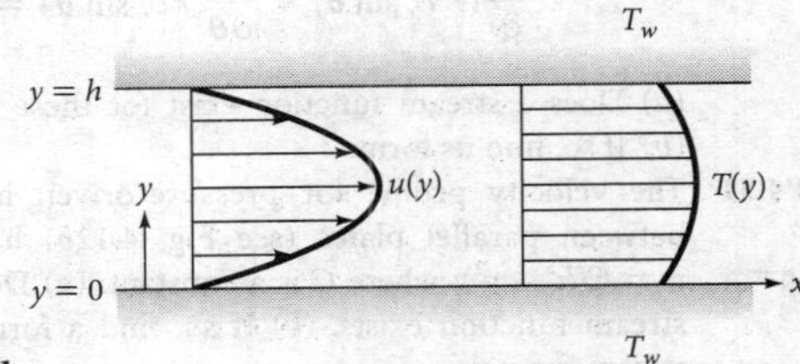

P4.41

Boundary conditions

P4.42 Suppose we wish to analyze the rotating, partly full cylinder of Fig. 2.23 as a *spin-up* problem, starting from rest and continuing until solid-body rotation is achieved. What are the appropriate boundary and initial conditions for this problem?

P4.43 For the draining liquid film of Fig. P4.36, what are the appropriate boundary conditions (*a*) at the bottom $y = 0$ and (*b*) at the surface $y = h$?

P4.44 Suppose that we wish to analyze the sudden pipe expansion flow of Fig. P3.59, using the full continuity and Navier-Stokes equations. What are the proper boundary conditions to handle this problem?

P4.45 For the sluice gate problem of Example 3.10, list all the boundary conditions needed to solve this flow exactly by, say, computational fluid dynamics.

P4.46 Fluid from a large reservoir at temperature T_0 flows into a circular pipe of radius R. The pipe walls are wound with an electric resistance coil that delivers heat to the fluid at a rate q_w (energy per unit wall area). If we wish to analyze this problem by using the full continuity, Navier-Stokes, and energy equations, what are the proper boundary conditions for the analysis?

Stream function

P4.47 A two-dimensional incompressible flow is given by the velocity field $\mathbf{V} = 3y\mathbf{i} + 2x\mathbf{j}$, in arbitrary units. Does this flow satisfy continuity? If so, find the stream function $\psi(x, y)$ and plot a few streamlines, with arrows.

P4.48 Consider the following two-dimensional incompressible flow, which clearly satisfies continuity:

$$u = U_0 = \text{constant}, \; v = V_0 = \text{constant}$$

Find the stream function $\psi(r, \theta)$ of this flow using *polar coordinates*.

P4.49 Investigate the stream function $\psi = K(x^2 - y^2)$, K = constant. Plot the streamlines in the full xy plane, find any stagnation points, and interpret what the flow could represent.

P4.50 In 1851, George Stokes (of Navier-Stokes fame) solved the problem of steady incompressible low-Reynolds-number flow past a sphere, using *spherical polar* coordinates (r, θ) [Ref. 5, page 168]. In these coordinates, the equation of continuity is

$$\frac{\partial}{\partial r}(r^2 \upsilon_r \sin\theta) + \frac{\partial}{\partial \theta}(r\upsilon_\theta \sin\theta) = 0$$

(*a*) Does a stream function exist for these coordinates? (*b*) If so, find its form.

P4.51 The velocity profile for pressure-driven laminar flow between parallel plates (see Fig. 4.12*b*) has the form $u = C(h^2 - y^2)$, where C is a constant. (*a*) Determine if a stream function exists. (*b*) If so, find a formula for the stream function.

P4.52 A two-dimensional, incompressible, frictionless fluid is guided by wedge-shaped walls into a small slot at the origin, as in Fig. P4.52. The width into the paper is b,

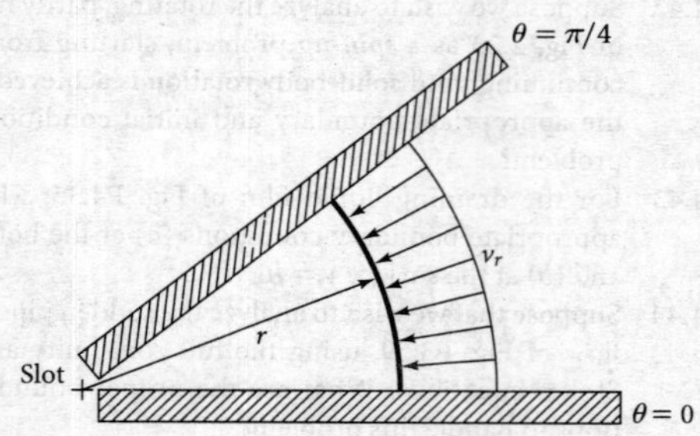

P4.52

and the volume flow rate is Q. At any given distance r from the slot, the flow is radial inward, with constant velocity. Find an expression for the polar coordinate stream function of this flow.

P4.53 For the fully developed laminar pipe flow solution of Eq. (4.137), find the axisymmetric stream function $\psi(r, z)$. Use this result to determine the average velocity $V = Q/A$ in the pipe as a ratio of u_{max}.

P4.54 An incompressible stream function is defined by

$$\psi(x, y) = \frac{U}{L^2}(3x^2y - y^3)$$

where U and L are (positive) constants. Where in this chapter are the streamlines of this flow plotted? Use this stream function to find the volume flow Q passing through the rectangular surface whose corners are defined by $(x, y, z) = (2L, 0, 0), (2L, 0, b), (0, L, b)$, and $(0, L, 0)$. Show the direction of Q.

P4.55 The proposed flow in Prob. P4.19 does indeed satisfy the equation of continuity. Determine the polar-coordinate stream function of this flow.

Velocity potential, vorticity

P4.56 Investigate the velocity potential $\phi = Kxy$, K = constant. Sketch the potential lines in the full xy plane, find any stagnation points, and sketch in by eye the orthogonal streamlines. What could the flow represent?

P4.57 A two-dimensional incompressible flow field is defined by the velocity components

$$u = 2V\left(\frac{x}{L} - \frac{y}{L}\right) \qquad \upsilon = -2V\frac{y}{L}$$

where V and L are constants. If they exist, find the stream function and velocity potential.

P4.58 Show that the incompressible velocity potential in plane polar coordinates $\phi(r, \theta)$ is such that

$$\upsilon_r = \frac{\partial \phi}{\partial r} \qquad \upsilon_\theta = \frac{1}{r}\frac{\partial \phi}{\partial \theta}$$

Finally show that ϕ as defined here satisfies Laplace's equation in polar coordinates for incompressible flow.

P4.59 Consider the two-dimensional incompressible velocity potential $\phi = xy + x^2 - y^2$. (*a*) Is it true that $\nabla^2\phi = 0$, and, if so, what does this mean? (*b*) If it exists, find the stream function $\psi(x, y)$ of this flow. (*c*) Find the equation of the streamline that passes through $(x, y) = (2, 1)$.

P4.60 Liquid drains from a small hole in a tank, as shown in Fig. P4.60, such that the velocity field set up is given by $\upsilon_r \approx 0$, $\upsilon_z \approx 0$, $\upsilon_\theta = KR^2/r$, where $z = H$ is the depth of the water far from the hole. Is this flow pattern rotational or irrotational? Find the depth z_C of the water at the radius $r = R$.

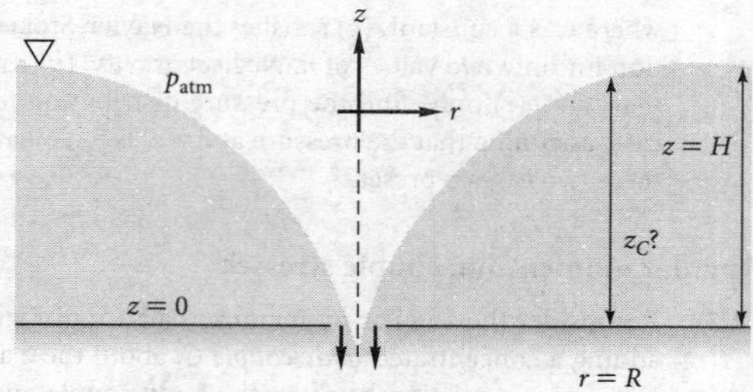

P4.60

P4.61 An incompressible stream function is given by $\psi = a\theta + br\sin\theta$. (*a*) Does this flow have a velocity potential? (*b*) If so, find it.

P4.62 Show that the linear Couette flow between plates in Fig. 1.7 has a stream function but no velocity potential. Why is this so?

P4.63 Find the two-dimensional velocity potential $\phi(r, \theta)$ for the polar coordinate flow pattern $v_r = Q/r$, $v_\theta = K/r$, where Q and K are constants.

P4.64 Show that the velocity potential $\phi(r, z)$ in axisymmetric cylindrical coordinates (see Fig. 4.2) is defined such that

$$v_r = \frac{\partial \phi}{\partial r} \qquad v_z = \frac{\partial \phi}{\partial z}$$

Further show that for incompressible flow this potential satisfies Laplace's equation in (r, z) coordinates.

P4.65 Consider the function $f = ay - by^3$. (*a*) Could this represent a realistic velocity potential? *Extra credit*: (*b*) Could it represent a stream function?

P4.66 A plane polar coordinate velocity potential is defined by

$$\phi = \frac{K \cos \theta}{r} \qquad K = \text{const}$$

Find the stream function for this flow, sketch some streamlines and potential lines, and interpret the flow pattern.

P4.67 A stream function for a plane, irrotational, polar coordinate flow is

$$\psi = C\theta - K \ln r \quad C \text{ and } K = \text{const}$$

Find the velocity potential for this flow. Sketch some streamlines and potential lines, and interpret the flow pattern.

Stream function *and* velocity potential

P4.68 For the velocity distribution of Prob. P4.4, (*a*) determine if a velocity potential exists, and (*b*), if it does, find an expression for $\phi(x, y)$ and sketch the potential line which passes through the point $(x, y) = (L/2, L/2)$.

P4.69 A steady, two-dimensional flow has the following polar-coordinate velocity potential:

$$\phi = C r \cos \theta + K \ln r$$

where C and K are constants. Determine the stream function $\psi(r, \theta)$ for this flow. For extra credit, let C be a velocity scale U and let $K = UL$, sketch what the flow might represent.

P4.70 A CFD model of steady two-dimensional incompressible flow has printed out the values of stream function $\psi(x, y)$, in m^2/s, at each of the four corners of a small 10-cm-by-10-cm cell, as shown in Fig. P4.70. Use these numbers to estimate the resultant velocity in the center of the cell and its angle α with respect to the x axis.

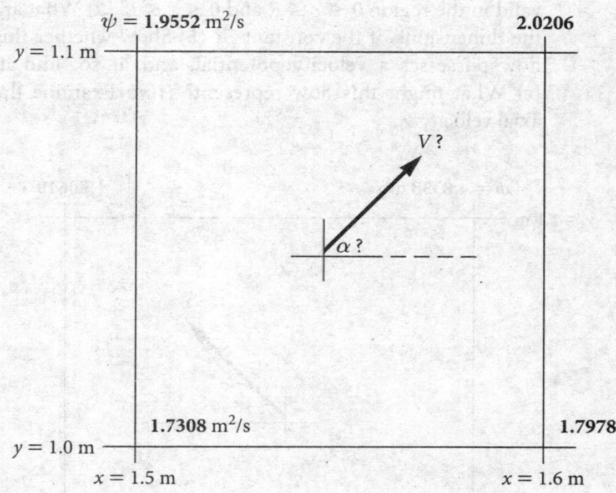

P4.70

P4.71 Consider the following two-dimensional function $f(x, y)$:

$$f = Ax^3 + Bxy^2 + Cx^2 + D \quad \text{where } A > 0$$

(*a*) Under what conditions, if any, on (A, B, C, D) can this function be a steady plane-flow velocity potential? (*b*) If you find a $\phi(x, y)$ to satisfy part (*a*), also find the associated stream function $\psi(x, y)$, if any, for this flow.

P4.72 Water flows through a two-dimensional narrowing wedge at 6×10^{-4} m^3/s per meter of width into the paper (Fig. P4.72). If this inward flow is purely radial, find an expression, in SI units, for (*a*) the stream function and (*b*) the velocity potential of the flow. Assume one-dimensional flow. The included angle of the wedge is 45°.

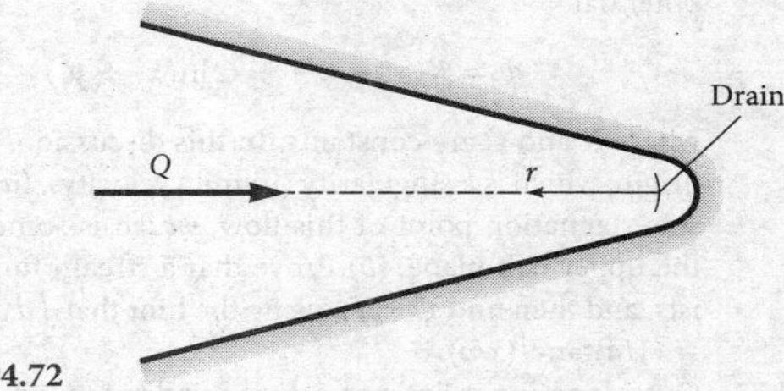

P4.72

P4.73 A CFD model of steady two-dimensional incompressible flow has printed out the values of velocity potential $\phi(x, y)$, in m^2/s, at each of the four corners of a small 10-cm-by-10-cm cell, as shown in Fig. P4.73. Use these numbers to estimate the resultant velocity in the center of the cell and its angle α with respect to the x axis.

P4.74 Consider the two-dimensional incompressible polar-coordinate velocity potential

$$\phi = Br \cos \theta + B L \theta$$

where B is a constant and L is a constant length scale. (*a*) What are the dimensions of B? (*b*) Locate the only stagnation point in this flow field. (*c*) Prove that a stream function exists and then find the function $\psi(r, \theta)$.

P4.75 Given the following steady *axisymmetric* stream function:

$$\psi = \frac{B}{2}\left(r^2 - \frac{r^4}{2R^2}\right) \quad \text{where } B \text{ and } R \text{ are constants}$$

valid in the region $0 \le r \le R$ and $0 \le z \le L$. (*a*) What are the dimensions of the constant B? (*b*) Show whether this flow possesses a velocity potential, and, if so, find it. (*c*) What might this flow represent? *Hint:* Examine the axial velocity v_z.

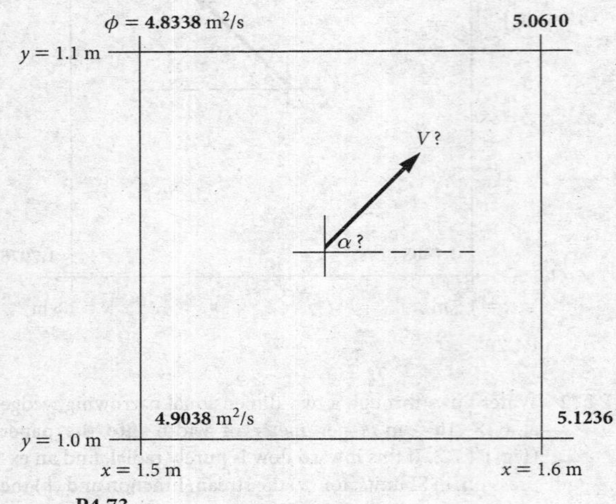

P4.73

***P4.76** A two-dimensional incompressible flow has the velocity potential

$$\phi = K(x^2 - y^2) + C\ln(x^2 + y^2)$$

where K and C are constants. In this discussion, avoid the origin, which is a singularity (infinite velocity). (*a*) Find the sole stagnation point of this flow, which is somewhere in the upper half plane. (*b*) Prove that a stream function exists, and then find $\psi(x, y)$, using the hint that $\int dx/(a^2 + x^2) = (1/a)\tan^{-1}(x/a)$.

P4.77 Outside an inner, intense-activity circle of radius R, a tropical storm can be simulated by a polar-coordinate velocity potential $\phi(r, \theta) = U_o R\,\theta$, where U_o is the wind velocity at radius R. (*a*) Determine the velocity components outside $r = R$. (*b*) If, at $R = 40$ km, the velocity is 45 m/s and the pressure 99 kPa, calculate the velocity and pressure at $r = 4R$.

P4.78 An incompressible, irrotational, two-dimensional flow has the following stream function in polar coordinates:

$$\psi = A\,r^n \sin(n\theta) \qquad \text{where } A \text{ and } n \text{ are constants.}$$

Find an expression for the velocity potential of this flow.

Incompressible viscous flows

***P4.79** Study the combined effect of the two viscous flows in Fig. 4.12. That is, find $u(y)$ when the upper plate moves at speed V and there is also a constant pressure gradient (dp/dx). Is superposition possible? If so, explain why. Plot representative velocity profiles for (*a*) zero, (*b*) positive, and (*c*) negative pressure gradients for the same upper-wall speed V.

***P4.80** Oil, of density ρ and viscosity μ, drains steadily down the side of a vertical plate, as in Fig. P4.80. After a development region near the top of the plate, the oil film will become independent of z and of constant thickness δ. Assume that $w = w(x)$ only and that the atmosphere offers no shear resistance to the surface of the film. (*a*) Solve the Navier-Stokes equation for $w(x)$, and sketch its approximate shape. (*b*) Suppose that film thickness δ and the slope of the velocity profile at the wall $[\partial w/\partial x]_{wall}$ are measured with a laser-Doppler anemometer (Chap. 6). Find an expression for oil viscosity μ as a function of $(\rho, \delta, g, [\partial w/\partial x]_{wall})$.

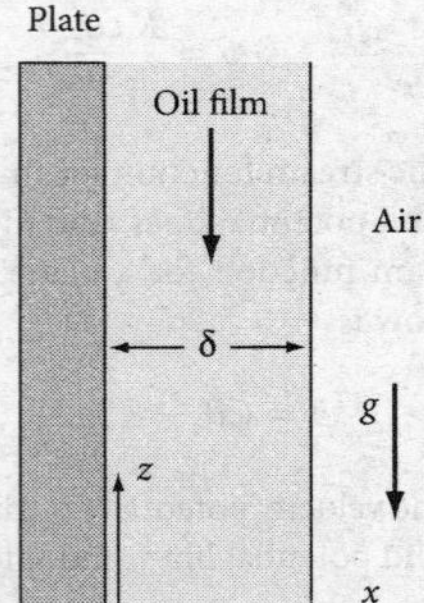

P4.80

P4.81 Modify the analysis of Fig. 4.13 to find the velocity u_θ when the inner cylinder is fixed and the outer cylinder rotates at angular velocity Ω_0. May this solution be *added* to Eq. (4.140) to represent the flow caused when both inner and outer cylinders rotate? Explain your conclusion.

***P4.82** A solid circular cylinder of radius R rotates at angular velocity Ω in a viscous incompressible fluid that is at rest far from the cylinder, as in Fig. P4.82. Make simplifying assumptions and derive the governing differential equation and boundary conditions for the velocity field v_θ in the fluid. Do not solve unless you are obsessed with this problem. What is the steady-state flow field for this problem?

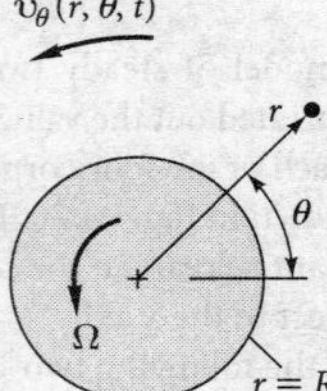

P4.82

P4.83 The flow pattern in bearing lubrication can be illustrated by Fig. P4.83, where a viscous oil (ρ, μ) is forced into the gap $h(x)$ between a fixed slipper block and a wall moving at velocity U. If the gap is thin, $h \ll L$, it can be shown that the pressure and velocity distributions are of the form $p = p(x)$, $u = u(y)$, $v = w = 0$. Neglecting gravity, reduce the

Navier-Stokes equations (4.38) to a single differential equation for $u(y)$. What are the proper boundary conditions? Integrate and show that

$$u = \frac{1}{2\mu}\frac{dp}{dx}(y^2 - yh) + U\left(1 - \frac{y}{h}\right)$$

where $h = h(x)$ may be an arbitrary, slowly varying gap width. (For further information on lubrication theory, see Ref. 16.)

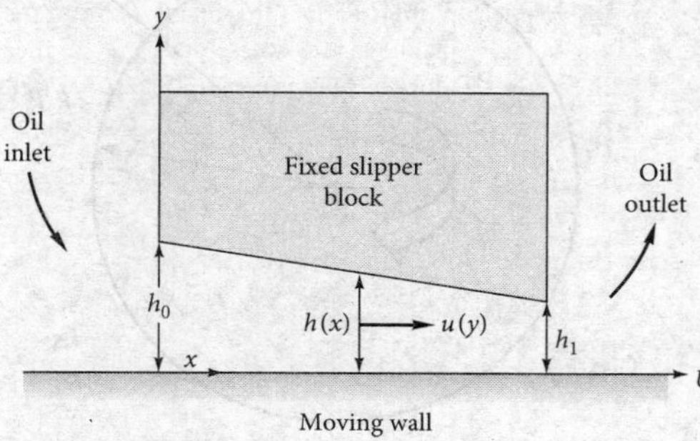

P4.83

***P4.84** Consider a viscous film of liquid draining uniformly down the side of a vertical rod of radius a, as in Fig. P4.84. At some distance down the rod the film will approach a terminal or *fully developed* draining flow of constant outer radius b, with $v_z = v_z(r)$, $v_\theta = v_r = 0$. Assume that the atmosphere offers no shear resistance to the film motion. Derive a differential equation for v_z, state the proper boundary conditions, and solve for the film velocity distribution. How does the film radius b relate to the total film volume flow rate Q?

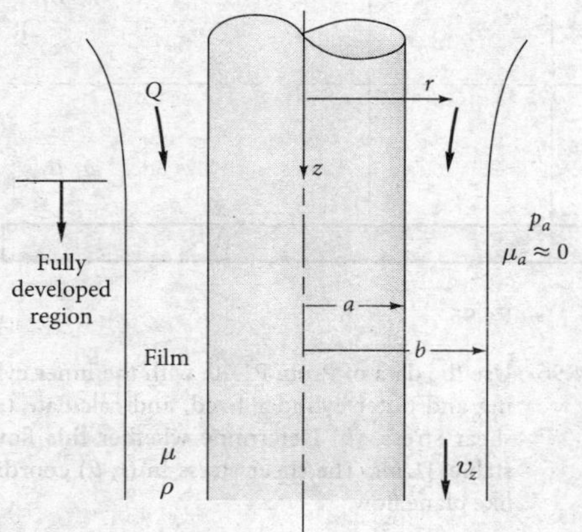

P4.84

P4.85 A flat plate of essentially infinite width and breadth oscillates sinusoidally in its own plane beneath a viscous fluid, as in Fig. P4.85. The fluid is at rest far above the plate. Making as many simplifying assumptions as you can, set up the governing differential equation and boundary conditions for finding the velocity field u in the fluid. Do not solve (if you *can* solve it immediately, you might be able to get exempted from the balance of this course with credit).

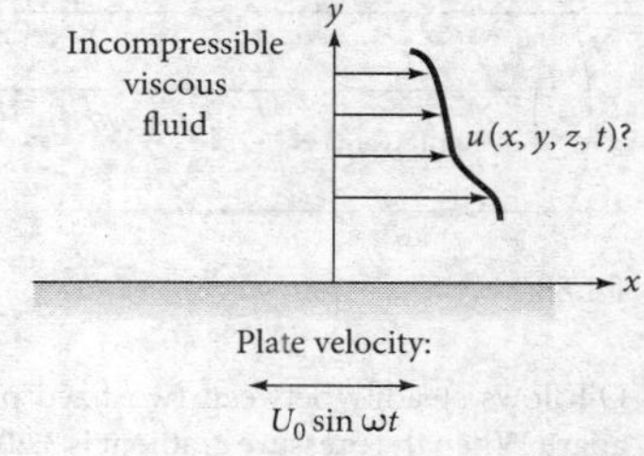

P4.85

P4.86 SAE 10 oil at 20°C flows between parallel plates 8 mm apart, as in Fig. P4.86. A mercury manometer, with wall pressure taps 1 m apart, registers a 6-cm height, as shown. Estimate the flow rate of oil for this condition.

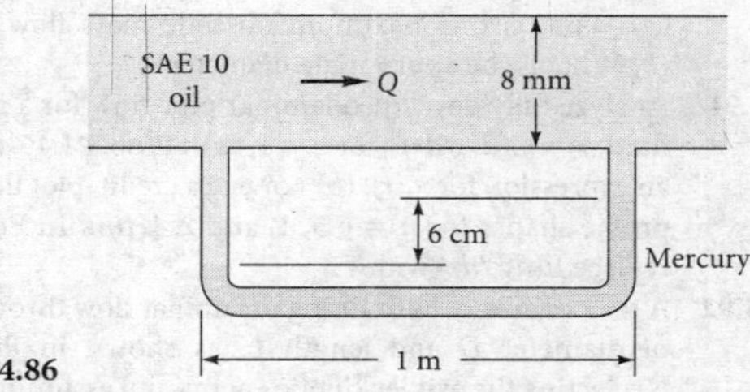

P4.86

P4.87 SAE 30W oil at 20°C flows through the 9-cm-diameter pipe in Fig. P4.87 at an average velocity of 4.3 m/s.

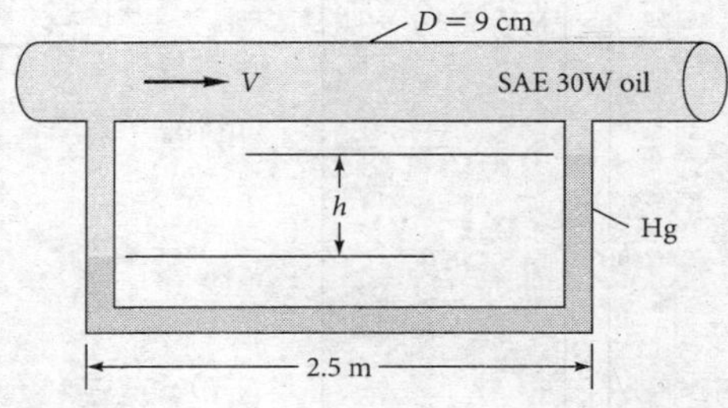

P4.87

(*a*) Verify that the flow is laminar. (*b*) Determine the volume flow rate in m^3/h. (*c*) Calculate the expected reading h of the mercury manometer, in cm.

P4.88 The viscous oil in Fig. P4.88 is set into steady motion by a concentric inner cylinder moving axially at velocity U inside a fixed outer cylinder. Assuming constant pressure and density and a purely axial fluid motion, solve Eqs. (4.38) for the fluid velocity distribution $v_z(r)$. What are the proper boundary conditions?

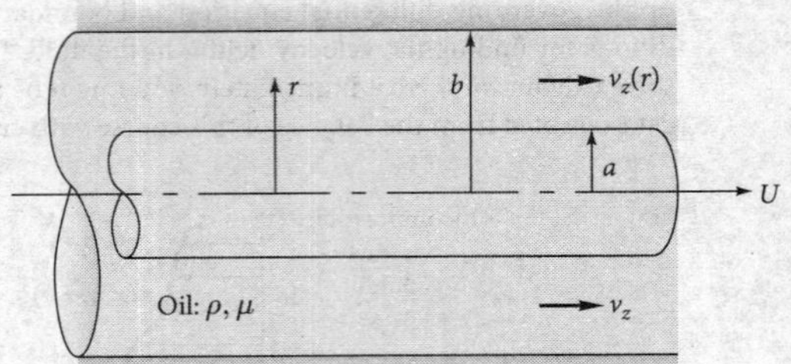

P4.88

P4.89 Oil flows steadily between two fixed plates that are 5 cm apart. When the pressure gradient is 3,200 pascals per meter, the average velocity is 0.8 m/s. (*a*) What is the flow rate per meter of width? (*b*) What oil in Table A.4 fits this data? (*c*) Can we be sure that the flow is laminar?

P4.90 It is desired to pump ethanol at 20°C through 25 m of straight smooth tubing under laminar-flow conditions, $Re_d = \rho V d/\mu < 2{,}300$. The available pressure drop is 10 kPa. (*a*) What is the maximum possible mass flow, in kg/h? (*b*) What is the appropriate diameter?

***P4.91** Analyze fully developed laminar pipe flow for a *power-law* fluid, $\tau = C(dv_z/dr)^n$, for $n \neq 1$, as in Prob. P1.46. (*a*) Derive an expression for $v_z(r)$. (*b*) For extra credit, plot the velocity profile shapes for $n = 0.5$, 1, and 2. [*Hint:* In Eq. (4.136), replace $\mu(dv_z/dr)$ with τ.]

P4.92 A tank of area A_0 is draining in laminar flow through a pipe of diameter D and length L, as shown in Fig. P4.92. Neglecting the exit jet kinetic energy and assuming the pipe flow is driven by the hydrostatic pressure at its entrance, derive a formula for the tank level $h(t)$ if its initial level is h_0.

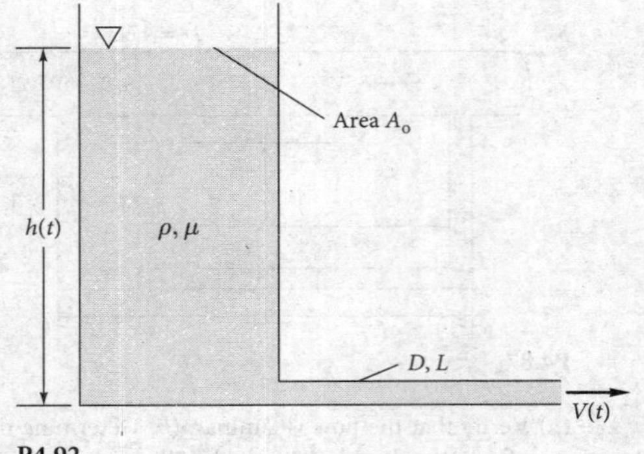

P4.92

P4.93 A number of straight 25-cm-long microtubes of diameter d are bundled together into a "honeycomb" whose total cross-sectional area is 0.0006 m². The pressure drop from entrance to exit is 1.5 kPa. It is desired that the total volume flow rate be 1 m³/h of water at 20°C. (*a*) What is the appropriate microtube diameter? (*b*) How many microtubes are in the bundle? (*c*) What is the Reynolds number of each microtube?

P4.94 A long, solid cylinder rotates steadily in a very viscous fluid, as in Fig. P4.94. Assuming laminar flow, solve the Navier-Stokes equation in polar coordinates to determine the resulting velocity distribution. The fluid is at rest far from the cylinder. [*Hint:* The cylinder does not induce any radial motion.]

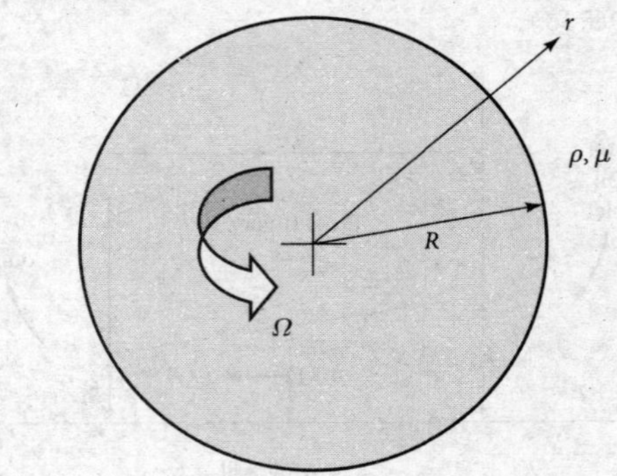

P4.94

***P4.95** Two immiscible liquids of equal thickness h are being sheared between a fixed and a moving plate, as in Fig. P4.95. Gravity is neglected, and there is no variation with x. Find an expression for (*a*) the velocity at the interface and (*b*) the shear stress in each fluid. Assume steady laminar flow.

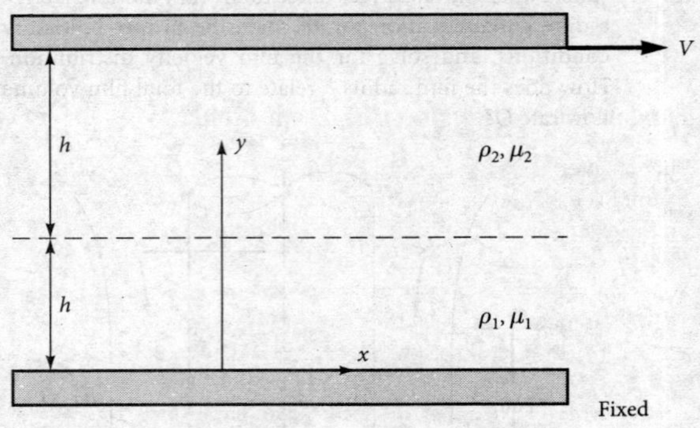

P4.95

P4.96 Use the data of Prob. P1.40, with the inner cylinder rotating and outer cylinder fixed, and calculate (*a*) the inner shear stress. (*b*) Determine whether this flow pattern is stable. [*Hint:* The shear stress in (r, θ) coordinates is *not* like plane flow.

Slip flows

P4.97 For Couette flow between a moving and a fixed plate, Fig. 4.12*a*, solve continuity and Navier-Stokes to find the velocity distribution when there is *slip* at both walls.

P4.98 For the pressure-gradient flow between two parallel plates of Fig. 4.12(*b*), reanalyze for the case of *slip flow* at both walls. Use the simple slip condition $u_{wall} = \ell(du/dy)_{wall}$, where ℓ is the mean free path of the fluid. (*a*) Sketch the expected velocity profile. (*b*) Find an expression for the shear stress at each wall. (*c*) Find the volume flow between the plates.

P4.99 For the pressure-gradient flow in a circular tube in Sec. 4.10, reanalyze for the case of *slip flow* at the wall. Use the simple slip condition $v_{z,wall} = \ell(dv_z/dr)_{wall}$, where ℓ is the mean free path of the fluid. (*a*) Sketch the expected velocity profile. (*b*) Find an expression for the shear stress at the wall. (*c*) Find the volume flow through the tube.

Word Problems

W4.1 The total acceleration of a fluid particle is given by Eq. (4.2) in the Eulerian[?] system, where **V** is a known function of space and time. Explain how we might evaluate particle acceleration in the Lagrangian[?] frame, where particle position **r** is a known function of time and initial position, $\mathbf{r} = \text{fcn}(\mathbf{r}_0, t)$. Can you give an illustrative example?

W4.2 Is it true that the continuity relation, Eq. (4.6), is valid for both viscous and inviscid, newtonian and nonnewtonian, compressible and incompressible flow? If so, are there *any* limitations on this equation?

W4.3 Consider a CD (compact disc) rotating at angular velocity Ω. Does it have *vorticity* in the sense of this chapter? If so, how much vorticity?

W4.4 How much acceleration can fluids endure? Are fluids like astronauts, who feel that 5*g* is severe? Perhaps use the flow pattern of Example 4.8, at $r = R$, to make some estimates of fluid acceleration magnitudes.

W4.5 State the conditions (there are more than one) under which the analysis of temperature distribution in a flow field can be completely uncoupled, so that a separate analysis for velocity and pressure is possible. Can we do this for both laminar and turbulent flow?

W4.6 Consider liquid flow over a dam or weir. How might the boundary conditions and the flow pattern change when we compare water flow over a large prototype to SAE 30 oil flow over a tiny scale model?

W4.7 What is the difference between the stream function ψ and our method of finding the streamlines from Sec. 1.11? Or are they essentially the same?

W4.8 Under what conditions do both the stream function ψ and the velocity potential ϕ exist for a flow field? When does one exist but not the other?

W4.9 How might the remarkable three-dimensional Taylor instability of Fig. 4.14 be predicted? Discuss a general procedure for examining the stability of a given flow pattern.

W4.10 Consider an irrotational, incompressible, axisymmetric ($\partial/\partial\theta = 0$) flow in (r, z) coordinates. Does a stream function exist? If so, does it satisfy Laplace's equation? Are lines of constant ψ equal to the flow streamlines? Does a velocity potential exist? If so, does it satisfy Laplace's equation? Are lines of constant ϕ everywhere perpendicular to the ψ lines?

Fundamentals of Engineering Exam Problems

This chapter is not a favorite of the people who prepare the FE Exam. Probably not a single problem from this chapter will appear on the exam, but if some did, they might be like these.

FE4.1 Given the steady, incompressible velocity distribution $\mathbf{V} = 3x\mathbf{i} + Cy\mathbf{j} + 0\mathbf{k}$, where C is a constant, if conservation of mass is satisfied, the value of C should be
(*a*) 3, (*b*) 3/2, (*c*) 0, (*d*) −3/2, (*e*) −3

FE4.2 Given the steady velocity distribution $\mathbf{V} = 3x\mathbf{i} + 0\mathbf{j} + Cy\mathbf{k}$, where C is a constant, if the flow is irrotational, the value of C should be
(*a*) 3, (*b*) 3/2, (*c*) 0, (*d*) −3/2, (*e*) −3

FE4.3 Given the steady, incompressible velocity distribution $\mathbf{V} = 3x\mathbf{i} + Cy\mathbf{j} + 0\mathbf{k}$, where C is a constant, the shear stress τ_{xy} at the point (x, y, z) is given by
(*a*) 3μ, (*b*) $(3x + Cy)\mu$, (*c*) 0, (*d*) $C\mu$, (*e*) $(3 + C)\mu$

FE4.4 Given the steady, incompressible velocity distribution $u = Ax$, $v = By$, and $w = Cxy$, where (A, B, C) are constants. This flow satisfies the equation of continuity if A equals
(*a*) B, (*b*) $B + C$, (*c*) $B - C$, (*d*) $-B$, (*e*) $-(B + C)$

FE4.5 For the velocity field in Prob. FE4.4, the convective acceleration in the x direction is
(*a*) Ax^2, (*b*) A^2x, (*c*) B^2y, (*d*) By^2, (*e*) Cx^2y

FE4.6 If, for laminar flow in a smooth, straight tube, the tube diameter and length both double, while everything else remains the same, the volume flow rate will increase by a factor of
(*a*) 2, (*b*) 4, (*c*) 8, (*d*) 12, (*e*) 16

Comprehensive Problems

C4.1 In a certain medical application, water at room temperature and pressure flows through a rectangular channel of length $L = 10$ cm, width $s = 1.0$ cm, and gap thickness $b = 0.30$ mm as in Fig. C4.1. The volume flow rate is sinusoidal with amplitude $\hat{Q} = 0.50$ mL/s and frequency $f = 20$ Hz, i.e., $Q = \hat{Q}\sin(2\pi ft)$.

(*a*) Calculate the maximum Reynolds number ($\mathrm{Re} = Vb/\upsilon$) based on maximum average velocity and gap thickness. Channel flow like this remains laminar for Re less than about 2,000. If Re is greater than about 2,000, the flow will be turbulent. Is this flow laminar or turbulent? (*b*) In this problem, the frequency is low enough that at any given time, the flow can be solved as if it were steady at the given flow rate. (This is called a *quasi-steady assumption.*) At any arbitrary instant of time, find an expression for streamwise velocity u as a function of y, μ, dp/dx, and b, where dp/dx is the pressure gradient required to push the flow through the channel at volume flow rate Q. In addition, estimate the maximum magnitude of velocity component u. (*c*) At any instant of time, find a relationship between volume flow rate Q and pressure gradient dp/dx. Your answer should be given as an expression for Q as a function of dp/dx, s, b, and viscosity μ. (*d*) Estimate the wall shear stress, τ_w as a function of $\hat{Q}$, f, μ, b, s, and time (t). (*e*) Finally, for the numbers given in the problem statement, estimate the amplitude of the wall shear stress, $\hat{\tau}_w$, in N/m^2.

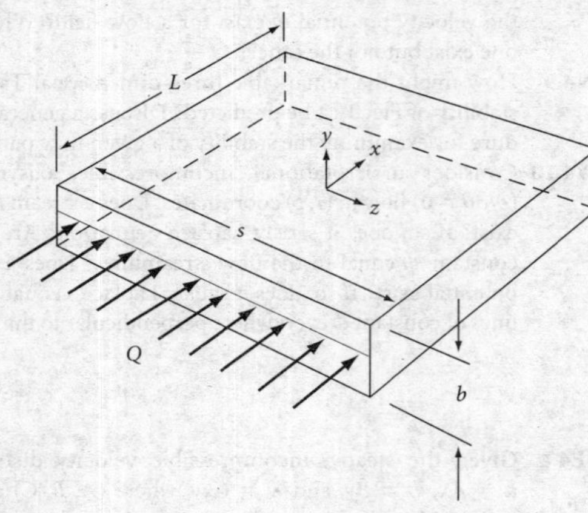

C4.1

C4.2 A belt moves upward at velocity V, dragging a film of viscous liquid of thickness h, as in Fig. C4.2. Near the belt, the film moves upward due to no slip. At its outer edge, the film moves downward due to gravity. Assuming that the only nonzero velocity is $\upsilon(x)$, with zero shear stress at the outer film edge, derive a formula for (*a*) $\upsilon(x)$, (*b*) the average velocity V_{avg} in the film, and (*c*) the velocity V_c for which there is no net flow either up or down. (*d*) Sketch $\upsilon(x)$ for case (*c*).

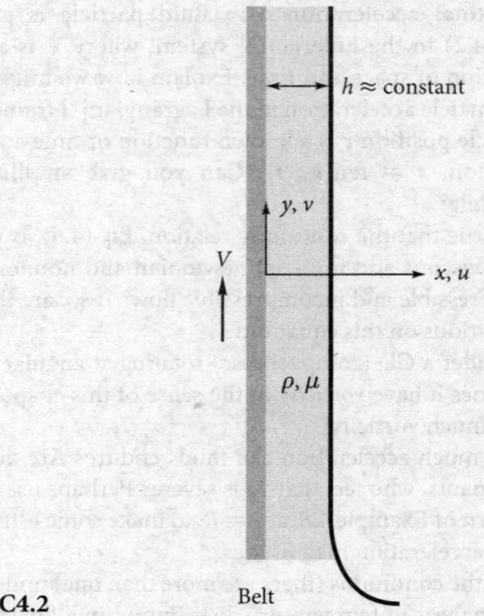

C4.2

References

1. J. D. Anderson, *Computational Fluid Dynamics: An Introduction*, 3d ed., Springer, New York, 2010.
2. C. E. Brennen, *Fundamentals of Multiphase Flow,* Cambridge University Press, New York, 2009. See also URL <http://caltechbook.library.caltech.edu/51/01/multiph.htm>
3. D. Zwillinger, *CRC Standard Mathematical Tables and Formulae,* 32d ed., CRC Press Inc., Cleveland, Ohio, 2011.
4. H. Schlichting and K. Gersten, *Boundary Layer Theory,* 8th ed., Springer, New York, 2000.
5. F. M. White, *Viscous Fluid Flow,* 3d ed., McGraw-Hill, New York, 2005.
6. E. B. Tadmor, R. E. Miller, and R. S. Elliott, *Continuum Mechanics and Thermodynamics,* Cambridge University Press, New York, 2012.
7. J. P. Holman, *Heat Transfer,* 10th ed., McGraw-Hill, New York, 2009.
8. W. M. Kays and M. E. Crawford, *Convective Heat and Mass Transfer,* 4th ed., McGraw-Hill, New York, 2004.
9. G. K. Batchelor, *An Introduction to Fluid Dynamics,* Cambridge University Press, Cambridge, England, 1967.
10. L. Prandtl and O. G. Tietjens, *Fundamentals of Hydro- and Aeromechanics,* Dover, New York, 1957.
11. D. Fleisch, *A Student's Guide to Vectors and Tensors,* Cambridge University Press, New York, 2011.
12. O. Gonzalez and A. M. Stuart, *A First Course in Continuum Mechanics,* Cambridge University Press, New York, 2008.
13. D. A. Danielson, *Vectors and Tensors in Engineering and Physics,* 2d ed., Westview (Perseus) Press, Boulder, CO, 2003.

14. R. I. Tanner, *Engineering Rheology,* 2d ed., Oxford University Press, New York, 2000.
15. H. Lamb, *Hydrodynamics,* 6th ed., Dover, New York, 1945.
16. J. P. Davin, *Tribology for Engineers: A Practical Guide*, Woodhead Publishing, Philadelphia, 2011.
17. G. I. Taylor, "Stability of a Viscous Liquid Contained between Two Rotating Cylinders," *Philos. Trans. Roy. Soc. London Ser. A*, vol. 223, 1923, pp. 289–343.
18. E. L. Koschmieder, "Turbulent Taylor Vortex Flow," *J. Fluid Mech.,* vol. 93, pt. 3, 1979, pp. 515–527.
19. M. T. Nair, T. K. Sengupta, and U. S. Chauhan, "Flow Past Rotating Cylinders at High Reynolds Numbers Using Higher Order Upwind Scheme," *Computers and Fluids,* vol. 27, no. 1, 1998, pp. 47–70.
20. M. Constanceau and C. Menard, "Influence of Rotation on the Near-Wake Development behind an Impulsively Started Circular Cylinder," *J. Fluid Mechanics,* vol. 1258, 1985, pp. 399–446.
21. J-Y. Hwang, K-S. Yang, and K. Bremhorst, "Direct Numerical Simulation of Turbulent Flow Around a Rotating Circular Cylinder." *Fluids Engineering*, vol. 129, Jan. 2007, pp. 40–47.

A full-scale NASA parachute, which helped lower the vehicle *Curiosity* to the Mars surface in 2012, was tested in the world's largest wind tunnel, at NASA Ames Research Center, Moffett Field, California. It is the largest disc-gap-band parachute [41] ever built, with a diameter of 15.5 m. In the Mars atmosphere it will generate up to 289 kN of drag, which leads to a problem assignment in Chapter 7. [*Image from NASA/JPL-Caltech.*]

Chapter 5
Dimensional Analysis and Similarity

Motivation. In this chapter we discuss the planning, presentation, and interpretation of experimental data. We shall try to convince you that such data are best presented in *dimensionless* form. Experiments that might result in tables of output, or even multiple volumes of tables, might be reduced to a single set of curves—or even a single curve—when suitably nondimensionalized. The technique for doing this is *dimensional analysis.* It is also effective in theoretical studies.

Chapter 3 presented large-scale control volume balances of mass, momentum, and energy, which led to global results: mass flow, force, torque, total work done, or heat transfer. Chapter 4 presented infinitesimal balances that led to the basic partial differential equations of fluid flow and some particular solutions for both inviscid and viscous (laminar) flow. These straight *analytical* techniques are limited to simple geometries and uniform boundary conditions. Only a fraction of engineering flow problems can be solved by direct analytical formulas.

Most practical fluid flow problems are too complex, both geometrically and physically, to be solved analytically. They must be tested by experiment or approximated by computational fluid dynamics (CFD) [2]. These results are typically reported as experimental or numerical data points and smoothed curves. Such data have much more generality if they are expressed in compact, economic form. This is the motivation for dimensional analysis. The technique is a mainstay of fluid mechanics and is also widely used in all engineering fields plus the physical, biological, medical, and social sciences. The present chapter shows how dimensional analysis improves the presentation of both data and theory.

5.1 Introduction

Basically, dimensional analysis is a method for reducing the number and complexity of experimental variables that affect a given physical phenomenon, by using a sort of compacting technique. If a phenomenon depends on n dimensional variables, dimensional analysis will reduce the problem to only k *dimensionless* variables, where the reduction $n - k = 1$, 2, 3, or 4, depending on the problem complexity. Generally, $n - k$ equals the number of different dimensions (sometimes called basic or primary or fundamental dimensions) that govern the problem. In fluid mechanics, the four basic dimensions are usually taken to be mass M, length L, time T, and temperature Θ, or an $MLT\Theta$ system for short. Alternatively, one uses an $FLT\Theta$ system, with force F replacing mass.

Although its purpose is to reduce variables and group them in dimensionless form, dimensional analysis has several side benefits. The first is enormous savings in time and money. Suppose one knew that the force F on a particular body shape immersed in a stream of fluid depended only on the body length L, stream velocity V, fluid density ρ, and fluid viscosity μ; that is,

$$F = f(L, V, \rho, \mu) \tag{5.1}$$

Suppose further that the geometry and flow conditions are so complicated that our integral theories (Chap. 3) and differential equations (Chap. 4) fail to yield the solution for the force. Then we must find the function $f(L, V, \rho, \mu)$ experimentally or numerically.

Generally speaking, it takes about 10 points to define a curve. To find the effect of body length in Eq. (5.1), we have to run the experiment for 10 lengths L. For each L we need 10 values of V, 10 values of ρ, and 10 values of μ, making a grand total of 10^4, or 10,000, experiments. At $100 per experiment—well, you see what we are getting into. However, with dimensional analysis, we can immediately reduce Eq. (5.1) to the equivalent form

$$\frac{F}{\rho V^2 L^2} = g\left(\frac{\rho V L}{\mu}\right)$$

or

$$C_F = g(\text{Re}) \tag{5.2}$$

That is, the dimensionless *force coefficient* $F/(\rho V^2 L^2)$ is a function only of the dimensionless *Reynolds number* $\rho VL/\mu$. We shall learn exactly how to make this reduction in Secs. 5.2 and 5.3. Equation (5.2) will be useful in Chap. 7.

Note that Eq. (5.2) is just an *example,* not the full story, of forces caused by fluid flows. Some fluid forces have a very weak or negligible Reynolds number dependence in wide regions (Fig. 5.3*a*). Other groups may also be important. The force coefficient may depend, in high-speed gas flow, on the *Mach number,* $\text{Ma} = V/a$, where a is the speed of sound. In free-surface flows, such as ship drag, C_F may depend upon *Froude number,* $\text{Fr} = V^2/(gL)$, where g is the acceleration of gravity. In turbulent flow, force may depend upon the *roughness ratio,* ϵ/L, where ϵ is the roughness height of the surface.

The function g is different mathematically from the original function f, but it contains all the same information. Nothing is lost in a dimensional analysis. And think of the savings: We can establish g by running the experiment for only 10 values of the single variable called the Reynolds number. We do not have to vary L, V, ρ, or μ separately but only the *grouping* $\rho VL/\mu$. This we do merely by varying velocity V in, say, a wind tunnel or drop test or water channel, and there is no need to build 10 different bodies or find 100 different fluids with 10 densities and 10 viscosities. The cost is now about $1,000, maybe less.

A second side benefit of dimensional analysis is that it helps our thinking and planning for an experiment or theory. It suggests dimensionless ways of writing equations before we spend money on computer analysis to find solutions. It suggests variables that can be discarded; sometimes dimensional analysis will immediately reject variables, and at other times it groups them off to the side, where a few simple tests will show them to be unimportant. Finally, dimensional analysis will often give a great deal of insight into the form of the physical relationship we are trying to study.

A third benefit is that dimensional analysis provides *scaling laws* that can convert data from a cheap, small *model* to design information for an expensive, large *prototype*. We do not build a million-dollar airplane and see whether it has enough lift force. We measure the lift on a small model and use a scaling law to predict the lift on the full-scale prototype airplane. There are rules we shall explain for finding scaling laws. When the scaling law is valid, we say that a condition of *similarity* exists between the model and the prototype. In the simple case of Eq. (5.1), similarity is achieved if the Reynolds number is the same for the model and prototype because the function g then requires the force coefficient to be the same also:

$$\text{If } \mathrm{Re}_m = \mathrm{Re}_p \quad \text{then} \quad C_{Fm} = C_{Fp} \tag{5.3}$$

where subscripts m and p mean model and prototype, respectively. From the definition of force coefficient, this means that

$$\frac{F_p}{F_m} = \frac{\rho_p}{\rho_m}\left(\frac{V_p}{V_m}\right)^2\left(\frac{L_p}{L_m}\right)^2 \tag{5.4}$$

for data taken where $\rho_p V_p L_p/\mu_p = \rho_m V_m L_m/\mu_m$. Equation (5.4) is a scaling law: If you measure the model force at the model Reynolds number, the prototype force at the same Reynolds number equals the model force times the density ratio times the velocity ratio squared times the length ratio squared. We shall give more examples later.

Do you understand these introductory explanations? Be careful; learning dimensional analysis is like learning to play tennis: There are levels of the game. We can establish some ground rules and do some fairly good work in this brief chapter, but dimensional analysis in the broad view has many subtleties and nuances that only time, practice, and maturity enable you to master. Although dimensional analysis has a firm physical and mathematical foundation, considerable art and skill are needed to use it effectively.

EXAMPLE 5.1

A copepod is a water crustacean approximately 1 mm in diameter. We want to know the drag force on the copepod when it moves slowly in fresh water. A scale model 100 times larger is made and tested in glycerin at $V = 30$ cm/s. The measured drag on the model is 1.3 N. For similar conditions, what are the velocity and drag of the actual copepod in water? Assume that Eq. (5.2) applies and the temperature is 20°C.

Solution

- *Property values:* From Table A.3, the densities and viscosities at 20°C are

Water (prototype):	$\mu_p = 0.001$ kg/(m-s)	$\rho_p = 998$ kg/m^3
Glycerin (model):	$\mu_m = 1.5$ kg/(m-s)	$\rho_m = 1{,}263$ kg/m^3

- *Assumptions:* Equation (5.2) is appropriate and *similarity* is achieved; that is, the model and prototype have the same Reynolds number and, therefore, the same force coefficient.
- *Approach:* The length scales are $L_m = 100$ mm and $L_p = 1$ mm. Calculate the Reynolds number and force coefficient of the model and set them equal to prototype values:

$$\mathrm{Re}_m = \frac{\rho_m V_m L_m}{\mu_m} = \frac{(1{,}263\ \mathrm{kg/m^3})(0.3\ \mathrm{m/s})(0.1\ \mathrm{m})}{1.5\ \mathrm{kg/(m\text{-}s)}} = 25.3 = \mathrm{Re}_p = \frac{(998\ \mathrm{kg/m^3})V_p(0.001\ \mathrm{m})}{0.001\ \mathrm{kg/(m\text{-}s)}}$$

$$\text{Solve for } V_p = 0.0253\ \mathrm{m/s} = 2.53\ \mathrm{cm/s} \qquad \textit{Ans.}$$

In like manner, using the prototype velocity just found, equate the force coefficients:

$$C_{Fm} = \frac{F_m}{\rho_m V_m^2 L_m^2} = \frac{1.3\ \text{N}}{(1{,}263\ \text{kg/m}^3)(0.3\ \text{m/s})^2(0.1\ \text{m})^2} = 1.14$$

$$= C_{Fp} = \frac{F_p}{(998\ \text{kg/m}^3)(0.0253\ \text{m/s})^2(0.001\ \text{m})^2}$$

Solve for $F_p = 7.3\text{E-}7\ \text{N}$ *Ans.*

- *Comments:* Assuming we modeled the Reynolds number correctly, the model test is a very good idea, as it would obviously be difficult to measure such a tiny copepod drag force.

Historically, the first person to write extensively about units and dimensional reasoning in physical relations was Euler in 1765. Euler's ideas were far ahead of his time, as were those of Joseph Fourier, whose 1822 book *Analytical Theory of Heat* outlined what is now called the *principle of dimensional homogeneity* and even developed some similarity rules for heat flow. There were no further significant advances until Lord Rayleigh's book in 1877, *Theory of Sound,* which proposed a "method of dimensions" and gave several examples of dimensional analysis. The final breakthrough that established the method as we know it today is generally credited to E. Buckingham in 1914 [1], whose paper outlined what is now called the *Buckingham Pi Theorem* for describing dimensionless parameters (see Sec. 5.3). However, it is now known that a Frenchman, A. Vaschy, in 1892 and a Russian, D. Riabouchinsky, in 1911 had independently published papers reporting results equivalent to the pi theorem. Following Buckingham's paper, P. W. Bridgman published a classic book in 1922 [3], outlining the general theory of dimensional analysis.

Dimensional analysis is so valuable and subtle, with both skill and art involved, that it has spawned a wide variety of textbooks and treatises. The writer is aware of more than 30 books on the subject, of which his engineering favorites are listed here [3–10]. Dimensional analysis is not confined to fluid mechanics, or even to engineering. Specialized books have been published on the application of dimensional analysis to metrology [11], astrophysics [12], economics [13], chemistry [14], hydrology [15], medications [16], clinical medicine [17], chemical processing pilot plants [18], social sciences [19], biomedical sciences [20], pharmacy [21], fractal geometry [22], and even the growth of plants [23]. Clearly this is a subject well worth learning for many career paths.

5.2 The Principle of Dimensional Homogeneity

In making the remarkable jump from the five-variable Eq. (5.1) to the two-variable Eq. (5.2), we were exploiting a rule that is almost a self-evident axiom in physics. This rule, the *principle of dimensional homogeneity* (PDH), can be stated as follows:

> If an equation truly expresses a proper relationship between variables in a physical process, it will be *dimensionally homogeneous;* that is, each of its additive terms will have the same dimensions.

All the equations that are derived from the theory of mechanics are of this form. For example, consider the relation that expresses the displacement of a falling body:

$$S = S_0 + V_0 t + \tfrac{1}{2} g t^2 \qquad (5.5)$$

Each term in this equation is a displacement, or length, and has dimensions $\{L\}$. The equation is dimensionally homogeneous. Note also that any consistent set of units can be used to calculate a result.

Consider Bernoulli's equation for incompressible flow:

$$\frac{p}{\rho} + \frac{1}{2}V^2 + gz = \text{const} \tag{5.6}$$

Each term, including the constant, has dimensions of velocity squared, or $\{L^2T^{-2}\}$. The equation is dimensionally homogeneous and gives proper results for any consistent set of units.

Students count on dimensional homogeneity and use it to check themselves when they cannot quite remember an equation during an exam. For example, which is it:

$$S = \tfrac{1}{2}gt^2? \quad \text{or} \quad S = \tfrac{1}{2}g^2t? \tag{5.7}$$

By checking the dimensions, we reject the second form and back up our faulty memory. We are exploiting the principle of dimensional homogeneity, and this chapter simply exploits it further.

Variables and Constants

Equations (5.5) and (5.6) also illustrate some other factors that often enter into a dimensional analysis:

Dimensional variables are the quantities that actually vary during a given case and would be plotted against each other to show the data. In Eq. (5.5), they are S and t; in Eq. (5.6) they are p, V, and z. All have dimensions, and all can be nondimensionalized as a dimensional analysis technique.

Dimensional constants may vary from case to case but are held constant during a given run. In Eq. (5.5) they are S_0, V_0, and g, and in Eq. (5.6) they are ρ, g, and C. They all have dimensions and conceivably could be nondimensionalized, but they are normally used to help nondimensionalize the variables in the problem.

Pure constants have no dimensions and never did. They arise from mathematical manipulations. In both Eqs. (5.5) and (5.6) they are $\frac{1}{2}$ and the exponent 2, both of which came from an integration: $\int t\,dt = \frac{1}{2}t^2$, $\int V\,dV = \frac{1}{2}V^2$. Other common dimensionless constants are π and e. Also, the argument of any mathematical function, such as ln, exp, cos, or J_0, is dimensionless.

Angles and *revolutions* are dimensionless. The preferred unit for an angle is the radian, which makes it clear that an angle is a ratio. In like manner, a revolution is 2π radians.

Counting numbers are dimensionless. For example, if we triple the energy E to $3E$, the coefficient 3 is dimensionless.

Note that integration and differentiation of an equation may change the dimensions but not the homogeneity of the equation. For example, integrate or differentiate Eq. (5.5):

$$\int S\,dt = S_0t + \tfrac{1}{2}V_0t^2 + \tfrac{1}{6}gt^3 \tag{5.8a}$$

$$\frac{dS}{dt} = V_0 + gt \tag{5.8b}$$

In the integrated form (5.8*a*) every term has dimensions of $\{LT\}$, while in the derivative form (5.8*b*) every term is a velocity $\{LT^{-1}\}$.

Finally, some physical variables are naturally dimensionless by virtue of their definition as ratios of dimensional quantities. Some examples are strain (change in length per unit length), Poisson's ratio (ratio of transverse strain to longitudinal strain), and specific gravity (ratio of density to standard water density).

The motive behind dimensional analysis is that any dimensionally homogeneous equation can be written in an entirely equivalent nondimensional form that is more compact. Usually, there are multiple methods of presenting one's dimensionless data or theory. Let us illustrate these concepts more thoroughly by using the falling-body relation (5.5) as an example.

Ambiguity: The Choice of Variables and Scaling Parameters[1]

Equation (5.5) is familiar and simple, yet it illustrates most of the concepts of dimensional analysis. It contains five terms (S, S_0, V_0, t, g), which we may divide, in our thinking, into variables and parameters. The *variables* are the things we wish to plot, the basic output of the experiment or theory: in this case, S versus t. The *parameters* are those quantities whose effect on the variables we wish to know: in this case S_0, V_0, and g. Almost any engineering study can be subdivided in this manner.

To nondimensionalize our results, we need to know how many dimensions are contained among our variables and parameters: in this case, only two, length $\{L\}$ and time $\{T\}$. Check each term to verify this:

$$\{S\} = \{S_0\} = \{L\} \qquad \{t\} = \{T\} \qquad \{V_0\} = \{LT^{-1}\} \qquad \{g\} = \{LT^{-2}\}$$

Among our parameters, we therefore select two to be *scaling parameters* (also called *repeating variables*), used to define dimensionless variables. What remains will be the "basic" parameter(s) whose effect we wish to show in our plot. These choices will not affect the content of our data, only the form of their presentation. Clearly there is ambiguity in these choices, something that often vexes the beginning experimenter. But, the ambiguity is deliberate. Its purpose is to show a particular effect, and the choice is yours to make.

For the falling-body problem, we select any two of the three parameters to be scaling parameters. Thus, we have three options. Let us discuss and display them in turn.

Option 1: Scaling parameters S_0 and V_0: the effect of gravity g.

First, use the scaling parameters (S_0, V_0) to define dimensionless (*) displacement and time. There is only one suitable definition for each:[2]

$$S^* = \frac{S}{S_0} \qquad t^* = \frac{V_0 t}{S_0} \tag{5.9}$$

Substitute these variables into Eq. (5.5) and clean everything up until each term is dimensionless. The result is our first option:

$$S^* = 1 + t^* + \frac{1}{2}\alpha t^{*2} \qquad \alpha = \frac{gS_0}{V_0^2} \tag{5.10}$$

This result is shown plotted in Fig. 5.1*a*. There is a single dimensionless parameter α, which shows here the effect of gravity. It cannot show the direct effects of S_0 and V_0, since these two are hidden in the ordinate and abscissa. We see that gravity increases the parabolic rate of fall for $t^* > 0$, but not the initial slope at $t^* = 0$. We would learn

[1] I am indebted to Prof. Jacques Lewalle of Syracuse University for suggesting, outlining, and clarifying this entire discussion.

[2] Make them *proportional* to S and t. Do not define dimensionless terms upside down: S_0/S or $S_0/(V_0 t)$. The plots will look funny, users of your data will be confused, and your supervisor will be angry. It is not a good idea.

the same from falling-body data, and the plot, within experimental accuracy, would look like Fig. 5.1*a*.

Option 2: Scaling parameters V_0 and g: the effect of initial displacement S_0.

Now use the new scaling parameters (V_0, g) to define dimensionless (**) displacement and time. Again there is only one suitable definition:

$$S^{**} = \frac{Sg}{V_0^2} \qquad t^{**} = t\frac{g}{V_0} \tag{5.11}$$

Substitute these variables into Eq. (5.5) and clean everything up again. The result is our second option:

$$S^{**} = \alpha + t^{**} + \frac{1}{2}t^{**2} \qquad \alpha = \frac{gS_0}{V_0^2} \tag{5.12}$$

This result is plotted in Fig. 5.1*b*. The same single parameter α again appears and here shows the effect of initial *displacement,* which merely moves the curves upward without changing their shape.

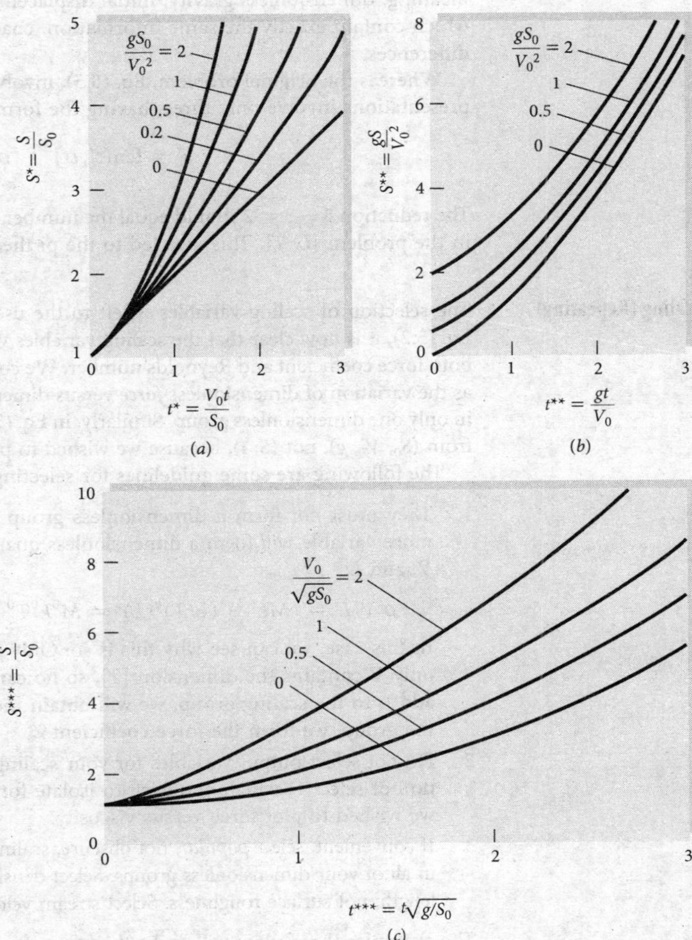

Fig. 5.1 Three entirely equivalent dimensionless presentations of the falling-body problem, Eq. (5.5): the effect of (*a*) gravity, (*b*) initial displacement, and (*c*) initial velocity. All plots contain the same information.

Option 3: Scaling parameters S_0 and g: the effect of initial speed V_0.

Finally, use the scaling parameters (S_0, g) to define dimensionless (***) displacement and time. Again, there is only one suitable definition:

$$S^{***} = \frac{S}{S_0} \qquad t^{***} = t\left(\frac{g}{S_0}\right)^{1/2} \tag{5.13}$$

Substitute these variables into Eq. (5.5) and clean everything up as usual. The result is our third and final option:

$$S^{***} = 1 + \beta t^{***} + \frac{1}{2}t^{***2} \qquad \beta = \frac{1}{\sqrt{\alpha}} = \frac{V_0}{\sqrt{gS_0}} \tag{5.14}$$

This final presentation is shown in Fig. 5.1*c*. Once again the parameter α appears, but we have redefined it upside down, $\beta = 1/\sqrt{\alpha}$, so that our display parameter V_0 is in the numerator and is linear. This is our free choice and simply improves the display. Figure 5.1*c* shows that initial *velocity* increases the falling displacement.

Note that, in all three options, the same parameter α appears but has a different meaning: dimensionless gravity, initial displacement, and initial velocity. The graphs, which contain exactly the same information, change their appearance to reflect these differences.

Whereas the original problem, Eq. (5.5), involved five quantities, the dimensionless presentations involve only three, having the form

$$S' = \text{fcn}(t', \alpha) \qquad \alpha = \frac{gS_0}{V_0^2} \tag{5.15}$$

The reduction 5 − 3 = 2 should equal the number of fundamental dimensions involved in the problem {L, T}. This idea led to the pi theorem (Sec. 5.3).

Selection of Scaling (Repeating) Variables

The selection of scaling variables is left to the user, but there are some guidelines. In Eq. (5.2), it is now clear that the scaling variables were ρ, V, and L, since they appear in both force coefficient and Reynolds number. We could then interpret data from Eq. (5.2) as the variation of dimensionless *force* versus dimensionless *viscosity,* since each appears in only one dimensionless group. Similarly, in Eq. (5.5) the scaling variables were selected from (S_0, V_0, g), not (S, t), because we wished to plot S versus t in the final result.

The following are some guidelines for selecting scaling variables:

1. They must *not* form a dimensionless group among themselves, but adding one more variable *will* form a dimensionless quantity. For example, test powers of ρ, V, and L:

$$\rho^a V^b L^c = (ML^{-3})^a (L/T)^b (L)^c = M^0 L^0 T^0 \text{ only if } a = 0, b = 0, c = 0$$

   In this case, we can see why this is so: Only ρ contains the dimension {M}, and only V contains the dimension {T}, so no cancellation is possible. If, now, we add μ to the scaling group, we will obtain the Reynolds number. If we add F to the group, we form the force coefficient.

2. Do not select output variables for your scaling parameters. In Eq. (5.1), certainly do not select F, which you wish to isolate for your plot. Nor was μ selected, for we wished to plot force versus viscosity.
3. If convenient, select *popular,* not obscure, scaling variables because they will appear in all of your dimensionless groups. Select density, not surface tension. Select body length, not surface roughness. Select stream velocity, not speed of sound.

The examples that follow will make this clear. Problem assignments might give hints.

Suppose we wish to study drag force versus *velocity*. Then we would not use V as a scaling parameter in Eq. (5.1). We would use (ρ, μ, L) instead, and the final dimensionless function would become

$$C'_F = \frac{\rho F}{\mu^2} = f(\text{Re}) \qquad \text{Re} = \frac{\rho V L}{\mu} \tag{5.16}$$

In plotting these data, we would not be able to discern the effect of ρ or μ, since they appear in both dimensionless groups. The grouping C'_F again would mean dimensionless force, and Re is now interpreted as either dimensionless velocity or size.[3] The plot would be quite different compared to Eq. (5.2), although it contains exactly the same information. The development of parameters such as C'_F and Re from the initial variables is the subject of the pi theorem (Sec. 5.3).

Some Peculiar Engineering Equations

The foundation of the dimensional analysis method rests on two assumptions: (1) The proposed physical relation is dimensionally homogeneous, and (2) all the relevant variables have been included in the proposed relation.

If a relevant variable is missing, dimensional analysis will fail, giving either algebraic difficulties or, worse, yielding a dimensionless formulation that does not resolve the process. A typical case is Manning's open-channel formula, discussed in Example 1.4 and Chap. 10.

$$V = \frac{1.49}{n} R^{2/3} S^{1/2} \tag{1}$$

Since V is velocity, R is a radius, and n and S are dimensionless, the formula is not dimensionally homogeneous. This should be a warning that (1) the formula changes if the *units* of V and R change and (2) if valid, it represents a very special case. Equation (1) in Example 1.4 predates the dimensional analysis technique and is valid only for water in rough channels at moderate velocities and large radii in BG units.

Such dimensionally inhomogeneous formulas abound in the hydraulics literature. Another example is the Hazen-Williams formula [24] for volume flow of water through a straight smooth pipe:

$$Q = 61.9 D^{2.63} \left(\frac{dp}{dx} \right)^{0.54} \tag{5.17}$$

where D is diameter and dp/dx is the pressure gradient. Some of these formulas arise because numbers have been inserted for fluid properties and other physical data into perfectly legitimate homogeneous formulas. We shall not give the units of Eq. (5.17) to avoid encouraging its use.

On the other hand, some formulas are "constructs" that cannot be made dimensionally homogeneous. The "variables" they relate cannot be analyzed by the dimensional analysis technique. Most of these formulas are raw empiricisms convenient to a small group of specialists. Here are three examples:

$$B = \frac{25{,}000}{100 - R} \tag{5.18}$$

$$S = \frac{140}{130 + \text{API}} \tag{5.19}$$

$$0.0147 D_E - \frac{3.74}{D_E} = 0.26 t_R - \frac{172}{t_R} \tag{5.20}$$

[3]We were lucky to achieve a size effect because in this case L, a scaling parameter, did not appear in the drag coefficient.

Equation (5.18) relates the Brinell hardness B of a metal to its Rockwell hardness R. Equation (5.19) relates the specific gravity S of an oil to its density in degrees API. Equation (5.20) relates the viscosity of a liquid in D_E, or degrees Engler, to its viscosity t_R in Saybolt seconds. Such formulas have a certain usefulness when communicated between fellow specialists, but we cannot handle them here. Variables like Brinell hardness and Saybolt viscosity are not suited to an $MLT\Theta$ dimensional system.

5.3 The Pi Theorem

There are several methods of reducing a number of dimensional variables into a smaller number of dimensionless groups. The first scheme given here was proposed in 1914 by Buckingham [1] and is now called the *Buckingham Pi Theorem.* The name *pi* comes from the mathematical notation Π, meaning a product of variables. The dimensionless groups found from the theorem are power products denoted by Π_1, Π_2, Π_3, etc. The method allows the pi groups to be found in sequential order without resorting to free exponents.

The first part of the pi theorem explains what reduction in variables to expect:

> If a physical process satisfies the PDH and involves n dimensional variables, it can be reduced to a relation between k dimensionless variables or Πs. The reduction $j = n - k$ equals the maximum number of variables that do not form a pi among themselves and is always less than or equal to the number of dimensions describing the variables.

Take the specific case of force on an immersed body: Eq. (5.1) contains five variables F, L, U, ρ, and μ described by three dimensions $\{MLT\}$. Thus $n = 5$ and $j \leq 3$. Therefore it is a good guess that we can reduce the problem to k pi groups, with $k = n - j \geq 5 - 3 = 2$. And this is exactly what we obtained: two dimensionless variables $\Pi_1 = C_F$ and $\Pi_2 = \text{Re}$. On rare occasions it may take more pi groups than this minimum (see Example 5.5).

The second part of the theorem shows how to find the pi groups one at a time:

> Find the reduction j, then select j scaling variables that do not form a pi among themselves.[4] Each desired pi group will be a power product of these j variables plus one additional variable, which is assigned any convenient nonzero exponent. Each pi group thus found is independent.

To be specific, suppose the process involves five variables:

$$v_1 = f(v_2, v_3, v_4, v_5)$$

Suppose there are three dimensions $\{MLT\}$ and we search around and find that indeed $j = 3$. Then, $k = 5 - 3 = 2$ and we expect, from the theorem, two and only two pi groups. Pick out three convenient variables that do *not* form a pi, and suppose these turn out to be v_2, v_3, and v_4. Then, the two pi groups are formed by power products of these three plus one additional variable, either v_1 or v_5:

$$\Pi_1 = (v_2)^a(v_3)^b(v_4)^c v_1 = M^0L^0T^0 \quad \Pi_2 = (v_2)^a(v_3)^b(v_4)^c v_5 = M^0L^0T^0$$

Here, we have arbitrarily chosen v_1 and v_5, the added variables, to have unit exponents. Equating exponents of the various dimensions is guaranteed by the theorem to give unique values of a, b, and c for each pi. And they are independent because only

[4]Make a clever choice here because all pi groups will contain these j variables in various groupings.

Π_1 contains υ_1 and only Π_2 contains υ_5. It is a very neat system once you get used to the procedure. We shall illustrate it with several examples.

Typically, six steps are involved:

1. List and count the n variables involved in the problem. If any important variables are missing, dimensional analysis will fail.
2. List the dimensions of each variable according to $\{MLT\Theta\}$ or $\{FLT\Theta\}$. A list is given in Table 5.1.
3. Find j. Initially guess j equal to the number of different dimensions present, and look for j variables that do not form a pi product. If no luck, reduce j by 1 and look again. With practice, you will find j rapidly.
4. Select j scaling parameters that do not form a pi product. Make sure they please you and have some generality if possible, because they will then appear in every one of your pi groups. Pick density or velocity or length. Do not pick surface tension, for example, or you will form six different independent Weber-number parameters and thoroughly annoy your colleagues.
5. Add one additional variable to your j repeating variables, and form a power product. Algebraically, find the exponents that make the product dimensionless. Try to arrange for your output or *dependent* variables (force, pressure drop, torque, power) to appear in the numerator, and your plots will look better. Do this sequentially, adding one new variable each time, and you will find all $n - j = k$ desired pi products.
6. Write the final dimensionless function, and check the terms to make sure all pi groups are dimensionless.

Table 5.1 Dimensions of Fluid-Mechanics Properties

Quantity	Symbol	Dimensions	
		***MLT*Θ**	***FLT*Θ**
Length	L	L	L
Area	A	L^2	L^2
Volume	$\mathcal{V}$	L^3	L^3
Velocity	V	LT^{-1}	LT^{-1}
Acceleration	dV/dt	LT^{-2}	LT^{-2}
Speed of sound	a	LT^{-1}	LT^{-1}
Volume flow	Q	L^3T^{-1}	L^3T^{-1}
Mass flow	$\dot{m}$	MT^{-1}	FTL^{-1}
Pressure, stress	p, σ, τ	$ML^{-1}T^{-2}$	FL^{-2}
Strain rate	$\dot{\varepsilon}$	T^{-1}	T^{-1}
Angle	θ	None	None
Angular velocity	ω, Ω	T^{-1}	T^{-1}
Viscosity	μ	$ML^{-1}T^{-1}$	FTL^{-2}
Kinematic viscosity	ν	L^2T^{-1}	L^2T^{-1}
Surface tension	Υ	MT^{-2}	FL^{-1}
Force	F	MLT^{-2}	F
Moment, torque	M	ML^2T^{-2}	FL
Power	P	ML^2T^{-3}	FLT^{-1}
Work, energy	W, E	ML^2T^{-2}	FL
Density	ρ	ML^{-3}	FT^2L^{-4}
Temperature	T	Θ	Θ
Specific heat	c_p, c_v	$L^2T^{-2}\Theta^{-1}$	$L^2T^{-2}\Theta^{-1}$
Specific weight	γ	$ML^{-2}T^{-2}$	FL^{-3}
Thermal conductivity	k	$MLT^{-3}\Theta^{-1}$	$FT^{-1}\Theta^{-1}$
Thermal expansion coefficient	β	Θ^{-1}	Θ^{-1}

EXAMPLE 5.2

Repeat the development of Eq. (5.2) from Eq. (5.1), using the pi theorem.

Solution

Step 1 Write the function and count variables:

$$F = f(L, U, \rho, \mu) \quad \text{there are five variables } (n = 5)$$

Step 2 List dimensions of each variable. From Table 5.1

F	L	U	ρ	μ
$\{MLT^{-2}\}$	$\{L\}$	$\{LT^{-1}\}$	$\{ML^{-3}\}$	$\{ML^{-1}T^{-1}\}$

Step 3 Find j. No variable contains the dimension Θ, and so j is less than or equal to 3 (MLT). We inspect the list and see that L, U, and ρ cannot form a pi group because only ρ contains mass and only U contains time. Therefore j does equal 3, and $n - j = 5 - 3 = 2 = k$. The pi theorem guarantees for this problem that there will be exactly two independent dimensionless groups.

Step 4 Select repeating j variables. The group L, U, ρ we found in step 3 will do fine.

Step 5 Combine L, U, ρ with one additional variable, in sequence, to find the two pi products.

First add force to find Π_1. You may select *any* exponent on this additional term as you please, to place it in the numerator or denominator to any power. Since F is the output, or dependent, variable, we select it to appear to the first power in the numerator:

$$\Pi_1 = L^a U^b \rho^c F = (L)^a (LT^{-1})^b (ML^{-3})^c (MLT^{-2}) = M^0 L^0 T^0$$

Equate exponents:

Length:	$a + b - 3c + 1 = 0$
Mass:	$c + 1 = 0$
Time:	$-b \quad -2 = 0$

We can solve explicitly for

$$a = -2 \qquad b = -2 \qquad c = -1$$

Therefore

$$\Pi_1 = L^{-2} U^{-2} \rho^{-1} F = \frac{F}{\rho U^2 L^2} = C_F \qquad \textit{Ans.}$$

This is exactly the right pi group as in Eq. (5.2). By varying the exponent on F, we could have found other equivalent groups such as $UL\rho^{1/2}/F^{1/2}$.

Finally, add viscosity to L, U, and ρ to find Π_2. Select any power you like for viscosity. By hindsight and custom, we select the power -1 to place it in the denominator:

$$\Pi_2 = L^a U^b \rho^c \mu^{-1} = L^a (LT^{-1})^b (ML^{-3})^c (ML^{-1}T^{-1})^{-1} = M^0 L^0 T^0$$

Equate exponents:

Length:	$a + b - 3c + 1 = 0$
Mass:	$c - 1 = 0$
Time:	$-b \quad +1 = 0$

from which we find

$$a = b = c = 1$$

Therefore $$\Pi_2 = L^1U^1\rho^1\mu^{-1} = \frac{\rho UL}{\mu} = \text{Re}$$ *Ans.*

Step 6 We know we are finished; this is the second and last pi group. The theorem guarantees that the functional relationship must be of the equivalent form

$$\frac{F}{\rho U^2L^2} = g\left(\frac{\rho UL}{\mu}\right)$$ *Ans.*

which is exactly Eq. (5.2).

EXAMPLE 5.3

The power input P to a centrifugal pump is a function of the volume flow Q, impeller diameter D, rotational rate Ω, and the density ρ and viscosity μ of the fluid:

$$P = f(Q, D, \Omega, \rho, \mu)$$

Rewrite this as a dimensionless relationship. *Hint:* Use Ω, ρ, and D as repeating variables. We will revisit this problem in Chap. 11.

Solution

Step 1 Count the variables. There are six (don't forget the one on the left, P).

Step 2 List the dimensions of each variable from Table 5.1. Use the $\{FLT\Theta\}$ system:

P	Q	D	Ω	ρ	μ
$\{FLT^{-1}\}$	$\{L^3T^{-1}\}$	$\{L\}$	$\{T^{-1}\}$	$\{FT^2L^{-4}\}$	$\{FTL^{-2}\}$

Step 3 Find j. Lucky us, we were told to use (Ω, ρ, D) as repeating variables, so surely $j = 3$, the number of dimensions (FLT)? Check that these three do *not* form a pi group:

$$\Omega^a\rho^bD^c = (T^{-1})^a(FT^2L^{-4})^b(L)^c = F^0L^0T^0 \text{ only if } \quad a = 0, b = 0, c = 0$$

Yes, $j = 3$. This was not as obvious as the scaling group (L, U, ρ) in Example 5.2, but it is true. We now know, from the theorem, that adding one more variable will indeed form a pi group.

Step 4a Combine (Ω, ρ, D) with power P to find the first pi group:

$$\Pi_1 = \Omega^a\rho^bD^cP = (T^{-1})^a(FT^2L^{-4})^b(L)^c(FLT^{-1}) = F^0L^0T^0$$

Equate exponents:

Force: $$b + 1 = 0$$

Length: $$-4b + c + 1 = 0$$

Time: $$-a + 2b - 1 = 0$$

Solve algebraically to obtain $a = -3$, $b = -1$, and $c = -5$. This first pi group, the output dimensionless variable, is called the *power coefficient* of a pump, C_P:

$$\Pi_1 = \Omega^{-3}\rho^{-1}D^{-5}P = \frac{P}{\rho\Omega^3D^5} = C_P$$

Step 4b Combine (Ω, ρ, D) with flow rate Q to find the second pi group:

$$\Pi_2 = \Omega^a \rho^b D^c Q = (T^{-1})^a (FT^2L^{-4})^b (L)^c (L^3T^{-1}) = F^0L^0T^0$$

After equating exponents, we now find $a = -1$, $b = 0$, and $c = -3$. This second pi group is called the *flow coefficient* of a pump, C_Q:

$$\Pi_2 = \Omega^{-1}\rho^0 D^{-3} Q = \frac{Q}{\Omega D^3} = C_Q$$

Step 4c Combine (Ω, ρ, D) with viscosity μ to find the third and last pi group:

$$\Pi_3 = \Omega^a \rho^b D^c \mu = (T^{-1})^a (FT^2L^{-4})^b (L)^c (FTL^{-2}) = F^0L^0T^0$$

This time, $a = -1$, $b = -1$, and $c = -2$; or $\Pi_3 = \mu/(\rho\Omega D^2)$, a sort of Reynolds number.

Step 5 The original relation between six variables is now reduced to three dimensionless groups:

$$\frac{P}{\rho\Omega^3D^5} = f\left(\frac{Q}{\Omega D^3}, \frac{\mu}{\rho\Omega D^2}\right) \qquad \textit{Ans.}$$

Comment: These three are the classical coefficients used to correlate pump power in Chap. 11.

EXAMPLE 5.4

At low velocities (laminar flow), the volume flow Q through a small-bore tube is a function only of the tube radius R, the fluid viscosity μ, and the pressure drop per unit tube length dp/dx. Using the pi theorem, find an appropriate dimensionless relationship.

Solution

Write the given relation and count variables:

$$Q = f\left(R, \mu, \frac{dp}{dx}\right) \qquad \text{four variables } (n = 4)$$

Make a list of the dimensions of these variables from Table 5.1 using the $\{MLT\}$ system:

Q	R	μ	dp/dx
$\{L^3T^{-1}\}$	$\{L\}$	$\{ML^{-1}T^{-1}\}$	$\{ML^{-2}T^{-2}\}$

There are three primary dimensions (M, L, T), hence $j \le 3$. By trial and error we determine that R, μ, and dp/dx cannot be combined into a pi group. Then $j = 3$, and $n - j = 4 - 3 = 1$. There is only *one* pi group, which we find by combining Q in a power product with the other three:

$$\Pi_1 = R^a \mu^b \left(\frac{dp}{dx}\right)^c Q^1 = (L)^a (ML^{-1}T^{-1})^b (ML^{-2}T^{-2})^c (L^3T^{-1})$$
$$= M^0L^0T^0$$

Equate exponents:

Mass: $\quad b + c = 0$

Length: $\quad a - b - 2c + 3 = 0$

Time: $\quad -b - 2c - 1 = 0$

Solving simultaneously, we obtain $a = -4$, $b = 1$, and $c = -1$. Then

$$\Pi_1 = R^{-4}\mu^1\left(\frac{dp}{dx}\right)^{-1}Q$$

or

$$\Pi_1 = \frac{Q\mu}{R^4(dp/dx)} = \text{const} \qquad \textit{Ans.}$$

Since there is only one pi group, it must equal a dimensionless constant. This is as far as dimensional analysis can take us. The laminar flow theory of Sec. 4.10 shows that the value of the constant is $-\frac{\pi}{8}$. This result is also useful in Chap. 6.

EXAMPLE 5.5

Assume that the tip deflection δ of a cantilever beam is a function of the tip load P, beam length L, area moment of inertia I, and material modulus of elasticity E; that is, $\delta = f(P, L, I, E)$. Rewrite this function in dimensionless form, and comment on its complexity and the peculiar value of j.

Solution

List the variables and their dimensions:

δ	P	L	I	E
$\{L\}$	$\{MLT^{-2}\}$	$\{L\}$	$\{L^4\}$	$\{ML^{-1}T^{-2}\}$

There are five variables ($n = 5$) and three primary dimensions (M, L, T), hence $j \leq 3$. But try as we may, we *cannot* find any combination of three variables that does not form a pi group. This is because $\{M\}$ and $\{T\}$ occur only in P and E and only in the same form, $\{MT^{-2}\}$. Thus we have encountered a special case of $j = 2$, which is less than the number of dimensions (M, L, T). To gain more insight into this peculiarity, you should rework the problem, using the (F, L, T) system of dimensions. You will find that only $\{F\}$ and $\{L\}$ occur in these variables, hence $j = 2$.

With $j = 2$, we select L and E as two variables that cannot form a pi group and then add other variables to form the three desired pis:

$$\Pi_1 = L^aE^bI^1 = (L)^a(ML^{-1}T^{-2})^b(L^4) = M^0L^0T^0$$

from which, after equating exponents, we find that $a = -4$, $b = 0$, or $\Pi_1 = I/L^4$. Then

$$\Pi_2 = L^aE^bP^1 = (L)^a(ML^{-1}T^{-2})^b(MLT^{-2}) = M^0L^0T^0$$

from which we find $a = -2$, $b = -1$, or $\Pi_2 = P/(EL^2)$, and

$$\Pi_3 = L^aE^b\delta^1 = (L)^a(ML^{-1}T^{-2})^b(L) = M^0L^0T^0$$

from which $a = -1$, $b = 0$, or $\Pi_3 = \delta/L$. The proper dimensionless function is $\Pi_3 = f(\Pi_2, \Pi_1)$, or

$$\frac{\delta}{L} = f\left(\frac{P}{EL^2}, \frac{I}{L^4}\right) \qquad \textit{Ans. (1)}$$

This is a complex three-variable function, but dimensional analysis alone can take us no further.

Comments: We can "improve" Eq. (1) by taking advantage of some physical reasoning, as Langhaar points out [4, p. 91]. For small elastic deflections, δ is proportional to load P and inversely proportional to moment of inertia I. Since P and I occur separately in Eq. (1), this means that Π_3 must be proportional to Π_2 and inversely proportional to Π_1. Thus, for these conditions,

$$\frac{\delta}{L} = (\text{const}) \frac{P}{EL^2} \frac{L^4}{I}$$

or
$$\delta = (\text{const}) \frac{PL^3}{EI} \qquad (2)$$

This could not be predicted by a pure dimensional analysis. Strength-of-materials theory predicts that the value of the constant is $\frac{1}{3}$.

An Alternate Step-by-Step Method by Ipsen (1960)[5]

The pi theorem method, just explained and illustrated, is often called the *repeating variable method* of dimensional analysis. Select the repeating variables, add one more, and you get a pi group. The writer likes it. This method is straightforward and systematically reveals all the desired pi groups. However, there are drawbacks: (1) All pi groups contain the same repeating variables and might lack variety or effectiveness, and (2) one must (sometimes laboriously) check that the selected repeating variables do *not* form a pi group among themselves (see Prob. P5.21).

Ipsen [5] suggests an entirely different procedure, a step-by-step method that obtains all of the pi groups at once, without any counting or checking. One simply successively eliminates each dimension in the desired function by division or multiplication. Let us illustrate with the same classical drag function proposed in Eq. (5.1). Underneath the variables, write out the dimensions of each quantity.

$$\underset{\{MLT^{-2}\}}{F} = \text{fcn}(\underset{\{L\}}{L}, \underset{\{LT^{-1}\}}{V}, \underset{\{ML^{-3}\}}{\rho}, \underset{\{ML^{-1}T^{-1}\}}{\mu}) \qquad (5.1)$$

There are three dimensions, $\{MLT\}$. Eliminate them successively by division or multiplication by a variable. Start with mass $\{M\}$. Pick a variable that contains mass and divide it into all the other variables with mass dimensions. We select ρ, divide, and rewrite the function (5.1):

$$\underset{\{L^4T^{-2}\}}{\frac{F}{\rho}} = \text{fcn}\left(\underset{\{L\}}{L}, \underset{\{LT^{-1}\}}{V}, \not{\rho}, \underset{\{L^2T^{-1}\}}{\frac{\mu}{\rho}}\right) \qquad (5.1a)$$

We did not divide into L or V, which do not contain $\{M\}$. Equation (5.1*a*) at first looks strange, but it contains five distinct variables and the same information as Eq. (5.1).

We see that ρ is no longer important. Thus, *discard* ρ, and now there are only four variables. Next, eliminate time $\{T\}$ by dividing the time-containing variables by suitable powers of, say, V. The result is

$$\underset{\{L^2\}}{\frac{F}{\rho V^2}} = \text{fcn}\left(\underset{\{L\}}{L}, \not{V}, \underset{\{L\}}{\frac{\mu}{\rho V}}\right) \qquad (5.1b)$$

Now, we see that V is no longer relevant. Finally, eliminate $\{L\}$ through division by, say, appropriate powers of L itself:

$$\underset{\{1\}}{\frac{F}{\rho V^2 L^2}} = \text{fcn}\left(\not{L}, \underset{\{1\}}{\frac{\mu}{\rho VL}}\right) \qquad (5.1c)$$

[5]This method may be omitted without loss of continuity.

Now, L by itself is no longer relevant, and so discard it also. The result is equivalent to Eq. (5.2):

$$\frac{F}{\rho V^2 L^2} = \text{fcn}\left(\frac{\mu}{\rho V L}\right) \tag{5.2}$$

In Ipsen's step-by-step method, we find the force coefficient is a function solely of the Reynolds number. We did no counting and did not find j. We just successively eliminated each primary dimension by division with the appropriate variables.

Recall Example 5.5, where we discovered, awkwardly, that the number of repeating variables was *less* than the number of primary dimensions. Ipsen's method avoids this preliminary check. Recall the beam-deflection problem proposed in Example 5.5 and the various dimensions:

$$\begin{matrix} \delta = f(P, & L, & I, & E) \\ \{L\} \quad \{MLT^{-2}\} & \{L\} & \{L^4\} & \{ML^{-1}T^{-2}\} \end{matrix}$$

For the first step, let us eliminate $\{M\}$ by dividing by E. We only have to divide into P:

$$\begin{matrix} \delta = f\left(\frac{P}{E}, \right. & L, & I, & \left. \cancel{E}\right) \\ \{L\} \quad \{L^2\} & \{L\} & \{L^4\} & \end{matrix}$$

We see that we may discard E as no longer relevant, and the dimension $\{T\}$ has vanished along with $\{M\}$. We need only eliminate $\{L\}$ by dividing by, say, powers of L itself:

$$\begin{matrix} \frac{\delta}{L} = \text{fcn}\left(\frac{P}{EL^2}, \cancel{L}, \frac{I}{L^4}\right) \\ \{1\} \qquad \{1\} \qquad \{1\} \end{matrix}$$

Discard L itself as now irrelevant, and we obtain *Answer* (1) to Example 5.5:

$$\frac{\delta}{L} = \text{fcn}\left(\frac{P}{EL^2}, \frac{I}{L^4}\right)$$

Ipsen's approach is again successful. The fact that $\{M\}$ and $\{T\}$ vanished in the same division is proof that there are only *two* repeating variables this time, not the three that would be inferred by the presence of $\{M\}$, $\{L\}$, and $\{T\}$.

EXAMPLE 5.6

The leading-edge aerodynamic moment M_{LE} on a supersonic airfoil is a function of its chord length C, angle of attack α, and several air parameters: approach velocity V, density ρ, speed of sound a, and specific heat ratio k (Fig. E5.6). There is a very weak effect of air viscosity, which is neglected here.

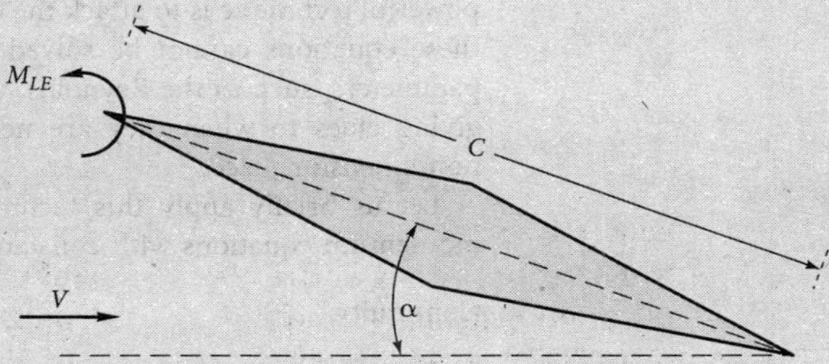

E5.6

Use Ipsen's method to rewrite this function in dimensionless form.

Solution

Write out the given function and list the variables' dimensions $\{MLT\}$ underneath:

$$\begin{array}{ccccccc} M_{LE} & = \text{fcn}(C, & \alpha, & V, & \rho, & a, & k) \\ \{ML^2/T^2\} & \{L\} & \{1\} & \{L/T\} & \{M/L^3\} & \{L/T\} & \{1\} \end{array}$$

Two of them, α and k, are already dimensionless. Leave them alone; they will be pi groups in the final function. You can eliminate any dimension. We choose mass $\{M\}$ and divide by ρ:

$$\begin{array}{ccccccc} \dfrac{M_{LE}}{\rho} & = \text{fcn}(C, & \alpha, & V, & \cancel{\rho,} & a, & k) \\ \{L^5/T^2\} & \{L\} & \{1\} & \{L/T\} & & \{L/T\} & \{1\} \end{array}$$

Recall Ipsen's rules: Only divide into variables containing mass, in this case only M_{LE}, and then discard the divisor, ρ. Now, eliminate time $\{T\}$ by dividing by appropriate powers of a:

$$\begin{array}{cccccc} \dfrac{M_{LE}}{\rho a^2} & = \text{fcn}\Bigg(C, & \alpha, & \dfrac{V}{a}, & \cancel{a,} & k\Bigg) \\ \{L^3\} & \{L\} & \{1\} & \{1\} & & \{1\} \end{array}$$

Finally, eliminate $\{L\}$ on the left side by dividing by C^3:

$$\begin{array}{ccccc} \dfrac{M_{LE}}{\rho a^2 C^3} & = \text{fcn}\Bigg(\cancel{C,} & \alpha, & \dfrac{V}{a}, & k\Bigg) \\ \{1\} & & \{1\} & \{1\} & \{1\} \end{array}$$

We end up with four pi groups and recognize V/a as the Mach number, Ma. In aerodynamics, the dimensionless moment is often called the *moment coefficient*, C_M. Thus, our final result could be written in the compact form

$$C_M = \text{fcn}(\alpha, \text{Ma}, k) \qquad \textit{Ans.}$$

Comments: Our analysis is fine, but experiment and theory and physical reasoning all indicate that M_{LE} varies more strongly with V than with a. Thus, aerodynamicists commonly define the moment coefficient as $C_M = M_{LE}/(\rho V^2 C^3)$ or something similar. We will study the analysis of supersonic forces and moments in Chap. 9.

5.4 Nondimensionalization of the Basic Equations

We could use the pi theorem method of the previous section to analyze problem after problem, finding the dimensionless parameters that govern in each case. Textbooks on dimensional analysis [for example, 5] do this. An alternative and very powerful technique is to attack the basic equations of flow from Chap. 4. Even though these equations cannot be solved in general, they will reveal basic dimensionless parameters, such as the Reynolds number, in their proper form and proper position, giving clues to when they are negligible. The boundary conditions must also be nondimensionalized.

Let us briefly apply this technique to the incompressible flow continuity and momentum equations with constant viscosity:

Continuity:
$$\nabla \cdot \mathbf{V} = 0 \tag{5.21a}$$

Navier-Stokes:
$$\rho \frac{d\mathbf{V}}{dt} = \rho \mathbf{g} - \nabla p + \mu \nabla^2 \mathbf{V} \tag{5.21b}$$

Typical boundary conditions for these two equations are (Sect. 4.6)

Fixed solid surface: $\mathbf{V} = 0$

Inlet or outlet: Known $\mathbf{V}, p$ (5.22)

Free surface, $z = \eta$: $w = \dfrac{d\eta}{dt}$ $\quad p = p_a - \Upsilon(R_x^{-1} + R_y^{-1})$

We omit the energy equation (4.75) and assign its dimensionless form in the problems (Prob. P5.43).

Equations (5.21) and (5.22) contain the three basic dimensions M, L, and T. All variables p, $\mathbf{V}$, x, y, z, and t can be nondimensionalized by using density and two reference constants that might be characteristic of the particular fluid flow:

$$\text{Reference velocity} = U \qquad \text{Reference length} = L$$

For example, U may be the inlet or upstream velocity and L the diameter of a body immersed in the stream.

Now, define all relevant dimensionless variables, denoting them by an asterisk:

$$\mathbf{V}^* = \frac{\mathbf{V}}{U} \qquad \nabla^* = L\nabla$$

$$x^* = \frac{x}{L} \quad y^* = \frac{y}{L} \quad z^* = \frac{z}{L} \quad R^* = \frac{R}{L} \tag{5.23}$$

$$t^* = \frac{tU}{L} \quad p^* = \frac{p + \rho g z}{\rho U^2}$$

All these are fairly obvious except for p^*, where we have introduced the piezometric pressure, assuming that z is up. This is a hindsight idea suggested by Bernoulli's equation (3.54).

Since ρ, U, and L are all constants, the derivatives in Eqs. (5.21) can all be handled in dimensionless form with dimensional coefficients. For example,

$$\frac{\partial u}{\partial x} = \frac{\partial (Uu^*)}{\partial (Lx^*)} = \frac{U}{L}\frac{\partial u^*}{\partial x^*}$$

Substitute the variables from Eqs. (5.23) into Eqs. (5.21) and (5.22) and divide through by the leading dimensional coefficient, in the same way as we handled Eq. (5.12). Here are the resulting dimensionless equations of motion:

Continuity: $$\boxed{\nabla^* \cdot \mathbf{V}^* = 0} \tag{5.24a}$$

Momentum: $$\boxed{\frac{d\mathbf{V}^*}{dt^*} = -\nabla^* p^* + \frac{\mu}{\rho U L}\nabla^{*2}(\mathbf{V}^*)} \tag{5.24b}$$

The dimensionless boundary conditions are:

Fixed solid surface: $\boxed{\mathbf{V}^* = 0}$

Inlet or outlet: $\boxed{\text{Known } \mathbf{V}^*, p^*}$

Free surface, $z^* = \eta^*$:

$$w^* = \frac{d\eta^*}{dt^*}$$

$$p^* = \frac{p_a}{\rho U^2} + \frac{gL}{U^2} z^* - \frac{\Upsilon}{\rho U^2 L}(R_x^{*-1} + R_y^{*-1}) \tag{5.25}$$

These equations reveal a total of four dimensionless parameters, one in the Navier-Stokes equation and three in the free-surface-pressure boundary condition.

Dimensionless Parameters

In the continuity equation, there are no parameters. The Navier-Stokes equation contains one, generally accepted as the most important parameter in fluid mechanics:

$$\text{Reynolds number Re} = \frac{\rho U L}{\mu}$$

It is named after Osborne Reynolds (1842–1912), a British engineer who first proposed it in 1883 (Ref. 4 of Chap. 6). The Reynolds number is always important, with or without a free surface, and can be neglected only in flow regions away from high-velocity gradients—for example, away from solid surfaces, jets, or wakes.

The no-slip and inlet-exit boundary conditions contain no parameters. The free-surface-pressure condition contains three:

$$\text{Euler number (pressure coefficient) Eu} = \frac{p_a}{\rho U^2}$$

This is named after Leonhard Euler (1707–1783) and is rarely important unless the pressure drops low enough to cause vapor formation (cavitation) in a liquid. The Euler number is often written in terms of pressure differences: $\text{Eu} = \Delta p/(\rho U^2)$. If Δp involves vapor pressure p_v, it is called the *cavitation number* $\text{Ca} = (p_a - p_v)/(\rho U^2)$. Cavitation problems are surprisingly common in many water flows.

The second free-surface parameter is much more important:

$$\text{Froude number Fr} = \frac{U^2}{gL}$$

It is named after William Froude (1810–1879), a British naval architect who, with his son Robert, developed the ship-model towing-tank concept and proposed similarity rules for free-surface flows (ship resistance, surface waves, open channels). The Froude number is the dominant effect in free-surface flows. It can also be important in *stratified flows,* where a strong density difference exists without a free surface. For example, see Ref. [42]. Chapter 10 investigates Froude number effects in detail.

The final free-surface parameter is

$$\text{Weber number We} = \frac{\rho U^2 L}{\Upsilon}$$

It is named after Moritz Weber (1871–1951) of the Polytechnic Institute of Berlin, who developed the laws of similitude in their modern form. It was Weber who named Re and Fr after Reynolds and Froude. The Weber number is important only if it is of order unity or less, which typically occurs when the surface curvature is comparable in size to the liquid depth, such as in droplets, capillary flows, ripple waves, and very small hydraulic models. If We is large, its effect may be neglected.

If there is no free surface, Fr, Eu, and We drop out entirely, except for the possibility of cavitation of a liquid at very small Eu. Thus, in low-speed viscous flows with no free surface, the Reynolds number is the only important dimensionless parameter.

Compressibility Parameters

In high-speed flow of a gas, there are significant changes in pressure, density, and temperature that must be related by an equation of state such as the perfect-gas law, Eq. (1.10). These thermodynamic changes introduce two additional dimensionless parameters mentioned briefly in earlier chapters:

$$\text{Mach number Ma} = \frac{U}{a} \qquad \text{Specific-heat ratio } k = \frac{c_p}{c_v}$$

The Mach number is named after Ernst Mach (1838–1916), an Austrian physicist. The effect of k is only slight to moderate, but Ma exerts a strong effect on compressible flow properties if it is greater than about 0.3. These effects are studied in Chap. 9.

Oscillating Flows

If the flow pattern is oscillating, a seventh parameter enters through the inlet boundary condition. For example, suppose that the inlet stream is of the form

$$u = U \cos \omega t$$

Nondimensionalization of this relation results in

$$\frac{u}{U} = u^* = \cos\left(\frac{\omega L}{U} t^*\right)$$

The argument of the cosine contains the new parameter

$$\text{Strouhal number St} = \frac{\omega L}{U}$$

The dimensionless forces and moments, friction, and heat transfer, and so on of such an oscillating flow would be a function of both Reynolds and Strouhal numbers. This parameter is named after V. Strouhal, a German physicist who experimented in 1878 with wires singing in the wind.

Some flows that you might guess to be perfectly steady actually have an oscillatory pattern that is dependent on the Reynolds number. An example is the periodic vortex shedding behind a blunt body immersed in a steady stream of velocity U. Figure 5.2*a* shows an array of alternating vortices shed from a circular cylinder immersed in a steady crossflow. This regular, periodic shedding is called a *Kármán vortex street,* after T. von Kármán, who explained it theoretically in 1912. The shedding occurs in the range $10^2 < \text{Re} < 10^7$, with an average Strouhal number $\omega d/(2\pi U) \approx 0.21$. Figure 5.2*b* shows measured shedding frequencies.

Resonance can occur if a vortex shedding frequency is near a body's structural vibration frequency. Electric transmission wires sing in the wind, undersea mooring lines gallop at certain current speeds, and slender structures flutter at critical wind or vehicle speeds. A striking example is the disastrous failure of the Tacoma Narrows suspension bridge in 1940, when wind-excited vortex shedding caused resonance with the natural torsional oscillations of the bridge. The problem was magnified by the bridge deck nonlinear stiffness, which occurred when the hangers went slack during the oscillation.

Other Dimensionless Parameters

We have discussed seven important parameters in fluid mechanics, and there are others. Four additional parameters arise from nondimensionalization of the energy equation (4.75) and its boundary conditions. These four (Prandtl number, Eckert number, Grashof number, and wall temperature ratio) are listed in Table 5.2 just in case you fail to solve Prob. P5.43. Another important and perhaps surprising parameter is the

(a)

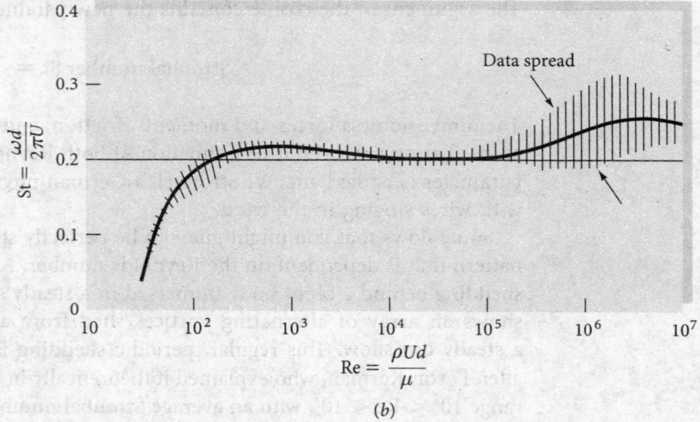

(b)

Fig. 5.2 Vortex shedding from a circular cylinder: (*a*) vortex street behind a circular cylinder *(Courtesy of U.S. Navy)*; (*b*) experimental shedding frequencies *(data from Refs. 25 and 26)*.

wall roughness ratio ε/L (in Table 5.2).[6] Slight changes in surface roughness have a striking effect in the turbulent flow or high-Reynolds-number range, as we shall see in Chap. 6 and in Fig. 5.3.

This book is primarily concerned with Reynolds-, Mach-, and Froude-number effects, which dominate most flows. Note that we discovered these parameters (except ε/L) simply by nondimensionalizing the basic equations without actually solving them.

If the reader is not satiated with the 19 parameters given in Table 5.2, Ref. 29 contains a list of over 1,200 dimensionless parameters in use in engineering and science.

A Successful Application

Dimensional analysis is fun, but does it work? Yes, if all important variables are included in the proposed function, the dimensionless function found by dimensional analysis will collapse all the data onto a single curve or set of curves.

[6]Roughness is easy to overlook because it is a slight geometric effect that does not appear in the equations of motion. It is a boundary condition that one might forget.

Table 5.2 Dimensionless Groups in Fluid Mechanics

Parameter	Definition	Qualitative ratio of effects	Importance
Reynolds number	$\text{Re} = \dfrac{\rho UL}{\mu}$	$\dfrac{\text{Inertia}}{\text{Viscosity}}$	Almost always
Mach number	$\text{Ma} = \dfrac{U}{a}$	$\dfrac{\text{Flow speed}}{\text{Sound speed}}$	Compressible flow
Froude number	$\text{Fr} = \dfrac{U^2}{gL}$	$\dfrac{\text{Inertia}}{\text{Gravity}}$	Free-surface flow
Weber number	$\text{We} = \dfrac{\rho U^2 L}{\Upsilon}$	$\dfrac{\text{Inertia}}{\text{Surface tension}}$	Free-surface flow
Rossby number	$\text{Ro} = \dfrac{U}{\Omega_{\text{earth}} L}$	$\dfrac{\text{Flow velocity}}{\text{Coriolis effect}}$	Geophysical flows
Cavitation number (Euler number)	$\text{Ca} = \dfrac{p - p_v}{\frac{1}{2}\rho U^2}$	$\dfrac{\text{Pressure}}{\text{Inertia}}$	Cavitation
Prandtl number	$\text{Pr} = \dfrac{\mu c_p}{k}$	$\dfrac{\text{Dissipation}}{\text{Conduction}}$	Heat convection
Eckert number	$\text{Ec} = \dfrac{U^2}{c_p T_0}$	$\dfrac{\text{Kinetic energy}}{\text{Enthalpy}}$	Dissipation
Specific-heat ratio	$k = \dfrac{c_p}{c_v}$	$\dfrac{\text{Enthalpy}}{\text{Internal energy}}$	Compressible flow
Strouhal number	$\text{St} = \dfrac{\omega L}{U}$	$\dfrac{\text{Oscillation}}{\text{Mean speed}}$	Oscillating flow
Roughness ratio	$\dfrac{\varepsilon}{L}$	$\dfrac{\text{Wall roughness}}{\text{Body length}}$	Turbulent, rough walls
Grashof number	$\text{Gr} = \dfrac{\beta \Delta T g L^3 \rho^2}{\mu^2}$	$\dfrac{\text{Buoyancy}}{\text{Viscosity}}$	Natural convection
Rayleigh number	$\text{Ra} = \dfrac{\beta \Delta T g L^3 \rho^2 c_p}{\mu k}$	$\dfrac{\text{Buoyancy}}{\text{Viscosity}}$	Natural convection
Temperature ratio	$\dfrac{T_w}{T_0}$	$\dfrac{\text{Wall temperature}}{\text{Stream temperature}}$	Heat transfer
Pressure coefficient	$C_p = \dfrac{p - p_\infty}{\frac{1}{2}\rho U^2}$	$\dfrac{\text{Static pressure}}{\text{Dynamic pressure}}$	Aerodynamics, hydrodynamics
Lift coefficient	$C_L = \dfrac{L}{\frac{1}{2}\rho U^2 A}$	$\dfrac{\text{Lift force}}{\text{Dynamic force}}$	Aerodynamics, hydrodynamics
Drag coefficient	$C_D = \dfrac{D}{\frac{1}{2}\rho U^2 A}$	$\dfrac{\text{Drag force}}{\text{Dynamic force}}$	Aerodynamics, hydrodynamics
Friction factor	$f = \dfrac{h_f}{(V^2/2g)(L/d)}$	$\dfrac{\text{Friction head loss}}{\text{Velocity head}}$	Pipe flow
Skin friction coefficient	$c_f = \dfrac{\tau_{\text{wall}}}{\rho V^2/2}$	$\dfrac{\text{Wall shear stress}}{\text{Dynamic pressure}}$	Boundary layer flow

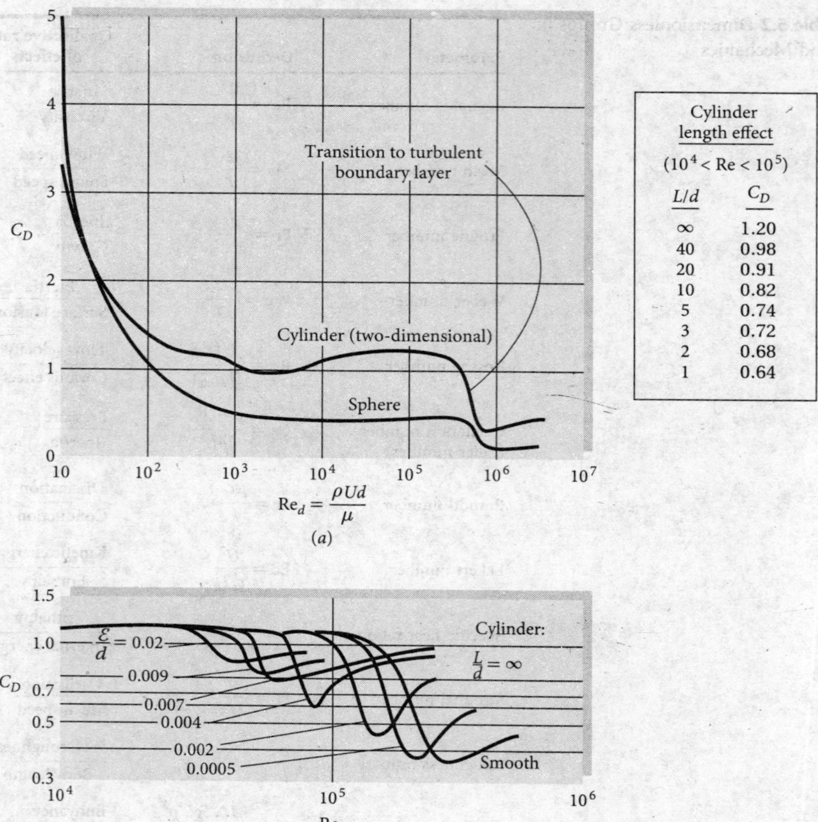

Fig. 5.3 The proof of practical dimensional analysis: drag coefficients of a cylinder and sphere: (*a*) drag coefficient of a smooth cylinder and sphere (data from many sources); (*b*) increased roughness causes earlier transition to a turbulent boundary layer.

An example of the success of dimensional analysis is given in Fig. 5.3 for the measured drag on smooth cylinders and spheres. The flow is normal to the axis of the cylinder, which is extremely long, $L/d \rightarrow \infty$. The data are from many sources, for both liquids and gases, and include bodies from several meters in diameter down to fine wires and balls less than 1 mm in size. Both curves in Fig. 5.3*a* are entirely experimental; the analysis of immersed body drag is one of the weakest areas of modern fluid mechanics theory. Except for digital computer calculations, there is little theory for cylinder and sphere drag except *creeping flow,* Re < 1.

The concept of a fluid-caused *drag force* on bodies is covered extensively in Chap. 7. Drag is the fluid force parallel to the oncoming stream—see Fig. 7.10 for details.

The Reynolds number of both bodies is based on diameter, hence the notation Re_d. But the drag coefficients are defined differently:

$$C_D = \begin{cases} \dfrac{\text{drag}}{\frac{1}{2}\rho U^2 L d} & \text{cylinder} \\[2ex] \dfrac{\text{drag}}{\frac{1}{2}\rho U^2 \frac{1}{4}\pi d^2} & \text{sphere} \end{cases} \tag{5.26}$$

They both have a factor $\frac{1}{2}$ because the term $\frac{1}{2}\rho U^2$ occurs in Bernoulli's equation, and both are based on the projected area—that is, the area one sees when looking toward the body from upstream. The usual definition of C_D is thus

$$C_D = \frac{\text{drag}}{\frac{1}{2}\rho U^2(\text{projected area})} \tag{5.27}$$

However, one should carefully check the definitions of C_D, Re, and the like before using data in the literature. Airfoils, for example, use the planform area.

Figure 5.3*a* is for long, smooth cylinders. If wall roughness and cylinder length are included as variables, we obtain from dimensional analysis a complex three-parameter function:

$$C_D = f\left(\text{Re}_d, \frac{\varepsilon}{d}, \frac{L}{d}\right) \tag{5.28}$$

To describe this function completely would require 1,000 or more experiments or CFD results. Therefore, it is customary to explore the length and roughness effects separately to establish trends.

The table with Fig. 5.3*a* shows the length effect with zero wall roughness. As length decreases, the drag decreases by up to 50 percent. Physically, the pressure is "relieved" at the ends as the flow is allowed to skirt around the tips instead of deflecting over and under the body.

Figure 5.3*b* shows the effect of wall roughness for an infinitely long cylinder. The sharp drop in drag occurs at lower Re_d as roughness causes an earlier transition to a turbulent boundary layer on the surface of the body. Roughness has the same effect on sphere drag, a fact that is exploited in sports by deliberate dimpling of golf balls to give them less drag at their flight $\text{Re}_d \approx 10^5$. See Fig. D5.2.

Figure 5.3 is a typical experimental study of a fluid mechanics problem, aided by dimensional analysis. As time and money and demand allow, the complete three-parameter relation (5.28) could be filled out by further experiments.

EXAMPLE 5.7

A smooth cylinder, 1 cm in diameter and 20 cm long, is tested in a wind tunnel for a crossflow of 45 m/s of air at 20°C and 1 atm. The measured drag is 2.2 ± 0.1 N. (*a*) Does this data point agree with the data in Fig. 5.3? (*b*) Can this data point be used to predict the drag of a chimney 1 m in diameter and 20 m high in winds at 20°C and 1 atm? If so, what is the recommended range of wind velocities and drag forces for this data point? (*c*) Why are the answers to part (*b*) always the same, regardless of the chimney height, as long as $L = 20d$?

Solution

(*a*) For air at 20°C and 1 atm, take $\rho = 1.2$ kg/m³ and $\mu = 1.8$ E−5 kg/(m-s). Since the test cylinder is short, $L/d = 20$, it should be compared with the tabulated value $C_D \approx 0.91$ in the table to the right of Fig. 5.3*a*. First calculate the Reynolds number of the test cylinder:

$$\text{Re}_d = \frac{\rho U d}{\mu} = \frac{(1.2\ \text{kg/m}^3)(45\ \text{m/s})(0.01\ \text{m})}{1.8\text{E}{-5}\ \text{kg/(m}-\text{s)}} = 30{,}000$$

Yes, this is in the range $10^4 < \text{Re} < 10^5$ listed in the table. Now calculate the test drag coefficient:

$$C_{D,\text{test}} = \frac{F}{(1/2)\rho U^2 L d} = \frac{2.2\ \text{N}}{(1/2)(1.2\ \text{kg/m}^3)(45\ \text{m/s})^2(0.2\ \text{m})(0.01\ \text{m})} = 0.905$$

Yes, this is close, and certainly within the range of ±5 percent stated by the test results. *Ans. (a)*

(*b*) Since the chimney has $L/d = 20$, we can use the data if the Reynolds number range is correct:

$$10^4 < \frac{(1.2\ \text{kg/m}^3)U_{\text{chimney}}(1\ \text{m})}{1.8\ \text{E}{-5}\ \text{kg/(m}\cdot\text{s)}} < 10^5 \quad \text{if} \quad 0.15\frac{\text{m}}{\text{s}} < U_{\text{chimney}} < 1.5\frac{\text{m}}{\text{s}}$$

These are negligible winds, so the test data point is not very useful *Ans. (b)*
The drag forces in this range are also negligibly small:

$$F_{\min} = C_D\frac{\rho}{2}U_{\min}^2 Ld = (0.91)\left(\frac{1.2\ \text{kg/m}^3}{2}\right)(0.15\ \text{m/s})^2(20\ \text{m})(1\ \text{m}) = 0.25\ \text{N}$$

$$F_{\max} = C_D\frac{\rho}{2}U_{\max}^2 Ld = (0.91)\left(\frac{1.2\ \text{kg/m}^3}{2}\right)(1.5\ \text{m/s})^2(20\ \text{m})(1\ \text{m}) = 25\ \text{N}$$

(*c*) Try this yourself. Choose any 20:1 size for the chimney, even something silly like 20 mm:1 mm. You will get the same results for U and F as in part (*b*) above. This is because the product Ud occurs in Re_d and, if $L = 20d$, the same product occurs in the drag force. For example, for $\text{Re} = 10^4$,

$$Ud = 10^4\frac{\mu}{\rho} \quad \text{then } F = C_D\frac{\rho}{2}U^2Ld = C_D\frac{\rho}{2}U^2(20d)d = 20C_D\frac{\rho}{2}(Ud)^2 = 20C_D\frac{\rho}{2}\left(\frac{10^4\mu}{\rho}\right)^2$$

The answer is always $F_{\min} = 0.25$ N. This is an algebraic quirk that seldom occurs.

EXAMPLE 5.8

Telephone wires are said to "sing" in the wind. Consider a wire of diameter 8 mm. At what sea-level wind velocity, if any, will the wire sing a middle C note?

Solution

For sea-level air, take $\nu \approx 1.5\ \text{E}{-5}\ \text{m}^2/\text{s}$. For nonmusical readers, middle C is 262 Hz. Measured shedding rates are plotted in Fig. 5.2*b*. Over a wide range, the Strouhal number is approximately 0.2, which we can take as a first guess. Note that $(\omega/2\pi) = f$, the shedding frequency. Thus

$$\text{St} = \frac{fd}{U} = \frac{(262\ \text{s}^{-1})(0.008\ \text{m})}{U} \approx 0.2$$

$$U \approx 10.5\ \frac{\text{m}}{\text{s}}$$

Now, check the Reynolds number to see if we fall into the appropriate range:

$$\text{Re}_d = \frac{Ud}{\nu} = \frac{(10.5\ \text{m/s})(0.008\ \text{m})}{1.5\ \text{E}{-5}\ \text{m}^2/\text{s}} \approx 5{,}600$$

In Fig. 5.2*b*, at $\text{Re} = 5{,}600$, maybe St is a little higher, at about 0.21. Thus a slightly improved estimate is

$$U_{\text{wind}} = (262)(0.008)/(0.21) \approx 10.0\ \text{m/s} \qquad \textit{Ans.}$$

5.5 Modeling and Similarity

So far we have learned about dimensional homogeneity and the pi theorem method, using power products, for converting a homogeneous physical relation to dimensionless form. This is straightforward mathematically, but certain engineering difficulties need to be discussed.

First, we have more or less taken for granted that the variables that affect the process can be listed and analyzed. Actually, selection of the important variables requires considerable judgment and experience. The engineer must decide, for example, whether viscosity can be neglected. Are there significant temperature effects? Is surface tension important? What about wall roughness? Each pi group that is retained increases the expense and effort required. Judgment in selecting variables will come through practice and maturity; this book should provide some of the necessary experience.

Once the variables are selected and the dimensional analysis is performed, the experimenter seeks to achieve *similarity* between the model tested and the prototype to be designed. With sufficient testing, the model data will reveal the desired dimensionless function between variables:

$$\Pi_1 = f(\Pi_2, \Pi_3, \ldots \Pi_k) \tag{5.29}$$

With Eq. (5.29) available in chart, graphical, or analytical form, we are in a position to ensure complete similarity between model and prototype. A formal statement would be as follows:

> Flow conditions for a model test are completely similar if all relevant dimensionless parameters have the same corresponding values for the model and the prototype.

This follows mathematically from Eq. (5.29). If $\Pi_{2m} = \Pi_{2p}$, $\Pi_{3m} = \Pi_{3p}$, and so forth, Eq. (5.29) guarantees that the desired output Π_{1m} will equal Π_{1p}. But, this is easier said than done, as we now discuss. There are specialized texts on model testing [30–32].

Instead of complete similarity, the engineering literature speaks of particular types of similarity, the most common being geometric, kinematic, dynamic, and thermal. Let us consider each separately.

Geometric Similarity

Geometric similarity concerns the length dimension $\{L\}$ and must be ensured before any sensible model testing can proceed. A formal definition is as follows:

> A model and prototype are *geometrically similar* if and only if all body dimensions in all three coordinates have the same linear scale ratio.

Note that *all* length scales must be the same. It is as if you took a photograph of the prototype and reduced it or enlarged it until it fitted the size of the model. If the model is to be made one-tenth the prototype size, its length, width, and height must each be one-tenth as large. Not only that, but also its entire shape must be one-tenth as large, and technically we speak of *homologous* points, which are points that have the same relative location. For example, the nose of the prototype is homologous to the nose of the model. The left wingtip of the prototype is homologous to the left wingtip of the model. Then geometric similarity requires that all homologous points be related by the same linear scale ratio. This applies to the fluid geometry as well as the model geometry.

> All angles are preserved in geometric similarity. All flow directions are preserved. The orientations of model and prototype with respect to the surroundings must be identical.

Figure 5.4 illustrates a prototype wing and a one-tenth-scale model. The model lengths are all one-tenth as large, but its angle of attack with respect to the free stream is the same for both model and prototype: 10° not 1°. All physical details on the model must be scaled, and some are rather subtle and sometimes overlooked:

1. The model nose radius must be one-tenth as large.
2. The model surface roughness must be one-tenth as large.
3. If the prototype has a 5-mm boundary layer trip wire 1.5 m from the leading edge, the model should have a 0.5-mm trip wire 0.15 m from its leading edge.
4. If the prototype is constructed with protruding fasteners, the model should have homologous protruding fasteners one-tenth as large.

And so on. Any departure from these details is a violation of geometric similarity and must be justified by experimental comparison to show that the prototype behavior was not significantly affected by the discrepancy.

Models that appear similar in shape but that clearly violate geometric similarity should not be compared except at your own risk. Figure 5.5 illustrates this point. The spheres in Fig. 5.5*a* are all geometrically similar and can be tested with a high expectation of success if the Reynolds number, Froude number, or the like is matched. But, the ellipsoids in Fig. 5.5*b* merely *look* similar. They actually have different linear scale ratios and therefore cannot be compared in a rational manner, even though they may

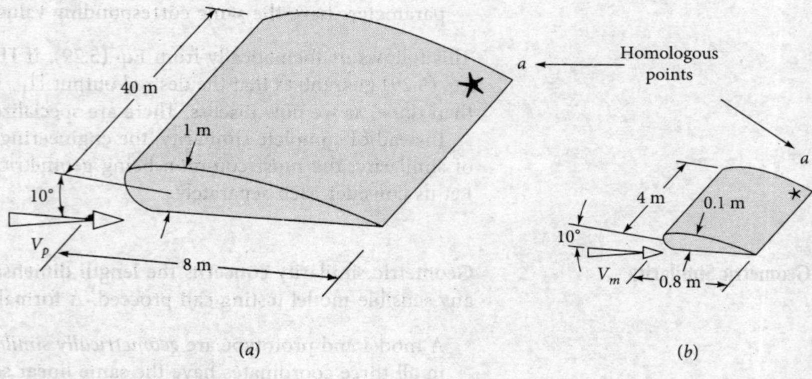

Fig. 5.4 Geometric similarity in model testing: (*a*) prototype; (*b*) one-tenth-scale model.

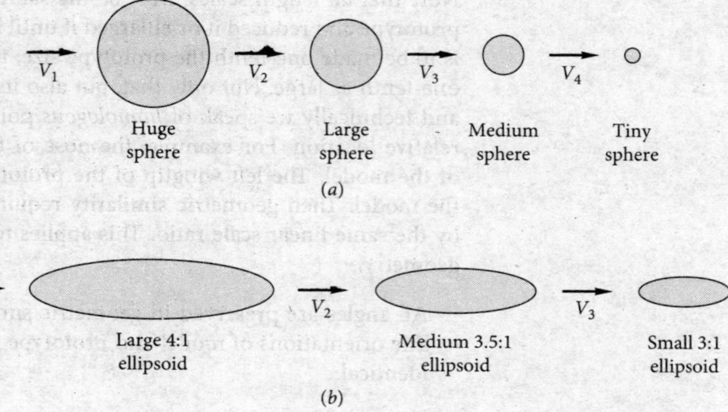

Fig. 5.5 Geometric similarity and dissimilarity of flows: (*a*) similar; (*b*) dissimilar.

have identical Reynolds and Froude numbers and so on. The data will not be the same for these ellipsoids, and any attempt to "compare" them is a matter of rough engineering judgment.

Kinematic Similarity

Kinematic similarity requires that the model and prototype have the same length scale ratio and the same time scale ratio. The result is that the velocity scale ratio will be the same for both. As Langhaar [4] states it:

> The motions of two systems are kinematically similar if homologous particles lie at homologous points at homologous times.

Length scale equivalence simply implies geometric similarity, but time scale equivalence may require additional dynamic considerations such as equivalence of the Reynolds and Mach numbers.

One special case is incompressible frictionless flow with no free surface, as sketched in Fig. 5.6*a*. These perfect-fluid flows are kinematically similar with independent length and time scales, and no additional parameters are necessary (see Chap. 8 for further details).

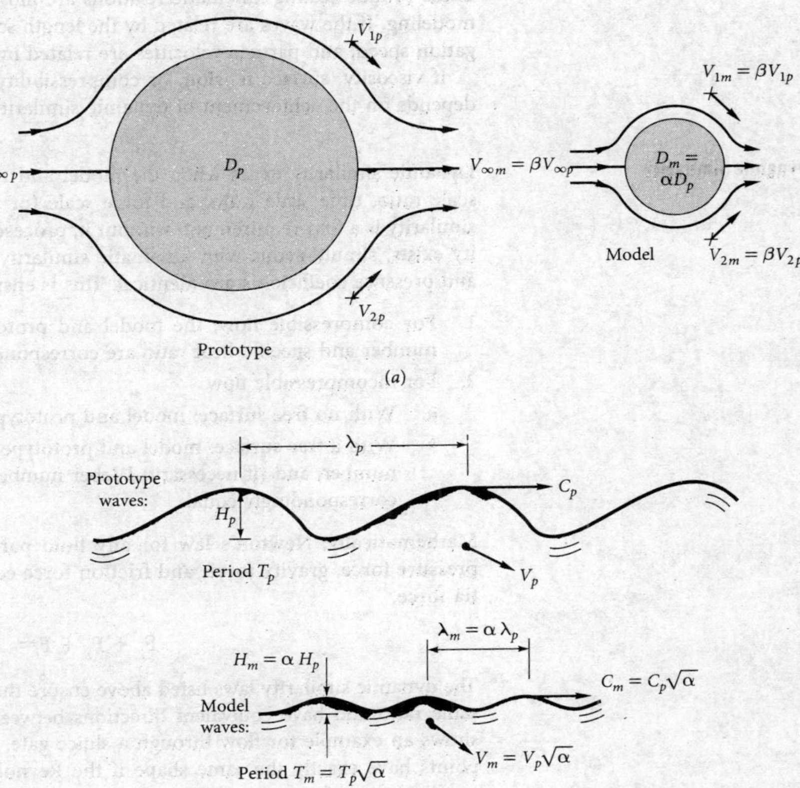

Fig. 5.6 Frictionless low-speed flows are kinematically similar: (*a*) Flows with no free surface are kinematically similar with independent length and time scale ratios; (*b*) free-surface flows are kinematically similar with length and time scales related by the Froude number.

Froude Scaling

Frictionless flows with a free surface, as in Fig. 5.6*b*, are kinematically similar if their Froude numbers are equal:

$$\mathrm{Fr}_m = \frac{V_m^2}{gL_m} = \frac{V_p^2}{gL_p} = Fr_p \tag{5.30}$$

Note that the Froude number contains only length and time dimensions and hence is a purely kinematic parameter that fixes the relation between length and time. From Eq. (5.30), if the length scale is

$$L_m = \alpha L_p \tag{5.31}$$

where α is a dimensionless ratio, the velocity scale is

$$\frac{V_m}{V_p} = \left(\frac{L_m}{L_p}\right)^{1/2} = \sqrt{\alpha} \tag{5.32}$$

and the time scale is

$$\frac{T_m}{T_p} = \frac{L_m/V_m}{L_p/V_p} = \sqrt{\alpha} \tag{5.33}$$

These Froude-scaling kinematic relations are illustrated in Fig. 5.6*b* for wave motion modeling. If the waves are related by the length scale α, then the wave period, propagation speed, and particle velocities are related by $\sqrt{\alpha}$.

If viscosity, surface tension, or compressibility is important, kinematic similarity depends on the achievement of dynamic similarity.

Dynamic Similarity

Dynamic similarity exists when the model and the prototype have the same length scale ratio, time scale ratio, and force scale (or mass scale) ratio. Again geometric similarity is a first requirement; without it, proceed no further. Then dynamic similarity exists, simultaneous with kinematic similarity, if the model and prototype force and pressure coefficients are identical. This is ensured if

1. For compressible flow, the model and prototype Reynolds number and Mach number and specific-heat ratio are correspondingly equal.
2. For incompressible flow
   a. With no free surface: model and prototype Reynolds numbers are equal.
   b. With a free surface: model and prototype Reynolds number, Froude number, and (if necessary) Weber number and cavitation number are correspondingly equal.

Mathematically, Newton's law for any fluid particle requires that the sum of the pressure force, gravity force, and friction force equal the acceleration term, or inertia force,

$$\mathbf{F}_p + \mathbf{F}_g + \mathbf{F}_f = \mathbf{F}_i$$

The dynamic similarity laws listed above ensure that each of these forces will be in the same ratio and have equivalent directions between model and prototype. Figure 5.7 shows an example for flow through a sluice gate. The force polygons at homologous points have exactly the same shape if the Reynolds and Froude numbers are equal (neglecting surface tension and cavitation, of course). Kinematic similarity is also ensured by these model laws.

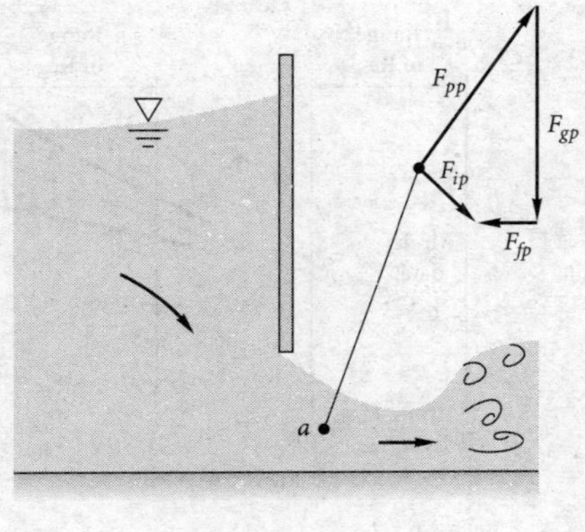

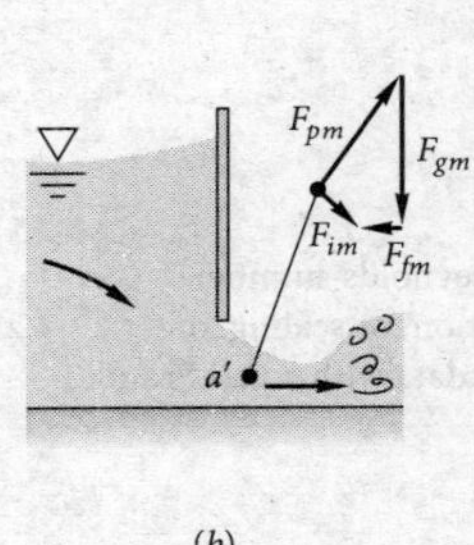

Fig. 5.7 Dynamic similarity in sluice gate flow. Model and prototype yield identical homologous force polygons if the Reynolds and Froude numbers are the same corresponding values: (*a*) prototype; (*b*) model.

Discrepancies in Water and Air Testing

The perfect dynamic similarity shown in Fig. 5.7 is more of a dream than a reality because true equivalence of Reynolds and Froude numbers can be achieved only by dramatic changes in fluid properties, whereas in fact most model testing is simply done with water or air, the cheapest fluids available.

First, consider hydraulic model testing with a free surface. Dynamic similarity requires equivalent Froude numbers, Eq. (5.30), *and* equivalent Reynolds numbers:

$$\frac{V_m L_m}{\nu_m} = \frac{V_p L_p}{\nu_p} \tag{5.34}$$

But, both velocity and length are constrained by the Froude number, Eqs. (5.31) and (5.32). Therefore, for a given length scale ratio α, Eq. (5.34) is true only if

$$\frac{\nu_m}{\nu_p} = \frac{L_m}{L_p}\frac{V_m}{V_p} = \alpha\sqrt{\alpha} = \alpha^{3/2} \tag{5.35}$$

For example, for a one-tenth-scale model, $\alpha = 0.1$ and $\alpha^{3/2} = 0.032$. Since ν_p is undoubtedly water, we need a fluid with only 0.032 times the kinematic viscosity of water to achieve dynamic similarity. Referring to Table 1.4, we see that this is impossible: Even mercury has only one-ninth the kinematic viscosity of water, and a mercury hydraulic model would be expensive and bad for your health. In practice, water is used for both the model and the prototype, and the Reynolds number similarity (5.34) is unavoidably violated. The Froude number is held constant since it is the dominant parameter in free-surface flows. Typically, the Reynolds number of the model flow is too small by a factor of 10 to 1,000. As shown in Fig. 5.8, the low-Reynolds-number model data are used to estimate by extrapolation the desired high-Reynolds-number prototype data. As the figure indicates, there is obviously considerable uncertainty in using such an extrapolation, but there is no other practical alternative in hydraulic model testing.

Second, consider aerodynamic model testing in air with no free surface. The important parameters are the Reynolds number and the Mach number. Equation (5.34) should be satisfied, plus the compressibility criterion

$$\frac{V_m}{a_m} = \frac{V_p}{a_p} \tag{5.36}$$

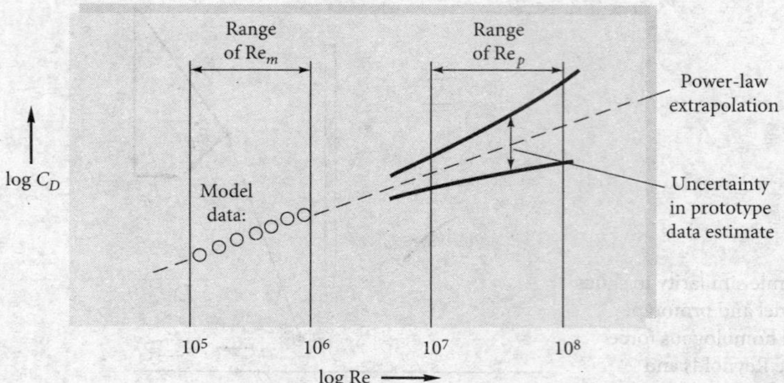

Fig. 5.8 Reynolds-number extrapolation, or scaling, of hydraulic data with equal Froude numbers.

Elimination of V_m/V_p between (5.34) and (5.36) gives

$$\frac{\nu_m}{\nu_p} = \frac{L_m}{L_p}\frac{a_m}{a_p} \tag{5.37}$$

Since the prototype is no doubt an air operation, we need a wind-tunnel fluid of low viscosity and high speed of sound. Hydrogen is the only practical example, but clearly it is too expensive and dangerous. Therefore, wind tunnels normally operate with air as the working fluid. Cooling and pressurizing the air will bring Eq. (5.37) into better agreement but not enough to satisfy a length scale reduction of, say, one-tenth. Therefore Reynolds number scaling is also commonly violated in aerodynamic testing, and an extrapolation like that in Fig. 5.8 is required here also.

There are specialized monographs devoted entirely to wind tunnel testing: low speed [38], high speed [39], and a detailed general discussion [40]. The following example illustrates modeling discrepancies in aeronautical testing.

EXAMPLE 5.9

A prototype airplane, with a chord length of 1.6 m, is to fly at Ma = 2 at 10 km standard altitude. A one-eighth scale model is to be tested in a helium wind tunnel at 100°C and 1 atm. Find the helium test section velocity that will match (*a*) the Mach number or (*b*) the Reynolds number of the prototype. In each case, criticize the lack of dynamic similarity. (*c*) What high pressure in the helium tunnel will match *both* the Mach and Reynolds numbers? (*d*) Why does part (*c*) *still* not achieve dynamic similarity?

Solution

For helium, from Table A.4, $R = 2{,}077\ \text{m}^2/(\text{s}^2\text{-K})$, $k = 1.66$, and estimate $\mu_{He} \approx 2.32$ E−5 kg/(m · s) from the power-law, $n = 0.67$, in the table. (*a*) Calculate the helium speed of sound and velocity:

$$a_{He} = \sqrt{(kRT)_{He}} = \sqrt{(1.66)(2{,}077\ \text{m}^2/\text{s}^2\text{K}) \times (373\ \text{K})} = 1{,}134\ \text{m/s}$$

$$\text{Ma}_{air} = \text{Ma}_{He} = 2.0 = \frac{V_{He}}{a_{He}} = \frac{V_{He}}{1{,}134\ \text{m/s}}$$

$$V_{He} = 2{,}268\ \frac{\text{m}}{\text{s}} \qquad \textit{Ans. (a)}$$

For dynamic similarity, the Reynolds numbers should also be equal. From Table A.6 at an altitude of 10,000 m, read $\rho_{air} = 0.4125\ kg/m^3$, $a_{air} = 299.5\ m/s$, and estimate $\mu_{air} \approx 1.48\ E{-5}\ kg/m \cdot s$ from the power-law, $n = 0.7$, in Table A.4. The air velocity is $V_{air} = (Ma)(a_{air}) = 2(299.5) = 599\ m/s$. The model chord length is (1.6 m)/8 = 0.2 m. The helium density is $\rho_{He} = (p/RT)_{He} = (101{,}350\ Pa)/[(2{,}077\ m^2/s^2\ K)(373\ K)] = 0.131\ kg/m^3$. Now calculate the two Reynolds numbers:

$$\mathrm{Re}_{C,air} = \left.\frac{\rho VC}{\mu}\right|_{air} = \frac{(0.4125\ kg/m^3)(599\ m/s)(1.6\ m)}{1.48\ E{-5}\ kg/(m \cdot s)} = 26.6\ E6$$

$$\mathrm{Re}_{C,He} = \left.\frac{\rho VC}{\mu}\right|_{He} = \frac{(0.131\ kg/m^3)(2{,}268\ m/s)(0.2\ m)}{2.32\ E{-5}\ kg/(m \cdot s)} = 2.56\ E6$$

The model Reynolds number is 10 times less than the prototype. This is typical when using small-scale models. The test results must be extrapolated for Reynolds number effects.
(*b*) Now ignore Mach number and let the model Reynolds number match the prototype:

$$\mathrm{Re}_{He} = \mathrm{Re}_{air} = 26.6\ E6 = \frac{(0.131\ kg/m^3)V_{He}(0.2\ m)}{2.32\ E{-5}\ kg/(m \cdot s)}$$

$$V_{He} = 23{,}600\ \frac{m}{s} \qquad \textit{Ans. (b)}$$

This is ridiculous: a hypersonic Mach number of 21, suitable for escaping from the earth's gravity. One should match the Mach numbers and correct for a lower Reynolds number.
(*c*) Match both Reynolds and Mach numbers by increasing the helium density:
Ma matches if

$$V_{He} = 2{,}268\ \frac{m}{s}$$

Then

$$\mathrm{Re}_{He} = 26.6\ E6 = \frac{\rho_{He}(2{,}268\ m/s)(0.2\ m)}{2.32\ E{-5}\ kg/(m \cdot s)}$$

Solve for

$$\rho_{He} = 1.36\ \frac{kg}{m^3} \quad p_{He} = \rho RT|_{He} = (1.36)(2{,}077)(373) = 1.05\ E6\ Pa \qquad \textit{Ans. (c)}$$

A match is possible if we increase the tunnel pressure by a factor of ten, a daunting task.
(*d*) Even with Ma and Re matched, we are *still* not dynamically similar because the two gases have different specific heat ratios: $k_{He} = 1.66$ and $k_{air} = 1.40$. This discrepancy will cause substantial differences in pressure, density, and temperature throughout supersonic flow.

Figure 5.9 shows a hydraulic model of the Bluestone Lake Dam in West Virginia. The model itself is located at the U.S. Army Waterways Experiment Station in Vicksburg, MS. The horizontal scale is 1:65, which is sufficient that the vertical scale can also be 1:65 without incurring significant surface tension (Weber number) effects. Velocities are scaled by the Froude number. However, the prototype Reynolds number, which is of order 1 E7, cannot be matched here. The engineers set the Reynolds number at about 2 E4, high enough for a reasonable approximation of prototype turbulent flow viscous

Fig. 5.9 Hydraulic model of the Bluestone Lake Dam on the New River near Hinton, West Virginia. The model scale is 1:65 both vertically and horizontally, and the Reynolds number, though far below the prototype value, is set high enough for the flow to be turbulent. *(Courtesy of the U.S. Army Corps of Engineers Waterways Experiment Station.)*

effects. Note the intense turbulence below the dam. The downstream bed, or *apron*, of a dam must be strengthened structurally to avoid bed erosion.

For hydraulic models of larger scale, such as harbors, estuaries, and embayments, geometric similarity may be violated of necessity. The vertical scale will be distorted to avoid Weber number effects. For example, the horizontal scale may be 1:1,000, while the vertical scale is only 1:100. Thus, the model channel may be *deeper* relative to its horizontal dimensions. Since deeper passages flow more efficiently, the model channel bottom may be deliberately roughened to create the friction level expected in the prototype.

EXAMPLE 5.10

The pressure drop due to friction for flow in a long, smooth pipe is a function of average flow velocity, density, viscosity, and pipe length and diameter: $\Delta p = \text{fcn}(V, \rho, \mu, L, D)$. We wish to know how Δp varies with V. (*a*) Use the pi theorem to rewrite this function in dimensionless form. (*b*) Then plot this function, using the following data for three pipes and three fluids:

D, cm	*L*, m	*Q*, m³/h	Δp, Pa	ρ, kg/m³	μ, kg/(m · s)	*V*, m/s*
1.0	5.0	0.3	4,680	680†	2.92 E-4†	1.06
1.0	7.0	0.6	22,300	680†	2.92 E-4†	2.12
1.0	9.0	1.0	70,800	680†	2.92 E-4†	3.54
2.0	4.0	1.0	2,080	998‡	0.0010‡	0.88
2.0	6.0	2.0	10,500	998‡	0.0010‡	1.77
2.0	8.0	3.1	30,400	998‡	0.0010‡	2.74
3.0	3.0	0.5	540	13,550§	1.56 E-3§	0.20
3.0	4.0	1.0	2,480	13,550§	1.56 E-3§	0.39
3.0	5.0	1.7	9,600	13,550§	1.56 E-3§	0.67

*$V = Q/A$, $A = \pi D^2/4$.
†Gasoline.
‡Water.
§Mercury.

(*c*) Suppose it is further known that Δp is proportional to *L* (which is quite true for long pipes with well-rounded entrances). Use this information to simplify and improve the pi theorem formulation. Plot the dimensionless data in this improved manner and comment on the results.

Solution

There are six variables with three primary dimensions involved {*MLT*}. Therefore, we expect that $j = 6 - 3 = 3$ pi groups. We are correct, for we can find three variables that do not form a pi product (e.g., ρ, *V*, *L*). Carefully select three (*j*) repeating variables, but not including Δp or *V*, which we plan to plot versus each other. We select (ρ, μ, *D*), and the pi theorem guarantees that three independent power-product groups will occur:

$$\Pi_1 = \rho^a \mu^b D^c \,\Delta p \qquad \Pi_2 = \rho^d \mu^e D^f V \qquad \Pi_3 = \rho^g \mu^h D^i L$$

or

$$\Pi_1 = \frac{\rho D^2 \Delta p}{\mu^2} \qquad \Pi_2 = \frac{\rho V D}{\mu} \qquad \Pi_3 = \frac{L}{D}$$

We have omitted the algebra of finding (*a*, *b*, *c*, *d*, *e*, *f*, *g*, *h*, *i*) by setting all exponents to zero M^0, L^0, T^0. Therefore, we wish to plot the dimensionless relation

$$\frac{\rho D^2 \,\Delta p}{\mu^2} = \text{fcn}\left(\frac{\rho V D}{\mu}, \frac{L}{D}\right) \qquad \textit{Ans. (a)}$$

We plot Π_1 versus Π_2 with Π_3 as a parameter. There will be nine data points. For example, the first row in the data here yields

$$\frac{\rho D^2 \,\Delta p}{\mu^2} = \frac{(680)(0.01)^2(4{,}680)}{(2.92\text{ E-}4)^2} = 3.73\text{ E}9$$

$$\frac{\rho V D}{\mu} = \frac{(680)(1.06)(0.01)}{2.92\text{ E-}4} = 24{,}700 \qquad \frac{L}{D} = 500$$

The nine data points are plotted as the open circles in Fig. 5.10. The values of *L*/*D* are listed for each point, and we see a significant length effect. In fact, if we connect the only two points that have the same *L*/*D* (= 200), we could see (and cross-plot to verify) that Δp increases linearly with *L*, as stated in the last part of the problem. Since *L* occurs only in $\Pi_3 = L/D$, the function $\Pi_1 = \text{fcn}(\Pi_2, \Pi_3)$ must reduce to $\Pi_1 = (L/D)\,\text{fcn}(\Pi_2)$, or simply a function involving only *two* parameters:

$$\frac{\rho D^3 \,\Delta p}{L\mu^2} = \text{fcn}\left(\frac{\rho V D}{\mu}\right) \quad \text{flow in a long pipe} \qquad \textit{Ans. (c)}$$

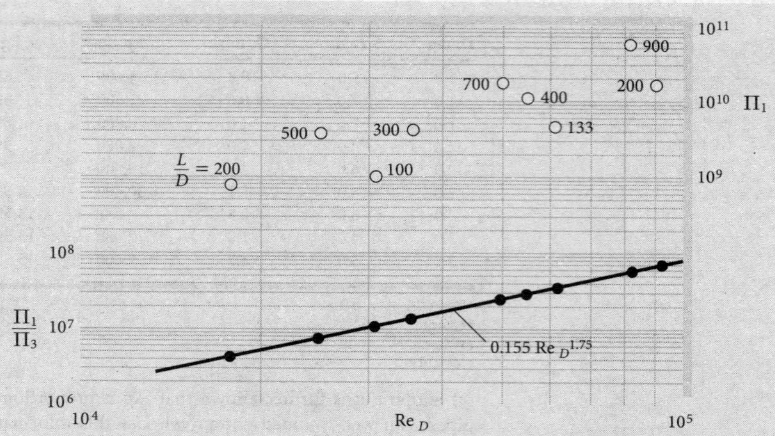

Fig. 5.10 Two different correlations of the data in Example 5.10: Open circles when plotting $\rho D^2\ \Delta p/\mu^2$ versus Re_D, L/D is a parameter; once it is known that Δp is proportional to L, a replot (solid circles) of $\rho D^3\ \Delta p/(L\mu^2)$ versus Re_D collapses into a single power-law curve.

We now modify each data point in Fig. 5.10 by dividing it by its L/D value. For example, for the first row of data, $\rho D^3\ \Delta p/(L\mu^2) = (3.73\ \mathrm{E9})/500 = 7.46\ \mathrm{E6}$. We replot these new data points as solid circles in Fig. 5.10. They correlate almost perfectly into a straight-line power-law function:

$$\frac{\rho D^3\ \Delta p}{L\mu^2} \approx 0.155\left(\frac{\rho VD}{\mu}\right)^{1.75} \qquad \textit{Ans. (c)}$$

All newtonian smooth pipe flows should correlate in this manner. This example is a variation of the first completely successful dimensional analysis, pipe-flow friction, performed by Prandtl's student Paul Blasius, who published a related plot in 1911. For this range of (turbulent flow) Reynolds numbers, the pressure drop increases approximately as $V^{1.75}$.

EXAMPLE 5.11

The smooth sphere data plotted in Fig. 5.3*a* represent dimensionless drag versus dimensionless *viscosity*, since (ρ, V, d) were selected as scaling or repeating variables. (*a*) Replot these data to display the effect of dimensionless *velocity* on the drag. (*b*) Use your new figure to predict the terminal (zero-acceleration) velocity of a 1-cm-diameter steel ball (SG = 7.86) falling through water at 20°C.

Solution

- *Assumptions:* Fig 5.3*a* is valid for any smooth sphere in that Reynolds number range.
- *Approach (a):* Form pi groups from the function $F = \mathrm{fcn}(d, V, \rho, \mu)$ in such a way that F is plotted versus V. The answer was already given as Eq. (5.16), but let us review the steps. The proper scaling variables are (ρ, μ, d), which do *not* form a pi. Therefore $j = 3$, and we expect $n - j = 5 - 3 = 2$ pi groups. Skipping the algebra, they arise as follows:

$$\Pi_1 = \rho^a\mu^b d^c\ F = \frac{\rho F}{\mu^2} \qquad \Pi_2 = \rho^a\mu^b d^c\ V = \frac{\rho V d}{\mu} \qquad \textit{Ans. (a)}$$

We may replot the data of Fig. 5.3*a* in this new form, noting that $\Pi_1 \equiv (\pi/8)(C_D)(\mathrm{Re})^2$. This replot is shown as Fig. 5.11. The drag increases rapidly with velocity up to transition, where there is a slight drop, after which it increases more than ever. If force is known, we may predict velocity from the figure, and vice versa.

- *Property values for part (b):* $\rho_{\text{water}} = 998 \text{ kg/m}^3$ $\mu_{\text{water}} = 0.001 \text{ kg/(m-s)}$

$$\rho_{\text{steel}} = 7.86\rho_{\text{water}} = 7{,}844 \text{ kg/m}^3.$$

- *Solution to part (b):* For terminal velocity, the drag force equals the net weight of the sphere in water:

$$F = W_{\text{net}} = (\rho_s - \rho_w)g\frac{\pi}{6}d^3 = (7{,}844 - 998)(9.81)\left(\frac{\pi}{6}\right)(0.01)^3 = 0.0351 \text{ N}$$

Therefore, the ordinate of Fig. 5.11 is known:

$$\text{Falling steel sphere:} \quad \frac{\rho F}{\mu^2} = \frac{(998 \text{ kg/m}^3)(0.0351 \text{ N})}{[0.001 \text{ kg/(m}\cdot\text{s)}]^2} \approx 3.5 \text{ E7}$$

From Fig. 5.11, at $\rho F/\mu^2 \approx 3.5$ E7, a magnifying glass reveals that $\text{Re}_d \approx 2$ E4. Then a crude estimate of the terminal fall velocity is

$$\frac{\rho V d}{\mu} \approx 20{,}000 \quad \text{or} \quad V \approx \frac{20{,}000[0.001 \text{ kg/(m}\cdot\text{s)}]}{(998 \text{ kg/m}^3)(0.01 \text{ m})} \approx 2.0 \frac{\text{m}}{\text{s}} \qquad \textit{Ans. (b)}$$

- *Comments:* Better accuracy could be obtained by expanding the scale of Fig. 5.11 in the region of the given force coefficient. However, there is considerable uncertainty in published drag data for spheres, so the predicted fall velocity is probably uncertain by at least ±10 percent.

  Note that we found the answer directly from Fig. 5.11. We could use Fig. 5.3*a* also but would have to iterate between the ordinate and abscissa to obtain the final result, since *V* is contained in both plotted variables.

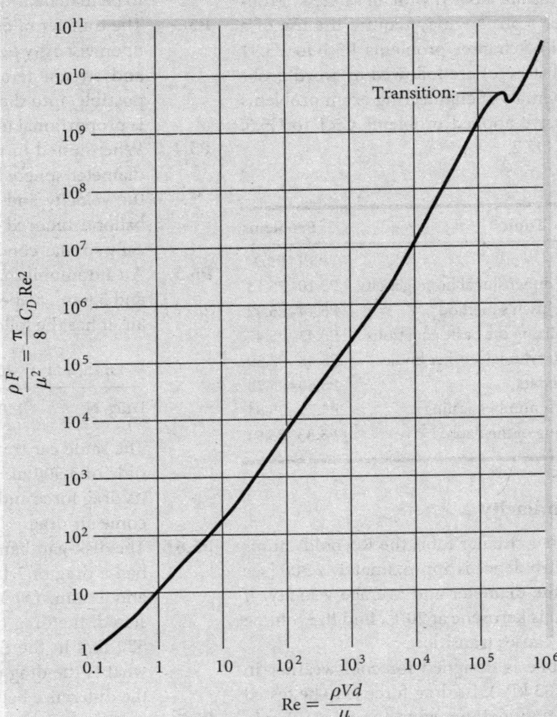

Fig. 5.11 Cross-plot of sphere drag data from Fig. 5.3*a* to show dimensionless force versus dimensionless velocity.

Summary

Chapters 3 and 4 presented integral and differential methods of mathematical analysis of fluid flow. This chapter introduces the third and final method: experimentation, as supplemented by the technique of dimensional analysis. Tests and experiments are used both to strengthen existing theories and to provide useful engineering results when theory is inadequate.

The chapter begins with a discussion of some familiar physical relations and how they can be recast in dimensionless form because they satisfy the principle of dimensional homogeneity. A general technique, the pi theorem, is then presented for systematically finding a set of dimensionless parameters by grouping a list of variables that govern any particular physical process. A second technique, Ipsen's method, is also described. Alternately, direct application of dimensional analysis to the basic equations of fluid mechanics yields the fundamental parameters governing flow patterns: Reynolds number, Froude number, Prandtl number, Mach number, and others.

It is shown that model testing in air and water often leads to scaling difficulties for which compromises must be made. Many model tests do not achieve true dynamic similarity. The chapter ends by pointing out that classic dimensionless charts and data can be manipulated and recast to provide direct solutions to problems that would otherwise be quite cumbersome and laboriously iterative.

Problems

Most of the problems herein are fairly straightforward. More difficult or open-ended assignments are labeled with an asterisk. Problems labeled with a computer icon may require the use of a computer. The standard end-of-chapter problems P5.1 to P5.91 (categorized in the problem list here) are followed by word problems W5.1 to W5.10, fundamentals of engineering exam problems FE5.1 to FE5.12, comprehensive applied problems C5.1 to C5.5, and design projects D5.1 and D5.2.

Problem Distribution

Section	Topic	Problems
5.1	Introduction	P5.1–P5.9
5.2	The principle of dimensional homogeneity	P5.10–P5.13
5.3	The pi theorem; Ipsen's method	P5.14–P5.42
5.4	Nondimensionalizing the basic equations	P5.43–P5.47
5.4	Data for spheres, cylinders, other bodies	P5.48–P5.59
5.5	Scaling of model data	P5.60–P5.74
5.5	Froude and Mach number scaling	P5.75–P5.84
5.5	Inventive rescaling of the data	P5.85–P5.91

Introduction; dynamic similarity

P5.1 For axial flow through a circular tube, the Reynolds number for transition to turbulence is approximately 2,300 [see Eq. (6.2)], based on the diameter and average velocity. If $d = 5$ cm and the fluid is kerosene at 20°C, find the volume flow rate in m^3/h that causes transition.

P5.2 A prototype automobile is designed for cold weather in Denver, CO (−10°C, 83 kPa). Its drag force is to be tested on a one-seventh-scale model in a wind tunnel at 70 m/s, 20°C, and 1 atm. If the model and prototype are to satisfy dynamic similarity, what prototype velocity, in m/s, needs to be matched? Comment on your result.

P5.3 The transfer of energy by viscous dissipation is dependent upon viscosity μ, thermal conductivity k, stream velocity U, and stream temperature T_0. Group these quantities, if possible, into the dimensionless *Brinkman number*, which is proportional to μ.

P5.4 When tested in water at 20°C flowing at 2 m/s, an 8-cm-diameter sphere has a measured drag of 5 N. What will be the velocity and drag force on a 1.5-m-diameter weather balloon moored in sea-level standard air under dynamically similar conditions?

P5.5 An automobile has a characteristic length and area of 2.5 m and 5.6 m^2, respectively. When tested in sea-level standard air, it has the following measured drag force versus speed:

V, km/h	30	65	95
Drag, N	140	510	1,110

The same car travels in Colorado at 105 km/hr at an altitude of 3,500 m. Using dimensional analysis, estimate (*a*) its drag force and (*b*) the power (in kW) required to overcome air drag.

P5.6 The disk-gap-band parachute in the chapter-opener photo had a drag of 7,100 N when tested at 7 m/s in air at 20°C and 1 atm. (*a*) What was its drag coefficient? (*b*) If, as stated, the drag on Mars is 289,000 N and the velocity is 580 m/s in the thin Mars atmosphere, $\rho \approx 0.020$ kg/m^3, what is the drag coefficient on Mars? (*c*) Can you explain the difference between (*a*) and (*b*)?

P5.7 A body is dropped on the moon ($g = 1.62$ m/s^2) with an initial velocity of 12 m/s. By using option 2 variables,

Eq. (5.11), the ground impact occurs at $t^{**} = 0.34$ and $S^{**} = 0.84$. Estimate (*a*) the initial displacement, (*b*) the final displacement, and (*c*) the time of impact.

P5.8 The Archimedes number, Ar, used in the flow of stratified fluids, is a dimensionless combination of gravity g, density difference $\Delta\rho$, fluid width L, and viscosity μ. Find the form of this number if it is proportional to g.

P5.9 The *Richardson number,* Ri, which correlates the production of turbulence by buoyancy, is a dimensionless combination of the acceleration of gravity g, the fluid temperature T_0, the local temperature gradient $\partial T/\partial z$, and the local velocity gradient $\partial u/\partial z$. Determine the form of the Richardson number if it is proportional to g.

The principle of dimensional homogeneity

P5.10 Determine the dimension $\{MLT\Theta\}$ of the following quantities:

$$(a)\ \rho u \frac{\partial u}{\partial x} \quad (b) \int_1^2 (p - p_0)\, dA \quad (c)\ \rho c_p \frac{\partial^2 T}{\partial x\, \partial y}$$

$$(d) \iiint \rho \frac{\partial u}{\partial t}\, dx\, dy\, dz$$

All quantities have their standard meanings; for example, ρ is density.

P5.11 During World War II, Sir Geoffrey Taylor, a British fluid dynamicist, used dimensional analysis to estimate the wave speed of an atomic bomb explosion. He assumed that the blast wave radius R was a function of energy released E, air density ρ, and time t. Use dimensional reasoning to show how wave radius must vary with time.

P5.12 The *Stokes number,* St, used in particle dynamics studies, is a dimensionless combination of *five* variables: acceleration of gravity g, viscosity μ, density ρ, particle velocity U, and particle diameter D. (*a*) If St is proportional to μ and inversely proportional to g, find its form. (*b*) Show that St is actually the quotient of two more traditional dimensionless groups.

P5.13 The speed of propagation C of a capillary wave in deep water is known to be a function only of density ρ, wavelength λ, and surface tension Υ. Find the proper functional relationship, completing it with a dimensionless constant. For a given density and wavelength, how does the propagation speed change if the surface tension is doubled?

The pi theorem or Ipsen's method

P5.14 Flow in a pipe is often measured with an orifice plate, as in Fig. P5.14. The volume flow Q is a function of the pressure drop Δp across the plate, the fluid density ρ, the pipe diameter D, and the orifice diameter d. Rewrite this functional relationship in dimensionless form.

P5.15 The wall shear stress τ_w in a boundary layer is assumed to be a function of stream velocity U, boundary layer thickness δ, local turbulence velocity u', density ρ, and local pressure gradient dp/dx. Using (ρ, U, δ) as repeating variables, rewrite this relationship as a dimensionless function.

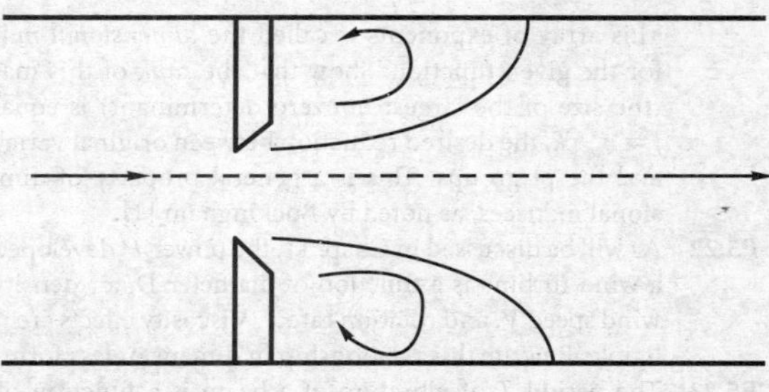

P5.14

P5.16 Convection heat transfer data are often reported as a *heat transfer coefficient h*, defined by

$$\dot{Q} = hA\,\Delta T$$

where $\dot{Q}$ = heat flow, J/s
A = surface area, m^2
ΔT = temperature difference, K

The dimensionless form of h, called the *Stanton number,* is a combination of h, fluid density ρ, specific heat c_p, and flow velocity V. Derive the Stanton number if it is proportional to h. What are the units of h?

P5.17 If you disturb a tank of length L and water depth h, the surface will oscillate back and forth at frequency Ω, assumed here to depend also upon water density ρ and the acceleration of gravity g. (*a*) Rewrite this as a dimensionless function. (*b*) If a tank of water sloshes at 2.0 Hz on earth, how fast would it oscillate on Mars ($g \approx 3.7$ m/s^2)?

P5.18 Under laminar conditions, the volume flow Q through a small triangular-section pore of side length b and length L is a function of viscosity μ, pressure drop per unit length $\Delta p/L$, and b. Using the pi theorem, rewrite this relation in dimensionless form. How does the volume flow change if the pore size b is doubled?

P5.19 The period of oscillation T of a water surface wave is assumed to be a function of density ρ, wavelength l, depth h, gravity g, and surface tension Υ. Rewrite this relationship in dimensionless form. What results if Υ is negligible? *Hint:* Take l, ρ, and g as repeating variables.

P5.20 A fixed cylinder of diameter D and length L, immersed in a stream flowing normal to its axis at velocity U, will experience zero average lift. However, if the cylinder is rotating at angular velocity Ω, a lift force F will arise. The fluid density ρ is important, but viscosity is secondary and can be neglected. Formulate this lift behavior as a dimensionless function.

P5.21 In Example 5.1 we used the pi theorem to develop Eq. (5.2) from Eq. (5.1). Instead of merely listing the primary dimensions of each variable, some workers list the *powers* of each primary dimension for each variable in an array:

$$\begin{array}{c} \\ M \\ L \\ T \end{array} \begin{array}{c} \begin{array}{ccccc} F & L & U & \rho & \mu \end{array} \\ \begin{bmatrix} 1 & 0 & 0 & 1 & 1 \\ 1 & 1 & 1 & -3 & -1 \\ -2 & 0 & -1 & 0 & -1 \end{bmatrix} \end{array}$$

This array of exponents is called the *dimensional matrix* for the given function. Show that the *rank* of this matrix (the size of the largest nonzero determinant) is equal to $j = n - k$, the desired reduction between original variables and the pi groups. This is a general property of dimensional matrices, as noted by Buckingham [1].

P5.22 As will be discussed in Chap. 11, the power P developed by a wind turbine is a function of diameter D, air density ρ, wind speed V, and rotation rate ω. Viscosity effects are negligible. Rewrite this relationship in dimensionless form.

P5.23 The period T of vibration of a beam is a function of its length L, area moment of inertia I, modulus of elasticity E, density ρ, and Poisson's ratio σ. Rewrite this relation in dimensionless form. What further reduction can we make if E and I can occur only in the product form EI? *Hint:* Take L, ρ, and E as repeating variables.

P5.24 The lift force F on a missile is a function of its length L, velocity V, diameter D, angle of attack α, density ρ, viscosity μ, and speed of sound a of the air. Write out the dimensional matrix of this function and determine its rank. (See Prob. P5.21 for an explanation of this concept.) Rewrite the function in terms of pi groups.

P5.25 The thrust F of a propeller is generally thought to be a function of its diameter D and angular velocity Ω, the forward speed V, and the density ρ and viscosity μ of the fluid. Rewrite this relationship as a dimensionless function.

P5.26 A pendulum has an oscillation period T which is assumed to depend on its length L, bob mass m, angle of swing θ, and the acceleration of gravity. A pendulum 1 m long, with a bob mass of 200 g, is tested on earth and found to have a period of 2.04 s when swinging at 20°. (*a*) What is its period when it swings at 45°? A similarly constructed pendulum, with $L = 30$ cm and $m = 100$ g, is to swing on the moon ($g = 1.62$ m/s^2) at $\theta = 20°$. (*b*) What will be its period?

P5.27 In studying sand transport by ocean waves, A. Shields in 1936 postulated that the threshold wave-induced bottom shear stress τ required to move particles depends on gravity g, particle size d and density ρ_p, and water density ρ and viscosity μ. Find suitable dimensionless groups of this problem, which resulted in 1936 in the celebrated Shields sand transport diagram.

P5.28 A simply supported beam of diameter D, length L, and modulus of elasticity E is subjected to a fluid crossflow of velocity V, density ρ, and viscosity μ. Its center deflection δ is assumed to be a function of all these variables. (*a*) Rewrite this proposed function in dimensionless form. (*b*) Suppose it is known that δ is independent of μ, inversely proportional to E, and dependent only on ρV^2, not ρ and V separately. Simplify the dimensionless function accordingly. *Hint:* Take L, ρ, and V as repeating variables.

P5.29 When fluid in a pipe is accelerated linearly from rest, it begins as laminar flow and then undergoes transition to turbulence at a time t_{tr} that depends on the pipe diameter D, fluid acceleration a, density ρ, and viscosity μ. Arrange this into a dimensionless relation between t_{tr} and D.

P5.30 When a large tank of high-pressure gas discharges through a nozzle, the exit mass flow $\dot{m}$ is a function of tank pressure p_0 and temperature T_0, gas constant R, specific heat c_p, and nozzle diameter D. Rewrite this as a dimensionless function. Check to see if you can use (p_0, T_0, R, D) as repeating variables.

P5.31 The pressure drop per unit length in horizontal pipe flow, $\Delta p/L$, depends on the fluid density ρ, viscosity μ, diameter D, and volume flow rate Q. Rewrite this function in terms of pi groups.

P5.32 A *weir* is an obstruction in a channel flow that can be calibrated to measure the flow rate, as in Fig. P5.32. The volume flow Q varies with gravity g, weir width b into the paper, and upstream water height H above the weir crest. If it is known that Q is proportional to b, use the pi theorem to find a unique functional relationship $Q(g, b, H)$.

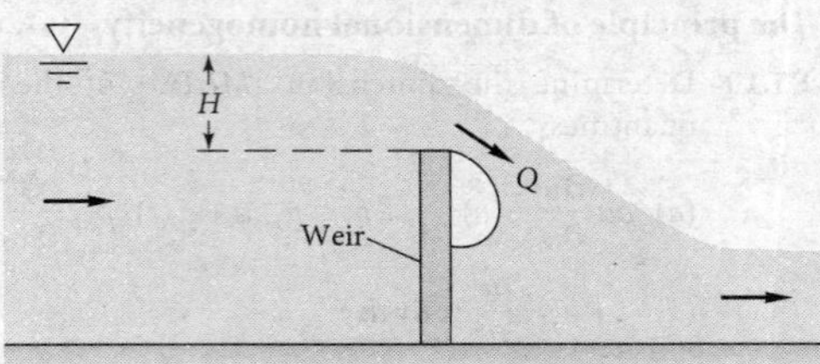

P5.32

P5.33 A spar buoy (see Prob. P2.113) has a period T of vertical (heave) oscillation that depends on the waterline cross-sectional area A, buoy mass m, and fluid specific weight γ. How does the period change due to doubling of (*a*) the mass and (*b*) the area? Instrument buoys should have long periods to avoid wave resonance. Sketch a possible long-period buoy design.

P5.34 To good approximation, the thermal conductivity k of a gas (see Ref. 21 of Chap. 1) depends only on the density ρ, mean free path l, gas constant R, and absolute temperature T. For air at 20°C and 1 atm, $k \approx 0.026$ W/(m · K) and $l \approx 6.5$ E-8 m. Use this information to determine k for hydrogen at 20°C and 1 atm if $l \approx 1.2$ E-7 m.

P5.35 The torque M required to turn the cone-plate viscometer in Fig. P5.35 depends on the radius R, rotation rate Ω, fluid viscosity μ, and cone angle θ. Rewrite this relation in dimensionless form. How does the relation simplify it if it is known that M is proportional to θ?

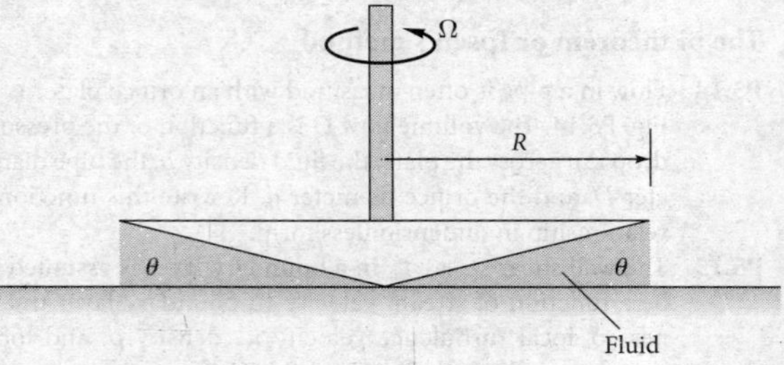

P5.35

P5.36 The rate of heat loss $\dot{Q}_{loss}$ through a window or wall is a function of the temperature difference between inside and outside ΔT, the window surface area A, and the R value of the window, which has units of ($m^2 \cdot s \cdot °C$)/ J. (*a*) Using the Buckingham Pi Theorem, find an expression for rate of heat loss as a function of the other three parameters in the problem. (*b*) If the temperature difference ΔT doubles, by what factor does the rate of heat loss increase?

P5.37 The volume flow Q through an orifice plate is a function of pipe diameter D, pressure drop Δp across the orifice, fluid density ρ and viscosity μ, and orifice diameter d. Using D, ρ, and Δp as repeating variables, express this relationship in dimensionless form.

P5.38 The size d of droplets produced by a liquid spray nozzle is thought to depend on the nozzle diameter D, jet velocity U, and the properties of the liquid ρ, μ, and Υ. Rewrite this relation in dimensionless form. *Hint:* Take D, ρ, and U as repeating variables.

P5.39 The volume flow Q over a certain dam is a function of dam width b, gravity g, and the upstream water depth H above the dam crest. It is known that Q is proportional to b. If b = 36 m and H = 38 cm, the flow rate is 17 m^3/s. What will be the flow rate if H = 1 m.

P5.40 The time t_d to drain a liquid from a hole in the bottom of a tank is a function of the hole diameter d, the initial fluid volume Υ_0, the initial liquid depth h_0, and the density ρ and viscosity μ of the fluid. Rewrite this relation as a dimensionless function, using Ipsen's method.

P5.41 A certain axial flow turbine has an output torque M that is proportional to the volume flow rate Q and also depends on the density ρ, rotor diameter D, and rotation rate Ω. How does the torque change due to a doubling of (*a*) D and (*b*) Ω?

P5.42 When disturbed, a floating buoy will bob up and down at frequency f. Assume that this frequency varies with buoy mass m, waterline diameter d, and the specific weight γ of the liquid. (*a*) Express this as a dimensionless function. (*b*) If d and γ are constant and the buoy mass is halved, how will the frequency change?

Nondimensionalizing the basic equations

P5.43 Nondimensionalize the energy equation (4.75) and its boundary conditions (4.62), (4.63), and (4.70) by defining $T^* = T/T_0$, where T_0 is the inlet temperature, assumed constant. Use other dimensionless variables as needed from Eqs. (5.23). Isolate all dimensionless parameters you find, and relate them to the list given in Table 5.2.

P5.44 The differential energy equation for incompressible two-dimensional flow through a "Darcy-type" porous medium is approximately

$$\rho c_p \frac{\sigma}{\mu}\frac{\partial p}{\partial x}\frac{\partial T}{\partial x} + \rho c_p \frac{\sigma}{\mu}\frac{\partial p}{\partial y}\frac{\partial T}{\partial y} + k\frac{\partial^2 T}{\partial y^2} = 0$$

where σ is the *permeability* of the porous medium. All other symbols have their usual meanings. (*a*) What are the appropriate dimensions for σ? (*b*) Nondimensionalize this equation, using (L, U, ρ, T_0) as scaling constants, and discuss any dimensionless parameters that arise.

P5.45 A model differential equation, for chemical reaction dynamics in a plug reactor, is as follows:

$$u\frac{\partial C}{\partial x} = D\frac{\partial^2 C}{\partial x^2} - kC - \frac{\partial C}{\partial t}$$

where u is the velocity, D is a diffusion coefficient, k is a reaction rate, x is distance along the reactor, and C is the (dimensionless) concentration of a given chemical in the reactor. (*a*) Determine the appropriate dimensions of D and k. (*b*) Using a characteristic length scale L and average velocity V as parameters, rewrite this equation in dimensionless form and comment on any pi groups appearing.

P5.46 If a vertical wall at temperature T_w is surrounded by a fluid at temperature T_0, a natural convection boundary layer flow will form. For laminar flow, the momentum equation is

$$\rho\left(u\frac{\partial u}{\partial x} + v\frac{\partial u}{\partial y}\right) = \rho\beta(T - T_0)g + \mu\frac{\partial^2 u}{\partial y^2}$$

to be solved, along with continuity and energy, for (u, v, T) with appropriate boundary conditions. The quantity β is the thermal expansion coefficient of the fluid. Use ρ, g, L, and ($T_w - T_0$) to nondimensionalize this equation. Note that there is no "stream" velocity in this type of flow.

P5.47 The differential equation for small-amplitude vibrations $y(x, t)$ of a simple beam is given by

$$\rho A\frac{\partial^2 y}{\partial t^2} + EI\frac{\partial^4 y}{\partial x^4} = 0$$

where ρ = beam material density
A = cross-sectional area
I = area moment of inertia
E = Young's modulus

Use only the quantities ρ, E, and A to nondimensionalize y, x, and t, and rewrite the differential equation in dimensionless form. Do any parameters remain? Could they be removed by further manipulation of the variables?

Data for spheres, cylinders, other bodies

P5.48 A smooth steel (SG = 7.86) sphere is immersed in a stream of ethanol at 20°C moving at 1.5 m/s. Estimate its drag in N from Fig. 5.3*a*. What stream velocity would quadruple its drag? Take D = 2.5 cm.

P5.49 The sphere in Prob. P5.48 is dropped in gasoline at 20°C. Ignoring its acceleration phase, what will its terminal (constant) fall velocity be, from Fig. 5.3*a*?

P5.50 The parachute in the chapter-opener photo is, of course, meant to decelerate the payload on Mars. The wind tunnel test gave a drag coefficient of about 1.1, based upon the projected area of the parachute. Suppose it was falling on *earth* and, at an altitude of 1,000 m, showed a steady descent rate of about 8 m/s. Estimate the weight of the payload.

P5.51 A ship is towing a sonar array that approximates a submerged cylinder 30 cm in diameter and 9 m long with its axis normal to the direction of tow. If the tow speed is 12 kn (1 kn = 0.515 m/s), estimate the horsepower required to tow this cylinder. What will be the frequency of vortices shed from the cylinder? Use Figs. 5.2 and 5.3.

P5.52 When fluid in a long pipe starts up from rest at a uniform acceleration a, the initial flow is laminar. The flow undergoes transition to turbulence at a time t^* which depends, to first approximation, only upon a, ρ, and μ. Experiments by P. J. Lefebvre, on water at 20°C starting from rest with 1-g acceleration in a 3-cm-diameter pipe, showed transition at $t^* = 1.02$ s. Use this data to estimate (*a*) the transition time and (*b*) the transition Reynolds number Re_D for water flow accelerating at 35 m/s^2 in a 5-cm-diameter pipe.

P5.53 Vortex shedding can be used to design a *vortex flowmeter* (Fig. 6.34). A blunt rod stretched across the pipe sheds vortices whose frequency is read by the sensor downstream. Suppose the pipe diameter is 5 cm and the rod is a cylinder of diameter 8 mm. If the sensor reads 5,400 counts per minute, estimate the volume flow rate of water in m^3/h. How might the meter react to other liquids?

P5.54 A fishnet is made of 1-mm-diameter strings knotted into 2 × 2 cm squares. Estimate the horsepower required to tow 30 m^2 of this netting at 1.5 m/s in seawater at 20°C. The net plane is normal to the flow direction.

P5.55 The radio antenna on a car begins to vibrate wildly at 8 Hz when the car is driven at 20 m/s over a rutted road that approximates a sine wave of amplitude 2 cm and wavelength $\lambda = 2.5$ m. The antenna diameter is 4 mm. Is the vibration due to the road or to vortex shedding?

P5.56 Flow past a long cylinder of square cross-section results in more drag than the comparable round cylinder. Here are data taken in a water tunnel for a square cylinder of side length b = 2 cm:

V, m/s	1.0	2.0	3.0	4.0
Drag, N/(m of depth)	21	85	191	335

(*a*) Use these data to predict the drag force per unit depth of wind blowing at 6 m/s, in air at 20°C, over a tall square chimney of side length b = 55 cm. (*b*) Is there any uncertainty in your estimate?

P5.57 The simply supported 1040 carbon-steel rod of Fig. P5.57 is subjected to a crossflow stream of air at 20°C and 1 atm. For what stream velocity U will the rod center deflection be approximately 1 cm?

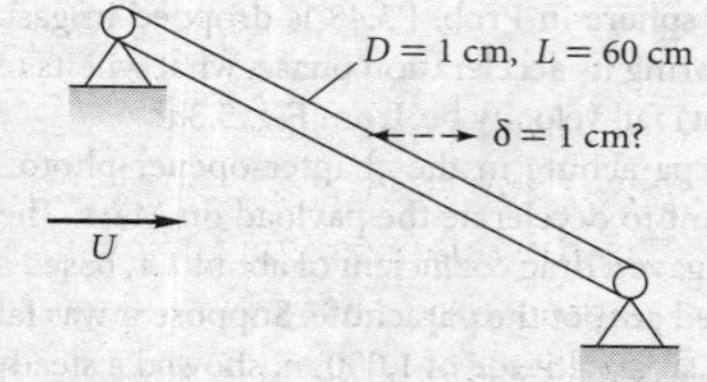

P5.57

P5.58 For the steel rod of Prob. P5.57, at what airstream velocity U will the rod begin to vibrate laterally in resonance in its first mode (a half sine wave)? *Hint:* Consult a vibration text [34,35] under "lateral beam vibration."

P5.59 A long, slender, smooth 3-cm-diameter flagpole bends alarmingly in 9 m/s sea-level winds, causing patriotic citizens to gasp. An engineer claims that the pole will bend less if its surface is deliberately roughened. Is she correct, at least qualitatively?

Scaling of model data

***P5.60** The thrust F of a free propeller, either aircraft or marine, depends upon density ρ, the rotation rate n in r/s, the diameter D, and the forward velocity V. Viscous effects are slight and neglected here. Tests of a 25-cm-diameter model aircraft propeller, in a sea-level wind tunnel, yield the following thrust data at a velocity of 20 m/s:

Rotation rate, r/min	4,800	6,000	8,000
Measured thrust, N	6.1	19	47

(*a*) Use this data to make a crude but effective dimensionless plot. (*b*) Use the dimensionless data to predict the thrust, in newtons, of a similar 1.6-m-diameter prototype propeller when rotating at 3,800 r/min and flying at 100 m/s at 4,000-m standard altitude.

P5.61 If viscosity is neglected, typical pump flow results from Example 5.3 are shown in Fig. P5.61 for a model pump tested in water. The pressure rise decreases and the power required increases with the dimensionless flow coefficient. Curve-fit expressions are given for the data. Suppose a similar pump of 12-cm diameter is built to move gasoline at 20°C and a flow rate of 25 m^3/h. If the pump rotation speed is 30 r/s, find (*a*) the pressure rise and (*b*) the power required.

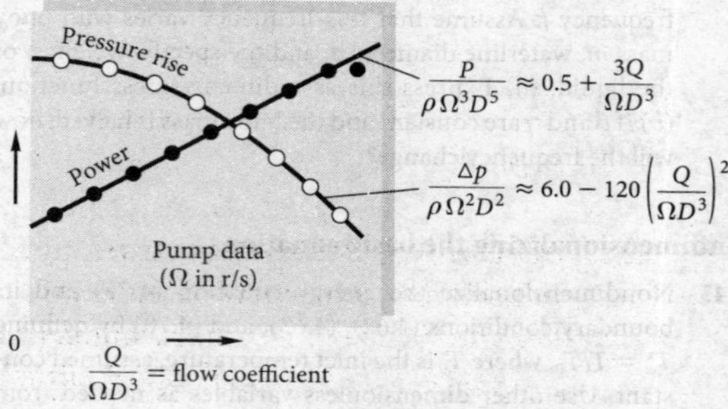

P5.61

P5.62 For the system of Prob. P5.22, assume that a small model wind turbine of diameter 90 cm, rotating at 1,200 r/min, delivers 280 watts when subjected to a wind of 12 m/s. The data is to be used for a prototype of diameter 50 m and winds of 8 m/s. For dynamic similarity, estimate (*a*) the rotation rate, and (*b*) the power delivered by the prototype. Assume sea-level air density.

P5.63 The Keystone Pipeline in the Chapter 6 opener photo has D = 0.9 m and an oil flow rate Q = 590,000 barrels per day

(1 barrel = 0.159 m^3). Its pressure drop per unit length, $\Delta p/L$, depends on the fluid density ρ, viscosity μ, diameter D, and flow rate Q. A water-flow model test, at 20°C, uses a 5-cm-diameter pipe and yields $\Delta p/L \approx$ 4,000 Pa/m. For dynamic similarity, estimate $\Delta p/L$ of the pipeline. For the oil take ρ = 860 kg/m^3 and μ = 0.005 kg/m · s.

P5.64 The natural frequency ω of vibration of a mass M attached to a rod, as in Fig. P5.64, depends only on M

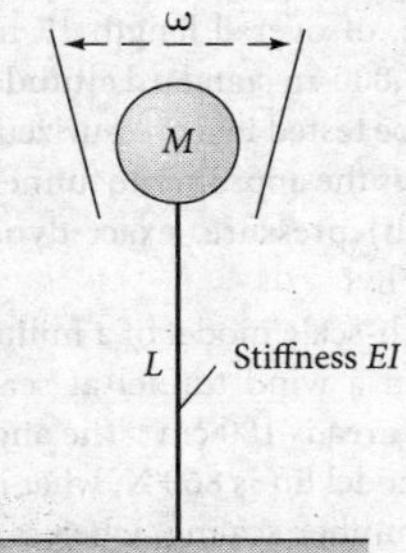

P5.64

and the stiffness EI and length L of the rod. Tests with a 2-kg mass attached to a 1040 carbon steel rod of diameter 12 mm and length 40 cm reveal a natural frequency of 0.9 Hz. Use these data to predict the natural frequency of a 1-kg mass attached to a 2024 aluminum alloy rod of the same size.

P5.65 In turbulent flow near a flat wall, the local velocity u varies only with distance y from the wall, wall shear stress τ_w, and fluid properties ρ and μ. The following data were taken in the University of Rhode Island wind tunnel for airflow, ρ = 1.185 kg/m^3, μ = 1.82E-5kg/m • s, and τ_w = 1.39 Pa.

y, mm	0.533	0.889	1.397	2.032	3.048	4.064
u, m/s	15.42	16.52	17.56	18.20	19.35	20.09

(*a*) Plot these data in the form of dimensionless u versus dimensionless y, and suggest a suitable power-law curve fit. (*b*) Suppose that the tunnel speed is increased until u = 27.5 m/s at y = 3 mm. Estimate the new wall shear stress, in Pa.

P5.66 A torpedo 8 m below the surface in 20°C seawater cavitates at a speed of 21 m/s when atmospheric pressure is 101 kPa. If Reynolds number and Froude number effects are negligible, at what speed will it cavitate when running at a depth of 20 m? At what depth should it be to avoid cavitation at 30 m/s?

P5.67 A student needs to measure the drag on a prototype of characteristic dimension d_p moving at velocity U_p in air at standard atmospheric conditions. He constructs a model of characteristic dimension d_m, such that the ratio d_p/d_m is some factor f. He then measures the drag on the model at dynamically similar conditions (also with air at standard atmospheric conditions). The student claims that the drag force on the prototype will be identical to that measured on the model. Is this claim correct? Explain.

P5.68 For the rotating-cylinder function of Prob. P5.20, if $L >> D$, the problem can be reduced to only two groups, $F/(\rho U^2 LD)$ versus $(\Omega D/U)$. Here are experimental data for a cylinder 30 cm in diameter and 2 m long, rotating in sea-level air, with U = 25 m/s.

Ω, rev/min	0	3,000	6,000	9,000	12,000	15,000
F, N	0	850	2,260	2,900	3,120	3,300

(*a*) Reduce this data to the two dimensionless groups and make a plot. (*b*) Use this plot to predict the lift of a cylinder with D = 5 cm, L = 80 cm, rotating at 3,800 rev/min in water at U = 4 m/s.

P5.69 A simple flow measurement device for streams and channels is a notch, of angle α, cut into the side of a dam, as shown in Fig. P5.69. The volume flow Q depends only on α, the acceleration of gravity g, and the height δ of the upstream water surface above the notch vertex. Tests of a model notch, of angle α = 55°, yield the following flow rate data:

δ, cm	10	20	30	40
Q, m^3/h	8	47	126	263

(*a*) Find a dimensionless correlation for the data. (*b*) Use the model data to predict the flow rate of a prototype notch, also of angle α = 55°, when the upstream height δ is 3.2 m.

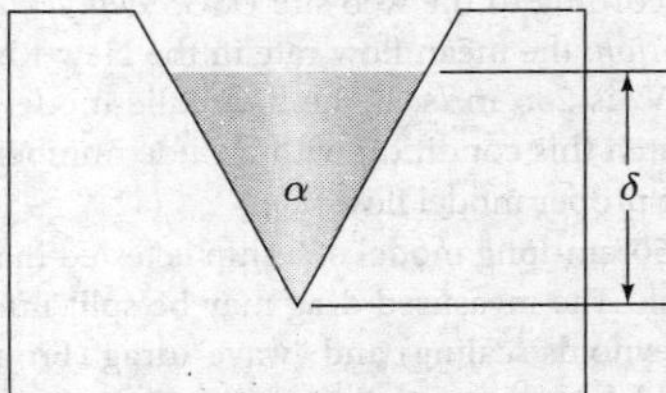

P5.69

P5.70 A diamond-shaped body, of characteristic length 23 cm, has the following measured drag forces when placed in a wind tunnel at sea-level standard conditions:

V, m/s	9	12	15	17	19
F, N	5.56	8.67	13.43	18.02	21.40

Use these data to predict the drag force of a similar 38-cm diamond placed at similar orientation in 20°C water flowing at 2.2 m/s.

P5.71 The pressure drop in a venturi meter (Fig. P3.128) varies only with the fluid density, pipe approach velocity, and diameter ratio of the meter. A model venturi meter tested in water at 20°C shows a 5-kPa drop when the approach velocity is 4 m/s. A geometrically similar prototype meter is used to measure gasoline at 20°C and a flow rate of 9 m^3/min. If the prototype pressure gage is most accurate at 15 kPa, what should the upstream pipe diameter be?

P5.72 A one-twelfth-scale model of a large commercial aircraft is tested in a wind tunnel at 20°C and 1 atm. The model chord length is 27 cm, and its wing area is 0.63 m^2. Test results for the drag of the model are as follows:

V, m/s	22.5	34	45	56
Drag, N	15	32	53	80

In the spirit of Fig. 5.8, use this data to estimate the drag of the full-scale aircraft when flying at 250 m/s, for the same angle of attack, at 10 km standard altitude.

P5.73 The power P generated by a certain windmill design depends on its diameter D, the air density ρ, the wind velocity V, the rotation rate Ω, and the number of blades n. (*a*) Write this relationship in dimensionless form. A model windmill, of diameter 50 cm, develops 2.7 kW at sea level when $V = 40$ m/s and when rotating at 4,800 r/min. (*b*) What power will be developed by a geometrically and dynamically similar prototype, of diameter 5 m, in winds of 12 m/s at 2,000 m standard altitude? (*c*) What is the appropriate rotation rate of the prototype?

P5.74 A one-tenth-scale model of a supersonic wing tested at 700 m/s in air at 20°C and 1 atm shows a pitching moment of 0.25 kN · m. If Reynolds number effects are negligible, what will the pitching moment of the prototype wing be if it is flying at the same Mach number at 8-km standard altitude?

Froude and Mach number scaling

P5.75 According to the web site *USGS Daily Water Data for the Nation*, the mean flow rate in the New River near Hinton, WV, is 286 m^3/s. If the hydraulic model in Fig. 5.9 is to match this condition with Froude number scaling, what is the proper model flow rate?

***P5.76** A 60-cm-long model of a ship is tested in a freshwater tow tank. The measured drag may be split into "friction" drag (Reynolds scaling) and "wave" drag (Froude scaling). The model data are as follows:

Tow speed, m/s	0.25	0.50	0.75	1	1.25	1.50
Friction drag, N	0.071	0.254	0.543	0.925	1.401	1.961
Wave drag, N	0.008	0.093	0.369	1.125	2.264	3.100

The prototype ship is 45 m long. Estimate its total drag when cruising at 8 m/s in seawater at 20°C.

P5.77 A dam 25 m wide, with a nominal flow rate of 7.5 m^3/s, is to be studied with a scale model 1 m wide, using Froude scaling. (*a*) What is the expected flow rate for the model? (*b*) What is the danger of only using Froude scaling for this test? (*c*) Derive a formula for a force on the model as compared to a force on the prototype.

P5.78 A prototype spillway has a characteristic velocity of 3 m/s and a characteristic length of 10 m. A small model is constructed by using Froude scaling. What is the minimum scale ratio of the model that will ensure that its minimum Weber number is 100? Both flows use water at 20°C.

P5.79 An East Coast estuary has a tidal period of 12.42 h (the semidiurnal lunar tide) and tidal currents of approximately 80 cm/s. If a one-five-hundredth-scale model is constructed with tides driven by a pump and storage apparatus, what should the period of the model tides be and what model current speeds are expected?

P5.80 A prototype ship is 35 m long and designed to cruise at 11 m/s. Its drag is to be simulated by a 1-m-long model pulled in a tow tank. For Froude scaling find (*a*) the tow speed, (*b*) the ratio of prototype to model drag, and (*c*) the ratio of prototype to model power.

P5.81 An airplane, of overall length 17 m, is designed to fly at 680 m/s at 8,000-m standard altitude. A one-thirtieth-scale model is to be tested in a pressurized helium wind tunnel at 20°C. What is the appropriate tunnel pressure in atm? Even at this (high) pressure, exact dynamic similarity is not achieved. Why?

P5.82 A one-fiftieth-scale model of a military airplane is tested at 1,020 m/s in a wind tunnel at sea-level conditions. The model wing area is 180 cm^2. The angle of attack is 3°. If the measured model lift is 860 N, what is the prototype lift, using Mach number scaling, when it flies at 10,000 m standard altitude under dynamically similar conditions? *Note:* Be careful with the area scaling.

P5.83 A one-fortieth-scale model of a ship's propeller is tested in a tow tank at 1,200 r/min and exhibits a power output of 2 watts. According to Froude scaling laws, what should the revolutions per minute and horsepower output of the prototype propeller be under dynamically similar conditions?

P5.84 A prototype ocean platform piling is expected to encounter currents of 150 cm/s and waves of 12-s period and 3-m height. If a one-fifteenth-scale model is tested in a wave channel, what current speed, wave period, and wave height should be encountered by the model?

Inventive rescaling of the data

***P5.85** As shown in Example 5.3, pump performance data can be nondimensionalized. Problem P5.61 gave typical dimensionless data for centrifugal pump "head," $H = \Delta p/\rho g$, as follows:

$$\frac{gH}{n^2D^2} \approx 6.0 - 120\left(\frac{Q}{nD^3}\right)^2$$

where Q is the volume flow rate, n the rotation rate in r/s, and D the impeller diameter. This type of correlation allows one to compute H when (ρ, Q, D) are known. (*a*) Show how to rearrange these pi groups so that one can *size* the pump, that is, compute D directly when (Q, H, n) are known. (*b*) Make a crude but effective plot of your new function. (*c*) Apply part (*b*) to the following example: Find D when $H = 37$ m, $Q = 0.14$ m^3/s, and $n = 35$ r/s. Find the pump diameter for this condition.

P5.86 Solve Prob. P5.49 for glycerin at 20°C, using the modified sphere-drag plot of Fig. 5.11.

P5.87 In Prob. P5.61 it would be difficult to solve for Ω because it appears in all three of the dimensionless pump coefficients. Suppose that, in Prob. 5.61, Ω is unknown but $D = 12$ cm and $Q = 25$ m^3/h. The fluid is gasoline at 20°C. Rescale the coefficients, using the data of Prob. P5.61, to make a plot of dimensionless power versus dimensionless rotation speed. Enter this plot to find the maximum rotation speed Ω for which the power will not exceed 300 W.

P5.88 Modify Prob. P5.61 as follows: Let $\Omega = 32$ r/s and $Q =$ 24 m^3/h for a geometrically similar pump. What is the maximum diameter if the power is not to exceed 340 W? Solve this problem by rescaling the data of Fig. P5.61 to make a plot of dimensionless power versus dimensionless diameter. Enter this plot directly to find the desired diameter.

P5.89 Wall friction τ_w, for turbulent flow at velocity U in a pipe of diameter D, was correlated, in 1911, with a dimensionless correlation by Ludwig Prandtl's student H. Blasius:

$$\frac{\tau_w}{\rho U^2} \approx \frac{0.632}{(\rho U D/\mu)^{1/4}}$$

Suppose that (ρ, U, μ, τ_w) were all known and it was desired to find the unknown velocity U. Rearrange and rewrite the formula so that U can be immediately calculated.

P5.90 Knowing that Δp is proportional to L, rescale the data of Example 5.10 to plot dimensionless Δp versus dimensionless *viscosity*. Use this plot to find the viscosity required in the first row of data in Example 5.10 if the pressure drop is increased to 10 kPa for the same flow rate, length, and density.

***P5.91** The traditional "Moody-type" pipe friction correlation in Chap. 6 is of the form

$$f = \frac{2\Delta p D}{\rho V^2 L} = \text{fcn}\left(\frac{\rho V D}{\mu}, \frac{\varepsilon}{D}\right)$$

where D is the pipe diameter, L the pipe length, and ε the wall roughness. Note that pipe average velocity V is used on both sides. This form is meant to find Δp when V is known. (*a*) Suppose that Δp is known, and we wish to find V. Rearrange the above function so that V is isolated on the left-hand side. Use the following data, for $\varepsilon/D = 0.005$, to make a plot of your new function, with your velocity parameter as the ordinate of the plot.

f	0.0356	0.0316	0.0308	0.0305	0.0304
$\rho V D/\mu$	15,000	75,000	250,000	900,000	3,330,000

(*b*) Use your plot to determine V, in m/s, for the following pipe flow: $D = 5$ cm, $\varepsilon = 0.025$ cm, $L = 10$ m, for water flow at 20°C and 1 atm. The pressure drop Δp is 110 kPa.

Word Problems

W5.1 In 98 percent of data analysis cases, the "reducing factor" j, which lowers the number n of dimensional variables to $n - j$ dimensionless groups, exactly equals the number of relevant dimensions (M, L, T, Θ). In one case (Example 5.5) this was not so. Explain in words why this situation happens.

W5.2 Consider the following equation: 1 dollar bill $\approx$ 15 cm. Is this relation dimensionally inconsistent? Does it satisfy the PDH? Why?

W5.3 In making a dimensional analysis, what rules do you follow for choosing your scaling variables?

W5.4 In an earlier edition, the writer asked the following question about Fig. 5.1: "Which of the three graphs is a more effective presentation?" Why was this a dumb question?

W5.5 This chapter discusses the difficulty of scaling Mach and Reynolds numbers together (an airplane) and Froude and Reynolds numbers together (a ship). Give an example of a flow that would combine Mach and Froude numbers. Would there be scaling problems for common fluids?

W5.6 What is different about a very *small* model of a weir or dam (Fig. P5.32) that would make the test results difficult to relate to the prototype?

W5.7 What else are you studying this term? Give an example of a popular equation or formula from another course (thermodynamics, strength of materials, or the like) that does not satisfy the principle of dimensional homogeneity. Explain what is wrong and whether it can be modified to be homogeneous.

W5.8 Some colleges (such as Colorado State University) have environmental wind tunnels that can be used to study phenomena like wind flow over city buildings. What details of scaling might be important in such studies?

W5.9 If the model scale ratio is $\alpha = L_m/L_p$, as in Eq. (5.31), and the Weber number is important, how must the model and prototype surface tension be related to α for dynamic similarity?

W5.10 For a typical incompressible velocity potential analysis in Chap. 8 we solve $\nabla^2\phi = 0$, subject to known values of $\partial\phi/\partial n$ on the boundaries. What dimensionless parameters govern this type of motion?

Fundamentals of Engineering Exam Problems

FE5.1 Given the parameters (U, L, g, ρ, μ) that affect a certain liquid flow problem, the ratio $V^2/(Lg)$ is usually known as the
(*a*) velocity head, (*b*) Bernoulli head, (*c*) Froude number, (*d*) kinetic energy, (*e*) impact energy

FE5.2 A ship 150 m long, designed to cruise at 9.3 m/s, is to be tested in a tow tank with a model 3 m long. The appropriate tow velocity is
(*a*) 0.19 m/s, (*b*) 0.35 m/s, (*c*) 1.31 m/s, (*d*) 2.55 m/s, (*e*) 8.35 m/s

FE5.3 A ship 150 m long, designed to cruise at 9.3 m/s, is to be tested in a tow tank with a model 3 m long. If the model wave drag is 2.2 N, the estimated full-size ship wave drag is
(*a*) 5,500 N, (*b*) 8,700 N, (*c*) 38,900 N, (*d*) 61,800 N, (*e*) 275,000 N

FE5.4 A tidal estuary is dominated by the semidiurnal lunar tide, with a period of 12.42 h. If a 1:500 model of the estuary is tested, what should be the model tidal period?
(*a*) 4.0 s, (*b*) 1.5 min, (*c*) 17 min, (*d*) 33 min, (*e*) 64 min

FE5.5 A football, meant to be thrown at 96.5 km/hr in sea-level air ($\rho = 1.22$ kg/m^3, $\mu = 1.78$ E-5 N · s/m^2), is to be tested using a one-quarter scale model in a water tunnel ($\rho = 998$ kg/m^3, $\mu = 0.0010$ N · s/m^2). For dynamic similarity, what is the proper model water velocity?
(*a*) 12.06 km/hr, (*b*) 24.14 km/hr, (*c*) 25.10 km/hr, (*d*) 26.55 km/hr, (*e*) 48.28 km/hr

FE5.6 A football, meant to be thrown at 96.5 km/hr in sea-level air ($\rho = 1.22$ kg/m^3, $\mu = 1.78$ E-5 N · m^2), is to be tested using a one-quarter scale model in a water tunnel ($\rho = 998$ kg/m^3, $\mu = 0.0010$ N · s/m^2). For dynamic similarity, what is the ratio of prototype force to model force?
(*a*) 3.86:1, (*b*) 16:1, (*c*) 32:1, (*d*) 56:1, (*e*) 64:1

FE5.7 Consider liquid flow of density ρ, viscosity μ, and velocity U over a very small model spillway of length scale L, such that the liquid surface tension coefficient Υ is important. The quantity $\rho U^2 L/\Upsilon$ in this case is important and is called the
(*a*) capillary rise, (*b*) Froude number, (*c*) Prandtl number, (*d*) Weber number, (*e*) Bond number

FE5.8 If a stream flowing at velocity U past a body of length L causes a force F on the body that depends only on U, L, and fluid viscosity μ, then F must be proportional to
(*a*) $\rho UL/\mu$, (*b*) ρU^2L^2, (*c*) $\mu U/L$, (*d*) μUL, (*e*) UL/μ

FE5.9 In supersonic wind tunnel testing, if different gases are used, dynamic similarity requires that the model and prototype have the same Mach number and the same
(*a*) Euler number, (*b*) speed of sound, (*c*) stagnation enthalpy, (*d*) Froude number, (*e*) specific-heat ratio

FE5.10 The Reynolds number for a 0.3 m-diameter sphere moving at 1.028 m/s through seawater (specific gravity 1.027, viscosity 1.07 E-3 N · s/m^2) is approximately
(*a*) 300, (*b*) 3,000, (*c*) 30,000, (*d*) 300,000, (*e*) 3,000,000

FE5.11 The Ekman number, important in physical oceanography, is a dimensionless combination of μ, L, ρ, and the earth's rotation rate Ω. If the Ekman number is proportional to Ω, it should take the form
(*a*) $\rho\Omega^2L^2/\mu$, (*b*) $\mu\Omega L/\rho$, (*c*) $\rho\Omega L/\mu$, (*d*) $\rho\Omega L^2/\mu$, (*e*) $\rho\Omega/L\mu$

FE5.12 A valid, but probably useless, dimensionless group is given by $(\mu T_0 g)/(\Upsilon L\alpha)$, where everything has its usual meaning, except α. What are the dimensions of α?
(*a*) $\Theta L^{-1}T^{-1}$, (*b*) $\Theta L^{-1}T^{-2}$, (*c*) ΘML^{-1}, (*d*) $\Theta^{-1}LT^{-1}$, (*e*) ΘLT^{-1}

Comprehensive Problems

C5.1 Estimating pipe wall friction is one of the most common tasks in fluids engineering. For long circular rough pipes in turbulent flow, wall shear τ_w is a function of density ρ, viscosity μ, average velocity V, pipe diameter d, and wall roughness height ε. Thus, functionally, we can write $\tau_w = \text{fcn}(\rho, \mu, V, d, \varepsilon)$. (*a*) Using dimensional analysis, rewrite this function in dimensionless form. (*b*) A certain pipe has $d = 5$ cm and $\varepsilon = 0.25$ mm. For flow of water at 20°C, measurements show the following values of wall shear stress:

Q, m^3/min	0.005	0.010	0.020	0.030	0.040	0.05
τ_w, Pa	0.05	0.18	0.37	0.64	0.86	1.25

Plot these data using the dimensionless form obtained in part (*a*) and suggest a curve-fit formula. Does your plot reveal the entire functional relation obtained in part (*a*)?

C5.2 When the fluid exiting a nozzle, as in Fig. P3.49, is a gas, instead of water, compressibility may be important, especially if upstream pressure p_1 is large and exit diameter d_2 is small. In this case, the difference $p_1 - p_2$ is no longer controlling, and the gas mass flow $\dot{m}$ reaches a maximum value that depends on p_1 and d_2 and also on the absolute upstream temperature T_1 and the gas constant R. Thus, functionally, $\dot{m} = \text{fcn}(p_1, d_2, T_1, R)$. (*a*) Using dimensional analysis, rewrite this function in dimensionless form. (*b*) A certain pipe has $d_2 = 1$ cm. For flow of air, measurements show the following values of mass flow through the nozzle:

T_1, K	300	300	300	500	800
p_1, kPa	200	250	300	300	300
$\dot{m}$, kg/s	0.037	0.046	0.055	0.043	0.034

Plot these data in the dimensionless form obtained in part (*a*). Does your plot reveal the entire functional relation obtained in part (*a*)?

C5.3 Reconsider the fully developed draining vertical oil film problem (see Fig. P4.80) as an exercise in dimensional analysis. Let the vertical velocity be a function only of distance from the plate, fluid properties, gravity, and film thickness. That is, $w = \text{fcn}(x, \rho, \mu, g, \delta)$. (*a*) Use the pi theorem to rewrite this function in terms of dimensionless parameters. (*b*) Verify that the exact solution from Prob. P4.80 is consistent with your result in part (*a*).

C5.4 The Taco Inc. model 4013 centrifugal pump has an impeller of diameter $D = 30$ cm. When pumping 20°C water at $\Omega = 1{,}160$ r/min, the measured flow rate Q and pressure rise Δp are given by the manufacturer as follows:

Q, m^3/min	0.75	1.1	1.5	1.90	2.25	2.65
Δp, kPa	248	241	234	220	200	160

(*a*) Assuming that $\Delta p = \text{fcn}(\rho, Q, D, \Omega)$, use the pi theorem to rewrite this function in terms of dimensionless parameters and then plot the given data in dimensionless form. (*b*) It is desired to use the same pump, running at 900 r/min, to pump 20°C gasoline at 1.5 m^3/min. According to your dimensionless correlation, what pressure rise Δp is expected, in kPa?

C5.5 Does an automobile radio antenna vibrate in resonance due to vortex shedding? Consider an antenna of length L and diameter D. According to beam vibration theory [see [34] or [35, p. 401]], the first mode natural frequency of a solid circular cantilever beam is $\omega_n = 3.516[EI/(\rho AL^4)]^{1/2}$, where E is the modulus of elasticity, I is the area moment of inertia, ρ is the beam material density, and A is the beam cross-section area. (*a*) Show that ω_n is proportional to the antenna radius R. (*b*) If the antenna is steel, with $L = 60$ cm and $D = 4$ mm, estimate the natural vibration frequency, in Hz. (*c*) Compare with the shedding frequency if the car moves at 30 m/s.

Design Projects

D5.1 We are given laboratory data, taken by Prof. Robert Kirchhoff and his students at the University of Massachusetts, for the spin rate of a 2-cup anemometer. The anemometer was made of ping-pong balls ($d = 38$ mm) split in half, facing in opposite directions, and glued to thin (6.5 mm) rods pegged to a center axle. (See Fig. P7.91 for a sketch.) There were four rods, of lengths $l = 65$, 98, 140, and 175 mm. The experimental data, for wind tunnel velocity U and rotation rate Ω, are as follows:

$l = 65$		$l = 98$		$l = 140$		$l = 175$	
U, m/s	**Ω, r/min**	**U, m/s**	**Ω, r/min**	**U, m/s**	**Ω, r/min**	**U, m/s**	**Ω, r/min**
5.776	435	5.776	225	6.126	140	7.074	115
6.767	545	7.068	290	8.159	215	8.412	145
7.894	650	8.885	370	9.562	260	9.775	175
9.126	760	9.994	425	10.99	295	10.99	195
11.72	970	11.72	495	11.90	327	12.07	215

Assume that the angular velocity Ω of the device is a function of wind speed U, air density ρ and viscosity μ, rod length l, and cup diameter d. For all data, assume air is at 1 atm and 20°C. Define appropriate pi groups for this problem, and plot the data in this dimensionless manner. Comment on the possible uncertainty of the results.

As a design application, suppose we are to use this anemometer geometry for a large-scale ($d = 30$ cm) airport wind anemometer. If wind speeds vary up to 25 m/s and we desire an average rotation rate $\Omega = 120$ r/min, what should be the proper rod length? What are possible limitations of your design? Predict the expected Ω (in r/min) of your design as affected by wind speeds from 0 to 25 m/s.

D5.2 By analogy with the cylinder drag data in Fig. 5.3*b*, spheres also show a strong roughness effect on drag, at least in the Reynolds number range 4 E4 < Re_D < 3 E5, which accounts for the dimpling of golf balls to increase their distance traveled. Some experimental data for roughened spheres [33] are given in Fig. D5.2. The figure also shows typical golf ball data. We see that some roughened spheres are better than golf balls in some regions. For the present study, let us neglect the ball's *spin*, which causes the very important side-force or *Magnus effect* (see Fig. 8.15) and assume that the ball is hit without spin and follows the equations of motion for plane motion (*x*, *z*):

$$m\ddot{x} = -F\cos\theta \qquad m\ddot{z} = -F\sin\theta - W$$

where

$$F = C_D\frac{\rho}{2}\frac{\pi}{4}D^2(\dot{x}^2 + \dot{z}^2) \qquad \theta = \tan^{-1}\frac{\dot{z}}{\dot{x}}$$

The ball has a particular $C_D(Re_D)$ curve from Fig. D5.2 and is struck with an initial velocity V_0 and angle θ_0. Take the ball's average mass to be 46 g and its diameter to be 4.3 cm. Assuming sea-level air and a modest but finite range of initial conditions, integrate the equations of motion to compare the trajectory of "roughened spheres" to actual golf ball calculations. Can the rough sphere outdrive a normal golf ball for any conditions? What roughness-effect differences occur between a low-impact duffer and, say, Tiger Woods?

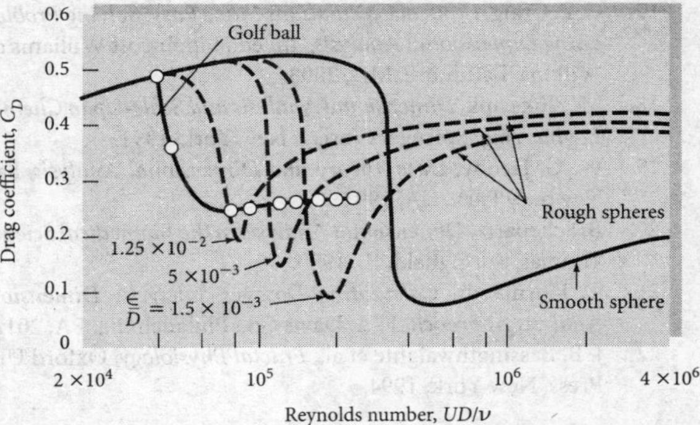

D5.2

References

1. E. Buckingham, "On Physically Similar Systems: Illustrations of the Use of Dimensional Equations," *Phys. Rev.*, vol. 4, no. 4, 1914, pp. 345–376.
2. J. D. Anderson, *Computational Fluid Dynamics: The Basics with Applications,* McGraw-Hill, New York, 1995.
3. P. W. Bridgman, *Dimensional Analysis,* Yale University Press, New Haven, CT, 1922, rev. ed., 1963.
4. H. L. Langhaar, *Dimensional Analysis and the Theory of Models,* Wiley, New York, 1951.
5. E. C. Ipsen, *Units, Dimensions, and Dimensionless Numbers,* McGraw-Hill, New York, 1960.
6. H. G. Hornung, *Dimensional Analysis: Examples of the Use of Symmetry,* Dover, New York, 2006.
7. E. S. Taylor, *Dimensional Analysis for Engineers,* Clarendon Press, Oxford, England, 1974.
8. G. I. Barenblatt, *Dimensional Analysis,* Gordon and Breach, New York, 1987.
9. A. C. Palmer, *Dimensional Analysis and Intelligent Experimentation,* World Scientific Publishing, Hackensack, NJ, 2008.
10. T. Szirtes, *Applied Dimensional Analysis and Modeling,* 2d ed., Butterworth-Heinemann, Burlington, MA, 2006.
11. R. Esnault-Pelterie, *Dimensional Analysis and Metrology,* F. Rouge, Lausanne, Switzerland, 1950.
12. R. Kurth, *Dimensional Analysis and Group Theory in Astrophysics,* Pergamon, New York, 1972.
13. R. Kimball and M. Ross, *The Data Warehouse Toolkit: The Complete Guide to Dimensional Modeling,* 2d ed., Wiley, New York, 2002.
14. R. Nakon, *Chemical Problem Solving Using Dimensional Analysis,* Prentice-Hall, Upper Saddle River, NJ, 1990.
15. D. R. Maidment (ed.), *Hydrologic and Hydraulic Modeling Support: With Geographic Information Systems,* Environmental Systems Research Institute, Redlands, CA, 2000.
16. A. M. Curren, *Dimensional Analysis for Meds,* 4th ed., Delmar Cengage Learning, Independence, KY, 2009.
17. G. P. Craig, *Clinical Calculations Made Easy: Solving Problems Using Dimensional Analysis,* 4th ed., Lippincott Williams and Wilkins, Baltimore, MD, 2008.
18. M. Zlokarnik, *Dimensional Analysis and Scale-Up in Chemical Engineering,* Springer-Verlag, New York, 1991.
19. W. G. Jacoby, *Data Theory and Dimensional Analysis,* Sage, Newbury Park, CA, 1991.
20. B. Schepartz, *Dimensional Analysis in the Biomedical Sciences,* Thomas, Springfield, IL, 1980.
21. T. Horntvedt, *Calculating Dosages Safely: A Dimensional Analysis Approach,* F. A. Davis Co., Philadelphia, PA, 2012.
22. J. B. Bassingthwaighte et al., *Fractal Physiology,* Oxford Univ. Press, New York, 1994.
23. K. J. Niklas, *Plant Allometry: The Scaling of Form and Process,* Univ. of Chicago Press, Chicago, 1994.
24. "*Flow of Fluids through Valves, Fittings, and Pipes,*" Crane Valve Group, Long Beach, CA, 1957 (now updated as a CD-ROM; see <http://www.cranevalves.com>).
25. A. Roshko, "On the Development of Turbulent Wakes from Vortex Streets," *NACA Rep.* 1191, 1954.
26. G. W. Jones, Jr., "Unsteady Lift Forces Generated by Vortex Shedding about a Large, Stationary, Oscillating Cylinder at High Reynolds Numbers," *ASME Symp. Unsteady Flow,* 1968.
27. O. M. Griffin and S. E. Ramberg, "The Vortex Street Wakes of Vibrating Cylinders," *J. Fluid Mech.*, vol. 66, pt. 3, 1974, pp. 553–576.
28. *Encyclopedia of Science and Technology,* 11th ed., McGraw-Hill, New York, 2012.
29. J. Kunes, *Dimensionless Physical Quantities in Science and Engineering,* Elsevier, New York, 2012.
30. V. P. Singh et al. (eds.), *Hydraulic Modeling,* Water Resources Publications LLC, Highlands Ranch, CO, 1999.
31. L. Armstrong, *Hydraulic Modeling and GIS,* ESRI Press, La Vergne, TN, 2011.
32. R. Ettema, *Hydraulic Modeling: Concepts and Practice,* American Society of Civil Engineers, Reston, VA, 2000.
33. R. D. Blevins, *Applied Fluid Dynamics Handbook,* van Nostrand Reinhold, New York, 1984.
34. W. J. Palm III, *Mechanical Vibration,* Wiley, New York, 2006.
35. S. S. Rao, *Mechanical Vibrations,* 5th ed., Prentice-Hall, Upper Saddle River, NJ, 2010.
36. G. I. Barenblatt, *Scaling,* Cambridge University Press, Cambridge, UK, 2003.
37. L. J. Fingersh, "Unsteady Aerodynamics Experiment," *Journal of Solar Energy Engineering,* vol. 123, Nov. 2001, p. 267.
38. J. B. Barlow, W. H. Rae, and A. Pope, *Low-Speed Wind Tunnel Testing,* Wiley, New York, 1999.
39. B. H. Goethert, *Transonic Wind Tunnel Testing,* Dover, New York, 2007.
40. American Institute of Aeronautics and Astronautics, *Recommended Practice: Wind Tunnel Testing,* 2 vols., Reston, VA, 2003.
41. P. N. Desai, J. T. Schofield, and M. E. Lisano, "Flight Reconstruction of the Mars Pathfinder Disk-Gap-Band Parachute Drag Coefficients," *J. Spacecraft and Rockets,* vol. 42, no. 4, July–August 2005, pp. 672–676.
42. K.-H. Kim, "Recent Advances in Cavitation Research," 14th International Symposium on Transport Phenomena, Honolulu, HI, March 2012.

This chapter is mostly about flow analysis. The photo shows the 0.9 m-diameter Keystone Pipeline, which has been operating since July 2010. This pipeline delivers heavy and light blends of oil from Hardisty, Alberta, Canada, to refineries in Texas. The pipeline is completely buried and emerges occasionally at delivery or pump stations; it can currently deliver up to 111,000 m^3 of oil per day. [*Image courtesy of TransCanada*]

Chapter 6
Viscous Flow in Ducts

Motivation. This chapter is completely devoted to an important practical fluids engineering problem: flow in ducts with various velocities, various fluids, and various duct shapes. Piping systems are encountered in almost every engineering design and thus have been studied extensively. There is a small amount of theory plus a large amount of experimentation.

The basic piping problem is this: Given the pipe geometry and its added components (such as fittings, valves, bends, and diffusers) plus the desired flow rate and fluid properties, what pressure drop is needed to drive the flow? Of course, it may be stated in alternative form: Given the pressure drop available from a pump, what flow rate will ensue? The correlations discussed in this chapter are adequate to solve most such piping problems.

This chapter is for incompressible flow; Chap. 9 treats compressible pipe flow.

6.1 Reynolds Number Regimes

Now that we have derived and studied the basic flow equations in Chap. 4, you would think that we could just whip off myriad beautiful solutions illustrating the full range of fluid behavior, of course expressing all these educational results in dimensionless form, using our new tool from Chap. 5, dimensional analysis.

The fact of the matter is that no general analysis of fluid motion yet exists. There are several dozen known particular solutions, there are many approximate digital computer solutions, and there are a great many experimental data. There is a lot of theory available if we neglect such important effects as viscosity and compressibility (Chap. 8), but there is no general theory and there may never be. The reason is that a profound and vexing change in fluid behavior occurs at moderate Reynolds numbers. The flow ceases being smooth and steady (*laminar*) and becomes fluctuating and agitated (*turbulent*). The changeover is called *transition* to turbulence. In Fig. 5.3*a*, we saw that transition on the cylinder and sphere occurred at about $\text{Re} = 3 \times 10^5$, where the sharp drop in the drag coefficient appeared. Transition depends on many effects, such as wall roughness (Fig. 5.3*b*) or fluctuations in the inlet stream, but the primary parameter is the Reynolds number. There are a great many data on transition but only a small amount of theory [1 to 3].

Turbulence can be detected from a measurement by a small, sensitive instrument such as a hot-wire anemometer (Fig. 6.29*e*) or a piezoelectric pressure transducer. The flow will appear steady on average but will reveal rapid, random fluctuations if

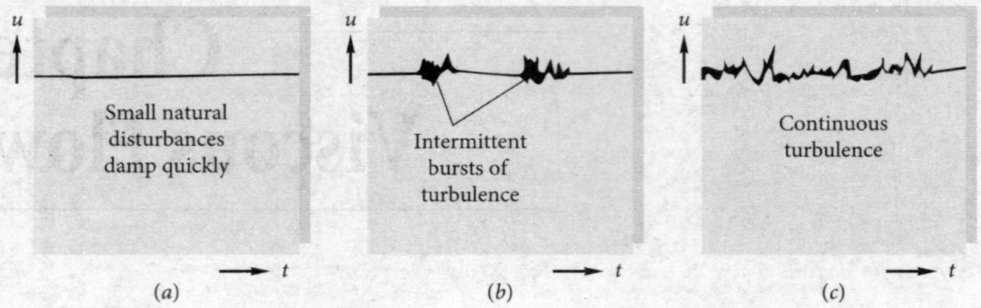

Fig. 6.1 The three regimes of viscous flow: (*a*) laminar flow at low Re; (*b*) transition at intermediate Re; (*c*) turbulent flow at high Re.

turbulence is present, as sketched in Fig. 6.1. If the flow is laminar, there may be occasional natural disturbances that damp out quickly (Fig. 6.1*a*). If transition is occurring, there will be sharp bursts of intermittent turbulent fluctuation (Fig. 6.1*b*) as the increasing Reynolds number causes a breakdown or instability of laminar motion. At sufficiently large Re, the flow will fluctuate continually (Fig. 6.1*c*) and is termed *fully turbulent*. The fluctuations, typically ranging from 1 to 20 percent of the average velocity, are not strictly periodic but are random and encompass a continuous range, or spectrum, of frequencies. In a typical wind tunnel flow at high Re, the turbulent frequency ranges from 1 to 10,000 Hz, and the wavelength ranges from about 0.01 to 400 cm.

EXAMPLE 6.1

The accepted transition Reynolds number for flow in a circular pipe is $Re_{d,crit} \approx 2{,}300$. For flow through a 5-cm-diameter pipe, at what velocity will this occur at 20°C for (*a*) airflow and (*b*) water flow?

Solution

Almost all pipe flow formulas are based on the *average* velocity $V = Q/A$, not centerline or any other point velocity. Thus, transition is specified at $\rho Vd/\mu \approx 2{,}300$. With d known, we introduce the appropriate fluid properties at 20°C from Tables A.3 and A.4:

(*a*) Air: $$\frac{\rho Vd}{\mu} = \frac{(1.205\ \text{kg/m}^3)\,V(0.05\ \text{m})}{1.80\ \text{E-5 kg/(m}\cdot\text{s)}} = 2{,}300 \quad \text{or} \quad V \approx 0.7\ \frac{\text{m}}{\text{s}}$$

(*b*) Water: $$\frac{\rho Vd}{\mu} = \frac{(998\ \text{kg/m}^3)\,V(0.05\ \text{m})}{0.001\ \text{kg/(m}\cdot\text{s)}} = 2{,}300 \quad \text{or} \quad V = 0.046\ \frac{\text{m}}{\text{s}}$$

These are very low velocities, so most engineering air and water pipe flows are turbulent, not laminar. We might expect laminar duct flow with more viscous fluids such as lubricating oils or glycerin.

In free-surface flows, turbulence can be observed directly. Figure 6.2 shows liquid flow issuing from the open end of a tube. The low-Reynolds-number jet (Fig. 6.2*a*) is smooth and laminar, with the fast center motion and slower wall flow forming different trajectories joined by a liquid sheet. The higher-Reynolds-number turbulent flow (Fig. 6.2*b*) is unsteady and irregular, but when averaged over time is steady and predictable.

How did turbulence form inside the pipe? The laminar parabolic flow profile, which is similar to Eq. (4.137), became unstable and, at $Re_d \approx 2{,}300$, began to form "slugs" or "puffs" of intense turbulence. A puff has a fast-moving front and a slow-moving

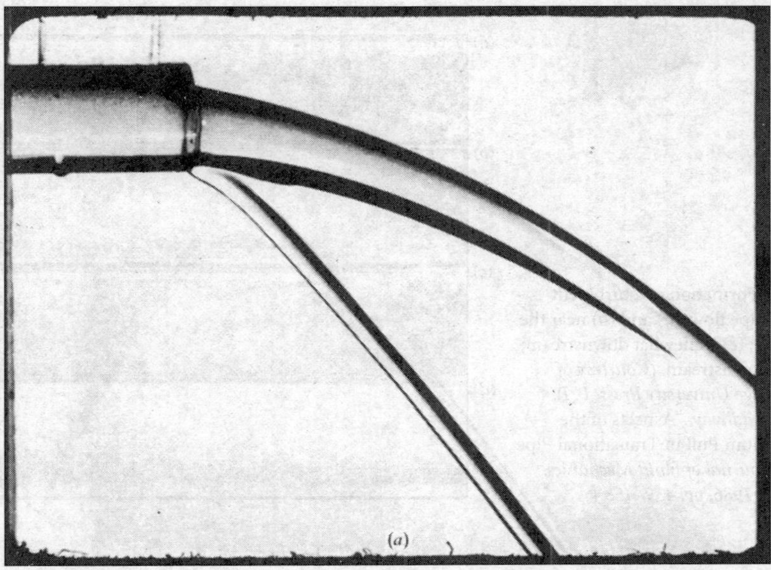

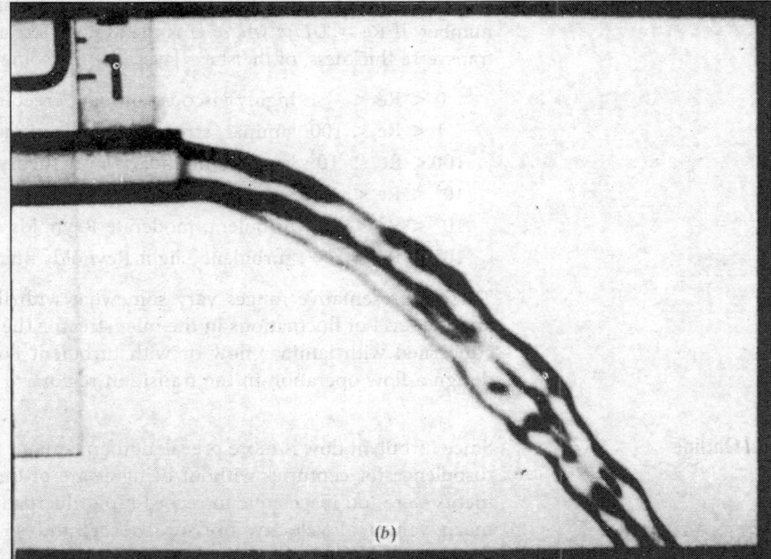

Fig. 6.2 Flow issuing at constant speed from a pipe: (*a*) high-viscosity, low-Reynolds-number, laminar flow; (*b*) low-viscosity, high-Reynolds-number, turbulent flow. Note the ragged, disorderly shape of the jet. *(National Committee for Fluid Mechanics Films, Education Development Center, Inc., © 1972.)*

rear and may be visualized by experimenting with glass tube flow. Figure 6.3 shows a puff as photographed by Bandyopadhyay [45]. Near the entrance (Fig. 6.3*a* and *b*) there is an irregular laminar–turbulent interface, and vortex roll-up is visible. Further downstream (Fig. 6.3*c*) the puff becomes fully turbulent and very active, with helical motions visible. Far downstream (Fig. 6.3*d*) the puff is cone-shaped and less active, with a fuzzy, ill-defined interface, sometimes called the "relaminarization" region.

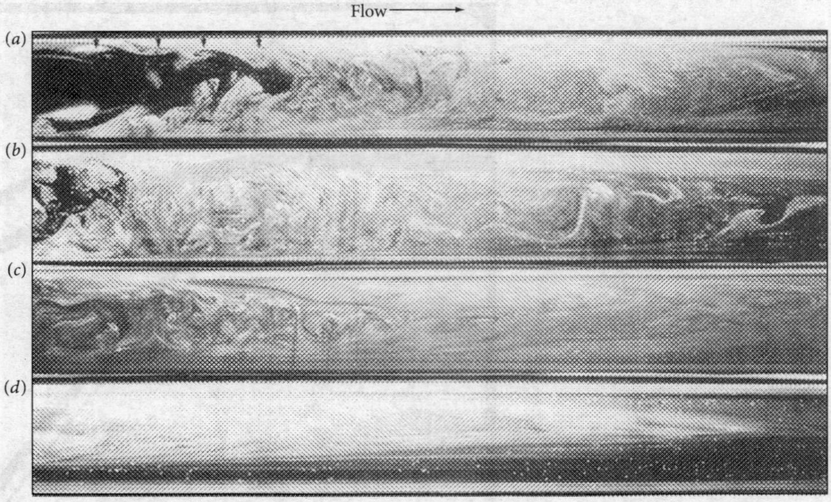

Fig. 6.3 Formation of a turbulent puff in pipe flow: (*a*) and (*b*) near the entrance; (*c*) somewhat downstream; (*d*) far downstream. *(Courtesy of Cambridge University Press–P. R. Bandyopadhyay,* "Aspects of the Equilibrium Puff in Transitional Pipe Flow," *Journal of Fluid Mechanics, vol. 163, 1986, pp. 439–458.)*

A complete description of the statistical aspects of turbulence is given in Ref. 1, while theory and data on transition effects are given in Refs. 2 and 3. At this introductory level we merely point out that the primary parameter affecting transition is the Reynolds number. If $\text{Re} = UL/\nu$, where U is the average stream velocity and L is the "width," or transverse thickness, of the shear layer, the following approximate ranges occur:

$0 < \text{Re} < 1$: highly viscous laminar "creeping" motion
$1 < \text{Re} < 100$: laminar, strong Reynolds number dependence
$100 < \text{Re} < 10^3$: laminar, boundary layer theory useful
$10^3 < \text{Re} < 10^4$: transition to turbulence
$10^4 < \text{Re} < 10^6$: turbulent, moderate Reynolds number dependence
$10^6 < \text{Re} < \infty$: turbulent, slight Reynolds number dependence

These representative ranges vary somewhat with flow geometry, surface roughness, and the level of fluctuations in the inlet stream. The great majority of our analyses are concerned with laminar flow or with turbulent flow, and one should not normally design a flow operation in the transition region.

Historical Outline

Since turbulent flow is more prevalent than laminar flow, experimenters have observed turbulence for centuries without being aware of the details. Before 1930 flow instruments were too insensitive to record rapid fluctuations, and workers simply reported mean values of velocity, pressure, force, and so on. But turbulence can change the mean values dramatically, as with the sharp drop in drag coefficient in Fig. 5.3. A German engineer named G. H. L. Hagen first reported in 1839 that there might be *two* regimes of viscous flow. He measured water flow in long brass pipes and deduced a pressure-drop law:

$$\Delta p = (\text{const}) \frac{LQ}{R^4} + \text{entrance effect} \tag{6.1}$$

This is exactly our laminar flow scaling law from Example 5.4, but Hagen did not realize that the constant was proportional to the fluid viscosity.

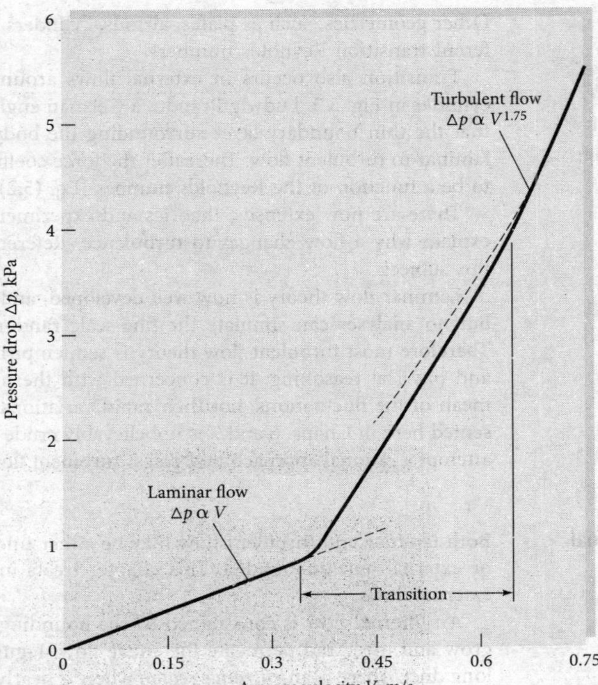

Fig. 6.4 Experimental evidence of transition for water flow in a 6 mm smooth pipe 3 m long.

The formula broke down as Hagen increased Q beyond a certain limit—that is, past the critical Reynolds number—and he stated in his paper that there must be a second mode of flow characterized by "strong movements of water for which Δp varies as the second power of the discharge. . . ." He admitted that he could not clarify the reasons for the change.

A typical example of Hagen's data is shown in Fig. 6.4. The pressure drop varies linearly with $V = Q/A$ up to about 0.35 m/s, where there is a sharp change. Above about $V = 0.7$ m/s the pressure drop is nearly quadratic with V. The actual power $\Delta p \propto V^{1.75}$ seems impossible on dimensional grounds, but is easily explained when the dimensionless pipe flow data (Fig. 5.10) are displayed.

In 1883 Osborne Reynolds, a British engineering professor, showed that the change depended on the parameter $\rho Vd/\mu$, now named in his honor. By introducing a dye streak into a pipe flow, Reynolds could observe transition and turbulence. His sketches [4] of the flow behavior are shown in Fig. 6.5.

If we examine Hagen's data and compute the Reynolds number at $V = 0.35$ m/s, we obtain $\text{Re}_d = 2{,}100$. The flow became fully turbulent, $V = 0.7$ m/s, at $\text{Re}_d = 4{,}200$. The accepted design value for pipe flow transition is now taken to be

$$\text{Re}_{d,\text{crit}} \approx 2{,}300 \tag{6.2}$$

This is accurate for commercial pipes (Fig. 6.13), although with special care in providing a rounded entrance, smooth walls, and a steady inlet stream, $\text{Re}_{d,\text{crit}}$ can be delayed until much higher values. The study of transition in pipe flow, both experimentally and theoretically, continues to be a fascinating topic for researchers, as discussed in a recent review article [55]. *Note:* The value of 2,300 is for transition in *pipes*.

Other geometries, such as plates, airfoils, cylinders, and spheres, have completely different transition Reynolds numbers.

Transition also occurs in external flows around bodies such as the sphere and cylinder in Fig. 5.3. Ludwig Prandtl, a German engineering professor, showed in 1914 that the thin boundary layer surrounding the body was undergoing transition from laminar to turbulent flow. Thereafter the force coefficient of a body was acknowledged to be a function of the Reynolds number [Eq. (5.2)].

There are now extensive theories and experiments of laminar flow instability that explain why a flow changes to turbulence. Reference 5 is an advanced textbook on this subject.

Laminar flow theory is now well developed, and many solutions are known [2, 3], but no analyses can simulate the fine-scale random fluctuations of turbulent flow.[1] Therefore most turbulent flow theory is semiempirical, based on dimensional analysis and physical reasoning; it is concerned with the mean flow properties only and the mean of the fluctuations, not their rapid variations. The turbulent flow "theory" presented here in Chaps. 6 and 7 is unbelievably crude yet surprisingly effective. We shall attempt a rational approach that places turbulent flow analysis on a firm physical basis.

6.2 Internal versus External Viscous Flows

Both laminar and turbulent flow may be either internal (that is, "bounded" by walls) or external and unbounded. This chapter treats internal flows, and Chap. 7 studies external flows.

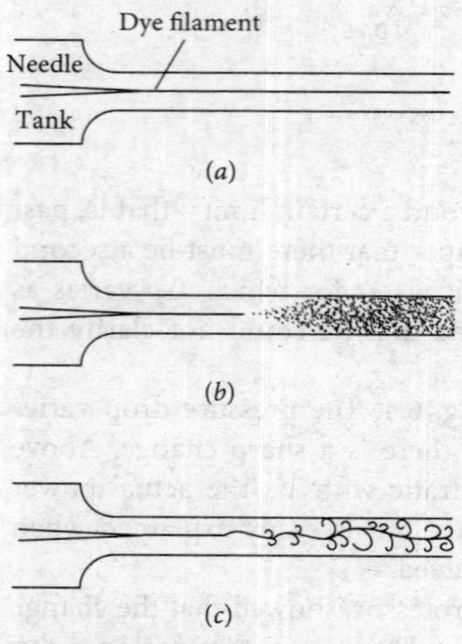

Fig. 6.5 Reynolds' sketches of pipe flow transition: (*a*) low-speed, laminar flow; (*b*) high-speed, turbulent flow; (*c*) spark photograph of condition (*b*).
Source: Reynolds, "An Experimental Investigation of the Circumstances which Determine Whether the Motion of Water Shall Be Direct or Sinuous and of the Law of Resistance in Parallel Channels," Phil. Trans. R. Soc., vol. 174, 1883, pp. 935–982.

An internal flow is constrained by the bounding walls, and the viscous effects will grow and meet and permeate the entire flow. Figure 6.6 shows an internal flow in a long duct. There is an *entrance region* where a nearly inviscid upstream flow converges and enters the tube. Viscous boundary layers grow downstream, retarding the axial flow $u(r, x)$ at the wall and thereby accelerating the center core flow to maintain the incompressible continuity requirement

$$Q = \int u \, dA = \text{const} \tag{6.3}$$

At a finite distance from the entrance, the boundary layers merge and the inviscid core disappears. The tube flow is then entirely viscous, and the axial velocity adjusts slightly further until at $x = L_e$ it no longer changes with x and is said to be *fully developed*, $u \approx u(r)$ only. Downstream of $x = L_e$ the velocity profile is constant, the wall shear is constant, and the pressure drops linearly with x, for either laminar or turbulent flow. All these details are shown in Fig. 6.6.

Dimensional analysis shows that the Reynolds number is the only parameter affecting entrance length. If

$$L_e = f(d, V, \rho, \mu) \qquad V = \frac{Q}{A}$$

then

$$\frac{L_e}{d} = g\left(\frac{\rho V d}{\mu}\right) = g(\text{Re}_d) \tag{6.4}$$

For laminar flow [2, 3], the accepted correlation is

$$\frac{L_e}{d} \approx 0.06 \, \text{Re}_d \qquad \text{laminar} \tag{6.5}$$

The maximum laminar entrance length, at $\text{Re}_{d,\text{crit}} = 2{,}300$, is $L_e = 138d$, which is the longest development length possible.

[1]However, direct numerical simulation (DNS) of low-Reynolds-number turbulence is now quite common [32].

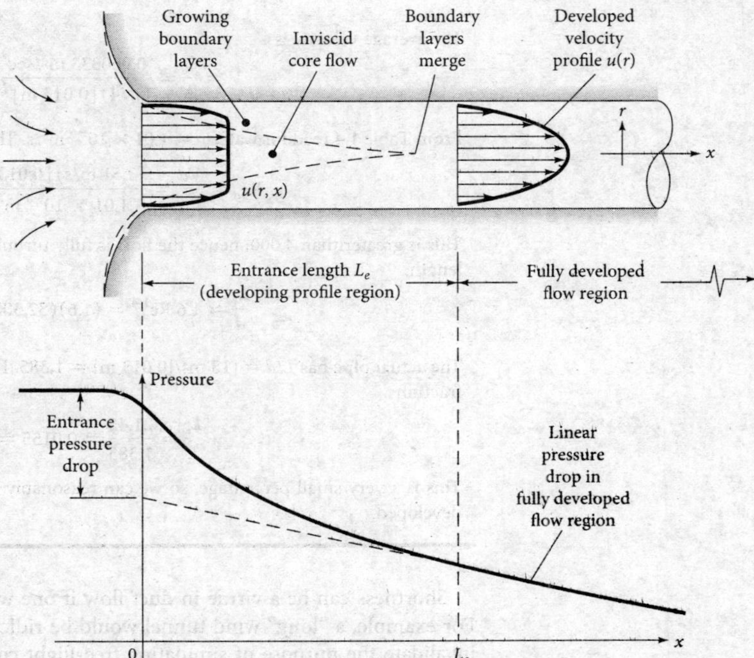

Fig. 6.6 Developing velocity profiles and pressure changes in the entrance of a duct flow.

In turbulent flow, the boundary layers grow faster, and L_e is relatively shorter. For decades, the writer has favored a sixth-power-law estimate, $L_e/d \approx 4.4\,\mathrm{Re}_d^{1/6}$, but recent CFD results, communicated by Fabien Anselmet, and separately by Sukanta Dash, indicate that a better turbulent entrance-length correlation is

$$\frac{L_e}{d} \approx 1.6\,\mathrm{Re}_d^{1/4} \quad \text{for} \quad \mathrm{Re}_d \leq 10^7 \tag{6.6}$$

Some computed turbulent entrance-length estimates are thus

Re_d	4,000	10^4	10^5	10^6	10^7
L_e/d	13	16	28	51	90

Now 90 diameters may seem "long," but typical pipe flow applications involve an L/d value of 1,000 or more, in which case the entrance effect may be neglected and a simple analysis made for fully developed flow. This is possible for both laminar and turbulent flows, including rough walls and noncircular cross sections.

EXAMPLE 6.2

A 13 mm-diameter water pipe is 18 m long and delivers water at 0.02 m^3/min at 20°C. What fraction of this pipe is taken up by the entrance region?

Solution

Convert

$$Q = (0.02\ \mathrm{m^3/min}) = 0.000333\ \mathrm{m^3/sec}$$

The average velocity is

$$V = \frac{Q}{A} = \frac{0.000333 \text{ m}^3/\text{sec}}{(\pi/4)[0.013 \text{ m}]^2} = 2.51 \text{ m/s}$$

From Table 1.4 read for water $\nu = 1.01 \times 10^{-6}$ m²/s. Then the pipe Reynolds number is

$$\text{Re}_d = \frac{Vd}{\nu} = \frac{(2.51 \text{ m/s})[0.013 \text{ m}]}{1.01 \times 10^{-6} \text{ m}^2/\text{s}} = 32{,}300$$

This is greater than 4,000; hence the flow is fully turbulent, and Eq. (6.6) applies for entrance length:

$$\frac{L_e}{d} \approx 1.6 \text{ Re}_d^{1/4} = (1.6)(32{,}300)^{1/4} = 21.45$$

The actual pipe has $L/d = (18 \text{ m})/[0.013 \text{ m}] = 1{,}385$. Hence the entrance region takes up the fraction

$$\frac{L_e}{L} = \frac{21.45}{1{,}385} = 0.0155 = 1.55\% \qquad \textit{Ans.}$$

This is a very small percentage, so we can reasonably treat this pipe flow as essentially fully developed.

Shortness can be a virtue in duct flow if one wishes to maintain the inviscid core. For example, a "long" wind tunnel would be ridiculous, since the viscous core would invalidate the purpose of simulating free-flight conditions. A typical laboratory low-speed wind tunnel test section is 1 m in diameter and 5 m long, with $V = 30$ m/s. If we take $\nu_{\text{air}} = 1.51 \times 10^{-5}$ m²/s from Table 1.4, then $\text{Re}_d = 1.99 \times 10^6$ and, from Eq. (6.6), $L_e/d \approx 49$. The test section has $L/d = 5$, which is much shorter than the development length. At the end of the section the wall boundary layers are only 10 cm thick, leaving 80 cm of inviscid core suitable for model testing.

An external flow has no restraining walls and is free to expand no matter how thick the viscous layers on the immersed body may become. Thus, far from the body the flow is nearly inviscid, and our analytical technique, treated in Chap. 7, is to patch an inviscid-flow solution onto a viscous boundary-layer solution computed for the wall region. There is no external equivalent of fully developed internal flow.

6.3 Head Loss—The Friction Factor

When applying pipe flow formulas to practical problems, it is customary to use a control volume analysis. Consider incompressible steady flow between sections 1 and 2 of the inclined constant-area pipe in Fig. 6.7. The one-dimensional continuity relation, Eq. (3.30), reduces to

$$Q_1 = Q_2 = \text{const} \quad \text{or} \quad V_1 = V_2 = V$$

since the pipe is of constant area. The steady flow energy equation (3.75) becomes

$$\left(\frac{p}{\rho g} + \alpha \frac{V^2}{2g} + z\right)_1 = \left(\frac{p}{\rho g} + \alpha \frac{V^2}{2g} + z\right)_2 + h_f \tag{6.7}$$

since there is no pump or turbine between 1 and 2. For fully developed flow, the velocity profile shape is the same at sections 1 and 2. Thus $\alpha_1 = \alpha_2$ and, since $V_1 = V_2$, Eq. (6.7) reduces to head loss versus pressure drop and elevation change:

$$h_f = (z_1 - z_2) + \left(\frac{p_1}{\rho g} - \frac{p_2}{\rho g}\right) = \Delta z + \frac{\Delta p}{\rho g} \tag{6.8}$$

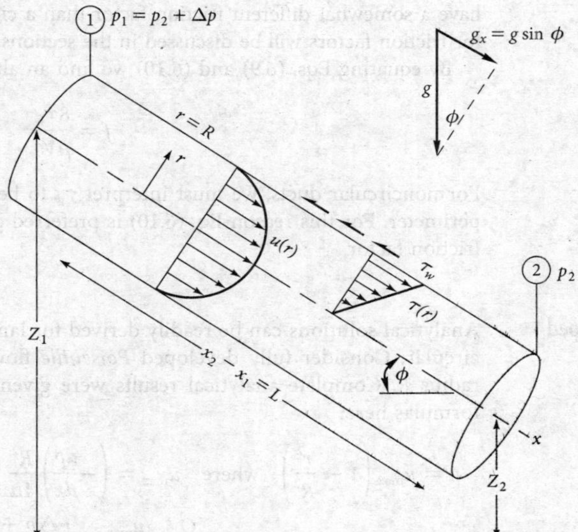

Fig. 6.7 Control volume, just inside the pipe wall, of steady, fully developed flow between two sections in an inclined pipe.

The pipe head loss equals the change in the sum of pressure and gravity head—that is, the change in height of the hydraulic grade line (HGL).

Finally, apply the momentum relation (3.40) to the control volume in Fig. 6.7, accounting for applied x-directed forces due to pressure, gravity, and shear:

$$\sum F_x = \Delta p(\pi R^2) + \rho g(\pi R^2)L \sin \phi - \tau_w(2\pi R)L = \dot{m}(V_2 - V_1) = 0 \quad (6.9a)$$

Rearrange this and we find that the head loss is also related to wall shear stress:

$$\Delta z + \frac{\Delta p}{\rho g} = h_f = \frac{2\tau_w}{\rho g}\frac{L}{R} = \frac{4\tau_w}{\rho g}\frac{L}{d} \quad (6.9b)$$

where we have substituted $\Delta z = L \sin \phi$ from the geometry of Fig. 6.7. Note that, regardless of whether the pipe is horizontal or tilted, the head loss is proportional to the wall shear stress.

How should we correlate the head loss for pipe flow problems? The answer was given a century and a half ago by Julius Weisbach, a German professor who in 1850 published the first modern textbook on hydrodynamics. Equation (6.9b) shows that h_f is proportional to (L/d), and data such as Hagen's in Fig. 6.6 show that, for turbulent flow, h_f is approximately proportional to V^2. The proposed correlation, still as effective today as in 1850, is

$$h_f = f\frac{L}{d}\frac{V^2}{2g} \quad \text{where} \quad f = \text{fcn}(\text{Re}_d, \frac{\varepsilon}{d}, \text{duct shape}) \quad (6.10)$$

The dimensionless parameter f is called the *Darcy friction factor,* after Henry Darcy (1803–1858), a French engineer whose pipe flow experiments in 1857 first established the effect of roughness on pipe resistance. The quantity ε is the wall roughness height, which is important in turbulent (but not laminar) pipe flow. We added the "duct shape" effect in Eq. (6.10) to remind us that square and triangular and other noncircular ducts have a somewhat different friction factor than a circular pipe. Actual data and theory for friction factors will be discussed in the sections that follow.

By equating Eqs. (6.9) and (6.10) we find an alternative form for friction factor:

$$f = \frac{8\tau_w}{\rho V^2} \tag{6.11}$$

For noncircular ducts, we must interpret τ_w to be an average value around the duct perimeter. For this reason Eq. (6.10) is preferred as a unified definition of the Darcy friction factor.

6.4 Laminar Fully Developed Pipe Flow

Analytical solutions can be readily derived for laminar flows, either circular or noncircular. Consider fully developed *Poiseuille* flow in a round pipe of diameter d, radius R. Complete analytical results were given in Sec. 4.10. Let us review those formulas here:

$$u = u_{\max}\left(1 - \frac{r^2}{R^2}\right) \quad \text{where} \quad u_{\max} = \left(-\frac{dp}{dx}\right)\frac{R^2}{4\mu} \quad \text{and} \quad \left(-\frac{dp}{dx}\right) = \left(\frac{\Delta p + \rho g \Delta z}{L}\right)$$

$$V = \frac{Q}{A} = \frac{u_{\max}}{2} = \left(\frac{\Delta p + \rho g \Delta z}{L}\right)\frac{R^2}{8\mu}$$

$$Q = \int u\,dA = \pi R^2 V = \frac{\pi R^4}{8\mu}\left(\frac{\Delta p + \rho g \Delta z}{L}\right) \tag{6.12}$$

$$\tau_w = \left|\mu \frac{du}{dr}\right|_{r=R} = \frac{4\mu V}{R} = \frac{8\mu V}{d} = \frac{R}{2}\left(\frac{\Delta p + \rho g \Delta z}{L}\right)$$

$$h_f = \frac{32\mu L V}{\rho g d^2} = \frac{128\mu L Q}{\pi \rho g d^4}$$

The paraboloid velocity profile has an average velocity V which is one-half of the maximum velocity. The quantity Δp is the pressure *drop* in a pipe of length L; that is, (dp/dx) is negative. These formulas are valid whenever the pipe Reynolds number, $\text{Re}_d = \rho V d/\mu$, is less than about 2,300. Note that τ_w is proportional to V (see Fig. 6.6) and is independent of density because the fluid acceleration is zero. Neither of these is true in turbulent flow.

With wall shear stress known, the Poiseuille flow friction factor is easily determined:

$$f_{\text{lam}} = \frac{8\tau_{w,\text{lam}}}{\rho V^2} = \frac{8(8\mu V/d)}{\rho V^2} = \frac{64}{\rho V d/\mu} = \frac{64}{\text{Re}_d} \tag{6.13}$$

In laminar flow, the pipe friction factor decreases inversely with Reynolds number. This famous formula is effective, but often the algebraic relations of Eqs. (6.12) are more direct for problems.

EXAMPLE 6.3

An oil with $\rho = 900$ kg/m³ and $\nu = 0.0002$ m²/s flows upward through an inclined pipe as shown in Fig. E6.3. The pressure and elevation are known at sections 1 and 2, 10 m apart. Assuming steady laminar flow, (*a*) verify that the flow is up, (*b*) compute h_f between 1 and 2, and compute (*c*) Q, (*d*) V, and (*e*) Re_d. Is the flow really laminar?

Solution

Part (a) For later use, calculate

$$\mu = \rho\nu = (900 \text{ kg/m}^3)(0.0002 \text{ m}^2/\text{s}) = 0.18 \text{ kg/(m} \cdot \text{s)}$$

$$z_2 = \Delta L \sin 40° = (10 \text{ m})(0.643) = 6.43 \text{ m}$$

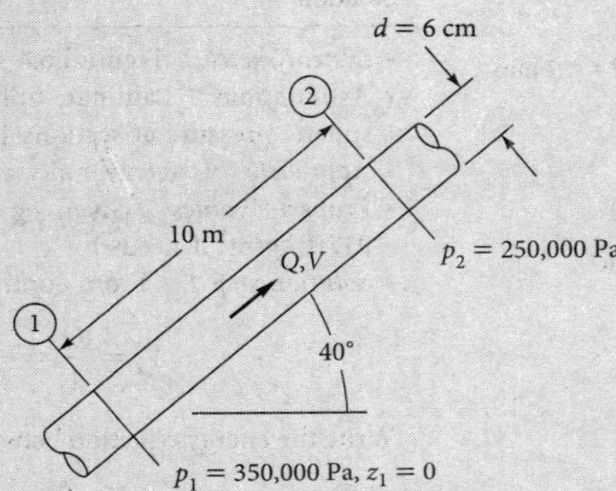

E6.3

The flow goes in the direction of falling HGL; therefore, compute the hydraulic grade-line height at each section:

$$\text{HGL}_1 = z_1 + \frac{p_1}{\rho g} = 0 + \frac{350{,}000}{900(9.807)} = 39.65 \text{ m}$$

$$\text{HGL}_2 = z_2 + \frac{p_2}{\rho g} = 6.43 + \frac{250{,}000}{900(9.807)} = 34.75 \text{ m}$$

The HGL is lower at section 2; hence the flow is up from 1 to 2 as assumed. *Ans. (a)*

Part (b) The head loss is the change in HGL:

$$h_f = \text{HGL}_1 - \text{HGL}_2 = 39.65 \text{ m} - 34.75 \text{ m} = 4.9 \text{ m}$$ *Ans. (b)*

Half the length of the pipe is quite a large head loss.

Part (c) We can compute Q from the various laminar flow formulas, notably Eq. (6.12):

$$Q = \frac{\pi\rho g d^4 h_f}{128\mu L} = \frac{\pi(900)(9.807)(0.06)^4(4.9)}{128(0.18)(10)} = 0.0076 \text{ m}^3/\text{s}$$ *Ans. (c)*

Part (d) Divide Q by the pipe area to get the average velocity:

$$V = \frac{Q}{\pi R^2} = \frac{0.0076}{\pi(0.03)^2} = 2.7 \text{ m/s}$$ *Ans. (d)*

Part (e) With V known, the Reynolds number is

$$\text{Re}_d = \frac{Vd}{\nu} = \frac{2.7(0.06)}{0.0002} = 810$$ *Ans. (e)*

This is well below the transition value $\text{Re}_d = 2{,}300$, so we are fairly certain the flow is laminar.

Notice that by sticking entirely to consistent SI units (meters, seconds, kilograms, newtons) for all variables we avoid the need for any conversion factors in the calculations.

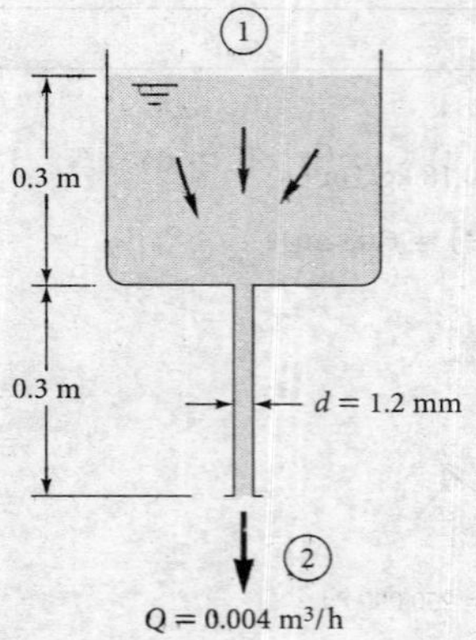

E6.4

EXAMPLE 6.4

A liquid of specific weight ρg = 9,000 N/m³ flows by gravity through a 0.3 m tank and a 0.3 m capillary tube at a rate of 0.004 m³/h, as shown in Fig. E6.4. Sections 1 and 2 are at atmospheric pressure. Neglecting entrance effects and friction in the large tank, compute the viscosity of the liquid.

Solution

- *System sketch:* Figure E6.4 shows L = 0.3 m, d = 1.2 mm, and Q =0.004 m³/h.
- *Assumptions:* Laminar, fully developed, incompressible (Poiseuille) pipe flow. Atmospheric pressure at sections 1 and 2. Negligible velocity at surface, $V_1 \approx 0$.
- *Approach:* Use continuity and energy to find the head loss and thence the viscosity.
- *Property values:* Given ρg = 9,000 N/m³, figure out ρ = 9,000 kg/m³/9.81 m/s² = 917.4 kg/m³ if needed.
- *Solution step 1:* From continuity and the known flow rate, determine V_2:

$$V_2 = \frac{Q}{A_2} = \frac{Q}{(\pi/4)d^2} = \frac{(1.111\text{ E-6})\text{m}^3/\text{sec}}{(\pi/4)(0.0012\text{ m})^2} = 0.982\text{ m/s}$$

Write the energy equation between 1 and 2, canceling terms, and find the head loss:

$$\frac{p_1}{\rho g} + \frac{\alpha_1 V_1^2}{2g} + z_1 = \frac{p_2}{\rho g} + \frac{\alpha_2 V_2^2}{2g} + z_2 + h_f$$

or

$$h_f = z_1 - z_2 - \frac{\alpha_2 V_2^2}{2g} = 0.6\text{ m} - 0\text{ m} - \frac{(2.0)(0.982\text{ m/s})^2}{2(9.81\text{ m/s}^2)} = 0.502\text{ m}$$

- *Comment:* We introduced α_2 = 2.0 for laminar pipe flow from Eq. (3.76). If we forgot α_2, we would have calculated h_f = 0.551 meters, a 10 percent error.
- *Solution step 2:* With head loss known, the viscosity follows from the laminar formula in Eqs. (6.12):

$$h_f = 0.502\text{ m} = \frac{32\,\mu L V}{(\rho g)d^2} = \frac{32\mu(0.3\text{ m})(0.982\text{ m/s})}{(9{,}000\text{ N/m}^3)(0.0012\text{ m})^2} \quad \textit{solve for } \mu = 6.90\text{ E-4 kg/m.s} \quad \textit{Ans.}$$

- *Comments:* We didn't need the value of ρ—the formula contains ρg, but who knew? Note also that L in this formula is the *pipe length* of 0.3 m, not the total elevation change.
- *Final check:* Calculate the Reynolds number to see if it is less than 2300 for laminar flow:

$$\text{Re}_d = \frac{\rho V d}{\mu} = \frac{(917.4\text{ kg/m}^3)(0.982\text{ m/s})(0.0012\text{ m})}{(6.90\text{ E-4 kg/m}\cdot\text{s})} \approx 1{,}567 \quad \text{Yes, laminar.}$$

- *Comments:* So we did need ρ after all to calculate Re_d.
- *Unexpected comment:* For this head loss, there is a *second* (turbulent) solution, as we shall see in Example 6.8.

6.5 Turbulence Modeling

Throughout this chapter we assume constant density and viscosity and no thermal interaction, so that only the continuity and momentum equations are to be solved for velocity and pressure

$$\text{Continuity:} \quad \frac{\partial u}{\partial x} + \frac{\partial v}{\partial y} + \frac{\partial w}{\partial z} = 0$$

$$\text{Momentum:} \quad \rho\frac{d\mathbf{V}}{dt} = -\nabla p + \rho\mathbf{g} + \mu\,\nabla^2\mathbf{V} \tag{6.14}$$

subject to no slip at the walls and known inlet and exit conditions. (We shall save our free-surface solutions for Chap. 10.)

We will not work with the differential energy relation, Eq. (4.53), in this chapter, but it is very important, both for heat transfer calculations and for general understanding of duct flow processes. There is work being done by pressure forces to drive the fluid through the duct. Where does this energy go? There is no work done by the wall shear stresses, because the velocity at the wall is zero. The answer is that pressure work is balanced by viscous dissipation in the interior of the flow. The integral of the dissipation function Φ, from Eq. (4.50), over the flow field will equal the pressure work. An example of this fundamental viscous flow energy balance is given in Problem C6.7.

Both laminar and turbulent flows satisfy Eqs. (6.14). For laminar flow, where there are no random fluctuations, we go right to the attack and solve them for a variety of geometries [2, 3], leaving many more, of course, for the problems.

Reynolds' Time-Averaging Concept

For turbulent flow, because of the fluctuations, every velocity and pressure term in Eqs. (6.14) is a rapidly varying random function of time and space. At present our mathematics cannot handle such instantaneous fluctuating variables. No single pair of random functions $\mathbf{V}(x, y, z, t)$ and $p(x, y, z, t)$ is known to be a solution to Eqs. (6.14). Moreover, our attention as engineers is toward the average or *mean* values of velocity, pressure, shear stress, and the like in a high-Reynolds-number (turbulent) flow. This approach led Osborne Reynolds in 1895 to rewrite Eqs. (6.14) in terms of mean or time-averaged turbulent variables.

The time mean $\overline{u}$ of a turbulent function $u(x, y, z, t)$ is defined by

$$\overline{u} = \frac{1}{T}\int_0^T u\, dt \tag{6.15}$$

where T is an averaging period taken to be longer than any significant period of the fluctuations themselves. The mean values of turbulent velocity and pressure are illustrated in Fig. 6.8. For turbulent gas and water flows, an averaging period $T \approx 5$ s is usually quite adequate.

The *fluctuation* u' is defined as the deviation of u from its average value

$$u' = u - \overline{u} \tag{6.16}$$

also shown in Fig. 6.8. It follows by definition that a fluctuation has zero mean value:

$$\overline{u'} = \frac{1}{T}\int_0^T (u - \overline{u})\, dt = \overline{u} - \overline{u} = 0 \tag{6.17}$$

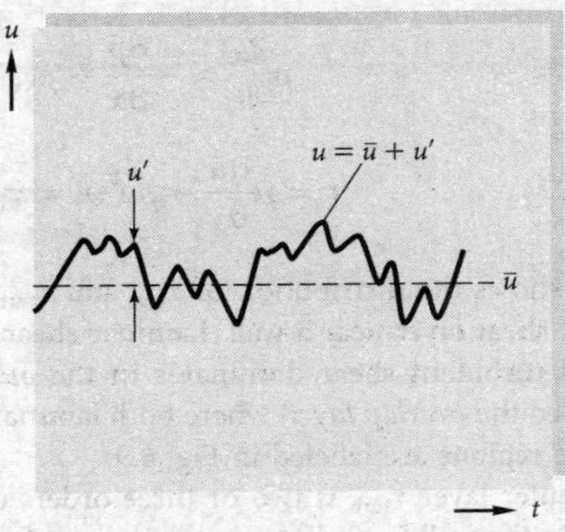

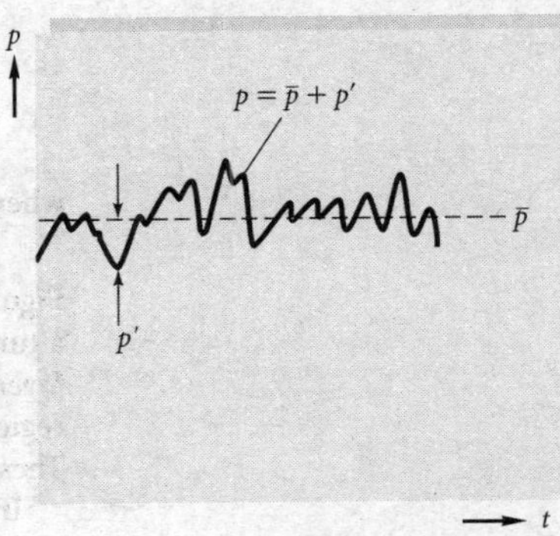

Fig. 6.8 Definition of mean and fluctuating turbulent variables: (*a*) velocity; (*b*) pressure.

However, the mean square of a fluctuation is not zero and is a measure of the *intensity* of the turbulence:

$$\overline{u'^2} = \frac{1}{T}\int_0^T u'^2\,dt \neq 0 \tag{6.18}$$

Nor in general are the mean fluctuation products such as $\overline{u'v'}$ and $\overline{u'p'}$ zero in a typical turbulent flow.

Reynolds' idea was to split each property into mean plus fluctuating variables:

$$u = \overline{u} + u' \quad v = \overline{v} + v' \quad w = \overline{w} + w' \quad p = \overline{p} + p' \tag{6.19}$$

Substitute these into Eqs. (6.14), and take the time mean of each equation. The continuity relation reduces to

$$\frac{\partial \overline{u}}{\partial x} + \frac{\partial \overline{v}}{\partial y} + \frac{\partial \overline{w}}{\partial z} = 0 \tag{6.20}$$

which is no different from a laminar continuity relation.

However, each component of the momentum equation (6.14*b*), after time averaging, will contain mean values plus three mean products, or *correlations,* of fluctuating velocities. The most important of these is the momentum relation in the mainstream, or *x*, direction, which takes the form

$$\rho\frac{d\overline{u}}{dt} = -\frac{\partial \overline{p}}{\partial x} + \rho g_x + \frac{\partial}{\partial x}\left(\mu\frac{\partial \overline{u}}{\partial x} - \rho\overline{u'^2}\right) + \frac{\partial}{\partial y}\left(\mu\frac{\partial \overline{u}}{\partial y} - \rho\overline{u'v'}\right) + \frac{\partial}{\partial z}\left(\mu\frac{\partial \overline{u}}{\partial z} - \rho\overline{u'w'}\right) \tag{6.21}$$

The three correlation terms $-\rho\overline{u'^2}$, $-\rho\overline{u'v'}$, and $-\rho\overline{u'w'}$ are called *turbulent stresses* because they have the same dimensions and occur right alongside the newtonian (laminar) stress terms $\mu(\partial\overline{u}/\partial x)$ and so on.

The turbulent stresses are unknown a priori and must be related by experiment to geometry and flow conditions, as detailed in Refs. 1 to 3. Fortunately, in duct and boundary layer flow, the stress $-\rho\overline{u'v'}$, associated with direction *y* normal to the wall is dominant, and we can approximate with excellent accuracy a simpler streamwise momentum equation

$$\rho\frac{d\overline{u}}{dt} \approx -\frac{\partial \overline{p}}{\partial x} + \rho g_x + \frac{\partial \tau}{\partial y} \tag{6.22}$$

where
$$\tau = \mu\frac{\partial \overline{u}}{\partial y} - \rho\overline{u'v'} = \tau_{\text{lam}} + \tau_{\text{turb}} \tag{6.23}$$

Figure 6.9 shows the distribution of τ_{lam} and τ_{turb} from typical measurements across a turbulent shear layer near a wall. Laminar shear is dominant near the wall (the *wall layer*), and turbulent shear dominates in the *outer layer.* There is an intermediate region, called the *overlap layer,* where both laminar and turbulent shear are important. These three regions are labeled in Fig. 6.9.

In the outer layer τ_{turb} is two or three orders of magnitude greater than τ_{lam}, and vice versa in the wall layer. These experimental facts enable us to use a crude but very effective model for the velocity distribution $\overline{u}(y)$ across a turbulent wall layer.

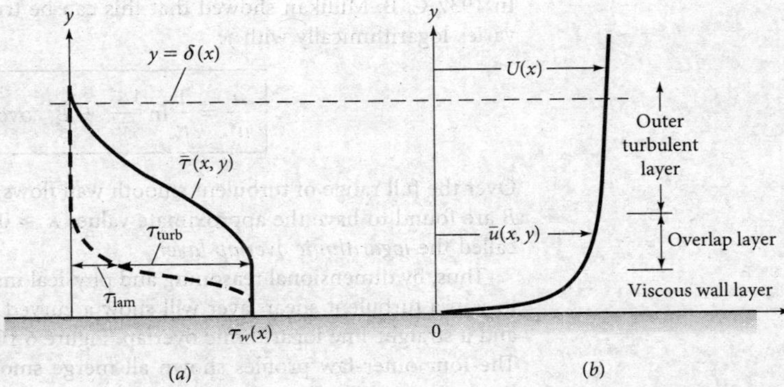

Fig. 6.9 Typical velocity and shear distributions in turbulent flow near a wall: (*a*) shear; (*b*) velocity.

The Logarithmic Overlap Law

We have seen in Fig. 6.9 that there are three regions in turbulent flow near a wall:

1. Wall layer: Viscous shear dominates.
2. Outer layer: Turbulent shear dominates.
3. Overlap layer: Both types of shear are important.

From now on let us agree to drop the overbar from velocity $\overline{u}$. Let τ_w be the wall shear stress, and let δ and U represent the thickness and velocity at the edge of the outer layer, $y = \delta$.

For the wall layer, Prandtl deduced in 1930 that u must be independent of the shear layer thickness:

$$u = f(\mu, \tau_w, \rho, y) \tag{6.24}$$

By dimensional analysis, this is equivalent to

$$u^+ = \frac{u}{u^*} = F\left(\frac{yu^*}{\nu}\right) \qquad u^* = \left(\frac{\tau_w}{\rho}\right)^{1/2} \tag{6.25}$$

Equation (6.25) is called the *law of the wall*, and the quantity u^* is termed the *friction velocity* because it has dimensions $\{LT^{-1}\}$, although it is not actually a flow velocity.

Subsequently, Kármán in 1933 deduced that u in the outer layer is independent of molecular viscosity, but its deviation from the stream velocity U must depend on the layer thickness δ and the other properties:

$$(U - u)_{\text{outer}} = g(\delta, \tau_w, \rho, y) \tag{6.26}$$

Again, by dimensional analysis we rewrite this as

$$\frac{U - u}{u^*} = G\left(\frac{y}{\delta}\right) \tag{6.27}$$

where u^* has the same meaning as in Eq. (6.25). Equation (6.27) is called the *velocity-defect law* for the outer layer.

Both the wall law (6.25) and the defect law (6.27) are found to be accurate for a wide variety of experimental turbulent duct and boundary layer flows [Refs. 1 to 3]. They are different in form, yet they must overlap smoothly in the intermediate layer.

In 1937 C. B. Millikan showed that this can be true only if the overlap layer velocity varies logarithmically with y:

$$\frac{u}{u^*} = \frac{1}{\kappa} \ln \frac{yu^*}{\nu} + B \quad \text{overlap layer} \tag{6.28}$$

Over the full range of turbulent smooth wall flows, the dimensionless constants κ and B are found to have the approximate values $\kappa \approx 0.41$ and $B \approx 5.0$. Equation (6.28) is called the *logarithmic overlap layer.*

Thus, by dimensional reasoning and physical insight we infer that a plot of u versus $\ln y$ in a turbulent shear layer will show a curved wall region, a curved outer region, and a straight-line logarithmic overlap. Figure 6.10 shows that this is exactly the case. The four outer-law profiles shown all merge smoothly with the logarithmic overlap law but have different magnitudes because they vary in external pressure gradient. The wall law is unique and follows the linear viscous relation

$$u^+ = \frac{u}{u^*} = \frac{yu^*}{\nu} = y^+ \tag{6.29}$$

from the wall to about $y^+ = 5$, thereafter curving over to merge with the logarithmic law at about $y^+ = 30$.

Believe it or not, Fig. 6.10, which is nothing more than a shrewd correlation of velocity profiles, is the basis for most existing "theory" of turbulent shear flows. Notice that we have not solved any equations at all, but have merely expressed the streamwise velocity in a neat form.

There is serendipity in Fig. 6.10: The logarithmic law (6.28), instead of just being a short overlapping link, actually approximates nearly the entire velocity profile, except

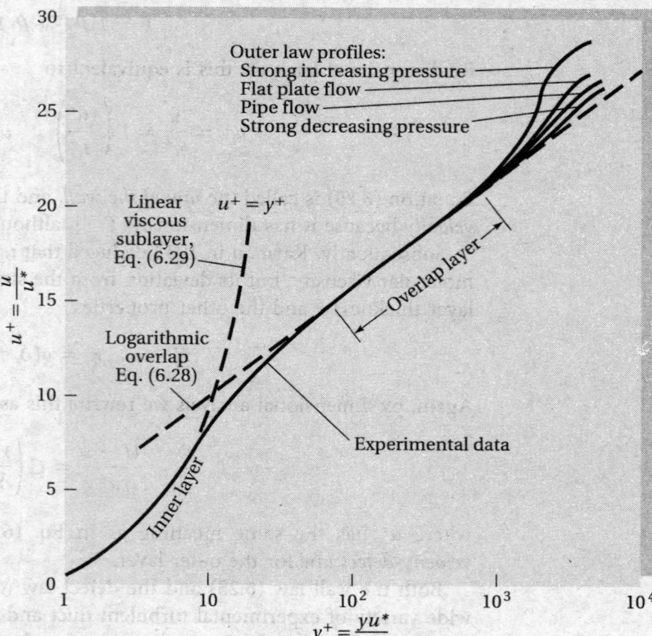

Fig. 6.10 Experimental verification of the inner, outer, and overlap layer laws relating velocity profiles in turbulent wall flow.

for the outer law when the pressure is increasing strongly downstream (as in a diffuser). The inner wall law typically extends over less than 2 percent of the profile and can be neglected. Thus, we can use Eq. (6.28) as an excellent approximation to solve nearly every turbulent flow problem presented in this and the next chapter. Many additional applications are given in Refs. 2 and 3.

Advanced Modeling Concepts

Turbulence modeling is a very active field. Scores of papers have been published to more accurately simulate the turbulent stresses in Eq. (6.21) and their y and z components. This research, now available in advanced texts [1, 13, 19], goes well beyond the present book, which is confined to the use of the logarithmic law (6.28) for pipe and boundary layer problems. For example, L. Prandtl, who invented boundary layer theory in 1904, later proposed an *eddy viscosity* model of the Reynolds stress term in Eq. (6.23):

$$-\rho\,\overline{u'v'} = \tau_{\text{turb}} \approx \mu_t \frac{du}{dy} \quad \text{where} \quad \mu_t \approx \rho\, l^2 \left| \frac{du}{dy} \right| \tag{6.30}$$

The term μ_t, which is a property of the *flow*, not the fluid, is called the *eddy viscosity* and can be modeled in various ways. The most popular form is Eq. (6.30), where l is called the *mixing length* of the turbulent eddies (analogous to mean free path in molecular theory). Near a solid wall, l is approximately proportional to distance from the wall, and Kármán suggested

$$l \approx \kappa y \quad \text{where} \quad \kappa = \text{Kármán's constant} \approx 0.41 \tag{6.31}$$

As a homework assignment, Prob. P6.40, you may show that Eqs. (6.30) and (6.31) lead to the logarithmic law (6.28) near a wall.

Modern turbulence models approximate three-dimensional turbulent flows and employ additional partial differential equations for such quantities as the turbulence kinetic energy, the turbulent dissipation, and the six Reynolds stresses. For details, see Refs. 1, 13, and 19.

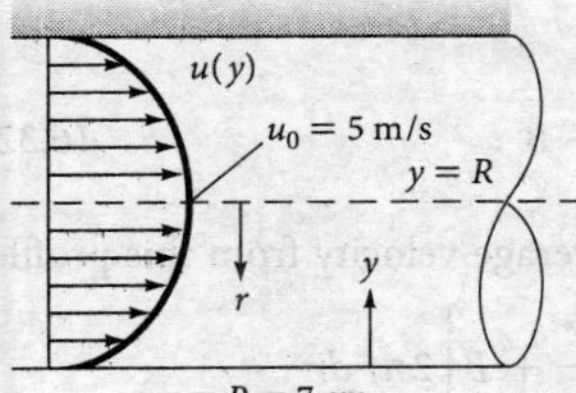

E6.5

EXAMPLE 6.5

Air at 20°C flows through a 14-cm-diameter tube under fully developed conditions. The centerline velocity is u_0 = 5 m/s. Estimate from Fig. 6.10 (*a*) the friction velocity u^* and (*b*) the wall shear stress τ_w.

Solution

- *System sketch:* Figure E6.5 shows turbulent pipe flow with u_0 = 5 m/s and R = 7 cm.
- *Assumptions:* Figure 6.10 shows that the logarithmic law, Eq. (6.28), is reasonable all the way to the center of the tube.
- *Approach:* Use Eq. (6.28) to estimate the unknown friction velocity u^*.
- *Property values:* For air at 20°C, ρ = 1.205 kg/m^3 and ν = 1.51 E-5 m^2/s.
- *Solution step:* Insert all the given data into Eq. (6.28) at $y = R$ (the centerline). The only unknown is u^*:

$$\frac{u_0}{u^*} = \frac{1}{\kappa}\ln\left(\frac{Ru^*}{\nu}\right) + B \quad \text{or} \quad \frac{5.0 \text{ m/s}}{u^*} = \frac{1}{0.41}\ln\left[\frac{(0.07\text{ m})u^*}{1.51\text{ E-5 m}^2/\text{s}}\right] + 5$$

Although the logarithm makes it awkward, one can solve this either by hand or by Excel iteration. There is an automatic iteration procedure in Excel—`File, Excel Options, Formulas, Enable iterative calculation`—but here we simply show how to iterate by repeat calculations, copied and pasted downward. For a single unknown, in this case u^*, we only need two columns, one for the unknown and one for the equation. The writer hopes that the following copy-and-iterate procedure is clear:

	A	B
	Here place the first guess for u*:	Here place the equation that solves for u*:
1	1.0	=(5.0/(1/0.41*ln(0.07*a1/1.51E-5)+5))
2	=B1 (the number, not the equation)	Copy B1 equation and place here
3	Copy A2 here	Copy B2 here
4	Keep copying down...	Keep copying down until convergence

Note that B2 uses the *cell* location for u^*, A1, not the notation u^*. Here are the actual numbers, not instructions or equations, for this problem:

A	B
1.0000	0.1954
0.1954	0.2314
0.2314	0.2271
0.2271	0.2275
0.2275	0.2275

The solution for u^* has converged to 0.2275. To three decimal places,

$$u^* \approx 0.228 \text{ m/s} \qquad \textit{Ans. (a)}$$

$$\tau_w = \rho u^{*2} = (1.205)(0.228)^2 \approx 0.062 \text{ Pa} \qquad \textit{Ans. (b)}$$

- *Comments:* The logarithmic law solved everything! This is a powerful technique, using an experimental velocity correlation to approximate general turbulent flows. You may check that the Reynolds number Re_d is about 40,000, definitely turbulent flow.

6.6 Turbulent Pipe Flow

For turbulent pipe flow, we need not solve a differential equation, but instead proceed with the logarithmic law, as in Example 6.5. Assume that Eq. (6.28) correlates the local mean velocity $u(r)$ all the way across the pipe

$$\frac{u(r)}{u^*} \approx \frac{1}{\kappa} \ln \frac{(R-r)u^*}{\nu} + B \tag{6.32}$$

where we have replaced y with $R - r$. Compute the average velocity from this profile:

$$V = \frac{Q}{A} = \frac{1}{\pi R^2} \int_0^R u^* \left[\frac{1}{\kappa} \ln \frac{(R-r)u^*}{\nu} + B \right] 2\pi r \, dr$$

$$= \frac{1}{2} u^* \left(\frac{2}{\kappa} \ln \frac{Ru^*}{\nu} + 2B - \frac{3}{\kappa} \right) \tag{6.33}$$

Introducing $\kappa = 0.41$ and $B = 5.0$, we obtain, numerically,

$$\frac{V}{u^*} \approx 2.44 \ln \frac{Ru^*}{\nu} + 1.34 \tag{6.34}$$

This looks only marginally interesting until we realize that V/u^* is directly related to the Darcy friction factor:

$$\frac{V}{u^*} = \left(\frac{\rho V^2}{\tau_w} \right)^{1/2} = \left(\frac{8}{f} \right)^{1/2} \tag{6.35}$$

Moreover, the argument of the logarithm in (6.34) is equivalent to

$$\frac{Ru^*}{\nu} = \frac{\frac{1}{2}Vd}{\nu}\frac{u^*}{V} = \frac{1}{2}\mathrm{Re}_d\left(\frac{f}{8}\right)^{1/2} \tag{6.36}$$

Introducing (6.35) and (6.36) into Eq. (6.34), changing to a base-10 logarithm, and rearranging, we obtain

$$\frac{1}{f^{1/2}} \approx 1.99 \log (\mathrm{Re}_d f^{1/2}) - 1.02 \tag{6.37}$$

In other words, by simply computing the mean velocity from the logarithmic law correlation, we obtain a relation between the friction factor and Reynolds number for turbulent pipe flow. Prandtl derived Eq. (6.37) in 1935 and then adjusted the constants slightly to fit friction data better:

$$\frac{1}{f^{1/2}} = 2.0 \log (\mathrm{Re}_d f^{1/2}) - 0.8 \tag{6.38}$$

This is the accepted formula for a smooth-walled pipe. Some numerical values may be listed as follows:

Re_d	4,000	10^4	10^5	10^6	10^7	10^8
f	0.0399	0.0309	0.0180	0.0116	0.0081	0.0059

Thus, f drops by only a factor of 5 over a 10,000-fold increase in Reynolds number. Equation (6.38) is cumbersome to solve if Re_d is known and f is wanted. There are many alternative approximations in the literature from which f can be computed explicitly from Re_d:

$$f = \begin{cases} 0.316\,\mathrm{Re}_d^{-1/4} & 4{,}000 < \mathrm{Re}_d < 10^5 \quad \text{H. Blasius (1911)} \\ \left(1.8 \log \dfrac{\mathrm{Re}_d}{6.9}\right)^{-2} & \text{Ref. 9, Colebrook} \end{cases} \tag{6.39}$$

However, Eq. (6.38) the preferred formula, is easily solved by computer iteration.

Blasius, a student of Prandtl, presented his formula in the first correlation ever made of pipe friction versus Reynolds number. Although his formula has a limited range, it illustrates what was happening in Fig. 6.4 to Hagen's 1839 pressure-drop data. For a horizontal pipe, from Eq. (6.39),

$$h_f = \frac{\Delta p}{\rho g} = f\frac{L}{d}\frac{V^2}{2g} \approx 0.316\left(\frac{\mu}{\rho V d}\right)^{1/4}\frac{L}{d}\frac{V^2}{2g}$$

or

$$\Delta p \approx 0.158\, L\rho^{3/4}\mu^{1/4}d^{-5/4}V^{7/4} \tag{6.40}$$

at low turbulent Reynolds numbers. This explains why Hagen's data for pressure drop begin to increase as the 1.75 power of the velocity, in Fig. 6.4. Note that Δp varies only slightly with viscosity, which is characteristic of turbulent flow. Introducing $Q = \frac{1}{4}\pi d^2 V$ into Eq. (6.40), we obtain the alternative form

$$\Delta p \approx 0.241 L\rho^{3/4}\mu^{1/4}d^{-4.75}Q^{1.75} \tag{6.41}$$

For a given flow rate Q, the turbulent pressure drop decreases with diameter even more sharply than the laminar formula (6.12). Thus, the quickest way to reduce required pumping pressure is to increase the pipe size, although, of course, the larger pipe is more expensive. Doubling the pipe size decreases Δp by a factor of about 27 for a given Q. Compare Eq. (6.40) with Example 5.7 and Fig. 5.10.

The maximum velocity in turbulent pipe flow is given by Eq. (6.32), evaluated at $r = 0$:

$$\frac{u_{max}}{u^*} \approx \frac{1}{\kappa} \ln \frac{Ru^*}{\nu} + B \tag{6.42}$$

Combining this with Eq. (6.33), we obtain the formula relating mean velocity to maximum velocity:

$$\frac{V}{u_{max}} \approx (1 + 1.3\sqrt{f})^{-1} \tag{6.43}$$

Some numerical values are

Re_d	4,000	10^4	10^5	10^6	10^7	10^8
V/u_{max}	0.794	0.814	0.852	0.877	0.895	0.909

The ratio varies with the Reynolds number and is much larger than the value of 0.5 predicted for all laminar pipe flow in Eq. (6.12). Thus a turbulent velocity profile, as shown in Fig. 6.11*b*, is very flat in the center and drops off sharply to zero at the wall.

Effect of Rough Walls

It was not known until experiments in 1800 by Coulomb [6] that surface roughness has an effect on friction resistance. It turns out that the effect is negligible for laminar pipe flow, and all the laminar formulas derived in this section are valid for rough walls also. But turbulent flow is strongly affected by roughness. In Fig. 6.10 the linear viscous sublayer extends out only to $y^+ = yu^*/\nu = 5$. Thus, compared with the diameter, the sublayer thickness y_s is only

$$\frac{y_s}{d} = \frac{5\nu/u^*}{d} = \frac{14.1}{Re_d f^{1/2}} \tag{6.44}$$

For example, at $Re_d = 10^5$, $f = 0.0180$, and $y_s/d = 0.001$, a wall roughness of about $0.001d$ will break up the sublayer and profoundly change the wall law in Fig. 6.10.

Measurements of $u(y)$ in turbulent rough-wall flow by Prandtl's student Nikuradse [7] show, as in Fig. 6.12*a*, that a roughness height ε will force the logarithm law profile outward on the abscissa by an amount approximately equal to $\ln \varepsilon^+$, where $\varepsilon^+ = \varepsilon u^*/\nu$. The slope of the logarithm law remains the same, $1/\kappa$, but the shift outward causes the constant B to be less by an amount $\Delta B \approx (1/\kappa) \ln \varepsilon^+$.

Nikuradse [7] simulated roughness by gluing uniform sand grains onto the inner walls of the pipes. He then measured the pressure drops and flow rates and correlated

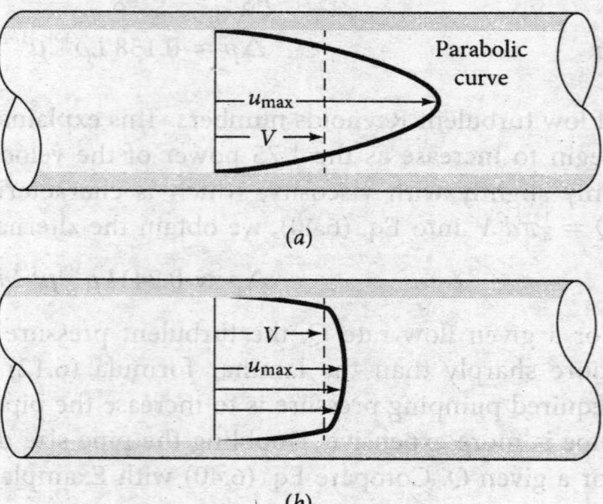

Fig. 6.11 Comparison of laminar and turbulent pipe flow velocity profiles for the same volume flow: (*a*) laminar flow; (*b*) turbulent flow.

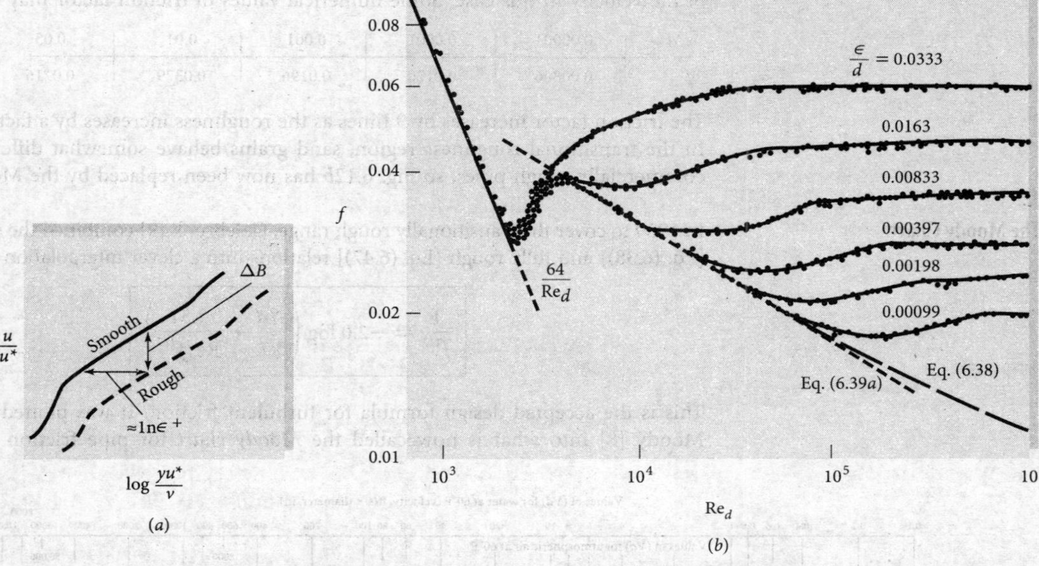

Fig. 6.12 Effect of wall roughness on turbulent pipe flow. (*a*) The logarithmic overlap velocity profile shifts down and to the right; (*b*) experiments with sand-grain roughness by Nikuradse [7] show a systematic increase of the turbulent friction factor with the roughness ratio.

friction factor versus Reynolds number in Fig. 6.12*b*. We see that laminar friction is unaffected, but turbulent friction, after an *onset* point, increases monotonically with the roughness ratio ε/d. For any given ε/d, the friction factor becomes constant *(fully rough)* at high Reynolds numbers. These points of change are certain values of $\varepsilon^+ = \varepsilon u^*/\nu$:

$\dfrac{\varepsilon u^*}{\nu} < 5$: *hydraulically smooth* walls, no effect of roughness on friction

$5 \le \dfrac{\varepsilon u^*}{\nu} \le 70$: *transitional* roughness, moderate Reynolds number effect

$\dfrac{\varepsilon u^*}{\nu} > 70$: *fully rough* flow, sublayer totally broken up and friction independent of Reynolds number

For fully rough flow, $\varepsilon^+ > 70$, the log law downshift ΔB in Fig. 6.12*a* is

$$\Delta B \approx \frac{1}{\kappa}\ln \varepsilon^+ - 3.5 \tag{6.45}$$

and the logarithm law modified for roughness becomes

$$u^+ = \frac{1}{\kappa}\ln y^+ + B - \Delta B = \frac{1}{\kappa}\ln\frac{y}{\varepsilon} + 8.5 \tag{6.46}$$

The viscosity vanishes, and hence fully rough flow is independent of the Reynolds number. If we integrate Eq. (6.46) to obtain the average velocity in the pipe, we obtain

$$\frac{V}{u^*} = 2.44\ln\frac{d}{\varepsilon} + 3.2$$

or

$$\frac{1}{f^{1/2}} = -2.0\log\frac{\varepsilon/d}{3.7} \qquad \text{fully rough flow} \tag{6.47}$$

There is no Reynolds number effect; hence the head loss varies exactly as the square of the velocity in this case. Some numerical values of friction factor may be listed:

ϵ/d	0.00001	0.0001	0.001	0.01	0.05
f	0.00806	0.0120	0.0196	0.0379	0.0716

The friction factor increases by 9 times as the roughness increases by a factor of 5,000. In the transitional roughness region, sand grains behave somewhat differently from commercially rough pipes, so Fig. 6.12*b* has now been replaced by the Moody chart.

The Moody Chart

In 1939 to cover the transitionally rough range, Colebrook [9] combined the smooth wall [Eq. (6.38)] and fully rough [Eq. (6.47)] relations into a clever interpolation formula:

$$\frac{1}{f^{1/2}} = -2.0 \log\left(\frac{\epsilon/d}{3.7} + \frac{2.51}{\mathrm{Re}_d f^{1/2}}\right) \tag{6.48}$$

This is the accepted design formula for turbulent friction. It was plotted in 1944 by Moody [8] into what is now called the *Moody chart* for pipe friction (Fig. 6.13).

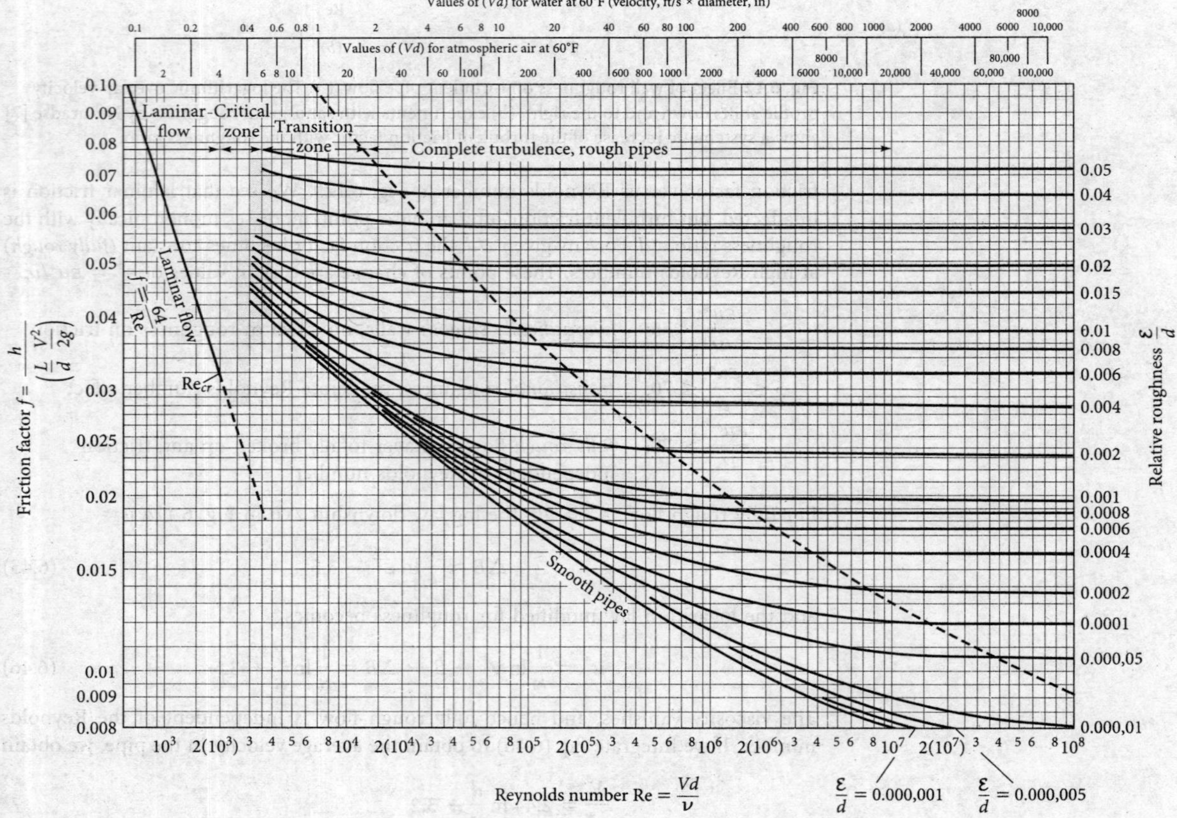

Fig. 6.13 The Moody chart for pipe friction with smooth and rough walls. This chart is identical to Eq. (6.48) for turbulent flow. *(From Ref. 8, Source: ASME.)*

Table 6.1 Recommended Roughness Values for Commercial Ducts

Material	Condition	ε ft	ε mm	Uncertainty, %
Steel	Sheet metal, new	0.00016	0.05	±60
	Stainless, new	0.000007	0.002	±50
	Commercial, new	0.00015	0.046	±30
	Riveted	0.01	3.0	±70
	Rusted	0.007	2.0	±50
Iron	Cast, new	0.00085	0.26	±50
	Wrought, new	0.00015	0.046	±20
	Galvanized, new	0.0005	0.15	±40
	Asphalted cast	0.0004	0.12	±50
Brass	Drawn, new	0.000007	0.002	±50
Plastic	Drawn tubing	0.000005	0.0015	±60
Glass	—	Smooth	Smooth	
Concrete	Smoothed	0.00013	0.04	±60
	Rough	0.007	2.0	±50
Rubber	Smoothed	0.000033	0.01	±60
Wood	Stave	0.0016	0.5	±40

The Moody chart is probably the most famous and useful figure in fluid mechanics. It is accurate to ±15 percent for design calculations over the full range shown in Fig. 6.13. It can be used for circular and noncircular (Sec. 6.6) pipe flows and for open-channel flows (Chap. 10). The data can even be adapted as an approximation to boundary layer flows (Chap. 7).

The Moody chart gives a good visual summary of laminar and turbulent pipe friction, including roughness effects. When the writer was in college, everyone solved problems by carefully reading this chart. Currently, though, Eq. (6.48), though implicit in f, is easily solved by iteration or a direct solver. If only a calculator is available, the clever explicit formula given by Haaland [33] as

$$\frac{1}{f^{1/2}} \approx -1.8 \log \left[\frac{6.9}{\mathrm{Re}_d} + \left(\frac{\varepsilon/d}{3.7} \right)^{1.11} \right] \tag{6.49}$$

varies less than 2 percent from Eq. (6.48).

The shaded area in the Moody chart indicates the range where transition from laminar to turbulent flow occurs. There are no reliable friction factors in this range, $2{,}000 < \mathrm{Re}_d < 4{,}000$. Notice that the roughness curves are nearly horizontal in the fully rough regime to the right of the dashed line.

From tests with commercial pipes, recommended values for average pipe roughness are listed in Table 6.1.

EXAMPLE 6.6[2]

Compute the loss of head and pressure drop in 60 m of horizontal 15 cm-diameter asphalted cast iron pipe carrying water with a mean velocity of 2 m/s.

Solution

- *System sketch:* See Fig. 6.7 for a horizontal pipe, with $\Delta z = 0$ and h_f proportional to Δp.
- *Assumptions:* Turbulent flow, asphalted horizontal cast iron pipe, $d = 0.15$ m, $L = 60$ m.
- *Approach:* Find Re_d and ε/d; enter the Moody chart, Fig. 6.13; find f, then h_f and Δp.

[2]This example was given by Moody in his 1944 paper [8].

- *Property values:* From Table A.3 for water, $\rho = 998 \text{ kg/m}^3$, $\mu = 0.001$ kg/m.s.
- *Solution step 1:* Calculate Re_d and the roughness ratio. As a crutch, Moody provided water and air values of "Vd" at the top of Fig. 6.13 to find Re_d. Instead, let's calculate it ourselves:

$$\text{Re}_d = \frac{\rho V d}{\mu} = \frac{(998 \text{ kg/m}^3)(2 \text{ m/s})(0.15 \text{ m})}{0.001 \text{ kg/m} \cdot \text{s}} \approx 299{,}400 \quad \text{(turbulent)}$$

From Table 6.1, for asphalted cast iron, $\varepsilon = 0.00012$ m. Then calculate

$$\varepsilon/d = (0.00012 \text{ m})/(0.15 \text{ m}) = 0.0008$$

- *Solution step 2:* Find the friction factor from the Moody chart or from Eq. (6.48). If you use the Moody chart, Fig. 6.13, you need practice. Find the line on the right side for $\varepsilon/d = 0.0008$ and follow it back to the left until it hits the vertical line for $\text{Re}_d \approx 2.99$ E5. Read, approximately, $f \approx 0.02$ [or compute $f = 0.0198$ from Eq. (6.48).]
- *Solution step 3:* Calculate h_f from Eq. (6.10) and Δp from Eq. (6.8) for a horizontal pipe:

$$h_f = f\frac{L}{d}\frac{V^2}{2g} = (0.02)\left(\frac{60 \text{ m}}{0.15 \text{ m}}\right)\frac{(2 \text{ m/s})^2}{2(9.81 \text{ m/s}^2)} \approx 1.63 \text{ m} \qquad \textit{Ans.}$$

$$\Delta p = \rho g h_f = (998 \text{ kg/m}^3)(9.81 \text{ m/s}^2)(1.63 \text{ m}) \approx 15{,}960 \text{ Pa} \qquad \textit{Ans.}$$

- *Comments:* In giving this example, Moody [8] stated that this estimate, even for clean new pipe, can be considered accurate only to about ±10 percent.

EXAMPLE 6.7

Oil, with $\rho = 900 \text{ kg/m}^3$ and $\nu = 0.00001 \text{ m}^2/\text{s}$, flows at $0.2 \text{ m}^3/\text{s}$ through 500 m of 200-mm-diameter cast iron pipe. Determine (*a*) the head loss and (*b*) the pressure drop if the pipe slopes down at 10° in the flow direction.

Solution

First compute the velocity from the known flow rate:

$$V = \frac{Q}{\pi R^2} = \frac{0.2 \text{ m}^3/\text{s}}{\pi(0.1 \text{ m})^2} = 6.4 \text{ m/s}$$

Then the Reynolds number is

$$\text{Re}_d = \frac{Vd}{\nu} = \frac{(6.4 \text{ m/s})(0.2 \text{ m})}{0.00001 \text{ m}^2/\text{s}} = 128{,}000$$

From Table 6.1, $\varepsilon = 0.26$ mm for cast iron pipe. Then

$$\frac{\varepsilon}{d} = \frac{0.26 \text{ mm}}{200 \text{ mm}} = 0.0013$$

Enter the Moody chart on the right at $\varepsilon/d = 0.0013$ (you will have to interpolate), and move to the left to intersect with Re = 128,000. Read $f \approx 0.0225$ [from Eq. (6.48) for these values we could compute $f = 0.0227$]. Then the head loss is

$$h_f = f\frac{L}{d}\frac{V^2}{2g} = (0.0225)\frac{500 \text{ m}}{0.2 \text{ m}}\frac{(6.4 \text{ m/s})^2}{2(9.81 \text{ m/s}^2)} = 117 \text{ m} \qquad \textit{Ans. (a)}$$

From Eq. (6.9) for the inclined pipe,

$$h_f = \frac{\Delta p}{\rho g} + z_1 - z_2 = \frac{\Delta p}{\rho g} + L \sin 10°$$

or $\Delta p = \rho g[h_f - (500 \text{ m}) \sin 10°] = \rho g(117 \text{ m} - 87 \text{ m})$

$$= (900 \text{ kg/m}^3)(9.81 \text{ m/s}^2)(30 \text{ m}) = 265{,}000 \text{ kg/(m} \cdot \text{s}^2) = 265{,}000 \text{ Pa} \quad \textit{Ans. (b)}$$

EXAMPLE 6.8

Repeat Example 6.4 to see whether there is any possible turbulent flow solution for a smooth-walled pipe.

Solution

In Example 6.4 we estimated a head loss $h_f \approx 0.502$ m, assuming laminar exit flow ($\alpha \approx 2.0$). For this condition the friction factor is

$$f = h_f \frac{d}{L}\frac{2g}{V^2} = (0.502 \text{ m}) \frac{(0.0012 \text{ m})(2)(9.81 \text{ m/s}^2)}{(0.3 \text{ m})(0.982 \text{ m/s})^2} \approx 0.04085$$

For laminar flow, $\text{Re}_d = 64/f = 64/0.04085 \approx 1{,}566$, as we showed in Example 6.4. However, from the Moody chart (Fig. 6.13), we see that $f = 0.041$ also corresponds to a *turbulent* smooth-wall condition, at $\text{Re}_d \approx 4{,}500$. If the flow actually were turbulent, we should change our kinetic energy factor to $\alpha \approx 1.06$ [Eq. (3.77)], whence the corrected $h_f \approx 0.548$ m and $f \approx 0.0446$. With f known, we can estimate the Reynolds number from our formulas:

$$\text{Re}_d \approx 2{,}770 \;\; [\text{Eq. } (6.38)] \quad \text{or} \quad \text{Re}_d \approx 2{,}950 \;\; [\text{Eq. } (6.39b)]$$

So the flow *might* have been turbulent, in which case the viscosity of the fluid would have been

$$\mu = \frac{\rho V d}{\text{Re}_d} = \frac{917 \text{ kg/m}^3 (0.982 \text{ m/s})(0.0012 \text{ m})}{2{,}850} = 3.79 \text{ E-4 kg/m} \cdot \text{s} \qquad \textit{Ans.}$$

This is about 55 percent less than our laminar estimate in Example 6.4. The moral is to keep the capillary-flow Reynolds number below about 1,000 to avoid such duplicate solutions.

6.7 Four Types of Pipe Flow Problems

The Moody chart (Fig. 6.13) can be used to solve almost any problem involving friction losses in long pipe flows. However, many such problems involve considerable iteration and repeated calculations using the chart because the standard Moody chart is essentially a *head loss chart.* One is supposed to know all other variables, compute Re_d, enter the chart, find f, and hence compute h_f. This is one of four fundamental problems which are commonly encountered in pipe flow calculations:

1. Given d, L, and V or Q, ρ, μ, and g, compute the head loss h_f (head loss problem).
2. Given d, L, h_f, ρ, μ, and g, compute the velocity V or flow rate Q (flow rate problem).
3. Given Q, L, h_f, ρ, μ, and g, compute the diameter d of the pipe (sizing problem).
4. Given Q, d, h_f, ρ, μ, and g, compute the pipe length L.

Problems 1 and 4 are well suited to the Moody chart. We have to iterate to compute velocity or diameter because both d and V are contained in the ordinate *and* the abscissa of the chart.

There are two alternatives to iteration for problems of type 2 and type 3: (*a*) preparation of a suitable new Moody-type formula (see Probs. P6.68 and P6.73); or (*b*) the use of *solver* software, like Excel. Examples 6.9 and 6.11 include the Excel approach to these problems.

Type 2 Problem: Find the Flow Rate

Even though velocity (or flow rate) appears in both the ordinate and the abscissa on the Moody chart, iteration for turbulent flow is nevertheless quite fast because f varies so slowly with Re_d. In earlier editions, the writer rescaled the Colebrook formula (6.48) into a relation where Q could be calculated directly. That idea is now downsized to Problem P6.68. Example 6.9, which follows, is illustrated both by iteration and by an Excel solution.

EXAMPLE 6.9

Oil, with $\rho = 950\ \mathrm{kg/m^3}$ and $\nu = 2\ \mathrm{E\text{-}5\ m^2/s}$, flows through a 30-cm-diameter pipe 100 m long with a head loss of 8 m. The roughness ratio is $\varepsilon/d = 0.0002$. Find the average velocity and flow rate.

Iterative Solution

By definition, the friction factor is known except for V:

$$f = h_f \frac{d}{L}\frac{2g}{V^2} = (8\ \mathrm{m})\left(\frac{0.3\ \mathrm{m}}{100\ \mathrm{m}}\right)\left[\frac{2(9.81\ \mathrm{m/s^2})}{V^2}\right] \quad \text{or} \quad fV^2 \approx 0.471 \quad \text{(SI units)}$$

To get started, we only need to guess f, compute $V = \sqrt{0.471/f}$, then get Re_d, compute a better f from the Moody chart, and repeat. The process converges fairly rapidly. A good first guess is the "fully rough" value for $\varepsilon/d = 0.0002$, or $f \approx 0.014$ from Fig. 6.13. The iteration would be as follows:

Guess $f \approx 0.014$, then $V = \sqrt{0.471/0.014} = 5.80$ m/s and $\mathrm{Re}_d = Vd/\nu \approx 87{,}000$. At $\mathrm{Re}_d = 87{,}000$ and $\varepsilon/d = 0.0002$, compute $f_{\text{new}} \approx 0.0195$ [Eq. (6.48)].

New $f \approx 0.0195$, $V = \sqrt{0.471/0.0195} = 4.91$ m/s and $\mathrm{Re}_d = Vd/\nu = 73{,}700$. At $\mathrm{Re}_d = 73{,}700$ and $\varepsilon/d = 0.0002$, compute $f_{\text{new}} \approx 0.0201$ [Eq. (6.48)].

Better $f \approx 0.0201$, $V = \sqrt{0.471/0.0201} = 4.84$ m/s and $\mathrm{Re}_d \approx 72{,}600$. At $\mathrm{Re}_d = 72{,}600$ and $\varepsilon/d = 0.0002$, compute $f_{\text{new}} \approx 0.0201$ [Eq. (6.48)].

We have converged to three significant figures. Thus our iterative solution is

$$V = 4.84\ \mathrm{m/s}$$

$$Q = V\left(\frac{\pi}{4}\right)d^2 = (4.84)\left(\frac{\pi}{4}\right)(0.3)^2 \approx 0.342\ \mathrm{m^3/s} \qquad \textit{Ans.}$$

The iterative approach is straightforward and not too onerous, so it is routinely used by engineers. Obviously this repetitive procedure is ideal for a personal computer.

Solution by Iteration with Excel

To iterate by repeated copying in Excel, we need five columns: velocity, flow rate, Reynolds number, an initial guess for f, and a calculation of f from $(\varepsilon/\mathrm{d}) = 0.0002$ and the current value of $\mathrm{Re_d}$. We modify our guess for f, in the next row, with the new value of f and calculate again, as shown in the following table. Since f is a slowly varying function, the process converges rapidly.

	V(m/s) = (0.471/E1)^0.5	Q(m³/s) = (π/4)A1*0.3^2	Re_d = A1*0.3/0.00002	f(Eq. 6.48)	f-guess
	A	B	C	D	E
1	5.8002	0.4100	87,004	0.02011	0.01400
2	4.8397	0.3421	72,596	0.02011	0.02011
3	**4.8397**	**0.3421**	72,596	0.02011	0.02011

As shown in the hand-iterated method, the proper solution is $V = 4.84$ m/s and $Q = 0.342\ \text{m}^2/\text{s}$.

Type 3 Problem: Find the Pipe Diameter

The Moody chart is especially awkward for finding the pipe size, since d occurs in all three parameters f, Re_d, and ε/d. Further, it depends on whether we know the velocity or the flow rate. We cannot know both, or else we could immediately compute $d = \sqrt{4Q/(\pi V)}$.

Let us assume that we know the flow rate Q. Note that this requires us to redefine the Reynolds number in terms of Q:

$$\text{Re}_d = \frac{Vd}{\nu} = \frac{4Q}{\pi d \nu} \tag{6.50}$$

If, instead, we knew the velocity V, we could use the first form for the Reynolds number. The writer finds it convenient to solve the Darcy friction factor correlation, Eq. (6.10), by solving for f:

$$f = h_f \frac{d}{L}\frac{2g}{V^2} = \frac{\pi^2}{8}\frac{g h_f d^5}{LQ^2} \tag{6.51}$$

The following two examples illustrate the iteration.

EXAMPLE 6.10

Work Example 6.9 backward, assuming that $Q = 0.342\ \text{m}^3/\text{s}$ and $\varepsilon = 0.06$ mm are known but that d (30 cm) is unknown. Recall $L = 100$ m, $\rho = 950\ \text{kg/m}^3$, $\nu = 2\ \text{E-5}\ \text{m}^2/\text{s}$, and $h_f = 8$ m.

Iterative Solution

First write the diameter in terms of the friction factor:

$$f = \frac{\pi^2}{8}\frac{(9.81\ \text{m/s}^2)(8\ \text{m})d^5}{(100\ \text{m})(0.342\ \text{m}^3/\text{s})^2} = 8.28d^5 \qquad \text{or} \qquad d \approx 0.655 f^{1/5} \tag{1}$$

in SI units. Also write the Reynolds number and roughness ratio in terms of the diameter:

$$\text{Re}_d = \frac{4(0.342\ \text{m}^3/\text{s})}{\pi(2\ \text{E-5}\ \text{m}^2/\text{s})d} = \frac{21{,}800}{d} \tag{2}$$

$$\frac{\varepsilon}{d} = \frac{6\ \text{E-5 m}}{d} \tag{3}$$

Guess f, compute d from (1), then compute Re_d from (2) and ε/d from (3), and compute a better f from the Moody chart or Eq. (6.48). Repeat until (fairly rapid) convergence. Having no initial estimate for f, the writer guesses $f \approx 0.03$ (about in the middle of the turbulent portion of the Moody chart). The following calculations result:

$$f \approx 0.03 \qquad d \approx 0.655(0.03)^{1/5} \approx 0.325 \text{ m}$$

$$Re_d \approx \frac{21{,}800}{0.325} \approx 67{,}000 \qquad \frac{\varepsilon}{d} \approx 1.85 \text{ E-4}$$

Eq. (6.48): $f_{new} \approx 0.0203$ then $d_{new} \approx 0.301$ m

$$Re_{d,new} \approx 72{,}500 \qquad \frac{\varepsilon}{d} \approx 2.0 \text{ E-4}$$

Eq. (6.48): $f_{better} \approx 0.0201$ and $d = 0.300$ m *Ans.*

The procedure has converged to the correct diameter of 30 cm given in Example 6.9.

Solution by Iteration with Excel

To iterate by repeated copying in Excel, we need five columns: ε/d, friction factor, Reynolds number, diameter d, and an initial guess for f. With the guess for f, we calculate $d \approx 0.655f^{1/5}$, $Re_d \approx 21{,}800/d$, and $\varepsilon/d = (0.00006 \text{ m})/d$. Replace the guessed f with the new f. Thus Excel is doing the work of our previous hand calculation:

	ε/d = 0.00006/d	f − Eq. (6.48)	Re_d = 21,800/d	d(meters) = 0.655f^0.2	f-guess
	A	B	C	D	E
1	0.000185	0.0196	67,111	0.325	0.0300
2	0.000201	0.0201	73,106	0.298	0.0196
3	0.000200	0.0201	72,677	0.300	0.0201
4	0.000200	0.0201	72,706	**0.300**	0.0201

As shown in our hand-iterated method, the proper solution is $d = 0.300$ m.

EXAMPLE 6.11

A smooth plastic pipe is to be designed to carry 0.25 m³/sec of water at 20°C through 300 m of horizontal pipe with an exit at 103,000 Pa. The pressure drop is to be approximately 1,700,000 Pa. Determine (*a*) the proper diameter for this pipe and (*b*) whether a Schedule 40 is suitable if the pipe material has an allowable stress of 55,000 kPa.

Solution by Excel Iteration

Assumptions: Steady turbulent flow, smooth walls. For water, take $\rho = 998$ kg/m³ and $\mu = 0.001$ kg/m · s. With d unknown, use Eq. (6.51):

$$f = \frac{\pi^2}{8}\frac{gh_f d^5}{LQ^2} = \frac{\pi^2}{8}\frac{\Delta p d^5}{\rho L Q^2} = \frac{\pi^2}{8}\frac{(1{,}700{,}000 \text{ Pa})d^5}{(998 \text{ kg/m}^3)(300 \text{ m})(0.25 \text{ m}^3/\text{s})^2} = 112.1\, d^5 \qquad (1)$$

We know neither d nor f, but they are related by the Prandtl formula, Eq. (6.38):

$$\frac{1}{f^{1/2}} \approx 2.0 \log(\mathrm{Re}_d f^{1/2}) - 0.8, \ \mathrm{Re}_d = \frac{\rho V d}{\mu} = \frac{4\rho Q}{\pi \mu d} = \frac{4(998\ \mathrm{kg/m^3})(0.25\ \mathrm{m^3/s})}{\pi(0.001\ \mathrm{kg/m \cdot s})d}$$

$$= \frac{317{,}700}{d} \quad (2)$$

Part (a) Equations (1) and (2) can be solved simultaneously for f and d. Using Excel iteration, we have four columns: a guessed $f = 0.02$, d from Eq. (1), Re_d from Eq. (2), and a better f from Eq. (6.38). The pipe is smooth, so we don't need roughness:

	f – Eq. (6.38)	Re_d = 311,700/C2	d = (D2/112.1)^0.2	f-guess
	A	B	C	D
1	0.01057	1,785,387	0.177945	0.02000
2	0.01035	2,028,259	0.156637	0.01057
3	0.01034	2,036,809	0.155979	0.01035
4	**0.01034**	**2,037,203**	**0.155949**	0.01034

The process converges rapidly to:

$$\mathrm{Re}_d \approx 2.04\,E6; \quad f \approx 0.01034; \quad d \approx 0.156\ \mathrm{m}$$

Take the next highest Schedule 40 diameter in Table 6.2: $d \approx 150$ mm *Ans. (a)*

Part (b) Check to see if Schedule 40 is strong enough. The maximum pressure occurs at the pipe entrance: $p_{max} = p_{exit} + \Delta p = 103{,}000\ \mathrm{Pa} + 1{,}700{,}000\ \mathrm{Pa} = 1{,}803{,}000\ \mathrm{Pa}$. The schedule number is thus

$$\text{Schedule number} = (1{,}000)\frac{(\text{maximum pressure})}{(\text{allowable stress})} = (1{,}000)\left(\frac{1{,}803{,}000\ \mathrm{Pa}}{55{,}000{,}000\ \mathrm{Pa}}\right) \approx 33$$

Schedule 30 is too weak for this pressure, so choose a **Schedule 40 pipe.** *Ans. (b)*

Commercial Pipe Sizes

In solving a problem to find the pipe diameter, we should note that commercial pipes are made only in certain sizes. Table 6.2 gives nominatl and actual sizes of pipes in the United States. The term *Schedule 40* is a measure of the pipe thickness and its resistance to stress caused by internal fluid pressure. If P is the internal fluid pressure and

Table 6.2 Nominal and Actual Sizes of Schedule 40 Pipe

Nominal size, mm	Actual ID, mm	Wall thickness, mm
3.2	6.8	1.7
6.4	9.2	2.2
9.5	12.5	2.3
12.7	15.8	2.8
19.1	20.9	2.9
25.4	26.6	3.4
38.1	40.9	3.7
50.8	52.5	3.9
63.5	62.7	5.2
76.2	77.9	5.5
101.6	102.3	6.0
127	128.2	6.6
152.4	154.1	7.1

S is the allowable stress of the pipe material, then the schedule number = (1000)(P/S). Commercial schedules vary from 5 to 160, but 40 and 80 are by far the most popular. Example 6.11 is a typical application.

Type 4 Problem: Find the Pipe Length

In designing piping systems, it is desirable to estimate the appropriate pipe length for a given pipe diameter, pump power, and flow rate. The pump head will match the piping head loss. If minor losses, Sec. 6.9, are neglected, the (horizontal) pipe length follows from Darcy's formula (6.10):

$$h_{pump} = \frac{\text{Power}}{\rho g Q} = h_f = f\frac{L}{d}\frac{V^2}{2g} \tag{6.52}$$

With Q, d, and ε known, we may compute Re_d and f, after which L is obtained from the formula. Note that pump efficiency varies strongly with flow rate (Chap. 11). Thus, it is important to match pipe length to the pump's region of maximum efficiency.

EXAMPLE 6.12

A pump delivers 0.6 hp to water at 20°C, flowing in a 15 cm-diameter asphalted cast iron horizontal pipe at $V = 2$ m/s. What is the proper pipe length to match these conditions?

Solution

- *Approach:* Find h_f from the known power and find f from Re_d and ε/d. Then find L.
- *Water properties:* For water at 20°C, Table A.3, $\rho = 998$ kg/m³ and $\mu = 0.001$ kg/m · s.
- *Pipe roughness:* From Table 6.1 for asphalted cast iron, $\varepsilon = 0.00012$ meter.
- *Solution step 1:* Find the pump head from the flow rate and the pump power:

$$Q = AV = \frac{\pi}{4}(0.15\text{ m})^2(2\text{ m/s}) = 0.0353\text{ m}^3/\text{s}$$

$$h_{pump} = \frac{\text{Power}}{\rho g Q} = \frac{(0.6\text{ hp})[746(\text{N}\cdot\text{m})/(\text{s}\cdot\text{hp})]}{(998\text{ kg/m}^3)(9.81\text{ m/s}^2)(0.0353\text{ m}^3/\text{s})} = 1.295\text{ m}$$

- *Solution step 2:* Compute the friction factor from the Colebrook formula, Eq. (6.48):

$$\text{Re}_d = \frac{\rho V d}{\mu} = \frac{(998\text{ kg/m}^3)(2\text{ m/s})(0.15\text{ m})}{0.001\text{ kg/(m}\cdot\text{s)}} = 299{,}400 \quad \frac{\varepsilon}{d} = \frac{0.00012\text{ m}}{0.15\text{ m}} = 0.0008$$

$$\frac{1}{\sqrt{f}} \approx -2.0\log_{10}\left(\frac{\varepsilon/d}{3.7} + \frac{2.51}{\text{Re}_d\sqrt{f}}\right) \quad \text{yields} \quad f = 0.0198$$

- *Solution step 3:* Find the pipe length from the Darcy formula (6.10):

$$h_p = h_f = 1.295\text{ m} = f\frac{L}{d}\frac{V^2}{2g} = (0.0198)\left(\frac{L}{0.15\text{ m}}\right)\frac{(2\text{ m/s})^2}{2(9.81\text{ m/s}^2)}$$

Solve for $L \approx 48.1$ m *Ans.*

- *Comment:* This is Moody's problem (Example 6.6) turned around so that the length is unknown.

6.8 Flow in Noncircular Ducts[3]

If the duct is noncircular, the analysis of fully developed flow follows that of the circular pipe but is more complicated algebraically. For laminar flow, one can solve the exact equations of continuity and momentum. For turbulent flow, the logarithm law velocity profile can be used, or (better and simpler) the hydraulic diameter is an excellent approximation.

The Hydraulic Diameter

For a noncircular duct, the control volume concept of Fig. 6.7 is still valid, but the cross-sectional area A does not equal πR^2 and the cross-sectional perimeter wetted by the shear stress $\mathcal{P}$ does not equal $2\pi R$. The momentum equation (6.9a) thus becomes

$$\Delta p\, A + \rho g A\, \Delta L \sin \phi - \overline{\tau}_w \mathcal{P} \Delta L = 0$$

or

$$h_f = \frac{\Delta p}{\rho g} + \Delta z = \frac{\overline{\tau}_w}{\rho g}\frac{\Delta L}{A/\mathcal{P}} \tag{6.53}$$

Comparing this to Eq. (6.9b), we see that $A/\mathcal{P}$ takes the place of one-fourth of the pipe diameter for a circular cross section. We define the friction factor in terms of average shear stress:

$$f_{\text{NCD}} = \frac{8\overline{\tau}_w}{\rho V^2} \tag{6.54}$$

where NCD stands for noncircular duct and $V = Q/A$ as usual, Eq. (6.53) becomes

$$h_f = f\frac{L}{D_h}\frac{V^2}{2g} \tag{6.55}$$

This is equivalent to Eq. (6.10) for pipe flow except that d is replaced by D_h. Therefore, we customarily define the *hydraulic diameter* as

$$D_h = \frac{4A}{\mathcal{P}} = \frac{4 \times \text{area}}{\text{wetted perimeter}} \tag{6.56}$$

We should stress that the wetted perimeter includes all surfaces acted upon by the shear stress. For example, in a circular annulus, both the outer and the inner perimeter should be added.

We would therefore expect by dimensional analysis that this friction factor f, based on hydraulic diameter as in Eq. (6.55), would correlate with the Reynolds number and roughness ratio based on the hydraulic diameter

$$f = F\left(\frac{VD_h}{\nu}, \frac{\varepsilon}{D_h}\right) \tag{6.57}$$

and this is the way the data are correlated. But we should not necessarily expect the Moody chart (Fig. 6.13) to hold exactly in terms of this new length scale. And it does not, but it is surprisingly accurate:

$$f \approx \begin{cases} \dfrac{64}{\text{Re}_{D_h}} & \pm 40\% \quad \text{laminar flow} \\[2ex] f_{\text{Moody}}\left(\text{Re}_{D_h}, \dfrac{\varepsilon}{D_h}\right) & \pm 15\% \quad \text{turbulent flow} \end{cases} \tag{6.58}$$

Now let us look at some particular cases.

[3]This section may be omitted without loss of continuity.

Flow between Parallel Plates

Probably the simplest noncircular duct flow is fully developed flow between parallel plates a distance $2h$ apart, as in Fig. 6.14. As noted in the figure, the width $b \gg h$, so the flow is essentially two-dimensional; that is, $u = u(y)$ only. The hydraulic diameter is

$$D_h = \frac{4A}{\mathcal{P}} = \lim_{b\to\infty} \frac{4(2bh)}{2b + 4h} = 4h \tag{6.59}$$

that is, twice the distance between the plates. The pressure gradient is constant, $(-dp/dx) = \Delta p/L$, where L is the length of the channel along the x axis.

Laminar Flow Solution

The laminar solution was given in Sec. 4.10, in connection with Fig. 4.16*b*. Let us review those results here:

$$\begin{aligned} u &= u_{\max}\left(1 - \frac{y^2}{h^2}\right) \quad \text{where } u_{\max} = \frac{h^2}{2\mu}\frac{\Delta p}{L} \\ Q &= \frac{2bh^3}{3\mu}\frac{\Delta p}{L} \\ V &= \frac{Q}{A} = \frac{h^2}{3\mu}\frac{\Delta p}{L} = \frac{2}{3}u_{\max} \\ \tau_w &= \mu\left|\frac{du}{dy}\right|_{y=h} = h\frac{\Delta p}{L} = \frac{3\mu V}{h} \\ h_f &= \frac{\Delta p}{\rho g} = \frac{3\mu L V}{\rho g h^2} \end{aligned} \tag{6.60}$$

Now use the head loss to establish the laminar friction factor:

$$f_{\text{lam}} = \frac{h_f}{(L/D_h)(V^2/2g)} = \frac{96\mu}{\rho V(4h)} = \frac{96}{\text{Re}_{D_h}} \tag{6.61}$$

Thus, if we could not work out the laminar theory and chose to use the approximation $f \approx 64/\text{Re}_{D_h}$, we would be 33 percent low. The hydraulic-diameter approximation is relatively crude in laminar flow, as Eq. (6.58) states.

Just as in circular-pipe flow, the laminar solution above becomes unstable at about $\text{Re}_{D_h} \approx 2{,}000$; transition occurs and turbulent flow results.

Turbulent Flow Solution

For turbulent flow between parallel plates, we can again use the logarithm law, Eq. (6.28), as an approximation across the entire channel, using not y but a wall coordinate Y, as shown in Fig. 6.14:

$$\frac{u(Y)}{u^*} \approx \frac{1}{\kappa}\ln\frac{Yu^*}{\nu} + B \qquad 0 < Y < h \tag{6.62}$$

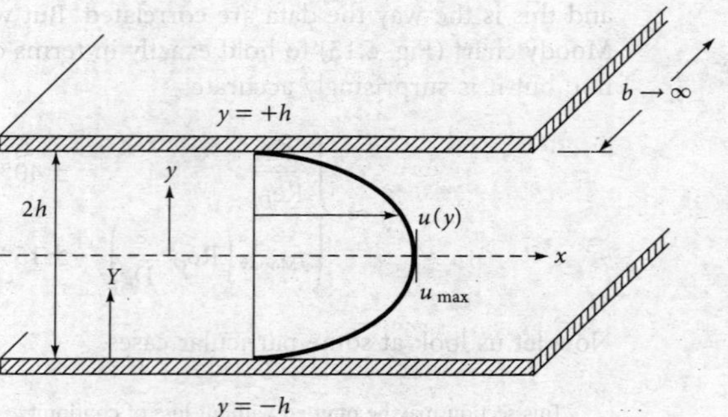

Fig. 6.14 Fully developed flow between parallel plates.

This distribution looks very much like the flat turbulent profile for pipe flow in Fig. 6.11*b*, and the mean velocity is

$$V = \frac{1}{h}\int_0^h u\,dY = u^*\left(\frac{1}{\kappa}\ln\frac{hu^*}{\nu} + B - \frac{1}{\kappa}\right) \tag{6.63}$$

Recalling that $V/u^* = (8/f)^{1/2}$, we see that Eq. (6.63) is equivalent to a parallel-plate friction law. Rearranging and cleaning up the constant terms, we obtain

$$\frac{1}{f^{1/2}} \approx 2.0 \log\,(\mathrm{Re}_{D_h} f^{1/2}) - 1.19 \tag{6.64}$$

where we have introduced the hydraulic diameter $D_h = 4h$. This is remarkably close to the smooth-wall pipe friction law, Eq. (6.38). Therefore we conclude that the use of the hydraulic diameter in this turbulent case is quite successful. That turns out to be true for other noncircular turbulent flows also.

Equation (6.64) can be brought into exact agreement with the pipe law by rewriting it in the form

$$\frac{1}{f^{1/2}} = 2.0 \log\,(0.64\,\mathrm{Re}_{D_h} f^{1/2}) - 0.8 \tag{6.65}$$

Thus the turbulent friction is predicted most accurately when we use an effective diameter D_{eff} equal to 0.64 times the hydraulic diameter. The effect on f itself is much less, about 10 percent at most. We can compare with Eq. (6.66) for laminar flow, which predicted

Parallel plates:
$$D_{eff} = \frac{64}{96}D_h = \frac{2}{3}D_h \tag{6.66}$$

This close resemblance ($0.64D_h$ versus $0.667D_h$) occurs so often in noncircular duct flow that we take it to be a general rule for computing turbulent friction in ducts:

$$D_{eff} = D_h = \frac{4A}{\mathcal{P}} \quad \text{reasonable accuracy}$$

$$D_{eff} = D_h\frac{64}{(f\,\mathrm{Re}_{D_h})\text{laminar theory}} \quad \text{better accuracy} \tag{6.67}$$

Jones [10] shows that the effective-laminar-diameter idea collapses all data for rectangular ducts of arbitrary height-to-width ratio onto the Moody chart for pipe flow. We recommend this idea for all noncircular ducts.

EXAMPLE 6.13

Fluid flows at an average velocity of 2 m/s between horizontal parallel plates a distance of 6 cm apart. Find the head loss and pressure drop for each 30 m of length for $\rho = 980\ \text{kg/m}^3$ and (*a*) $\nu = 2$ E-6 m^2/s and (*b*) $\nu = 0.0002\ \text{m}^2/\text{s}$. Assume smooth walls.

Solution

Part (a) The viscosity $\mu = \rho\nu = 0.00196$ kg/m · s. The spacing is $2h = 0.06$ m, and $D_h = 4h = 0.12$ m. The Reynolds number is

$$\mathrm{Re}_{D_h} = \frac{VD_h}{\nu} = \frac{(2\ \text{m/s})(0.12\ \text{m})}{2\ \text{E-6 m}^2/\text{s}} = 120{,}000$$

The flow is therefore turbulent. For reasonable accuracy, simply look on the Moody chart (Fig. 6.13) for smooth walls:

$$f \approx 0.0173 \quad h_f \approx f\frac{L}{D_h}\frac{V^2}{2g} = 0.0173\frac{30\text{ m}}{0.12\text{ m}}\frac{(2\text{ m/s})^2}{2(9.81\text{ m/s}^2)} \approx 0.882\text{ m} \qquad \textit{Ans. (a)}$$

Since there is no change in elevation,

$$\Delta p = \rho g h_f = 980\text{ kg/m}^3(9.81\text{ m/s}^2)(0.882\text{ m}) = 8{,}480\text{ Pa} \qquad \textit{Ans. (a)}$$

This is the head loss and pressure drop per 30 m of channel. For more accuracy, take $D_{\text{eff}} = \frac{64}{96}D_h$ from laminar theory; then

$$\text{Re}_{\text{eff}} = \tfrac{64}{96}(120{,}000) = 80{,}000$$

and from the Moody chart read $f \approx 0.0189$ for smooth walls. Thus a better estimate is

$$h_f = 0.0189\frac{30\text{ m}}{0.12\text{ m}}\frac{(2\text{ m/s})^2}{2(9.81\text{ m/s}^2)} = 0.963\text{ m}$$

and $$\Delta p = 980\text{ kg/m}^3(9.81\text{ m/s}^2)(0.963\text{ m}) = 9{,}260\text{ Pa} \qquad \textit{Better ans. (a)}$$

The more accurate formula predicts friction about 9 percent higher.

Part (a) Compute $\mu = \rho\nu = 0.196$ kg/m · s. The Reynolds number is 2 m/s(0.12 m)/0.0002 m^2/s = 1,200; therefore the flow is laminar, since Re is less than 2,300.

You could use the laminar flow friction factor, Eq. (6.61)

$$f_{\text{lam}} = \frac{96}{\text{Re}_{D_h}} = \frac{96}{1{,}200} = 0.08$$

from which $$h_f = 0.08\frac{30\text{ m}}{0.12\text{ m}}\frac{(2\text{ m/s})^2}{2(9.81\text{ m/s}^2)} = 4.08\text{ m}$$

and $$\Delta p = 980\text{ kg/m}^3(9.81\text{ m/s}^2)(4.08\text{ m}) = 39{,}200\text{ Pa} \qquad \textit{Ans. (b)}$$

Alternately you can finesse the Reynolds number and go directly to the appropriate laminar flow formula, Eq. (6.60):

$$V = \frac{h^2}{3\mu}\frac{\Delta p}{L}$$

or $$\Delta p = \frac{3(2\text{ m/s})\,[0.196\text{ kg/m}\cdot\text{s}]\,(30\text{ m})}{(0.03\text{ m})^2} = 39{,}200\text{ Pa}$$

and $$h_f = \frac{\Delta p}{\rho g} = \frac{39{,}200\text{ Pa}}{980\text{ kg/m}^3(9.81\text{ m/s}^2)} = 4.08\text{ m}$$

Flow through a Concentric Annulus

Consider steady axial laminar flow in the annular space between two concentric cylinders, as in Fig. 6.15. There is no slip at the inner ($r = b$) and outer radius ($r = a$). For $u = u(r)$ only, the governing relation is Eq. (D.7) in Appendix D:

$$\frac{d}{dr}\left(r\mu\frac{du}{dr}\right) = Kr \qquad K = \frac{d}{dx}(p + \rho g z) \tag{6.68}$$

Integrate this twice:

$$u = \frac{1}{4}r^2\frac{K}{\mu} + C_1 \ln r + C_2$$

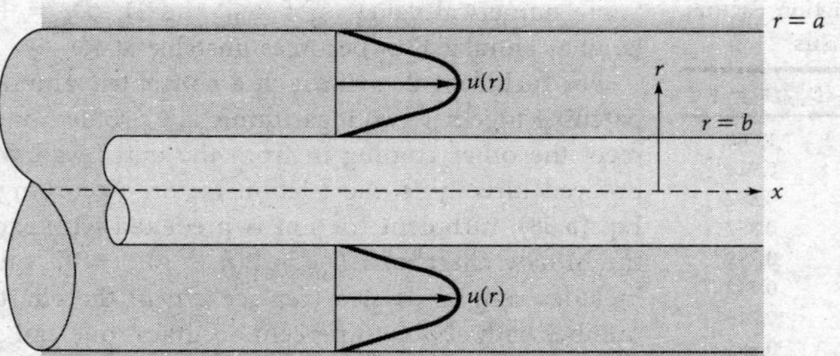

Fig. 6.15 Fully developed flow through a concentric annulus.

The constants are found from the two no-slip conditions:

$$u(r = a) = 0 = \frac{1}{4}a^2\frac{K}{\mu} + C_1 \ln a + C_2$$

$$u(r = b) = 0 = \frac{1}{4}b^2\frac{K}{\mu} + C_1 \ln b + C_2$$

The final solution for the velocity profile is

$$u = \frac{1}{4\mu}\left[-\frac{d}{dx}(p + \rho gz)\right]\left[a^2 - r^2 + \frac{a^2 - b^2}{\ln(b/a)}\ln\frac{a}{r}\right] \tag{6.69}$$

The volume flow is given by

$$Q = \int_b^a u\, 2\pi r\, dr = \frac{\pi}{8\mu}\left[-\frac{d}{dx}(p + \rho gz)\right]\left[a^4 - b^4 - \frac{(a^2 - b^2)^2}{\ln(a/b)}\right] \tag{6.70}$$

The velocity profile $u(r)$ resembles a parabola wrapped around in a circle to form a split doughnut, as in Fig. 6.15.

It is confusing to base the friction factor on the wall shear because there are two shear stresses, the inner stress being greater than the outer. It is better to define f with respect to the head loss, as in Eq. (6.55),

$$f = h_f\frac{D_h}{L}\frac{2g}{V^2} \quad \text{where } V = \frac{Q}{\pi(a^2 - b^2)} \tag{6.71}$$

The hydraulic diameter for an annulus is

$$D_h = \frac{4\pi(a^2 - b^2)}{2\pi(a + b)} = 2(a - b) \tag{6.72}$$

It is twice the clearance, rather like the parallel-plate result of twice the distance between plates [Eq. (6.59)].

Substituting h_f, D_h, and V into Eq. (6.71), we find that the friction factor for laminar flow in a concentric annulus is of the form

$$f = \frac{64\zeta}{\mathrm{Re}_{D_h}} \qquad \zeta = \frac{(a - b)^2(a^2 - b^2)}{a^4 - b^4 - (a^2 - b^2)^2/\ln(a/b)} \tag{6.73}$$

The dimensionless term ζ is a sort of correction factor for the hydraulic diameter. We could rewrite Eq. (6.73) as

$$\text{Concentric annulus:} \qquad f = \frac{64}{\mathrm{Re}_{\mathrm{eff}}} \qquad \mathrm{Re}_{\mathrm{eff}} = \frac{1}{\zeta}\mathrm{Re}_{D_h} \tag{6.74}$$

Table 6.3 Laminar Friction Factors for a Concentric Annulus

b/a	$f\,\mathrm{Re}_{D_h}$	$D_{\mathrm{eff}}/D_h = 1/\zeta$
0.0	64.0	1.000
0.00001	70.09	0.913
0.0001	71.78	0.892
0.001	74.68	0.857
0.01	80.11	0.799
0.05	86.27	0.742
0.1	89.37	0.716
0.2	92.35	0.693
0.4	94.71	0.676
0.6	95.59	0.670
0.8	95.92	0.667
1.0	96.0	0.667

Some numerical values of $f\,\mathrm{Re}_{D_h}$ and $D_{\mathrm{eff}}/D_h = 1/\zeta$ are given in Table 6.3. Again, laminar annular flow becomes unstable at $\mathrm{Re}_{D_h} \approx 2{,}000$.

For turbulent flow through a concentric annulus, the analysis might proceed by patching together two logarithmic law profiles, one going out from the inner wall to meet the other coming in from the outer wall. We omit such a scheme here and proceed directly to the friction factor. According to the general rule proposed in Eq. (6.58), turbulent friction is predicted with excellent accuracy by replacing d in the Moody chart with $D_{\mathrm{eff}} = 2(a - b)/\zeta$, with values listed in Table 6.3.[4] This idea includes roughness also (replace ε/d in the chart with $\varepsilon/D_{\mathrm{eff}}$). For a quick design number with about 10 percent accuracy, one can simply use the hydraulic diameter $D_h = 2(a - b)$.

EXAMPLE 6.14

What should the reservoir level h be to maintain a flow of 0.01 m³/s through the commercial steel annulus 30 m long shown in Fig. E6.14? Neglect entrance effects and take $\rho = 1{,}000$ kg/m³ and $\nu = 1.02 \times 10^{-6}$ m²/s for water.

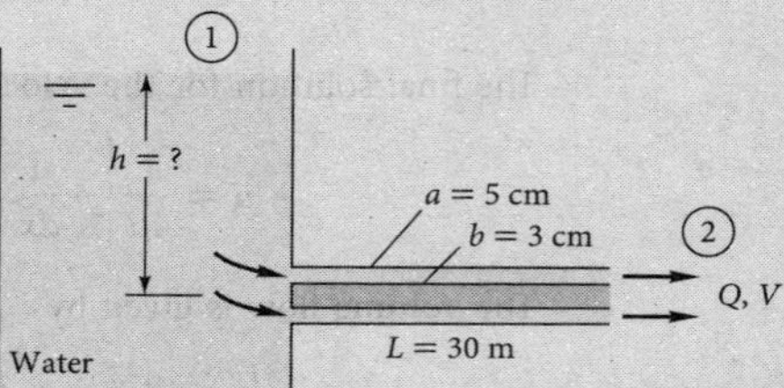

E6.14

Solution

- *Assumptions:* Fully developed annulus flow, minor losses neglected.
- *Approach:* Determine the Reynolds number, then find f and h_f and thence h.
- *Property values:* Given $\rho = 1{,}000$ kg/m³ and $\nu = 1.02$ E-6 m²/s.
- *Solution step 1:* Calculate the velocity, hydraulic diameter, and Reynolds number:

$$V = \frac{Q}{A} = \frac{0.01\ \mathrm{m^3/s}}{\pi[(0.05\ \mathrm{m})^2 - (0.03\ \mathrm{m})^2]} = 1.99\ \frac{\mathrm{m}}{\mathrm{s}}$$

$$D_h = 2(a - b) = 2(0.05\ \mathrm{m} - 0.03\ \mathrm{m}) = 0.04\ \mathrm{m}$$

$$\mathrm{Re}_{D_h} = \frac{VD_h}{\nu} = \frac{(1.99\ \mathrm{m/s})(0.04\ \mathrm{m})}{1.02\ \mathrm{E\text{-}6\ m^2/s}} = 78{,}000 \quad \text{(turbulent flow)}$$

- *Solution step 2:* Apply the steady flow energy equation between sections 1 and 2:

$$\frac{p_1}{\rho g} + \frac{\alpha_1 V_1^2}{2g} + z_1 = \frac{p_2}{\rho g} + \frac{\alpha_2 V_2^2}{2g} + z_2 + h_f$$

or

$$h = \frac{\alpha_2 V_2^2}{2g} + h_f = \frac{V_2^2}{2g}\left(\alpha_2 + f\frac{L}{D_h}\right) \tag{1}$$

Note that $z_1 = h$. For turbulent flow, from Eq. (3.43*c*), we estimate $\alpha_2 \approx 1.03$

[4] Jones and Leung [44] show that data for annular flow also satisfy the effective-laminar-diameter idea.

- *Solution step 3:* Determine the roughness ratio and the friction factor. From Table 6.1, for (new) commercial steel pipe, $\varepsilon = 0.046$ mm. Then

$$\frac{\varepsilon}{D_h} = \frac{0.046 \text{ mm}}{40 \text{ mm}} = 0.00115$$

For a reasonable estimate, use Re_{D_h} to estimate the friction factor from Eq. (6.48):

$$\frac{1}{\sqrt{f}} \approx -2.0 \log_{10}\left(\frac{0.00115}{3.7} + \frac{2.51}{78{,}000\sqrt{f}}\right) \quad \text{solve for } f \approx 0.0232$$

For slightly better accuracy, we could use $D_{\text{eff}} = D_h/\zeta$. From Table 6.3, for $b/a = 3/5$, $1/\zeta = 0.67$. Then $D_{\text{eff}} = 0.67(40 \text{ mm}) = 26.8$ mm, whence $\text{Re}_{D_{\text{eff}}} = 52{,}300$, $\varepsilon/D_{\text{eff}} = 0.00172$, and $f_{\text{eff}} \approx 0.0257$. Using the latter estimate, we find the required reservoir level from Eq. (1):

$$h = \frac{V_2^2}{2g}\left(\alpha_2 + f_{\text{eff}}\frac{L}{D_h}\right) = \frac{(1.99 \text{ m/s})^2}{2(9.81 \text{ m/s})^2}\left[1.03 + 0.0257\frac{30 \text{ m}}{0.04 \text{ m}}\right] \approx 4.1 \text{ m} \qquad \textit{Ans.}$$

- *Comments:* Note that we do *not* replace D_h with D_{eff} in the head loss term fL/D_h, which comes from a momentum balance and *requires* hydraulic diameter. If we used the simpler friction estimate, $f \approx 0.0232$, we would obtain $h \approx 3.72$ m, or about 9 percent lower.

Other Noncircular Cross Sections

In principle, any duct cross section can be solved analytically for the laminar flow velocity distribution, volume flow, and friction factor. This is because any cross section can be mapped onto a circle by the methods of complex variables, and other powerful analytical techniques are also available. Many examples are given by White [3, pp. 112–115], Berker [11], and Olson [12]. Reference 34 is devoted entirely to laminar duct flow.

In general, however, most unusual duct sections have strictly academic and not commercial value. We list here only the rectangular and isosceles-triangular sections, in Table 6.4, leaving other cross sections for you to find in the references.

For turbulent flow in a duct of unusual cross section, one should replace d with D_h on the Moody chart if no laminar theory is available. If laminar results are known, such as Table 6.4, replace d with $D_{\text{eff}} = [64/(f\text{Re})]D_h$ for the particular geometry of the duct.

For laminar flow in rectangles and triangles, the wall friction varies greatly, being largest near the midpoints of the sides and zero in the corners. In turbulent flow through the same sections, the shear is nearly constant along the sides, dropping off sharply to zero in the corners. This is because of the phenomenon of turbulent *secondary flow,* in which there are nonzero mean velocities v and w in the plane of the cross section. Some measurements of axial velocity and secondary flow patterns are shown in Fig. 6.16, as sketched by Nikuradse in his 1926 dissertation. The secondary flow "cells" drive the mean flow toward the corners, so that the axial velocity contours are similar to the cross section and the wall shear is nearly constant. This is why the hydraulic-diameter concept is so successful for turbulent flow. Laminar flow in a straight noncircular duct has no secondary flow. An accurate theoretical prediction of turbulent secondary flow has yet to be achieved, although numerical models are often successful [36].

Table 6.4 Laminar Friction Constants $f\text{Re}$ for Rectangular and Triangular Ducts

Rectangular		Isosceles triangle	
b, a		2θ	
b/a	$f\text{Re}_{D_h}$	θ, deg	$f\text{Re}_{D_h}$
0.0	96.00	0	48.0
0.05	89.91	10	51.6
0.1	84.68	20	52.9
0.125	82.34	30	53.3
0.167	78.81	40	52.9
0.25	72.93	50	52.0
0.4	65.47	60	51.1
0.5	62.19	70	49.5
0.75	57.89	80	48.3
1.0	56.91	90	48.0

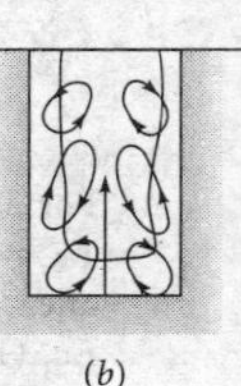

Fig. 6.16 Illustration of secondary turbulent flow in noncircular ducts: (*a*) axial mean velocity contours; (*b*) secondary flow in-plane cellular motions. *(After J. Nikuradse, dissertation, Göttingen, 1926.)*

EXAMPLE 6.15

Air, with $\rho = 1.22$ kg/m^3 and $\nu = 1.45$ E-5 m^2/s, is forced through a horizontal square 25 cm by 25 cm duct 30 m long at 0.7 m^3/s. Find the pressure drop if $\varepsilon = 1.0\text{e-}4$ m.

Solution

Compute the mean velocity and hydraulic diameter:

$$V = \frac{0.7\ \text{m}^3/\text{s}}{(0.25\ \text{m})^2} = 11.2\ \text{m/s}$$

$$D_h = \frac{4A}{\mathcal{P}} = \frac{4(0.0625\ \text{m}^2)}{1\ \text{m}} = 0.25\ \text{m}$$

From Table 6.4, for $b/a = 1.0$, the effective diameter is

$$D_{\text{eff}} = \frac{64}{56.91} D_h = 0.281\ \text{m}$$

whence

$$\text{Re}_{\text{eff}} = \frac{V D_{\text{eff}}}{\nu} = \frac{11.2\ \text{m/s}(0.281\ \text{m})}{1.45\ \text{E-5}\ \text{m}^2/\text{s}} = 217{,}000$$

$$\frac{\varepsilon}{D_{\text{eff}}} = \frac{1.0\text{e-}4\ \text{m}}{0.281\ \text{m}} = 0.000356$$

From the Moody chart, read $f = 0.0177$. Then the pressure drop is

$$\Delta p = \rho g h_f = \rho g \left(f \frac{L}{D_h} \frac{V^2}{2g} \right) = 1.22\ \text{kg/m}^3 (9.81\ \text{m/s}^2) \left[0.0177 \frac{30\ \text{m}}{0.25\ \text{m}} \frac{(11.2\ \text{m/s})^2}{2(9.81\ \text{m/s}^2)} \right]$$

or

$$\Delta p = 163\ \text{Pa} \qquad \textit{Ans.}$$

Pressure drop in air ducts is usually small because of the low density.

6.9 Minor or Local Losses in Pipe Systems

For any pipe system, in addition to the Moody-type friction loss computed for the length of pipe, there are additional so-called *minor losses* or *local losses* due to

1. Pipe entrance or exit.
2. Sudden expansion or contraction.

3. Bends, elbows, tees, and other fittings.
4. Valves, open or partially closed.
5. Gradual expansions or contractions.

The losses may not be so minor; for example, a partially closed valve can cause a greater pressure drop than a long pipe.

Since the flow pattern in fittings and valves is quite complex, the theory is very weak. The losses are commonly measured experimentally and correlated with the pipe flow parameters. The data, especially for valves, are somewhat dependent on the particular manufacturer's design, so that the values listed here must be taken as average design estimates [15, 16, 35, 43, 46].

The measured minor loss is usually given as a ratio of the head loss $h_m = \Delta p/(\rho g)$ through the device to the velocity head $V^2/(2g)$ of the associated piping system:

$$\text{Loss coefficient } K = \frac{h_m}{V^2/(2g)} = \frac{\Delta p}{\frac{1}{2}\rho V^2} \tag{6.75}$$

Although K is dimensionless, it often is not correlated in the literature with the Reynolds number and roughness ratio but rather simply with the raw size of the pipe in, say, inches. Almost all data are reported for turbulent flow conditions.

A single pipe system may have many minor losses. Since all are correlated with $V^2/(2g)$, they can be summed into a single total system loss if the pipe has constant diameter:

$$\Delta h_{\text{tot}} = h_f + \sum h_m = \frac{V^2}{2g}\left(\frac{fL}{d} + \sum K\right) \tag{6.76}$$

Note, however, that we must sum the losses separately if the pipe size changes so that V^2 changes. The length L in Eq. (6.76) is the total length of the pipe axis.

There are many different valve designs in commercial use. Figure 6.17 shows five typical designs: (*a*) the *gate*, which slides down across the section; (*b*) the *globe*, which closes a hole in a special insert; (*c*) the *angle*, similar to a globe but with a 90° turn; (*d*) the *swing-check* valve, which allows only one-way flow; and (*e*) the *disk*, which closes the section with a circular gate. The globe, with its tortuous flow path, has the highest losses when fully open. Many excellent details about these and other valves are given in the handbooks by Skousen [35] and Crane Co. [52].

Table 6.5 lists loss coefficients K for four types of valve, three angles of elbow fitting, and two tee connections. Fittings may be connected by either internal screws or flanges, hence the two listings. We see that K generally decreases with pipe size, which is consistent with the higher Reynolds number and decreased roughness ratio of large pipes. We stress that Table 6.5 represents losses *averaged among various manufacturers*, so there is an uncertainty as high as ±50 percent.

In addition, most of the data in Table 6.5 are relatively old [15, 16] and therefore based on fittings manufactured in the 1950s. Modern forged and molded fittings may yield somewhat different loss factors, often less than those listed in Table 6.5. An example, shown in Fig. 6.18*a*, gives recent data [48] for fairly short (bend-radius/elbow-diameter = 1.2) flanged 90° elbows. The elbow diameter was 4.3 cm. Notice first that K is plotted versus Reynolds number, rather than versus the raw (dimensional) pipe diameters in Table 6.5, and therefore Fig. 6.18*a* has more generality. Then notice that the K values of 0.23 ± 0.05 are significantly less than the values for 90° elbows in Table 6.5, indicating smoother walls and/or better design. One may conclude that (1) Table 6.5 data are probably conservative and (2) loss factors are highly dependent on actual design and manufacturing factors, with Table 6.5 serving only as a rough guide.

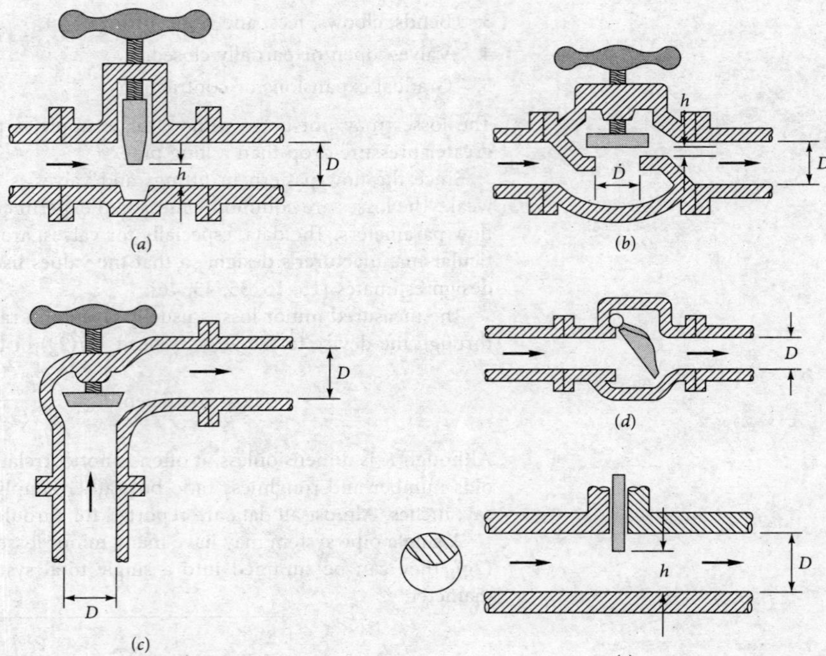

Fig. 6.17 Typical commercial valve geometries: (*a*) gate valve; (*b*) globe valve; (*c*) angle valve; (*d*) swing-check valve; (*e*) disk-type gate valve.

Table 6.5 Resistance Coefficients $K = h_m/[V^2/(2g)]$ for Open Valves, Elbows, and Tees

	Nominal diameter, mm								
	Screwed				Flanged				
	13	25	50	100	25	50	100	200	500
Valves (fully open):									
Globe	14	8.2	6.9	5.7	13	8.5	6.0	5.8	5.5
Gate	0.30	0.24	0.16	0.11	0.80	0.35	0.16	0.07	0.03
Swing check	5.1	2.9	2.1	2.0	2.0	2.0	2.0	2.0	2.0
Angle	9.0	4.7	2.0	1.0	4.5	2.4	2.0	2.0	2.0
Elbows:									
45° regular	0.39	0.32	0.30	0.29					
45° long radius					0.21	0.20	0.19	0.16	0.14
90° regular	2.0	1.5	0.95	0.64	0.50	0.39	0.30	0.26	0.21
90° long radius	1.0	0.72	0.41	0.23	0.40	0.30	0.19	0.15	0.10
180° regular	2.0	1.5	0.95	0.64	0.41	0.35	0.30	0.25	0.20
180° long radius					0.40	0.30	0.21	0.15	0.10
Tees:									
Line flow	0.90	0.90	0.90	0.90	0.24	0.19	0.14	0.10	0.07
Branch flow	2.4	1.8	1.4	1.1	1.0	0.80	0.64	0.58	0.41

The valve losses in Table 6.5 are for the fully open condition. Losses can be much higher for a partially open valve. Figure 6.18*b* gives average losses for three valves as a function of "percentage open," as defined by the opening-distance ratio h/D (see Fig. 6.17 for the geometries). Again we should warn of a possible uncertainty of ± 50 percent. Of all minor losses, valves, because of their complex geometry, are most sensitive to manufacturers' design details. For more accuracy, the particular design and manufacturer should be consulted [35].

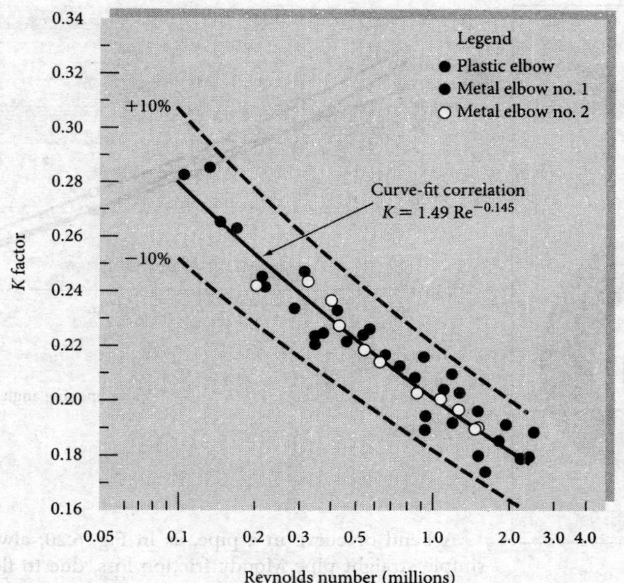

Fig. 6.18a Recent measured loss coefficients for 90° elbows. These values are less than those reported in Table 6.5. *(From Ref. 48, Source of Data R. D. Coffield.)*

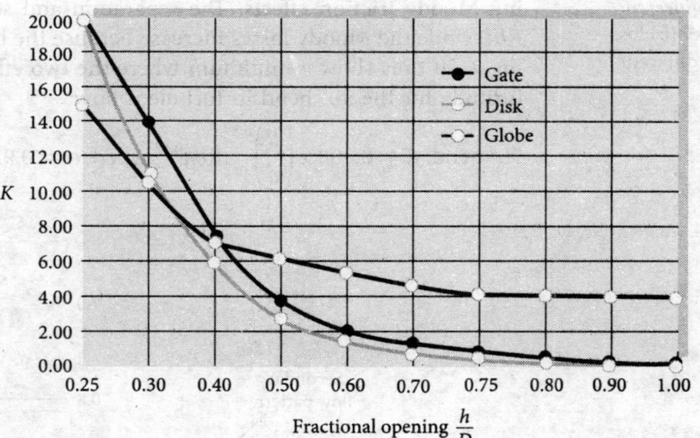

Fig. 6.18b Average loss coefficients for partially open valves (see sketches in Fig. 6.17).

The *butterfly* valve of Fig. 6.19*a* is a stem-mounted disk that, when closed, seats against an O-ring or compliant seal near the pipe surface. A single 90° turn opens the valve completely, hence the design is ideal for controllable quick-opening and quick-closing situations such as occur in fire protection and the electric power industry. However, considerable dynamic torque is needed to close these valves, and losses are high when the valves are nearly closed.

Figure 6.19*b* shows butterfly-valve loss coefficients as a function of the opening angle θ for turbulent flow conditions ($\theta = 0$ is closed). The losses are huge when the opening is small, and K drops off nearly exponentially with the opening angle. There is a factor of 2 spread among the various manufacturers. Note that K in Fig. 6.19*b* is, as usual, based on the average *pipe* velocity $V = Q/A$, not on the increased velocity of the flow as it passes through the narrow valve passage.

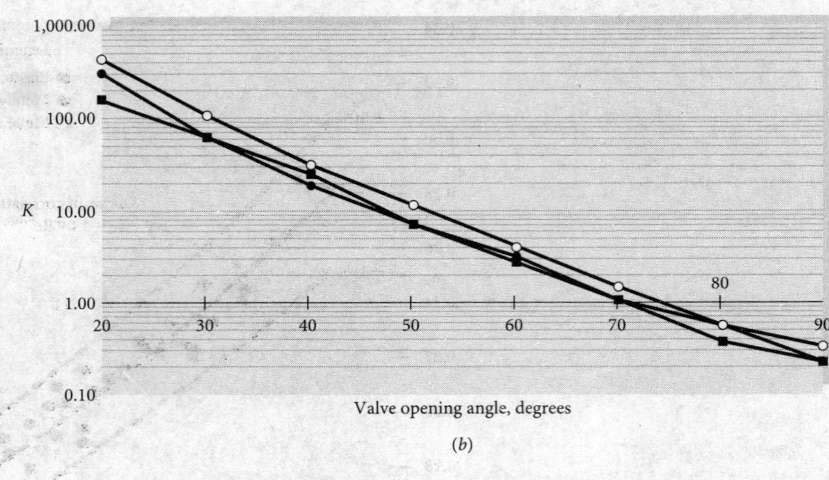

(b)

(a)

Fig. 6.19 Performance of butterfly valves: (*a*) typical geometry (*Courtesy of Tyco Engineered Products and Services*); (*b*) loss coefficients for three different manufacturers.

A bend or curve in a pipe, as in Fig. 6.20, always induces a loss larger than the simple straight-pipe Moody friction loss, due to flow separation on the curved walls and a swirling secondary flow arising from the centripetal acceleration. The smooth-wall loss coefficients K in Fig. 6.20, from the data of Ito [49], are for *total* loss, including Moody friction effects. The separation and secondary flow losses decrease with R/d, while the Moody losses increase because the bend length increases. The curves in Fig. 6.20 thus show a minimum where the two effects cross. Ito [49] gives a curve-fit formula for the 90° bend in turbulent flow:

$$90^\circ \text{ bend: } K \approx 0.388\alpha \left(\frac{R}{d}\right)^{0.84} \mathrm{Re}_D^{-0.17} \quad \text{where } \alpha = 0.95 + 4.42\left(\frac{R}{d}\right)^{-1.96} \geq 1 \qquad (6.80a)$$

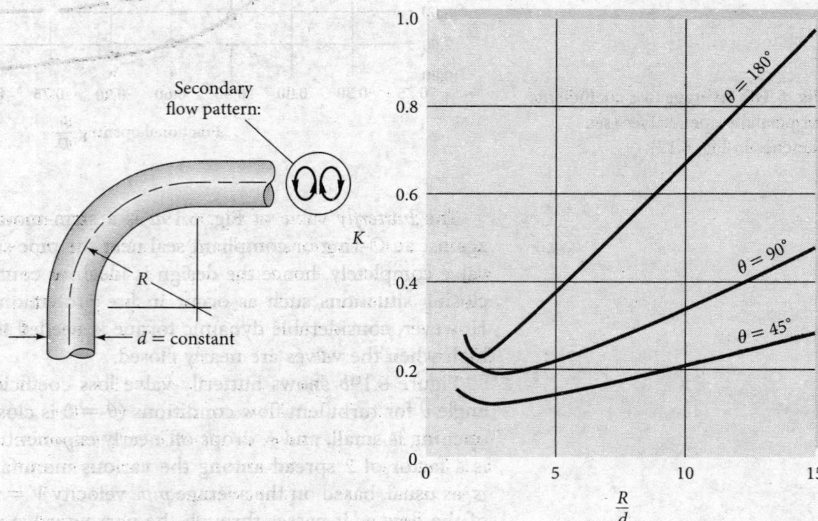

Fig. 6.20 Resistance coefficients for smooth-walled 45°, 90°, and 180° bends, at $\mathrm{Re}_d = 200{,}000$, after Ito [49].
Source: After H. Ito, "Pressure Losses in Smooth Pipe Bends," Journal of Basic Engineering, March 1960, pp. 131–143.

The formula accounts for Reynolds number, which equals 200,000 in Fig. 6.20. Comprehensive reviews of curved-pipe flow, for both laminar and turbulent flow, are given by Berger et al. [53] and for 90° bends by Spedding et al. [54].

As shown in Fig. 6.21, entrance losses are highly dependent on entrance geometry, but exit losses are not. Sharp edges or protrusions in the entrance cause large zones of flow separation and large losses. A little rounding goes a long way, and a well-rounded entrance ($r = 0.2d$) has a nearly negligible loss $K = 0.05$. At a submerged exit, on the other hand, the flow simply passes out of the pipe into the large downstream reservoir and loses all its velocity head due to viscous dissipation. Therefore $K = 1.0$ for all *submerged exits,* no matter how well rounded.

If the entrance is from a finite reservoir, it is termed a *sudden contraction* (SC) between two sizes of pipe. If the exit is to finite-sized pipe, it is termed a *sudden expansion* (SE). The losses for both are graphed in Fig. 6.22. For the sudden expansion, the shear stress in the corner separated flow, or deadwater region, is negligible, so that a control volume analysis between the expansion section and the end of the separation zone gives a theoretical loss:

$$K_{SE} = \left(1 - \frac{d^2}{D^2}\right)^2 = \frac{h_m}{V^2/(2g)} \tag{6.77}$$

Note that K is based on the velocity head in the small pipe. Equation (6.77) is in excellent agreement with experiment.

For the sudden contraction, however, flow separation in the downstream pipe causes the main stream to contract through a minimum diameter d_{min}, called the *vena contracta,* as sketched in Fig. 6.22. Because the theory of the vena contracta is not well

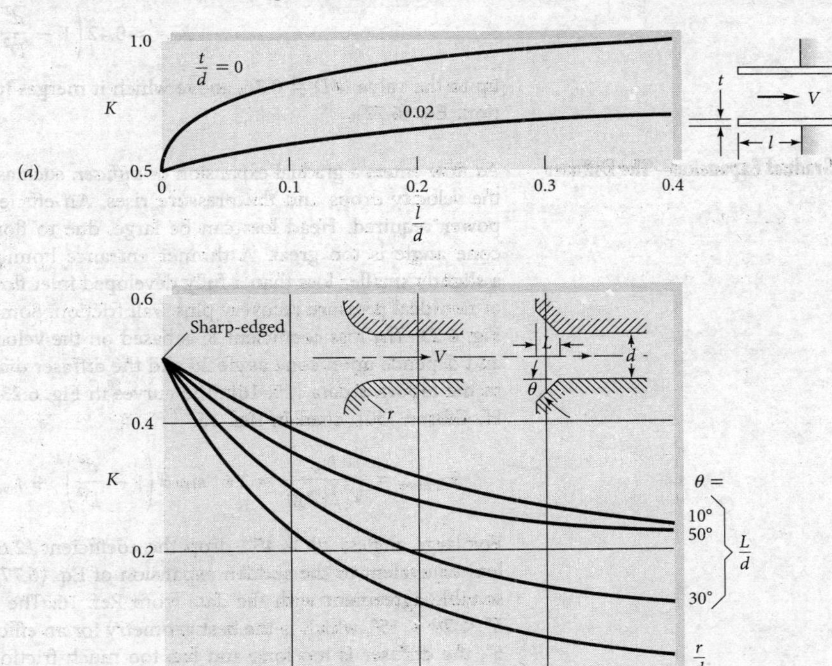

Fig. 6.21 Entrance and exit loss coefficients: (*a*) reentrant inlets; (*b*) rounded and beveled inlets. Exit losses are $K \approx 1.0$ for all shapes of exit (reentrant, sharp, beveled, or rounded).

Source: From ASHRAE Handbook-2012 Fundamentals, ASHRAE, Atlanta, GA, 2012.

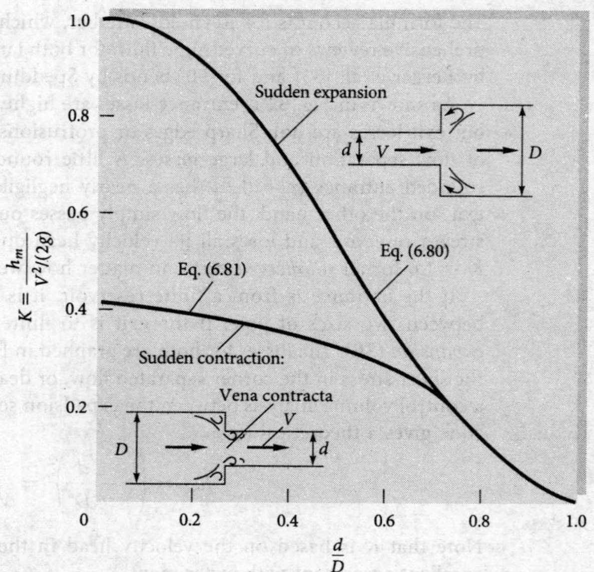

Fig. 6.22 Sudden expansion and contraction losses. Note that the loss is based on velocity head in the small pipe.

developed, the loss coefficient in the figure for sudden contraction is experimental. It fits the empirical formula

$$K_{SC} \approx 0.42\left(1 - \frac{d^2}{D^2}\right) \tag{6.78}$$

up to the value $d/D = 0.76$, above which it merges into the sudden-expansion prediction, Eq. (6.77).

Gradual Expansion—The Diffuser

As flow enters a gradual expansion or *diffuser,* such as the conical geometry of Fig. 6.23, the velocity drops and the pressure rises. An efficient diffuser reduces the pumping power required. Head loss can be large, due to flow separation on the walls, if the cone angle is too great. A thinner entrance boundary layer, as in Fig. 6.6, causes a slightly smaller loss than a fully developed inlet flow. The flow loss is a combination of nonideal pressure recovery plus wall friction. Some correlating curves are shown in Fig. 6.23. The loss coefficient K is based on the velocity head in the inlet (small) pipe and depends upon cone angle 2θ and the diffuser diameter ratio d_1/d_2. There is scatter in the reported data [15, 16]. The curves in Fig. 6.23 are based on a correlation by A. H. Gibson [50], cited in Ref. 15:

$$K_{\text{diffuser}} = \frac{h_m}{V_1^2/(2g)} \approx 2.61 \sin\theta \left(1 - \frac{d^2}{D^2}\right)^2 + f_{\text{avg}}\frac{L}{d_{\text{avg}}} \quad \text{for} \quad 2\theta \le 45° \tag{6.79}$$

For large angles, $2\theta > 45°$, drop the coefficient ($2.61 \sin\theta$), which leaves us with a loss equivalent to the sudden expansion of Eq. (6.77). As seen, the formula is in reasonable agreement with the data from Ref. 16. The minimum loss lies in the region $5° < 2\theta < 15°$, which is the best geometry for an efficient diffuser. For angles less than 5°, the diffuser is too long and has too much friction. Angles greater than 15° cause flow separation, resulting in poor pressure recovery. Professor Gordon Holloway provided the writer a recent example, where an improved diffuser design reduced the power requirement of a wind tunnel by 40 percent (100 hp decrease!). We shall look again at diffusers in Sec. 6.11, using the data of Ref. 14.

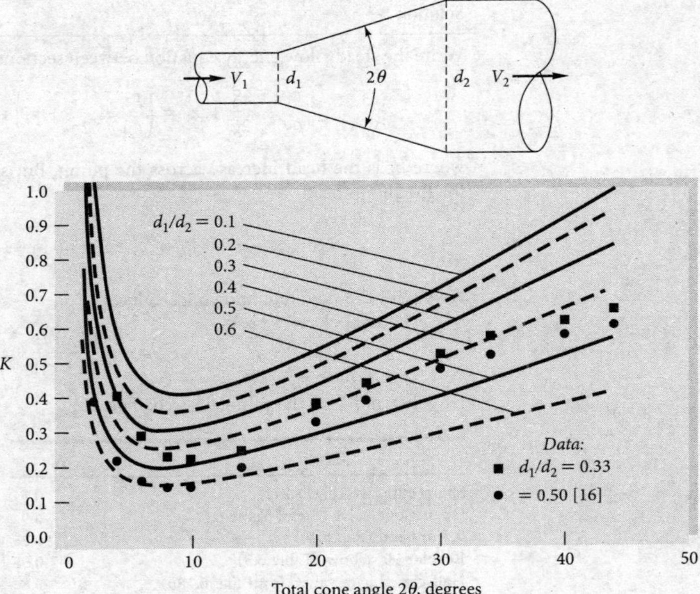

Fig. 6.23 Flow losses in a gradual conical expansion region, as calculated from Gibson's suggestion [15, 50], Eq. (6.79), for a smooth wall.

For a gradual *contraction*, the loss is very small, as seen from the following experimental values [15]:

Contraction cone angle 2θ, deg	30	45	60
K for gradual contraction	0.02	0.04	0.07

References 15, 16, 43, and 46 contain additional data on minor losses.

EXAMPLE 6.16

Water, $\rho = 998$ kg/m^3 and $\nu = 1$ E-6 m^2/s, is pumped between two reservoirs at 0.006 m^3/s through 120 m of 5 cm-diameter pipe and several minor losses, as shown in Fig. E6.16. The roughness ratio is $\varepsilon/d = 0.001$. Compute the pump horsepower required.

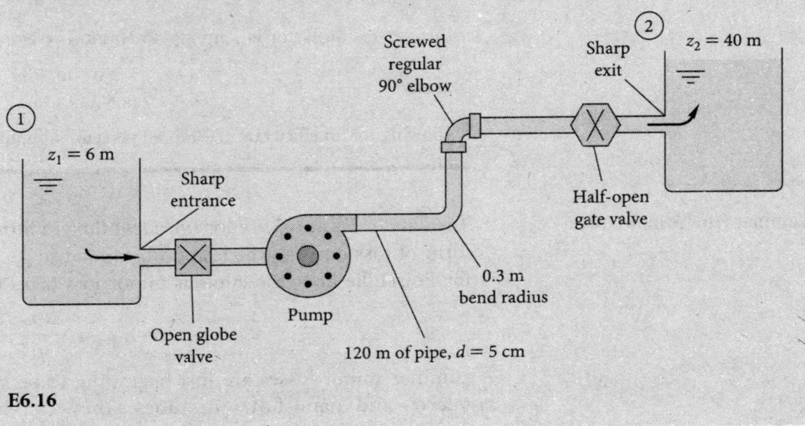

E6.16

Solution

Write the steady flow energy equation between sections 1 and 2, the two reservoir surfaces:

$$\frac{p_1}{\rho g} + \frac{V_1^2}{2g} + z_1 = \left(\frac{p_2}{\rho g} + \frac{V_2^2}{2g} + z_2\right) + h_f + \sum h_m - h_p$$

where h_p is the head increase across the pump. But since $p_1 = p_2$ and $V_1 = V_2 \approx 0$, solve for the pump head:

$$h_p = z_2 - z_1 + h_f + \sum h_m = 40 \text{ m} - 6 \text{ m} + \frac{V^2}{2g}\left(\frac{fL}{d} + \sum K\right) \quad (1)$$

Now with the flow rate known, calculate

$$V = \frac{Q}{A} = \frac{0.006 \text{ m}^3/\text{s}}{\frac{1}{4}\pi(0.05 \text{ m})^2} = 3.06 \text{ m/s}$$

Now list and sum the minor loss coefficients:

Loss	K
Sharp entrance (Fig. 6.21)	0.5
Open globe valve (5 cm, Table 6.5)	6.9
0.3 m bend (Fig. 6.20)	0.25
Regular 90° elbow (Table 6.5)	0.95
Half-closed gate valve (from Fig. 6.18*b*)	3.8
Sharp exit (Fig. 6.21)	1.0
	$\sum K = 13.4$

Calculate the Reynolds number and pipe friction factor:

$$\text{Re}_d = \frac{Vd}{\nu} = \frac{3.06 \text{ m/s}(0.05 \text{ m})}{1 \text{ E-6 m}^2/\text{s}} = 153{,}000$$

For $\varepsilon/d = 0.001$, from the Moody chart read $f = 0.0216$. Substitute into Eq. (1):

$$h_p = 34 \text{ m} + \frac{(3.06 \text{ m/s})^2}{2(9.81 \text{ m/s}^2)}\left[\frac{0.0216(120 \text{ m})}{0.05 \text{ m}} + 13.4\right]$$

$$= 34 \text{ m} + 31.14 \text{ m} = 65.14 \text{ m pump head}$$

The pump must provide a power to the water of

$$P = \rho g Q h_p = [998 \text{ kg/m}^3(9.81 \text{ m/s}^2)](0.006 \text{ m}^3/\text{s})(65.14 \text{ m}) \approx 3{,}830 \text{ Nm/s}$$

The conversion factor is 1 hp = 746 Nm/s. Therefore

$$P = \frac{3{,}830 \text{ Nm/s}}{746 \text{ Nm/s} \cdot \text{hp}} = 5.13 \text{ hp} \qquad \textit{Ans.}$$

Allowing for an efficiency of 70 to 80 percent, a pump is needed with an input of about 7 hp.

Laminar Flow Minor Losses

The data in Table 6.5 are for *turbulent* flow in fittings. If the flow is laminar, a different form of loss occurs, which is proportional to V, not V^2. By analogy with Eqs. (6.12) for Poiseuille flow, the laminar minor loss takes the form

$$K_{\text{lam}} = \frac{\Delta p_{\text{loss}}\, d}{\mu V}$$

Laminar minor losses are just beginning to be studied, due to increased interest in micro- and nano-flows in tubes. They can be substantial, comparable to the

Poiseuille loss. Professor Bruce Finlayson, of the University of Washington, kindly provided the writer with new data in the following table:

Laminar Minor Loss Coefficients K_{lam} in Tube Fittings for $1 \leq Re_d \leq 10$ [60]

Type of fitting	K_{lam}
45° bend, long radius	0.2
90° bend, short radius	0.5
90° bend, long radius	0.36
2:1 pipe contraction	7.3
3:1 pipe contraction	8.6
4:1 pipe contraction	9.0
2:1 pipe expansion	3.1
3:1 pipe expansion	4.1
4:1 pipe expansion	4.5

For the bends in the table, K_{lam} is the excess loss after calculating Poiseuille flow around the centerline of the bend. For the contractions and expansion, K_{lam} is based upon the velocity in the smaller section.

6.10 Multiple-Pipe Systems[5]

If you can solve the equations for one-pipe systems, you can solve them all; but when systems contain two or more pipes, certain basic rules make the calculations very smooth. Any resemblance between these rules and the rules for handling electric circuits is not coincidental.

Figure 6.24 shows three examples of multiple-pipe systems.

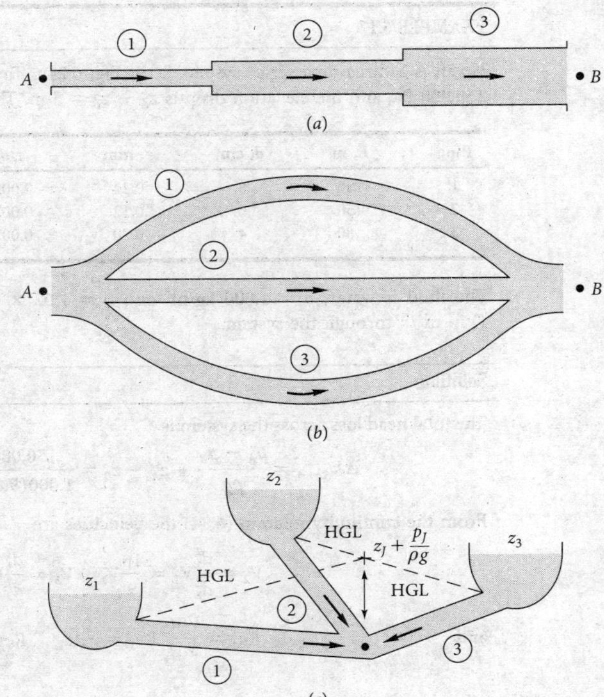

Fig. 6.24 Examples of multiple-pipe systems: (*a*) pipes in series; (*b*) pipes in parallel; (*c*) the three-reservoir junction problem.

Pipes in Series The first is a set of three (or more) pipes in series. Rule 1 is that the flow rate is the same in all pipes:

$$Q_1 = Q_2 = Q_3 = \text{const} \tag{6.80}$$

or

$$V_1 d_1^2 = V_2 d_2^2 = V_3 d_3^2 \tag{6.81}$$

Rule 2 is that the total head loss through the system equals the sum of the head loss in each pipe:

$$\Delta h_{A\to B} = \Delta h_1 + \Delta h_2 + \Delta h_3 \tag{6.82}$$

In terms of the friction and minor losses in each pipe, we could rewrite this as

$$\Delta h_{A\to B} = \frac{V_1^2}{2g}\left(\frac{f_1 L_1}{d_1} + \sum K_1\right) + \frac{V_2^2}{2g}\left(\frac{f_2 L_2}{d_2} + \sum K_2\right) + \frac{V_3^2}{2g}\left(\frac{f_3 L_3}{d_3} + \sum K_3\right) \tag{6.83}$$

and so on for any number of pipes in the series. Since V_2 and V_3 are proportional to V_1 from Eq. (6.81), Eq. (6.83) is of the form

$$\Delta h_{A\to B} = \frac{V_1^2}{2g}(\alpha_0 + \alpha_1 f_1 + \alpha_2 f_2 + \alpha_3 f_3) \tag{6.84}$$

where the α_i are dimensionless constants. If the flow rate is given, we can evaluate the right-hand side and hence the total head loss. If the head loss is given, a little iteration is needed, since f_1, f_2, and f_3 all depend on V_1 through the Reynolds number. Begin by calculating f_1, f_2, and f_3, assuming fully rough flow, and the solution for V_1 will converge with one or two iterations.

EXAMPLE 6.17

Given is a three-pipe series system, as in Fig. 6.24*a*. The total pressure drop is $p_A - p_B =$ 150,000 Pa, and the elevation drop is $z_A - z_B = 5$ m. The pipe data are

Pipe	L, m	d, cm	ε, mm	ε/d
1	100	8	0.24	0.003
2	150	6	0.12	0.002
3	80	4	0.20	0.005

The fluid is water, $\rho = 1{,}000$ kg/m^3 and $\nu = 1.02 \times 10^{-6}$ m^2/s. Calculate the flow rate Q in m^3/h through the system.

Solution

The total head loss across the system is

$$\Delta h_{A\to B} = \frac{p_A - p_B}{\rho g} + z_A - z_B = \frac{150{,}000}{1{,}000(9.81)} + 5\text{ m} = 20.3\text{ m}$$

From the continuity relation (6.84) the velocities are

$$V_2 = \frac{d_1^2}{d_2^2}V_1 = \frac{16}{9}V_1 \qquad V_3 = \frac{d_1^2}{d_3^2}V_1 = 4V_1$$

and

$$\text{Re}_2 = \frac{V_2 d_2}{V_1 d_1}\text{Re}_1 = \frac{4}{3}\text{Re}_1 \qquad \text{Re}_3 = 2\text{Re}_1$$

[5]This section may be omitted without loss of continuity.

Neglecting minor losses and substituting into Eq. (6.83), we obtain

$$\Delta h_{A\to B} = \frac{V_1^2}{2g}\left[1{,}250f_1 + 2{,}500\left(\frac{16}{9}\right)^2 f_2 + 2{,}000(4)^2 f_3\right]$$

or

$$20.3 \text{ m} = \frac{V_1^2}{2g}(1{,}250f_1 + 7{,}900f_2 + 32{,}000f_3) \tag{1}$$

This is the form that was hinted at in Eq. (6.84). It seems to be dominated by the third pipe loss $32{,}000f_3$. Begin by estimating f_1, f_2, and f_3 from the Moody-chart fully rough regime:

$$f_1 = 0.0262 \qquad f_2 = 0.0234 \qquad f_3 = 0.0304$$

Substitute in Eq. (1) to find $V_1^2 \approx 2g(20.3)/(33 + 185 + 973)$. The first estimate thus is $V_1 = 0.58$ m/s, from which

$$\text{Re}_1 \approx 45{,}400 \qquad \text{Re}_2 = 60{,}500 \qquad \text{Re}_3 = 90{,}800$$

Hence, from the Moody chart,

$$f_1 = 0.0288 \qquad f_2 = 0.0260 \qquad f_3 = 0.0314$$

Substitution into Eq. (1) gives the better estimate

$$V_1 = 0.565 \text{ m/s} \qquad Q = \tfrac{1}{4}\pi d_1^2 V_1 = 2.84 \times 10^{-3} \text{ m}^3\text{/s}$$

or

$$Q = 10.2 \text{ m}^3\text{/h} \qquad \textit{Ans.}$$

A second iteration gives $Q = 10.22$ m³/h, a negligible change.

Pipes in Parallel

The second multiple-pipe system is the *parallel* flow case shown in Fig. 6.24*b*. Here the pressure drop is the same in each pipe, and the total flow is the sum of the individual flows:

$$\Delta h_{A\to B} = \Delta h_1 = \Delta h_2 = \Delta h_3 \tag{6.85a}$$

$$Q = Q_1 + Q_2 + Q_3 \tag{6.85b}$$

If the total head loss is known, it is straightforward to solve for Q_i in each pipe and sum them, as will be seen in Example 6.18. The reverse problem, of determining $\sum Q_i$ when h_f is known, requires iteration. Each pipe is related to h_f by the Moody relation $h_f = f(L/d)(V^2/2g) = fQ^2/C$, where $C = \pi^2 gd^5/8L$. Thus each pipe has nearly quadratic nonlinear parallel resistance, and head loss is related to total flow rate by

$$h_f = \frac{Q^2}{(\sum\sqrt{C_i/f_i})^2} \quad \text{where } C_i = \frac{\pi^2 g d_i^5}{8L_i} \tag{6.86}$$

Since the f_i vary with Reynolds number and roughness ratio, one begins Eq. (6.86) by guessing values of f_i (fully rough values are recommended) and calculating a first estimate of h_f. Then each pipe yields a flow-rate estimate $Q_i \approx (C_i h_f/f_i)^{1/2}$ and hence a new Reynolds number and a better estimate of f_i. Then repeat Eq. (6.86) to convergence.

It should be noted that both of these parallel-pipe cases—finding either $\sum Q$ or h_f—are easily solved by Excel if reasonable guesses are given.

EXAMPLE 6.18

Assume that the same three pipes in Example 6.17 are now in parallel with the same total head loss of 20.3 m. Compute the total flow rate Q, neglecting minor losses.

Solution

From Eq. (6.85*a*) we can solve for each V separately:

$$20.3 \text{ m} = \frac{V_1^2}{2g} 1{,}250 f_1 = \frac{V_2^2}{2g} 2{,}500 f_2 = \frac{V_3^2}{2g} 2{,}000 f_3 \tag{1}$$

Guess fully rough flow in pipe 1: $f_1 = 0.0262$, $V_1 = 3.49$ m/s; hence $\text{Re}_1 = V_1 d_1/\nu = 273{,}000$. From the Moody chart read $f_1 = 0.0267$; recompute $V_1 = 3.46$ m/s, $Q_1 = 62.5$ m³/h.

Next guess for pipe 2: $f_2 \approx 0.0234$, $V_2 \approx 2.61$ m/s; then $\text{Re}_2 = 153{,}000$, and hence $f_2 = 0.0246$, $V_2 = 2.55$ m/s, $Q_2 = 25.9$ m³/h.

Finally guess for pipe 3: $f_3 \approx 0.0304$, $V_3 \approx 2.56$ m/s; then $\text{Re}_3 = 100{,}000$, and hence $f_3 = 0.0313$, $V_3 = 2.52$ m/s, $Q_3 = 11.4$ m³/h.

This is satisfactory convergence. The total flow rate is

$$Q = Q_1 + Q_2 + Q_3 = 62.5 + 25.9 + 11.4 = 99.8 \text{ m}^3/\text{h} \qquad \textit{Ans.}$$

These three pipes carry 10 times more flow in parallel than they do in series.

This example may be solved by Excel iteration using the Colebrook-formula procedure outlined in Ex. 6.9. Each pipe is a separate iteration of friction factor, Reynolds number, and flow rate. The pipes are rough, so only one iteration is needed. Here are the Excel results:

	A	B	C	D	E	F
			Ex. 6.18 – Pipe 1			
	Re_1	**$(\varepsilon/d)_1$**	**V_1 – m/s**	**Q_1 – m³/h**	**f_1**	**f_1-guess**
1	313,053	0.003	3.991	72.2	0.0267	0.0200
2	271,100	0.003	3.457	**62.5**	0.0267	0.0267
			Ex. 6.18 – Pipe 2			
	Re_2	**$(\varepsilon/d)_2$**	**V_2 – m/s**	**Q_2 – m³/h**	**f_2**	**f_2-guess**
1	166,021	0.002	2.822	28.7	0.0246	0.0200
2	149,739	0.002	2.546	**25.9**	0.0246	0.0246
			Ex. 6.18 – Pipe 3			
	Re_3	**$(\varepsilon/d)_3$**	**V_3 – m/s**	**Q_3 – m³/h**	**f_3**	**f_3-guess**
1	123,745	0.005	3.155	14.3	0.0313	0.0200
2	98,891	0.005	2.522	**11.4**	0.0313	0.0313

Thus, as in the hand calculations, the total flow rate = 62.5 + 25.9 + 11.4 = **99.8** m³/h. *Ans.*

Three-Reservoir Junction

Consider the third example of a *three-reservoir pipe junction,* as in Fig. 6.24*c*. If all flows are considered positive toward the junction, then

$$Q_1 + Q_2 + Q_3 = 0 \tag{6.87}$$

which obviously implies that one or two of the flows must be away from the junction. The pressure must change through each pipe so as to give the same static pressure p_J at the junction. In other words, let the HGL at the junction have the elevation

$$h_J = z_J + \frac{p_J}{\rho g}$$

where p_J is in gage pressure for simplicity. Then the head loss through each, assuming $p_1 = p_2 = p_3 = 0$ (gage) at each reservoir surface, must be such that

$$\Delta h_1 = \frac{V_1^2}{2g}\frac{f_1 L_1}{d_1} = z_1 - h_J$$

$$\Delta h_2 = \frac{V_2^2}{2g}\frac{f_2 L_2}{d_2} = z_2 - h_J \qquad (6.88)$$

$$\Delta h_3 = \frac{V_3^2}{2g}\frac{f_3 L_3}{d_3} = z_3 - h_J$$

We guess the position h_J and solve Eqs. (6.88) for V_1, V_2, and V_3 and hence Q_1, Q_2, and Q_3, iterating until the flow rates balance at the junction according to Eq. (6.87). If we guess h_J too *high*, the sum $Q_1 + Q_2 + Q_3$ will be *negative* and the remedy is to reduce h_J, and vice versa.

EXAMPLE 6.19

Take the same three pipes as in Example 6.17, and assume that they connect three reservoirs at these surface elevations

$$z_1 = 20 \text{ m} \quad z_2 = 100 \text{ m} \quad z_3 = 40 \text{ m}$$

Find the resulting flow rates in each pipe, neglecting minor losses.

Solution

As a first guess, take h_J equal to the middle reservoir height, $z_3 = h_J = 40$ m. This saves one calculation ($Q_3 = 0$) and enables us to get the lay of the land:

Reservoir	h_J, m	$z_i - h_J$, m	f_i	V_i, m/s	Q_i, m³/h	L_i/d_i
1	40	−20	0.0267	−3.43	−62.1	1,250
2	40	60	0.0241	4.42	45.0	2,500
3	40	0		0	0	2,000
					$\sum Q = -17.1$	

Since the sum of the flow rates toward the junction is negative, we guessed h_J too high. Reduce h_J to 30 m and repeat:

Reservoir	h_J, m	$z_i - h_J$, m	f_i	V_i, m/s	Q_i, m³/h
1	30	−10	0.0269	−2.42	−43.7
2	30	70	0.0241	4.78	48.6
3	30	10	0.0317	1.76	8.0
					$\sum Q = 12.9$

This is positive $\sum Q$, and so we can linearly interpolate to get an accurate guess: $h_J \approx 34.3$ m. Make one final list:

Reservoir	h_J, m	$z_i - h_J$, m	f_i	V_i, m/s	Q_i, m³/h
1	34.3	−14.3	0.0268	−2.90	−52.4
2	34.3	65.7	0.0241	4.63	47.1
3	34.3	5.7	0.0321	1.32	6.0
					$\sum Q = 0.7$

This is close enough; hence we calculate that the flow rate is 52.4 m^3/h toward reservoir 3, balanced by 47.1 m^3/h away from reservoir 1 and 6.0 m^3/h away from reservoir 3.

One further iteration with this problem would give $h_J = 34.53$ m, resulting in $Q_1 = -52.8$, $Q_2 = 47.0$, and $Q_3 = 5.8$ m^3/h, so that $\sum Q = 0$ to three-place accuracy. Pedagogically speaking, we would then be exhausted.

Pipe Networks

The ultimate case of a multipipe system is the *piping network* illustrated in Fig. 6.25. This might represent a water supply system for an apartment or subdivision or even a city. This network is quite complex algebraically but follows the same basic rules:

1. The net flow into any junction must be zero.
2. The net pressure change around any closed loop must be zero. In other words, the HGL at each junction must have one and only one elevation.
3. All pressure changes must satisfy the Moody and minor-loss friction correlations.

By supplying these rules to each junction and independent loop in the network, one obtains a set of simultaneous equations for the flow rates in each pipe leg and the HGL (or pressure) at each junction. Solution may then be obtained by numerical iteration, as first developed in a hand calculation technique by Prof. Hardy Cross in 1936 [17]. Computer solution of pipe network problems is now quite common and is covered in at least one specialized text [18]. Network analysis is quite useful for real water distribution systems if well calibrated with the actual system head loss data.

6.11 Experimental Duct Flows: Diffuser Performance[6]

The Moody chart is such a great correlation for tubes of any cross section with any roughness or flow rate that we may be deluded into thinking that the world of internal flow prediction is at our feet. Not so. The theory is reliable only for ducts of constant cross section. As soon as the section varies, we must rely principally on experiment to determine the flow properties. As mentioned many times before, experimentation is a vital part of fluid mechanics.

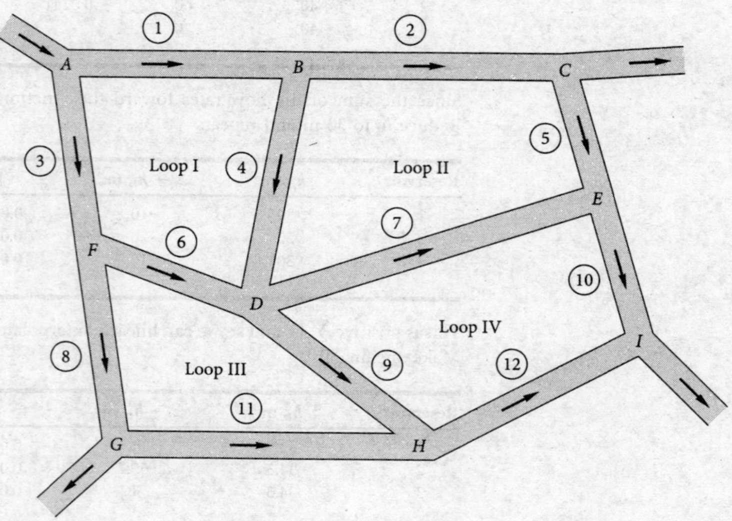

Fig. 6.25 Schematic of a piping network.

[6]This section may be omitted without loss of continuity.

Literally thousands of papers in the literature report experimental data for specific internal and external viscous flows. We have already seen several examples:

1. Vortex shedding from a cylinder (Fig. 5.2).
2. Drag of a sphere and a cylinder (Fig. 5.3).
3. Hydraulic model of a dam spillway (Fig. 5.9).
4. Rough-wall pipe flows (Fig. 6.12).
5. Secondary flow in ducts (Fig. 6.16).
6. Minor duct loss coefficients (Sec. 6.9).

Chapter 7 will treat a great many more external flow experiments, especially in Sec. 7.6. Here we shall show data for one type of internal flow, the diffuser.

Diffuser Performance

A diffuser, shown in Fig. 6.26*a* and *b*, is an expansion or area increase intended to reduce velocity in order to recover the pressure head of the flow. Rouse and Ince [6] relate that it may have been invented by customers of the early Roman (about 100 A.D.) water supply system, where water flowed continuously and was billed according to pipe size. The ingenious customers discovered that they could increase the flow rate at no extra cost by flaring the outlet section of the pipe.

Engineers have always designed diffusers to increase pressure and reduce kinetic energy of ducted flows, but until about 1950, diffuser design was a combination of art, luck, and vast amounts of empiricism. Small changes in design parameters caused large changes in performance. The Bernoulli equation seemed highly suspect as a useful tool.

Neglecting losses and gravity effects, the incompressible Bernoulli equation predicts that

$$p + \tfrac{1}{2}\rho V^2 = p_0 = \text{const} \tag{6.89}$$

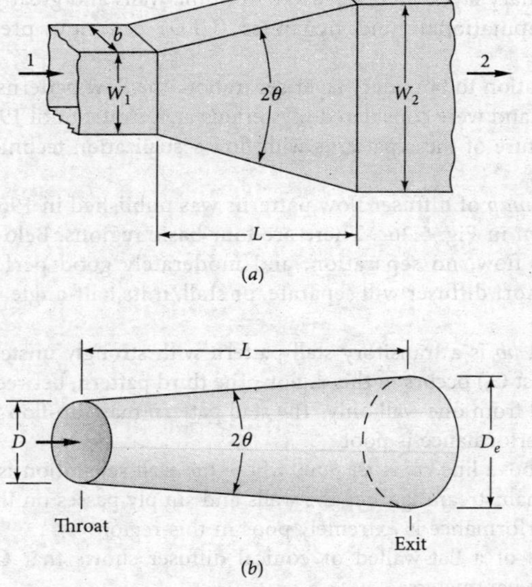

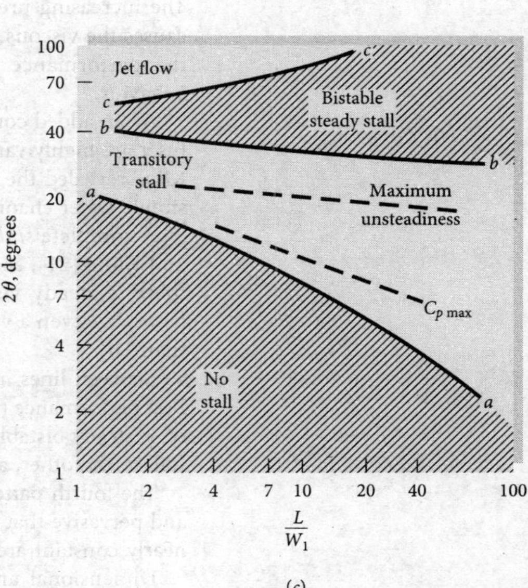

Fig. 6.26 Diffuser geometry and typical flow regimes: (*a*) geometry of a flat-walled diffuser; (*b*) geometry of a conical diffuser; (*c*) flat diffuser stability map. *(From Ref. 14, by permission of Creare, Inc.)*

where p_0 is the stagnation pressure the fluid would achieve if the fluid were slowed to rest ($V = 0$) without losses.

The basic output of a diffuser is the *pressure-recovery coefficient* C_p, defined as

$$C_p = \frac{p_e - p_t}{p_{0t} - p_t} \tag{6.90}$$

where subscripts e and t mean the exit and the throat (or inlet), respectively. Higher C_p means better performance.

Consider the flat-walled diffuser in Fig. 6.26a, where section 1 is the inlet and section 2 the exit. Application of Bernoulli's equation (6.89) to this diffuser predicts that

$$p_{01} = p_1 + \tfrac{1}{2}\rho V_1^2 = p_2 + \tfrac{1}{2}\rho V_2^2 = p_{02}$$

or

$$C_{p,\text{frictionless}} = 1 - \left(\frac{V_2}{V_1}\right)^2 \tag{6.91}$$

Meanwhile, steady one-dimensional continuity would require that

$$Q = V_1 A_1 = V_2 A_2 \tag{6.92}$$

Combining (6.91) and (6.92), we can write the performance in terms of the *area ratio* $\text{AR} = A_2/A_1$, which is a basic parameter in diffuser design:

$$C_{p,\text{frictionless}} = 1 - (\text{AR})^{-2} \tag{6.93}$$

A typical design would have AR = 5:1, for which Eq. (6.93) predicts $C_p = 0.96$, or nearly full recovery. But, in fact, measured values of C_p for this area ratio [14] are only as high as 0.86 and can be as low as 0.24.

The basic reason for the discrepancy is flow separation, as sketched in Fig. 6.27b. The increasing pressure in the diffuser is an unfavorable gradient (Sec. 7.5), which causes the viscous boundary layers to break away from the walls and greatly reduces the performance. Computational fluid dynamics (CFD) can now predict this behavior.

As an added complication to boundary layer separation, the flow patterns in a diffuser are highly variable and were considered mysterious and erratic until 1955, when Kline revealed the structure of these patterns with flow visualization techniques in a simple water channel.

A complete *stability map* of diffuser flow patterns was published in 1962 by Fox and Kline [21], as shown in Fig. 6.26c. There are four basic regions. Below line aa there is steady viscous flow, no separation, and moderately good performance. Note that even a very short diffuser will separate, or stall, if its half-angle is greater than 10°.

Between lines aa and bb is a transitory stall pattern with strongly unsteady flow. Best performance (highest C_p) occurs in this region. The third pattern, between bb and cc, is steady bistable stall from one wall only. The stall pattern may flip-flop from one wall to the other, and performance is poor.

The fourth pattern, above line cc, is *jet flow,* where the wall separation is so gross and pervasive that the mainstream ignores the walls and simply passes on through at nearly constant area. Performance is extremely poor in this region.

Dimensional analysis of a flat-walled or conical diffuser shows that C_p should depend on the following parameters:

1. Any two of the following geometric parameters:
   a. Area ratio $\text{AR} = A_2/A_1$ or $(D_e/D)^2$

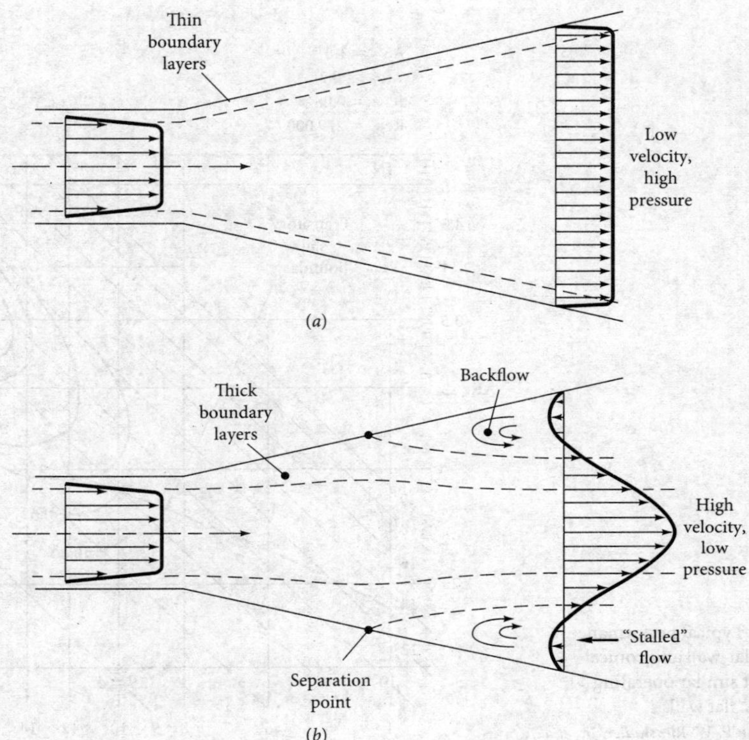

Fig. 6.27 Diffuser performance: (*a*) ideal pattern with good performance; (*b*) actual measured pattern with boundary layer separation and resultant poor performance.

  b. Divergence angle 2θ
  c. Slenderness L/W_1 or L/D
2. Inlet Reynolds number $\text{Re}_t = V_1W_1/\nu$ or $\text{Re}_t = V_1D/\nu$
3. Inlet Mach number $\text{Ma}_t = V_1/a_1$
4. Inlet boundary layer *blockage factor* $B_t = A_{\text{BL}}/A_1$, where A_{BL} is the wall area blocked, or displaced, by the retarded boundary layer flow in the inlet (typically B_t varies from 0.03 to 0.12)

A flat-walled diffuser would require an additional shape parameter to describe its cross section:

5. Aspect ratio $\text{AS} = b/W_1$

Even with this formidable list, we have omitted five possible important effects: inlet turbulence, inlet swirl, inlet profile vorticity, superimposed pulsations, and downstream obstruction, all of which occur in practical machinery applications.

The three most important parameters are AR, θ, and B. Typical performance maps for diffusers are shown in Fig. 6.28. For this case of 8 to 9 percent blockage, both the flat-walled and conical types give about the same maximum performance, $C_p = 0.70$, but at different divergence angles (9° flat versus 4.5° conical). Both types fall far short of the Bernoulli estimates of $C_p = 0.93$ (flat) and 0.99 (conical), primarily because of the blockage effect.

From the data of Ref. 14 we can determine that, in general, performance decreases with blockage and is approximately the same for both flat-walled and conical

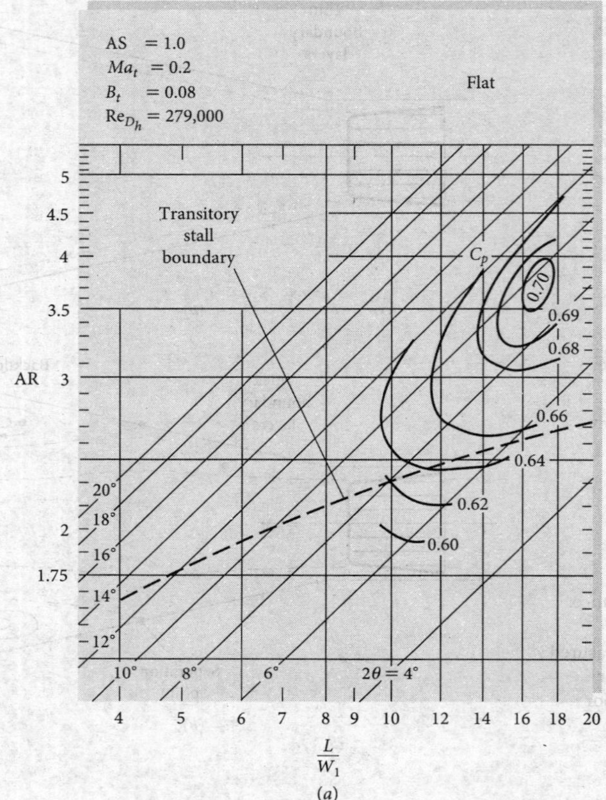

Fig. 6.28a Typical performance maps for flat-wall and conical diffusers at similar operating conditions: flat wall.
Source: From P. W. Runstadler, Jr., et al., "Diffuser Data Book," Crème Inc. Tech. Note 186, Hanover, NH, 1975., by permission of Creare, Inc.

diffusers, as shown in Table 6.6. In all cases, the best conical diffuser is 10 to 80 percent longer than the best flat-walled design. Therefore, if length is limited in the design, the flat-walled design will give the better performance depending on duct cross section.

The experimental design of a diffuser is an excellent example of a successful attempt to minimize the undesirable effects of adverse pressure gradient and flow separation.

Table 6.6 Maximum Diffuser Performance Data [14]
Source: From P. W. Runstadler, Jr., et al., "Diffuser Data Book," Crème Inc. Tech. Note 186, Hanover, NH, 1975.

Inlet blockage	Flat-walled		Conical	
B_t	C_p,max	L/W_1	C_p,max	L/d
0.02	0.86	18	0.83	20
0.04	0.80	18	0.78	22
0.06	0.75	19	0.74	24
0.08	0.70	20	0.71	26
0.10	0.66	18	0.68	28
0.12	0.63	16	0.65	30

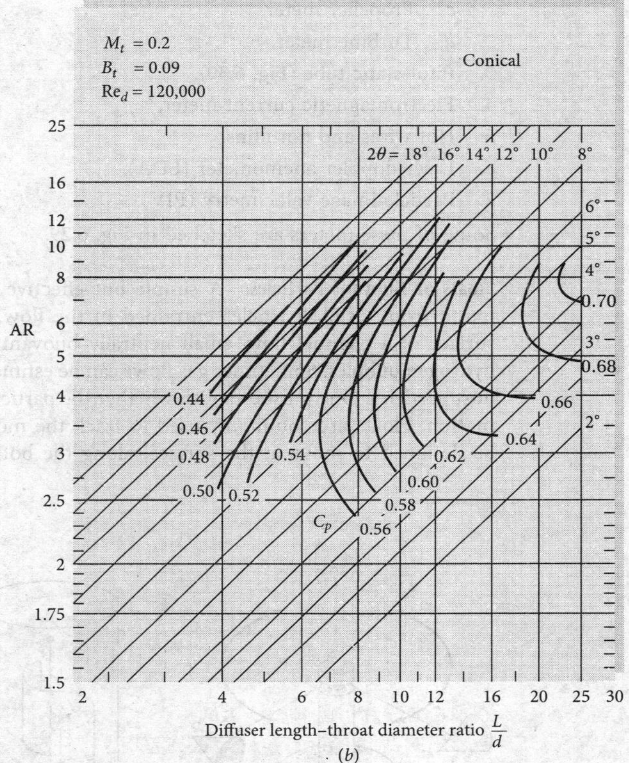

Fig. 6.28b Typical performance maps for flat-wall and conical diffusers at similar operating conditions: conical wall. *(From Ref. 14, by permission of Creare, Inc.)*

6.12 Fluid Meters

Almost all practical fluid engineering problems are associated with the need for an accurate flow measurement. There is a need to measure *local* properties (velocity, pressure, temperature, density, viscosity, turbulent intensity), *integrated* properties (mass flow and volume flow), and *global* properties (visualization of the entire flow field). We shall concentrate in this section on velocity and volume flow measurements.

We have discussed pressure measurement in Sec. 2.10. Measurement of other thermodynamic properties, such as density, temperature, and viscosity, is beyond the scope of this text and is treated in specialized books such as Refs. 22 and 23. Global visualization techniques were discussed in Sec. 1.11 for low-speed flows, and the special optical techniques used in high-speed flows are treated in Ref. 34 of Chap. 1. Flow measurement schemes suitable for open-channel and other free-surface flows are treated in Chap. 10.

Local Velocity Measurements

Velocity averaged over a small region, or point, can be measured by several different physical principles, listed in order of increasing complexity and sophistication:

1. Trajectory of floats or neutrally buoyant particles.
2. Rotating mechanical devices:
   a. Cup anemometer.
   b. Savonius rotor.

   c. Propeller meter.
   d. Turbine meter.
3. Pitot-static tube (Fig. 6.30).
4. Electromagnetic current meter.
5. Hot wires and hot films.
6. Laser-doppler anemometer (LDA).
7. Particle image velocimetry (PIV).

Some of these meters are sketched in Fig. 6.29.

Floats or Buoyant Particles. A simple but effective estimate of flow velocity can be found from visible particles entrained in the flow. Examples include flakes on the surface of a channel flow, small neutrally buoyant spheres mixed with a liquid, or hydrogen bubbles. Sometimes gas flows can be estimated from the motion of entrained dust particles. One must establish whether the particle motion truly simulates the fluid motion. Floats are commonly used to track the movement of ocean waters and can be designed to move at the surface, along the bottom, or at any given depth [24].

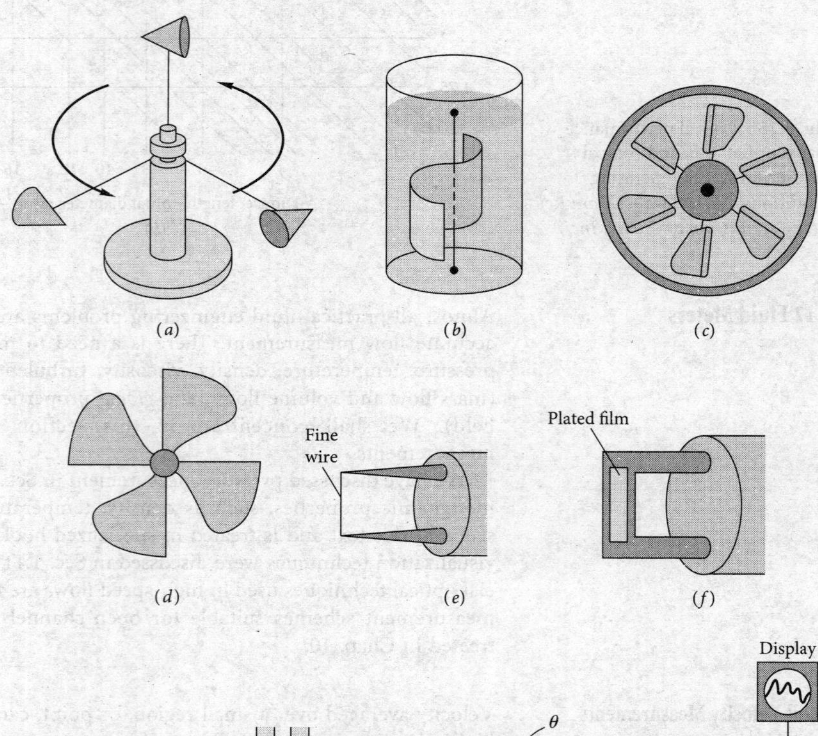

Fig. 6.29 Eight common velocity meters: (a) three-cup anemometer; (b) Savonius rotor; (c) turbine mounted in a duct; (d) free-propeller meter; (e) hot-wire anemometer; (f) hot-film anemometer; (g) pitot-static tube; (h) laser-doppler anemometer.

Many official tidal current charts [25] were obtained by releasing and timing a floating spar attached to a length of string. One can release whole groups of spars to determine a flow pattern.

Rotating Sensors. The rotating devices of Fig. 6.29*a* to *d* can be used in either gases or liquids, and their rotation rate is approximately proportional to the flow velocity. The cup anemometer (Fig. 6.29*a*) and Savonius rotor (Fig. 6.29*b*) always rotate the same way, regardless of flow direction. They are popular in atmospheric and oceanographic applications and can be fitted with a direction vane to align themselves with the flow. The ducted-propeller (Fig. 6.29*c*) and free-propeller (Fig. 6.29*d*) meters must be aligned with the flow parallel to their axis of rotation. They can sense reverse flow because they will then rotate in the opposite direction. All these rotating sensors can be attached to counters or sensed by electromagnetic or slip-ring devices for either a continuous or a digital reading of flow velocity. All have the disadvantage of being relatively large and thus not representing a "point."

Pitot-Static Tube. A slender tube aligned with the flow (Figs. 6.29*g* and 6.30) can measure local velocity by means of a pressure difference. It has sidewall holes to measure the static pressure p_s in the moving stream and a hole in the front to measure the *stagnation* pressure p_0, where the stream is decelerated to zero velocity. Instead of measuring p_0 or p_s separately, it is customary to measure their difference with, say, a transducer, as in Fig. 6.30.

If $\mathrm{Re}_D > 1{,}000$, where D is the probe diameter, the flow around the probe is nearly frictionless and Bernoulli's relation, Eq. (3.54), applies with good accuracy. For incompressible flow

$$p_s + \tfrac{1}{2}\rho V^2 + \rho g z_s \approx p_0 + \tfrac{1}{2}\rho(0)^2 + \rho g z_0$$

Assuming that the elevation pressure difference $\rho g(z_s - z_0)$ is negligible, this reduces to

$$V \approx \left[2\frac{(p_0 - p_s)}{\rho}\right]^{1/2} \tag{6.94}$$

This is the *Pitot formula,* named after the French engineer, Henri de Pitot, who designed the device in 1732.

The primary disadvantage of the pitot tube is that it must be aligned with the flow direction, which may be unknown. For yaw angles greater than 5°, there are substantial

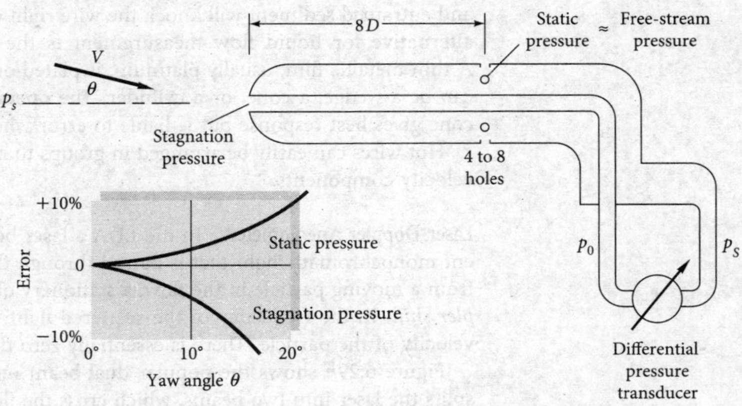

Fig. 6.30 Pitot-static tube for combined measurement of static and stagnation pressure in a moving stream.

errors in both the p_0 and p_s measurements, as shown in Fig. 6.30. The pitot-static tube is useful in liquids and gases; for gases a compressibility correction is necessary if the stream Mach number is high (Chap. 9). Because of the slow response of the fluid-filled tubes leading to the pressure sensors, it is not useful for unsteady flow measurements. It does resemble a point and can be made small enough to measure, for example, blood flow in arteries and veins. It is not suitable for low-velocity measurement in gases because of the small pressure differences developed. For example, if $V = 0.3$ m/s in standard air, from Eq. (6.94) we compute $p_0 - p$ equal to only 0.05 Pa. This is beyond the resolution of most pressure gages.

Electromagnetic Meter. If a magnetic field is applied across a conducting fluid, the fluid motion will induce a voltage across two electrodes placed in or near the flow. The electrodes can be streamlined or built into the wall, and they cause little or no flow resistance. The output is very strong for highly conducting fluids such as liquid metals. Seawater also gives good output, and electromagnetic current meters are in common use in oceanography. Even low-conductivity fresh water can be measured by amplifying the output and insulating the electrodes. Commercial instruments are available for most liquid flows but are relatively costly. Electromagnetic flowmeters are treated in Ref. 26.

Hot-Wire Anemometer. A very fine wire ($d = 0.01$ mm or less) heated between two small probes, as in Fig. 6.29*e*, is ideally suited to measure rapidly fluctuating flows such as the turbulent boundary layer. The idea dates back to work by L. V. King in 1914 on heat loss from long, thin cylinders. If electric power is supplied to heat the cylinder, the loss varies with flow velocity across the cylinder according to *King's law*

$$q = I^2R \approx a + b(\rho V)^n \tag{6.95}$$

where $n \approx \frac{1}{3}$ at very low Reynolds numbers and equals $\frac{1}{2}$ at high Reynolds numbers. The hot wire normally operates in the high-Reynolds-number range but should be calibrated in each situation to find the best-fit a, b, and n. The wire can be operated either at constant current I, so that resistance R is a measure of V, or at constant resistance R (constant temperature), with I a measure of velocity. In either case, the output is a nonlinear function of V, and the equipment should contain a *linearizer* to produce convenient velocity data. Many varieties of commercial hot-wire equipment are available, as are do-it-yourself designs [27]. Excellent detailed discussions of the hot wire are given in Ref. 28.

Because of its frailty, the hot wire is not suited to liquid flows, whose high density and entrained sediment will knock the wire right off. A more stable yet quite sensitive alternative for liquid flow measurement is the hot-film anemometer (Fig. 6.29*f*). A thin metallic film, usually platinum, is plated onto a relatively thick support, which can be a wedge, a cone, or a cylinder. The operation is similar to the hot wire. The cone gives best response but is liable to error when the flow is yawed to its axis.

Hot wires can easily be arranged in groups to measure two- and three-dimensional velocity components.

Laser-Doppler Anemometer. In the LDA a laser beam provides highly focused, coherent monochromatic light that is passed through the flow. When this light is scattered from a moving particle in the flow, a stationary observer can detect a change, or *doppler shift,* in the frequency of the scattered light. The shift Δf is proportional to the velocity of the particle. There is essentially zero disturbance of the flow by the laser.

Figure 6.29*h* shows the popular dual-beam mode of the LDA. A focusing device splits the laser into two beams, which cross the flow at an angle θ. Their intersection, which is the measuring volume or resolution of the measurement, resembles an ellipsoid about 0.5 mm wide and 0.1 mm in diameter. Particles passing through this measuring volume scatter the beams; they then pass through receiving optics to a photodetector, which converts the light to an electric signal. A signal processor then converts electric frequency to a voltage that can be either displayed or stored. If l is the wavelength of the laser light, the measured velocity is given by

$$V = \frac{\lambda \, \Delta f}{2 \sin(\theta/2)} \tag{6.96}$$

Multiple components of velocity can be detected by using more than one photodetector and other operating modes. Either liquids or gases can be measured as long as scattering particles are present. In liquids, normal impurities serve as scatterers, but gases may have to be seeded. The particles may be as small as the wavelength of the light. Although the measuring volume is not as small as with a hot wire, the LDA is capable of measuring turbulent fluctuations.

The advantages of the LDA are as follows:

1. No disturbance of the flow.
2. High spatial resolution of the flow field.
3. Velocity data that are independent of the fluid thermodynamic properties.
4. An output voltage that is linear with velocity.
5. No need for calibration.

The disadvantages are that both the apparatus and the fluid must be transparent to light and that the cost is high (a basic system shown in Fig. 6.29*h* begins at about $50,000).

Once installed, an LDA can map the entire flow field in minutest detail. To truly appreciate the power of the LDA, one should examine, for instance, the amazingly detailed three-dimensional flow profiles measured by Eckardt [29] in a high-speed centrifugal compressor impeller. Extensive discussions of laser velocimetry are given in Refs. 38 and 39.

Particle Image Velocimetry. This popular new idea, called PIV for short, measures not just a single point but instead maps the entire field of flow. An illustration was shown in Fig. 1.18*b*. The flow is seeded with neutrally buoyant particles. A planar laser light sheet across the flow is pulsed twice and photographed twice. If $\Delta\mathbf{r}$ is the particle displacement vector over a short time Δt, an estimate of its velocity is $\mathbf{V} \approx \Delta\mathbf{r}/\Delta t$. A dedicated computer applies this formula to a whole cloud of particles and thus maps the flow field. One can also use the data to calculate velocity gradient and vorticity fields. Since the particles all look alike, other cameras may be needed to identify them. Three-dimensional velocities can be measured by two cameras in a stereoscopic arrangement. The PIV method is not limited to stop-action. New high-speed cameras (up to 10,000 frames per second) can record movies of unsteady flow fields. For further details, see the monograph by M. Raffel [51].

EXAMPLE 6.20

The pitot-static tube of Fig. 6.30 uses mercury as a manometer fluid. When it is placed in a water flow, the manometer height reading is h = 20 cm. Neglecting yaw and other errors, what is the flow velocity V in m/s?

Solution

From the two-fluid manometer relation (2.23b), with $z_A = z_2$, the pressure difference is related to h by

$$p_0 - p_s = (\gamma_M - \gamma_w)h$$

Taking the specific weights of mercury and water from Table 2.1, we have

$$p_0 - p_s = (133{,}100 \text{ N/m}^3 - 9{,}790 \text{ N/m}^3)\ 0.2 \text{ m} = 24{,}660 \text{ Pa}$$

The density of water is $(9790 \text{ N/m}^3)/(9.81 \text{ m/s}^2) = 998 \text{ kg/m}^3$. Introducing these values into the pitot-static formula (6.97), we obtain

$$V = \left[\frac{2(24{,}660 \text{ Pa})}{998 \text{ kg/m}^3}\right]^{1/2} = 7.03 \text{ m/s} \qquad \textit{Ans.}$$

Since this is a low-speed flow, no compressibility correction is needed.

Volume Flow Measurements

It is often desirable to measure the integrated mass, or volume flow, passing through a duct. Accurate measurement of flow is vital in billing customers for a given amount of liquid or gas passing through a duct. The different devices available to make these measurements are discussed in great detail in the ASME text on fluid meters [30]. These devices split into two classes: mechanical instruments and head loss instruments.

The mechanical instruments measure actual mass or volume of fluid by trapping it and counting it. The various types of measurement are

1. Mass measurement
   a. Weighing tanks
   b. Tilting traps
2. Volume measurement
   a. Volume tanks
   b. Reciprocating pistons
   c. Rotating slotted rings
   d. Nutating disc
   e. Sliding vanes
   f. Gear or lobed impellers
   g. Reciprocating bellows
   h. Sealed-drum compartments

The last three of these are suitable for gas flow measurement.

The head loss devices obstruct the flow and cause a pressure drop, which is a measure of flux:

1. Bernoulli-type devices
   a. Thin-plate orifice
   b. Flow nozzle
   c. Venturi tube
2. Friction loss devices
   a. Capillary tube
   b. Porous plug

The friction loss meters cause a large nonrecoverable head loss and obstruct the flow too much to be generally useful.

Six other widely used meters operate on different physical principles:

1. Turbine meter
2. Vortex meter
3. Ultrasonic flowmeter
4. Rotameter
5. Coriolis mass flowmeter
6. Laminar flow element

Nutating Disc Meter. For measuring liquid *volumes,* as opposed to volume rates, the most common devices are the nutating disc and the turbine meter. Figure 6.31 shows a cutaway sketch of *a nutating disc meter,* widely used in both water and gasoline delivery systems. The mechanism is clever and perhaps beyond the writer's capability to explain. The metering chamber is a slice of a sphere and contains a rotating disc set at an angle to the incoming flow. The fluid causes the disc to *nutate* (spin eccentrically), and one revolution corresponds to a certain fluid volume passing through. Total volume is obtained by counting the number of revolutions.

Turbine Meter. The turbine meter, sometimes called a *propeller meter,* is a freely rotating propeller that can be installed in a pipeline. A typical design is shown in Fig. 6.32*a*. There are flow straighteners upstream of the rotor, and the rotation is measured by electric or magnetic pickup of pulses caused by passage of a point on the rotor. The rotor rotation is approximately proportional to the volume flow in the pipe.

Like the nutating disc, a major advantage of the turbine meter is that each pulse corresponds to a finite incremental volume of fluid, and the pulses are digital and can be summed easily. Liquid flow turbine meters have as few as two blades and produce a constant number of pulses per unit fluid volume over a 5:1 flow rate range with ±0.25 percent accuracy. Gas meters need many blades to produce sufficient torque and are accurate to ±1 percent.

Since turbine meters are very individualistic, flow calibration is an absolute necessity. A typical liquid meter calibration curve is shown in Fig. 6.32*b*. Researchers attempting to establish universal calibration curves have met with little practical success as a result of manufacturing variabilities.

Turbine meters can also be used in unconfined flow situations, such as winds or ocean currents. They can be compact, even microsize with two or three component directions. Figure 6.33 illustrates a handheld wind velocity meter that uses a seven-bladed turbine with a calibrated digital output. The accuracy of this device is quoted at ±2 percent.

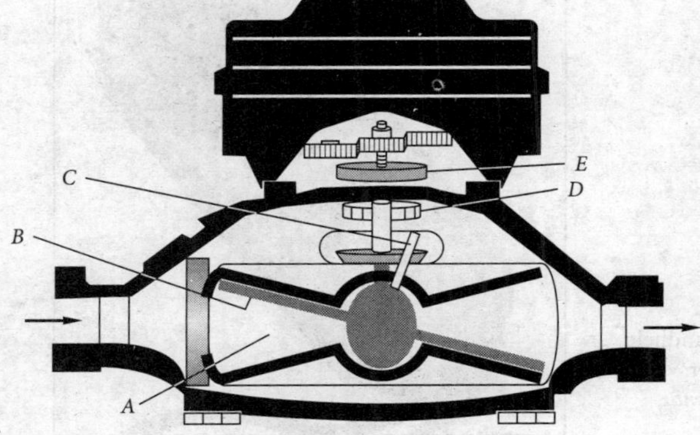

Fig. 6.31 Cutaway sketch of a nutating disc fluid meter. *A*: metered-volume chamber; *B*: nutating disc; *C*: rotating spindle; *D*: drive magnet; *E*: magnetic counter sensor.
Source: Courtesy of Badger Meter, Inc., Milwaukee, Wisconsin.

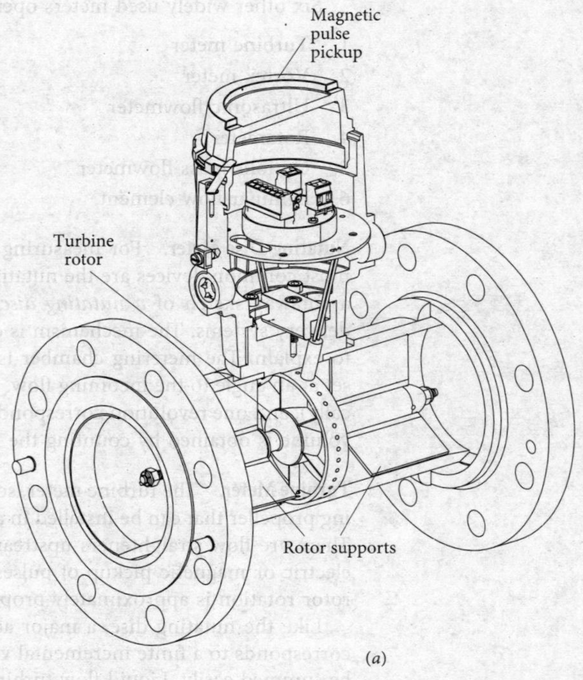

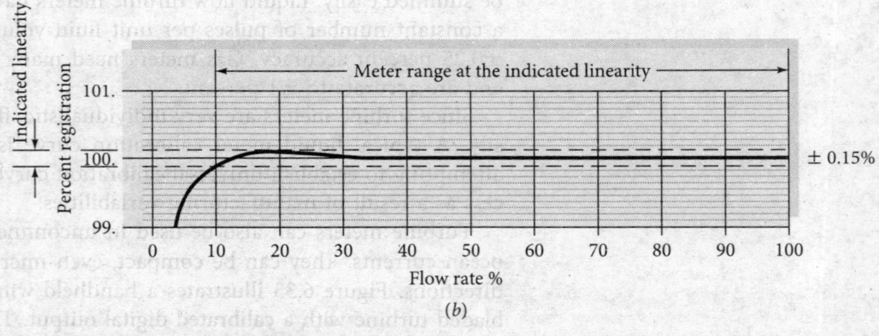

Fig. 6.32 The turbine meter widely used in the oil and gas industry: (*a*) basic design; (*b*) the linearity curve is the measure of variation in the signal output across the 10% to 100% nominal flow range of the meter. *(Daniel Measurement and Control, Houston, TX.)*
Source: (a) Daniel Industries of Fluke Calibration, Houston, TX.

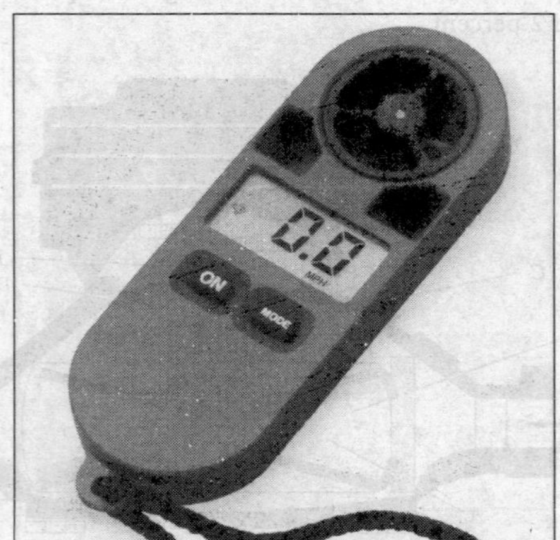

Fig. 6.33 A Commercial handheld wind velocity turbine meter. *(Courtesy of Nielsen-Kellerman Company.)*

Vortex Flowmeters. Recall from Fig. 5.2 that a bluff body placed in a uniform crossflow sheds alternating vortices at a nearly uniform Strouhal number St $= fL/U$, where U is the approach velocity and L is a characteristic body width. Since L and St are constant, this means that the shedding frequency is proportional to velocity:

$$f = (\text{const})(U) \tag{6.97}$$

The vortex meter introduces a shedding element across a pipe flow and picks up the shedding frequency downstream with a pressure, ultrasonic, or heat transfer type of sensor. A typical design is shown in Fig. 6.34.

The advantages of a vortex meter are as follows:

1. Absence of moving parts.
2. Accuracy to ± 1 percent over a wide flow rate range (up to 100:1).
3. Ability to handle very hot or very cold fluids.
4. Requirement of only a short pipe length.
5. Calibration insensitive to fluid density or viscosity.

For further details see Ref. 40.

Ultrasonic Flowmeters. The sound-wave analog of the laser velocimeter of Fig. 6.29*h* is the ultrasonic flowmeter. Two examples are shown in Fig. 6.35. The pulse-type flowmeter is shown in Fig. 6.35*a*. Upstream piezoelectric transducer A is excited with a short sonic pulse that propagates across the flow to downstream transducer B. The arrival at B triggers another pulse to be created at A, resulting in a regular pulse frequency f_A. The same process is duplicated in the reverse direction from B to A, creating frequency f_B. The difference $f_A - f_B$ is proportional to the flow rate. Figure 6.35*b* shows a doppler-type arrangement, where sound waves from transmitter T are

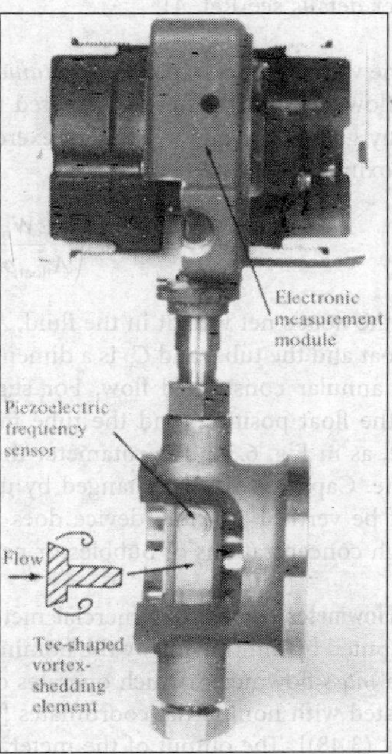

Fig. 6.34 A vortex flowmeter. *(Courtesy of Invensys p/c.)*

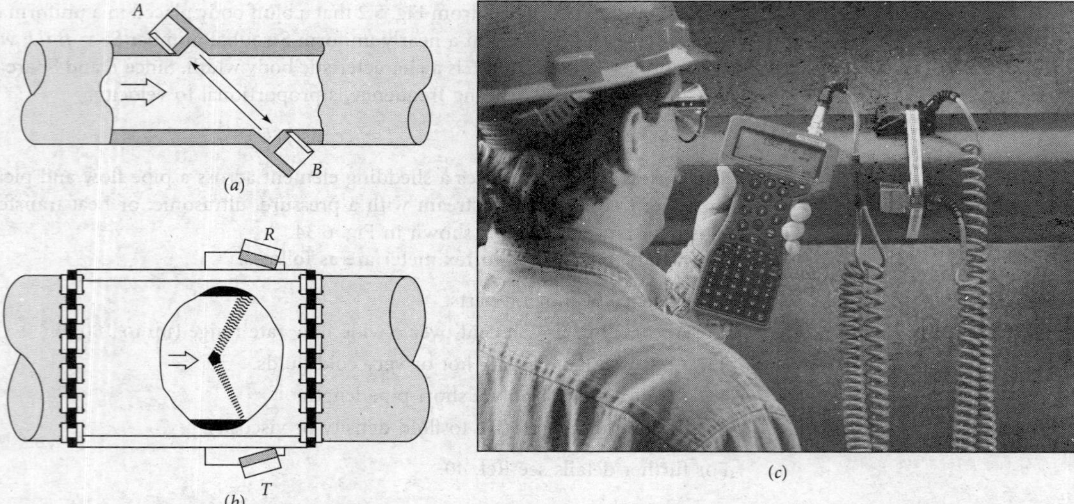

Fig. 6.35 Ultrasonic flowmeters: (*a*) pulse type; (*b*) doppler-shift type *(from Ref. 41)*; (*c*) a portable noninvasive installation *(Courtesy of Thermo Polysonics, Houston, TX.)*

scattered by particles or contaminants in the flow to receiver R. Comparison of the two signals reveals a doppler frequency shift that is proportional to the flow rate. Ultrasonic meters are nonintrusive and can be directly attached to pipe flows in the field (Fig. 6.35*c*). Their quoted uncertainty of ±1 to 2 percent can rise to ±5 percent or more due to irregularities in velocity profile, fluid temperature, or Reynolds number. For further details see Ref. 41.

Rotameter. The variable-area transparent *rotameter* of Fig. 6.36 has a float that, under the action of flow, rises in the vertical tapered tube and takes a certain equilibrium position for any given flow rate. A student exercise for the forces on the float would yield the approximate relation

$$Q = C_d A_a \left(\frac{2 W_{\text{net}}}{A_{\text{float}} \rho_{\text{fluid}}} \right)^{1/2} \tag{6.98}$$

where W_{net} is the float's net weight in the fluid, $A_a = A_{\text{tube}} - A_{\text{float}}$ is the annular area between the float and the tube, and C_d is a dimensionless discharge coefficient of order unity, for the annular constricted flow. For slightly tapered tubes, A_a varies nearly linearly with the float position, and the tube may be calibrated and marked with a flow rate scale, as in Fig. 6.36. The rotameter thus provides a readily visible measure of the flow rate. Capacity may be changed by using different-sized floats. Obviously the tube must be vertical, and the device does not give accurate readings for fluids containing high concentrations of bubbles or particles.

Coriolis Mass Flowmeter. Most commercial meters measure *volume* flow, with mass flow then computed by multiplying by the nominal fluid density. An attractive modern alternative is a *mass* flowmeter, which operates on the principle of the Coriolis acceleration associated with noninertial coordinates [recall Fig. 3.11 and the Coriolis term $2\Omega \times V$ in Eq. (3.48)]. The output of the meter is directly proportional to mass flow.

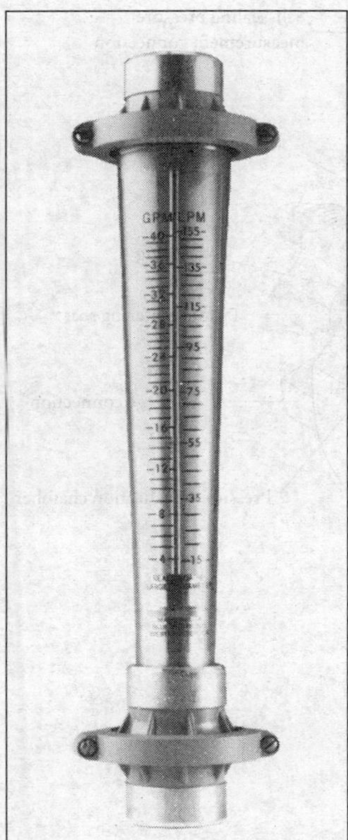

Fig. 6.36 A commercial rotameter. The float rises in the tapered tube to an equilibrium position, which is a measure of the fluid flow rate. *(Courtesy of Blue White Industries, Huntington Beach, CA.)*

Figure 6.37 is a schematic of a Coriolis device, to be inserted into a piping system. The flow enters a loop arrangement, which is electromagnetically vibrated at a high natural frequency (amplitude < 1 mm and frequency > 100 Hz). The Coriolis effect induces a downward force on the loop entrance and an upward force on the loop exit, as shown. The loop twists, and the twist angle can be measured and is proportional to the mass flow through the tube. Accuracy is typically less than 1 percent of full scale.

Laminar Flow Element. In many, perhaps most, commercial flowmeters, the flow through the meter is turbulent and the variation of flow rate with pressure drop is nonlinear. In laminar duct flow, however, Q is linearly proportional to Δp, as in Eq. (6.12): $Q = [\pi R^4/(8\mu L)]\ \Delta p$. Thus a *laminar* flow sensing element is attractive, since its calibration will be linear. To ensure laminar flow for what otherwise would be a turbulent condition, all or part of the fluid is directed into small passages, each of which has a low (laminar) Reynolds number. A honeycomb is a popular design.

Figure 6.38 uses axial flow through a narrow annulus to create laminar flow. The theory again predicts $Q \propto \Delta p$, as in Eq. (6.70). However, the flow is very sensitive to passage size; for example, halving the annulus clearance increases Δp more than eight times. Careful calibration is thus necessary. In Fig. 6.38 the laminar flow concept has been synthesized into a complete mass flow system, with temperature control, differential pressure measurement, and a microprocessor all self-contained. The accuracy of this device is rated at ± 0.2 percent.

Bernoulli Obstruction Theory. Consider the generalized flow obstruction shown in Fig. 6.39. The flow in the basic duct of diameter D is forced through an obstruction of diameter d; the β ratio of the device is a key parameter:

$$\beta = \frac{d}{D} \tag{6.99}$$

After leaving the obstruction, the flow may neck down even more through a vena contracta of diameter $D_2 < d$, as shown. Apply the Bernoulli and continuity equations for incompressible steady frictionless flow to estimate the pressure change:

Continuity: $$Q = \frac{\pi}{4} D^2 V_1 = \frac{\pi}{4} D_2^2 V_2$$

Bernoulli: $$p_0 = p_1 + \tfrac{1}{2}\rho V_1^2 = p_2 + \tfrac{1}{2}\rho V_2^2$$

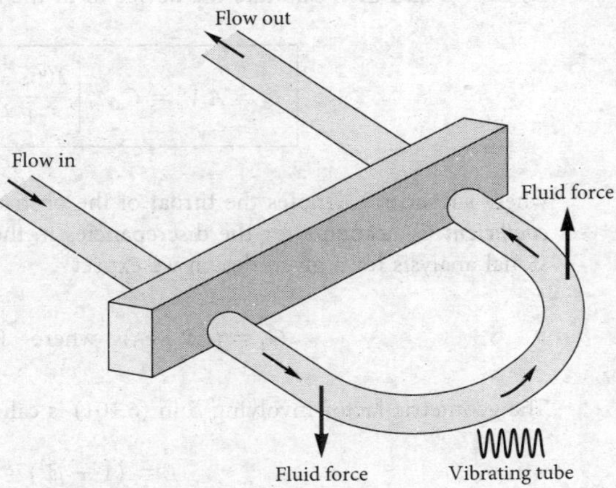

Fig. 6.37 Schematic of a Coriolis mass flowmeter.

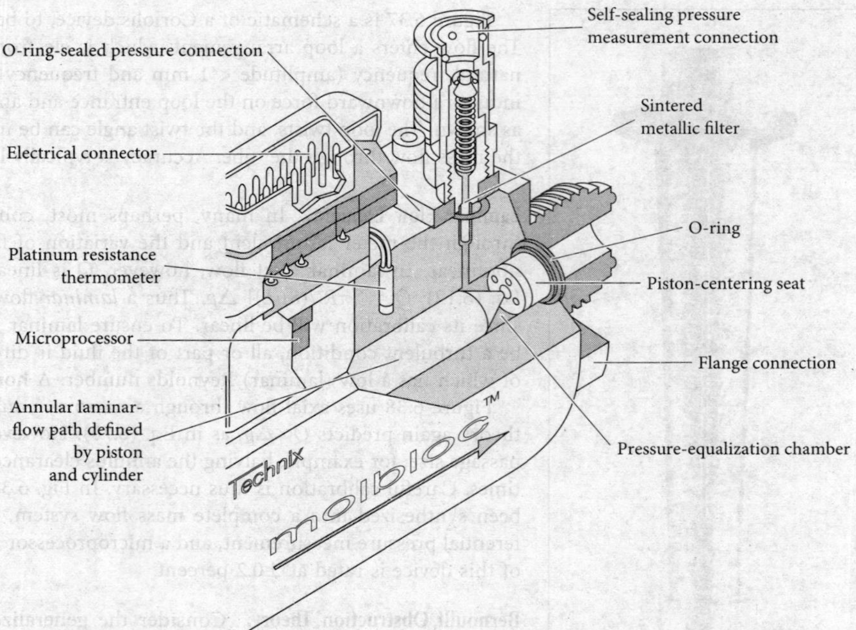

Fig. 6.38 A complete flowmeter system using a laminar flow element (in this case a narrow annulus). The flow rate is linearly proportional to the pressure drop.
Source: Courtesy of Martin Girard, DH Instruments, Inc.

Eliminating V_1, we solve these for V_2 or Q in terms of the pressure change $p_1 - p_2$:

$$\frac{Q}{A_2} = V_2 \approx \left[\frac{2(p_1 - p_2)}{\rho(1 - D_2^4/D^4)}\right]^{1/2} \tag{6.100}$$

But, this is surely inaccurate because we have neglected friction in a duct flow, where we know friction will be very important. Nor do we want to get into the business of measuring vena contracta ratios D_2/d for use in (6.100). Therefore, we assume that $D_2/D \approx \beta$ and then calibrate the device to fit the relation

$$Q = A_t V_t = C_d A_t \left[\frac{2(p_1 - p_2)/\rho}{1 - \beta^4}\right]^{1/2} \tag{6.101}$$

where subscript t denotes the throat of the obstruction. The dimensionless *discharge coefficient* C_d accounts for the discrepancies in the approximate analysis. By dimensional analysis for a given design we expect

$$C_d = f(\beta, \mathrm{Re}_D) \quad \text{where} \quad \mathrm{Re}_D = \frac{V_1 D}{\nu} \tag{6.102}$$

The geometric factor involving β in (6.101) is called the *velocity-of-approach factor*:

$$E = (1 - \beta^4)^{-1/2} \tag{6.103}$$

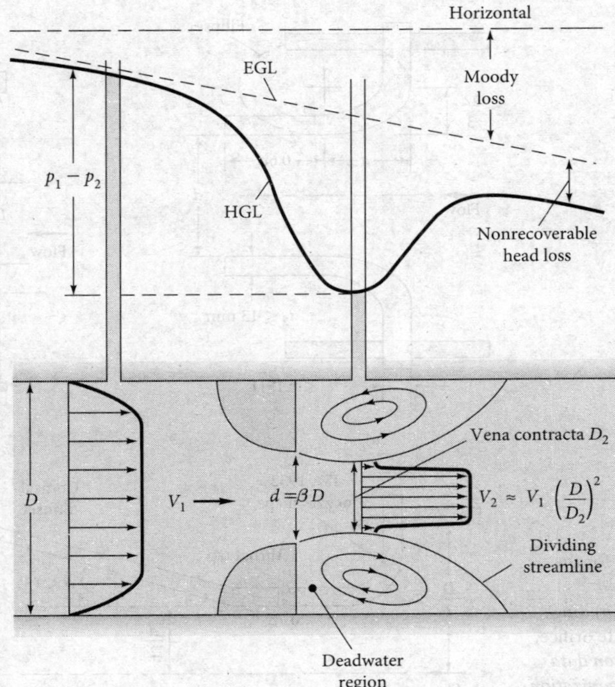

Fig. 6.39 Velocity and pressure change through a generalized Bernoulli obstruction meter.

One can also group C_d and E in Eq. (6.101) to form the dimensionless *flow coefficient* α:

$$\alpha = C_d E = \frac{C_d}{(1 - \beta^4)^{1/2}} \tag{6.104}$$

Thus, Eq. (6.101) can be written in the equivalent form

$$Q = \alpha A_t \left[\frac{2(p_1 - p_2)}{\rho}\right]^{1/2} \tag{6.105}$$

Obviously, the flow coefficient is correlated in the same manner:

$$\alpha = f(\beta, \mathrm{Re}_D) \tag{6.106}$$

Occasionally, one uses the throat Reynolds number instead of the approach Reynolds number:

$$\mathrm{Re}_d = \frac{V_t d}{\nu} = \frac{\mathrm{Re}_D}{\beta} \tag{6.107}$$

Since the design parameters are assumed known, the correlation of α from Eq. (6.106) or of C_d from Eq. (6.102) is the desired solution to the fluid metering problem.

The mass flow is related to Q by

$$\dot{m} = \rho Q \tag{6.108}$$

and is thus correlated by exactly the same formulas.

Figure 6.40 shows the three basic devices recommended for use by the International Organization for Standardization (ISO) [31]: the orifice, nozzle, and venturi tube.

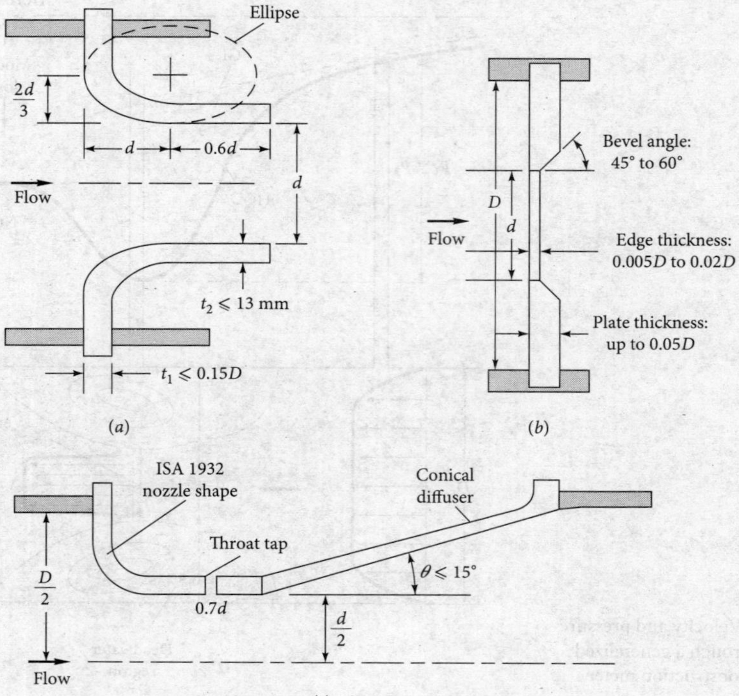

Fig. 6.40 Standard shapes for the three primary Bernoulli obstruction-type meters: (*a*) long-radius nozzle; (*b*) thin-plate orifice; (*c*) venturi nozzle. *(Based on data from the International Organization for Standardization.)*

Thin-Plate Orifice. The thin-plate orifice, Fig. 6.40*b*, can be made with β in the range of 0.2 to 0.8, except that the hole diameter d should not be less than 12.5 mm. To measure p_1 and p_2, three types of tappings are commonly used:

1. Corner taps where the plate meets the pipe wall.
2. D: $\frac{1}{2}D$ taps: pipe-wall taps at D upstream and $\frac{1}{2}D$ downstream.
3. Flange taps: 25 mm (1 inch) upstream and 25 mm (1 inch) downstream of the plate, regardless of the size D.

Types 1 and 2 approximate geometric similarity, but since the flange taps 3 do not, they must be correlated separately for every single size of pipe in which a flange-tap plate is used [30, 31].

Figure 6.41 shows the discharge coefficient of an orifice with D: $\frac{1}{2}D$ or type 2 taps in the Reynolds number range $\text{Re}_D = 10^4$ to 10^7 of normal use. Although detailed charts such as Fig. 6.41 are available for designers [30], the ASME recommends use of the curve-fit formulas developed by the ISO [31]. The basic form of the curve fit is [42]

$$C_d = f(\beta) + 91.71\beta^{2.5}\text{Re}_D^{-0.75} + \frac{0.09\beta^4}{1-\beta^4}F_1 - 0.0337\beta^3F_2 \qquad (6.109)$$

where
$$f(\beta) = 0.5959 + 0.0312\beta^{2.1} - 0.184\beta^8$$

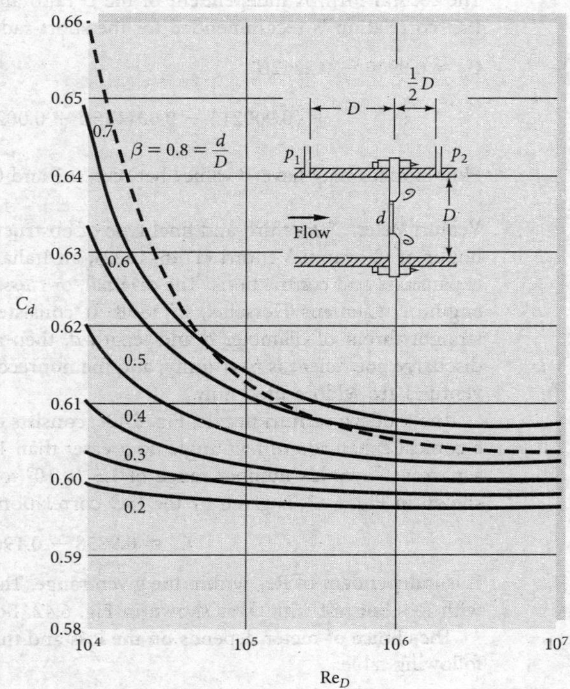

Fig. 6.41 Discharge coefficient for a thin-plate orifice with D: $\frac{1}{2}D$ taps, plotted from Eqs. (6.109) and (6.110*b*).

The correlation factors F_1 and F_2 vary with tap position:

Corner taps: $$F_1 = 0 \quad F_2 = 0 \tag{6.110a}$$

D: $\frac{1}{2}D$ taps: $$F_1 = 0.4333 \quad F_2 = 0.47 \tag{6.110b}$$

Flange taps: $$F_2 = \frac{25.4}{D\,(\mathrm{mm})} \quad F_1 = \begin{cases} \dfrac{25.4}{D\,(\mathrm{mm})} & D > 58.5\ \mathrm{mm} \\ 0.4333 & 51 \le D \le 58.5\ \mathrm{mm} \end{cases} \tag{6.110c}$$

Note that the flange taps (6.110*c*), not being geometrically similar, use raw diameter in mm in the formula. The constants will change if other diameter units are used. We cautioned against such dimensional formulas in Example 1.4 and Eq. (5.17) and give Eq. (6.110*c*) only because flange taps are widely used in the United States.

Flow Nozzle. The flow nozzle comes in two types, a long-radius type shown in Fig. 6.40*a* and a short-radius type (not shown) called the ISA 1932 nozzle [30, 31]. The flow nozzle, with its smooth, rounded entrance convergence, practically eliminates the vena contracta and gives discharge coefficients near unity. The nonrecoverable loss is still large because there is no diffuser provided for gradual expansion.

The ISO recommended correlation for long-radius-nozzle discharge coefficient is

$$C_d \approx 0.9965 - 0.00653\beta^{1/2}\left(\frac{10^6}{\mathrm{Re}_D}\right)^{1/2} = 0.9965 - 0.00653\left(\frac{10^6}{\mathrm{Re}_d}\right)^{1/2} \tag{6.111}$$

The second form is independent of the β ratio and is plotted in Fig. 6.42. A similar ISO correlation is recommended for the short-radius ISA 1932 flow nozzle:

$$C_d \approx 0.9900 - 0.2262\beta^{4.1} + (0.000215 - 0.001125\beta + 0.00249\beta^{4.7})\left(\frac{10^6}{\text{Re}_D}\right)^{1.15} \tag{6.112}$$

Flow nozzles may have β values between 0.2 and 0.8.

Venturi Meter. The third and final type of obstruction meter is the venturi, named in honor of Giovanni Venturi (1746–1822), an Italian physicist who first tested conical expansions and contractions. The original, or *classical,* venturi was invented by a U.S. engineer, Clemens Herschel, in 1898. It consisted of a 21° conical contraction, a straight throat of diameter d and length d, then a 7° to 15° conical expansion. The discharge coefficient is near unity, and the nonrecoverable loss is very small. Herschel venturis are seldom used now.

The modern venturi nozzle, Fig. 6.40*c*, consists of an ISA 1932 nozzle entrance and a conical expansion of half-angle no greater than 15°. It is intended to be operated in a narrow Reynolds number range of 1.5×10^5 to 2×10^6. Its discharge coefficient, shown in Fig. 6.43, is given by the ISO correlation formula

$$C_d \approx 0.9858 - 0.196\beta^{4.5} \tag{6.113}$$

It is independent of Re_D within the given range. The Herschel venturi discharge varies with Re_D but not with β, as shown in Fig. 6.42. Both have very low net losses.

The choice of meter depends on the loss and the cost and can be illustrated by the following table:

Type of meter	Net head loss	Cost
Orifice	Large	Small
Nozzle	Medium	Medium
Venturi	Small	Large

As so often happens, the product of inefficiency and initial cost is approximately constant.

The average nonrecoverable head losses for the three types of meters, expressed as a fraction of the throat velocity head $V_t^2/(2g)$, are shown in Fig. 6.44. The orifice has

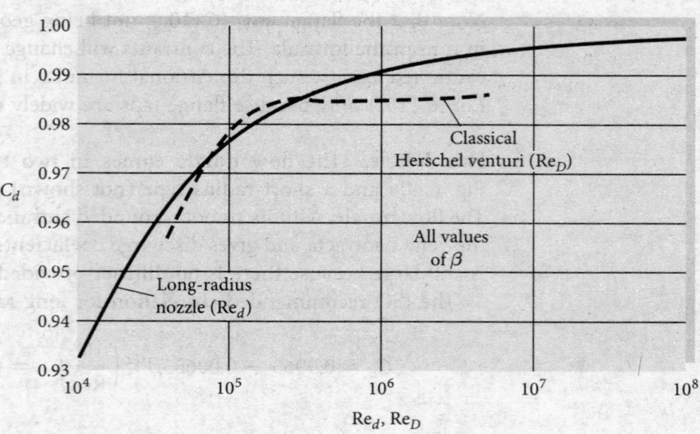

Fig. 6.42 Discharge coefficient for long-radius nozzle and classical Herschel-type venturi.

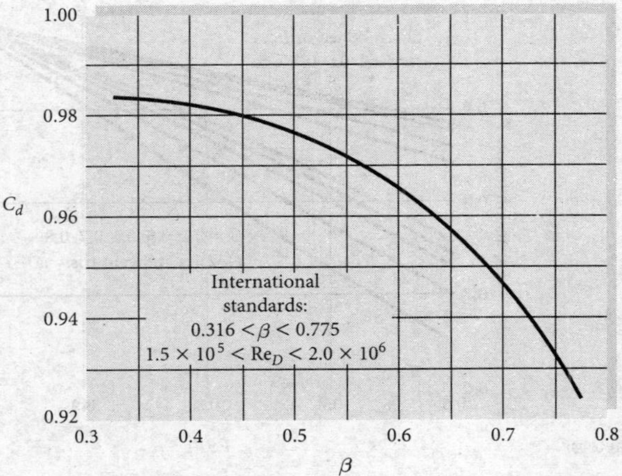

Fig. 6.43 Discharge coefficient for a venturi nozzle.

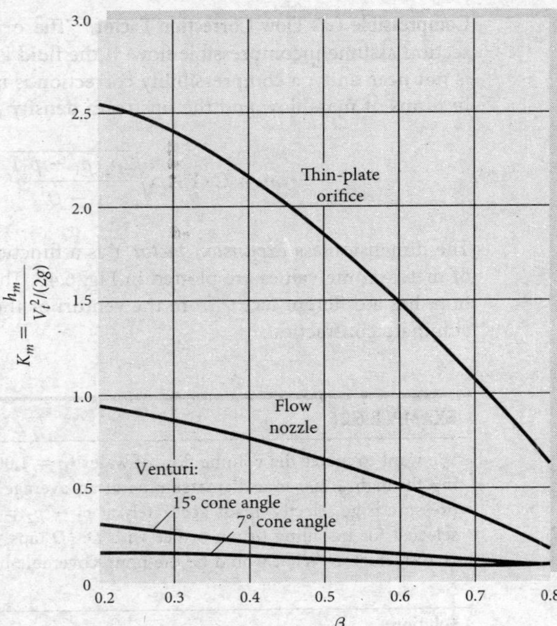

Fig. 6.44 Nonrecoverable head loss in Bernoulli obstruction meters. *(Adapted from Ref. 30.)*

the greatest loss and the venturi the least, as discussed. The orifice and nozzle simulate partially closed valves as in Fig. 6.18*b*, while the venturi is a very minor loss. When the loss is given as a fraction of the measured *pressure drop*, the orifice and nozzle have nearly equal losses, as Example 6.21 will illustrate.

The other types of instruments discussed earlier in this section can also serve as flowmeters if properly constructed. For example, a hot wire mounted in a tube can be calibrated to read volume flow rather than point velocity. Such hot-wire meters are commercially available, as are other meters modified to use velocity instruments. For further details see Ref. 30.

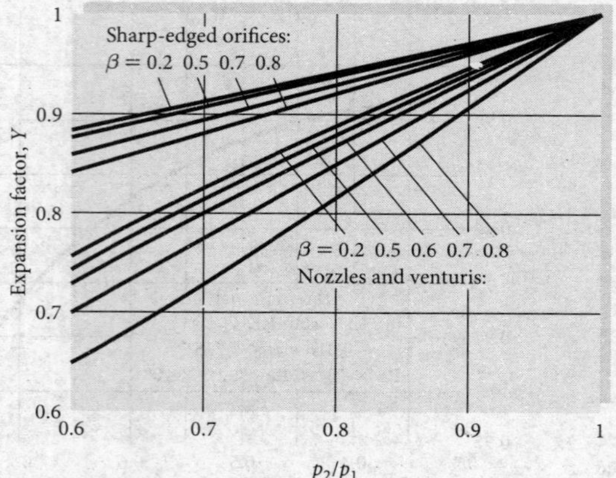

Fig. 6.45 Compressible flow expansion factor Y for flowmeters.

Compressible Gas Flow Correction Factor. The orifice/nozzle/venturi formulas in this section assume incompressible flow. If the fluid is a gas, and the pressure ratio (p_2/p_1) is not near unity, a compressibility correction is needed. Equation (6.101) is rewritten in terms of mass flow and the upstream density ρ_1:

$$\dot{m} = C_d\, Y A_t \sqrt{\frac{2\rho_1(p_1 - p_2)}{1 - \beta^4}} \quad \text{where} \quad \beta = \frac{d}{D} \tag{6.114}$$

The dimensionless *expansion factor* Y is a function of pressure ratio, β, and the type of meter. Some values are plotted in Fig. 6.45. The orifice, with its strong jet contraction, has a different factor from the venturi or the flow nozzle, which are designed to eliminate contraction.

EXAMPLE 6.21

We want to meter the volume flow of water ($\rho = 1{,}000$ kg/m^3, $\nu = 1.02 \times 10^{-6}$ m^2/s) moving through a 200-mm-diameter pipe at an average velocity of 2.0 m/s. If the differential pressure gage selected reads accurately at $p_1 - p_2 = 50{,}000$ Pa, what size meter should be selected for installing (*a*) an orifice with D: $\frac{1}{2}D$ taps, (*b*) a long-radius flow nozzle, or (*c*) a venturi nozzle? What would be the nonrecoverable head loss for each design?

Solution

Here the unknown is the β ratio of the meter. Since the discharge coefficient is a complicated function of β, iteration will be necessary. We are given $D = 0.2$ m and $V_1 = 2.0$ m/s. The pipe-approach Reynolds number is thus

$$\mathrm{Re}_D = \frac{V_1 D}{\nu} = \frac{(2.0)(0.2)}{1.02 \times 10^{-6}} = 392{,}000$$

For all three cases [(*a*) to (*c*)] the generalized formula (6.105) holds:

$$V_t = \frac{V_1}{\beta^2} = \alpha\left[\frac{2(p_1 - p_2)}{\rho}\right]^{1/2} \qquad \alpha = \frac{C_d}{(1 - \beta^4)^{1/2}} \tag{1}$$

where the given data are $V_1 = 2.0$ m/s, $\rho = 1{,}000$ kg/m^3, and $\Delta p = 50{,}000$ Pa. Inserting these known values into Eq. (1) gives a relation between β and α:

$$\frac{2.0}{\beta^2} = \alpha\left[\frac{2(50{,}000)}{1{,}000}\right]^{1/2} \quad \text{or} \quad \beta^2 = \frac{0.2}{\alpha} \tag{2}$$

The unknowns are β (or α) and C_d. Parts (*a*) to (*c*) depend on the particular chart or formula needed for $C_d = \text{fcn}(\text{Re}_D, \beta)$. We can make an initial guess $\beta \approx 0.5$ and iterate to convergence.

Part (a) For the orifice with D: $\frac{1}{2}D$ taps, use Eq. (6.109) or Fig. 6.41. The iterative sequence is

$$\beta_1 \approx 0.5,\ C_{d1} \approx 0.604,\ \alpha_1 \approx 0.624,\ \beta_2 \approx 0.566,\ C_{d2} \approx 0.606,\ \alpha_2 \approx 0.640,\ \beta_3 = 0.559$$

We have converged to three figures. The proper orifice diameter is

$$d = \beta D = 112 \text{ mm} \qquad \textit{Ans. (a)}$$

Part (b) For the long-radius flow nozzle, use Eq. (6.111) or Fig. 6.42. The iterative sequence is

$$\beta_1 \approx 0.5,\ C_{d1} \approx 0.9891,\ \alpha_1 \approx 1.022,\ \beta_2 \approx 0.442,\ C_{d2} \approx 0.9896,\ \alpha_2 \approx 1.009,\ \beta_3 = 0.445$$

We have converged to three figures. The proper nozzle diameter is

$$d = \beta D = 89 \text{ mm} \qquad \textit{Ans. (b)}$$

Part (c) For the venturi nozzle, use Eq. (6.113) or Fig. 6.43. The iterative sequence is

$$\beta_1 \approx 0.5,\ C_{d1} \approx 0.977,\ \alpha_1 \approx 1.009,\ \beta_2 \approx 0.445,\ C_{d2} \approx 0.9807,\ \alpha_2 \approx 1.0004,\ \beta_3 = 0.447$$

We have converged to three figures. The proper venturi diameter is

$$d = \beta D = 89 \text{ mm} \qquad \textit{Ans. (c)}$$

Comments: These meters are of similar size, but their head losses are not the same. From Fig. 6.44 for the three different shapes we may read the three K factors and compute

$$h_{m,\text{orifice}} \approx 3.5 \text{ m} \qquad h_{m,\text{nozzle}} \approx 3.6 \text{ m} \qquad h_{m,\text{venturi}} \approx 0.8 \text{ m}$$

The venturi loss is only about 22 percent of the orifice and nozzle losses.

Solution by Excel Iteration for the Flow Nozzle

Parts (*a, b, c*) were solved by hand, but Excel is ideal for these calculations. You may review this procedure from the instructions in Example 6.5. We need five columns: C_d, calculated from Eq. (6.111), throat velocity V_t calculated from Δp, α as calculated from Eq. (6.104), and β calculated from the velocity ratio (V/V_t). The fifth column is an initial guess for β, which is replaced in its next row by the newly computed β. Any initial $\beta < 1$ will do. Here we chose $\beta = 0.5$ as in part (*b*) for the flow nozzle. Remember to use *cell* names, not symbols: in row 1, C_d = A1, V_t = B1, α = C1, and β = D1. The process converges rapidly, in only two or three iterations:

	C_d from Eq.(6.114)	$V_t = \alpha(2\Delta p/r)$	$\alpha = C_d/(1 - \beta\text{^}4)\text{^}0.5$	$\beta = (V/V_t)\text{^}0.5$	β-guess
	A	B	C	D	E
1	0.9891	10.216	1.0216	0.4425	0.5000
2	0.9896	10.091	1.0091	0.4452	0.4425
3	0.9895	10.096	1.0096	0.4451	0.4452
4	0.9895	10.096	1.0096	0.4451	0.4451

The final answers for the long-radius flow nozzle are:

$$\alpha = 1.0096 \qquad C_d = 0.9895 \qquad \beta = 0.4451 \qquad \textit{Ans. (b)}$$

EXAMPLE 6.22

A long-radius nozzle of diameter 6 cm is used to meter airflow in a 10-cm-diameter pipe. Upstream conditions are $p_1 = 200$ kPa and $T_1 = 100°C$. If the pressure drop through the nozzle is 60 kPa, estimate the flow rate in m^3/s.

Solution

- *Assumptions:* The pressure drops 30 percent, so we need the compressibility factor Y, and Eq. (6.114) is applicable to this problem.
- *Approach:* Find ρ_1 and C_d and apply Eq. (6.114) with $\beta = 6/10 = 0.6$.
- *Property values:* Given p_1 and T_1, $\rho_1 = p_1/RT_1 = (200{,}000)/[287(100 + 273)] = 1.87\ \text{kg/m}^3$. The downstream pressure is $p_2 = 200 - 60 = 140$ kPa, hence $p_2/p_1 = 0.7$. At 100°C, from Table A.2, the viscosity of air is 2.17 E-5 kg/m-s.
- *Solution steps:* Initially apply Eq. (6.114) by guessing, from Fig. 6.42, that $C_d \approx 0.98$. From Fig. 6.45, for a nozzle with $p_2/p_1 = 0.7$ and $\beta = 0.6$, read $Y \approx 0.80$. Then

$$\dot{m} = C_d Y A_t \sqrt{\frac{2\rho_1(p_1 - p_2)}{1 - \beta^4}} \approx (0.98)(0.80)\frac{\pi}{4}(0.06\ \text{m})^2\sqrt{\frac{2(1.87\ \text{kg/m}^3)(60{,}000\ \text{Pa})}{1 - (0.6)}}$$

$$\approx 1.13\ \frac{\text{kg}}{\text{s}}$$

Now estimate Re_d, putting it in the convenient mass flow form:

$$\text{Re}_d = \frac{\rho V d}{\mu} = \frac{4\dot{m}}{\pi \mu d} = \frac{4(1.13\ \text{kg/s})}{\pi(2.17\ \text{E-5 kg/m} - \text{s})(0.06\ \text{m})} \approx 1.11\ \text{E6}$$

Returning to Fig. 6.42, we could read a slightly better $C_d \approx 0.99$. Thus our final estimate is

$$\dot{m} \approx 1.14\ \text{kg/s} \qquad \textit{Ans.}$$

- *Comments:* Figure 6.45 is not just a "chart" for engineers to use casually. It is based on the compressible flow theory of Chap. 9. There, we may reassign this example as a *theory*.

Summary

This chapter has been concerned with internal pipe and duct flows, which are probably the most common problems encountered in engineering fluid mechanics. Such flows are very sensitive to the Reynolds number and change from laminar to transitional to turbulent flow as the Reynolds number increases.

The various Reynolds number regimes were outlined, and a semiempirical approach to turbulent flow modeling was presented. The chapter then made a detailed analysis of flow through a straight circular pipe, leading to the famous Moody chart (Fig. 6.13) for the friction factor. Possible uses of the Moody chart were discussed for flow rate and sizing problems, as well as the application of the Moody chart to noncircular ducts using an equivalent duct "diameter." The addition of minor losses due to valves, elbows, fittings, and other devices was presented in the form of loss coefficients to be incorporated along with Moody-type friction losses. Multiple-pipe systems were discussed briefly and were seen to be quite complex algebraically and appropriate for computer solution.

Diffusers are added to ducts to increase pressure recovery at the exit of a system. Their behavior was presented as experimental data, since the theory of real diffusers is still not well developed. The chapter ended with a discussion of flowmeters, especially the pitot-static tube and the Bernoulli obstruction type of meter. Flowmeters also require careful experimental calibration.

Problems

Most of the problems herein are fairly straightforward. More difficult or open-ended assignments are labeled with an asterisk. Problems labeled with a computer icon may require the use of a computer. The standard end-of-chapter problems P6.1 to P6.163 (categorized in the problem list here) are followed by word problems W6.1 to W6.4, fundamentals of engineering exam problems FE6.1 to FE6.15, comprehensive problems C6.1 to C6.9, and design projects D6.1 and D6.2.

Problem Distribution

Section	Topic	Problems
6.1	Reynolds number regimes	P6.1–P6.5
6.2	Internal and external flow	P6.6–P6.8
6.3	Head loss—friction factor	P6.9–P6.11
6.4	Laminar pipe flow	P6.12–P6.33
6.5	Turbulence modeling	P6.34–P6.40
6.6	Turbulent pipe flow	P6.41–P6.62
6.7	Flow rate and sizing problems	P6.63–P6.85
6.8	Noncircular ducts	P6.86–P6.98
6.9	Minor or local losses	P6.99–P6.110
6.10	Series and parallel pipe systems	P6.111–P6.120
6.10	Three-reservoir and pipe network systems	P6.121–P6.130
6.11	Diffuser performance	P6.131–P6.134
6.12	The pitot-static tube	P6.135–P6.139
6.12	Flowmeters: the orifice plate	P6.140–P6.148
6.12	Flowmeters: the flow nozzle	P6.149–P6.153
6.12	Flowmeters: the venturi meter	P6.154–P6.159
6.12	Flowmeters: other designs	P6.160–P6.161
6.12	Flowmeters: compressibility correction	P6.162–P6.163

Reynolds number regimes

P6.1 An engineer claims that the flow of SAE 30W oil, at 20°C, through a 5-cm-diameter smooth pipe at 1 million N/h, is laminar. Do you agree? A million newtons is a lot, so this sounds like an awfully high flow rate.

P6.2 The present pumping rate of crude oil through the Alaska Pipeline, with an ID of 1.22 m, is 1 m^3/s. (*a*) Is this a turbulent flow? (*b*) What would be the maximum rate if the flow were constrained to be laminar? Assume that Alaskan oil fits Fig. A.1 of the Appendix at 60°C.

P6.3 The Keystone Pipeline in the chapter opener photo has a maximum proposed flow rate of 206,700 m^3 of crude oil per day. Estimate the Reynolds number and whether the flow is laminar. Assume that Keystone crude oil fits Fig. A.1 of the Appendix at 40°C.

P6.4 For flow of SAE 30 oil through a 5-cm-diameter pipe, from Fig. A.1, for what flow rate in m^3/h would we expect transition to turbulence at (*a*) 20°C and (*b*) 100°C?

P6.5 In flow past a body or wall, early transition to turbulence can be induced by placing a trip wire on the wall across the flow, as in Fig. P6.5. If the trip wire in Fig. P6.5 is placed where the local velocity is U, it will trigger turbulence if $Ud/\nu = 850$, where d is the wire diameter [3, p. 388]. If the sphere diameter is 20 cm and transition is observed at $Re_D = 90,000$, what is the diameter of the trip wire in mm?

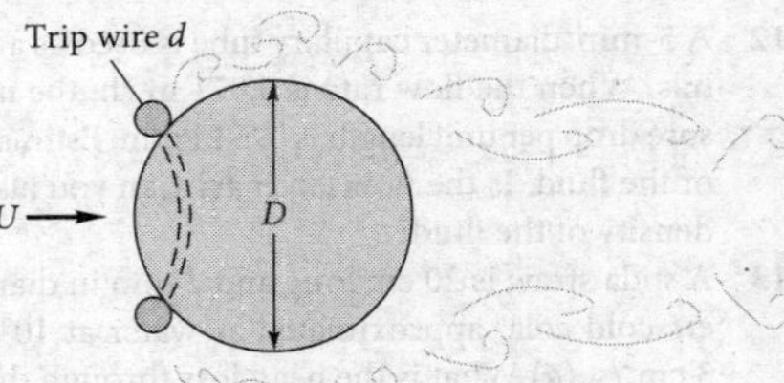

P6.5

Internal and external flow

P6.6 For flow of a uniform stream parallel to a sharp flat plate, transition to a turbulent boundary layer on the plate may occur at $Re_x = \rho U x/\mu \approx 1$ E6, where U is the approach velocity and x is distance along the plate. If $U = 2.5$ m/s, determine the distance x for the following fluids at 20°C and 1 atm: (*a*) hydrogen, (*b*) air, (*c*) gasoline, (*d*) water, (*e*) mercury, and (*f*) glycerin.

P6.7 SAE 10W30 oil at 20°C flows from a tank into a 2-cm-diameter tube 40 cm long. The flow rate is 1.1 m^3/hr. Is the entrance length region a significant part of this tube flow?

P6.8 When water at 20°C is in steady turbulent flow through an 8-cm-diameter pipe, the wall shear stress is 72 Pa. What is the axial pressure gradient ($\partial p/\partial x$) if the pipe is (*a*) horizontal and (*b*) vertical with the flow up?

Head loss—friction factor

P6.9 A light liquid ($\rho \approx 950$ kg/m^3) flows at an average velocity of 10 m/s through a horizontal smooth tube of diameter 5 cm. The fluid pressure is measured at 1-m intervals along the pipe, as follows:

x, m	0	1	2	3	4	5	6
p, kPa	304	273	255	240	226	213	200

Estimate (*a*) the total head loss, in meters; (*b*) the wall shear stress in the fully developed section of the pipe; and (*c*) the overall friction factor.

P6.10 Water at 20°C flows through an inclined 8-cm-diameter pipe. At sections A and B the following data are taken: $p_A = 186$ kPa, $V_A = 3.2$ m/s, $z_A = 24.5$ m, and $p_B = 260$ kPa, $V_B = 3.2$ m/s, $z_B = 9.1$ m. Which way is the flow going? What is the head loss in meters?

P6.11 Water at 20°C flows upward at 4 m/s in a 6-cm-diameter pipe. The pipe length between points 1 and 2 is 5 m, and point 2 is 3 m higher. A mercury manometer, connected between 1 and 2, has a reading $h = 135$ mm, with p_1 higher. (*a*) What is the pressure change ($p_1 - p_2$)? (*b*) What is the head loss, in meters? (*c*) Is the manometer reading proportional to head loss? Explain. (*d*) What is the friction factor of the flow?

In Probs. 6.12 to 6.99, neglect minor losses.

Laminar pipe flow—no minor losses

P6.12 A 5-mm-diameter capillary tube is used as a viscometer for oils. When the flow rate is 0.071 m^3/h, the measured pressure drop per unit length is 375 kPa/m. Estimate the viscosity of the fluid. Is the flow laminar? Can you also estimate the density of the fluid?

P6.13 A soda straw is 20 cm long and 2 mm in diameter. It delivers cold cola, approximated as water at 10°C, at a rate of 3 cm^3/s. (*a*) What is the head loss through the straw? What is the axial pressure gradient $\partial p/\partial x$ if the flow is (*b*) vertically up or (*c*) horizontal? Can the human lung deliver this much flow?

P6.14 Water at 20°C is to be siphoned through a tube 1 m long and 2 mm in diameter, as in Fig. P6.14. Is there any height H for which the flow might not be laminar? What is the flow rate if $H = 50$ cm? Neglect the tube curvature.

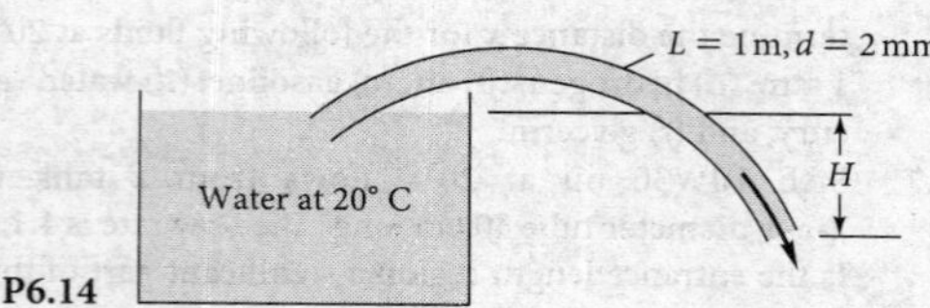

P6.14

P6.15 Professor Gordon Holloway and his students at the University of New Brunswick went to a fast-food emporium and tried to drink chocolate shakes ($\rho \approx 1{,}200$ kg/m^3, $\mu \approx$ 6 kg/m-s) through fat straws 8 mm in diameter and 30 cm long. (*a*) Verify that their human lungs, which can develop approximately 3,000 Pa of vacuum pressure, would be unable to drink the milkshake through the vertical straw. (*b*) A student cut 15 cm from his straw and proceeded to drink happily. What rate of milkshake flow was produced by this strategy?

P6.16 Fluid flows steadily, at volume rate Q, through a large pipe and then divides into two small pipes, the larger of which has an inside diameter of 25 mm and carries three times the flow of the smaller pipe. Both small pipes have the same length and pressure drop. If all flows are laminar, estimate the diameter of the smaller pipe.

P6.17 A *capillary viscometer* measures the time required for a specified volume υ of liquid to flow through a small-bore glass tube, as in Fig. P6.17. This transit time is then correlated with fluid viscosity. For the system shown, (*a*) derive an approximate formula for the time required, assuming laminar flow with no entrance and exit losses. (*b*) If $L = 12$ cm, $l = 2$ cm, $\upsilon = 8$ cm^3, and the fluid is water at 20°C, what capillary diameter D will result in a transit time t of 6 seconds?

P6.18 SAE 50W oil at 20°C flows from one tank to another through a tube 160 cm long and 5 cm in diameter. Estimate the flow rate in m^3/hr if $z_1 = 2$ m and $z_2 = 0.8$ m.

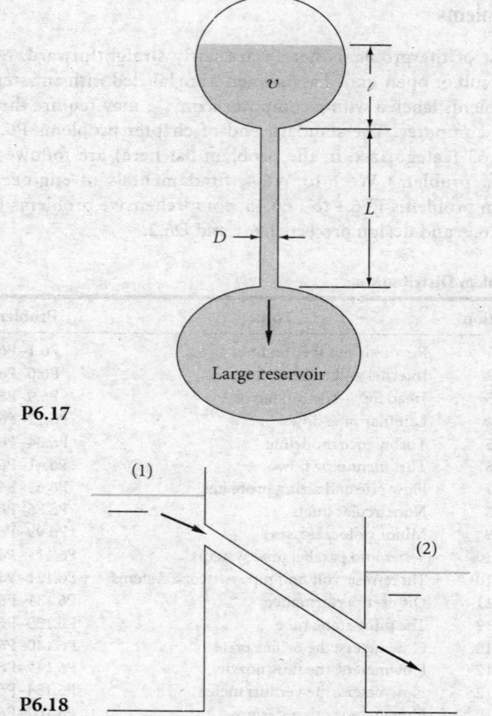

P6.17

P6.18

P6.19 An oil (SG = 0.9) issues from the pipe in Fig. P6.19 at $Q = 1$ m^3/h. What is the kinematic viscosity of the oil in m^2/s? Is the flow laminar?

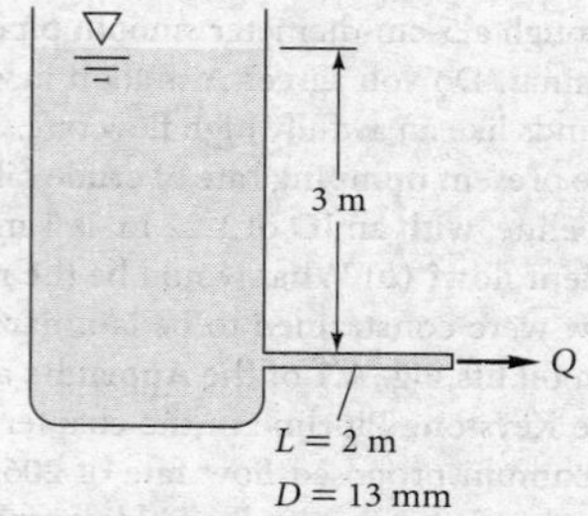

P6.19

P6.20 The oil tanks in Tinyland are only 160 cm high, and they discharge to the Tinyland oil truck through a smooth tube 4 mm in diameter and 55 cm long. The tube exit is open to the atmosphere and 145 cm below the tank surface. The fluid is medium fuel oil, $\rho = 850$ kg/m^3 and $\mu =$ 0.11 kg/(m · s). Estimate the oil flow rate in cm^3/h.

P6.21 In Tinyland, houses are less than 30 cm high! The rainfall is laminar! The drainpipe in Fig. P6.21 is only 2 mm in diameter. (*a*) When the gutter is full, what is the rate of draining? (*b*) The gutter is designed for a sudden rainstorm of up to 5 mm per hour. For this condition, what is the maximum roof area that can be drained successfully? (*c*) What is Re_d?

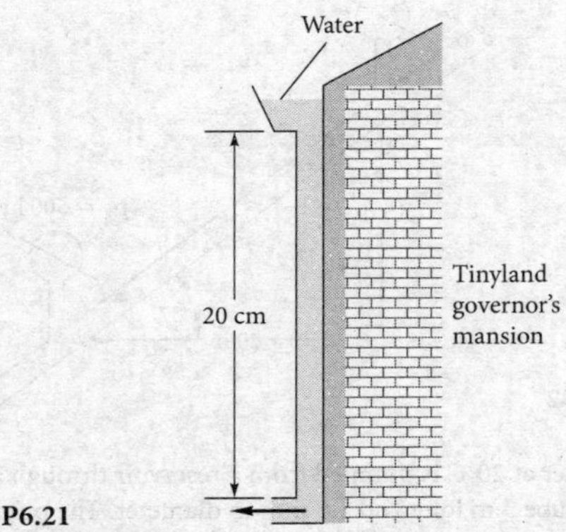

P6.21

P6.22 A steady push on the piston in Fig. P6.22 causes a flow rate $Q = 0.15$ cm^3/s through the needle. The fluid has $\rho = 900$ kg/m^3 and $\mu = 0.002$ kg/(m · s). What force *F* is required to maintain the flow?

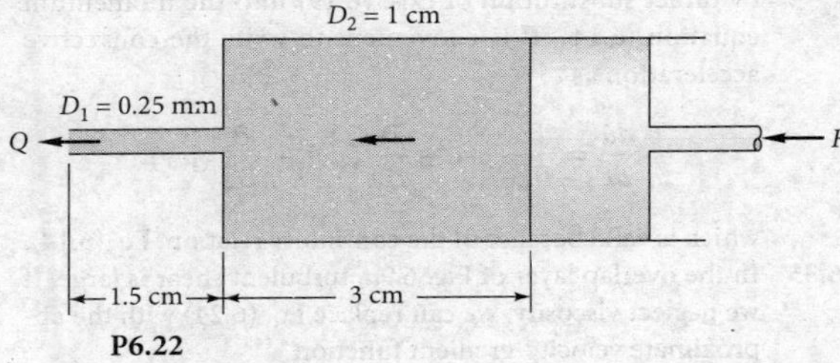

P6.22

P6.23 SAE 10 oil at 20°C flows in a vertical pipe of diameter 2.5 cm. It is found that the pressure is constant throughout the fluid. What is the oil flow rate in m^3/h? Is the flow up or down?

P6.24 Two tanks of water at 20°C are connected by a capillary tube 4 mm in diameter and 3.5 m long. The surface of tank 1 is 30 cm higher than the surface of tank 2. (*a*) Estimate the flow rate in m^3/h. Is the flow laminar? (*b*) For what tube diameter will Re_d be 500?

P6.25 For the configuration shown in Fig. P6.25, the fluid is ethyl alcohol at 20°C, and the tanks are very wide. Find the flow rate which occurs in m^3/h. Is the flow laminar?

P6.26 Two oil tanks are connected by two 9-m-long pipes, as in Fig. P6.26. Pipe 1 is 5 cm in diameter and is 6 m higher than pipe 2. It is found that the flow rate in pipe 2 is twice as large as the flow in pipe 1. (*a*) What is the diameter of pipe 2? (*b*) Are both pipe flows laminar? (*c*) What is the flow rate in pipe 2 (m^3/s)? Neglect minor losses.

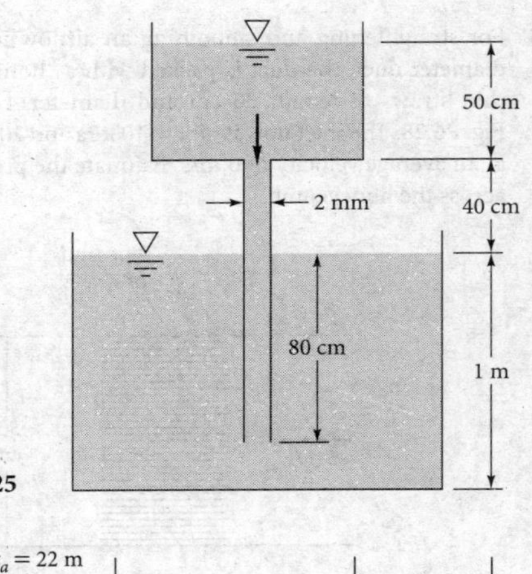

P6.25

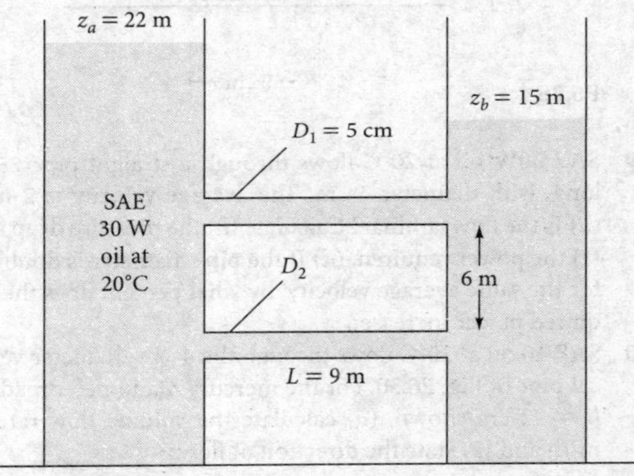

P6.26

***P6.27** Let us attack Prob. P6.25 in symbolic fashion, using Fig. P6.27. All parameters are constant except the upper tank depth $Z(t)$. Find an expression for the flow rate $Q(t)$ as a function of $Z(t)$. Set up a differential equation, and solve for the time t_0 to drain the upper tank completely. Assume quasi-steady laminar flow.

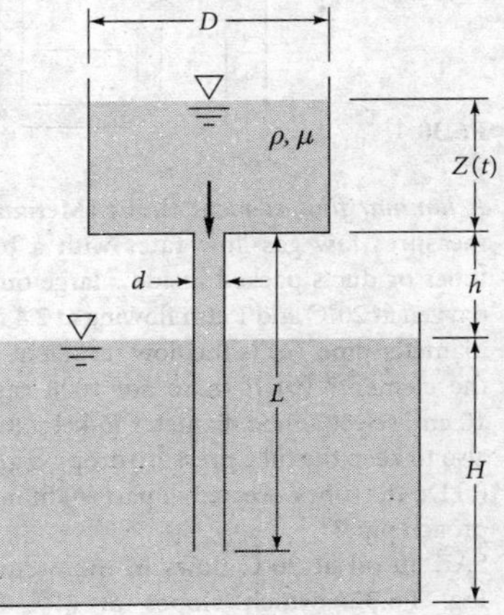

P6.27

P6.28 For straightening and smoothing an airflow in a 50-cm-diameter duct, the duct is packed with a "honeycomb" of thin straws of length 30 cm and diameter 4 mm, as in Fig. P6.28. The inlet flow is air at 110 kPa and 20°C, moving at an average velocity of 6 m/s. Estimate the pressure drop across the honeycomb.

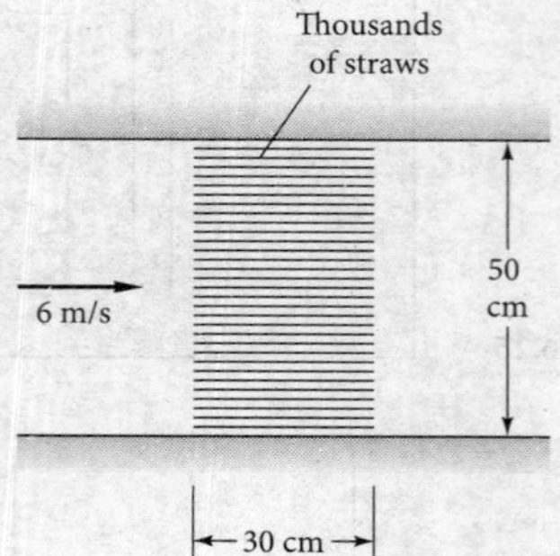

P6.28

P6.29 SAE 30W oil at 20°C flows through a straight pipe 25 m long, with diameter 4 cm. The average velocity is 2 m/s. (*a*) Is the flow laminar? Calculate (*b*) the pressure drop and (*c*) the power required. (*d*) If the pipe diameter is doubled, for the same average velocity, by what percent does the required power increase?

P6.30 SAE 10 oil at 20°C flows through the 4-cm-diameter vertical pipe of Fig. P6.30. For the mercury manometer reading h = 42 cm shown, (*a*) calculate the volume flow rate in m^3/h and (*b*) state the direction of flow.

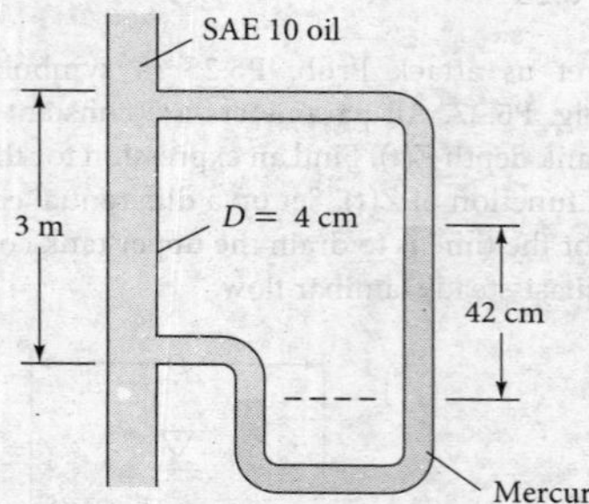

P6.30

P6.31 A *laminar flow element* (LFE) (Meriam Instrument Co.) measures low gas-flow rates with a bundle of capillary tubes or ducts packed inside a large outer tube. Consider oxygen at 20°C and 1 atm flowing at 2.4 m^3/min in a 10 cm-diameter pipe. (*a*) Is the flow turbulent when approaching the element? (*b*) If there are 1000 capillary tubes, L = 10 cm, select a tube diameter to keep Re_d below 1,500 and also to keep the tube pressure drop no greater than 3.5 kPa. (*c*) Do the tubes selected in part (*b*) fit nicely within the approach pipe?

P6.32 SAE 30 oil at 20°C flows in the 3-cm-diameter pipe in Fig. P6.32, which slopes at 37°. For the pressure measurements shown, determine (*a*) whether the flow is up or down and (*b*) the flow rate in m^3/h.

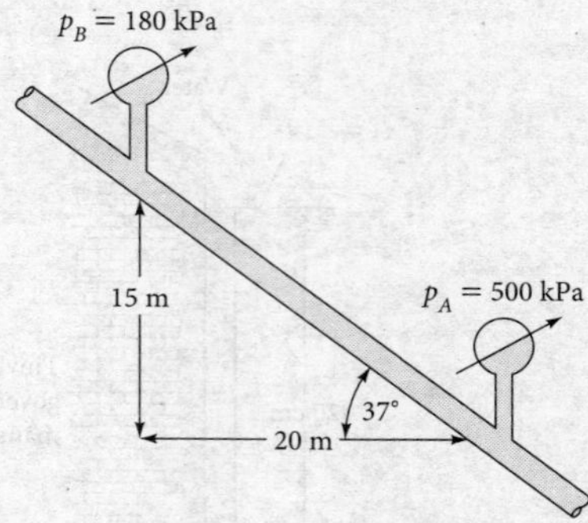

P6.32

P6.33 Water at 20°C is pumped from a reservoir through a vertical tube 3 m long and 1.6 mm in diameter. The pump provides a pressure rise of 76 kPa to the flow. Neglect entrance losses. (*a*) Calculate the exit velocity. (*b*) Approximately how high will the exit water jet rise? (*c*) Verify that the flow is laminar.

Turbulence modeling

P6.34 Derive the time-averaged x-momentum equation (6.21) by direct substitution of Eqs. (6.19) into the momentum equation (6.14). It is convenient to write the convective acceleration as

$$\frac{du}{dt} = \frac{\partial}{\partial x}(u^2) + \frac{\partial}{\partial y}(uv) + \frac{\partial}{\partial z}(uw)$$

which is valid because of the continuity relation, Eq. (6.14).

P6.35 In the overlap layer of Fig. 6.9*a*, turbulent shear is large. If we neglect viscosity, we can replace Eq. (6.24) with the approximate velocity-gradient function

$$\frac{du}{dy} = fcn(y, \tau_w, \rho)$$

Show by dimensional analysis that this leads to the logarithmic overlap relation (6.28).

P6.36 The following turbulent flow velocity data $u(y)$, for air at 24°C and 1 atm near a smooth flat wall were taken in the University of Rhode Island wind tunnel:

y, mm	0.64	0.89	1.19	1.40	1.65
u, m/s	15.6	16.5	17.3	17.6	18.0

Estimate (*a*) the wall shear stress and (*b*) the velocity u at y = 5.6 mm.

P6.37 Two infinite plates a distance h apart are parallel to the xz plane with the upper plate moving at speed V, as in Fig. P6.37. There is a fluid of viscosity μ and constant pressure between the plates. Neglecting gravity and assuming incompressible turbulent flow $u(y)$ between the plates, use the logarithmic law and appropriate boundary conditions to derive a formula for dimensionless wall shear stress versus dimensionless plate velocity. Sketch a typical shape of the profile $u(y)$.

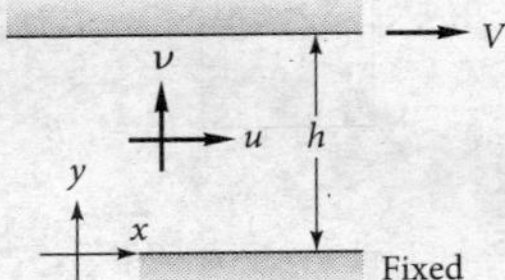

P6.37

P6.38 Suppose in Fig. P6.37 that $h = 3$ cm, the fluid in water at 20°C, and the flow is turbulent, so that the logarithmic law is valid. If the shear stress in the fluid is 15 Pa, what is V in m/s?

P6.39 By analogy with laminar shear, $\tau = \mu\ du/dy$, T. V. Boussinesq in 1877 postulated that turbulent shear could also be related to the mean velocity gradient $\tau_{turb} = \varepsilon\ du/dy$, where ε is called the *eddy viscosity* and is much larger than μ. If the logarithmic overlap law, Eq. (6.28), is valid with $\tau_{turb} \approx \tau_w$, show that $\varepsilon \approx \kappa\rho u^* y$.

P6.40 Theodore von Kármán in 1930 theorized that turbulent shear could be represented by $\tau_{turb} = \varepsilon\ du/dy$, where $\varepsilon = \rho\kappa^2 y^2 |du/dy|$ is called the *mixing-length eddy viscosity* and $\kappa \approx 0.41$ is Kármán's dimensionless *mixing-length constant* [2, 3]. Assuming that $\tau_{turb} \approx \tau_w$ near the wall, show that this expression can be integrated to yield the logarithmic overlap law, Eq. (6.28).

Turbulent pipe flow—no minor losses

P6.41 Two reservoirs, which differ in surface elevation by 40 m, are connected by 350 m of new pipe of diameter 8 cm. If the desired flow rate is at least 130 N/s of water at 20°C, can the pipe material be made of (*a*) galvanized iron, (*b*) commercial steel, or (*c*) cast iron? Neglect minor losses.

P6.42 Fluid flows steadily, at volume rate Q, through a large horizontal pipe and then divides into two small pipes, the larger of which has an inside diameter of 25 mm and carries three times the flow of the smaller pipe. Both small pipes have the same length and pressure drop. If all flows are turbulent, at Re_D near 10^4, estimate the diameter of the smaller pipe.

P6.43 A reservoir supplies water through 100 m of 30-cm-diameter cast iron pipe to a turbine that extracts 80 hp from the flow. The water then exhausts to the atmosphere.

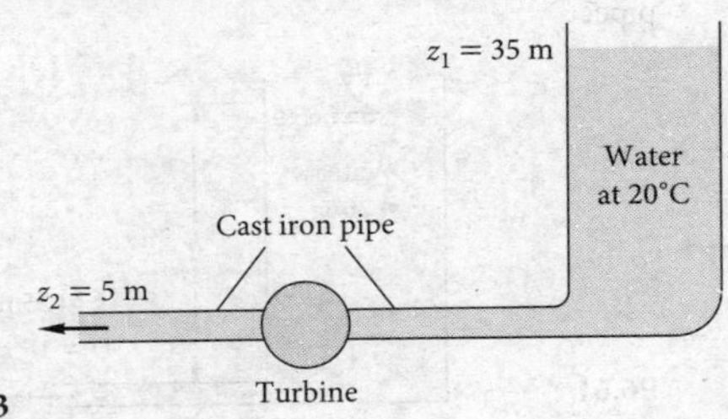

P6.43

Neglect minor losses. (*a*) Assuming that $f \approx 0.019$, find the flow rate (which results in a cubic polynomial). Explain why there are *two* legitimate solutions. (*b*) For extra credit, solve for the flow rates using the actual friction factors.

P6.44 Mercury at 20°C flows through 4 m of 7-mm-diameter glass tubing at an average velocity of 5 m/s. Estimate the head loss in m and the pressure drop in kPa.

P6.45 Oil, SG = 0.88 and ν = 4 E-5 m²/s, flows at 25 L/s through a 15 cm asphalted cast iron pipe. The pipe is 800 m long and slopes upward at 8° in the flow direction. Compute the head loss in m and the pressure change.

P6.46 The Keystone Pipeline in the chapter opener photo has a diameter of 0.9 m and a design flow rate of 94,000 m³ per day of crude oil at 40°C. If the pipe material is new steel, estimate the pump horsepower required per km of pipe.

P6.47 The gutter and smooth drainpipe in Fig. P6.47 remove rainwater from the roof of a building. The smooth drainpipe is 7 cm in diameter. (*a*) When the gutter is full, estimate the rate of draining. (*b*) The gutter is designed for a sudden rainstorm of up to 130 mm per hour. For this condition, what is the maximum roof area that can be drained successfully?

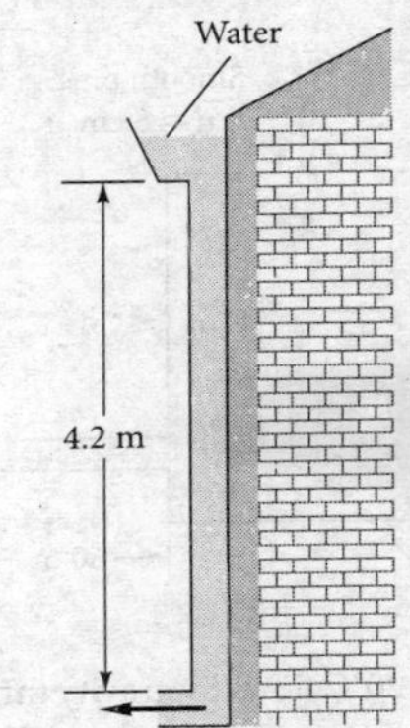

P6.47

P6.48 Follow up Prob. P6.46 with the following question. If the total Keystone pipeline length, from Alberta to Texas, is 3,454 km, how much flow, in m³ per minute, will result if the total available pumping power is 8,000 hp?

P6.49 The tank–pipe system of Fig. P6.49 is to deliver at least 11 m³/h of water at 20°C to the reservoir. What is the maximum roughness height ε allowable for the pipe?

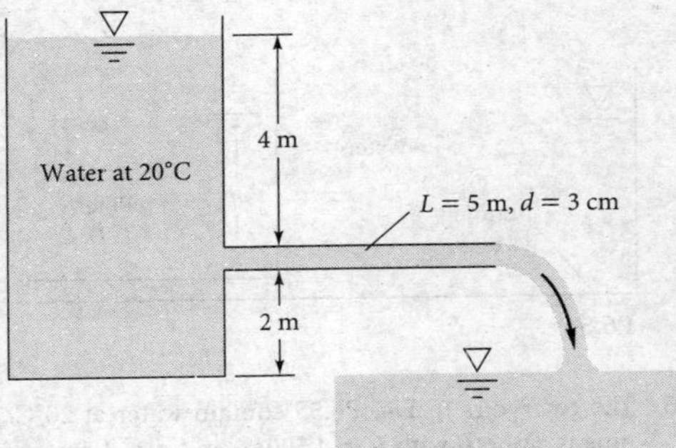

P6.49

P6.50 Ethanol at 20°C flows at 8 L/s through a horizontal cast iron pipe with $L = 12$ m and $d = 5$ cm. Neglecting entrance effects, estimate (*a*) the pressure gradient dp/dx, (*b*) the wall shear stress τ_w, and (*c*) the percentage reduction in friction factor if the pipe walls are polished to a smooth surface.

P6.51 The viscous sublayer (Fig. 6.9) is normally less than 1 percent of the pipe diameter and therefore very difficult to probe with a finite-sized instrument. In an effort to generate a thick sublayer for probing, Pennsylvania State University in 1964 built a pipe with a flow of glycerin. Assume a smooth 0.3 m-diameter pipe with $V = 18$ m/s and glycerin at 20°C. Compute the sublayer thickness in mm and the pumping horsepower required at 75 percent efficiency if $L = 12$ m.

P6.52 The pipe flow in Fig. P6.52 is driven by pressurized air in the tank. What gage pressure p_1 is needed to provide a 20°C water flow rate $Q = 60$ m^3/h?

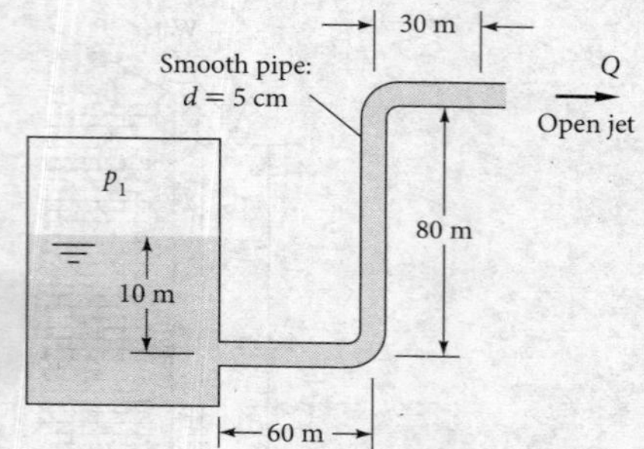

P6.52

P6.53 Water at 20°C flows by gravity through a smooth pipe from one reservoir to a lower one. The elevation difference is 60 m. The pipe is 360 m long, with a diameter of 12 cm. Calculate the expected flow rate in m^3/h. Neglect minor losses.

***P6.54** A swimming pool W by Y by h deep is to be emptied by gravity through the long pipe shown in Fig. P6.54. Assuming an average pipe friction factor f_{av} and neglecting minor losses, derive a formula for the time to empty the tank from an initial level h_o.

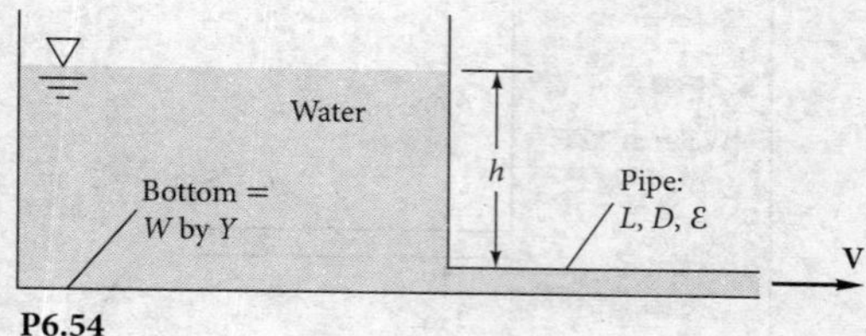

P6.54

P6.55 The reservoirs in Fig. P6.55 contain water at 20°C. If the pipe is smooth with $L = 4{,}500$ m and $d = 4$ cm, what will the flow rate in m^3/h be for $\Delta z = 100$ m?

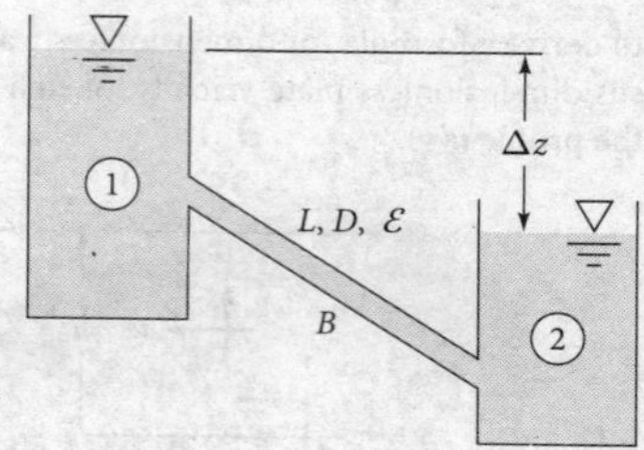

P6.55

P6.56 The Alaska Pipeline in the chapter opener photo has a design flow rate of 1.7 E5 m^3 per day of crude oil at 60°C (see Fig. A.1). (*a*) Assuming a galvanized-iron wall, estimate the total pressure drop required for the 1,287 km trip. (*b*) If there are nine equally spaced pumps, estimate the horsepower each pump must deliver.

P6.57 Apply the analysis of Prob. P6.54 to the following data. Let $W = 5$ m, $Y = 8$ m, $h_o = 2$ m, $L = 15$ m, $D = 5$ cm, and $\varepsilon = 0$. (*a*) By letting $h = 1.5$ m and 0.5 m as representative depths, estimate the average friction factor. Then (*b*) estimate the time to drain the pool.

P6.58 For the system in Prob. 6.53, a pump is used at night to drive water back to the upper reservoir. If the pump delivers 15,000 W to the water, estimate the flow rate.

P6.59 The following data were obtained for flow of 20°C water at 20 m^3/h through a badly corroded 5-cm-diameter pipe that slopes downward at an angle of 8°: $p_1 = 420$ kPa, $z_1 = 12$ m, $p_2 = 250$ kPa, $z_2 = 3$ m. Estimate (*a*) the roughness ratio of the pipe and (*b*) the percentage change in head loss if the pipe were smooth and the flow rate the same.

P6.60 In the spirit of Haaland's explicit pipe friction factor approximation, Eq. (6.49), Jeppson [20] proposed the following explicit formula:

$$\frac{1}{\sqrt{f}} \approx -2.0 \log_{10}\left(\frac{\varepsilon/d}{3.7} + \frac{5.74}{\mathrm{Re}_d^{0.9}}\right)$$

(*a*) Is this identical to Haaland's formula with just a simple rearrangement? Explain. (*b*) Compare Jeppson's formula to Haaland's for a few representative values of (turbulent) Re_d and ε/d and their errors compared to the Colebrook formula (6.48). Discuss briefly.

P6.61 What level h must be maintained in Fig. P6.61 to deliver a flow rate of 0.4 L/s through the 15 mm commercial steel pipe?

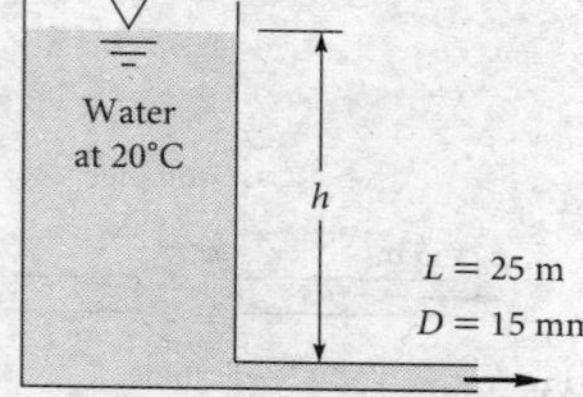

P6.61

P6.62 Water at 20°C is to be pumped through 600 m of pipe from reservoir 1 to 2 at a rate of 0.1 m^3/s, as shown in

Fig. P6.62. If the pipe is cast iron of diameter 15 cm and the pump is 75 percent efficient, what horsepower pump is needed?

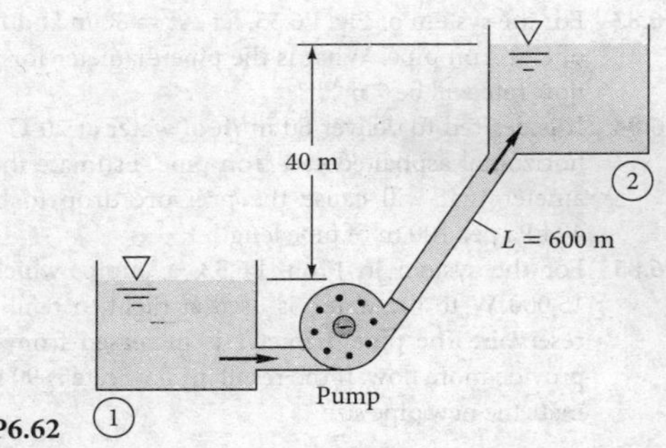

P6.62

Flow rate and sizing problems

P6.63 A tank contains 1 m^3 of water at 20°C and has a drawn-capillary outlet tube at the bottom, as in Fig. P6.63. Find the outlet volume flux Q in m^3/h at this instant.

P6.64 For the system in Fig. P6.63, solve for the flow rate in m^3/h if the fluid is SAE 10 oil at 20°C. Is the flow laminar or turbulent?

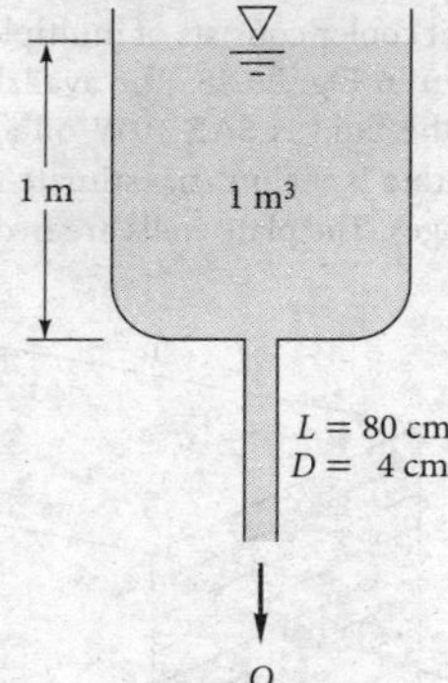

P6.63

P6.65 In Prob. P6.63 the initial flow is turbulent. As the water drains out of the tank, will the flow revert to laminar motion as the tank becomes nearly empty? If so, at what tank depth? Estimate the time, in h, to drain the tank completely.

P6.66 Ethyl alcohol at 20°C flows through a 10-cm horizontal drawn tube 100 m long. The fully developed wall shear stress is 14 Pa. Estimate (*a*) the pressure drop, (*b*) the volume flow rate, and (*c*) the velocity u at $r = 1$ cm.

P6.67 A straight 10-cm commercial-steel pipe is 1 km long and is laid on a constant slope of 5°. Water at 20°C flows downward, due to gravity only. Estimate the flow rate in m^3/h. What happens if the pipe length is 2 km?

***P6.68** The Moody chart cannot find V directly, since V appears in both ordinate and abscissa. (*a*) Arrange the variables (h_f, d, g, L, ν) into a single dimensionless group, with $h_f d^3$ in the numerator, denoted as ξ, which equals ($f \mathrm{Re}_d^2/2$). (*b*) Rearrange the Colebrook formula (6.48) to solve for Re_d in terms of ξ. (*c*) For extra credit, solve Example 6.9 with this new formula.

P6.69 For Prob. P6.62 suppose the only pump available can deliver 80 hp to the fluid. What is the proper pipe size in meters to maintain the 0.1 m^3/s flow rate?

P6.70 Ethylene glycol at 20°C flows through 80 m of cast iron pipe of diameter 6 cm. The measured pressure drop is 250 kPa. Neglect minor losses. Using a noniterative formulation, estimate the flow rate in m^3/h.

***P6.71** It is desired to solve Prob. 6.62 for the most economical pump and cast iron pipe system. If the pump costs \$125 per horsepower delivered to the fluid and the pipe costs \$2,500 per centimeter of diameter, what are the minimum cost and the pipe and pump size to maintain the 0.1 m^3/s flow rate? Make some simplifying assumptions.

P6.72 Modify Prob. P6.57 by letting the diameter be unknown. Find the proper pipe diameter for which the pool will drain in about two hours flat.

P6.73 For 20°C water flow in a smooth, horizontal 10-cm pipe, with $\Delta p/L = 1{,}000$ Pa/m, the writer computed a flow rate of 0.030 m^3/s. (*a*) Verify, or disprove, the writer's answer. (*b*) If verified, use the power-law friction factor relation, Eq. (6.41), to estimate the pipe diameter that will triple this flow rate. (*c*) For extra credit, use the more exact friction factor relation, Eq. (6.38), to solve part (*b*).

P6.74 Two reservoirs, which differ in surface elevation by 40 m, are connected by a new commercial steel pipe of diameter 8 cm. If the desired flow rate is 200 N/s of water at 20°C, what is the proper length of the pipe?

P6.75 You wish to water your garden with 30 m of 15 mm-diameter hose whose roughness is 0.3 mm. What will be the delivery, in L/s, if the gage pressure at the faucet is 400 kPa? If there is no nozzle (just an open hose exit), what is the maximum horizontal distance the exit jet will carry?

P6.76 The small turbine in Fig. P6.76 extracts 400 W of power from the water flow. Both pipes are wrought iron. Compute the flow rate Q in m^3/h. Why are there two solutions? Which is better?

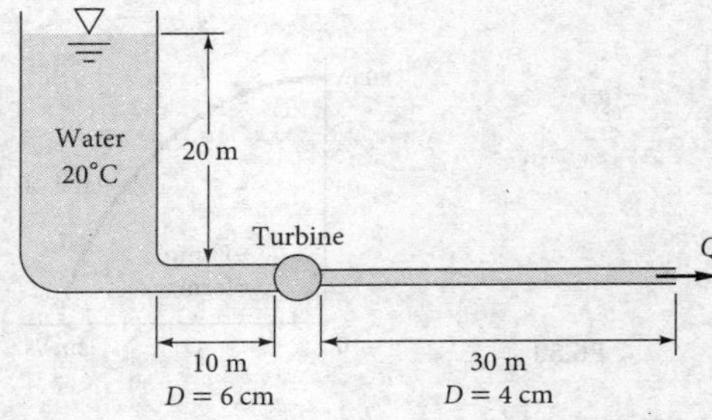

P6.76

***P6.77** Modify Prob. P6.76 into an economic analysis, as follows: Let the 40 m of wrought iron pipe have a uniform diameter d. Let the steady water flow available be $Q = 30\ m^3/h$. The cost of the turbine is $4 per watt developed, and the cost of the piping is $75 per centimeter of diameter. The power generated may be sold for $0.08 per kilowatt-hour. Find the proper pipe diameter for minimum *payback time*—that is, the minimum time for which the power sales will equal the initial cost of the system.

P6.78 In Fig. P6.78 the connecting pipe is commercial steel 6 cm in diameter. Estimate the flow rate, in m^3/h, if the fluid is water at 20°C. Which way is the flow?

P6.79 A garden hose is to be used as the return line in a waterfall display at a mall. In order to select the proper pump, you need to know the roughness height inside the garden hose. Unfortunately, roughness information is not supplied by the hose manufacturer. So you devise a simple experiment to measure the roughness. The hose is attached to the drain of an above-ground swimming pool, the surface of which is 3.0 m above the hose outlet. You estimate the minor loss coefficient of the entrance region as 0.5, and the drain valve has a minor loss equivalent length of 200 diameters when fully open. Using a bucket and stopwatch, you open the valve and measure the flow rate to be $2.0 \times 10^{-4}\ m^3/s$ for a hose that is 10.0 m long and has an inside diameter of 1.50 cm. Estimate the roughness height in mm inside the hose.

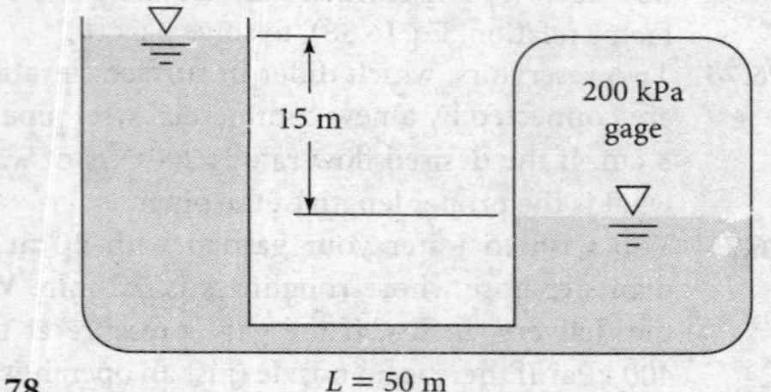

P6.78

P6.80 The head-versus-flow-rate characteristics of a centrifugal pump are shown in Fig. P6.80. If this pump drives water at 20°C through 120 m of 30-cm-diameter cast iron pipe, what will be the resulting flow rate, in m^3/s?

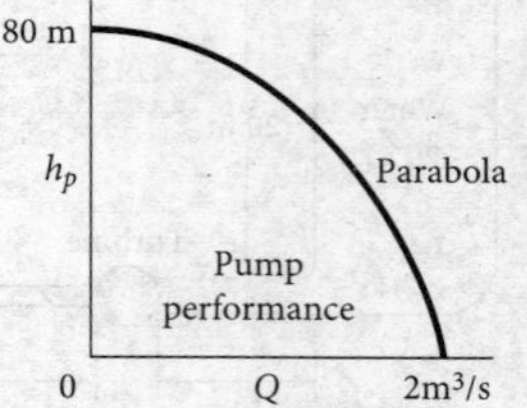

P6.80

P6.81 The pump in Fig. P6.80 is used to deliver gasoline at 20°C through 350 m of 30-cm-diameter galvanized iron pipe. Estimate the resulting flow rate, in m^3/s. (Note that the pump head is now in meters of gasoline.)

P6.82 Fluid at 20°C flows through a horizontal galvanized-iron pipe 20 m long and 8 cm in diameter. The wall shear stress is 90 Pa. Calculate the flow rate in m^3/h if the fluid is (*a*) glycerin and (*b*) water.

P6.83 For the system of Fig. P6.55, let $\Delta z = 80$ m and $L = 185$ m of cast iron pipe. What is the pipe diameter for which the flow rate will be 7 m^3/h?

P6.84 It is desired to deliver 60 m^3/h of water at 20°C through a horizontal asphalted cast iron pipe. Estimate the pipe diameter that will cause the pressure drop to be exactly 40 kPa per 100 m of pipe length.

P6.85 For the system in Prob. P6.53, a pump, which delivers 15,000 W to the water, is used at night to refill the upper reservoir. The pipe diameter is increased from 12 cm to provide more flow. If the resultant flow rate is 90 m^3/h, estimate the new pipe size.

Noncircular ducts

P6.86 SAE 10 oil at 20°C flows at an average velocity of 2 m/s between two smooth parallel horizontal plates 3 cm apart. Estimate (*a*) the centerline velocity, (*b*) the head loss per meter, and (*c*) the pressure drop per meter.

P6.87 A commercial steel annulus 12 m long, with $a = 20$ mm and $b = 10$ mm, connects two reservoirs that differ in surface height by 5 m. Compute the flow rate in L/s through the annulus if the fluid is water at 20°C.

P6.88 An oil cooler consists of multiple parallel-plate passages, as shown in Fig. P6.88. The available pressure drop is 6 kPa, and the fluid is SAE 10W oil at 20°C. If the desired total flow rate is 900 m^3/h, estimate the appropriate number of passages. The plate walls are hydraulically smooth.

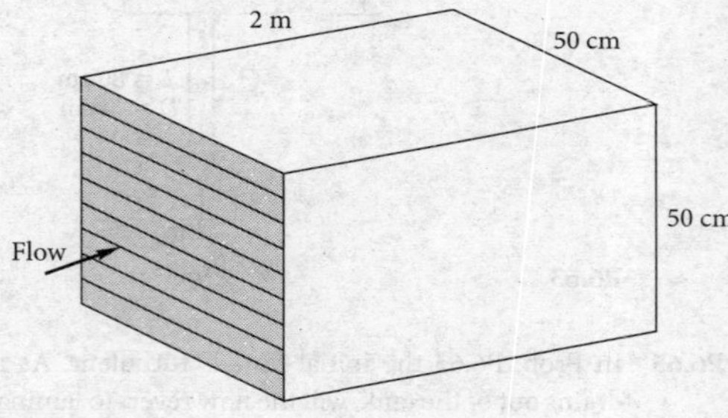

P6.88

P6.89 An annulus of narrow clearance causes a very large pressure drop and is useful as an accurate measurement of viscosity. If a smooth annulus 1 m long with $a = 50$ mm and $b = 49$ mm carries an oil flow at 0.001 m^3/s, what is the oil viscosity if the pressure drop is 250 kPa?

P6.90 A rectangular sheet-metal duct is 60 m long and has a fixed height $H = 15$ cm. The width B, however, may vary from 15 to 90 cm. A blower provides a pressure drop of 80 Pa of air at 20°C and 1 atm. What is the optimum width B that will provide the most airflow in m^3/s?

P6.91 Heat exchangers often consist of many triangular passages. Typical is Fig. P6.91, with $L = 60$ cm and an isosceles-triangle cross section of side length $a = 2$ cm and included angle $\beta = 80°$. If the average velocity is $V = 2$ m/s and the fluid is SAE 10 oil at 20°C, estimate the pressure drop.

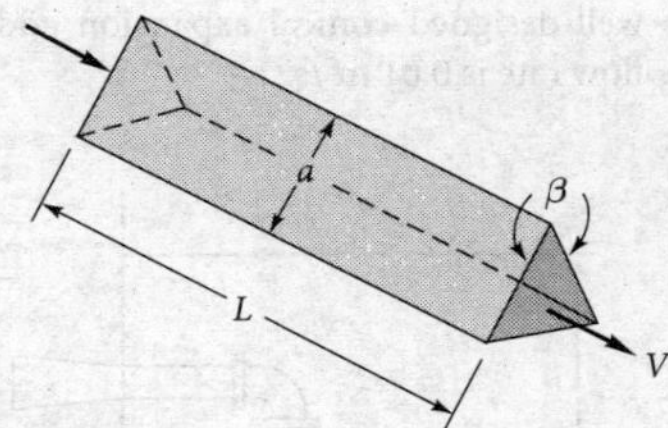

P6.91

P6.92 A large room uses a fan to draw in atmospheric air at 20°C through a 30-cm by 30-cm commercial-steel duct 12 m long, as in Fig. P6.92. Estimate (*a*) the airflow rate in m³/h if the room pressure is 10 Pa vacuum and (*b*) the room pressure if the flow rate is 1,200 m³/h. Neglect minor losses.

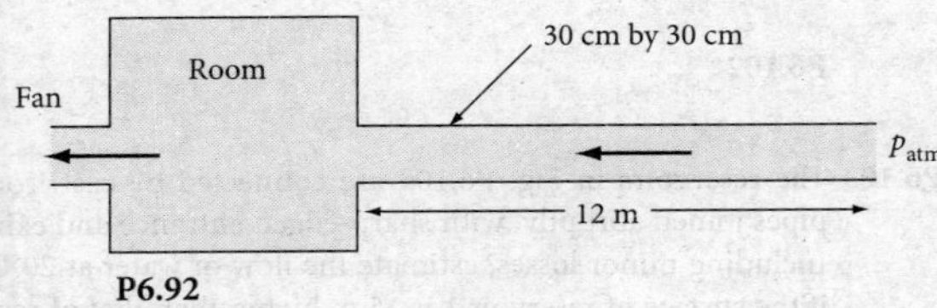

P6.92

P6.93 In Moody's Example 6.6, the 15 cm diameter, 60 m-long asphalted cast iron pipe has a pressure drop of about 13.4 kPa when the average water velocity is 2 m/s. Compare this to an *annular* cast iron pipe with an inner diameter of 15 cm and the same annular average velocity of 2 m/s. (*a*) What outer diameter would cause the flow to have the same pressure drop of 13.4 kPa? (*b*) How do the cross-section areas compare, and why? Use the hydraulic diameter approximation.

P6.94 Air at 20°C flows through a smooth duct of diameter 20 cm at an average velocity of 5 m/s. It then flows into a smooth square duct of side length *a*. Find the square duct size *a* for which the pressure drop per meter will be exactly the same as the circular duct.

P6.95 Although analytical solutions are available for laminar flow in many duct shapes [34], what do we do about ducts of arbitrary shape? Bahrami et al. [57] propose that a better approach to the pipe result, $f\text{Re} = 64$, is achieved by replacing the hydraulic diameter D_h with $\sqrt{A}$, where A is the area of the cross section. Test this idea for the isosceles triangles of Table 6.4. If time is short, at least try 10°, 50°, and 80°. What do you conclude about this idea?

P6.96 A fuel cell [59] consists of air (or oxygen) and hydrogen micro ducts, separated by a membrane that promotes proton exchange for an electric current, as in Fig. P6.96. Suppose that the air side, at 20°C and approximately 1 atm, has five 1 mm by 1 mm ducts, each 1 m long. The total flow rate is 1.5 E-4 kg/s. (*a*) Determine if the flow is laminar or turbulent. (*b*) Estimate the pressure drop. (*Problem courtesy of Dr. Pezhman Shirvanian.*)

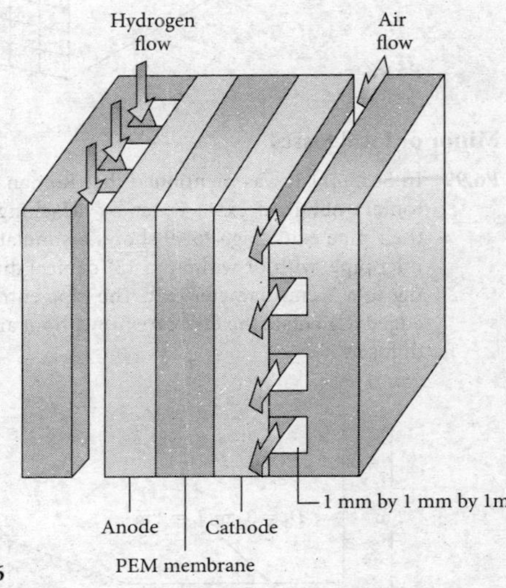

P6.96

P6.97 A heat exchanger consists of multiple parallel-plate passages, as shown in Fig. P6.97. The available pressure drop is 2 kPa, and the fluid is water at 20°C. If the desired total flow rate is 900 m³/h, estimate the appropriate number of passages. The plate walls are hydraulically smooth.

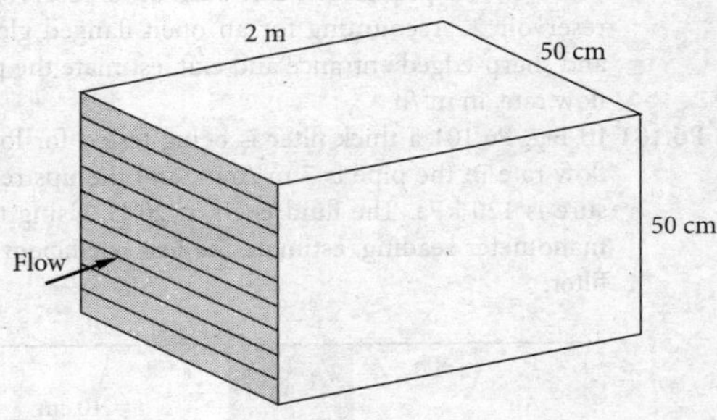

P6.97

P6.98 A rectangular heat exchanger is to be divided into smaller sections using sheets of commercial steel 0.4 mm thick, as sketched in Fig. P6.98. The flow rate is 20 kg/s of water at 20°C. Basic dimensions are $L = 1$ m, $W = 20$ cm, and $H = 10$ cm. What is the proper number of *square* sections if the overall pressure drop is to be no more than 1,600 Pa?

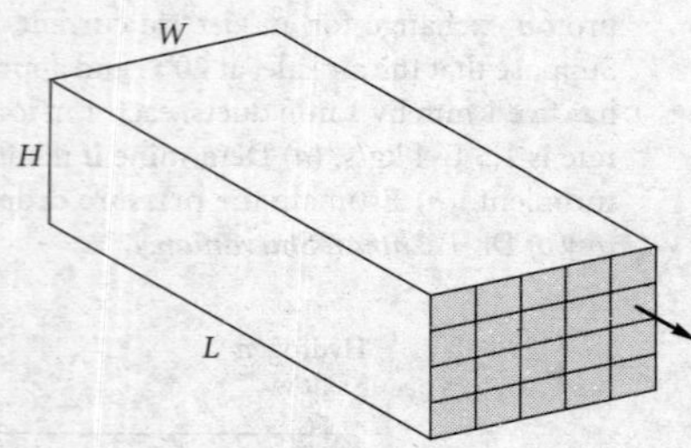

P6.98

Minor or local losses

P6.99 In Sec. 6.11 it was mentioned that Roman aqueduct customers obtained extra water by attaching a diffuser to their pipe exits. Fig. P6.99 shows a simulation: a smooth inlet pipe, with or without a 15° conical diffuser expanding to a 5-cm-diameter exit. The pipe entrance is sharp-edged. Calculate the flow rate (*a*) without and (*b*) with the diffuser.

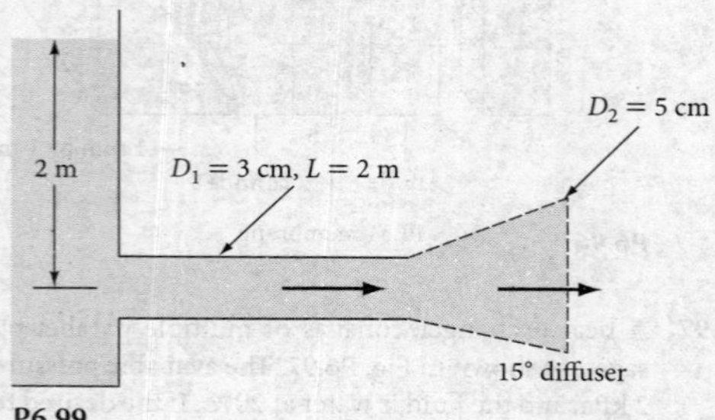

P6.99

***P6.100** Modify Prob. P6.55 as follows: Assume a pump can deliver 3 kW to pump the water back up to reservoir 1 from reservoir 2. Accounting for an open flanged globe valve and sharp-edged entrance and exit, estimate the predicted flow rate, in m^3/h.

P6.101 In Fig. P6.101 a thick filter is being tested for losses. The flow rate in the pipe is 7 m^3/min, and the upstream pressure is 120 kPa. The fluid is air at 20°C. Using the water manometer reading, estimate the loss coefficient *K* of the filter.

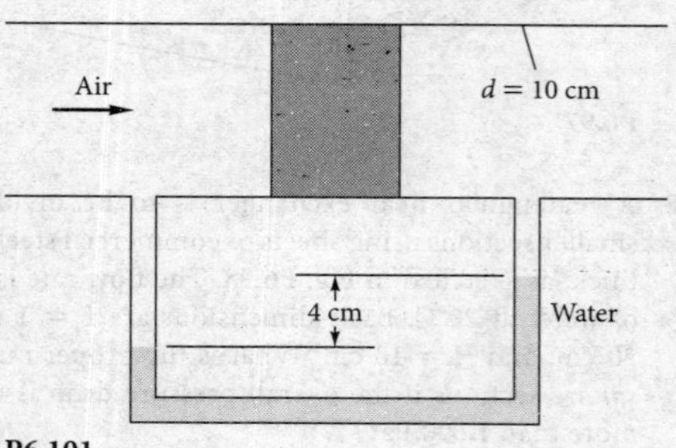

P6.101

***P6.102** A 70 percent efficient pump delivers water at 20°C from one reservoir to another 6 m higher, as in Fig. P6.102. The piping system consists of 20 m of galvanized iron 5 cm pipe, a reentrant entrance, two screwed 90° long-radius elbows, a screwed-open gate valve, and a sharp exit. What is the input power required in horsepower with and without a 6° well-designed conical expansion added to the exit? The flow rate is 0.01 m^3/s.

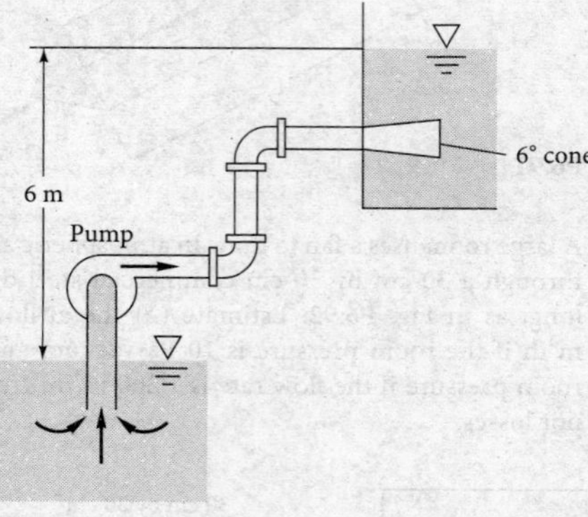

P6.102

P6.103 The reservoirs in Fig. P6.103 are connected by cast iron pipes joined abruptly, with sharp-edged entrance and exit. Including minor losses, estimate the flow of water at 20°C if the surface of reservoir 1 is 15 m higher than that of reservoir 2.

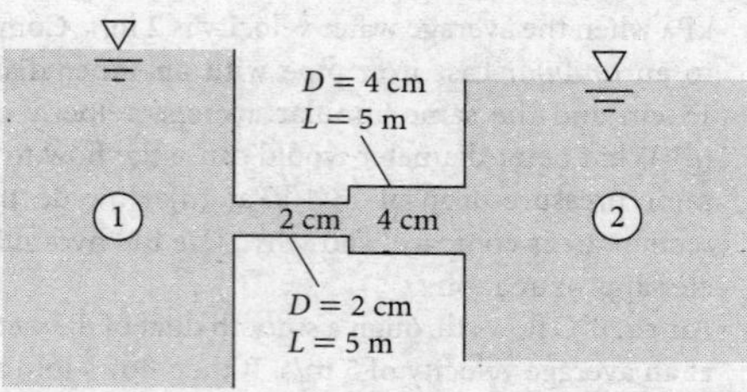

P6.103

P6.104 Consider a 20°C flow at 2 m/s through a smooth 3-mm diameter microtube which consists of a straight run of 10 cm, a long radius bend, and another straight run of 10 cm. Compute the total pressure drop if the fluid is (*a*) water; and (*b*) ethylene glycol.

P6.105 The system in Fig. P6.105 consists of 1,200 m of 5 cm cast iron pipe, two 45° and four 90° flanged long-radius elbows, a fully open flanged globe valve, and a sharp exit into a reservoir. If the elevation at point 1 is 400 m, what gage pressure is required at point 1 to deliver 0.005 m^3/s of water at 20°C into the reservoir?

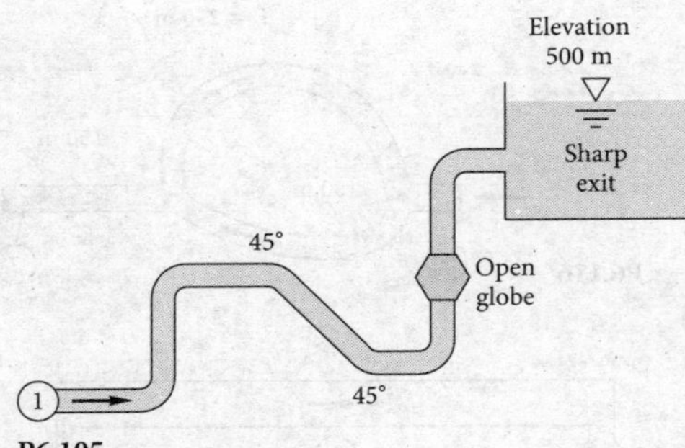

P6.105

P6.106 The water pipe in Fig. P6.106 slopes upward at 30°. The pipe has a 25 mm diameter and is smooth. The flanged globe valve is fully open. If the mercury manometer shows a 20 cm deflection, what is the flow rate in m^3/s?

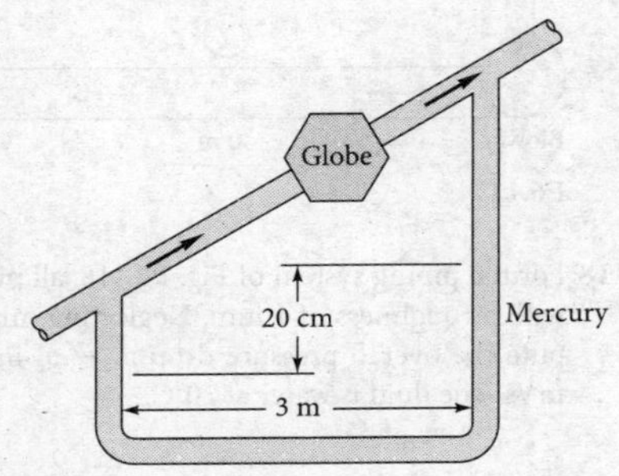

P6.106

***P6.107** A tank of water 4 m in diameter and 7 m deep is to be drained by a 5-cm-diameter exit pipe at the bottom, as in Fig. P6.107. In design (1), the pipe extends out for 1 m and into the tank for 10 cm. In design (2), the interior pipe is removed and the entrance beveled, Fig. 6.21, so that $K \approx 0.1$ in the entrance. (*a*) An engineer claims that design (2) will drain 25 percent faster than design (1). Is this claim true? (*b*) Estimate the time to drain of design (2), assuming $f \approx 0.020$.

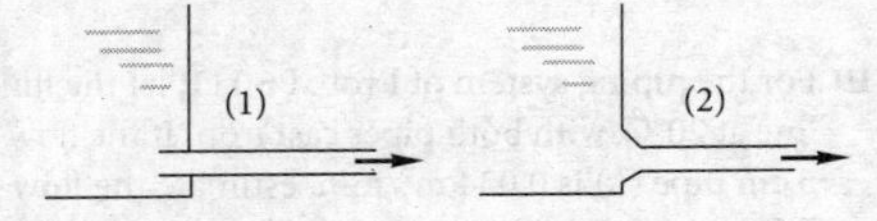

P6.107

P6.108 The water pump in Fig. P6.108 maintains a pressure of 45 kPa at point 1. There is a filter, a half-open disk valve, and two regular screwed elbows. There are 25 m of 10 cm diameter commercial steel pipe. (*a*) If the flow rate is 0.01 m^3/s, what is the loss coefficient of the filter? (*b*) If the disk valve is wide open and $K_{filter} = 7$, what is the resulting flow rate?

P6.109 In Fig. P6.109 there are 40 m of 5 cm pipe, 25 m of 15 cm pipe, and 45 m of 7.5 cm pipe, all cast iron. There are two 90° elbows and an open globe valve, all flanged. If the exit elevation is zero, what horsepower is extracted by the turbine when the flow rate is 0.005 m^3/s of water at 20°C?

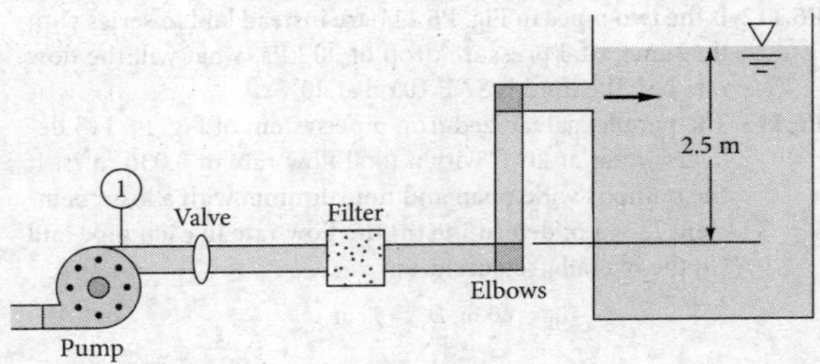

P6.108

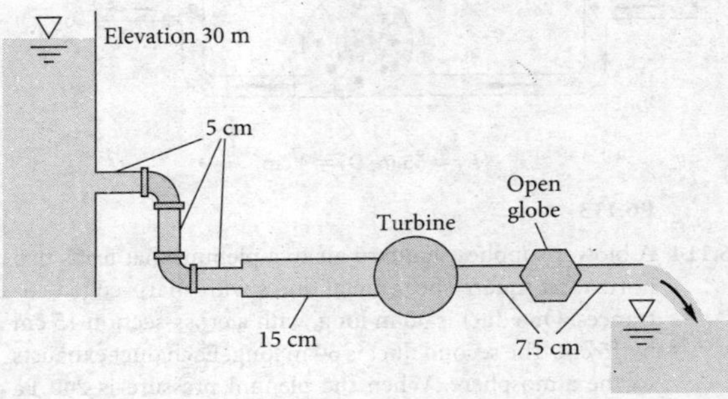

P6.109

P6.110 In Fig. P6.110 the pipe entrance is sharp-edged. If the flow rate is 0.004 m^3/s, what power, in W, is extracted by the turbine?

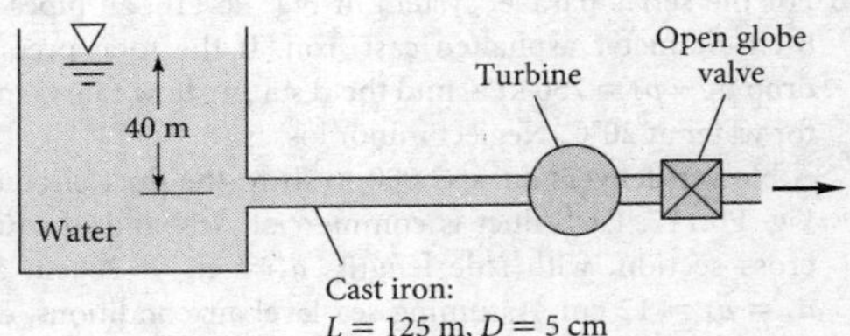

P6.110

Series and parallel pipe systems

P6.111 For the parallel-pipe system of Fig. P6.111, each pipe is cast iron, and the pressure drop $p_1 - p_2 = 20$ kPa. Compute the total flow rate between 1 and 2 if the fluid is SAE 10 oil at 20°C.

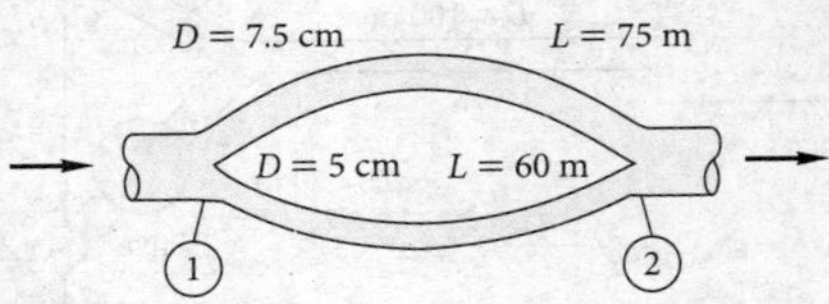

P6.111

P6.112 If the two pipes in Fig. P6.111 are instead laid in series with the same total pressure drop of 20 kPa, what will the flow rate be? The fluid is SAE 10 oil at 20°C.

P6.113 The parallel galvanized iron pipe system of Fig. P6.113 delivers water at 20°C with a total flow rate of 0.036 m^3/s. If the pump is wide open and not running, with a loss coefficient $K = 1.5$, determine (*a*) the flow rate in each pipe and (*b*) the overall pressure drop.

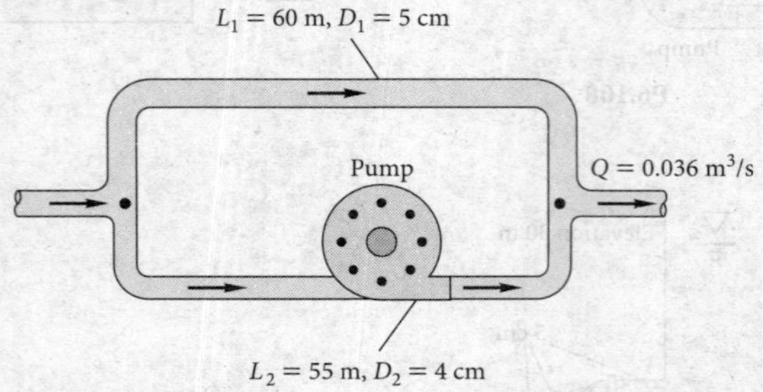

P6.113

***P6.114** A blower supplies standard air to a plenum that feeds two horizontal square sheet-metal ducts with sharp-edged entrances. One duct is 30 m long, with a cross-section 15 cm by 15 cm. The second duct is 60 m long. Each duct exhausts to the atmosphere. When the plenum pressure is 240 Pa (gage) the volume flow in the longer duct is three times the flow in the shorter duct. Estimate both volume flows and the cross-section size of the longer duct.

P6.115 In Fig. P6.115 all pipes are 8-cm-diameter cast iron. Determine the flow rate from reservoir 1 if valve *C* is (*a*) closed and (*b*) open, $K = 0.5$.

P6.116 For the series-parallel system of Fig. P6.116, all pipes are 8-cm-diameter asphalted cast iron. If the total pressure drop $p_1 - p_2 = 750$ kPa, find the resulting flow rate Q m^3/h for water at 20°C. Neglect minor losses.

P6.117 A blower delivers air at 3,000 m^3/h to the duct circuit in Fig. P6.117. Each duct is commercial steel and of square cross section, with side lengths $a_1 = a_3 = 20$ cm and $a_2 = a_4 = 12$ cm. Assuming sea-level air conditions, estimate the power required if the blower has an efficiency of 75 percent. Neglect minor losses.

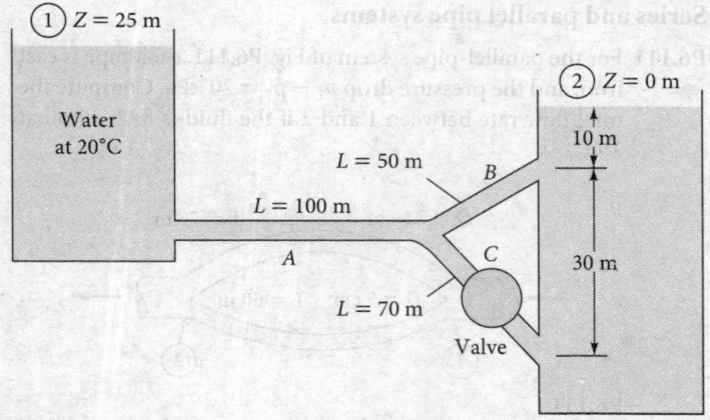

P6.115

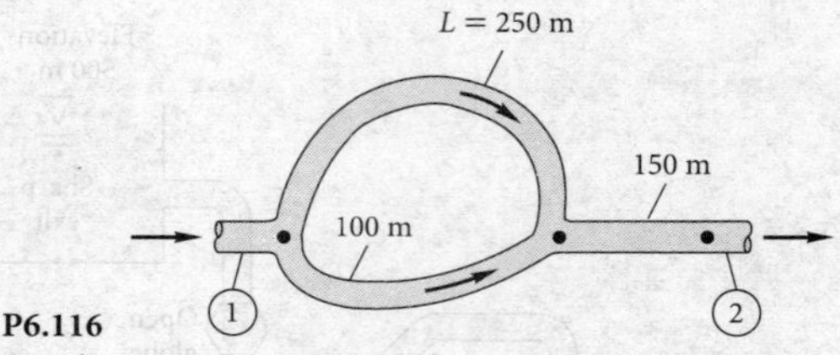

P6.116

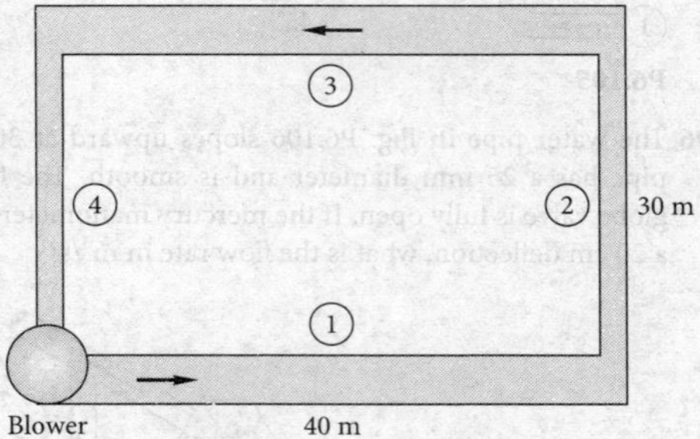

P6.117

P6.118 For the piping system of Fig. P6.118, all pipes are concrete with a roughness of 1 mm. Neglecting minor losses, compute the overall pressure drop $p_1 - p_2$ in kPa if $Q = 0.6$ m^3/s. The fluid is water at 20°C.

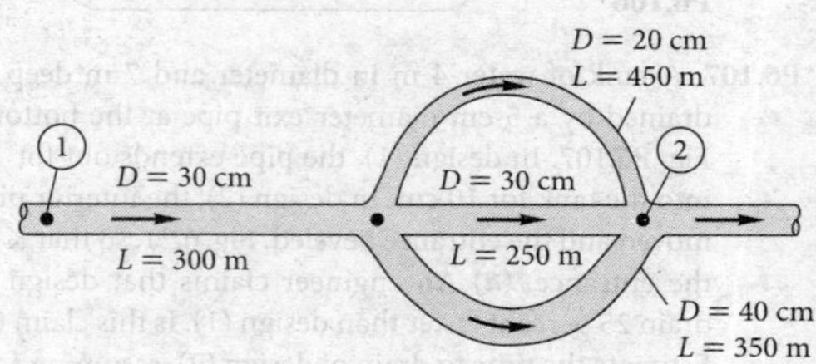

P6.118

P6.119 For the piping system of Prob. P6.111, let the fluid be gasoline at 20°C, with both pipes cast iron. If the flow rate in the 5 cm pipe (*b*) is 0.034 m^3/min, estimate the flow rate in the 7.5 cm pipe (*a*), in m^3/min.

P6.120 Three cast iron pipes are laid in parallel with these dimensions:

Pipe	Length, m	Diameter, cm
1	800	12
2	600	8
3	900	10

The total flow rate is 200 m^3/h of water at 20°C. Determine (*a*) the flow rate in each pipe and (*b*) the pressure drop across the system.

Three-reservoir and pipe network systems

P6.121 Consider the three-reservoir system of Fig. P6.121 with the following data:

$$L_1 = 95\text{m} \quad L_2 = 125 \text{ m} \quad L_3 = 160 \text{ m}$$

$$z_1 = 25 \text{ m} \quad z_2 = 115 \text{ m} \quad z_3 = 85 \text{ m}$$

All pipes are 28-cm-diameter unfinished concrete (ε = 1 mm). Compute the steady flow rate in all pipes for water at 20°C.

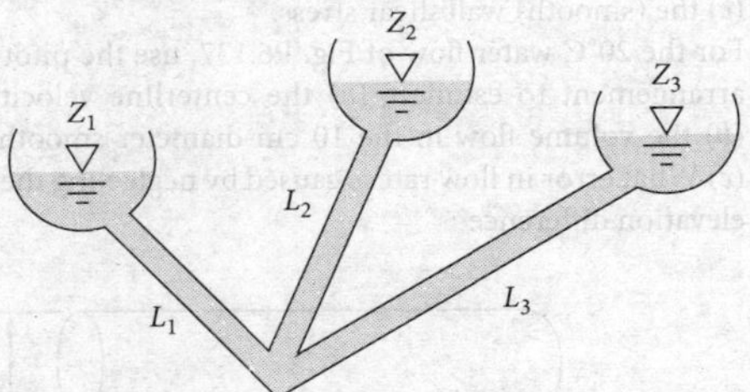

P6.121

P6.122 Modify Prob. P6.121 as follows: Reduce the diameter to 15 cm (with ε = 1 mm), and compute the flow rates for water at 20°C. These flow rates distribute in nearly the same manner as in Prob. P6.121 but are about 5.2 times lower. Can you explain this difference?

P6.123 Modify Prob. P6.121 as follows: All data are the same except that z_3 is unknown. Find the value of z_3 for which the flow rate in pipe 3 is 0.2 m³/s toward the junction. (This problem requires iteration and is best suited to a computer.)

P6.124 The three-reservoir system in Fig. P6.124 delivers water at 20°C. The system data are as follows:

$$D_1 = 20 \text{ cm} \quad D_2 = 15 \text{ cm} \quad D_3 = 23 \text{ cm}$$

$$L_1 = 540 \text{ m} \quad L_2 = 360 \text{ m} \quad L_3 = 480 \text{ m}$$

All pipes are galvanized iron. Compute the flow rate in all pipes.

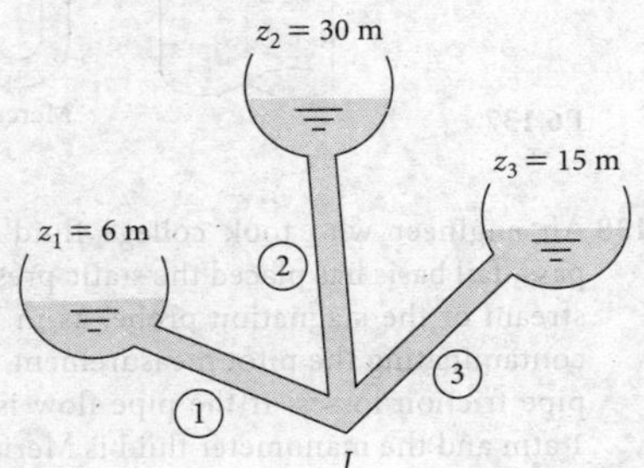

P6.124

P6.125 Suppose that the three cast iron pipes in Prob. P6.120 are instead connected to meet smoothly at a point *B*, as shown in Fig. P6.125. The inlet pressures in each pipe are

$$p_1 = 200 \text{ kPa} \quad p_2 = 160 \text{ kPa} \quad p_3 = 100 \text{ kPa}.$$

The fluid is water at 20°C. Neglect minor losses. Estimate the flow rate in each pipe and whether it is toward or away from point *B*.

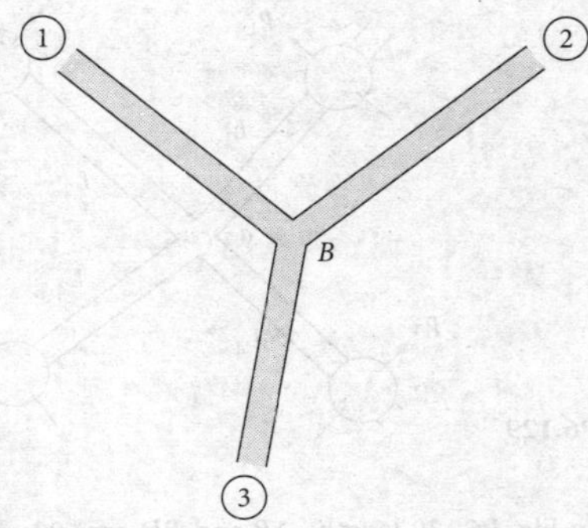

P6.125

P6.126 Modify Prob. P6.124 as follows: Let all data be the same except that pipe 1 is fitted with a butterfly valve (Fig. 6.19*b*). Estimate the proper valve opening angle (in degrees) for the flow rate through pipe 1 to be reduced to 0.04 m³/s toward reservoir 1. (This problem requires iteration and is best suited to a computer.)

P6.127 In the five-pipe horizontal network of Fig. P6.127, assume that all pipes have a friction factor f = 0.025. For the given inlet and exit flow rate of 0.05 m³/s of water at 20°C, determine the flow rate and direction in all pipes. If p_A = 800 kPa gage, determine the pressures at points *B*, *C*, and *D*.

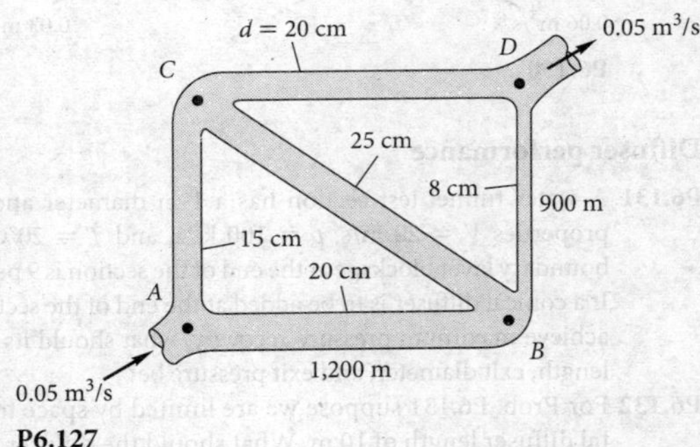

P6.127

P6.128 Modify Prob. P6.127 as follows: Let the inlet flow rate at *A* and the exit flow at *D* be unknown. Let $p_A - p_B$ = 700 kPa. Compute the flow rate in all five pipes.

P6.129 In Fig. P6.129 all four horizontal cast iron pipes are 45 m long and 8 cm in diameter and meet at junction *a*, delivering water at 20°C. The pressures are known at four points as shown:

$$p_1 = 950 \text{ kPa} \quad p_2 = 350 \text{ kPa}$$

$$p_3 = 675 \text{ kPa} \quad p_4 = 100 \text{ kPa}$$

Neglecting minor losses, determine the flow rate in each pipe.

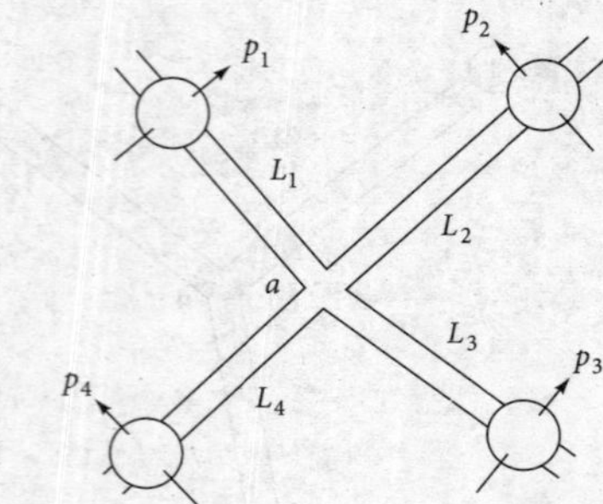

P6.129

P6.130 In Fig. P6.130 lengths AB and BD are 600 and 450 m, respectively. The friction factor is 0.022 everywhere, and p_A = 620 kPa gage. All pipes have a diameter of 15 cm. For water at 20°C, determine the flow rate in all pipes and the pressures at points B, C, and D.

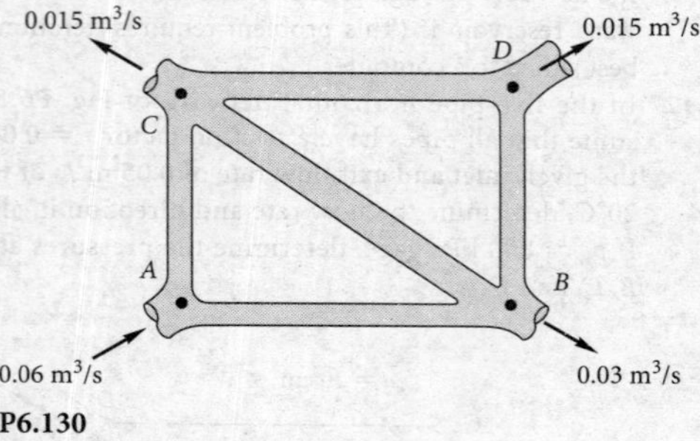

P6.130

Diffuser performance

P6.131 A water tunnel test section has a 1-m diameter and flow properties V = 20 m/s, p = 100 kPa, and T = 20°C. The boundary layer blockage at the end of the section is 9 percent. If a conical diffuser is to be added at the end of the section to achieve maximum pressure recovery, what should its angle, length, exit diameter, and exit pressure be?

P6.132 For Prob. P6.131 suppose we are limited by space to a total diffuser length of 10 m. What should the diffuser angle, exit diameter, and exit pressure be for maximum recovery?

P6.133 A wind tunnel test section is 1 m square with flow properties V = 45 m/s, p = 103 kPa absolute, and T = 20°C. Boundary layer blockage at the end of the test section is 8 percent. Find the angle, length, exit height, and exit pressure of a flat-walled diffuser added onto the section to achieve maximum pressure recovery.

P6.134 For Prob. P6.133 suppose we are limited by space to a total diffuser length of 10 m. What should the diffuser angle, exit height, and exit pressure be for maximum recovery?

The pitot-static tube

P6.135 An airplane uses a pitot-static tube as a velocimeter. The measurements, with their uncertainties, are a static temperature of (−11 ± 3)°C, a static pressure of 60 ± 2 kPa, and a pressure difference ($p_o - p_s$) = 3,200 ± 60 Pa. (a) Estimate the airplane's velocity and its uncertainty. (b) Is a compressibility correction needed?

P6.136 For the pitot-static pressure arrangement of Fig. P6.136, the manometer fluid is (colored) water at 20°C. Estimate (a) the centerline velocity, (b) the pipe volume flow, and (c) the (smooth) wall shear stress.

P6.137 For the 20°C water flow of Fig. P6.137, use the pitot-static arrangement to estimate (a) the centerline velocity and (b) the volume flow in the 10 cm-diameter smooth pipe. (c) What error in flow rate is caused by neglecting the 0.3 m elevation difference?

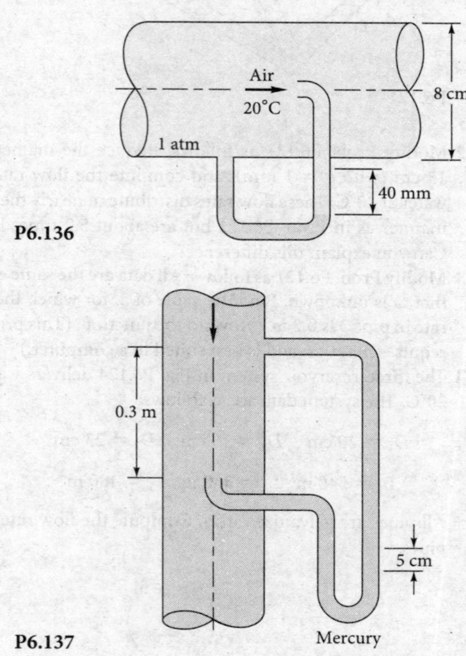

P6.136

P6.137

P6.138 An engineer who took college fluid mechanics on a pass–fail basis has placed the static pressure hole far upstream of the stagnation probe, as in Fig. P6.138, thus contaminating the pitot measurement ridiculously with pipe friction losses. If the pipe flow is air at 20°C and 1 atm and the manometer fluid is Meriam red oil (SG = 0.827), estimate the air centerline velocity for the given manometer reading of 16 cm. Assume a smooth-walled tube.

P6.139 Professor Walter Tunnel needs to measure the flow velocity in a water tunnel. Due to budgetary restrictions, he cannot afford a pitot-static probe, but instead inserts a total head probe and a static pressure probe, as shown in Fig. P6.139, a distance h_1 apart from each other. Both probes are in the main free stream of the water tunnel, unaffected by the thin boundary layers on the sidewalls. The two probes are connected as shown to a U-tube manometer. The densities and vertical distances are shown in Fig. P6.139. (*a*) Write an expression for velocity V in terms of the parameters in the problem. (*b*) Is it critical that distance h_1 be measured accurately? (*c*) How does the expression for velocity V differ from that which would be obtained if a pitot-static probe had been available and used with the same U-tube manometer?

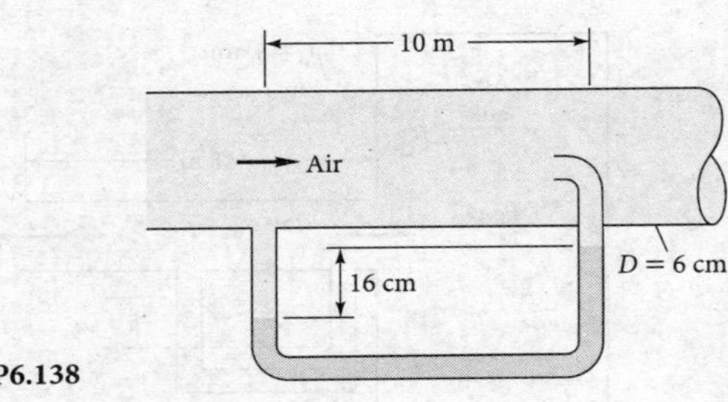

P6.138

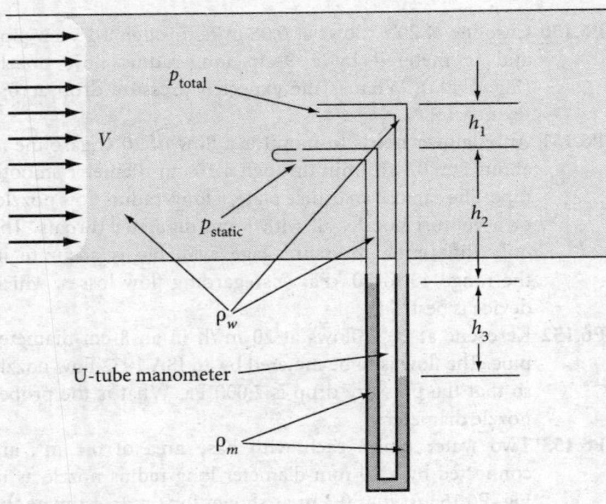

P6.139

Flowmeters: the orifice plate

P6.140 Gasoline at 20°C flows at 3 m³/h in a 6-cm-diameter pipe. A 4-cm-diameter thin-plate orifice with corner taps is installed. Estimate the measured pressure drop, in Pa.

P6.141 Gasoline at 20°C flows at 105 m³/h in a 10-cm-diameter pipe. We wish to meter the flow with a thin-plate orifice and a differential pressure transducer that reads best at about 55 kPa. What is the proper β ratio for the orifice?

P6.142 The shower head in Fig. P6.142 delivers water at 50°C. An orifice-type flow reducer is to be installed. The upstream pressure is constant at 400 kPa. What flow rate, in m³/min, results without the reducer? What reducer orifice diameter would decrease the flow by 40 percent?

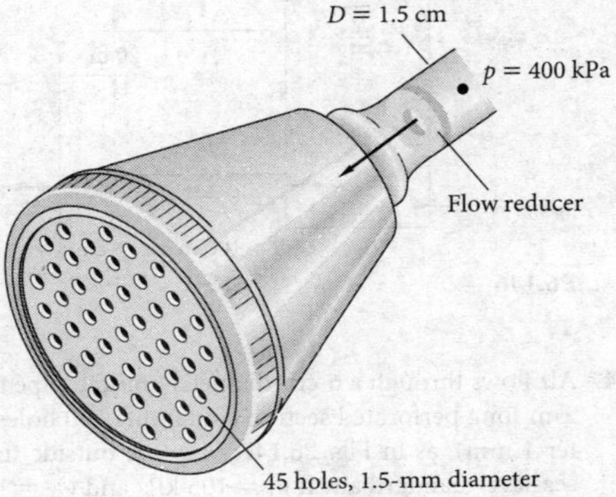

P6.142

P6.143 A 10-cm-diameter smooth pipe contains an orifice plate with D: $\frac{1}{2}$D taps and $\beta = 0.5$. The measured orifice pressure drop is 75 kPa for water flow at 20°C. Estimate the flow rate, in m³/h. What is the nonrecoverable head loss?

***P6.144** Water at 20°C flows through the orifice in Fig. P6.154, which is monitored by a mercury manometer. If $d = 3$ cm, (*a*) what is h when the flow rate is 20 m³/h and (*b*) what is Q in m³/h when $h = 58$ cm?

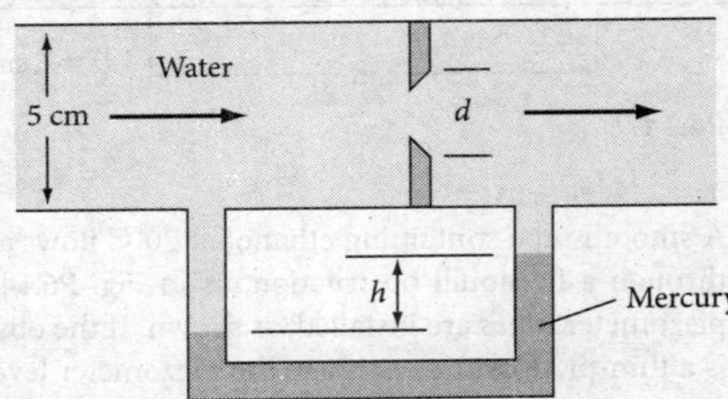

P6.144

P6.145 The 1-m-diameter tank in Fig. P6.145 is initially filled with gasoline at 20°C. There is a 2-cm-diameter orifice in the bottom. If the orifice is suddenly opened, estimate the time for the fluid level $h(t)$ to drop from 2.0 to 1.6 m.

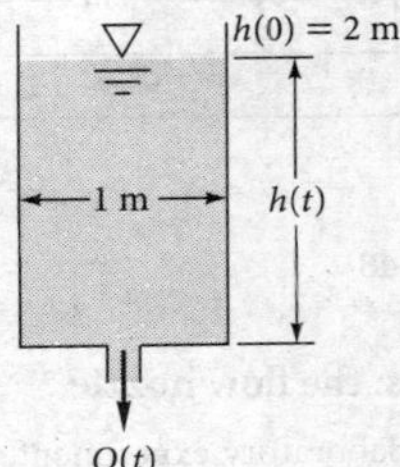

P6.145

P6.146 A pipe connecting two reservoirs, as in Fig. P6.146, contains a thin-plate orifice. For water flow at 20°C, estimate (*a*) the volume flow through the pipe and (*b*) the pressure drop across the orifice plate.

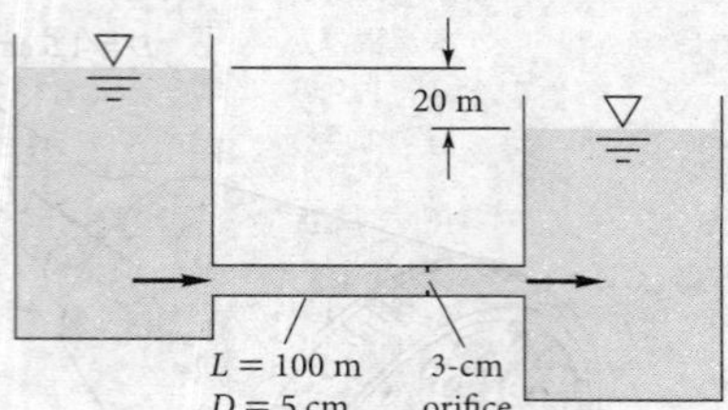

P6.146

P6.147 Air flows through a 6-cm-diameter smooth pipe that has a 2-m-long perforated section containing 500 holes (diameter 1 mm), as in Fig. P6.147. Pressure outside the pipe is sea-level standard air. If $p_1 = 105$ kPa and $Q_1 = 110$ m^3/h, estimate p_2 and Q_2, assuming that the holes are approximated by thin-plate orifices. (*Hint:* A momentum control volume may be very useful.)

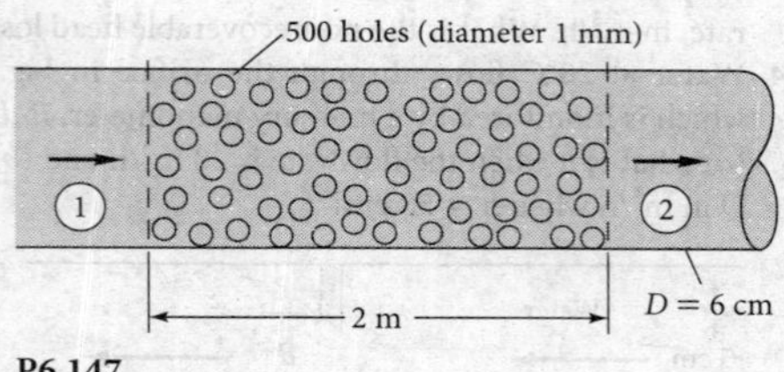

P6.147

P6.148 A smooth pipe containing ethanol at 20°C flows at 7 m^3/h through a Bernoulli obstruction, as in Fig. P6.148. Three piezometer tubes are installed, as shown. If the obstruction is a thin-plate orifice, estimate the piezometer levels (*a*) h_2 and (*b*) h_3.

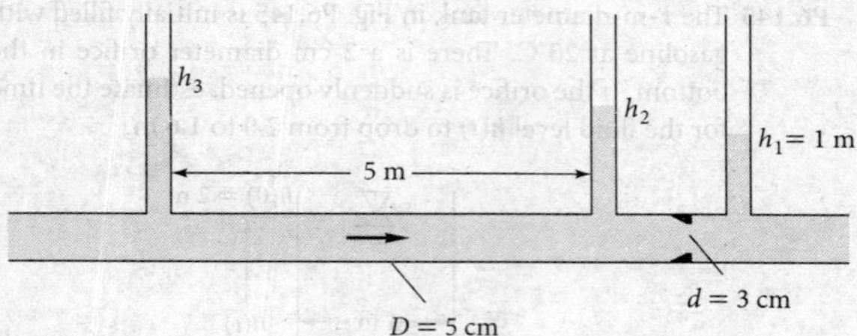

P6.148

Flowmeters: the flow nozzle

P6.149 In a laboratory experiment, air at 20°C flows from a large tank through a 2-cm-diameter smooth pipe into a sea-level atmosphere, as in Fig. P6.149. The flow is metered by a long-radius nozzle of 1-cm diameter, using a manometer with Meriam red oil (SG = 0.827). The pipe is 8 m long. The measurements of tank pressure and manometer height are as follows:

p_{tank}, Pa(gage):	60	320	1,200	2,050	2,470	3,500	4,900
h_{mano}, mm:	6	38	160	295	380	575	820

Use these data to calculate the flow rates Q and Reynolds numbers Re_D and make a plot of measured flow rate versus tank pressure. Is the flow laminar or turbulent? Compare the data with theoretical results obtained from the Moody chart, including minor losses. Discuss.

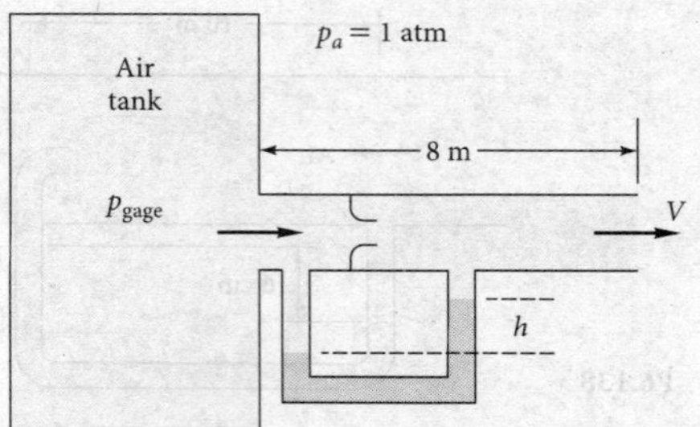

P6.149

P6.150 Gasoline at 20°C flows at 0.06 m^3/s through a 15-cm pipe and is metered by a 9-cm long-radius flow nozzle (Fig. 6.40*a*). What is the expected pressure drop across the nozzle?

P6.151 An engineer needs to monitor a flow of 20°C gasoline at about 1 ± 0.1 m^3/min through a 10 cm-diameter smooth pipe. She can use an orifice plate, a long-radius flow nozzle, or a venturi nozzle, all with 5 cm-diameter throats. The only differential pressure gage available is accurate in the range 40 to 70 kPa. Disregarding flow losses, which device is best?

P6.152 Kerosene at 20°C flows at 20 m^3/h in an 8-cm-diameter pipe. The flow is to be metered by an ISA 1932 flow nozzle so that the pressure drop is 7,000 Pa. What is the proper nozzle diameter?

P6.153 Two water tanks, each with base area of 0.1 m^2, are connected by a 10 mm-diameter long-radius nozzle as in Fig. P6.153. If $h = 0.3$ m as shown for $t = 0$, estimate the time for $h(t)$ to drop to 0.08 m.

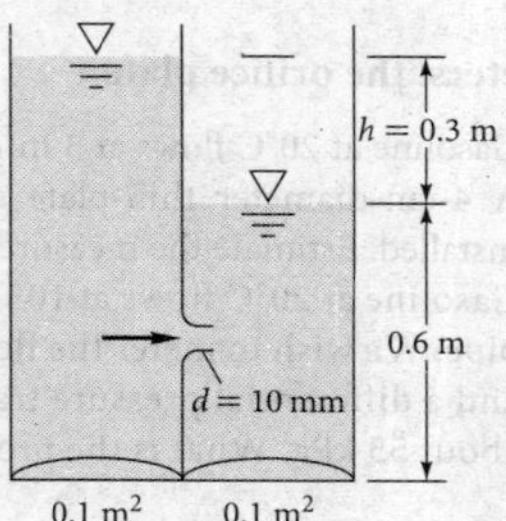

P6.153

Flowmeters: the venturi meter

P6.154 Gasoline at 20°C flows through a 6-cm-diameter pipe. It is metered by a modern venturi nozzle with $d = 4$ cm. The measured pressure drop is 8.5 kPa. Estimate the flow rate in m^3/min.

P6.155 It is desired to meter methanol at 20°C flowing through a 15 cm-diameter pipe. The expected flow rate is about 20 L/s. Two flowmeters are available: a venturi nozzle and a thinplate orifice, each with $d = 5$ cm. The differential pressure gage on hand is most accurate at about 80–105 kPa. Which meter is better for this job?

P6.156 Ethanol at 20°C flows down through a modern venturi nozzle as in Fig. P6.156. If the mercury manometer reading is 10 cm, as shown, estimate the flow rate, in m^3/min.

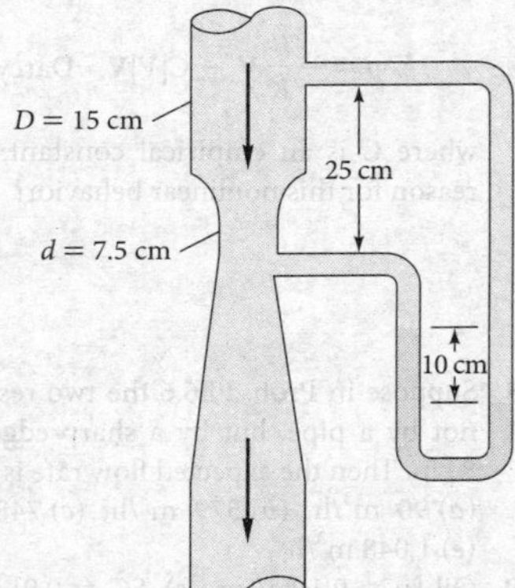

P6.156

P6.157 Modify Prob. P6.156 if the fluid is air at 20°C, entering the venturi at a pressure of 124 kPa. Should a compressibility correction be used?

P6.158 Water at 20°C flows in a long horizontal commercial steel 6-cm-diameter pipe that contains a classical Herschel venturi with a 4-cm throat. The venturi is connected to a mercury manometer whose reading is $h = 40$ cm. Estimate (*a*) the flow rate, in m^3/h, and (*b*) the total pressure difference between points 50 cm upstream and 50 cm downstream of the venturi.

P6.159 A modern venturi nozzle is tested in a laboratory flow with water at 20°C. The pipe diameter is 5.5 cm, and the venturi throat diameter is 3.5 cm. The flow rate is measured by a weigh tank and the pressure drop by a water–mercury manometer. The mass flow rate and manometer readings are as follows:

$\dot{m}$, kg/s	0.95	1.98	2.99	5.06	8.15
h, mm	3.7	15.9	36.2	102.4	264.4

Use these data to plot a calibration curve of venturi discharge coefficient versus Reynolds number. Compare with the accepted correlation, Eq. (6.114).

Flowmeters: other designs

P6.160 An instrument popular in the beverage industry is the *target flowmeter* in Fig. P6.160. A small flat disk is mounted in the center of the pipe, supported by a strong but thin rod. (*a*) Explain how the flowmeter works. (*b*) If the bending moment M of the rod is measured at the wall, derive a formula for the estimated velocity of the flow. (*c*) List a few advantages and disadvantages of such an instrument.

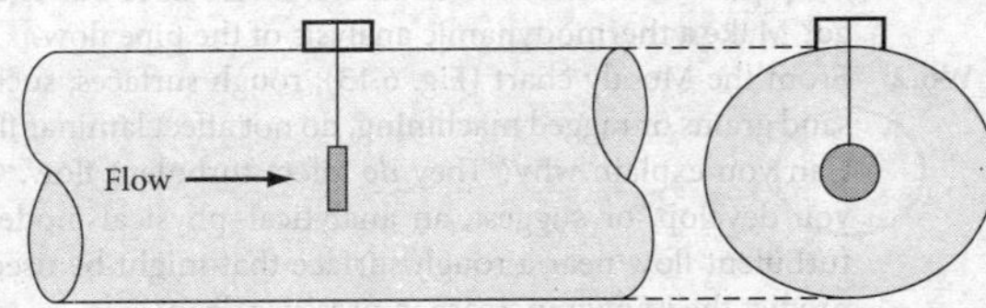

P6.160

P6.161 An instrument popular in the water supply industry, sketched in Fig. P6.161, is the single jet water meter. (*a*) How does it work? (*b*) What do you think a typical calibration curve would look like? (*c*) Can you cite further details, for example, reliability, head loss, cost [58]?

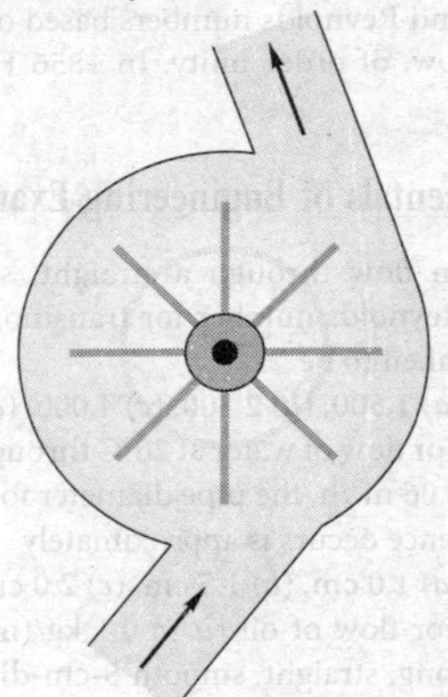

P6.161

Flowmeters: compressibility correction

P6.162 Air flows at high speed through a Herschel venturi monitored by a mercury manometer, as shown in Fig. P6.162. The upstream conditions are 150 kPa and 80°C. If $h = 37$ cm, estimate the mass flow in kg/s. (*Hint:* The flow is compressible.)

P6.163 Modify Prob. P6.162 as follows: Find the manometer reading h for which the mass flow through the venturi is approximately 0.4 kg/s. (*Hint:* The flow is compressible.)

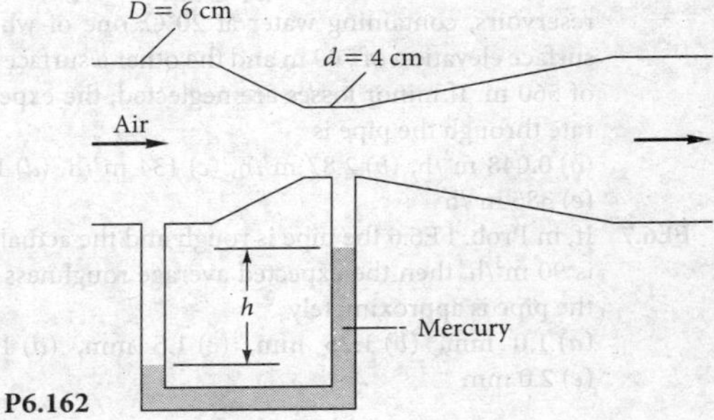

P6.162

Word Problems

W6.1 In fully developed straight-duct flow, the velocity profiles do not change (why?), but the pressure drops along the pipe axis. Thus there is pressure work done on the fluid. If, say, the pipe is insulated from heat loss, where does this energy go? Make a thermodynamic analysis of the pipe flow.

W6.2 From the Moody chart (Fig. 6.13), rough surfaces, such as sand grains or ragged machining, do not affect laminar flow. Can you explain why? They *do* affect turbulent flow. Can you develop, or suggest, an analytical–physical model of turbulent flow near a rough surface that might be used to predict the known increase in pressure drop?

W6.3 Differentiation of the laminar pipe flow solution, Eq. (6.40), shows that the fluid shear stress $\tau(r)$ varies linearly from zero at the axis to τ_w at the wall. It is claimed that this is also true, at least in the time mean, for fully developed *turbulent* flow. Can you verify this claim analytically?

W6.4 A porous medium consists of many tiny tortuous passages, and Reynolds numbers based on pore size are usually very low, of order unity. In 1856 H. Darcy proposed that the pressure gradient in a porous medium was directly proportional to the volume-averaged velocity **V** of the fluid:

$$\nabla p = -\frac{\mu}{K}\mathbf{V}$$

where K is termed the *permeability* of the medium. This is now called *Darcy's law* of porous flow. Can you make a Poiseuille flow model of porous-media flow that verifies Darcy's law? Meanwhile, as the Reynolds number increases, so that $VK^{1/2}/\nu > 1$, the pressure drop becomes nonlinear, as was shown experimentally by P. H. Forscheimer as early as 1782. The flow is still decidedly laminar, yet the pressure gradient is quadratic:

$$\nabla p = -\frac{\mu}{K}\mathbf{V} - C|V|\mathbf{V} \quad \text{Darcy-Forscheimer law}$$

where C is an empirical constant. Can you explain the reason for this nonlinear behavior?

Fundamentals of Engineering Exam Problems

FE6.1 In flow through a straight, smooth pipe, the diameter Reynolds number for transition to turbulence is generally taken to be
(*a*) 1,500, (*b*) 2,300, (*c*) 4,000, (*d*) 250,000, (*e*) 500,000

FE6.2 For flow of water at 20°C through a straight, smooth pipe at 0.06 m^3/h, the pipe diameter for which transition to turbulence occurs is approximately
(*a*) 1.0 cm, (*b*) 1.5 cm, (*c*) 2.0 cm, (*d*) 2.5 cm, (*e*) 3.0 cm

FE6.3 For flow of oil [$\mu = 0.1$ kg/(m · s), SG = 0.9] through a long, straight, smooth 5-cm-diameter pipe at 14 m^3/h, the pressure drop per meter is approximately
(*a*) 2,200 Pa, (*b*) 2,500 Pa, (*c*) 10,000 Pa, (*d*) 160 Pa, (*e*) 2,800 Pa

FE6.4 For flow of water at a Reynolds number of 1.03 E6 through a 5-cm-diameter pipe of roughness height 0.5 mm, the approximate Moody friction factor is
(*a*) 0.012, (*b*) 0.018, (*c*) 0.038, (*d*) 0.049, (*e*) 0.102

FE6.5 Minor losses through valves, fittings, bends, contractions, and the like are commonly modeled as proportional to
(*a*) total head, (*b*) static head, (*c*) velocity head, (*d*) pressure drop, (*e*) velocity

FE6.6 A smooth 8-cm-diameter pipe, 200 m long, connects two reservoirs, containing water at 20°C, one of which has a surface elevation of 700 m and the other a surface elevation of 560 m. If minor losses are neglected, the expected flow rate through the pipe is
(*a*) 0.048 m^3/h, (*b*) 2.87 m^3/h, (*c*) 134 m^3/h, (*d*) 172 m^3/h, (*e*) 385 m^3/h

FE6.7 If, in Prob. FE6.6 the pipe is rough and the actual flow rate is 90 m^3/h, then the expected average roughness height of the pipe is approximately
(*a*) 1.0 mm, (*b*) 1.25 mm, (*c*) 1.5 mm, (*d*) 1.75 mm, (*e*) 2.0 mm

FE6.8 Suppose in Prob. FE6.6 the two reservoirs are connected, not by a pipe, but by a sharp-edged orifice of diameter 8 cm. Then the expected flow rate is approximately
(*a*) 90 m^3/h, (*b*) 579 m^3/h, (*c*) 748 m^3/h, (*d*) 949 m^3/h, (*e*) 1,048 m^3/h

FE6.9 Oil [$\mu = 0.1$ kg/(m · s), SG = 0.9] flows through a 50-m-long smooth 8-cm-diameter pipe. The maximum pressure drop for which laminar flow is expected is approximately
(*a*) 30 kPa, (*b*) 40 kPa, (*c*) 50 kPa, (*d*) 60 kPa, (*e*) 70 kPa

FE6.10 Air at 20°C and approximately 1 atm flows through a smooth 30-cm-square duct at 42.5 m^3/min. The expected pressure drop per meter of duct length is
(*a*) 1.0 Pa, (*b*) 2.0 Pa, (*c*) 3.0 Pa, (*d*) 4.0 Pa, (*e*) 5.0 Pa

FE6.11 Water at 20°C flows at 3 m^3/h through a sharp-edged 3-cm-diameter orifice in a 6-cm-diameter pipe. Estimate the expected pressure drop across the orifice.
(*a*) 440 Pa, (*b*) 680 Pa, (*c*) 875 Pa, (*d*) 1,750 Pa, (*e*) 1,870 Pa

FE6.12 Water flows through a straight 10-cm-diameter pipe at a diameter Reynolds number of 250,000. If the pipe roughness is 0.06 mm, what is the approximate Moody friction factor?
(*a*) 0.015, (*b*) 0.017, (*c*) 0.019, (*d*) 0.026, (*e*) 0.032

FE6.13 What is the hydraulic diameter of a rectangular air-ventilation duct whose cross section is 1 m by 25 cm?
(*a*) 25 cm, (*b*) 40 cm, (*c*) 50 cm, (*d*) 75 cm, (*e*) 100 cm

FE6.14 Water at 20°C flows through a pipe at 19 L/s with a friction head loss of 13.5 m. What is the power required to drive this flow?
(*a*) 0.16 kW, (*b*) 1.88 kW, (*c*) 2.54 kW, (*d*) 3.41 kW, (*e*) 4.24 kW

FE6.15 Water at 20°C flows at 12 L/s through a pipe 150 m long and 8 cm in diameter. If the friction head loss is 12 m, what is the Moody friction factor?
(*a*) 0.010, (*b*) 0.015, (*c*) 0.020, (*d*) 0.025, (*e*) 0.030

Comprehensive Problems

C6.1 A pitot-static probe will be used to measure the velocity distribution in a water tunnel at 20°C. The two pressure lines from the probe will be connected to a U-tube manometer that uses a liquid of specific gravity 1.7. The maximum velocity expected in the water tunnel is 2.3 m/s. Your job is to select an appropriate U-tube from a manufacturer that supplies manometers of heights 20, 30, 40, 60, and 90 cm. The cost increases significantly with manometer height. Which of these should you purchase?

***C6.2** A pump delivers a steady flow of water (ρ, μ) from a large tank to two other higher-elevation tanks, as shown in Fig. C6.2. The same pipe of diameter d and roughness ε is used throughout. All minor losses *except through the valve* are neglected, and the partially closed valve has a loss coefficient K_{valve}. Turbulent flow may be assumed with all kinetic energy flux correction coefficients equal to 1.06. The pump net head H is a known function of Q_A and hence also of $V_A = Q_A/A_{pipe}$; for example, $H = a - bV_A^2$, where a and b are constants. Subscript J refers to the junction point at the tee where branch A splits into B and C. Pipe length L_C is much longer than L_B. It is desired to predict the pressure at J, the three pipe velocities and friction factors, and the pump head. Thus there are eight variables: H, V_A, V_B, V_C, f_A, f_B, f_C, p_J. Write down the eight equations needed to resolve this problem, but *do not solve,* since an elaborate iteration procedure would be required.

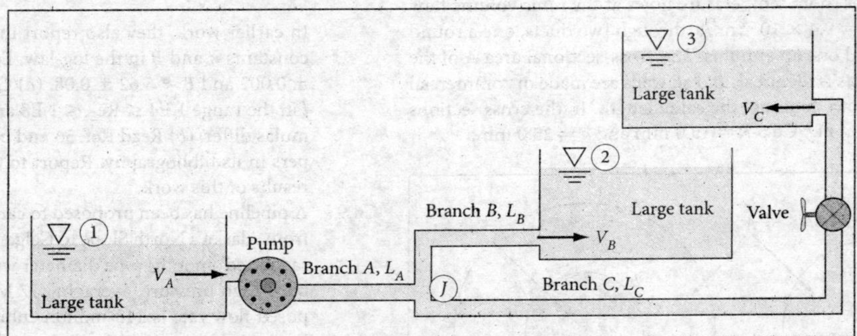

C6.2

C6.3 A small water slide is to be installed inside a swimming pool. See Fig. C6.3. The slide manufacturer recommends a continuous water flow rate Q of 1.39×10^{-3} m^3/s down the slide, to ensure that the customers do not burn their bottoms. A pump is to be installed under the slide, with a 5.00-m-long, 4.00-cm-diameter hose supplying swimming pool water for the slide. The pump is 80 percent efficient and will rest fully submerged 1.00 m below the water surface. The roughness inside the hose is about 0.0080 cm. The hose discharges the water at the top of the slide as a free jet open to the atmosphere. The hose outlet is 4.00 m above the water surface. For fully developed turbulent pipe flow, the kinetic energy flux correction factor is about 1.06. Ignore any minor losses here. Assume that $\rho = 998$ kg/m^3 and $\nu = 1.00 \times 10^{-6}$ m^2/s for this water. Find the brake horsepower (that is, the actual shaft power in watts) required to drive the pump.

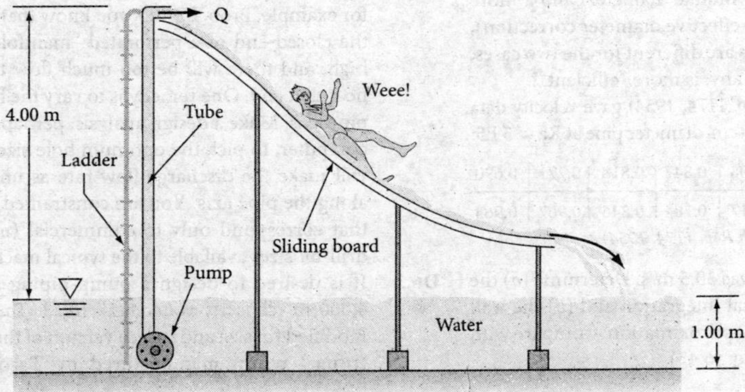

C6.3

***C6.4** Suppose you build a rural house where you need to run a pipe to the nearest water supply, which is fortunately at an elevation of about 1,000 m above that of your house. The pipe will be 6.0 km long (the distance to the water supply), and the gage pressure at the water supply is 1,000 kPa. You require a minimum of 0.2 L/s of water when the end of your pipe is open to the atmosphere. To minimize cost, you want to buy the smallest-diameter pipe possible. The pipe you will use is extremely smooth. (*a*) Find the total head loss from the pipe inlet to its exit. Neglect any minor losses due to valves, elbows, entrance lengths, and so on, since the length is so long here and major losses dominate. Assume the outlet of the pipe is open to the atmosphere. (*b*) Which is more important in this problem, the head loss due to elevation difference or the head loss due to pressure drop in the pipe? (*c*) Find the minimum required pipe diameter.

C6.5 Water at room temperature flows at the *same* volume flow rate, $Q = 9.4 \times 10^{-4}$ m³/s, through two ducts, one a round pipe and one an annulus. The cross-sectional area A of the two ducts is identical, and all walls are made of commercial steel. Both ducts are the same length. In the cross sections shown in Fig. C6.5 $R = 15.0$ mm and $a = 25.0$ mm.

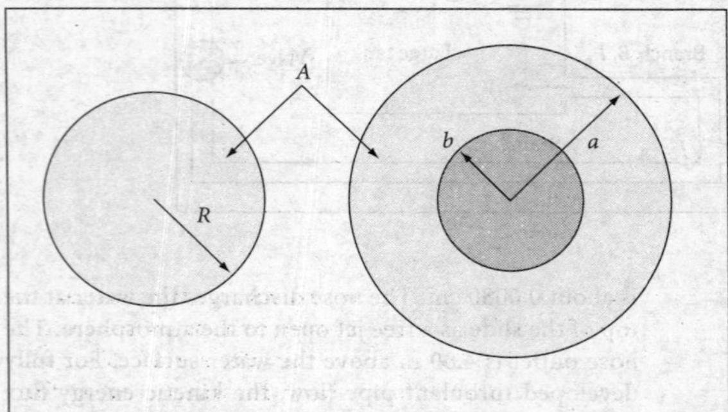

C6.5

(*a*) What is the radius b such that the cross-sectional areas of the two ducts are identical? (*b*) Compare the frictional head loss h_f per unit length of pipe for the two cases, assuming fully developed flow. For the annulus, do both a quick estimate (using the hydraulic diameter) and a more accurate estimate (using the effective diameter correction), and compare. (*c*) If the losses are different for the two cases, explain why. Which duct, if any, is more "efficient"?

C6.6 John Laufer (*NACA Tech Rep.* 1174, 1954) gave velocity data 20°C airflow in a smooth 24.7-cm-diameter pipe at Re ≈ 5 E5:

u/u_{CL}	1.0	0.997	0.988	0.959	0.908	0.847	0.818	0.771	0.690
r/R	0.0	0.102	0.206	0.412	0.617	0.784	0.846	0.907	0.963

Source: John Laufer (NASA Tech Rep. 1174, 1954)

The centerline velocity u_{CL} was 30.5 m/s. Determine (*a*) the average velocity by numerical integration and (*b*) the wall shear stress from the log law approximation. Compare with the Moody chart and with Eq. (6.43).

C6.7 Consider energy exchange in fully developed laminar flow between parallel plates, as in Eqs. (6.60). Let the pressure drop over a length L be Δp. Calculate the rate of work done by this pressure drop on the fluid in the region ($0 < x < L$, $-h < y < +h$) and compare with the integrated energy dissipated due to the viscous function Φ from Eq. (4.50) over this same region. The two should be equal. Explain why this is so. Can you relate the viscous drag force and the wall shear stress to this energy result?

C6.8 This text has presented the traditional correlations for the turbulent smooth-wall friction factor, Eq. (6.38), and the law of the wall, Eq. (6.28). Recently, groups at Princeton and Oregon [56] have made new friction measurements and suggest the following smooth-wall friction law:

$$\frac{1}{\sqrt{f}} = 1.930 \log_{10}(\mathrm{Re}_D\sqrt{f}) - 0.537$$

In earlier work, they also report that better values for the constants κ and B in the log-law, Eq. (6.28), are $\kappa \approx 0.421 \pm 0.002$ and $B \approx 5.62 \pm 0.08$. (*a*) Calculate a few values of f in the range $1\text{ E4} \le \mathrm{Re}_D \le 1\text{ E8}$ and see how the two formulas differ. (*b*) Read Ref. 56 and briefly check the five papers in its bibliography. Report to the class on the general results of this work.

C6.9 A pipeline has been proposed to carry natural gas 2,760 km from Alaska's North Slope to Calgary, Alberta, Canada. The (assumed smooth) pipe diameter will be 1.3 m. The gas will be at high pressure, averaging 17 MPa. (*a*) Why? The proposed flow rate is 113 million cubic meter per day at sea-level conditions. (*b*) What volume flow rate, at 20°C, would carry the same mass at the high pressure? (*c*) If natural gas is assumed to be methane (CH_4), what is the total pressure drop? (*d*) If each pumping station can deliver 12,000 hp to the flow, how many stations are needed?

Design Projects

D6.1 A hydroponic garden uses the 10-m-long perforated-pipe system in Fig. D6.1 to deliver water at 20°C. The pipe is 5 cm in diameter and contains a circular hole every 20 cm. A pump delivers water at 75 kPa (gage) at the entrance, while the other end of the pipe is closed. If you attempted, for example, Prob. P3.125, you know that the pressure near the closed end of a perforated "manifold" is surprisingly high, and there will be too much flow through the holes near that end. One remedy is to vary the hole size along the pipe axis. Make a design analysis, perhaps using a personal computer, to pick the optimum hole size distribution that will make the discharge flow rate as uniform as possible along the pipe axis. You are constrained to pick hole sizes that correspond only to commercial (numbered) metric drill-bit sizes available to the typical machine shop.

D6.2 It is desired to design a pump-piping system to keep a 4,000 m³ capacity water tank filled. The plan is to use a modified (in size and speed) version of the model 1206 centrifugal pump manufactured by Taco Inc., Cranston,

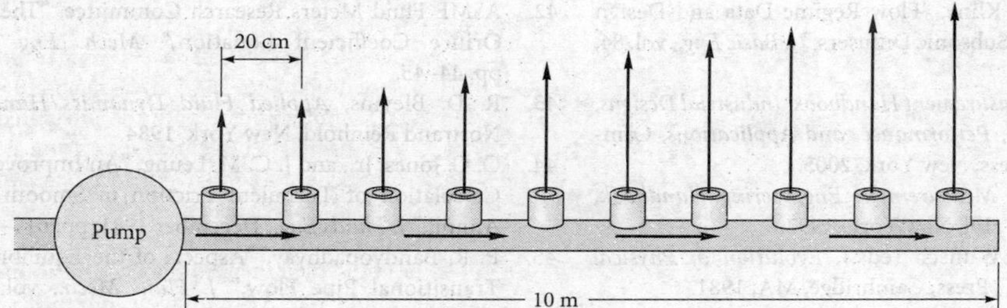

D6.1

Rhode Island. Test data have been provided to us by Taco Inc. for a small model of this pump: $D = 14$ cm, $\Omega = 1{,}760$ r/min, tested with water at 20°C:

Q, L/s	0	0.3	0.6	0.9	1.2	1.5	1.8	2.1	2.4	2.7	3.0	3.3	3.6
H, m	8.5	8.5	8.8	8.8	8.5	8.5	8.2	7.9	7.6	7.0	6.4	5.5	4.6
Efficiency, %	0	13	25	35	44	48	51	53	54	55	53	50	45

The tank is to be filled daily with rather chilly (10°C) groundwater from an aquifer, which is 1,300 m from the tank and 45 m lower than the tank. Estimated daily water use is 5,700 m^3/day. Filling time should not exceed 8 h per day. The piping system should have four "butterfly" valves with variable openings (see Fig. 6.19), 10 elbows of various angles, and galvanized iron pipe of a size to be selected in the design. The design should be economical—both in capital costs and operating expense. Taco Inc. has provided the following cost estimates for system components:

Pump and motor	\$3,500 plus \$1,500 per 25 mm of impeller size
Pump speed	Between 900 and 1,800 r/min
Valves	\$300 + \$200 per 25 mm of pipe size
Elbows	\$50 plus \$50 per 25 mm of pipe size
Pipes	\$1 per 25 mm of diameter per meter of length
Electricity cost	10¢ per kilowatt-hour

Your design task is to select an economical pipe size and pump impeller size and speed for this task, using the pump test data in nondimensional form (see Prob. P5.61) as design data. Write a brief report (five to six pages) showing your calculations and graphs.

References

1. P. S. Bernard and J. M. Wallace, *Turbulent Flow: Analysis, Measurement, and Prediction,* Wiley, New York, 2002.
2. H. Schlichting et al., *Boundary Layer Theory,* Springer, New York, 2000.
3. F. M. White, *Viscous Fluid Flow,* 3d ed., McGraw-Hill, New York, 2005.
4. O. Reynolds, "An Experimental Investigation of the Circumstances which Determine Whether the Motion of Water Shall Be Direct or Sinuous and of the Law of Resistance in Parallel Channels," *Phil. Trans. R. Soc.*, vol. 174, 1883, pp. 935–982.
5. P. G. Drazin and W. H. Reid, *Hydrodynamic Stability,* 2d ed., Cambridge University Press, New York, 2004.
6. H. Rouse and S. Ince, *History of Hydraulics,* Iowa Institute of Hydraulic Research, State University of Iowa, Iowa City, 1957.
7. J. Nikuradse, "Strömungsgesetze in Rauhen Rohren," *VDI Forschungsh.* 361, 1933; English trans., *NACA Tech. Mem.*1292.
8. L. F. Moody, "Friction Factors for Pipe Flow," *ASME Trans.*, vol. 66, pp. 671–684, 1944.
9. C. F. Colebrook, "Turbulent Flow in Pipes, with Particular Reference to the Transition between the Smooth and Rough Pipe Laws," *J. Inst. Civ. Eng. Lond.*, vol. 11, 1938–1939, pp. 133–156.
10. O. C. Jones, Jr., "An Improvement in the Calculations of Turbulent Friction in Rectangular Ducts," *J. Fluids Eng.*, June 1976, pp. 173–181.
11. R. Berker, *Handbuch der Physik,* vol. 7, no. 2, pp. 1–384, Springer-Verlag, Berlin, 1963.
12. R. M. Olson, *Essentials of Engineering Fluid Mechanics,* Literary Licensing LLC, Whitefish, MT, 2012.
13. P. A. Durbin and B. A. Pettersson, *Statistical Theory and Modeling for Turbulent Flows,* 2d ed., Wiley, New York, 2010.
14. P. W. Runstadler, Jr., et al., "Diffuser Data Book," *Creare Inc. Tech. Note* 186, Hanover, NH, 1975.
15. "*Flow of Fluids through Valves, Fittings, and Pipes,*" Tech. Paper 410, Crane Valve Group, Long Beach, CA, 1957 (now updated as a CD-ROM; see < http://www.cranevalves.com >).
16. E. F. Brater, H. W. King, J. E. Lindell, and C. Y. Wei, *Handbook of Hydraulics,* 7th ed., McGraw-Hill, New York, 1996.
17. H. Cross, "Analysis of Flow in Networks of Conduits or Conductors," *Univ. Ill. Bull.* 286, November 1936.
18. P. K. Swamee and A. K. Sharma, *Design of Water Supply Pipe Networks,* Wiley-Interscience, New York, 2008.
19. D. C. Wilcox, *Turbulence Modeling for CFD,* 3d ed., DCW Industries, La Cañada, CA, 2006.
20. R. W. Jeppson, *Analysis of Flow in Pipe Networks,* Butterworth-Heinemann, Woburn, MA, 1976.

21. R. W. Fox and S. J. Kline, "Flow Regime Data and Design Methods for Curved Subsonic Diffusers," *J. Basic Eng.*, vol. 84, 1962, pp. 303–312.
22. R. C. Baker, *Flow Measurement Handbook: Industrial Designs, Operating Principles, Performance, and Applications*, Cambridge University Press, New York, 2005.
23. R. W. Miller, *Flow Measurement Engineering Handbook*, 3d edition, McGraw-Hill, New York, 1997.
24. B. Warren and C. Wunsch (eds.), *Evolution of Physical Oceanography*, M.I.T. Press, Cambridge, MA, 1981.
25. U.S. Department of Commerce, *Tidal Current Tables*, National Oceanographic and Atmospheric Administration, Washington, DC, 1971.
26. J. A. Shercliff, *Electromagnetic Flow Measurement*, Cambridge University Press, New York, 1962.
27. J. A. Miller, "A Simple Linearized Hot-Wire Anemometer," *J. Fluids Eng.*, December 1976, pp. 749–752.
28. R. J. Goldstein (ed.), *Fluid Mechanics Measurements*, 2d ed., Hemisphere, New York, 1996.
29. D. Eckardt, "Detailed Flow Investigations within a High Speed Centrifugal Compressor Impeller," *J. Fluids Eng.*, September 1976, pp. 390–402.
30. H. S. Bean (ed.), *Fluid Meters: Their Theory and Application*, 6th ed., American Society of Mechanical Engineers, New York, 1971.
31. "Measurement of Fluid Flow by Means of Orifice Plates, Nozzles, and Venturi Tubes Inserted in Circular Cross Section Conduits Running Full," *Int. Organ. Stand. Rep.* DIS-5167, Geneva, April 1976.
32. P. Sagaut and C. Meneveau, *Large Eddy Simulation for Incompressible Flows: An Introduction*, 3d ed., Springer, New York, 2006.
33. S. E. Haaland, "Simple and Explicit Formulas for the Friction Factor in Turbulent Pipe Flow," *J. Fluids Eng.*, March 1983, pp. 89–90.
34. R. K. Shah and A. L. London, *Laminar Flow Forced Convection in Ducts*, Academic, New York, 1979.
35. P. L. Skousen, *Valve Handbook*, 3d ed. McGraw-Hill, New York, 2011.
36. W. Li, W.-X. Chen, and S.-Z. Xie, "Numerical Simulation of Stress-Induced Secondary Flows with Hybrid Finite Analytic Method," *Journal of Hydrodynamics*, vol. 14, no. 4, December 2002, pp. 24–30.
37. *ASHRAE Handbook—2012 Fundamentals*, ASHRAE, Atlanta, GA, 2012.
38. F. Durst, A. Melling, and J. H. Whitelaw, *Principles and Practice of Laser-Doppler Anemometry*, 2d ed., Academic, New York, 1981.
39. A. P. Lisitsyn et al., *Laser Doppler and Phase Doppler Measurement Techniques*, Springer-Verlag, New York, 2003.
40. J. E. Amadi-Echendu, H. Zhu, and E. H. Higham, "Analysis of Signals from Vortex Flowmeters," *Flow Measurement and Instrumentation*, vol. 4, no. 4, Oct. 1993, pp. 225–231.
41. G. Vass, "Ultrasonic Flowmeter Basics," *Sensors*, vol. 14, no. 10, Oct. 1997, pp. 73–78.
42. ASME Fluid Meters Research Committee, "The ISO-ASME Orifice Coefficient Equation," *Mech. Eng.* July 1981, pp. 44–45.
43. R. D. Blevins, *Applied Fluid Dynamics Handbook*, Van Nostrand Reinhold, New York, 1984.
44. O. C. Jones, Jr., and J. C. M. Leung, "An Improvement in the Calculation of Turbulent Friction in Smooth Concentric Annuli," *J. Fluids Eng.*, December 1981, pp. 615–623.
45. P. R. Bandyopadhyay, "Aspects of the Equilibrium Puff in Transitional Pipe Flow," *J. Fluid Mech.*, vol. 163, 1986, pp. 439–458.
46. I. E. Idelchik, *Handbook of Hydraulic Resistance*, 3d ed., CRC Press, Boca Raton, FL, 1993.
47. S. Klein and W. Beckman, *Engineering Equation Solver (EES)*, University of Wisconsin, Madison, WI, 2014.
48. R D. Coffield, P. T. McKeown, and R. B. Hammond, "Irrecoverable Pressure Loss Coefficients for Two Elbows in Series with Various Orientation Angles and Separation Distances," *Report WAPD-T-3117*, Bettis Atomic Power Laboratory, West Mifflin, PA, 1997.
49. H. Ito, "Pressure Losses in Smooth Pipe Bends," *Journal of Basic Engineering*, March 1960, pp. 131–143.
50. A. H. Gibson, "On the Flow of Water through Pipes and Passages," *Proc. Roy. Soc. London*, Ser. A, vol. 83, 1910, pp. 366–378.
51. M. Raffel et al., *Particle Image Velocimetry: A Practical Guide*, 2d ed., Springer, New York, 2007.
52. Crane Co., *Flow of Fluids through Valves, Fittings, and Pipe*, Crane, Stanford, CT, 2009.
53. S. A. Berger, L. Talbot, and L.-S. Yao, "Flow in Curved Pipes," *Annual Review of Fluid Mechanics*, vol. 15, 1983, pp. 461–512.
54. P. L. Spedding, E. Benard, and G. M. McNally, "Fluid Flow through 90° Bends," *Developments in Chemical Engineering and Mineral Processing*, vol. 12, nos. 1–2, 2004, pp. 107–128.
55. R. R. Kerswell, "Recent Progress in Understanding the Transition to Turbulence in a Pipe," *Nonlinearity*, vol. 18, 2005, pp. R17–R44.
56. B. J. McKeon et al., "Friction Factors for Smooth Pipe Flow," *J. Fluid Mech.*, vol. 511, 2004, pp. 41–44.
57. M. Bahrami, M. M. Yovanovich, and J. R. Culham, "Pressure Drop of Fully-Developed Laminar Flow in Microchannels of Arbitrary Cross-Section," *J. Fluids Engineering*, vol. 128, Sept. 2006, pp. 1036–1044.
58. G. S. Larraona, A. Rivas, and J. C. Ramos, "Computational Modeling and Simulation of a Single-Jet Water Meter," *J. Fluids Engineering*, vol. 130, May 2008, pp. 0511021–05110212.
59. C. Spiegel, *Designing and Building Fuel Cells*, McGraw-Hill, New York, 2007.
60. B. A. Finlayson et al., *Microcomponent Flow Characterization*, Chap. 8 of *Micro Instrumentation*, M. V. Koch (Ed.), John Wiley, Hoboken, NJ, 2007.

© Solar Impulse | Revillard | Rezo

This chapter is devoted to lift and drag of various bodies immersed in an approaching stream of fluid. Pictured is the Swiss solar-powered aircraft, Solar Impulse, over the Golden Gate Bridge. Earlier solar aircraft needed to be towed aloft before flying and were not able to fly at night. The Solar Impulse is the first solar airplane to fly day and night, approaching the notion of perpetual flight. The long, high-aspect-ratio wings have more lift, and less drag, than a short wing of the same area. Its first international flight was from Switzerland to Brussels, on May 14, 2011. In the summer of 2013, as shown, it flew from San Francisco to New York City, in five legs. The pilots were Bertrand Piccard and André Borschberg.

Chapter 7 Flow Past Immersed Bodies

Motivation. This chapter is devoted to "external" flows around bodies immersed in a fluid stream. Such a flow will have viscous (shear and no-slip) effects near the body surfaces and in its wake, but will typically be nearly inviscid far from the body. These are unconfined *boundary layer* flows.

Chapter 6 considered "internal" flows confined by the walls of a duct. In that case the viscous boundary layers grow from the sidewalls, meet downstream, and fill the entire duct. Viscous shear is the dominant effect. For example, the Moody chart of Fig. 6.13 is essentially a correlation of wall shear stress for long ducts of constant cross section.

External flows are unconfined, free to expand no matter how thick the viscous layers grow. Although boundary layer theory (Sec. 7.3) and computational fluid dynamics (CFD) [4] are helpful in understanding external flows, complex body geometries usually require experimental data on the forces and moments caused by the flow. Such immersed-body flows are commonly encountered in engineering studies: *aerodynamics* (airplanes, rockets, projectiles), *hydrodynamics* (ships, submarines, torpedos), *transportation* (automobiles, trucks, cycles), *wind engineering* (buildings, bridges, water towers, wind turbines), and *ocean engineering* (buoys, breakwaters, pilings, cables, moored instruments). This chapter provides data and analysis to assist in such studies.

7.1 Reynolds Number and Geometry Effects

The technique of boundary layer (BL) analysis can be used to compute viscous effects near solid walls and to "patch" these onto the outer inviscid motion. This patching is more successful as the body Reynolds number becomes larger, as shown in Fig. 7.1.

In Fig. 7.1 a uniform stream U moves parallel to a sharp flat plate of length L. If the Reynolds number UL/ν is low (Fig. 7.1*a*), the viscous region is very broad and extends far ahead and to the sides of the plate. The plate retards the oncoming stream greatly, and small changes in flow parameters cause large changes in the pressure distribution along the plate. Thus although in principle it should be possible to patch the viscous and inviscid layers in a mathematical analysis, their

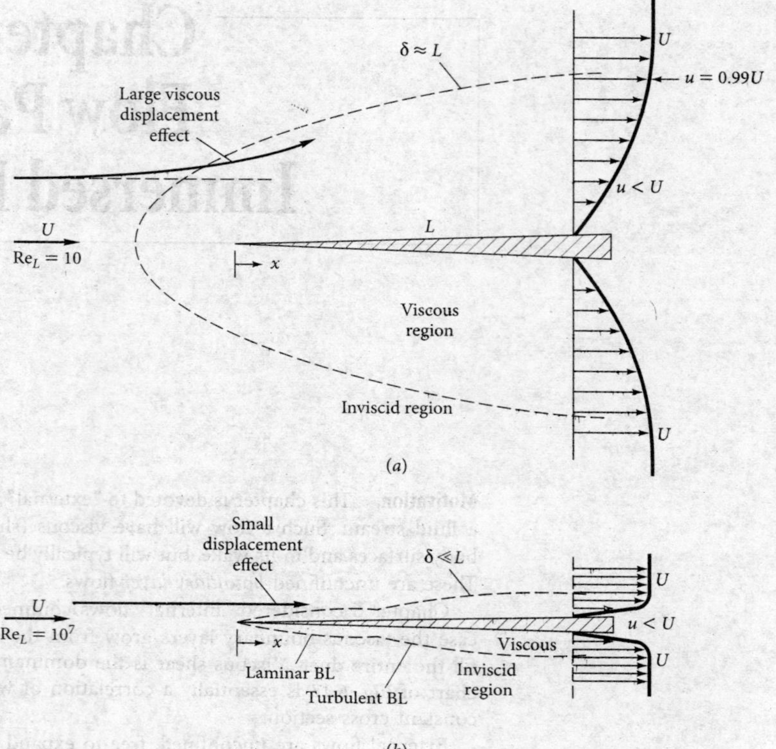

Fig. 7.1 Comparison of flow past a sharp flat plate at low and high Reynolds numbers: (*a*) laminar, low-Re flow; (*b*) high-Re flow.

interaction is strong and nonlinear [1 to 3]. There is no existing simple theory for external flow analysis at Reynolds numbers from 1 to about 1,000. Such thick-shear-layer flows are typically studied by experiment or by numerical modeling of the flow field on a computer [4].

A high-Reynolds-number flow (Fig. 7.1*b*) is much more amenable to boundary layer patching, as first pointed out by Prandtl in 1904. The viscous layers, either laminar or turbulent, are very thin, thinner even than the drawing shows. We define the boundary layer thickness δ as the locus of points where the velocity u parallel to the plate reaches 99 percent of the external velocity U. As we shall see in Sec. 7.4, the accepted formulas for flat-plate flow, and their approximate ranges, are

$$\frac{\delta}{x} \approx \begin{cases} \dfrac{5.0}{\mathrm{Re}_x^{1/2}} & \text{laminar} \quad 10^3 < \mathrm{Re}_x < 10^6 \quad (7.1a) \\[2ex] \dfrac{0.16}{\mathrm{Re}_x^{1/7}} & \text{turbulent} \quad 10^6 < \mathrm{Re}_x \quad (7.1b) \end{cases}$$

where $\mathrm{Re}_x = Ux/\nu$ is called the *local Reynolds number* of the flow along the plate surface. The turbulent flow formula applies for Re_x greater than approximately 10^6.

Some computed values from Eq. (7.1) are

Re_x	10^4	10^5	10^6	10^7	10^8
$(\delta/x)_{\text{lam}}$	0.050	0.016	0.005		
$(\delta/x)_{\text{turb}}$			0.022	0.016	0.011

The blanks indicate that the formula is not applicable. In all cases, these boundary layers are so thin that their displacement effect on the outer inviscid layer is negligible. Thus, the pressure distribution along the plate can be computed from inviscid theory as if the boundary layer were not even there. This external pressure field then "drives" the boundary layer flow, acting as a forcing function in the momentum equation along the surface. We shall explain this boundary layer theory in Secs. 7.4 and 7.5.

For slender bodies, such as plates and airfoils parallel to the oncoming stream, we conclude that this assumption of negligible interaction between the boundary layer and the outer pressure distribution is an excellent approximation.

For a blunt-body flow, however, even at very high Reynolds numbers, there is a discrepancy in the viscous–inviscid patching concept. Figure 7.2 shows two sketches of flow past a two- or three-dimensional blunt body. In the idealized sketch (7.2*a*), there is a thin film of boundary layer about the body and a narrow sheet of viscous wake in the rear. The patching theory would be glorious for this picture, but it is false. In the actual flow (Fig. 7.2*b*), the boundary layer is thin on the front, or windward, side of the body, where the pressure decreases along the surface (*favorable* pressure gradient). But in the rear the boundary layer encounters increasing pressure (*adverse* pressure gradient) and breaks off, or separates, into a broad, pulsating wake. (See Fig. 5.2*a* for a photograph of a specific example.) The mainstream is deflected by this wake, so that the external flow is quite different from the prediction from inviscid theory with the addition of a thin boundary layer.

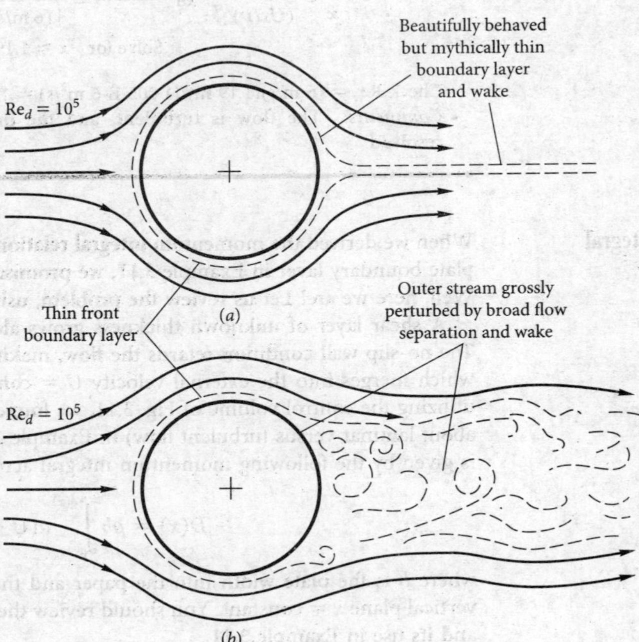

Fig. 7.2 Illustration of the strong interaction between viscous and inviscid regions in the rear of blunt-body flow: (*a*) idealized and definitely false picture of blunt-body flow; (*b*) actual picture of blunt-body flow.

The theory of strong interaction between blunt-body viscous and inviscid layers is not well developed. Flows like that of Fig. 7.2*b* are normally studied experimentally or with CFD [4]. Reference 5 is an example of efforts to improve the theory of separated flows. Reference 6 is another textbook devoted to separated flow.

EXAMPLE 7.1

A long, thin, flat plate is placed parallel to a 6 m/s stream of water at 20°C. At what distance x from the leading edge will the boundary layer thickness be 2 cm

Solution

- *Assumptions:* Flat-plate flow, with Eqs. (7.1) applying in their appropriate ranges.
- *Approach:* Guess laminar flow first. If contradictory, try turbulent flow.
- *Property values:* From Table A.1 for water at 20°C, $\nu \approx 1.005$ E-6 m²/s.
- *Solution step 1:* With $\delta = 0.02$ m, try laminar flow, Eq. (7.1*a*):

$$\frac{\delta}{x}\bigg|_{\text{lam}} = \frac{5}{(Ux/\nu)^{1/2}} \quad \text{or} \quad \frac{0.02\text{ m}}{x} = \frac{5}{[(6\text{ m/s})x/(1.005\text{ E-6 m}^2/\text{s})]^{1/2}}$$

$$\text{Solve for} \quad x \approx 95.5\text{ m}$$

Pretty long plate! This does not sound right. Check the local Reynolds number:

$$\text{Re}_x = \frac{Ux}{\nu} = \frac{(6\text{ m/s})(95.5\text{ m})}{1.005\text{ E-6 m}^2/\text{s}} = 5.7\text{ E8} \quad (!)$$

This is impossible, since laminar boundary layer flow only persists up to about 10^6 (or, with special care to avoid disturbances, up to 3×10^6).

- *Solution step 2:* Try turbulent flow, Eq. (7.1*b*):

$$\frac{\delta}{x} = \frac{0.16}{(Ux/\nu)^{1/7}} \quad \text{or} \quad \frac{0.02\text{ m}}{x} = \frac{0.16}{[(6\text{ m/s})x/(1.005\text{ E-6 m}^2/\text{s})]^{1/7}}$$

$$\text{Solve for} \quad x \approx 1.19\text{ m} \qquad \textit{Ans.}$$

Check $\text{Re}_x = (6\text{ m/s})(1.19\text{ m})/(1.005\text{ E-6 m}^2/\text{s}) = 7.1\text{ E6} > 10^6$. OK, turbulent flow.

- *Comments:* The flow is turbulent, and the inherent ambiguity of the theory is resolved.

7.2 Momentum Integral Estimates

When we derived the momentum integral relation, Eq. (3.37), and applied it to a flat-plate boundary layer in Example 3.11, we promised to consider it further in Chap. 7. Well, here we are! Let us review the problem, using Fig. 7.3.

A shear layer of unknown thickness grows along the sharp flat plate in Fig. 7.3. The no-slip wall condition retards the flow, making it into a rounded profile $u(x, y)$, which merges into the external velocity U = constant at a "thickness" $y = \delta(x)$. By utilizing the control volume of Fig. 3.11, we found (without making any assumptions about laminar versus turbulent flow) in Example 3.11 that the drag force on the plate is given by the following momentum integral across the exit plane:

$$D(x) = \rho b \int_0^{\delta(x)} u(U - u)\, dy \qquad (7.2)$$

where b is the plate width into the paper and the integration is carried out along a vertical plane x = constant. You should review the momentum integral relation (3.37) and its use in Example 3.11.

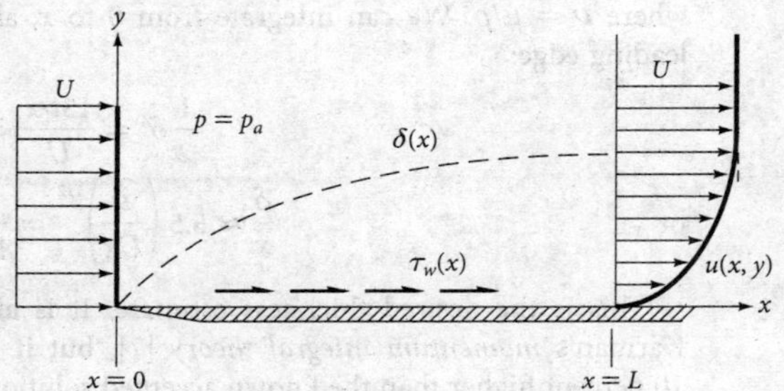

Fig. 7.3 Growth of a boundary layer on a flat plate. The thickness is exaggerated.

Kármán's Analysis of the Flat Plate

Equation (7.2) was derived in 1921 by Kármán [7], who wrote it in the convenient form of the *momentum thickness* θ:

$$D(x) = \rho b U^2 \theta \qquad \theta = \int_0^\delta \frac{u}{U}\left(1 - \frac{u}{U}\right) dy \tag{7.3}$$

Momentum thickness is thus a measure of total plate drag. Kármán then noted that the drag also equals the integrated wall shear stress along the plate:

$$D(x) = b \int_0^x \tau_w(x)\, dx$$

or

$$\frac{dD}{dx} = b\tau_w \tag{7.4}$$

Meanwhile, the derivative of Eq. (7.3), with $U =$ constant, is

$$\frac{dD}{dx} = \rho b U^2 \frac{d\theta}{dx}$$

By comparing this with Eq. (7.4) Kármán arrived at what is now called the *momentum integral relation* for flat-plate boundary layer flow:

$$\boxed{\tau_w = \rho U^2 \frac{d\theta}{dx}} \tag{7.5}$$

It is valid for either laminar or turbulent flat-plate flow.

To get a numerical result for laminar flow, Kármán assumed that the velocity profiles had an approximately parabolic shape

$$u(x, y) \approx U\left(\frac{2y}{\delta} - \frac{y^2}{\delta^2}\right) \qquad 0 \le y \le \delta(x) \tag{7.6}$$

which makes it possible to estimate both momentum thickness and wall shear:

$$\theta = \int_0^\delta \left(\frac{2y}{\delta} - \frac{y^2}{\delta^2}\right)\left(1 - \frac{2y}{\delta} + \frac{y^2}{\delta^2}\right) dy \approx \frac{2}{15}\delta$$

$$\tau_w = \mu \left.\frac{\partial u}{\partial y}\right|_{y=0} \approx \frac{2\mu U}{\delta} \tag{7.7}$$

By substituting (7.7) into (7.5) and rearranging, we obtain

$$\delta\, d\delta \approx 15 \frac{\nu}{U} dx \tag{7.8}$$

where $\nu = \mu/\rho$. We can integrate from 0 to x, assuming that $\delta = 0$ at $x = 0$, the leading edge:

$$\frac{1}{2}\delta^2 = \frac{15\nu x}{U}$$

or

$$\frac{\delta}{x} \approx 5.5\left(\frac{\nu}{Ux}\right)^{1/2} = \frac{5.5}{\mathrm{Re}_x^{1/2}} \tag{7.9}$$

This is the desired thickness estimate. It is all approximate, of course, part of Kármán's *momentum integral theory* [7], but it is startlingly accurate, being only 10 percent higher than the known accepted solution for laminar flat-plate flow, which we gave as Eq. (7.1*a*).

By combining Eqs. (7.9) and (7.7) we also obtain a shear stress estimate along the plate:

$$c_f = \frac{2\tau_w}{\rho U^2} \approx \left(\frac{\frac{8}{15}}{\mathrm{Re}_x}\right)^{1/2} = \frac{0.73}{\mathrm{Re}_x^{1/2}} \tag{7.10}$$

Again this estimate, in spite of the crudeness of the profile assumption [Eq. (7.6)] is only 10 percent higher than the known exact laminar-plate-flow solution $c_f = 0.664/\mathrm{Re}_x^{1/2}$, treated in Sec. 7.4. The dimensionless quantity c_f, called the *skin friction coefficient,* is analogous to the friction factor f in ducts.

A boundary layer can be judged as "thin" if, say, the ratio δ/x is less than about 0.1. This occurs at $\delta/x = 0.1 = 5.0/\mathrm{Re}_x^{1/2}$ or at $\mathrm{Re}_x = 2{,}500$. For Re_x less than 2,500 we can estimate that boundary layer theory fails because the thick layer has a significant effect on the outer inviscid flow. The upper limit on Re_x for laminar flow is about 3×10^6, where measurements on a smooth flat plate [8] show that the flow undergoes transition to a turbulent boundary layer. From 3×10^6 upward the turbulent Reynolds number may be arbitrarily large, and a practical limit at present is 5×10^{10} for oil supertankers.

Displacement Thickness

Another interesting effect of a boundary layer is its small but finite displacement of the outer streamlines. As shown in Fig. 7.4, outer streamlines must deflect outward a distance $\delta^*(x)$ to satisfy conservation of mass between the inlet and outlet:

$$\int_0^h \rho U b\, dy = \int_0^\delta \rho u b\, dy \qquad \delta = h + \delta^* \tag{7.11}$$

The quantity δ^* is called the *displacement thickness* of the boundary layer. To relate it to $u(y)$, cancel ρ and b from Eq. (7.11), evaluate the left integral, and slyly add and subtract U from the right integrand:

$$Uh = \int_0^\delta (U + u - U)\, dy = U(h + \delta^*) + \int_0^\delta (u - U)\, dy$$

or

$$\delta^* = \int_0^\delta \left(1 - \frac{u}{U}\right) dy \tag{7.12}$$

Thus, the ratio of δ^*/δ varies only with the dimensionless velocity profile shape u/U.

Introducing our profile approximation (7.6) into (7.12), we obtain by integration this approximate result:

$$\delta^* \approx \frac{1}{3}\delta \qquad \frac{\delta^*}{x} \approx \frac{1.83}{\mathrm{Re}_x^{1/2}} \tag{7.13}$$

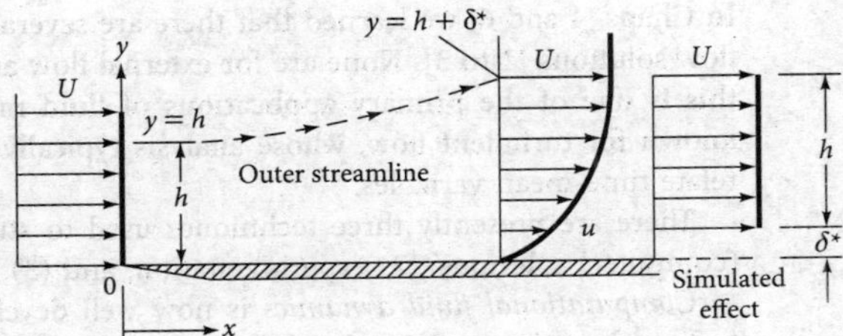

Fig. 7.4 Displacement effect of a boundary layer.

These estimates are only 6 percent away from the exact solutions for laminar flat-plate flow given in Sec. 7.4: $\delta^* = 0.344\delta = 1.721x/\mathrm{Re}_x^{1/2}$. Since δ^* is much smaller than x for large Re_x and the outer streamline slope V/U is proportional to δ^*, we conclude that the velocity normal to the wall is much smaller than the velocity parallel to the wall. This is a key assumption in boundary layer theory (Sec. 7.3).

We also conclude from the success of these simple parabolic estimates that Kármán's momentum integral theory is effective and useful. Many details of this theory are given in Refs. 1 to 3.

EXAMPLE 7.2

Are low-speed, small-scale air and water boundary layers really thin? Consider flow at $U = 0.3$ m/s past a flat plate 0.3 m long. Compute the boundary layer thickness at the trailing edge for (*a*) air and (*b*) water at 20°C.

Solution

Part (a) From Table A.2, $\nu_{\text{air}} \approx 1.5$ E-5 m²/s. The trailing-edge Reynolds number thus is

$$\mathrm{Re}_L = \frac{UL}{\nu} = \frac{(0.3\ \text{m/s})(0.3\ \text{m})}{1.5\ \text{E-5 m}^2/\text{s}} = 6{,}000$$

Since this is less than 10^6, the flow is presumed laminar, and since it is greater than 2,500, the boundary layer is reasonably thin. From Eq. (7.1*a*), the predicted laminar thickness is

$$\frac{\delta}{x} = \frac{5.0}{\sqrt{6{,}000}} = 0.0645$$

or, at $x = 0.3$ m $\qquad \delta = 0.01935\ \text{m} \approx 0.02\ \text{m}$ *Ans. (a)*

Part (b) From Table A.1, $\nu_{\text{water}} \approx 1.005$ E-6 m²/s. The trailing-edge Reynolds number is

$$\mathrm{Re}_L = \frac{(3\ \text{m/s})(0.3\ \text{m})}{1.005\ \text{E-6 m}^2/\text{s}} \approx 89{,}600$$

This again satisfies the laminar and thinness conditions. The boundary layer thickness is

$$\frac{\delta}{x} \approx \frac{5.0}{\sqrt{89{,}600}} = 0.0167$$

or, at $x = 0.3$ m $\qquad \delta = \approx 0.005\ \text{m}$ *Ans. (b)*

Thus, even at such low velocities and short lengths, both airflows and water flows satisfy the boundary layer approximations.

7.3 The Boundary Layer Equations

In Chaps. 4 and 6, we learned that there are several dozen known analytical laminar flow solutions [1 to 3]. None are for external flow around immersed bodies, although this is one of the primary applications of fluid mechanics. No exact solutions are known for turbulent flow, whose analysis typically uses empirical modeling laws to relate time-mean variables.

There are presently three techniques used to study external flows: (1) numerical (computer) solutions, (2) experimentation, and (3) boundary layer theory.

Computational fluid dynamics is now well developed and described in advanced texts such as that by Anderson [4]. Thousands of computer solutions and models have been published; execution times, mesh sizes, and graphical presentations are improving each year. Both laminar and turbulent flow solutions have been published, and turbulence modeling is a current research topic [9]. Except for a brief discussion of computer analysis in Chap. 8, the topic of CFD is beyond our scope here.

Experimentation is the most common method of studying external flows. Chapter 5 outlined the technique of dimensional analysis, and we shall give many nondimensional experimental data for external flows in Sec. 7.6.

The third tool is boundary layer theory, first formulated by Ludwig Prandtl in 1904. We shall follow Prandtl's ideas here and make certain order-of-magnitude assumptions to greatly simplify the Navier-Stokes equations (4.38) into boundary layer equations that are solved relatively easily and patched onto the outer inviscid flow field.

One of the great achievements of boundary layer theory is its ability to predict the flow separation that occurs in adverse (positive) pressure gradients, as illustrated in Fig. 7.2*b*. Before 1904, when Prandtl published his pioneering paper, no one realized that such thin shear layers could cause such a gross effect as flow separation. Even today, however, boundary layer theory cannot accurately predict the behavior of the separated-flow region and its interaction with the outer flow. Modern research [4, 9] has focused on detailed CFD simulations of separated flow, and the resultant wakes, to gain further insight.

Derivation for Two-Dimensional Flow

We consider only steady two-dimensional incompressible viscous flow with the x direction along the wall and y normal to the wall, as in Fig. 7.3.[1] We neglect gravity, which is important only in boundary layers where fluid buoyancy is dominant [2, sec. 4.14]. From Chap. 4, the complete equations of motion consist of continuity and the x- and y-momentum relations:

$$\frac{\partial u}{\partial x} + \frac{\partial v}{\partial y} = 0 \tag{7.14a}$$

$$\rho\left(u\frac{\partial u}{\partial x} + v\frac{\partial u}{\partial y}\right) = -\frac{\partial p}{\partial x} + \mu\left(\frac{\partial^2 u}{\partial x^2} + \frac{\partial^2 u}{\partial y^2}\right) \tag{7.14b}$$

$$\rho\left(u\frac{\partial v}{\partial x} + v\frac{\partial v}{\partial y}\right) = -\frac{\partial p}{\partial y} + \mu\left(\frac{\partial^2 v}{\partial x^2} + \frac{\partial^2 v}{\partial y^2}\right) \tag{7.14c}$$

These should be solved for u, v, and p subject to typical no-slip, inlet, and exit boundary conditions, but in fact they are too difficult to handle for most external flows except with CFD.

[1]For a curved wall, x can represent the arc length along the wall and y can be everywhere normal to x with negligible change in the boundary layer equations as long as the radius of curvature of the wall is large compared with the boundary layer thickness [1 to 3].

In 1904 Prandtl correctly deduced that a shear layer must be very thin if the Reynolds number is large, so that the following approximations apply:

Velocities: $$v \ll u \tag{7.15a}$$

Rates of change: $$\frac{\partial u}{\partial x} \ll \frac{\partial u}{\partial y} \qquad \frac{\partial v}{\partial x} \ll \frac{\partial v}{\partial y} \tag{7.15b}$$

Reynolds number: $$\mathrm{Re}_x = \frac{Ux}{\nu} \gg 1 \tag{7.15c}$$

Our discussion of displacement thickness in the previous section was intended to justify these assumptions.

Applying these approximations to Eq. (7.14*c*) results in a powerful simplification:

$$\underset{\text{small}}{\rho\left(u\frac{\partial v}{\partial x}\right)} + \underset{\text{small}}{\rho\left(v\frac{\partial v}{\partial y}\right)} = -\frac{\partial p}{\partial y} + \underset{\text{very small}}{\mu\left(\frac{\partial^2 v}{\partial x^2}\right)} + \underset{\text{small}}{\mu\left(\frac{\partial^2 v}{\partial y^2}\right)}$$

$$\frac{\partial p}{\partial y} \approx 0 \quad \text{or} \quad p \approx p(x) \quad \text{only} \tag{7.16}$$

In other words, the *y*-momentum equation can be neglected entirely, and the pressure varies only *along* the boundary layer, not through it. The pressure gradient term in Eq. (7.14*b*) is assumed to be known in advance from Bernoulli's equation applied to the outer inviscid flow:

$$\frac{\partial p}{\partial x} = \frac{dp}{dx} = -\rho U \frac{dU}{dx} \tag{7.17}$$

Presumably we have already made the inviscid analysis and know the distribution of $U(x)$ along the wall (Chap. 8).

Meanwhile, one term in Eq. (7.14*b*) is negligible due to Eqs. (7.15):

$$\frac{\partial^2 u}{\partial x^2} \ll \frac{\partial^2 u}{\partial y^2} \tag{7.18}$$

However, neither term in the continuity relation (7.14*a*) can be neglected—another warning that continuity is always a vital part of any fluid flow analysis.

The net result is that the three full equations of motion (7.14) are reduced to Prandtl's two boundary layer equations for two-dimensional incompressible flow:

Continuity: $$\frac{\partial u}{\partial x} + \frac{\partial v}{\partial y} = 0 \tag{7.19a}$$

Momentum along wall: $$u\frac{\partial u}{\partial x} + v\frac{\partial u}{\partial y} \approx U\frac{dU}{dx} + \frac{1}{\rho}\frac{\partial \tau}{\partial y} \tag{7.19b}$$

where $$\tau = \begin{cases} \mu \dfrac{\partial u}{\partial y} & \text{laminar flow} \\ \mu \dfrac{\partial u}{\partial y} - \rho\overline{u'v'} & \text{turbulent flow} \end{cases}$$

These are to be solved for $u(x, y)$ and $v(x, y)$, with $U(x)$ assumed to be a known function from the outer inviscid flow analysis. There are two boundary conditions on u and one on v:

At $y = 0$ (wall): $$u = v = 0 \quad \text{(no slip)} \tag{7.20a}$$

As $y = \delta(x)$ (other stream): $$u = U(x) \quad \text{(patching)} \tag{7.20b}$$

Unlike the Navier-Stokes equations (7.14), which are mathematically elliptic and must be solved simultaneously over the entire flow field, the boundary layer equations (7.19) are mathematically parabolic and are solved by beginning at the leading edge and marching downstream as far as you like, stopping at the separation point or earlier if you prefer.[2]

The boundary layer equations have been solved for scores of interesting cases of internal and external flow for both laminar and turbulent flow, utilizing the inviscid distribution $U(x)$ appropriate to each flow. Full details of boundary layer theory and results and comparison with experiment are given in Refs. 1 to 3. Here we shall confine ourselves primarily to flat-plate solutions (Sec. 7.4).

7.4 The Flat-Plate Boundary Layer

The classic and most often used solution of boundary layer theory is for flat-plate flow, as in Fig. 7.3, which can represent either laminar or turbulent flow.

Laminar Flow

For laminar flow past the plate, the boundary layer equations (7.19) can be solved exactly for u and v, assuming that the free-stream velocity U is constant ($dU/dx = 0$). The solution was given by Prandtl's student Blasius, in his 1908 dissertation from Göttingen. With an ingenious coordinate transformation, Blasius showed that the dimensionless velocity profile u/U is a function only of the single composite dimensionless variable $(y)[U/(\nu x)]^{1/2}$:

$$\frac{u}{U} = f'(\eta) \qquad \eta = y\left(\frac{U}{\nu x}\right)^{1/2} \tag{7.21}$$

where the prime denotes differentiation with respect to η. Substitution of (7.21) into the boundary layer equations (7.19) reduces the problem, after much algebra, to a single third-order nonlinear ordinary differential equation for f [1–3]:

$$f''' + \tfrac{1}{2} f f'' = 0 \tag{7.22}$$

The boundary conditions (7.20) become

$$\text{At } y = 0: \qquad f(0) = f'(0) = 0 \tag{7.23a}$$

$$\text{As } y \to \infty: \qquad f'(\infty) \to 1.0 \tag{7.23b}$$

This is the *Blasius equation*, for which accurate solutions have been obtained only by numerical integration. Some tabulated values of the velocity profile shape $f'(\eta) = u/U$ are given in Table 7.1.

Since u/U approaches 1.0 only as $y \to \infty$, it is customary to select the boundary layer thickness δ as that point where $u/U = 0.99$. From the table, this occurs at $\eta \approx 5.0$:

$$\delta_{99\%}\left(\frac{U}{\nu x}\right)^{1/2} \approx 5.0$$

or

$$\boxed{\frac{\delta}{x} \approx \frac{5.0}{\mathrm{Re}_x^{1/2}} \qquad \text{Blasius (1908)}} \tag{7.24}$$

With the profile known, Blasius, of course, could also compute the wall shear and displacement thickness:

$$\boxed{c_f = \frac{0.664}{\mathrm{Re}_x^{1/2}} \qquad \frac{\delta^*}{x} = \frac{1.721}{\mathrm{Re}_x^{1/2}}} \tag{7.25}$$

[2]For further mathematical details, see Ref. 2, Sec. 2.8.

Table 7.1 The Blasius Velocity Profile [1 to 3]

$y[U/(\nu x)]^{1/2}$	u/U	$y[U/(\nu x)]^{1/2}$	u/U
0.0	0.0	2.8	0.81152
0.2	0.06641	3.0	0.84605
0.4	0.13277	3.2	0.87609
0.6	0.19894	3.4	0.90177
0.8	0.26471	3.6	0.92333
1.0	0.32979	3.8	0.94112
1.2	0.39378	4.0	0.95552
1.4	0.45627	4.2	0.96696
1.6	0.51676	4.4	0.97587
1.8	0.57477	4.6	0.98269
2.0	0.62977	4.8	0.98779
2.2	0.68132	5.0	0.99155
2.4	0.72899	∞	1.00000
2.6	0.77246		

Notice how close these are to our integral estimates, Eqs. (7.9), (7.10), and (7.13). When c_f is converted to dimensional form, we have

$$\tau_w(x) = \frac{0.332\rho^{1/2}\mu^{1/2}U^{1.5}}{x^{1/2}}$$

The wall shear drops off with $x^{1/2}$ because of boundary layer growth and varies as velocity to the 1.5 power. This is in contrast to laminar pipe flow, where $\tau_w \propto U$ and is independent of x.

If $\tau_w(x)$ is substituted into Eq. (7.4), we compute the total drag force:

$$D(x) = b\int_0^x \tau_w(x)\,dx = 0.664b\rho^{1/2}\mu^{1/2}U^{1.5}x^{1/2} \tag{7.26}$$

The drag increases only as the square root of the plate length. The nondimensional *drag coefficient* is defined as

$$C_D = \frac{2D(L)}{\rho U^2 bL} = \frac{1.328}{\mathrm{Re}_L^{1/2}} = 2c_f(L) \tag{7.27}$$

Thus, for laminar plate flow, C_D equals twice the value of the skin friction coefficient at the trailing edge. This is the drag on one side of the plate.

Kármán pointed out that the drag could also be computed from the momentum relation (7.2). In dimensionless form, Eq. (7.2) becomes

$$C_D = \frac{2}{L}\int_0^\delta \frac{u}{U}\left(1 - \frac{u}{U}\right) dy \tag{7.28}$$

This can be rewritten in terms of the momentum thickness at the trailing edge:

$$C_D = \frac{2\theta(L)}{L} \tag{7.29}$$

Computation of θ from the profile u/U or from C_D gives

$$\frac{\theta}{x} = \frac{0.664}{\mathrm{Re}_x^{1/2}} \qquad \text{laminar flat plate} \tag{7.30}$$

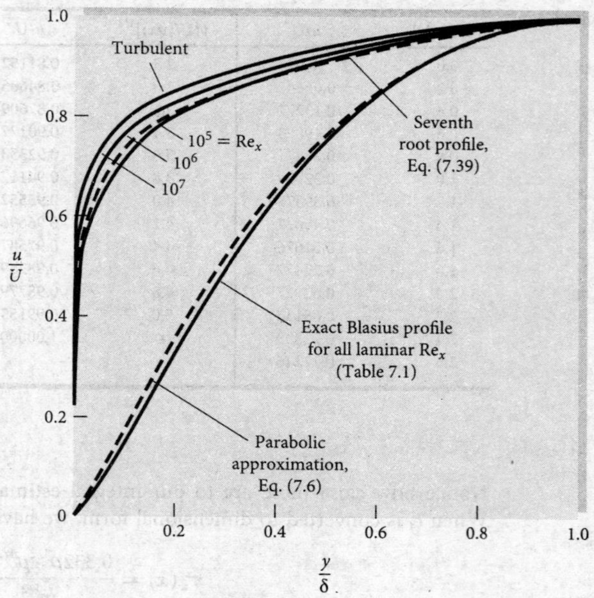

Fig. 7.5 Comparison of dimensionless laminar and turbulent flat-plate velocity profiles.

Since δ is so ill defined, the momentum thickness, being definite, is often used to correlate data taken for a variety of boundary layers under differing conditions. The ratio of displacement to momentum thickness, called the dimensionless-profile *shape factor,* is also useful in integral theories. For laminar flat-plate flow

$$H = \frac{\delta^*}{\theta} = \frac{1.721}{0.664} = 2.59 \tag{7.31}$$

A large shape factor then implies that boundary layer separation is about to occur.

If we plot the Blasius velocity profile from Table 7.1 in the form of u/U versus y/δ, we can see why the simple integral theory guess, Eq. (7.6), was such a great success. This is done in Fig. 7.5. The simple parabolic approximation is not far from the true Blasius profile; hence its momentum thickness is within 10 percent of the true value. Also shown in Fig. 7.5 are three typical turbulent flat-plate velocity profiles. Notice how strikingly different in shape they are from the laminar profiles. Instead of decreasing parabolically to zero, the turbulent profiles are very flat and then drop off sharply at the wall. As you might guess, they follow the logarithmic law shape and thus can be analyzed by momentum integral theory if this shape is properly represented.

Transition to Turbulence

The laminar flat-plate boundary layer eventually becomes turbulent, but there is no unique value for this change to occur. With care in polishing the wall and keeping the free stream quiet, one can delay the transition Reynolds number to $\mathrm{Re}_{x,\mathrm{tr}} \approx$ 3 E6 [8]. However, for typical commercial surfaces and gusty free streams, a more realistic value is

$$\mathrm{Re}_{x,\mathrm{tr}} \approx 5\ \mathrm{E}5.$$

EXAMPLE 7.3

A sharp flat plate with $L = 50$ cm and $b = 3$ m is parallel to a stream of velocity 2.5 m/s. Find the drag on *one side* of the plate, and the boundary thickness δ at the trailing edge, for (*a*) air and (*b*) water at 20°C and 1 atm.

Solution

- *Assumptions:* Laminar flat-plate flow, but we should check the Reynolds numbers.
- *Approach:* Find the Reynolds number and use the appropriate boundary layer formulas.
- *Property values:* From Table A.2 for air at 20°C, $\rho = 1.2$ kg/m³, $\nu = 1.5$ E-5 m²/s. From Table A.1 for water at 20°C, $\rho = 998$ kg/m³, $\nu = 1.005$ E-6 m²/s.
- *(a) Solution for air:* Calculate the Reynolds number at the trailing edge:

$$\mathrm{Re}_L = \frac{UL}{\nu_{\mathrm{air}}} = \frac{(2.5\ \mathrm{m/s})(0.5\ \mathrm{m})}{1.5\ \mathrm{E\text{-}5\ m^2/s}} = 83{,}300 < 5\ \mathrm{E5}\ \text{therefore assuredly laminar}$$

The appropriate thickness relation is Eq. (7.24):

$$\frac{\delta}{L} = \frac{5}{\mathrm{Re}_L^{1/2}} = \frac{5}{(83{,}300)^{1/2}} = 0.0173, \text{ or } \delta_{x=L} = 0.0173(0.5\ \mathrm{m}) \cong 0.0087\ \mathrm{m} \qquad \textit{Ans. (a)}$$

The laminar boundary layer is only 8.7 mm thick. The drag coefficient follows from Eq. (7.27):

$$C_D = \frac{1.328}{\mathrm{Re}_L^{1/2}} = \frac{1.328}{(83{,}300)^{1/2}} = 0.0046$$

$$\text{or } D_{\text{one side}} = C_D\frac{\rho}{2}U^2bL = (0.0046)\frac{1.2\ \mathrm{kg/m^3}}{2}(2.5\ \mathrm{m/s})^2(3\ \mathrm{m})(0.5\ \mathrm{m}) \approx 0.026\ \mathrm{N} \qquad \textit{Ans. (a)}$$

- *Comment (a):* This is purely *friction* drag and is very small for gases at low velocities.
- *(b) Solution for water:* Again calculate the Reynolds number at the trailing edge:

$$\mathrm{Re}_L = \frac{UL}{\nu_{\mathrm{water}}} = \frac{(2.5\ \mathrm{m/s})(0.5\ \mathrm{m})}{1.005\ \mathrm{E\text{-}6\ m^2/s}} = 1.24\ \mathrm{E6} > 5\ \mathrm{E5}\ \text{therefore it might be turbulent}$$

This is a quandary. If the plate is rough or encounters disturbances, the flow at the trailing edge will be turbulent. Let us assume a smooth, undisturbed plate, which will remain laminar. Then again the appropriate thickness relation is Eq. (7.24):

$$\frac{\delta}{L} = \frac{5}{\mathrm{Re}_L^{1/2}} = \frac{5}{(1.24\ \mathrm{E6})^{1/2}} = 0.00448 \text{ or } \delta_{x=L} = 0.00448(0.5\ \mathrm{m}) \cong 0.0022\ \mathrm{m} \qquad \textit{Ans. (b)}$$

This is four times thinner than the air result in part (*a*), due to the high laminar Reynolds number. Again the drag coefficient follows from Eq. (7.27):

$$C_D = \frac{1.328}{\mathrm{Re}_L^{1/2}} = \frac{1.328}{(1.24\ \mathrm{E6})^{1/2}} = 0.0012$$

$$\text{or } D_{\text{one side}} = C_D\frac{\rho}{2}U^2bL = (0.0012)\frac{998\ \mathrm{kg/m^3}}{2}(2.5\ \mathrm{m/s})^2(3\ \mathrm{m})(0.5\ \mathrm{m}) \approx 5.6\ \mathrm{N} \qquad \textit{Ans. (b)}$$

- *Comment (b):* The drag is 215 times larger for water, although C_D is lower, reflecting that water is 56 times more viscous and 830 times denser than air. From Eq. (7.26), for the same U and x, the water drag should be $(56)^{1/2}(830)^{1/2} \approx 215$ times higher. *Note:* If transition to turbulence had occurred at $\mathrm{Re}_x = 5$ E5 (at about $x = 20$ cm), the drag would be about 2.5 times higher, and the trailing edge thickness about four times higher than for fully laminar flow.

Turbulent Flow

There is no exact theory for turbulent flat-plate flow, although there are many elegant computer solutions of the boundary layer equations using various empirical models for the turbulent eddy viscosity [9]. The most widely accepted result is simply an integral analysis similar to our study of the laminar profile approximation (7.6).

We begin with Eq. (7.5), which is valid for laminar or turbulent flow. We write it here for convenient reference:

$$\tau_w(x) = \rho U^2 \frac{d\theta}{dx} \tag{7.32}$$

From the definition of c_f, Eq. (7.10), this can be rewritten as

$$c_f = 2\frac{d\theta}{dx} \tag{7.33}$$

Now, recall from Fig. 7.5 that the turbulent profiles are nowhere near parabolic. Going back to Fig. 6.10, we see that flat-plate flow is very nearly logarithmic, with a slight outer wake and a thin viscous sublayer. Therefore, just as in turbulent pipe flow, we assume that the logarithmic law (6.28) holds all the way across the boundary layer

$$\frac{u}{u^*} \approx \frac{1}{\kappa}\ln\frac{yu^*}{\nu} + B \qquad u^* = \left(\frac{\tau_w}{\rho}\right)^{1/2} \tag{7.34}$$

with, as usual, $\kappa = 0.41$ and $B = 5.0$. At the outer edge of the boundary layer, $y = \delta$ and $u = U$, and Eq. (7.34) becomes

$$\frac{U}{u^*} = \frac{1}{\kappa}\ln\frac{\delta u^*}{\nu} + \text{B} \tag{7.35}$$

But the definition of the skin friction coefficient, Eq. (7.10), is such that the following identities hold:

$$\frac{U}{u^*} \equiv \left(\frac{2}{c_f}\right)^{1/2} \qquad \frac{\delta u^*}{\nu} \equiv \text{Re}_\delta\left(\frac{c_f}{2}\right)^{1/2} \tag{7.36}$$

Therefore, Eq. (7.35) is a *skin friction law* for turbulent flat-plate flow:

$$\left(\frac{2}{c_f}\right)^{1/2} \approx 2.44\ln\left[\text{Re}_\delta\left(\frac{c_f}{2}\right)^{1/2}\right] + 5.0 \tag{7.37}$$

It is a complicated law, but we can at least solve for a few values and list them:

Re_δ	10^4	10^5	10^6	10^7
c_f	0.00493	0.00315	0.00217	0.00158

Following a suggestion of Prandtl, we can forget the complex log friction law (7.37) and simply fit the numbers in the table to a power-law approximation:

$$c_f \approx 0.02\,\text{Re}_\delta^{-1/6} \tag{7.38}$$

This we shall use as the left-hand side of Eq. (7.33). For the right-hand side, we need an estimate for $\theta(x)$ in terms of $\delta(x)$. If we use the logarithmic law profile (7.34), we shall be up to our hips in logarithmic integrations for the momentum thickness. Instead we follow another suggestion of Prandtl, who pointed out that the turbulent profiles in Fig. 7.5 can be approximated by a one-seventh-power law:

$$\left(\frac{u}{U}\right)_{\text{turb}} \approx \left(\frac{y}{\delta}\right)^{1/7} \tag{7.39}$$

This is shown as a dashed line in Fig. 7.5. It is an excellent fit to the low-Reynolds-number turbulent data, which were all that were available to Prandtl at the time. With this simple approximation, the momentum thickness (7.28) can easily be evaluated:

$$\theta \approx \int_0^\delta \left(\frac{y}{\delta}\right)^{1/7}\left[1-\left(\frac{y}{\delta}\right)^{1/7}\right] dy = \frac{7}{72}\,\delta \tag{7.40}$$

We accept this result and substitute Eqs. (7.38) and (7.40) into Kármán's momentum law (7.33):

$$c_f = 0.02\,\mathrm{Re}_\delta^{-1/6} = 2\frac{d}{dx}\left(\frac{7}{72}\,\delta\right)$$

or

$$\mathrm{Re}_\delta^{-1/6} = 9.72\,\frac{d\delta}{dx} = 9.72\,\frac{d(\mathrm{Re}_\delta)}{d(\mathrm{Re}_x)} \tag{7.41}$$

Separate the variables and integrate, assuming $\delta = 0$ at $x = 0$:

$$\mathrm{Re}_\delta \approx 0.16\,\mathrm{Re}_x^{6/7} \quad \text{or} \quad \frac{\delta}{x} \approx \frac{0.16}{\mathrm{Re}_x^{1/7}} \tag{7.42}$$

Thus, the thickness of a turbulent boundary layer increases as $x^{6/7}$, far more rapidly than the laminar increase $x^{1/2}$. Equation (7.42) is the solution to the problem, because all other parameters are now available. For example, combining Eqs. (7.42) and (7.38), we obtain the friction variation

$$c_f \approx \frac{0.027}{\mathrm{Re}_x^{1/7}} \tag{7.43}$$

Writing this out in dimensional form, we have

$$\tau_{w,\mathrm{turb}} \approx \frac{0.0135\mu^{1/7}\rho^{6/7}U^{13/7}}{x^{1/7}} \tag{7.44}$$

Turbulent plate friction drops slowly with x, increases nearly as ρ and U^2, and is rather insensitive to viscosity.

We can evaluate the drag coefficient by integrating the wall friction:

$$D = \int_0^L \tau_w b\,dx$$

or

$$C_D = \frac{2D}{\rho U^2 bL} = \int_0^1 c_f\,d\left(\frac{x}{L}\right)$$

$$C_D = \frac{0.031}{\mathrm{Re}_L^{1/7}} = \frac{7}{6}\,c_f\,(L) \tag{7.45}$$

Then C_D is only 16 percent greater than the trailing-edge skin friction coefficient [compare with Eq. (7.27) for laminar flow].

The displacement thickness can be estimated from the one-seventh-power law and Eq. (7.12):

$$\delta^* \approx \int_0^\delta \left[1-\left(\frac{y}{\delta}\right)^{1/7}\right] dy = \frac{1}{8}\,\delta \tag{7.46}$$

The turbulent flat-plate shape factor is approximately

$$H = \frac{\delta^*}{\theta} = \frac{\frac{1}{8}}{\frac{7}{72}} = 1.3 \tag{7.47}$$

These are the basic results of turbulent flat-plate theory.

Figure 7.6 shows flat-plate drag coefficients for both laminar and turbulent flow conditions. The smooth-wall relations (7.27) and (7.45) are shown, along with the effect of wall roughness, which is quite strong. The proper roughness parameter here is x/ε or L/ε, by analogy with the pipe parameter ε/d. In the fully rough regime, C_D is independent of the Reynolds number, so that the drag varies exactly as U^2 and is independent of μ. Reference 2 presents a theory of rough flat-plate flow, and Ref. 1 gives a curve fit for skin friction and drag in the fully rough regime:

$$c_f \approx \left(2.87 + 1.58 \log \frac{x}{\varepsilon}\right)^{-2.5} \tag{7.48a}$$

$$C_D \approx \left(1.89 + 1.62 \log \frac{L}{\varepsilon}\right)^{-2.5} \tag{7.48b}$$

Equation (7.48*b*) is plotted to the right of the dashed line in Fig. 7.6. The figure also shows the behavior of the drag coefficient in the transition region $5 \times 10^5 < \mathrm{Re}_L < 8 \times 10^7$, where the laminar drag at the leading edge is an appreciable fraction of the

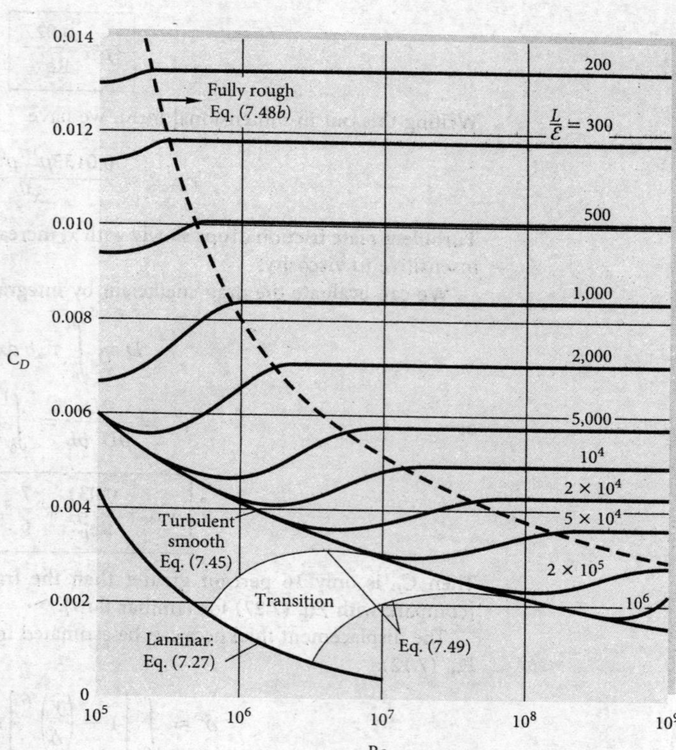

Fig. 7.6 Drag coefficient of laminar and turbulent boundary layers on smooth and rough flat plates. This chart is the flat-plate analog of the Moody diagram of Fig. 6.13.

total drag. Schlichting [1] suggests the following curve fits for these transition drag curves, depending on the Reynolds number $\mathrm{Re}_{\mathrm{trans}}$ where transition begins:

$$C_D \approx \begin{cases} \dfrac{0.031}{\mathrm{Re}_L^{1/7}} - \dfrac{1{,}440}{\mathrm{Re}_L} & \mathrm{Re}_{\mathrm{trans}} = 5 \times 10^5 \quad (7.49a) \\ \dfrac{0.031}{\mathrm{Re}_L^{1/7}} - \dfrac{8{,}700}{\mathrm{Re}_L} & \mathrm{Re}_{\mathrm{trans}} = 3 \times 10^6 \quad (7.49b) \end{cases}$$

EXAMPLE 7.4

A hydrofoil 0.4 m long and 2 m wide is placed in a seawater flow of 12 m/s, with $\rho = 1{,}025\ \mathrm{kg/m^3}$ and $\nu = 1.05\ \mathrm{E}\text{-}6\ \mathrm{m^2/s}$ (*a*) Estimate the boundary layer thickness at the end of the plate. Estimate the friction drag for (*b*) turbulent smooth-wall flow from the leading edge, (*c*) laminar turbulent flow with $\mathrm{Re}_{\mathrm{trans}} = 5 \times 10^5$, and (*d*) turbulent rough-wall flow with $\varepsilon = 0.00012$ m.

Solution

Part (a) The Reynolds number is

$$\mathrm{Re}_L = \frac{UL}{\nu} = \frac{(12\ \mathrm{m/s})(0.4\ \mathrm{m})}{1.05\ \mathrm{E6\ m^2/s}} = 4.57\ \mathrm{E6}$$

Thus the trailing-edge flow is certainly turbulent. The maximum boundary layer thickness would occur for turbulent flow starting at the leading edge. From Eq. (7.42),

$$\frac{\delta(L)}{L} = \frac{0.16}{(4.57\ \mathrm{E6})^{1/7}} = 0.0178$$

or $$\delta = 0.0178(0.4\ \mathrm{m}) = 0.00712\ \mathrm{m} \qquad \textit{Ans. (a)}$$

This is 7.5 times thicker than a fully laminar boundary layer at the same Reynolds number.

Part (b) For fully turbulent smooth-wall flow, the drag coefficient on one side of the plate is, from Eq. (7.45),

$$C_D = \frac{0.031}{(4.57\ \mathrm{E6})^{1/7}} = 0.00347$$

Then the drag on both sides of the foil is approximately

$$D = 2C_D(\tfrac{1}{2}\rho U^2)bL = 2(0.00347)(\tfrac{1}{2})(1{,}025\ \mathrm{kg/m^3})(12\ \mathrm{m/s})^2(2\ \mathrm{m})(0.4\ m) = 410\ \mathrm{N} \qquad \textit{Ans. (b)}$$

Part (c) With a laminar leading edge and $\mathrm{Re}_{\mathrm{trans}} = 5 \times 10^5$, Eq. (7.49*a*) applies:

$$C_D = 0.00347 - \frac{1{,}440}{4.57\ \mathrm{E6}} = 0.00315$$

The drag can be recomputed for this lower drag coefficient:

$$D = 2C_D(\tfrac{1}{2}\rho U^2)bL = 372\ \mathrm{N} \qquad \textit{Ans. (c)}$$

Part (d) Finally, for the rough wall, we calculate

$$\frac{L}{\varepsilon} = \frac{0.4\ \mathrm{m}}{0.00012\ \mathrm{m}} = 3{,}333$$

From Fig. 7.6 at $\mathrm{Re}_L = 4.57$ E6 this condition is just inside the fully rough regime. Equation (7.48*b*) applies:

$$C_D = (1.89 + 1.62 \log 3{,}335)^{-2.5} = 0.00629$$

and the drag estimate is

$$D = 2C_D(\tfrac{1}{2}\rho U^2)bL = 743\text{ N} \qquad \textit{Ans. (d)}$$

This small roughness nearly doubles the drag. It is probable that the total hydrofoil drag is still another factor of 2 larger because of trailing-edge flow separation effects.

7.5 Boundary Layers with Pressure Gradient[3]

The flat-plate analysis of the previous section should give us a good feeling for the behavior of both laminar and turbulent boundary layers, except for one important effect: flow separation. Prandtl showed that separation like that in Fig. 7.2*b* is caused by excessive momentum loss near the wall in a boundary layer trying to move downstream against increasing pressure, $dp/dx > 0$, which is called an *adverse pressure gradient*. The opposite case of decreasing pressure, $dp/dx < 0$, is called a *favorable gradient*, where flow separation can never occur. In a typical immersed-body flow, such as in Fig. 7.2*b*, the favorable gradient is on the front of the body and the adverse gradient is in the rear, as discussed in detail in Chap. 8.

We can explain flow separation with a geometric argument about the second derivative of velocity u at the wall. From the momentum equation (7.19*b*) at the wall, where $u = v = 0$, we obtain

$$\left.\frac{\partial \tau}{\partial y}\right|_{\text{wall}} = \mu \left.\frac{\partial^2 u}{\partial y^2}\right|_{\text{wall}} = -\rho U \frac{dU}{dx} = \frac{dp}{dx}$$

or

$$\left.\frac{\partial^2 u}{\partial y^2}\right|_{\text{wall}} = \frac{1}{\mu}\frac{dp}{dx} \qquad (7.50)$$

for either laminar or turbulent flow. Thus in an adverse gradient the second derivative of velocity is positive at the wall; yet it must be negative at the outer layer ($y = \delta$) to merge smoothly with the mainstream flow $U(x)$. It follows that the second derivative must pass through zero somewhere in between, at a point of inflection, and any boundary layer profile in an adverse gradient must exhibit a characteristic S shape.

Figure 7.7 illustrates the general case. In a favorable gradient (Fig. 7.7*a*) the profile is very rounded, there is no point of inflection, there can be no separation, and laminar profiles of this type are very resistant to a transition to turbulence [1 to 3].

In a zero pressure gradient (Fig. 7.7*b*), such as a flat-plate flow, the point of inflection is at the wall itself. There can be no separation, and the flow will undergo transition at Re_x no greater than about 3×10^6, as discussed earlier.

In an adverse gradient (Fig. 7.7*c* to *e*), a point of inflection (PI) occurs in the boundary layer, its distance from the wall increasing with the strength of the adverse gradient. For a weak gradient (Fig. 7.7*c*) the flow does not actually separate, but it is vulnerable to transition to turbulence at Re_x as low as 10^5 [1, 2]. At a moderate gradient, a critical condition (Fig. 7.7*d*) is reached where the wall shear is exactly zero ($\partial u/\partial y = 0$). This is defined as the *separation point* ($\tau_w = 0$), because any stronger gradient will actually cause backflow at the wall (Fig. 7.7*e*): the boundary layer thickens greatly, and the main flow breaks away, or separates, from the wall (Fig. 7.2*b*).

The flow profiles of Fig. 7.7 usually occur in sequence as the boundary layer progresses along the wall of a body. For example, in Fig. 7.2*a*, a favorable gradient occurs on the front of the body, zero pressure gradient occurs just upstream of the shoulder, and an adverse gradient occurs successively as we move around the rear of the body.

[3]This section may be omitted without loss of continuity.

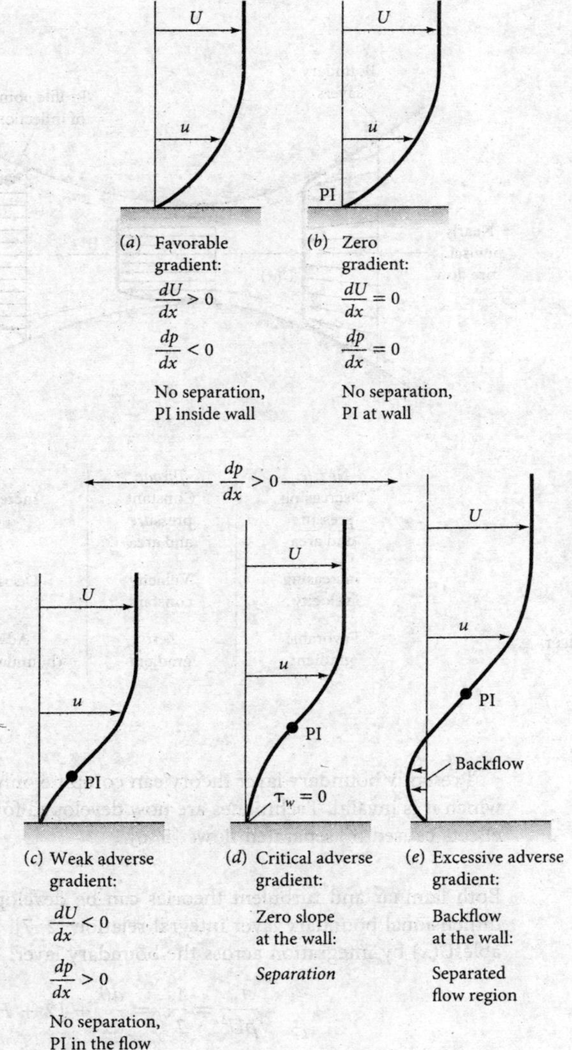

Fig. 7.7 Effect of pressure gradient on boundary layer profiles; PI = point of inflection.

A second practical example is the flow in a duct consisting of a nozzle, throat, and diffuser, as in Fig. 7.8. The nozzle flow is a favorable gradient and never separates, nor does the throat flow where the pressure gradient is approximately zero. But the expanding-area diffuser produces low velocity and increasing pressure, an adverse gradient. If the diffuser angle is too large, the adverse gradient is excessive. and the boundary layer will separate at one or both walls, with backflow, increased losses, and poor pressure recovery. In the diffuser literature, [10] this condition is called *diffuser stall,* a term used also in airfoil aerodynamics (Sec. 7.6) to denote airfoil boundary layer separation. Thus, the boundary layer behavior explains why a large-angle diffuser has heavy flow losses (Fig. 6.23) and poor performance (Fig. 6.28).

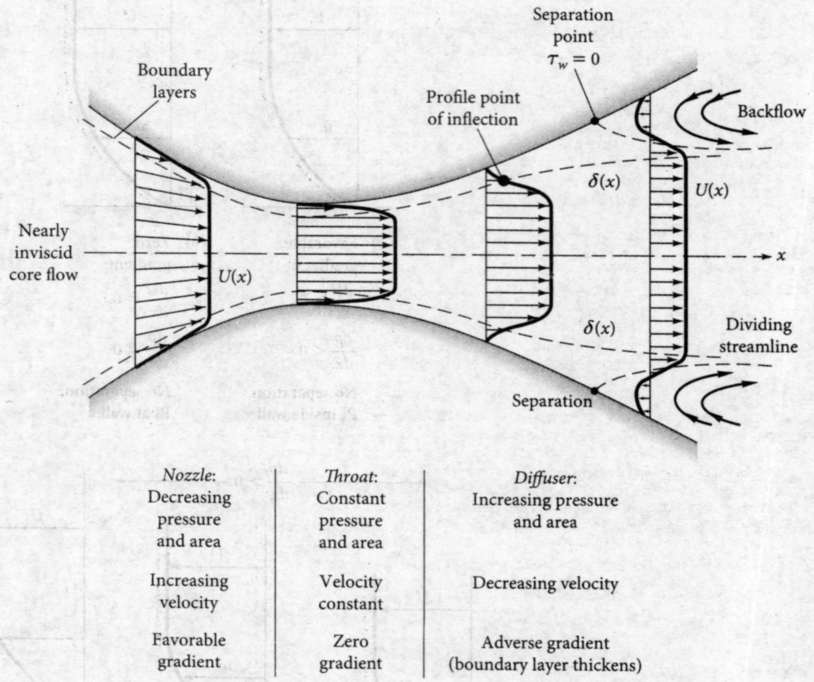

Fig. 7.8 Boundary layer growth and separation in a nozzle–diffuser configuration.

Presently boundary layer theory can compute only up to the separation point, after which it is invalid. Techniques are now developed for analyzing the strong interaction effects caused by separated flows [5, 6].

Laminar Integral Theory[4]

Both laminar and turbulent theories can be developed from Kármán's general two-dimensional boundary layer integral relation [2, 7], which extends Eq. (7.33) to variable $U(x)$ by integration across the boundary layer:

$$\frac{\tau_w}{\rho U^2} = \frac{1}{2} c_f = \frac{d\theta}{dx} + (2 + H)\frac{\theta}{U}\frac{dU}{dx} \tag{7.51}$$

where $\theta(x)$ is the momentum thickness and $H(x) = \delta^*(x)/\theta(x)$ is the shape factor. From Eq. (7.17) negative dU/dx is equivalent to positive dp/dx—that is, an adverse gradient.

We can integrate Eq. (7.51) to determine $\theta(x)$ for a given $U(x)$ if we correlate c_f and H with the momentum thickness. This has been done by examining typical velocity profiles of laminar and turbulent boundary layer flows for various pressure gradients. Some examples are given in Fig. 7.9, showing that the shape factor H is a good indicator of the pressure gradient. The higher the H, the stronger the adverse gradient, and separation occurs approximately at

$$H \approx \begin{cases} 3.5 & \text{laminar flow} \\ 2.4 & \text{turbulent flow} \end{cases} \tag{7.52}$$

[4]This section may be omitted without loss of continuity.

The laminar profiles (Fig. 7.9*a*) clearly exhibit the S shape and a point of inflection with an adverse gradient. But in the turbulent profiles (Fig. 7.9*b*) the points of inflection are typically buried deep within the thin viscous sublayer, which can hardly be seen on the scale of the figure.

There are scores of turbulent theories in the literature, but they are all complicated algebraically and will be omitted here. The reader is referred to advanced texts [1–3, 9].

For laminar flow, a simple and effective method was developed by Thwaites [11], who found that Eq. (7.51) can be correlated by a single dimensionless momentum thickness variable λ, defined as

$$\lambda = \frac{\theta^2}{\nu}\frac{dU}{dx} \tag{7.53}$$

Using a straight-line fit to his correlation, Thwaites was able to integrate Eq. (7.51) in closed form, with the result

$$\theta^2 = \theta_0^2\left(\frac{U_0}{U}\right)^6 + \frac{0.45\nu}{U^6}\int_0^x U^5\,dx \tag{7.54}$$

where θ_0 is the momentum thickness at $x = 0$ (usually taken to be zero). Separation ($c_f = 0$) was found to occur at a particular value of λ:

Separation: $$\lambda = -0.09 \tag{7.55}$$

Finally, Thwaites correlated values of the dimensionless shear stress $S = \tau_w\theta/(\mu U)$ with λ, and his graphed result can be curve-fitted as follows:

$$S(\lambda) = \frac{\tau_w\theta}{\mu U} \approx (\lambda + 0.09)^{0.62} \tag{7.56}$$

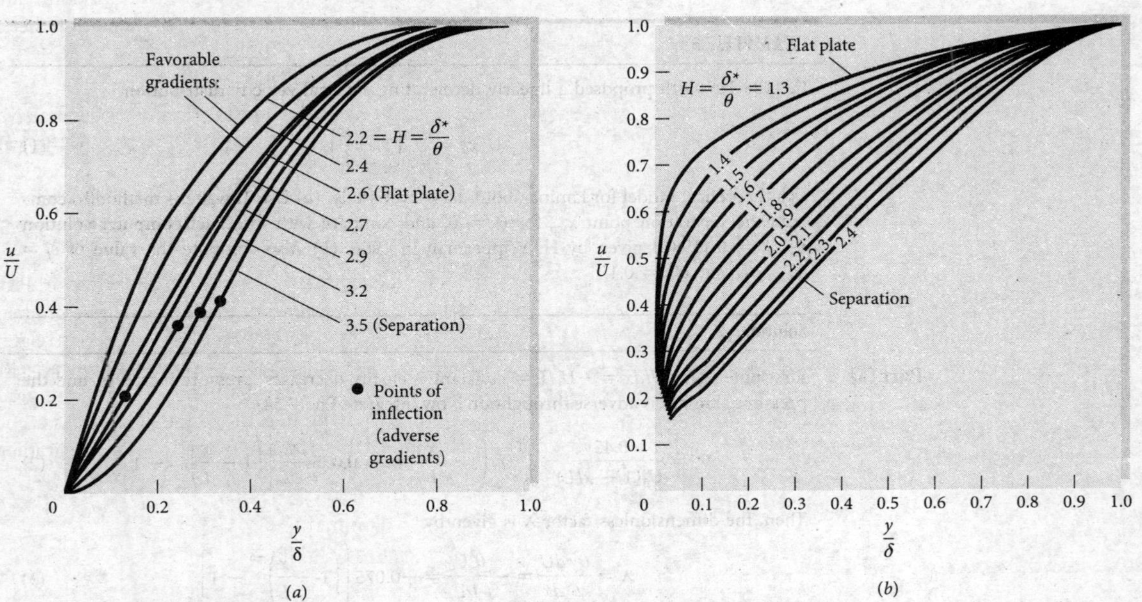

Fig. 7.9 Velocity profiles with pressure gradient: (*a*) laminar flow; (*b*) turbulent flow with adverse gradients.

This parameter is related to the skin friction by the identity

$$S \equiv \tfrac{1}{2} c_f \, \mathrm{Re}_\theta \tag{7.57}$$

Equations (7.54) to (7.56) constitute a complete theory for the laminar boundary layer with variable $U(x)$, with an accuracy of ± 10 percent compared with computer solutions of the laminar-boundary-layer equations (7.19). Complete details of Thwaites's and other laminar theories are given in Ref. 2.

As a demonstration of Thwaites's method, take a flat plate, where U = constant, $\lambda = 0$, and $\theta_0 = 0$. Equation (7.54) integrates to

$$\theta^2 = \frac{0.45\nu x}{U}$$

or

$$\frac{\theta}{x} = \frac{0.671}{\mathrm{Re}_x^{1/2}} \tag{7.58}$$

This is within 1 percent of Blasius's numerical solution, Eq. (7.30).

With $\lambda = 0$, Eq. (7.56) predicts the flat-plate shear to be

$$\frac{\tau_w \theta}{\mu U} = (0.09)^{0.62} = 0.225$$

or

$$c_f = \frac{2\tau_w}{\rho U^2} = \frac{0.671}{\mathrm{Re}_x^{1/2}} \tag{7.59}$$

This is also within 1 percent of the Blasius result, Eq. (7.25). However, the general accuracy of this method is poorer than 1 percent because Thwaites actually "tuned" his correlation constants to make them agree with exact flat-plate theory.

We shall not compute any more boundary layer details here; but as we go along, investigating various immersed-body flows, especially in Chap. 8, we shall use Thwaites's method to make qualitative assessments of the boundary layer behavior.

EXAMPLE 7.5

In 1938 Howarth proposed a linearly decelerating external velocity distribution

$$U(x) = U_0\left(1 - \frac{x}{L}\right) \tag{1}$$

as a theoretical model for laminar-boundary-layer study. (*a*) Use Thwaites's method to compute the separation point x_{sep} for $\theta_0 = 0$, and compare with the exact computer solution $x_{sep}/L = 0.119863$ given by H. Wipperman in 1966. (*b*) Also compute the value of $c_f = 2\tau_w/(\rho U^2)$ at $x/L = 0.1$.

Solution

Part (a) First note that $dU/dx = -U_0/L$ = constant: Velocity decreases, pressure increases, and the pressure gradient is adverse throughout. Now integrate Eq. (7.54):

$$\theta^2 = \frac{0.45\nu}{U_0^6(1 - x/L)^6}\int_0^x U_0^5\left(1 - \frac{x}{L}\right)^5 dx = 0.075\frac{\nu L}{U_0}\left[\left(1 - \frac{x}{L}\right)^{-6} - 1\right] \tag{2}$$

Then, the dimensionless factor λ is given by

$$\lambda = \frac{\theta^2}{\nu}\frac{dU}{dx} = -\frac{\theta^2 U_0}{\nu L} = -0.075\left[\left(1 - \frac{x}{L}\right)^{-6} - 1\right] \tag{3}$$

From Eq. (7.55) we set this equal to −0.09 for separation:

$$\lambda_{sep} = -0.09 = -0.075\left[\left(1 - \frac{x_{sep}}{L}\right)^{-6} - 1\right]$$

or $$\frac{x_{sep}}{L} = 1 - (2.2)^{-1/6} = 0.123 \qquad \textit{Ans. (a)}$$

This is less than 3 percent higher than Wipperman's exact solution, and the computational effort is very modest.

Part (b) To compute c_f at $x/L = 0.1$ (just before separation), we first compute λ at this point, using Eq. (3):

$$\lambda(x = 0.1L) = -0.075[(1 - 0.1)^{-6} - 1] = -0.0661$$

Then from Eq. (7.56) the shear parameter is

$$S(x = 0.1L) = (-0.0661 + 0.09)^{0.62} = 0.099 = \tfrac{1}{2}c_f \mathrm{Re}_\theta \qquad (4)$$

We can compute Re_θ in terms of Re_L from Eq. (2) or (3):

$$\frac{\theta^2}{L^2} = \frac{0.0661}{UL/\nu} = \frac{0.0661}{\mathrm{Re}_L}$$

or $$\mathrm{Re}_\theta = 0.257\,\mathrm{Re}_L^{1/2} \quad \text{at} \quad \frac{x}{L} = 0.1$$

Substitute into Eq. (4):

$$0.099 = \tfrac{1}{2}c_f(0.257\,\mathrm{Re}_L^{1/2})$$

or $$c_f = \frac{0.77}{\mathrm{Re}_L^{1/2}} \qquad \mathrm{Re}_L = \frac{UL}{\nu} \qquad \textit{Ans. (b)}$$

We cannot actually compute c_f without the value of, say, U_0L/ν.

7.6 Experimental External Flows

Boundary layer theory is very interesting and illuminating and gives us a great qualitative grasp of viscous flow behavior; but, because of flow separation, the theory does not generally allow a quantitative computation of the complete flow field. In particular, there is at present no satisfactory theory, except CFD results, for the forces on an arbitrary body immersed in a stream flowing at an arbitrary Reynolds number. Therefore, experimentation is the key to treating external flows.

Literally thousands of papers in the literature report experimental data on specific external viscous flows. This section gives a brief description of the following external flow problems:

1. Drag of two- and three-dimensional bodies:
   a. Blunt bodies.
   b. Streamlined shapes.
2. Performance of lifting bodies:
   a. Airfoils and aircraft.
   b. Projectiles and finned bodies.
   c. Birds and insects.

For further reading see the goldmine of data compiled in Hoerner [12]. In later chapters we shall study data on supersonic airfoils (Chap. 9), open-channel friction (Chap. 10), and turbomachinery performance (Chap. 11).

Drag of Immersed Bodies

Any body of any shape when immersed in a fluid stream will experience forces and moments from the flow. If the body has arbitrary shape and orientation, the flow will exert forces and moments about all three coordinate axes, as shown in Fig. 7.10. It is customary to choose one axis parallel to the free stream and positive downstream. The force on the body along this axis is called *drag,* and the moment about that axis the *rolling moment.* The drag is essentially a flow loss and must be overcome if the body is to move against the stream.

A second and very important force is perpendicular to the drag and usually performs a useful job, such as bearing the weight of the body. It is called the *lift.* The moment about the lift axis is called *yaw.*

The third component, neither a loss nor a gain, is the *side force,* and about this axis is the *pitching moment.* To deal with this three-dimensional force-moment situation is more properly the role of a textbook on aerodynamics [for example, 13]. We shall limit the discussion here to lift and drag.

When the body has symmetry about the lift–drag axis, as with airplanes, ships, and cars moving directly into a stream, the side force, yaw, and roll vanish, and the problem reduces to a two-dimensional case: two forces, lift and drag, and one moment, pitch.

A final simplification often occurs when the body has two planes of symmetry, as in Fig. 7.11. A wide variety of shapes such as cylinders, wings, and all bodies of revolution satisfy this requirement. If the free stream is parallel to the intersection of these two planes, called the *principal chord line of the body,* the body experiences drag only, with no lift, side force, or moments.[5] This type of degenerate one-force drag data is

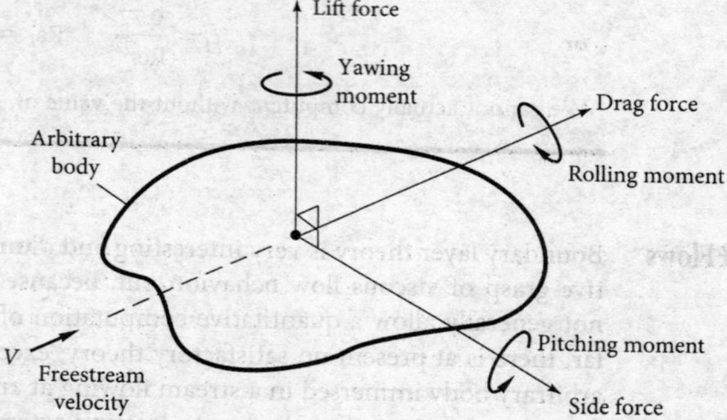

Fig. 7.10 Definition of forces and moments on a body immersed in a uniform flow.

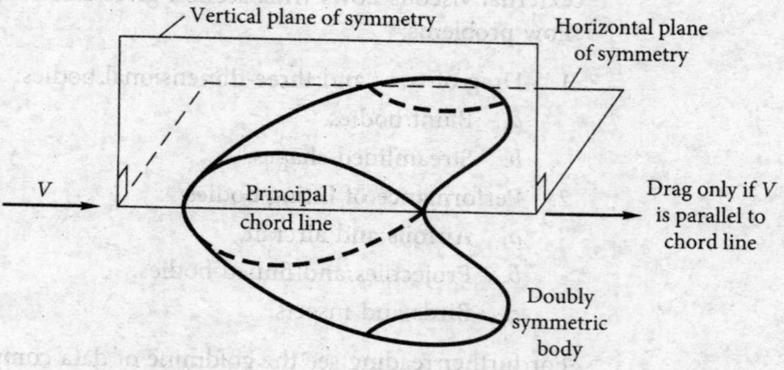

Fig. 7.11 Only the drag force occurs if the flow is parallel to both planes of symmetry.

[5]In bodies with shed vortices, such as the cylinder in Fig. 5.2, there may be *oscillating* lift, side force, and moments, but their mean value is zero.

what is most commonly reported in the literature, but if the free stream is not parallel to the chord line, the body will have an unsymmetric orientation and all three forces and three moments can arise in principle.

In low-speed flow past geometrically similar bodies with identical orientation and relative roughness, the drag coefficient should be a function of the body Reynolds number:

$$C_D = f(\text{Re}) \tag{7.60}$$

The Reynolds number is based upon the free-stream velocity V and a characteristic length L of the body, usually the chord or body length parallel to the stream:

$$\text{Re} = \frac{VL}{\nu} \tag{7.61}$$

For cylinders, spheres, and disks, the characteristic length is the diameter D.

Characteristic Area

Drag coefficients are defined by using a characteristic area A, which may differ depending on the body shape:

$$C_D = \frac{\text{drag}}{\frac{1}{2}\rho V^2 A} \tag{7.62}$$

The factor $\frac{1}{2}$ is our traditional tribute to Euler and Bernoulli. The area A is usually one of three types:

1. *Frontal area,* the body as seen from the stream; suitable for thick, stubby bodies, such as spheres, cylinders, cars, trucks, missiles, projectiles, and torpedoes.
2. *Planform area,* the body area as seen from above; suitable for wide, flat bodies such as wings and hydrofoils.
3. *Wetted area,* customary for surface ships and barges.

In using drag or other fluid force data, it is important to note what length and area are being used to scale the measured coefficients.

Friction Drag and Pressure Drag

As we have mentioned, the theory of drag is weak and inadequate, except for the flat plate. This is because of flow separation. Boundary layer theory can predict the separation point, but cannot accurately estimate the (usually low) pressure distribution in the separated region. The difference between the high pressure in the front stagnation region and the low pressure in the rear separated region causes a large drag contribution called *pressure drag.* This is added to the integrated shear stress or *friction drag* of the body, which it often exceeds:

$$C_D = C_{D,\text{press}} + C_{D,\text{fric}} \tag{7.63}$$

The relative contribution of friction and pressure drag depends upon the body's shape, especially its thickness. Figure 7.12 shows drag data for a streamlined cylinder of very large depth into the paper. At zero thickness the body is a flat plate and exhibits 100 percent friction drag. At thickness equal to the chord length, simulating a circular cylinder, the friction drag is only about 3 percent. Friction and pressure drag are about equal at thickness $t/c = 0.25$. Note that C_D in Fig. 7.12b looks quite

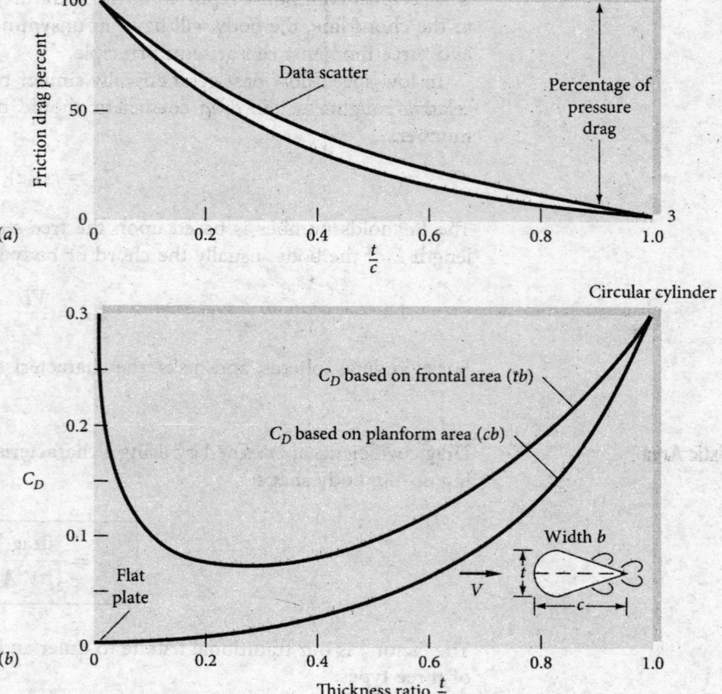

Fig. 7.12 Drag of a streamlined two-dimensional cylinder at $\mathrm{Re}_c = 10^6$: (*a*) effect of thickness ratio on percentage of friction drag; (*b*) total drag versus thickness when based on two different areas.

different when based on frontal area instead of planform area, planform being the usual choice for this body shape. The two curves in Fig. 7.12*b* represent exactly the same drag data.

Figure 7.13 illustrates the dramatic effect of separated flow and the subsequent failure of boundary layer theory. The theoretical inviscid pressure distribution on a circular cylinder (Chap. 8) is shown as the dashed line in Fig. 7.13*c*:

$$C_p = \frac{p - p_\infty}{\frac{1}{2}\rho V^2} = 1 - 4\sin^2\theta$$

where P_∞ and V are the pressure and velocity, respectively, in the free stream. The actual laminar and turbulent boundary layer pressure distributions in Fig. 7.13*c* are startlingly different from those predicted by theory. Laminar flow is very vulnerable to the adverse gradient on the rear of the cylinder, and separation occurs at $\theta = 82°$, which certainly could not have been predicted from inviscid theory. The broad wake and very low pressure in the separated laminar region cause the large drag $C_D = 1.2$.

The turbulent boundary layer in Fig. 7.13*b* is more resistant, and separation is delayed until $\theta = 120°$, with a resulting smaller wake, higher pressure on the rear, and 75 percent less drag, $C_D = 0.3$. This explains the sharp drop in drag at transition in Fig. 5.3.

The same sharp difference between vulnerable laminar separation and resistant turbulent separation can be seen for a sphere in Fig. 7.14. The laminar flow (Fig. 7.14*a*)

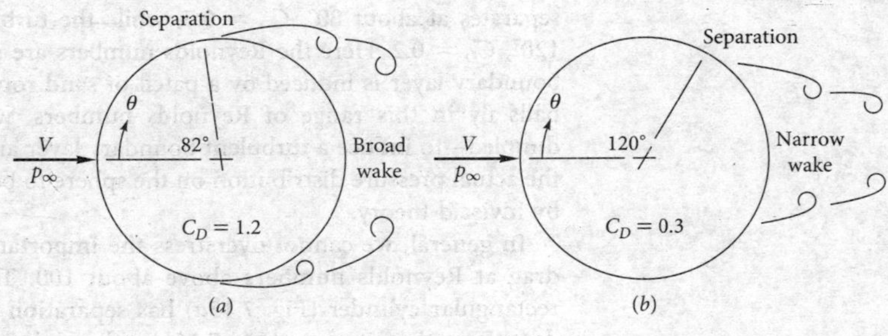

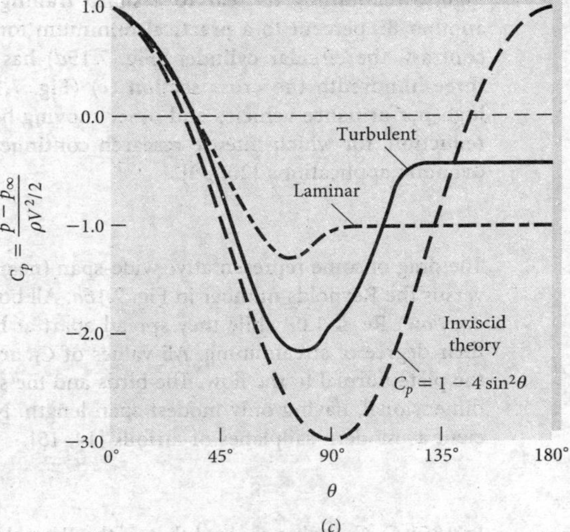

Fig. 7.13 Flow past a circular cylinder: (*a*) laminar separation; (*b*) turbulent separation; (*c*) theoretical and actual surface pressure distributions.

(*a*) (*b*)

Fig. 7.14 Strong differences in laminar and turbulent separation on an 0.22 m bowling ball entering water at 7.6 m/s (*a*) smooth ball, laminar boundary layer; (*b*) same entry, turbulent flow induced by patch of nose-sand roughness. *(NAVAIR Weapons Division Historical Archives.)*

separates at about 80°, $C_D = 0.5$, while the turbulent flow (Fig. 7.14*b*) separates at 120°, $C_D = 0.2$. Here the Reynolds numbers are exactly the same, and the turbulent boundary layer is induced by a patch of sand roughness at the nose of the ball. Golf balls fly in this range of Reynolds numbers, which is why they are deliberately dimpled—to induce a turbulent boundary layer and lower drag. Again we would find the actual pressure distribution on the sphere to be quite different from that predicted by inviscid theory.

In general, we cannot overstress the importance of body streamlining to reduce drag at Reynolds numbers above about 100. This is illustrated in Fig. 7.15. The rectangular cylinder (Fig. 7.15*a*) has separation at all sharp corners and very high drag. Rounding its nose (Fig. 7.15*b*) reduces drag by about 45 percent, but C_D is still high. Streamlining its rear to a sharp trailing edge (Fig. 7.15*c*) reduces its drag another 85 percent to a practical minimum for the given thickness. As a dramatic contrast, the circular cylinder (Fig. 7.15*d*) has one-eighth the thickness and one-three-hundredth the cross section (*c*) (Fig. 7.15*c*), yet it has the same drag. For high-performance vehicles and other moving bodies, the name of the game is drag reduction, for which intense research continues for both aerodynamic and hydrodynamic applications [20, 39].

Two-Dimensional Bodies

The drag of some representative wide-span (nearly two-dimensional) bodies is shown versus the Reynolds number in Fig. 7.16*a*. All bodies have high C_D at very low (*creeping flow*) Re $\leq$ 1.0, while they spread apart at high Reynolds numbers according to their degree of streamlining. All values of C_D are based on the planform area except the plate normal to the flow. The birds and the sailplane are, of course, not very two-dimensional, having only modest span length. Note that birds are not nearly as efficient as modern sailplanes or airfoils [14, 15].

Creeping Flow

In 1851 G. G. Stokes showed that, if the Reynolds number is very small, Re $\ll$ 1, the acceleration terms in the Navier-Stokes equations (7.14*b*, *c*) are negligible. The flow is termed *creeping flow*, or Stokes flow, and is a balance between pressure gradient and viscous stresses. Continuity and momentum reduce to two linear equations for velocity and pressure:

$$\text{Re} \ll 1: \qquad \nabla \cdot \mathbf{V} = 0 \qquad \text{and} \qquad \nabla p \approx \mu \nabla^2 \mathbf{V}$$

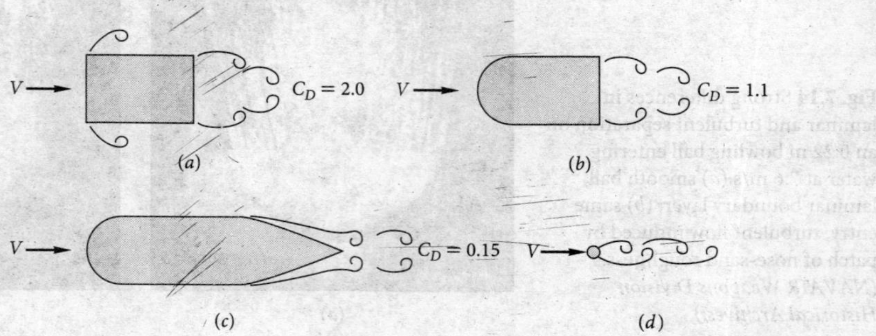

Fig. 7.15 The importance of streamlining in reducing drag of a body (C_D based on frontal area): (*a*) rectangular cylinder; (*b*) rounded nose; (*c*) rounded nose and streamlined sharp trailing edge; (*d*) circular cylinder with the same drag as case (*c*).

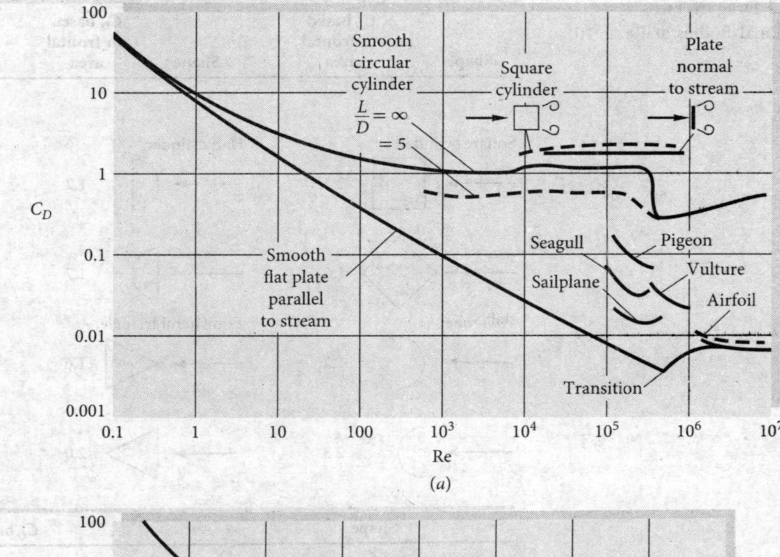

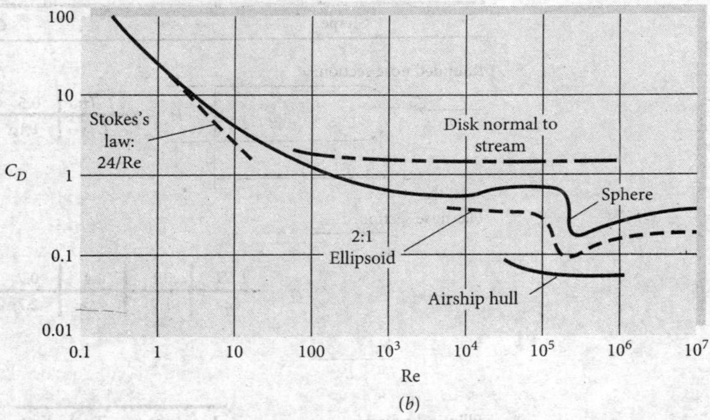

Fig. 7.16 Drag coefficients of smooth bodies at low Mach numbers: (*a*) two-dimensional bodies; (*b*) three-dimensional bodies. Note the Reynolds number independence of blunt bodies at high Re.

If the geometry is simple (for example, a sphere or disk), closed-form solutions can be found and the body drag can be computed [2]. Stokes himself provided the sphere drag formula:

$$F_{\text{sphere}} = 3\pi\,\mu U d$$

or

$$C_D = \frac{F}{\frac{1}{2}\rho U^2 \frac{\pi}{4} d^2} = \frac{24}{\rho U d/\mu} = \frac{24}{\text{Re}_d} \tag{7.64}$$

This relation is plotted in Fig. 7.16*b* and is seen to be accurate for about $\text{Re}_d \leq 1$.

Table 7.2 gives a few data on drag, based on frontal area, of two-dimensional bodies of various cross section, at $\text{Re} \geq 10^4$. The sharp-edged bodies, which tend to cause flow separation regardless of the character of the boundary layer, are insensitive to the Reynolds number. The elliptic cylinders, being smoothly rounded, have the laminar-to-turbulent transition effect of Figs. 7.13 and 7.14 and are therefore quite sensitive to whether the boundary layer is laminar or turbulent.

Table 7.2 Drag of Two-Dimensional Bodies at Re $\geq 10^4$

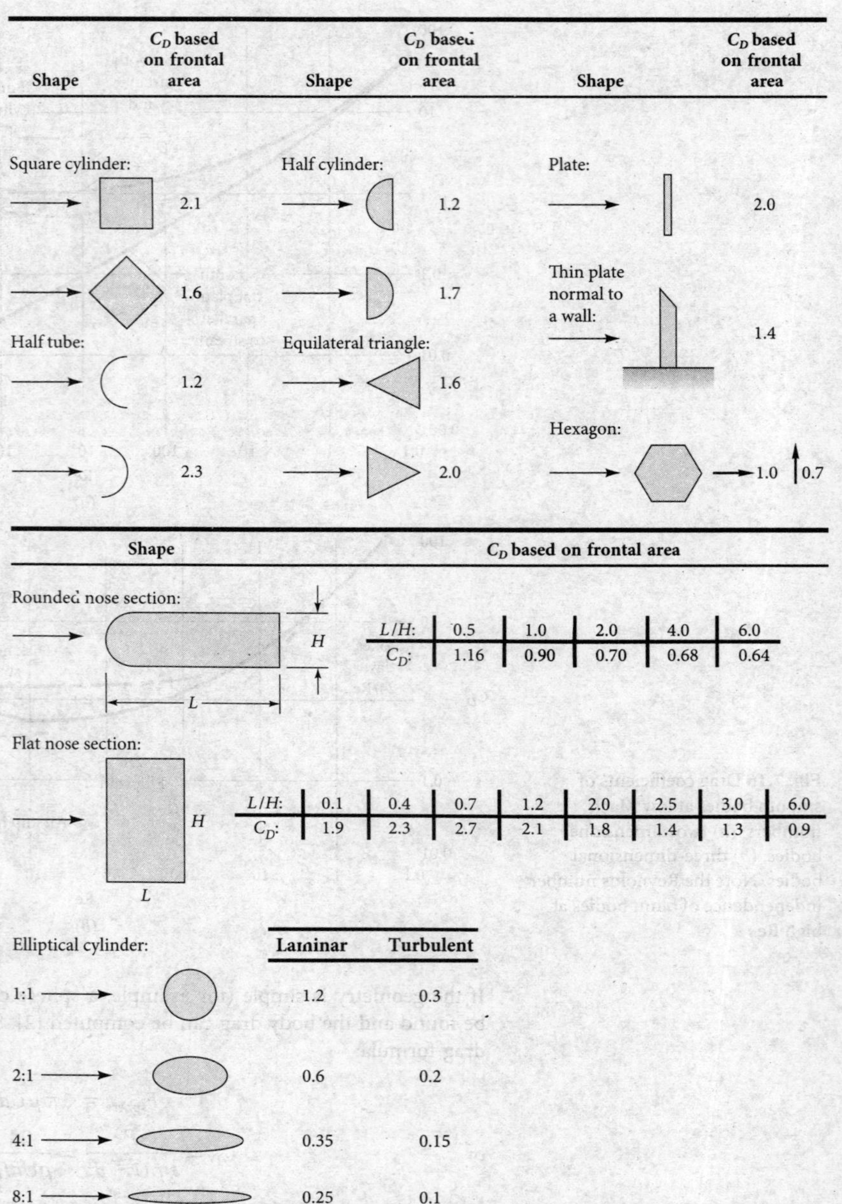

Shape	C_D based on frontal area	Shape	C_D based on frontal area	Shape	C_D based on frontal area
Square cylinder: (flat face to flow)	2.1	Half cylinder: (flat face downstream)	1.2	Plate:	2.0
Square cylinder (edge to flow)	1.6	Half cylinder (flat face to flow)	1.7	Thin plate normal to a wall:	1.4
Half tube: (convex side to flow)	1.2	Equilateral triangle: (point to flow)	1.6	Hexagon:	→ 1.0 ↑ 0.7
Half tube (concave side to flow)	2.3	Equilateral triangle (flat face to flow)	2.0		

Shape	C_D based on frontal area							
Rounded nose section:	L/H:	0.5	1.0	2.0	4.0	6.0		
	C_D:	1.16	0.90	0.70	0.68	0.64		
Flat nose section:	L/H:	0.1	0.4	0.7	1.2	2.0	2.5	3.0	6.0
	C_D:	1.9	2.3	2.7	2.1	1.8	1.4	1.3	0.9

Elliptical cylinder:	Laminar	Turbulent
1:1	1.2	0.3
2:1	0.6	0.2
4:1	0.35	0.15
8:1	0.25	0.1

EXAMPLE 7.6

A square 0.15 m piling is acted on by a water flow of 1.5 m/s that is 6 m deep, as shown in Fig. E7.6. Estimate the maximum bending exerted by the flow on the bottom of the piling.

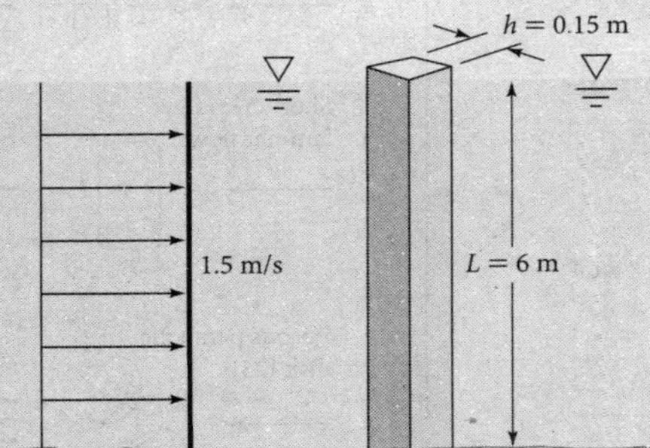

E7.6

Solution

Assume seawater with $\rho = 1{,}025\ \text{kg/m}^3$ and kinematic viscosity $\nu = 1.05\text{ E-6 m}^2/\text{s}$. With a piling width of 0.15 m, we have

$$\text{Re}_h = \frac{(1.5\ \text{m/s})(0.15\ \text{m})}{1.05\ \text{E-6 m}^2/\text{s}} = 2.14\ \text{E5}$$

This is the range where Table 7.2 applies. The worst case occurs when the flow strikes the flat side of the piling, $C_D \approx 2.1$. The frontal area is $A = Lh = (6\ \text{m})(0.15\ \text{m}) = 0.9\ \text{m}^2$. The drag is estimated by

$$F = C_D(\tfrac{1}{2}\rho V^2 A) \approx 2.1(\tfrac{1}{2})(1.025\ \text{kg/m}^3)(1.5\ \text{m/s})^2(0.9\ \text{m}^2) = 2{,}180\ \text{N}$$

If the flow is uniform, the center of this force should be at approximately middepth. Therefore, the bottom bending moment is

$$M_0 \approx \frac{FL}{2} = 2{,}180\ \text{N}(3\ \text{m}) = 6{,}540\ \text{N}\cdot\text{m} \qquad \textit{Ans.}$$

According to the flexure formula from strength of materials, the bending stress at the bottom would be

$$S = \frac{M_0 y}{I} = \frac{(6{,}540\ \text{N}\cdot\text{m})(0.075\ \text{m})}{\frac{1}{12}(0.15\ \text{m})^4} = 11.63\ \text{MPa}$$

to be multiplied, of course, by the stress concentration factor due to the built-in end conditions.

Three-Dimensional Bodies

Some drag coefficients of three-dimensional bodies are listed in Table 7.3 and Fig. 7.16*b*. Again we can conclude that sharp edges always cause flow separation and high drag that is insensitive to the Reynolds number. Rounded bodies like the ellipsoid have drag that depends on the point of separation, so both the Reynolds number and the character of the boundary layer are important. Body length will generally decrease pressure drag by making the body relatively more slender, but sooner or later the friction drag will catch up. For the flat-faced cylinder in Table 7.3, pressure drag decreases with L/d but friction increases, so minimum drag occurs at about $L/d = 2$.

Table 7.3 Drag of Three-Dimensional Bodies at Re ≥ 10^4

Body	C_D based on frontal area
Cube: (face to flow)	1.07
Cube: (edge to flow)	0.81
Cup: (open side to flow)	1.4
Cup: (closed side to flow)	0.4
Disk:	1.17
Parachute (Low porosity):	1.2
Streamlined train (approximately 5 cars):	$C_DA = 8.5\ \text{m}^2$
Bicycle:	Upright: $C_DA = 0.51\ \text{m}^2$; Racing: $C_DA = 0.30\ \text{m}^2$

Cone: [60]

θ:	10°	20°	30°	40°	60°	75°	90°
C_D:	0.30	0.40	0.55	0.65	0.80	1.05	1.15

Short cylinder, laminar flow:

L/D:	1	2	3	5	10	20	40	∞
C_D:	0.64	0.68	0.72	0.74	0.82	0.91	0.98	1.20

Porous parabolic dish [23]:

Porosity:	0	0.1	0.2	0.3	0.4	0.5
← C_D:	1.42	1.33	1.20	1.05	0.95	0.82
→ C_D:	0.95	0.92	0.90	0.86	0.83	0.80

Average person: → $C_DA = 0.84\ \text{m}^2$ ↑ $C_DA = 0.11\ \text{m}^2$

Pine and spruce trees [24]:

U, m/s:	10	20	30	40
C_D:	1.2 ± 0.2	1.0 ± 0.2	0.7 ± 0.2	0.5 ± 0.2

Tractor-trailer truck: Without deflector: 0.96; with deflector: 0.76

Body	Ratio	C_D based on frontal area
Rectangular plate:	b/h 1	1.18
	5	1.2
	10	1.3
	20	1.5
	∞	2.0

Body	Ratio	C_D based on frontal area
Flat-faced cylinder:	L/d 0.5	1.15
	1	0.90
	2	0.85
	4	0.87
	8	0.99

Body	Ratio	Laminar	Turbulent
Ellipsoid:	L/d 0.75	0.5	0.2
	1	0.47	0.2
	2	0.27	0.13
	4	0.25	0.1
	8	0.2	0.08

Buoyant rising sphere [50], $135 < \text{Re}_d < 1\text{E}5$: $C_D \approx 0.95$

Buoyant Rising Light Spheres

The sphere data in Fig. 7.16*b* are for fixed models in wind tunnels and from falling sphere tests and indicate a drag coefficient of about 0.5 in the range 1 E3 < Re_d < 1 E5. It was pointed out [50] that this is *not* the case for a freely rising buoyant sphere or bubble. If the sphere is light, $\rho_{sphere} < 0.8\ \rho_{fluid}$, a wake instability arises in the range 135 < Re_d < 1 E5. The sphere then spirals upward at an angle of about 60° from the horizontal. The drag coefficient is approximately doubled, to an average value $C_D \approx$ 0.95, as listed in Table 7.3 [50]. For a heavier body, $\rho_{sphere} \approx \rho_{fluid}$, the buoyant sphere rises vertically and the drag coefficient follows the standard curve in Fig. 7.16*b*.

EXAMPLE 7.7

According to Ref. 12, the drag coefficient of a blimp, based on surface area, is approximately 0.006 if $Re_L > 10^6$. A certain blimp is 75 m long and has a surface area of 3,400 m^2. Estimate the power required to propel this blimp at 18 m/s at a standard altitude of 1,000 m.

Solution

- *Assumptions:* We hope the Reynolds number will be high enough that the given data are valid.
- *Approach:* Determine if $Re_L > 10^6$ and, if so, compute the drag and the power required.
- *Property values:* Table A.6 at z = 1,000 m: ρ = 1.112 kg/m^3, T = 282 K, thus $\mu \approx$ 1.75 E-5 kg/m · s.
- *Solution steps:* Determine the Reynolds number of the blimp:

$$Re_L = \frac{\rho U L}{\mu} = \frac{(1.112\ \text{kg/m}^3)(18\ \text{m/s})(75\ \text{m})}{1.75\ \text{E-5 kg/m}\cdot\text{s}} = 8.6\ \text{E7} > 10^6 \quad \text{OK}$$

The given drag coefficient is valid. Compute the blimp drag and the power = (drag) × (velocity):

$$F = C_D \frac{\rho}{2} U^2 A_{wet} = (0.006)\frac{1.112\ \text{kg/m}^3}{2}(18\ \text{m/s})^2\,(3{,}400\ \text{m}^2) = 3{,}675\ \text{N}$$

$$\text{Power} = FV = (3{,}675\ \text{N})(18\ \text{m/s}) = 66{,}000\ \text{W}\ \ (89\ \text{hp}) \qquad \textit{Ans.}$$

- *Comments:* These are nominal estimates. Drag is highly dependent on both shape and Reynolds number, and the coefficient C_D = 0.006 has considerable uncertainty.

Aerodynamic Forces on Road Vehicles

Automobiles and trucks are now the subject of much research on aerodynamic forces, both lift and drag [21]. At least one textbook is devoted to the subject [22]. A very readable description of race car drag is given by Katz [51]. Consumer, manufacturer, and government interest has cycled between high speed/high horsepower and lower speed/lower drag. Better streamlining of car shapes has resulted over the years in a large decrease in the automobile drag coefficient, as shown in Fig. 7.17*a*. Modern cars have an average drag coefficient of about 0.25, based on the frontal area. Since the frontal area has also decreased sharply, the actual raw drag *force* on cars has dropped even more than indicated in Fig. 7.17*a*. The theoretical minimum shown in the figure, $C_D \approx$ 0.15, is about right for a commercial automobile, but lower values are possible for experimental vehicles; see Prob. P7.109. Note that basing C_D on the frontal area is awkward, since one would need an accurate drawing of the automobile to estimate its frontal area. For this reason, some technical articles simply report the raw drag in newtons or pound-force, or the product $C_D A$.

Many companies and laboratories have automotive wind tunnels, some full-scale and/or with moving floors to approximate actual kinematic similarity. The blunt

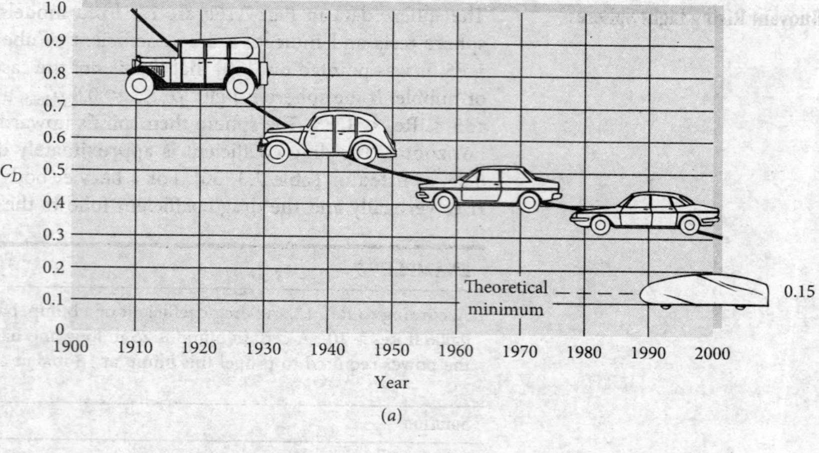

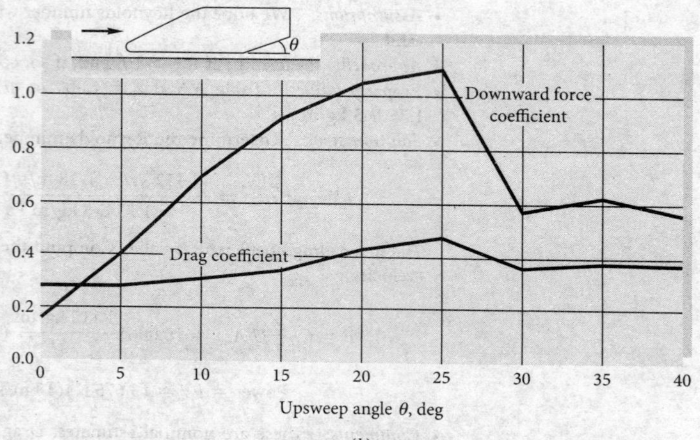

Fig. 7.17 Aerodynamics of automobiles: (*a*) the historical trend for drag coefficients (from Ref. 21); (*b*) effect of bottom rear upsweep angle on drag and downward lift force (from Ref. 25).

shapes of most automobiles, together with their proximity to the ground, cause a wide variety of flow and geometric effects. Simple changes in part of the shape can have a large influence on aerodynamic forces. Figure 7.17*b* shows force data by Bearman et al. [25] for an idealized smooth automobile shape with upsweep in the rear of the bottom section. We see that by simply adding an upsweep angle of 25°, we can quadruple the downward force, gaining tire traction at the expense of doubling the drag. For this study, the effect of a moving floor was small—about a 10 percent increase in both drag and lift compared to a fixed floor.

It is difficult to quantify the exact effect of geometric changes on automotive forces, since, for example, changes in a windshield shape might interact with downstream flow over the roof and trunk. Nevertheless, based on correlation of many model and full-scale tests, Ref. 26 proposes a formula for automobile drag that adds separate effects such as front ends, cowls, fenders, windshield, roofs, and rear ends.

Figure 7.18*a* illustrates the power required to drive a typical tractor-trailer truck. An approximation is that the rolling resistance increases linearly and the drag quadratically with speed. The two are about equal at 90 km/h. Figure 7.18*b* shows that air drag can be reduced by attaching a deflector to the top of the cab. When the deflector

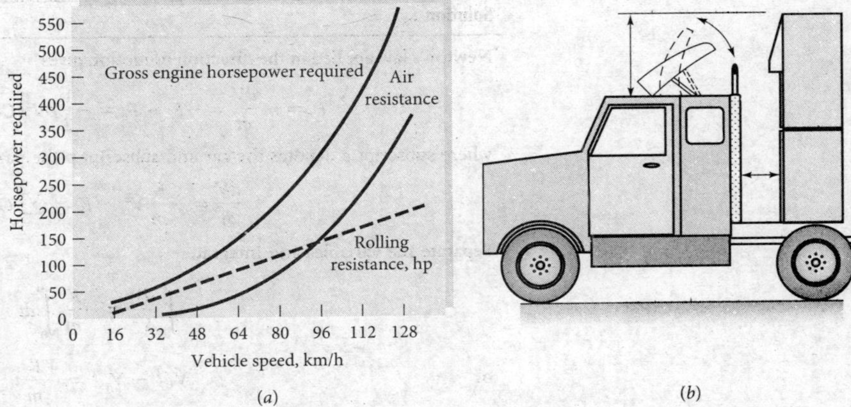

Fig. 7.18 Drag reduction of a tractor-trailer truck: (*a*) horsepower required to overcome resistance; (*b*) deflector added to cab reduces air drag by 20 percent.

angle is adjusted to carry the flow smoothly over the top and sides of the trailer, the reduction in C_D is about 20 percent. This type of applied fluids engineering is very important for modern transportation problems [58].

The velocity effect on rolling resistance is mostly due to the engine-transmission-wheel-bearing system. The tires generally have a nearly constant *rolling resistance coefficient,*

$$C_{rr} = \frac{F_{rr}}{N}$$

where F_{rr} is the resistance force and N is the normal force on the tires [61]. This coefficient C_{rr} is analogous to a solid friction factor but is much smaller: about 0.01 to 0.04 for passenger car tires and 0.006 to 0.01 for truck tires.

Progress in computational fluid dynamics means that complicated vehicle flow fields can be predicted fairly well. Reference 42 compares one-equation and two-equation turbulence models [9] with NASA data for a simplified tractor-trailer model. Even with two million mesh points, the predicted vehicle drag is from 20 to 50 percent higher than the measurements. The turbulence models do not reproduce the pressures and wake structure in the rear of the vehicle. Newer models, such as large eddy simulation (LES) and direct numerical simulation (DNS) will no doubt improve the calculations.

EXAMPLE 7.8

A high-speed car with $m = 2{,}000$ kg, $C_D = 0.3$, and $A = 1\ \text{m}^2$ deploys a 2-m parachute to slow down from an initial velocity of 100 m/s (Fig. E7.8). Assuming constant C_D, brakes free, and no rolling resistance, calculate the distance and velocity of the car after 1, 10, 100, and 1,000 s. For air assume $\rho = 1.2\ \text{kg/m}^3$, and neglect interference between the wake of the car and the parachute.

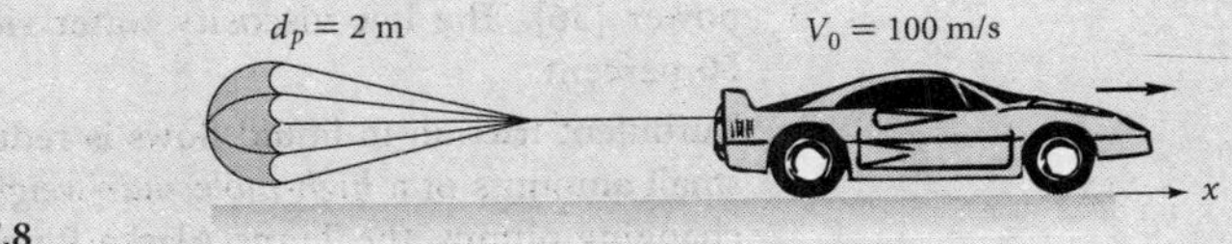

E7.8

Solution

Newton's law applied in the direction of motion gives

$$F_x = m\frac{dV}{dt} = -F_c - F_p = -\frac{1}{2}\rho V^2(C_{Dc}A_c + C_{Dp}A_p)$$

where subscript c denotes the car and subscript p the parachute. This is of the form

$$\frac{dV}{dt} = -\frac{K}{m}V^2 \qquad K = \sum C_D A\frac{\rho}{2}$$

Separate the variables and integrate:

$$\int_{V_0}^{V}\frac{dV}{V^2} = -\frac{K}{m}\int_0^t dt$$

or

$$V_0^{-1} - V^{-1} = -\frac{K}{m}t$$

Rearrange and solve for the velocity V:

$$V = \frac{V_0}{1 + (K/m)V_0 t} \qquad K = \frac{(C_{Dc}A_c + C_{Dp}A_p)\rho}{2} \tag{1}$$

We can integrate this to find the distance traveled:

$$S = \frac{V_0}{\alpha}\ln(1 + \alpha t) \qquad \alpha = \frac{K}{m}V_0 \tag{2}$$

Now work out some numbers. From Table 7.3, $C_{Dp} \approx 1.2$; hence

$$C_{Dc}A_c + C_{Dp}A_p = 0.3(1\text{ m}^2) + 1.2\frac{\pi}{4}(2\text{ m})^2 = 4.07\text{ m}^2$$

Then

$$\frac{K}{m}V_0 = \frac{\frac{1}{2}(4.07\text{ m}^2)(1.2\text{ kg/m}^3)(100\text{ m/s})}{2{,}000\text{ kg}} = 0.122\text{ s}^{-1} = \alpha$$

Now make a table of the results for V and S from Eqs. (1) and (2):

t, s	1	10	100	1,000
V, m/s	89	45	7.6	0.8
S, m	94	654	2,110	3,940

Air resistance alone will not stop a body completely. If you don't apply the brakes, you'll be halfway to the Yukon Territory and still going.

Other Methods of Drag Reduction

Sometimes drag is good, for example, when using a parachute. Do not jump out of an airplane holding a flat plate parallel to your motion (see Prob. P7.81). Mostly, though, drag is bad and should be reduced. The classical method of drag reduction is *streamlining* (Figs. 7.15 and 7.18). For example, nose fairings and body panels have produced motorcycles that can travel over 300 km/h. More recent research has uncovered other methods that hold great promise, especially for turbulent flows.

1. Oil pipelines introduce an *annular strip* of water to reduce the pumping power [36]. The low-viscosity water rides the wall and reduces friction up to 60 percent.
2. Turbulent friction in liquid flows is reduced up to 60 percent by dissolving small amounts of a *high-molecular-weight polymer additive* [37]. Without changing pumps, the Trans-Alaska Pipeline System (TAPS) increased oil flow 50 percent by injecting small amounts of polymer dissolved in kerosene.

3. Stream-oriented surface *vee-groove microriblets* can reduce turbulent friction up to 8 percent [38]. Riblet heights are of order 1 mm and were used on the Stars and Stripes yacht hull in the Americas Cup races. Riblets are also effective on aircraft skins.
4. Small, near-wall *large-eddy breakup devices* (LEBUs) reduce local turbulent friction up to 10 percent [39]. However, one must add these small structures to the surface, and LEBU drag may be significant.
5. Air *microbubbles* injected at the wall of a water flow create a low-shear bubble blanket [40]. At high void fractions, drag reduction can be 80 percent.
6. Spanwise (transverse) *wall oscillation* may reduce turbulent friction up to 30 percent [41].
7. *Active flow control*, especially of turbulent flows, is the wave of the future, as reviewed in Ref. 47. These methods generally require expenditure of energy but can be worth it. For example, tangential blowing at the rear of an auto [48] evokes the *Coanda effect*, in which the separated near-wake flow attaches itself to the body surface and reduces auto drag up to 10 percent.

Drag reduction is presently an area of intense and fruitful research and applies to many types of airflows [39, 53] and water flows for both vehicles and conduits.

Drag of Surface Ships

The drag data given so far, such as Tables 7.2 and 7.3, are for bodies "fully immersed" in a free stream—that is, with no free surface. If, however, the body moves at or near a free liquid surface, *wave-making drag* becomes important and is dependent on both the Reynolds number and the Froude number. To move through a water surface, a ship must create waves on both sides. This implies putting energy into the water surface and requires a finite drag force to keep the ship moving, even in a frictionless fluid. The total drag of a ship can then be approximated as the sum of friction drag and wave-making drag:

$$F \approx F_{\text{fric}} + F_{\text{wave}} \quad \text{or} \quad C_D \approx C_{D,\text{fric}} + C_{D,\text{wave}}$$

The friction drag can be estimated by the (turbulent) flat-plate formula, Eq. (7.45), based on the below-water or *wetted area* of the ship.

Reference 27 is an interesting review of both theory and experiment for wake-making surface ship drag. Generally speaking, the bow of the ship creates a wave system whose wavelength is related to the ship speed but not necessarily to the ship length. If the stern of the ship is a wave *trough*, the ship is essentially climbing uphill and has high wave drag. If the stern is a wave crest, the ship is nearly level and has lower drag. The criterion for these two conditions results in certain approximate Froude numbers [27]:

$$\text{Fr} = \frac{V}{\sqrt{gL}} \approx \frac{0.53}{\sqrt{N}} \quad \begin{array}{l}\text{high drag if } N = 1, 3, 5, 7, \ldots; \\ \text{low drag if } N = 2, 4, 6, 8, \ldots\end{array} \tag{7.65}$$

where V is the ship's speed, L is the ship's length along the centerline, and N is the number of half-lengths, from bow to stern, of the drag-making wave system. The wave drag will increase with the Froude number and oscillate between lower drag ($\text{Fr} \approx 0.38, 0.27, 0.22, \ldots$) and higher drag ($\text{Fr} \approx 0.53, 0.31, 0.24, \ldots$) with negligible variation for $\text{Fr} < 0.2$. Thus it is best to design a ship to cruise at $N = 2, 4, 6, 8$. Shaping the bow and stern can further reduce wave-making drag.

Figure 7.19 shows the data of Inui [27] for a model ship. The main hull, curve A, shows peaks and valleys in wave drag at the appropriate Froude numbers > 0.2.

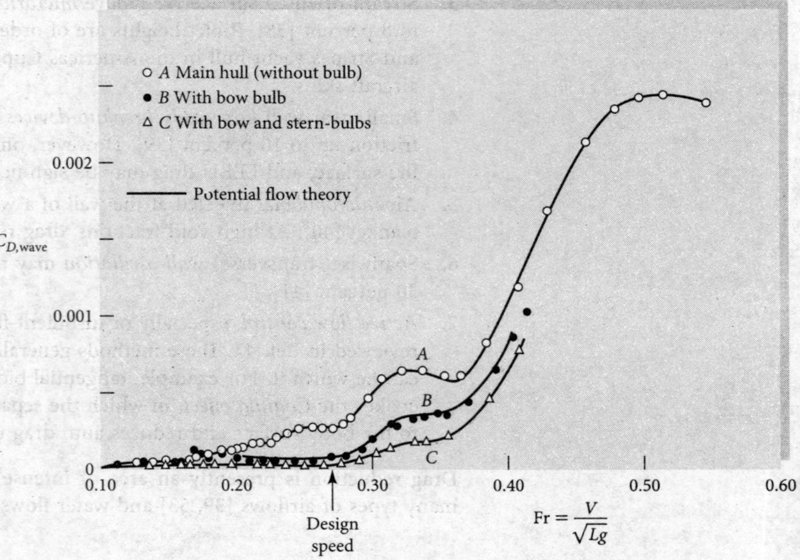

Fig. 7.19 Wave-making drag on a ship model. (*After Inui [27].*) *Note:* The drag coefficient is defined as $C_{DW} = 2F/(\rho V^2 L^2)$.

Introduction of a *bulb* protrusion on the bow, curve *B*, greatly reduces the drag. Adding a second bulb to the stern, curve *C*, is still better, and Inui recommends that the design speed of this two-bulb ship be at $N = 4$, $\text{Fr} \approx 0.27$, which is a nearly "waveless" condition. In this figure $C_{D,\text{wave}}$ is defined as $2F_{\text{wave}}/(\rho V^2 L^2)$ instead of using the wetted area.

The solid curves in Fig. 7.19 are based on potential flow theory for the below-water hull shape. Chapter 8 is an introduction to potential flow theory. Modern computers can be programmed for numerical CFD solutions of potential flow over the hulls of ships, submarines, yachts, and sailboats, including boundary layer effects driven by the potential flow [28]. Thus theoretical prediction of flow past surface ships is now at a fairly high level. See also Ref. 15.

Body Drag at High Mach Numbers

All the data presented to this point are for nearly incompressible flows, with Mach numbers assumed less than about 0.3. Beyond this value compressibility can be very important, with $C_D = \text{fcn}(\text{Re, Ma})$. As the stream Mach number increases, at some subsonic value $M_{\text{crit}} < 1$ that depends on the body's bluntness and thickness, the local velocity at some point near the body surface will become sonic. If Ma increases beyond Ma_{crit}, shock waves form, intensify, and spread, raising surface pressures near the front of the body and therefore increasing the pressure drag. The effect can be dramatic with C_D increasing tenfold, and 70 years ago this sharp increase was called the *sonic barrier*, implying that it could not be surmounted. Of course, it can be—the rise in C_D is finite, as supersonic bullets have proved for centuries.

Figure 7.20 shows the effect of the Mach number on the drag coefficient of various body shapes tested in air.[6] We see that compressibility affects blunt bodies earlier, with Ma_{crit} equal to 0.4 for cylinders, 0.6 for spheres, and 0.7 for airfoils and pointed projectiles. Also the Reynolds number (laminar versus turbulent boundary layer flow) has a large effect below Ma_{crit} for spheres and cylinders but becomes unimportant above

[6]There is a slight effect of the specific-heat ratio k, which would appear if other gases were tested.

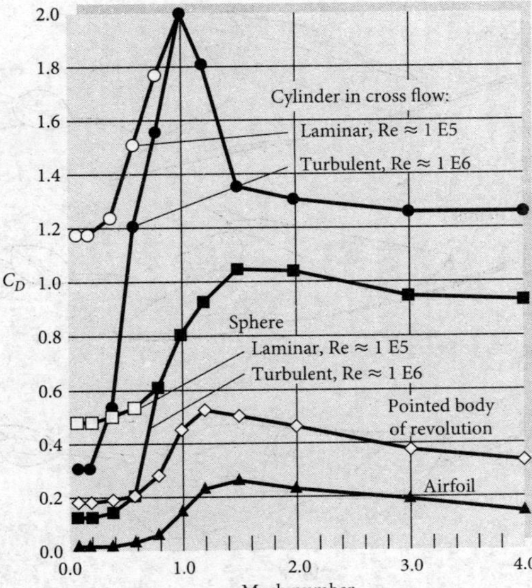

Fig. 7.20 Effect of the Mach number on the drag of various body shapes. (*Data from Refs. 23 and 29.*)

Ma ≈ 1. In contrast, the effect of the Reynolds number is small for airfoils and projectiles and is not shown in Fig. 7.20. A general statement might divide Reynolds and Mach number effects as follows:

$$\text{Ma} \le 0.3\text{: Reynolds number important, Mach number unimportant}$$

$$0.3 < \text{Ma} < 1\text{: both Reynolds and Mach numbers important}$$

$$\text{Ma} > 1.0\text{: Reynolds number unimportant, Mach number important}$$

At supersonic speeds, a broad *bow shock wave* forms in front of the body (see Figs. 9.10*b* and 9.19), and the drag is mainly due to high shock-induced pressures on the front. Making the bow a sharp point can sharply reduce the drag (Fig. 9.28) but does not eliminate the bow shock. Chapter 9 gives a brief treatment of compressible flow. References 30 and 31 are more advanced textbooks devoted entirely to compressible flow.

Biological Drag Reduction

A great deal of engineering effort goes into designing immersed bodies to reduce their drag. Most such effort concentrates on rigid-body shapes. A different process occurs in nature, as organisms adapt to survive high winds or currents, as reported in a series of papers by S. Vogel [33, 34]. A good example is a tree, whose flexible structure allows it to reconfigure in high winds and thus reduce drag and damage. Tree root systems have evolved in several ways to resist wind-induced bending moments, and trunk cross sections have become resistant to bending but relatively easy to twist and reconfigure. We saw this in Table 7.3, where tree drag coefficients [24] reduced by 60 percent as wind velocity increased. The shape of the tree changes to offer less resistance.

The individual branches and leaves of a tree also curl and cluster to reduce drag. Figure 7.21 shows the results of wind tunnel experiments by Vogel [33]. A tulip tree leaf, Fig. 7.21*a*, broad and open in low wind, curls into a conical low-drag shape as

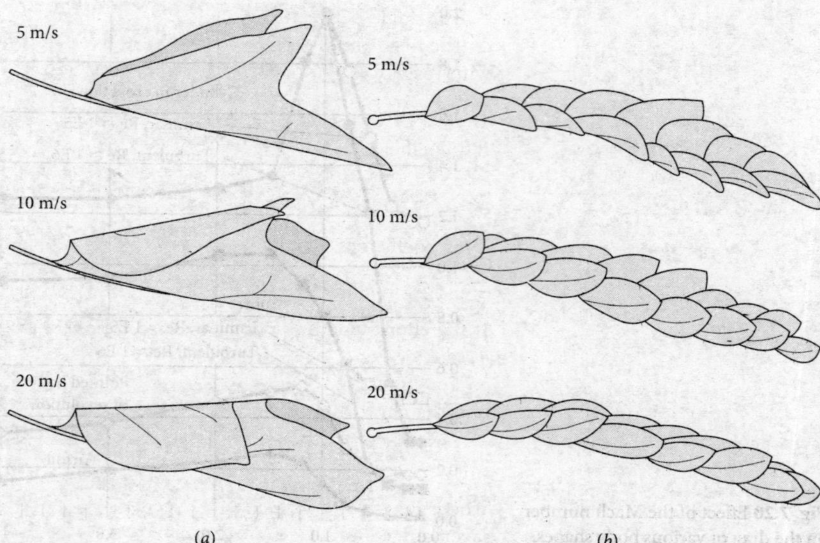

Fig. 7.21 Biological adaptation to wind forces: (*a*) a tulip tree leaf curls into a conical shape at high velocity; (*b*) black walnut leaves cluster into a low-drag shape as wind increases. (*From Vogel, Ref. 33.*)

wind increases. A compound black walnut leaf group, Fig. 7.21*b*, clusters into a low-drag shape at high wind speed. Although drag coefficients were reduced up to 50 percent by flexibility, Vogel points out that rigid structures are sometimes just as effective. An interesting recent symposium [35] was devoted entirely to the solid mechanics and fluid mechanics of biological organisms.

Forces on Lifting Bodies

Lifting bodies (airfoils, hydrofoils, or vanes) are intended to provide a large force normal to the free stream and as little drag as possible. Conventional design practice has evolved a shape not unlike a bird's wing—that is, relatively thin ($t/c \leq 0.24$) with a rounded leading edge and a sharp trailing edge. A typical shape is sketched in Fig. 7.22.

For our purposes we consider the body to be symmetric, as in Fig. 7.11, with the free-stream velocity in the vertical plane. If the chord line between the leading and trailing edge is not a line of symmetry, the airfoil is said to be *cambered.* The camber line is the line midway between the upper and lower surfaces of the vane.

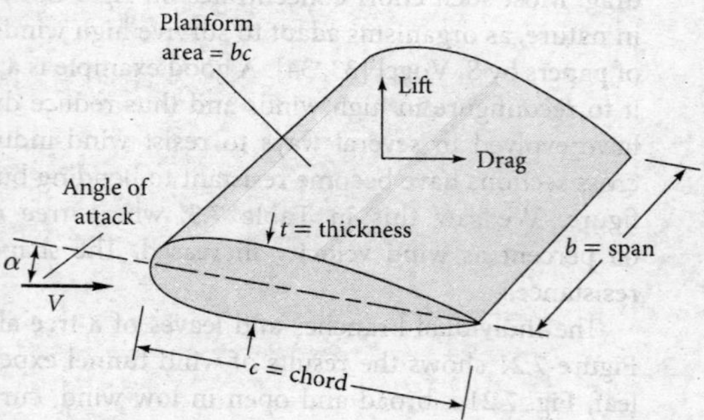

Fig. 7.22 Definition sketch for a lifting vane.

The angle between the free stream and the chord line is called the *angle of attack* α. The lift L and the drag D vary with this angle. The dimensionless forces are defined with respect to the planform area $A_p = bc$:

$$\text{Lift coefficient:} \qquad C_L = \frac{L}{\frac{1}{2}\rho V^2 A_p} \tag{7.66a}$$

$$\text{Drag coefficient:} \qquad C_D = \frac{D}{\frac{1}{2}\rho V^2 A_p} \tag{7.66b}$$

If the chord length is not constant, as in the tapered wings of modern aircraft, $A_p = \int c\, db$.

For low-speed flow with a given roughness ratio, C_L and C_D should vary with α and the chord Reynolds number:

$$C_L = f(\alpha, \text{Re}_c) \qquad \text{or} \qquad C_D = f(\alpha, \text{Re}_c)$$

where $\text{Re}_c = Vc/\nu$. The Reynolds numbers are commonly in the turbulent boundary layer range and have a modest effect.

The rounded leading edge prevents flow separation there, but the sharp trailing edge causes a tangential wake motion that generates the lift. Figure 7.23 shows what happens when a flow starts up past a lifting vane or an airfoil.

Just after start-up in Fig. 7.23*a* the streamline motion is irrotational and inviscid. The rear stagnation point, assuming a positive angle of attack, is on the upper surface, and there is no lift; but, the flow cannot long negotiate the sharp turn at the trailing edge: it separates, and a *starting vortex* forms in Fig. 7.23*b*. This starting vortex is shed downstream in Figs. 7.23*c* and *d*, and a smooth streamline flow develops over the wing, leaving the foil in a direction approximately parallel to the chord line. Lift at this time is fully developed, and the starting vortex is gone. Should the flow now cease, a *stopping vortex* of opposite (clockwise) sense will form and be shed. During flight, increases or decreases in lift will cause incremental starting or stopping vortices, always with the effect of maintaining a smooth parallel flow at the trailing edge. We pursue this idea mathematically in Chap. 8.

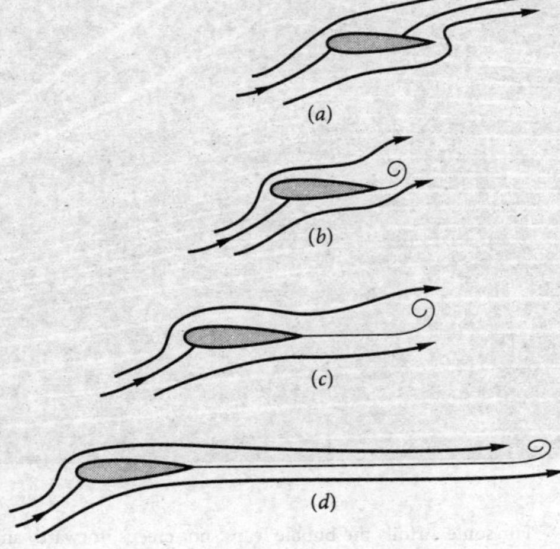

Fig. 7.23 Transient stages in the development of lift: (*a*) start-up: rear stagnation point on the upper surface: no lift; (*b*) sharp trailing edge induces separation, and a starting vortex forms: slight lift; (*c*) starting vortex is shed; and streamlines flow smoothly from trailing edge: lift is now 80 percent developed; (*d*) starting vortex now shed far behind, trailing edge now very smooth: lift fully developed.

At a low angle of attack, the rear surfaces have an adverse pressure gradient but not enough to cause significant boundary layer separation. The flow pattern is smooth, as in Fig. 7.23*d*, and drag is small and lift excellent. As the angle of attack is increased, the upper-surface adverse gradient becomes stronger, and generally a *separation bubble* begins to creep forward on the upper surface.[7] At a certain angle $\alpha = 15$ to $20°$, the flow is separated completely from the upper surface, as in Fig. 7.24. The airfoil is said to be *stalled*: Lift drops off markedly, drag increases markedly, and the foil is no longer flyable.

Early airfoils were thin, modeled after birds' wings. The German engineer Otto Lilienthal (1848–1896) experimented with flat and cambered plates on a rotating arm. He and his brother Gustav flew the world's first glider in 1891. Horatio Frederick Phillips (1845–1912) built the first wind tunnel in 1884 and measured the lift and drag of cambered vanes. The first theory of lift was proposed by Frederick W. Lanchester shortly afterward. Modern airfoil theory dates from 1905, when the Russian hydrodynamicist N. E. Joukowsky (1847–1921) developed a circulation theorem (Chap. 8) for computing airfoil lift for arbitrary camber and thickness. With this basic theory, as extended and developed by Prandtl and Kármán and their students, it is now possible to design a low-speed airfoil to satisfy particular surface pressure distributions and boundary layer characteristics. There are whole families of airfoil designs, notably those developed in the United States under the sponsorship of the NACA (now NASA). Extensive theory and data on these airfoils are contained in Ref. 16. We shall discuss this further in Chap. 8. The history of aeronautics is a rich and engaging topic and highly recommended to the reader [43, 44].

Figure 7.25 shows the lift and drag on a symmetric airfoil denoted as the NACA 0009 foil, the last digit indicating the thickness of 9 percent. With no flap extended,

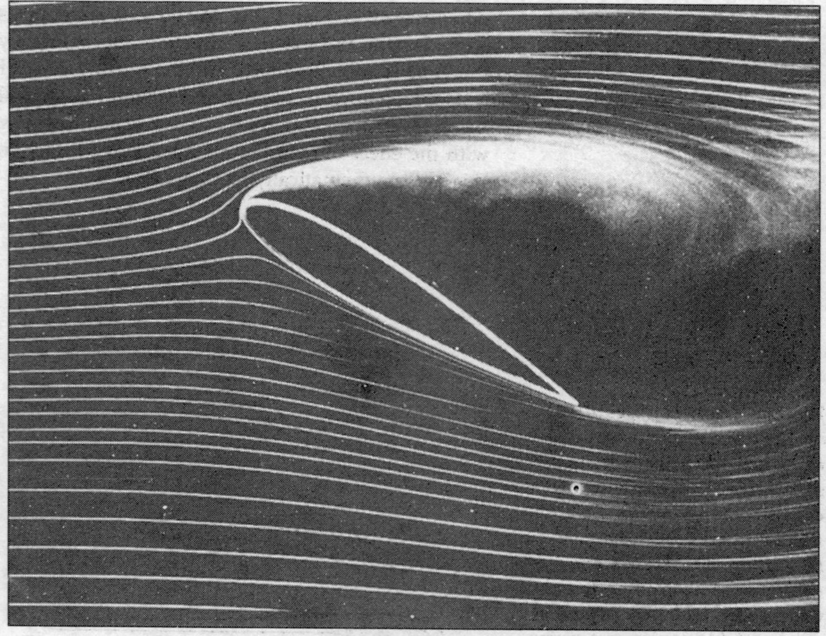

Fig. 7.24 At high angle of attack, smoke flow visualization shows stalled flow on the upper surface of a lifting vane. (*National Committee for Fluid Mechanics Films, Education Development Center, Inc., © 1972.*)

[7]For some airfoils the bubble leaps, not creeps, forward, and stall occurs rapidly and dangerously.

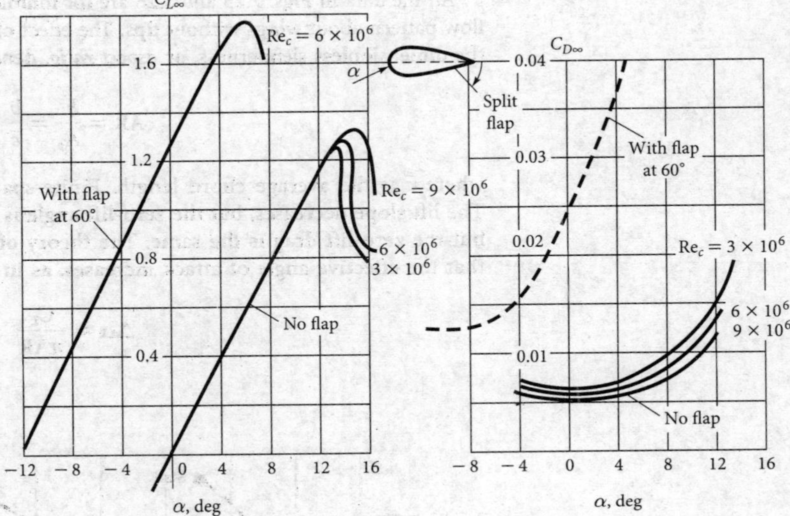

Fig. 7.25 Lift and drag of a symmetric NACA 0009 airfoil of infinite span, including effect of a split-flap deflection. Note that roughness can increase C_D from 100 to 300 percent.

this airfoil, as expected, has zero lift at zero angle of attack. Up to about 12° the lift coefficient increases linearly with a slope of 0.1 per degree, or 6.0 per radian. This is in agreement with the theory outlined in Chap. 8:

$$C_{L,\text{theory}} \approx 2\pi \sin\left(\alpha + \frac{2h}{c}\right) \tag{7.67}$$

where h/c is the maximum camber expressed as a fraction of the chord. The NACA 0009 has zero camber; hence $C_L = 2\pi \sin \alpha \approx 0.11\alpha$, where α is in degrees. This is excellent agreement.

The drag coefficient of the smooth-model airfoils in Fig. 7.25 is as low as 0.005, which is actually lower than both sides of a flat plate in turbulent flow. This is misleading inasmuch as a commercial foil will have roughness effects; for example, a paint job will double the drag coefficient.

The effect of increasing Reynolds number in Fig. 7.25 is to increase the maximum lift and stall angle (without changing the slope appreciably) and to reduce the drag coefficient. This is a salutary effect since the prototype will probably be at a higher Reynolds number than the model (10^7 or more).

For takeoff and landing, the lift is greatly increased by deflecting a split flap, as shown in Fig. 7.25. This makes the airfoil unsymmetric (or effectively cambered) and changes the zero-lift point to $\alpha = -12°$. The drag is also greatly increased by the flap, but the reduction in takeoff and landing distance is worth the extra power needed.

A lifting craft cruises at low angle of attack, where the lift is much larger than the drag. Maximum lift-to-drag ratios for the common airfoils lie between 20 and 50.

Some airfoils, such as the NACA 6 series, are shaped to provide favorable gradients over much of the upper surface at low angles. Thus separation is small, and transition to turbulence is delayed; the airfoil retains a good length of laminar flow even at high Reynolds numbers. The lift-drag *polar plot* in Fig. 7.26 shows the NACA 0009 data from Fig. 7.25 and a laminar flow airfoil, NACA 63-009, of the same thickness. The laminar flow airfoil has a low-drag bucket at small angles but also suffers lower stall angle and lower maximum lift coefficient. The drag is 30 percent less in the bucket, but the bucket disappears if there is significant surface roughness.

All the data in Figs. 7.25 and 7.26 are for infinite span—that is, a two-dimensional flow pattern about wings without tips. The effect of finite span can be correlated with the dimensionless slenderness, or *aspect ratio*, denoted (AR):

$$\mathrm{AR} = \frac{b^2}{A_p} = \frac{b}{\bar{c}} \tag{7.68}$$

where $\bar{c}$ is the average chord length. Finite-span effects are shown in Fig. 7.27. The lift slope decreases, but the zero-lift angle is the same; and the drag increases, but the zero-lift drag is the same. The theory of finite-span airfoils [16] predicts that the effective angle of attack increases, as in Fig. 7.27, by the amount

$$\Delta\alpha \approx \frac{C_L}{\pi \mathrm{AR}} \tag{7.69}$$

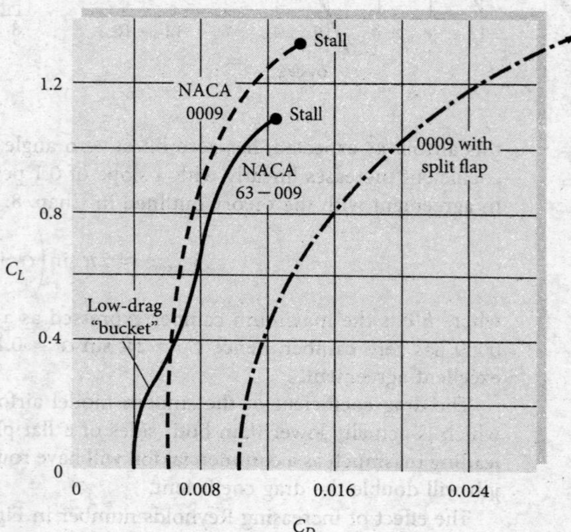

Fig. 7.26 Lift-drag polar plot for standard (0009) and a laminar flow (63-009) NACA airfoil.

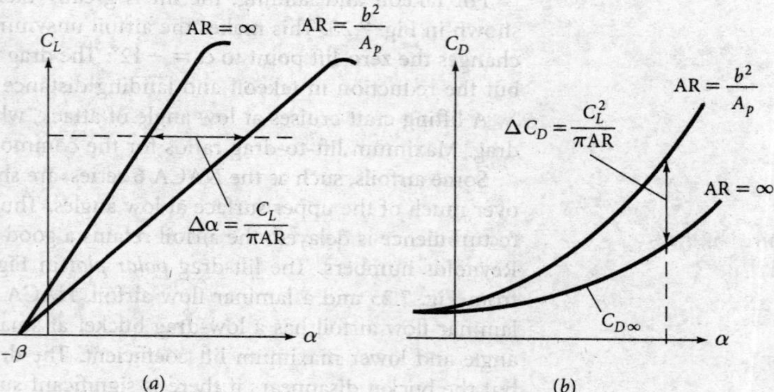

Fig. 7.27 Effect of finite aspect ratio on lift and drag of an airfoil: (*a*) effective angle increase; (*b*) induced drag increase.

When applied to Eq. (7.67), the finite-span lift becomes

$$C_L \approx \frac{2\pi \sin (\alpha + 2h/c)}{1 + 2/\text{AR}} \tag{7.70}$$

The associated drag increase is $\Delta C_D \approx C_L \sin \Delta\alpha \approx C_L \, \Delta\alpha$, or

$$C_D \approx C_{D\infty} + \frac{C_L^2}{\pi \text{AR}} \tag{7.71}$$

where $C_{D\infty}$ is the drag of the infinite-span airfoil, as sketched in Fig. 7.25. These correlations are in good agreement with experiments on finite-span wings [16].

The existence of a maximum lift coefficient implies the existence of a minimum speed, or *stall speed*, for a craft whose lift supports its weight:

$$L = W = C_{L,\max}(\tfrac{1}{2}\rho V_s^2 A_p)$$

or

$$V_s = \left(\frac{2W}{C_{L,\max}\rho A_p}\right)^{1/2} \tag{7.72}$$

The stall speed of typical aircraft varies between 20 m/s and 60 m/s, depending on the weight and value of $C_{L,\max}$. The pilot must hold the speed greater than about $1.2V_s$ to avoid the instability associated with complete stall.

The split flap in Fig. 7.25 is only one of many devices used to secure high lift at low speeds. Figure 7.28*a* shows six such devices whose lift performance is given in Fig. 7.28*b* along with a standard (*A*) and laminar flow (*B*) airfoil. The double-slotted flap achieves $C_{L,\max} \approx 3.4$, and a combination of this plus a leading-edge slat can achieve $C_{L,\max} \approx 4.0$. These are not scientific curiosities; for instance, the Boeing 727 commercial jet aircraft uses a triple-slotted flap plus a leading-edge slat during landing.

A violation of conventional aerodynamic wisdom is that military aircraft are beginning to fly, briefly, *above the stall point.* Fighter pilots are learning to make quick maneuvers in the stalled region as detailed in Ref. 32. Some planes can even *fly continuously* while stalled—the Grumman X-29 experimental aircraft recently set a record by flying at $\alpha = 67°$.

The Kline-Fogelman Airfoil

Traditionally, an airfoil is a thin teardrop shape, with a rounded leading edge and a sharp trailing edge, Fig. 7.28*a*. It provides low drag but stalls at low $\alpha \approx 10$ to 15°. In 1972 R. F. Kline and F. F. Fogelman designed an airfoil with a sharp leading edge and a rear cut-out [17]. When tested in a wind tunnel, Fig. 7.28*b*, it did not stall until $\alpha \approx 45°$, but the drag was very high. Fertis [55] rounded the leading edge and reduced the drag. Finaish and Witherspoon [56] made further improvements, but the drag is still too high for full-scale commercial applications. The KF airfoil, though, is extremely popular for radio-controlled model aircraft.

A Wing Inspired by the Humpback Whale

Biologists have long noticed the high maneuverability of the humpback whale when it seeks prey. Unlike most whales, the humpback has tubercles, or bumps, on the leading edge of its flippers. Miklosovic et al. [57] tested this idea, using a standard wing with periodic bumps glued to its leading edge, as in Fig. 7.29. They report a 40 percent increase in stall angle, compared to the same wing without bumps, plus higher lift and higher lift-to-drag ratios. The concept has promise for commercial applications such

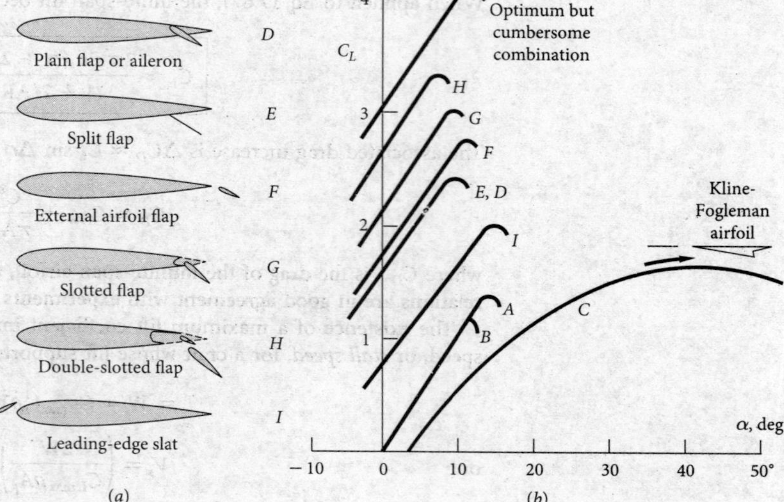

Fig. 7.28 Performance of airfoils with and without high-lift devices: A = NACA 0009; B = NACA 63-009; C = Kline-Fogleman airfoil (*from Ref. 17*); D to I shown in (*a*): (*a*) types of high-lift devices; (*b*) lift coefficients for various devices (Ref. 62).

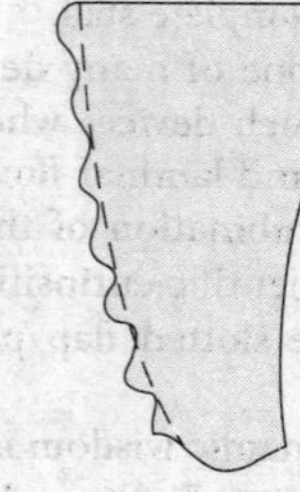

Fig. 7.29 New experimental airfoils: plan view of a wing modeled on the humpback whale flipper [57].
Source: D. S. Miklosovic et al., "Leading Edge Tubercles Delay Stall on Humpback Whale," Physics of Fluids, vol. 16, no. 5, May 2004, pp. L39–L42.

as wind turbine blades. Flow visualization shows that the bumps create energetic streamwise vortices along the wing surface, helping to delay separation.

A Combination Car and Airplane

Engineers have long dreamed of a viable car that can fly. Hop in at home, drive to the airport, fly somewhere, then drive to the motel. Designer efforts date back to the Glenn Curtiss 1917 Autoplane, with other projects in the 1930s and 1940s. Perhaps the most famous was Moulton Taylor's Aerocar in 1947. Only five Aerocars were built. The year 2008 seems to have been the Year of the Car-Plane, with at least five different companies working on designs. Engineers can now use lightweight materials, better engines, and guidance systems. The writer's favorite is the *Transition*®, made by Terrafugia, Inc., shown in Fig. 7.30.

The Transition® has wings that fold out to a span of 8.4 m. The twin tails do not fold. The front canard wing doubles as a bumper for the highway. The 100-hp engine drives both the rear propeller for flight and also the front wheels for the highway. Gross takeoff weight is 6,360 N. Operators need only a Light Sport Airplane license. The Transition® made a successful maiden flight on Mar. 5, 2009. This craft's data will clearly be useful for setting end-of-chapter problems.

Further information on the performance of lifting craft can be found in Refs. 12, 13, and 16. We discuss this matter again briefly in Chap. 8.

Fig. 7.30 The Transition® car-plane in flight on March 23, 2012. It has a gross weight of 6,360 N and a cruise velocity of 170 km/h. (*Image from Terrafugia, Inc., http:// www.terrafugia.com*)

EXAMPLE 7.9

An aircraft weighs 333,600 N has a planform area of 230 m^2 and can deliver a constant thrust of 53,400 N. It has an aspect ratio of 7, and $C_{D\infty} \approx 0.02$. Neglecting rolling resistance, estimate the takeoff distance at sea level if takeoff speed equals 1.2 times stall speed. Take $C_{L,\max} = 2.0$.

Solution

The stall speed from Eq. (7.72), with sea-level density $\rho = 1.225$ kg/m^3, is

$$V_s = \left(\frac{2W}{C_{L,\max}\rho A_p}\right)^{1/2} = \left[\frac{2(333{,}600 \text{ N})}{2.0(1.225 \text{ kg/m}^3)(230 \text{ m}^2)}\right]^{1/2} = 34.4 \text{ m/s}$$

Hence takeoff speed $V_0 = 1.2V_s = 41.3$ m/s. The drag is estimated from Eq. (7.71) for AR = 7 as

$$C_D \approx 0.02 + \frac{C_L^2}{7\pi} = 0.02 + 0.0455C_L^2$$

A force balance in the direction of takeoff gives

$$F_s = m\frac{dV}{dt} = \text{thrust} - \text{drag} = T - kV^2 \qquad k = \tfrac{1}{2}C_D\rho A_p \tag{1}$$

Since we are looking for distance, not time, we introduce $dV/dt = V\, dV/ds$ into Eq. (1), separate variables, and integrate:

$$\int_0^{S_0} dS = \frac{m}{2}\int_0^{V_0} \frac{d(V^2)}{T - kV^2} \qquad k \approx \text{const}$$

or

$$S_0 = \frac{m}{2k}\ln\frac{T}{T - kV_0^2} = \frac{m}{2k}\ln\frac{T}{T - D_0} \tag{2}$$

where $D_0 = kV_0^2$ is the takeoff drag. Equation (2) is the desired theoretical relation for takeoff distance. For the particular numerical values, take

$$m = \frac{333{,}600 \text{ N}}{9.81 \text{ m/s}^2} = 34{,}000 \text{ kg}$$

$$C_{L_o} = \frac{W}{\frac{1}{2}\rho V_0^2 A_p} = \frac{333{,}600\ \text{N}}{\frac{1}{2}(1.225\ \text{kg/m}^3)(41.3\ \text{m/s})^2(230\ \text{m}^2)} = 1.39$$

$$C_{D_o} = 0.02 + 0.0455(C_{L_o})^2 = 0.108$$

$$k \approx \tfrac{1}{2} C_{D_o} \rho A_p = (\tfrac{1}{2})(0.108)(1.225\ \text{kg/m}^3)(230\ \text{m}^2) = 15.21\ \text{kg/m}$$

$$D_0 = kV_0^2 = 25{,}900\ \text{N}$$

Then Eq. (2) predicts that

$$S_0 = \frac{34{,}000\ \text{kg}}{2(15.21\ \text{kg/m})} \ln \frac{53{,}400\ \text{N}}{53{,}400\ \text{N} - 25{,}900\ \text{N}} = 1{,}118\ \text{m}\ (\ln 1.94) = 741\ \text{m} \qquad \textit{Ans.}$$

A more exact analysis accounting for variable k [13] gives the same result to within 1 percent.

EXAMPLE 7.10

For the aircraft of Example 7.9, if maximum thrust is applied during flight at 6,000 m standard altitude, estimate the resulting velocity of the plane, in km/h.

Solution

- *Assumptions:* Given $W = 333{,}600$ N, $A_p = 230\ \text{m}^2$, $T = 53{,}400$ N, $AR = 7$, $C_{D\infty} = 0.02$.
- *Approach:* Set lift equal to weight and drag equal to thrust and solve for the velocity.
- *Property values.* From Table A.6, at $z = 6{,}000$ m, $\rho = 0.6596\ \text{kg/m}^3$.
- *Solution steps:* Write out the formulas for lift and drag. The unknowns will be C_L and V.

$$W = 333{,}600\ \text{N} = \text{lift} = C_L \frac{\rho}{2} V^2 A_p = C_L \frac{0.6596\ \text{kg/m}^3}{2} V^2 (230\ \text{m}^2)$$

$$T = 53{,}400\ \text{N} = \text{drag} = \left(C_{D\infty} + \frac{C_L^2}{\pi AR} \right) \frac{\rho}{2} V^2 A_p$$

$$= \left[0.02 + \frac{C_L^2}{\pi(7)} \right] \frac{0.6596\ \text{kg/m}^3}{2} V^2 (230\ \text{m}^2)$$

Some clever manipulation (dividing W by T) would reveal a quadratic equation for C_L. The final solution is

$$C_L = 0.13 \qquad V \approx 184\ \text{m/s} = 662\ \text{km/h} \qquad \textit{Ans.}$$

- *Comments:* These are *preliminary design* estimates that do not depend on airfoil shape.

Summary

This chapter has dealt with viscous effects in external flow past bodies immersed in a stream. When the Reynolds number is large, viscous forces are confined to a thin boundary layer and wake in the vicinity of the body. Flow outside these "shear layers" is essentially inviscid and can be predicted by potential theory and Bernoulli's equation.

The chapter began with a discussion of the flat-plate boundary layer and the use of momentum integral estimates to predict the wall shear, friction drag, and thickness of such layers. These approximations suggest how to eliminate certain small terms in the Navier-Stokes equations, resulting in Prandtl's boundary layer equations for laminar and turbulent flow. Section 7.4 then solved the boundary layer equations to give very accurate formulas for flat-plate flow at high Reynolds numbers. Rough-wall effects were included, and Sec. 7.5 gave a brief introduction to pressure gradient effects. An adverse (decelerating) gradient was seen to cause flow separation, where the boundary layer breaks away from the surface and forms a broad, low-pressure wake.

Boundary layer theory fails in separated flows, which are commonly studied by experiment or CFD. Section 7.6 gave data on drag coefficients of various two- and three-dimensional body shapes. The chapter ended with a brief discussion of lift forces generated by lifting bodies such as airfoils and hydrofoils. Airfoils also suffer flow separation or *stall* at high angles of incidence.

Problems

Most of the problems herein are fairly straightforward. More difficult or open-ended assignments are labeled with an asterisk. Problems labeled with a computer icon may require the use of a computer. The standard end-of-chapter problems P7.1 to P7.127 (categorized in the problem list here) are followed by word problems W7.1 to W7.12, fundamentals of engineering exam problems FE7.1 to FE7.10, comprehensive problems C7.1 to C7.5, and design project D7.1.

Problem Distribution

Section	Topic	Problems
7.1	Reynolds number and geometry	P7.1–P7.5
7.2	Momentum integral estimates	P7.6–P7.12
7.3	The boundary layer equations	P7.13–P7.15
7.4	Laminar flat-plate flow	P7.16–P7.29
7.4	Turbulent flat-plate flow	P7.30–P7.47
7.5	Boundary layers with pressure gradient	P7.48–P7.50
7.6	Drag of bodies	P7.51–P7.114
7.6	Lifting bodies—airfoils	P7.115–P7.127

Reynolds number and geometry

P7.1 An ideal gas, at 20°C and 1 atm, flows at 12 m/s past a thin flat plate. At a position 60 cm downstream of the leading edge, the boundary layer thickness is 5 mm. Which of the 13 gases in Table A.4 is this likely to be?

P7.2 A gas at 20°C and 1 atm flows at 2 m/s past a thin flat plate. At $x = 1$ m, the boundary layer thickness is 0.016 m. Assuming laminar flow, which of the gases in Table A.4 is this likely to be?

P7.3 Equation (7.1*b*) assumes that the boundary layer on the plate is turbulent from the leading edge onward. Devise a scheme for determining the boundary layer thickness more accurately when the flow is laminar up to a point $\mathrm{Re}_{x,\mathrm{crit}}$ and turbulent thereafter. Apply this scheme to computation of the boundary layer thickness at $x = 1.5$ m in 40 m/s flow of air at 20°C and 1 atm past a flat plate. Compare your result with Eq. (7.1*b*). Assume $\mathrm{Re}_{x,\mathrm{crit}} \approx 1.2$ E6.

P7.4 A smooth ceramic sphere (SG = 2.6) is immersed in a flow of water at 20°C and 25 cm/s. What is the sphere diameter if it is encountering (*a*) creeping motion, $\mathrm{Re}_d = 1$ or (*b*) transition to turbulence, $\mathrm{Re}_d = 250{,}000$?

P7.5 SAE 30 oil at 20°C flows at 0.05 m³/s from a reservoir into a 15-cm-diameter pipe. Use flat-plate theory to estimate the position x where the pipe wall boundary layers meet in the center. Compare with Eq. (6.5), and give some explanations for the discrepancy.

Momentum integral estimates

P7.6 For the laminar parabolic boundary layer profile of Eq. (7.6), compute the shape factor H and compare with the exact Blasius result, Eq. (7.31).

P7.7 Air at 20°C and 1 atm enters a 40-cm-square duct as in Fig. P7.7. Using the "displacement thickness" concept of Fig. 7.4, estimate (*a*) the mean velocity and (*b*) the mean pressure in the core of the flow at the position $x = 3$ m. (*c*) What is the average gradient, in Pa/m, in this section?

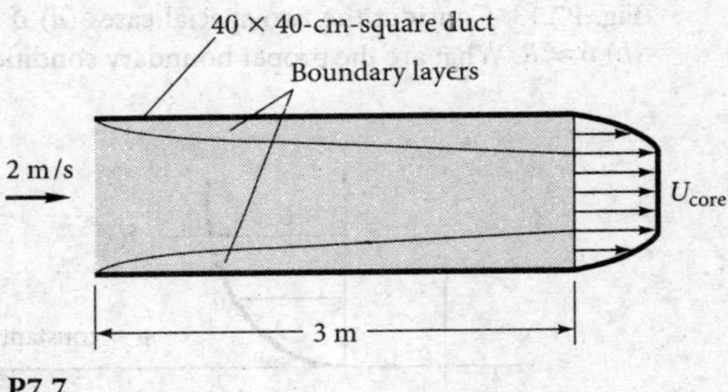

P7.7

P7.8 Air, $\rho = 1.2$ kg/m³ and $\mu = 1.8$ E-5 kg/(m · s), flows at 10 m/s past a flat plate. At the trailing edge of the plate, the following velocity profile data are measured:

y, mm	0	0.5	1.0	2.0	3.0	4.0	5.0	6.0
u, m/s	0	1.75	3.47	6.58	8.70	9.68	10.0	10.0

If the upper surface has an area of 0.6 m², estimate, using momentum concepts, the friction drag, in N, on the upper surface.

P7.9 Repeat the flat-plate momentum analysis of Sec. 7.2 by replacing Eq. (7.6) with the simple but unrealistic linear velocity profile suggested by Schlichting [1]:

$$\frac{u}{U} \approx \frac{y}{\delta} \quad \text{for} \quad 0 \le y \le \delta$$

Compute momentum–integral estimates of c_f, θ/x, δ^*/x, and H.

P7.10 Repeat Prob. P7.9, using a trigonometric profile approximation:

$$\frac{u}{U} \approx \sin\left(\frac{\pi y}{2\delta}\right)$$

Does this profile satisfy the conditions of laminar flat-plate flow?

P7.11 Air at 20°C and 1 atm flows at 2 m/s past a sharp flat plate. Assuming that Kármán's parabolic-profile analysis, Eqs. (7.6–7.10), is accurate, estimate (*a*) the local velocity u and (*b*) the local shear stress τ at the position $(x, y) =$ (50 cm, 5 mm).

P7.12 The velocity profile shape $u/U \approx 1 - \exp(-4.605y/\delta)$ is a smooth curve with $u = 0$ at $y = 0$ and $u = 0.99U$ at $y = \delta$ and thus would seem to be a reasonable substitute for the parabolic flat-plate profile of Eq. (7.3). Yet when this new profile is used in the integral analysis of Sec. 7.3, we get the lousy result $\delta/x \approx 9.2/\mathrm{Re}_x^{1/2}$, which is 80 percent high. What is the reason for the inaccuracy? [*Hint:* The answer lies in evaluating the laminar boundary layer momentum equation (7.19*b*) at the wall, $y = 0$.]

The boundary layer equations

P7.13 Derive modified forms of the laminar boundary layer equations (7.19) for the case of axisymmetric flow along the outside of a circular cylinder of constant radius R, as in Fig. P7.13. Consider the two special cases (*a*) $\delta \ll R$ and (*b*) $\delta \approx R$. What are the proper boundary conditions?

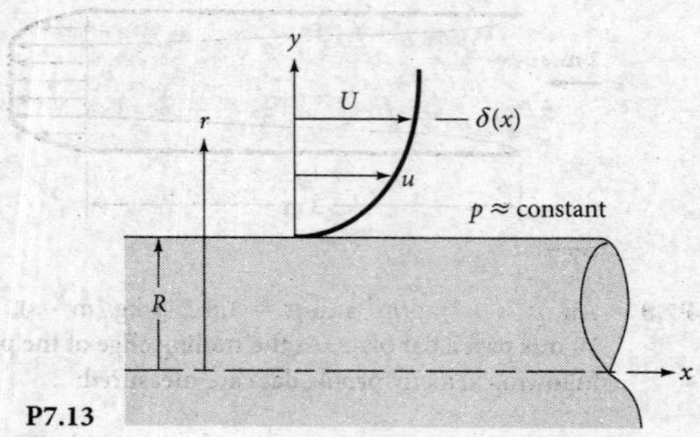

P7.13

P7.14 Show that the two-dimensional laminar flow pattern with $dp/dx = 0$

$$u = U_0(1 - e^{Cy}) \qquad v = v_0 < 0$$

is an exact solution to the boundary layer equations (7.19). Find the value of the constant C in terms of the flow parameters. Are the boundary conditions satisfied? What might this flow represent?

P7.15 Discuss whether fully developed laminar incompressible flow between parallel plates, Eq. (4.134) and Fig. 4.14*b*, represents an exact solution to the boundary layer equations (7.19) and the boundary conditions (7.20). In what sense, if any, are duct flows also boundary layer flows?

Laminar flat-plate flow

P7.16 A thin flat plate 55 by 110 cm is immersed in a 6-m/s stream of SAE 10 oil at 20°C. Compute the total friction drag if the stream is parallel to (*a*) the long side and (*b*) the short side.

P7.17 Consider laminar flow past a sharp flat plate of width b and length L. What percentage of the friction drag on the plate is carried by the rear half of the plate?

P7.18 Air at 20°C and 1 atm flows at 5 m/s past a flat plate. At $x =$ 60 cm and $y = 2.95$ mm, use the Blasius solution, Table 7.1, to find (*a*) the velocity u; and (*b*) the wall shear stress. (*c*) For extra credit, find a Blasius formula for the shear stress away from the wall.

P7.19 Air at 20°C and 1 atm flows at 15 m/s past a thin flat plate whose area (bL) is 2.2 m². If the total friction drag is 1.3 N, what are the length and width of the plate?

P7.20 Air at 20°C and 1 atm flows at 20 m/s past the flat plate in Fig. P7.20. A pitot stagnation tube, placed 2 mm from the wall, develops a manometer head h = 16 mm of Meriam red oil, SG = 0.827. Use this information to estimate the downstream position x of the pitot tube. Assume laminar flow.

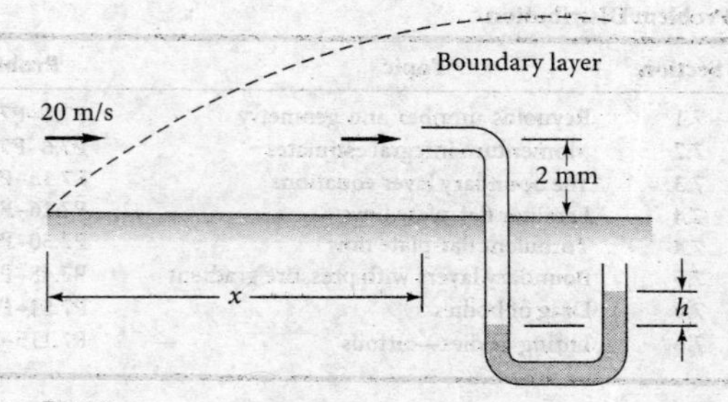

P7.20

P7.21 For the experimental setup of Fig. P7.20, suppose the stream velocity is unknown and the pitot stagnation tube is traversed across the boundary layer of air at 1 atm and 20°C. The manometer fluid is Meriam red oil, and the following readings are made:

y, mm	0.5	1.0	1.5	2.0	2.5	3.0	3.5	4.0	4.5	5.0
h, mm	1.2	4.6	9.8	15.8	21.2	25.3	27.8	29.0	29.7	29.7

Using these data only (not the Blasius theory) estimate (*a*) the stream velocity, (*b*) the boundary layer thickness, (*c*) the wall shear stress, and (*d*) the total friction drag between the leading edge and the position of the pitot tube.

P7.22 In the Blasius equation (7.22), f is a dimensionless plane stream function:

$$f(\eta) = \frac{\psi(x, y)}{\sqrt{\nu U x}}$$

Values of f are not given in Table 7.1, but one published value is $f(2.0) = 0.6500$. Consider airflow at 6 m/s, 20°C, and

1 atm past a flat plate. At $x = 1$ m, estimate (*a*) the height *y*; (*b*) the velocity, and (*c*) the stream function at $\eta = 2.0$.

P7.23 Suppose you buy a 120- by 240-cm sheet of plywood and put it on your roof rack. (See Fig. P7.23.) You drive home at 56 km/h. (*a*) Assuming the board is perfectly aligned with the airflow, how thick is the boundary layer at the end of the board? (*b*) Estimate the drag on the sheet of plywood if the boundary layer remains laminar. (*c*) Estimate the drag on the sheet of plywood if the boundary layer is turbulent (assume the wood is smooth), and compare the result to that of the laminar boundary layer case.

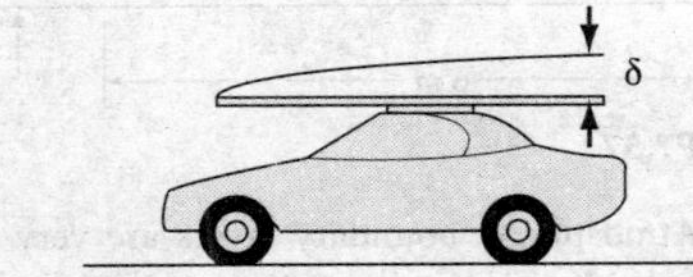

P7.23

***P7.24** Air at 20°C and 1 atm flows past the flat plate in Fig. P7.24 under laminar conditions. There are two equally spaced pitot stagnation tubes, each placed 2 mm from the wall. The manometer fluid is water at 20°C. If $U = 15$ m/s and $L = 50$ cm, determine the values of the manometer readings h_1 and h_2, in mm.

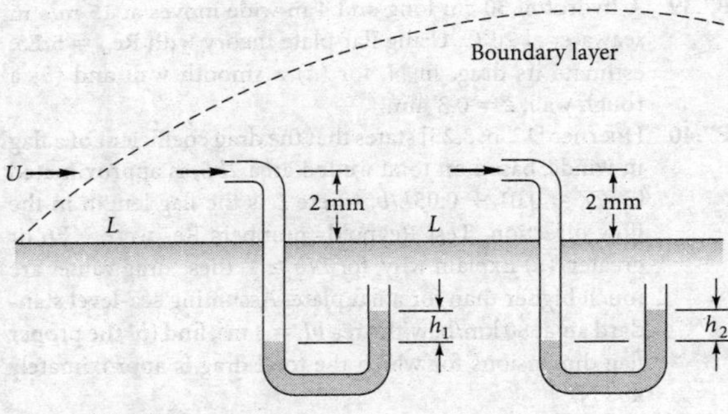

P7.24

P7.25 Consider the smooth square 10-cm-by-10-cm duct in Fig. P7.25. The fluid is air at 20°C and 1 atm, flowing at $V_{avg} = 24$ m/s. It is desired to increase the pressure drop over the 1-m length by adding sharp 8-mm-long flat plates across the duct, as shown. (*a*) Estimate the pressure drop if there are no plates. (*b*) Estimate how many plates are needed to generate an additional 100 Pa of pressure drop.

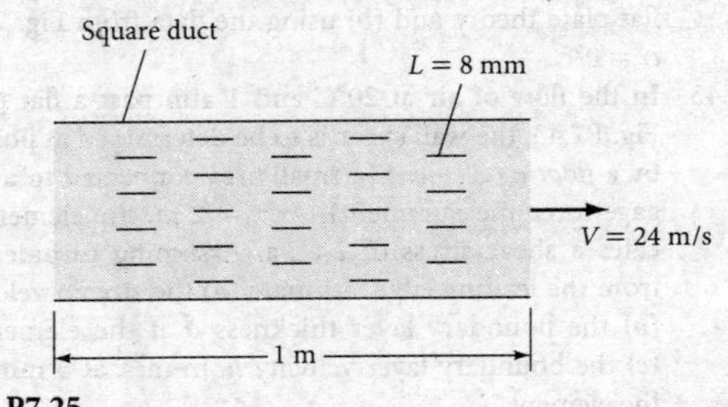

P7.25

P7.26 Consider laminar boundary layer flow past the square-plate arrangements in Fig. P7.26. Compared to the friction drag of a single plate 1, how much larger is the drag of four plates together as in configurations (*a*) and (*b*)? Explain your results.

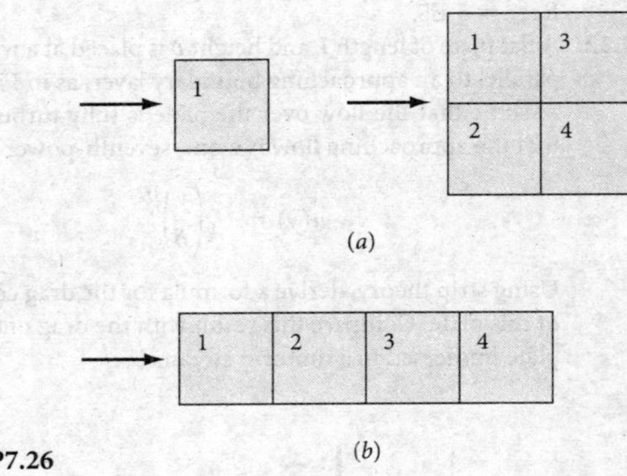

P7.26

P7.27 Air at 20°C and 1 atm flows at 3 m/s past a sharp flat plate 2 m wide and 1 m long. (*a*) What is the wall shear stress at the end of the plate? (*b*) What is the air velocity at a point 4.5 mm normal to the end of the plate? (*c*) What is the total friction drag on the plate?

P7.28 Flow straighteners are arrays of narrow ducts placed in wind tunnels to remove swirl and other in-plane secondary velocities. They can be idealized as square boxes constructed by vertical and horizontal plates, as in Fig. P7.28. The cross section is *a* by *a*, and the box length is *L*. Assuming laminar flat-plate flow and an array of $N \times N$ boxes, derive a formula for (*a*) the total drag on the bundle of boxes and (*b*) the effective pressure drop across the bundle.

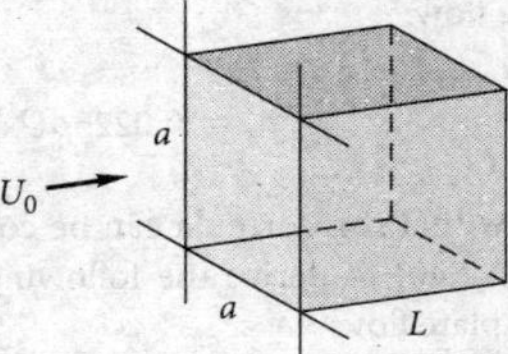

P7.28

P7.29 Let the flow straighteners in Fig. P7.28 form an array of 20 × 20 boxes of size $a = 4$ cm and $L = 25$ cm. If the approach velocity is $U_0 = 12$ m/s and the fluid is sea-level standard air, estimate (*a*) the total array drag and (*b*) the pressure drop across the array. Compare with Sec. 6.8.

Turbulent flat-plate flow

P7.30 In Ref. 56 of Chap. 6, McKeon et al. propose new, more accurate values for the turbulent log-law constants, $\kappa = 0.421$ and $B = 5.62$. Use these constants, and the one-seventh power-law, to repeat the analysis that led to the formula for turbulent boundary layer thickness, Eq. (7.42).

By what percent is δ/x in your new formula different from that in Eq. (7.42)? Comment.

P7.31 The centerboard on a sailboat is 1 m long parallel to the flow and protrudes 2 m down below the hull into seawater at 20°C. Using flat-plate theory for a smooth surface, estimate its drag if the boat moves at 5 m/s. Assume $Re_{x,tr} = 5$ E5.

P7.32 A flat plate of length L and height δ is placed at a wall and is parallel to an approaching boundary layer, as in Fig. P7.32. Assume that the flow over the plate is fully turbulent and that the approaching flow is a one-seventh-power law:

$$u(y) = U_0\left(\frac{y}{\delta}\right)^{1/7}$$

Using strip theory, derive a formula for the drag coefficient of this plate. Compare this result with the drag of the same plate immersed in a uniform stream U_0.

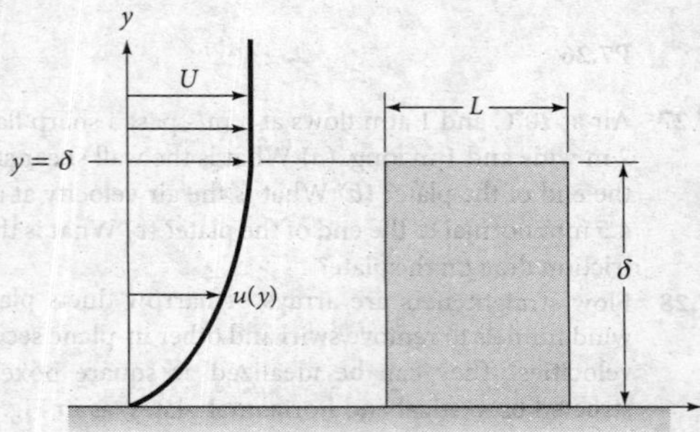

P7.32

P7.33 An alternate analysis of turbulent flat-plate flow was given by Prandtl in 1927, using a wall shear stress formula from pipe flow:

$$\tau_w = 0.0225\rho U^2\left(\frac{\nu}{U\delta}\right)^{1/4}$$

Show that this formula can be combined with Eqs. (7.33) and (7.40) to derive the following relations for turbulent flat-plate flow:

$$\frac{\delta}{x} = \frac{0.37}{Re_x^{1/5}} \quad c_f = \frac{0.0577}{Re_x^{1/5}} \quad C_D = \frac{0.072}{Re_L^{1/5}}$$

These formulas are limited to Re_x between 5×10^5 and 10^7.

P7.34 Consider turbulent flow past a sharp, smooth flat plate of width b and length L. What percentage of the friction drag on the plate is carried by the rear half of the plate?

P7.35 Water at 20°C flows at 5 m/s past a 2-m-wide sharp flat plate. (*a*) Estimate the boundary layer thickness at $x =$ 1.2 m. (*b*) If the total drag (on both sides of the plate) is 310 N, estimate the length of the plate using, for simplicity, Eq. (7.45).

P7.36 A ship is 125 m long and has a wetted area of 3,500 m^2. Its propellers can deliver a maximum power of 1.1 MW to seawater at 20°C. If all drag is due to friction, estimate the maximum ship speed, in m/s.

P7.37 Air at 20°C and 1 atm flows past a long flat plate, at the end of which is placed a narrow scoop, as shown in Fig. P7.37. (*a*) Estimate the height h of the scoop if it is to extract 4 kg/s per meter of width into the paper. (*b*) Find the drag on the plate up to the inlet of the scoop, per meter of width.

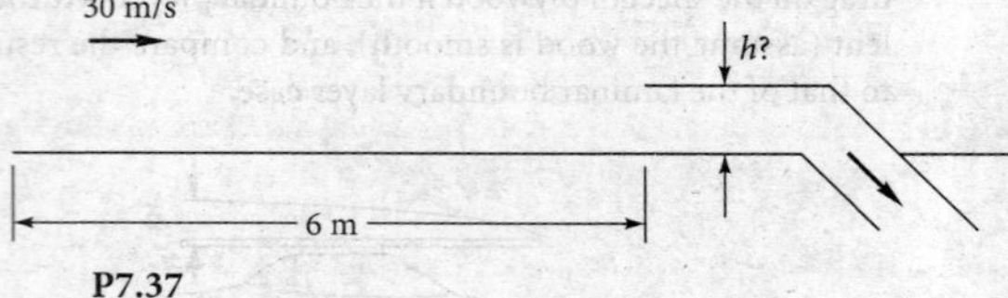

P7.37

P7.38 Atmospheric boundary layers are very thick but follow formulas very similar to those of flat-plate theory. Consider wind blowing at 10 m/s at a height of 80 m above a smooth beach. Estimate the wind shear stress, in Pa, on the beach if the air is standard sea-level conditions. What will the wind velocity striking your nose be if (*a*) you are standing up and your nose is 170 cm off the ground and (*b*) you are lying on the beach and your nose is 17 cm off the ground?

P7.39 A hydrofoil 50 cm long and 4 m wide moves at 15 m/s in seawater at 20°C. Using flat-plate theory with $Re_{tr} = 5$ E5, estimate its drag, in N, for (*a*) a smooth wall and (*b*) a rough wall, $\varepsilon = 0.3$ mm.

P7.40 Hoerner [12, p. 3.25] states that the drag coefficient of a flag in winds, based on total wetted area $2bL$, is approximated by $C_D \approx 0.01 + 0.05L/b$, where L is the flag length in the flow direction. Test Reynolds numbers Re_L were 1 E6 or greater. (*a*) Explain why, for $L/b \geq 1$, these drag values are much higher than for a flat plate. Assuming sea-level standard air at 80 km/h, with area $bL = 4$ m^2, find (*b*) the proper flag dimensions for which the total drag is approximately 400 N.

P7.41 Repeat Prob. P7.20 with the sole change that the pitot probe is now 10 mm from the wall (5 times higher). Show that the flow there cannot possibly be laminar, and use smooth-wall turbulent flow theory to estimate the position x of the probe, in m.

P7.42 A light aircraft flies at 30 m/s in air at 20°C and 1 atm. Its wing is an NACA 0009 airfoil, with a chord length of 150 cm and a very wide span (neglect aspect ratio effects). Estimate the drag of this wing, per unit span length, (*a*) by flat plate theory and (*b*) using the data from Fig. 7.25 for $\alpha = 0°$.

P7.43 In the flow of air at 20°C and 1 atm past a flat plate in Fig. P7.43, the wall shear is to be determined at position x by a *floating element* (a small area connected to a strain-gage force measurement). At $x = 2$ m, the element indicates a shear stress of 2.1 Pa. Assuming turbulent flow from the leading edge, estimate (*a*) the stream velocity U, (*b*) the boundary layer thickness δ at the element, and (*c*) the boundary layer velocity u, in m/s, at 5 mm above the element.

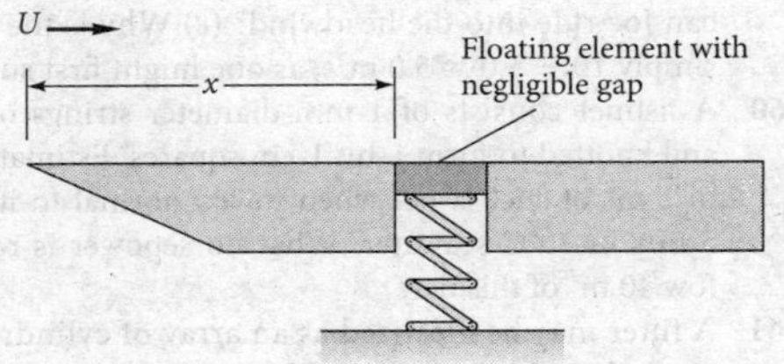

P7.43

P7.44 Extensive measurements of wall shear stress and local velocity for turbulent airflow on the flat surface of the University of Rhode Island wind tunnel have led to the following proposed correlation:

$$\frac{\rho y^2 \tau_w}{\mu^2} \approx 0.0207 \left(\frac{uy}{\nu}\right)^{1.77}$$

Thus, if y and $u(y)$ are known at a point in a flat-plate boundary layer, the wall shear may be computed directly. If the answer to part (c) of Prob. P7.43 is $u \approx 26.3$ m/s, determine the shear stress and compare with Prob. P7.43. Discuss.

P7.45 A thin sheet of fiberboard weighs 90 N and lies on a rooftop, as shown in Fig. P7.45. Assume ambient air at 20°C and 1 atm. If the coefficient of solid friction between board and roof is $\sigma \approx 0.12$, what wind velocity will generate enough fluid friction to dislodge the board?

P7.46 A ship is 150 m long and has a wetted area of 5,000 m^2. If it is encrusted with barnacles, the ship requires 7,000 hp to overcome friction drag when moving in seawater at 8 m/s and 20°C. What is the average roughness of the barnacles? How fast would the ship move with the same power if the surface were smooth? Neglect wave drag.

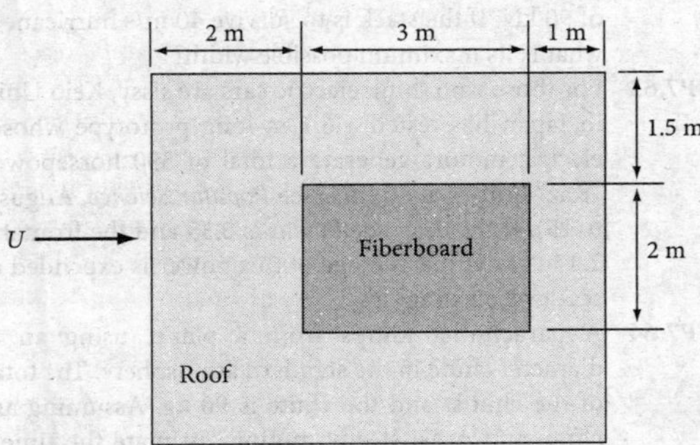

P7.45

P7.47 Local boundary layer effects, such as shear stress and heat transfer, are best correlated with local variables, rather using distance x from the leading edge. The momentum thickness θ is often used as a length scale. Use the analysis of turbulent flat-plate flow to write local wall shear stress τ_w in terms of dimensionless θ and compare with the formula recommended by Schlichting [1]: $C_f \approx 0.033\,\mathrm{Re}_\theta^{-0.268}$.

Boundary layers with pressure gradient

P7.48 In 1957 H. Görtler proposed the adverse gradient test cases

$$U = \frac{U_0}{(1 + x/L)^n}$$

and computed separation for laminar flow at $n = 1$ to be $x_{sep}/L = 0.159$. Compare with Thwaites's method, assuming $\theta_0 = 0$.

P7.49 Based strictly on your understanding of flat-plate theory plus adverse and favorable pressure gradients, explain the direction (left or right) for which airflow past the slender airfoil shape in Fig. P7.49 will have lower total (friction + pressure) drag.

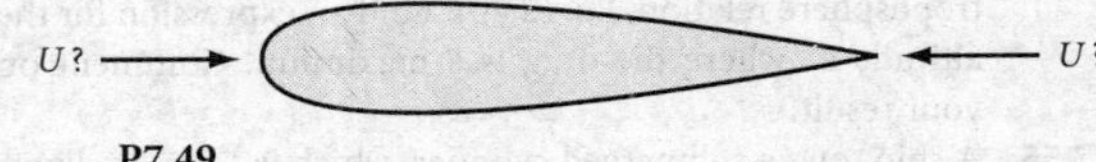

P7.49

P7.50 Consider the flat-walled diffuser in Fig. P7.50, which is similar to that of Fig. 6.26*a* with constant width b. If x is measured from the inlet and the wall boundary layers are thin, show that the core velocity $U(x)$ in the diffuser is given approximately by

$$U = \frac{U_0}{1 + (2x \tan \theta)/W}$$

where W is the inlet height. Use this velocity distribution with Thwaites's method to compute the wall angle θ for which laminar separation will occur in the exit plane when diffuser length $L = 2W$. Note that the result is independent of the Reynolds number.

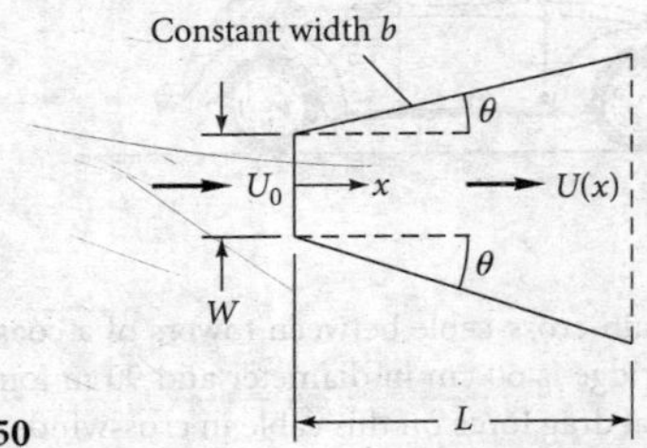

P7.50

Drag of bodies

P7.51 A 2-cm-diameter solid metal sphere falls steadily at about 1 m/s in 20°C fresh water. If we use Table 7.3 for a drag estimate, is the sphere made of steel, aluminum, or copper?

P7.52 Clift et al. [46] give the formula $F \approx (6\pi/5)(4 + a/b)\mu U b$ for the drag of a prolate spheroid in *creeping motion*, as shown in Fig. P7.52. The half-thickness b is 4 mm. If the fluid is SAE 50W oil at 20°C, (*a*) check that $\mathrm{Re}_b < 1$ and (*b*) estimate the spheroid length if the drag is 0.02 N.

P7.53 From Table 7.2, the drag coefficient of a wide plate normal to a stream is approximately 2.0. Let the stream conditions be U_∞ and p_∞. If the average pressure on the front of the plate is approximately equal to the free-stream stagnation pressure, what is the average pressure on the rear?

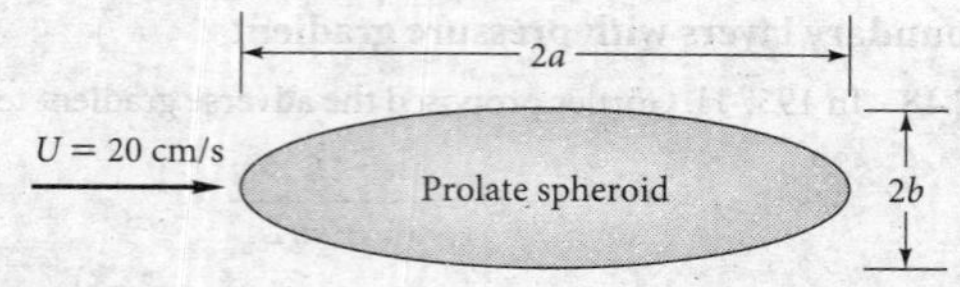

P7.52

***P7.54** If a missile takes off vertically from sea level and leaves the atmosphere, it has zero drag when it starts and zero drag when it finishes. It follows that the drag must be a maximum somewhere in between. To simplify the analysis, assume a constant drag coefficient, C_D, and a constant vertical acceleration, a. Let the density variation be modeled by the troposphere relation, Eq. (2.20). Find an expression for the altitude z^* where the drag is a maximum. Comment on your result.

P7.55 A ship tows a submerged cylinder, which is 1.5 m in diameter and 22 m long, at 5 m/s in fresh water at 20°C. Estimate the towing power, in kW, required if the cylinder is (*a*) parallel and (*b*) normal to the tow direction.

P7.56 A delivery vehicle carries a long sign on top, as in Fig. P7.56. If the sign is very thin and the vehicle moves at 105 km/h, (*a*) estimate the force on the sign with no crosswind and (*b*) discuss the effect of a crosswind.

P7.56

P7.57 The main cross-cable between towers of a coastal suspension bridge is 60 cm in diameter and 90 m long. Estimate the total drag force on this cable in crosswinds of 80 km/h. Are these laminar flow conditions?

***P7.58** Modify Prob. P7.54 to be more realistic by accounting for missile drag during ascent. Assume constant thrust T and missile weight W. Neglect the variation of g with altitude. Solve for the altitude z^* in the standard atmosphere where the drag is a maximum, for $T = 16{,}000$ N, $W = 8{,}000$ N, and $C_DA = 0.4$ m^2. The writer does not believe an analytic solution is possible.

***P7.59** Joe can pedal his bike at 10 m/s on a straight level road with no wind. The rolling resistance of his bike is 0.80 N · s/m—that is, 0.80 N of force per m/s of speed. The drag area (C_DA) of Joe and his bike is 0.422 m^2. Joe's mass is 80 kg and that of the bike is 15 kg. He now encounters a headwind of 5.0 m/s. (*a*) Develop an equation for the speed at which Joe can pedal into the wind. *Hint:* A cubic equation for V will result. (*b*) Solve for V; that is, how fast can Joe ride into the headwind? (*c*) Why is the result not simply $10 - 5.0 = 5.0$ m/s, as one might first suspect?

P7.60 A fishnet consists of 1-mm-diameter strings overlapped and knotted to form 1-by-1-cm squares. Estimate the drag of 1 m^2 of such a net when towed normal to its plane at 3 m/s in 20°C seawater. What horsepower is required to tow 40 m^2 of this net?

P7.61 A filter may be idealized as an array of cylindrical fibers normal to the flow, as in Fig. P7.61. Assuming that the fibers are uniformly distributed and have drag coefficients given by Fig. 7.16*a*, derive an approximate expression for the pressure drop Δp through a filter of thickness L.

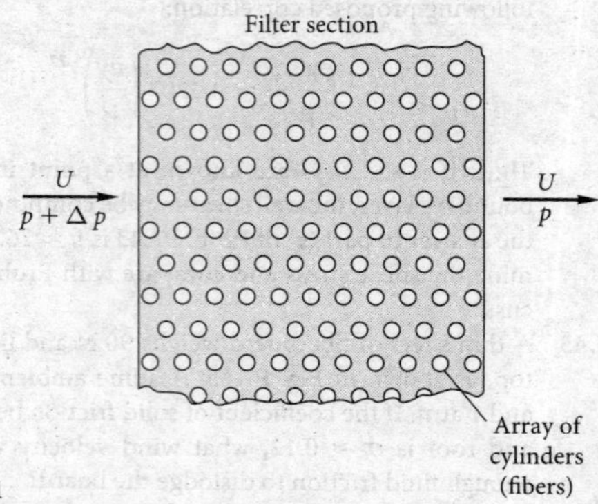

P7.61

P7.62 A sea-level smokestack is 52 m high and has a square cross section. Its supports can withstand a maximum side force of 90 kN. If the stack is to survive 40 m/s hurricane winds, what is its maximum possible width?

P7.63 For those who think electric cars are sissy, Keio University in Japan has tested a 6.7 m-long prototype whose eight electric motors generate a total of 590 horsepower. The "Kaz" cruises at 80 m/s (see *Popular Science*, August 2001, p. 15). If the drag coefficient is 0.35 and the frontal area is 2.4 m^2, what percentage of this power is expended against sea-level air drag?

P7.64 A parachutist jumps from a plane, using an 8.5-m-diameter chute in the standard atmosphere. The total mass of the chutist and the chute is 90 kg. Assuming an open chute and quasi-steady motion, estimate the time to fall from 2,000- to 1,000-m altitude.

P7.65 As soldiers get bigger and packs get heavier, a parachutist and load can weigh as much as 1,800 N. The standard 8.5 m parachute may descend too fast for safety. For heavier loads, the U.S. Army Natick Center has developed a 8.5 m, higher-drag, less porous XT-11 parachute (see <http://www.natick.army.mil>). This parachute has a sea-level descent speed of 5 m/s with a 1,800 N load. (*a*) What is the drag coefficient of the XT-11? (*b*) How fast would the standard chute descend at sea level with such a load?

P7.66 A sphere of density ρ_s and diameter D is dropped from rest in a fluid of density ρ and viscosity μ. Assuming a constant drag coefficient C_{d_0}, derive a differential equation for the fall velocity $V(t)$ and show that the solution is

$$V = \left[\frac{4gD(S-1)}{3C_{d_0}}\right]^{1/2} \tanh Ct$$

$$C = \left[\frac{3gC_{d_0}(S-1)}{4S^2D}\right]^{1/2}$$

where $S = \rho_s/\rho$ is the specific gravity of the sphere material.

P7.67 The Toyota Prius has a drag coefficient of 0.25, a frontal area of 2.2 m^2, and an empty weight of 13,500 N. Its rolling resistance coefficient is $C_{rr} = 0.03$, that is, the rolling resistance is 3 percent of the normal force on the tires. If rolling freely down a slope of 8° at an altitude of 500 m, calculate its maximum velocity, in km/h.

P7.68 The Mars roving-laboratory parachute, in the Chap. 5 opener photo, is a 15.5-m-diameter disk-gap-band chute, with a measured drag coefficient of 1.12 [59]. Mars has very low density, about 0.015 kg/m^3, and its gravity is only 38 percent of earth gravity. If the mass of payload and chute is 2,400 kg, estimate the terminal fall velocity of the parachute.

P7.69 Two baseballs of diameter 7.35 cm are connected to a rod 7 mm in diameter and 56 cm long, as in Fig. P7.69. What power, in W, is required to keep the system spinning at 400 r/min? Include the drag of the rod, and assume sea-level standard air.

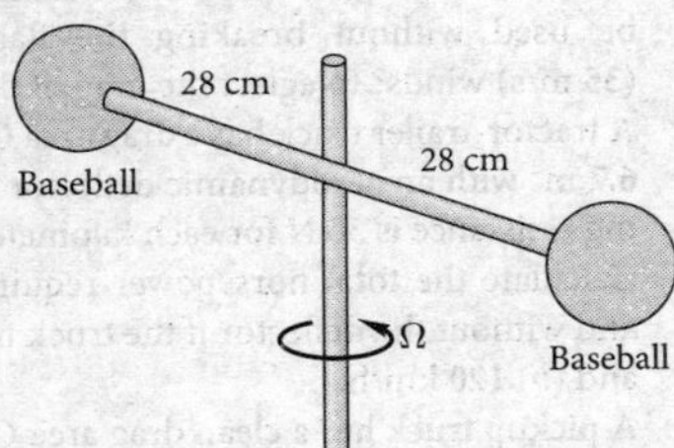

P7.69

P7.70 The Army's new ATPS personnel parachute is said to be able to bring a 1,800 N load, trooper plus pack, to ground at 5 m/s in "mile-high" Denver, Colorado. If we assume that Table 7.3 is valid, what is the approximate diameter of this new parachute?

P7.71 The 2013 Toyota Camry has an empty weight of 14,200 N, a frontal area of 2.05 m^2, and a drag coefficient of 0.28. Its rolling resistance is $C_{rr} \approx 0.035$. Estimate the maximum velocity, in km/h, this car can attain when rolling freely at sea level down a 4° slope.

P7.72 A settling tank for a municipal water supply is 2.5 m deep, and 20°C water flows through continuously at 35 cm/s. Estimate the minimum length of the tank that will ensure that all sediment (SG = 2.55) will fall to the bottom for particle diameters greater than (*a*) 1 mm and (*b*) 100 μm.

P7.73 A balloon is 4 m in diameter and contains helium at 125 kPa and 15°C. Balloon material and payload weigh 200 N, not including the helium. Estimate (*a*) the terminal ascent velocity in sea-level standard air, (*b*) the final standard altitude (neglecting winds) at which the balloon will come to rest, and (*c*) the minimum diameter (< 4 m) for which the balloon will just barely begin to rise in sea-level standard air.

P7.74 It is difficult to define the "frontal area" of a motorcycle due to its complex shape. One then measures the *drag area* (that is, C_DA) in area units. Hoerner [12] reports the drag area of a typical motorcycle, including the (upright) driver, as about 0.5 m^2. Rolling friction is typically about 1.93 N per km/h of speed. If that is the case, estimate the maximum sea-level speed (in km/h) of the Harley-Davidson V-Rod[TM] cycle, whose liquid-cooled engine produces 115 hp.

P7.75 The helium-filled balloon in Fig. P7.75 is tethered at 20°C and 1 atm with a string of negligible weight and drag. The diameter is 50 cm, and the balloon material weighs 0.2 N, not including the helium. The helium pressure is 120 kPa. Estimate the tilt angle θ if the airstream velocity U is (*a*) 5 m/s or (*b*) 20 m/s.

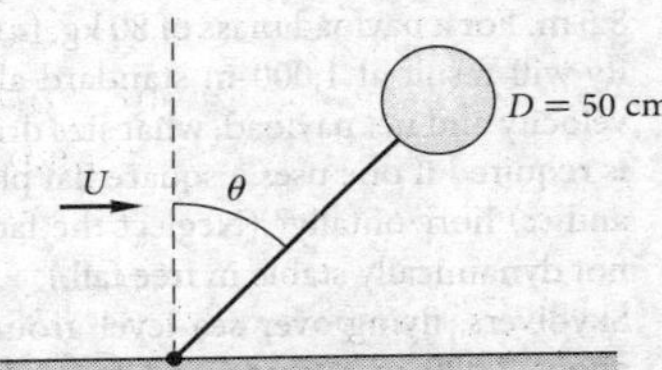

P7.75

P7.76 The movie *The World's Fastest Indian* tells the story of Burt Munro, a New Zealander who, in 1967, set a motorcycle record of 323 km/h on the Bonneville Salt Flats. Using the data of Prob. P7.74, (*a*) estimate the horsepower needed to drive this fast. (*b*) What horsepower would have gotten Burt up to 400 km/h?

P7.77 To measure the drag of an upright person, without violating human subject protocols, a life-sized mannequin is attached to the end of a 6-m rod and rotated at Ω = 80 rev/min, as in Fig. P7.77. The power required to maintain the rotation is 60 kW. By including rod drag power, which is significant, estimate the drag area C_DA of the mannequin, in m^2.

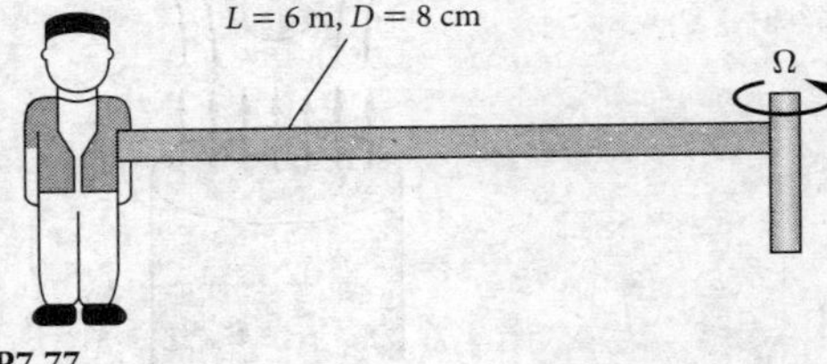

P7.77

P7.78 On April 24, 2007, a French bullet train set a new speed record, for rail-driven trains, of 575 km/h, beating the old record by 12 percent. Using the data in Table 7.3, estimate the sea-level horsepower required to drive this train at such a speed.

P7.79 Assume that a radioactive dust particle approximates a sphere of density 2,400 kg/m^3. How long, in days, will it take such a particle to settle to sea level from an altitude of 12 km if the particle diameter is (*a*) 1 μm or (*b*) 20 μm?

P7.80 A heavy sphere attached to a string should hang at an angle θ when immersed in a stream of velocity U, as in Fig. P7.80. Derive an expression for θ as a function of the sphere and flow properties. What is θ if the sphere is steel (SG = 7.86) of diameter 3 cm and the flow is sea-level standard air at $U = 40$ m/s? Neglect the string drag.

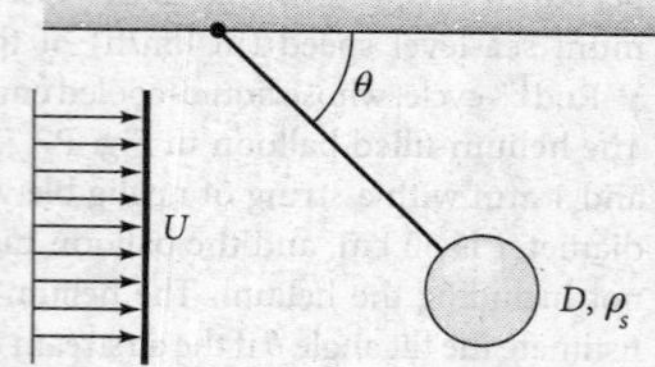

P7.80

P7.81 A typical U.S. Army parachute has a projected diameter of 8.5 m. For a payload mass of 80 kg, (*a*) what terminal velocity will result at 1,000-m standard altitude? For the same velocity and net payload, what size drag-producing "chute" is required if one uses a square flat plate held (*b*) vertically and (*c*) horizontally? (Neglect the fact that flat shapes are not dynamically stable in free fall.)

P7.82 Skydivers, flying over sea-level ground, typically jump at about 2,400 m altitude and free-fall spread-eagled until they open their chutes at about 600 m. They take about 10 s to reach terminal velocity. Estimate how many seconds of free-fall they enjoy if (*a*) they fall spread-eagled or (*b*) they fall feet first? Assume a total skydiver weight of 1,000 N.

P7.83 A blimp approximates a 4:1 spheroid that is 60 m long. It is powered by two 150 hp ducted fans. Estimate the maximum speed attainable, in km/h, at an altitude of 2,500 m.

P7.84 A Ping-Pong ball weighs 2.6 g and has a diameter of 3.8 cm. It can be supported by an air jet from a vacuum cleaner outlet, as in Fig. P7.84. For sea-level standard air, what jet velocity is required?

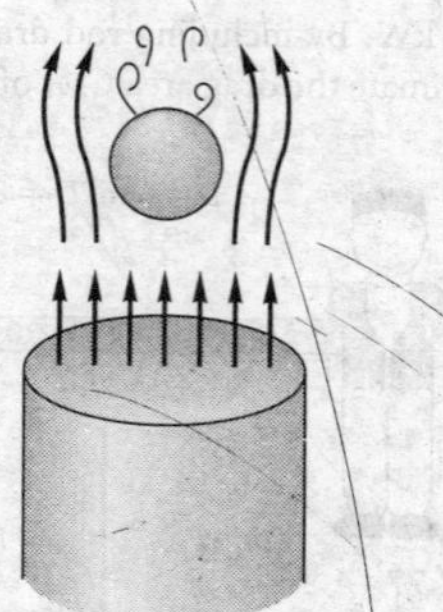

P7.84

P7.85 In this era of expensive fossil fuels, many alternatives have been pursued. One idea from SkySails, Inc., shown in Fig. P7.85, is the assisted propulsion of a ship by a large tethered kite. The tow force of the kite assists the ship's propeller and is said to reduce annual fuel consumption by 10–35 percent. For a typical example, let the ship be 120 m long, with a wetted area of 2,800 m^2. The kite area is 330 m^2 and has a force coefficient of 0.8. The kite cable makes an angle of 25° with the horizontal. Let $V_{wind} = 13.5$ m/s. Neglect ship wave drag. Estimate the ship speed (*a*) due to the kite only and (*b*) if the propeller delivers 1,250 hp to the water. [*Hint:* The kite sees the *relative* velocity of the wind.]

P7.85 Ship propulsion assisted by a large kite. *(Courtesy of SkySails, Inc.)*

P7.86 Hoerner [Ref. 12, pp. 3–25] states that the drag coefficient of a flag of 2:1 aspect ratio is 0.11 based on planform area. The University of Rhode Island has an aluminum flagpole 25 m high and 14 cm in diameter. It flies equal-sized national and state flags together. If the fracture stress of aluminum is 210 MPa, what is the maximum flag size that can be used without breaking the flagpole in hurricane (35 m/s) winds? (Neglect the drag of the flagpole.)

P7.87 A tractor-trailer truck has a drag area $C_DA = 8$ m^2 bare and 6.7 m^2 with an aerodynamic deflector (Fig. 7.18*b*). Its rolling resistance is 30 N for each kilometer per hour of speed. Calculate the total horsepower required at sea level with and without the deflector if the truck moves at (*a*) 90 km/h and (*b*) 120 km/h.

P7.88 A pickup truck has a clean drag area C_DA of 3.25 m^2. Estimate the horsepower required to drive the truck at 90 km/h (*a*) clean and (*b*) with the 0.9- by 1.8-m sign in Fig. P7.88 installed if the rolling resistance is 700 N at sea level.

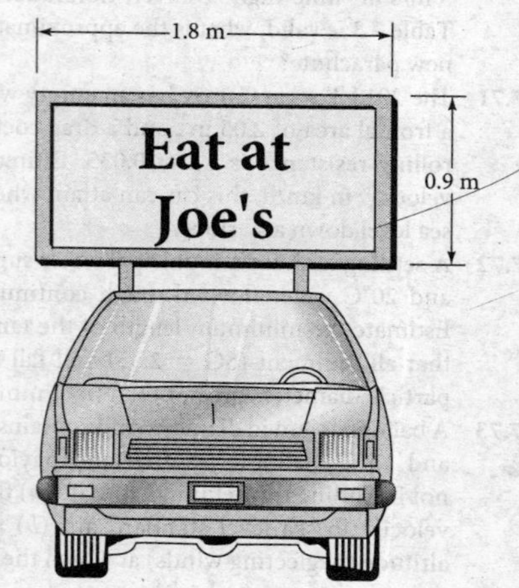

P7.88

P7.89 The AMTRAK Acela train passes through Kingston, RI, at 210 km/h, scaring all the villagers daily. Its total weight is 570 kN, with a rolling resistance $C_{rr} \approx 0.0024$. Estimate the horsepower required to drive the train this fast.

P7.90 In the great hurricane of 1938, winds of 38 m/s blew over a boxcar in Providence, Rhode Island. The boxcar was 3 m high, 12 m long, and 1.8 m wide, with a 0.9 m clearance above tracks 1.5 m apart. What wind speed would topple a boxcar weighing 180 kN?

***P7.91** A cup anemometer uses two 5-cm-diameter hollow hemispheres connected to 15-cm rods, as in Fig. P7.91. Rod drag is negligible, and the central bearing has a retarding torque of 0.004 N · m. Making simplifying assumptions to average out the time-varying geometry, estimate and plot the variation of anemometer rotation rate Ω with wind velocity U in the range $0 < U < 25$ m/s for sea-level standard air.

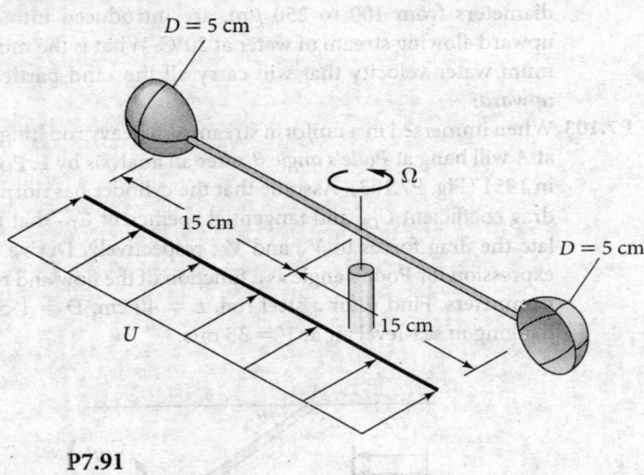

P7.91

P7.92 A 1,500-kg automobile uses its drag area $C_DA = 0.4$ m^2, plus brakes and a parachute, to slow down from 50 m/s. Its brakes apply 5,000 N of resistance. Assume sea-level standard air. If the automobile must stop in 8 s, what diameter parachute is appropriate?

P7.93 A hot-film probe is mounted on a cone-and-rod system in a sea-level airstream of 45 m/s, as in Fig. P7.93. Estimate the maximum cone vertex angle allowable if the flow-induced bending moment at the root of the rod is not to exceed 30 N · cm.

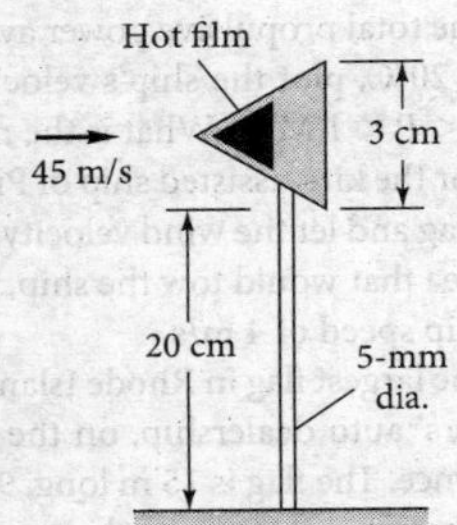

P7.93

P7.94 Baseball drag data from the University of Texas are shown in Fig. P7.94. A baseball weighs approximately 145 grams and has a diameter of 7.4 cm. Hall-of-Famer Nolan Ryan, in a 1974 game, threw the fastest pitch ever recorded: 174 km/h. If it is 18 m from Nolan's hand to the catcher's mitt, estimate the sea-level ball velocity which the catcher experiences for (*a*) a normal baseball, and (*b*) a perfectly smooth baseball.

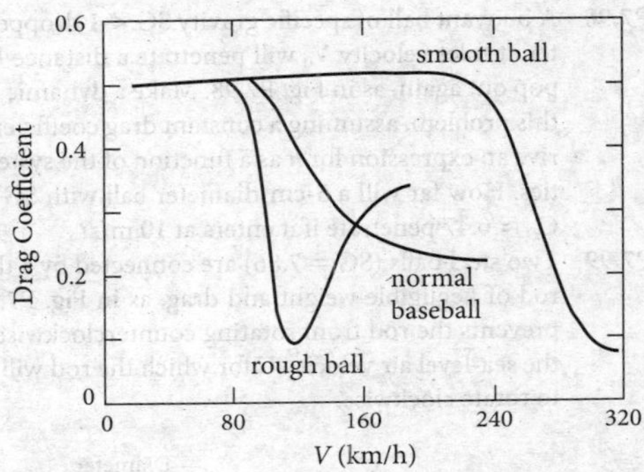

Fig. P7.94

P7.95 An airplane weighing 28 kN, with a drag area $C_DA \approx 5$ m^2, lands at sea level at 55 m/s and deploys a drag parachute 3 m in diameter. No other brakes are applied. (*a*) How long will it take the plane to slow down to 20 m/s? (*b*) How far will it have traveled in that time?

***P7.96** A Savonius rotor (Fig. 6.29*b*) can be approximated by the two open half-tubes in Fig. P7.96 mounted on a central axis. If the drag of each tube is similar to that in Table 7.2, derive an approximate formula for the rotation rate Ω as a function of U, D, L, and the fluid properties (ρ, μ).

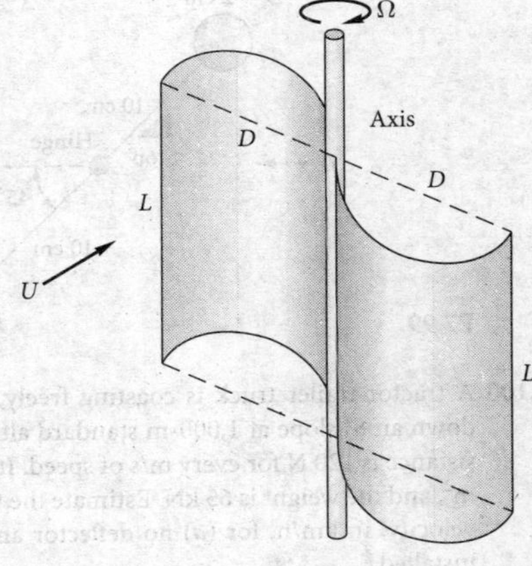

P7.96

P7.97 A simple measurement of automobile drag can be found by an unpowered *coastdown* on a level road with no wind. Assume rolling resistance proportional to velocity. For an automobile of mass 1,500 kg and frontal area 2 m^2, the following velocity-versus-time data are obtained during a coastdown:

t, s	0	10	20	30	40
V, m/s	27.0	24.2	21.8	19.7	17.9

Estimate (*a*) the rolling resistance and (*b*) the drag coefficient. This problem is well suited for computer analysis but can be done by hand also.

***P7.98** A buoyant ball of specific gravity SG < 1 dropped into water at inlet velocity V_0 will penetrate a distance h and then pop out again, as in Fig. P7.98. Make a dynamic analysis of this problem, assuming a constant drag coefficient, and derive an expression for h as a function of the system properties. How far will a 5-cm-diameter ball with SG $= 0.5$ and $C_D \approx 0.47$ penetrate if it enters at 10 m/s?

P7.99 Two steel balls (SG $= 7.86$) are connected by a thin hinged rod of negligible weight and drag, as in Fig. P7.99. A stop prevents the rod from rotating counterclockwise. Estimate the sea-level air velocity U for which the rod will first begin to rotate clockwise.

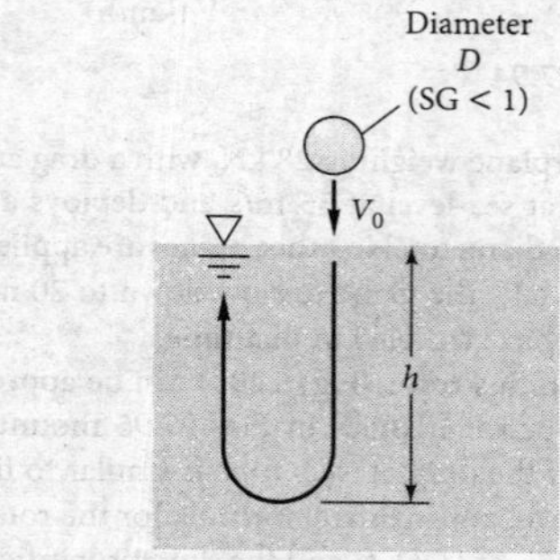

P7.98

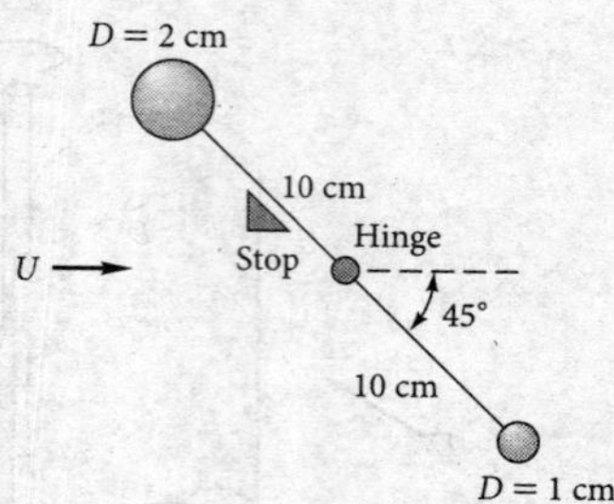

P7.99

P7.100 A tractor-trailer truck is coasting freely, with no brakes, down an 8° slope at 1,000-m standard altitude. Rolling resistance is 120 N for every m/s of speed. Its frontal area is 9 m^2, and the weight is 65 kN. Estimate the terminal coasting velocity, in km/h, for (*a*) no deflector and (*b*) a deflector installed.

P7.101 Icebergs can be driven at substantial speeds by the wind. Let the iceberg be idealized as a large, flat cylinder, $D \gg L$, with one-eighth of its bulk exposed, as in Fig. P7.101. Let the seawater be at rest. If the upper and lower drag forces depend on relative velocities between the iceberg and the fluid, derive an approximate expression for the steady iceberg speed V when driven by wind velocity U.

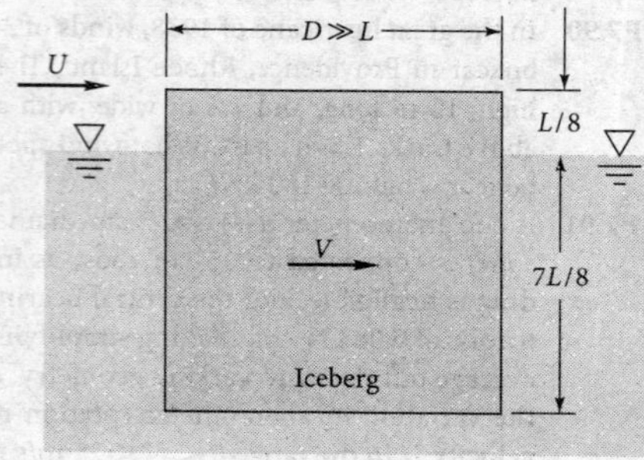

P7.101

P7.102 Sand particles (SG $= 2.7$), approximately spherical with diameters from 100 to 250 μm, are introduced into an upward-flowing stream of water at 20°C. What is the minimum water velocity that will carry all the sand particles *upward?*

P7.103 When immersed in a uniform stream V, a heavy rod hinged at A will hang at *Pode's angle* θ, after an analysis by L. Pode in 1951 (Fig. P7.103). Assume that the cylinder has normal drag coefficient C_{DN} and tangential coefficient C_{DT} that relate the drag forces to V_N and V_T, respectively. Derive an expression for Pode's angle as a function of the flow and rod parameters. Find θ for a steel rod, $L = 40$ cm, $D = 1$ cm, hanging in sea-level air at $V = 35$ m/s.

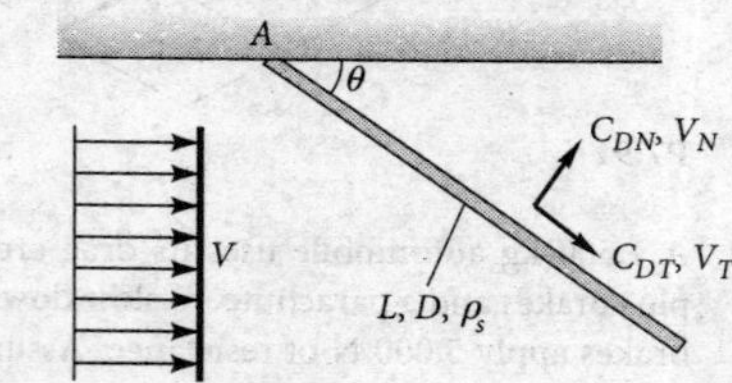

P7.103

P7.104 The Russian Typhoon-class submarine is 170 m long, with a maximum diameter of 23 m. Its propulsor can deliver up to 80,000 hp to the seawater. Model the submarine as an 8:1 ellipsoid and estimate the maximum speed, in m/s, of this ship.

P7.105 A ship 50 m long, with a wetted area of 800 m^2, has the hull shape tested in Fig. 7.19. There are no bow or stern bulbs. The total propulsive power available is 1 MW. For seawater at 20°C, plot the ship's velocity V m/s versus power P for $0 < P < 1$ MW. What is the most efficient setting?

P7.106 For the kite-assisted ship of Prob. P7.85, again neglect wave drag and let the wind velocity be 50 km/h. Estimate the kite area that would tow the ship, unaided by the propeller, at a ship speed of 4 m/s.

P7.107 The largest flag in Rhode Island stands outside Herb Chambers' auto dealership, on the edge of Route I-95 in Providence. The flag is 15 m long, 9 m wide, weighs 1,100 N, and takes four strong people to raise it or lower it. Using Prob.

P7.40 for input, estimate (*a*) the wind speed, in km/h, for which the flag drag is 4,500 N and (*b*) the flag drag when the wind is a low-end category 1 hurricane, 120 km/h. [*Hint:* Providence is at sea level.]

P7.108 The data in Fig. P7.108 are for the lift and drag of a spinning sphere from Ref. 45. Suppose that a tennis ball ($W \approx 0.56$ N, $D \approx 6.35$ cm) is struck at sea level with initial velocity $V_0 = 30$ m/s and "topspin" (front of the ball rotating downward) of 120 r/s. If the initial height of the ball is 1.5 m, estimate the horizontal distance traveled before it strikes the ground.

P7.109 The world record for automobile mileage, 5,384 kilometers per liter, was set in 2005 by the PAC-CAR II in Fig. P7.109, built by students at the Swiss Federal Institute of Technology in Zurich [52]. This little car, with an empty weight of 285 N and a height of only 0.8 m, traveled a 21-km course at 30 km/h to set the record. It has a reported drag coefficient of 0.075 (comparable to an airfoil), based upon a frontal area of 0.3 m². (*a*) What is the drag of this little car when on the course? (*b*) What horsepower is required to propel it? (*c*) Do a bit of research and explain why a value of kilometers per liter is completely misleading in this particular case.

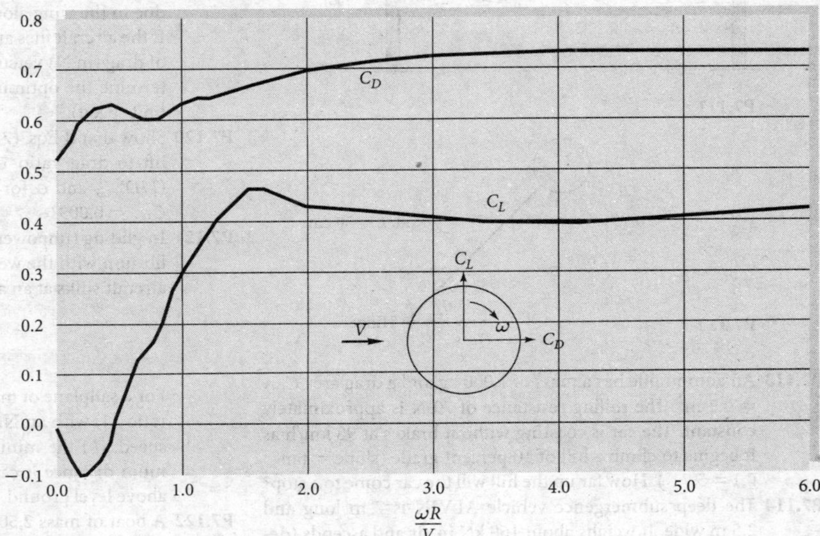

P7.108 Drag and lift coefficients for a rotating sphere at $\text{Re}_D \approx 10^5$, from Ref. 45. *(Reproduced by permission of the American Society of Mechanical Engineers.)*

P7.109 The world's best mileage set by PAC-Car II of ETH Zurich.

P7.110 A baseball pitcher throws a curveball with an initial velocity of 105 km/h and a spin of 6,500 r/min about a vertical axis. A baseball weighs 1.42 N and has a diameter of 7.5 cm. Using the data of Fig. P7.108 for turbulent flow, estimate how far such a curveball will have deviated from its straight-line path when it reaches home plate 18.5 m away.

***P7.111** A table tennis ball has a mass of 2.6 g and a diameter of 3.81 cm. It is struck horizontally at an initial velocity of 20 m/s while it is 50 cm above the table, as in Fig. P7.111. For sea-level air, what spin, in r/min, will cause the ball to strike the opposite edge of the table, 4 m away? Make an analytical estimate, using Fig. P7.108, and account for the fact that the ball decelerates during flight.

P7.112 A smooth wooden sphere (SG = 0.65) is connected by a thin rigid rod to a hinge in a wind tunnel, as in Fig. P7.112. Air at 20°C and 1 atm flows and levitates the sphere. (*a*) Plot the angle θ versus sphere diameter d in the range 1 cm ≤ d ≤ 15 cm. (*b*) Comment on the feasibility of this configuration. Neglect rod drag.

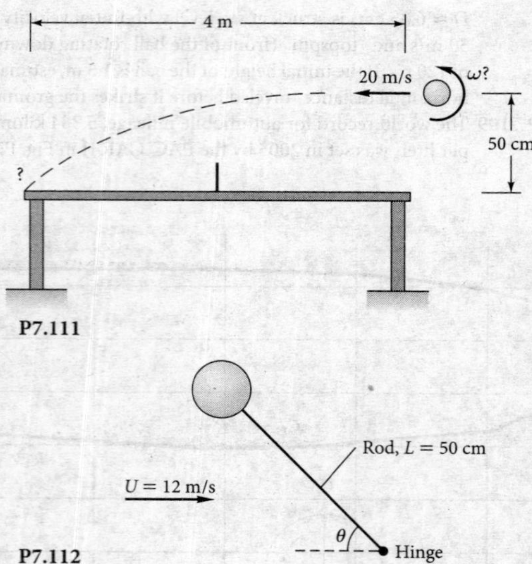

P7.111

P7.112

P7.113 An automobile has a mass of 1,000 kg and a drag area $C_D A = 0.7\ \text{m}^2$. The rolling resistance of 70 N is approximately constant. The car is coasting without brakes at 90 km/h as it begins to climb a hill of 10 percent grade (slope = $\tan^{-1} 0.1 = 5.71°$). How far up the hill will the car come to a stop?

P7.114 The deep submergence vehicle ALVIN is 7 m long and 2.5 m wide. It weighs about 160 kN in air and ascends (descends) in the seawater due to about 1,600 N of positive (negative) buoyancy. Noting that the front face of the ship is quite different for ascent and descent, (*a*) estimate the velocity for each direction, in meters per minute. (*b*) How long does it take to ascend from its maximum depth of 4,500 m?

Lifting bodies—airfoils

P7.115 The Cessna Citation executive jet weighs 67 kN and has a wing area of 32 m². It cruises at 10 km standard altitude with a lift coefficient of 0.21 and a drag coefficient of 0.015. Estimate (*a*) the cruise speed in km/h and (*b*) the horsepower required to maintain cruise velocity.

P7.116 An airplane weighs 180 kN and has a wing area of 160 m² and a mean chord of 4 m. The airfoil properties are given by Fig. 7.25. If the plane is designed to land at $V_0 = 1.2V_{stall}$, using a split flap set at 60°, (*a*) what is the proper landing speed in km/h? (*b*) What power is required for takeoff at the same speed?

P7.117 The Transition* auto-car in Fig. 7.30 has a weight of 5,300 N, a wingspan of 8.5 m, and a wing area of 14 m², with a symmetrical airfoil, $C_{D\infty} \approx 0.02$. Assume that the fuselage and tail section have a drag-area comparable to the Toyota *Prius* [21], $C_D A \approx 6.24\ \text{t}^2$. If the pusher propeller provides a thrust of 1,100 N, how fast, in km/h, can this car-plane fly at an altitude of 2,500 m?

***P7.118** Suppose that the airplane of Prob. P7.116 is fitted with all the best high-lift devices of Fig. 7.28. What is its minimum stall speed in km/h? Estimate the stopping distance if the plane lands at $V_0 = 1.25V_{stall}$ with constant $C_L = 3.0$ and $C_D = 0.2$ and the braking force is 20 percent of the weight on the wheels.

P7.119 A transport plane has a mass of 45,000 kg, a wing area of 160 m², and an aspect ratio of 7. Assume all lift and drag due to the wing alone, with $C_{D\infty} = 0.020$ and $C_{L,max} = 1.5$. If the aircraft flies at 9,000 m standard altitude, make a plot of drag (in N) versus speed (from stall to 240 m/s) and determine the optimum cruise velocity (minimum drag per unit speed).

P7.120 Show that if Eqs. (7.70) and (7.71) are valid, the maximum lift-to-drag ratio occurs when $C_D = 2C_{D\infty}$. What are $(L/D)_{max}$ and α for a symmetric wing when AR = 5 and $C_{D\infty} = 0.009$?

P7.121 In gliding (unpowered) flight, the lift and drag are in equilibrium with the weight. Show that if there is no wind, the aircraft sinks at an angle

$$\tan\theta \approx \frac{\text{drag}}{\text{lift}}$$

For a sailplane of mass 200 kg, wing area 12 m², and aspect ratio 11, with an NACA 0009 airfoil, estimate (*a*) the stall speed, (*b*) the minimum gliding angle, and (*c*) the maximum distance it can glide in still air when it is 1,2200 m above level ground.

P7.122 A boat of mass 2,500 kg has two hydrofoils, each of chord 30 cm and span 1.5 m, with $C_{L,max} = 1.2$ and $C_{D\infty} = 0.08$. Its engine can deliver 130 kW to the water. For seawater at 20°C, estimate (*a*) the minimum speed for which the foils support the boat and (*b*) the maximum speed attainable.

P7.123 In prewar days there was a controversy, perhaps apocryphal, about whether the bumblebee has a legitimate aerodynamic right to fly. The average bumblebee (*Bombus terrestris*) weighs 0.88 g, with a wingspan of 1.73 cm and a wing area of 1.26 cm². It can indeed fly at 10 m/s. Using fixed-wing theory, what is the lift coefficient of the bee at this speed? Is this reasonable for typical airfoils?

***P7.124** The bumblebee can hover at zero speed by flapping its wings. Using the data of Prob. P7.123, devise a theory for flapping wings where the downstroke approximates a short flat plate normal to the flow (Table 7.3) and the upstroke is feathered at nearly zero drag. How many flaps per second of such a model wing are needed to support the bee's weight? (Actual measurements of bees show a flapping rate of 194 Hz.)

***P7.125** The Solar Impulse aircraft in the chapter-opener photo has a wingspan of 63.5 m, a wing area of 199 m², and a weight of 15,700 N. Its propellers deliver an average of 24 hp to the air at a cruising altitude of 8.5 km. Assuming an NACA 0009 airfoil, and neglecting the drag of the fuselage and tail, estimate (*a*) the wing aspect ratio, (*b*) the cruise speed, in km/h, and (*c*) the wing angle of attack. [*Hint:* Simplify by using Fig. 7.25 to estimate lift and drag.]

P7.126 Using the data for the Transition® auto-car from Prob. P7.117, and a maximum lift coefficient of 1.3, estimate the distance for the vehicle to take off at a speed of $1.2V_{stall}$. Note that we have to add the car-body drag to the wing drag.

P7.127 The so-called Rocket Man, Yves Rossy, flew across the Alps in 2008, wearing a rocket-propelled wing-suit with the following data: thrust = 890 N, altitude = 2,500 m, and wingspan = 2.5 m (http://en.wikipedia.org/wiki/Yves_Rossy). Further assume a wing area of 1 m^2, total weight of 1,200 N, $C_{D\infty}$ = 0.08 for the wing, and a drag area of 0.16 m^2 for Rocket Man. Estimate the maximum velocity possible for this condition, in km/h.

Word Problems

W7.1 How do you *recognize* a boundary layer? Cite some physical properties and some measurements that reveal appropriate characteristics.

W7.2 In Chap. 6 the Reynolds number for transition to turbulence in pipe flow was about $Re_{tr} \approx$ 2,300, whereas in flat-plate flow $Re_{tr} \approx$ 1 E6, nearly three orders of magnitude higher. What accounts for the difference?

W7.3 Without writing any equations, give a verbal description of boundary layer displacement thickness.

W7.4 Describe, in words only, the basic ideas behind the "boundary layer approximations."

W7.5 What is an *adverse* pressure gradient? Give three examples of flow regimes where such gradients occur.

W7.6 What is a *favorable* pressure gradient? Give three examples of flow regimes where such gradients occur.

W7.7 The drag of an airfoil (Fig. 7.12) increases considerably if you turn the sharp edge around 180° to face the stream. Can you explain this?

W7.8 In Table 7.3, the drag coefficient of a spruce tree decreases sharply with wind velocity. Can you explain this?

W7.9 Thrust is required to propel an airplane at a finite forward velocity. Does this imply an energy loss to the system? Explain the concepts of thrust and drag in terms of the first law of thermodynamics.

W7.10 How does the concept of *drafting,* in automobile and bicycle racing, apply to the material studied in this chapter?

W7.11 The circular cylinder of Fig. 7.13 is doubly symmetric and therefore should have no lift. Yet a lift sensor would definitely reveal a finite root-mean-square value of lift. Can you explain this behavior?

W7.12 Explain in words why a thrown spinning ball moves in a curved trajectory. Give some physical reasons why a side force is developed in addition to the drag.

Fundamentals of Engineering Exam Problems

FE7.1 A smooth 12-cm-diameter sphere is immersed in a stream of 20°C water moving at 6 m/s. The appropriate Reynolds number of this sphere is approximately
(*a*) 2.3 E5, (*b*) 7.2 E5, (*c*) 2.3 E6, (*d*) 7.2 E6, (*e*) 7.2 E7

FE7.2 If, in Prob. FE7.1, the drag coefficient based on frontal area is 0.5, what is the drag force on the sphere?
(*a*) 17 N, (*b*) 51 N, (*c*) 102 N, (*d*) 130 N, (*e*) 203 N

FE7.3 If, in Prob. FE7.1, the drag coefficient based on frontal area is 0.5, at what terminal velocity will an aluminum sphere (SG = 2.7) fall in still water?
(*a*) 2.3 m/s, (*b*) 2.9 m/s, (*c*) 4.6 m/s, (*d*) 6.5 m/s, (*e*) 8.2 m/s

FE7.4 For flow of sea-level standard air at 4 m/s parallel to a thin flat plate, estimate the boundary layer thickness at x = 60 cm from the leading edge:
(*a*) 1.0 mm, (*b*) 2.6 mm, (*c*) 5.3 mm, (*d*) 7.5 mm, (*e*) 20.2 mm

FE7.5 In Prob. FE7.4, for the same flow conditions, what is the wall shear stress at x = 60 cm from the leading edge?
(*a*) 0.053 Pa, (*b*) 0.11 Pa, (*c*) 0.16 Pa, (*d*) 0.32 Pa, (*e*) 0.64 Pa

FE7.6 Wind at 20°C and 1 atm blows at 75 km/h past a flagpole 18 m high and 20 cm in diameter. The drag coefficient, based on frontal area, is 1.15. Estimate the wind-induced bending moment at the base of the pole.
(*a*) 9.7 kN · m, (*b*) 15.2 kN · m, (*c*) 19.4 kN · m, (*d*) 30.5 kN · m, (*e*) 61.0 kN · m

FE7.7 Consider wind at 20°C and 1 atm blowing past a chimney 30 m high and 80 cm in diameter. If the chimney may fracture at a base bending moment of 486 kN · m, and its drag coefficient based on frontal area is 0.5, what is the approximate maximum allowable wind velocity to avoid fracture?
(*a*) 80 km/h, (*b*) 120 km/h, (*c*) 160 km/h, (*d*) 200 km/h, (*e*) 240 km/h

FE7.8 A dust particle of density 2,600 kg/m^3, small enough to satisfy Stokes's drag law, settles at 1.5 mm/s in air at 20°C and 1 atm. What is its approximate diameter?
(*a*) 1.8 μm, (*b*) 2.9 μm, (*c*) 4.4 μm, (*d*) 16.8 μm, (*e*) 234 μm

FE7.9 An airplane has a mass of 19,550 kg, a wingspan of 20 m, and an average wing chord of 3 m. When flying in air of density 0.5 kg/m^3, its engines provide a thrust of 12 kN

against an overall drag coefficient of 0.025. What is its approximate velocity?
(*a*) 112 m/s, (*b*) 134 m/s, (*c*) 156 m/s, (*d*) 179 m/s, (*e*) 201 m/s

FE7.10 For the flight conditions of the airplane in Prob. FE7.9 above, what is its approximate lift coefficient?
(*a*) 0.1, (*b*) 0.2, (*c*) 0.3, (*d*) 0.4, (*e*) 0.5

Comprehensive Problems

C7.1 Jane wants to estimate the drag coefficient of herself on her bicycle. She measures the projected frontal area to be 0.40 m^2 and the rolling resistance to be 0.80 N · s/m. The mass of the bike is 15 kg, while the mass of Jane is 80 kg. Jane coasts down a long hill that has a constant 4° slope. (See Fig. C7.1.) She reaches a terminal (steady state) speed of 14 m/s down the hill. Estimate the aerodynamic drag coefficient C_D of the rider and bicycle combination.

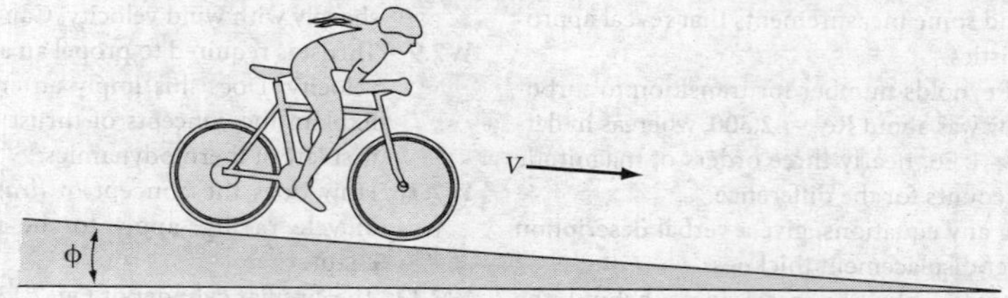

C7.1

C7.2 Air at 20°C and 1 atm flows at V_{avg} = 5 m/s between long, smooth parallel heat exchanger plates 10 cm apart, as in Fig. C7.2. It is proposed to add a number of widely spaced 1-cm-long interrupter plates to increase the heat transfer, as shown. Although the flow in the channel is turbulent, the boundary layers over the interrupter plates are essentially laminar. Assume all plates are 1 m wide into the paper. Find (*a*) the pressure drop in Pa/m without the small plates present. Then find (*b*) the number of small plates per meter of channel length that will cause the pressure drop to rise to 10.0 Pa/m.

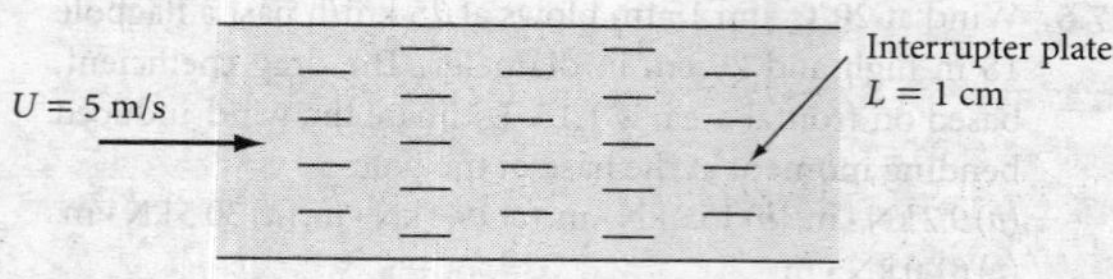

C7.2

C7.3 A new pizza store is planning to open. It will, of course, offer free delivery, and therefore need a small delivery car with a large sign attached. The sign (a flat plate) is 0.45 m high and 1.5 m long. The boss (having no feel for fluid mechanics) mounts the sign bluntly facing the wind. One of his drivers is taking fluid mechanics and tells his boss he can save lots of money by mounting the sign parallel to the wind. (See Fig. C7.3.) (*a*) Calculate the drag (in N) on the *sign alone* at 60 km/h (16.7 m/s) in *both orientations*. (*b*) Suppose the car without any sign has a drag coefficient of 0.4 and a frontal area of 3.75 m^2. For $V = 60$ km/h, calculate the *total* drag of the car–sign combination for both orientations. (*c*) If the car has a rolling resistance of 180 N at 60 km/h, calculate the horsepower required by the engine to drive the car at 60 km/h in both orientations. (*d*) Finally, if the engine can deliver 3 hp for 1 h on a liter of gasoline, calculate the fuel efficiency in km/L for both orientations at 60 km/h.

C7.3

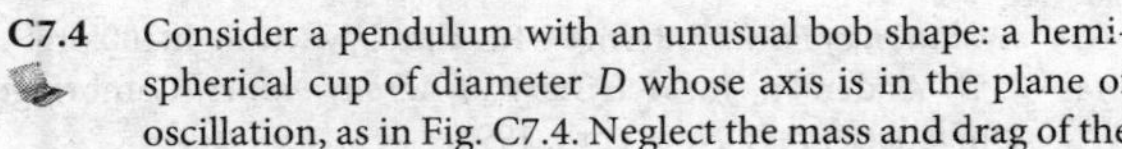

C7.4 Consider a pendulum with an unusual bob shape: a hemispherical cup of diameter D whose axis is in the plane of oscillation, as in Fig. C7.4. Neglect the mass and drag of the rod L. (*a*) Set up the differential equation for the oscillation $\theta(t)$, including different cup drag (air density ρ) in each direction, and (*b*) nondimensionalize this equation. (*c*) Determine the natural frequency of oscillation for small $\theta \ll 1$ rad. (*d*) For the special case $L = 1$ m, $D = 10$ cm, $m = 50$ g, and air at 20°C and 1 atm, with $\theta(0) = 30°$, find (numerically) the time required for the oscillation amplitude to drop to 1°.

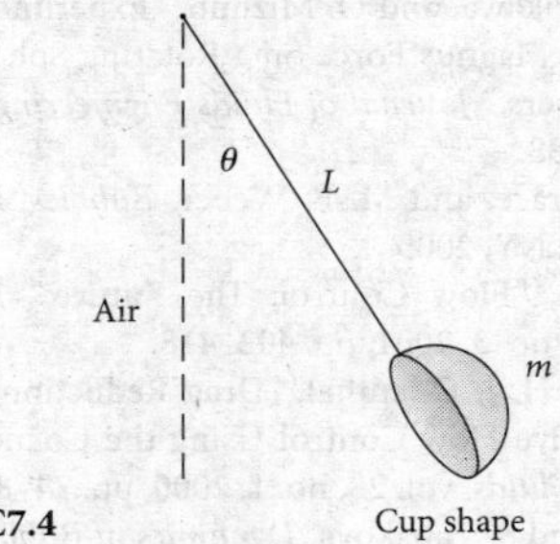

C7.4

C7.5 Program a method of numerical solution of the Blasius flat-plate relation, Eq. (7.22), subject to the conditions in Eqs. (7.23). You will find that you cannot get started without knowing the initial second derivative $f''(0)$, which lies between 0.2 and 0.5. Devise an iteration scheme that starts at $f''(0) \approx 0.2$ and converges to the correct value. Print out $u/U = f'(\eta)$ and compare with Table 7.1.

Design Project

D7.1 It is desired to design a cup anemometer for wind speed, similar to Fig. P7.91, with a more sophisticated approach than the "average-torque" method of Prob. P7.91. The design should achieve an approximately linear relation between wind velocity and rotation rate in the range $30 < U < 60$ km/h, and the anemometer should rotate at about 6 r/s at $U = 45$ km/h. All specifications—cup diameter D, rod length L, rod diameter d, the bearing type, and all materials—are to be selected through your analysis. Make suitable assumptions about the instantaneous drag of the cups and rods at any given angle $\theta(t)$ of the system. Compute the instantaneous torque $T(t)$, and find and integrate the instantaneous angular acceleration of the device. Develop a complete theory for rotation rate versus wind speed in the range $0 < U < 75$ km/h. Try to include actual commercial bearing friction properties.

References

1. H. Schlichting and K. Gersten, *Boundary Layer Theory*, 8th ed., Springer, New York, 2000.
2. F. M. White, *Viscous Fluid Flow*, 3d ed., McGraw-Hill, New York, 2005.
3. J. Cousteix, *Modeling and Computation of Boundary-Layer Flows*, 2d ed., Springer-Verlag, New York, 2005.
4. J. D. Anderson, *Computational Fluid Dynamics: An Introduction*, 3d ed., Springer, New York, 2010.
5. V. V. Sychev et al., *Asymptotic Theory of Separated Flows*, Cambridge University Press, New York, 2008.
6. I. J. Sobey, *Introduction to Interactive Boundary Layer Theory*, Oxford University Press, New York, 2001.
7. T. von Kármán, "On Laminar and Turbulent Friction," *Z. Angew. Math. Mech.*, vol. 1, 1921, pp. 235–236.
8. G. B. Schubauer and H. K. Skramstad, "Laminar Boundary Layer Oscillations and Stability of Laminar Flow," *Natl. Bur. Stand. Res. Pap.* 1772, April 1943 (see also *J. Aero. Sci.*, vol. 14, 1947, pp. 69–78, and *NACA Rep.* 909, 1947).
9. P. S. Bernard and J. M. Wallace, *Turbulent Flow: Analysis, Measurement, and Prediction*, Wiley, New York, 2002.
10. P. W. Runstadler, Jr., et al., "Diffuser Data Book," Creare Inc., *Tech. Note* 186, Hanover, NH, May 1975.
11. B. Thwaites, "Approximate Calculation of the Laminar Boundary Layer," *Aeronaut. Q.*, vol. 1, 1949, pp. 245–280.
12. S. F. Hoerner, *Fluid Dynamic Drag*, published by the author, Midland Park, NJ, 1965.
13. J. D. Anderson, *Fundamentals of Aerodynamics*, 5th ed., McGraw-Hill, New York, 2010.
14. V. Tucker and G. C. Parrott, "Aerodynamics of Gliding Flight of Falcons and Other Birds," *J. Exp. Biol.*, vol. 52, 1970, pp. 345–368.
15. E. C. Tupper, *Introduction to Naval Architecture*, 5th ed., Butterworth-Heinemann, Burlington, MA, 2013.
16. I. H. Abbott and A. E. von Doenhoff, *Theory of Wing Sections*, Dover, New York, 1981.
17. R. L. Kline and F. F. Fogelman, "Airfoil for Aircraft," U. S. Patent 3,706,430, Dec. 19, 1972.
18. A. Azuma, *The Biokinetics of Swimming and Flying*, AIAA, Reston, VA, 2006.
19. National Committee for Fluid Mechanics Films, *Illustrated Experiments in Fluid Mechanics*, M.I.T. Press, Cambridge, MA, 1972.
20. D. M. Bushnell and J. Hefner (Eds.), *Viscous Drag Reduction in Boundary Layers*, American Institute of Aeronautics & Astronautics, Reston, VA, 1990.

21. "Automobile Drag Coefficient," URL <http://en.wikipedia.org/wiki/Automobile_drag_coefficients>.
22. R. H. Barnard, *Road Vehicle Aerodynamic Design*, 3d ed., Mechaero Publishing, St. Albans, U.K., 2010.
23. R. D. Blevins, *Applied Fluid Dynamics Handbook*, BBS, New York, 2009.
24. R. C. Johnson, Jr., G. E. Ramey, and D. S. O'Hagen, "Wind Induced Forces on Trees," *J. Fluids Eng.*, vol. 104, March 1983, pp. 25–30.
25. P. W. Bearman et al., "The Effect of a Moving Floor on Wind-Tunnel Simulation of Road Vehicles," Paper No. 880245, SAE Transactions, *J. Passenger Cars,* vol. 97, sec. 4, 1988, pp. 4.200–4.214.
26. *CRC Handbook of Tables for Applied Engineering Science,* 2d ed., CRC Press, Boca Raton, FL, 1973.
27. T. Inui, "Wavemaking Resistance of Ships," *Trans. Soc. Nav. Arch. Marine Engrs.*, vol. 70, 1962, pp. 283–326.
28. L. Larsson, "CFD in Ship Design—Prospects and Limitations," *Ship Technology Research,* vol. 44, no. 3, July 1997, pp. 133–154.
29. R. L. Street, G. Z. Watters, and J. K. Vennard, *Elementary Fluid Mechanics,* 7th ed., Wiley, New York, 1995.
30. J. D. Anderson, Jr., *Modern Compressible Flow: with Historical Perspective,* 3d ed., McGraw-Hill, New York, 2002.
31. J. D. Anderson, Jr., *Hypersonic and High Temperature Gas Dynamics,* AIAA, Reston, VA, 2000.
32. J. Rom, *High Angle of Attack Aerodynamics: Subsonic, Transonic, and Supersonic Flows,* Springer-Verlag, New York, 2011.
33. S. Vogel, "Drag and Reconfiguration of Broad Leaves in High Winds," *J. Exp. Bot.*, vol. 40, no. 217, August 1989, pp. 941–948.
34. S. Vogel, *Life in Moving Fluids,* Princeton University Press 2d ed., Princeton, NJ, 1996.
35. J. A. C. Humphrey (ed.), *Proceedings 2d International Symposium on Mechanics of Plants, Animals, and Their Environment,* Engineering Foundation, New York, January 2000.
36. D. D. Joseph, R. Bai, K. P. Chen, and Y. Y. Renardy, "Core-Annular Flows," *Annu. Rev. Fluid Mech.*, vol. 29, 1997, pp. 65–90.
37. J. W. Hoyt and R. H. J. Sellin, "Scale Effects in Polymer Solution Pipe Flow," *Experiments in Fluids,* vol. 15, no. 1, June 1993, pp. 70–74.
38. S. Nakao, "Application of V-Shape Riblets to Pipe Flows," *J. Fluids Eng.*, vol. 113, December 1991, pp. 587–590.
39. P. Thiede (ed.), *Aerodynamic Drag Reduction Technologies,* Springer, New York, 2001.
40. C. L. Merkle and S. Deutsch, "Microbubble Drag Reduction in Liquid Turbulent Boundary Layers," *Applied Mechanics Reviews,* vol. 45, no. 3 part 1, March 1992, pp. 103–127.
41. K. S. Choi and G. E. Karniadakis, "Mechanisms on Transverse Motions in Turbulent Wall Flows," *Annual Review of Fluid Mechanics,* vol. 35, 2003, pp. 45–62.
42. C. J. Roy, J. Payne, and M. McWherter-Payne, "RANS Simulations of a Simplified Tractor-Trailer Geometry," *J. Fluids Engineering,* vol. 128, Sept. 2006, pp. 1083–1089.
43. *Evolution of Flight,* Internet URL <http://www.flight100.org>.
44. J. D. Anderson, Jr., *A History of Aerodynamics,* Cambridge University Press, New York, 1999.
45. Y. Tsuji, Y. Morikawa, and O. Mizuno, "Experimental Measurement of the Magnus Force on a Rotating Sphere at Low Reynolds Numbers," *Journal of Fluids Engineering,* vol. 107, 1985, pp. 484–488.
46. R. Clift, J. R. Grace, and M. E. Weber, *Bubbles, Drops and Particles,* Dover, NY, 2005.
47. M. Gad-el-Hak, "Flow Control: The Future," *Journal of Aircraft,* vol. 38, no. 3, 2001, pp. 402–418.
48. D. Geropp and H. J. Odenthal, "Drag Reduction of Motor Vehicles by Active Flow Control Using the Coanda Effect," *Experiments in Fluids,* vol. 28, no. 1, 2000, pp. 74–85.
49. Z. Zapryanov and S. Tabakova, *Dynamics of Bubbles, Drops, and Rigid Particles,* Kluwer Academic Pub., New York, 1998.
50. D. G. Karamanev, and L. N. Nikolov, "Freely Rising Spheres Do Not Obey Newton's Law for Free Settling," *AIChE Journal,* vol. 38, no. 1, Nov. 1992, pp. 1843–1846.
51. Katz J., *Race-Car Aerodynamics,* Robert Bentley Inc., Cambridge, MA, 2003.
52. A. S. Brown, "More than 12,000 Miles to the Gallon," *Mechanical Engineering,* January 2006, p. 64.
53. D. M. Bushnell, "Aircraft Drag Reduction: A Review," *Proceedings of the Institution of Mechanical Engineers, Part G: Journal of Aerospace Engineering,* vol. 217, no. 1, 2003, pp. 1–18.
54. D. B. Spalding, "A Single Formula for the Law of the Wall," *J. Appl. Mechanics,* vol. 28, no. 3, 1961, pp. 444–458.
55. D. G. Fertis, "New Airfoil-Design Concept with Improved Aerodynamic Characteristics," *J. Aerospace Engineering,* vol. 7, no. 3, July 1994, pp. 328–339.
56. F. Finaish and S. Witherspoon, "Aerodynamic Performance of an Airfoil with Step-Induced Vortex for Lift Augmentation," *J. Aerospace Engineering,* vol. 11, no. 1, Jan. 1998, pp. 9–16.
57. D. S. Miklosovic et al., "Leading Edge Tubercles Delay Stall on Humpback Whale," *Physics of Fluids,* vol. 16, no. 5, May 2004, pp. L39–L42.
58. R. McCallen, J. Ross, and F. Browand, *The Aerodynamics of Heavy Vehicles: Trucks, Buses, and Trains,* Springer-Verlag, New York, 2005.
59. J. R. Cruz et al., "Wind Tunnel Testing of Various Disk-Gap-Band Parachutes," AIAA Paper 2003-2129, 17th AIAA Aerodynamic Decelerator Systems Conference, May 2003.
60. B. de Gomars, "Drag of Cones at Zero Incidence," URL http://perso.numericable.fr/fbouquetbe63/gomars/cx_cones.
61. National Highway Traffic Safety Administration, "NHTSA Tire Fuel Efficiency," Report DOT HS 811 154, August 2009.
62. J. F. Cahill, "Summary of Section Data on Trailing-Edge High-Lift Devices," National Advisory Committee for Aeronautics, Report 938, 1949.

Until they reach shallow water and feel the bottom, ocean waves are almost frictionless. Waves are created by winds, especially storms. Long waves—large distances between crests—travel the fastest and decay the slowest. Short waves decay more quickly but are still nearly frictionless. The pictured long waves, breaking on the beach in Narragansett, Rhode Island, might have been formed from a storm off the coast of Africa. The theory of ocean waves [21] is based almost entirely upon frictionless flow. *(Photo courtesy of Ellen Emerson White.)*

Chapter 8 Potential Flow and Computational Fluid Dynamics

Motivation. The basic partial differential equations of mass, momentum, and energy were discussed in Chap. 4. A few solutions were then given for incompressible *viscous* flow in Sec. 4.10. The viscous solutions were limited to simple geometries and unidirectional flows, where the difficult nonlinear convective terms were neglected. Potential flows are not limited by such nonlinear terms. Then, in Chap. 7, we found an approximation: patching *boundary layer flows* onto an outer inviscid flow pattern. For more complicated viscous flows, we found no theory or solutions, just experimental data or computer solutions.

The purposes of the present chapter are (1) to explore examples of potential theory and (2) to indicate some flows that can be approximated by computational fluid dynamics (CFD). The combination of these two gives us a good picture of incompressible-flow theory and its relation to experiment. One of the most important applications of potential-flow theory is to aerodynamics and marine hydrodynamics. First, however, we will review and extend the concepts of Chap. 4.

8.1 Introduction and Review

Figure 8.1 reminds us of the problems to be faced. A free stream approaches two closely spaced bodies, creating an "internal" flow between them and "external" flows above and below them. The fronts of the bodies are regions of favorable gradient (decreasing pressure along the surface), and the boundary layers will be attached and thin: Inviscid theory will give excellent results for the outer flow if $\text{Re} > 10^4$. For the internal flow between bodies, the boundary layers will grow and eventually meet, and the inviscid core vanishes. Inviscid theory works well in a "short" duct $L/D < 10$, such as the nozzle of a wind tunnel. For longer ducts we must estimate boundary layer growth and be cautious about using inviscid theory.

For the external flows above and below the bodies in Fig. 8.1, inviscid theory should work well for the outer flows, until the surface pressure gradient becomes adverse (increasing pressure) and the boundary layer separates or stalls. After the separation point, boundary layer theory becomes inaccurate, and the outer flow streamlines are

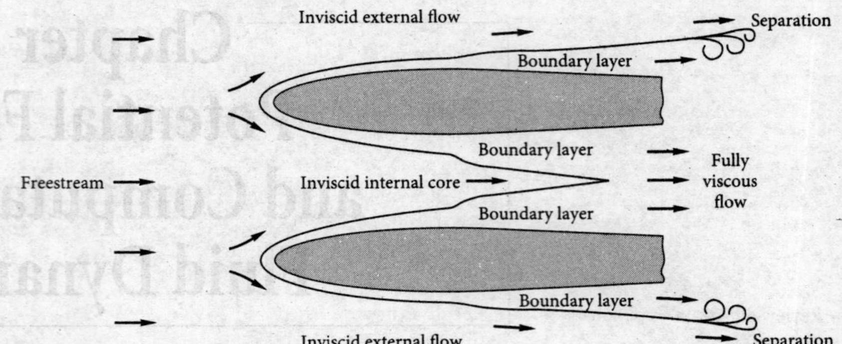

Fig. 8.1 Patching viscous and inviscid flow regions. Potential theory in this chapter does not apply to the boundary layer regions.

deflected and have a strong interaction with the viscous near-wall regions. The theoretical analysis of separated-flow regions is an active research area at present.

Review of Velocity Potential Concepts

Recall from Sec. 4.9 that if viscous effects are neglected, low-speed flows are irrotational, $\nabla \times \mathbf{V} = 0$, and the velocity potential ϕ exists, such that

$$\mathbf{V} = \nabla\phi \quad \text{or} \quad u = \frac{\partial\phi}{\partial x} \quad v = \frac{\partial\phi}{\partial y} \quad w = \frac{\partial\phi}{\partial z} \tag{8.1}$$

The continuity equation (4.73), $\nabla \cdot \mathbf{V} = 0$, reduces to Laplace's equation for ϕ:

$$\nabla^2\phi = \frac{\partial^2\phi}{\partial x^2} + \frac{\partial^2\phi}{\partial y^2} + \frac{\partial^2\phi}{\partial z^2} = 0 \tag{8.2}$$

and the momentum equation (4.74) reduces to Bernoulli's equation:

$$\frac{\partial\phi}{\partial t} + \frac{p}{\rho} + \frac{1}{2}V^2 + gz = \text{const} \quad \text{where } V = |\nabla\phi| \tag{8.3}$$

Typical boundary conditions are known free-stream conditions

$$\text{Outer boundaries:} \qquad \text{Known } \frac{\partial\phi}{\partial x}, \frac{\partial\phi}{\partial y}, \frac{\partial\phi}{\partial z} \tag{8.4}$$

and no velocity normal to the boundary at the body surface:

$$\text{Solid surfaces:} \quad \frac{\partial\phi}{\partial n} = 0 \quad \text{where } n \text{ is perpendicular to body} \tag{8.5}$$

Unlike the no-slip condition in viscous flow, here there is *no* condition on the tangential surface velocity $V_s = \partial\phi/\partial s$, where s is the coordinate along the surface. This velocity is determined as part of the solution to the problem.

Occasionally the problem involves a free surface, for which the boundary pressure is known and equal to p_a, usually a constant. The Bernoulli equation (8.3) then supplies a relation at the surface between V and the elevation z of the surface. For steady flow,

$$\text{Free surface:} \quad V^2 = |\nabla\phi|^2 = \text{const} - 2gz_{\text{surf}} \tag{8.6}$$

It should be clear to the reader that this use of Laplace's equation, with known values of the derivative of ϕ along the boundaries, is much easier than a direct attack using the fully viscous Navier-Stokes equations. The analysis of Laplace's equation is very well developed and is termed *potential theory*, with whole books written about its application to fluid mechanics [1 to 4]. There are many analytical techniques, including superposition of elementary functions, conformal mapping [4], numerical finite differences [5], numerical finite elements [6], numerical boundary elements [7], and electric or mechanical analogs [8] that are now outdated. Having found $\phi(x, y, z, t)$ from such an analysis, we then compute **V** by direct differentiation in Eq. (8.1), after which we compute p from Eq. (8.3). The procedure is quite straightforward, and many interesting albeit idealized results can be obtained. A beautiful collection of computer-generated potential flow sketches is given by Kirchhoff [43].

Review of Stream Function Concepts

Recall from Sec. 4.7 that if a flow is described by only two coordinates, the stream function ψ also exists as an alternate approach. For plane incompressible flow in xy coordinates, the correct form is

$$u = \frac{\partial \psi}{\partial y} \qquad v = -\frac{\partial \psi}{\partial x} \tag{8.7}$$

The condition of irrotationality reduces to Laplace's equation for ψ also:

$$2\omega_z = 0 = \frac{\partial v}{\partial x} - \frac{\partial u}{\partial y} = \frac{\partial}{\partial x}\left(-\frac{\partial \psi}{\partial x}\right) - \frac{\partial}{\partial y}\left(\frac{\partial \psi}{\partial y}\right)$$

or

$$\frac{\partial^2 \psi}{\partial x^2} + \frac{\partial^2 \psi}{\partial y^2} = 0 \tag{8.8}$$

The boundary conditions again are known velocity in the stream and no flow through any solid surface:

$$\text{Free stream:} \qquad \text{Known } \frac{\partial \psi}{\partial x}, \frac{\partial \psi}{\partial y} \tag{8.9a}$$

$$\text{Solid surface:} \qquad \psi_{\text{body}} = \text{const} \tag{8.9b}$$

Equation (8.9*b*) is particularly interesting because *any* line of constant ψ in a flow can therefore be interpreted as a body shape and may lead to interesting applications.

For the applications in this chapter, we may compute either ϕ or ψ or both, and the solution will be an *orthogonal flow net* as in Fig. 8.2. Once found, either set of lines may be considered the ϕ lines, and the other set will be the ψ lines. Both sets of lines are laplacian and could be useful.

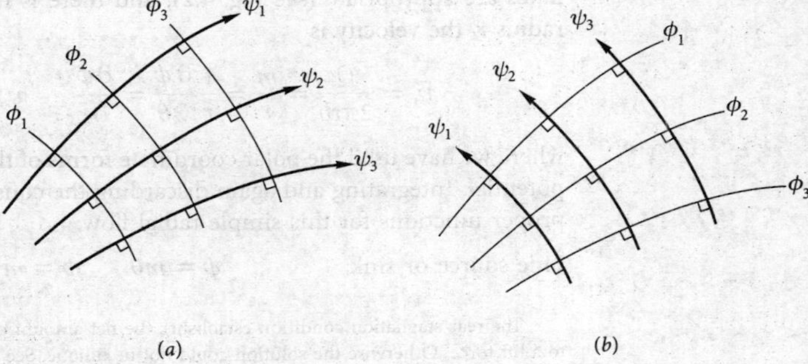

Fig. 8.2 Streamlines and potential lines are orthogonal and may reverse roles if results are useful: (*a*) typical inviscid flow pattern; (*b*) same as (*a*) with roles reversed.

Plane Polar Coordinates

Many solutions in this chapter are conveniently expressed in polar coordinates (r, θ). Both the velocity components and the differential relations for ϕ and ψ are then changed, as follows:

$$v_r = \frac{\partial \phi}{\partial r} = \frac{1}{r}\frac{\partial \psi}{\partial \theta} \qquad v_\theta = \frac{1}{r}\frac{\partial \phi}{\partial \theta} = -\frac{\partial \psi}{\partial r} \tag{8.10}$$

Laplace's equation takes the form

$$\frac{1}{r}\frac{\partial}{\partial r}\left(r\frac{\partial \phi}{\partial r}\right) + \frac{1}{r^2}\frac{\partial^2 \phi}{\partial \theta^2} = 0 \tag{8.11}$$

Exactly the same equation holds for the polar-coordinate form of $\psi(r, \theta)$.

An intriguing facet of potential flow with no free surface is that the governing equations (8.2) and (8.8) contain no parameters, nor do the boundary conditions. Therefore, the solutions are purely geometric, depending only on the body shape, the free-stream orientation, and—surprisingly—the position of the rear stagnation point.[1] There is no Reynolds, Froude, or Mach number to complicate the dynamic similarity. Inviscid flows are kinematically similar without additional parameters—recall Fig. 5.6*a*.

8.2 Elementary Plane Flow Solutions

The present chapter is a detailed introductory study of inviscid incompressible flows, especially those that possess both a stream function and a velocity potential. Many solutions make use of the superposition principle, so we begin with the three elementary building blocks illustrated in Fig. 8.3: (*a*) a uniform stream in the x direction, (*b*) a line source or sink at the origin, and (*c*) a line vortex at the origin.

Uniform Stream in the *x* Direction

A uniform stream $\mathbf{V} = \mathbf{i}U$, as in Fig. 8.3*a*, possesses both a stream function and a velocity potential, which may be found as follows:

$$u = U = \frac{\partial \phi}{\partial x} = \frac{\partial \psi}{\partial y} \qquad v = 0 = \frac{\partial \phi}{\partial y} = -\frac{\partial \psi}{\partial x}$$

We may integrate each expression and discard the constants of integration, which do not affect the velocities in the flow. The results are

$$\text{Uniform stream } \mathbf{i}U: \qquad \psi = Uy \qquad \phi = Ux \tag{8.12}$$

The streamlines are horizontal straight lines (y = const), and the potential lines are vertical (x = const)—that is, orthogonal to the streamlines, as expected.

Line Source or Sink at the Origin

Suppose that the z axis were a sort of thin pipe manifold through which fluid issued at total rate Q uniformly along its length b. Looking at the xy plane, we would see a cylindrical radial outflow or *line source,* as sketched in Fig. 8.3*b*. Plane polar coordinates are appropriate (see Fig. 4.2), and there is no circumferential velocity. At any radius r, the velocity is

$$v_r = \frac{Q}{2\pi r b} = \frac{m}{r} = \frac{1}{r}\frac{\partial \psi}{\partial \theta} = \frac{\partial \phi}{\partial r} \qquad v_\theta = 0 = -\frac{\partial \psi}{\partial r} = \frac{1}{r}\frac{\partial \phi}{\partial \theta}$$

where we have used the polar coordinate forms of the stream function and the velocity potential. Integrating and again discarding the constants of integration, we obtain the proper functions for this simple radial flow:

$$\text{Line source or sink:} \qquad \psi = m\theta \qquad \phi = m \ln r \tag{8.13}$$

[1]The rear stagnation condition establishes the net amount of "circulation" about the body, giving rise to a lift force. Otherwise the solution could not be unique. See Sec. 8.4.

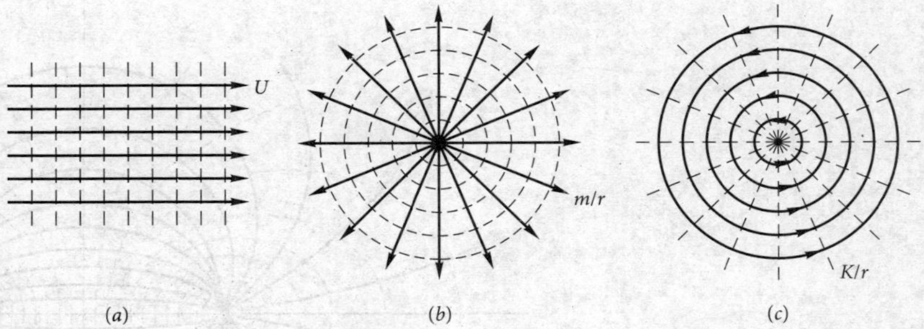

Fig. 8.3 Three elementary plane potential flows. Solid lines are streamlines; dashed lines are potential lines. (*a*) uniform stream; (*b*) line sink; (*c*) line vortex.

where $m = Q/(2\pi b)$ is a constant, positive for a source, negative for a sink. As shown in Fig. 8.3*b*, the streamlines are radial spokes (constant θ), and the potential lines are circles (constant r).

Line Irrotational Vortex

A (two-dimensional) line vortex is a purely circulating steady motion, $\upsilon_\theta = f(r)$ only, $\upsilon_r = 0$. This satisfies the continuity equation identically, as may be checked from Eq. (4.12*b*). We may also note that a variety of velocity distributions $\upsilon_\theta(r)$ satisfy the θ momentum equation of a viscous fluid, Eq. (D.6). We may show, as a problem exercise, that only one function $\upsilon_\theta(r)$ is *irrotational;* that is, curl $\mathbf{V} = 0$, and $\upsilon_\theta = K/r$, where K is a constant. This is sometimes called a *free vortex*, for which the stream function and velocity may be found:

$$\upsilon_r = 0 = \frac{1}{r}\frac{\partial \psi}{\partial \theta} = \frac{\partial \phi}{\partial r} \qquad \upsilon_\theta = \frac{K}{r} = -\frac{\partial \psi}{\partial r} = \frac{1}{r}\frac{\partial \phi}{\partial \theta}$$

We may again integrate to determine the appropriate functions:

$$\psi = -K \ln r \qquad \phi = K\theta \tag{8.14}$$

where K is a constant called the *strength* of the vortex. As shown in Fig. 8.3*c*, the streamlines are circles (constant r), and the potential lines are radial spokes (constant θ). Note the similarity between Eqs. (8.13) and (8.14). A free vortex is a sort of reversed image of a source. The "bathtub vortex," formed when water drains through a bottom hole in a tank, is a good approximation to the free-vortex pattern.

Superposition: Source Plus an Equal Sink

Each of the three elementary flow patterns in Fig. 8.3 is an incompressible irrotational flow and therefore satisfies both plane "potential flow" equations $\nabla^2\psi = 0$ and $\nabla^2\phi = 0$. Since these are linear partial differential equations, any *sum* of such basic solutions is also a solution. Some of these composite solutions are quite interesting and useful.

For example, consider a source $+m$ at $(x, y) = (-a, 0)$, combined with a sink of equal strength $-m$, placed at $(+a, 0)$, as in Fig. 8.4. The resulting stream function is simply the sum of the two. In cartesian coordinates,

$$\psi = \psi_{\text{source}} + \psi_{\text{sink}} = m \tan^{-1}\frac{y}{x+a} - m \tan^{-1}\frac{y}{x-a}$$

Similarly, the composite velocity potential is

$$\phi = \phi_{\text{source}} + \phi_{\text{sink}} = \frac{1}{2} m \ln\left[(x+a)^2 + y^2\right] - \frac{1}{2} m \ln\left[(x-a)^2 + y^2\right]$$

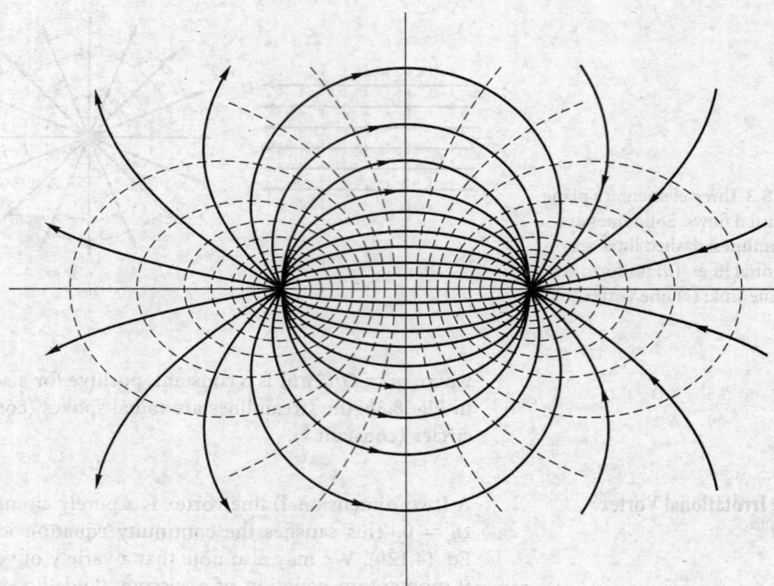

Fig. 8.4 Potential flow due to a line source plus an equal line sink, from Eq. (8.15). Solid lines are streamlines; dashed lines are potential lines.

By using trigonometric and logarithmic identities, these may be simplified to

$$\text{Source plus sink:} \qquad \psi = -m \tan^{-1} \frac{2ay}{x^2 + y^2 - a^2}$$

$$\phi = \frac{1}{2} m \ln \frac{(x + a)^2 + y^2}{(x - a)^2 + y^2} \qquad (8.15)$$

These lines are plotted in Fig. 8.4 and are seen to be two families of orthogonal circles, with the streamlines passing through the source and sink and the potential lines encircling them. They are harmonic (laplacian) functions that are exactly analogous in electromagnetic theory to the electric current and electric potential patterns of a magnet with poles at $(\pm a, 0)$.

Sink Plus a Vortex at the Origin

An interesting flow pattern, approximated in nature, occurs by superposition of a sink and a vortex, both centered at the origin. The composite stream function and velocity potential are

$$\text{Sink plus vortex:} \qquad \psi = m\theta - K \ln r \qquad \phi = m \ln r + K\theta \qquad (8.16)$$

When plotted, these form two orthogonal families of logarithmic spirals, as shown in Fig. 8.5. This is a fairly realistic simulation of a tornado (where the sink flow moves up the z axis into the atmosphere) or a rapidly draining bathtub vortex. At the center of a real (viscous) vortex, where Eq. (8.16) predicts infinite velocity, the actual circulating flow is highly *rotational* and approximates solid-body rotation $v_\theta \approx Cr$.

Uniform Stream Plus a Source at the Origin: The Rankine Half-Body

If we superimpose a uniform x-directed stream against an isolated source, a half-body shape appears. If the source is at the origin, the combined stream function is, in polar coordinates,

$$\text{Uniform stream plus source:} \qquad \psi = Ur \sin\theta + m\theta \qquad (8.17)$$

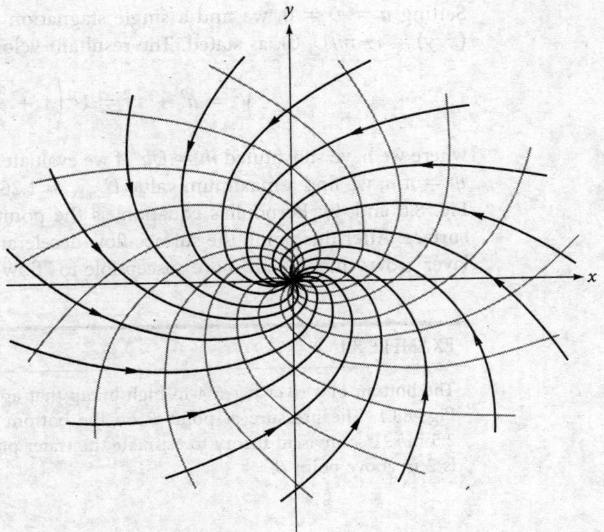

Fig. 8.5 Superposition of a sink plus a vortex, Eq. (8.16), simulates a tornado.

We can set this equal to various constants and plot the streamlines, as shown in Fig. 8.6. A curved, roughly elliptical, *half-body* shape appears, which separates the source flow from the stream flow. The body shape, which is named after the Scottish engineer W. J. M. Rankine (1820–1872), is formed by the particular streamlines $\psi = \pm\pi m$. The half-width of the body far downstream is $\pi m/U$. The upper surface may be plotted from the relation

$$r = \frac{m(\pi - \theta)}{U \sin \theta} \tag{8.18}$$

It is not a true ellipse. The nose of the body, which is a "stagnation" point where $V = 0$, stands at $(x, y) = (-a, 0)$, where $a = m/U$. The streamline $\psi = 0$ also crosses this point—recall that streamlines can cross only at a stagnation point.

The cartesian velocity components are found by differentiation:

$$u = \frac{\partial \psi}{\partial y} = U + \frac{m}{r}\cos\theta \qquad v = -\frac{\partial \psi}{\partial x} = \frac{m}{r}\sin\theta \tag{8.19}$$

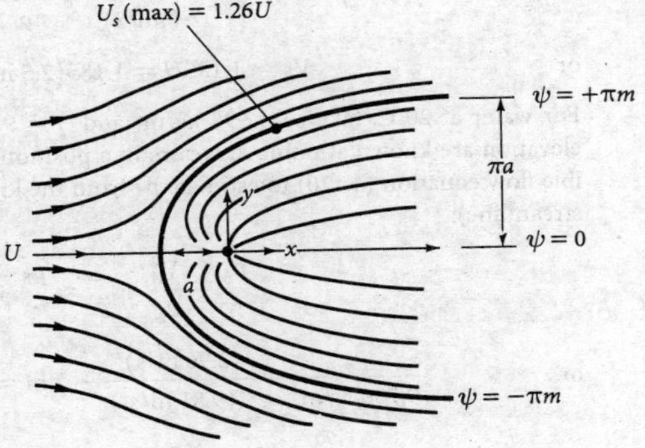

Fig. 8.6 Superposition of a source plus a uniform stream forms a Rankine half-body.

Setting $u = v = 0$, we find a single stagnation point at $\theta = 180°$ and $r = m/U$, or $(x, y) = (-m/U, 0)$, as stated. The resultant velocity at any point is

$$V^2 = u^2 + v^2 = U^2\left(1 + \frac{a^2}{r^2} + \frac{2a}{r}\cos\theta\right) \tag{8.20}$$

where we have substituted $m = Ua$. If we evaluate the velocities along the upper surface $\psi = \pi m$, we find a maximum value $U_{s,max} \approx 1.26U$ at $\theta = 63°$. This point is labeled in Fig. 8.6 and, by Bernoulli's equation, is the point of minimum pressure on the body surface. After this point, the surface flow decelerates, the pressure rises, and the viscous layer grows thicker and more susceptible to "flow separation," as we saw in Chap. 7.

EXAMPLE 8.1

The bottom of a river has a 4-m-high bump that approximates a Rankine half-body, as in Fig. E8.1. The pressure at point B on the bottom is 130 kPa, and the river velocity is 2.5 m/s. Use inviscid theory to estimate the water pressure at point A on the bump, which is 2 m above point B.

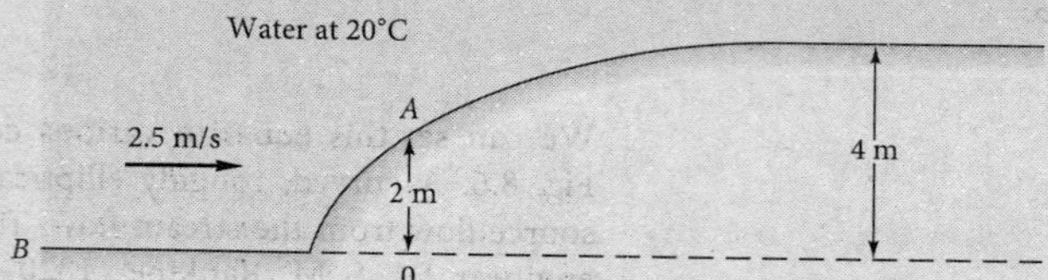

E8.1

Solution

As in all inviscid theories, we ignore the low-velocity boundary layers that form on solid surfaces due to the no-slip condition. From Eq. (8.18) and Fig. 8.6, the downstream bump half-height equals πa. Therefore, for our case, $a = (4 \text{ m})/\pi = 1.27$ m. We have to find the spot where the bump height is half that much, $h = 2 \text{ m} = \pi a/2$. From Eq. (8.18) we may compute

$$r = h_A = \frac{a(\pi - \theta)}{\sin\theta} = \frac{\pi}{2}a \quad \text{or} \quad \theta = \frac{\pi}{2} = 90°$$

Thus point A in Fig. E8.1 is directly above the (initially unknown) origin of coordinates (labeled O in Fig. E8.1) and is 1.27 m to the right of the nose of the bump. With $r = \pi a/2$ and $\theta = \pi/2$ known, we compute the velocity at point A from Eq. (8.20):

$$V_A^2 = U^2\left[1 + \frac{a^2}{(\pi a/2)^2} + \frac{2a}{\pi a/2}\cos\frac{\pi}{2}\right] = 1.405U^2$$

or

$$V_A \approx 1.185U = 1.185(2.5 \text{ m/s}) = 2.96 \text{ m/s}$$

For water at 20°C, take $\rho = 998 \text{ kg/m}^2$ and $\gamma = 9{,}790 \text{ N/m}^3$. Now, since the velocity and elevation are known at point A, we are in a position to use Bernoulli's inviscid, incompressible flow equation (4.120) to estimate p_A from the known properties at point B (on the same streamline):

$$\frac{p_A}{\gamma} + \frac{V_A^2}{2g} + z_A \approx \frac{p_B}{\gamma} + \frac{V_B^2}{2g} + z_B$$

or

$$\frac{p_A}{9{,}790 \text{ N/m}^3} + \frac{(2.96 \text{ m/s})^2}{2(9.81 \text{ m/s}^2)} + 2 \text{ m} \approx \frac{130{,}000}{9{,}790} + \frac{(2.5)^2}{2(9.81)} + 0$$

Solving, we find

$$p_A = (13.60 - 2.45)(9{,}790) \approx 109{,}200 \text{ Pa} \qquad \textit{Ans.}$$

If the approach velocity is uniform, this should be a pretty good approximation, since water is relatively inviscid and its boundary layers are thin.

Uniform Stream at an Angle α

If the uniform stream is written in plane polar coordinates, it becomes

$$\text{Uniform stream } \mathbf{i}U: \qquad \psi = Ur\sin\theta \qquad \phi = Ur\cos\theta \tag{8.21}$$

This makes it easier to superimpose, say, a stream and a source or vortex by using the same coordinates. If the uniform stream is moving at angle α with respect to the x axis—that is,

$$u = U\cos\alpha = \frac{\partial\psi}{\partial y} = \frac{\partial\phi}{\partial x} \qquad v = U\sin\alpha = -\frac{\partial\psi}{\partial x} = \frac{\partial\phi}{\partial y}$$

then by integration we obtain the correct functions for flow at an angle:

$$\psi = U(y\cos\alpha - x\sin\alpha) \qquad \phi = U(x\cos\alpha + y\sin\alpha) \tag{8.22}$$

These expressions are useful in airfoil angle-of-attack problems (Sec. 8.7).

Circulation

The line vortex flow is irrotational everywhere except at the origin, where the vorticity $\nabla \times \mathbf{V}$ is infinite. This means that a certain line integral called the *fluid circulation* Γ does not vanish when taken around a vortex center.

With reference to Fig. 8.7, the circulation is defined as the counterclockwise line integral, around a closed curve C, of arc length ds times the velocity component tangent to the curve:

$$\Gamma = \oint_C V\cos\alpha\, ds = \int_C \mathbf{V}\cdot d\mathbf{s} = \int_C (u\,dx + v\,dy + w\,dz) \tag{8.23}$$

From the definition of ϕ, $\mathbf{V}\cdot d\mathbf{s} = \nabla\phi \cdot d\mathbf{s} = d\phi$ for an irrotational flow; hence normally Γ in an irrotational flow would equal the final value of ϕ minus the initial value of ϕ. Since we start and end at the same point, we compute $\Gamma = 0$, but not for vortex flow: With $\phi = K\theta$ from Eq. (8.14) there is a change in ϕ of amount $2\pi K$ as we make one complete circle:

$$\text{Path enclosing a vortex:} \qquad \Gamma = 2\pi K \tag{8.24}$$

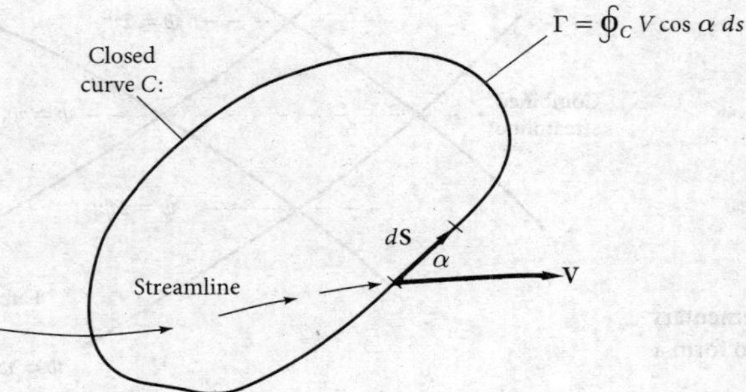

Fig. 8.7 Definition of the fluid circulation Γ.

Alternatively the calculation can be made by defining a circular path of radius r around the vortex center, from Eq. (8.23):

$$\Gamma = \int_C \upsilon_\theta \, ds = \int_0^{2\pi} \frac{K}{r} r \, d\theta = 2\pi K$$

In general, Γ denotes the net algebraic strength of all the vortex filaments contained within the closed curve. In the next section we shall see that a region of finite circulation within a flowing stream will be subjected to a lift force proportional to both U_∞ and Γ.

One can show, by using Eq. (8.23), that a source or sink creates no circulation. If there are no vortices present, the circulation will be zero for any path enclosing any number of sources and sinks.

8.3 Superposition of Plane Flow Solutions

We can now form a variety of interesting potential flows by summing the velocity potential and stream functions of a uniform stream, source or sink, and vortex. Most of the results are classic, of course, needing only a brief treatment here. Superposition is valid because the basic equations, (8.2) and (8.8), are linear.

Graphical Method of Superposition

A simple means of accomplishing $\psi_{tot} = \Sigma\,\psi_i$ graphically is to plot the individual stream functions separately and then look at their intersections. The value of ψ_{tot} at each intersection is the sum of the individual values ψ_i that cross there. Connecting intersections with the same value of ψ_{tot} creates the desired superimposed flow streamlines.

A simple example is shown in Fig. 8.8, summing two families of streamlines ψ_a and ψ_b. The individual components are plotted separately, and four typical intersections are shown. Dashed lines are then drawn through intersections representing the same sum of $\psi_a + \psi_b$. These dashed lines are the desired solution. Often this graphical method is a quick means of evaluating the proposed superposition before a full-blown numerical plot routine is executed.

Boundary Layer Separation on a Half-Body

Although the inviscid flow patterns seen in Figs. 8.9*a* and *c* are mirror images, their viscous (boundary layer) behavior is different. The body shape and the velocity along the surface are

$$V^2 = U_\infty^2\left(1 + \frac{a^2}{r^2} + \frac{2a}{r}\cos\theta\right) \quad \text{along} \quad r = \frac{m(\pi - \theta)}{U_\infty \sin\theta} \tag{8.25}$$

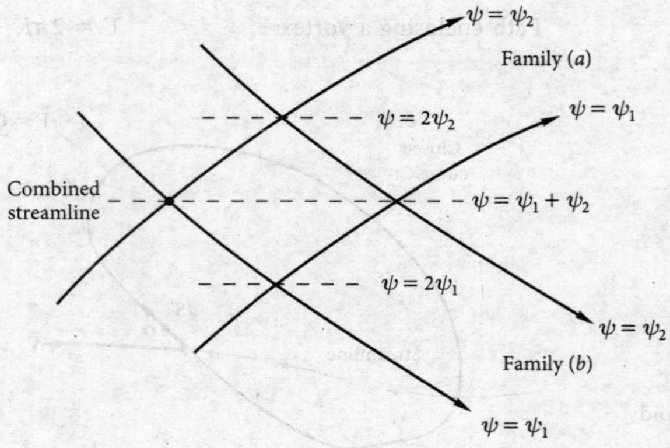

Fig. 8.8 Intersections of elementary streamlines can be joined to form a combined streamline.

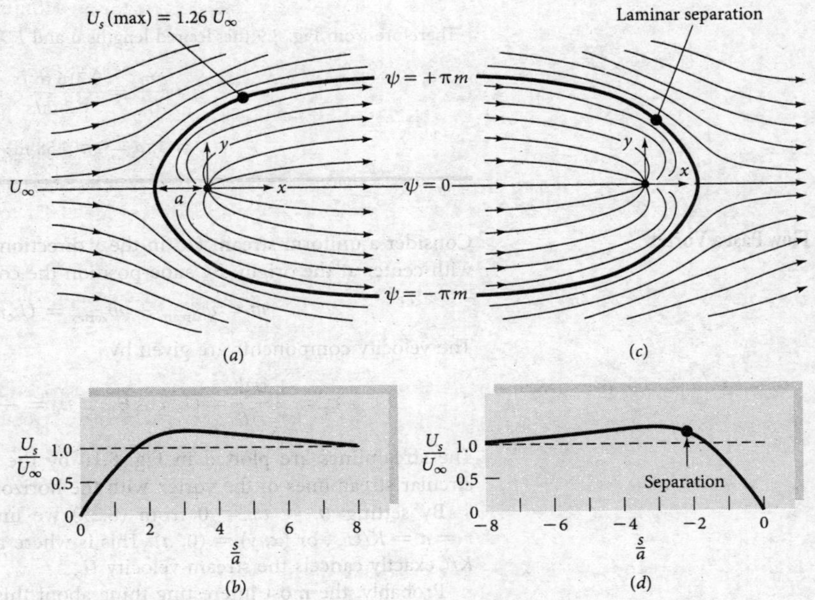

Fig. 8.9 The Rankine half-body; pattern (*c*) is not found in a real fluid because of boundary layer separation. (*a*) Uniform stream plus a source equals a half-body; stagnation point at $x = -a = -m/U_\infty$. (*b*) Slight adverse gradient for s/a greater than 3.0: no separation. (*c*) Uniform stream plus a sink equals the rear of a half-body; stagnation point at $x = a = m/U_\infty$. (*d*) Strong adverse gradient for $s/a > -3.0$: separation.

The computed surface velocities are plotted along the half-body contours in Fig. 8.9*b* and *d* as a function of arc length s/a measured from the stagnation point. These plots are also mirror images. However, if the nose is in front, Fig. 8.9*b*, the pressure gradient there is *favorable* (decreasing pressure along the surface). In contrast, the pressure gradient is *adverse* (increasing pressure along the surface) when the nose is in the rear, Fig. 8.9*d*, and boundary layer separation may occur.

Application to Fig. 8.9*b* of Thwaites's laminar boundary method from Eqs. (7.54) and (7.56) reveals that separation does not occur on the front nose of the half-body. Therefore, Fig. 8.9*a* is a very realistic picture of streamlines past a half-body nose. In contrast, when applied to the tail, Fig. 8.9*c*, Thwaites's method predicts separation at about $s/a \approx -2.2$, or $\theta \approx 110°$. Thus, if a half-body is a solid surface, Fig. 8.9*c* is *not* realistic and a broad separated wake will form. However, if the half-body tail is a *fluid line* separating the sink-directed flow from the outer stream, as in Example 8.2, then Fig. 8.9*c* is quite realistic and useful. Computations for turbulent boundary layer theory would be similar: separation on the tail, no separation on the nose.

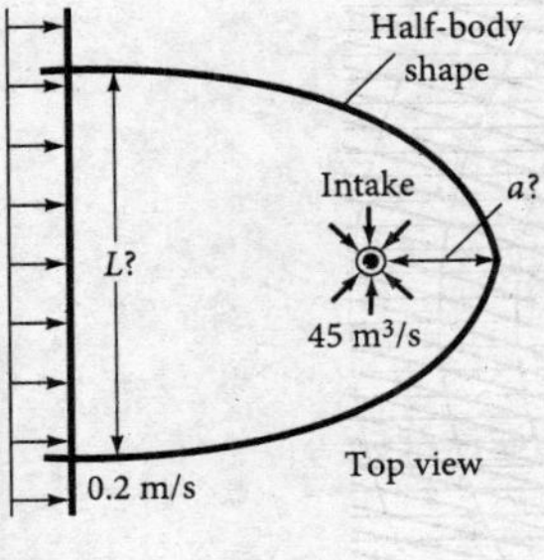

E8.2

EXAMPLE 8.2

An offshore power plant cooling-water intake sucks in 45 m³/s in water 10 m deep, as in Fig. E8.2. If the tidal velocity approaching the intake is 0.2 m/s, (*a*) how far downstream does the intake effect extend and (*b*) how much width *L* of tidal flow is entrained into the intake?

Solution

Recall from Eq. (8.13) that the sink strength *m* is related to the volume flow *Q* and the depth *b* into the paper:

$$m = \frac{Q}{2\pi b} = \frac{45\ \text{m}^3/\text{s}}{2\pi(10\ \text{m})} = 0.716\ \text{m}^2/\text{s}$$

Therefore from Fig. 8.9 the desired lengths a and L are

$$a = \frac{m}{U_\infty} = \frac{0.716\ \text{m}^2/\text{s}}{0.2\ \text{m/s}} = 3.58\ \text{m} \qquad \textit{Ans. (a)}$$

$$L = 2\pi a = 2\pi(3.58\ \text{m}) = 22.5\ \text{m} \qquad \textit{Ans. (b)}$$

Flow Past a Vortex

Consider a uniform stream U_∞ in the x direction flowing past a vortex of strength K with center at the origin. By superposition the combined stream function is

$$\psi = \psi_{\text{stream}} + \psi_{\text{vortex}} = U_\infty r \sin\theta - K \ln r \tag{8.26}$$

The velocity components are given by

$$v_r = \frac{1}{r}\frac{\partial \psi}{\partial \theta} = U_\infty \cos\theta \qquad v_\theta = -\frac{\partial \psi}{\partial r} = -U_\infty \sin\theta + \frac{K}{r} \tag{8.27}$$

The streamlines are plotted in Fig. 8.10 by the graphical method, intersecting the circular streamlines of the vortex with the horizontal lines of the uniform stream.

By setting $v_r = v_\theta = 0$ from (8.27) we find a stagnation point at $\theta = 90°$, $r = a = K/U_\infty$, or $(x, y) = (0, a)$. This is where the counterclockwise vortex velocity K/r exactly cancels the stream velocity U_∞.

Probably, the most interesting thing about this example is that there is a nonzero lift force normal to the stream on the surface of any region enclosing the vortex, but we postpone this discussion until the next section.

An Infinite Row of Vortices

Consider an infinite row of vortices of equal strength K and equal spacing a, as in Fig. 8.11a. This case is included here to illustrate the interesting concept of a *vortex sheet*.

From Eq. (8.14), the ith vortex in Fig. 8.11a has a stream function $\psi_i = -K \ln r_i$, so that the total infinite row has a combined stream function

$$\psi = -K \sum_{i=1}^{\infty} \ln r_i$$

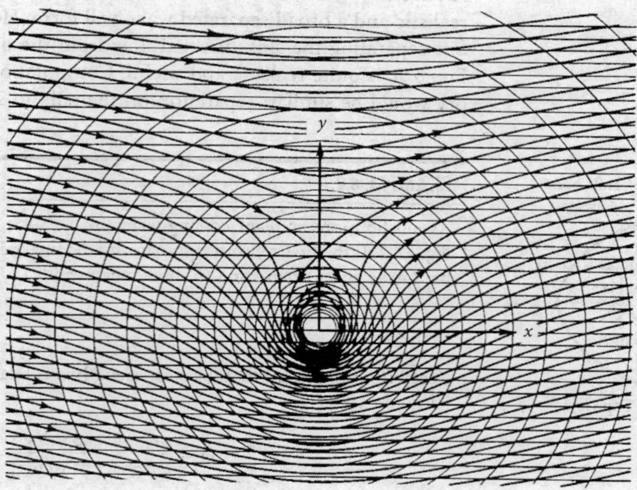

Fig. 8.10 Flow of a uniform stream past a vortex constructed by the graphical method.

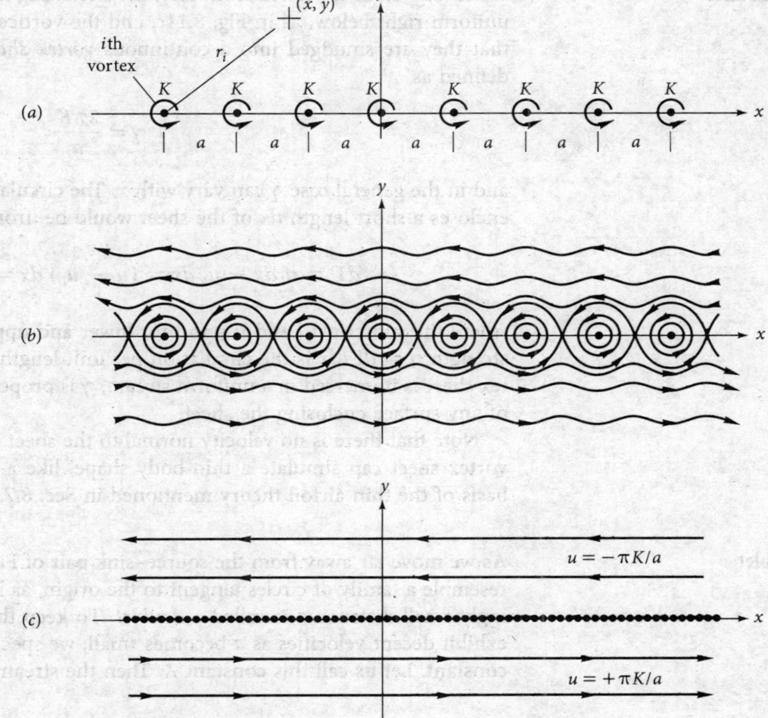

Fig. 8.11 Superposition of vortices: (*a*) an infinite row of equal strength; (*b*) streamline pattern for part (*a*); (*c*) vortex sheet: part (*b*) viewed from afar.

It can be shown [2, Sec. 4.51] that this infinite sum of logarithms is equivalent to a closed-form function:

$$\psi = -\tfrac{1}{2}K \ln\left[\frac{1}{2}\left(\cosh\frac{2\pi y}{a} - \cos\frac{2\pi x}{a}\right)\right] \tag{8.28}$$

Since the proof uses the complex variable $z = x + iy$, $i = (-1)^{1/2}$, we are not going to show the details here.

The streamlines from Eq. (8.28) are plotted in Fig. 8.11*b*, showing what is called a *cat's-eye* pattern of enclosed flow cells surrounding the individual vortices. Above the cat's eyes the flow is entirely to the left, and below the cat's eyes the flow is to the right. Moreover, these left and right flows are uniform if $|y| \gg a$, which follows by differentiating Eq. (8.28):

$$u = \frac{\partial \psi}{\partial y}\bigg|_{|y| \gg a} = \pm\frac{\pi K}{a}$$

where the plus sign applies below the row and the minus sign above the row. This uniform left and right streaming is sketched in Fig. 8.11*c*. We stress that this effect is induced by the row of vortices: There is no uniform stream approaching the row in this example.

The Vortex Sheet

When Fig. 8.11*b* is viewed from afar, the streaming motion is uniform left above and uniform right below, as in Fig. 8.11*c*, and the vortices are packed so closely together that they are smudged into a continuous *vortex sheet*. The strength of the sheet is defined as

$$\gamma = \frac{2\pi K}{a} \tag{8.29}$$

and in the general case γ can vary with x. The circulation about any closed curve that encloses a short length dx of the sheet would be, from Eqs. (8.23) and (8.29),

$$d\Gamma = u_l\,dx - u_u\,dx = (u_l - u_u)\,dx = \frac{2\pi K}{a}\,dx = \gamma\,dx \tag{8.30}$$

where the subscripts l and u stand for lower and upper, respectively. Thus, the sheet strength $\gamma = d\Gamma/dx$ is the circulation per unit length of the sheet. Thus, when a vortex sheet is immersed in a uniform stream, γ is proportional to the lift per unit length of any surface enclosing the sheet.

Note that there is no velocity normal to the sheet at the sheet surface. Therefore, a vortex sheet can simulate a thin-body shape, like a plate or thin airfoil. This is the basis of the thin airfoil theory mentioned in Sec. 8.7.

The Doublet

As we move far away from the source–sink pair of Fig. 8.4, the flow pattern begins to resemble a family of circles tangent to the origin, as in Fig. 8.12. This limit of vanishingly small distance a is called a *doublet*. To keep the flow strength large enough to exhibit decent velocities as a becomes small, we specify that the product $2am$ remain constant. Let us call this constant λ. Then the stream function of a doublet is

$$\psi = \lim_{\substack{a\to 0\\ 2am=\lambda}} \left(-m\tan^{-1}\frac{2ay}{x^2 + y^2 - a^2}\right) = -\frac{2amy}{x^2 + y^2} = -\frac{\lambda y}{x^2 + y^2} \tag{8.31}$$

We have used the fact that $\tan^{-1}\alpha \approx \alpha$ as α becomes small. The quantity λ is called the *strength* of the doublet.

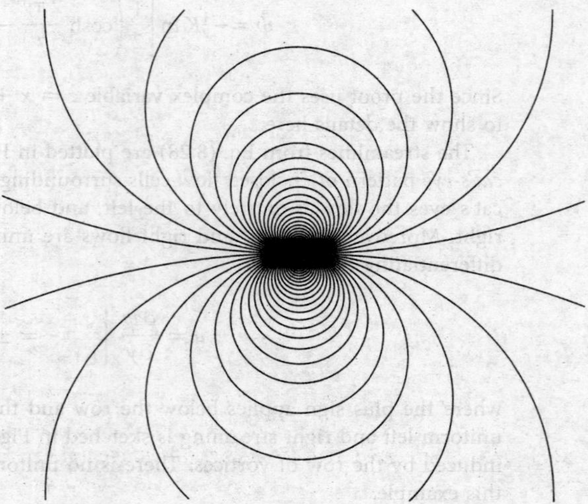

Fig. 8.12 A doublet, or source–sink pair, is the limiting case of Fig. 8.4 viewed from afar. Streamlines are circles tangent to the x axis at the origin. This figure was prepared using the *contour* feature of MATLAB [34, 35].

Equation (8.31) can be rearranged to yield

$$x^2 + \left(y + \frac{\lambda}{2\psi}\right)^2 = \left(\frac{\lambda}{2\psi}\right)^2$$

so that, as advertised, the streamlines are circles tangent to the origin with centers on the y axis. This pattern is sketched in Fig. 8.12.

Although the author has in the past laboriously sketched streamlines by hand, this is no longer necessary. Figure 8.12 was computer-drawn, using the *contour* feature of the student version of MATLAB [34]. Simply set up a grid of points, spell out the stream function, and call for a contour. For Fig. 8.12, the actual statements were

```
[X, Y] = meshgrid (-1 : .02 : 1);
PSI = -Y. / (X. ^2 + Y. ^2);
contour (X, Y, PSI, 100)
```

This would produce 100 contour lines of ψ from Eq. (8.31), with $\lambda = 1$ for convenience. The plot would include grid lines, scale markings, and a surrounding box, and the circles might look a bit elliptical. These blemishes can be eliminated with three statements of cosmetic improvement:

```
axis square
grid off
axis off
```

The final plot, Fig. 8.12, has no markings, but the streamlines themselves. MATLAB is thus a recommended tool and, in addition, has scores of other uses. All this chapter's problem assignments that call for "sketch the streamlines/potential lines" can be completed using this contour feature. For further details, consult Ref. 34.

In a similar manner the velocity potential of a doublet is found by taking the limit of Eq. (8.15) as $a \rightarrow 0$ and $2am = \lambda$:

$$\phi_{\text{doublet}} = \frac{\lambda x}{x^2 + y^2}$$

or

$$\left(x - \frac{\lambda}{2\phi}\right)^2 + y^2 = \left(\frac{\lambda}{2\phi}\right)^2 \tag{8.32}$$

The potential lines are circles tangent to the origin with centers on the x axis. Simply turn Fig. 8.12 clockwise 90° to visualize the ϕ lines, which are everywhere normal to the streamlines.

The doublet functions can also be written in polar coordinates:

$$\psi = \frac{\lambda \sin \theta}{r} \qquad \phi = \frac{\lambda \cos \theta}{r} \tag{8.33}$$

These forms are convenient for the cylinder flows of the next section.

8.4 Plane Flow Past Closed-Body Shapes

A variety of closed-body external flows can be constructed by superimposing a uniform stream with sources, sinks, and vortices. The body shape will be closed only if the net source outflow equals the net sink inflow.

The Rankine Oval

A cylindrical shape called a *Rankine oval,* which is long compared with its height, is formed by a source–sink pair aligned parallel to a uniform stream, as in Fig. 8.13*a*.

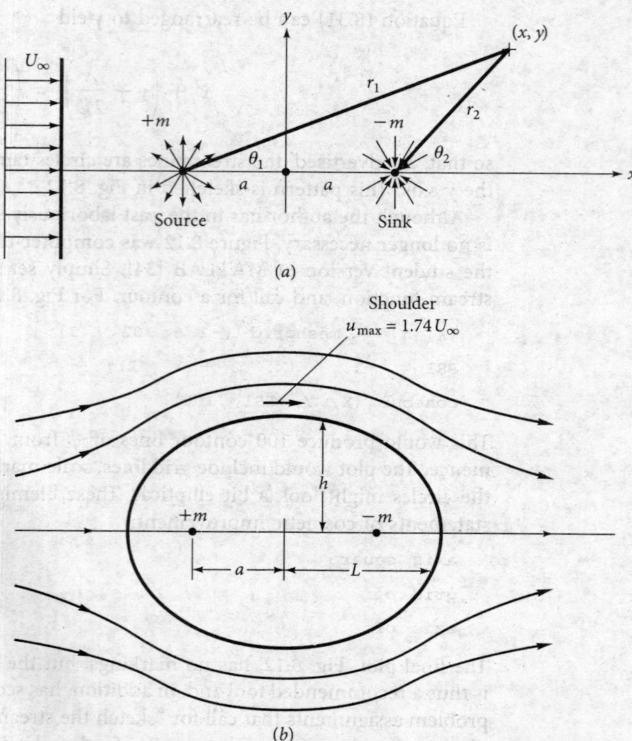

Fig. 8.13 Flow past a Rankine oval: (*a*) uniform stream plus a source–sink pair; (*b*) oval shape and streamlines for $m/(U_\infty a) = 1.0$.

From Eqs. (8.12) and (8.15) the combined stream function is

$$\psi = U_\infty y - m \tan^{-1} \frac{2ay}{x^2 + y^2 - a^2} = U_\infty r \sin \theta + m(\theta_1 - \theta_2) \qquad (8.34)$$

When streamlines of constant ψ are plotted from Eq. (8.34), an oval body shape appears, as in Fig. 8.13*b*. The half-length L and half-height h of the oval depend on the relative strength of source and stream—that is, the ratio $m/(U_\infty a)$, which equals 1.0 in Fig. 8.13*b*. The circulating streamlines inside the oval are uninteresting and not usually shown. The oval is the line $\psi = 0$.

There are stagnation points at the front and rear, $x = \pm L$, and points of maximum velocity and minimum pressure at the shoulders, $y = \pm h$, of the oval. All these parameters are a function of the basic dimensionless parameter $m/(U_\infty a)$, which we can determine from Eq. (8.34):

$$\frac{h}{a} = \cot \frac{h/a}{2m/(U_\infty a)} \qquad \frac{L}{a} = \left(1 + \frac{2m}{U_\infty a}\right)^{1/2}$$
$$\frac{u_{max}}{U_\infty} = 1 + \frac{2m/(U_\infty a)}{1 + h^2/a^2} \qquad (8.35)$$

As we increase $m/(U_\infty a)$ from zero to large values, the oval shape increases in size and thickness from a flat plate of length $2a$ to a huge, nearly circular cylinder. This is shown in Table 8.1. In the limit as $m/(U_\infty a) \to \infty$, $L/h \to 1.0$ and $u_{max}/U_\infty \to 2.0$, which is equivalent to flow past a circular cylinder.

Table 8.1 Rankine Oval Parameters from Eq. (8.30)

$m/(U_\infty a)$	h/a	L/a	L/h	u_{max}/U_∞
0.0	0.0	1.0	∞	1.0
0.01	0.031	1.010	32.79	1.020
0.1	0.263	1.095	4.169	1.187
1.0	1.307	1.732	1.326	1.739
10.0	4.435	4.583	1.033	1.968
100.0	14.130	14.177	1.003	1.997
∞	∞	∞	1.000	2.000

All the Rankine ovals except very thin ones have a large adverse pressure gradient on their leeward surface. Thus, boundary layer separation will occur in the rear with a broad wake flow, and the inviscid pattern is unrealistic in that region.

Flow Past a Circular Cylinder with Circulation

From Table 8.1 at large source strength the Rankine oval becomes a large circle, much greater in diameter than the source–sink spacing $2a$. Viewed on the scale of the cylinder, this is equivalent to a uniform stream plus a doublet. We also throw in a vortex at the doublet center, which does not change the shape of the cylinder.

Thus, the stream function for flow past a circular cylinder with circulation, centered at the origin, is a uniform stream plus a doublet plus a vortex:

$$\psi = U_\infty r \sin\theta - \frac{\lambda \sin\theta}{r} - K \ln r + \text{const} \tag{8.36}$$

The doublet strength λ has units of velocity times length squared. For convenience, let $\lambda = U_\infty a^2$, where a is a length, and let the arbitrary constant in Eq. (8.36) equal $K \ln a$. Then the stream function becomes

$$\psi = U_\infty \sin\theta \left(r - \frac{a^2}{r}\right) - K \ln \frac{r}{a} \tag{8.37}$$

The streamlines are plotted in Fig. 8.14 for four different values of the dimensionless vortex strength $K/(U_\infty a)$. For all cases the line $\psi = 0$ corresponds to the circle $r = a$—that is, the shape of the cylindrical body. As circulation $\Gamma = 2\pi K$ increases, the velocity becomes faster and faster below the cylinder and slower and slower above it. The velocity components in the flow are given by

$$\begin{aligned} v_r &= \frac{1}{r}\frac{\partial \psi}{\partial \theta} = U_\infty \cos\theta \left(1 - \frac{a^2}{r^2}\right) \\ v_\theta &= -\frac{\partial \psi}{\partial r} = -U_\infty \sin\theta \left(1 + \frac{a^2}{r^2}\right) + \frac{K}{r} \end{aligned} \tag{8.38}$$

The velocity at the cylinder surface $r = a$ is purely tangential, as expected:

$$v_r(r = a) = 0 \quad v_\theta(r = a) = -2U_\infty \sin\theta + \frac{K}{a} \tag{8.39}$$

For small K, two stagnation points appear on the surface at angles θ_s where $v_\theta = 0$; or, from Eq. (8.39),

$$\sin\theta_s = \frac{K}{2U_\infty a} \tag{8.40}$$

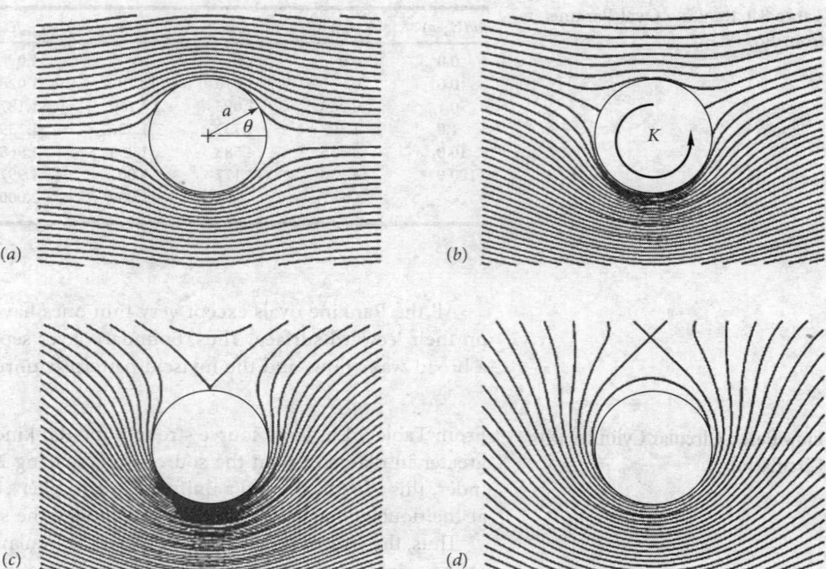

Fig. 8.14 Flow past a circular cylinder with circulation for values of $K/(U_\infty a)$ of (*a*) 0, (*b*) 1.0, (*c*) 2.0, and (*d*) 3.0.

Figure 8.14*a* is for $K = 0$, $\theta_s = 0$ and 180°, or doubly symmetric inviscid flow past a cylinder with no circulation. Figure 8.14*b* is for $K/(U_\infty a) = 1$, $\theta_s = 30$ and 150°; and Fig. 8.14*c* is the limiting case where the two stagnation points meet at the top, $K/(U_\infty a) = 2$, $\theta_s = 90°$.

For $K > 2U_\infty a$, Eq. (8.40) is invalid, and the single stagnation point is above the cylinder, as in Fig. 8.14*d*, at a point $y = h$ given by

$$\frac{h}{a} = \frac{1}{2}[\beta + (\beta^2 - 4)^{1/2}] \qquad \beta = \frac{K}{U_\infty a} > 2$$

In Fig. 8.14*d*, $K/(U_\infty a) = 3.0$, and $h/a = 2.6$.

The Kutta-Joukowski Lift Theorem

The cylinder flows with circulation, Figs. 8.14*b* to *d*, develop an inviscid downward *lift* normal to the free stream, called the *Magnus-Robins force*. This lift is proportional to stream velocity and vortex strength. Its discovery, by experiment, has long been attributed to the German physicist Gustav Magnus, who observed it in 1853. It is now known [40, 45] that the brilliant British engineer Benjamin Robins first reported a lift force on a spinning body in 1761. We see from the streamline patterns that the velocity on top of the cylinder is less, and, thus, from Bernoulli's equation, the pressure is higher. On the bottom, we see tightly packed streamlines, high velocity, and low pressure; viscosity is neglected. Inviscid theory predicts this force.

The surface velocity is given by Eq. (8.39). From Bernoulli's equation (8.3), neglecting gravity, the surface pressure p_s is given by

$$p_\infty + \frac{1}{2}\rho U_\infty^2 = p_s + \frac{1}{2}\rho\left(-2U_\infty \sin\theta + \frac{K}{a}\right)^2$$

or

$$p_s = p_\infty + \tfrac{1}{2}\rho U_\infty^2(1 - 4\sin^2\theta + 4\beta\sin\theta - \beta^2) \tag{8.41}$$

where $\beta = K/(U_\infty a)$ and p_∞ is the free-stream pressure. If b is the cylinder depth into the paper, the drag D is the integral over the surface of the horizontal component of pressure force:

$$D = -\int_0^{2\pi} (p_s - p_\infty) \cos\theta \, ba \, d\theta$$

where $p_s - p_\infty$ is substituted from Eq. (8.41). But the integral of $\cos\theta$ times any power of $\sin\theta$ over a full cycle 2π is identically zero. Thus we obtain the (perhaps surprising) result

$$D(\text{cylinder with circulation}) = 0 \tag{8.42}$$

This is a special case of d'Alembert's paradox, mentioned in Sec. 1.2:

> According to inviscid theory, the drag of any body of any shape immersed in a uniform stream is identically zero.

D'Alembert published this result in 1752 and pointed out himself that it did not square with the facts for real fluid flows. This unfortunate paradox caused everyone to overreact and reject all inviscid theory until 1904, when Prandtl first pointed out the profound effect of the thin viscous boundary layer on the flow pattern in the rear, as in Fig. 7.2*b*, for example.

The lift force L normal to the stream, taken positive upward, is given by summation of vertical pressure forces:

$$L = -\int_0^{2\pi} (p_s - p_\infty) \sin\theta \, ba \, d\theta$$

Since the integral over 2π of any odd power of $\sin\theta$ is zero, only the third term in the parentheses in Eq. (8.41) contributes to the lift:

$$L = -\frac{1}{2}\rho U_\infty^2 \frac{4K}{aU_\infty} ba \int_0^{2\pi} \sin^2\theta \, d\theta = -\rho U_\infty (2\pi K) b$$

or

$$\boxed{\frac{L}{b} = -\rho U_\infty \Gamma} \tag{8.43}$$

Notice that the lift seems independent of the radius a of the cylinder. Actually, though, as we shall see in Sec. 8.7, the circulation Γ depends on the body size and orientation through a physical requirement.

Equation (8.43) was generalized by W. M. Kutta in 1902 and independently by N. Joukowski in 1906 as follows:

> According to inviscid theory, the lift per unit depth of any cylinder of any shape immersed in a uniform stream equals $\rho u_\infty \Gamma$, where Γ is the total net circulation contained within the body shape. The direction of the lift is 90° from the stream direction, rotating opposite to the circulation.

The problem in airfoil analysis, Sec. 8.7, is thus to determine the circulation Γ as a function of airfoil shape and orientation.

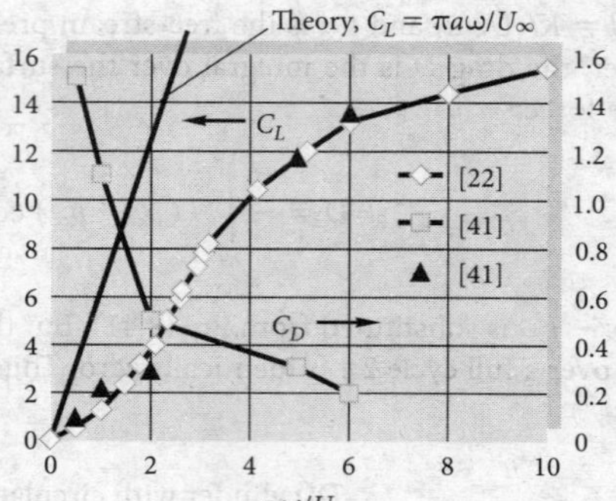

Fig. 8.15 Drag and lift of a rotating cylinder of large aspect ratio at $\mathrm{Re}_D = 3{,}800$, after Tokumaru and Dimotakis [22] and Sengupta et al. [41].

Lift and Drag of Rotating Cylinders[2]

The flows in Fig. 8.14 are mathematical: a doublet plus a vortex plus a uniform stream. The physical realization could be a rotating cylinder in a free stream. The no-slip condition would cause the fluid in contact with the cylinder to move tangentially at velocity $v_\theta = a\omega$, setting up a net circulation Γ. Measurement of forces on a spinning cylinder is very difficult, and no reliable drag data are known to the author. However, Tokumaru and Dimotakis [22] used a clever auxiliary scheme to measure lift forces at $\mathrm{Re}_D = 3{,}800$.

Figure 8.15 shows lift and drag coefficients, based on frontal area ($2ab$), for a rotating cylinder at $\mathrm{Re}_D = 3{,}800$. The drag curve is from CFD calculations [41]. Reported CFD drag results, from several different authors, are quite controversial because they do not agree, even qualitatively. The writer feels that Ref. 41 gives the most reliable results. Note that the experimental C_L increases to a value of 15.3 at $a\omega/U_\infty = 10$. This contradicts an early theory of Prandtl, in 1926, that the maximum possible value of C_L would be $4\pi \approx 12.6$, corresponding to the flow conditions in Fig. 8.14*c*. The inviscid theory for lift would be:

$$C_L = \frac{L}{\frac{1}{2}\rho U_\infty^2(2ba)} = \frac{2\pi\rho U_\infty K b}{\rho U_\infty^2 ba} = \frac{2\pi \upsilon_{\theta s}}{U_\infty} \tag{8.44}$$

where $\upsilon_{\theta s} = K/a$ is the peripheral speed of the cylinder.

Figure 8.15 shows that the theoretical lift from Eq. (8.44) is much too high, but the measured lift is quite respectable, much larger in fact than a typical airfoil of the same chord length, as in Fig. 7.25. Thus rotating cylinders have practical possibilities. The Flettner rotor ship built in Germany in 1924 employed rotating vertical cylinders that developed a thrust due to any winds blowing past the ship. The Flettner design did not gain any popularity, but such inventions may be more attractive in this era of high energy costs.

EXAMPLE 8.3

The experimental Flettner rotor sailboat at the University of Rhode Island is shown in Fig. E8.3. The rotor is 0.75 m in diameter and 3 m long and rotates at 220 r/min. It is driven by a small lawnmower engine. If the wind is a steady 5 m/s and boat relative motion is neglected, what is the maximum thrust expected for the rotor? Assume standard air and water density.

[2] The writer is indebted to Prof. T. K. Sengupta of I.I.T. Kanpur for data and discussion for this subsection.

E8.3 *(Courtesy of R. C. Lessmann, University of Rhode Island.)*

Solution

Convert the rotation rate to $\omega = 2\pi(220)/60 = 23.04$ rad/s. The wind velocity is 5 m/s, so the velocity ratio is

$$\frac{a\omega}{U_\infty} = \frac{(0.375\text{ m})(23.04\text{ rad/s})}{5\text{ m}} = 1.728$$

Using Fig. 8.15, we read $C_D \approx 0.7$ and $C_L \approx 2.5$. From Table A.6, standard air density is 1.2255 kg/m^3. Then the estimated rotor lift and drag are

$$L = C_L \frac{1}{2}\rho U_\infty^2\, 2ba = (2.5)\frac{1}{2}(1.226\text{ kg/m}^3)(5\text{ m/s})^2 2(3\text{ m})(0.375\text{ m}) = 86.2\text{ N}$$

$$D = C_D \frac{1}{2}\rho U_\infty^2\, 2ba = (0.7)\frac{1}{2}(1.226\text{ kg/m}^3)(5\text{ m/s})^2 2(3\text{ m})(0.375\text{ m}) = 24.1\text{ N}$$

The maximum thrust available to the sailboat is the resultant of these two:

$$F = [(86.2)^2 + (24.1)^2]^{1/2} = 89.5\text{ N} \qquad \textit{Ans.}$$

Note that water density did not enter into this calculation, which is a force due to *air*. If aligned along the boat's keel, this thrust will drive the boat through the water at a speed of about 2 m/s.
Comment: For the sake of a numerical example, we have done something improper here. We have used data for $\text{Re}_D = 3{,}800$ to estimate forces when the rotor $\text{Re}_D \approx 260{,}000$. Do not do this in your real job after you graduate!

The Kelvin Oval

A family of body shapes taller than they are wide can be formed by letting a uniform stream flow normal to a vortex pair. If U_∞ is to the right, the negative vortex $-K$ is placed at $y = +a$ and the counterclockwise vortex $+K$ placed at $y = -a$, as in Fig. 8.16. The combined stream function is

$$\psi = U_\infty y - \frac{1}{2}K \ln \frac{x^2 + (y+a)^2}{x^2 + (y-a)^2} \tag{8.45}$$

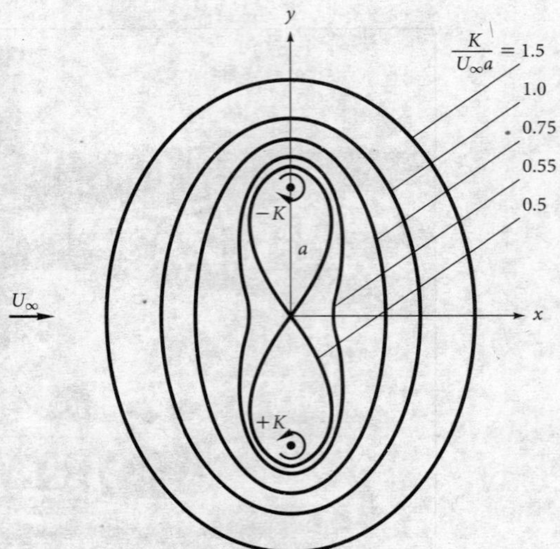

Fig. 8.16 Kelvin oval body shapes as a function of the vortex strength parameter $K/(U_\infty a)$; outer streamlines not shown.

The body shape is the line $\psi = 0$, and some of these shapes are shown in Fig. 8.16. For $K/(U_\infty a) > 10$ the shape is within 1 percent of a Rankine oval (Fig. 8.13) turned 90°, but for small $K/(U_\infty a)$ the waist becomes pinched in, and a figure-eight shape occurs at 0.5. For $K/(U_\infty a) < 0.5$ the stream blasts right between the vortices and isolates two more or less circular body shapes, one surrounding each vortex.

A closed body of practically any shape can be constructed by proper superposition of sources, sinks, and vortices. See the advanced work in Refs. 1 to 3 for further details. A summary of elementary potential flows is given in Table 8.2.

Potential Flow Analogs

For complicated potential flow geometries, one can resort to other methods than superposition of sources, sinks, and vortices. A variety of devices simulate solutions to Laplace's equation.

From 1897 to 1900 Hele-Shaw [9] developed a technique whereby laminar flow between very closely spaced parallel plates simulated potential flow when viewed from above the plates. Obstructions simulate body shapes, and dye streaks represent the streamlines. The Hele-Shaw apparatus makes an excellent laboratory demonstration of potential flow [10, pp. 197–198, 219–220]. Figure 8.17*a* illustrates Hele-Shaw

Table 8.2 Summary of Plane Incompressible Potential Flows

Type of flow	Potential functions	Remarks
Stream $\mathbf{i}U$	$\psi = Uy \quad \phi = Ux$	See Fig. 8.3*a*
Line source ($m > 0$) or sink ($m < 0$)	$\psi = m\theta \quad \phi = m \ln r$	See Fig. 8.3*b*
Line vortex	$\psi = -K \ln r \quad \phi = K\theta$	See Fig. 8.3*c*
Half-body	$\psi = Ur \sin\theta + m\theta$ $\phi = Ur\cos\theta + m \ln r$	See Fig. 8.9
Doublet	$\psi = \dfrac{-\lambda \sin\theta}{r} \quad \phi = \dfrac{\lambda\cos\theta}{r}$	See Fig. 8.12
Rankine oval	$\psi = Ur\sin\theta + m(\theta_1 - \theta_2)$	See Fig. 8.13
Cylinder with circulation	$\psi = U\sin\theta\left(r - \dfrac{a^2}{r}\right) - K\ln\dfrac{r}{a}$	See Fig. 8.14

(potential) flow through an array of cylinders, a flow pattern that would be difficult to analyze just using Laplace's equation. However beautiful this array pattern may be, it is not a good approximation to real (laminar viscous) array flow. Figure 8.17*b* shows experimental streakline patterns for a similar staggered-array flow at Re $\approx$ 6,400. We see that the interacting wakes of the real flow (Fig. 8.17*b*) cause intensive mixing and transverse motion, not the smooth streaming passage of the potential flow model (Fig. 8.17*a*). The moral is that this is an internal flow with multiple bodies and, therefore, not a good candidate for a realistic potential flow model.

Other flow-mapping techniques are discussed in Ref. 8. Electromagnetic fields also satisfy Laplace's equation, with voltage analogous to velocity potential and current lines analogous to streamlines. At one time commercial analog field plotters were available, using thin conducting paper cut to the shape of the flow geometry. Potential lines (voltage contours) were plotted by probing the paper with a potentiometer pointer. Hand-sketching "curvilinear square" techniques were also popular. The availability and the simplicity of computer potential flow methods [5 to 7] have made analog models obsolete.

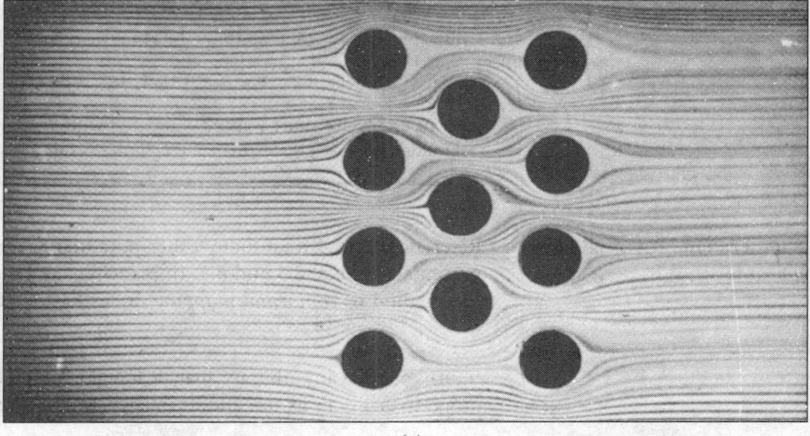

(*a*)

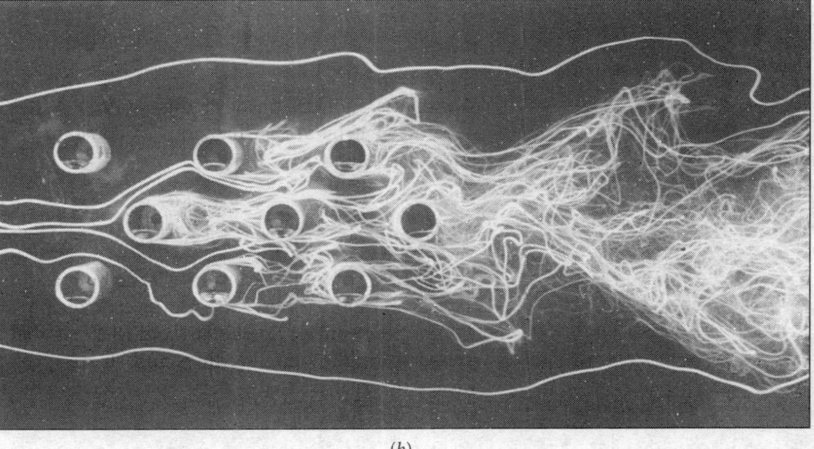

(*b*)

Fig. 8.17 Flow past a staggered array of cylinders: (*a*) potential flow model using the Hele-Shaw apparatus *(TQ Education and Training Ltd.)*; (*b*) experimental streaklines for actual staggered-array flow at $Re_D \approx 6{,}400$. *(From Ref. 36, Courtesy of Jack Hoyt, with the permission of the American Society of Mechanical Engineers.)*

EXAMPLE 8.4

A Kelvin oval from Fig. 8.16 has $K/(U_\infty a) = 1.0$. Compute the velocity at the top shoulder of the oval in terms of U_∞.

Solution

We must locate the shoulder $y = h$ from Eq. (8.45) for $\psi = 0$ and then compute the velocity by differentiation. At $\psi = 0$ and $y = h$ and $x = 0$, Eq. (8.45) becomes

$$\frac{h}{a} = \frac{K}{U_\infty a} \ln \frac{h/a + 1}{h/a - 1}$$

With $K/(U_\infty a) = 1.0$ and the initial guess $h/a \approx 1.5$ from Fig. 8.16, we iterate and find the location $h/a = 1.5434$.

By inspection $v = 0$ at the shoulder because the streamline is horizontal. Therefore the shoulder velocity is, from Eq. (8.45),

$$u\Big|_{y=h} = \frac{\partial \psi}{\partial y}\Big|_{y=h} = U_\infty + \frac{K}{h-a} - \frac{K}{h+a}$$

Introducing $K = U_\infty a$ and $h = 1.5434a$, we obtain

$$u_{\text{shoulder}} = U_\infty(1.0 + 1.84 - 0.39) = 2.45U_\infty \qquad \textit{Ans.}$$

Because they are short-waisted compared with a circular cylinder, all the Kelvin ovals have shoulder velocity greater than the cylinder result $2.0U_\infty$ from Eq. (8.39).

8.5 Other Plane Potential Flows[3]

References 2 to 4 treat many other potential flows of interest in addition to the cases presented in Secs. 8.3 and 8.4. In principle, any plane potential flow can be solved by the method of *conformal mapping,* by using the complex variable

$$z = x + iy \qquad i = (-1)^{1/2}$$

It turns out that any arbitrary analytic function of this complex variable z has the remarkable property that both its real and its imaginary parts are solutions of Laplace's equation. If

$$f(z) = f(x + iy) = f_1(x, y) + i f_2(x, y)$$

then

$$\frac{\partial^2 f_1}{\partial x^2} + \frac{\partial^2 f_1}{\partial y^2} = 0 = \frac{\partial^2 f_2}{\partial x^2} + \frac{\partial^2 f_2}{\partial y^2} \tag{8.46}$$

We shall assign the proof of this as Prob. W8.4. Even more remarkable if you have never seen it before is that lines of constant f_1 will be everywhere perpendicular to lines of constant f_2:

$$\left(\frac{dy}{dx}\right)_{f_1=C} = -\frac{1}{(dy/dx)_{f_2=C}} \tag{8.47}$$

This is true for totally arbitrary $f(z)$ as long as this function is analytic; that is, it must have a unique derivative df/dz at every point in the region.

The net result of Eqs. (8.46) and (8.47) is that the functions f_1 and f_2 can be interpreted to be the potential lines and streamlines of an inviscid flow. By long custom we let the real part of $f(z)$ be the velocity potential and the imaginary part be the stream function:

$$f(z) = \phi(x, y) + i\psi(x, y) \tag{8.48}$$

We try various functions $f(z)$ and see whether any interesting flow pattern results. Of course, most of them have already been found, and we simply report on them here.

We shall not go into the details, but there are excellent treatments of this complex-variable technique on both an introductory [4] and a more advanced [2, 3] level. The method is less important now because of the popularity of computer techniques.

As a simple example, consider the linear function

$$f(z) = U_\infty z = U_\infty x + iU_\infty y$$

It follows from Eq. (8.48) that $\phi = U_\infty x$ and $\psi = U_\infty y$, which, we recall from Eq. (8.12), represents a uniform stream in the x direction. Once you get used to the complex variable, the solution practically falls in your lap.

To find the velocities, you may either separate ϕ and ψ from $f(z)$ and differentiate, or differentiate f directly:

$$\frac{df}{dz} = \frac{\partial \phi}{\partial x} + i\frac{\partial \psi}{\partial x} = -i\frac{\partial \phi}{\partial y} + \frac{\partial \psi}{\partial y} = u - iv \tag{8.49}$$

Thus the real part of df/dz equals $u(x, y)$, and the imaginary part equals $-v(x, y)$. To get a practical result, the derivative df/dz must exist and be unique, hence the requirement that f be an analytic function. For $f(z) = U_\infty z$, $df/dz = U_\infty = u$, since it is real, and $v = 0$, as expected.

Sometimes it is convenient to use the polar coordinate form of the complex variable

$$z = x + iy = re^{i\theta} = r\cos\theta + ir\sin\theta$$

where $\quad r = (x^2 + y^2)^{1/2} \qquad \theta = \tan^{-1}\frac{y}{x}$

This form is especially convenient when powers of z occur.

Uniform Stream at an Angle of Attack

All the elementary plane flows of Sec. 8.2 have a complex-variable formulation. The uniform stream U_∞ at an angle of attack α has the complex potential

$$f(z) = U_\infty z e^{-i\alpha} \tag{8.50}$$

Compare this form with Eq. (8.22).

Line Source at a Point z_0

Consider a line source of strength m placed off the origin at a point $z_0 = x_0 + iy_0$. Its complex potential is

$$f(z) = m\ln(z - z_0) \tag{8.51}$$

This can be compared with Eq. (8.13), which is valid only for the source at the origin. For a line sink, the strength m is negative.

Line Vortex at a Point z_0

If a line vortex of strength K is placed at point z_0, its complex potential is

$$f(z) = -iK\ln(z - z_0) \tag{8.52}$$

to be compared with Eq. (8.14). Also compare to Eq. (8.51) to see that we reverse the meaning of ϕ and ψ simply by multiplying the complex potential by $-i$.

[3]This section may be omitted without loss of continuity.

Flow around a Corner of Arbitrary Angle

Corner flow is an example of a pattern that cannot be conveniently produced by superimposing sources, sinks, and vortices. It has a strikingly simple complex representation:

$$f(z) = Az^n = Ar^n e^{in\theta} = Ar^n \cos n\theta + iAr^n \sin n\theta$$

where A and n are constants.

It follows from Eq. (8.48) that for this pattern

$$\phi = Ar^n \cos n\theta \qquad \psi = Ar^n \sin n\theta \tag{8.53}$$

Streamlines from Eq. (8.53) are plotted in Fig. 8.18 for five different values of n. The flow is seen to represent a stream turning through an angle $\beta = \pi/n$. Patterns in Fig. 8.18*d* and *e* are not realistic on the downstream side of the corner, where separation will occur due to the adverse pressure gradient and sudden change of direction. In general, separation always occurs downstream of salient, or protruding corners, except in creeping flows at low Reynolds number Re < 1.

Since 360° = 2π is the largest possible corner, the patterns for $n < \frac{1}{2}$ do not represent corner flow.

If we expand the plot of Fig. 8.18*a* to *c* to double size, we can represent stagnation flow toward a corner of angle $2\beta = 2\pi/n$. This is done in Fig. 8.19 for n = 3, 2, and 1.5. These are very realistic flows; although they slip at the wall, they can be patched to boundary layer theories very successfully. We took a brief look at corner flows before, in Examples 4.5 and 4.9 and in Probs. P4.49 to P4.51.

Flow Normal to a Flat Plate

We treat this case separately because the Kelvin ovals of Fig. 8.16 failed to degenerate into a flat plate as K became small. The flat plate normal to a uniform stream is an extreme case worthy of our attention.

Although the result is quite simple, the derivation is very complicated and is given, for example, in Ref. 2, Sec. 9.3. There are three changes of complex variable, or *mappings,* beginning with the basic cylinder flow solution of Fig. 8.14*a*. First the uniform stream is rotated to be vertical upward, then the cylinder is squeezed down into a plate

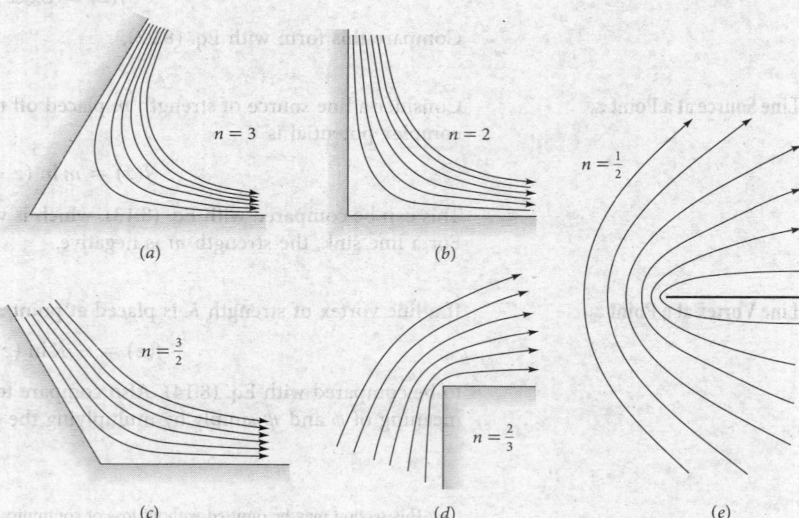

Fig. 8.18 Streamlines for corner flow, Eq. (8.53), for corner angle β of (*a*) 60°, (*b*) 90°, (*c*) 120°, (*d*) 270°, and (*e*) 360°.

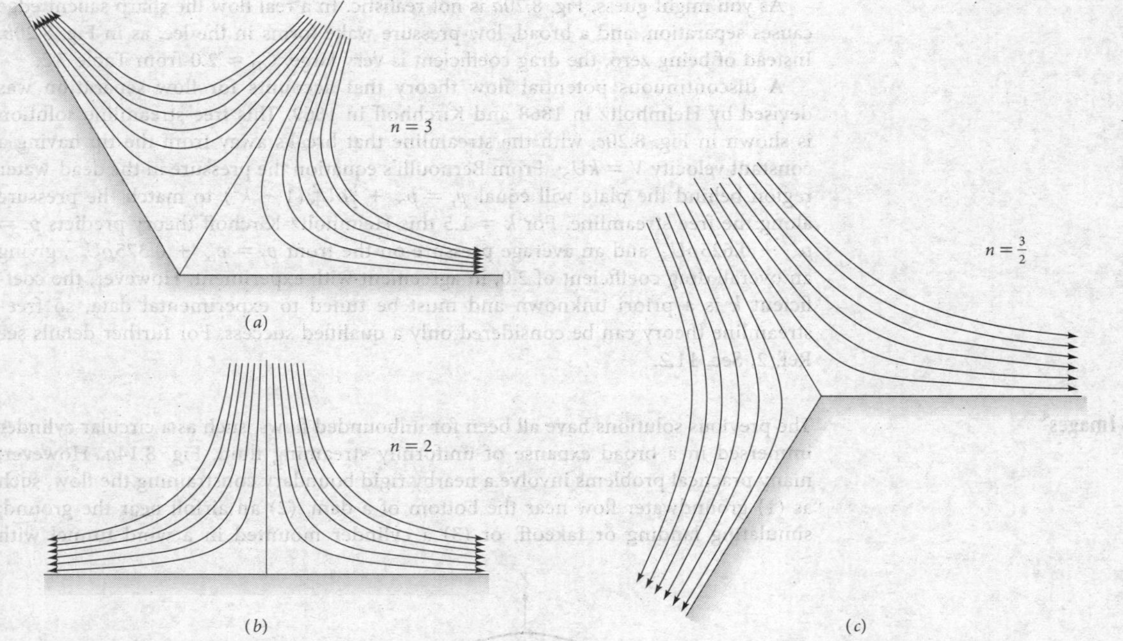

Fig. 8.19 Streamlines for stagnation flow from Eq. (8.53) for corner angle 2β of (*a*) 120°, (*b*) 180°, and (*c*) 240°.

shape, and finally the free stream is rotated back to the horizontal. The final result for complex potential is

$$f(z) = \phi + i\psi = U_\infty(z^2 + a^2)^{1/2} \tag{8.54}$$

where $2a$ is the height of the plate. To isolate ϕ or ψ, square both sides and separate real and imaginary parts:

$$\phi^2 - \psi^2 = U_\infty^2(x^2 - y^2 + a^2) \qquad \phi\psi = U_\infty^2 xy$$

We can solve for ψ to determine the streamlines

$$\psi^4 + \psi^2 U_\infty^2(x^2 - y^2 + a^2) = U_\infty^4 x^2 y^2 \tag{8.55}$$

Equation (8.55) is plotted in Fig. 8.20*a*, revealing a doubly symmetric pattern of streamlines that approach very closely to the plate and then deflect up and over, with very high velocities and low pressures near the plate tips.

The velocity υ_s along the plate surface is found by computing df/dz from Eq. (8.54) and isolating the imaginary part:

$$\left.\frac{\upsilon_s}{U_\infty}\right|_{\text{plate surface}} = \frac{y/a}{(1 - y^2/a^2)^{1/2}} \tag{8.56}$$

Some values of surface velocity can be tabulated as follows:

y/a	0.0	0.2	0.4	0.6	0.707	0.8	0.9	1.0
υ_s/U_∞	0.0	0.204	0.436	0.750	1.00	1.33	2.07	∞

The origin is a stagnation point; then the velocity grows linearly at first and very rapidly near the tip, with both velocity and acceleration being infinite at the tip.

As you might guess, Fig. 8.20*a* is not realistic. In a real flow the sharp salient edge causes separation, and a broad, low-pressure wake forms in the lee, as in Fig. 8.20*b*. Instead of being zero, the drag coefficient is very large, $C_D \approx 2.0$ from Table 7.2.

A discontinuous potential flow theory that accounts for flow separation was devised by Helmholtz in 1868 and Kirchhoff in 1869. This free-streamline solution is shown in Fig. 8.20*c*, with the streamline that breaks away from the tip having a constant velocity $V = kU_\infty$. From Bernoulli's equation the pressure in the dead-water region behind the plate will equal $p_r = p_\infty + \frac{1}{2}\rho U_\infty^2(1 - k^2)$ to match the pressure along the free streamline. For $k = 1.5$ this Helmholtz-Kirchoff theory predicts $p_r = p_\infty - 0.625\rho U_\infty^2$ and an average pressure on the front $p_f = p_\infty + 0.375\rho U_\infty^2$, giving an overall drag coefficient of 2.0, in agreement with experiment. However, the coefficient k is a priori unknown and must be tuned to experimental data, so free-streamline theory can be considered only a qualified success. For further details see Ref. 2, Sec. 11.2.

8.6 Images[4]

The previous solutions have all been for unbounded flows, such as a circular cylinder immersed in a broad expanse of uniformly streaming fluid, Fig. 8.14*a*. However, many practical problems involve a nearby rigid boundary constraining the flow, such as (1) groundwater flow near the bottom of a dam, (2) an airfoil near the ground, simulating landing or takeoff, or (3) a cylinder mounted in a wind tunnel with

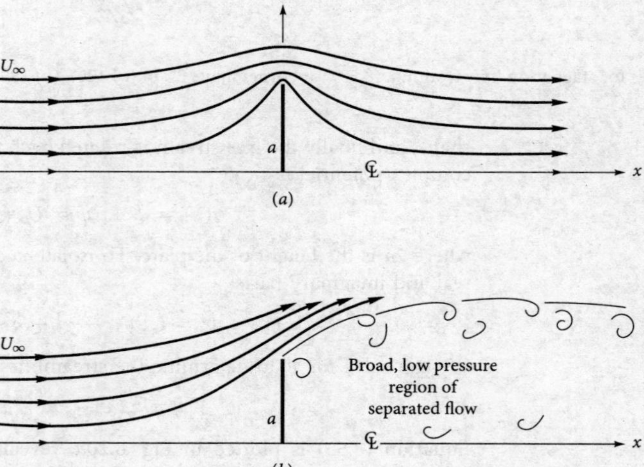

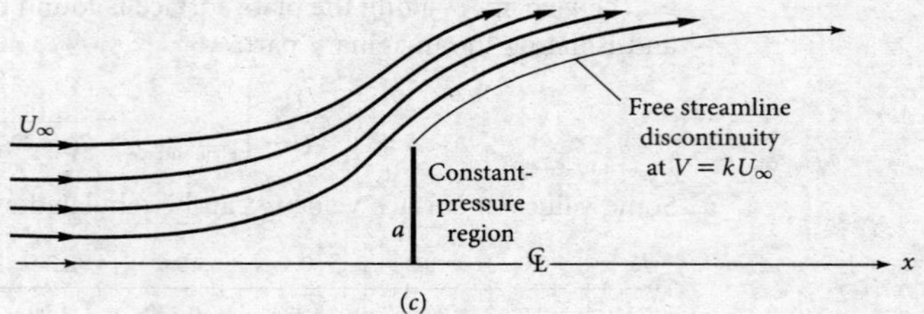

Fig. 8.20 Streamlines in upper half-plane for flow normal to a flat plate of height $2a$: (*a*) continuous potential flow theory, Eq. (8.55); (*b*) actual measured flow pattern; (*c*) discontinuous potential theory with $k \approx 1.5$.

[4] This section may be omitted without loss of continuity.

narrow walls. In such cases the basic unbounded potential flow solutions can be modified for wall effects by the method of *images*.

Consider a line source placed a distance *a* from a wall, as in Fig. 8.21*a*. To create the desired wall, an image source of identical strength is placed the same distance below the wall. By symmetry the two sources create a plane surface streamline between them, which is taken to be the wall.

In Fig. 8.21*b*, a vortex near a wall requires an image vortex the same distance below but of *opposite* rotation. We have shaded in the wall, but of course the pattern could also be interpreted as the flow near a vortex pair in an unbounded fluid.

In Fig. 8.21*c*, an airfoil in a uniform stream near the ground is created by an image airfoil below the ground of opposite circulation and lift. This looks easy, but actually it is not because the airfoils are so close together that they interact and distort each other's shapes. A rule of thumb is that nonnegligible shape distortion occurs if the body shape is within two chord lengths of the wall. To eliminate distortion, a whole series of "corrective" images must be added to the flow to recapture the shape of the original

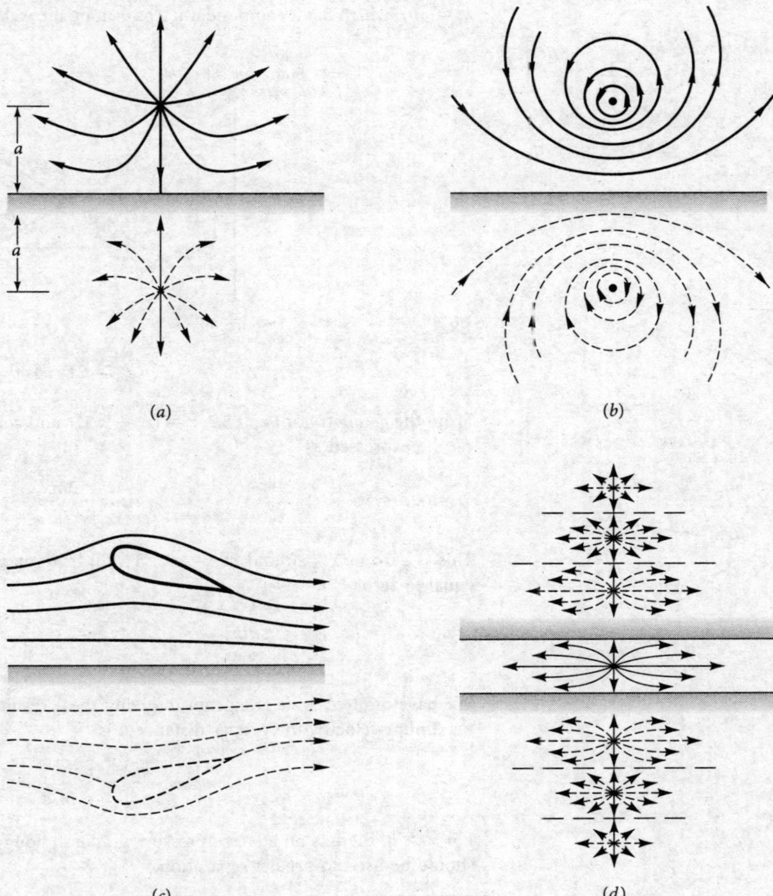

Fig. 8.21 Constraining walls can be created by image flows: (*a*) source near a wall with identical image source; (*b*) vortex near a wall with image vortex of opposite sense; (*c*) airfoil in ground effect with image airfoil of opposite circulation; (*d*) source between two walls requiring an infinite row of images.

isolated airfoil. Reference 2, Sec. 7.75, has a good discussion of this procedure, which usually requires computer summation of the multiple images needed.

Figure 8.21*d* shows a source constrained between two walls. One wall required only one image in Fig. 8.21*a*, but *two* walls require an infinite array of image sources above and below the desired pattern, as shown. Usually computer summation is necessary, but sometimes a closed-form summation can be achieved, as in the infinite vortex row of Eq. (8.28).

EXAMPLE 8.5

For the source near a wall as in Fig. 8.21*a*, the wall velocity is zero between the sources, rises to a maximum moving out along the wall, and then drops to zero far from the sources. If the source strength is 8 m²/s, how far from the wall should the source be to ensure that the maximum velocity along the wall will be 5 m/s?

Solution

At any point x along the wall, as in Fig. E8.5, each source induces a radial outward velocity $v_r = m/r$, which has a component $v_r \cos\theta$ along the wall. The total wall velocity is thus

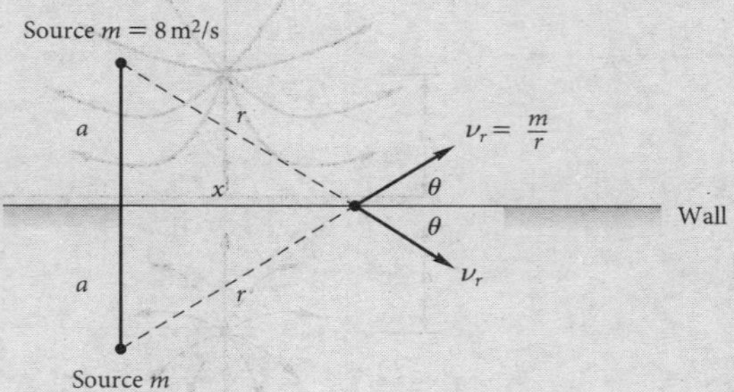

E8.5

$$u_{\text{wall}} = 2v_r \cos\theta$$

From the geometry of Fig. E8.5, $r = (x^2 + a^2)^{1/2}$ and $\cos\theta = x/r$. Then the total wall velocity can be expressed as

$$u = \frac{2mx}{x^2 + a^2}$$

This is zero at $x = 0$ and at $x \rightarrow \infty$. To find the maximum velocity, differentiate and set equal to zero:

$$\frac{du}{dx} = 0 \quad \text{at} \quad x = a \qquad \text{and} \qquad u_{\max} = \frac{m}{a}$$

We have omitted a bit of algebra in giving these results. For the given source strength and maximum velocity, the proper distance a is

$$a = \frac{m}{u_{\max}} = \frac{8\ \text{m}^2/\text{s}}{5\ \text{m/s}} = 1.6\ \text{m} \qquad \textit{Ans.}$$

For $x > a$, there is an adverse pressure gradient along the wall, and boundary layer theory should be used to predict separation.

8.7 Airfoil Theory[5]

As mentioned in conjunction with the Kutta-Joukowski lift theorem, Eq. (8.43), the problem in airfoil theory is to determine the net circulation Γ as a function of airfoil shape and free-stream angle of attack α.

The Kutta Condition

Even if the airfoil shape and free-stream angle of attack are specified, the potential flow theory solution is nonunique: An infinite family of solutions can be found corresponding to different values of circulation Γ. Four examples of this nonuniqueness were shown for the cylinder flows in Fig. 8.14. The same is true of the airfoil, and Fig. 8.22 shows three mathematically acceptable "solutions" to a given airfoil flow for small (Fig. 8.22*a*), large (Fig. 8.22*b*), and medium (Fig. 8.22*c*) net circulation. You can guess which case best simulates a real airfoil from the earlier discussion of transient lift development in Fig. 7.23. It is the case (Fig. 8.22*c*) where the upper and lower flows meet and leave the trailing edge smoothly. If the trailing edge is rounded slightly, there will be a stagnation point there. If the trailing edge is sharp, approximating most airfoil designs, the upper- and lower-surface flow velocities will be equal as they meet and leave the airfoil.

This statement of the physically proper value of Γ is generally attributed to W. M. Kutta, hence the name *Kutta condition,* although some texts give credit to Joukowski and/or Chaplygin. All airfoil theories use the Kutta condition, which is in good agreement with experiment. It turns out that the correct circulation Γ_{Kutta} depends on flow velocity, angle of attack, and airfoil shape.

Potential Theory for Thick Cambered Airfoils

The theory of thick cambered airfoils is covered in advanced texts [for example, 2 to 4]; Ref. 13 has a thorough and comprehensive review of both inviscid and viscous aspects of airfoil behavior.

Basically, the theory uses a complex-variable mapping that transforms the flow about a cylinder with circulation in Fig. 8.14 into flow about a foil shape with

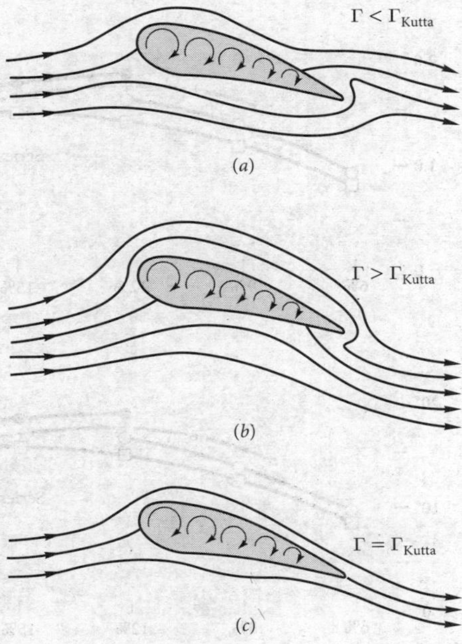

Fig. 8.22 The Kutta condition properly simulates the flow about an airfoil; (*a*) too little circulation, stagnation point on rear upper surface; (*b*) too much, stagnation point on rear lower surface; (*c*) just right, Kutta condition requires smooth flow at trailing edge.

[5]This section may be omitted without loss of continuity.

circulation. The circulation is then adjusted to match the Kutta condition of smooth exit flow from the trailing edge.

Regardless of the exact airfoil shape, the inviscid mapping theory predicts that the correct circulation for any thick cambered airfoil is

$$\Gamma_{\text{Kutta}} = \pi C U_\infty \left(1 + 0.77\frac{t}{C}\right) \sin(\alpha + \beta) \tag{8.57}$$

where $\beta = \tan^{-1}(2h/C)$ and h is the maximum camber, or maximum deviation of the airfoil midline from its chord line, as in Fig. 8.24*a*.

The lift coefficient of the infinite-span airfoil is thus

$$C_L = \frac{\rho U_\infty \Gamma}{\frac{1}{2}\rho U_\infty^2 bC} = 2\pi\left(1 + 0.77\frac{t}{C}\right)\sin(\alpha + \beta) \tag{8.58}$$

Figure 8.23 shows that the thickness effect $1 + 0.77t/C$ is not verified by experiment. Some airfoils increase lift with thickness, others decrease, and none approach the

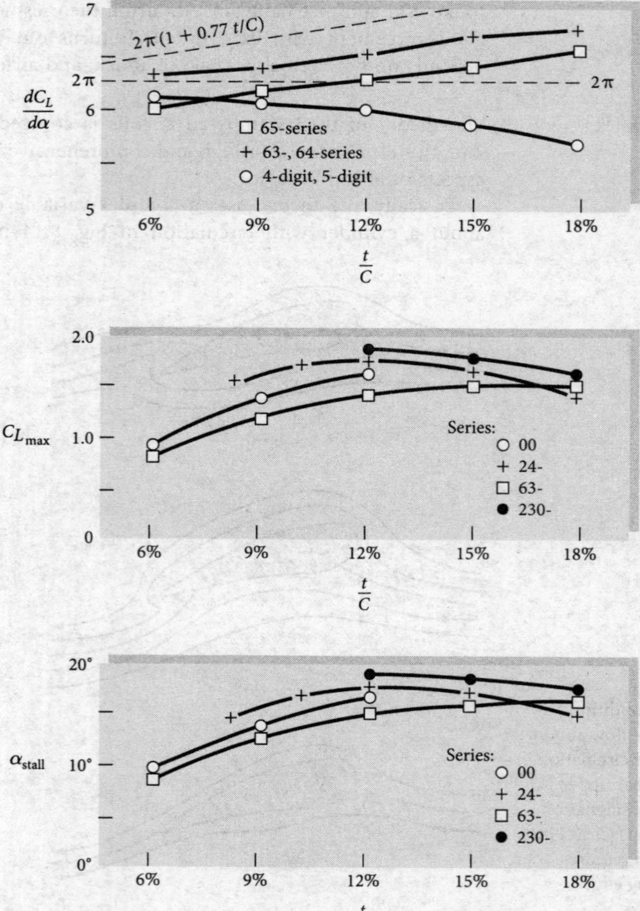

Fig. 8.23 Lift characteristics of smooth NACA airfoils as a function of thickness ratio, for infinite aspect ratio. *(From Ref. 12.)*

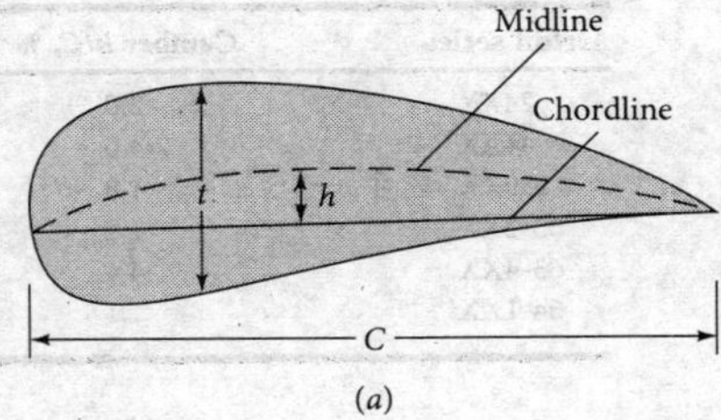

(a)

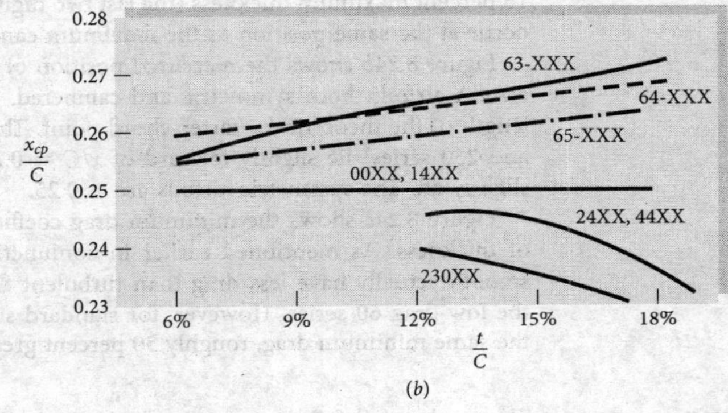

(b)

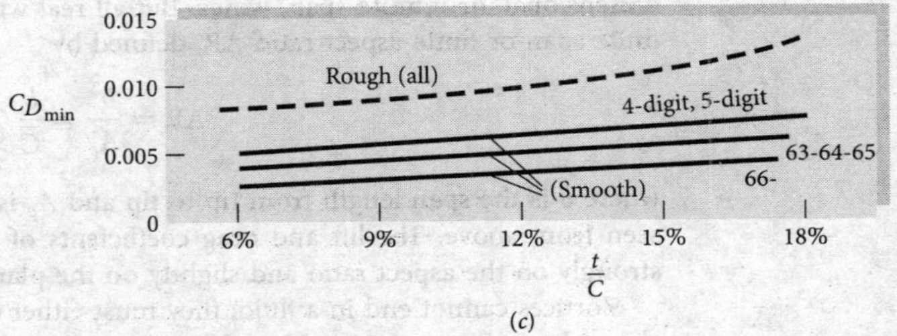

(c)

Fig. 8.24 Characteristics of NACA airfoils: (*a*) typical thick cambered airfoil; (*b*) center-of-pressure data; and (*c*) minimum drag coefficient.

theory very closely, the primary reason being the boundary layer growth on the upper surface affecting the airfoil "shape." Thus, it is customary to drop the thickness effect from the theory:

$$C_L \approx 2\pi \sin (\alpha + \beta) \tag{8.59}$$

The theory correctly predicts that a cambered airfoil will have finite lift at zero angle of attack and zero lift (ZL) at an angle

$$\alpha_{ZL} = -\beta = -\tan^{-1} \frac{2h}{C} \tag{8.60}$$

Equation (8.60) overpredicts the measured zero-lift angle by 1° or so, as shown in Table 8.3. The measured values are essentially independent of thickness. The designation XX in the NACA series indicates the thickness in percent, and the other digits refer to camber and other details. For example, the 2415 airfoil has 2 percent maximum camber (the first digit) occurring at 40 percent chord (the second digit) with

Table 8.3 Zero-Lift Angle of NACA Airfoils

Airfoil series	Camber h/C, %	Measured α_{ZL}, deg	Theory $-\beta$, deg
24XX	2.0	−2.1	−2.3
44XX	4.0	−4.0	−4.6
230XX	1.8	−1.3	−2.1
63-2XX	2.2	−1.8	−2.5
63-4XX	4.4	−3.1	−5.0
64-1XX	1.1	−0.8	−1.2

15 percent maximum thickness (the last two digits). The maximum thickness need not occur at the same position as the maximum camber.

Figure 8.24*b* shows the measured position of the center of pressure of the various NACA airfoils, both symmetric and cambered. In all cases x_{CP} is within 0.02 chord length of the theoretical quarter-chord point. The standard cambered airfoils (24, 44, and 230 series) lie slightly forward of $x/C = 0.25$ and the low-drag (60 series) foils slightly aft. The symmetric airfoils are at 0.25.

Figure 8.24*c* shows the minimum drag coefficient of NACA airfoils as a function of thickness. As mentioned earlier in conjunction with Fig. 7.25, these foils when smooth actually have less drag than turbulent flow parallel to a flat plate, especially the low-drag 60 series. However, for standard surface roughness all foils have about the same minimum drag, roughly 30 percent greater than that for a smooth flat plate.

Wings of Finite Span

The results of airfoil theory and experiment in the previous subsection were for two-dimensional, or infinite-span, wings. But all real wings have tips and are therefore of finite span or finite aspect ratio AR, defined by

$$\mathrm{AR} = \frac{b^2}{A_p} = \frac{b}{\bar{C}} \tag{8.61}$$

where b is the span length from tip to tip and A_p is the planform area of the wing as seen from above. The lift and drag coefficients of a finite-aspect-ratio wing depend strongly on the aspect ratio and slightly on the planform shape of the wing.

Vortices cannot end in a fluid; they must either extend to the boundary or form a closed loop. Figure 8.25*a* shows how the vortices that provide the wing circulation bend downstream at finite wing tips and extend far behind the wing to join the starting vortex (Fig. 7.23) downstream. The strongest vortices are shed from the tips, but some are shed from the body of the wing, as sketched schematically in Fig. 8.25*b*. The effective circulation $\Gamma(y)$ of these trailing shed vortices is zero at the tips and usually a maximum at the center plane, or root, of the wing. In 1918 Prandtl successfully modeled this flow by replacing the wing with a single lifting line and a continuous sheet of semi-infinite trailing vortices of strength $\gamma(y) = d\Gamma/dy$, as in Fig. 8.25*c*. Each elemental piece of trailing sheet $\gamma(\eta)\, d\eta$ induces a downwash, or downward velocity, $dw(y)$, given by

$$dw(y) = \frac{\gamma(\eta)\, d\eta}{4\pi(y - \eta)}$$

at position y on the lifting line. Note the denominator term 4π rather than 2π because the trailing vortex extends only from 0 to ∞ rather than from $-\infty$ to $+\infty$.

The total downwash $w(y)$ induced by the entire trailing vortex system is thus

$$w(y) = \frac{1}{4\pi} \int_{-(1/2)b}^{(1/2)b} \frac{\gamma(\eta)\, d\eta}{y - \eta} \tag{8.62}$$

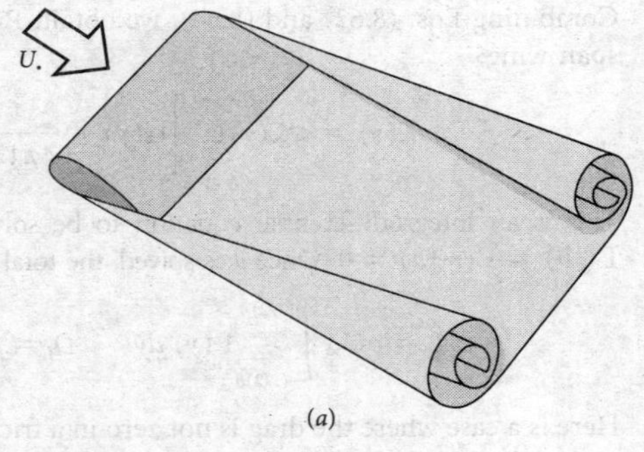

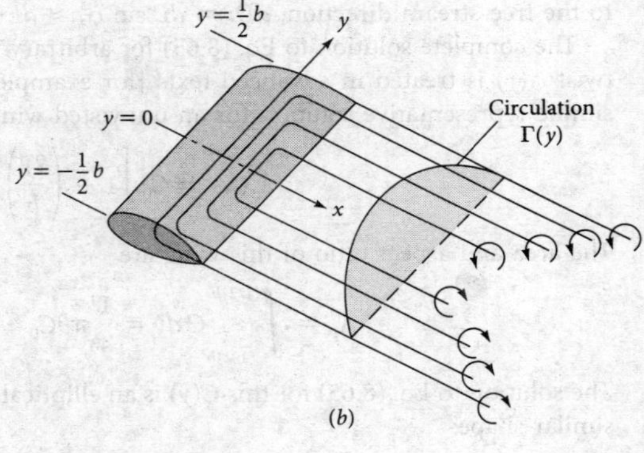

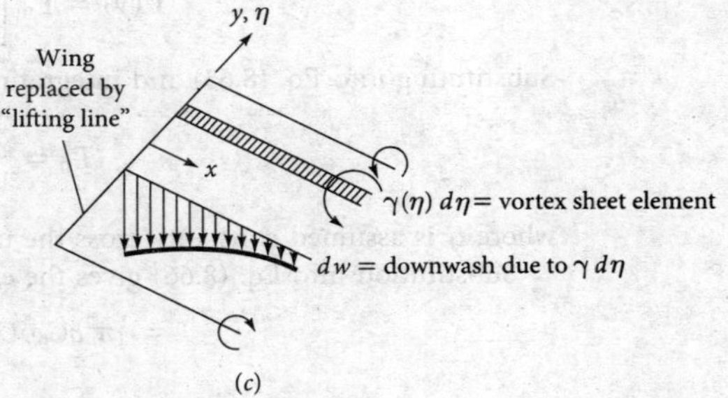

Fig. 8.25 Lifting-line theory for a finite wing: (*a*) actual trailing-vortex system behind a wing; (*b*) simulation by vortex system "bound" to the wing; (*c*) downwash on the wing due to an element of the trailing-vortex system.

When the downwash is vectorially added to the approaching free stream U_∞, the effective angle of attack at this section of the wing is reduced to

$$\alpha_{\text{eff}} = \alpha - \alpha_i \qquad \alpha_i = \tan^{-1}\frac{w}{U_\infty} \approx \frac{w}{U_\infty} \tag{8.63}$$

where we have used a small-amplitude approximation $w \ll U_\infty$.

The final step is to assume that the local circulation $\Gamma(y)$ is equal to that of a two-dimensional wing of the same shape and same effective angle of attack. From thin-airfoil theory we have the estimate

$$C_L = \frac{\rho U_\infty \Gamma b}{\frac{1}{2}\rho U_\infty^2 bC} \approx 2\pi\alpha_{\text{eff}}$$

or

$$\Gamma \approx \pi C U_\infty \alpha_{\text{eff}} \tag{8.64}$$

Combining Eqs. (8.62) and (8.64), we obtain Prandtl's lifting-line theory for a finite-span wing:

$$\Gamma(y) = \pi C(y) U_\infty \left[\alpha(y) - \frac{1}{4\pi U_\infty} \int_{-(1/2)b}^{(1/2)b} \frac{(d\Gamma/d\eta)\, d\eta}{y - \eta} \right] \tag{8.65}$$

This is an integrodifferential equation to be solved for $\Gamma(y)$ subject to the conditions $\Gamma(\frac{1}{2}b) = \Gamma(-\frac{1}{2}b) = 0$. Once it is solved, the total wing lift and induced drag are given by

$$L = \rho U_\infty \int_{-(1/2)b}^{(1/2)b} \Gamma(y)\, dy \qquad D_i = \rho U_\infty \int_{-(1/2)b}^{(1/2)b} \Gamma(y)\alpha_i(y)\, dy \tag{8.66}$$

Here is a case where the drag is not zero in a frictionless theory because the downwash causes the lift to slant backward by angle α_i so that it has a drag component parallel to the free-stream direction, $dDi = dL \sin \alpha_i \approx dL\, \alpha_i$.

The complete solution to Eq. (8.65) for arbitrary wing planform $C(y)$ and arbitrary twist $\alpha(y)$ is treated in advanced texts [for example, 11]. It turns out that there is a simple representative solution for an untwisted wing of elliptical planform:

$$C(y) = C_0 \left[1 - \left(\frac{2y}{b} \right)^2 \right]^{1/2}$$

The area and aspect ratio of this wing are

$$A_p = \int_{-(1/2)b}^{(1/2)b} C\, dy = \frac{1}{4}\pi b C_0 \qquad \text{AR} = \frac{4b}{\pi C_0} \tag{8.67}$$

The solution to Eq. (8.65) for this $C(y)$ is an elliptical circulation distribution of exactly similar shape:

$$\Gamma(y) = \Gamma_0 \left[1 - \left(\frac{2y}{b} \right)^2 \right]^{1/2}$$

Substituting into Eq. (8.65) and integrating give a relation between Γ_0 and C_0:

$$\Gamma_0 = \frac{\pi C_0 U_\infty \alpha}{1 + 2/\text{AR}}$$

where α is assumed constant across the untwisted wing.

Substitution into Eq. (8.66) gives the elliptical wing lift:

$$L = \tfrac{1}{4}\pi^2 b C_0 \rho U_\infty^2 \alpha / (1 + 2/\text{AR})$$

or

$$C_L = \frac{2\pi\alpha}{1 + 2/\text{AR}} \tag{8.68}$$

If we generalize this to a thick cambered finite wing of approximately elliptical planform, we obtain

$$\boxed{C_L = \frac{2\pi \sin(\alpha + \beta)}{1 + 2/\text{AR}} = \frac{2L}{\rho U_\infty^2 A_p}} \tag{8.69}$$

This result was given without proof as Eq. (7.70). From Eq. (8.62) the computed downwash for the elliptical wing is constant:

$$w(y) = \frac{2U_\infty \alpha}{2 + \text{AR}} = \text{const} \tag{8.70}$$

Finally, the induced drag coefficient from Eq. (8.63) is

$$C_{Di} = C_L \frac{w}{U_\infty} = \frac{C_L^2}{\pi \text{AR}} \tag{8.71}$$

This was given without proof as Eq. (7.71).

Figure 8.26 shows the effectiveness of this theory when tested against a nonelliptical cambered wing by Prandtl in 1921 [14]. Figures 8.26*a* and *b* show the measured lift curves and drag polars for five different aspect ratios. Note the increase in stall angle and drag and the decrease in lift slope as the aspect ratio decreases.

Figure 8.26*c* shows the lift data replotted against effective angle of attack $\alpha_{\text{eff}} = (\alpha + \beta)/(1 + 2/\text{AR})$, as predicted by Eq. (8.69). These curves should be equivalent to an infinite-aspect-ratio wing, and they do collapse together except near stall. Their common slope $dC_L/d\alpha$ is about 10 percent less than the theoretical value 2π, but this is consistent with the thickness and shape effects noted in Fig. 8.23.

Figure 8.26*d* shows the drag data replotted with the theoretical induced drag $C_{Di} = C_L^2/(\pi\text{AR})$ subtracted out. Again, except near stall, the data collapse onto a single line of nearly constant infinite-aspect-ratio drag $C_{D0} \approx 0.01$. We conclude that the finite-wing theory is very effective and may be used for design calculations.

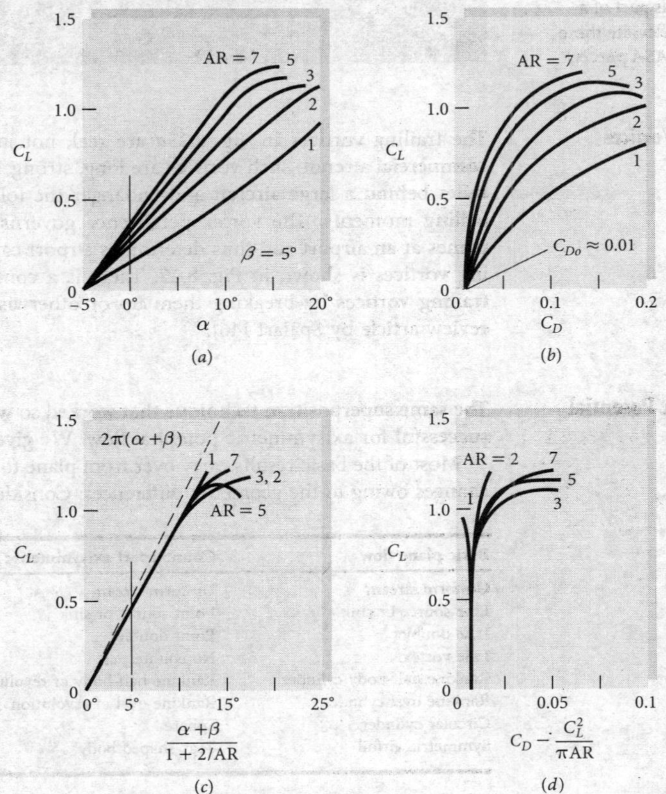

Fig. 8.26 Comparison of theory and experiment for a finite wing: (*a*) measured lift [14]; (*b*) measured drag polar [14]; (*c*) lift reduced to infinite aspect ratio; (*d*) drag polar reduced to infinite aspect ratio.

Fig. 8.27 Wingtip vortices from a smoke-visualization test of a Boeing 737. Vortices from large airplanes can be extremely dangerous to any following aircraft, especially small planes. This test was part of a research effort to alleviate these swirling wakes. *(NASA photo.)*

Aircraft Trailing Vortices

The trailing vortices in Fig. 8.25*a* are real, not just mathematical abstractions. On commercial aircraft, such vortices are long, strong, and lingering. They can stretch for miles behind a large aircraft and endanger the following planes by inducing drastic rolling moments. The vortex persistence governs the separation distance between planes at an airport and thus determines airport capacity. An example of strong trailing vortices is shown in Fig. 8.27. There is a continuing research effort to alleviate trailing vortices by breaking them up or otherwise causing them to decay. See the review article by Spalart [46].

8.8 Axisymmetric Potential Flow[6]

The same superposition technique that worked so well for plane flow in Sec. 8.3 is also successful for axisymmetric potential flow. We give some brief examples here.

Most of the basic results carry over from plane to axisymmetric flow with only slight changes owing to the geometric differences. Consider the following related flows:

Basic plane flow	Counterpart axisymmetric flow
Uniform stream	Uniform stream
Line source or sink	Point source or sink
Line doublet	Point doublet
Line vortex	No counterpart
Rankine half-body cylinder	Rankine half-body of revolution
Rankine oval cylinder	Rankine oval of revolution
Circular cylinder	Sphere
Symmetric airfoil	Tear-shaped body

[6] This section may be omitted without loss of continuity.

Since there is no such thing as a point vortex, we must forgo the pleasure of studying circulation effects in axisymmetric flow. However, as any smoker knows, there is an axisymmetric ring vortex, and there are also ring sources and ring sinks, which we leave to advanced texts [for example, 3].

Spherical Polar Coordinates

Axisymmetric potential flows are conveniently treated in the spherical polar coordinates of Fig. 8.28. There are only two coordinates (r, θ), and flow properties are constant on a circle of radius $r \sin\theta$ about the x axis.

The equation of continuity for incompressible flow in these coordinates is

$$\frac{\partial}{\partial r}(r^2 v_r \sin\theta) + \frac{\partial}{\partial\theta}(r v_\theta \sin\theta) = 0 \tag{8.72}$$

where v_r and v_θ are radial and tangential velocity as shown. Thus a spherical polar stream function[7] exists such that

$$v_r = -\frac{1}{r^2 \sin\theta}\frac{\partial\psi}{\partial\theta} \qquad v_\theta = \frac{1}{r\sin\theta}\frac{\partial\psi}{\partial r} \tag{8.73}$$

In like manner a velocity potential $\phi(r, \theta)$ exists such that

$$v_r = \frac{\partial\phi}{\partial r} \qquad v_\theta = \frac{1}{r}\frac{\partial\phi}{\partial\theta} \tag{8.74}$$

These formulas serve to deduce the ψ and ϕ functions for various elementary axisymmetric potential flows.

Uniform Stream in the x Direction

A stream U_∞ in the x direction has components

$$v_r = U_\infty \cos\theta \qquad v_\theta = -U_\infty \sin\theta$$

Substitution into Eqs. (8.73) and (8.74) and integrating give

$$\text{Uniform stream:} \qquad \psi = -\tfrac{1}{2}U_\infty r^2 \sin^2\theta \qquad \phi = U_\infty r\cos\theta \tag{8.75}$$

As usual, arbitrary constants of integration have been neglected.

Point Source or Sink

Consider a volume flux Q issuing from a point source. The flow will spread out radially and at radius r will equal Q divided by the area $4\pi r^2$ of a sphere. Thus

$$v_r = \frac{Q}{4\pi r^2} = \frac{m}{r^2} \qquad v_\theta = 0 \tag{8.76}$$

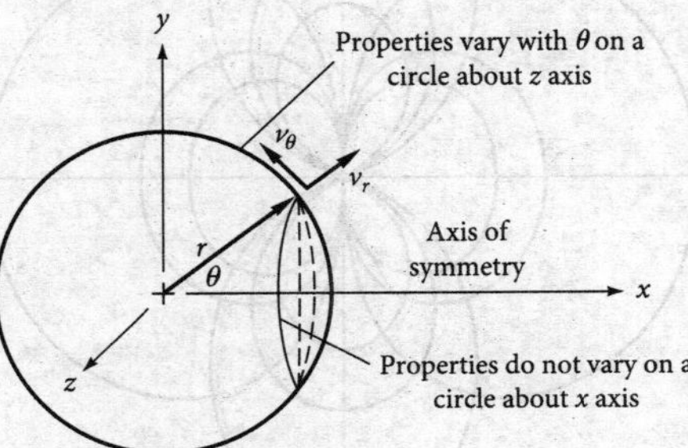

Fig. 8.28 Spherical polar coordinates for axisymmetric flow.

[7]It is often called *Stokes's stream function*, having been used in a paper Stokes wrote in 1851 on viscous sphere flow.

with $m = Q/(4\pi)$ for convenience. Integrating (8.73) and (8.74) gives

$$\text{Point source} \qquad \psi = m\cos\theta \qquad \phi = -\frac{m}{r} \tag{8.77}$$

For a point sink, change m to $-m$ in Eq. (8.77).

Point Doublet

Exactly as in Fig. 8.12, place a source at $(x, y) = (-a, 0)$ and an equal sink at $(+a, 0)$, taking the limit as a becomes small with the product $2am = \lambda$ held constant:

$$\psi_{\text{doublet}} = \lim_{\substack{a\to 0\\ 2am=\lambda}} (m\cos\theta_{\text{source}} - m\cos\theta_{\text{sink}}) = \frac{\lambda\sin^2\theta}{r} \tag{8.78}$$

We leave the proof of this limit as a problem. The point-doublet velocity potential is

$$\phi_{\text{doublet}} = \lim_{\substack{a\to 0\\ 2am=\lambda}} \left(-\frac{m}{r_{\text{source}}} + \frac{m}{r_{\text{sink}}}\right) = \frac{\lambda\cos\theta}{r^2} \tag{8.79}$$

The streamlines and potential lines are shown in Fig. 8.29. Unlike the plane doublet flow of Fig. 8.12, neither set of lines represents perfect circles.

Uniform Stream plus a Point Source

By combining Eqs. (8.75) and (8.77), we obtain the stream function for a uniform stream plus a point source at the origin:

$$\psi = -\tfrac{1}{2}U_\infty r^2\sin^2\theta + m\cos\theta \tag{8.80}$$

From Eq. (8.73) the velocity components are, by differentiation,

$$v_r = U_\infty\cos\theta + \frac{m}{r^2} \qquad v_\theta = -U_\infty\sin\theta \tag{8.81}$$

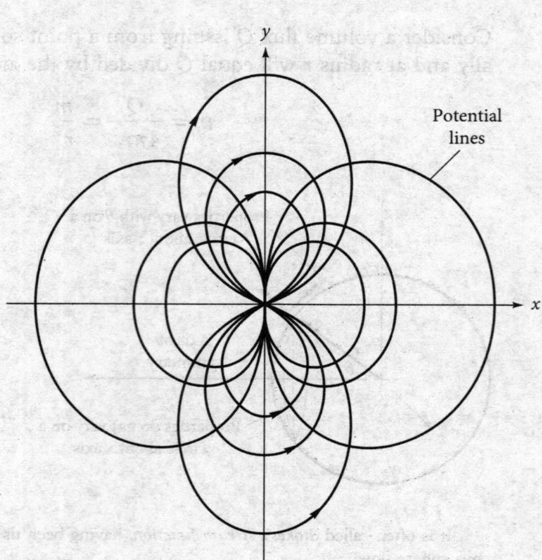

Fig. 8.29 Streamlines and potential lines due to a point doublet at the origin, from Eqs. (8.78) and (8.79).

Setting these equal to zero reveals a stagnation point at $\theta = 180°$ and $r = a = (m/U_\infty)^{1/2}$, as shown in Fig. 8.30. If we let $m = U_\infty a^2$, the stream function can be rewritten as

$$\frac{\psi}{U_\infty a^2} = \cos\theta - \frac{1}{2}\left(\frac{r}{a}\right)^2 \sin^2\theta \tag{8.82}$$

The stream surface that passes through the stagnation point $(r, \theta) = (a, \pi)$ has the value $\psi = -U_\infty a^2$ and forms a half-body of revolution enclosing the point source, as shown in Fig. 8.30. This half-body can be used to simulate a pitot tube. Far downstream the half-body approaches the constant radius $R = 2a$ about the x axis. The maximum velocity and minimum pressure along the half-body surface occur at $\theta = 70.5°$, $r = a\sqrt{3}$, $V_s = 1.155U_\infty$. Downstream of this point there is an adverse gradient as V_s slowly decelerates to U_∞, but boundary layer theory indicates no flow separation. Thus, Eq. (8.82) is a very realistic simulation of a real half-body flow. But when the uniform stream is added to a sink to form a half-body rear surface, similar to Fig. 8.9*c*, separation is predicted and the rear inviscid pattern is not realistic.

Uniform Stream plus a Point Doublet

From Eqs. (8.75) and (8.78), combination of a uniform stream and a point doublet at the origin gives

$$\psi = -\frac{1}{2} U_\infty r^2 \sin^2\theta + \frac{\lambda}{r}\sin^2\theta \tag{8.83}$$

Examination of this relation reveals that the stream surface $\psi = 0$ corresponds to the sphere of radius

$$r = a = \left(\frac{2\lambda}{U_\infty}\right)^{1/3} \tag{8.84}$$

This is exactly analogous to the cylinder flow of Fig. 8.14*a* formed by combining a uniform stream and a line doublet.

Letting $\lambda = \frac{1}{2}U_\infty a^3$ for convenience, we rewrite Eq. (8.83) as

$$\frac{\psi}{\frac{1}{2}U_\infty a^2} = -\sin^2\theta\left(\frac{r^2}{a^2} - \frac{a}{r}\right) \tag{8.85}$$

The streamlines for this sphere flow are plotted in Fig. 8.31. By differentiation from Eq. (8.73) the velocity components are

$$v_r = U_\infty \cos\theta\left(1 - \frac{a^3}{r^3}\right) \qquad v_\theta = -\frac{1}{2}U_\infty \sin\theta\left(2 + \frac{a^3}{r^3}\right) \tag{8.86}$$

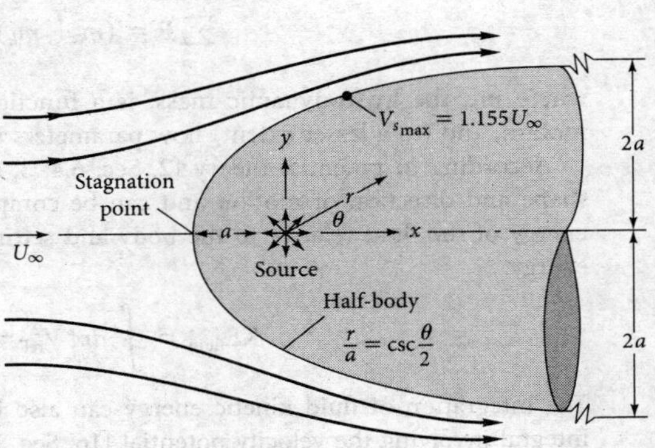

Fig. 8.30 Streamlines for a Rankine half-body of revolution.

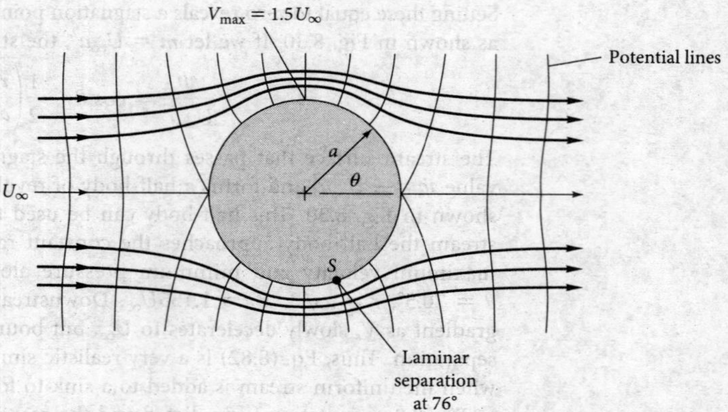

Fig. 8.31 Streamlines and potential lines for inviscid flow past a sphere.

We see that the radial velocity vanishes at the sphere surface $r = a$, as expected. There is a stagnation point at the front (a, π) and the rear $(a, 0)$ of the sphere. The maximum velocity occurs at the shoulder $(a, \pm\frac{1}{2}\pi)$, where $\upsilon_r = 0$ and $\upsilon_\theta = \mp 1.5U_\infty$. The surface velocity distribution is

$$V_s = -\upsilon_\theta|_{r=a} = \tfrac{3}{2}U_\infty \sin\theta \tag{8.87}$$

Note the similarity to the cylinder surface velocity equal to $2U_\infty \sin\theta$ from Eq. (8.39) with zero circulation.

Equation (8.87) predicts, as expected, an adverse pressure gradient on the rear $(\theta < 90°)$ of the sphere. If we use this distribution with laminar boundary layer theory [for example, 15, p. 294], separation is computed to occur at about $\theta = 76°$, so that in the actual flow pattern of Fig. 7.14 a broad wake forms in the rear. This wake interacts with the free stream and causes Eq. (8.87) to be inaccurate even in the front of the sphere. The measured maximum surface velocity is equal only to about $1.3U_\infty$ and occurs at about $\theta = 107°$ (see Ref. 15, Sec. 4.10.4, for further details).

The Concept of Hydrodynamic Mass

When a body moves through a fluid, it must push a finite mass of fluid out of the way. If the body is accelerated, the surrounding fluid must also be accelerated. The body behaves as if it were heavier by an amount called the *hydrodynamic mass* (also called the *added* or *virtual mass*) of the fluid. If the instantaneous body velocity is $\mathbf{U}(t)$, the summation of forces must include this effect:

$$\sum \mathbf{F} = (m + m_h)\frac{d\mathbf{U}}{dt} \tag{8.88}$$

where m_h, the hydrodynamic mass, is a function of body shape, the direction of motion, and (to a lesser extent) flow parameters such as the Reynolds number.

According to potential theory [2, Sec. 6.4; 3, Sec. 9.22], m_h depends only on the shape and direction of motion and can be computed by summing the total kinetic energy of the fluid relative to the body and setting this equal to an equivalent body energy:

$$KE_{fluid} = \int \tfrac{1}{2}dm\, V_{rel}^2 = \tfrac{1}{2}m_h U^2 \tag{8.89}$$

The integration of fluid kinetic energy can also be accomplished by a body-surface integral involving the velocity potential [16, Sec. 11].

Consider the previous example of a sphere immersed in a uniform stream. By subtracting out the stream velocity we can replot the flow as in Fig. 8.32, showing the streamlines relative to the moving sphere. Note the similarity to the doublet flow in Fig. 8.29. The relative velocity components are found by subtracting U from Eqs. (8.86):

$$v_r = -\frac{Ua^3 \cos\theta}{r^3} \qquad v_\theta = -\frac{Ua^3 \sin\theta}{2r^3}$$

The element of fluid mass, in spherical polar coordinates, is

$$dm = \rho(2\pi r \sin\theta) r\, dr\, d\theta$$

When dm and $V_{rel}^2 = v_r^2 + v_\theta^2$ are substituted into Eq. (8.89), the integral can be evaluated:

$$\mathrm{KE}_{\mathrm{fluid}} = \tfrac{1}{3}\rho\pi a^3 U^2$$

or

$$m_h(\text{sphere}) = \tfrac{2}{3}\rho\pi a^3 \tag{8.90}$$

Thus, according to potential theory, the hydrodynamic mass of a sphere equals one-half of its displaced mass, independent of the direction of motion.

A similar result for a cylinder moving normal to its axis can be computed from Eqs. (8.38) after subtracting out the stream velocity. The result is

$$m_h(\text{cylinder}) = \rho\pi a^2 L \tag{8.91}$$

for a cylinder of length L, assuming two-dimensional motion. The cylinder's hydrodynamic mass equals its displaced mass.

Tables of hydrodynamic mass for various body shapes and directions of motion are given by Patton [17]. See also Ref. 21.

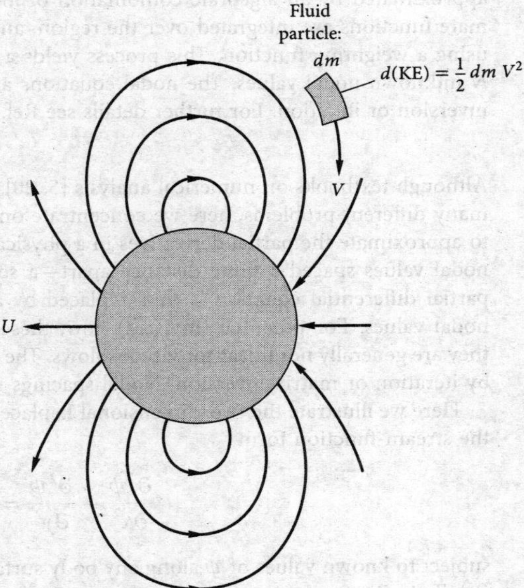

Fig. 8.32 Potential flow streamlines relative to a moving sphere. Compare with Figs. 8.29 and 8.31.

8.9 Numerical Analysis

When potential flow involves complicated geometries or unusual stream conditions, the classical superposition scheme of Secs. 8.3 and 8.4 becomes less attractive. Conformal mapping of body shapes, by using the complex-variable technique of Sec. 8.5, is no longer popular. Numerical analysis is the appropriate modern approach, and at least three different approaches are in use:

1. The finite element method (FEM) [6, 19]
2. The finite difference method (FDM) [5, 20, 23–26], or its close sibling, the finite volume method [27].
3. *a.* Integral methods with distributed singularities [18]
   b. The boundary element method (BEM) [7, 38]

Methods 3*a* and 3*b* are closely related, having first been developed on an ad hoc basis by aerodynamicists in the 1960s [18] and then generalized into a multipurpose applied mechanics technique in the 1970s [7].

Methods 1 (or FEM) and 2 (or FDM), though strikingly different in concept, are comparable in scope, mesh size, and general accuracy. We concentrate here on the latter method for illustration purposes.

All three of these methods—FEM, FDM, and BEM—are popular in present-day computational fluid dynamics. Although simplified online CFD codes are available—sometimes free—the writer believes that CFD, as a serious method of flow analysis, should wait until one studies the professional software available. This subject is more appropriate for advanced electives or graduate school. The discussion here will be brief and descriptive, with only nominal illustrations for an FDM method.

The Finite Element Method

The finite element method [19] is applicable to all types of linear and nonlinear partial differential equations in physics and engineering. The computational domain is divided into small regions, usually triangular or quadrilateral. These regions are delineated with a finite number of *nodes* where the field variables—temperature, velocity, pressure, stream function, and so on—are to be calculated. The solution in each region is approximated by an algebraic combination of local nodal values. Then the approximate functions are integrated over the region, and their error is minimized, often by using a weighting function. This process yields a set of N algebraic equations for the N unknown nodal values. The nodal equations are solved simultaneously, by matrix inversion or iteration. For further details see Ref. 6 or 19.

The Finite Difference Method

Although textbooks on numerical analysis [5, 20] apply finite difference techniques to many different problems, here we concentrate on potential flow. The idea of FDM is to approximate the partial derivatives in a physical equation by "differences" between nodal values spaced a finite distance apart—a sort of numerical calculus. The basic partial differential equation is thus replaced by a set of algebraic equations for the nodal values. For potential (inviscid) flow, these algebraic equations are linear, but they are generally nonlinear for viscous flows. The solution for nodal values is obtained by iteration or matrix inversion. Nodal spacings need not be equal.

Here we illustrate the two-dimensional Laplace equation, choosing for convenience the stream-function form

$$\frac{\partial^2 \psi}{\partial x^2} + \frac{\partial^2 \psi}{\partial y^2} = 0 \tag{8.92}$$

subject to known values of ψ along any body surface and known values of $\partial\psi/\partial x$ and $\partial\psi/\partial y$ in the free stream.

Our finite difference technique divides the flow field into equally spaced nodes, as shown in Fig. 8.33. To economize on the use of parentheses or functional notation, subscripts i and j denote the position of an arbitrary, equally spaced node, and $\psi_{i,j}$ denotes the value of the stream function at that node:

$$\psi_{i,j} = \psi(x_0 + i\,\Delta x, y_0 + j\,\Delta y)$$

Thus, $\psi_{i+1,j}$ is just to the right of $\psi_{i,j}$, and $\psi_{i,j+1}$ is just above.

An algebraic approximation for the derivative $\partial\psi/\partial x$ is

$$\frac{\partial \psi}{\partial x} \approx \frac{\psi(x + \Delta x, y) - \psi(x, y)}{\Delta x}$$

A similar approximation for the second derivative is

$$\frac{\partial^2 \psi}{\partial x^2} \approx \frac{1}{\Delta x}\left[\frac{\psi(x + \Delta x, y) - \psi(x, y)}{\Delta x} - \frac{\psi(x, y) - \psi(x - \Delta x, y)}{\Delta x}\right]$$

The subscript notation makes these expressions more compact:

$$\frac{\partial \psi}{\partial x} \approx \frac{1}{\Delta x}(\psi_{i+1,j} - \psi_{i,j})$$
$$\frac{\partial^2 \psi}{\partial x^2} \approx \frac{1}{\Delta x^2}(\psi_{i+1,j} - 2\psi_{i,j} + \psi_{i-1,j}) \tag{8.93}$$

These formulas are exact in the calculus limit as $\Delta x \to 0$, but in numerical analysis we keep Δx and Δy finite, hence the term *finite differences.*

In an exactly similar manner, we can derive the equivalent difference expressions for the y direction:

$$\frac{\partial \psi}{\partial y} \approx \frac{1}{\Delta y}(\psi_{i,j+1} - \psi_{i,j})$$
$$\frac{\partial^2 \psi}{\partial y^2} \approx \frac{1}{\Delta y^2}(\psi_{i,j+1} - 2\psi_{i,j} + \psi_{i,j-1}) \tag{8.94}$$

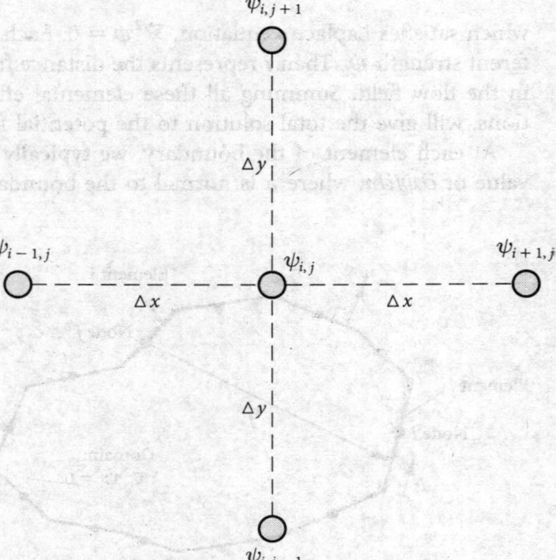

Fig. 8.33 Definition sketch for a two-dimensional rectangular finite difference grid.

The use of subscript notation allows these expressions to be programmed directly into a scientific computer language.

When (8.93) and (8.94) are substituted into Laplace's equation (8.92), the result is the algebraic formula

$$2(1 + \beta)\psi_{i,j} \approx \psi_{i-1,j} + \psi_{i+1,j} + \beta(\psi_{i,j-1} + \psi_{i,j+1}) \tag{8.95}$$

where $\beta = (\Delta x/\Delta y)^2$ depends on the mesh size selected. This finite difference model of Laplace's equation states that every nodal stream-function value $\psi_{i,j}$ is a linear combination of its four nearest neighbors.

The most commonly programmed case is a square mesh ($\beta = 1$), for which Eq. (8.95) reduces to

$$\boxed{\psi_{i,j} \approx \tfrac{1}{4}(\psi_{i,j+1} + \psi_{i,j-1} + \psi_{i+1,j} + \psi_{i-1,j})} \tag{8.96}$$

Thus, for a square mesh, each nodal value equals the arithmetic average of the four neighbors shown in Fig. 8.33. The formula is easily remembered and easily programmed. The formula is applied in iterative fashion sweeping over each of the internal nodes (I, J), with known values of P specified at each of the surrounding boundary nodes. Any initial guesses can be specified for the internal nodes, and the iteration process will converge to the final algebraic solution in a finite number of sweeps. The numerical error, compared with the exact solution of Laplace's equation, is proportional to the square of the mesh size.

The Boundary Element Method

A relatively new technique for numerical solution of partial differential equations is the *boundary element method* (BEM). Reference 7 is an introductory textbook outlining the concepts of BEM. There are no interior elements. Rather, all nodes are placed on the boundary of the domain, as in Fig. 8.34. The "element" is a small piece of the boundary surrounding the node. The "strength" of the element can be either constant or variable.

For plane potential flow, the method takes advantage of the particular solution

$$\psi^* = \frac{1}{2\pi} \ln \frac{1}{r} \tag{8.97}$$

which satisfies Laplace's equation, $\nabla^2\psi = 0$. Each element i is assumed to have a different strength ψ_i. Then r represents the distance from that element to any other point in the flow field. Summing all these elemental effects, with proper boundary conditions, will give the total solution to the potential flow problem.

At each element of the boundary, we typically know either the value of ψ or the value of $\partial\psi/\partial n$, where n is normal to the boundary. (Mixed combinations of ψ and

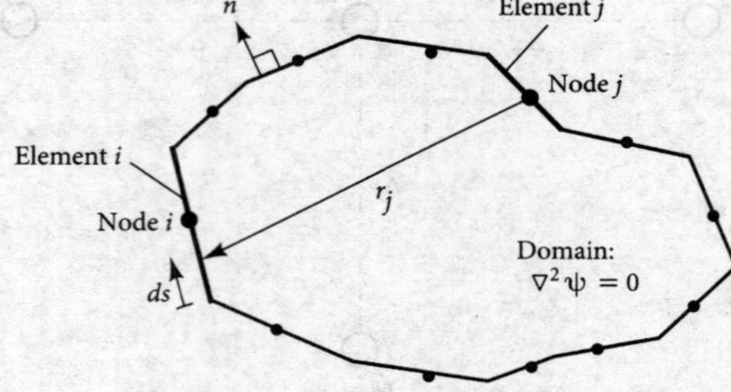

Fig. 8.34 Boundary elements of constant strength in plane potential flow.

$\partial\psi/\partial n$ are also possible but are not discussed here.) The correct strengths ψ_i are such that these boundary conditions are satisfied at every element. Summing these effects over N elements requires integration by parts plus a careful evaluation of the (singular) effect of element i upon itself. The mathematical details are given in Ref. 7. The result is a set of N algebraic equations for the unknown boundary values. In the case of elements of constant strength, the final expression is

$$\frac{1}{2}\psi_i + \sum_{j=1}^{N}\psi_j\left(\int_j \frac{\partial\psi^*}{\partial n}\,ds\right) = \sum_{j=1}^{N}\left(\frac{\partial\psi}{\partial n}\right)_j\left(\int_j \psi^*\,ds\right) \qquad i = 1 \text{ to } N \tag{8.98}$$

The integrals, which involve the logarithmic particular solution ψ^* from Eq. (8.97), are evaluated numerically for each element.

Reference 7 is a general introduction to boundary elements, while Ref. 38 emphasizes programming methods. Meanwhile, research continues. Dargush and Grigoriev [42] have developed a multilevel boundary element method for steady Stokes or *creeping* flows (see Sec. 7.6) in irregular geometries. Their scheme avoids the heavy memory and CPU-time requirements of most boundary element methods. They estimate that CPU time is reduced by a factor of 700,000 and required memory is reduced by a factor of 16,000.

Viscous Flow Computer Models

Our previous finite difference model of Laplace's equation, as in Eq. (8.96), was very well behaved and converged nicely with or without overrelaxation. Much more care is needed to model the full Navier-Stokes equations. The challenges are quite different, and they have been met to a large extent, so there are now many textbooks [5, 20, 23 to 27] on (fully viscous) CFD. This is not a textbook on CFD, but we will address some of the issues in this section.

One-Dimensional Unsteady Flow

We consider a simplified problem, showing that even a single viscous term introduces new effects and possible instabilities. Recall (or review) Prob. P4.85, where a wall moves and drives a viscous fluid parallel to itself. Gravity is neglected. Let the wall be the plane $y = 0$, moving at a speed $U_0(t)$, as in Fig. 8.35. A uniform vertical grid, of spacing Δy, has nodes n at which the local velocity u_n^j is to be calculated, where superscript j denotes the time-step $j\Delta t$. The wall is $n = 1$. If $u = u(y, t)$ only and $v = w = 0$, continuity, $\nabla \cdot \mathbf{V} = 0$, is satisfied, and we need only solve the x-momentum Navier-Stokes equation:

$$\frac{\partial u}{\partial t} = \nu\frac{\partial^2 u}{\partial y^2} \tag{8.99}$$

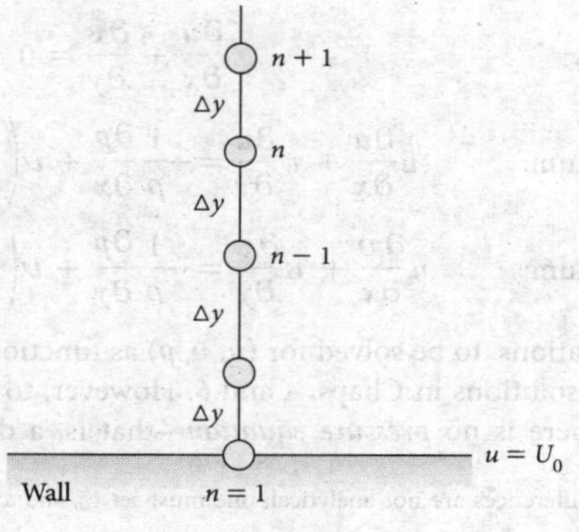

n + 1
Δy
n
Δy
n − 1
Δy
Δy
Wall
n = 1
u = U₀

Fig. 8.35 An equally spaced finite difference mesh for one-dimensional viscous flow [Eq. (8.99)].

where $\nu = \mu/\rho$. Utilizing the same finite difference approximations as in Eq. (8.93), we may model Eq. (8.99) algebraically as a forward time difference and a central spatial difference:

$$\frac{u_n^{j+1} - u_n^j}{\Delta t} \approx \nu \frac{u_{n+1}^j - 2u_n^j + u_{n-1}^j}{\Delta y^2}$$

Rearrange and find that we can solve explicitly for u_n at the next time-step $j + 1$:

$$u_n^{j+1} \approx (1 - 2\sigma)\, u_n^j + \sigma(u_{n-1}^j + u_{n+1}^j) \qquad \sigma = \frac{\nu \Delta t}{\Delta y^2} \tag{8.100}$$

Thus, u at node n at the next time-step $j + 1$ is a weighted average of three previous values, similar to the "four-nearest-neighbors" average in the laplacian model of Eq. (8.96). Since the new velocity is calculated immediately, Eq. (8.100) is called an *explicit* model. It differs from the well-behaved laplacian model, however, because it may be *unstable*. The weighting coefficients in Eq. (8.100) must all be positive to avoid divergence. Now σ is positive, but $(1 - 2\sigma)$ may not be. Therefore, our explicit viscous flow model has a stability requirement:

$$\sigma = \frac{\nu \Delta t}{\Delta y^2} \leq \frac{1}{2} \tag{8.101}$$

Normally, one would first set up the mesh size Δy in Fig. 8.35, after which Eq. (8.101) would limit the time-step Δt. The solutions for nodal values would then be stable, but not necessarily that accurate. The mesh sizes Δy and Δt could be reduced to increase accuracy, similar to the case of the potential flow laplacian model (8.96).

For example, to solve Prob. P4.85 numerically, one sets up a mesh with plenty of nodes (30 or more Δy within the expected viscous layer); selects Δt according to Eq. (8.101); and sets two boundary conditions[8] for all j: $u_1 = U_0 \sin \omega t$ and $u_N = 0$, where N is the outermost node. For initial conditions, perhaps assume the fluid initially at rest: $u_n^1 = 0$ for $2 \leq n \leq N - 1$. Sweeping the nodes $2 \leq n \leq N - 1$ using Eq. (8.100) (an Excel spreadsheet is excellent for this), one generates numerical values of u_n^j for as long as one desires. After an initial transient, the final "steady" fluid oscillation will approach the classical solution in viscous flow textbooks [15]. Try Prob. P8.115 to demonstrate this.

Steady Two-Dimensional Laminar Flow

The previous example, unsteady one-dimensional flow, had only one viscous term and no convective accelerations. Let us look briefly at incompressible two-dimensional steady flow, which has four of each type of term, plus a nontrivial continuity equation:

$$\text{Continuity:} \qquad \frac{\partial u}{\partial x} + \frac{\partial v}{\partial y} = 0 \tag{8.102a}$$

$$x \text{ momentum:} \qquad u\frac{\partial u}{\partial x} + v\frac{\partial u}{\partial y} = -\frac{1}{\rho}\frac{\partial p}{\partial x} + \nu\left(\frac{\partial^2 u}{\partial x^2} + \frac{\partial^2 u}{\partial y^2}\right) \tag{8.102b}$$

$$y \text{ momentum:} \qquad u\frac{\partial v}{\partial x} + v\frac{\partial v}{\partial y} = -\frac{1}{\rho}\frac{\partial p}{\partial y} + \nu\left(\frac{\partial^2 v}{\partial x^2} + \frac{\partial^2 v}{\partial y^2}\right) \tag{8.102c}$$

These equations, to be solved for (u, v, p) as functions of (x, y), are familiar to us from analytical solutions in Chaps. 4 and 6. However, to a numerical analyst, they are odd, because there is no *pressure equation*—that is, a differential equation for which the

[8] Finite differences are not analytical; one must set U_0 and ω equal to numerical values.

dominant derivatives involve p. This situation has led to several different "pressure adjustment" schemes in the literature [20, 23 to 27], most of which manipulate the continuity equation to insert a pressure correction.

A second difficulty in Eqs. (8.102b and c) is the presence of nonlinear convective accelerations such as $u(\partial u/\partial x)$, which creates asymmetry in viscous flows. Early attempts, which modeled such terms with a central difference, led to numerical instability. The remedy is to relate convection finite differences solely to the *upwind* flow entering the cell, ignoring the downwind cell. For example, the derivative $\partial u/\partial x$ could be modeled, for a given cell, as $(u_{\text{upwind}} - u_{\text{cell}})/\Delta x$. Such improvements have made fully viscous CFD an effective tool, with various commercial user-friendly codes available. For details beyond our scope, see Refs. 20 and 23 to 27.

Mesh generation and gridding have also become quite refined in modern CFD. Figure 8.36 illustrates a CFD solution of two-dimensional flow past an NACA 66(MOD) hydrofoil [28]. The gridding in Fig. 8.36a is of the C type, which wraps around the leading edge and trails off behind the foil, thus capturing the important near-wall and wake details without wasting nodes in front or to the sides. The grid size is 262 by 91.

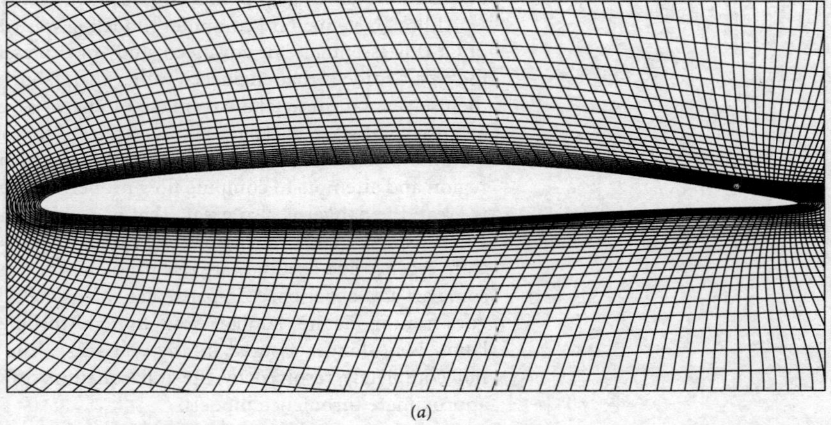

(*a*)

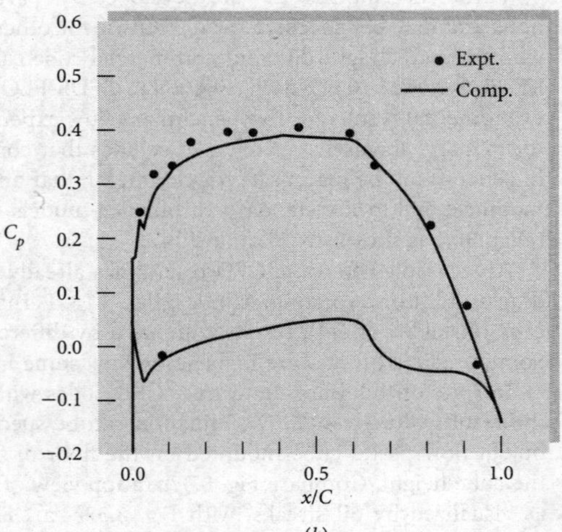

(*b*)

Fig. 8.36 CFD results for water flow past an NACA 66(MOD) hydrofoil *(from Ref. 28, with permission of the American Society of Mechanical Engineers)*: (*a*) C gridding, 262 by 91 nodes; (*b*) surface pressures at $\alpha = 1°$.

The CFD model for this hydrofoil flow is also quite sophisticated: a full Navier-Stokes solver with turbulence modeling [29] and allowance for cavitation bubble formation when surface pressures drop below the local vapor pressure. Figure 8.36*b* compares computed and experimental surface pressure coefficients for an angle of attack of 1°. The dimensionless pressure coefficient is defined as $C_p = (p_{\text{surface}} - p_\infty)/(\rho V_\infty^2/2)$. The agreement is excellent, as indeed it is also for cases where the hydrofoil cavitates [28]. Clearly, when properly implemented for the proper flow cases, CFD can be an extremely effective tool for engineers.

Commercial CFD Codes

The arrival of the third millennium has seen an enormous emphasis on computer applications in nearly every field, fluid mechanics being a prime example. It is now possible, at least for moderately complex geometries and flow patterns, to model on a computer, approximately, the equations of motion of fluid flow, with dedicated CFD textbooks available [5, 20, 23 to 27]. The flow region is broken into a fine grid of elements and nodes, which algebraically simulate the basic partial differential equations of flow. While simple two-dimensional flow simulations have long been reported and can be programmed as student exercises, three-dimensional flows, involving thousands or even millions of grid points, are now solvable with the modern supercomputer.

Although elementary computer modeling was treated briefly here, the general topic of CFD is essentially for advanced study or professional practice. The big change over the past decade is that engineers, rather than laboriously programming CFD problems themselves, can now take advantage of any of several commercial CFD codes. These extensive software packages allow engineers to construct a geometry and boundary conditions to simulate a given viscous flow problem. The software then grids the flow region and attempts to compute flow properties at each grid element. The convenience is great; the danger is also great. That is, computations are not merely automatic, like when using a hand calculator, but rather require care and concern from the user. Convergence and accuracy are real problems for the modeler. Use of the codes requires some art and experience. In particular, when the flow Reynolds number, $\text{Re} = \rho VL/\mu$, goes from moderate (laminar flow) to high (turbulent flow), the accuracy of the simulation is no longer assured in any real sense. The reason is that turbulent flows are not completely resolved by the full equations of motion, and one resorts to using approximate turbulence models.

Turbulence models [29] are developed for particular geometries and flow conditions and may be inaccurate or unrealistic for others. This is discussed by Freitas [30], who compared eight different commercial code calculations (FLOW-3D, FLOTRAN, STAR-CD, N3S, CFD-ACE, FLUENT, CFDS-FLOW3D, and NISA/3D-FLUID) with experimental results for five benchmark flow experiments. Calculations were made by the vendors themselves. Freitas concludes that commercial codes, though promising in general, can be inaccurate for certain laminar and turbulent flow situations. Recent modifications to the standard turbulence models have improved their accuracy and reliability, as shown by Elkhoury [47].

An example of erratic CFD results has already been mentioned here, namely, the drag and lift of a rotating cylinder, Fig. 8.15. Perhaps because the flow itself is physically unstable [41, 44], results computed by different workers are strikingly different: Some predicted forces are high, some low, some increase, some decrease.

In spite of this warning to treat CFD codes with care, one should also realize that the results of a given CFD simulation can be spectacular. Figure 8.37 illustrates turbulent flow past a cube mounted on the floor of a channel whose clearance is twice the cube height. Compare Fig. 8.37*a*, a top view of the experimental surface flow [31] as visualized by oil streaks, with Fig. 8.37*b*, a CFD supercomputer result using the method of large-eddy simulation [32, 33]. The agreement is remarkable. The C-shaped flow pattern in front of the cube is caused by formation of a horseshoe vortex, as seen in a side view of the experiment [31] in Fig. 8.37*c*. Horseshoe vortices commonly result when surface shear flows meet an obstacle. We conclude that CFD has a tremendous potential for flow prediction.

(*a*)

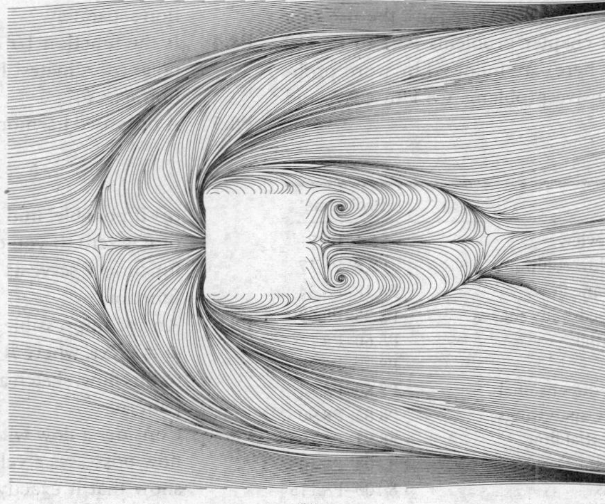

(*b*)

(*c*)

Fig. 8.37 Flow over a surface-mounted cube creates a complex and perhaps unexpected pattern: (*a*) experimental oil-streak visualization of surface flow at Re = 40,000 (based on cube height) *(Courtesy of Robert Martinuzzi with the permission of the American Society of Mechanical Engineers)*; (*b*) computational large-eddy simulation of the surface flow in *(a) (from Ref. 32, courtesy of Kishan Shah, Stanford University)*; and (*c*) a side view of the flow in (*a*) visualized by smoke generation and a laser light sheet *(Courtesy of Robert Martinuzzi with the permission of the American Society of Mechanical Engineers)*.

Summary

This chapter has analyzed a highly idealized but very useful type of flow: inviscid, incompressible, irrotational flow, for which Laplace's equation holds for the velocity potential (8.1) and for the plane stream function (8.7). The mathematics is well developed, and solutions of potential flows can be obtained for practically any body shape.

Some solution techniques outlined here are (1) superposition of elementary line or point solutions in both plane and axisymmetric flow, (2) the analytic functions of a complex variable, and (3) numerical analysis on a computer. Potential theory is especially useful and accurate for thin bodies such as airfoils. The only requirement is that the boundary layer be thin—in other words, that the Reynolds number be large.

For blunt bodies or highly divergent flows, potential theory serves as a first approximation, to be used as input to a boundary layer analysis. The reader should consult the advanced texts [for example, 2 to 4, 11 to 13] for further applications of potential theory. Section 8.9 discussed computational methods for viscous (nonpotential) flows.

Problems

Most of the problems herein are fairly straightforward. More difficult or open-ended assignments are labeled with an asterisk. Problems labeled with a computer icon may require the use of a computer. The standard end-of-chapter problems P8.1 to P8.115 (categorized in the problem list here) are followed by word problems W8.1 to W8.7, comprehensive problems C8.1 to C8.7, and design projects D8.1 to D8.3.

Problem Distribution

Section	Topic	Problems
8.1	Introduction and review	P8.1–P8.7
8.2	Elementary plane flow solutions	P8.8–P8.17
8.3	Superposition of plane flows	P8.18–P8.34
8.4	Plane flow past closed-body shapes	P8.35–P8.59
8.5	The complex potential	P8.60–P8.71
8.6	Images	P8.72–P8.79
8.7	Airfoil theory: two-dimensional	P8.80–P8.84
8.7	Airfoil theory: finite-span wings	P8.85–P8.90
8.8	Axisymmetric potential flow	P8.91–P8.103
8.8	Hydrodynamic mass	P8.104–P8.105
8.9	Numerical methods	P8.106–P8.115

Introduction and review

P8.1 Prove that the streamlines $\psi(r, \theta)$ in polar coordinates from Eqs. (8.10) are orthogonal to the potential lines $\phi(r, \theta)$.

P8.2 The steady plane flow in Fig. P8.2 has the polar velocity components $v_\theta = \Omega r$ and $v_r = 0$. Determine the circulation Γ around the path shown.

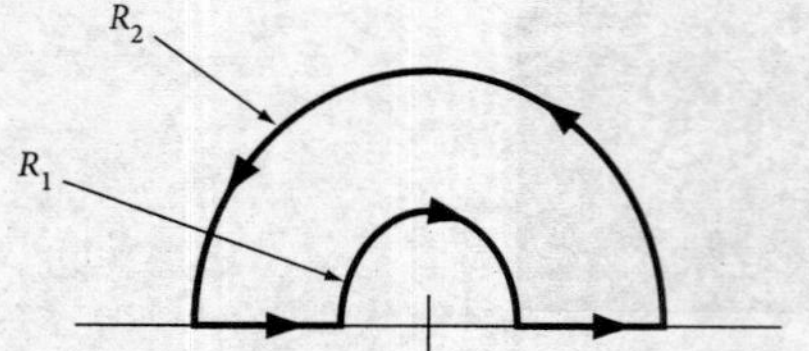

P8.2

P8.3 Using cartesian coordinates, show that each velocity component (u, v, w) of a potential flow satisfies Laplace's equation separately.

P8.4 Is the function $1/r$ a legitimate velocity potential in plane polar coordinates? If so, what is the associated stream function $\psi(r, \theta)$?

P8.5 A proposed harmonic function $F(x, y, z)$ is given by

$$F = 2x^2 + y^3 - 4xz + f(y)$$

(*a*) If possible, find a function $f(y)$ for which the laplacian of F is zero. If you do indeed solve part (*a*), can your final function F serve as (*b*) a velocity potential or (*c*) a stream function?

P8.6 An incompressible plane flow has the velocity potential $\phi = 2Kxy$, where B is a constant. Find the stream function of this flow, sketch a few streamlines, and interpret the flow pattern.

P8.7 Consider a flow with constant density and viscosity. If the flow possesses a velocity potential as defined by Eq. (8.1), show that it exactly satisfies the full Navier-Stokes equations (4.38). If this is so, why for inviscid theory do we back away from the full Navier-Stokes equations?

Elementary plane flow solutions

P8.8 For the velocity distribution $u = -By$, $v = +Bx$, $w = 0$, evaluate the circulation Γ about the rectangular closed curve defined by $(x, y) = (1,1), (3,1), (3,2)$, and $(1,2)$. Interpret your result, especially vis-ã -vis the velocity potential.

P8.9 Consider the two-dimensional flow $u = -Ax$, $v = Ay$, where A is a constant. Evaluate the circulation Γ around the rectangular closed curve defined by $(x, y) = (1, 1), (4, 1), (4, 3)$, and $(1, 3)$. Interpret your result, especially vis-ã -vis the velocity potential.

P8.10 A two-dimensional Rankine half-body, 8 cm thick, is placed in a water tunnel at 20°C. The water pressure far upstream along the body centerline is 105 kPa. What is the nose radius of the half-body? At what tunnel flow velocity will cavitation bubbles begin to form on the surface of the body?

P8.11 A power plant discharges cooling water through the manifold in Fig. P8.11, which is 55 cm in diameter and 8 m high and is perforated with 25,000 holes 1 cm in diameter. Does this manifold simulate a line source? If so, what is the equivalent source strength m?

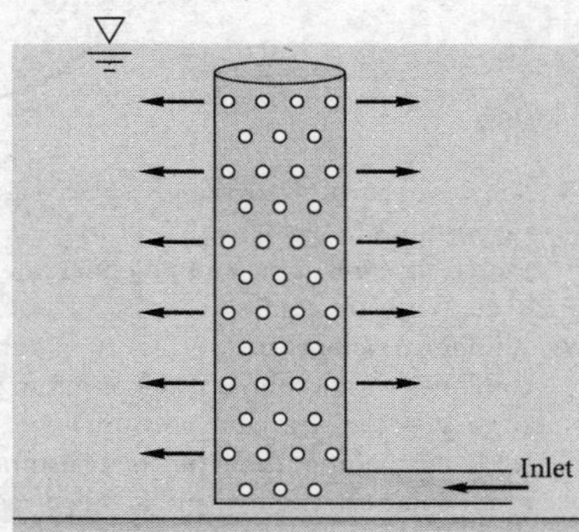

P8.11

P8.12 Consider the flow due to a vortex of strength K at the origin. Evaluate the circulation from Eq. (8.23) about the clockwise path from $(r, \theta) = (a, 0)$ to $(2a, 0)$ to $(2a, 3\pi/2)$ to $(a, 3\pi/2)$ and back to $(a, 0)$. Interpret the result.

P8.13 Starting at the stagnation point in Fig. 8.6, the fluid *acceleration* along the half-body surface rises to a maximum and eventually drops off to zero far downstream. (*a*) Does this maximum occur at the point in Fig. 8.6 where $U_{max} = 1.26U$? (*b*) If not, does the maximum acceleration occur before or after that point? Explain.

P8.14 A tornado may be modeled as the circulating flow shown in Fig. P8.14, with $\upsilon_r = \upsilon_z = 0$ and $\upsilon_\theta(r)$ such that

$$\upsilon_\theta = \begin{cases} \omega r & r \le R \\ \dfrac{\omega R^2}{r} & r > R \end{cases}$$

Determine whether this flow pattern is irrotational in either the inner or outer region. Using the r-momentum equation (D.5) of App. D, determine the pressure distribution $p(r)$ in the tornado, assuming $p = p_\infty$ as $r \to \infty$. Find the location and magnitude of the lowest pressure.

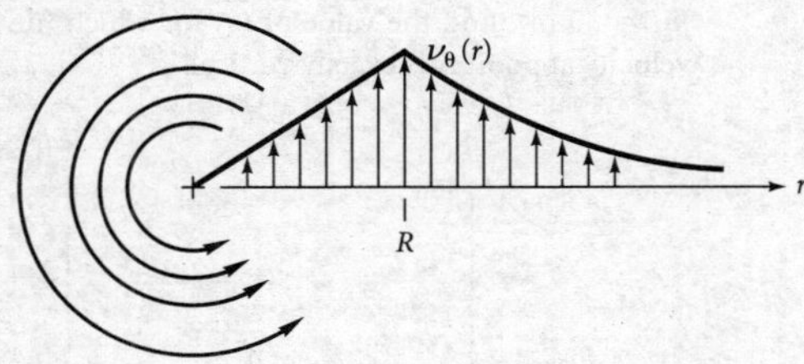

P8.14

P8.15 Hurricane Sandy, which hit the New Jersey coast on Oct. 29, 2012, was extremely broad, with wind velocities of 65 km/h at 650 km from its center. Its maximum velocity was 145 km/h. Using the model of Fig. P8.14, at 20°C with a pressure of 100 kPa far from the center, estimate (*a*) the radius R of maximum velocity, in mi; and (*b*) the pressure at $r = R$.

P8.16 Air flows at 1.2 m/s along a flat surface when it encounters a jet of air issuing from the horizontal wall at point A, as in Fig. P8.16. The jet volume flow is 0.4 m³/s per unit depth into the paper. If the jet is approximated as an inviscid line source, (*a*) locate the stagnation point S on the wall. (*b*) How far vertically will the jet flow extend into the stream?

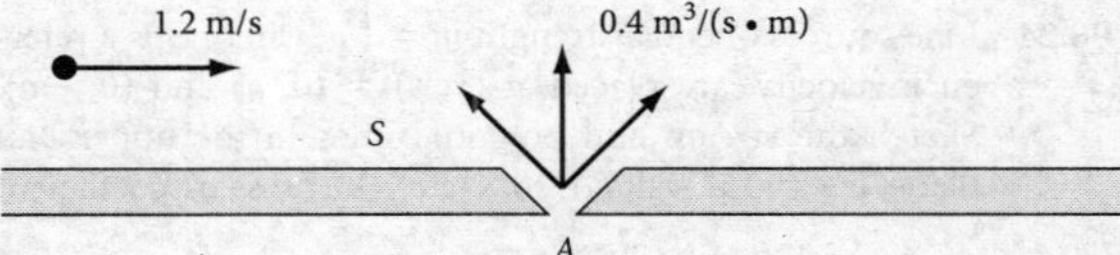

P8.16

P8.17 Find the position (x, y) on the upper surface of the half-body in Fig. 8.9*a* for which the local velocity equals the uniform stream velocity. What should be the pressure at this point?

Superposition of plane flows

P8.18 Plot the streamlines and potential lines of the flow due to a line source of strength m at $(a, 0)$ plus a source $3m$ at $(-a, 0)$. What is the flow pattern viewed from afar?

P8.19 Plot the streamlines and potential lines of the flow due to a line source of strength $3m$ at $(a, 0)$ plus a sink $-m$ at $(-a, 0)$. What is the pattern viewed from afar?

P8.20 Plot the streamlines of the flow due to a line vortex $+K$ at $(0, +a)$ and a vortex $-K$ at $(0, -a)$. What is the pattern viewed from afar?

P8.21 At point A in Fig. P8.21 is a clockwise line vortex of strength $K = 12$ m²/s. At point B is a line source of strength $m = 25$ m²/s. Determine the resultant velocity induced by these two at point C.

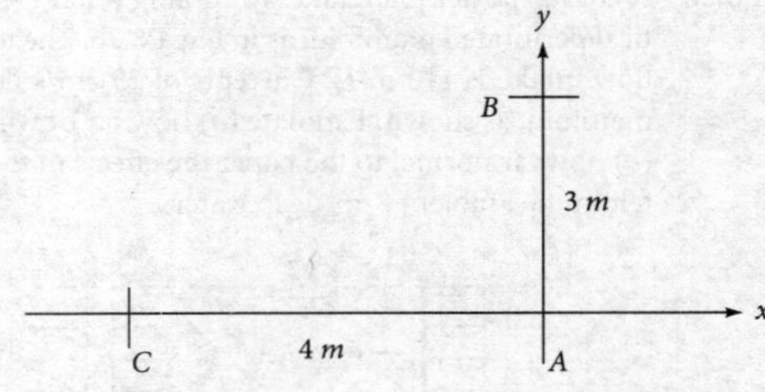

P8.21

P8.22 Consider inviscid stagnation flow, $\psi = Kxy$ (see Fig. 8.19*b*), superimposed with a source at the origin of strength m. Plot the resulting streamlines in the upper half-plane, using the length scale $a = (m/K)^{1/2}$. Give a physical interpretation of the flow pattern.

P8.23 Sources of strength $m = 10\ m^2/s$ are placed at points A and B in Fig. P8.23. At what height h should source B be placed so that the net induced horizontal velocity component at the origin is 8 m/s to the left?

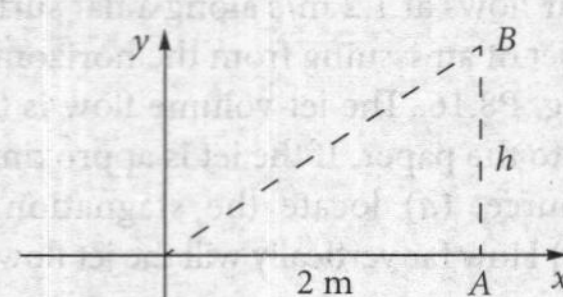

P8.23

P8.24 Line sources of equal strength $m = Ua$, where U is a reference velocity, are placed at $(x, y) = (0, a)$ and $(0, -a)$. Sketch the stream and potential lines in the upper half plane. Is $y = 0$ a "wall"? If so, sketch the pressure coefficient

$$C_p = \frac{p - p_0}{\frac{1}{2}\rho U^2}$$

along the wall, where p_0 is the pressure at (0, 0). Find the minimum pressure point and indicate where flow separation might occur in the boundary layer.

P8.25 Let the vortex/sink flow of Eq. (8.16) simulate a tornado as in Fig. P8.25. Suppose that the circulation about the tornado is $\Gamma = 8{,}500\ m^2/s$ and that the pressure at $r = 40$ m is 2,200 Pa less than the far-field pressure. Assuming inviscid flow at sea-level density, estimate (*a*) the appropriate sink strength $-m$, (*b*) the pressure at $r = 15$ m, and (*c*) the angle β at which the streamlines cross the circle at $r = 40$ m (see Fig. P8.25).

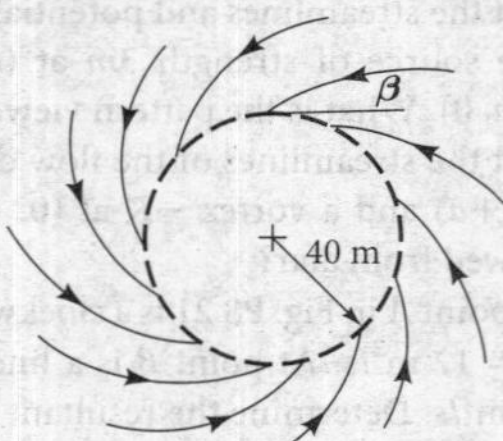

P8.25

P8.26 A coastal power plant takes in cooling water through a vertical perforated manifold, as in Fig. P8.26. The total volume flow intake is 110 m³/s. Currents of 25 cm/s flow past the manifold, as shown. Estimate (*a*) how far downstream and (*b*) how far normal to the paper the effects of the intake are felt in the ambient 8-m-deep waters.

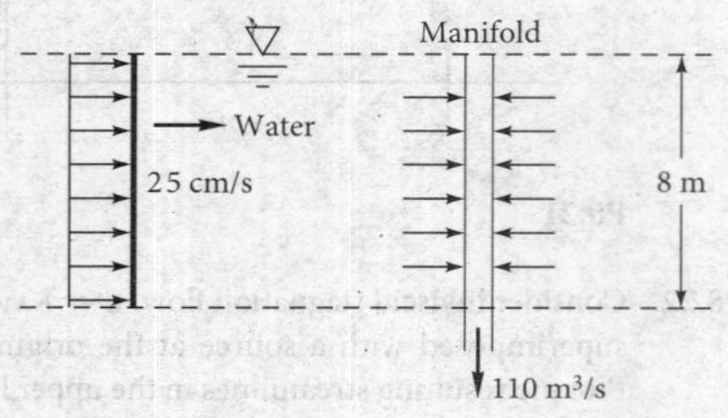

P8.26

P8.27 Water at 20°C flows past a half-body as shown in Fig. P8.27. Measured pressures at points A and B are 160 kPa and 90 kPa, respectively, with uncertainties of 3 kPa each. Estimate the stream velocity and its uncertainty.

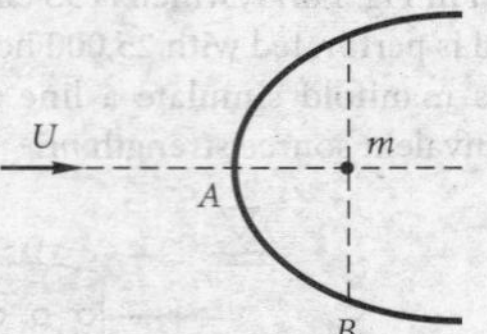

P8.27

P8.28 Sources of equal strength m are placed at the four symmetric positions $(x, y) = (a, a), (-a, a), (-a, -a)$, and $(a, -a)$. Sketch the streamline and potential line patterns. Do any plane "walls" appear?

P8.29 A uniform water stream, $U_\infty = 20$ m/s and $\rho = 998\ kg/m^3$, combines with a source at the origin to form a half-body. At $(x, y) = (0, 1.2\ m)$, the pressure is 12.5 kPa less than p_∞. (*a*) Is this point outside the body? Estimate (*b*) the appropriate source strength m and (*c*) the pressure at the nose of the body.

P8.30 A tornado is simulated by a line sink $m = -1{,}000\ m^2/s$ plus a line vortex $K = 1{,}600\ m^2/s$. Find the angle between any streamline and a radial line, and show that it is independent of both r and θ. If this tornado forms in sea-level standard air, at what radius will the local pressure be equivalent to 740 mmHg?

P8.31 A Rankine half-body is formed as shown in Fig. P8.31. For the stream velocity and body dimension shown, compute (*a*) the source strength m in m²/s, (*b*) the distance a, (*c*) the distance h, and (*d*) the total velocity at point A.

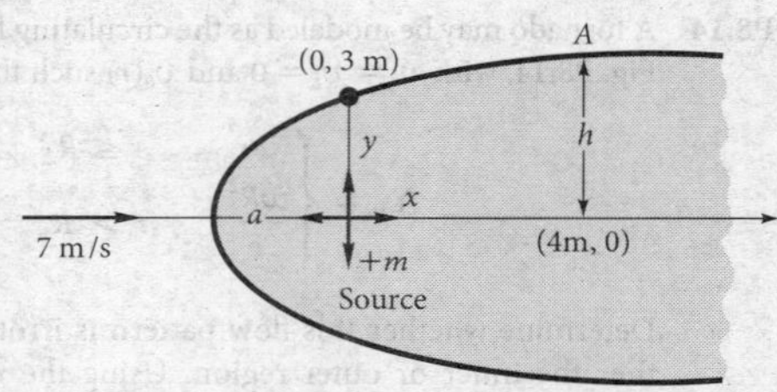

P8.31

P8.32 Line sources m_1 and m_2 are near point A, as in Fig. P8.32. If $m_1 = 30\ m^2$, find the value of m_2 for which the resultant velocity at point A is exactly vertical.

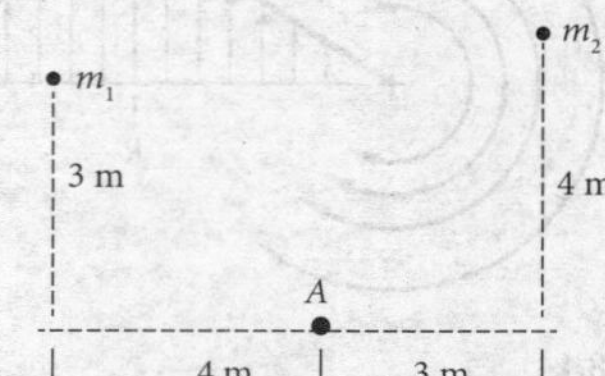

P8.32

P8.33 Sketch the streamlines, especially the body shape, due to equal line sources $+m$ at $(0, +a)$ and $(0, -a)$ plus a uniform stream $U_\infty = ma$.

P8.34 Consider three equally spaced sources of strength m placed at $(x, y) = (+a, 0)$, $(0, 0)$, and $(-a, 0)$. Sketch the resulting streamlines, noting the position of any stagnation points. What would the pattern look like from afar?

Plane flow past closed-body shapes

P8.35 A uniform stream, $U_\infty = 4$ m/s, approaches a Rankine oval as in Fig. 8.13, with $a = 50$ cm. Find the strength m of the source–sink pair, in m^2/s, which will cause the total length of the oval to be 250 cm. What is the maximum width of this oval?

P8.36 When a line source–sink pair with $m = 2$ m^2/s is combined with a uniform stream, it forms a Rankine oval whose minimum dimension is 40 cm. If $a = 15$ cm, what are the stream velocity and the velocity at the shoulder? What is the maximum dimension?

P8.37 A Rankine oval 2 m long and 1 m high is immersed in a stream $U_\infty = 10$ m/s, as in Fig. P8.37. Estimate (*a*) the velocity at point *A* and (*b*) the location of point *B* where a particle approaching the stagnation point achieves its maximum deceleration.

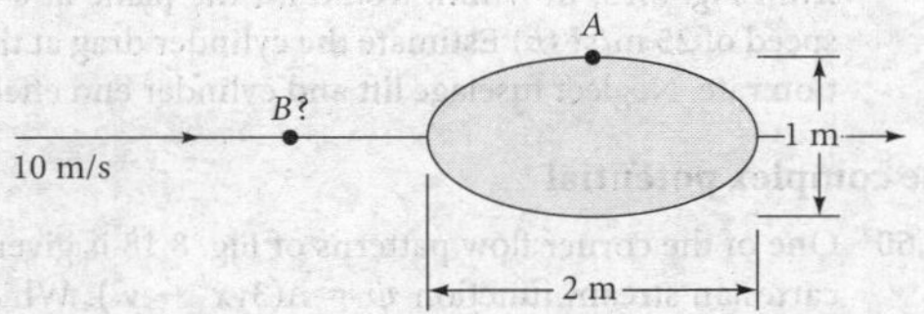

P8.37

P8.38 Consider potential flow of a uniform stream in the *x* direction plus two equal sources, one at $(x, y) = (0, +a)$ and the other at $(x, y) = (0, -a)$. Sketch your ideas of the body contours that would arise if the sources were (*a*) very weak and (*b*) very strong.

P8.39 A large Rankine oval, with $a = 1$ m and $h = 1$ m, is immersed in 20°C water flowing at 10 m/s. The upstream pressure on the oval centerline is 200 kPa. Calculate (*a*) the value of *m*; and (*b*) the pressure on the top of the oval (analogous to point A in Fig. P8.37).

P8.40 Modify the Rankine oval in Fig. P8.37 so that the stream velocity and body length are the same but the thickness is unknown (not 1 m). The fluid is water at 30°C and the pressure far upstream along the body centerline is 108 kPa. Find the body thickness for which cavitation will occur at point *A*.

P8.41 A Kelvin oval is formed by a line–vortex pair with $K = 9$ m^2/s, $a = 1$ m, and $U = 10$ m/s. What are the height, width, and shoulder velocity of this oval?

P8.42 The vertical keel of a sailboat approximates a Rankine oval 125 cm long and 30 cm thick. The boat sails in seawater in standard atmosphere at 7 m/s, parallel to the keel. At a section 2 m below the surface, estimate the lowest pressure on the surface of the keel.

P8.43 Water at 20°C flows past a 1-m-diameter circular cylinder. The upstream centerline pressure is 128,500 Pa. If the lowest pressure on the cylinder surface is exactly the vapor pressure, estimate, by potential theory, the stream velocity.

P8.44 Suppose that circulation is added to the cylinder flow of Prob. P8.43 sufficient to place the stagnation points at θ equal to 35° and 145°. What is the required vortex strength *K* in m^2/s? Compute the resulting pressure and surface velocity at (*a*) the stagnation points and (*b*) the upper and lower shoulders. What will the lift per meter of cylinder width be?

P8.45 If circulation *K* is added to the cylinder flow in Prob. P8.43, (*a*) for what value of *K* will the flow begin to cavitate at the surface? (*b*) Where on the surface will cavitation begin? (*c*) For this condition, where will the stagnation points lie?

P8.46 A cylinder is formed by bolting two semicylindrical channels together on the inside, as shown in Fig. P8.46. There are 10 bolts per meter of width on each side, and the inside pressure is 50 kPa (gage). Using potential theory for the outside pressure, compute the tension force in each bolt if the fluid outside is sea-level air.

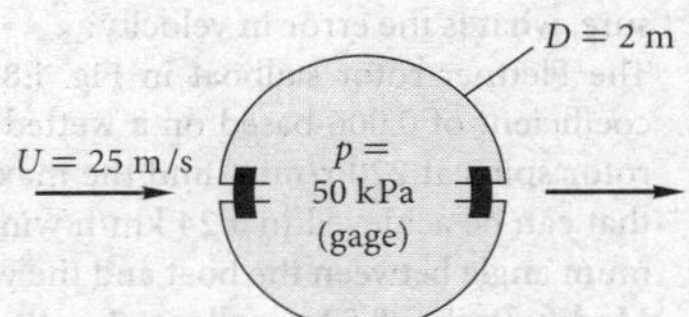

P8.46

P8.47 A circular cylinder is fitted with two surface-mounted pressure sensors, to measure p_a at $\theta = 180°$ and p_b at $\theta = 105°$. The intention is to use the cylinder as a stream velocimeter. Using inviscid theory, derive a formula for estimating U_∞ in terms of p_a, p_b, ρ, and the cylinder radius *a*.

***P8.48** Wind at U_∞ and p_∞ flows past a Quonset hut which is a half-cylinder of radius *a* and length *L* (Fig. P8.48). The internal pressure is p_i. Using inviscid theory, derive an expression for the upward force on the hut due to the difference between p_i and p_s.

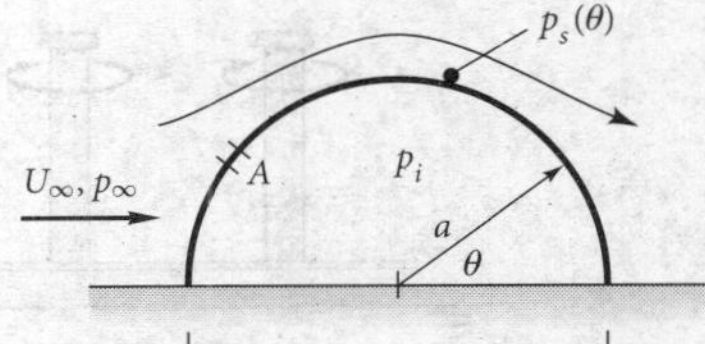

P8.48

P8.49 In strong winds the force in Prob. P8.48 can be quite large. Suppose that a hole is introduced in the hut roof at point *A* to make p_i equal to the surface pressure there. At what angle θ should hole *A* be placed to make the net wind force zero?

P8.50 It is desired to simulate flow past a two-dimensional ridge or bump by using a streamline that passes above the flow over a cylinder, as in Fig. P8.50. The bump is to be $a/2$ high, where a is the cylinder radius. What is the elevation h of this streamline? What is U_{max} on the bump compared with stream velocity U?

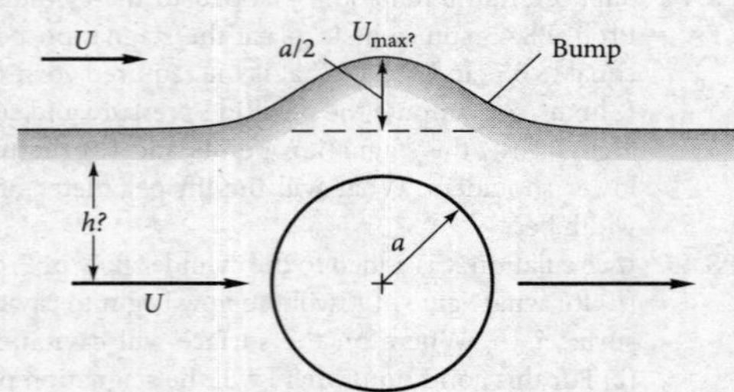

P8.50

P8.51 A hole is placed in the front of a cylinder to measure the stream velocity of sea-level fresh water. The measured pressure at the hole is 136 kPa. If the hole is misaligned by 12° from the stream, and misinterpreted as stagnation pressure, what is the error in velocity?

P8.52 The Flettner rotor sailboat in Fig. E8.3 has a water drag coefficient of 0.006 based on a wetted area of 4 m^2. If the rotor spins at 220 r/min, find the maximum boat velocity that can be achieved in a 24 km/h wind. What is the optimum angle between the boat and the wind?

P8.53 Modify Prob. P8.52 as follows. For the same sailboat data, find the wind velocity, in km/h, that will drive the boat at an optimum speed of 4 m/s parallel to its keel.

P8.54 The original Flettner rotor ship was approximately 30 m long, displaced 800 tons, and had a wetted area of 325 m^2. As sketched in Fig. P8.54, it had two rotors 15 m high and 3 m in diameter rotating at 750 r/min, which is far outside the range of Fig. 8.15. The measured lift and drag coefficients for each rotor were about 10 and 4, respectively. If the ship is moored and subjected to a crosswind of 8 m/s, as in Fig. P8.54, what will the wind force parallel and normal to the ship centerline be? Estimate the power required to drive the rotors.

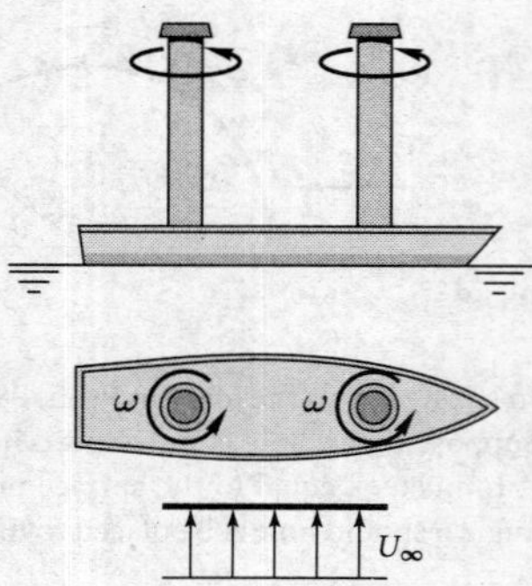

P8.54

P8.55 Assume that the Flettner rotor ship of Fig. P8.54 has a water resistance coefficient of 0.005. How fast will the ship sail in seawater at 20°C in a 6 m/s wind if the keel aligns itself with the resultant force on the rotors? [*Hint:* This is a problem in relative velocities.]

P8.56 A proposed free-stream velocimeter would use a cylinder with pressure taps at $\theta = 180°$ and at 150°. The pressure difference would be a measure of stream velocity U_∞. However, the cylinder must be aligned so that one tap exactly faces the free stream. Let the misalignment angle be δ; that is, the two taps are at $(180° + \delta)$ and $(150° + \delta)$. Make a plot of the percentage error in velocity measurement in the range $-20° < \delta < +20°$ and comment on the idea.

P8.57 In principle, it is possible to use rotating cylinders as aircraft wings. Consider a cylinder 30 cm in diameter, rotating at 2,400 r/min. It is to lift a 55-kN airplane cruising at 100 m/s. What should the cylinder length be? How much power is required to maintain this speed? Neglect end effects on the rotating wing.

P8.58 Plot the streamlines due to the combined flow of a line sink $-m$ at the origin plus line sources $+m$ at $(a, 0)$ and $(4a, 0)$. [*Hint:* A cylinder of radius $2a$ will appear.]

P8.59 The Transition® car-plane in Fig. 7.30 has a gross weight of 6,360 N. Suppose we replace the wing with a 0.3 m diameter rotating cylinder 6 m long. (*a*) What rotation rate from Fig. 8.15, in r/min, would lift the plane at a take-off speed of 25 m/s? (*b*) Estimate the cylinder drag at this rotation rate. Neglect fuselage lift and cylinder end effects.

The complex potential

P8.60 One of the corner flow patterns of Fig. 8.18 is given by the cartesian stream function $\psi = A(3yx^2 - y^3)$. Which one? Can the correspondence be proved from Eq. (8.53)?

P8.61 Plot the streamlines of Eq. (8.53) in the upper right quadrant for $n = 4$. How does the velocity increase with x outward along the x axis from the origin? For what corner angle and value of n would this increase be linear in x? For what corner angle and n would the increase be as x^5?

P8.62 Combine stagnation flow, Fig. 8.19*b*, with a source at the origin:

$$f(z) = Az^2 + m \ln z$$

Plot the streamlines for $m = AL^2$, where L is a length scale. Interpret.

P8.63 The superposition in Prob. P8.62 leads to stagnation flow near a curved bump, in contrast to the flat wall of Fig. 8.19*b*. Determine the maximum height H of the bump as a function of the constants A and m.

P8.64 Consider the polar-coordinate stream function $\psi = Br^{1.2} \sin(1.2\,\theta)$, with B equal, for convenience, to 0.3 m$^{0.8}$/s. (*a*) Plot the streamline $\psi = 0$ in the upper half plane. (*b*) Plot the streamline $\psi = 1.0$ and interpret the flow pattern. (*c*) Find the locus of points above $\psi = 0$ for which the resultant velocity = 0.3 m/s.

P8.65 Potential flow past a wedge of half-angle θ leads to an important application of laminar boundary layer theory called the *Falkner-Skan flows* [15, pp. 239–245]. Let x

denote distance along the wedge wall, as in Fig. P8.65, and let $\theta = 10°$. Use Eq. (8.53) to find the variation of surface velocity $U(x)$ along the wall. Is the pressure gradient adverse or favorable?

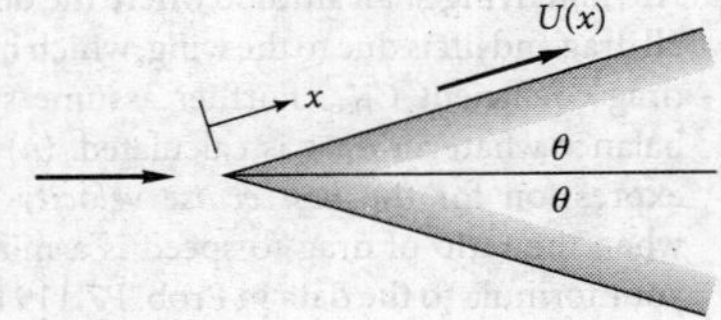

P8.65

***P8.66** The inviscid velocity along the wedge in Prob. P8.65 has the analytic form $U(x) = Cx^m$, where $m = n - 1$ and n is the exponent in Eq. (8.53). Show that, for any C and n, computation of the boundary layer by Thwaites's method, Eqs. (7.53) and (7.54), leads to a unique value of the Thwaites parameter λ. Thus wedge flows are called *similar* [15, p. 241].

P8.67 Investigate the complex potential function $f(z) = U_\infty(z + a^2/z)$ and interpret the flow pattern.

P8.68 Investigate the complex potential function $f(z) = U_\infty z + m \ln [(z + a)/(z - a)]$ and interpret the flow pattern.

P8.69 Investigate the complex potential $f(z) = A \cosh [\pi(z/a)]$, and plot the streamlines inside the region shown in Fig. P8.69. What hyphenated word (originally French) might describe such a flow pattern?

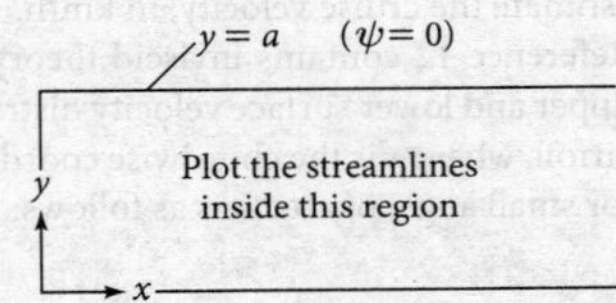

P8.69

P8.70 Show that the complex potential $f = U_\infty\{z + \frac{1}{4}a \coth [\pi(z/a)]\}$ represents flow past an oval shape placed midway between two parallel walls $y = \pm\frac{1}{2}a$. What is a practical application?

P8.71 Figure P8.71 shows the streamlines and potential lines of flow over a thin-plate weir as computed by the complex potential method. Compare qualitatively with Fig. 10.16*a*. State the proper boundary conditions at all boundaries. The velocity potential has equally spaced values. Why do the flow-net "squares" become smaller in the overflow jet?

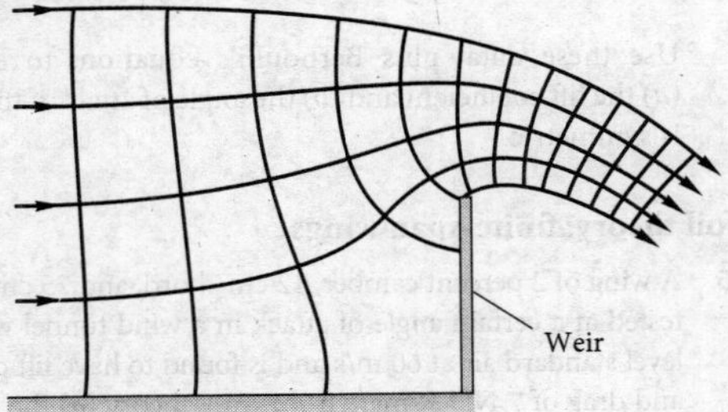

P8.71

Images

P8.72 Use the method of images to construct the flow pattern for a source $+m$ near two walls, as shown in Fig. P8.72. Sketch the velocity distribution along the lower wall ($y = 0$). Is there any danger of flow separation along this wall?

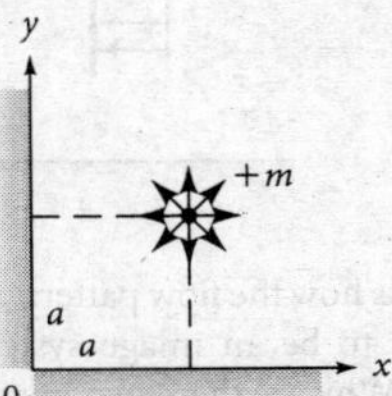

P8.72

P8.73 Set up an image system to compute the flow of a source at unequal distances from two walls, as in Fig. P8.73. Find the point of maximum velocity on the y axis.

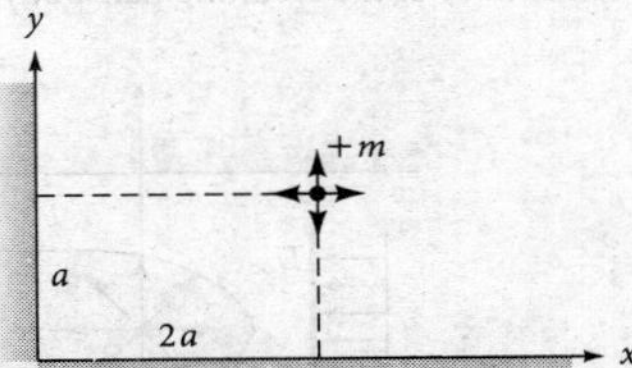

P8.73

P8.74 A positive line vortex K is trapped in a corner, as in Fig. P8.74. Compute the total induced velocity vector at point B, $(x, y) = (2a, a)$, and compare with the induced velocity when no walls are present.

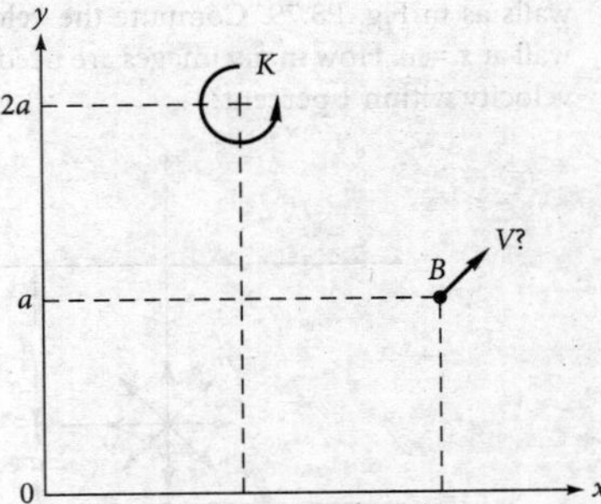

P8.74

P8.75 Using the four-source image pattern needed to construct the flow near a corner in Fig. P8.72, find the value of the source strength m that will induce a wall velocity of 4.0 m/s at the point $(x, y) = (a, 0)$ just below the source shown, if $a = 50$ cm.

P8.76 Use the method of images to approximate the flow pattern past a cylinder a distance $4a$ from a single wall, as in Fig. P8.76. To illustrate the effect of the wall, compute the velocities at corresponding points A, B, C, and D, comparing with a cylinder flow in an infinite expanse of fluid.

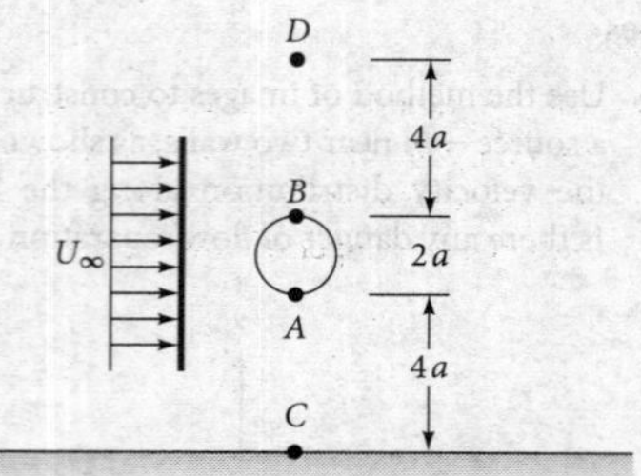

P8.76

P8.77 Discuss how the flow pattern of Prob. P8.58 might be interpreted to be an image system construction for circular walls. Why are there two images instead of one?

***P8.78** Indicate the system of images needed to construct the flow of a uniform stream past a Rankine half-body constrained between two parallel walls, as in Fig. P8.78. For the particular dimensions shown in this figure, estimate the position of the nose of the resulting half-body.

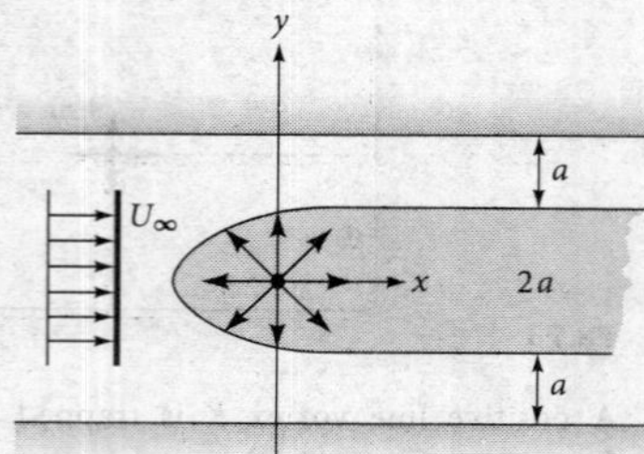

P8.78

P8.79 Explain the system of images needed to simulate the flow of a line source placed unsymmetrically between two parallel walls as in Fig. P8.79. Compute the velocity on the lower wall at $x = a$. How many images are needed to estimate this velocity within 1 percent?

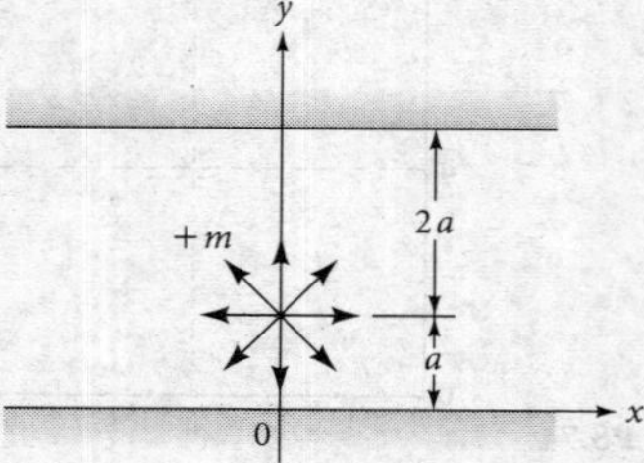

P8.79

Airfoil theory: two-dimensional

***P8.80** The beautiful expression for lift of a two-dimensional airfoil, Eq. (8.59), arose from applying the Joukowski transformation, $\zeta = z + a^2/z$, where $z = x + iy$ and $\zeta = \eta + i\beta$. The constant a is a length scale. The theory transforms a certain circle in the z plane into an airfoil in the ζ plane. Taking $a = 1$ unit for convenience, show that (*a*) a circle with center at the origin and radius > 1 will become an ellipse in the ζ plane and (*b*) a circle with center at $x = -\varepsilon \ll 1$, $y = 0$, and radius $(1 + \varepsilon)$ will become an airfoil shape in the ζ plane. [*Hint:* The Excel spreadsheet is excellent for solving this problem.]

***P8.81** Given an airplane of weight W, wing area A, aspect ratio AR, and flying at an altitude where the density is ρ. Assume all drag and lift is due to the wing, which has an infinite-span drag coefficient $C_{D\infty}$. Further assume sufficient thrust to balance whatever drag is calculated. (*a*) Find an algebraic expression for the *best cruise velocity* V_b, which occurs when the ratio of drag to speed is a minimum. (*b*) Apply your formula to the data in Prob. P7.119 for which a laborious graphing procedure gave an answer $V_b \approx 180$ m/s.

P8.82 The ultralight plane *Gossamer Condor* in 1977 was the first to complete the Kremer Prize figure-eight course under human power. Its wingspan was 29 m, with $C_{av} = 2.3$ m and a total mass of 95 kg. The drag coefficient was approximately 0.05. The pilot was able to deliver $\frac{1}{4}$ hp to propel the plane. Assuming two-dimensional flow at sea level, estimate (*a*) the cruise speed attained, (*b*) the lift coefficient, and (*c*) the horsepower required to achieve a speed of 8 m/s.

P8.83 The world's largest airplane, the Airbus A380, has a maximum weight of 5,338,000 N, wing area of 845 m^2, wingspan of 80 m, and $C_{Do} = 0.026$. When cruising at maximum weight at 10,700 m, the four engines each provide 311,000 N of thrust. Assuming all lift and drag are due to the wing, estimate the cruise velocity, in km/h.

P8.84 Reference 12 contains inviscid theory calculations for the upper and lower surface velocity distributions $V(x)$ over an airfoil, where x is the chordwise coordinate. A typical result for small angle of attack is as follows:

x/c	V/U_∞(upper)	V/U_∞(lower)
0.0	0.0	0.0
0.025	0.97	0.82
0.05	1.23	0.98
0.1	1.28	1.05
0.2	1.29	1.13
0.3	1.29	1.16
0.4	1.24	1.16
0.6	1.14	1.08
0.8	0.99	0.95
1.0	0.82	0.82

Use these data, plus Bernoulli's equation, to estimate (*a*) the lift coefficient and (*b*) the angle of attack if the airfoil is symmetric.

Airfoil theory: finite-span wings

P8.85 A wing of 2 percent camber, 12 cm chord, and 75 cm span is tested at a certain angle of attack in a wind tunnel with sea-level standard air at 60 m/s and is found to have lift of 135 N and drag of 7 N. Estimate from wing theory (*a*) the angle of attack, (*b*) the minimum drag of the wing and the angle of attack at which it occurs, and (*c*) the maximum lift-to-drag ratio.

P8.86 An airplane has a mass of 20,000 kg and flies at 175 m/s at 5,000-m standard altitude. Its rectangular wing has a 3-m chord and a symmetric airfoil at 2.5° angle of attack. Estimate (*a*) the wing span, (*b*) the aspect ratio, and (*c*) the induced drag.

P8.87 A freshwater boat of mass 400 kg is supported by a rectangular hydrofoil of aspect ratio 8, 2 percent camber, and 12 percent thickness. If the boat travels at 7 m/s and $\alpha = 2.5°$, estimate (*a*) the chord length, (*b*) the power required if $C_{D_\infty} = 0.01$, and (*c*) the top speed if the boat is refitted with an engine that delivers 20 hp to the water.

P8.88 The Boeing 787-8 Dreamliner has a maximum weight of 2,235,000 N, a wingspan of 60 m, a wing area of 325 m^2, and cruises at 250 m/s at 10,700 m altitude. When cruising, its overall drag coefficient is about 0.027. Estimate (*a*) the aspect ratio, (*b*) the lift coefficient, (*c*) the cruise Mach number, and (*d*) the engine thrust needed when cruising.

P8.89 The Beechcraft T-34C aircraft has a gross weight of 24,500 N and a wing area of 5.6 m^2 and flies at 144 m/s at 3,0000 m standard altitude. It is driven by a propeller that delivers 300 hp to the air. Assume for this problem that its airfoil is the NACA 2412 section described in Figs. 8.23 and 8.24, and neglect all drag except the wing. What is the appropriate aspect ratio for the wing?

P8.90 NASA is developing a swing-wing airplane called the Bird of Prey [37]. As shown in Fig. P8.90, the wings pivot like a pocketknife blade: forward (*a*), straight (*b*), or backward (*c*). Discuss a possible advantage for each of these wing positions. If you can't think of one, read the article [37] and report to the class.

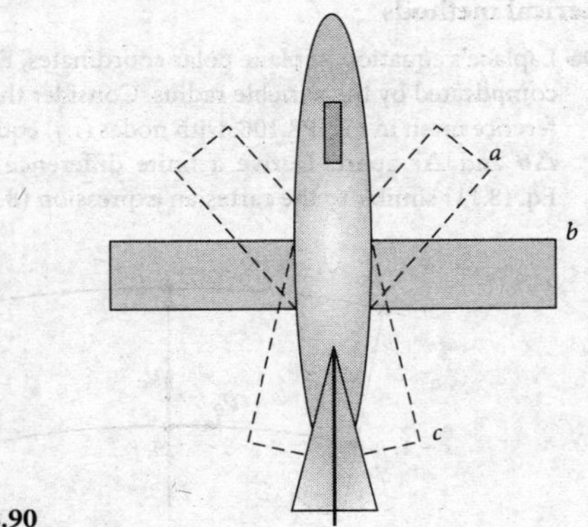

P8.90

Axisymmetric potential flow

P8.91 If $\phi(r, \theta)$ in axisymmetric flow is defined by Eq. (8.72) and the coordinates are given in Fig. 8.28, determine what partial differential equation is satisfied by ϕ.

P8.92 A point source with volume flow $Q = 30$ m^3/s is immersed in a uniform stream of speed 4 m/s. A Rankine half-body of revolution results. Compute (*a*) the distance from source to the stagnation point and (*b*) the two points (r, θ) on the body surface where the local velocity equals 4.5 m/s.

P8.93 The Rankine half-body of revolution (Fig. 8.30) could simulate the shape of a pitot-static tube (Fig. 6.30). According to inviscid theory, how far downstream from the nose should the static pressure holes be placed so that the local velocity is within ±0.5 percent of U_∞? Compare your answer with the recommendation $x \approx 8D$ in Fig. 6.30.

P8.94 Determine whether the Stokes streamlines from Eq. (8.73) are everywhere orthogonal to the Stokes potential lines from Eq. (8.74), as is the case for Cartesian and plane polar coordinates.

P8.95 Show that the axisymmetric potential flow formed by superposition of a point source $+m$ at $(x, y) = (-a, 0)$, a point sink $-m$ at $(+a, 0)$, and a stream U_∞ in the x direction forms a Rankine body of revolution as in Fig. P8.95. Find analytic expressions for determining the length $2L$ and maximum diameter $2R$ of the body in terms of m, U_∞, and a.

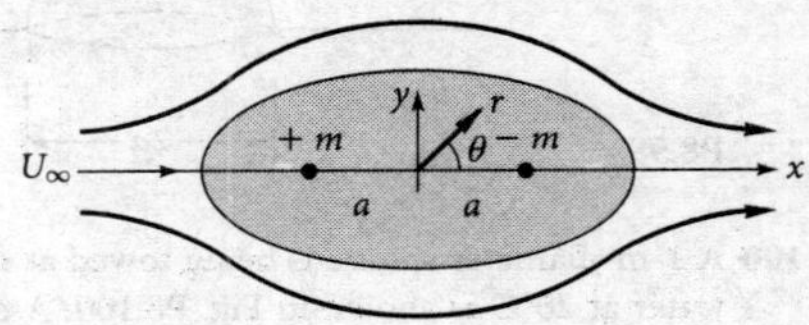

P8.95

P8.96 Consider inviscid flow along the streamline approaching the front stagnation point of a sphere, as in Fig. 8.31. Find (*a*) the maximum fluid deceleration along this streamline and (*b*) its position.

P8.97 The Rankine body of revolution in Fig. P8.97 is 60 cm long and 30 cm in diameter. When it is immersed in the low-pressure water tunnel as shown, cavitation may appear at point *A*. Compute the stream velocity *U*, neglecting surface wave formation, for which cavitation occurs.

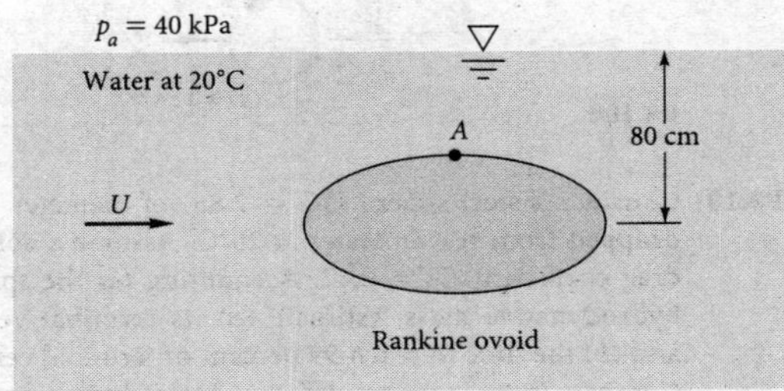

P8.97

P8.98 We have studied the point source (sink) and the line source (sink) of infinite depth into the paper. Does it make any sense to define a finite-length line sink (source) as in Fig. P8.98? If so, how would you establish the mathematical properties of such a finite line sink? When combined with a uniform stream and a point source of equivalent strength as in Fig. P8.98, should a closed-body shape be formed? Make a guess and sketch some of these possible shapes for various values of the dimensionless parameter $m/(U_\infty L^2)$.

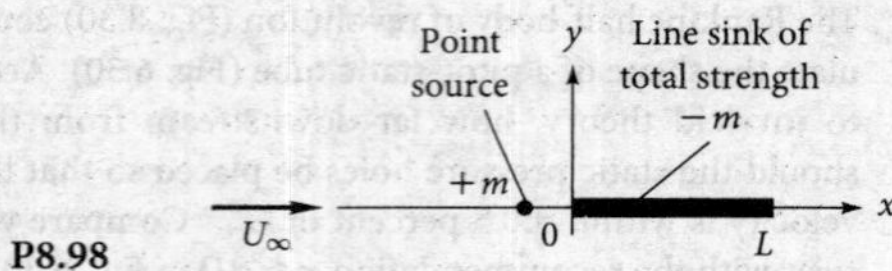

P8.98

*P8.99 Consider air flowing past a hemisphere resting on a flat surface, as in Fig. P8.99. If the internal pressure is p_i, find an expression for the pressure force on the hemisphere. By analogy with Prob. P8.49, at what point A on the hemisphere should a hole be cut so that the pressure force will be zero according to inviscid theory?

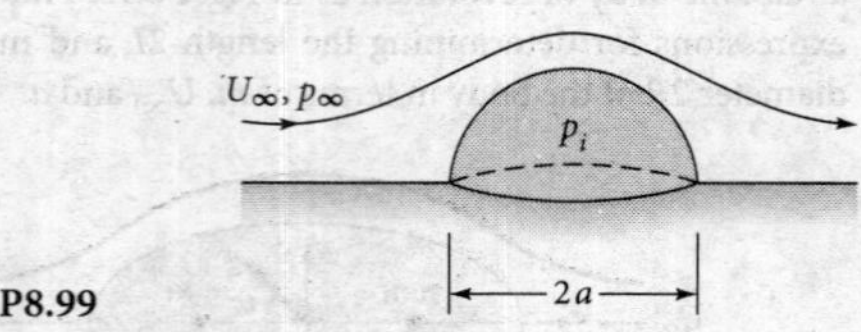

P8.99

P8.100 A 1-m-diameter sphere is being towed at speed V in fresh water at 20°C as shown in Fig. P8.100. Assuming inviscid theory with an undistorted free surface, estimate the speed V in m/s at which cavitation will first appear on the sphere surface. Where will cavitation appear? For this condition, what will be the pressure at point A on the sphere, which is 45° up from the direction of travel?

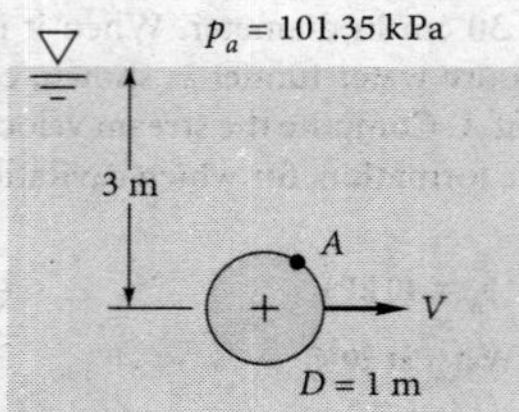

P8.100

P8.101 Consider a steel sphere (SG = 7.85) of diameter 2 cm, dropped from rest in water at 20°C. Assume a constant drag coefficient $C_D = 0.47$. Accounting for the sphere's hydrodynamic mass, estimate (a) its terminal velocity and (b) the time to reach 99 percent of terminal velocity. Compare these to the results when hydrodynamic mass is neglected, $V_{terminal} \approx 1.95$ m/s and $t_{99\%} \approx 0.605$ s, and discuss.

P8.102 A golf ball weighs 0.454 N and has a diameter of 38 mm. A professional golfer strikes the ball at an initial velocity of 75 m/s, an upward angle of 20°, and a backspin (front of the ball rotating upward). Assume that the lift coefficient on the ball (based on frontal area) follows Fig. P7.108. If the ground is level and drag is neglected, make a simple analysis to predict the impact point (a) without spin and (b) with backspin of 7,500 r/min.

P8.103 Consider inviscid flow past a sphere, as in Fig. 8.31. Find (a) the point on the front surface where the fluid acceleration a_{max} is maximum and (b) the magnitude of a_{max}. (c) If the stream velocity is 1 m/s, find the sphere diameter for which a_{max} is 10 times the acceleration of gravity. Comment.

Hydrodynamic mass

P8.104 Consider a cylinder of radius a moving at speed U_∞ through a still fluid, as in Fig. P8.104. Plot the streamlines relative to the cylinder by modifying Eq. (8.32) to give the relative flow with $K = 0$. Integrate to find the total relative kinetic energy, and verify the hydrodynamic mass of a cylinder from Eq. (8.91).

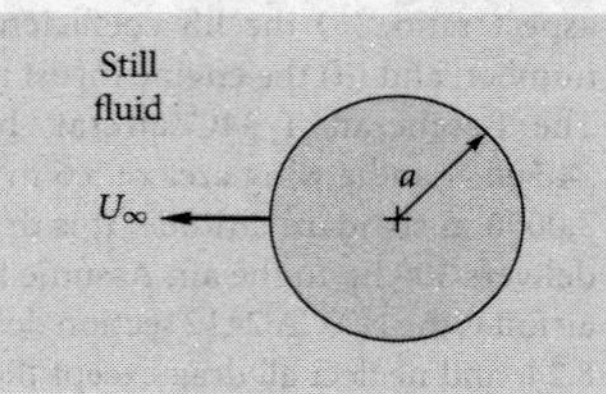

P8.104

P8.105 A 22-cm-diameter solid aluminum sphere (SG = 2.7) is accelerating at 12 m/s² in water at 20°C. (a) According to potential theory, what is the hydrodynamic mass of the sphere? (b) Estimate the force being applied to the sphere at this instant.

Numerical methods

P8.106 Laplace's equation in plane polar coordinates, Eq. (8.11), is complicated by the variable radius. Consider the finite difference mesh in Fig. P8.106, with nodes (i, j) equally spaced $\Delta\theta$ and Δr apart. Derive a finite difference model for Eq. (8.11) similar to the cartesian expression (8.96).

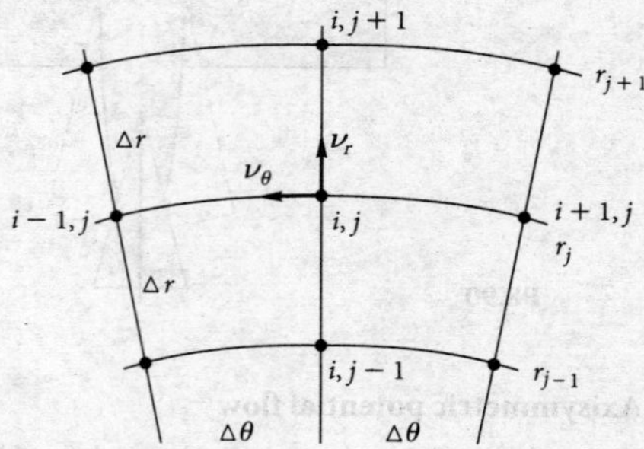

P8.106

P8.107 SAE 10W30 oil at 20°C is at rest near a wall when the wall suddenly begins moving at a constant 1 m/s. (a) Use Δy = 1 cm and $\Delta t = 0.2$ s and check the stability criterion (8.101). (b) Carry out Eq. (8.100) to $t = 2$ s and report the velocity u at $y = 4$ cm.

P8.108 Consider two-dimensional potential flow into a step contraction as in Fig. P8.108. The inlet velocity $U_1 = 7$ m/s, and the outlet velocity U_2 is uniform. The nodes (i, j) are labeled in the figure. Set up the complete finite difference algebraic relations for all nodes. Solve, if possible, on a computer and plot the streamlines in the flow.

P8.109 Consider inviscid flow through a two-dimensional 90° bend with a contraction, as in Fig. P8.109. Assume uniform flow at the entrance and exit. Make a finite difference computer analysis for small grid size (at least 150 nodes), determine the dimensionless pressure distribution along the walls, and sketch the streamlines. (You may use either square or rectangular grids.)

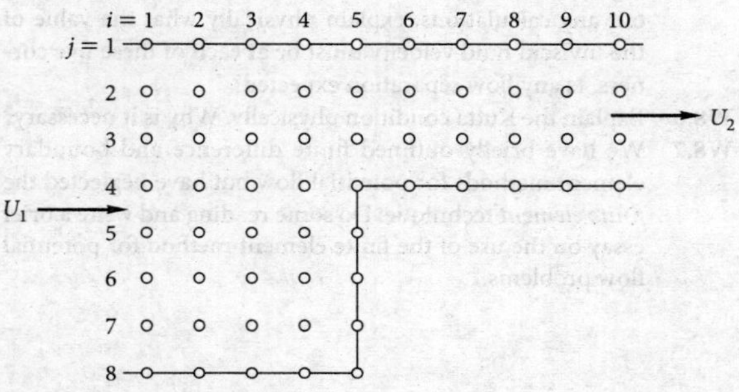

P8.108

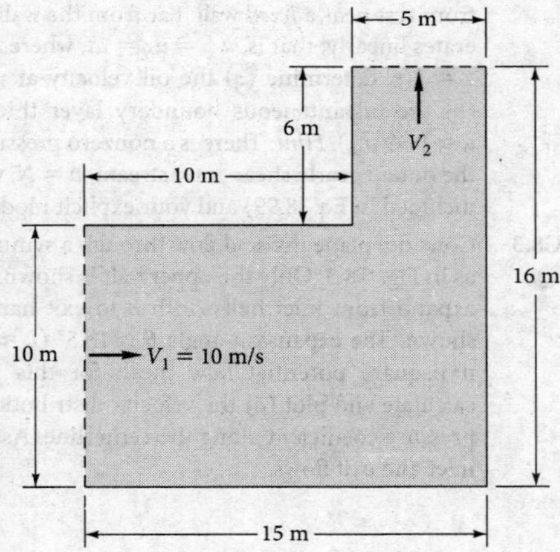

P8.109

P8.110 For fully developed laminar incompressible flow through a straight noncircular duct, as in Sec. 6.8, the Navier-Stokes equations (4.38) reduce to

$$\frac{\partial^2 u}{\partial y^2} + \frac{\partial^2 u}{\partial z^2} = \frac{1}{\mu}\frac{dp}{dx} = \text{const} < 0$$

where (y, z) is the plane of the duct cross section and x is along the duct axis. Gravity is neglected. Using a nonsquare rectangular grid $(\Delta x, \Delta y)$, develop a finite difference model for this equation, and indicate how it may be applied to solve for flow in a rectangular duct of side lengths a and b.

P8.111 Solve Prob. P8.110 numerically for a rectangular duct of side length b by $2b$, using at least 100 nodal points. Evaluate the volume flow rate and the friction factor, and compare with the results in Table 6.4:

$$Q \approx 0.1143\frac{b^4}{\mu}\left(-\frac{dp}{dx}\right) \qquad f\,\text{Re}_{D_h} \approx 62.19$$

where $D_h = 4A/P = 4b/3$ for this case. Comment on the possible truncation errors of your model.

P8.112 In CFD textbooks [5, 23–27], one often replaces the left-hand sides of Eqs. (8.102*b* and *c*) with the following two expressions, respectively:

$$\frac{\partial}{\partial x}(u^2) + \frac{\partial}{\partial y}(vu) \quad \text{and} \quad \frac{\partial}{\partial x}(uv) + \frac{\partial}{\partial y}(v^2)$$

Are these equivalent expressions, or are they merely simplified approximations? Either way, why might these forms be better for finite difference purposes?

P8.113 Formulate a numerical model for Eq. (8.99), which has no instability, by evaluating the second derivative at the *next* time step, $j + 1$. Solve for the center velocity at the next time step and comment on the result. This is called an *implicit model* and requires iteration.

P8.114 If your institution has an online potential flow boundary element computer code, consider flow past a symmetric airfoil, as in Fig. P8.114. The basic shape of an NACA symmetric airfoil is defined by the function [12]

$$\frac{2y}{t_{max}} \approx 1.4845\zeta^{1/2} - 0.63\zeta - 1.758\zeta^2 + 1.4215\zeta^3 - 0.5075\zeta^4$$

where $\zeta = x/C$ and the maximum thickness t_{max} occurs at $\zeta = 0.3$. Use this shape as part of the lower boundary for zero angle of attack. Let the thickness be fairly large, say, $t_{max} = 0.12$, 0.15, or 0.18. Choose a generous number of nodes (≥ 60), and calculate and plot the velocity distribution V/U_∞ along the airfoil surface. Compare with the

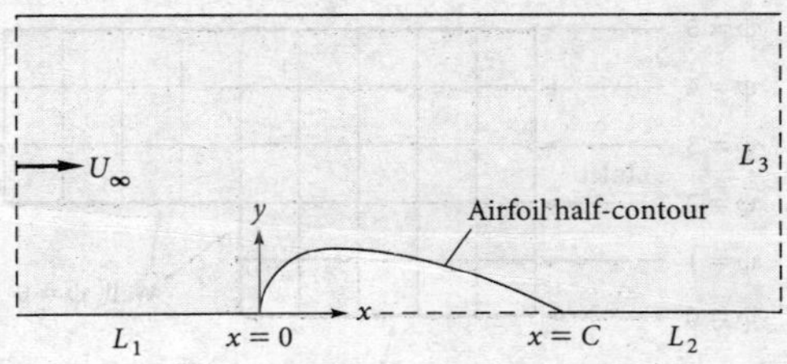

P8.114

theoretical results in Ref. 12 for NACA 0012, 0015, or 0018 airfoils. If time permits, investigate the effect of the boundary lengths L_1, L_2, and L_3, which can initially be set equal to the chord length C.

P8.115 Use the explicit method of Eq. (8.100) to solve Prob. P4.85 numerically for SAE 30 oil at 20°C with U_0 = 1 m/s and $\omega = M$ rad/s, where M is the number of letters in your surname. (This author will solve the problem for M = 5.) When steady oscillation is reached, plot the oil velocity versus time at y = 2 cm.

Word Problems

W8.1 What simplifications have been made, in the potential flow theory of this chapter, which result in the elimination of the Reynolds number, Froude number, and Mach number as important parameters?

W8.2 In this chapter we superimpose many basic solutions, a concept associated with *linear* equations. Yet Bernoulli's equation (8.3) is *nonlinear*, being proportional to the square of the velocity. How, then, do we justify the use of superposition in inviscid flow analysis?

W8.3 Give a physical explanation of circulation Γ as it relates to the lift force on an immersed body. If the line integral defined by Eq. (8.23) is zero, it means that the integrand is a perfect differential—but of what variable?

W8.4 Give a simple proof of Eq. (8.46)—namely, that both the real and imaginary parts of a function $f(z)$ are laplacian if $z = x + iy$. What is the secret of this remarkable behavior?

W8.5 Figure 8.18 contains five body corners. Without carrying out any calculations, explain physically what the value of the inviscid fluid velocity must be at each of these five corners. Is any flow separation expected?

W8.6 Explain the Kutta condition physically. Why is it necessary?

W8.7 We have briefly outlined finite difference and boundary element methods for potential flow but have neglected the *finite element* technique. Do some reading and write a brief essay on the use of the finite element method for potential flow problems.

Comprehensive Problems

C8.1 Did you know that you can solve simple fluid mechanics problems with Microsoft Excel? The successive relaxation technique for solving the Laplace equation for potential flow problems is easily set up on a spreadsheet, since the stream function at each interior cell is simply the average of its four neighbors. As an example, solve for the irrotational potential flow through a contraction, as given in Fig. C8.1. *Note:* To avoid the "circular reference" error, you must turn on the iteration option. Use the help index for more information. For full credit, attach a printout of your spreadsheet, with stream function converged and the value of the stream function at each node displayed to four digits of accuracy.

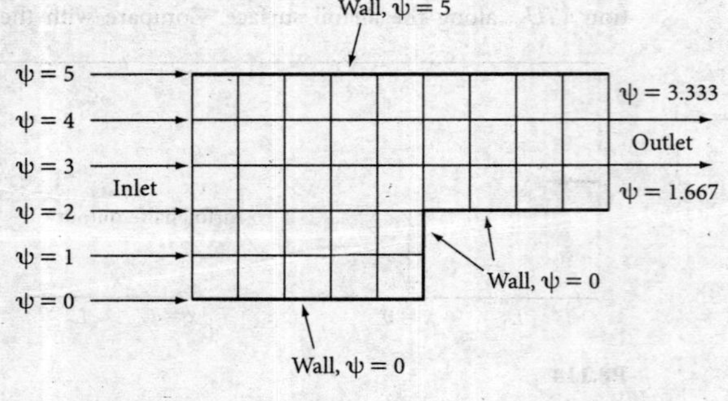

C8.1

C8.2 Use an explicit method, similar to but not identical to Eq. (8.100), to solve the case of SAE 30 oil at 20°C starting from rest near a *fixed* wall. Far from the wall, the oil accelerates linearly; that is, $u_\infty = u_N = at$, where a = 9 m/s². At t = 1 s, determine (*a*) the oil velocity at y = 1 cm and (*b*) the instantaneous boundary layer thickness (where $u \approx 0.99\, u_\infty$). *Hint:* There is a nonzero pressure gradient in the outer (nearly shear-free) stream, $n = N$, which must be included in Eq. (8.99) and your explicit model.

C8.3 Consider plane inviscid flow through a symmetric diffuser, as in Fig. C8.3. Only the upper half is shown. The flow is to expand from inlet half-width h to exit half-width $2h$, as shown. The expansion angle θ is 18.5° ($L \approx 3h$). Set up a nonsquare potential flow mesh for this problem, and calculate and plot (*a*) the velocity distribution and (*b*) the pressure coefficient along the centerline. Assume uniform inlet and exit flows.

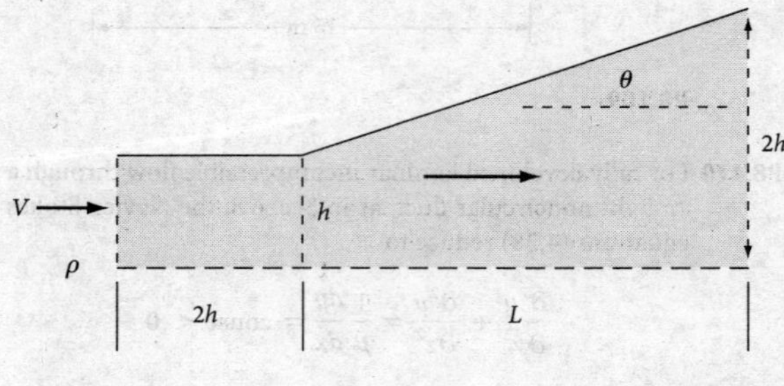

C8.3

C8.4 Use potential flow to approximate the flow of air being sucked up into a vacuum cleaner through a two-dimensional slit attachment, as in Fig. C8.4. In the xy plane through the centerline of the attachment, model the flow as a line sink of strength $(-m)$, with its axis in the z direction at height a above the floor. (*a*) Sketch the streamlines and locate any stagnation points in the flow. (*b*) Find the magnitude of velocity $V(x)$ along the floor in terms of the parameters a and m. (*c*) Let the pressure far away be p_∞, where velocity is zero. Define a velocity scale $U = m/a$. Determine the variation of dimensionless pressure coefficient, $C_p = (p - p_\infty)/(\rho U^2/2)$, along the floor. (*d*) The vacuum cleaner is most effective where C_p is a minimum—that is, where velocity is maximum. Find the locations of minimum pressure coefficient along the x axis. (*e*) At which points along the x axis do you expect the vacuum cleaner to work most effectively? Is it best at $x = 0$ directly beneath the slit, or at some other x location along the floor? Conduct a scientific experiment at home with a vacuum cleaner and some small pieces of dust or dirt to test your prediction. Report your results and discuss the agreement with prediction. Give reasons for any disagreements.

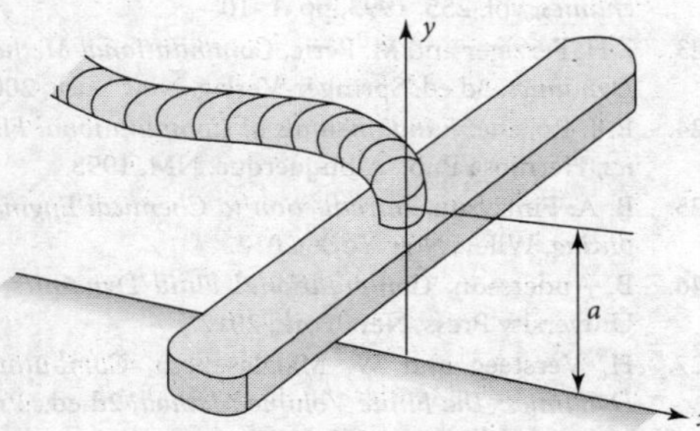

C8.4

C8.5 Consider a three-dimensional, incompressible, irrotational flow. Use the following two methods to prove that the viscous term in the Navier-Stokes equation is identically zero: (*a*) using vector notation; and (*b*) expanding out the scalar terms and substituting terms from the definition of irrotationality.

C8.6 Find, either on-line or in Ref. 12, lift-drag data for the NACA 4412 airfoil. (*a*) Draw the polar lift–drag plot and compare qualitatively with Fig. 7.26. (*b*) Find the maximum value of the lift-to-drag ratio. (*c*) Demonstrate a straight-line construction on the polar plot that will immediately yield the maximum L/D in (*b*). (*d*) If an aircraft could use this two-dimensional wing in actual flight (no induced drag) and had a perfect pilot, estimate how far (in kilometers) this aircraft could glide to a sea-level runway if it lost power at 7,600 m altitude.

C8.7 Find a formula for the stream function for flow of a doublet of strength λ a distance a from a wall, as in Fig. C8.7. (*a*) Sketch the streamlines. (*b*) Are there any stagnation points? (*c*) Find the maximum velocity along the wall and its position.

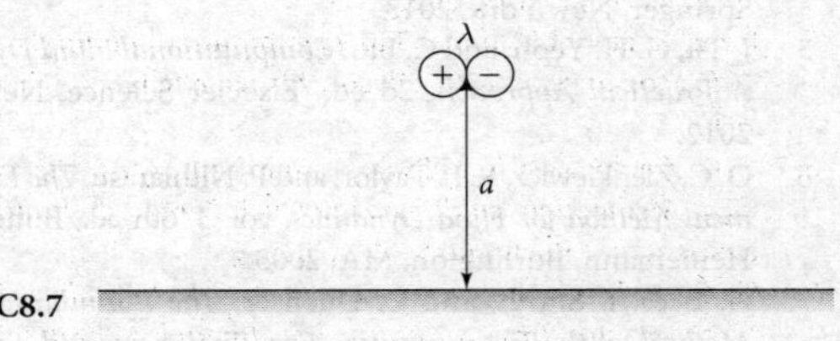

C8.7

Design Projects

D8.1 In 1927, Theodore von Kármán developed a scheme to use a uniform stream, plus a row of sources and sinks, to generate an arbitrary closed-body shape. A schematic of the idea is sketched in Fig. D8.1. The body is symmetric and at zero angle of attack. A total of N sources and sinks are distributed along the axis within the body, with strengths m_i at positions x_i, for $i = 1$ to N. The object is to find the correct distribution of strengths that approximates a given body shape $y(x)$ at a finite number of surface locations and then to compute the approximate surface velocity and pressure. The technique should work for either two-dimensional bodies (distributed line sources) or bodies of revolution (distributed point sources).

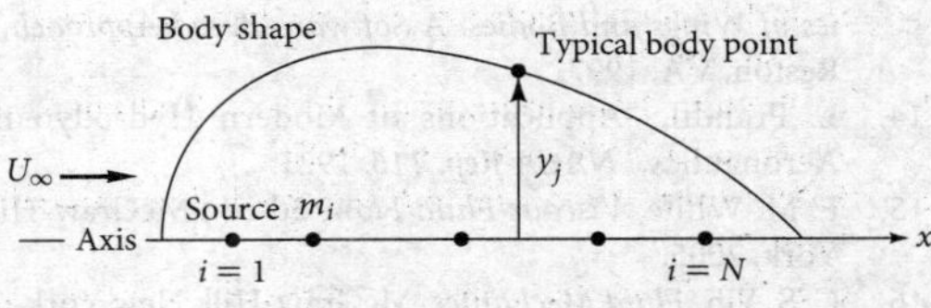

D8.1

For our body shape let us select the NACA 0018 airfoil, given by the formula in Prob. P8.114 with $t_{max} = 0.18$. Develop the ideas stated here into N simultaneous algebraic equations that can be used to solve for the N unknown line source/sink strengths. Then program your equations for a computer, with $N \geq 20$; solve for m_i; compute the surface velocities; and compare with the theoretical velocities for this shape in Ref. 12. Your goal should be to achieve accuracy within ± 1 percent of the classic results. If necessary, you should adjust N and the locations of the sources.

D8.2 Modify Prob. D8.1 to solve for the point-source distribution that approximates an "0018" body-of-revolution shape. Since no theoretical results are published, simply make sure that your results converge to ± 1 percent.

D8.3 Consider water at 20°C flowing at 12 m/s in a water channel. A Rankine oval cylinder, 40 cm long, is to be placed parallel to the flow, where the water static pressure is 120 kPa. The oval's thickness is a design parameter. Prepare a plot of the minimum pressure on the oval's surface as a function of body thickness. Especially note the thicknesses where (*a*) the local pressure is 50 kPa and (*b*) cavitation first occurs on the surface.

References

1. J. Wermer, *Potential Theory*, Springer-Verlag, New York, 2008.
2. J. M. Robertson, *Hydrodynamics in Theory and Application*, Prentice-Hall, Englewood Cliffs, NJ, 1965.
3. L. M. Milne-Thomson, *Theoretical Hydrodynamics*, 4th ed., Dover, New York, 1996.
4. D. H. Armitage and S. J. Gardiner, *Classical Potential Theory*, Springer, New York, 2013.
5. J. Tu, G. H. Yeoh, and C. Liu, *Computational Fluid Dynamics: A Practical Approach*, 2d ed., Elsevier Science, New York, 2012.
6. O. C. Zienkiewicz, R. L. Taylor, and P. Nithiarasu, *The Finite Element Method for Fluid Dynamics*, vol. 3, 6th ed., Butterworth-Heinemann, Burlington, MA, 2005.
7. G. Beer, I. Smith, and C. Duenser, *The Boundary Element Method with Programming: For Engineers and Scientists*, Springer-Verlag, New York, 2010.
8. A. D. Moore, "Fields from Fluid Flow Mappers," *J. Appl. Phys.*, vol. 20, 1949, pp. 790–804.
9. H. J. S. Hele-Shaw, "Investigation of the Nature of the Surface Resistance of Water and of Streamline Motion under Certain Experimental Conditions," *Trans. Inst. Nav. Archit.*, vol. 40, 1898, p. 25.
10. S. W. Churchill, *Viscous Flows: The Practical Use of Theory*, Butterworth, Stoneham, MA, 1988.
11. J. D. Anderson, Jr., *Fundamentals of Aerodynamics*, 5th ed., McGraw-Hill, New York, 2010.
12. I. H. Abbott and A. E. von Doenhoff, *Theory of Wing Sections*, Dover, New York, 1981.
13. F. O. Smetana, *Introductory Aerodynamics and Hydrodynamics of Wings and Bodies: A Software-Based Approach*, AIAA, Reston, VA, 1997.
14. L. Prandtl, "Applications of Modern Hydrodynamics to Aeronautics," *NACA Rep. 116*, 1921.
15. F. M. White, *Viscous Fluid Flow*, 3d ed., McGraw-Hill, New York, 2005.
16. C. S. Yih, *Fluid Mechanics*, McGraw-Hill, New York, 1969.
17. K. T. Patton, "Tables of Hydrodynamic Mass Factors for Translational Motion," *ASME Winter Annual Meeting*, Paper 65-WA/UNT-2, 1965.
18. J. L. Hess and A. M. O. Smith, "Calculation of Nonlifting Potential Flow about Arbitrary Three-Dimensional Bodies," *J. Ship Res.*, vol. 8, 1964, pp. 22–44.
19. K. H. Huebner, *The Finite Element Method for Engineers*, 4th ed., Wiley, New York, 2001.
20. J. C. Tannehill, D. A. Anderson, and R. H. Pletcher, *Computational Fluid Mechanics and Heat Transfer*, 3d ed., Taylor and Francis, Bristol, PA, 2011.
21. J. N. Newman, *Marine Hydrodynamics*, M.I.T. Press, Cambridge, MA, 1977.
22. P. T. Tokumaru and P. E. Dimotakis, "The Lift of a Cylinder Executing Rotary Motions in a Uniform Flow," *J. Fluid Mechanics*, vol. 255, 1993, pp. 1–10.
23. J. H. Ferziger and M. Peric, *Computational Methods for Fluid Dynamics*, 3d ed. Springer-Verlag, New York, 2002.
24. P. J. Roache, *Fundamentals of Computational Fluid Dynamics*, Hermosa Pub., Albuquerque, NM, 1998.
25. B. A. Finlayson, *Introduction to Chemical Engineering Computing*, Wiley, New York, 2012.
26. B. Andersson, *Computational Fluid Dynamics*, Cambridge University Press, New York, 2012.
27. H. Versteeg and W. Malalasekera, *Computational Fluid Dynamics: The Finite Volume Method*, 2d ed., Prentice-Hall, Upper Saddle River, NJ, 2007.
28. M. Deshpande, J. Feng, and C. L. Merkle, "Numerical Modeling of the Thermodynamic Effects of Cavitation," *J. Fluids Eng.*, June 1997, pp. 420–427.
29. P. A. Durbin and R. B. A. Pettersson, *Statistical Theory and Modeling for Turbulent Flows*, Wiley, New York, 2001.
30. C. J. Freitas, "Perspective: Selected Benchmarks from Commercial CFD Codes," *J. Fluids Eng.*, vol. 117, June 1995, pp. 208–218.
31. R. Martinuzzi and C. Tropea, "The Flow around Surface-Mounted, Prismatic Obstacles in a Fully Developed Channel Flow," *J. Fluids Eng.*, vol. 115, March 1993, pp. 85–92.
32. K. B. Shah and J. H. Ferzier, "Fluid Mechanicians View of Wind Engineering: Large Eddy Simulation of Flow Past a Cubic Obstacle," *J. Wind Engineering and Industrial Aerodynamics*, vol. 67–68, 1997, pp. 221–224.
33. P. Sagaut, *Large Eddy Simulation for Incompressible Flows: An Introduction*, 3rd ed., Springer, New York, 2010.
34. W. J. Palm, *Introduction to MATLAB 7 for Engineers*, 3d ed. McGraw-Hill, New York, 2010.
35. A. Gilat, *MATLAB: An Introduction with Applications*, 4th ed., Wiley, New York, 2010.

36. J. W. Hoyt and R. H. J. Sellin, "Flow over Tube Banks—A Visualization Study," *J. Fluids Eng.,* vol. 119, June 1997, pp. 480–483.
37. S. Douglass, "Switchblade Fighter Bomber," *Popular Science,* Nov. 2000, pp. 52–55.
38. G. Beer, I. Smith, and C. Duenser, The Boundary Element Method with Programming: For Engineers and Scientists, Springer, New York, 2010.
39. J. D. Anderson, *A History of Aerodynamics and Its Impact on Flying Machines,* Cambridge University Press, Cambridge, UK, 1999.
40. B. Robins, *Mathematical Tracts 1 & 2,* J. Nourse, London, 1761.
41. T. K. Sengupta, A. Kasliwal, S. De, and M. Nair, "Temporal Flow Instability for Magnus-Robins Effect at High Rotation Rates," *J. Fluids and Structures,* vol. 17, 2003, pp. 941–953.
42. G. F. Dargush and M. M. Grigoriev, "Fast and Accurate Solutions of Steady Stokes Flows Using Multilevel Boundary Element Methods," *J. Fluids Eng.,* vol. 127, July 2005, pp. 640–646.
43. R. H. Kirchhoff, *Potential Flows: Computer Graphic Solutions,* Marcel Dekker, New York, 2001.
44. H. Werle, "Hydrodynamic Visualization of the Flow around a Streamlined Cylinder with Suction: Cousteau-Malavard Turbine Sail Model," *Le Recherche Aerospatiale,* vol. 4, 1984, pp. 29–38.
45. T. K. Sengupta and S. R. Talla, "Robins-Magnus Effect: A Continuing Saga," *Current Science,* vol. 86, no. 7, 2004, pp. 1033–1036.
46. P. R. Spalart, "Airplane Trailing Vortices," *Annual Review Fluid Mechanics,* vol. 30, 1998, pp. 107–138.
47. M. Elkhoury, "Assessment and Modification of One-Equation Models of Turbulence for Wall-Bounded Flows," *J. Fluids Eng.,* vol. 129, July 2007, pp. 921–928.

Ever since the demise of the Concorde, engineers have been working on the design of an overland supersonic airplane. For such airplanes to be practical, sonic booms must be reduced to an acceptable level. Theory, though useful, cannot solve this problem without extensive testing. Shown here is Boeing's *Lynx* design, being tested in NASA's Glenn Research Center in Cleveland. Sensors capture both the forces on the plane and the pressures far from the vehicle. The goal is to generate sonic booms so low that they barely register on the ground. [*Photo courtesy of NASA*]

Chapter 9
Compressible Flow

Motivation. All eight of our previous chapters have been concerned with "low-speed" or "incompressible" flow, where the fluid velocity is much less than its speed of sound. In fact, we did not even develop an expression for the speed of sound of a fluid. That is done in this chapter.

When a fluid moves at speeds comparable to its speed of sound, density changes become significant and the flow is termed *compressible*. Such flows are difficult to obtain in liquids, since high pressures on the order of 1,000 atm are needed to generate sonic velocities. In gases, however, a pressure ratio of only 2:1 will likely cause sonic flow. Thus compressible gas flow is quite common, and this subject is often called *gas dynamics*. The most important parameter is the Mach number.

Probably, the two most important and distinctive effects of compressibility on flow are (1) *choking*, wherein the duct flow rate is sharply limited by the sonic condition, and (2) *shock waves*, which are nearly discontinuous property changes in a supersonic flow. The purpose of this chapter is to explain such striking phenomena and to familiarize the reader with engineering calculations of compressible flow.

Speaking of calculations, the present chapter can use the help of Excel. Compressible flow analysis is filled with scores of complicated algebraic equations, many of which are difficult to manipulate or invert. Consequently, for nearly a century, compressible flow textbooks have relied on extensive tables of Mach number relations (see App. B) for numerical work. With Excel, however, any equation in this chapter can be typed into a cell and iterated to solve for any variables—see part (b) of Example 9.13 for an especially intricate example. With such a tool, App. B serves only as a backup, for initial estimates, and may soon vanish from textbooks.

9.1 Introduction: Review of Thermodynamics

We took a brief look in Chap. 4 [Eqs. (4.13) to (4.17)] to see when we might safely neglect the compressibility inherent in every real fluid. We found that the proper criterion for a nearly incompressible flow was a small Mach number

$$\mathrm{Ma} = \frac{V}{a} \ll 1$$

where V is the flow velocity and a is the speed of sound of the fluid. Under small Mach number conditions, changes in fluid density are everywhere small in the flow field. The energy equation becomes uncoupled, and temperature effects can be either ignored or put aside for later study. The equation of state degenerates into the simple statement that density is nearly constant. This means that an incompressible flow requires only a momentum and continuity analysis, as we showed with many examples in Chaps. 7 and 8.

This chapter treats compressible flows, which have Mach numbers greater than about 0.3 and thus exhibit nonnegligible density changes. If the density change is significant, it follows from the equation of state that the temperature and pressure changes are also substantial. Large temperature changes imply that the energy equation can no longer be neglected. Therefore, the work is doubled from two basic equations to four

1. Continuity equation
2. Momentum equation
3. Energy equation
4. Equation of state

to be solved simultaneously for four unknowns: pressure, density, temperature, and flow velocity (p, ρ, T, V). Thus, the general theory of compressible flow is quite complicated, and we try here to make further simplifications, especially by assuming a reversible adiabatic or *isentropic* flow.

We note in passing that at least two flow patterns depend strongly on very small density differences, acoustics, and natural convection. Acoustics [7, 9] is the study of sound wave propagation, which is accompanied by extremely small changes in density, pressure, and temperature. Natural convection is the gentle circulating pattern set up by buoyancy forces in a fluid stratified by uneven heating or uneven concentration of dissolved materials. Here we are concerned only with steady compressible flow where the fluid velocity is of magnitude comparable to that of the speed of sound.

The Mach Number

The Mach number is the dominant parameter in compressible flow analysis, with different effects depending on its magnitude. Aerodynamicists especially make a distinction between the various ranges of Mach number, and the following rough classifications are commonly used:

$\text{Ma} < 0.3$: *incompressible flow,* where density effects are negligible.

$0.3 < \text{Ma} < 0.8$: *subsonic flow,* where density effects are important but no shock waves appear.

$0.8 < \text{Ma} < 1.2$: *transonic flow,* where shock waves first appear, dividing subsonic and supersonic regions of the flow. Powered flight in the transonic region is difficult because of the mixed character of the flow field.

$1.2 < \text{Ma} < 3.0$: *supersonic flow,* where shock waves are present but there are no subsonic regions.

$3.0 < \text{Ma}$: *hypersonic flow* [11], where shock waves and other flow changes are especially strong.

The numerical values listed are only rough guides. These five categories of flow are appropriate to external high-speed aerodynamics. For internal (duct) flows, the most important question is simply whether the flow is subsonic ($\text{Ma} < 1$) or supersonic ($\text{Ma} > 1$), because the effect of area changes reverses, as we show in Sec. 9.4. Since supersonic flow effects may go against intuition, you should study these differences carefully.

The Specific-Heat Ratio

In addition to geometry and Mach number, compressible flow calculations also depend on a second dimensionless parameter, the *specific-heat ratio* of the gas:

$$k = \frac{c_p}{c_v} \tag{9.1}$$

Earlier, in Chaps. 1 and 4, we used the same symbol k to denote the thermal conductivity of a fluid. We apologize for the duplication; thermal conductivity does not appear in these later chapters of the text.

Recall from Fig. 1.4 that k for the common gases decreases slowly with temperature and lies between 1.0 and 1.7. Variations in k have only a slight effect on compressible flow computations, and air, $k \approx 1.40$, is the dominant fluid of interest. Therefore, although we assign some problems involving other gases like steam and CO_2 and helium, the compressible flow tables in App. B are based solely on the single value $k = 1.40$ for air.

This text contains only a single chapter on compressible flow, but, as usual, whole books have been written on the subject. Here we list only certain recent or classic texts. References 1 to 4 are introductory or intermediate treatments, while Refs. 5 to 10 are advanced books. One can also become specialized within this specialty of compressible flow. Reference 11 concerns *hypersonic flow*—that is, at very high Mach numbers. Reference 12 explains the exciting new technique of direct simulation of gas flows with a *molecular dynamics model.* Compressible flow is also well suited for computational fluid dynamics (CFD), as described in Ref. 13. Finally, a short, thoroughly readable (no calculus) Ref. 14 describes the principles and promise of high-speed (supersonic) flight. From time to time we shall defer some specialized topic to these other texts.

The Perfect Gas

In principle, compressible flow calculations can be made for any fluid equation of state, and we shall assign a few problems involving the steam tables [15], the gas tables [16], and liquids [Eq. (1.19)]. But in fact most elementary treatments are confined to the perfect gas with constant specific heats:

$$p = \rho R T \qquad R = c_p - c_v = \text{const} \qquad k = \frac{c_p}{c_v} = \text{const} \tag{9.2}$$

For all real gases, c_p, c_v, and k vary with temperature but only moderately; for example, c_p of air increases 30 percent as temperature increases from −20 to 2,800°C. Since we rarely deal with such large temperature changes, it is quite reasonable to assume constant specific heats.

Recall from Sec. 1.8 that the gas constant is related to a universal constant Λ divided by the gas molecular weight:

$$R_{\text{gas}} = \frac{\Lambda}{M_{\text{gas}}} \tag{9.3}$$

where $\Lambda = 49{,}720 \text{ ft-lbf/(lbmol}\cdot{}^\circ\text{R)} = 8{,}314 \text{ J/(kmol}\cdot\text{K)}$

For air, $M = 28.97$, and we shall adopt the following property values for air throughout this chapter:

$$\begin{aligned} R &= 1{,}716 \text{ ft}^2/(\text{s}^2\cdot{}^\circ\text{R}) = 287 \text{ m}^2/(\text{s}^2\cdot\text{K}) \qquad k = 1.400 \\ c_v &= \frac{R}{k-1} = 4{,}293 \text{ ft}^2/(\text{s}^2\cdot{}^\circ\text{R}) = 718 \text{ m}^2/(\text{s}^2\cdot\text{K}) \\ c_p &= \frac{kR}{k-1} = 6{,}009 \text{ ft}^2/(\text{s}^2\cdot{}^\circ\text{R}) = 1{,}005 \text{ m}^2/(\text{s}^2\cdot\text{K}) \end{aligned} \tag{9.4}$$

Experimental values of k for eight common gases were shown in Fig. 1.4. From this figure and the molecular weight, the other properties can be computed, as in Eqs. (9.4).

The changes in the internal energy $\hat{u}$ and enthalpy h of a perfect gas are computed for constant specific heats as

$$\hat{u}_2 - \hat{u}_1 = c_v(T_2 - T_1) \qquad h_2 - h_1 = c_p(T_2 - T_1) \tag{9.5}$$

For variable specific heats one must integrate $\hat{u} = \int c_v dT$ and $h = \int c_p dT$ or use the gas tables [16]. Most modern thermodynamics texts now contain software for evaluating properties of nonideal gases [17].

Isentropic Process

The isentropic approximation is common in compressible flow theory. We compute the entropy change from the first and second laws of thermodynamics for a pure substance [17 or 18]:

$$T\,ds = dh - \frac{dp}{\rho} \tag{9.6}$$

Introducing $dh = c_p dT$ for a perfect gas and solving for ds, we substitute $\rho T = p/R$ from the perfect-gas law and obtain

$$\int_1^2 ds = \int_1^2 c_p \frac{dT}{T} - R\int_1^2 \frac{dp}{p} \tag{9.7}$$

If c_p is variable, the gas tables will be needed, but for constant c_p we obtain the analytic results

$$s_2 - s_1 = c_p \ln\frac{T_2}{T_1} - R\ln\frac{p_2}{p_1} = c_v \ln\frac{T_2}{T_1} - R\ln\frac{\rho_2}{\rho_1} \tag{9.8}$$

Equations (9.8) are used to compute the entropy change across a shock wave (Sec. 9.5), which is an irreversible process.

For isentropic flow, we set $s_2 = s_1$ and obtain these interesting power-law relations for an isentropic perfect gas:

$$\frac{p_2}{p_1} = \left(\frac{T_2}{T_1}\right)^{k/(k-1)} = \left(\frac{\rho_2}{\rho_1}\right)^k \tag{9.9}$$

These relations are used in Sec. 9.3.

EXAMPLE 9.1

Argon flows through a tube such that its initial condition is $p_1 = 1.7$ MPa and $\rho_1 = 18$ kg/m^3 and its final condition is $p_2 = 248$ kPa and $T_2 = 400$ K. Estimate (*a*) the initial temperature, (*b*) the final density, (*c*) the change in enthalpy, and (*d*) the change in entropy of the gas.

Solution

From Table A.4 for argon, $R = 208$ m^2/(s$^2 \cdot$ K) and $k = 1.67$. Therefore estimate its specific heat at constant pressure from Eq. (9.4):

$$c_p = \frac{kR}{k-1} = \frac{1.67(208)}{1.67 - 1} \approx 519 \text{ m}^2/(\text{s}^2 \cdot \text{K})$$

The initial temperature and final density are estimated from the ideal-gas law, Eq. (9.2):

$$T_1 = \frac{p_1}{\rho_1 R} = \frac{1.7\text{ E6 N/m}^2}{(18\text{ kg/m}^3)[208\text{ m}^2/(\text{s}^2 \cdot \text{K})]} = 454 \text{ K} \qquad \textit{Ans. (a)}$$

$$\rho_2 = \frac{p_2}{T_2 R} = \frac{248\text{ E3 N/m}^2}{(400\text{ K})[208\text{ m}^2/(\text{s}^2 \cdot \text{K})]} = 2.98 \text{ kg/m}^3 \qquad \textit{Ans. (b)}$$

From Eq. (9.5) the enthalpy change is

$$h_2 - h_1 = c_p(T_2 - T_1) = 519(400 - 454) \approx -28{,}000 \text{ J/kg (or m}^2\text{/s}^2) \qquad \textit{Ans. (c)}$$

The argon temperature and enthalpy decrease as we move down the tube. Actually, there may not be any external cooling; that is, the fluid enthalpy may be converted by friction to increased kinetic energy (Sec. 9.7).

Finally, the entropy change is computed from Eq. (9.8):

$$s_2 - s_1 = c_p \ln \frac{T_2}{T_1} - R \ln \frac{p_2}{p_1}$$

$$= 519 \ln \frac{400}{454} - 208 \ln \frac{0.248 \text{ E6}}{1.7 \text{ E6}}$$

$$= -66 + 400 \approx 334 \text{ m}^2/(\text{s}^2 \cdot \text{K}) \qquad \textit{Ans. (d)}$$

The fluid entropy has increased. If there is no heat transfer, this indicates an irreversible process. Note that entropy has the same units as the gas constant and specific heat.

This problem is not just arbitrary numbers. It correctly simulates the behavior of argon moving subsonically through a tube with large frictional effects (Sec. 9.7).

9.2 The Speed of Sound

The so-called speed of sound is the rate of propagation of a pressure pulse of infinitesimal strength through a still fluid. It is a thermodynamic property of a fluid. Let us analyze it by first considering a pulse of finite strength, as in Fig. 9.1. In Fig. 9.1*a* the pulse, or pressure wave, moves at speed C toward the still fluid ($p, \rho, T, V = 0$) at the left, leaving behind at the right a fluid of increased properties ($p + \Delta p, \rho + \Delta\rho, T + \Delta T$) and a fluid velocity ΔV toward the left following the wave but much slower. We can determine these effects by making a control volume analysis across the wave. To avoid the unsteady terms that would be necessary in Fig. 9.1*a*, we adopt instead the control volume of Fig. 9.1*b*, which moves at wave speed C to the left. The wave appears fixed from this viewpoint, and the fluid appears to have velocity C on the left

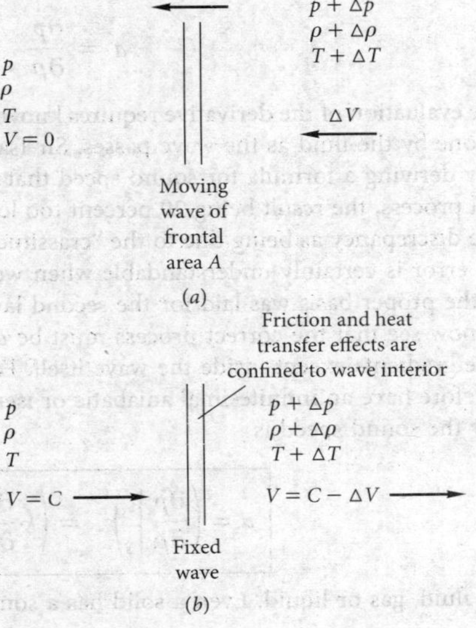

Fig. 9.1 Control volume analysis of a finite-strength pressure wave: (*a*) control volume fixed to still fluid at left; (*b*) control volume moving left at wave speed *C*.

and $C - \Delta V$ on the right. The thermodynamic properties p, ρ, and T are not affected by this change of viewpoint.

The flow in Fig. 9.1*b* is steady and one-dimensional across the wave. The continuity equation is thus, from Eq. (3.24),

$$\rho AC = (\rho + \Delta\rho)(A)(C - \Delta V)$$

or

$$\Delta V = C\frac{\Delta\rho}{\rho + \Delta\rho} \tag{9.10}$$

This proves our contention that the induced fluid velocity on the right is much smaller than the wave speed C. In the limit of infinitesimal wave strength (sound wave) this speed is itself infinitesimal.

Notice that there are no velocity gradients on either side of the wave. Therefore, even if fluid viscosity is large, frictional effects are confined to the interior of the wave. Advanced texts [for example, 9] show that the thickness of pressure waves in gases is of order 10^{-7} m at atmospheric pressure. Thus, we can safely neglect friction and apply the one-dimensional momentum equation (3.40) across the wave:

$$\sum F_{\text{right}} = \dot{m}(V_{\text{out}} - V_{\text{in}})$$

or

$$pA - (p + \Delta p)A = (\rho AC)(C - \Delta V - C) \tag{9.11}$$

Again, the area cancels, and we can solve for the pressure change:

$$\Delta p = \rho C\,\Delta V \tag{9.12}$$

If the wave strength is very small, the pressure change is small.

Finally, combine Eqs. (9.10) and (9.12) to give an expression for the wave speed:

$$C^2 = \frac{\Delta p}{\Delta\rho}\left(1 + \frac{\Delta\rho}{\rho}\right) \tag{9.13}$$

The larger the strength $\Delta\rho/\rho$ of the wave, the faster the wave speed; that is, powerful explosion waves move much more quickly than sound waves. In the limit of infinitesimal strength $\Delta\rho \rightarrow 0$, we have what is defined to be the speed of sound a of a fluid:

$$a^2 = \frac{\partial p}{\partial \rho} \tag{9.14}$$

But, the evaluation of the derivative requires knowledge of the thermodynamic process undergone by the fluid as the wave passes. Sir Isaac Newton in 1686 made a famous error by deriving a formula for sound speed that was equivalent to assuming an isothermal process, the result being 20 percent too low for air, for example. He rationalized the discrepancy as being due to the "crassitude" (dust particles and so on) in the air; the error is certainly understandable when we reflect that it was made 180 years before the proper basis was laid for the second law of thermodynamics.

We now see that the correct process must be *adiabatic* because there are no temperature gradients except inside the wave itself. For vanishing-strength sound waves, we therefore have an infinitesimal adiabatic or isentropic process. The correct expression for the sound speed is

$$a = \left(\left.\frac{\partial p}{\partial \rho}\right|_s\right)^{1/2} = \left(k\left.\frac{\partial p}{\partial \rho}\right|_T\right)^{1/2} \tag{9.15}$$

for any fluid, gas or liquid. Even a solid has a sound speed.

Table 9.1 Sound Speed of Various Materials at 60°F (15.5°C) and 1 atm

Material	*a*, ft/s	*a*, m/s
Gases:		
H_2	4,246	1,294
He	3,281	1,000
Air	1,117	340
Ar	1,040	317
CO_2	873	266
CH_4	607	185
$^{238}UF_6$	297	91
Liquids:		
Glycerin	6,100	1,860
Water	4,890	1,490
Mercury	4,760	1,450
Ethyl alcohol	3,940	1,200
Solids:*		
Aluminum	16,900	5,150
Steel	16,600	5,060
Hickory	13,200	4,020
Ice	10,500	3,200

*Plane waves. Solids also have a *shear-wave speed.*

For a perfect gas, from Eq. (9.2) or (9.9), we deduce that the speed of sound is

$$a = \left(\frac{kp}{\rho}\right)^{1/2} = (kRT)^{1/2} \tag{9.16}$$

The speed of sound increases as the square root of the absolute temperature. For air, with $k = 1.4$, an easily memorized dimensional formula is

$$a(\text{ft/s}) \approx 49[T(°\text{R})]^{1/2} \qquad a(\text{m/s}) \approx 20[T(\text{K})]^{1/2} \tag{9.17}$$

At sea-level standard temperature, 15°C = 288.2 K, $a = 340.3$ m/s. This decreases in the upper atmosphere, which is cooler; at 15,000 m standard altitude, $T = -56.5°\text{C} = 216.7$ K and $a = 20(216.7\ \text{K})^{1/2} = 295.1$ m/s, or 13 percent less.

Some representative values of sound speed in various materials are given in Table 9.1. For liquids and solids it is common to define the *bulk modulus K* of the material:

$$K = -\upsilon\frac{\partial p}{\partial \upsilon}\bigg|_s = \rho\frac{\partial p}{\partial \rho}\bigg|_s \tag{9.18}$$

In terms of bulk modulus, then, $a = (K/\rho)^{1/2}$. For example, at standard conditions, the bulk modulus of liquid carbon tetrachloride is 1.32 GPa absolute, and its density is 1,590 kg/m³. Its speed of sound is therefore $a = (1.3\ \text{E9 Pa}/1{,}590\ \text{kg/m}^3)^{1/2} = 911$ m/s. Steel has a bulk modulus of about 2 E11 Pa and water about 2.2 E9 Pa (see Table A.3), or 90 times less than steel.

For solids, it is sometimes assumed that the bulk modulus is approximately equivalent to Young's modulus of elasticity *E*, but in fact their ratio depends on Poisson's ratio σ:

$$\frac{E}{K} = 3(1 - 2\sigma) \tag{9.19}$$

The two are equal for $\sigma = \frac{1}{3}$, which is approximately the case for many common metals such as steel and aluminum.

EXAMPLE 9.2

Estimate the speed of sound of carbon monoxide at 200-kPa pressure and 300°C in m/s.

Solution

From Table A.4, for CO, the molecular weight is 28.01 and $k \approx 1.40$. Thus from Eq. (9.3) $R_{CO} = 8314/28.01 = 297\ \text{m}^2/(\text{s}^2 \cdot \text{K})$, and the given temperature is 300°C + 273 = 573 K. Thus from Eq. (9.16) we estimate

$$a_{CO} = (kRT)^{1/2} = [1.40(297)(573)]^{1/2} = 488\ \text{m/s} \qquad \textit{Ans.}$$

9.3 Adiabatic and Isentropic Steady Flow

As mentioned in Sec. 9.1, the isentropic approximation greatly simplifies a compressible flow calculation. So does the assumption of adiabatic flow, even if nonisentropic.

Consider high-speed flow of a gas past an insulated wall, as in Fig. 9.2. There is no shaft work delivered to any part of the fluid. Therefore, every streamtube in the flow satisfies the steady flow energy equation in the form of Eq. (3.70):

$$h_1 + \tfrac{1}{2}V_1^2 + gz_1 = h_2 + \tfrac{1}{2}V_2^2 + gz_2 - q + w_v \tag{9.20}$$

where point 1 is upstream of point 2. You may wish to review the details of Eq. (3.70) and its development. We saw in Example 3.20 that potential energy changes of a gas are extremely small compared with kinetic energy and enthalpy terms. We shall neglect the terms gz_1 and gz_2 in all gas dynamic analyses.

Inside the thermal and velocity boundary layers in Fig. 9.2, the heat transfer and viscous work terms q and w_v are not zero. But outside the boundary layer q and w_v are zero by definition, so that the outer flow satisfies the simple relation

$$h_1 + \tfrac{1}{2}V_1^2 = h_2 + \tfrac{1}{2}V_2^2 = \text{const} \tag{9.21}$$

The constant in Eq. (9.21) is equal to the maximum enthalpy that the fluid would achieve if brought to rest adiabatically. We call this value h_0, the *stagnation enthalpy* of the flow. Thus, we rewrite Eq. (9.21) in the form

$$h + \tfrac{1}{2}V^2 = h_0 = \text{const} \tag{9.22}$$

This should hold for steady adiabatic flow of any compressible fluid outside the boundary layer. The wall in Fig. 9.2 could be either the surface of an immersed body or the wall of a duct. We have shown the details of Fig. 9.2; typically the thermal layer thickness δ_T is greater than the velocity layer thickness δ_V because most gases have a dimensionless Prandtl number Pr less than unity (see, for example, Ref. 19, Sec. 4-3.2). Note that the stagnation enthalpy varies inside the thermal boundary layer, but its average value is the same as that at the outer layer due to the insulated wall.

For nonperfect gases we may have to use the steam tables [15] or the gas tables [16] to implement Eq. (9.22). But, for a perfect gas $h = c_pT$, and Eq. (9.22) becomes

$$c_pT + \tfrac{1}{2}V^2 = c_pT_0 \tag{9.23}$$

This establishes the stagnation temperature T_0 of an adiabatic perfect-gas flow—that is, the temperature it achieves when decelerated to rest adiabatically.

An alternate interpretation of Eq. (9.22) occurs when the enthalpy and temperature drop to (absolute) zero, so the velocity achieves a maximum value:

$$V_{\max} = (2h_0)^{1/2} = (2c_pT_0)^{1/2} \tag{9.24}$$

No higher flow velocity can occur unless additional energy is added to the fluid through shaft work or heat transfer (Sec. 9.8).

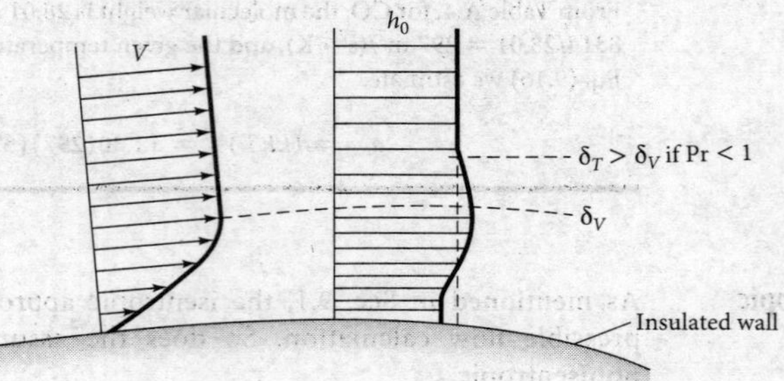

Fig. 9.2 Velocity and stagnation enthalpy distributions near an insulated wall in a typical high-speed gas flow.

Mach Number Relations

The dimensionless form of Eq. (9.23) brings in the Mach number Ma as a parameter, by using Eq. (9.16) for the speed of sound of a perfect gas. Divide through by c_pT to obtain

$$1 + \frac{V^2}{2c_pT} = \frac{T_0}{T} \tag{9.25}$$

But, from the perfect-gas law, $c_pT = [kR/(k-1)]T = a^2/(k-1)$, so that Eq. (9.25) becomes

$$1 + \frac{(k-1)V^2}{2a^2} = \frac{T_0}{T}$$

or

$$\frac{T_0}{T} = 1 + \frac{k-1}{2}\text{Ma}^2 \qquad \text{Ma} = \frac{V}{a} \tag{9.26}$$

This relation is plotted in Fig. 9.3 versus the Mach number for $k = 1.4$. At Ma = 5 the temperature has dropped to $\frac{1}{6}\,T_0$.

Since $a \propto T^{1/2}$, the ratio a_0/a is the square root of (9.26):

$$\frac{a_0}{a} = \left(\frac{T_0}{T}\right)^{1/2} = \left[1 + \frac{1}{2}(k-1)\text{Ma}^2\right]^{1/2} \tag{9.27}$$

Equation (9.27) is also plotted in Fig. 9.3. At Ma = 5 the speed of sound has dropped to 41 percent of the stagnation value.

Isentropic Pressure and Density Relations

Note that Eqs. (9.26) and (9.27) require only adiabatic flow and hold even in the presence of irreversibilities such as friction losses or shock waves.

If the flow is also *isentropic,* then for a perfect gas the pressure and density ratios can be computed from Eq. (9.9) as a power of the temperature ratio:

$$\frac{p_0}{p} = \left(\frac{T_0}{T}\right)^{k/(k-1)} = \left[1 + \frac{1}{2}(k-1)\text{Ma}^2\right]^{k/(k-1)} \tag{9.28a}$$

$$\frac{\rho_0}{\rho} = \left(\frac{T_0}{T}\right)^{1/(k-1)} = \left[1 + \frac{1}{2}(k-1)\text{Ma}^2\right]^{1/(k-1)} \tag{9.28b}$$

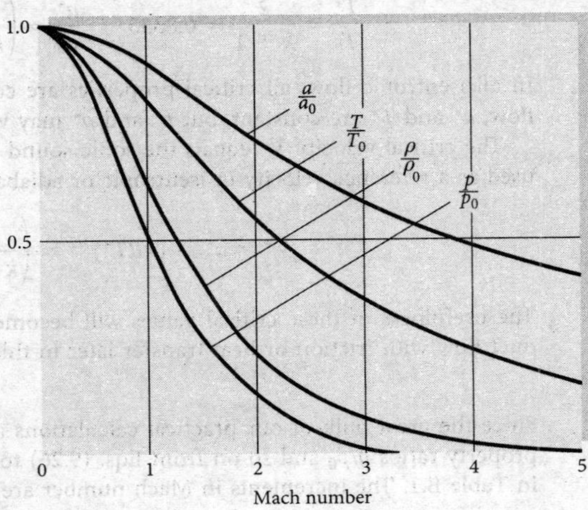

Fig. 9.3 Adiabatic (T/T_0 and a/a_0) and isentropic (p/p_0 and ρ/ρ_0) properties versus Mach number for $k = 1.4$.

These relations are also plotted in Fig. 9.3; at Ma = 5 the density is 1.13 percent of its stagnation value, and the pressure is only 0.19 percent of stagnation pressure.

The quantities p_0 and ρ_0 are the isentropic stagnation pressure and density, respectively that is, the pressure and density that the flow would achieve if brought isentropically to rest. In an adiabatic nonisentropic flow, p_0 and ρ_0 retain their local meaning, but they vary throughout the flow as the entropy changes due to friction or shock waves. The quantities h_0, T_0, and a_0 are constant in an adiabatic nonisentropic flow (see Sec. 9.7 for further details).

Relationship to Bernoulli's Equation

The isentropic assumptions (9.28) are effective, but are they realistic? Yes. To see why, take the differential of Eq. (9.22):

Adiabatic:
$$dh + V\,dV = 0 \tag{9.29}$$

Meanwhile, from Eq. (9.6), if $ds = 0$ (isentropic process),

$$dh = \frac{dp}{\rho} \tag{9.30}$$

Combining (9.29) and (9.30), we find that an isentropic streamtube flow must be

$$\frac{dp}{\rho} + V\,dV = 0 \tag{9.31}$$

But this is exactly the Bernoulli relation, Eq. (3.54), for steady frictionless flow with negligible gravity terms. Thus we see that the isentropic flow assumption is equivalent to use of the Bernoulli or streamline form of the frictionless momentum equation.

Critical Values at the Sonic Point

The stagnation values (a_0, T_0, p_0, ρ_0) are useful reference conditions in a compressible flow, but of comparable usefulness are the conditions where the flow is sonic, Ma = 1.0. These sonic, or *critical*, properties are denoted by asterisks: p^*, ρ^*, a^*, and T^*. They are certain ratios of the stagnation properties as given by Eqs. (9.26) to (9.28) when Ma = 1.0; for $k = 1.4$

$$\frac{p^*}{p_0} = \left(\frac{2}{k+1}\right)^{k/(k-1)} = 0.5283 \qquad \frac{\rho^*}{\rho_0} = \left(\frac{2}{k+1}\right)^{1/(k-1)} = 0.6339$$
$$\frac{T^*}{T_0} = \frac{2}{k+1} = 0.8333 \qquad \frac{a^*}{a_0} = \left(\frac{2}{k+1}\right)^{1/2} = 0.9129 \tag{9.32}$$

In all isentropic flow, all critical properties are constant; in adiabatic nonisentropic flow, a^* and T^* are constant, but p^* and ρ^* may vary.

The critical velocity V^* equals the sonic sound speed a^* by definition and is often used as a reference velocity in isentropic or adiabatic flow:

$$V^* = a^* = (kRT^*)^{1/2} = \left(\frac{2k}{k+1}RT_0\right)^{1/2} \tag{9.33}$$

The usefulness of these critical values will become clearer as we study compressible duct flow with friction or heat transfer later in this chapter.

Some Useful Numbers for Air

Since the great bulk of our practical calculations are for air, $k = 1.4$, the stagnation property ratios p/p_0 and so on from Eqs. (9.26) to (9.28) are tabulated for this value in Table B.1. The increments in Mach number are rather coarse in this table because the values are meant as only a guide; these equations are now a trivial matter to manipulate on a hand calculator. Thirty years ago every text had extensive compressible flow tables with Mach number spacings of about 0.01, so that accurate values could be interpolated. Even today, reference books are available [20, 21, 29] with tables and charts and computer programs for a wide variety of compressible flow situations. Reference 22 contains formulas and charts applying to the thermodynamics of *real* (nonperfect) gas flows.

For $k = 1.4$, the following numerical versions of the isentropic and adiabatic flow formulas are obtained:

$$\frac{T_0}{T} = 1 + 0.2\,\text{Ma}^2 \qquad \frac{\rho_0}{\rho} = (1 + 0.2\,\text{Ma}^2)^{2.5}$$
$$\frac{p_0}{p} = (1 + 0.2\,\text{Ma}^2)^{3.5} \tag{9.34}$$

Or, if we are given the properties, it is equally easy to solve for the Mach number (again with $k = 1.4$):

$$\text{Ma}^2 = 5\left(\frac{T_0}{T} - 1\right) = 5\left[\left(\frac{\rho_0}{\rho}\right)^{2/5} - 1\right] = 5\left[\left(\frac{p_0}{p}\right)^{2/7} - 1\right] \tag{9.35}$$

Note that these isentropic flow formulas serve as the equivalent of the frictionless adiabatic momentum and energy equations. They relate velocity to physical properties for a perfect gas, but they are *not* the "solution" to a gas dynamics problem. The complete solution is not obtained until the continuity equation has also been satisfied, for either one-dimensional (Sec. 9.4) or multidimensional (Sec. 9.9) flow.

One final note: These isentropic-ratio–versus–Mach-number formulas are seductive, tempting one to solve all problems by jumping right into the tables. Actually, many problems involving (dimensional) velocity and temperature can be solved more easily from the original raw dimensional energy equation (9.23) plus the perfect-gas law (9.2), as the next example will illustrate.

EXAMPLE 9.3

Air flows adiabatically through a duct. At point 1 the velocity is 240 m/s, with $T_1 = 320$ K and $p_1 = 170$ kPa. Compute (*a*) T_0, (*b*) p_0, (*c*) ρ_0, (*d*) Ma, (*e*) V_{max}, and (*f*) V^*. At point 2 further downstream $V_2 = 290$ m/s and $p_2 = 135$ kPa. (*g*) What is the stagnation pressure p_{02}?

Solution

- *Assumptions:* Let air be approximated as an ideal gas with constant k. The flow is adiabatic but *not* isentropic. Isentropic formulas are used *only* to compute local p_0 and ρ_0, which vary.
- *Approach:* Use adiabatic and isentropic formulas to find the various properties.
- *Ideal gas parameters:* For air, $R = 287\ \text{m}^2/(\text{s}^2 \cdot \text{K})$, $k = 1.40$, and $c_p = 1{,}005\ \text{m}^2/(\text{s}^2 \cdot \text{K})$.
- *Solution steps (a, b, c, d):* With T_1, p_1, and V_1 known, other properties at point 1 follow:

$$T_{01} = T_1 + \frac{V_1^2}{2c_p} = 320 + \frac{(240\ \text{m/s})^2}{2\,[1{,}005\ \text{m}^2/(\text{s}^2 \cdot \text{K})]} = 320 + 29 = 349\ \text{K} \qquad \textit{Ans. (a)}$$

Once the Mach number is found from Eq. (9.35), local stagnation pressure and density follow:

$$\text{Ma}_1 = \sqrt{5\left(\frac{T_{01}}{T_1} - 1\right)} = \sqrt{5\left(\frac{349\text{ K}}{320\text{ K}} - 1\right)} = \sqrt{0.448} = 0.67 \qquad \textit{Ans. (b)}$$

$$p_{01} = p_1(1 + 0.2\,\text{Ma}_1^2)^{3.5} = (170\text{ kPa})\,[1 + 0.2(0.67)^2]^{3.5} = 230\text{ kPa} \qquad \textit{Ans. (d)}$$

$$\rho_{01} = \frac{p_{01}}{RT_{01}} = \frac{230{,}000\text{ N/m}^2}{[287\text{ m}^2/(\text{s}^2\cdot\text{K})]\,(349\text{ K})} = 2.29\,\frac{\text{N}\cdot\text{s}^2/\text{m}}{\text{m}^3} = 2.29\,\frac{\text{kg}}{\text{m}^3} \qquad \textit{Ans. (c)}$$

- *Comment:* Note that we used dimensional (non-Mach-number) formulas where convenient.
- *Solution steps (e, f):* Both V_{max} and V^* are directly related to stagnation temperature from Eqs. (9.24) and (9.33):

$$V_{max} = \sqrt{2c_pT_0} = \sqrt{2[1{,}005\text{ m}^2/(\text{s}^2\cdot\text{K})]\,(349\text{ K})} = 837\,\frac{\text{m}}{\text{s}} \qquad \textit{Ans. (e)}$$

$$V^* = \sqrt{\frac{2k}{k+1}RT_0} = \sqrt{\frac{2(1.4)}{(1.4+1)}\left(287\,\frac{\text{m}^2}{\text{s}^2\cdot\text{K}}\right)(349\text{ K})} = 342\,\frac{\text{m}}{\text{s}} \qquad \textit{Ans. (f)}$$

- At point 2 downstream, the temperature is unknown, but since the flow is adiabatic, the stagnation temperature is constant: $T_{01} = T_{02} = 349$ K. Thus, from Eq. (9.23),

$$T_2 = T_{02} - \frac{V_2^2}{2c_p} = 349 - \frac{(290\text{ m/s})^2}{2[1{,}006\text{ m}^2/(\text{s}^2\cdot\text{K})]} = 307\text{ K}$$

Hence, from Eq. (9.28*a*), the isentropic stagnation pressure at point 2 is

$$p_{02} = p_2\left(\frac{T_{02}}{T_2}\right)^{k/(k-1)} = (135\text{ kPa})\left(\frac{349\text{ K}}{307\text{ K}}\right)^{1.4/0.4} = 211\text{ kPa} \qquad \textit{Ans. (g)}$$

- *Comments:* Part (*g*), a ratio-type ideal-gas formula, is more direct than finding the Mach number, which turns out to be $\text{Ma}_2 = 0.83$, and using the Mach number formula, Eq. (9.34) for p_{02}. Note that p_{02} is 8 percent less than p_{01}. The flow is nonisentropic: Entropy rises downstream, and stagnation pressure and density drop, due in this case to frictional losses.

9.4 Isentropic Flow with Area Changes

By combining the isentropic and/or adiabatic flow relations with the equation of continuity we can study practical compressible flow problems. This section treats the one-dimensional flow approximation.

Figure 9.4 illustrates the one-dimensional flow assumption. A real flow, Fig. 9.4*a*, has no slip at the walls and a velocity profile $V(x, y)$ that varies across the duct section (compare with Fig. 7.8). If, however, the area change is small and the wall radius of curvature large

$$\frac{d\mathbf{b}}{dx} \ll 1 \qquad \mathbf{b}(x) \ll R(x) \qquad (9.36)$$

then the flow is approximately one-dimensional, as in Fig. 9.4*b*, with $V \approx V(x)$ reacting to area change $A(x)$. Compressible flow nozzles and diffusers do not always satisfy conditions (9.36), but we use the one-dimensional theory anyway because of its simplicity.

For steady one-dimensional flow the equation of continuity is, from Eq. (3.24),

$$\rho(x)V(x)A(x) = \dot{m} = \text{const} \qquad (9.37)$$

y
x
V(x, y)
b(x)
Wall radius of
curvature R(x)
(a)
y
x
Area
A(x)
V(x)
(b)

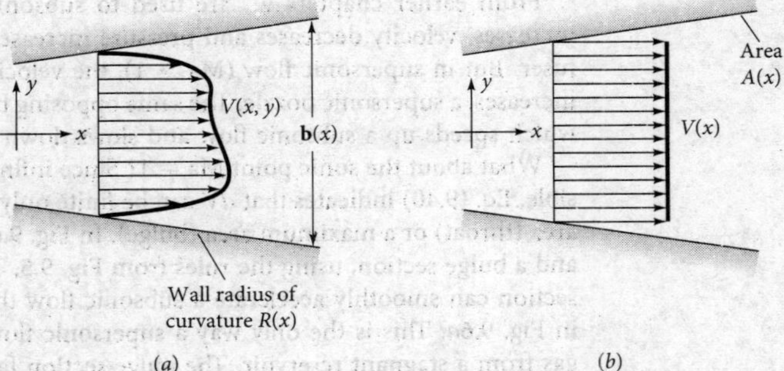

Fig. 9.4 Compressible flow through a duct: (*a*) real-fluid velocity profile; (*b*) one-dimensional approximation.

Before applying this to duct theory, we can learn a lot from the differential form of Eq. (9.37):

$$\frac{d\rho}{\rho} + \frac{dV}{V} + \frac{dA}{A} = 0 \tag{9.38}$$

The differential forms of the frictionless momentum equation (9.31) and the sound-speed relation (9.15) are recalled here for convenience:

$$\text{Momentum} \qquad \frac{dp}{\rho} + V\,dV = 0$$
$$\text{Sound speed:} \qquad dp = a^2\,d\rho \tag{9.39}$$

Now eliminate dp and $d\rho$ between Eqs. (9.38) and (9.39) to obtain the following relation between velocity change and area change in isentropic duct flow:

$$\frac{dV}{V} = \frac{dA}{A}\,\frac{1}{\mathrm{Ma}^2 - 1} = -\frac{dp}{\rho V^2} \tag{9.40}$$

Inspection of this equation, without actually solving it, reveals a fascinating aspect of compressible flow: Property changes are of opposite sign for subsonic and supersonic flow because of the term $\mathrm{Ma}^2 - 1$. There are four combinations of area change and Mach number, summarized in Fig. 9.5.

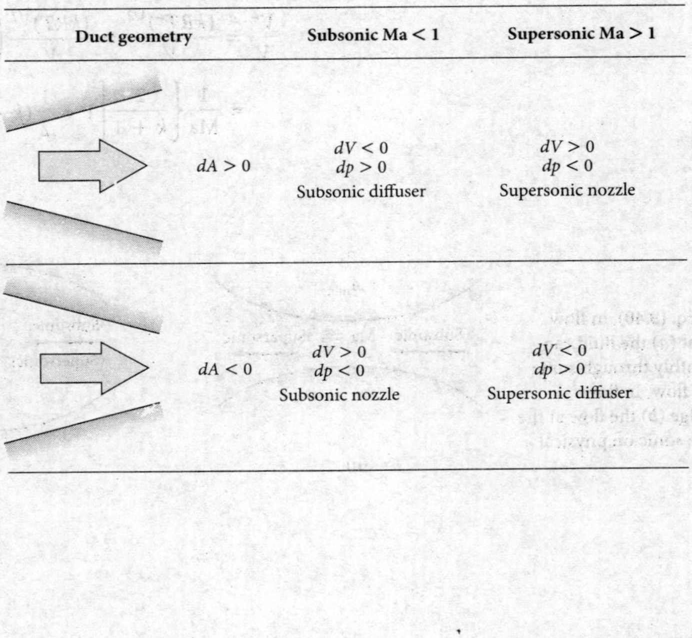

Duct geometry
Subsonic Ma < 1
Supersonic Ma > 1
dA > 0
dV < 0
dp > 0
Subsonic diffuser
dV > 0
dp < 0
Supersonic nozzle
dA < 0
dV > 0
dp < 0
Subsonic nozzle
dV < 0
dp > 0
Supersonic diffuser

Fig. 9.5 Effect of Mach number on property changes with area change in duct flow.

From earlier chapters w are used to subsonic behavior (Ma < 1): When area increases, velocity decreases and pressure increases, which is denoted a subsonic diffuser. But in supersonic flow (Ma > 1), the velocity actually increases when the area increases, a supersonic nozzle. The same opposing behavior occurs for an area decrease, which speeds up a subsonic flow and slows down a supersonic flow.

What about the sonic point Ma = 1? Since infinite acceleration is physically impossible, Eq. (9.40) indicates that dV can be finite only when $dA = 0$—that is, a minimum area (throat) or a maximum area (bulge). In Fig. 9.6 we patch together a throat section and a bulge section, using the rules from Fig. 9.5. The throat or converging–diverging section can smoothly accelerate a subsonic flow through sonic to supersonic flow, as in Fig. 9.6*a*. This is the only way a supersonic flow can be created by expanding the gas from a stagnant reservoir. The bulge section fails; the bulge Mach number moves away from a sonic condition rather than toward it.

Although supersonic flow downstream of a nozzle requires a sonic throat, the opposite is not true: A compressible gas can pass through a throat section without becoming sonic.

Perfect-Gas Area Change

We can use the perfect-gas and isentropic flow relations to convert the continuity relation (9.37) into an algebraic expression involving only area and Mach number, as follows. Equate the mass flow at any section to the mass flow under sonic conditions (which may not actually occur in the duct):

$$\rho V A = \rho^* V^* A^*$$

or

$$\frac{A}{A^*} = \frac{\rho^*}{\rho}\frac{V^*}{V} \tag{9.41}$$

Both terms on the right are functions only of Mach number for isentropic flow. From Eqs. (9.28) and (9.32)

$$\frac{\rho^*}{\rho} = \frac{\rho^*}{\rho_0}\frac{\rho_0}{\rho} = \left\{\frac{2}{k+1}\left[1 + \frac{1}{2}(k-1)\mathrm{Ma}^2\right]\right\}^{1/(k-1)} \tag{9.42}$$

From Eqs. (9.26) and (9.32) we obtain

$$\frac{V^*}{V} = \frac{(kRT^*)^{1/2}}{V} = \frac{(kRT)^{1/2}}{V}\left(\frac{T^*}{T_0}\right)^{1/2}\left(\frac{T_0}{T}\right)^{1/2}$$

$$= \frac{1}{\mathrm{Ma}}\left\{\frac{2}{k+1}\left[1 + \frac{1}{2}(k-1)\mathrm{Ma}^2\right]\right\}^{1/2} \tag{9.43}$$

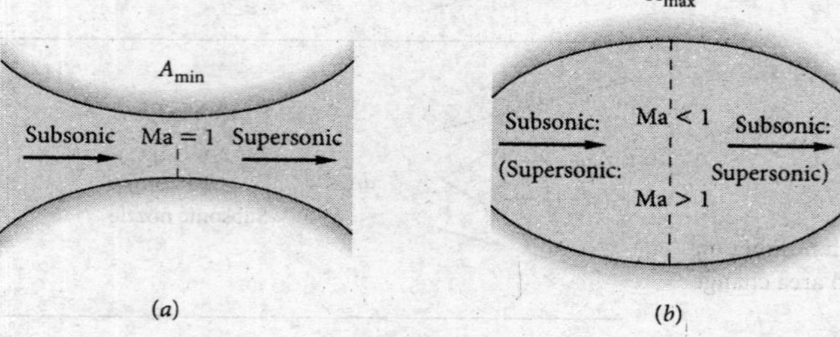

Fig. 9.6 From Eq. (9.40), in flow through a throat (*a*) the fluid can accelerate smoothly through sonic and supersonic flow. In flow through the bulge (*b*) the flow at the bulge cannot be sonic on physical grounds.

Combining Eqs. (9.41) to (9.43), we get the desired result:

$$\frac{A}{A^*} = \frac{1}{\text{Ma}}\left[\frac{1 + \frac{1}{2}(k-1)\,\text{Ma}^2}{\frac{1}{2}(k+1)}\right]^{(1/2)(k+1)/(k-1)} \quad (9.44)$$

For $k = 1.4$, Eq. (9.44) takes the numerical form

$$\frac{A}{A^*} = \frac{1}{\text{Ma}}\frac{(1 + 0.2\,\text{Ma}^2)^3}{1.728} \quad (9.45)$$

which is plotted in Fig. 9.7. Equations (9.45) and (9.34) enable us to solve any one-dimensional isentropic airflow problem given, say, the shape of the duct $A(x)$ and the stagnation conditions and assuming that there are no shock waves in the duct.

Figure 9.7 shows that the minimum area that can occur in a given isentropic duct flow is the sonic, or critical, throat area. All other duct sections must have A greater than A^*. In many flows a critical sonic throat is not actually present, and the flow in the duct is either entirely subsonic or, more rarely, entirely supersonic.

Choking

From Eq. (9.41) the inverse ratio A^*/A equals $\rho V/(\rho^* V^*)$, the mass flow per unit area at any section compared with the critical mass flow per unit area. From Fig. 9.7 this inverse ratio rises from zero at Ma = 0 to unity at Ma = 1 and back down to zero at large Ma. Thus, for given stagnation conditions, the maximum possible mass flow passes through a duct when its throat is at the critical or sonic condition. The duct is then said to be *choked* and can carry no additional mass flow unless the throat is widened. If the throat is constricted further, the mass flow through the duct must decrease.

From Eqs. (9.32) and (9.33) the maximum mass flow is

$$\dot{m}_{\text{max}} = \rho^* A^* V^* = \rho_0\left(\frac{2}{k+1}\right)^{1/(k-1)} A^*\left(\frac{2k}{k+1}RT_0\right)^{1/2}$$

$$= k^{1/2}\left(\frac{2}{k+1}\right)^{(1/2)(k+1)/(k-1)} A^*\rho_0\,(RT_0)^{1/2} \quad (9.46a)$$

For $k = 1.4$ this reduces to

$$\dot{m}_{\text{max}} = 0.6847 A^*\rho_0(RT_0)^{1/2} = \frac{0.6847 p_0\, A^*}{(RT_0)^{1/2}} \quad (9.46b)$$

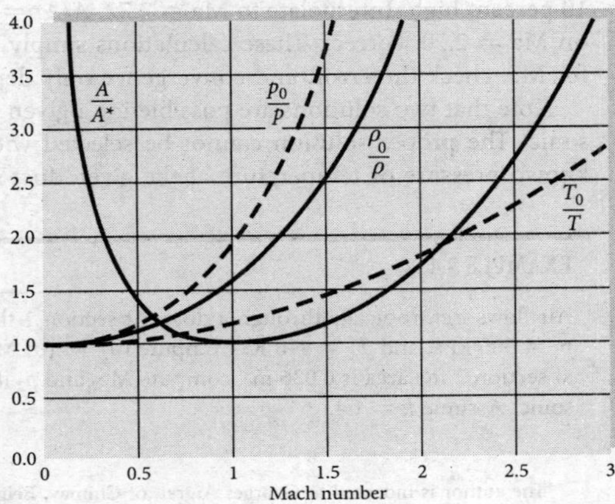

Fig. 9.7 Area ratio and fluid properties versus Mach number for isentropic flow of a perfect gas with $k = 1.4$.

For isentropic flow through a duct, the maximum mass flow possible is proportional to the throat area and stagnation pressure and inversely proportional to the square root of the stagnation temperature. These are somewhat abstract facts, so let us illustrate with some examples.

The Local Mass Flow Function

Equations (9.46) give the *maximum* mass flow, which occurs at the choking condition (sonic exit). They can be modified to predict the actual (nonmaximum) mass flow at any section where local area A and pressure p are known.[1] The algebra is convoluted, so here we give only the final result, expressed in dimensionless form:

$$\text{Mass flow function} = \frac{\dot{m}\sqrt{RT_0}}{A\,p_0} = \sqrt{\frac{2k}{k-1}\left(\frac{p}{p_0}\right)^{2/k}\left[1-\left(\frac{p}{p_0}\right)^{(k-1)/k}\right]} \tag{9.47}$$

We stress that p and A in this relation are the *local* values at position x. As p/p_0 falls, this function rises rapidly and then levels out at the maximum of Eqs. (9.46). A few values may be tabulated here for $k = 1.4$:

p/p_0	1.0	0.98	0.95	0.9	0.8	0.7	0.6	≤0.5283
Function	0.0	0.1978	0.3076	0.4226	0.5607	0.6383	0.6769	0.6847

Equation (9.47) is handy if stagnation conditions are known and the flow is not choked.

When A/A^* is known and the Mach number is unknown, no algebraic solution of Eq. (9.44) is known to the writer. One could interpolate in Table B.1 or simply iterate Eq. (9.44) with a calculator. But Excel can iterate Eq. (9.44) for subsonic flow in its direct form:

$$\text{Subsonic flow: Ma} = \frac{A^*}{A}\left[\frac{1+0.5(k-1)\text{Ma}^2}{0.5(k+1)}\right]^{0.5(k+1)/(k-1)} \tag{9.48}$$

Make a subsonic guess for Ma on the right side and then replace it with the value calculated on the left side. For example, suppose $A/A^* = 2.035$, corresponding to Ma = 0.300. A poor guess of Ma = 0.5 in Eq. (9.44) leads to a better Ma = 0.329, then 0.303, then 0.300.

For supersonic flow, iteration of Eq. (9.44) diverges. Instead, simply try different Mach numbers in Eq. (9.44) until the proper area is achieved. For example, suppose $A/A^* = 3.183$, corresponding to Ma = 2.70. A poor guess of Ma = 2.4 yields $A/A^* = 2.403$, 24 percent low. Improve the guess to Ma = 2.8 to give $A/A^* = 3.500$, or 10 percent high. Interpolate to Ma = 2.72, $A/A^* = 3.244$, 2 percent high. Finally settle on Ma = 2.70, correct. These calculations simply require that you retype your guess for Ma, check the error, and convergence only depends upon your cleverness.

Note that two solutions are possible for a given A/A^*, one subsonic and one supersonic. The proper solution cannot be selected without further information, such as known pressure or temperature at the given duct section.

EXAMPLE 9.4

Air flows isentropically through a duct. At section 1 the area is 0.05 m^2 and $V_1 = 180$ m/s, $p_1 = 500$ kPa, and $T_1 = 470$ K. Compute (*a*) T_0, (*b*) Ma_1, (*c*) p_0, and (*d*) both A^* and $\dot{m}$. If at section 2 the area is 0.036 m^2, compute Ma_2 and p_2 if the flow is (*e*) subsonic or (*f*) supersonic. Assume $k = 1.4$.

[1]The author is indebted to Georges Aigret, of Chimay, Belgium, for suggesting this useful function.

Solution

Part (a) A general sketch of the problem is shown in Fig. E9.4. With V_1 and T_1 known, the energy equation (9.23) gives

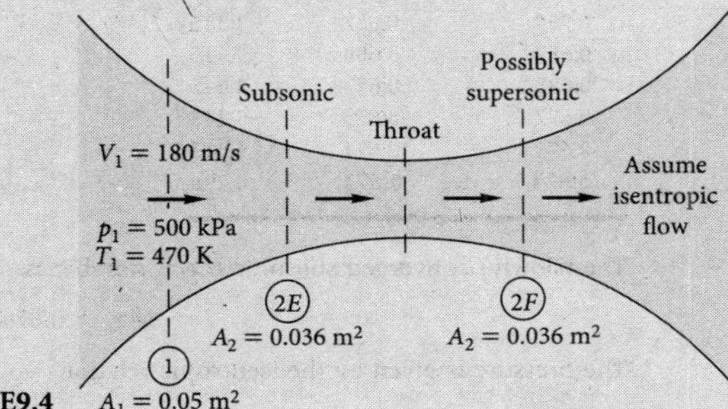

E9.4

$$T_0 = T_1 + \frac{V_1^2}{2c_p} = 470 + \frac{(180)^2}{2(1{,}005)} = 486 \text{ K} \qquad \textit{Ans. (a)}$$

Part (b) The local sound speed $a_1 = \sqrt{kRT_1} = [(1.4)(287)(470)]^{1/2} = 435$ m/s. Hence

$$\text{Ma}_1 = \frac{V_1}{a_1} = \frac{180}{435} = 0.414 \qquad \textit{Ans. (b)}$$

Part (c) With Ma_1 known, the stagnation pressure follows from Eq. (9.34):

$$p_0 = p_1(1 + 0.2\,\text{Ma}_1^2)^{3.5} = (500 \text{ kPa})[1 + 0.2(0.414)^2]^{3.5} = 563 \text{ kPa} \qquad \textit{Ans. (c)}$$

Part (d) Similarly, from Eq. (9.45), the critical sonic throat area is

$$\frac{A_1}{A^*} = \frac{(1 + 0.2\,\text{Ma}_1^2)^3}{1.728\,\text{Ma}_1} = \frac{[1 + 0.2(0.414)^2]^3}{1.728(0.414)} = 1.547$$

or

$$A^* = \frac{A_1}{1.547} = \frac{0.05 \text{ m}^2}{1.547} = 0.0323 \text{ m}^2 \qquad \textit{Ans. (d)}$$

This throat must *actually be present* in the duct if the flow is to become supersonic.

We now know A^*. So, to compute the mass flow we can use Eqs. (9.46), which remain valid, based on the numerical value of A^*, whether or not a throat actually exists:

$$\dot{m} = 0.6847 \frac{p_0 A^*}{\sqrt{RT_0}} = 0.6847 \frac{(563{,}000)(0.0323)}{\sqrt{(287)(486)}} = 33.4 \text{ kg/s} \qquad \textit{Ans. (d)}$$

Or, we could fare equally well with our new "local mass flow" formula, Eq. (9.47), using, say, the pressure and area at section 1. Given $p_1/p_0 = 500/563 = 0.889$, Eq. (9.47) yields

$$\dot{m}\frac{\sqrt{287(486)}}{563{,}000(0.05)} = \sqrt{\frac{2(1.4)}{0.4}(0.889)^{2/1.4}[1 - (0.889)^{0.4/1.4}]} = 0.444 \qquad \dot{m} = 33.4 \frac{\text{kg}}{\text{s}}$$

Ans. (d)

Part (e) For subsonic flow upstream of the throat at section 2E, the area ratio is $A_2/A^* = 0.036/0.0323 = 1.115$, corresponding to the left side of Fig. 9.7 or the subsonic numbers in Table B.1, neither of which is very accurate. Guess Ma_2 at section 2*E*, from Fig. 9.7, at about 0.70. Enter this guess into Eq. (9.48) and repeat. The Excel table is:

Ma – guess	Ma – Eq. (9.48)	A/A^*
0.700	0.687	1.115
0.687	0.680	1.115
0.680	0.677	1.115
0.677	0.675	1.115
0.675	0.674	1.115
0.674	**0.674**	1.115

The (slowly) converged subsonic Mach number is

$$Ma_2 = 0.674 \qquad \textit{Ans. (e)}$$

The pressure is given by the isentropic relation

$$p_2 = \frac{p_o}{[1 + 0.2(0.674)^2]^{3.5}} = \frac{563 \text{ kPa}}{1.356} = 415 \text{ kPa} \qquad \textit{Ans. (e)}$$

Part (*e*) does not require a throat, sonic or otherwise; the flow could simply be contracting subsonically from A_1 to A_2.

Part (f) For supersonic flow at section 2*F*, again the area ratio is 0.036/0.0323 = 1.115. On the right side of Fig. 9.7 we estimate $Ma_2 \approx 1.5$. The table from Eq. (9.44) is

Ma – guess	A/A^* – Eq. (9.44)	A/A^*
1.5000	1.1762	1.1150
1.4000	1.1149	1.1150
1.4001	**1.1150**	**1.1150**

We were lucky that this Mach number is easy to guess:

$$Ma_2 = 1.4001 \qquad \textit{Ans. (f)}$$

Again the pressure is given by the isentropic relation at this new Mach number:

$$p_2 = \frac{p_o}{[1 + 0.2(1.4001)^2]^{3.5}} = \frac{563 \text{ kPa}}{3.183} = 177 \text{ kPa} \qquad \textit{Ans. (f)}$$

Note that the supersonic-flow pressure level is much less than p_2 in part (*e*), and a sonic throat *must* have occurred between sections 1 and 2*F*.

EXAMPLE 9.5

It is desired to expand air from $p_0 = 200$ kPa and $T_0 = 500$ K through a throat to an exit Mach number of 2.5. If the desired mass flow is 3 kg/s, compute (*a*) the throat area and the exit (*b*) pressure, (*c*) temperature, (*d*) velocity, and (*e*) area, assuming isentropic flow, with $k = 1.4$.

Solution

The throat area follows from Eq. (9.47), because the throat flow must be sonic to produce a supersonic exit:

$$A^* = \frac{\dot{m}(RT_0)^{1/2}}{0.6847 p_0} = \frac{3.0[287(500)]^{1/2}}{0.6847(200{,}000)} = 0.00830 \text{ m}^2 = \frac{1}{4}\pi D^{*2}$$

or

$$D_{\text{throat}} = 10.3 \text{ cm} \qquad \textit{Ans. (a)}$$

With the exit Mach number known, the isentropic flow relations give the pressure and temperature:

$$p_e = \frac{p_0}{[1 + 0.2(2.5)^2]^{3.5}} = \frac{200{,}000}{17.08} = 11{,}700 \text{ Pa} \qquad \textit{Ans. (b)}$$

$$T_e = \frac{T_0}{1 + 0.2(2.5)^2} = \frac{500}{2.25} = 222 \text{ K} \qquad \textit{Ans. (c)}$$

The exit velocity follows from the known Mach number and temperature:

$$V_e = \text{Ma}_e(kRT_e)^{1/2} = 2.5[1.4(287)(222)]^{1/2} = 2.5(299 \text{ m/s}) = 747 \text{ m/s} \qquad \textit{Ans. (d)}$$

The exit area follows from the known throat area and exit Mach number and Eq. (9.45):

$$\frac{A_e}{A^*} = \frac{[1 + 0.2(2.5)^2]^3}{1.728(2.5)} = 2.64$$

or

$$A_e = 2.64A^* = 2.64(0.0083 \text{ m}^2) = 0.0219 \text{ m}^2 = \tfrac{1}{4}\pi D_e^2$$

or

$$D_e = 16.7 \text{ cm} \qquad \textit{Ans. (e)}$$

One point might be noted: The computation of the throat area A^* did not depend in any way on the numerical value of the exit Mach number. The exit was supersonic; therefore the throat is sonic and choked, and no further information is needed.

9.5 The Normal Shock Wave

Shock waves are nearly discontinuous changes in a supersonic flow. They can occur due to a higher downstream pressure, a sudden change in flow direction, blockage by a downstream body, or the result of an explosion. The simplest algebraically is a one-dimensional change, or *normal shock wave*, shown in Fig. 9.8. We select a control volume just before and after the wave.

The analysis is identical to that of Fig. 9.1; that is, a shock wave is a fixed strong pressure wave. To compute all property changes rather than just the wave speed, we use all our basic one-dimensional steady flow relations, letting section 1 be upstream and section 2 be downstream:

Continuity: $$\rho_1 V_1 = \rho_2 V_2 = G = \text{const} \qquad (9.49a)$$

Momentum: $$p_1 - p_2 = \rho_2 V_2^2 - \rho_1 V_1^2 \qquad (9.49b)$$

Energy: $$h_1 + \tfrac{1}{2}V_1^2 = h_2 + \tfrac{1}{2}V_2^2 = h_0 = \text{const} \qquad (9.49c)$$

Perfect gas: $$\frac{p_1}{\rho_1 T_1} = \frac{p_2}{\rho_2 T_2} \qquad (9.49d)$$

Constant c_p: $$h = c_p T \qquad k = \text{const} \qquad (9.49e)$$

Note that we have canceled out the areas $A_1 \approx A_2$, which is justified even in a variable duct section because of the thinness of the wave. The first successful analyses of these normal shock relations are credited to W. J. M. Rankine (1870) and A. Hugoniot (1887), hence the modern term *Rankine-Hugoniot relations*. If we assume that the upstream conditions (p_1, V_1, ρ_1, h_1, T_1) are known, Eqs. (9.49) are five algebraic relations in the five unknowns (p_2, V_2, ρ_2, h_2, T_2). Because of the velocity-squared term, two solutions are found, and the correct one is determined from the second law of thermodynamics, which requires that $s_2 > s_1$.

The velocities V_1 and V_2 can be eliminated from Eqs. (9.49*a*) to (9.49*c*) to obtain the Rankine-Hugoniot relation:

$$h_2 - h_1 = \frac{1}{2}(p_2 - p_1)\left(\frac{1}{\rho_2} + \frac{1}{\rho_1}\right) \qquad (9.50)$$

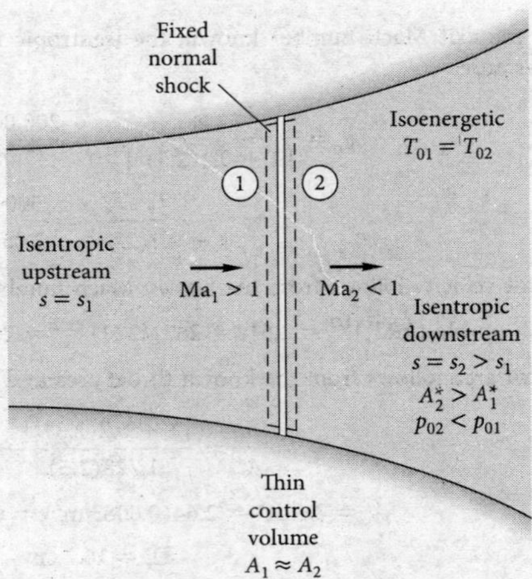

Fig. 9.8 Flow through a fixed normal shock wave.

This contains only thermodynamic properties and is independent of the equation of state. Introducing the perfect-gas law $h = c_pT = kp/[(k-1)\rho]$, we can rewrite this as

$$\frac{\rho_2}{\rho_1} = \frac{1 + \beta p_2/p_1}{\beta + p_2/p_1} \qquad \beta = \frac{k+1}{k-1} \tag{9.51}$$

We can compare this with the isentropic flow relation for a very weak pressure wave in a perfect gas:

$$\frac{\rho_2}{\rho_1} = \left(\frac{p_2}{p_1}\right)^{1/k} \tag{9.52}$$

Also, the actual change in entropy across the shock can be computed from the perfect-gas relation:

$$\frac{s_2 - s_1}{c_v} = \ln\left[\frac{p_2}{p_1}\left(\frac{\rho_1}{\rho_2}\right)^k\right] \tag{9.53}$$

Assuming a given wave strength p_2/p_1, we can compute the density ratio and the entropy change and list them as follows for $k = 1.4$:

$\frac{p_2}{p_1}$	ρ_2/ρ_1 Eq. (9.51)	ρ_2/ρ_1 Isentropic	$\frac{s_2 - s_1}{c_v}$
0.5	0.6154	0.6095	−0.0134
0.9	0.9275	0.9275	−0.00005
1.0	1.0	1.0	0.0
1.1	1.00704	1.00705	0.00004
1.5	1.3333	1.3359	0.0027
2.0	1.6250	1.6407	0.0134

We see that the entropy change is negative if the pressure decreases across the shock, which violates the second law. Thus a rarefaction shock is impossible in a perfect gas.[2] We see also that weak shock waves ($p_2/p_1 \leq 2.0$) are very nearly isentropic.

[2]This is true also for most real gases; see Ref. 9, Sec. 7.3.

Mach Number Relations

For a perfect gas all the property ratios across the normal shock are unique functions of k and the upstream Mach number Ma_1. For example, if we eliminate ρ_2 and V_2 from Eqs. (9.49*a*) to (9.49*c*) and introduce $h = kp/[(k-1)\rho]$, we obtain

$$\frac{p_2}{p_1} = \frac{1}{k+1}\left[\frac{2\rho_1 V_1^2}{p_1} - (k-1)\right] \tag{9.54}$$

But for a perfect gas $\rho_1 V_1^2/p_1 = kV_1^2/(kRT_1) = k\,\mathrm{Ma}_1^2$, so that Eq. (9.54) is equivalent to

$$\frac{p_2}{p_1} = \frac{1}{k+1}\left[2k\,\mathrm{Ma}_1^2 - (k-1)\right] \tag{9.55}$$

From this equation we see that, for any k, $p_2 > p_1$ only if $\mathrm{Ma}_1 > 1.0$. Thus for flow through a normal shock wave, the upstream Mach number must be supersonic to satisfy the second law of thermodynamics.

What about the downstream Mach number? From the perfect-gas identity $\rho V^2 = kp\,\mathrm{Ma}^2$, we can rewrite Eq. (9.49*b*) as

$$\frac{p_2}{p_1} = \frac{1 + k\,\mathrm{Ma}_1^2}{1 + k\,\mathrm{Ma}_2^2} \tag{9.56}$$

which relates the pressure ratio to both Mach numbers. By equating Eqs. (9.55) and (9.56) we can solve for

$$\mathrm{Ma}_2^2 = \frac{(k-1)\,\mathrm{Ma}_1^2 + 2}{2k\,\mathrm{Ma}_1^2 - (k-1)} \tag{9.57}$$

Since Ma_1 must be supersonic, this equation predicts for all $k > 1$ that Ma_2 must be subsonic. Thus, a normal shock wave decelerates a flow almost discontinuously from supersonic to subsonic conditions.

Further manipulation of the basic relations (9.49) for a perfect gas gives additional equations relating the change in properties across a normal shock wave in a perfect gas:

$$\frac{\rho_2}{\rho_1} = \frac{(k+1)\mathrm{Ma}_1^2}{(k-1)\mathrm{Ma}_1^2 + 2} = \frac{V_1}{V_2}$$

$$\frac{T_2}{T_1} = [2 + (k-1)\mathrm{Ma}_1^2]\,\frac{2k\,\mathrm{Ma}_1^2 - (k-1)}{(k+1)^2\,\mathrm{Ma}_1^2} \tag{9.58}$$

$$T_{02} = T_{01}$$

$$\frac{p_{02}}{p_{01}} = \frac{\rho_{02}}{\rho_{01}} = \left[\frac{(k+1)\,\mathrm{Ma}_1^2}{2 + (k-1)\,\mathrm{Ma}_1^2}\right]^{k/(k-1)}\left[\frac{k+1}{2k\,\mathrm{Ma}_1^2 - (k-1)}\right]^{1/(k-1)}$$

Of additional interest is the fact that the critical, or sonic, throat area A^* in a duct increases across a normal shock:

$$\frac{A_2^*}{A_1^*} = \frac{\mathrm{Ma}_2}{\mathrm{Ma}_1}\left[\frac{2 + (k-1)\,\mathrm{Ma}_1^2}{2 + (k-1)\,\mathrm{Ma}_2^2}\right]^{(1/2)(k+1)/(k-1)} \tag{9.59}$$

All these relations are given in Table B.2 and plotted versus upstream Mach number Ma_1 in Fig. 9.9 for $k = 1.4$. We see that pressure increases greatly while temperature and density increase moderately. The effective throat area A^* increases slowly at first and then rapidly. The failure of students to account for this change in A^* is a common source of error in shock calculations.

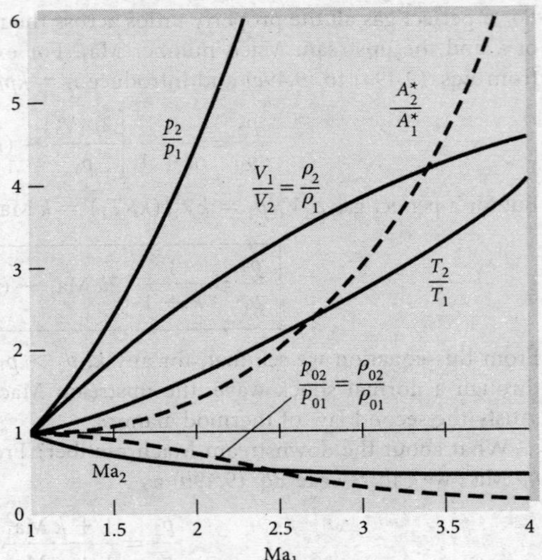

Fig. 9.9 Change in flow properties across a normal shock wave for $k = 1.4$.

The stagnation temperature remains the same, but the stagnation pressure and density decrease in the same ratio; in other words, the flow across the shock is adiabatic but nonisentropic. Other basic principles governing the behavior of shock waves can be summarized as follows:

1. The upstream flow is supersonic, and the downstream flow is subsonic.
2. For perfect gases (and also for real fluids except under bizarre thermodynamic conditions) rarefaction shocks are impossible, and only a compression shock can exist.
3. The entropy increases across a shock with consequent decreases in stagnation pressure and stagnation density and an increase in the effective sonic throat area.
4. Weak shock waves are very nearly isentropic.

Normal shock waves form in ducts under transient conditions, such as in shock tubes, and in steady flow for certain ranges of the downstream pressure. Figure 9.10*a* shows a normal shock in a supersonic nozzle. Flow is from left to right. The oblique wave pattern to the left is formed by roughness elements on the nozzle walls and indicates that the upstream flow is supersonic. Note the absence of these Mach waves (see Sec. 9.10) in the subsonic flow downstream.

Normal shock waves occur not only in supersonic duct flows but also in a variety of supersonic external flows. An example is the supersonic flow past a blunt body shown in Fig. 9.10*b*. The bow shock is curved, with a portion in front of the body that is essentially normal to the oncoming flow. This normal portion of the bow shock satisfies the property change conditions just as outlined in this section. The flow inside the shock near the body nose is thus subsonic and at relatively high temperature $T_2 > T_1$, and convective heat transfer is especially high in this region.

Each nonnormal portion of the bow shock in Fig. 9.10*b* satisfies the oblique shock relations to be outlined in Sec. 9.9. Note also the oblique recompression shock on the sides of the body. What has happened is that the subsonic nose flow has accelerated around the corners back to supersonic flow at low pressure, which must then pass through the second shock to match the higher downstream pressure conditions.

Note the fine-grained turbulent wake structure in the rear of the body in Fig. 9.10*b*. The turbulent boundary layer along the sides of the body is also clearly visible.

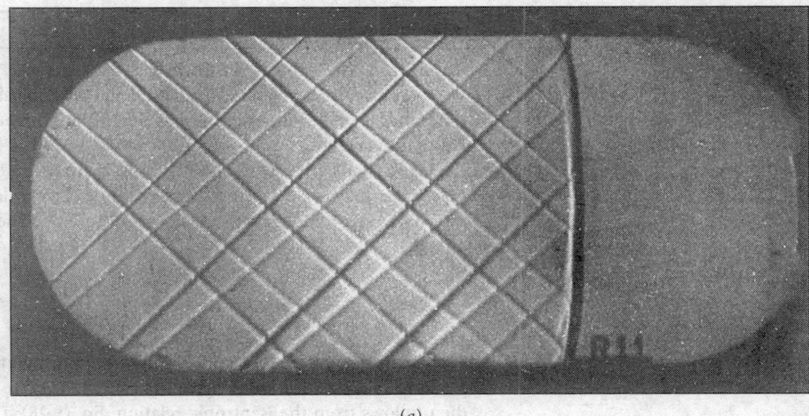

(a)

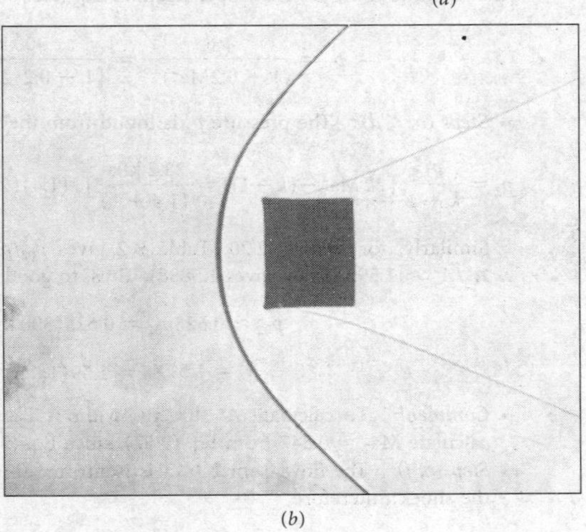

(b)

Fig. 9.10 Normal shocks form in both internal and external flows. (*a*) Normal shock in a duct; note the Mach wave pattern to the left (upstream), indicating supersonic flow. *(Courtesy of U.S. Air Force Arnold Engineering Development Center.)* (*b*) Supersonic flow past a blunt body creates a normal shock at the nose; the apparent shock thickness and body-corner curvature are optical distortions. *(Courtesy of U.S. Army Ballistic Research Laboratory, Aberdeen Proving Ground.)*

The analysis of a complex multidimensional supersonic flow such as in Fig. 9.10 is beyond the scope of this book. For further information see, for example, Ref. 9, Chap. 9, or Ref. 5, Chap. 16.

Moving Normal Shocks

The preceding analysis of the fixed shock applies equally well to the moving shock if we reverse the transformation used in Fig. 9.1. To make the upstream conditions simulate a still fluid, we move the shock of Fig. 9.8 to the left at speed V_1; that is, we fix our coordinates to a control volume moving with the shock. The downstream flow then appears to move to the left at a slower speed $V_1 - V_2$ following the shock. The thermodynamic properties are not changed by this transformation, so that all our Eqs. (9.50) to (9.59) are still valid.

EXAMPLE 9.6

Air flows from a reservoir where $p = 300$ kPa and $T = 500$ K through a throat to section 1 in Fig. E9.6, where there is a normal shock wave. Compute (*a*) p_1, (*b*) p_2, (*c*) p_{02}, (*d*) A_2^*, (*e*) p_{03}, (*f*) A_3^*, (*g*) p_3, and (*h*) T_{03}.

1 | 2 | 3

1 m²

2 m²

3 m²

E9.6

Solution

- *System sketch:* This is shown in Fig. E9.6. Between sections 1 and 2 is a normal shock.
- *Assumptions:* Isentropic flow before and after the shock. Lower p_0 and ρ_0 after the shock.
- *Approach:* After first noting that the throat is *sonic,* work your way from 1 to 2 to 3.
- *Property values:* For air, $R = 287\ \text{m}^2/(\text{s}^2 \cdot \text{K})$, $k = 1.40$, and $c_p = 1{,}005\ \text{m}^2/(\text{s}^2 \cdot \text{K})$. The inlet stagnation pressure of 300 kPa is constant up to point 1.
- *Solution step (a):* A shock wave cannot exist unless Ma_1 is supersonic. Therefore the throat is *sonic* and choked: $A_{\text{throat}} = A_1^* = 1\ \text{m}^2$. The area ratio gives Ma_1 from Eq. (9.45) for $k = 1.4$:

$$\frac{A_1}{A_1^*} = \frac{2\ \text{m}^2}{1\ \text{m}^2} = 2.0 = \frac{1}{\text{Ma}_1}\frac{(1 + 0.2\ \text{Ma}_1^2)^3}{1.728} \qquad \text{solve for} \qquad \text{Ma}_1 = 2.1972$$

Such four-decimal-place accuracy might require iteration or the use of Excel. Linear interpolation in Table B.1 would give $\text{Ma}_1 \approx 2.194$, quite good also. The pressure at section 1 then follows from the isentropic relation, Eq. (9.28):

$$p_1 = \frac{p_{01}}{(1 + 0.2\text{Ma}_1^2)^{3.5}} = \frac{300\ \text{kPa}}{[1 + 0.2(2.194)^2]^{3.5}} = 28.2\ \text{kPa} \qquad \textit{Ans. (a)}$$

- *Steps (b, c, d):* The pressure p_2 is found from the normal shock Eq. (9.55) or Table B.2:

$$p_2 = \frac{p_1}{k+1}[2k\,\text{Ma}_1^2 - (k-1)] = \frac{28.2\ \text{kPa}}{(1.4+1)}[2(1.4)(2.194)^2 - (1.4-1)] = 154\ \text{kPa} \quad \textit{Ans. (b)}$$

Similarly, for $\text{Ma}_1 \approx 2.20$, Table B.2 gives $p_{02}/p_{01} \approx 0.628$ (Excel gives 0.6294) and $A_2^*/A_1^* \approx 1.592$ (Excel gives 1.5888). Thus, to good accuracy,

$$p_{02} \approx 0.628 p_{01} = 0.628(300\ \text{kPa}) \approx 188\ \text{kPa} \qquad \textit{Ans. (c)}$$

$$A_2^* = 1.59 A_1^* = 1.59(1.0\ \text{m}^2) \approx 1.59\ \text{m}^2 \qquad \textit{Ans. (d)}$$

- *Comment:* To calculate A_2^* directly, without Table B.2, you would need to pause and calculate $\text{Ma}_2 \approx 0.547$ from Eq. (9.57), since Eq. (9.59) involves both Ma_1 and Ma_2.
- *Step (e, f):* The flow from 2 to 3 is isentropic (but at higher entropy than upstream of the shock); therefore

$$p_{03} = p_{02} \approx 188\ \text{kPa} \qquad \textit{Ans. (e)}$$

$$A_3^* = A_2^* \approx 1.59\ \text{m}^2 \qquad \textit{Ans. (f)}$$

- *Steps (g, h):* The flow is adiabatic throughout, so the stagnation temperature is constant:

$$T_{03} = T_{02} = T_{01} = 500\ \text{K} \qquad \textit{Ans. (h)}$$

Next, the area ratio, using the *new* sonic area, gives the Mach number at section 3:

$$\frac{A_3}{A_3^*} = \frac{3\ \text{m}^2}{1.59\ \text{m}^2} = 1.89 = \frac{1}{\text{Ma}_3}\frac{(1 + 0.2\ \text{Ma}_3^2)^3}{1.728} \qquad \text{solve for} \qquad \text{Ma}_3 \approx 0.33$$

Excel would yield $\text{Ma}_3 = 0.327$. Finally, with p_{02} known, Eq. (9.28) yields p_3:

$$p_3 = \frac{p_{02}}{(1 + 0.2\ \text{Ma}_3^2)^{3.5}} \approx \frac{188\ \text{kPa}}{[1 + 0.2(0.33)^2]^{3.5}} \approx 174\ \text{kPa} \qquad \textit{Ans. (g)}$$

- *Comments:* Excel would give $p_3 = 175$ kPa, so we see that Table B.2 is satisfactory for this type of problem. A duct flow with a normal shock wave requires straightforward application of algebraic perfect-gas relations, coupled with a little thought as to which formula is appropriate for the given property.

EXAMPLE 9.7

An explosion in air, $k = 1.4$, creates a spherical shock wave propagating radially into still air at standard conditions. At the instant shown in Fig. E9.7, the pressure just inside the shock is 1,400 kPa absolute. Estimate (*a*) the shock speed C and (*b*) the air velocity V just inside the shock.

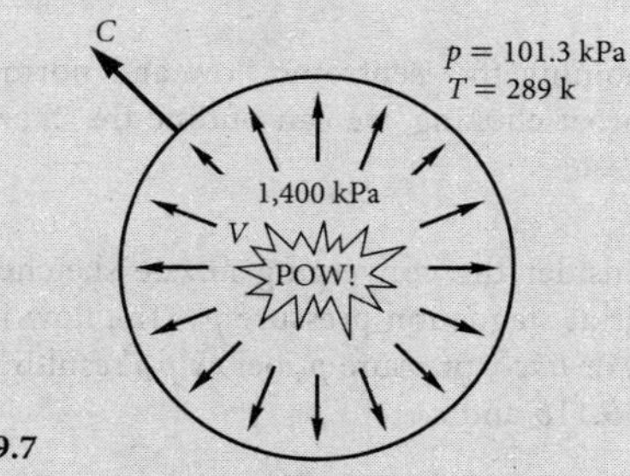

E9.7

Solution

Part (a)

In spite of the spherical geometry, the flow across the shock moves normal to the spherical wave front; hence the normal shock relations (9.50) to (9.59) apply. Fixing our control volume to the moving shock, we find that the proper conditions to use in Fig. 9.8 are

$$C = V_1 \qquad p_1 = 101.3\text{ kPa absolute} \qquad T_1 = 289\text{ K}$$

$$V = V_1 - V_2 \qquad p_2 = 1{,}400\text{ kPa absolute}$$

The speed of sound outside the shock is $a_1 \approx 20T_1^{1/2} = 340.8$ m/s. We can find Ma_1 from the known pressure ratio across the shock:

$$\frac{p_2}{p_1} = \frac{1{,}400\text{ kPa absolute}}{101.3\text{ kPa absolute}} = 13.82$$

From Eq. (9.55) or Table B.2

$$13.82 = \frac{1}{2.4}(2.8\,\text{Ma}_1^2 - 0.4) \qquad \text{or} \qquad \text{Ma}_1 = 3.462$$

Then, by definition of the Mach number,

$$C = V_1 = \text{Ma}_1\, a_1 = 3.462(340.8\text{ m/s}) = 1{,}180\text{ m/s} \qquad \textit{Ans. (a)}$$

Part (b)

To find V_2, we need the temperature or sound speed inside the shock. Since Ma_1 is known, from Eq. (9.58) or Table B.2 for $\text{Ma}_1 = 3.462$ we compute $T_2/T_1 = 3.263$. Then

$$T_2 = 3.263T_1 = 3.263(289\text{ K}) = 943\text{ K}$$

At such a high temperature we should account for non-perfect-gas effects or at least use the gas tables [16], but we won't. Here just estimate from the perfect-gas energy equation (9.23) that

$$V_2^2 = 2c_p(T_1 - T_2) + V_1^2 = 2(1{,}005\text{ (m}^2)/(\text{s}^2\cdot\text{K}))(289\text{ K} - 943\text{ K}) + (1{,}180\text{ m/s})^2$$
$$= 77{,}860\text{ m}^2/\text{s}^2$$

or

$$V_2 \approx 279\text{ m/s}$$

Notice that we did this without bothering to compute Ma_2, which equals 0.453, or $a_2 \approx 20T_2^{1/2} = 614$ m/s.

Finally, the air velocity behind the shock is

$$V = V_1 - V_2 = 1{,}180 \text{ m/s} - 279 \text{ m/s} \approx 901 \text{ m/s} \qquad \textit{Ans. (b)}$$

Thus, a powerful explosion creates a brief but intense blast wind as it passes.[3]

9.6 Operation of Converging and Diverging Nozzles

By combining the isentropic flow and normal shock relations plus the concept of sonic throat choking, we can outline the characteristics of converging and diverging nozzles.

Converging Nozzle

First consider the converging nozzle sketched in Fig. 9.11*a*. There is an upstream reservoir at stagnation pressure p_0. The flow is induced by lowering the downstream outside, or *back,* pressure p_b below p_0, resulting in the sequence of states *a* to *e* shown in Figs. 9.11*b* and *c*.

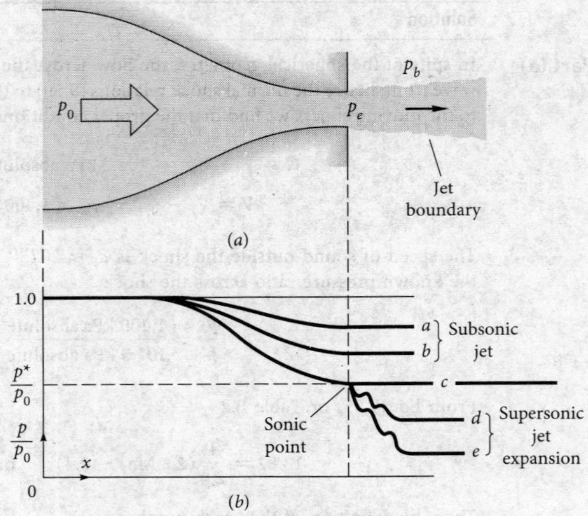

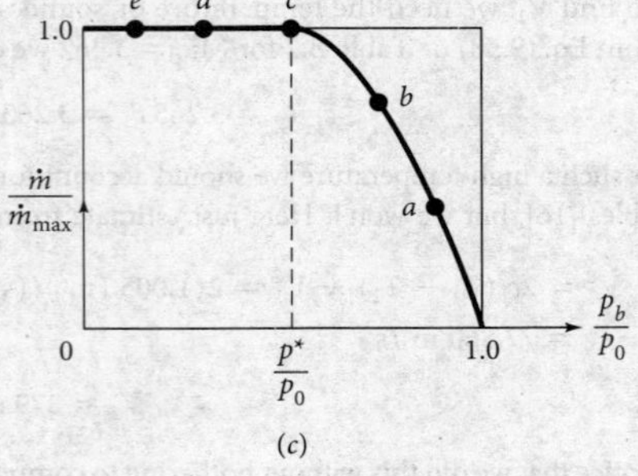

Fig. 9.11 Operation of a converging nozzle: (*a*) nozzle geometry showing characteristic pressures; (*b*) pressure distribution caused by various back pressures; (*c*) mass flow versus back pressure.

[3]This is the principle of the *shock tube wind tunnel,* in which a controlled explosion creates a brief flow at very high Mach number, with data taken by fast-response instruments. See, for example, Ref. 5.

For a moderate drop in p_b to states a and b, the throat pressure is higher than the critical value p^* that would make the throat sonic. The flow in the nozzle is subsonic throughout, and the jet exit pressure p_e equals the back pressure p_b. The mass flow is predicted by subsonic isentropic theory and is less than the critical value $\dot{m}_{max}$, as shown in Fig. 9.11*c*.

For condition *c*, the back pressure exactly equals the critical pressure p^* of the throat. The throat becomes sonic, the jet exit flow is sonic, $p_e = p_b$, and the mass flow equals its maximum value from Eqs. (9.46). The flow upstream of the throat is subsonic everywhere and predicted by isentropic theory based on the local area ratio $A(x)/A^*$ and Table B.1.

Finally, if p_b is lowered further to conditions d or e below p^*, the nozzle cannot respond further because it is choked at its maximum throat mass flow. The throat remains sonic with $p_e = p^*$, and the nozzle pressure distribution is the same as in state *c*, as sketched in Fig. 9.11*b*. The exit jet expands supersonically so that the jet pressure can be reduced from p^* down to p_b. The jet structure is complex and multidimensional and is not shown here. Being supersonic, the jet cannot send any signal upstream to influence the choked flow conditions in the nozzle.

If the stagnation plenum chamber is large or supplemented by a compressor, and if the discharge chamber is larger or supplemented by a vacuum pump, the converging nozzle flow will be steady or nearly so. Otherwise, the nozzle will be blowing down, with p_0 decreasing and p_b increasing, and the flow states will be changing from, say, state *e* backward to state *a*. Blowdown calculations are usually made by a quasi-steady analysis based on isentropic steady flow theory for the instantaneous pressures $p_0(t)$ and $p_b(t)$.

EXAMPLE 9.8

A converging nozzle has a throat area of 6 cm² and stagnation air conditions of 120 kPa and 400 K. Compute the exit pressure and mass flow if the back pressure is (*a*) 90 kPa and (*b*) 45 kPa. Assume $k = 1.4$.

Solution

From Eq. (9.32) for $k = 1.4$ the critical (sonic) throat pressure is

$$\frac{p^*}{p_0} = 0.5283 \quad \text{or} \quad p^* = (0.5283)(120\ \text{kPa}) = 63.4\ \text{kPa}$$

If the back pressure is less than this amount, the nozzle flow is choked.

Part (a) For $p_b = 90\ \text{kPa} > p^*$, the flow is subsonic, not choked. The exit pressure is $p_e = p_b$. The throat Mach number is found from the isentropic relation (9.35) or Table B.1:

$$\text{Ma}_e^2 = 5\left[\left(\frac{p_0}{p_e}\right)^{2/7} - 1\right] = 5\left[\left(\frac{120}{90}\right)^{2/7} - 1\right] = 0.4283 \quad \text{Ma}_e = 0.654$$

To find the mass flow, we could proceed with a serial attack on Ma_e, T_e, a_e, V_e, and ρ_e, hence to compute $\rho_e A_e V_e$. However, since the local pressure is known, this part is ideally suited for the dimensionless mass flow function in Eq. (9.47). With $p_e/p_0 = 90/120 = 0.75$, compute

$$\frac{\dot{m}\sqrt{RT_0}}{Ap_0} = \sqrt{\frac{2(1.4)}{0.4}(0.75)^{2/1.4}[1 - (0.75)^{0.4/1.4}]} = 0.6052$$

hence $$\dot{m} = 0.6052\frac{(0.0006)(120{,}000)}{\sqrt{287(400)}} = 0.129\ \text{kg/s} \qquad \textit{Ans. (a)}$$

for $$p_e = p_b = 90\ \text{kPa} \qquad \textit{Ans. (a)}$$

Part (b) For $p_b = 45$ kPa $< p^*$, the flow is choked, similar to condition d in Fig. 9.11b. The exit pressure is sonic:

$$p_e = p^* = 63.4 \text{ kPa} \qquad \textit{Ans. (b)}$$

The (choked) mass flow is a maximum from Eq. (9.46b):

$$\dot{m} = \dot{m}_{\max} = \frac{0.6847 p_0 A_e}{(RT_0)^{1/2}} = \frac{0.6847(120{,}000)(0.0006)}{[287(400)]^{1/2}} = 0.145 \text{ kg/s} \qquad \textit{Ans. (b)}$$

Any back pressure less than 63.4 kPa would cause this same choked mass flow. Note that the 50 percent increase in exit Mach number, from 0.654 to 1.0, has increased the mass flow only 12 percent, from 0.128 to 0.145 kg/s.

Converging–Diverging Nozzle

Now consider the converging–diverging nozzle sketched in Fig. 9.12a. If the back pressure p_b is low enough, there will be supersonic flow in the diverging portion and a variety of shock wave conditions may occur, which are sketched in Fig. 9.12b. Let the back pressure be gradually decreased.

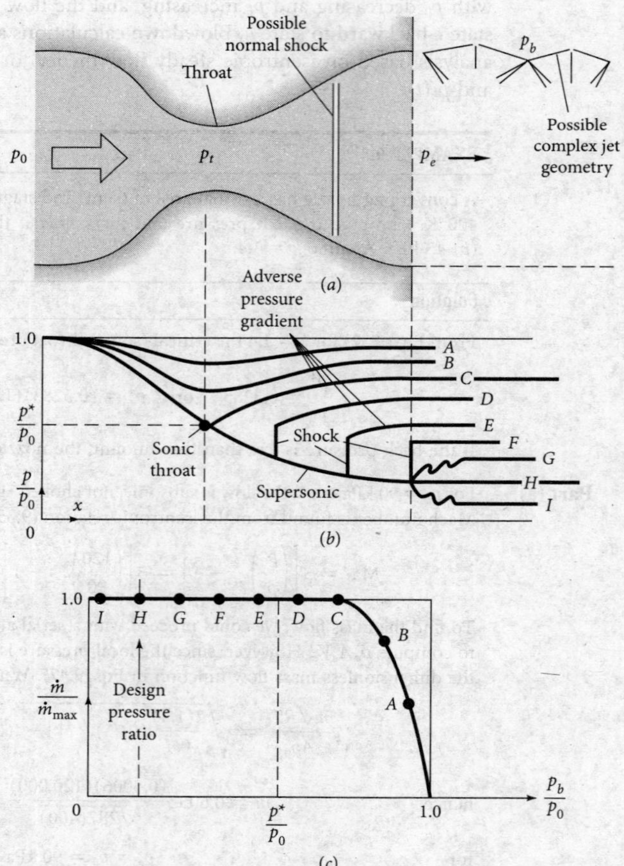

Fig. 9.12 Operation of a converging–diverging nozzle: (a) nozzle geometry with possible flow configurations; (b) pressure distribution caused by various back pressures; (c) mass flow versus back pressure.

For curves A and B in Fig. 9.12*b* the back pressure is not low enough to induce sonic flow in the throat, and the flow in the nozzle is subsonic throughout. The pressure distribution is computed from subsonic isentropic area-change relations, such as in Table B.1. The exit pressure $p_e = p_b$, and the jet is subsonic.

For curve C, the area ratio A_e/A_t exactly equals the critical ratio A_e/A^* for a subsonic Ma_e in Table B.1. The throat becomes sonic, and the mass flux reaches a maximum in Fig. 9.12*c*. The remainder of the nozzle flow is subsonic, including the exit jet, and $p_e = p_b$.

Now jump for a moment to curve H. Here p_b is such that p_b/p_0 exactly corresponds to the critical area ratio A_e/A^* for a *supersonic* Ma_e in Table B.1. The diverging flow is entirely supersonic, including the jet flow, and $p_e = p_b$. This is called the *design pressure ratio* of the nozzle and is the back pressure suitable for operating a supersonic wind tunnel or an efficient rocket exhaust.

Now, back up and suppose that p_b lies between curves C and H, which is impossible according to purely isentropic flow calculations. Then, back pressures D to F occur in Fig. 9.12*b*. The throat remains choked at the sonic value, and we can match $p_e = p_b$ by placing a normal shock at just the right place in the diverging section to cause a *subsonic diffuser* flow back to the back-pressure condition. The mass flow remains at maximum in Fig. 9.12*c*. At back pressure F the required normal shock stands in the duct exit. At back pressure G no single normal shock can do the job, and so the flow compresses outside the exit in a complex series of oblique shocks until it matches p_b.

Finally, at back pressure I, p_b is lower than the design pressure H, but the nozzle is choked and cannot respond. The exit flow expands in a complex series of supersonic wave motions until it matches the low back pressure. See, Ref. 7, Sec. 5.4, for further details of these off-design jet flow configurations.

Note that for p_b less than back pressure C, there is supersonic flow in the nozzle, and the throat can receive no signal from the exit behavior. The flow remains choked, and the throat has no idea what the exit conditions are.

Note also that the normal shock-patching idea is idealized. Downstream of the shock, the nozzle flow has an adverse pressure gradient, usually leading to wall boundary layer separation. Blockage by the greatly thickened separated layer interacts strongly with the core flow (recall Fig. 6.27) and usually induces a series of weak two-dimensional compression shocks rather than a single one-dimensional normal shock (see, Ref. 9, pp. 292 and 293, for further details).

EXAMPLE 9.9

A converging–diverging nozzle (Fig. 9.12*a*) has a throat area of 0.002 m^2 and an exit area of 0.008 m^2. Air stagnation conditions are $p_0 = 1{,}000$ kPa and $T_0 = 500$ K. Compute the exit pressure and mass flow for (*a*) design condition and the exit pressure and mass flow if (*b*) $p_b \approx 300$ kPa and (*c*) $p_b \approx 900$ kPa. Assume $k = 1.4$.

Solution

Part (a) The design condition corresponds to supersonic isentropic flow at the given area ratio $A_e/A_t = 0.008/0.002 = 4.0$. We can find the design Mach number by iteration of the area ratio formula (9.45):

$$\text{Ma}_{e,\text{design}} \approx 2.95$$

The design pressure ratio follows from Eq. (9.34):

$$\frac{p_0}{p_e} = [1 + 0.2(2.95)^2]^{3.5} = 34.1$$

or $$p_{e,\text{design}} = \frac{1{,}000 \text{ kPa}}{34.1} = 29.3 \text{ kPa} \qquad \textit{Ans. (a)}$$

Since the throat is clearly sonic at design conditions, Eq. (9.46*b*) applies:

$$\dot{m}_{\text{design}} = \dot{m}_{\max} = \frac{0.6847 p_0 A_t}{(RT_0)^{1/2}} = \frac{0.6847(10^6 \text{ Pa})(0.002 \text{ m}^2)}{[287(500)]^{1/2}} = 3.61 \text{ kg/s} \qquad \textit{Ans. (a)}$$

Part (b) For $p_b = 300$ kPa we are definitely far below the subsonic isentropic condition *C* in Fig. 9.12*b*, but we may even be below condition *F* with a normal shock in the exit—that is, in condition *G*, where oblique shocks occur outside the exit plane. If it is condition *G*, then $p_e = p_{e,\text{design}} =$ 29.3 kPa because no shock has yet occurred. To find out, compute condition *F* by assuming an exit normal shock with $\text{Ma}_1 = 2.95$—that is, the design Mach number just upstream of the shock. From Eq. (9.55)

$$\frac{p_2}{p_1} = \frac{1}{2.4}[2.8(2.95)^2 - 0.4] = 9.99$$

or $$p_2 = 9.99 p_1 = 9.99 p_{e,\text{design}} = 293 \text{ kPa}$$

Since this is less than the given $p_b = 300$ kPa, there is a normal shock just upstream of the exit plane (condition *E*). The exit flow is subsonic and equals the back pressure:

$$p_e = p_b = 300 \text{ kPa} \qquad \textit{Ans. (b)}$$

Also $$\dot{m} = \dot{m}_{\max} = 3.61 \text{ kg/s} \qquad \textit{Ans. (b)}$$

The throat is still sonic and choked at its maximum mass flow.

Part (c) Finally, for $p_b = 900$ kPa, which is up near condition *C*, we compute Ma_e and p_e for condition *C* as a comparison. Again $A_e/A_t = 4.0$ for this condition, with a subsonic Ma_e estimated from Eq. (9.48):

$$\text{Ma}_e(C) \approx 0.147 \qquad (\text{exact} = 0.14655)$$

Then, the isentropic exit pressure ratio for this condition is

$$\frac{p_0}{p_e} = [1 + 0.2(0.147)^2]^{3.5} = 1.0152$$

or $$p_e = \frac{1{,}000}{1.0152} = 985 \text{ kPa}$$

The given back pressure of 900 kPa is less than this value, corresponding roughly to condition *D* in Fig. 9.12*b*. Thus for this case there is a normal shock just downstream of the throat, and the throat is choked:

$$p_e = p_b = 900 \text{ kPa} \qquad \dot{m} = \dot{m}_{\max} = 3.61 \text{ kg/s} \qquad \textit{Ans. (c)}$$

For this large exit area ratio, the exit pressure would have to be larger than 985 kPa to cause a subsonic flow in the throat and a mass flow less than maximum.

9.7 Compressible Duct Flow with Friction[4]

Section 9.4 showed the effect of area change on a compressible flow while neglecting friction and heat transfer. We could now add friction and heat transfer to the area change and consider coupled effects, which is done in advanced texts [for example, Ref. 5, Chap. 8]. Instead, as an elementary introduction, this section treats only the effect of friction, neglecting area change and heat transfer. The basic assumptions are

1. Steady one-dimensional adiabatic flow.
2. Perfect gas with constant specific heats.
3. Constant-area straight duct.
4. Negligible shaft work and potential energy changes.
5. Wall shear stress correlated by a Darcy friction factor.

In effect, we are studying a Moody-type pipe friction problem but with large changes in kinetic energy, enthalpy, and pressure in the flow.

This type of duct flow—constant area, constant stagnation enthalpy, constant mass flow, but variable momentum (due to friction)—is often termed *Fanno flow,* after Gino Fanno, an Italian engineer born in 1882, who first studied this flow. For a given mass flow and stagnation enthalpy, a plot of enthalpy versus entropy for all possible flow states, subsonic or supersonic, is called a *Fanno line.* See Probs. P9.94 and P9.111 for examples of a Fanno line.

Consider the elemental duct control volume of area A and length dx in Fig. 9.13. The area is constant, but other flow properties (p, ρ, T, h, V) may vary with x. Application of the three conservation laws to this control volume gives three differential equations:

Continuity:
$$\rho V = \frac{\dot{m}}{A} = G = \text{const}$$

or
$$\frac{d\rho}{\rho} + \frac{dV}{V} = 0 \tag{9.60a}$$

x momentum:
$$pA - (p + dp)A - \tau_w \pi D\, dx = \dot{m}(V + dV - V)$$

or
$$dp + \frac{4\tau_w dx}{D} + \rho V\, dV = 0 \tag{9.60b}$$

Energy:
$$h + \tfrac{1}{2}V^2 = h_0 = c_p T_0 = c_p T + \tfrac{1}{2}V^2$$

or
$$c_p\, dT + V\, dV = 0 \tag{9.60c}$$

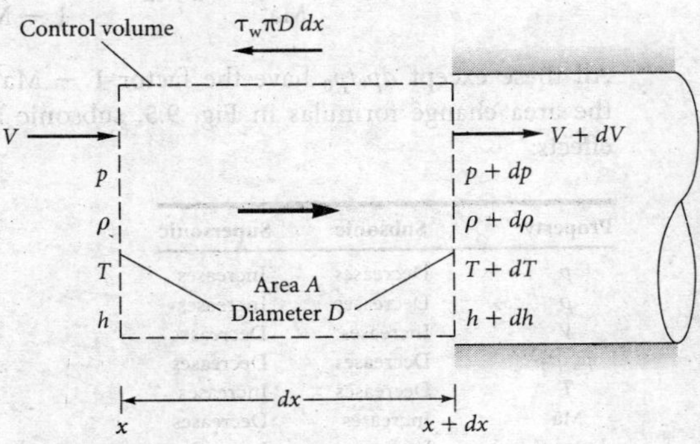

Fig. 9.13 Elemental control volume for flow in a constant-area duct with friction.

[4]This section may be omitted without loss of continuity.

Since these three equations have five unknowns—p, ρ, T, V, and τ_w—we need two additional relations. One is the perfect-gas law:

$$p = \rho RT \quad \text{or} \quad \frac{dp}{p} = \frac{d\rho}{\rho} + \frac{dT}{T} \tag{9.61}$$

To eliminate τ_w as an unknown, it is assumed that wall shear is correlated by a local Darcy friction factor f

$$\tau_w = \tfrac{1}{8} f \rho V^2 = \tfrac{1}{8} f k p \,\text{Ma}^2 \tag{9.62}$$

where the last form follows from the perfect-gas speed-of-sound expression $a^2 = kp/\rho$. In practice, f can be related to the local Reynolds number and wall roughness from, say, the Moody chart, Fig. 6.13.

Equations (9.60) and (9.61) are first-order differential equations and can be integrated, by using friction factor data, from any inlet section 1, where p_1, T_1, V_1, and so on are known, to determine $p(x)$, $T(x)$, and other properties along the duct. It is practically impossible to eliminate all but one variable to give, say, a single differential equation for $p(x)$, but all equations can be written in terms of the Mach number Ma(x) and the friction factor, by using this definition of Mach number:

$$V^2 = \text{Ma}^2\, kRT$$

$$\text{or} \quad \frac{2\,dV}{V} = \frac{2\,d\text{Ma}}{\text{Ma}} + \frac{dT}{T} \tag{9.63}$$

Adiabatic Flow

By eliminating variables between Eqs. (9.60) to (9.63), we obtain the working relations

$$\frac{dp}{p} = -k\,\text{Ma}^2 \frac{1 + (k-1)\text{Ma}^2}{2(1 - \text{Ma}^2)} f \frac{dx}{D} \tag{9.64a}$$

$$\frac{d\rho}{\rho} = -\frac{k\,\text{Ma}^2}{2(1 - \text{Ma}^2)} f \frac{dx}{D} = -\frac{dV}{V} \tag{9.64b}$$

$$\frac{dp_0}{p_0} = \frac{d\rho_0}{\rho_0} = -\frac{1}{2} k\,\text{Ma}^2 f \frac{dx}{D} \tag{9.64c}$$

$$\frac{dT}{T} = -\frac{k(k-1)\text{Ma}^4}{2(1 - \text{Ma}^2)} f \frac{dx}{D} \tag{9.64d}$$

$$\frac{d\,\text{Ma}^2}{\text{Ma}^2} = k\,\text{Ma}^2 \frac{1 + \frac{1}{2}(k-1)\text{Ma}^2}{1 - \text{Ma}^2} f \frac{dx}{D} \tag{9.64e}$$

All these except dp_0/p_0 have the factor $1 - \text{Ma}^2$ in the denominator, so that, like the area change formulas in Fig. 9.5, subsonic and supersonic flow have opposite effects:

Property	Subsonic	Supersonic
p	Decreases	Increases
ρ	Decreases	Increases
V	Increases	Decreases
p_0, ρ_0	Decreases	Decreases
T	Decreases	Increases
Ma	Increases	Decreases
Entropy	Increases	Increases

We have added to this list that entropy must increase along the duct for either subsonic or supersonic flow as a consequence of the second law for adiabatic flow. For the same reason, stagnation pressure and density must both decrease.

The key parameter in this discussion is the Mach number. Whether the inlet flow is subsonic or supersonic, the duct Mach number always tends downstream toward $\text{Ma} = 1$ because this is the path along which the entropy increases. If the pressure and density are computed from Eqs. (9.64*a*) and (9.64*b*) and the entropy from Eq. (9.53), the result can be plotted in Fig. 9.14 versus Mach number for $k = 1.4$. The maximum entropy occurs at $\text{Ma} = 1$, so the second law requires that the duct flow properties continually approach the sonic point. Since p_0 and ρ_0 continually decrease along the duct due to the frictional (nonisentropic) losses, they are not useful as reference properties. Instead, the sonic properties p^*, ρ^*, T^*, p_0^*, and ρ_0^* are the appropriate constant reference quantities in adiabatic duct flow. The theory then computes the ratios p/p^*, T/T^*, and so forth as a function of local Mach number and the integrated friction effect.

To derive working formulas, we first attack Eq. (9.64*e*), which relates the Mach number to friction. Separate the variables and integrate:

$$\int_0^{L^*} f\frac{dx}{D} = \int_{\text{Ma}^2}^{1.0} \frac{1 - \text{Ma}^2}{k\,\text{Ma}^4[1 + \frac{1}{2}(k-1)\text{Ma}^2]}\, d\,\text{Ma}^2 \qquad (9.65)$$

The upper limit is the sonic point, whether or not it is actually reached in the duct flow. The lower limit is arbitrarily placed at the position $x = 0$, where the Mach number is Ma. The result of the integration is

$$\frac{f\bar{L}^*}{D} = \frac{1 - \text{Ma}^2}{k\,\text{Ma}^2} + \frac{k+1}{2k}\ln\frac{(k+1)\text{Ma}^2}{2 + (k-1)\text{Ma}^2} \qquad (9.66)$$

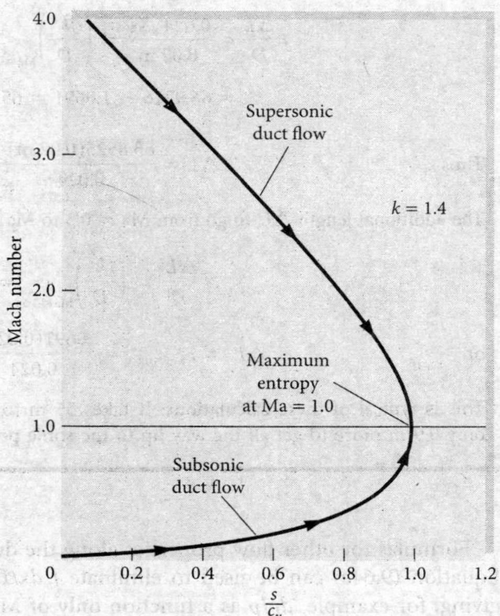

Fig. 9.14 Adiabatic frictional flow in a constant-area duct always approaches $\text{Ma} = 1$ to satisfy the second law of thermodynamics. The computed curve is independent of the value of the friction factor.

where $\bar{f}$ is the average friction factor between 0 and L^*. In practice, an average f is always assumed, and no attempt is made to account for the slight changes in Reynolds number along the duct. For noncircular ducts, D is replaced by the hydraulic diameter $D_h = (4 \times \text{area})/\text{perimeter}$ as in Eq. 6.56.

Equation (9.66) is tabulated versus Mach number in Table B.3. The length L^* is the length of duct required to develop a duct flow from Mach number Ma to the sonic point. Many problems involve short ducts that never become sonic, for which the solution uses the differences in the tabulated "maximum," or sonic, length. For example, the length ΔL required to develop from Ma_1 to Ma_2 is given by

$$\bar{f}\frac{\Delta L}{D} = \left(\frac{\bar{f}L^*}{D}\right)_1 - \left(\frac{\bar{f}L^*}{D}\right)_2 \tag{9.67}$$

This avoids the need for separate tabulations for short ducts.

It is recommended that the friction factor $\bar{f}$ be estimated from the Moody chart (Fig. 6.13) for the average Reynolds number and wall roughness ratio of the duct. Available data [23] on duct friction for compressible flow show good agreement with the Moody chart for subsonic flow, but the measured data in supersonic duct flow are up to 50 percent less than the equivalent Moody friction factor.

EXAMPLE 9.10

Air flows subsonically in an adiabatic 2-cm-diameter duct. The average friction factor is 0.024. What length of duct is necessary to accelerate the flow from $Ma_1 = 0.1$ to $Ma_2 = 0.5$? What additional length will accelerate it to $Ma_3 = 1.0$? Assume $k = 1.4$.

Solution

Equation (9.67) applies with values of $\bar{f}L^*/D$ computed from Eq. (9.66) or read from Table B.3:

$$\bar{f}\frac{\Delta L}{D} = \frac{0.024\,\Delta L}{0.02\text{ m}} = \left(\frac{\bar{f}L^*}{D}\right)_{Ma=0.1} - \left(\frac{\bar{f}L^*}{D}\right)_{Ma=0.5}$$

$$= 66.9216 - 1.0691 = 65.8525$$

Thus $$\Delta L = \frac{65.8525(0.02\text{ m})}{0.024} = 55\text{ m} \qquad \textit{Ans. (a)}$$

The additional length $\Delta L'$ to go from Ma = 0.5 to Ma = 1.0 is taken directly from Table B.2:

$$\bar{f}\frac{\Delta L'}{D} = \left(\frac{\bar{f}L^*}{D}\right)_{Ma=0.5} = 1.0691$$

or $$\Delta L' = L^*_{Ma=0.5} = \frac{1.0691(0.02\text{ m})}{0.024} = 0.9\text{ m} \qquad \textit{Ans. (b)}$$

This is typical of these calculations: It takes 55 m to accelerate up to Ma = 0.5 and then only 0.9 m more to get all the way up to the sonic point.

Formulas for other flow properties along the duct can be derived from Eqs. (9.64). Equation (9.64*e*) can be used to eliminate $f\,dx/D$ from each of the other relations, giving, for example, dp/p as a function only of Ma and $d\,Ma^2/Ma^2$. For convenience in tabulating the results, each expression is then integrated all the way from (p, Ma) to the sonic point (p^*, 1.0). The integrated results are

$$\frac{p}{p^*} = \frac{1}{\text{Ma}}\left[\frac{k+1}{2+(k-1)\,\text{Ma}^2}\right]^{1/2} \tag{9.68a}$$

$$\frac{\rho}{\rho^*} = \frac{V^*}{V} = \frac{1}{\text{Ma}}\left[\frac{2+(k-1)\,\text{Ma}^2}{k+1}\right]^{1/2} \tag{9.68b}$$

$$\frac{T}{T^*} = \frac{a^2}{a^{*2}} = \frac{k+1}{2+(k-1)\,\text{Ma}^2} \tag{9.68c}$$

$$\frac{p_0}{p_0^*} = \frac{\rho_0}{\rho_0^*} = \frac{1}{\text{Ma}}\left[\frac{2+(k-1)\,\text{Ma}^2}{k+1}\right]^{(1/2)(k+1)/(k-1)} \tag{9.68d}$$

All these ratios are also tabulated in Table B.3. For finding changes between points Ma_1 and Ma_2 that are not sonic, products of these ratios are used. For example,

$$\frac{p_2}{p_1} = \frac{p_2}{p^*}\frac{p^*}{p_1} \tag{9.69}$$

since p^* is a constant reference value for the flow.

EXAMPLE 9.11

For the duct flow of Example 9.10 assume that, at $\text{Ma}_1 = 0.1$, we have $p_1 = 600$ kPa and $T_1 = 450$ K. At section 2 farther downstream, $\text{Ma}_2 = 0.5$. Compute (*a*) p_2, (*b*) T_2, (*c*) V_2, and (*d*) p_{02}.

Solution

As preliminary information, we can compute V_1 and p_{01} from the given data:

$$V_1 = \text{Ma}_1\, a_1 = 0.1[(1.4)(287)(450)]^{1/2} = 0.1(425\text{ m/s}) = 42.5\text{ m/s}$$

$$p_{01} = p_1(1 + 0.2\,\text{Ma}_1^2)^{3.5} = (600\text{ kPa})[1 + 0.2(0.1)^2]^{3.5} = 604\text{ kPa}$$

Now enter Table B.3 or Eqs. (9.68) to find the following property ratios:

Section	Ma	p/p^*	T/T^*	V/V^*	p_0/p_0^*
1	0.1	10.9435	1.1976	0.1094	5.8218
2	0.5	2.1381	1.1429	0.5345	1.3399

Use these ratios to compute all properties downstream:

$$p_2 = p_1\frac{p_2/p^*}{p_1/p^*} = (600\text{ kPa})\frac{2.1381}{10.9435} = 117\text{ kPa} \qquad \textit{Ans. (a)}$$

$$T_2 = T_1\frac{T_2/T^*}{T_1/T^*} = (450\text{ K})\frac{1.1429}{1.1976} = 429\text{ K} \qquad \textit{Ans. (b)}$$

$$V_2 = V_1\frac{V_2/V^*}{V_1/V^*} = (42.5\text{ m/s})\frac{0.5345}{0.1094} = 208\ \frac{\text{m}}{\text{s}} \qquad \textit{Ans. (c)}$$

$$p_{02} = p_{01}\frac{p_{02}/p_0^*}{p_{01}/p_0^*} = (604\text{ kPa})\frac{1.3399}{5.8218} = 139\text{ kPa} \qquad \textit{Ans. (d)}$$

Note the 77 percent reduction in stagnation pressure due to friction. The formulas are seductive, so check your work by other means. For example, check $p_{02} = p_2(1 + 0.2\,\text{Ma}_2^2)^{3.5}$.

Choking Due to Friction

The theory here predicts that for adiabatic frictional flow in a constant-area duct, no matter what the inlet Mach number Ma_1 is, the flow downstream tends toward the sonic point. There is a certain duct length $L^*(Ma_1)$ for which the exit Mach number will be exactly unity. The duct is then choked.

But what if the actual length L is greater than the predicted "maximum" length L^*? Then the flow conditions must change, and there are two classifications.

Subsonic Inlet. If $L > L^*(Ma_1)$, the flow slows down until an inlet Mach number Ma_2 is reached such that $L = L^*(Ma_2)$. The exit flow is sonic, and the mass flow has been reduced by *frictional choking*. Further increases in duct length will continue to decrease the inlet Ma and mass flow.

Supersonic Inlet. From Table B.3, we see that friction has a very large effect on supersonic duct flow. Even an infinite inlet Mach number will be reduced to sonic conditions in only 41 diameters for $\bar{f} = 0.02$. Some typical numerical values are shown in Fig. 9.15, assuming an inlet Ma = 3.0 and $\bar{f} = 0.02$. For this condition $L^* = 26$ diameters. If L is increased beyond $26D$, the flow will not choke, but a normal shock will form at just the right place for the subsequent subsonic frictional flow to become sonic exactly at the exit. Figure 9.15 shows two examples, for $L/D =$ 40 and 53. As the length increases, the required normal shock moves upstream until, for Fig. 9.15, the shock is at the inlet for $L/D = 63$. Further increase in L causes the shock to move upstream of the inlet into the supersonic nozzle feeding the duct. Yet the mass flow is still the same as for the very short duct, because presumably the feed nozzle still has a sonic throat. Eventually, a very long duct will cause the feed-nozzle throat to become choked, thus reducing the duct mass flow. Thus supersonic friction changes the flow pattern if $L > L^*$ but does not choke the flow until L is much larger than L^*.

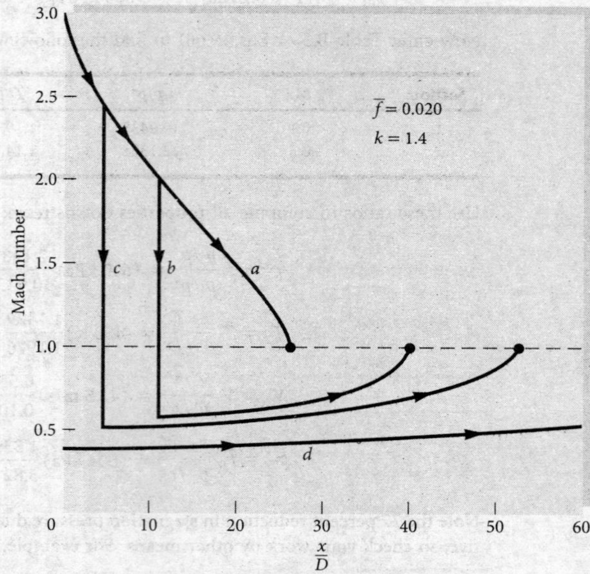

Fig. 9.15 Behavior of duct flow with a nominal supersonic inlet condition Ma = 3.0: (*a*) $L/D \le 26$, flow is supersonic throughout duct; (*b*) $L/D = 40 > L^*/D$, normal shock at Ma = 2.0 with subsonic flow then accelerating to sonic exit point; (*c*) $L/D = 53$, shock must now occur at Ma = 2.5; (*d*) $L/D >$ 63, flow must be entirely subsonic and choked at exit.

EXAMPLE 9.12

Air enters a 3-cm-diameter duct at $p_0 = 200$ kPa, $T_0 = 500$ K, and $V_1 = 100$ m/s. The friction factor is 0.02. Compute (*a*) the maximum duct length for these conditions, (*b*) the mass flow if the duct length is 15 m, and (*c*) the reduced mass flow if $L = 30$ m.

Solution

Part (a) First compute

$$T_1 = T_0 - \frac{\frac{1}{2}V_1^2}{c_p} = 500 - \frac{\frac{1}{2}(100\ \text{m(s)})^2}{1{,}005\ \text{m}^2/\text{s}^2 \cdot \text{K}} = 500 - 5 = 495\ \text{K}$$

$$a_1 = (kRT_1)^{1/2} \approx 20(495)^{1/2} = 445\ \text{m/s}$$

Thus $$\text{Ma}_1 = \frac{V_1}{a_1} = \frac{100}{445} = 0.225$$

For this Ma_1, from Eq. (9.66) or interpolation in Table B.3,

$$\frac{\bar{f}L^*}{D} = 11.0$$

The maximum duct length possible for these inlet conditions is

$$L^* = \frac{(\bar{f}L^*/D)D}{\bar{f}} = \frac{11.0(0.03\ \text{m})}{0.02} = 16.5\ \text{m} \qquad \textit{Ans. (a)}$$

Part (b) The given $L = 15$ m is less than L^*, and so the duct is not choked and the mass flow follows from the inlet conditions:

$$\rho_{01} = \frac{p_{01}}{RT_0} = \frac{200{,}000\ \text{Pa}}{287(500\ \text{K})} = 1.394\ \text{kg/m}^3$$

$$\rho_1 = \frac{\rho_{01}}{[1 + 0.2(0.225)^2]^{2.5}} = \frac{1.394}{1.0255} = 1.359\ \text{kg/m}^3$$

whence $$\dot{m} = \rho_1 A V_1 = (1.359\ \text{kg/m}^3)\left[\frac{\pi}{4}(0.03\ \text{m})^2\right](100\ \text{m/s})$$

$$= 0.0961\ \text{kg/s} \qquad \textit{Ans. (b)}$$

Part (c) Since $L = 30$ m is greater than L^*, the duct must choke back until $L = L^*$, corresponding to a lower inlet Ma_1:

$$L^* = L = 30\ \text{m}$$

$$\frac{\bar{f}L^*}{D} = \frac{0.02(30\ \text{m})}{0.03\ \text{m}} = 20.0$$

Although it is difficult to interpolate in the coarse Table B.3 for $fL/D = 20$, it is a simple matter for Excel to iterate to find this subsonic Mach number. Program Eq. (9.66) into a cell, guess a subsonic Mach number, and calculate fL/D. Adjust Ma until you approximate $fL/D = 20$. The writer's effort took five guesses, as in the following Excel table:

	A	B
	Ma_1	$(fL/D)_1$ – Eq. (9.66)
1	0.2	14.533
2	0.15	27.932
3	0.17	21.115
4	0.175	19.772
5	**0.1741**	**20.005**

An accurate solution for Ma is thus $\quad \text{Ma}_{\text{choked}} \approx 0.174 \quad$ (23 percent less)

$$T_{1,\text{new}} = \frac{T_0}{1 + 0.2(0.174)^2} = 497\ \text{K}$$

$$a_{1,\text{new}} \approx 20(497\ \text{K})^{1/2} = 446\ \text{m/s}$$

$$V_{1,\text{new}} = \text{Ma}_1\, a_1 = 0.174(446) = 77.6\ \text{m/s}$$

$$\rho_{1,\text{new}} = \frac{\rho_{01}}{[1 + 0.2(0.174)^2]^{2.5}} = 1.373\ \text{kg/m}^3$$

$$\dot{m}_{\text{new}} = \rho_1 A V_1 = 1.373\left[\frac{\pi}{4}(0.03)^2\right](77.6)$$

$$= 0.0753\ \text{kg/s} \quad \text{(22 percent less)}$$

Ans. (*c*)

Minor Losses in Compressible Flow

For incompressible pipe flow, as in Eq. (6.78), the loss coefficient K is the ratio of pressure head loss ($\Delta p/\rho g$) to the velocity head ($V^2/2g$) in the pipe. This is inappropriate in compressible pipe flow, where ρ and V are not constant. Benedict [24] suggests that the static pressure loss ($p_1 - p_2$) be related to downstream conditions and a *static loss coefficient* K_s:

$$K_s = \frac{2(p_1 - p_2)}{\rho_2 V_2^2} \tag{9.70}$$

Benedict [24] gives examples of compressible losses in sudden contractions and expansions. If data are unavailable, a first approximation would be to use $K_s \approx K$ from Sec. 6.9.

Isothermal Flow with Friction: Long Pipelines

The adiabatic frictional flow assumption is appropriate to high-speed flow in short ducts. For flow in long ducts, such as natural gas pipelines, the gas state more closely approximates an isothermal flow. The analysis is the same except that the isoenergetic energy equation (9.60*c*) is replaced by the simple relation

$$T = \text{const} \quad dT = 0$$

Again, it is possible to write all property changes in terms of the Mach number. Integration of the Mach number–friction relation yields

$$\frac{\bar{f}L_{\max}}{D} = \frac{1 - k\,\text{Ma}^2}{k\,\text{Ma}^2} + \ln(k\,\text{Ma}^2) \tag{9.71}$$

which is the isothermal analog of Eq. (9.66) for adiabatic flow.

This friction relation has the interesting result that $L_{\max}$ becomes zero not at the sonic point but at $\text{Ma}_{\text{crit}} = 1/k^{1/2} = 0.845$ if $k = 1.4$. The inlet flow, whether subsonic or supersonic, tends downstream toward this limiting Mach number $1/k^{1/2}$. If the tube length L is greater than $L_{\max}$ from Eq. (9.71), a subsonic flow will choke back to a smaller Ma_1 and mass flow and a supersonic flow will experience a normal shock adjustment similar to Fig. 9.15.

The exit isothermal choked flow is not sonic, and so the use of the asterisk is inappropriate. Let p', ρ', and V' represent properties at the choking point $L = L_{\max}$. Then the isothermal analysis leads to the following Mach number relations for the flow properties:

$$\frac{p}{p'} = \frac{1}{\text{Ma}\,k^{1/2}} \qquad \frac{V}{V'} = \frac{\rho'}{\rho} = \text{Ma}\,k^{1/2} \tag{9.72}$$

The complete analysis and some examples are given in advanced texts [for example, Ref. 5, Sec. 6.4].

Mass Flow for a Given Pressure Drop

An interesting by-product of the isothermal analysis is an explicit relation between the pressure drop and duct mass flow. This common problem requires numerical iteration for adiabatic flow, as outlined here. In isothermal flow, we may substitute $dV/V = -dp/p$ and $V^2 = G^2/[p/(RT)]^2$ in Eq. (9.63) to obtain

$$\frac{2p\,dp}{G^2RT} + f\frac{dx}{D} - \frac{2\,dp}{p} = 0$$

Since G^2RT is constant for isothermal flow, this may be integrated in closed form between $(x, p) = (0, p_1)$ and (L, p_2):

$$G^2 = \left(\frac{\dot{m}}{A}\right)^2 = \frac{p_1^2 - p_2^2}{RT\,[\bar{f}L/D + 2\ln(p_1/p_2)]} \tag{9.73}$$

Thus, mass flow follows directly from the known end pressures, without any use of Mach numbers or tables.

The writer does not know of any direct analogy to Eq. (9.73) for adiabatic flow. However, a useful adiabatic relation, involving velocities instead of pressures, is

$$V_1^2 = \frac{a_0^2[1 - (V_1/V_2)^2]}{k\bar{f}L/D + (k+1)\ln(V_2/V_1)} \tag{9.74}$$

where $a_0 = (kRT_0)^{1/2}$ is the stagnation speed of sound, constant for adiabatic flow. This may be combined with continuity for constant duct area $V_1/V_2 = \rho_2/\rho_1$, plus the following combination of adiabatic energy and the perfect-gas relation:

$$\frac{V_1}{V_2} = \frac{p_2}{p_1}\frac{T_1}{T_2} = \frac{p_2}{p_1}\left[\frac{2a_0^2 - (k-1)V_1^2}{2a_0^2 - (k-1)V_2^2}\right] \tag{9.75}$$

If we are given the end pressures, neither V_1 nor V_2 will likely be known in advance. We suggest only the following simple procedure: Begin with $a_0 \approx a_1$ and the bracketed term in Eq. (9.75) approximately equal to 1.0. Solve Eq. (9.75) for a first estimate of V_1/V_2, and use this value in Eq. (9.74) to get a better estimate of V_1. Use V_1 to improve your estimate of a_0, and repeat the procedure. The process should converge in a few iterations.

Equations (9.73) and (9.74) have one flaw: With the Mach number eliminated, the frictional choking phenomenon is not directly evident. Therefore, assuming a subsonic inlet flow, one should check the exit Mach number Ma_2 to ensure that it is not greater than $1/k^{1/2}$ for isothermal flow or greater than 1.0 for adiabatic flow. We illustrate both adiabatic and isothermal flow with the following example.

EXAMPLE 9.13

Air enters a pipe of 1-cm diameter and 1.2-m length at $p_1 = 220$ kPa and $T_1 = 300$ K. If $\bar{f} = 0.025$ and the exit pressure is $p_2 = 140$ kPa, estimate the mass flow for (*a*) isothermal flow and (*b*) adiabatic flow.

Solution

Part (a) For isothermal flow Eq. (9.73) applies without iteration:

$$\frac{\bar{f}L}{D} + 2\ln\frac{p_1}{p_2} = \frac{(0.025)(1.2\text{ m})}{0.01\text{ m}} + 2\ln\frac{220}{140} = 3.904$$

$$G^2 = \frac{(220{,}000\text{ Pa})^2 - (140{,}000\text{ Pa})^2}{[287\text{ m}^2/(\text{s}^2\cdot\text{K})](300\text{ K})(3.904)} = 85{,}700 \quad \text{or} \quad G = 293\text{ kg}/(\text{s}\cdot\text{m}^2)$$

Since $A = (\pi/4)(0.01\text{ m})^2 = 7.85\text{ E-5 m}^2$, the isothermal mass flow estimate is

$$\dot{m} = GA = (293)(7.85\text{ E-5}) \approx 0.0230\text{ kg/s} \qquad \textit{Ans. (a)}$$

Check that the exit Mach number is not choked:

$$\rho_2 = \frac{p_2}{RT} = \frac{140{,}000}{(287)(300)} = 1.626\text{ kg/m}^3 \qquad V_2 = \frac{G}{\rho_2} = \frac{293}{1.626} = 180\text{ m/s}$$

or

$$\text{Ma}_2 = \frac{V_2}{\sqrt{kRT}} = \frac{180}{[1.4(287)(300)]^{1/2}} = \frac{180}{347} \approx 0.52$$

This is well below choking, and the isothermal solution is accurate.

Part (b) For adiabatic flow, we can iterate by hand, in the time-honored fashion, using Eqs. (9.74) and (9.75), plus the definition of stagnation speed of sound, $a_o = (kRT_o)^{1/2}$. A few years ago the writer would have done just that, laboriously. However, these equations can be iterated and manipulated by Excel. First, list the given data and requirements:

$$k = 1.4;\ p_1 = 220{,}000\text{ Pa};\ p_2 = 140{,}000\text{ Pa};\ T_1 = 300\text{ K};\ \frac{\bar{f}\Delta L}{D} = \frac{(0.025)(1.2)}{0.01} = 3.0,\ p_1^* = p_2^*$$

Now, use Excel to apply Eqs. (9.66), for $\bar{f}L/D$, and (9.68*a*), for p/p^*, to points 1 and 2 in the pipe. We are iterating to find the inlet Mach number for which (*a*) $\bar{f}\Delta L/D = 3.0$ and (*b*) $p_1^* = p_2^*$. Even with Excel doing all the work, the iteration is exasperating. We guess Ma_1 and adjust Ma_2 until $\Delta(fL/D) = 3.0$, after which we check for equal values of p^*. The tabulated results are below. It took four guesses of Ma_1 to arrive at $p^* \approx 67{,}900$ Pa.

	A	B	C	D	E	F	G	H	I
	Ma_1	$(fL/D)_1$	p_1/p^*	Ma_2	$(fL/D)_2$	p_2/p^*	$\Delta(fL/D)$	p_1^*	p_2^*
1	0.2	14.533	5.455	0.221	11.533	4.944	3.000	40,327	28,317
2	0.3	5.299	3.619	0.401	2.299	2.692	3.000	60,789	51,999
3	0.34	3.752	3.185	0.546	0.752	1.950	3.000	69,068	71,802
4	**0.3343**	**3.936**	**3.241**	**0.518**	**0.935**	**2.062**	**3.001**	**67,884**	**67,886**

The mass flow follows from the inlet Mach number and density and is quite close to part (*a*):

$$V_1 = Ma_1\sqrt{kRT_1} = (0.3343)\sqrt{1.4(287)(300)} = 116\text{ m/s}$$

$$\rho_1 = p_1/(RT_1) = (220{,}000)/[287(300)] = 2.56\text{ kg/m}^3$$

$$\dot{m} = \rho_1 A V_1 = (2.56)(\pi/4)(0.01)^2(116) = 0.0233\text{ kg/s} \qquad \textit{Ans. (b)}$$

9.8 Frictionless Duct Flow with Heat Transfer[5]

Heat addition or removal has an interesting effect on a compressible flow. Advanced texts [for example, Ref. 5, Chap. 8] consider the combined effect of heat transfer coupled with friction and area change in a duct. Here, we confine the analysis to heat transfer with no friction in a constant-area duct.

[5]This section may be omitted without loss of continuity.

This type of duct flow—constant area, constant momentum, constant mass flow, but variable stagnation enthalpy (due to heat transfer)—is often termed *Rayleigh flow* after John William Strutt, Lord Rayleigh (1842–1919), a famous physicist and engineer. For a given mass flow and momentum, a plot of enthalpy versus entropy for all possible flow states, subsonic or supersonic, forms a *Rayleigh line*. See Probs. P9.110 and P9.111 for examples of a Rayleigh line.

Consider the elemental duct control volume in Fig. 9.16. Between sections 1 and 2 an amount of heat δQ is added (or removed) to each incremental mass δm passing through. With no friction or area change, the control volume conservation relations are quite simple:

Continuity: $$\rho_1 V_1 = \rho_2 V_2 = G = \text{const} \tag{9.76a}$$

x momentum: $$p_1 - p_2 = G(V_2 - V_1) \tag{9.76b}$$

Energy: $$\dot{Q} = \dot{m}(h_2 + \tfrac{1}{2}V_2^2 - h_1 - \tfrac{1}{2}V_1^2)$$

or $$q = \frac{\dot{Q}}{\dot{m}} = \frac{\delta Q}{\delta m} = h_{02} - h_{01} \tag{9.76c}$$

The heat transfer results in a change in stagnation enthalpy of the flow. We shall not specify exactly how the heat is transferred—combustion, nuclear reaction, evaporation, condensation, or wall heat exchange—but simply that it happened in amount q between 1 and 2. We remark, however, that wall heat exchange is not a good candidate for the theory because wall convection is inevitably coupled with wall friction, which we neglected.

To complete the analysis, we use the perfect-gas and Mach number relations:

$$\frac{p_2}{\rho_2 T_2} = \frac{p_1}{\rho_1 T_1} \qquad h_{02} - h_{01} = c_p(T_{02} - T_{01})$$
$$\frac{V_2}{V_1} = \frac{\text{Ma}_2\, a_2}{\text{Ma}_1\, a_1} = \frac{\text{Ma}_2}{\text{Ma}_1}\left(\frac{T_2}{T_1}\right)^{1/2} \tag{9.77}$$

For a given heat transfer $q = \delta Q/\delta m$ or, equivalently, a given change $h_{02} - h_{01}$, Eqs. (9.76) and (9.77) can be solved algebraically for the property ratios p_2/p_1, Ma_2/Ma_1, and so on between inlet and outlet. Note that because the heat transfer allows the entropy to either increase or decrease, the second law imposes no restrictions on these solutions.

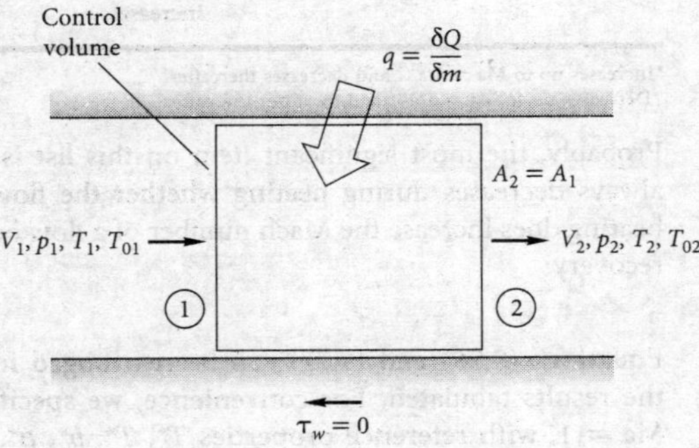

Fig. 9.16 Elemental control volume for frictionless flow in a constant-area duct with heat transfer. The length of the element is indeterminate in this simplified theory.

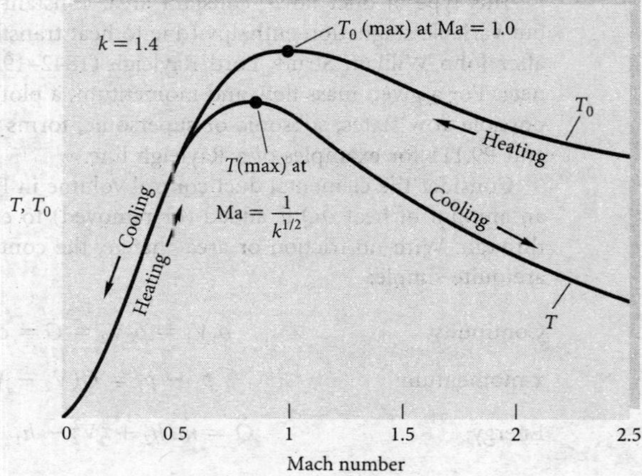

Fig. 9.17 Effect of heat transfer on Mach number.

Before writing down these property ratio functions, we illustrate the effect of heat transfer in Fig. 9.17, which shows T_0 and T versus Mach number in the duct. Heating increases T_0, and cooling decreases it. The maximum possible T_0 occurs at Ma = 1.0, and we see that heating, whether the inlet is subsonic or supersonic, drives the duct Mach number toward unity. This is analogous to the effect of friction in the previous section. The temperature of a perfect gas increases from Ma = 0 up to Ma = $1/k^{1/2}$ and then decreases. Thus there is a peculiar—or at least unexpected—region where heating (increasing T_0) actually decreases the gas temperature, the difference being reflected in a large increase of the gas kinetic energy. For $k = 1.4$ this peculiar area lies between Ma = 0.845 and Ma = 1.0 (interesting but not very useful information).

The complete list of the effects of simple T_0 change on duct flow properties is as follows:

	Heating		Cooling	
	Subsonic	Supersonic	Subsonic	Supersonic
T_0	Increases	Increases	Decreases	Decreases
Ma	Increases	Decreases	Decreases	Increases
p	Decreases	Increases	Increases	Decreases
ρ	Decreases	Increases	Increases	Decreases
V	Increases	Decreases	Decreases	Increases
p_0	Decreases	Decreases	Increases	Increases
s	Increases	Increases	Decreases	Decreases
T	*	Increases	†	Decreases

*Increases up to Ma = $1/k^{1/2}$ and decreases thereafter.
†Decreases up to Ma = $1/k^{1/2}$ and increases thereafter.

Probably, the most significant item on this list is the stagnation pressure p_0, which always decreases during heating whether the flow is subsonic or supersonic. Thus, heating does increase the Mach number of a flow but entails a loss in effective pressure recovery.

Mach Number Relations

Equations (9.76) and (9.77) can be rearranged in terms of the Mach number and the results tabulated. For convenience, we specify that the outlet section is sonic, Ma = 1, with reference properties T_0^*, T^*, p^*, ρ^*, V^*, and p_0^*. The inlet is assumed to be at arbitrary Mach number Ma. Equations (9.76) and (9.77) then take the following form:

$$\frac{T_0}{T_0^*} = \frac{(k+1)\,\text{Ma}^2\,[2+(k-1)\text{Ma}^2]}{(1+k\,\text{Ma}^2)^2} \tag{9.78a}$$

$$\frac{T}{T^*} = \frac{(k+1)^2\,\text{Ma}^2}{(1+k\,\text{Ma}^2)^2} \tag{9.78b}$$

$$\frac{p}{p^*} = \frac{k+1}{1+k\,\text{Ma}^2} \tag{9.78c}$$

$$\frac{V}{V^*} = \frac{\rho^*}{\rho} = \frac{(k+1)\text{Ma}^2}{1+k\,\text{Ma}^2} \tag{9.78d}$$

$$\frac{p_0}{p_0^*} = \frac{k+1}{1+k\,\text{Ma}^2}\left[\frac{2+(k-1)\text{Ma}^2}{k+1}\right]^{k/(k-1)} \tag{9.78e}$$

These formulas are all tabulated versus Mach number in Table B.4. The tables are very convenient if inlet properties Ma_1, V_1, and the like are given but are somewhat cumbersome if the given information centers on T_{01} and T_{02}. Let us illustrate with an example.

EXAMPLE 9.14

A fuel–air mixture, approximated as air with $k = 1.4$, enters a duct combustion chamber at $V_1 = 75$ m/s, $p_1 = 150$ kPa, and $T_1 = 300$ K. The heat addition by combustion is 900 kJ/kg of mixture. Compute (*a*) the exit properties V_2, p_2, and T_2 and (*b*) the total heat addition that would have caused a sonic exit flow.

Solution

Part (a) First, compute $T_{01} = T_1 + V_1^2/(2c_p) = 300 + (75)^2/[2(1{,}005)] = 303$ K. Then compute the change in stagnation temperature of the gas:

$$q = c_p(T_{02} - T_{01})$$

or $$T_{02} = T_{01} + \frac{q}{c_p} = 303\text{ K} + \frac{900{,}000\text{ J/kg}}{1{,}005\text{ J/(kg}\cdot\text{K)}} = 1{,}199\text{ K}$$

We have enough information to compute the initial Mach number:

$$a_1 = \sqrt{kRT_1} = [1.4(287)(300)]^{1/2} = 347\text{ m/s} \qquad \text{Ma}_1 = \frac{V_1}{a_1} = \frac{75}{347} = 0.216$$

For this Mach number, use Eq. (9.78*a*) or Table B.4 to find the sonic value T_0^*:

$$\text{At Ma}_1 = 0.216: \qquad \frac{T_{01}}{T_0^*} \approx 0.1992 \qquad \text{or} \qquad T_0^* = \frac{303\text{ K}}{0.1992} \approx 1{,}521\text{ K}$$

Then, the stagnation temperature ratio at section 2 is $T_{02}/T_0^* = 1{,}199/1{,}521 = 0.788$, which corresponds in Table B.4 to a Mach number $\text{Ma}_2 \approx 0.573$.

Now, use Eqs. (9.78) at Ma_1 and Ma_2 to tabulate the desired property ratios.

Section	Ma	V/V^*	p/p^*	T/T^*
1	0.216	0.1051	2.2528	0.2368
2	0.573	0.5398	1.6442	0.8876

The exit properties are computed by using these ratios to find state 2 from state 1:

$$V_2 = V_1 \frac{V_2/V^*}{V_1/V^*} = (75 \text{ m/s}) \frac{0.5398}{0.1051} = 385 \text{ m/s} \qquad \textit{Ans. (a)}$$

$$p_2 = p_1 \frac{p_2/p^*}{p_1/p^*} = (150 \text{ kPa}) \frac{1.6442}{2.2528} = 109 \text{ kPa} \qquad \textit{Ans. (a)}$$

$$T_2 = T_1 \frac{T_2/T^*}{T_1/T^*} = (300 \text{ K}) \frac{0.8876}{0.2368} = 1124 \text{ K} \qquad \textit{Ans. (a)}$$

Part (b) The maximum allowable heat addition would drive the exit Mach number to unity:

$$T_{02} = T_0^* = 1{,}521 \text{ K}$$

$$q_{max} = c_p(T_0^* - T_{01}) = [1{,}005 \text{ J/(kg} \cdot \text{K)}](1{,}521 - 303 \text{ K}) \approx 1.22 \text{ E6 J/kg} \qquad \textit{Ans. (b)}$$

Choking Effects Due to Simple Heating

Equation (9.78*a*) and Table B.4 indicate that the maximum possible stagnation temperature in simple heating corresponds to T_0^*, or the sonic exit Mach number. Thus, for given inlet conditions, only a certain maximum amount of heat can be added to the flow—for example, 1.22 MJ/kg in Example 9.14. For a subsonic inlet there is no theoretical limit on heat addition: The flow chokes more and more as we add more heat, with the inlet velocity approaching zero. For supersonic flow, even if Ma_1 is infinite, there is a finite ratio $T_{01}/T_0^* = 0.4898$ for $k = 1.4$. Thus if heat is added without limit to a supersonic flow, a normal shock wave adjustment is required to accommodate the required property changes.

In subsonic flow, there is no theoretical limit to the amount of cooling allowed: The exit flow just becomes slower and slower, and the temperature approaches zero. In supersonic flow only a finite amount of cooling can be allowed before the exit flow approaches infinite Mach number, with $T_{02}/T_0^* = 0.4898$ and the exit temperature equal to zero. There are very few practical applications for supersonic cooling.

EXAMPLE 9.15

What happens to the inlet flow in Example 9.14 if the heat addition is increased to 1,400 kJ/kg and the inlet pressure and stagnation temperature are fixed? What will be the subsequent decrease in mass flow?

Solution

For $q = 1{,}400$ kJ/kg, the exit will be choked at the stagnation temperature:

$$T_0^* = T_{01} + \frac{q}{c_p} = 303 + \frac{1.4 \text{ E6 J/kg}}{1{,}005 \text{ J/(kg} \cdot \text{K)}} \approx 1{,}696 \text{ K}$$

This is higher than the value $T_0^* = 1{,}521$ K in Example 9.14, so we know that condition 1 will have to choke down to a lower Mach number. The proper value is found from the ratio $T_{01}/T_0^* = 303/1{,}696 = 0.1787$. From Table B.4 or Eq. (9.78*a*) for this condition, we read the new, lowered entrance Mach number: $\text{Ma}_{1,\text{new}} \approx 0.203$. With T_{01} and p_1 known, the other inlet properties follow from this Mach number:

$$T_1 = \frac{T_{01}}{1 + 0.2\,\text{Ma}_1^2} = \frac{303}{1 + 0.2(0.203)^2} = 301 \text{ K}$$

$$a_1 = \sqrt{kRT_1} = [1.4(287)(301)]^{1/2} = 348 \text{ m/s}$$

$$V_1 = \text{Ma}_1 a_1 = (0.203)(348 \text{ m/s}) = 71 \text{ m/s}$$

$$\rho_1 = \frac{p_1}{RT_1} = \frac{150{,}000}{(287)(301)} = 1.74 \text{ kg/m}^3$$

Finally, the new lowered mass flow per unit area is

$$\frac{\dot{m}_{\text{new}}}{A} = \rho_1 V_1 = (1.74 \text{ kg/m}^3)(71 \text{ m/s}) = 123 \text{ kg/(s} \cdot \text{m}^2)$$

This is 7 percent less than in Example 9.14, due to choking by excess heat addition.

Relationship to the Normal Shock Wave

The normal shock wave relations of Sec. 9.5 actually lurk within the simple heating relations as a special case. From Table B.4 or Fig. 9.17 we see that for a given stagnation temperature less than T_0^* two flow states satisfy the simple heating relations, one subsonic and the other supersonic. These two states have (1) the same value of T_0, (2) the same mass flow per unit area, and (3) the same value of $p + \rho V^2$. Therefore these two states are exactly equivalent to the conditions on each side of a normal shock wave. The second law would again require that the upstream flow Ma_1 be supersonic.

To illustrate this point, take $\text{Ma}_1 = 3.0$ and from Table B.4 read $T_{01}/T_0^* = 0.6540$ and $p_1/p^* = 0.1765$. Now, for the same value $T_{02}/T_0^* = 0.6540$, use Table B.4 or Eq. (9.78*a*) to compute $\text{Ma}_2 = 0.4752$ and $p_2/p^* = 1.8235$. The value of Ma_2 is exactly what we read in the shock table, Table B.2, as the downstream Mach number when $\text{Ma}_1 = 3.0$. The pressure ratio for these two states is $p_2/p_1 = (p_2/p^*)/(p_1/p^*) = 1.8235/0.1765 = 10.33$, which again is just what we read in Table B.2 for $\text{Ma}_1 = 3.0$. This illustration is meant only to show the physical background of the simple heating relations; it would be silly to make a practice of computing normal shock waves in this manner.

9.9 Mach Waves and Oblique Shock Waves

Up to this point we have considered only one-dimensional compressible flow theories. This illustrated many important effects, but a one-dimensional world completely loses sight of the wave motions that are so characteristic of supersonic flow. The only "wave motion" we could muster in a one-dimensional theory was the normal shock wave, which amounted only to a flow discontinuity in the duct.

Mach Waves

When we add a second dimension to the flow, wave motions immediately become apparent if the flow is supersonic. Figure 9.18 shows a celebrated graphical construction that appears in every fluid mechanics textbook and was first presented by Ernst Mach in 1887. The figure shows the pattern of pressure disturbances (sound waves) sent out by a small particle moving at speed U through a still fluid whose sound velocity is a.

As the particle moves, it continually crashes against fluid particles and sends out spherical sound waves emanating from every point along its path. A few of these spherical disturbance fronts are shown in Fig. 9.18. The behavior of these fronts is quite different according to whether the particle speed is subsonic or supersonic.

In Fig. 9.18*a*, the particle moves subsonically, $U < a$, $\text{Ma} = U/a < 1$. The spherical disturbances move out in all directions and do not catch up with one another. They move well out in front of the particle also, because they travel a distance $a\,\delta t$ during the time interval δt in which the particle has moved only $U\,\delta t$. Therefore a subsonic body motion makes its presence felt everywhere in the flow field: You can "hear" or "feel" the pressure rise of an oncoming body before it reaches you. This is apparently

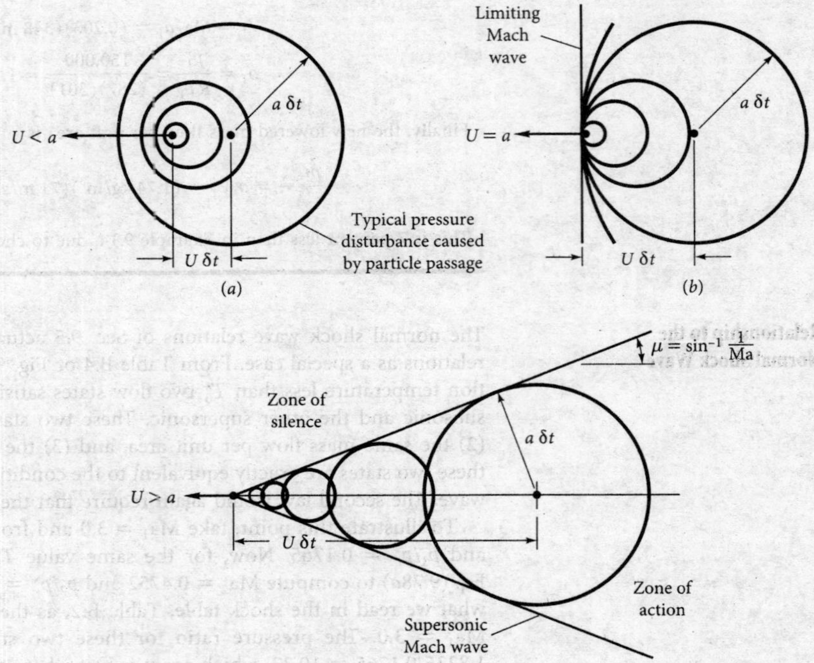

Fig. 9.18 Wave patterns set up by a particle moving at speed U into still fluid of sound velocity a: (*a*) subsonic, (*b*) sonic, and (*c*) supersonic motion.

why that pigeon in the road, without turning around to look at you, takes to the air and avoids being hit by your car.

At sonic speed, $U = a$, Fig. 9.18*b*, the pressure disturbances move at exactly the speed of the particle and thus pile up on the left at the position of the particle into a sort of "front locus," which is now called a *Mach wave,* after Ernst Mach. No disturbance reaches beyond the particle. If you are stationed to the left of the particle, you cannot "hear" the oncoming motion. If the particle blew its horn, you couldn't hear that either: A sonic car can sneak up on a pigeon.

In supersonic motion, $U > a$, the lack of advance warning is even more pronounced. The disturbance spheres cannot catch up with the fast-moving particle that created them. They all trail behind the particle and are tangent to a conical locus called the *Mach cone.* From the geometry of Fig. 9.18*c* the angle of the Mach cone is seen to be

$$\mu = \sin^{-1}\frac{a\,\delta t}{U\,\delta t} = \sin^{-1}\frac{a}{U} = \sin^{-1}\frac{1}{\text{Ma}} \tag{9.79}$$

The higher the particle Mach number, the more slender the Mach cone; for example, μ is 30° at Ma = 2.0 and 11.5° at Ma = 5.0. For the limiting case of sonic flow, Ma = 1, $\mu = 90°$; the Mach cone becomes a plane front moving with the particle, in agreement with Fig. 9.18*b*.

You cannot "hear" the disturbance caused by the supersonic particle in Fig. 9.18*c* until you are in the *zone of action* inside the Mach cone. No warning can reach your ears if you are in the *zone of silence* outside the cone. Thus an observer on the ground beneath a supersonic airplane does not hear the *sonic boom* of the passing cone until the plane is well past.

Fig. 9.19 The wave pattern around a model of the X-15 fighter, moving at about Ma = 1.7. The heavy lines are oblique shock waves, caused by sharp turns, and the light lines are Mach waves, caused by gentle turns. [*Photo Courtesy of NASA*]

The Mach wave need not be a cone: Similar waves are formed by a small disturbance of any shape moving supersonically with respect to the ambient fluid. For example, the "particle" in Fig. 9.18*c* could be the leading edge of a sharp flat plate, which would form a Mach wedge of exactly the same angle μ. Mach waves are formed by small roughnesses or boundary layer irregularities in a supersonic wind tunnel or at the surface of a supersonic body. Look again at Fig. 9.10: Mach waves are clearly visible along the body surface downstream of the recompression shock, especially at the rear corner. Their angle is about 30°, indicating a Mach number of about 2.0 along this surface.

A more complicated system of waves, seen in Fig. 9.19, emanates from a model of the supersonic X-15 fighter plane, catapulted at Ma $\approx$ 1.7 through a wind tunnel. The Mach and shock waves are visualized by the knife-edge *schlieren* photographic technique [31]. Note the supersonic turbulent wake.

EXAMPLE 9.16

An observer on the ground does not hear the sonic boom caused by an airplane moving at 5-km altitude until it is 9 km past her. What is the approximate Mach number of the plane? Assume a small disturbance, and neglect the variation of sound speed with altitude.

Solution

A finite disturbance like an airplane will create a finite-strength oblique shock wave whose angle will be somewhat larger than the Mach wave angle μ and will curve downward due to the variation in atmospheric sound speed. If we neglect these effects, the altitude and distance are a measure of μ, as seen in Fig. E9.16.

Thus, $$\tan \mu = \frac{5 \text{ km}}{9 \text{ km}} = 0.5556 \quad \text{or} \quad \mu = 29.05°$$

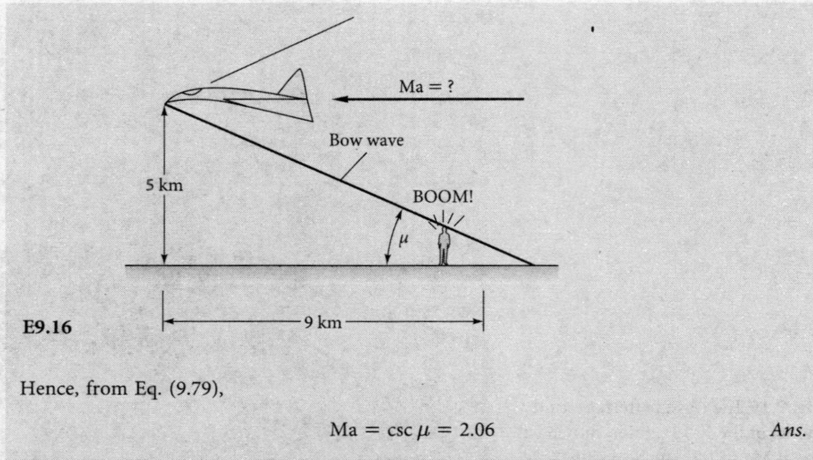

E9.16

Hence, from Eq. (9.79),

$$\text{Ma} = \csc \mu = 2.06 \qquad \textit{Ans.}$$

The Oblique Shock Wave

Figures 9.10 and 9.19 and our earlier discussion all indicate that a shock wave can form at an oblique angle to the oncoming supersonic stream. Such a wave will deflect the stream through an angle θ, unlike the normal shock wave, for which the downstream flow is in the same direction. In essence, an oblique shock is caused by the necessity for a supersonic stream to turn through such an angle. Examples could be a finite wedge at the leading edge of a body and a ramp in the wall of a supersonic wind tunnel.

The flow geometry of an oblique shock is shown in Fig. 9.20. As for the normal shock of Fig. 9.8, state 1 denotes the upstream conditions and state 2 is downstream. The shock angle has an arbitrary value β, and the downstream flow V_2 turns at an angle θ which is a function of β and state 1 conditions. The upstream flow is always supersonic, but the downstream Mach number $\text{Ma}_2 = V_2/a_2$ may be subsonic, sonic, or supersonic, depending on the conditions.

It is convenient to analyze the flow by breaking it up into normal and tangential components with respect to the wave, as shown in Fig. 9.20. For a thin control volume just encompassing the wave, we can then derive the following integral relations, canceling out $A_1 = A_2$ on each side of the wave:

Continuity: $$\rho_1 V_{n1} = \rho_2 V_{n2} \tag{9.80a}$$

Normal momentum: $$p_1 - p_2 = \rho_2 V_{n2}^2 - \rho_1 V_{n1}^2 \tag{9.80b}$$

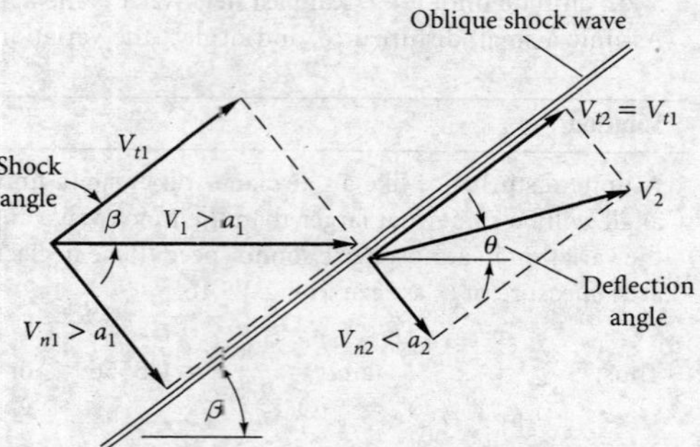

Fig. 9.20 Geometry of flow through an oblique shock wave.

Tangential momentum: $$0 = \rho_1 V_{n1}(V_{t2} - V_{t1}) \tag{9.80c}$$

Energy: $$h_1 + \tfrac{1}{2}V_{n1}^2 + \tfrac{1}{2}V_{t1}^2 = h_2 + \tfrac{1}{2}V_{n2}^2 + \tfrac{1}{2}V_{t2}^2 = h_0 \tag{9.80d}$$

We see from Eq. (9.80*c*) that there is no change in tangential velocity across an oblique shock:

$$V_{t2} = V_{t1} = V_t = \text{const} \tag{9.81}$$

Thus, tangential velocity has as its only effect the addition of a constant kinetic energy $\frac{1}{2}V_t^2$ to each side of the energy equation (9.80*d*). We conclude that Eqs. (9.80) are identical to the normal shock relations (9.49), with V_1 and V_2 replaced by the normal components V_{n1} and V_{n2}. All the various relations from Sec. 9.5 can be used to compute properties of an oblique shock wave. The trick is to use the "normal" Mach numbers in place of Ma_1 and Ma_2:

$$\boxed{Ma_{n1} = \frac{V_{n1}}{a_1} = Ma_1 \sin \beta}$$
$$\boxed{Ma_{n2} = \frac{V_{n2}}{a_2} = Ma_2 \sin (\beta - \theta)} \tag{9.82}$$

Then, for a perfect gas with constant specific heats, the property ratios across the oblique shock are the analogs of Eqs. (9.55) to (9.58) with Ma_1 replaced by Ma_{n1}:

$$\frac{p_2}{p_1} = \frac{1}{k+1}\left[2k\, Ma_1^2 \sin^2 \beta - (k-1)\right] \tag{9.83a}$$

$$\frac{\rho_2}{\rho_1} = \frac{\tan \beta}{\tan (\beta - \theta)} = \frac{(k+1)\, Ma_1^2 \sin^2 \beta}{(k-1)\, Ma_1^2 \sin^2 \beta + 2} = \frac{V_{n1}}{V_{n2}} \tag{9.83b}$$

$$\frac{T_2}{T_1} = \left[2 + (k-1)\, Ma_1^2 \sin^2 \beta\right] \frac{2k\, Ma_1^2 \sin^2 \beta - (k-1)}{(k+1)^2\, Ma_1^2 \sin^2 \beta} \tag{9.83c}$$

$$T_{02} = T_{01} \tag{9.83d}$$

$$\frac{p_{02}}{p_{01}} = \left[\frac{(k+1)\, Ma_1^2 \sin^2 \beta}{2 + (k-1)\, Ma_1^2 \sin^2 \beta}\right]^{k/(k-1)} \left[\frac{k+1}{2k\, Ma_1^2 \sin^2 \beta - (k-1)}\right]^{1/(k-1)} \tag{9.83e}$$

$$Ma_{n2}^2 = \frac{(k-1)\, Ma_{n1}^2 + 2}{2k\, Ma_{n1}^2 - (k-1)} \tag{9.83f}$$

All these are tabulated in the normal shock Table B.2. If you wondered why that table listed the Mach numbers as Ma_{n1} and Ma_{n2}, it should be clear now that the table is also valid for the oblique shock wave.

Thinking all this over, we realize with hindsight that an oblique shock wave is the flow pattern one would observe by running along a normal shock wave (Fig. 9.8) at a constant tangential speed V_t. Thus the normal and oblique shocks are related by a galilean, or inertial, velocity transformation and therefore satisfy the same basic equations.

If we continue with this run-along-the-shock analogy, we find that the deflection angle θ increases with speed V_t up to a maximum and then decreases. From the geometry of Fig. 9.20, the deflection angle is given by

$$\theta = \tan^{-1}\frac{V_t}{V_{n2}} - \tan^{-1}\frac{V_t}{V_{n1}} \tag{9.84}$$

If we differentiate θ with respect to V_t and set the result equal to zero, we find that the maximum deflection occurs when $V_t/V_{n1} = (V_{n2}/V_{n1})^{1/2}$. We can substitute this back into Eq. (9.84) to compute

$$\theta_{max} = \tan^{-1} r^{1/2} - \tan^{-1} r^{-1/2} \qquad r = \frac{V_{n1}}{V_{n2}} \tag{9.85}$$

For example, if $Ma_{n1} = 3.0$, from Table B.2 we find that $V_{n1}/V_{n2} = 3.8571$, the square root of which is 1.9640. Then Eq. (9.85) predicts a maximum deflection of $\tan^{-1} 1.9640 - \tan^{-1} (1/1.9640) = 36.03°$. The deflection is quite limited even for infinite Ma_{n1}: From Table B.2 for this case $V_{n1}/V_{n2} = 6.0$, and we compute from Eq. (9.85) that $\theta_{max} = 45.58°$.

This limited-deflection idea and other facts become more evident if we plot some of the solutions of Eqs. (9.83). For given values of V_1 and a_1, assuming as usual that $k = 1.4$, we can plot all possible solutions for V_2 downstream of the shock. Figure 9.21 does this in velocity-component coordinates V_x and V_y, with x parallel to V_1. Such a plot is called a *hodograph*. The heavy dark line that looks like a fat airfoil is the locus, or *shock polar,* of all physically possible solutions for the given Ma_1. The two dashed-line fishtails are solutions that increase V_2; they are physically impossible because they violate the second law.

Examining the shock polar in Fig. 9.21, we see that a given deflection line of small angle θ crosses the polar at two possible solutions: the *strong* shock, which greatly decelerates the flow, and the *weak* shock, which causes a much milder deceleration. The flow downstream of the strong shock is always subsonic, while that of the weak shock is usually supersonic but occasionally subsonic if the deflection is large. Both types of shock occur in practice. The weak shock is more prevalent, but the strong shock will occur if there is a blockage or high-pressure condition downstream.

Since the shock polar is only of finite size, there is a maximum deflection θ_{max}, shown in Fig. 9.21, that just grazes the upper edge of the polar curve. This verifies the kinematic discussion that led to Eq. (9.85). What happens if a supersonic flow is forced to deflect through an angle greater than θ_{max}? The answer is illustrated in Fig. 9.22 for flow past a wedge-shaped body.

In Fig. 9.22*a*, the wedge half-angle θ is less than θ_{max}, and thus an oblique shock forms at the nose of wave angle β just sufficient to cause the oncoming supersonic stream to deflect through the wedge angle θ. Except for the usually small effect of boundary layer growth (see, for example, Ref. 19, Sec. 7–5.2), the Mach number Ma_2 is constant along the wedge surface and is given by the solution of Eqs. (9.83). The

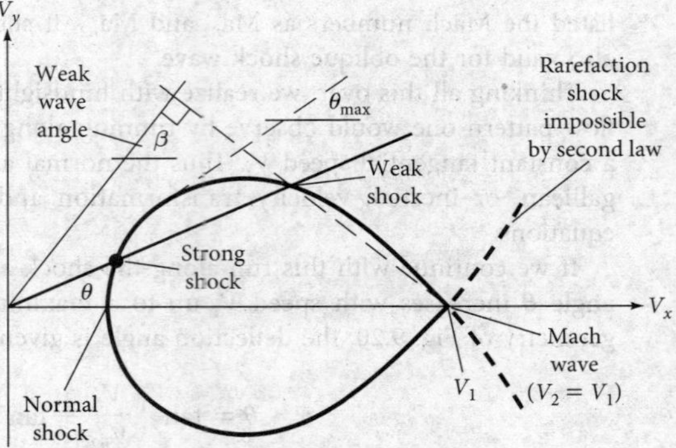

Fig. 9.21 The oblique shock polar hodograph, showing double solutions (strong and weak) for small deflection angle and no solutions at all for large deflection.

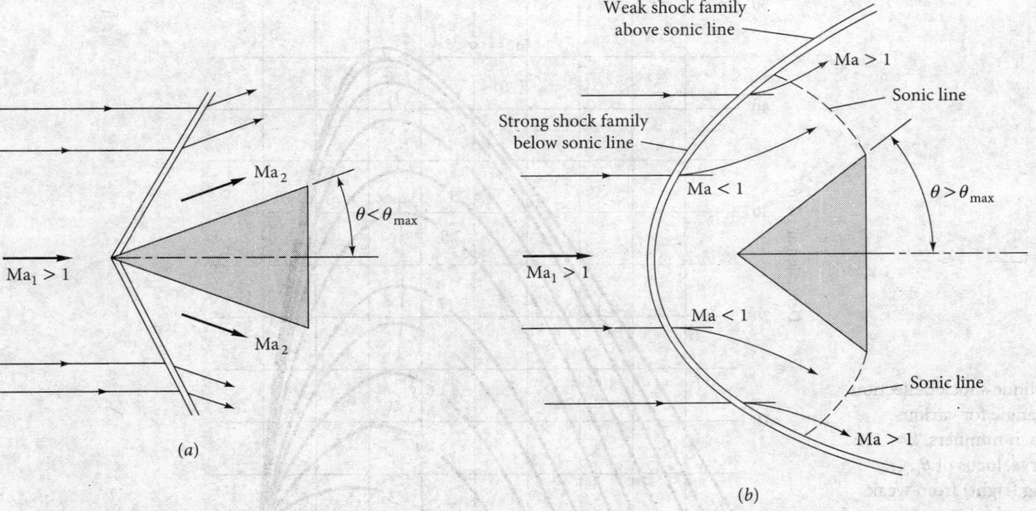

Fig. 9.22 Supersonic flow past a wedge: (*a*) small wedge angle, attached oblique shock forms; (*b*) large wedge angle, attached shock not possible, broad curved detached shock forms.

pressure, density, and temperature along the surface are also nearly constant, as predicted by Eqs. (9.83). When the flow reaches the corner of the wedge, it expands to higher Mach number and forms a wake (not shown) similar to that in Fig. 9.10.

In Fig. 9.22*b*, the wedge half-angle is greater than θ_{max}, and an attached oblique shock is impossible. The flow cannot deflect at once through the entire angle θ_{max}, yet somehow the flow must get around the wedge. A detached curve shock wave forms in front of the body, discontinuously deflecting the flow through angles smaller than θ_{max}. The flow then curves, expands, and deflects subsonically around the wedge, becoming sonic and then supersonic as it passes the corner region. The flow just inside each point on the curved shock exactly satisfies the oblique shock relations (9.83) for that particular value of β and the given Ma_1. Every condition along the curved shock is a point on the shock polar of Fig. 9.21. Points near the front of the wedge are in the strong shock family, and points aft of the sonic line are in the weak shock family. The analysis of detached shock waves is extremely complex [13], and experimentation is usually needed, such as the shadowgraph optical technique of Fig. 9.10.

The complete family of oblique shock solutions can be plotted or computed from Eqs. (9.83). For a given k, the wave angle β varies with Ma_1 and θ, from Eq. (9.83*b*). By using a trigonometric identity for tan $(\beta - \theta)$ this can be rewritten in the more convenient form

$$\tan\theta = \frac{2\cot\beta\,(Ma_1^2\sin^2\beta - 1)}{Ma_1^2\,(k + \cos 2\beta) + 2} \tag{9.86}$$

All possible solutions of Eq. (9.86) for $k = 1.4$ are shown in Fig. 9.23. For deflections $\theta < \theta_{max}$ there are two solutions: a weak shock (small β) and a strong shock (large β), as expected. All points along the dash–dot line for θ_{max} satisfy Eq. (9.85). A dashed line has been added to show where Ma_2 is exactly sonic. We see that there is a narrow region near maximum deflection where the weak shock downstream flow is subsonic.

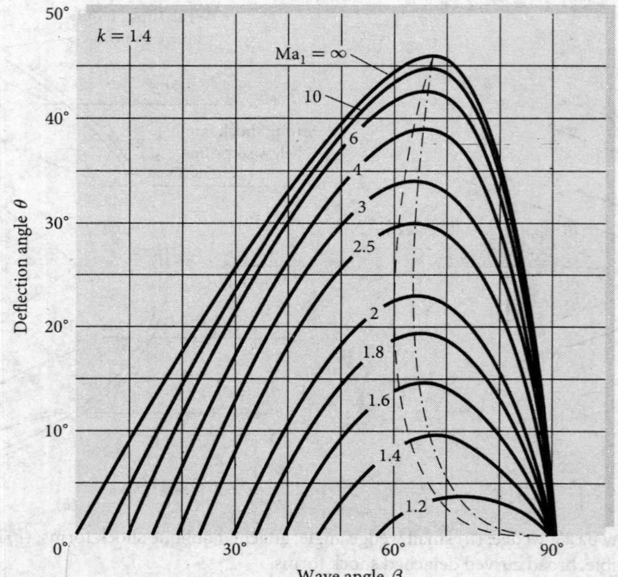

Fig. 9.23 Oblique shock deflection versus wave angle for various upstream Mach numbers, $k = 1.4$: dash–dot curve, locus of θ_{max}, divides strong (right) from weak (left) shocks; dashed curve, locus of sonic points, divides subsonic Ma_2 (right) from supersonic Ma_2 (left).

For zero deflections ($\theta = 0$) the weak shock family satisfies the wave angle relation

$$\beta = \mu = \sin^{-1}\frac{1}{Ma_1} \tag{9.87}$$

Thus weak shocks of vanishing deflection are equivalent to Mach waves. Meanwhile the strong shocks all converge at zero deflection to the normal shock condition $\beta = 90°$.

Two additional oblique shock charts are given in App. B for $k = 1.4$, where Fig. B.1 gives the downstream Mach number Ma_2 and Fig. B.2 the pressure ratio p_2/p_1, each plotted as a function of Ma_1 and θ. Additional graphs, tables, and computer programs are given in Refs. 20 and 21.

Very Weak Shock Waves

For any finite θ the wave angle β for a weak shock is greater than the Mach angle μ. For small θ Eq. (9.86) can be expanded in a power series in $\tan\theta$ with the following linearized result for the wave angle:

$$\sin\beta = \sin\mu + \frac{k+1}{4\cos\mu}\tan\theta + \cdots + \mathcal{O}(\tan^2\theta) + \cdots \tag{9.88}$$

For Ma_1 between 1.4 and 20.0 and deflections less than 6° this relation predicts β to within 1° for a weak shock. For larger deflections it can be used as a useful initial guess for iterative solution of Eq. (9.86).

Other property changes across the oblique shock can also be expanded in a power series for small deflection angles. Of particular interest is the pressure change from Eq. (9.83*a*), for which the linearized result for a weak shock is

$$\frac{p_2 - p_1}{p_1} = \frac{k\,Ma_1^2}{(Ma_1^2 - 1)^{1/2}}\tan\theta + \cdots + \mathcal{O}(\tan^2\theta) + \cdots \tag{9.89}$$

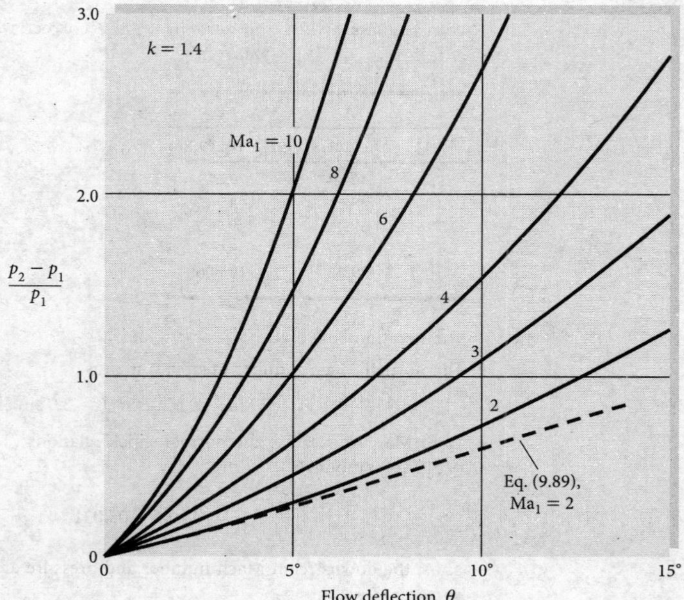

Fig. 9.24 Pressure jump across a weak oblique shock wave from Eq. (9.83*a*) for $k = 1.4$. For very small deflections Eq. (9.89) applies.

The differential form of this relation is used in the next section to develop a theory for supersonic expansion turns. Figure 9.24 shows the exact weak shock pressure jump computed from Eq. (9.83*a*). At very small deflections the curves are linear with slopes given by Eq. (9.89).

Finally, it is educational to examine the entropy change across a very weak shock. Using the same power series expansion technique, we can obtain the following result for small flow deflections:

$$\frac{s_2 - s_1}{c_p} = \frac{(k^2 - 1)\text{Ma}_1^6}{12(\text{Ma}_1^2 - 1)^{3/2}} \tan^3\theta + \cdots + \mathcal{O}(\tan^4\theta) + \cdots \tag{9.90}$$

The entropy change is cubic in the deflection angle θ. Thus, weak shock waves are very nearly isentropic, a fact that is also used in the next section.

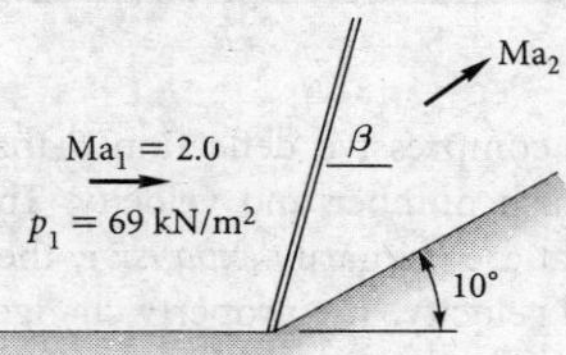

E9.17

EXAMPLE 9.17

Air at Ma = 2.0 and p = 69 kPa absolute is forced to turn through 10° by a ramp at the body surface. A weak oblique shock forms as in Fig. E9.17. For $k = 1.4$ compute from exact oblique shock theory (*a*) the wave angle β, (*b*) Ma_2, and (*c*) p_2. Also use the linearized theory to estimate (*d*) β and (*e*) p_2.

Solution using Excel

With $\text{Ma}_1 = 2.0$ and $\theta = 10°$ known, we can estimate $\beta \approx 40° \pm 2°$ from Fig. 9.23. For more accuracy, we can set an Excel iteration, using improved guesses for β. Begin with a guess of 40° and zero in on the correct wave angle. The writer's guesses are as follows:

	A	B
	β – guess	θ – Eq. (9.86)
1	40.00	10.623
2	38.00	8.767
3	39.00	9.710
4	39.30	9.987
5	**39.32**	**10.006**

The iteration converges to $\beta = 39.32°$ *Ans. (a)*

The normal Mach number upstream is thus

$$\text{Ma}_{n1} = \text{Ma}_1 \sin\beta = 2.0 \sin 39.32° = 1.267$$

With Ma_{n1} we can use the normal shock relations (Table B.2) or Fig. 9.9 or Eqs. (9.56) to (9.58) to compute

$$\text{Ma}_{n2} = 0.8031 \qquad \frac{p_2}{p_1} = 1.707$$

Thus, the downstream Mach number and pressure are

$$\text{Ma}_2 = \frac{\text{Ma}_{n2}}{\sin(\beta - \theta)} = \frac{0.8031}{\sin(39.32° - 10°)} = 1.64 \qquad \textit{Ans. (b)}$$

$$p_2 = (69 \text{ kPa absolute})(1.707) = 117.8 \text{ kPa absolute} \qquad \textit{Ans. (c)}$$

Notice that the computed pressure ratio agrees with Figs. 9.24 and B.2.

For the linearized theory the Mach angle is $\mu = \sin^{-1}(1/2.0) = 30°$. Equation (9.88) then estimates that

$$\sin\beta \approx \sin 30° + \frac{2.4 \tan 10°}{4 \cos 30°} = 0.622$$

or

$$\beta \approx 38.5° \qquad \textit{Ans. (d)}$$

Equation (9.89) estimates that

$$\frac{p_2}{p_1} \approx 1 + \frac{1.4(2)^2 \tan 10°}{(2^2 - 1)^{1/2}} = 1.57$$

or

$$p_2 \approx 1.57(69 \text{ kPa absolute}) \approx 108.3 \text{ kPa absolute} \qquad \textit{Ans. (e)}$$

These are reasonable estimates in spite of the fact that 10° is really not a "small" flow deflection.

9.10 Prandtl-Meyer Expansion Waves

The oblique shock solution of Sec. 9.9 is for a finite compressive deflection θ that obstructs a supersonic flow and thus decreases its Mach number and velocity. The present section treats gradual changes in flow angle that are primarily *expansive;* they widen the flow area and increase the Mach number and velocity. The property changes accumulate in infinitesimal increments, and the linearized relations (9.88) and (9.89) are used. The local flow deflections are infinitesimal, so the flow is nearly isentropic according to Eq. (9.90).

Figure 9.25 shows four examples, one of which (Fig. 9.25*c*) fails the test for gradual changes. The gradual compression of Fig. 9.25*a* is essentially isentropic, with a smooth

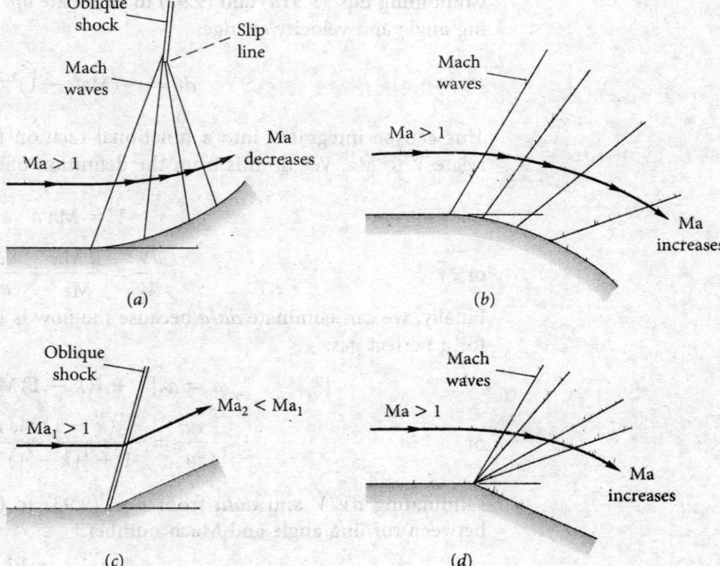

Fig. 9.25 Some examples of supersonic expansion and compression: (*a*) gradual isentropic compression on a concave surface, Mach waves coalesce farther out to form oblique shock; (*b*) gradual isentropic expansion on convex surface, Mach waves diverge; (*c*) sudden compression, nonisentropic shock forms; (*d*) sudden expansion, centered isentropic fan of Mach waves forms.

increase in pressure along the surface, but the Mach angle increases along the surface and the waves tend to coalesce farther out into an oblique shock wave. The gradual expansion of Fig. 9.25*b* causes a smooth isentropic increase of Mach number and velocity along the surface, with diverging Mach waves formed.

The sudden compression of Fig. 9.25*c* cannot be accomplished by Mach waves: An oblique shock forms, and the flow is nonisentropic. This could be what you would see if you looked at Fig. 9.25*a* from far away. Finally, the sudden expansion of Fig. 9.25*d* is isentropic and forms a fan of centered Mach waves emanating from the corner. Note that the flow on any streamline passing through the fan changes smoothly to higher Mach number and velocity. In the limit as we near the corner the flow expands almost discontinuously at the surface. The cases in Figs. 9.25*a*, *b*, and *d* can all be handled by the Prandtl-Meyer supersonic wave theory of this section, first formulated by Ludwig Prandtl and his student Theodor Meyer in 1907 to 1908.

Note that none of this discussion makes sense if the upstream Mach number is subsonic, since Mach wave and shock wave patterns cannot exist in subsonic flow.

The Prandtl-Meyer Perfect-Gas Function

Consider a small, nearly infinitesimal flow deflection $d\theta$ such as occurs between the first two Mach waves in Fig. 9.25*a*. From Eqs. (9.88) and (9.89) we have, in the limit,

$$\beta \approx \mu = \sin^{-1}\frac{1}{\text{Ma}} \tag{9.91a}$$

$$\frac{dp}{p} \approx \frac{k\,\text{Ma}^2}{(\text{Ma}^2 - 1)^{1/2}}\,d\theta \tag{9.91b}$$

Since the flow is nearly isentropic, we have the frictionless differential momentum equation for a perfect gas:

$$dp = -\rho V\,dV = -kp\,\text{Ma}^2\frac{dV}{V} \tag{9.92}$$

Combining Eqs. (9.91*a*) and (9.92) to eliminate *dp*, we obtain a relation between turning angle and velocity change:

$$d\theta = -(\text{Ma}^2 - 1)^{1/2} \frac{dV}{V} \tag{9.93}$$

This can be integrated into a functional relation for finite turning angles if we can relate *V* to Ma. We do this from the definition of Mach number:

$$V = \text{Ma}\, a$$

or

$$\frac{dV}{V} = \frac{d\,\text{Ma}}{\text{Ma}} + \frac{da}{a} \tag{9.94}$$

Finally, we can eliminate *da*/*a* because the flow is isentropic and hence a_0 is constant for a perfect gas:

$$a = a_0[1 + \tfrac{1}{2}(k - 1)\text{Ma}^2]^{-1/2}$$

or

$$\frac{da}{a} = \frac{-\frac{1}{2}(k - 1)\,\text{Ma}\,d\,\text{Ma}}{1 + \frac{1}{2}(k - 1)\,\text{Ma}^2} \tag{9.95}$$

Eliminating *dV*/*V* and *da*/*a* from Eqs. (9.93) to (9.95), we obtain a relation solely between turning angle and Mach number:

$$d\theta = -\frac{(\text{Ma}^2 - 1)^{1/2}}{1 + \frac{1}{2}(k - 1)\,\text{Ma}^2}\frac{d\,\text{Ma}}{\text{Ma}} \tag{9.96}$$

Before integrating this expression, we note that the primary application is to expansions: increasing Ma and decreasing θ. Therefore, for convenience, we define the Prandtl-Meyer angle $\omega(\text{Ma})$, which increases when θ decreases and is zero at the sonic point:

$$d\omega = -d\theta \qquad \omega = 0 \quad \text{at} \quad \text{Ma} = 1 \tag{9.97}$$

Thus, we integrate Eq. (9.96) from the sonic point to any value of Ma:

$$\int_0^{\omega} d\omega = \int_1^{\text{Ma}} \frac{(\text{Ma}^2 - 1)^{1/2}}{1 + \frac{1}{2}(k - 1)\,\text{Ma}^2}\frac{d\,\text{Ma}}{\text{Ma}} \tag{9.98}$$

The integrals are evaluated in closed form, with the result, in radians,

$$\omega(\text{Ma}) = K^{1/2} \tan^{-1}\left(\frac{\text{Ma}^2 - 1}{K}\right)^{1/2} - \tan^{-1}(\text{Ma}^2 - 1)^{1/2} \tag{9.99}$$

where

$$K = \frac{k + 1}{k - 1}$$

This is the *Prandtl-Meyer supersonic expansion function,* which is plotted in Fig. 9.26 and tabulated in Table B.5 for $k = 1.4$, $K = 6$. The angle ω changes rapidly at first and then levels off at high Mach number to a limiting value as $\text{Ma} \to \infty$:

$$\omega_{\max} = \frac{\pi}{2}(K^{1/2} - 1) = 130.45° \qquad \text{if} \qquad k = 1.4 \tag{9.100}$$

Thus, a supersonic flow can expand only through a finite turning angle before it reaches infinite Mach number, maximum velocity, and zero temperature.

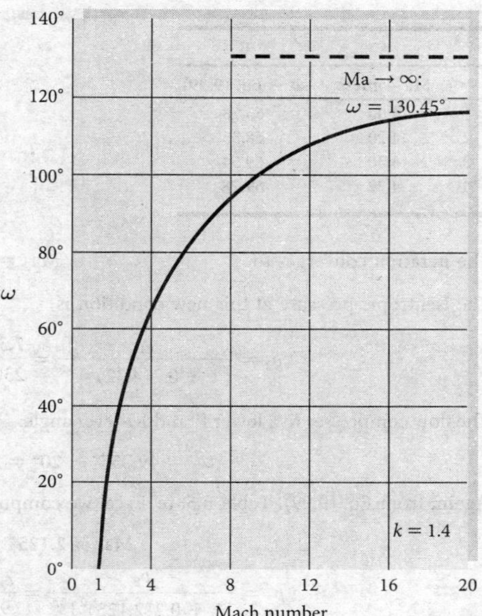

Fig. 9.26 The Prandtl-Meyer supersonic expansion from Eq. (9.99) for $k = 1.4$.

Gradual expansion or compression between finite Mach numbers Ma_1 and Ma_2, neither of which is unity, is computed by relating the turning angle $\Delta\omega$ to the difference in Prandtl-Meyer angles for the two conditions

$$\Delta\omega_{1\to 2} = \omega(Ma_2) - \omega(Ma_1) \tag{9.101}$$

The change $\Delta\omega$ may be either positive (expansion) or negative (compression) as long as the end conditions lie in the supersonic range. Let us illustrate with an example.

EXAMPLE 9.18

Air ($k = 1.4$) flows at $Ma_1 = 3.0$ and $p_1 = 200$ kPa. Compute the final downstream Mach number and pressure for *(a)* an expansion turn of 20° and *(b)* a gradual compression turn of 20°.

Solution using Excel

Part (a) The isentropic stagnation pressure is

$$p_0 = p_1[1 + 0.2(3.0)^2]^{3.5} = 7{,}347 \text{ kPa}$$

and this will be the same at the downstream point. For $Ma_1 = 3.0$ we find from Table B.5 or Eq. (9.99) that $\omega_1 = 49.757°$. The flow expands to a new condition such that

$$\omega_2 = \omega_1 + \Delta\omega = 49.757° + 20° = 69.757°$$

Inversion of Eq. (9.99), to find Ma when ω is given, requires iteration, and Excel is well suited for this job. Hard to read, but Fig. 9.26 indicates $\omega \approx 4$. Make a guess of $\omega = 4$ and program Eq. (9.99) into an Excel cell. The writer's improved guesses are shown.

	A	B
	Ma – guess	ω – Eq. (9.99)
1	4.00	65.78
2	4.20	68.33
3	4.30	69.54
4	**4.32**	**69.78**

The iteration converges to $\mathrm{Ma}_2 = 4.32$ *Ans. (a)*

The isentropic pressure at this new condition is

$$p_2 = \frac{p_0}{[1 + 0.2(4.32)^2]^{3.5}} = \frac{7{,}347}{230.1} = 31.9 \text{ kPa} \qquad \textit{Ans. (a)}$$

Part (b) The flow compresses to a lower Prandtl-Meyer angle:

$$\omega_2 = 49.757° - 20° = 29.757°$$

Again, from Eq. (9.99), Table B.5, or Excel we compute that

$$\mathrm{Ma}_2 = 2.125 \qquad \textit{Ans. (b)}$$

$$p_2 = \frac{p_0}{[1 + 0.2(2.125)^2]^{3.5}} = \frac{7{,}347}{9.51} = 773 \text{ kPa} \qquad \textit{Ans. (b)}$$

Similarly, we compute density and temperature changes by noticing that T_0 and ρ_0 are constant for isentropic flow.

Application to Supersonic Airfoils

The oblique shock and Prandtl-Meyer expansion theories can be used to patch together a number of interesting and practical supersonic flow fields. This marriage, called *shock expansion theory,* is limited by two conditions: (1) Except in rare instances the flow must be supersonic throughout, and (2) the wave pattern must not suffer interference from waves formed in other parts of the flow field.

A very successful application of shock expansion theory is to supersonic airfoils. Figure 9.27 shows two examples, a flat plate and a diamond-shaped foil. In contrast to subsonic flow designs (Fig. 8.21), these airfoils must have sharp leading edges, which form attached oblique shocks or expansion fans. Rounded supersonic leading edges would cause detached bow shocks, as in Fig. 9.19 or 9.22*b*, greatly increasing the drag and lowering the lift.

In applying shock expansion theory, one examines each surface turning angle to see whether it is an expansion ("opening up") or compression (obstruction) to the surface flow. Figure 9.27*a* shows a flat-plate foil at an angle of attack. There is a leading-edge shock on the lower edge with flow deflection $\theta = \alpha$, while the upper edge has an expansion fan with increasing Prandtl-Meyer angle $\Delta\omega = \alpha$. We compute p_3 with expansion theory and p_2 with oblique shock theory. The force on the plate is thus $F = (p_2 - p_3)Cb$, where C is the chord length and b the span width (assuming no wingtip effects). This force is normal to the plate, and thus the lift force normal to the stream is $L = F \cos \alpha$, and the drag parallel to the stream is $D = F \sin \alpha$. The dimensionless coefficients C_L and C_D have the same definitions as in low-speed flow, Eqs. (7.66), except that the perfect-gas law identity $\frac{1}{2}\rho V^2 \equiv \frac{1}{2}kp\,\mathrm{Ma}^2$ is very useful here:

$$C_L = \frac{L}{\frac{1}{2}kp_\infty \mathrm{Ma}_\infty^2 bC} \qquad C_D = \frac{D}{\frac{1}{2}kp_\infty \mathrm{Ma}_\infty^2 bC} \tag{9.102}$$

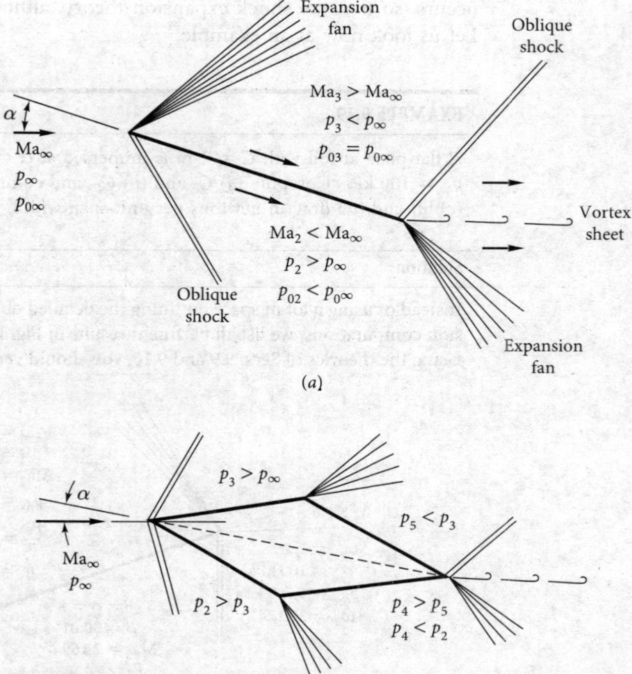

Fig. 9.27 Supersonic airfoils: (*a*) flat plate, higher pressure on lower surface, drag due to small downstream component of net pressure force; (*b*) diamond foil, higher pressures on both lower surfaces, additional drag due to body thickness.

The typical supersonic lift coefficient is much smaller than the subsonic value $C_L \approx 2\pi\alpha$, but the lift can be very large because of the large value of $\frac{1}{2}\rho V^2$ at supersonic speeds.

At the trailing edge in Fig. 9.27*a*, a shock and fan appear in reversed positions and bend the two flows back so that they are parallel in the wake and have the same pressure. They do not have quite the same velocity because of the unequal shock strengths on the upper and lower surfaces; hence a vortex sheet trails behind the wing. This is very interesting, but in the theory you ignore the trailing-edge pattern entirely, since it does not affect the surface pressures: The supersonic surface flow cannot "hear" the wake disturbances.

The diamond foil in Fig. 9.27*b* adds two more wave patterns to the flow. At this particular α less than the diamond half-angle, there are leading-edge shocks on both surfaces, the upper shock being much weaker. Then there are expansion fans on each shoulder of the diamond: The Prandtl-Meyer angle change $\Delta\omega$ equals the sum of the leading-edge and trailing-edge diamond half-angles. Finally, the trailing-edge pattern is similar to that of the flat plate (9.27*a*) and can be ignored in the calculation. Both lower-surface pressures p_2 and p_4 are greater than their upper counterparts, and the lift is nearly that of the flat plate. There is an additional drag due to thickness because p_4 and p_5 on the trailing surfaces are lower than their counterparts p_2 and p_3. The diamond drag is greater than the flat-plate drag, but this must be endured in practice to achieve a wing structure strong enough to support these forces.

The theory sketched in Fig. 9.27 is in good agreement with measured supersonic lift and drag as long as the Reynolds number is not too small (thick boundary layers) and the Mach number not too large (hypersonic flow). It turns out that for large Re_C and moderate supersonic Ma_∞ the boundary layers are thin and separation seldom occurs, so that the shock expansion theory, although frictionless, is quite successful. Let us look now at an example.

EXAMPLE 9.19

A flat-plate airfoil with $C = 2$ m is immersed at $\alpha = 8°$ in a stream with $\text{Ma}_\infty = 2.5$ and $p_\infty = 100$ kPa. Compute *(a)* C_L and *(b)* C_D, and compare with low-speed airfoils. Compute *(c)* lift and *(d)* drag in newtons per unit span width.

Solution

Instead of using a lot of space outlining the detailed oblique shock and Prandtl-Meyer expansion computations, we list all pertinent results in Fig. E9.19 on the upper and lower surfaces. Using the theories of Secs. 9.9 and 9.10, you should verify every single one of the calculations

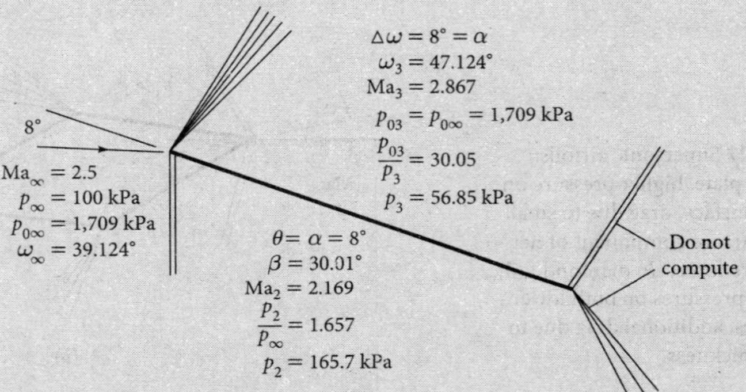

E9.19

in Fig. E9.19 to make sure that all details of shock expansion theory are well understood.

The important final results are p_2 and p_3, from which the total force per unit width on the plate is

$$F = (p_2 - p_3)bC = (165.7 - 56.85)(\text{kPa})(1\text{ m})(2\text{ m}) = 218\text{ kN}$$

The lift and drag per meter width are thus

$$L = F\cos 8° = 216\text{ kN} \qquad \textit{Ans. (c)}$$

$$D = F\sin 8° = 30\text{ kN} \qquad \textit{Ans. (d)}$$

These are very large forces for only 2 m^2 of wing area.

From Eq. (9.102) the lift coefficient is

$$C_L = \frac{216\text{ kN}}{\frac{1}{2}(1.4)(100\text{ kPa})(2.5)^2(2\text{ m}^2)} = 0.246 \qquad \textit{Ans. (a)}$$

The comparable low-speed coefficient from Eq. (8.67) is $C_L = 2\pi\sin 8° = 0.874$, which is 3.5 times larger.

From Eq. (9.102) the drag coefficient is

$$C_D = \frac{30\text{ kN}}{\frac{1}{2}(1.4)(100\text{ kPa})(2.5)^2(2\text{ m}^2)} = 0.035 \qquad \textit{Ans. (b)}$$

From Fig. 7.25 for the NACA 0009 airfoil, C_D at $\alpha = 8°$ is about 0.009, or about 4 times smaller.

Notice that this supersonic theory predicts a finite drag in spite of assuming frictionless flow with infinite wing aspect ratio. This is called *wave drag,* and we see that the d'Alembert paradox of zero body drag does not occur in supersonic flow.

Thin-Airfoil Theory

In spite of the simplicity of the flat-plate geometry, the calculations in Example 9.19 were laborious. In 1925 Ackeret [28] developed simple yet effective expressions for the lift, drag, and center of pressure of supersonic airfoils, assuming small thickness and angle of attack.

The theory is based on the linearized expression (9.89), where $\tan\theta \approx$ surface deflection relative to the free stream and condition 1 is the free stream, $Ma_1 = Ma_\infty$. For the flat-plate airfoil, the total force F is based on

$$\frac{p_2 - p_3}{p_\infty} = \frac{p_2 - p_\infty}{p_\infty} - \frac{p_3 - p_\infty}{p_\infty}$$

$$= \frac{k\,Ma_\infty^2}{(Ma_\infty^2 - 1)^{1/2}}[\alpha - (-\alpha)] \tag{9.103}$$

Substitution into Eq. (9.102) gives the linearized lift coefficient for a supersonic flat-plate airfoil:

$$C_L \approx \frac{(p_2 - p_3)bC}{\frac{1}{2}kp_\infty\,Ma_\infty^2\,bC} \approx \frac{4\alpha}{(Ma_\infty^2 - 1)^{1/2}} \tag{9.104}$$

Computations for diamond and other finite-thickness airfoils show no first-order effect of thickness on lift. Therefore, Eq. (9.104) is valid for any sharp-edged supersonic thin airfoil at a small angle of attack.

The flat-plate drag coefficient is

$$C_D = C_L \tan\alpha \approx C_L\alpha \approx \frac{4\alpha^2}{(Ma_\infty^2 - 1)^{1/2}} \tag{9.105}$$

However, the thicker airfoils have additional thickness drag. Let the chord line of the airfoil be the x axis, and let the upper-surface shape be denoted by $y_u(x)$ and the lower profile by $y_l(x)$. Then, the complete Ackeret drag theory (see Ref. 5, Sec. 14.6, for details) shows that the additional drag depends on the mean square of the slopes of the upper and lower surfaces, defined by

$$\overline{y'^2} = \frac{1}{C}\int_0^C \left(\frac{dy}{dx}\right)^2 dx \tag{9.106}$$

The final expression for drag [5, p. 442] is

$$C_D \approx \frac{4}{(Ma_\infty^2 - 1)^{1/2}}\left[\alpha^2 + \frac{1}{2}(\overline{y_u'^2} + \overline{y_l'^2})\right] \tag{9.107}$$

These are all in reasonable agreement with more exact computations, and their extreme simplicity makes them attractive alternatives to the laborious, but accurate shock expansion theory. Consider the following example.

EXAMPLE 9.20

Repeat parts (*a*) and (*b*) of Example 9.19, using the linearized Ackeret theory.

Solution

From Eqs. (9.104) and (9.105) we have, for $Ma_\infty = 2.5$ and $\alpha = 8° = 0.1396$ rad,

$$C_L \approx \frac{4(0.1396)}{(2.5^2 - 1)^{1/2}} = 0.244 \qquad C_D = \frac{4(0.1396)^2}{(2.5^2 - 1)^{1/2}} = 0.034 \qquad \textit{Ans.}$$

These are less than 3 percent lower than the more exact computations of Example 9.19.

A further result of the Ackeret linearized theory is an expression for the position x_{CP} of the center of pressure (CP) of the force distribution on the wing:

$$\frac{x_{CP}}{C} = 0.5 + \frac{S_u - S_l}{2\alpha C^2} \tag{9.108}$$

where S_u is the cross-sectional area between the upper surface and the chord and S_l is the area between the chord and the lower surface. For a symmetric airfoil ($S_l = S_u$) we obtain x_{CP} at the half-chord point, in contrast with the low-speed airfoil result, where x_{CP} is at the quarter-chord.

The difference in difficulty between the simple Ackeret theory and shock expansion theory is even greater for a thick airfoil, as the following example shows.

EXAMPLE 9.21

By analogy with Example 9.19 analyze a diamond, or double-wedge, airfoil of 2° half-angle and $C = 2$ m at $\alpha = 8°$ and $Ma_\infty = 2.5$. Compute C_L and C_D by (*a*) shock expansion theory and (*b*) Ackeret theory. Pinpoint the difference from Example 9.19.

Solution

Part (a) Again, we omit the details of shock expansion theory and simply list the properties computed on each of the four airfoil surfaces in Fig. E9.21. Assume $p_\infty = 100$ kPa. There are both a force F normal to the chord line and a force P parallel to the chord. For the normal force the pressure difference on the front half is $p_2 - p_3 = 186.4 - 65.9 = 120.5$ kPa, and on the rear half it is $p_4 - p_5 = 146.9 - 48.8 = 98.1$ kPa. The average pressure difference is $\frac{1}{2}(120.5 + 98.1) = 109.3$ kPa, so that the normal force is

$$F = (109.3 \text{ kPa})(2 \text{ m}^2) = 218.6 \text{ kN}$$

For the chordwise force P the pressure difference on the top half is $p_3 - p_5 = 65.9 - 48.8 = 17.1$ kPa, and on the bottom half it is $p_2 - p_4 = 186.4 - 146.9 = 39.5$ kPa. The average difference is $\frac{1}{2}(17.1 + 39.5) = 28.3$ kPa, which when multiplied by the frontal area (maximum thickness times 1-m width) gives

$$P = (28.3 \text{ kPa})(0.07 \text{ m})(1 \text{ m}) = 2.0 \text{ kN}$$

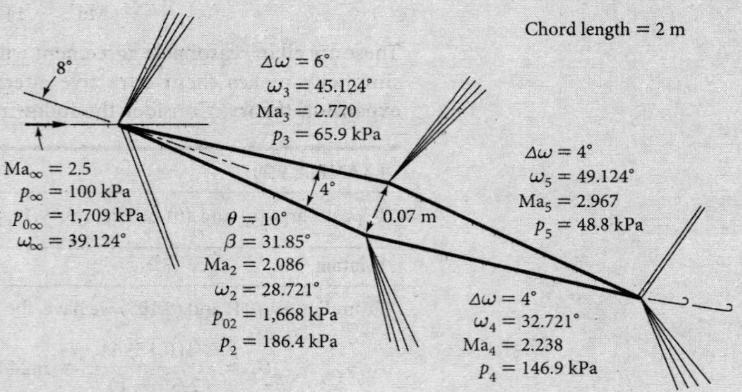

E9.21

Both F and P have components in the lift and drag directions. The lift force normal to the free stream is

$$L = F\cos 8° - P\sin 8° = 216.2\text{ kN}$$

and

$$D = F\sin 8° + P\cos 8° = 32.4\text{ kN}$$

For computing the coefficients, the denominator of Eq. (9.102) is the same as in Example 9.19: $\frac{1}{2}kp_\infty \text{Ma}_\infty^2 bC = \frac{1}{2}(1.4)(100\text{ kPa})(2.5)^2(2\text{ m}^2) = 875\text{ kN}$. Thus, finally, shock expansion theory predicts

$$C_L = \frac{216.2\text{ kN}}{875\text{ kN}} = 0.247 \qquad C_D = \frac{32.4\text{ kN}}{875\text{ kN}} = 0.0370 \qquad \textit{Ans. (a)}$$

Part (b) Meanwhile, by Ackeret theory, C_L is the same as in Example 9.20:

$$C_L = \frac{4(0.1396)}{(2.5^2 - 1)^{1/2}} = 0.244 \qquad \textit{Ans. (b)}$$

This is 1 percent less than the shock expansion result above. For the drag we need the mean-square slopes from Eq. (9.106):

$$\overline{y_u'^2} = \overline{y_l'^2} = \tan^2 2° = 0.00122$$

Then Eq. (9.107) predicts this linearized result:

$$C_D = \frac{4}{(2.5^2 - 1)^{1/2}}\left[(0.1396)^2 + \tfrac{1}{2}(0.00122 + 0.00122)\right] = 0.0362 \qquad \textit{Ans. (b)}$$

This is 2 percent lower than shock expansion theory predicts. We could judge Ackeret theory to be "satisfactory." Ackeret theory predicts $p_2 = 167$ kPa (-11 percent), $p_3 = 60$ kPa (-9 percent), $p_4 = 140$ kPa (-5 percent), and $p_5 = 33$ kPa (-6 percent).

Three-Dimensional Supersonic Flow

We have gone about as far as we can go in an introductory treatment of compressible flow. Of course, there is much more, and you are invited to study further in the references at the end of the chapter.

Three-dimensional supersonic flows are highly complex, especially if they concern blunt bodies, which therefore contain embedded regions of subsonic and transonic flow, as in Fig. 9.10. Some flows, however, yield to accurate theoretical treatment such as flow past a cone at zero incidence, as shown in Fig. 9.28. The exact theory of cone flow is discussed in advanced texts [for example, Ref. 5, Chap. 17], and extensive tables of such solutions have been published [25]. There are similarities between cone flow and the wedge flows illustrated in Fig. 9.22: an attached oblique shock, a thin turbulent boundary layer, and an expansion fan at the rear corner. However, the conical shock deflects the flow through an angle less than the cone half-angle, unlike the wedge shock. As in the wedge flow, there is a maximum cone angle above which the shock must detach, as in Fig. 9.22*b*. For $k = 1.4$ and $\text{Ma}_\infty = \infty$, the maximum cone half-angle for an attached shock is about 57°, compared with the maximum wedge angle of 45.6° (see Ref. 25).

The use of computational fluid dynamics (CFD) is now very popular and successful in compressible flow studies [13]. For example, a supersonic cone flow such as Fig. 9.28, even at an angle of attack, can be solved by numerical simulation of the full three-dimensional (viscous) Navier-Stokes equations [26].

For more complicated body shapes one usually resorts to experimentation in a supersonic wind tunnel. Figure 9.29 shows a wind tunnel study of supersonic flow past a model of an interceptor aircraft. The many junctions and wingtips and shape

Fig. 9.28 Shadowgraph of flow past an 8° half-angle cone at $Ma_\infty = 2.0$. The turbulent boundary layer is clearly visible. The Mach lines curve slightly, and the Mach number varies from 1.98 just inside the shock to 1.90 at the surface. *(Courtesy of U.S. Army Ballistic Research Laboratory, Aberdeen Proving Ground.)*

Fig. 9.29 Wind tunnel test of the Cobra P-530 supersonic interceptor. The surface flow patterns are visualized by the smearing of oil droplets. *(Courtesy of Northrop Grumman.)*

changes make theoretical analysis very difficult. Here the surface flow patterns, which indicate boundary layer development and regions of flow separation, have been visualized by the smearing of oil drops placed on the model surface before the test.

As we shall see in the next chapter, there is an interesting analogy between gas dynamic shock waves and the surface water waves that form in an open-channel flow. Chapter 11 of Ref. 9 explains how a water channel can be used in an inexpensive simulation of supersonic flow experiments.

New Trends in Aeronautics

The previous edition of this text discussed NASA's proposed hypersonic scramjet aircraft, the X-43A [30], which set a new world speed record, in 2004, of Mach 9.6, or nearly 11,300 km/h. This is hardly a design for a hypersonic airliner, though, since it has to be launched at high altitude from a B-52 bomber.

Also discussed earlier was the Air Force X-35 Joint Strike Fighter, whose wind tunnel test is shown here in Figure 9.19. This design is now operational, designated as the F-35, seen in Fig. 9.30, and it has been ordered by the U.S. military and also

Fig. 9.30 The F-35 Joint Strike Fighter is planned to become the standard supersonic fighter plane for the countries aligned with the United States. [*Lockheed Martin photo provided by F-35 Lightning II Program Office.*]

by Australia and seven NATO countries. A special version, for the U.S. Marine Corps, takes off and lands vertically. It reaches a speed of Mach 1.6 at 12,000 meters altitude. Its shortcomings are the present poor world economy and the fact that the price of one F-35 has risen to 220 million dollars.

Summary

This chapter briefly introduced a very broad subject, compressible flow, sometimes called *gas dynamics*. The primary parameter is the Mach number Ma $= V/a$, which is large and causes the fluid density to vary significantly. This means that the continuity and momentum equations must be coupled to the energy relation and the equation of state to solve for the four unknowns (p, ρ, T, V).

The chapter reviewed the thermodynamic properties of an ideal gas and derived a formula for the speed of sound of a fluid. The analysis was then simplified to one-dimensional steady adiabatic flow without shaft work, for which the stagnation enthalpy of the gas is constant. A further simplification to isentropic flow enables formulas to be derived for high-speed gas flow in a variable-area duct. This reveals the phenomenon of sonic-flow *choking* (maximum mass flow) in the throat of a nozzle. At supersonic velocities there is the possibility of a normal shock wave, where the gas discontinuously reverts to subsonic conditions. The normal shock explains the effect of back pressure on the performance of converging–diverging nozzles.

To illustrate nonisentropic flow conditions, the chapter briefly focused on constant-area duct flow with friction and with heat transfer, both of which lead to choking of the exit flow.

The chapter ended with a discussion of two-dimensional supersonic flow, where obliaue shock waves and Prandtl-Meyer (isentropic) expansion waves appear. With a proper combination of shocks and expansions one can analyze supersonic airfoils.

Problems

Most of the problems herein are fairly straightforward. More difficult or open-ended assignments are labeled with an asterisk. Problems labeled with a computer icon may require the use of a computer. The standard end-of-chapter problems P9.1 to P9.157 (categorized in the problem list here) are followed by word problems W9.1 to W9.8, fundamentals of engineering exam problems FE9.1 to FE9.10, comprehensive problems C9.1 to C9.8, and design projects D9.1 and D9.2.

Problem Distribution

Section	Topic	Problems
9.1	Introduction	P9.1–P9.9
9.2	The speed of sound	P9.10–P9.18
9.3	Adiabatic and isentropic flow	P9.19–P9.33
9.4	Isentropic flow with area changes	P9.34–P9.53
9.5	The normal shock wave	P9.54–P9.62
9.6	Converging and diverging nozzles	P9.63–P9.85
9.7	Duct flow with friction	P9.86–P9.106
9.8	Frictionless duct flow with heat transfer	P9.107–P9.115
9.9	Mach waves	P9.116–P9.121
9.9	The oblique shock wave	P9.122–P9.139
9.10	Prandtl-Meyer expansion waves	P9.140–P9.148
9.10	Supersonic airfoils	P9.149–P9.157

Introduction

P9.1 An ideal gas flows adiabatically through a duct. At section 1, $p_1 = 140$ kPa, $T_1 = 260°C$, and $V_1 = 75$ m/s. Farther downstream, $p_2 = 30$ kPa and $T_2 = 207°C$. Calculate V_2 in m/s and $s_2 - s_1$ in J/(kg · K) if the gas is (*a*) air, $k = 1.4$, and (*b*) argon, $k = 1.67$.

P9.2 Solve Prob. P9.1 if the gas is steam. Use two approaches: (*a*) an ideal gas from Table A.4 and (*b*) real gas data from the steam tables [15].

P9.3 If 8 kg of oxygen in a closed tank at 200°C and 300 kPa is heated until the pressure rises to 400 kPa, calculate (*a*) the new temperature, (*b*) the total heat transfer, and (*c*) the change in entropy.

P9.4 Consider steady adiabatic airflow in a duct. At section B, the pressure is 600 kPa and the temperature is 177°C. At section D, the density is 1.13 kg/m^3 and the temperature is 156°C. (*a*) Find the entropy change, if any. (*b*) Which way is the air flowing?

P9.5 Steam enters a nozzle at 377°C, 1.6 MPa, and a steady speed of 200 m/s and accelerates isentropically until it exits at saturation conditions. Estimate the exit velocity and temperature.

P9.6 Methane, approximated as a perfect gas, is compressed adiabatically from 101 kPa and 20°C to 300 kPa. Estimate (*a*) the final temperature, and (*b*) the final density.

P9.7 Air flows through a variable-area duct. At section 1, $A_1 = 20\ \text{cm}^2$, $p_1 = 300$ kPa, $\rho_1 = 1.75\ \text{kg/m}^3$, and $V_1 = 122.5$ m/s. At section 2, the area is exactly the same, but the density is much lower: $\rho_2 = 0.266\ \text{kg/m}^3$ and $T_2 = 281$ K. There is no transfer of work or heat. Assume one-dimensional steady flow. (*a*) How can you reconcile these differences? (*b*) Find the mass flow at section 2. Calculate (*c*) V_2, (*d*) p_2, and (*e*) $s_2 - s_1$. [*Hint:* This problem requires the continuity equation.]

P9.8 Atmospheric air at 20°C enters and fills an insulated tank that is initially evacuated. Using a control volume analysis from Eq. (3.67), compute the tank air temperature when it is full.

P9.9 Liquid hydrogen and oxygen are burned in a combustion chamber and fed through a rocket nozzle that exhausts at $V_{exit} = 1{,}600$ m/s to an ambient pressure of 54 kPa. The nozzle exit diameter is 45 cm, and the jet exit density is $0.15\ \text{kg/m}^3$. If the exhaust gas has a molecular weight of 18, estimate (*a*) the exit gas temperature, (*b*) the mass flow, and (*c*) the thrust developed by the rocket.

The speed of sound

P9.10 A certain aircraft flies at 980 km/h at standard sea level. (*a*) What is its Mach number? (*b*) If it flies at the same Mach number at 10,000 m altitude, how much slower (or faster) is it flying, in km/h?

P9.11 At 300°C and 1 atm, estimate the speed of sound of (*a*) nitrogen, (*b*) hydrogen, (*c*) helium, (*d*) steam, and (*e*) $^{238}UF_6$ ($k \approx 1.06$).

P9.12 Assume that water follows Eq. (1.19) with $n \approx 7$ and $B \approx 3{,}000$. Compute the bulk modulus (in kPa) and the speed of sound (in m/s) at (*a*) 1 atm and (*b*) 1,100 atm (the deepest part of the ocean). (*c*) Compute the speed of sound at 20°C and 9,000 atm and compare with the measured value of 2,650 m/s (A. H. Smith and A. W. Lawson, *J. Chem. Phys.*, vol. 22, 1954, p. 351).

P9.13 Consider steam at 500 K and 200 kPa. Estimate its speed of sound by two different methods: (*a*) assuming an ideal gas from Table B.4, or (*b*) using finite differences for isentropic densities between 210 kPa and 190 kPa.

P9.14 Benzene, listed in Table A.3, has a measured density of $925\ \text{kg/m}^3$ at a pressure of 70 MPa. Use this data to estimate the speed of sound of benzene.

P9.15 The pressure-density relation for ethanol is approximated by Eq. (1.19) with $B = 1{,}600$ and $n = 7$. Use this relation to estimate the speed of sound of ethanol at 200 MPa.

P9.16 A weak pressure pulse Δp propagates through still air. Discuss the type of reflected pulse that occurs and the boundary conditions that must be satisfied when the wave strikes normal to, and is reflected from, (*a*) a solid wall and (*b*) a free liquid surface.

P9.17 A submarine at a depth of 800 m sends a sonar signal and receives the reflected wave back from a similar submerged object in 15 s. Using Prob. P9.12 as a guide, estimate the distance to the other object.

P9.18 Race cars at the Indianapolis Speedway average speeds of 300 km/h. After determining the altitude of Indianapolis, find the Mach number of these cars and estimate whether compressibility might affect their aerodynamics.

Adiabatic and isentropic flow

P9.19 In 1976, the SR-71A, flying at 20 km standard altitude, set a jet-powered aircraft speed record of 3,326 km/h. Estimate the temperature, in °C, at its front stagnation point. At what Mach number would it have a front stagnation-point temperature of 500°C?

P9.20 Air flows isentropically in a channel. Properties at section 1 are $V_1 = 250$ m/s, $T_1 = 330$ K, and $p_1 = 80$ kPa. At section 2 downstream, the temperature has dropped to 0°C. Find (*a*) the pressure, (*b*) velocity, and (*c*) Mach number at section 2.

P9.21 N_2O expands isentropically through a duct from $p_1 = 200$ kPa and $T_1 = 250$°C to a downstream section where $p_2 = 26$ kPa and $V_2 = 594$ m/s. Compute (*a*) T_2; (*b*) Ma_2; (*c*) T_o; (*d*) p_o; (*e*) V_1; and (*f*) Ma_1.

P9.22 Given the pitot stagnation temperature and pressure and the static pressure measurements in Fig. P9.22, estimate the air velocity *V*, assuming (*a*) incompressible flow and (*b*) compressible flow.

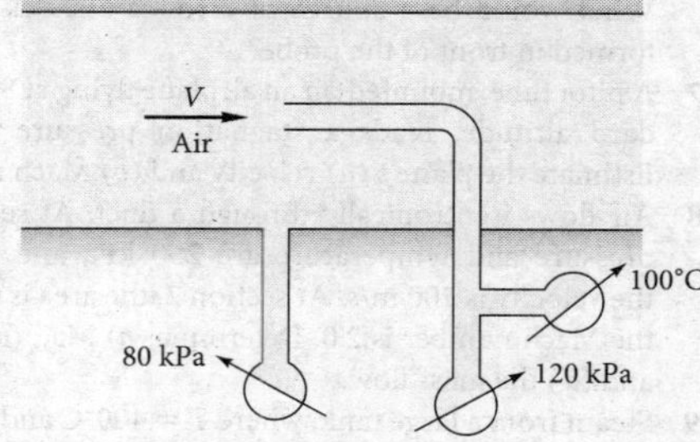

P9.22

P9.23 A gas, assumed ideal, flows isentropically from point 1, where the velocity is negligible, the pressure is 200 kPa, and the temperature is 300°C, to point 2, where the pressure is 40 kPa. What is the Mach number Ma_2 if the gas is (*a*) air, (*b*) argon, or (*c*) CH_4? (*d*) Can you tell, without calculating, which gas will be the coldest at point 2?

P9.24 For low-speed (nearly incompressible) gas flow, the stagnation pressure can be computed from Bernoulli's equation:

$$p_0 = p + \frac{1}{2}\rho V^2$$

(*a*) For higher subsonic speeds, show that the isentropic relation (9.28*a*) can be expanded in a power series as follows:

$$p_0 \approx p + \frac{1}{2}\rho V^2\left(1 + \frac{1}{4}\text{Ma}^2 + \frac{2-k}{24}\text{Ma}^4 + \cdots\right)$$

(*b*) Suppose that a pitot-static tube in air measures the pressure difference $p_0 - p$ and uses the Bernoulli relation, with stagnation density, to estimate the gas velocity. At what Mach number will the error be 4 percent?

P9.25 If it is known that the air velocity in the duct is 230 m/s, use the mercury manometer measurement in Fig. P9.25 to estimate the static pressure in the duct in kPa absolute.

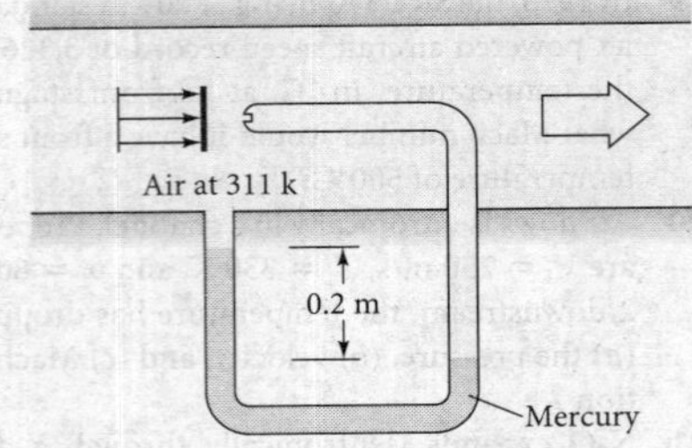

P9.25

P9.26 Show that for isentropic flow of a perfect gas if a pitot-static probe measures p_0, p, and T_0, the gas velocity can be calculated from

$$V^2 = 2c_pT_0\left[1 - \left(\frac{p}{p_0}\right)^{(k-1)/k}\right]$$

What would be a source of error if a shock wave were formed in front of the probe?

P9.27 A pitot tube, mounted on an airplane flying at 8000 m standard altitude, reads a stagnation pressure of 57 kPa. Estimate the plane's (*a*) velocity and (*b*) Mach number.

P9.28 Air flows isentropically through a duct. At section 1, the pressure and temperature are 250 kPa and 125°C, and the velocity is 200 m/s. At section 2, the area is 0.25 m^2 and the Mach number is 2.0. Determine (*a*) Ma_1; (*b*) T_2; (*c*) V_2; and (*d*) the mass flow.

P9.29 Steam from a large tank, where $T = 400$°C and $p = 1$ MPa, expands isentropically through a nozzle until, at a section of 2-cm diameter, the pressure is 500 kPa. Using the steam tables [15], estimate (*a*) the temperature, (*b*) the velocity, and (*c*) the mass flow at this section. Is the flow subsonic?

P9.30 When does the incompressible-flow assumption begin to fail for pressures? Construct a graph of p_0/p for incompressible flow of a perfect gas as compared to Eq. (9.28*a*). Plot both versus Mach number for $0 \leq \text{Ma} \leq 0.6$ and decide for yourself where the deviation is too great.

P9.31 Air flows adiabatically through a duct. At one section $V_1 = 120$ m/s, $T_1 = 370$ K, and $p_1 = 250$ kPa absolute, while farther downstream $V_2 = 350$ m/s and $p_2 = 120$ kPa absolute. Compute (*a*) Ma_2, (*b*) U_{max}, and (*c*) p_{02}/p_{01}.

P9.32 The large compressed-air tank in Fig. P9.32 exhausts from a nozzle at an exit velocity of 235 m/s. The mercury manometer reads $h = 30$ cm. Assuming isentropic flow, compute the pressure (*a*) in the tank and (*b*) in the atmosphere. (*c*) What is the exit Mach number?

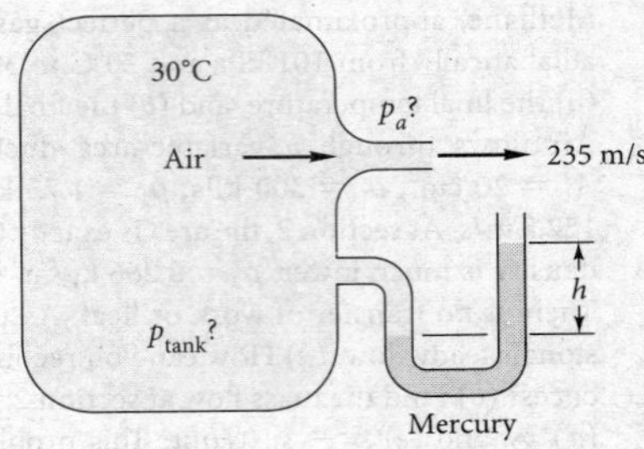

P9.32

P9.33 Air flows isentropically from a reservoir, where $p = 300$ kPa and $T = 500$ K, to section 1 in a duct, where $A_1 = 0.2$ m^2 and $V_1 = 550$ m/s. Compute (*a*) Ma_1, (*b*) T_1, (*c*) p_1, (*d*) $\dot{m}$, and (*e*) A^*. Is the flow choked?

Isentropic flow with area changes

P9.34 Air in a large tank, at 300°C and 400 kPa, flows through a converging-diverging nozzle with throat diameter 2 cm. It exits smoothly at a Mach number of 2.8. According to one-dimensional isentropic theory, what is (*a*) the exit diameter, and (*b*) the mass flow?

P9.35 Helium, at $T_0 = 400$ K, enters a nozzle isentropically. At section 1, where $A_1 = 0.1$ m^2, a pitot-static arrangement (see Fig. P9.25) measures stagnation pressure of 150 kPa and static pressure of 123 kPa. Estimate (*a*) Ma_1, (*b*) mass flow $\dot{m}$, (*c*) T_1, and (*d*) A^*.

P9.36 An air tank of volume 1.5 m^3 is initially at 800 kPa and 20°C. At $t = 0$, it begins exhausting through a converging nozzle to sea-level conditions. The throat area is 0.75 cm^2. Estimate (*a*) the initial mass flow in kg/s, (*b*) the time required to blow down to 500 kPa, and (*c*) the time at which the nozzle ceases being choked.

P9.37 Make an exact control volume analysis of the blowdown process in Fig. P9.37, assuming an insulated tank with negligible kinetic and potential energy within. Assume critical flow at the exit, and show that both p_0 and T_0 decrease during blowdown. Set up first-order differential equations for $p_0(t)$ and $T_0(t)$, and reduce and solve as far as you can.

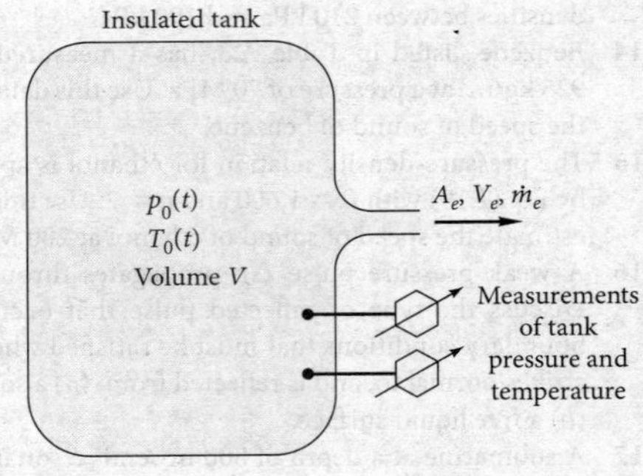

P9.37

P9.38 Prob. P9.37 makes an ideal senior project or combined laboratory and computer problem, as described in Ref. 27, Sec. 8.6. In Bober and Kenyon's lab experiment, the tank had a volume of 0.001 m^3 and was initially filled with air at 345 kPa gage and 295 K. Atmospheric pressure was 100 kPa absolute, and the nozzle exit diameter was 0.13 cm. After 2 s of blowdown, the measured tank pressure was 138 kPa gage and the tank temperature was 253 K. Compare these values with the theoretical analysis of Prob. P9.37.

P9.39 Consider isentropic flow in a channel of varying area, from section 1 to section 2. We know that $\text{Ma}_1 = 2.0$ and desire that the velocity ratio V_2/V_1 be 1.2. Estimate (*a*) Ma_2 and (*b*) A_2/A_1. (*c*) Sketch what this channel looks like. For example, does it converge or diverge? Is there a throat?

P9.40 Steam, in a tank at 300 kPa and 600 K, discharges isentropically to a low-pressure atmosphere through a converging nozzle with exit area 5 cm^2. (*a*) Using an ideal gas approximation from Table B.4, estimate the mass flow. (*b*) Without actual calculations, indicate how you would use real properties of steam to find the mass flow.

P9.41 Air, with a stagnation pressure of 100 kPa, flows through the nozzle in Fig. P9.41, which is 2 m long and has an area variation approximated by

$$A \approx 20 - 20x + 10x^2$$

with A in cm^2 and x in m. It is desired to plot the complete family of isentropic pressures $p(x)$ in this nozzle, for the range of inlet pressures $1 < p(0) < 100$ kPa. Indicate which inlet pressures are not physically possible and discuss briefly. If your computer has an online graphics routine, plot at least 15 pressure profiles; otherwise just hit the highlights and explain.

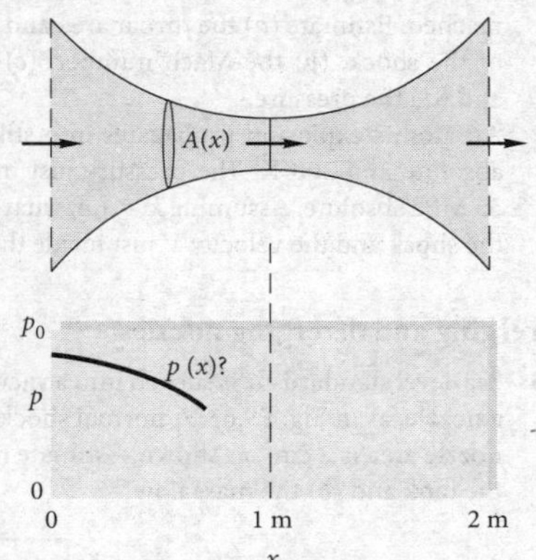

P9.41

P9.42 A bicycle tire is filled with air at an absolute pressure of 169.12 kPa, and the temperature inside is 30.0°C. Suppose the valve breaks, and air starts to exhaust out of the tire into the atmosphere (p_a = 100 kPa absolute and T_a = 20.0°C). The valve exit is 2.00 mm in diameter and is the smallest cross-sectional area of the entire system. Frictional losses can be ignored here; one-dimensional isentropic flow is a reasonable assumption. (*a*) Find the Mach number, velocity, and temperature at the exit plane of the valve (initially). (*b*) Find the initial mass flow rate out of the tire. (*c*) Estimate the velocity at the exit plane using the incompressible Bernoulli equation. How well does this estimate agree with the "exact" answer of part (*a*)? Explain.

P9.43 Air flows isentropically through a variable-area duct. At section 1, A_1= 20 cm^2, p_1= 300 kPa, ρ_1= 1.75 kg/m^3, and Ma_1= 0.25. At section 2, the area is exactly the same, but the flow is much faster. Compute (*a*) V_2, (*b*) Ma_2, (*c*) T_2, and (*d*) the mass flow. (*e*) Is there a sonic throat between sections 1 and 2? If so, find its area.

P9.44 In Prob. P3.34 we knew nothing about compressible flow at the time, so we merely assumed exit conditions p_2 and T_2 and computed V_2 as an application of the continuity equation. Suppose that the throat diameter is 7.5 cm. For the given stagnation conditions in the rocket chamber in Fig. P3.34 and assuming $k = 1.4$ and a molecular weight of 26, compute the actual exit velocity, pressure, and temperature according to one-dimensional theory. If p_a = 101.3 kPa absolute, compute the thrust from the analysis of Prob. P3.68. This thrust is entirely independent of the stagnation temperature (check this by changing T_0 to 1,100 K if you like). Why?

P9.45 It is desired to have an isentropic airflow achieve a velocity of 550 m/s at a 6-cm-diameter section where the pressure is 87 kPa and the density 1.3 kg/m^3. (*a*) Is a sonic throat needed? (*b*) If so, estimate its diameter, and compute (*c*) the stagnation temperature and (*d*) the mass flow.

P9.46 A one-dimensional isentropic airflow has the following properties at one section where the area is 53 cm^2: p = 12 kPa, ρ = 0.182 kg/m^3, and V = 760 m/s. Determine (*a*) the throat area, (*b*) the stagnation temperature, and (*c*) the mass flow.

P9.47 In wind tunnel testing near Mach 1, a small area decrease caused by model blockage can be important. Suppose the test section area is 1 m^2, with unblocked test conditions Ma = 1.10 and T = 20°C. What model area will first cause the test section to choke? If the model cross section is 0.004 m^2 (0.4 percent blockage), what percentage change in test section velocity results?

P9.48 A force F = 1,100 N pushes a piston of diameter 12 cm through an insulated cylinder containing air at 20°C, as in Fig. P9.48. The exit diameter is 3 mm, and p_a = 1 atm. Estimate (*a*) V_e, (*b*) V_p, and (*c*) $\dot{m}_e$.

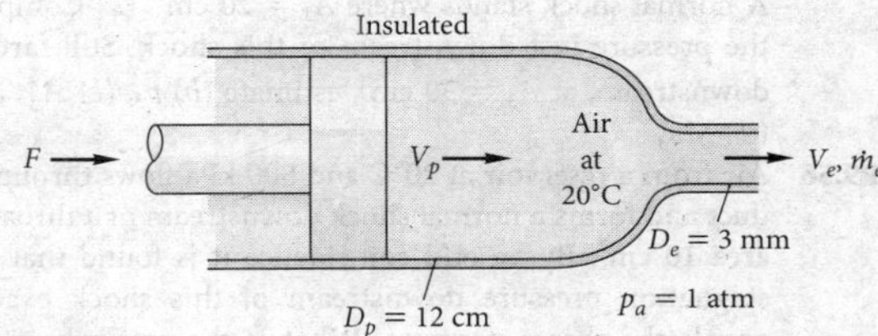

P9.48

P9.49 Consider the venturi nozzle of Fig. 6.40*c*, with $D = 5$ cm and $d = 3$ cm. Stagnation temperature is 300 K, and the upstream velocity $V_1 = 72$ m/s. If the throat pressure is 124 kPa, estimate, with isentropic flow theory, (*a*) p_1, (*b*) Ma_2, and (*c*) the mass flow.

P9.50 Methane is stored in a tank at 120 kPa and 330 K. It discharges to a second tank through a converging nozzle whose exit area is 5 cm². What is the initial mass flow rate if the second tank has a pressure of (*a*) 70 kPa or (*b*) 40 kPa?

P9.51 The scramjet engine is supersonic throughout. A sketch is shown in Fig. C9.8. Test the following design. The flow enters at Ma = 7 and air properties for 10,000 m altitude. Inlet area is 1 m², the minimum area is 0.1 m², and the exit area is 0.8 m². If there is no combustion, (*a*) will the flow still be supersonic in the throat? Also, determine (*b*) the exit Mach number, (*c*) exit velocity, and (*d*) exit pressure.

P9.52 A converging–diverging nozzle exits smoothly to sea-level standard atmosphere. It is supplied by a 40-m³ tank initially at 800 kPa and 100°C. Assuming isentropic flow in the nozzle, estimate (*a*) the throat area and (*b*) the tank pressure after 10 s of operation. The exit area is 10 cm².

P9.53 Air flows steadily from a reservoir at 20°C through a nozzle of exit area 20 cm² and strikes a vertical plate as in Fig. P9.53. The flow is subsonic throughout. A force of 135 N is required to hold the plate stationary. Compute (*a*) V_e, (*b*) Ma_e, and (*c*) p_0 if $p_a = 101$ kPa.

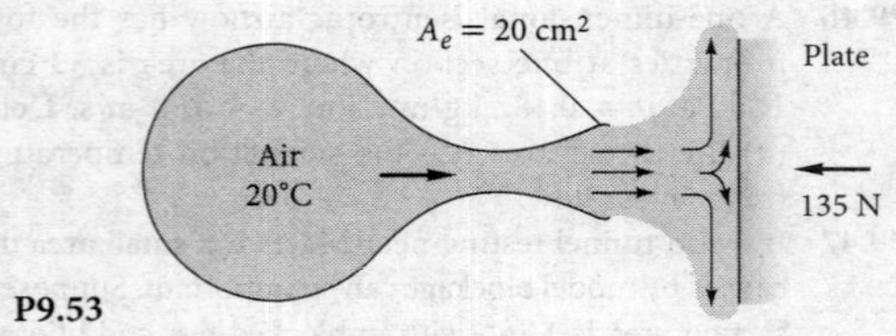

P9.53

The normal shock wave

P9.54 The airflow in Prob. P9.46 undergoes a normal shock just past the section where data was given. Determine the (*a*) Mach number, (*b*) pressure, and (*c*) velocity just downstream of the shock.

P9.55 Air, supplied by a reservoir at 450 kPa, flows through a converging–diverging nozzle whose throat area is 12 cm². A normal shock stands where $A_1 = 20$ cm². (*a*) Compute the pressure just downstream of this shock. Still farther downstream, at $A_3 = 30$ cm², estimate (*b*) p_3, (*c*) A_3^*, and (*d*) Ma_3.

P9.56 Air from a reservoir at 20°C and 500 kPa flows through a duct and forms a normal shock downstream of a throat of area 10 cm². By an odd coincidence it is found that the stagnation pressure downstream of this shock exactly equals the throat pressure. What is the area where the shock wave stands?

P9.57 Air flows from a tank through a nozzle into the standard atmosphere, as in Fig. P9.57. A normal shock stands in the exit of the nozzle, as shown. Estimate (*a*) the pressure in the tank and (*b*) the mass flow.

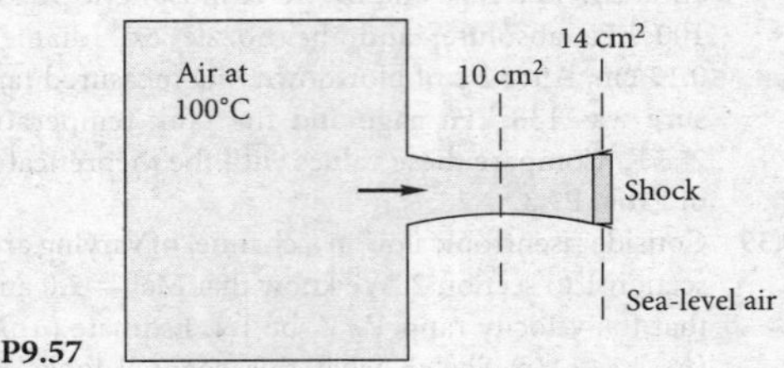

P9.57

P9.58 Downstream of a normal shock wave, in airflow, the conditions are $T_2 = 603$ K, $V_2 = 222$ m/s, and $p_2 = 900$ kPa. Estimate the following conditions just upstream of the shock: (*a*) Ma_1; (*b*) T_1; (*c*) p_1; (*d*) p_{01}; and (*e*) T_{01}.

P9.59 Air, at stagnation conditions of 450 K and 250 kPa, flows through a nozzle. At section 1, where the area is 15 cm², there is a normal shock wave. If the mass flow is 0.4 kg/s, estimate (*a*) the Mach number and (*b*) the stagnation pressure just downstream of the shock.

P9.60 When a pitot tube such as in Fig. 6.30 is placed in a supersonic flow, a normal shock will stand in front of the probe. Suppose the probe reads $p_0 = 190$ kPa and $p = 150$ kPa. If the stagnation temperature is 400 K, estimate the (supersonic) Mach number and velocity upstream of the shock.

P9.61 Air flows from a large tank, where $T = 376$ K and $p = 360$ kPa, to a design condition where the pressure is 9800 Pa. The mass flow is 0.9 kg/s. However, there is a normal shock in the exit plane just after this condition is reached. Estimate (*a*) the throat area and, just downstream of the shock, (*b*) the Mach number, (*c*) the temperature, and (*d*) the pressure.

P9.62 An atomic explosion propagates into still air at 101.3 kPa absolute and 300 K. The pressure just inside the shock is 35 MPa absolute. Assuming $k = 1.4$, what are the speed C of the shock and the velocity V just inside the shock?

Converging and diverging nozzles

P9.63 Sea-level standard air is sucked into a vacuum tank through a nozzle, as in Fig. P9.63. A normal shock stands where the nozzle area is 2 cm², as shown. Estimate (*a*) the pressure in the tank and (*b*) the mass flow.

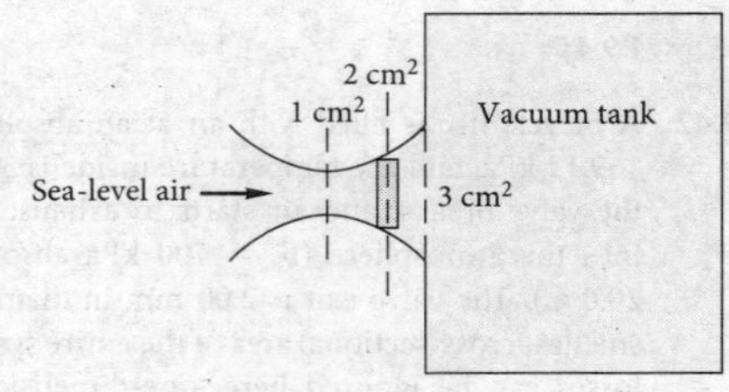

P9.63

P9.64 Air, from a reservoir at 350 K and 500 kPa, flows through a converging–diverging nozzle. The throat area is 3 cm^2. A normal shock appears, for which the downstream Mach number is 0.6405. (*a*) What is the area where the shock appears? Calculate (*b*) the pressure and (*c*) the temperature downstream of the shock.

P9.65 Air flows through a converging–diverging nozzle between two large reservoirs, as shown in Fig. P9.65. A mercury manometer between the throat and the downstream reservoir reads h = 15 cm. Estimate the downstream reservoir pressure. Is there a normal shock in the flow? If so, does it stand in the exit plane or farther upstream?

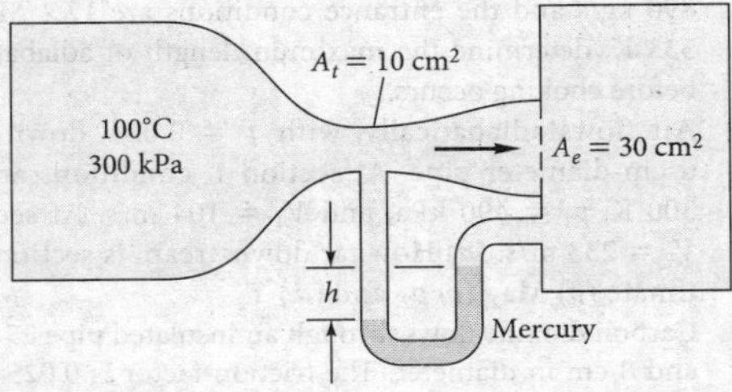

P9.65

P9.66 In Prob. P9.65 what would be the mercury manometer reading h if the nozzle were operating exactly at supersonic design conditions?

P9.67 A supply tank at 500 kPa and 400 K feeds air to a converging–diverging nozzle whose throat area is 9 cm^2. The exit area is 46 cm^2. State the conditions in the nozzle if the pressure outside the exit plane is (*a*) 400 kPa, (*b*) 120 kPa, and (*c*) 9 kPa. (*d*) In each of these cases, find the mass flow.

P9.68 Air in a tank at 120 kPa and 300 K exhausts to the atmosphere through a 5-cm^2-throat converging nozzle at a rate of 0.12 kg/s. What is the atmospheric pressure? What is the maximum mass flow possible at low atmospheric pressure?

P9.69 With reference to Prob. P3.68, show that the thrust of a rocket engine exhausting into a vacuum is given by

$$F = \frac{p_0 A_e(1 + k\,\mathrm{Ma}_e^2)}{\left(1 + \frac{k-1}{2}\mathrm{Ma}_e^2\right)^{k/(k-1)}}$$

where A_e = exit area
Ma_e = exit Mach number
p_0 = stagnation pressure in combustion chamber

Note that stagnation temperature does not enter into the thrust.

P9.70 Air, with p_o = 500 kPa and T_o = 600 K, flows through a converging–diverging nozzle. The exit area is 51.2 cm^2, and mass flow is 0.825 kg/s. What is the highest possible back pressure that will still maintain supersonic flow inside the diverging section?

P9.71 A converging-diverging nozzle has a throat area of 10 cm^2 and an exit area of 28.96 cm^2. A normal shock stands in the exit when the back pressure is sea-level standard. If the upstream tank temperature is 400 K, estimate (*a*) the tank pressure and (*b*) the mass flow.

P9.72 A large tank at 500 K and 165 kPa feeds air to a converging nozzle. The back pressure outside the nozzle exit is sea-level standard. What is the appropriate exit diameter if the desired mass flow is 72 kg/h?

P9.73 Air flows isentropically in a converging–diverging nozzle with a throat area of 3 cm^2. At section 1, the pressure is 101 kPa, the temperature is 300 K, and the velocity is 868 m/s. (*a*) Is the nozzle choked? Determine (*b*) A_1 and (*c*) the mass flow. Suppose, without changing stagnation conditions or A_1, the (flexible) throat is reduced to 2 cm^2. Assuming shock-free flow, will there be any change in the gas properties at section 1? If so, compute new p_1, V_1, and T_1 and explain.

P9.74 Use your strategic ideas, from part (*b*) of Prob. P9.40, to actually carry out the calculations for mass flow of steam, with p_o = 300 kPa and T_o = 600 K, discharging through a converging nozzle of choked exit area 5 cm^2.

***P9.75** A double-tank system in Fig. P9.75 has two identical converging nozzles of 6.5 cm^2 throat area. Tank 1 is very large, and tank 2 is small enough to be in steady-flow equilibrium with the jet from tank 1. Nozzle flow is isentropic, but entropy changes between 1 and 3 due to jet dissipation in tank 2. Compute the mass flow. (If you give up, Ref. 9, pp. 288–290, has a good discussion.)

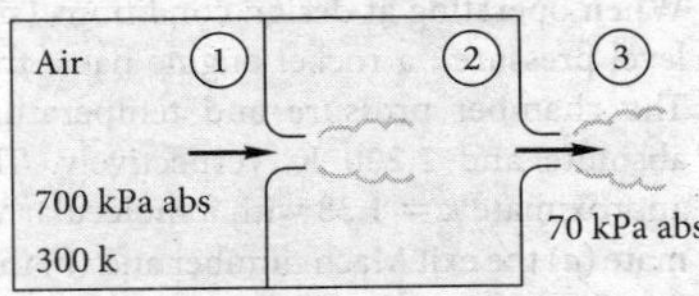

P9.75

P9.76 A large reservoir at 20°C and 800 kPa is used to fill a small insulated tank through a converging–diverging nozzle with 1-cm^2 throat area and 1.66-cm^2 exit area. The small tank has a volume of 1 m^3 and is initially at 20°C and 100 kPa. Estimate the elapsed time when (*a*) shock waves begin to appear inside the nozzle and (*b*) the mass flow begins to drop below its maximum value.

P9.77 A perfect gas (not air) expands isentropically through a supersonic nozzle with an exit area 5 times its throat area. The exit Mach number is 3.8. What is the specific-heat ratio of the gas? What might this gas be? If p_0 = 300 kPa, what is the exit pressure of the gas?

P9.78 The orientation of a hole can make a difference. Consider holes A and B in Fig. P9.78, which are identical but reversed. For the given air properties on either side, compute the mass flow through each hole and explain why they are different.

P9.79 A large tank, at 400 kPa and 450 K, supplies air to a converging-diverging nozzle of throat area 4 cm^2 and exit area 5 cm^2. For what range of back pressures will the flow (*a*) be entirely subsonic; (*b*) have a shock wave inside the nozzle; (*c*) have oblique shocks outside the exit; and (*d*) have supersonic expansion waves outside the exit?

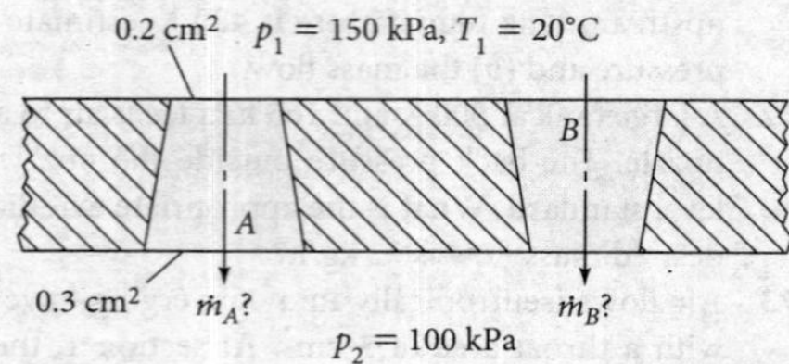

P9.78

P9.80 A sea-level automobile tire is initially at 220 kPa gage pressure and 297 K. When it is punctured with a hole that resembles a converging nozzle, its pressure drops to 103 kPa gage in 12 min. Estimate the size of the hole, in mm. The tire volume is 0.07 m^3.

P9.81 Air, at p_o = 1,100 kPa and T_o = 422 K, flows isentropically through a converging–diverging nozzle. At section 1, where A_1 = 0.2 m^2, the velocity is V_1 = 630 m/s. Calculate (*a*) Ma_1, (*b*) A^*, (*c*) p_1, and (*d*) the mass flow, in kg/s.

P9.82 Air at 500 K flows through a converging–diverging nozzle with throat area of 1 cm^2 and exit area of 2.7 cm^2. When the mass flow is 182.2 kg/h, a pitot-static probe placed in the exit plane reads p_0 = 250.6 kPa and p = 240.1 kPa. Estimate the exit velocity. Is there a normal shock wave in the duct? If so, compute the Mach number just downstream of this shock.

P9.83 When operating at design conditions (smooth exit to sea-level pressure), a rocket engine has a thrust of 4,500 kN. The chamber pressure and temperature are 4,100 kPa absolute and 2,200 K, respectively. The exhaust gases approximate k = 1.38 with a molecular weight of 26. Estimate (*a*) the exit Mach number and (*b*) the throat diameter.

P9.84 Air flows through a duct as in Fig. P9.84, where A_1 = 24 cm^2, A_2 = 18 cm^2, and A_3 = 32 cm^2. A normal shock stands at section 2. Compute (*a*) the mass flow, (*b*) the Mach number, and (*c*) the stagnation pressure at section 3.

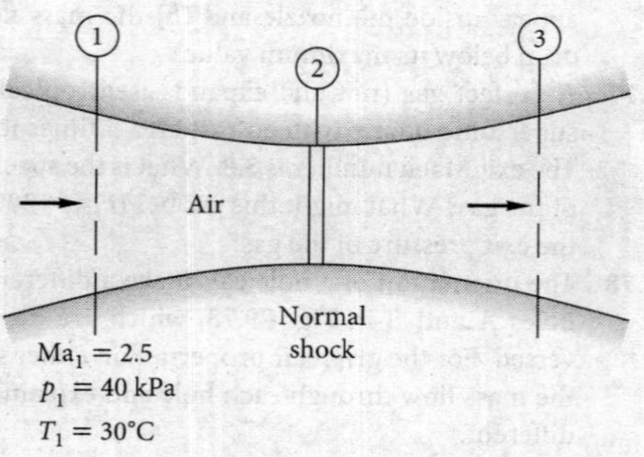

P9.84

P9.85 A typical carbon dioxide tank for a paintball gun holds about 340 g of liquid CO_2. The tank is filled no more than one-third with liquid, which, at room temperature, maintains the gaseous phase at about 5.86 MPa, abs.. (*a*) If a valve is opened that simulates a converging nozzle with an exit diameter of 1.3 mm, what mass flow and exit velocity results? (*b*) Repeat the calculations for helium.

Duct flow with friction

P9.86 Air enters a 3-cm-diameter pipe 15 m long at V_1 = 73 m/s, p_1 = 550 kPa, and T_1 = 60°C. The friction factor is 0.018. Compute V_2, p_2, T_2, and p_{02} at the end of the pipe. How much additional pipe length would cause the exit flow to be sonic?

P9.87 Problem C6.9 gives data for a proposed Alaska-to-Canada natural gas (assume CH_4) pipeline. If the design flow rate is 890 kg/s and the entrance conditions are 17.2 MPa and 333 K, determine the maximum length of adiabatic pipe before choking occurs.

P9.88 Air flows adiabatically, with $\bar{f}$ = 0.024, down a long 6-cm-diameter pipe. At section 1, conditions are T_1 = 300 K, p_1 = 400 kPa, and V_1 = 104 m/s. At section 2, V_2 = 233 m/s. (*a*) How far downstream is section 2? Estimate (*b*) Ma_2, (*c*) p_2, and (*d*) T_2.

P9.89 Carbon dioxide flows through an insulated pipe 25 m long and 8 cm in diameter. The friction factor is 0.025. At the entrance, p = 300 kPa and T = 400 K. The mass flow is 1.5 kg/s. Estimate the pressure drop by (*a*) compressible and (*b*) incompressible (Sec. 6.6) flow theory. (*c*) For what pipe length will the exit flow be choked?

P9.90 Air flows through a rough pipe 35 m long and 8 cm in diameter. Entrance conditions are p = 620 kPa, T = 293 K, and V = 70 m/s. The flow chokes at the end of the pipe. (*a*) What is the average friction factor? (*b*) What is the pressure at the end of the pipe?

P9.91 Air flows steadily from a tank through the pipe in Fig. P9.91. There is a converging nozzle on the end. If the mass flow is 3 kg/s and the nozzle is choked, estimate (*a*) the Mach number at section 1 and (*b*) the pressure inside the tank.

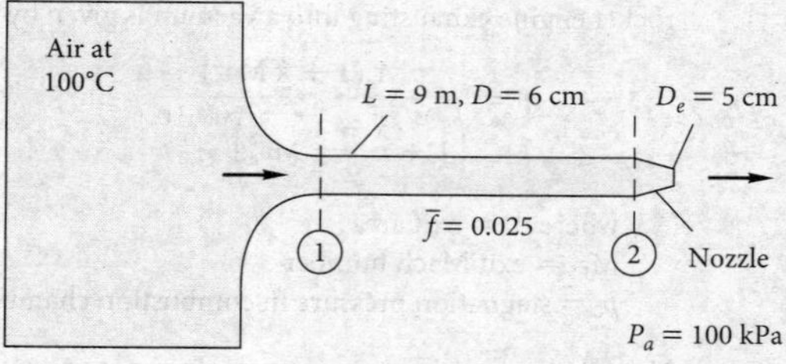

P9.91

P9.92 Air enters a 5-cm-diameter pipe at 380 kPa, 3.3 kg/m^3, and 120 m/s. The friction factor is 0.017. Find the pipe length for which the velocity (*a*) doubles, (*b*) triples, and (*c*) quadruples.

P9.93 Air flows adiabatically in a 3-cm-diameter duct, with $\bar{f}$ = 0.018. At the entrance, T_1 = 323 K, p_1 = 200 kPa, and V_1 = 72 m/s. (*a*) What is the mass flow? (*b*) For what tube length will the flow choke? (*c*) If the tube length is increased to 112 m, with the same inlet pressure and temperature, what will be the new mass flow?

P9.94 Compressible pipe flow with friction, Sec. 9.7, assumes constant stagnation enthalpy and mass flow but variable momentum. Such a flow is often called *Fanno flow*, and a line representing all possible property changes on a temperature–entropy chart is called a *Fanno line*. Assuming a perfect gas with $k = 1.4$ and the data of Prob. P9.86, draw a Fanno curve of the flow for a range of velocities from very low (Ma $\ll$ 1) to very high (Ma $\gg$ 1). Comment on the meaning of the maximum-entropy point on this curve.

P9.95 Helium (Table A.4) enters a 5-cm-diameter pipe at $p_1 =$ 550 kPa, $V_1 = 312$ m/s, and $T_1 = 40°C$. The friction factor is 0.025. If the flow is choked, determine (*a*) the length of the duct and (*b*) the exit pressure.

P9.96 Methane (CH_4) flows through an insulated 15-cm-diameter pipe with $f = 0.023$. Entrance conditions are 600 kPa, 100°C, and a mass flow of 5 kg/s. What lengths of pipe will (*a*) choke the flow, (*b*) raise the velocity by 50 percent, or (*c*) decrease the pressure by 50 percent?

P9.97 By making a few algebraic substitutions, show that Eq. (9.74) may be written in the density form

$$\rho_1^2 = \rho_2^2 + \rho^{*2}\left(\frac{2k}{k+1}\frac{\bar{f}L}{D} + 2\ln\frac{\rho_1}{\rho_2}\right)$$

Why is this formula awkward if one is trying to solve for the mass flow when the pressures are given at sections 1 and 2?

P9.98 Compressible *laminar* flow, $f \approx 64/\text{Re}$, may occur in capillary tubes. Consider air, at stagnation conditions of 100°C and 200 kPa, entering a tube 3 cm long and 0.1 mm in diameter. If the receiver pressure is near vacuum, estimate (*a*) the average Reynolds number, (*b*) the Mach number at the entrance, and (*c*) the mass flow in kg/h.

P9.99 A compressor forces air through a smooth pipe 20 m long and 4 cm in diameter, as in Fig. P9.99. The air leaves at 101 kPa and 200°C. The compressor data for pressure rise versus mass flow are shown in the figure. Using the Moody chart to estimate $\bar{f}$, compute the resulting mass flow.

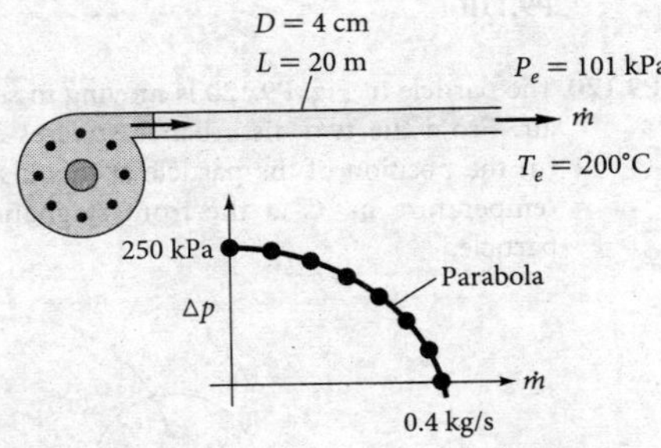

P9.99

P9.100 Natural gas, approximated as CH_4, flows through a Schedule 40 150 mm pipe from Providence to Narragansett, RI, a distance of 50 km. Gas companies use the *barg* as a pressure unit, meaning a bar of pressure *gage*, above ambient pressure. Assuming isothermal flow at 293 K, with $f \approx 0.019$, estimate the mass flow if the pressure is 5 bargs in Providence and 1 barg in Narragansett.

P9.101 How do the compressible pipe flow formulas behave for small pressure drops? Let air at 20°C enter a tube of diameter 1 cm and length 3 m. If $\bar{f} = 0.028$ with $p_1 = 102$ kPa and $p_2 = 100$ kPa, estimate the mass flow in kg/h for (*a*) isothermal flow, (*b*) adiabatic flow, and (*c*) incompressible flow (Chap. 6) at the entrance density.

P9.102 Air at 550 kPa and 100°C enters a smooth 1-m-long pipe and then passes through a second smooth pipe to a 30-kPa reservoir, as in Fig. P9.102. Using the Moody chart to compute $\bar{f}$, estimate the mass flow through this system. Is the flow choked?

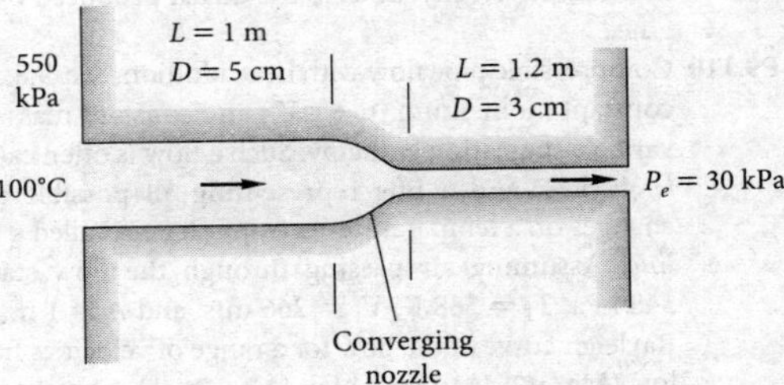

P9.102

P9.103 Natural gas, with $k \approx 1.3$ and a molecular weight of 16, is to be pumped through 100 km of 81-cm-diameter pipeline. The downstream pressure is 150 kPa. If the gas enters at 60°C, the mass flow is 20 kg/s, and $\bar{f} = 0.024$, estimate the required entrance pressure for (*a*) isothermal flow and (*b*) adiabatic flow.

P9.104 A tank of oxygen (Table A.4) at 20°C is to supply an astronaut through an umbilical tube 12 m long and 1.5 cm in diameter. The exit pressure in the tube is 40 kPa. If the desired mass flow is 90 kg/h and $\bar{f} = 0.025$, what should be the pressure in the tank?

P9.105 Modify Prob. P9.87 as follows: The pipeline will not be allowed to choke. It will have pumping stations about every 320 km. (*a*) Find the length of pipe for wł ich the pressure has dropped to 13.8 MPa. (*b*) What is the temperature at that point?

P9.106 Air, from a 3 cubic meter tank initially at 300 kPa and 200°C, blows down adiabatically through a smooth pipe 1 cm in diameter and 2.5 m long. Estimate the time required to reduce the tank pressure to 200 kPa. For simplicity, assume constant tank temperature and $f \approx 0.020$.

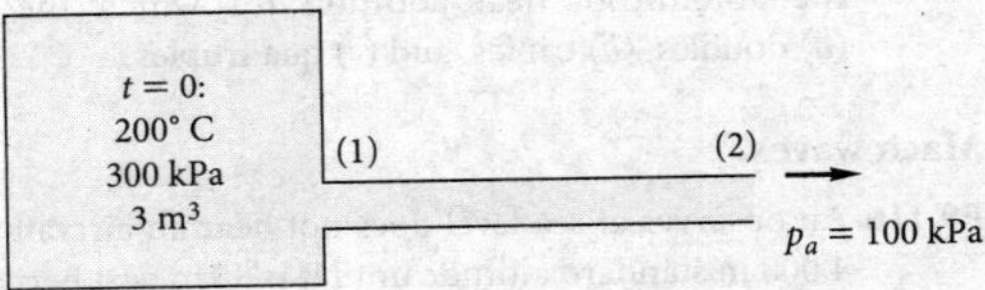

P9.106

Frictionless flow with heat transfer

P9.107 A fuel–air mixture, assumed equivalent to air, enters a duct combustion chamber at $V_1 = 104$ m/s and $T_1 = 300$ K. What amount of heat addition in kJ/kg will cause the exit flow to be choked? What will be the exit Mach number and temperature if 504 kJ/kg are added during combustion?

P9.108 What happens to the inlet flow of Prob. P9.107 if the combustion yields 1,500 kJ/kg heat addition and p_{01} and T_{01} remain the same? How much is the mass flow reduced?

P9.109 A jet engine at 7,000-m altitude takes in 45 kg/s of air and adds 550 kJ/kg in the combustion chamber. The chamber cross section is 0.5 m^2, and the air enters the chamber at 80 kPa and 5°C. After combustion the air expands through an isentropic converging nozzle to exit at atmospheric pressure. Estimate (*a*) the nozzle throat diameter, (*b*) the nozzle exit velocity, and (*c*) the thrust produced by the engine.

P9.110 Compressible pipe flow with heat addition, Sec. 9.8, assumes constant momentum ($p + \rho V^2$) and constant mass flow but variable stagnation enthalpy. Such a flow is often called *Rayleigh flow*, and a line representing all possible property changes on a temperature–entropy chart is called a *Rayleigh line*. Assuming air passing through the flow state $p_1 = 548$ kPa, $T_1 = 588$ K, $V_1 = 266$ m/s, and $A = 1$ m^2, draw a Rayleigh curve of the flow for a range of velocities from very low (Ma $\ll$ 1) to very high (Ma $\gg$ 1). Comment on the meaning of the maximum-entropy point on this curve.

P9.111 Add to your Rayleigh line of Prob. P9.110 a Fanno line (see Prob. P9.94) for stagnation enthalpy equal to the value associated with state 1 in Prob. P9.110. The two curves will intersect at state 1, which is subsonic, and at a certain state 2, which is supersonic. Interpret these two states vis-ā-vis Table B.2.

P9.112 Air enters a duct at $V_1 = 144$ m/s, $p_1 = 200$ kPa, and $T_1 = 323$ K. Assuming frictionless heat addition, estimate (*a*) the heat addition needed to raise the velocity to 372 m/s; and (*b*) the pressure at this new section 2.

P9.113 Air enters a constant-area duct at $p_1 = 90$ kPa, $V_1 = 520$ m/s, and $T_1 = 558$°C. It is then cooled with negligible friction until it exits at $p_2 = 160$ kPa. Estimate (*a*) V_2, (*b*) T_2, and (*c*) the total amount of cooling in kJ/kg.

P9.114 The scramjet of Fig. C9.8 operates with supersonic flow throughout. Assume that the heat addition of 500 kJ/kg, between sections 2 and 3, is frictionless and at constant area of 0.2 m^2. Given Ma$_2$ = 4.0, $p_2 = 260$ kPa, and $T_2 = 420$ K. Assume airflow at $k = 1.40$. At the combustion section exit, find (*a*) Ma$_3$, (*b*) p_3, and (*c*) T_3.

P9.115 Air enters a 5-cm-diameter pipe at 380 kPa, 3.3 kg/m^3, and 120 m/s. Assume frictionless flow with heat addition. Find the amount of heat addition for which the velocity (*a*) doubles, (*b*) triples, and (*c*) quadruples.

Mach waves

P9.116 An observer at sea level does not hear an aircraft flying at 4,000 m standard altitude until it is 8 km past her. Estimate the aircraft speed in m/s.

P9.117 A tiny scratch in the side of a supersonic wind tunnel creates a very weak wave of angle 17°, as shown in Fig. P9.117, after which a normal shock occurs. The air temperature in region (1) is 250 K. Estimate the temperature in region (2).

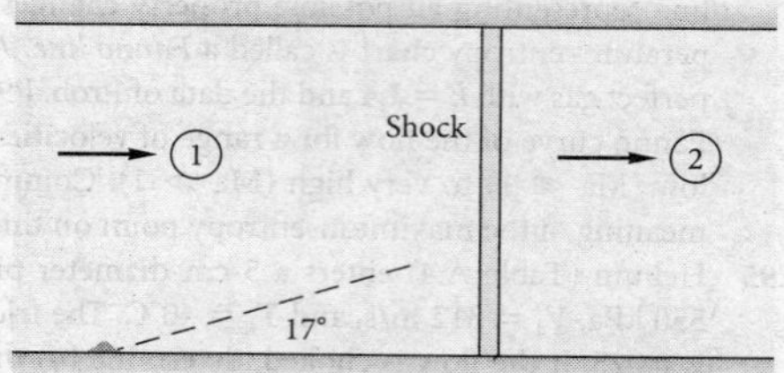

P9.117

P9.118 A particle moving at uniform velocity in sea-level standard air creates the two disturbance spheres shown in Fig. P9.118. Compute the particle velocity and Mach number.

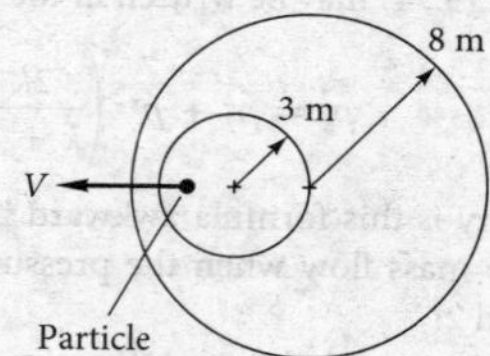

P9.118

P9.119 The particle in Fig. P9.119 is moving supersonically in sea-level standard air. From the two given disturbance spheres, compute the particle Mach number, velocity, and Mach angle.

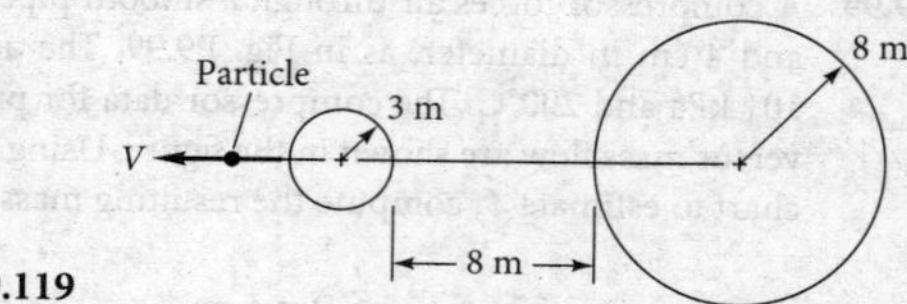

P9.119

P9.120 The particle in Fig. P9.120 is moving in sea-level standard air. From the two disturbance spheres shown, estimate (*a*) the position of the particle at this instant and (*b*) the temperature in °C at the front stagnation point of the particle.

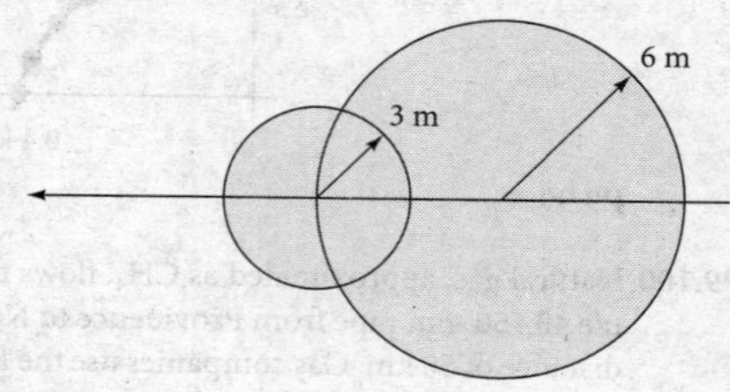

P9.120

P9.121 A thermistor probe, in the shape of a needle parallel to the flow, reads a static temperature of −25°C when inserted into a supersonic airstream. A conical disturbance cone of half-angle 17° is created. Estimate (*a*) the Mach number, (*b*) the velocity, and (*c*) the stagnation temperature of the stream.

The oblique shock wave

P9.122 Supersonic air takes a 5° compression turn, as in Fig. P9.122. Compute the downstream pressure and Mach number and the wave angle, and compare with small-disturbance theory.

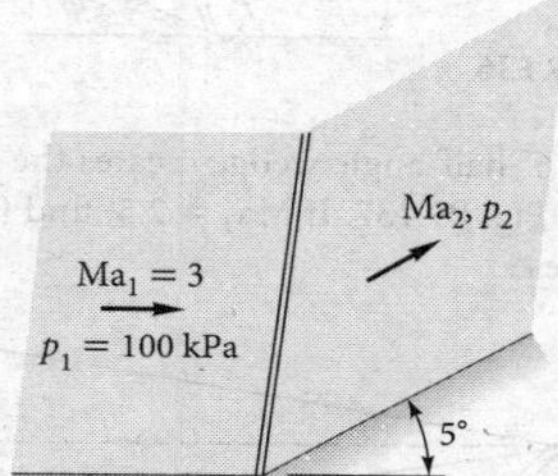

P9.122

P9.123 The 10° deflection in Example 9.17 caused a final Mach number of 1.641 and a pressure ratio of 1.707. Compare this with the case of the flow passing through two 5° deflections. Comment on the results and why they might be higher or lower in the second case.

P9.124 When a sea-level flow approaches a ramp of angle 20°, an oblique shock wave forms as in Figure P9.124. Calculate (*a*) Ma_1, (*b*) p_2, (*c*) T_2, and (*d*) V_2.

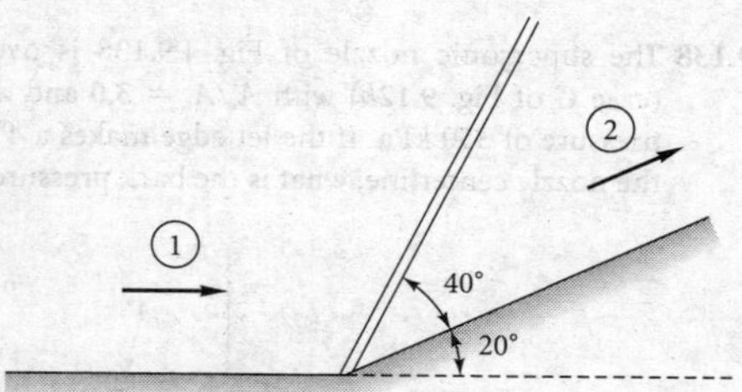

P9.124

P9.125 We saw in the text that, for $k = 1.40$, the maximum possible deflection caused by an oblique shock wave occurs at infinite approach Mach number and is $\theta_{max} = 45.58°$. Assuming an ideal gas, what is θ_{max} for (*a*) argon and (*b*) carbon dioxide?

P9.126 Airflow at Ma = 2.8, $p = 80$ kPa, and $T = 280$ K undergoes a 15° compression turn. Find the downstream values of (*a*) Mach number, (*b*) pressure, and (*c*) temperature.

P9.127 Do the Mach waves upstream of an oblique shock wave intersect with the shock? Assuming supersonic downstream flow, do the downstream Mach waves intersect the shock? Show that for small deflections the shock wave angle β lies halfway between μ_1 and $\mu_2 + \theta$ for any Mach number.

P9.128 Air flows past a two-dimensional wedge-nosed body as in Fig. P9.128. Determine the wedge half-angle δ for which the horizontal component of the total pressure force on the nose is 35 kN/m of depth into the paper.

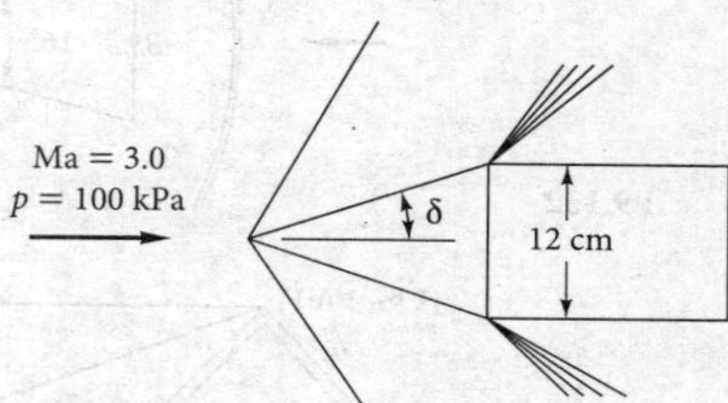

P9.128

P9.129 Air flows at supersonic speed toward a compression ramp, as in Fig. P9.129. A scratch on the wall at point *a* creates a wave of 30° angle, while the oblique shock created has a 50° angle. What is (*a*) the ramp angle θ and (*b*) the wave angle ϕ caused by a scratch at *b*?

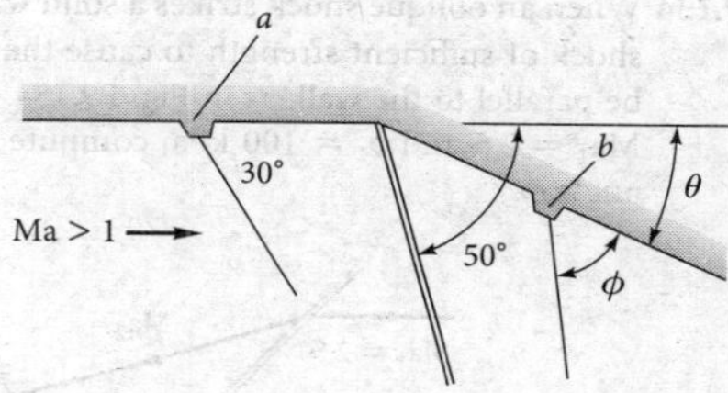

P9.129

P9.130 A supersonic airflow, at a temperature of 300 K, strikes a wedge and is deflected 12°. If the resulting shock wave is attached, and the temperature after the shock is 450 K, (*a*) estimate the approach Mach number and wave angle. (*b*) Why are there two solutions?

P9.131 The following formula has been suggested as an alternate to Eq. (9.86) to relate upstream Mach number to the oblique shock wave angle β and turning angle θ:

$$\sin^2 \beta = \frac{1}{Ma_1^2} + \frac{(k+1)\sin\beta\sin\theta}{2\cos(\beta-\theta)}$$

Can you prove or disprove this relation? If not, try a few numerical values and compare with the results from Eq. (9.86).

P9.132 Air flows at Ma = 3 and $p = 70$ kPa absolute toward a wedge of 16° angle at zero incidence in Fig. P9.132. If the pointed edge is forward, what will be the pressure at point *A*? If the blunt edge is forward, what will be the pressure at point *B*?

P9.133 Air flows supersonically toward the double-wedge system in Fig. P9.133. The (*x*, *y*) coordinates of the tips are given. The shock wave of the forward wedge strikes the tip of the aft wedge. Both wedges have 15° deflection angles. What is the free-stream Mach number?

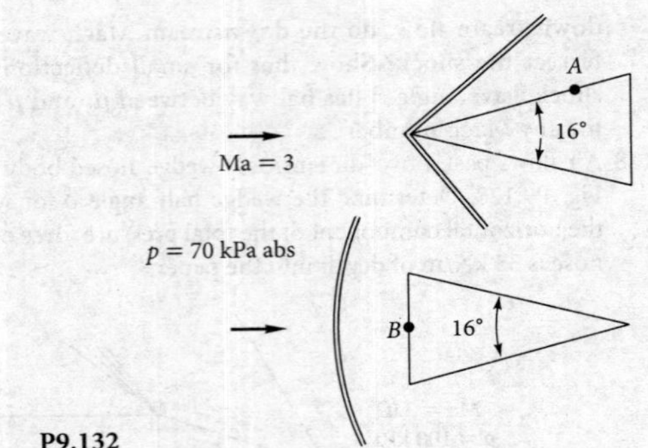

P9.132

(1 m, 1 m)
Shocks
Ma_∞
(0, 0)

P9.133

P9.134 When an oblique shock strikes a solid wall, it reflects as a shock of sufficient strength to cause the exit flow Ma_3 to be parallel to the wall, as in Fig. P9.134. For airflow with $Ma_1 = 2.5$ and $p_1 = 100$ kPa, compute Ma_3, p_3, and the angle ϕ.

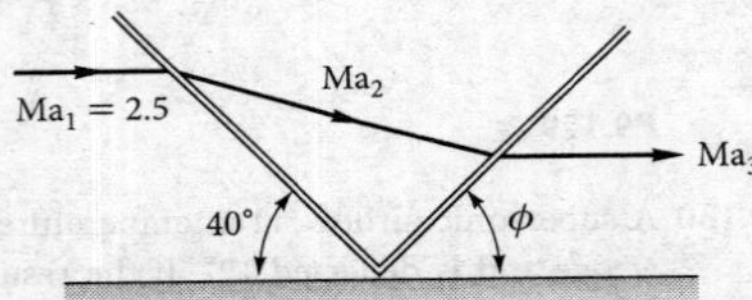

P9.134

P9.135 A bend in the bottom of a supersonic duct flow induces a shock wave that reflects from the upper wall, as in Fig. P9.135. Compute the Mach number and pressure in region 3.

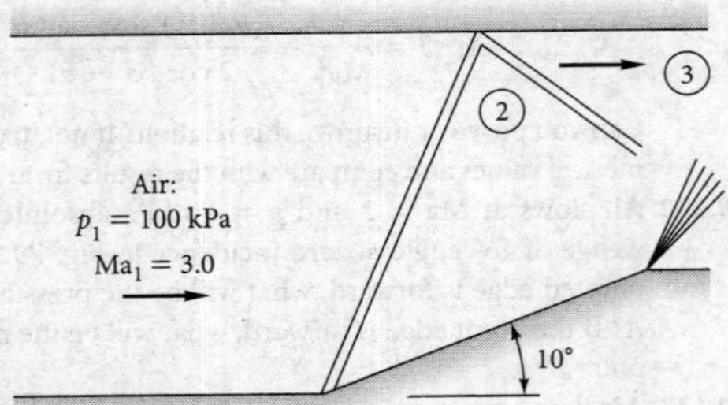

P9.135

P9.136 Figure P9.136 is a special application of Prob. P9.135. With careful design, one can orient the bend on the lower wall so that the reflected wave is exactly canceled by the return bend, as shown. This is a method of reducing the Mach number in a channel (a supersonic diffuser). If the bend angle is $\phi = 10°$, find (*a*) the downstream width *h* and (*b*) the downstream Mach number. Assume a weak shock wave.

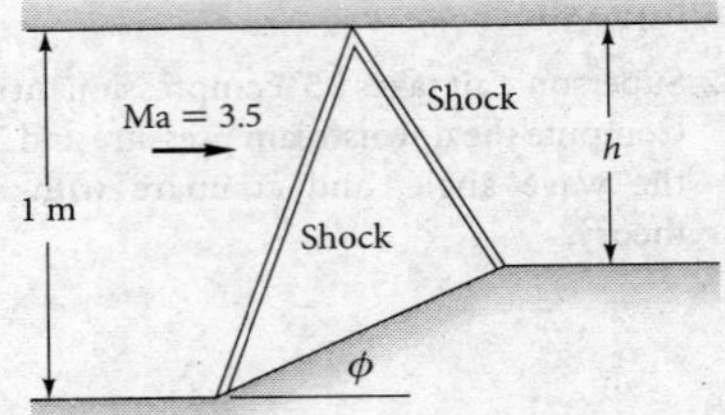

P9.136

P9.137 A 6° half-angle wedge creates the reflected shock system in Fig. P9.137. If $Ma_3 = 2.5$, find (*a*) Ma_1 and (*b*) the angle α.

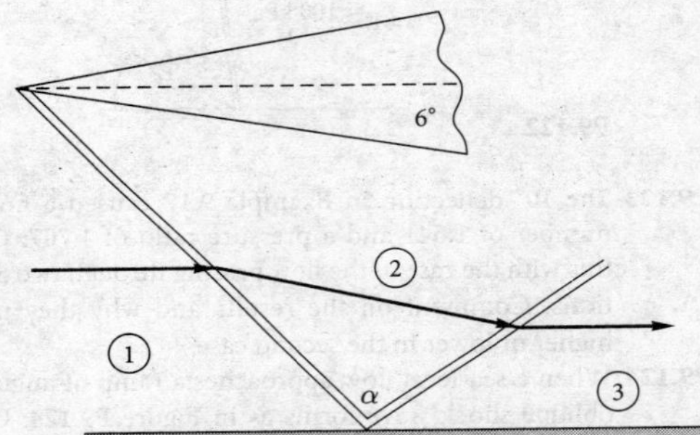

P9.137

P9.138 The supersonic nozzle of Fig. P9.138 is overexpanded (case *G* of Fig. 9.12*b*) with $A_e/A_t = 3.0$ and a stagnation pressure of 350 kPa. If the jet edge makes a 4° angle with the nozzle centerline, what is the back pressure p_r in kPa?

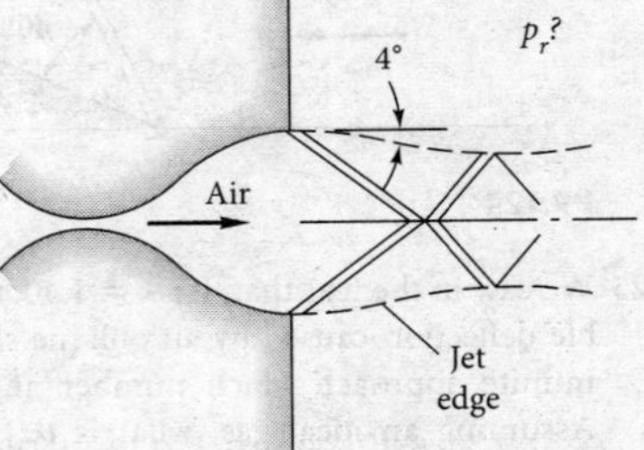

P9.138

P9.139 Airflow at Ma = 2.2 takes a compression turn of 12° and then another turn of angle θ in Fig. P9.139. What is the maximum value of θ for the second shock to be attached? Will the two shocks intersect for any θ less than θ_{max}?

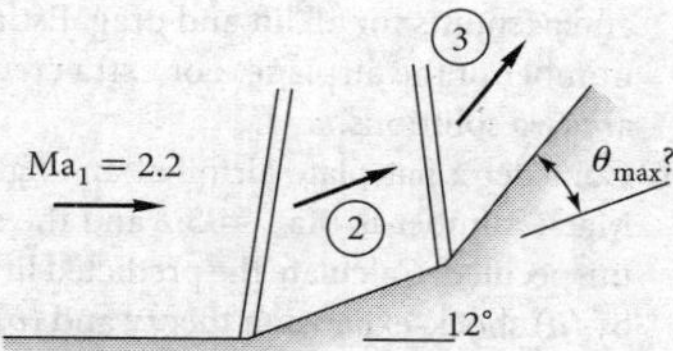

P9.139

Prandtl-Meyer expansion waves

P9.140 The solution to Prob. P9.122 is $Ma_2 = 2.750$ and $p_2 = 145.5$ kPa. Compare these results with an isentropic compression turn of 5°, using Prandtl-Meyer theory.

P9.141 Supersonic airflow takes a 5° expansion turn, as in Fig. P9.141. Compute the downstream Mach number and pressure, and compare with small-disturbance theory.

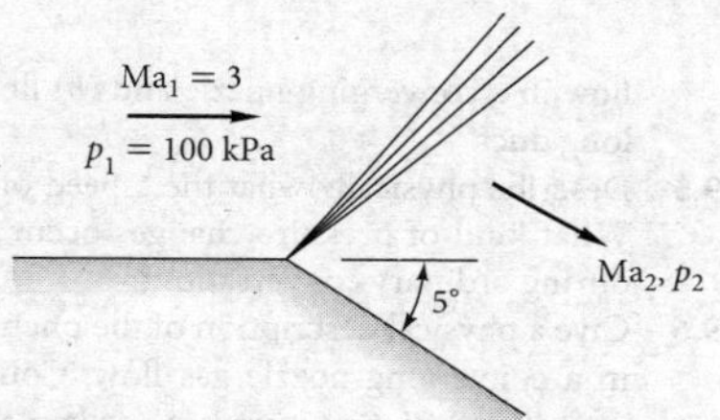

P9.141

P9.142 A supersonic airflow at $Ma_1 = 3.2$ and $p_1 = 50$ kPa undergoes a compression shock followed by an isentropic expansion turn. The flow deflection is 30° for each turn. Compute Ma_2 and p_2 if (*a*) the shock is followed by the expansion and (*b*) the expansion is followed by the shock.

P9.143 Airflow at Ma = 3.4 and 300 K encounters a 28° oblique shock turn. What subsequent isentropic expansion turn will bring the temperature back to 300 K?

P9.144 The 10° deflection in Example 9.17 caused the Mach number to drop to 1.64. (*a*) What turn angle will create a Prandtl-Meyer fan and bring the Mach number back up to 2.0? (*b*) What will be the final pressure?

P9.145 Air at $Ma_1 = 2.0$ and $p_1 = 100$ kPa undergoes an isentropic expansion to a downstream pressure of 50 kPa. What is the desired turn angle in degrees?

P9.146 Air flows supersonically over a surface that changes direction twice, as in Fig. P9.146. Calculate (*a*) Ma_2 and (*b*) p_3.

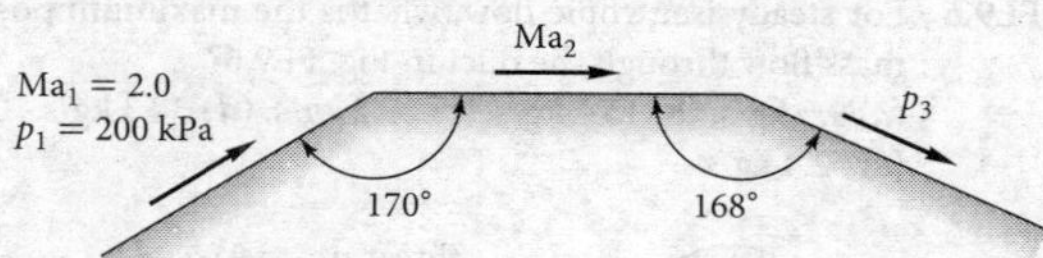

P9.146

P9.147 A converging–diverging nozzle with a 4:1 exit-area ratio and $p_0 = 500$ kPa operates in an underexpanded condition (case *I* of Fig. 9.12*b*) as in Fig. P9.147. The receiver pressure is $p_a = 10$ kPa, which is less than the exit pressure, so that expansion waves form outside the exit. For the given conditions, what will the Mach number Ma_2 and the angle ϕ of the edge of the jet be? Assume $k = 1.4$ as usual.

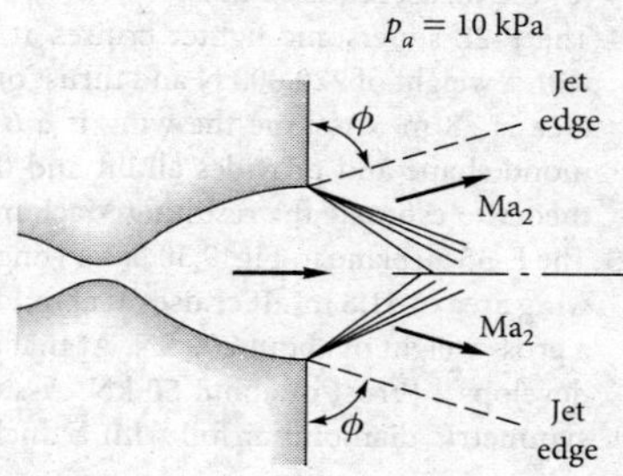

P9.147

P9.148 Air flows supersonically over a circular-arc surface as in Fig. P9.148. Estimate (*a*) the Mach number Ma_2 and (*b*) the pressure p_2 as the flow leaves the circular surface.

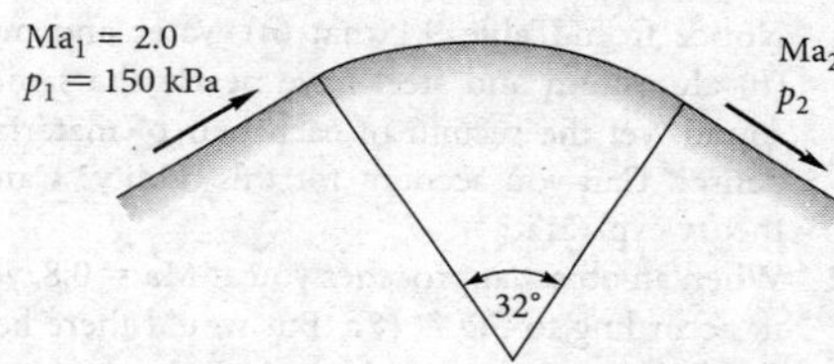

P9.148

Supersonic airfoils

P9.149 Air flows at $Ma_\infty = 3.0$ past a doubly symmetric diamond airfoil whose front and rear included angles are both 24°. For zero angle of attack, compute the drag coefficient obtained using shock-expansion theory and compare with Ackeret theory.

P9.150 A flat-plate airfoil with $C = 1.2$ m is to have a lift of 30 kN/m when flying at 5,000-m standard altitude with $U_\infty = 641$ m/s. Using Ackeret theory, estimate (*a*) the angle of attack and (*b*) the drag force in N/m.

P9.151 Air flows at Ma = 2.5 past a half-wedge airfoil whose angles are 4°, as in Fig. P9.151. Compute the lift and drag coefficient at α equal to (*a*) 0° and (*b*) 6°.

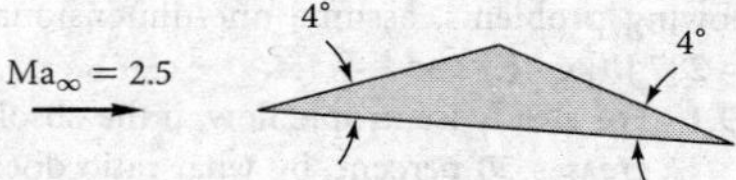

P9.151

P9.152 The X-43 model A scramjet aircraft in Fig. C9.8 is small W = 13,000 N, and unmanned, only 3.75 m long and 1.67 m wide. The aerodynamics of a slender arrowhead-shaped hypersonic vehicle is beyond our scope. Instead, let us assume it is a flat plate airfoil of area 2.0 m^2. Let Ma = 7 at 12,000 m standard altitude. Estimate the drag, by shock-expansion theory. *Hint:* Use Ackeret theory to estimate the angle of attack.

P9.153 A supersonic transport has a mass of 65,000 kg and cruises at 11-km standard altitude at a Mach number of 2.25. If the angle of attack is 2° and its wings can be approximated by flat plates, estimate (*a*) the required wing area in m² and (*b*) the thrust required in N.

P9.154 The F-22 supersonic fighter cruises at 11,000 m altitude, with a weight of 220,000 N and thrust of 45,000 N. Its wing area is 78 m². Assume the wing is a 6-percent-thick diamond shape and provides all lift and thrust. Use Ackeret theory to estimate the resulting Mach number.

***P9.155** The F-35 airplane in Fig. 9.30 has a wingspan of 10 m and a wing area of 41.8 m². It cruises at about 10 km altitude with a gross weight of about 200 kN. At that altitude, the engine develops a thrust of about 50 kN. Assume the wing has a symmetric diamond airfoil with a thickness of 8 percent, and accounts for all lift and drag. Estimate the cruise Mach number of the airplane. For extra credit, explain why there are *two* solutions.

P9.156 Consider a flat-plate airfoil at an angle of attack of 6°. The Mach number is $Ma_\infty = 3.2$ and the stream pressure p_∞ is unspecified. Calculate the predicted lift and drag coefficients by (*a*) shock-expansion theory and (*b*) Ackeret theory.

P9.157 The Ackeret airfoil theory of Eq. (9.104) is meant for *moderate* supersonic speeds, $1.2 < Ma < 4$. How does it fare for *hypersonic* speeds? To illustrate, calculate (*a*) C_L and (*b*) C_D for a flat-plate airfoil at $a = 5°$ and $Ma_\infty = 8.0$, using shock-expansion theory, and compare with Ackeret theory. Comment.

Word Problems

W9.1 Notice from Table 9.1 that (*a*) water and mercury and (*b*) aluminum and steel have nearly the same speeds of sound, yet the second of each pair of materials is much denser. Can you account for this oddity? Can molecular theory explain it?

W9.2 When an object approaches you at Ma = 0.8, you can hear it, according to Fig. 9.18*a*. But would there be a Doppler shift? For example, would a musical tone seem to you to have a higher or a lower pitch?

W9.3 The subject of this chapter is commonly called *gas dynamics*. But can liquids not perform in this manner? Using water as an example, make a rule-of-thumb estimate of the pressure level needed to drive a water flow at velocities comparable to the sound speed.

W9.4 Suppose a gas is driven at compressible subsonic speeds by a large pressure drop, p_1 to p_2. Describe its behavior on an appropriately labeled Mollier chart for (*a*) frictionless flow in a converging nozzle and (*b*) flow with friction in a long duct.

W9.5 Describe physically what the "speed of sound" represents. What kind of pressure changes occur in air sound waves during ordinary conversation?

W9.6 Give a physical description of the phenomenon of choking in a converging-nozzle gas flow. Could choking happen even if wall friction were not negligible?

W9.7 Shock waves are treated as discontinuities here, but they actually have a very small finite thickness. After giving it some thought, sketch your idea of the distribution of gas velocity, pressure, temperature, and entropy through the inside of a shock wave.

W9.8 Describe how an observer, running along a normal shock wave at finite speed *V*, will see what appears to be an oblique shock wave. Is there any limit to the running speed?

Fundamentals of Engineering Exam Problems

One-dimensional compressible flow problems have become quite popular on the FE Exam, especially in the afternoon sessions. In the following problems, assume one-dimensional flow of ideal air, $R = 287$ J/(kg · K) and $k = 1.4$.

FE9.1 For steady isentropic flow, if the absolute temperature increases 50 percent, by what ratio does the static pressure increase?
(*a*) 1.12, (*b*) 1.22, (*c*) 2.25, (*d*) 2.76, (*e*) 4.13

FE9.2 For steady isentropic flow, if the density doubles, by what ratio does the static pressure increase?
(*a*) 1.22, (*b*) 1.32, (*c*) 1.44, (*d*) 2.64, (*e*) 5.66

FE9.3 A large tank, at 500 K and 200 kPa, supplies isentropic airflow to a nozzle. At section 1, the pressure is only 120 kPa. What is the Mach number at this section?
(*a*) 0.63, (*b*) 0.78, (*c*) 0.89, (*d*) 1.00, (*e*) 1.83

FE9.4 In Prob. FE9.3 what is the temperature at section 1?
(*a*) 300 K, (*b*) 408 K, (*c*) 417 K, (*d*) 432 K, (*e*) 500 K

FE9.5 In Prob. FE9.3, if the area at section 1 is 0.15 m², what is the mass flow?
(*a*) 38.1 kg/s, (*b*) 53.6 kg/s, (*c*) 57.8 kg/s, (*d*) 67.8 kg/s, (*e*) 77.2 kg/s

FE9.6 For steady isentropic flow, what is the maximum possible mass flow through the duct in Fig. FE9.6?
(*a*) 9.5 kg/s, (*b*) 15.1 kg/s, (*c*) 26.2 kg/s, (*d*) 30.3 kg/s, (*e*) 52.4 kg/s

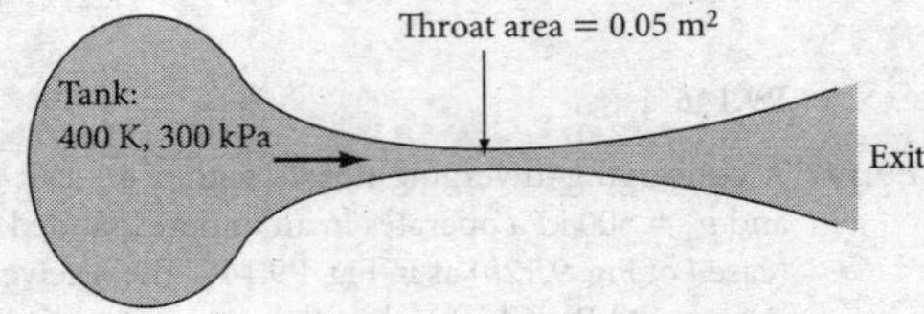

FE9.6

FE9.7 If the exit Mach number in Fig. FE9.6 is 2.2, what is the exit area?
(*a*) 0.10 m^2, (*b*) 0.12 m^2, (*c*) 0.15 m^2, (*d*) 0.18 m^2, (*e*) 0.22 m^2

FE9.8 If there are no shock waves and the pressure at one duct section in Fig. FE9.6 is 55.5 kPa, what is the velocity at that section?
(*a*) 166 m/s, (*b*) 232 m/s, (*c*) 554 m/s, (*d*) 706 m/s, (*e*) 774 m/s

FE9.9 If, in Fig. FE9.6, there is a normal shock wave at a section where the area is 0.07 m^2, what is the air density just upstream of that shock?
(*a*) 0.48 kg/m^3, (*b*) 0.78 kg/m^3, (*c*) 1.35 kg/m^3, (*d*) 1.61 kg/m^3, (*e*) 2.61 kg/m^3

FE9.10 In Prob. FE9.9, what is the Mach number just downstream of the shock wave?
(*a*) 0.42, (*b*) 0.55, (*c*) 0.63, (*d*) 1.00, (*e*) 1.76

Comprehensive Problems

C9.1 The converging–diverging nozzle sketched in Fig. C9.1 is designed to have a Mach number of 2.00 at the exit plane (assuming the flow remains nearly isentropic). The flow travels from tank *a* to tank *b*, where tank *a* is much larger than tank *b*. (*a*) Find the area at the exit A_e and the back pressure p_b that will allow the system to operate at design conditions. (*b*) As time goes on, the back pressure will grow, since the second tank slowly fills up with more air. Since tank *a* is huge, the flow in the nozzle will remain the same, however, until a normal shock wave appears at the exit plane. At what back pressure will this occur? (*c*) If tank *b* is held at constant temperature, $T = 20°C$, estimate how long it will take for the flow to go from design conditions to the condition of part (*b*)—that is, with a shock wave at the exit plane.

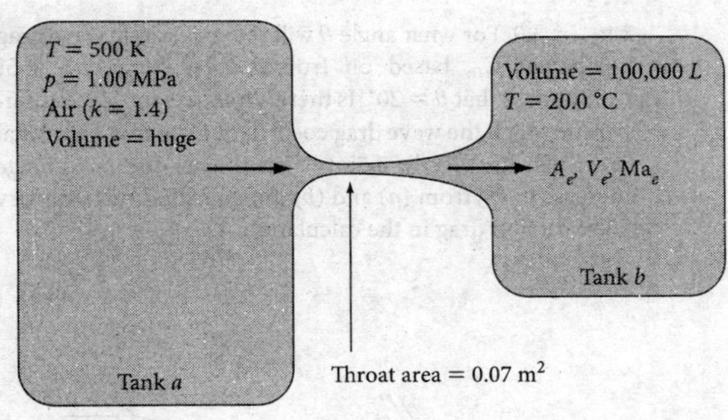

C9.1

C9.2 Two large air tanks, one at 400 K and 300 kPa and the other at 300 K and 100 kPa, are connected by a straight tube 6 m long and 5 cm in diameter. The average friction factor is 0.0225. Assuming adiabatic flow, estimate the mass flow through the tube.

***C9.3** Figure C9.3 shows the exit of a converging–diverging nozzle, where an oblique shock pattern is formed. In the exit plane, which has an area of 15 cm^2, the air pressure is 16 kPa and the temperature is 250 K. Just outside the exit shock, which makes an angle of 50° with the exit plane, the temperature is 430 K. Estimate (*a*) the mass flow, (*b*) the throat area, (*c*) the turning angle of the exit flow, and, in the tank supplying the air, (*d*) the pressure and (*e*) the temperature.

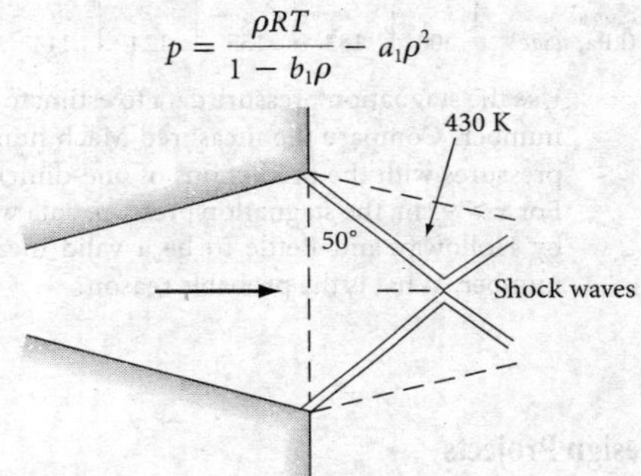

C9.3

C9.4 The properties of a dense gas (high pressure and low temperature) are often approximated by van der Waals's equation of state [17, 18]:

$$p = \frac{\rho RT}{1 - b_1\rho} - a_1\rho^2$$

where constants a_1 and b_1 can be found from the critical temperature and pressure

$$a_1 = \frac{27R^2T_c^2}{64p_c} = 162 \text{ N} \cdot \text{m}^4/\text{kg}^2$$

for air, and

$$b_1 = \frac{RT_c}{8p_c} = 0.00126 \text{ m}^3/\text{kg}$$

for air. Find an analytic expression for the speed of sound of a van der Waals gas. Assuming $k = 1.4$, compute the speed of sound of air in m/s at 310 K and 2 MPa for (*a*) a perfect gas and (*b*) a van der Waals gas. What percentage higher density does the van der Waals relation predict?

C9.5 Consider one-dimensional steady flow of a nonideal gas, steam, in a converging nozzle. Stagnation conditions are $p_0 = 100$ kPa and $T_0 = 200°C$. The nozzle exit diameter is 2 cm. (*a*) If the nozzle exit pressure is 70 kPa, calculate the mass flow and the exit temperature for real steam from the steam tables. (As a first estimate, assume steam to be an ideal gas from Table A.4.) Is the flow choked? (*b*) Find the nozzle exit pressure and mass flow for which the steam flow *is* choked, using the steam tables.

C9.6 Extend Prob. C9.5 as follows: Let the nozzle be converging–diverging, with an exit diameter of 3 cm. Assume isentropic flow. Find the exit Mach number, pressure, and temperature for an ideal gas from Table A.4. Does the mass flow agree with the value of 0.0452 kg/s in Prob. C9.5?

C9.7 Professor Gordon Holloway and his student, Jason Bettle, of the University of New Brunswick obtained the following tabulated data for blow-down airflow through a converging–diverging nozzle similar in shape to Fig. P3.22. The supply tank pressure and temperature were 200 kPa, gage and 296 K, respectively. Atmospheric pressure was 101.3 kPa, abs. Wall pressures and centerline stagnation pressures were measured in the expansion section, which was a frustrum of a cone. The nozzle throat is at $x = 0$.

x(cm)	0	1.5	3	4.5	6	7.5	9
Diameter (cm)	1.00	1.098	1.195	1.293	1.390	1.488	1.585
p_{wall} (kPa, gage)	53	−18	−34	−50	−45	−72	−51
$p_{stagnation}$ (kPa, gage)	200	183	155	124	114	96	69

Use the stagnation pressure data to estimate the local Mach number. Compare the measured Mach numbers and wall pressures with the predictions of one-dimensional theory. For $x > 9$ cm, the stagnation pressure data was not thought by Holloway and Bettle to be a valid measure of Mach number. What is the probable reason?

C9.8 Engineers call the supersonic combustion in a scramjet engine almost miraculous, "like lighting a match in a hurricane." Figure C9.8 is a crude idealization of the engine. Air enters, burns fuel in the narrow section, then exits, all at supersonic speeds. There are no shock waves. Assume areas of 1 m^2 at sections 1 and 4 and 0.2 m^2 at sections 2 and 3. Let the entrance conditions be $\text{Ma}_1 = 6$, at 10,000 m standard altitude. Assume isentropic flow from 1 to 2, frictionless heat transfer from 2 to 3 with $Q = 500$ kJ/kg, and isentropic flow from 3 to 4. Calculate the exit conditions and the thrust produced.

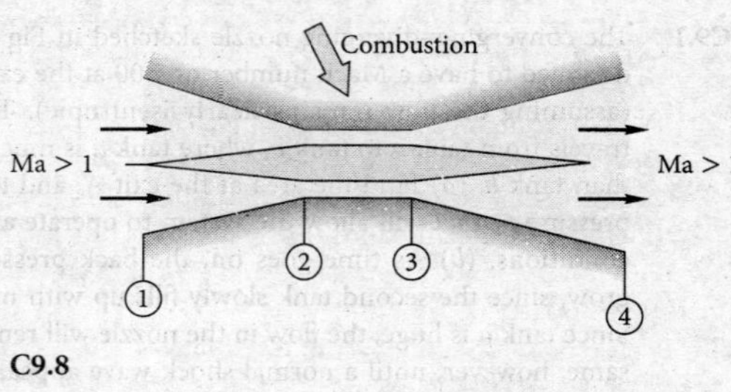

C9.8

Design Projects

D9.1 It is desired to select a rectangular wing for a fighter aircraft. The plane must be able (*a*) to take off and land on a 1,400 m-long sea-level runway and (*b*) to cruise supersonically at Ma = 2.3 at 8,500 m altitude. For simplicity, assume a wing with zero sweepback. Let the aircraft maximum weight equal (30 + *n*)(4,500) N, where *n* is the number of letters in your surname. Let the available sea-level maximum thrust be one-third of the maximum weight, decreasing at altitude proportional to ambient density. Making suitable assumptions about the effect of finite aspect ratio on wing lift and drag for both subsonic and supersonic flight, select a wing of minimum area sufficient to perform these takeoff/landing and cruise requirements. Some thought should be given to analyzing the wingtips and wing roots in supersonic flight, where Mach cones form and the flow is not two-dimensional. If no satisfactory solution is possible, gradually increase the available thrust to converge to an acceptable design.

D9.2 Consider supersonic flow of air at sea-level conditions past a wedge of half-angle θ, as shown in Fig. D9.2. Assume that the pressure on the back of the wedge equals the fluid pressure as it exits the Prandtl-Meyer fan. (*a*) Suppose $\text{Ma}_\infty = 3.0$. For what angle θ will the supersonic wave drag coefficient C_D, based on frontal area, be exactly 0.5? (*b*) Suppose that $\theta = 20°$. Is there a free-stream Mach number for which the wave drag coefficient C_D, based on frontal area, will be exactly 0.5? (*c*) Investigate the percentage increase in C_D from (*a*) and (*b*) due to including boundary layer friction drag in the calculation.

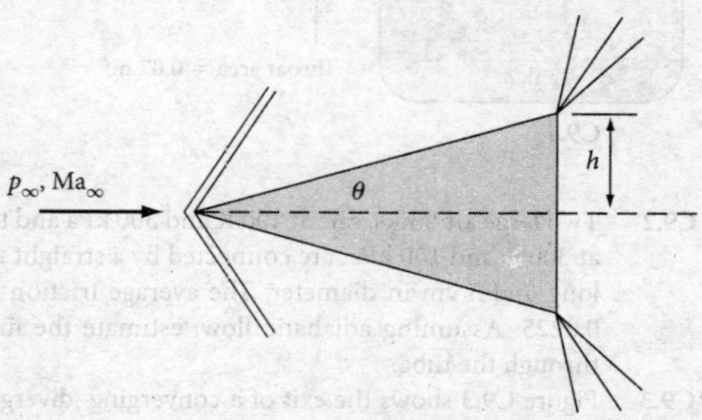

D9.2

References

1. J. E. A. John and T. G. Keith, *Gas Dynamics,* 3d ed., Pearson Education, Upper Saddle River, NJ, 2005.
2. B. K. Hodge and K. Koenig, *Compressible Fluid Dynamics: With Personal Computer Applications,* Pearson Prentice-Hall, Upper Saddle River, NJ, 1995.
3. R. D. Zucker and O. Biblarz, *Fundamentals of Gas Dynamics,* 2d ed., Wiley, New York, 2002.
4. J. D. Anderson, *Modern Compressible Flow: with Historical Perspective,* 3d ed., McGraw-Hill, New York, 2002.
5. A. H. Shapiro, *The Dynamics and Thermodynamics of Compressible Fluid Flow,* 2 vols., Wiley, New York, 1953.
6. C. Cercignani, *Rarefied Gas Dynamics,* Cambridge University Press, New York, 2000.
7. H. W. Liepmann and A. Roshko, *Elements of Gas Dynamics,* Dover, New York, 2001.
8. I. Straskraba, *Introduction to the Mathematical Theory of Compressible Flow,* Oxford University Press, New York, 2004.
9. P. A. Thompson, *Compressible Fluid Dynamics,* McGraw-Hill, New York, 1972.
10. P. H. Oosthuizen and W. E. Carscallen, *Compressible Fluid Flow,* McGraw-Hill, New York, 2003.
11. J. D. Anderson, *Hypersonic and High Temperature Gas Dynamics,* 2d ed., AIAA, Reston, VA, 2006.
12. G. A. Bird, *Molecular Gas Dynamics and the Direct Simulation of Gas Flows,* Clarendon Press, Oxford, 1994.
13. D. D. Knight, *Elements of Numerical Methods for Compressible Flows,* Cambridge University Press, New York, 2012.
14. L. W. Reithmaier, *Mach 1 and Beyond: The Illustrated Guide to High-Speed Flight,* McGraw-Hill, 1994.
15. W. T. Parry, *ASME International Steam Tables for Industrial Use,* 2d ed., ASME, New York, 2009.
16. J. H. Keenan et al., *Gas Tables: International Version,* Krieger Publishing, Melbourne, FL, 1992.
17. Y. A. Cengel and M. A. Boles, *Thermodynamics: An Engineering Approach,* 7th ed., McGraw-Hill, New York, 2010.
18. M. J. Moran and H. A. Shapiro, *Fundamentals of Engineering Thermodynamics,* 7th ed., Wiley, New York, 2010.
19. F. M. White, *Viscous Fluid Flow,* 3d ed., McGraw-Hill, New York, 2005.
20. J. Palmer, K. Ramsden, and E. Goodger, *Compressible Flow Tables for Engineers: With Appropriate Computer Programs,* Scholium Intl., Port Washington, NY, 1989.
21. M. R. Lindeburg, *Consolidated Gas Dynamics Tables,* Professional Publications, Inc., Belmont, CA, 1994.
22. A. M. Shektman, *Gasdynamic Functions of Real Gases,* Taylor and Francis, New York, 1991.
23. J. H. Keenan and E. P. Neumann, "Measurements of Friction in a Pipe for Subsonic and Supersonic Flow of Air," *J. Applied Mechanics,* vol. 13, no. 2, 1946, p. A-91.
24. R. P. Benedict, *Fundamentals of Pipe Flow,* John Wiley, New York, 1980.
25. J. L. Sims, *Tables for Supersonic Flow around Right Circular Cones at Zero Angle of Attack,* NASA SP-3004, 1964 (see also NASA SP-3007).
26. J. L. Thomas, "Reynolds Number Effects on Supersonic Asymmetrical Flows over a Cone," *J. Aircraft,* vol. 30, no. 4, 1993, pp. 488–495.
27. W. Bober and R. A. Kenyon, *Fluid Mechanics,* Wiley, New York, 1980.
28. J. Ackeret, "Air Forces on Airfoils Moving Faster than Sound Velocity," *NACA Tech. Memo.* 317, 1925.
29. W. B. Brower, *Theory, Tables and Data for Compressible Flow,* Taylor & Francis, New York, 1990.
30. M. Belfiore, "The Hypersonic Age is Near," *Popular Science,* January 2008, pp. 36–41.
31. G. S. Settles, *Schlieren and Shadowgraph Techniques: Visualizing Phenomena in Transparent Media,* Springer-Verlag, Berlin, 2001.

In March of 2010, the normally placid Saugatucket River, in South Kingstown, Rhode Island, was flooded by heavy rains. Instead of its usual gentle trickle over the Main Street dam, the flow rate was enormously increased and flooded the medical building in the background, ruining their offices and X-ray machines. The open-channel flow analysis methods in this chapter can handle both the trickle and the flood. [*Photo courtesy of Independent Newspapers*]

Chapter 10
Open-Channel Flow

Motivation. An *open-channel flow* denotes a flow with a free surface touching an atmosphere, like a river or a canal or a flume. Closed-duct flows (Chap. 6) are full of fluid, either liquid or gas, have no free surface within, and are driven by a pressure gradient along the duct axis. The open-channel flows here are driven by gravity alone, and the pressure gradient at the atmospheric interface is negligible. The basic force balance in an open channel is between gravity and friction.

Open-channel flows are an especially important mode of fluid mechanics for civil and environmental engineers. One needs to predict the flow rates and water depths that result from a given channel geometry, whether natural or artificial, and a given wet-surface roughness. Water is almost always the relevant fluid, and the channel size is usually large. Thus open-channel flows are generally turbulent, three-dimensional, sometimes unsteady, and often quite complex. This chapter presents some simple engineering theories and experimental correlations for steady flow in straight channels of regular geometry. We can borrow and use some concepts from duct flow analysis: hydraulic radius, friction factor, and head losses.

10.1 Introduction

Simply stated, open-channel flow is the flow of a liquid in a conduit with a free surface. There are many practical examples, both artificial (flumes, spillways, canals, weirs, drainage ditches, culverts) and natural (streams, rivers, estuaries, floodplains). This chapter introduces the elementary analysis of such flows, which are dominated by the effects of gravity.

The presence of the free surface, which is essentially at atmospheric pressure, both helps and hurts the analysis. It helps because the pressure can be taken as constant along the free surface, which therefore is equivalent to the *hydraulic grade line* (HGL) of the flow. Unlike flow in closed ducts, the pressure gradient is not a direct factor in open-channel flow, where the balance of forces is confined to gravity and friction.[1] But, the free surface complicates the analysis because its shape is a priori unknown: The depth profile changes with conditions and must be computed as part of the problem, especially in unsteady problems involving wave motion.

Before proceeding, we remark, as usual, that whole books have been written on open-channel hydraulics [1 to 7, 32]. There are also specialized texts devoted to wave motion [8 to 10] and to engineering aspects of coastal free-surface flows [11 to 13].

[1]Surface tension is rarely important because open channels are normally quite large and have a very large Weber number. Surface tension affects small models of large channels.

This chapter is only an introduction to broader and more detailed treatments. The writer recommends, as an occasional break from free-surface flow analysis, Ref. 31, which is an enchanting and spectacular gallery of ocean wave photographs.

The One-Dimensional Approximation

An open channel always has two sides and a bottom, where the flow satisfies the no-slip condition. Therefore, even a straight channel has a three-dimensional velocity distribution. Some measurements of straight-channel velocity contours are shown in Fig. 10.1. The profiles are quite complex, with maximum velocity typically occurring in the midplane about 20 percent below the surface. In very broad shallow channels the maximum velocity is near the surface, and the velocity profile is nearly logarithmic from the bottom to the free surface, as in Eq. (6.62). In noncircular channels, there are also secondary motions similar to Fig. 6.16 for closed-duct flows. If the channel curves or meanders, the secondary motion intensifies due to centrifugal effects, with high velocity occurring near the outer radius of the bend. Curved natural channels are subject to strong bottom erosion and deposition effects.

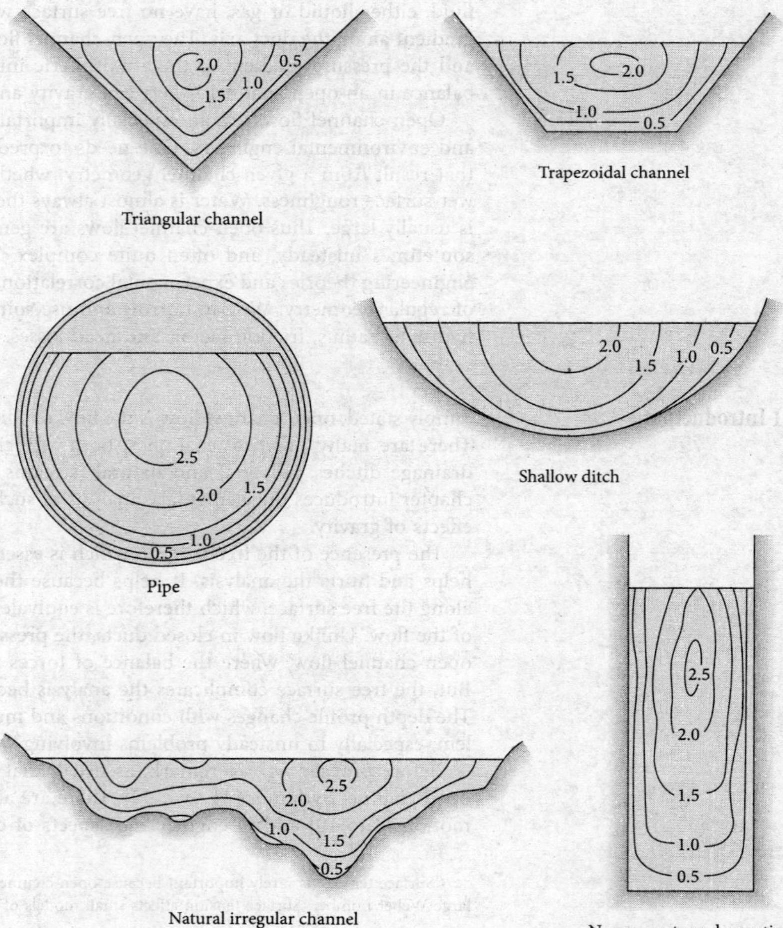

Fig. 10.1 Measured isovelocity contours in typical straight open-channel flows. *(From Ref. 2.)*
Source: From V. T. Chow, Open Channel Hydraulics, Blackburn Press, Caldwell, NJ, 2009.

With the advent of the supercomputer, it is possible to make numerical simulations of complex flow patterns such as those in Fig. 10.1 [27, 28]. However, the practical engineering approach, used here, is to make a one-dimensional flow approximation, as in Fig. 10.2. Since the liquid density is nearly constant, the steady flow continuity equation reduces to constant-volume flow Q along the channel

$$Q = V(x)A(x) = \text{const} \tag{10.1}$$

where V is average velocity and A the local cross-sectional area, as sketched in Fig. 10.2.

A second one-dimensional relation between velocity and channel geometry is the energy equation, including friction losses. If points 1 (upstream) and 2 (downstream) are on the free surface, $p_1 = p_2 = p_a$, and we have, for steady flow,

$$\frac{V_1^2}{2g} + z_1 = \frac{V_2^2}{2g} + z_2 + h_f \tag{10.2}$$

where z denotes the total elevation of the free surface, which includes the water depth y (see Fig. 10.2*a*) plus the height of the (sloping) bottom. The friction head loss h_f is analogous to head loss in duct flow from Eq. (6.10):

$$h_f \approx f\frac{x_2 - x_1}{D_h}\frac{V_{\text{av}}^2}{2g} \qquad D_h = \text{hydraulic diameter} = \frac{4A}{P} \tag{10.3}$$

where f is the average friction factor (Fig. 6.13) between sections 1 and 2. Since channels are irregular in shape, their "size" is taken to be the hydraulic *radius*:

$$\boxed{R_h = \frac{1}{4}D_h = \frac{A}{P}} \tag{10.4}$$

The local Reynolds number of the channel would be $\text{Re} = VR_h/\nu$, which is usually highly turbulent (>1 E5). The only commonly occurring laminar channel flows are the thin sheets that form as rainwater drains from crowned streets and airport runways.

The wetted perimeter P (see Fig. 10.2*b*) includes the sides and bottom of the channel but, not the free surface and, of course, not the parts of the sides above the water level. For example, if a rectangular channel is b wide and h high and contains water to depth y, its wetted perimeter is

$$P = b + 2y$$

not $2b + 2h$.

Although the Moody chart (Fig. 6.13) would give a good estimate of the friction factor in channel flow, in practice it is seldom used. An alternative correlation due to Robert Manning, discussed in Sec. 10.2, is the formula of choice in open-channel hydraulics.

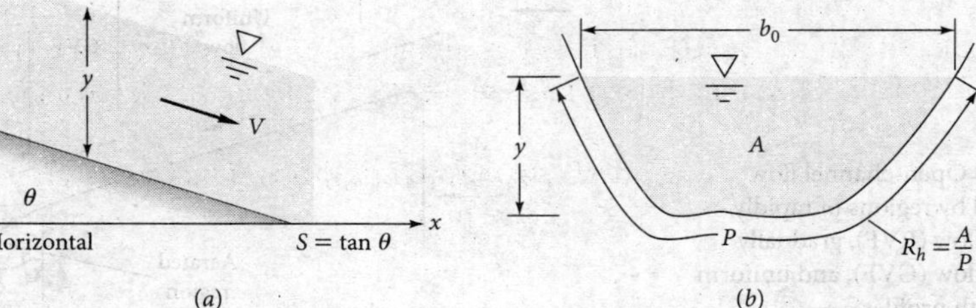

Fig. 10.2 Geometry and notation for open-channel flow: (*a*) side view; (*b*) cross section. All these parameters are constant in uniform flow.

Flow Classification by Depth Variation

The most common method of classifying open-channel flows is by the rate of change of the free-surface depth. The simplest and most widely analyzed case is *uniform flow*, where the depth and area (hence the velocity in steady flow) remain constant. Uniform flow conditions are approximated by long, straight runs of constant-slope and constant-area channel. A channel in uniform flow is said to be moving at its *normal depth* y_n, which is an important design parameter.

If the channel slope or cross section changes or there is an obstruction in the flow, then the depth changes and the flow is said to be *varied*. The flow is *gradually varying* if the one-dimensional approximation is valid and *rapidly varying* if not. Some examples of this method of classification are shown in Fig. 10.3. The classes can be summarized as follows:

1. Uniform flow (constant depth and slope)
2. Varied flow:
   a. Gradually varied (one-dimensional)
   b. Rapidly varied (multidimensional)

Typically, uniform flow is separated from rapidly varying flow by a region of gradually varied flow. Gradually varied flow can be analyzed by a first-order differential equation (Sec. 10.6), but rapidly varying flow usually requires experimentation or three-dimensional computational fluid dynamics [14, 27, 28].

Flow Classification by Froude Number

A second and very useful classification of open-channel flow is by the dimensionless Froude number, Fr, which is the ratio of channel velocity to the speed of propagation of a small-disturbance wave in the channel. For a rectangular or very wide constant-depth channel, this takes the form

$$\text{Fr} = \frac{\text{flow velocity}}{\text{surface wave speed}} = \frac{V}{\sqrt{gy}} \tag{10.5}$$

where y is the water depth. The flow behaves differently depending on these three flow regimes:

$$\begin{aligned} \text{Fr} &< 1.0 \quad \text{subcritical flow} \\ \text{Fr} &= 1.0 \quad \text{critical flow} \\ \text{Fr} &> 1.0 \quad \text{supercritical flow} \end{aligned} \tag{10.6}$$

The Froude number for irregular channels is defined in Sec. 10.4. As mentioned in Sec. 9.10, there is a strong analogy here with the three compressible flow regimes of

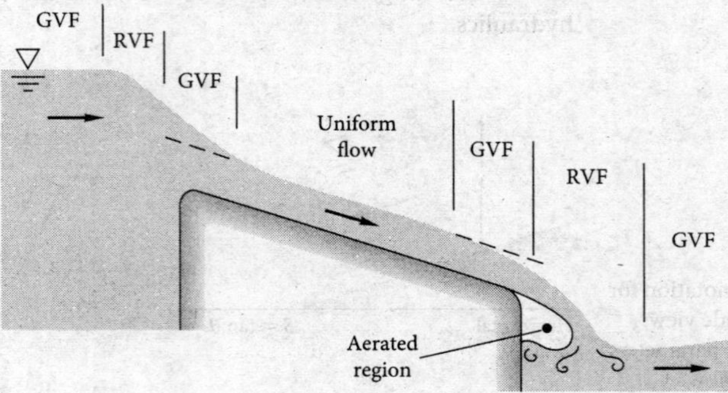

Fig. 10.3 Open-channel flow classified by regions of rapidly varying flow (RVF), gradually varying flow (GVF), and uniform flow depth profiles.

the Mach number: subsonic (Ma < 1), sonic (Ma = 1), and supersonic (Ma > 1). We shall use the analogy in Secs. 10.4 and 10.5. The analogy is pursued in Ref. 21.

Surface Wave Speed

The Froude number denominator $(gy)^{1/2}$ is the speed of an infinitesimal shallow-water surface wave. We can derive this with reference to Fig. 10.4*a*, which shows a wave of height δy propagating at speed c into still liquid. To achieve a steady flow inertial frame of reference, we fix the coordinates on the wave as in Fig. 10.4*b*, so that the still water moves to the right at velocity c. Figure 10.4 is exactly analogous to Fig. 9.1, which analyzed the speed of sound in a fluid. It can be used to analyze tidal bores, Fig. P10.86, which are described by Chanson [34].

For the control volume of Fig. 10.4*b*, the one-dimensional continuity relation is, for channel width b,

$$\rho cby = \rho(c - \delta V)(y + \delta y)b$$

or

$$\delta V = c\frac{\delta y}{y + \delta y} \tag{10.7}$$

This is analogous to Eq. (9.10); the velocity change δV induced by a surface wave is small if the wave is "weak," $\delta y \ll y$. If we neglect bottom friction in the short distance across the wave in Fig. 10.4*b*, the momentum relation is a balance between the net hydrostatic pressure force and momentum:

$$-\tfrac{1}{2}\rho gb[(y + \delta y)^2 - y^2] = \rho cby(c - \delta V - c)$$

or

$$g\left(1 + \frac{\tfrac{1}{2}\delta y}{y}\right)\delta y = c\,\delta V \tag{10.8}$$

This is analogous to Eq. (9.12). By eliminating δV between Eqs. (10.7) and (10.8) we obtain the desired expression for wave propagation speed:

$$c^2 = gy\left(1 + \frac{\delta y}{y}\right)\left(1 + \frac{\tfrac{1}{2}\delta y}{y}\right) \tag{10.9}$$

The "stronger" the wave height δy, the faster the wave speed c, by analogy with Eq. (9.13). In the limit of an infinitesimal wave height $\delta y \rightarrow 0$, the speed becomes

$$\boxed{c_0^2 = gy} \tag{10.10}$$

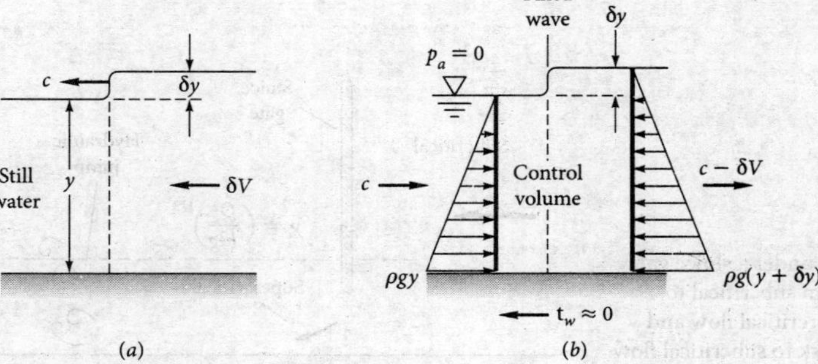

Fig. 10.4 Analysis of a small surface wave propagating into still shallow water; (*a*) moving wave, nonsteady frame; (*b*) fixed wave, inertial frame of reference.

This is the surface-wave equivalent of fluid sound speed a, and thus the Froude number in channel flow $Fr = V/c_0$ is the analog of the Mach number. For $y = 1$ m, $c_o = 3.1$ m/s.

As in gas dynamics, a channel flow can accelerate from subcritical to critical to supercritical flow and then return to subcritical flow through a sort of normal shock called a *hydraulic jump* (Sec. 10.5). This is illustrated in Fig. 10.5. The flow upstream of the sluice gate is subcritical. It then accelerates to critical and supercritical flow as it passes under the gate, which serves as a sort of "nozzle." Further downstream the flow "shocks" back to subcritical flow because the downstream "receiver" height is too high to maintain supercritical flow. Note the similarity with the nozzle gas flows of Fig. 9.12.

The critical depth $y_c = [Q^2/(b^2g)]^{1/3}$ is sketched as a dashed line in Fig. 10.5 for reference. Like the normal depth y_n, y_c is an important parameter in characterizing open-channel flow (see Sec. 10.4).

An excellent discussion of the various regimes of open-channel flow is given in Ref. 15.

10.2 Uniform Flow; The Chézy Formula

Uniform flow can occur in long, straight runs of constant slope and constant channel cross section. The water depth is constant at $y = y_n$, and the velocity is constant at $V = V_0$. Let the slope be $S_0 = \tan\theta$, where θ is the angle the bottom makes with the horizontal, considered positive for downhill flow. Then, Eq. (10.2), with $V_1 = V_2 = V_0$, becomes

$$h_f = z_1 - z_2 = S_0 L \tag{10.11}$$

where L is the horizontal distance between sections 1 and 2. The head loss thus balances the loss in height of the channel. The flow is essentially fully developed, so the Darcy-Weisbach relation, Eq. (6.10), holds

$$h_f = f\frac{L}{D_h}\frac{V_0^2}{2g} \qquad D_h = 4R_h \tag{10.12}$$

with $D_h = 4A/P$ used to accommodate noncircular channels. The geometry and notation for open-channel flow analysis are shown in Fig. 10.2.

By combining Eqs. (10.11) and (10.12) we obtain an expression for flow velocity in uniform channel flow:

$$V_0 = \left(\frac{8g}{f}\right)^{1/2} R_h^{1/2} S_0^{1/2} \tag{10.13}$$

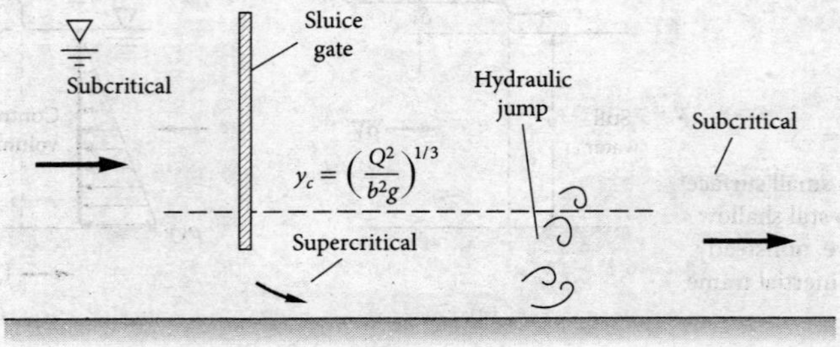

Fig. 10.5 Flow under a sluice gate accelerates from subcritical to critical to supercritical flow and then jumps back to subcritical flow.

For a given channel shape and bottom roughness, the quantity $(8g/f)^{1/2}$ is constant and can be denoted by C. Equation (10.13) becomes

$$V_0 = C(R_h S_0)^{1/2} \quad Q = CA(R_h S_0)^{1/2} \quad \tau_{avg} = \rho g R_h S_0 \tag{10.14}$$

These are called the *Chézy formulas,* first developed by the French engineer Antoine Chézy in conjunction with his experiments on the Seine River and the Courpalet Canal in 1769. The quantity C, called the *Chézy coefficient,* varies from about 30 $m^{1/2}/s$ for small, rough channels to 90 $m^{1/2}/s$ for large, smooth channels.

Over the past century a great deal of hydraulics research [16] has been devoted to the correlation of the Chézy coefficient with the roughness, shape, and slope of various open channels. Correlations are due to Ganguillet and Kutter in 1869, Manning in 1889, Bazin in 1897, and Powell in 1950 [16]. All these formulations are discussed in delicious detail in Ref. 2, Chap. 5. Here, we confine our treatment to Manning's correlation, the most popular.

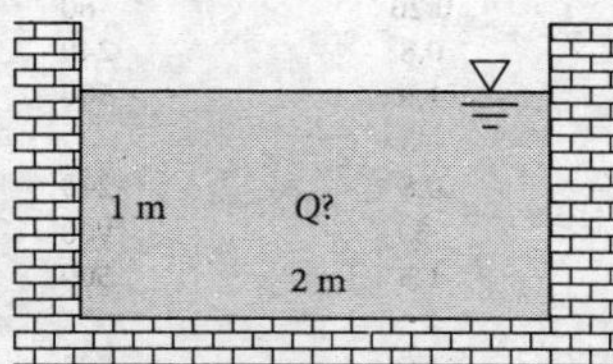

E10.1

EXAMPLE 10.1

A straight rectangular channel is 2 m wide and 1 m deep and laid on a slope of 2°. The friction factor is 0.022. Estimate the uniform flow rate in cubic meters per second.

Solution

- *System sketch:* The channel cross section is shown in Fig. E10.1.
- *Assumptions:* Steady, uniform channel flow with $\theta = 2°$.
- *Approach:* Evaluate the Chézy formula, Eq. (10.13) or (10.14).
- *Property values:* Please note that there are *no* fluid physical properties involved in the Chézy formula. Can you explain this?
- *Solution step:* Simply evaluate each term in the Chézy formula, Eq. (10.13):

$$C = \sqrt{\frac{8g}{f}} = \sqrt{\frac{8(9.81\ \text{m/s}^2)}{0.022}} = 60\ \text{m}^{1/2}/\text{s} \qquad A = by = (2\ \text{m})(1\ \text{m}) = 2\ \text{m}^2$$

$$R_h = \frac{A}{P_{wet}} = \frac{2\ \text{m}^2}{(1\ \text{m} + 2\ \text{m} + 1\ \text{m})\ \text{meters}} = 0.5\ \text{m} \qquad S_0 = \tan(\theta) = \tan(2°)$$

Then $Q = CAR_h^{1/2}S_0^{1/2} = (60\ \text{m}^{1/2}/\text{s})(2\ \text{m}^2)(0.5\ \text{m})^{1/2}(\tan 2°)^{1/2} \approx 16\ \text{m}^3/\text{s}$ *Ans.*

- *Comments:* Uniform flow estimates are straightforward if the geometry is simple. Results are independent of water density and viscosity because the flow is fully rough and driven by gravity. Note the high flow rate, larger than some rivers. Two degrees is a substantial channel slope.

The Manning Roughness Correlation

The most fundamentally sound approach to the Chézy formula is to use Eq. (10.13) with f estimated from the Moody friction factor chart, Fig. 6.13. Indeed, the open-channel research establishment [18] strongly recommends use of the friction factor in all calculations. Since typical channels are large and rough, we would generally use the fully rough turbulent flow limit of Eq. (6.48)

$$f \approx \left(2.0 \log \frac{14.8 R_h}{\varepsilon}\right)^{-2} \tag{10.15}$$

where ε is the roughness height, with typical values listed in Table 10.1.

Table 10.1 Experimental Values of Manning's n Factor*

	n	Average roughness height ϵ ft	Average roughness height ϵ mm
Artificial lined channels:			
Glass	0.010 ± 0.002	0.0011	0.3
Brass	0.011 ± 0.002	0.0019	0.6
Steel, smooth	0.012 ± 0.002	0.0032	1.0
Painted	0.014 ± 0.003	0.0080	2.4
Riveted	0.015 ± 0.002	0.012	3.7
Cast iron	0.013 ± 0.003	0.0051	1.6
Concrete, finished	0.012 ± 0.002	0.0032	1.0
Unfinished	0.014 ± 0.002	0.0080	2.4
Planed wood	0.012 ± 0.002	0.0032	1.0
Clay tile	0.014 ± 0.003	0.0080	2.4
Brickwork	0.015 ± 0.002	0.012	3.7
Asphalt	0.016 ± 0.003	0.018	5.4
Corrugated metal	0.022 ± 0.005	0.12	37
Rubble masonry	0.025 ± 0.005	0.26	80
Excavated earth channels:			
Clean	0.022 ± 0.004	0.12	37
Gravelly	0.025 ± 0.005	0.26	80
Weedy	0.030 ± 0.005	0.8	240
Stony, cobbles	0.035 ± 0.010	1.5	500
Natural channels:			
Clean and straight	0.030 ± 0.005	0.8	240
Sluggish, deep pools	0.040 ± 0.010	3	900
Major rivers	0.035 ± 0.010	1.5	500
Floodplains:			
Pasture, farmland	0.035 ± 0.010	1.5	500
Light brush	0.05 ± 0.02	6	2,000
Heavy brush	0.075 ± 0.025	15	5,000
Trees	0.15 ± 0.05	?	?

*A more complete list is given in Ref. 2, pp. 110–113.

In spite of the attractiveness of this friction factor approach, most engineers prefer to use a simple (dimensional) correlation published in 1891 by Robert Manning [17], an Irish engineer. In tests with real channels, Manning found that the Chézy coefficient C increased approximately as the sixth root of the channel size. He proposed the simple formula

$$C = \left(\frac{8g}{f}\right)^{1/2} \approx \alpha \frac{R_h^{1/6}}{n} \tag{10.16}$$

where n is a roughness parameter. Since the formula is clearly not dimensionally consistent, it requires a conversion factor α that changes with the system of units used:

$$\alpha = 1.0 \quad \text{SI units} \qquad \alpha = 1.486 \quad \text{BG units} \tag{10.17}$$

Recall that we warned about this awkwardness in Example 1.4. You may verify that α is the cube root of the conversion factor between the meter and your chosen length scale: In BG units, $\alpha = (3.2808 \text{ ft/m})^{1/3} = 1.486$.[2]

[2]An interesting discussion of the history and "dimensionality" of Manning's formula is given in Ref. 2, pp. 98–99.

The Manning formula for uniform flow velocity is thus

$$V_0\,(\text{m/s}) \approx \frac{1.0}{n}\,[R_h(\text{m})]^{2/3}S_0^{1/2}$$
$$V_0\,(\text{ft/s}) \approx \frac{1.486}{n}\,[R_h(\text{ft})]^{2/3}S_0^{1/2} \tag{10.18}$$

The channel slope S_0 is dimensionless, and n is taken to be the same in both systems. The volume flow rate simply multiplies this result by the area:

Uniform flow: $$Q = V_0 A \approx \frac{\alpha}{n} A R_h^{2/3} S_0^{1/2} \tag{10.19}$$

Experimental values of n (and the corresponding roughness height) are listed in Table 10.1 for various channel surfaces. There is a factor-of-15 variation from a smooth glass surface ($n \approx 0.01$) to a tree-lined floodplain ($n \approx 0.15$). Due to the irregularity of typical channel shapes and roughness, the scatter bands in Table 10.1 should be taken seriously. For routine calculations, always use the average roughness in Table 10.1.

Since Manning's sixth-root size variation is not exact, real channels can have a variable n depending on the water depth. The Mississippi River near Memphis, Tennessee, has $n \approx 0.032$ at 12 m flood depth, 0.030 at normal 6 m depth, and 0.040 at 1.5 m low-stage depth. Seasonal vegetative growth and factors such as bottom erosion can also affect the value of n. Even nearly identical man-made channels can vary. Brater et al. [19] report that U.S. Bureau of Reclamation tests, on large concrete-lined canals, yielded values of n ranging from 0.012 to 0.017.

EXAMPLE 10.2

Engineers find that the most efficient rectangular channel (maximum uniform flow for a given area) flows at a depth equal to half the bottom width. Consider a rectangular brickwork channel laid on a slope of 0.006. What is the best bottom width for a flow rate of 3 m³/s?

Solution

- *Assumptions:* Uniform flow in a straight channel of constant of slope $S = 0.006$.
- *Approach:* Use the Manning formula in SI units, Eq. (10.19), to predict the flow rate.
- *Property values:* For brickwork, from Table 10.1, the roughness factor $n \approx 0.015$.
- *Solution:* For bottom width b, take the water depth to be $y = b/2$. Equation (10.19) becomes

$$A = by = b(b/2) = \frac{b^2}{2} \qquad R_h = \frac{A}{P} = \frac{by}{b+2y} = \frac{b^2/2}{b+2(b/2)} = \frac{b}{4}$$

$$Q = \frac{\alpha}{n} A R_h^{2/3} S^{1/2} = \frac{1}{0.015}\left(\frac{b^2}{2}\right)\left(\frac{b}{4}\right)^{2/3}(0.006)^{1/2} = 3\ \text{m}^3/\text{s}$$

Clean this up: $b^{8/3} = 2.928$ solve for $b \approx 1.5$ m *Ans.*

- *Comments:* The Manning approach is simple and effective. The Moody friction factor method, Eq. (10.14), requires laborious iteration and leads to a result $b \approx 1.5$ m.

Normal Depth Estimates

With water depth y known, the computation of Q is quite straightforward. However, if Q is given, the computation of the normal depth y_n may require iteration. Since the normal depth is a characteristic flow parameter, this is an important type of problem.

The *normal depth*, y_n, is the depth, in uniform flow, of the water in a straight, constant-area, constant-slope channel. It varies with the flow rate and is a useful reference depth, calculated by solving Eq. (10.19) when Q is given.

EXAMPLE 10.3

The asphalt-lined trapezoidal channel in Fig. E10.3 carries 9 m^3/s of water under uniform flow conditions when $S = 0.0015$. What is the normal depth y_n?

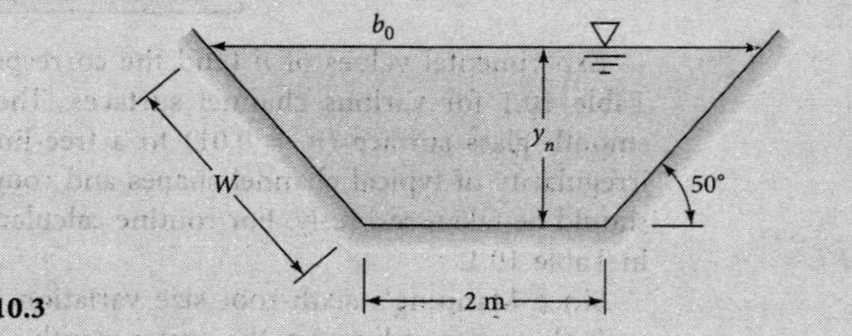

Note: See Fig. 10.7 for generalized trapezoid notation.

E10.3

Solution using Excel

From Table 10.1, for asphalt, $n \approx 0.016$. The area and hydraulic radius are functions of y_n, which is unknown:

$$b_0 = 2\text{ m} + 2y_n \cot 50° \qquad A = \tfrac{1}{2}(2 + b_0)y_n = 2y_n + y_n^2 \cot 50°$$

$$P = 2 + 2W = 2 + 2y_n \csc 50°$$

From Manning's formula (10.19) with a known $Q = 9$ m^3/s we have

$$9 = \frac{1}{0.016}(2y_n + y_n^2 \cot 50°)\left(\frac{2y_n + y_n^2 \cot 50°}{2 + 2y_n \csc 50°}\right)^{2/3}(0.0015)^{1/2}$$

or

$$(2y_n + y_n^2 \cot 50°)^{5/3} = 3.718(2 + 2y_n \csc 50°)^{2/3} \qquad (1)$$

One can iterate Eq. (1) by hand laboriously and eventually find $y_n \approx 1.387$ m. However, it is a good candidate for iteration in Excel. One could iterate the Chézy formula directly and find the value of y_n that gives a flow rate of 9 m^3/s. The writer chose to deal with the rearranged version, Eq. (1), guessing values of y_n until the left-hand side equals the right-hand side. The writer's first guess was $y_n = 3$ m, which was much too deep. Four more guesses achieved quite accurate convergence, in the following table:

	y_n	Eq. (1) LHS	Eq. (1) RHS	LHS − RHS
	A	B	C	D
1	3	77.03	17.06	59.97
2	2	27.83	13.89	13.93
3	1.5	14.08	12.16	1.92
4	1.4	12.02	11.80	0.21
5	**1.387**	**11.76**	**11.75**	0.01

At 9 m^3/s, the normal depth for this channel is $\quad y_n = 1.387$ m $\quad$ *Ans.*

For later use, we might also list the other properties of this channel:

$$b_o = 4.328 \text{ m}; \quad P = 5.621 \text{ m}; \quad A = 4.388 \text{ m}^2; \quad R_h = 0.781 \text{ m}$$

Hand calculation is too cumbersome for open-channel problems where the depth is unknown.

Uniform Flow in a Partly Full Circular Pipe

Consider the partially full pipe of Fig. 10.6*a* in uniform flow. The maximum velocity and flow rate actually occur before the pipe is completely full. In terms of the pipe radius R and the angle θ up to the free surface, the geometric properties are

$$A = R^2\left(\theta - \frac{\sin 2\theta}{2}\right) \quad P = 2R\theta \quad R_h = \frac{R}{2}\left(1 - \frac{\sin 2\theta}{2\theta}\right)$$

The Manning formulas (10.19) predict a uniform flow as follows:

$$V_0 \approx \frac{\alpha}{n}\left[\frac{R}{2}\left(1 - \frac{\sin 2\theta}{2\theta}\right)\right]^{2/3} S_0^{1/2} \quad Q = V_0 R^2\left(\theta - \frac{\sin 2\theta}{2}\right) \tag{10.20}$$

For a given n and slope S_0, we may plot these two relations versus y/D in Fig. 10.6*b*. There are two different maxima, as follows:

$$\begin{aligned} V_{\max} &= 0.718\frac{\alpha}{n}R^{2/3}S_0^{1/2} \quad \text{at} \quad \theta = 128.73° \quad \text{and} \quad y = 0.813D \\ Q_{\max} &= 2.129\frac{\alpha}{n}R^{8/3}S_0^{1/2} \quad \text{at} \quad \theta = 151.21° \quad \text{and} \quad y = 0.938D \end{aligned} \tag{10.21}$$

As shown in Fig. 10.6*b*, the maximum velocity is 14 percent more than the velocity when running full, and similarly the maximum discharge is 8 percent more. Since real pipes running nearly full tend to have somewhat unstable flow, these differences are not that significant.

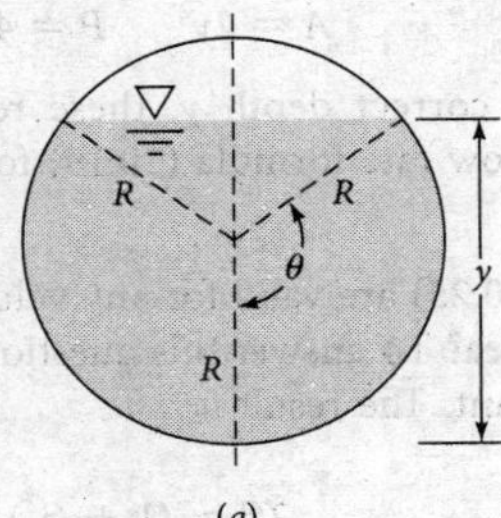

(*a*)

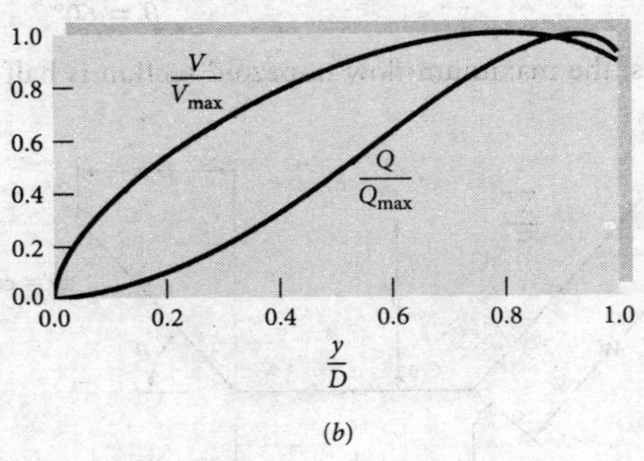

(*b*)

Fig. 10.6 Uniform flow in a partly full circular channel: (*a*) geometry; (*b*) velocity and flow rate versus depth.

10.3 Efficient Uniform-Flow Channels

The engineering design of an open channel has many parameters. If the channel surface can erode or scour, a low-velocity design might be sought. A dirt channel could be planted with grass to minimize erosion. For nonerodible surfaces, construction and lining costs might dominate, suggesting a cross section of minimum wetted perimeter. Nonerodible channels can be designed for maximum flow.

The simplicity of Manning's formulation (10.19) enables us to analyze channel flows to determine the most efficient low-resistance sections for given conditions. The most common problem is that of maximizing R_h for a given flow area and discharge. Since $R_h = A/P$, maximizing R_h for given A is the same as minimizing the wetted perimeter P. There is no general solution for arbitrary cross sections, but an analysis of the trapezoid section will show the basic results.

Consider the generalized symmetric trapezoid of angle θ in Fig. 10.7. For a given side angle θ, the flow area is

$$A = by + \beta y^2 \qquad \beta = \cot\theta \tag{10.22}$$

The wetted perimeter is

$$P = b + 2W = b + 2y(1 + \beta^2)^{1/2} \tag{10.23}$$

Eliminating b between (10.22) and (10.23) gives

$$P = \frac{A}{y} - \beta y + 2y(1 + \beta^2)^{1/2} \tag{10.24}$$

To minimize P, evaluate dP/dy for constant A and β and set equal to zero. The result is

$$A = y^2[2(1 + \beta^2)^{1/2} - \beta] \quad P = 4y(1 + \beta^2)^{1/2} - 2\beta y \quad R_h = \tfrac{1}{2}y \tag{10.25}$$

The last result is very interesting: For any angle θ, the most efficient cross section for uniform flow occurs when the hydraulic radius is half the depth.

Since a rectangle is a trapezoid with $\beta = 0$, the most efficient rectangular section is such that

$$A = 2y^2 \qquad P = 4y \qquad R_h = \tfrac{1}{2}y \qquad b = 2y \tag{10.26}$$

To find the correct depth y, these relations must be solved in conjunction with Manning's flow rate formula (10.19) for the given discharge Q.

Best Trapezoid Angle

Equations (10.25) are valid for any value of β. What is the best value of β for a given depth and area? To answer this question, evaluate $dP/d\beta$ from Eq. (10.24) with A and y held constant. The result is

$$2\beta = (1 + \beta^2)^{1/2} \qquad \beta = \cot\theta = \frac{1}{3^{1/2}}$$

or

$$\theta = 60^\circ \tag{10.27}$$

Thus, the maximum-flow trapezoid section is half a hexagon.

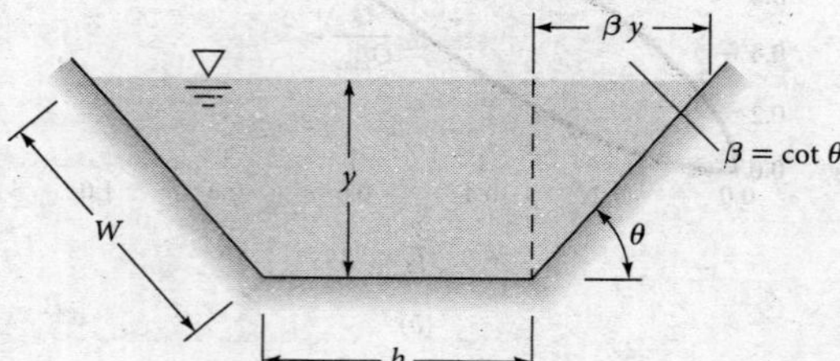

Fig. 10.7 Geometry of a trapezoidal channel section.

Similar calculations with a circular channel section running partially full show best efficiency for a semicircle, $y = \frac{1}{2}D$. In fact, the semicircle is the best of all possible channel sections (minimum wetted perimeter for a given flow area). The percentage improvement over, say, half a hexagon is very slight, however.

EXAMPLE 10.4

(*a*) What are the best dimensions y and b for a rectangular brick channel designed to carry 5 m³/s of water in uniform flow with $S_0 = 0.001$? (*b*) Compare results with a half-hexagon and semicircle.

Solution

Part (a) From Eq. (10.26), $A = 2y^2$ and $R_h = \frac{1}{2}y$. Manning's formula (10.19) in SI units gives, with $n \approx 0.015$ from Table 10.1,

$$Q = \frac{1.0}{n} A R_h^{2/3} S_0^{1/2} \quad \text{or} \quad 5\ \text{m}^3/\text{s} = \frac{1.0}{0.015}(2y^2)\left(\frac{1}{2}y\right)^{2/3}(0.001)^{1/2}$$

which can be solved for

$$y^{8/3} = 1.882\ \text{m}^{8/3}$$

$$y = 1.27\ \text{m} \qquad \textit{Ans.}$$

The proper area and width are

$$A = 2y^2 = 3.21\ \text{m}^2 \qquad b = \frac{A}{y} = 2.53\ \text{m} \qquad \textit{Ans.}$$

Part (b) It is constructive to see what flow rate a half-hexagon and semicircle would carry for the same area of 3.214 m².

For the half-hexagon (HH), with $\beta = 1/3^{1/2} = 0.577$, Eq. (10.25) predicts

$$A = y_{HH}^2[2(1 + 0.577^2)^{1/2} - 0.577] = 1.732 y_{HH}^2 = 3.214$$

or $y_{HH} = 1.362$ m, whence $R_h = \frac{1}{2}y = 0.681$ m. The half-hexagon flow rate is thus

$$Q = \frac{1.0}{0.015}(3.214)(0.681)^{2/3}(0.001)^{1/2} = 5.25\ \text{m}^3/\text{s}$$

or about 5 percent more than that for the rectangle.

For a semicircle, $A = 3.214\ \text{m}^2 = \pi D^2/8$, or $D = 2.861$ m, whence $P = \frac{1}{2}\pi D = 4.494$ m and $R_h = A/P = 3.214/4.494 = 0.715$ m. The semicircle flow rate will thus be

$$Q = \frac{1.0}{0.015}(3.214)(0.715)^{2/3}(0.001)^{1/2} = 5.42\ \text{m}^3/\text{s}$$

or about 8 percent more than that of the rectangle and 3 percent more than that of the half-hexagon.

10.4 Specific Energy; Critical Depth

The total head of any incompressible flow is the sum of its velocity head $\alpha V^2/(2g)$, pressure head p/γ, and potential head z. For open-channel flow, surface pressure is everywhere atmospheric, so channel energy is a balance between velocity and elevation head only. Since the flow is turbulent, we assume that $\alpha \approx 1$—recall Eq. (3.77). The final result is the quantity called *specific energy E*, as suggested by Bakhmeteff [1] in 1913:

$$E = y + \frac{V^2}{2g} \tag{10.28}$$

where y is the water depth. It is seen from Fig. 10.8 that E is the height of the *energy grade line* (EGL) above the channel bottom. For a given flow rate, there are usually two states possible, called *alternate states,* for the same specific energy. There is also a minimum energy, $E_{\min}$, which corresponds to a Froude number of unity.

Rectangular Channels

Consider the possible states at a given location. Let $q = Q/b = Vy$ be the discharge per unit width of a rectangular channel. Then, with q constant, Eq. (10.28) becomes

$$E = y + \frac{q^2}{2gy^2} \qquad q = \frac{Q}{b} \tag{10.29}$$

Figure 10.8 is a plot of y versus E for constant q from Eq. (10.29). There is a minimum value of E at a certain value of y called the *critical depth.* By setting $dE/dy = 0$ at constant q, we find that $E_{\min}$ occurs at

$$y = y_c = \left(\frac{q^2}{g}\right)^{1/3} = \left(\frac{Q^2}{b^2 g}\right)^{1/3} \tag{10.30}$$

The associated minimum energy is

$$E_{\min} = E(y_c) = \tfrac{3}{2}y_c \tag{10.31}$$

The depth y_c corresponds to channel velocity equal to the shallow-water wave propagation speed C_0 from Eq. (10.10). To see this, rewrite Eq. (10.30) as

$$q^2 = gy_c^3 = (gy_c)y_c^2 = V_c^2 y_c^2 \tag{10.32}$$

By comparison, it follows that the critical channel velocity is

$$V_c = (gy_c)^{1/2} = C_0 \qquad \text{Fr} = 1 \tag{10.33}$$

For $E < E_{\min}$ no solution exists in Fig. 10.8, and thus such a flow is impossible physically. For $E > E_{\min}$ two solutions are possible: (1) large depth with $V < V_c$, called *subcritical,* and (2) small depth with $V > V_c$, called *supercritical.* In subcritical flow, disturbances can propagate upstream because wave speed $C_0 > V$. In supercritical flow, waves are swept downstream: Upstream is a zone of silence, and a small

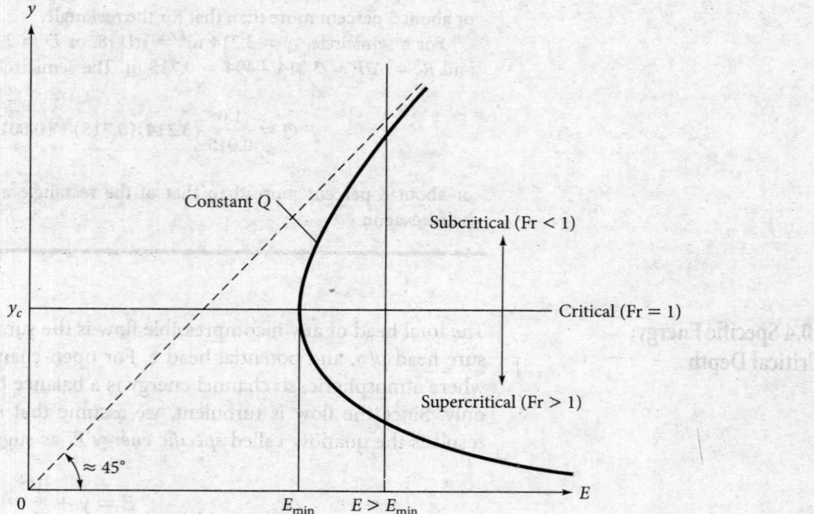

Fig. 10.8 Illustration of a specific energy curve. The curve for each flow rate Q has a minimum energy corresponding to critical flow. For energy greater than minimum, there are two *alternate* flow states, one subcritical and one supercritical.

obstruction in the flow will create a wedge-shaped wave exactly analogous to the Mach waves in Fig. 9.18*c*. The angle of these waves must be

$$\mu = \sin^{-1}\frac{c_0}{V} = \sin^{-1}\frac{(gy)^{1/2}}{V} = \sin^{-1}\left(\frac{1}{\text{Fr}}\right) \quad (10.34)$$

The wave angle and the depth can thus be used as a simple measurement of supercritical flow velocity.

Note from Fig. 10.8 that small changes in E near $E_{\min}$ cause a large change in the depth y, by analogy with small changes in duct area near the sonic point in Fig. 9.7. Thus, critical flow is neutrally stable and is often accompanied by waves and undulations in the free surface. Channel designers should avoid long runs of near-critical flow.

The Water Channel Compressible Flow Analogy

The simple water wave expression of Eq. (10.34), $\sin \mu = 1/\text{Fr}$, can be extended into a general, often qualitative, analogy between gas dynamics and water channel flow [21, Chap. 11]. The analogy is derived from small-disturbance theory and has the results

Froude number is analogous to Mach number.
Water depth is analogous to gas density.
Water depth squared is analogous to gas pressure.
Hydraulic jumps are analogous to gas shock waves.

Both experiments and CFD results have exploited this analogy [37, 38]. A further result is that small-disturbance channel flow is equivalent to a fictitious gas with a specific heat ratio $k = 2.0$. However, for larger disturbances, such as shock waves, the numerical results differ and, oddly, are more accurate for $k = 1.4$ than for $k = 2.0$. A few examples are assigned here as Probs. P10.85, P10.88, and P10.91.

EXAMPLE 10.5

A wide rectangular clean-earth channel has a flow rate $q = 20\ \text{m}^3/(\text{s}\cdot\text{m})$. (*a*) What is the critical depth? (*b*) What type of flow exists if $y = 2$ m?

Solution

Part (a) The critical depth is independent of channel roughness and simply follows from Eq. (10.30):

$$y_c = \left(\frac{q^2}{g}\right)^{1/3} = \left(\frac{20^2}{9.81}\right)^{1/3} = 3.44 \text{ m} \qquad \textit{Ans. (a)}$$

Part (b) If the actual depth is 2 m, which is less than y_c, the flow must be *supercritical*. *Ans. (b)*

Nonrectangular Channels

If the channel width varies with y, the specific energy must be written in the form

$$E = y + \frac{Q^2}{2gA^2} \quad (10.35)$$

The critical point of minimum energy occurs where $dE/dy = 0$ at constant Q. Since $A = A(y)$, Eq. (10.35) yields, for $E = E_{\min}$,

$$\frac{dA}{dy} = \frac{gA^3}{Q^2} \quad (10.36)$$

But $dA = b_0\,dy$, where b_0 is the channel width at the free surface. Therefore, Eq. (10.36) is equivalent to

$$A_c = \left(\frac{b_0 Q^2}{g}\right)^{1/3} \tag{10.37a}$$

$$V_c = \frac{Q}{A_c} = \left(\frac{gA_c}{b_0}\right)^{1/2} \tag{10.37b}$$

For a given channel shape $A(y)$ and $b_0(y)$ and a given Q, Eqs. (10.37) have to be solved by trial and error or by Excel iteration to find the critical area A_c, from which V_c can be computed.

By comparing the actual depth and velocity with the critical values, we can determine the local flow condition.

$y > y_c$, $V < V_c$: subcritical flow (Fr < 1)

$y = y_c$, $V = V_c$: critical flow (Fr $= 1$)

$y < y_c$, $V > V_c$: supercritical flow (Fr < 1)

Note that V_c is equal to the speed of propagation c of a shallow-water wave in the channel and is dependent upon the depth, as in Fig. 10.4*a*. For a rectangular channel, $c = (gy)^{1/2}$.

Critical Uniform Flow: The Critical Slope

If a critical channel flow is also moving uniformly (at constant depth), it must correspond to a *critical slope* S_c, with $y_n = y_c$. This condition is analyzed by equating Eq. (10.37a) to the Chézy (or Manning) formula:

$$Q^2 = \frac{gA_c^3}{b_0} = C^2 A_c^2 R_h S_c = \frac{\alpha^2}{n^2} A_c^2 R_h^{4/3} S_c$$

or
$$S_c = \frac{n^2 g A_c}{\alpha^2 b_0 R_{hc}^{4/3}} = \frac{n^2 V_c^2}{\alpha^2 R_{hc}^{4/3}} = \frac{n^2 g}{\alpha^2 R_{hc}^{1/3}} \frac{P}{b_0} = \frac{f}{8} \frac{P}{b_0} \tag{10.38}$$

where α^2 equals 1.0 for SI units and 2.208 for BG units. Equation (10.38) is valid for any channel shape. For a wide rectangular channel, $b_0 \gg y_c$, the formula reduces to

Wide rectangular channel: $$S_c \approx \frac{n^2 g}{\alpha^2 y_c^{1/3}} \approx \frac{f}{8}$$

This is a special case, a reference point. In most channel flows $y_n \neq y_c$. For fully rough turbulent flow, the critical slope varies between 0.002 and 0.008.

EXAMPLE 10.6

The 50° triangular channel in Fig. E10.6 has a flow rate $Q = 16\ \text{m}^3/\text{s}$. Compute (*a*) y_c, (*b*) V_c, and (*c*) S_c if $n = 0.018$.

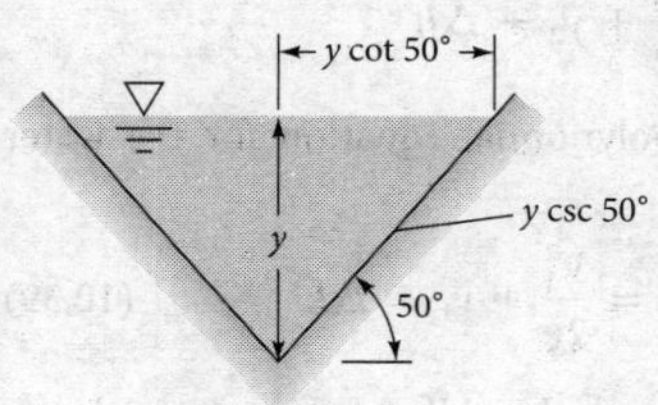

E10.6

Solution

Part (a) This is an easy cross section because all geometric quantities can be written directly in terms of depth y:

$$P = 2y \csc 50° \qquad A = y^2 \cot 50°$$
$$R_h = \tfrac{1}{2} y \cos 50° \qquad b_0 = 2y \cot 50° \tag{1}$$

The critical flow condition satisfies Eq. (10.37*a*):

$$gA_c^3 = b_0 Q^2$$

or

$$g(y_c^2 \cot 50°)^3 = (2y_c \cot 50°)Q^2$$

$$y_c = \left(\frac{2Q^2}{g \cot^2 50°}\right)^{1/5} = \left[\frac{2(16)^2}{9.81(0.839)^2}\right]^{1/5} = 2.37 \text{ m} \qquad \textit{Ans. (a)}$$

Part (b) With y_c known, from Eqs. (1) we compute $P_c = 6.18$ m, $R_{hc} = 0.760$ m, $A_c = 4.70$ m^2, and $b_{0c} =$ 3.97 m. The critical velocity from Eq. (10.37*b*) is

$$V_c = \frac{Q}{A_c} = \frac{16 \text{ m}^3/\text{s}}{4.70 \text{ m}^2} = 3.41 \text{ m/s} \qquad \textit{Ans. (b)}$$

Part (c) With $n = 0.018$, we compute from Eq. (10.38) a critical slope:

$$S_c = \frac{gn^2 P}{\alpha^2 R_h^{1/3} b_0} = \frac{9.81(0.018)^2(6.18)}{1.0(0.760)^{1/3}(3.97)} = 0.00542 \qquad \textit{Ans. (c)}$$

Frictionless Flow over a Bump

A rough analogy to compressible gas flow in a nozzle (Fig. 9.12) is open-channel flow over a bump, as in Fig. 10.9*a*. The behavior of the free surface is sharply different according to whether the approach flow is subcritical or supercritical. The height of

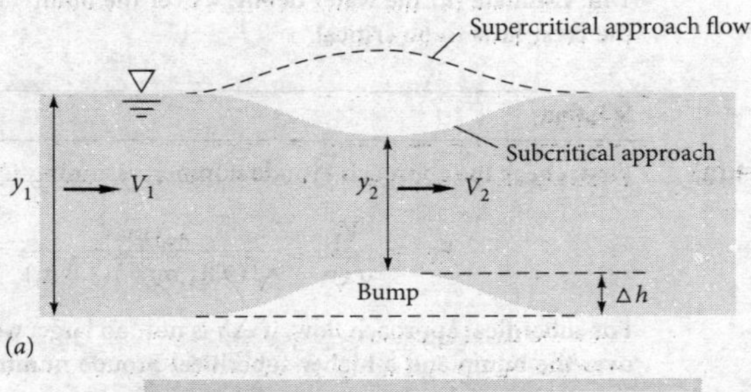

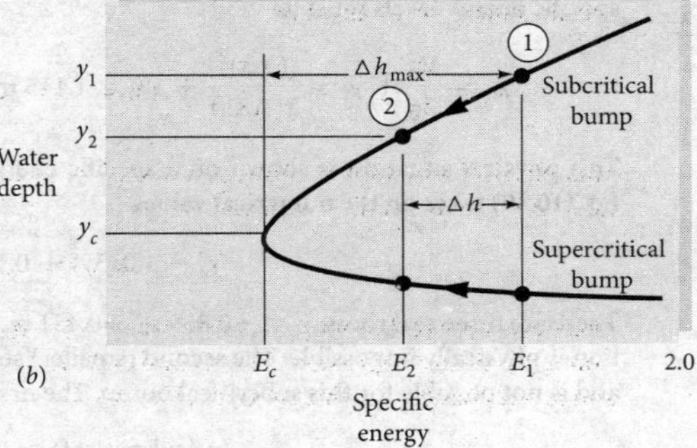

Fig. 10.9 Frictionless two-dimensional flow over a bump: (*a*) definition sketch showing Froude number dependence; (*b*) specific energy plot showing bump size and water depths.

the bump also can change the character of the results. For frictionless two-dimensional flow, sections 1 and 2 in Fig. 10.9*a* are related by continuity and momentum:

$$V_1 y_1 = V_2 y_2 \qquad \frac{V_1^2}{2g} + y_1 = \frac{V_2^2}{2g} + y_2 + \Delta h$$

Eliminating V_2 between these two gives a cubic polynomial equation for the water depth y_2 over the bump:

$$y_2^3 - E_2 y_2^2 + \frac{V_1^2 y_1^2}{2g} = 0 \qquad \text{where } E_2 = \frac{V_1^2}{2g} + y_1 - \Delta h \tag{10.39}$$

This equation has one negative and two positive solutions if Δh is not too large. Its behavior is illustrated in Fig. 10.9*b* and depends on whether condition 1 is on the upper or lower leg of the energy curve. The specific energy E_2 is exactly Δh less than the approach energy E_1, and point 2 will lie on the same leg of the curve as E_1. A subcritical approach, $Fr_1 < 1$, will cause the water level to decrease at the bump. Supercritical approach flow, $Fr_1 > 1$, causes a water level increase over the bump.

If the bump height reaches $\Delta h_{max} = E_1 - E_c$, as illustrated in Fig. 10.9*b*, the flow at the crest will be exactly critical ($Fr = 1$). If $\Delta h > \Delta h_{max}$, there are no physically correct solutions to Eq. (10.39). That is, a bump too large will "choke" the channel and cause frictional effects, typically a hydraulic jump (Sec. 10.5).

These bump arguments are reversed if the channel has a *depression* ($\Delta h < 0$): Subcritical approach flow will cause a water level rise and supercritical flow a fall in depth. Point 2 will be $|\Delta h|$ to the right of point 1, and critical flow cannot occur.

EXAMPLE 10.7

Water flow in a wide channel approaches a 10-cm-high bump at 1.5 m/s and a depth of 1 m. Estimate (*a*) the water depth y_2 over the bump and (*b*) the bump height that will cause the crest flow to be critical.

Solution

Part (a) First, check the approach Froude number, assuming $C_0 = \sqrt{gy}$:

$$Fr_1 = \frac{V_1}{\sqrt{gy_1}} = \frac{1.5 \text{ m/s}}{\sqrt{(9.81 \text{ m/s}^2)(1.0 \text{ m})}} = 0.479 \qquad \text{(subcritical)}$$

For subcritical approach flow, if Δh is not too large, we expect a depression in the water level over the bump and a higher subcritical Froude number at the crest. With $\Delta h = 0.1$ m, the specific energy levels must be

$$E_1 = \frac{V_1^2}{2g} + y_1 = \frac{(1.5)^2}{2(9.81)} + 1.0 = 1.115 \text{ m} \qquad E_2 = E_1 - \Delta h = 1.015 \text{ m}$$

This physical situation is shown on a specific energy plot in Fig. E10.7. With y_1 in meters, Eq. (10.39) takes on the numerical values

$$y_2^3 - 1.015 y_2^2 + 0.115 = 0$$

There are three real roots: $y_2 = +0.859$ m, $+0.451$ m, and -0.296 m. The third (negative) solution is physically impossible. The second (smaller) solution is the *supercritical* condition for E_2 and is not possible for this subcritical bump. The first solution is correct:

$$y_2(\text{subcritical}) \approx 0.859 \text{ m} \qquad \textit{Ans. (a)}$$

The surface level has dropped by $y_1 - y_2 - \Delta h = 1.0 - 0.859 - 0.1 = 0.041$ m. The crest velocity is $V_2 = V_1 y_1 / y_2 = 1.745$ m/s. The Froude number at the crest is $Fr_2 = 0.601$. Flow downstream of the bump is subcritical. These flow conditions are shown in Fig. E10.7.

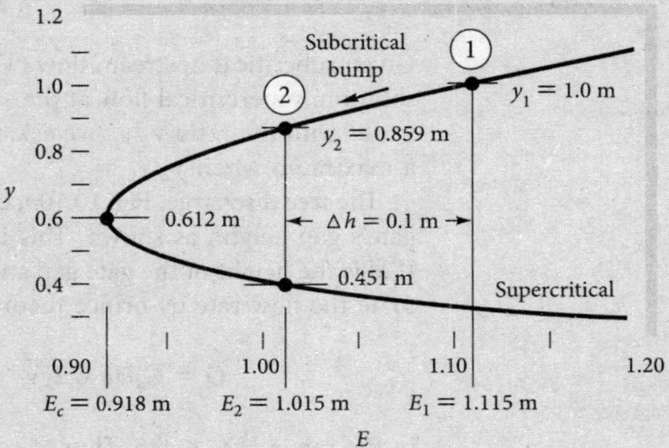

E10.7

Part (b) For critical flow in a wide channel, with $q = Vy = 1.5\ \text{m}^2/\text{s}$, from Eq. (10.31),

$$E_{2,\min} = E_c = \frac{3}{2} y_c = \frac{3}{2}\left(\frac{q^2}{g}\right)^{1/3} = \frac{3}{2}\left[\frac{(1.5\ \text{m}^2/\text{s})^2}{9.81\ \text{m/s}^2}\right]^{1/3} = 0.918\ \text{m}$$

Therefore, the maximum height for frictionless flow over this particular bump is

$$\Delta h_{\max} = E_1 - E_{2,\min} = 1.115 - 0.918 = 0.197\ \text{m} \qquad \textit{Ans. (b)}$$

For this bump, the solution of Eq. (10.39) is $y_2 = y_c = 0.612$ m, and the Froude number is unity at the crest. At critical flow the surface level has dropped by $y_1 - y_2 - \Delta h = 0.191$ m.

Flow under a Sluice Gate

A sluice gate is a bottom opening in a wall, as sketched in Fig. 10.10*a*, commonly used in control of rivers and channel flows. If the flow is allowed free discharge through the gap, as in Fig. 10.10*a*, the flow smoothly accelerates from subcritical (upstream) to critical (near the gap) to supercritical (downstream). The gate is then analogous to a converging–diverging nozzle in gas dynamics, as in Fig. 9.12, operating at its *design condition* (similar to point *H* in Fig. 9.12*b*).

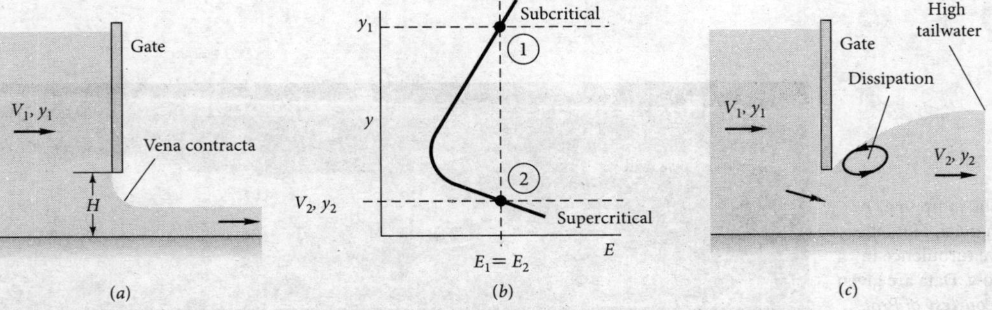

Fig. 10.10 Flow under a sluice gate passes through critical flow: (*a*) free discharge with vena contracta; (*b*) specific energy for free discharge; (*c*) dissipative flow under a drowned gate.

For free discharge, friction may be neglected, and since there is no bump ($\Delta h = 0$), Eq. (10.39) applies with $E_1 = E_2$:

$$y_2^3 - \left(\frac{V_1^2}{2g} + y_1\right)y_2^2 + \frac{V_1^2 y_1^2}{2g} = 0 \tag{10.40}$$

Given subcritical upstream flow (V_1, y_1), this cubic equation has only one positive real solution: supercritical flow at the same specific energy, as in Fig. 10.10*b*. The flow rate varies with the ratio y_2/y_1; we ask, as a problem exercise, to show that the flow rate is a maximum when $y_2/y_1 = \frac{2}{3}$.

The free discharge, Fig. 10.10*a*, contracts to a depth y_2 about 40 percent less than the gate's gap height, as shown. This is similar to a free *orifice* discharge, as in Fig. 6.39. If H is the height of the gate gap and b is the gap width into the paper, we can approximate the flow rate by orifice theory:

$$Q = C_d H b \sqrt{2 g y_1} \qquad \text{where} \qquad C_d \approx \frac{0.61}{\sqrt{1 + 0.61 H/y_1}} \tag{10.41}$$

in the range $H/y_1 < 0.5$. Thus, a continuous variation in flow rate is accomplished by raising the gate.

If the tailwater is high, as in Fig. 10.10*c*, free discharge is not possible. The sluice gate is said to be *drowned* or partially drowned. There will be energy dissipation in the exit flow, probably in the form of a drowned hydraulic jump, and the downstream flow will return to subcritical. Equations (10.40) and (10.41) do not apply to this situation, and experimental discharge correlations are necessary [3, 19]. See Prob. P10.77.

10.5 The Hydraulic Jump

In open-channel flow, a supercritical flow can change quickly back to a subcritical flow by passing through a hydraulic jump, as in Fig. 10.5. The upstream flow is fast and shallow, and the downstream flow is slow and deep, analogous to the normal shock wave of Fig. 9.8. Unlike the infinitesimally thin normal shock, the hydraulic jump is quite thick, ranging in length from 4 to 6 times the downstream depth y_2 [20].

Being extremely turbulent and agitated, the hydraulic jump is a very effective energy dissipator and is a feature of stilling-basin and spillway applications [20]. Figure 10.11 shows the jump formed in a laboratory open channel. It is very important that such jumps be located on specially designed aprons; otherwise the channel bottom will be badly scoured by the agitation. Jumps also mix fluids very effectively and have application to sewage and water treatment designs.

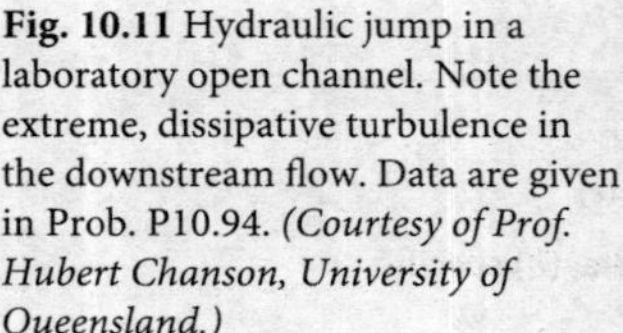

Fig. 10.11 Hydraulic jump in a laboratory open channel. Note the extreme, dissipative turbulence in the downstream flow. Data are given in Prob. P10.94. *(Courtesy of Prof. Hubert Chanson, University of Queensland.)*

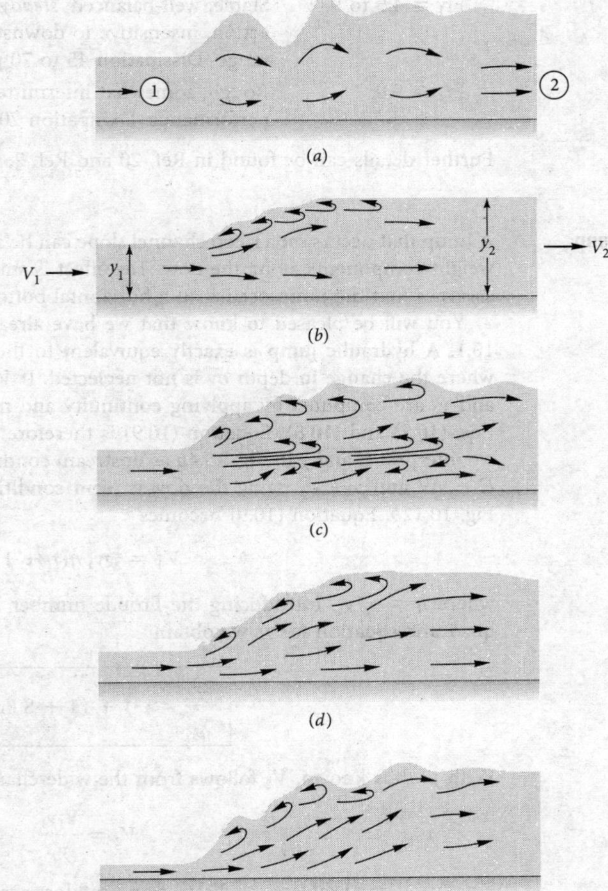

Fig. 10.12 Classification of hydraulic jumps: (*a*) Fr = 1.0 to 1.7: undular jump; (*b*) Fr = 1.7 to 2.5: weak jump; (*c*) Fr = 2.5 to 4.5: oscillating jump; (*d*) Fr = 4.5 to 9.0: steady jump; (*e*) Fr > 9.0: strong jump.
Source: Adapted from U.S. Bureau of Reclamation, "Research studies on stilling Basins, Energy Dissipators, and Associated Appurtenances," Hydraulic Lab, Rep., Hyd-399, June 1, 1995.

Classification

The principal parameter affecting hydraulic jump performance is the upstream Froude number $Fr_1 = V_1/(gy_1)^{1/2}$. The Reynolds number and channel geometry have only a secondary effect. As detailed in Ref. 20, the following ranges of operation can be outlined, as illustrated in Fig. 10.12:

$Fr_1 < 1.0$: Jump impossible, violates second law of thermodynamics.

$Fr_1 = 1.0$ to 1.7: Standing-wave or *undular jump* about $4y_2$ long; low dissipation, less than 5 percent.

$Fr_1 = 1.7$ to 2.5: Smooth surface rise with small rollers, known as a *weak jump;* dissipation 5 to 15 percent.

$Fr_1 = 2.5$ to 4.5: Unstable, *oscillating jump;* each irregular pulsation creates a large wave that can travel downstream for miles, damaging earth banks and other structures. Not recommended for design conditions. Dissipation 15 to 45 percent.

$Fr_1 = 4.5$ to 9.0: Stable, well-balanced, *steady jump;* best performance and action, insensitive to downstream conditions. Best design range. Dissipation 45 to 70 percent.

$Fr_1 > 9.0$: Rough, somewhat intermittent *strong jump,* but good performance. Dissipation 70 to 85 percent.

Further details can be found in Ref. 20 and Ref. 2, Chap. 15.

Theory for a Horizontal Jump

A jump that occurs on a steep channel slope can be affected by the difference in water-weight components along the flow. The effect is small, however, so the classic theory assumes that the jump occurs on a horizontal bottom.

You will be pleased to know that we have already analyzed this problem in Sec. 10.1. A hydraulic jump is exactly equivalent to the strong fixed wave in Fig. 10.4*b*, where the change in depth δy is not neglected. If V_1 and y_1 upstream are known, V_2 and y_2 are computed by applying continuity and momentum across the wave, as in Eqs. (10.7) and (10.8). Equation (10.9) is therefore the correct solution for a jump if we interpret C and y in Fig. 10.4*b* as upstream conditions V_1 and y_1, respectively, with $C - \delta V$ and $y + \delta y$ being the downstream conditions V_2 and y_2, respectively, as in Fig. 10.12*b*. Equation (10.9) becomes

$$V_1^2 = \tfrac{1}{2} g y_1 \eta(\eta + 1) \tag{10.42}$$

where $\eta = y_2/y_1$. Introducing the Froude number $Fr_1 = V_1/(gy_1)^{1/2}$ and solving this quadratic equation for η, we obtain

$$\boxed{\frac{2y_2}{y_1} = -1 + (1 + 8\,Fr_1^2)^{1/2}} \tag{10.43}$$

With y_2 thus known, V_2 follows from the wide-channel continuity relation:

$$V_2 = \frac{V_1 y_1}{y_2} \tag{10.44}$$

Finally, we can evaluate the dissipation head loss across the jump from the steady flow energy equation:

$$h_f = E_1 - E_2 = \left(y_1 + \frac{V_1^2}{2g}\right) - \left(y_2 + \frac{V_2^2}{2g}\right)$$

Introducing y_2 and V_2 from Eqs. (10.43) and (10.44), we find after considerable algebraic manipulation that

$$\boxed{h_f = \frac{(y_2 - y_1)^3}{4 y_1 y_2}} \tag{10.45}$$

Equation (10.45) shows that the dissipation loss is positive only if $y_2 > y_1$, which is a requirement of the second law of thermodynamics. Equation (10.43) then requires that $Fr_1 > 1.0$; that is, the upstream flow must be supercritical. Finally, Eq. (10.44) shows that $V_2 < V_1$ and the downstream flow is subcritical. All these results agree with our previous experience analyzing the normal shock wave.

The present theory is for hydraulic jumps in wide or rectangular horizontal channels. For the theory of prismatic or sloping channels see advanced texts [for example, 2, Chaps. 15 and 16].

EXAMPLE 10.8

Water flows in a wide channel at $q = 10\ \text{m}^3/(\text{s} \cdot \text{m})$ and $y_1 = 1.25$ m. If the flow undergoes a hydraulic jump, compute (*a*) y_2, (*b*) V_2, (*c*) Fr_2, (*d*) h_f, (*e*) the percentage dissipation, (*f*) the power dissipated per unit width, and (*g*) the temperature rise due to dissipation if $c_p = 4{,}200\ \text{J/(kg} \cdot \text{K)}$.

Solution

Part (a) The upstream velocity is

$$V_1 = \frac{q}{y_1} = \frac{10\ \text{m}^3/(\text{s} \cdot \text{m})}{1.25\ \text{m}} = 8.0\ \text{m/s}$$

The upstream Froude number is therefore

$$\text{Fr}_1 = \frac{V_1}{(gy_1)^{1/2}} = \frac{8.0}{[9.81(1.25)]^{1/2}} = 2.285$$

From Fig. 10.12 this is a weak jump. The depth y_2 is obtained from Eq. (10.43):

$$\frac{2y_2}{y_1} = -1 + [1 + 8(2.285)^2]^{1/2} = 5.54$$

or

$$y_2 = \tfrac{1}{2}y_1(5.54) = \tfrac{1}{2}(1.25)(5.54) = 3.46\ \text{m} \qquad \textit{Ans. (a)}$$

Part (b) From Eq. (10.44) the downstream velocity is

$$V_2 = \frac{V_1 y_1}{y_2} = \frac{8.0(1.25)}{3.46} = 2.89\ \text{m/s} \qquad \textit{Ans. (b)}$$

Part (c) The downstream Froude number is

$$\text{Fr}_2 = \frac{V_2}{(gy_2)^{1/2}} = \frac{2.89}{[9.81(3.46)]^{1/2}} = 0.496 \qquad \textit{Ans. (c)}$$

Part (d) As expected, Fr_2 is subcritical. From Eq. (10.45) the dissipation loss is

$$h_f = \frac{(3.46 - 1.25)^3}{4(3.46)(1.25)} = 0.625\ \text{m} \qquad \textit{Ans. (d)}$$

Part (e) The percentage dissipation relates h_f to upstream energy:

$$E_1 = y_1 + \frac{V_1^2}{2g} = 1.25 + \frac{(8.0)^2}{2(9.81)} = 4.51\ \text{m}$$

Hence

$$\text{Percentage loss} = (100)\frac{h_f}{E_1} = \frac{100(0.625)}{4.51} = 14\ \text{percent} \qquad \textit{Ans. (e)}$$

Part (f) The power dissipated per unit width is

$$\text{Power} = \rho g q h_f = (9{,}800\ \text{N/m}^3)[10\ \text{m}^3/(\text{s} \cdot \text{m})](0.625\ \text{m})$$

$$= 61.3\ \text{kW/m} \qquad \textit{Ans. (f)}$$

Part (g) Finally, the mass flow rate is $\dot{m} = \rho q = (1{,}000\ \text{kg/m}^3)[10\ \text{m}^3/(\text{s} \cdot \text{m})] = 10{,}000\ \text{kg/(s} \cdot \text{m)}$, and the temperature rise from the steady flow energy equation is

$$\text{Power dissipated} = \dot{m} c_p\, \Delta T$$

or

$$61{,}300\ \text{W/m} = [10{,}000\ \text{kg/(s} \cdot \text{m)}][4{,}200\ \text{J/(kg} \cdot \text{K)}]\Delta T$$

from which

$$\Delta T = 0.0015\ \text{K} \qquad \textit{Ans. (g)}$$

The dissipation is large, but the temperature rise is negligible.

10.6 Gradually Varied Flow[3]

In practical channel flows both the bottom slope and the water depth change with position, as in Fig. 10.3. An approximate analysis is possible if the flow is gradually varied, as is the case if the slopes are small and changes not too sudden. The basic assumptions are

1. Slowly changing bottom slope.
2. Slowly changing water depth (no hydraulic jumps).
3. Slowly changing cross section.
4. One-dimensional velocity distribution.
5. Pressure distribution approximately hydrostatic.

The flow then satisfies the continuity relation (10.1) plus the energy equation with bottom friction losses included. The two unknowns for steady flow are velocity $V(x)$ and water depth $y(x)$, where x is distance along the channel.

Basic Differential Equation

Consider the length of channel dx illustrated in Fig. 10.13. All the terms that enter the steady flow energy equation are shown, and the balance between x and $x + dx$ is

$$\frac{V^2}{2g} + y + S_0\,dx = S\,dx + \frac{V^2}{2g} + d\left(\frac{V^2}{2g}\right) + y + dy$$

or

$$\frac{dy}{dx} + \frac{d}{dx}\left(\frac{V^2}{2g}\right) = S_0 - S \tag{10.46}$$

where S_0 is the slope of the channel bottom (positive as shown in Fig. 10.13) and S is the slope of the EGL (which drops due to wall friction losses).

To eliminate the velocity derivative, differentiate the continuity relation:

$$\frac{dQ}{dx} = 0 = A\frac{dV}{dx} + V\frac{dA}{dx} \tag{10.47}$$

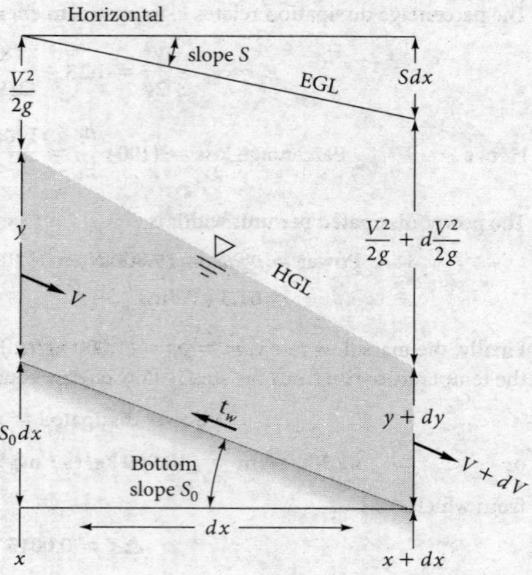

Fig. 10.13 Energy balance between two sections in a gradually varied open-channel flow.

[3]This section may be omitted without loss of continuity.

But $dA = b_0\, dy$, where b_0 is the channel width at the surface. Eliminating dV/dx between Eqs. (10.46) and (10.47), we obtain

$$\frac{dy}{dx}\left(1 - \frac{V^2 b_0}{gA}\right) = S_0 - S \tag{10.48}$$

Finally, recall from Eq. (10.37) that $V^2 b_0/(gA)$ is the square of the Froude number of the local channel flow. The final desired form of the gradually varied flow equation is

$$\frac{dy}{dx} = \frac{S_0 - S}{1 - \mathrm{Fr}^2} \tag{10.49}$$

This equation changes sign according to whether the Froude number is subcritical or supercritical and is analogous to the one-dimensional gas dynamic area-change formula (9.40).

The numerator of Eq. (10.49) changes sign according to whether S_0 is greater or less than S, which is the slope equivalent to uniform flow at the same discharge Q:

$$S = S_{0n} = \frac{f}{D_h}\frac{V^2}{2g} = \frac{V^2}{R_h C^2} = \frac{n^2 V^2}{\alpha^2 R_h^{4/3}} \tag{10.50}$$

where C is the Chézy coefficient. The behavior of Eq. (10.49) thus depends on the relative magnitude of the local bottom slope $S_0(x)$, compared with (1) uniform flow, $y = y_n$, and (2) critical flow, $y = y_c$. As in Eq. (10.38), the dimensional parameter α^2 equals 1.0 for SI units and 2.208 for BG units.

Classification of Solutions

It is customary to compare the actual channel slope S_0 with the critical slope S_c for the same Q from Eq. (10.38). There are five classes for S_0, giving rise to 12 distinct types of solution curves, all of which are illustrated in Fig. 10.14:

Slope class	Slope notation	Depth class	Solution curves
$S_0 > S_c$	Steep	$y_c > y_n$	S-1, S-2, S-3
$S_0 = S_c$	Critical	$y_c = y_n$	C-1, C-3
$S_0 < S_c$	Mild	$y_c < y_n$	M-1, M-2, M-3
$S_0 = 0$	Horizontal	$y_n = \infty$	H-2, H-3
$S_0 < 0$	Adverse	$y_n =$ imaginary	A-2, A-3

The solution letters S, C, M, H, and A obviously denote the names of the five types of slopes. The numbers 1, 2, 3 relate to the position of the initial point on the solution curve with respect to the normal depth y_n and the critical depth y_c. In type 1 solutions, the initial point is above both y_n and y_c, and in all cases the water depth solution $y(x)$ becomes even deeper and farther away from y_n and y_c. In type 2 solutions, the initial point lies between y_n and y_c, and if there is no change in S_0 or roughness, the solution tends asymptotically toward the lower of y_n or y_c. In type 3 cases, the initial point lies below both y_n and y_c, and the solution curve tends asymptotically toward the lower of these.

Figure 10.14 shows the basic character of the local solutions, but in practice, of course, S_0 varies with x, and the overall solution patches together the various cases to form a continuous depth profile $y(x)$ compatible with a given initial condition and a given discharge Q. There is a fine discussion of various composite solutions in Ref. 2, Chap. 9; see also Ref. 22, Sec. 12.7.

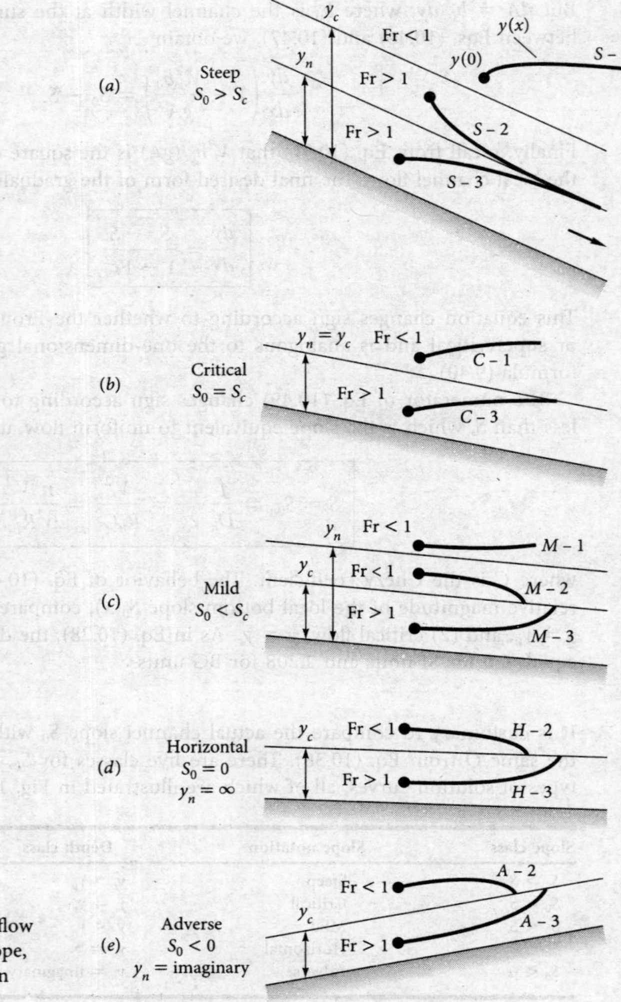

Fig. 10.14 Gradually varied flow for five classes of channel slope, showing the 12 basic solution curves.

Numerical Solution

The basic relation for gradually varied flow, Eq. (10.49), is a first-order ordinary differential equation that can be easily solved numerically. For a given constant-volume flow rate Q, it may be written in the form

$$\frac{dy}{dx} = \frac{S_0 - n^2Q^2/(\alpha^2 A^2 R_h^{4/3})}{1 - Q^2 b_0/(gA^3)} \tag{10.51}$$

subject to an initial condition $y = y_0$ at $x = x_0$. It is assumed that the bottom slope $S_0(x)$ and the cross-sectional shape parameters (b_0, P, A) are known everywhere along the channel. Then one may solve Eq. (10.51) for local water depth $y(x)$ by any standard numerical method. The author uses an Excel spreadsheet for a personal computer. Step sizes Δx may be selected so that each change Δy is limited to no greater than, say, 1 percent. The solution curves are generally well behaved unless there are discontinuous changes in channel parameters. Note that if one approaches the critical depth y_c, the denominator of Eq. (10.51) approaches zero, so small step sizes are required. It helps physically to know what type of solution curve (M-1, S-2, or the like) you are proceeding along, but this is not mathematically necessary.

EXAMPLE 10.9

Let us extend the data of Example 10.5 to compute a portion of the profile shape. Given is a wide channel with $n = 0.022$, $S_0 = 0.0040$, and $q = 20\ \text{m}^3/(\text{s} \cdot \text{m})$. If $y_0 = 2$ m at $x = 0$, how far along the channel $x = L$ does it take the depth to rise to $y_L = 3$ m? Is the 3 m depth position upstream or downstream in Fig. E10.9*a*?

Solution

In Example 10.5 we computed $y_c = 3.44$ m. Since our initial depth $y = 2$ m is less than y_c, we know the flow is supercritical. Let us also compute the normal depth for the given slope S_0 by setting $q = 20\ \text{m}^3/(\text{s} \cdot \text{m})$ in the Chézy formula (10.19) with $R_h = y_n$:

$$q = \frac{\alpha}{n} A R_h^{2/3} S_0^{1/2} = \frac{1}{0.022}\,[y_n(1\ \text{m})]\,y_n^{2/3}(0.0040)^{1/2} = 20\ \text{m}^3/(\text{s} \cdot \text{m})$$

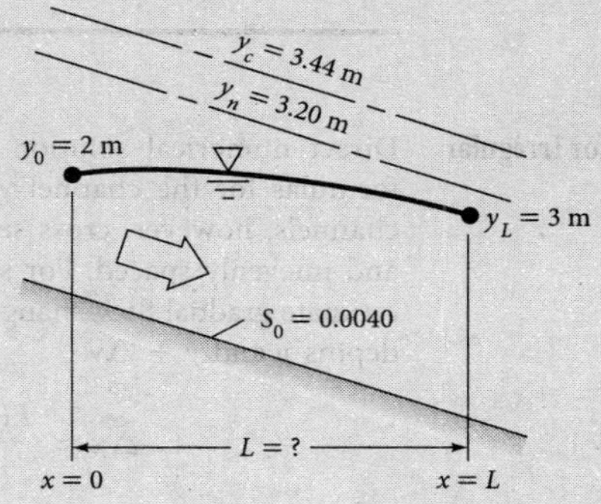

E10.9a

Solve for: $y_n \approx 3.20$ m

Thus both $y(0) = 2$ m and $y(L) = 3$ m are less than y_n, which is less than y_c, so we *must* be on an S-3 curve, as in Fig. 10.14*a*. For a wide channel, Eq. (10.51) reduces to

$$\frac{dy}{dx} = \frac{S_0 - n^2 q^2/(\alpha^2 y^{10/3})}{1 - q^2/(g y^3)}$$

$$\approx \frac{0.0040 - (0.022)^2(20)^2/(1 y^{10/3})}{1 - (20)^2/(9.81 y^3)} \qquad \text{with } y(0) = 2\ \text{m}$$

The initial slope is $y'(0) \approx 0.00371$, and a step size $\Delta x = 3$ m would cause a change $\Delta y \approx (0.00371)(3\ \text{m}) \approx 0.0111$ m, less than 1 percent. We therefore integrate numerically with $\Delta x = 3$ m to determine when the depth $y = 3$ m is achieved. Tabulate some values:

x, m	0	60	120	180	240	333
y, m	2.00	2.22	2.42	2.61	2.78	3.00

The water depth, still supercritical, reaches $y = 3$ m at

$$x \approx 333\ \text{m downstream} \qquad \textit{Ans.}$$

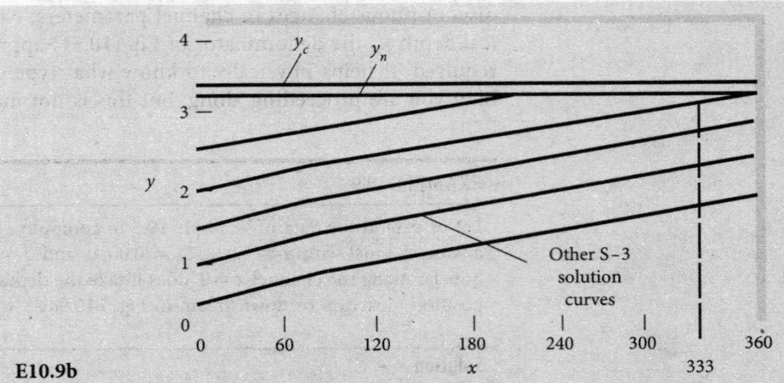

E10.9b

We verify from Fig. 10.14*a* that water depth does increase downstream on an S-3 curve. The solution curve $y(x)$ is shown as the bold line in Fig. E10.9*b*.

For little extra effort we can investigate the entire family of S-3 solution curves for this problem. Figure E10.9*b* also shows what happens if the initial depth is varied from 0.5 to 2.5 m in increments of 0.5 m. All S-3 solutions smoothly rise and asymptotically approach the uniform flow condition $y = y_n = 3.20$ m.

Approximate Solution for Irregular Channels

Direct numerical solution of Eq. (10.51) is appropriate when we have analytical formulas for the channel variations $A(x)$, $S_0(x)$, $n(x)$, $b_0(x)$, and $R_h(x)$. For natural channels, however, cross sections are often highly irregular, and data can be sparse and unevenly spaced. For such cases, civil engineers use an approximate method to estimate gradual flow changes. Write Eq. (10.46) in finite-difference form between two depths y and $y + \Delta y$:

$$\Delta x \approx \frac{E(y + \Delta y) - E(y)}{(S_0 - S_{avg})} \quad \text{where } E = y + \frac{V^2}{2g} \tag{10.52}$$

Average values of velocity, slope, and hydraulic radius are estimated between the two sections. For example,

$$V_{avg} \approx \frac{1}{2}[V(y) + V(y + \Delta y)];\ R_{h,\,avg} \approx \frac{1}{2}[R_h(y) + R_h(y + \Delta y)];\ S_{avg} \approx \frac{n^2 V_{avg}^2}{\alpha^2 R_{h,\,avg}^{4/3}}$$

Again, computation can proceed either upstream or downstream, using small values of Δy. Further details of such computations are given in Chap. 10 of Ref. 2.

EXAMPLE 10.10

Repeat Example 10.9 using the approximate method of Eq. (10.52) with a 0.25 meter increment in Δy. Find the distance required for y to rise from 2 m to 3 m.

Solution

Recall from Example 10.9 that $n = 0.022$, $S_0 = 0.0040$, and $q = 20\ m^3/(s \cdot m)$. Note that $R_h = y$ for a wide channel. Make a table with y varying from 2 to 3 m in increments of 0.25 m, computing $V = q/y$, $E = y + V^2/(2g)$, and $S_{avg} = [n^2V^2/(1y^{4/3})]_{avg}$:

y, m	V (m/s) = $20/y$	$E = y + V^2/(2g)$	S	S_{avg}	$\Delta x = \Delta E/(S_0 - S)_{avg}$	$x = \sum \Delta x$
2	10.00	7.097	0.01921	—	—	0
2.25	8.889	6.277	0.01297	0.01580	69.5	69.5
2.5	8.000	5.762	0.00913	0.01089	74.7	144.2
2.75	7.273	5.446	0.00665	0.00779	83.3	227.5
3.0 m	6.667 m/s	5.265	0.00497	0.00575	103.1 m	330.6 m

Comment: The accuracy is excellent, giving the almost same result, $x = 333$ m, as the Excel spreadsheet numerical integration in Example 10.9. Much of this accuracy is due to the smooth, slowly varying nature of the profile. Less precision is expected when the channel is irregular and given as uneven cross sections.

Some Illustrative Composite-Flow Transitions

The solution curves in Fig. 10.14 are somewhat simplistic, since they postulate constant-bottom slopes. In practice, channel slopes can vary greatly, $S_0 = S_0(x)$, and the solution curves can cross between two regimes. Other parameter changes, such as $A(x)$, $b_0(x)$, and $n(x)$, can cause interesting composite-flow profiles. Some examples are shown in Fig. 10.15.[4]

Figure 10.15*a* shows a transition from a mild slope to a steep slope in a constant-width channel. The initial M-2 curve must change to an S-2 curve farther down the steep slope. The only way this can happen physically is for the solution curve to pass smoothly through the critical depth, as shown. The critical point is mathematically *singular* [2, Sec. 9.6], and the flow near this point is generally *rapidly*, not gradually, varied. The flow pattern, accelerating from subcritical to supercritical, is similar to a converging–diverging nozzle in gas dynamics. Other scenarios for Fig. 10.15*a* are impossible. For example, the upstream curve cannot be M-1, for the break in slope would cause an S-1 curve that would move away from uniform steep flow.

Figure 10.15*b* shows a mild slope that suddenly changes to an even milder slope. The approach flow is assumed uniform, and the break in slope makes its presence known upstream. The water depth moves smoothly along an M-1 curve until it exactly merges, at the break point, with a uniform flow at the new (milder) depth y_{n2}.

Figure 10.15*c* shows a steep slope that suddenly changes to a less steep slope. Note for both slopes that $y_n < y_c$. Because of the supercritical ($V > V_c$) approach flow, the break in slope cannot make its presence known upstream. Thus, not until the break point does an S-3 curve form, and then this profile proceeds smoothly to uniform flow at the new (higher) normal depth.

Figure 10.15*d* shows a steep slope that suddenly changes to a mild slope. Various cases may occur, possibly beyond the ability of this author to describe. The two cases shown depend on the relative magnitude of the mild slope. If the downstream depth y_{n2} is shallow, an M-3 curve will start at the break and develop until the local supercritical flow is just sufficient to form a hydraulic jump up to the new normal depth. As y_{n2} increases, the jump moves upstream until, for the "high" case shown, it forms on the steep side, followed by an S-1 curve that merges into normal depth y_{n2} at the break point.

Figure 10.15*e* illustrates a *free overfall* with a mild slope. This acts as a *control section* to the upstream flow, which then forms an M-2 curve and accelerates to critical flow near the overfall. The falling stream will be supercritical. The overfall "controls"

[4]The author is indebted to Prof. Bruce Larock for clarification of these transition profiles.

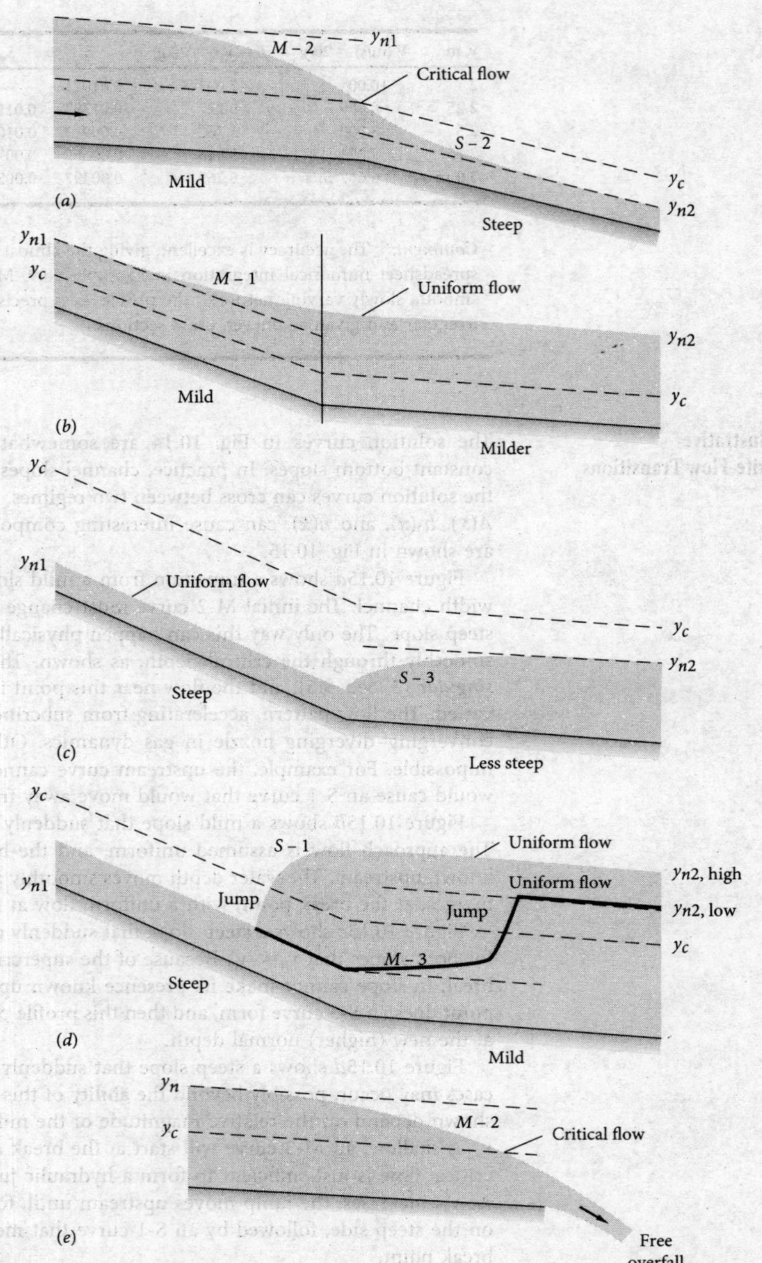

Fig. 10.15 Some examples of composite-flow transition profiles.

the water depths upstream and can serve as an initial condition for computation of $y(x)$. This is the type of flow that occurs in a weir or waterfall, Sec. 10.7.

The examples in Fig. 10.15 show that changing conditions in open-channel flow can result in complex flow patterns. Many more examples of composite-flow profiles are given in Ref. 2, pp. 229–233.

10.7 Flow Measurement and Control by Weirs

A *weir*, of which the ordinary dam is an example, is a channel obstruction over which the flow must deflect. For simple geometries, the channel discharge Q correlates with gravity and with the blockage height H to which the upstream flow is backed up above the weir elevation (see Fig. 10.16). Thus a weir is a simple but effective open-channel flowmeter. We used a weir as an example of dimensional analysis in Prob. P5.32.

Figure 10.16 shows two common weirs, sharp-crested and broad-crested, assumed to be very wide. In both cases, the flow upstream is subcritical, accelerates to critical near the top of the weir, and spills over into a supercritical *nappe*. For both weirs the discharge q per unit width is proportional to $g^{1/2}H^{3/2}$, but with somewhat different coefficients. The short-crested (or thin-plate) weir nappe should be *ventilated* to the atmosphere; that is, it should spring clear of the weir crest. Unventilated or drowned nappes are more difficult to correlate and depend on tailwater conditions. (The spillway of Fig. 10.11 is a sort of unventilated weir.)

A very complete discussion of weirs, including other designs such as the polygonal "Crump" weir and various contracting flumes, is given in the text by Ackers et al. [23]. See Prob. P10.122.

Analysis of Sharp-Crested Weirs

It is possible to analyze weir flow by inviscid potential theory with an unknown (but solvable) free surface, as in Fig. P8.71. Here, however, we simply use one-dimensional flow theory plus dimensional analysis to develop suitable weir flow rate correlations.

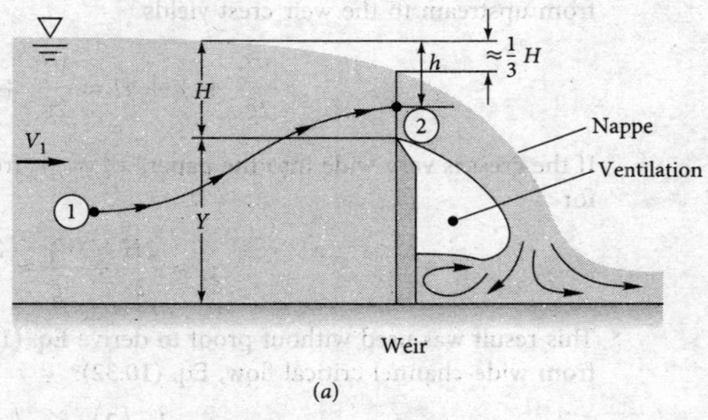

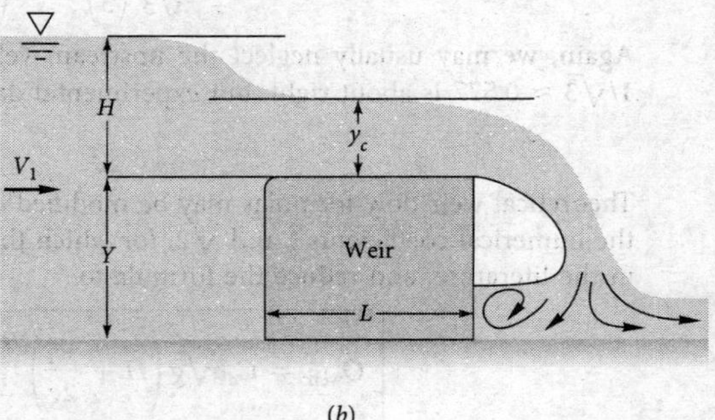

Fig. 10.16 Flow over wide, well-ventilated weirs: (*a*) sharp-crested; (*b*) broad-crested.

A very early theoretical approach is credited to J. Weisbach in 1855. The velocity head at any point 2 above the weir crest is assumed to equal the total head upstream; in other words, Bernoulli's equation is used with no losses:

$$\frac{V_2^2}{2g} + H - h \approx \frac{V_1^2}{2g} + H \quad \text{or} \quad V_2(h) \approx \sqrt{2gh + V_1^2}$$

where h is the vertical distance down to point 2, as shown in Fig. 10.16*a*. If we accept for the moment, without proof, that the flow over the crest draws down to $h_{min} \approx H/3$, the volume flow $q = Q/b$ over the crest is approximately

$$q = \int_{crest} V_2\, dh \approx \int_{H/3}^{H} (2gh + V_1^2)^{1/2}\, dh$$

$$= \frac{2}{3}\sqrt{2g}\left[\left(H + \frac{V_1^2}{2g}\right)^{3/2} - \left(\frac{H}{3} + \frac{V_1^2}{2g}\right)^{3/2}\right]$$

Normally the upstream velocity head $V_1^2/(2g)$ is neglected, so this expression reduces to

Sharp-crested theory: $$q \approx 0.81\left(\tfrac{2}{3}\right)(2g)^{1/2}H^{3/2} \quad (10.53)$$

This formula is functionally correct, but the coefficient 0.81 is too high and should be replaced by an experimentally determined discharge coefficient.

Analysis of Broad-Crested Weirs

The broad-crested weir of Fig. 10.16*b* can be analyzed more accurately because it creates a short run of nearly one-dimensional critical flow, as shown. Bernoulli's equation from upstream to the weir crest yields

$$\frac{V_1^2}{2g} + Y + H \approx \frac{V_c^2}{2g} + Y + y_c$$

If the crest is very wide into the paper, $V_c^2 = gy_c$ from Eq. (10.33). Thus, we can solve for

$$y_c \approx \frac{2H}{3} + \frac{V_1^2}{3g} \approx \frac{2H}{3}$$

This result was used without proof to derive Eq. (10.53). Finally, the flow rate follows from wide-channel critical flow, Eq. (10.32):

Broad-crested theory: $$q = \sqrt{gy_c^3} \approx \frac{1}{\sqrt{3}}\left(\frac{2}{3}\right)\sqrt{2g}\left(H + \frac{V_1^2}{2g}\right)^{3/2} \quad (10.54)$$

Again, we may usually neglect the upstream velocity head $V_1^2/(2g)$. The coefficient $1/\sqrt{3} \approx 0.577$ is about right, but experimental data are preferred.

Experimental Weir Discharge Coefficients

Theoretical weir flow formulas may be modified experimentally as follows: Eliminate the numerical coefficients $\frac{2}{3}$ and $\sqrt{2}$, for which there is much sentimental attachment in the literature, and reduce the formula to

$$Q_{weir} = C_d b\sqrt{g}\left(H + \frac{V_1^2}{2g}\right)^{3/2} \approx C_d b\sqrt{g}H^{3/2} \quad (10.55)$$

where b is the crest width and C_d is a dimensionless, experimentally determined *weir discharge coefficient,* which may vary with the weir geometry, Reynolds number, and Weber number. Many data for many different weirs have been reported in the literature, as detailed in Ref. 23.

An accurate (±2 percent) composite correlation for wide ventilated sharp crests is recommended as follows [23]:

$$\text{Wide sharp-crested weir: } C_d \approx 0.564 + 0.0846\frac{H}{Y} \quad \text{for} \quad \frac{H}{Y} \le 2 \tag{10.56}$$

The Reynolds numbers V_1H/ν for these data vary from 1 E4 to 2 E6, but the formula should apply to higher Re, such as large dams on rivers.

The broad-crested weir of Fig. 10.16*b* is considerably more sensitive to geometric parameters, including the surface roughness ε of the crest. If the leading-edge nose is rounded, $R/L \ge 0.05$, available data [23, Chap. 7] may be correlated as follows:

$$\text{Round-nosed broad-crested weir: } C_d \approx 0.544\left(1 - \frac{\delta^*/L}{H/L}\right)^{3/2} \tag{10.57}$$

$$\text{where} \quad \frac{\delta^*}{L} \approx 0.001 + 0.2\sqrt{\varepsilon/L}$$

The chief effect is due to turbulent boundary layer displacement-thickness growth δ^* on the crest as compared to upstream head H. The formula is limited to $H/L < 0.7$, $\varepsilon/L \le 0.002$, and $V_1H/\nu > 3$ E5. If the nose is round, there is no significant effect of weir height Y, at least if $H/Y < 2.4$.

If the broad-crested weir has a sharp leading edge, thus commonly called a *rectangular* weir, the discharge may depend on the weir height Y. However, in a certain range of weir height and length, C_d is nearly constant:

$$\text{Sharp-nosed broad-crested weir: } \quad C_d \approx 0.462 \quad \text{for} \quad 0.08 < \frac{H}{L} < 0.33 \quad \text{and} \quad 0.22 < \frac{H}{Y} < 0.56 \tag{10.58}$$

Surface roughness is not a significant factor here. For $H/L < 0.08$ there is large scatter (±10 percent) in the data. For $H/L > 0.33$ and $H/Y > 0.56$, C_d increases up to 10 percent due to each parameter, and complex charts are needed for the discharge coefficient [19, Chap. 5].

EXAMPLE 10.11

A weir in a horizontal channel is 1 m high and 4 m wide. The water depth upstream is 1.6 m. Estimate the discharge if the weir is (*a*) sharp-crested and (*b*) round-nosed with an unfinished concrete broad crest 1.2 m long. Neglect $V_1^2/(2g)$.

Solution

Part (a) We are given $Y = 1$ m and $H + Y \approx 1.6$ m, hence $H \approx 0.6$ m. Since $H \ll b$, we assume that the weir is "wide." For a sharp crest, Eq. (10.56) applies:

$$C_d \approx 0.564 + 0.0846\frac{0.6 \text{ m}}{1 \text{ m}} \approx 0.615$$

Then, the discharge is given by the basic correlation, Eq. (10.55):

$$Q = C_d b\sqrt{g}H^{3/2} = (0.615)(4 \text{ m})\sqrt{(9.81 \text{ m/s}^2)}(0.6 \text{ m})^{3/2} \approx 3.58 \text{ m}^3/\text{s} \qquad \textit{Ans. (a)}$$

We check that $H/Y = 0.6 < 2.0$ for Eq. (10.56) to be valid. From continuity, $V_1 = Q/(by_1) = 3.58/[(4.0)(1.6)] = 0.56$ m/s, giving a Reynolds number $V_1H/\nu \approx 3.4$ E5.

Part (b) For a round-nosed broad-crested weir, Eq. (10.57) applies. For an unfinished concrete surface, read $\varepsilon \approx 2.4$ mm from Table 10.1. Then the displacement thickness is

$$\frac{\delta^*}{L} \approx 0.001 + 0.2\sqrt{\varepsilon/L} = 0.001 + 0.2\left(\frac{0.0024 \text{ m}}{1.2 \text{ m}}\right)^{1/2} \approx 0.00994$$

Then, Eq. (10.57) predicts the discharge coefficient:

$$C_d \approx 0.544\left(1 - \frac{0.00994}{0.6 \text{ m}/1.2 \text{ m}}\right)^{3/2} \approx 0.528$$

The estimated flow rate is thus

$$Q = C_d b\sqrt{g}H^{3/2} = 0.528(4 \text{ m})\sqrt{(9.81 \text{ m}^2/\text{s})}(0.6 \text{ m})^{3/2} \approx 3.07 \text{ m}^3/\text{s} \qquad \textit{Ans. (b)}$$

Check that $H/L = 0.5 < 0.7$ as required. The approach Reynolds number is $V_1H/\nu \approx 2.9$ E5, just barely below the recommended limit in Eq. (10.57).

Since $V_1 \approx 0.5$ m/s, $V_1^2/(2g) \approx 0.012$ m, so the error in taking total head equal to 0.6 m is about 2 percent. We could correct this for upstream velocity head if desired.

Other Thin-Plate Weir Designs

Weirs are often used for flow measurement and control of artificial channels. The two most common shapes are a rectangle and a V notch, as shown in Table 10.2. All should be fully ventilated and not drowned.

Table 10.2*a* shows a full-width rectangle, which will have slight end-boundary-layer effects but no end contractions. For a thin-plate design, the top is approximately sharp-crested, and Eq. (10.56) should give adequate accuracy, as shown in the table. Since the overfall spans the entire channel, artificial ventilation may be needed, such as holes in the channel walls.

Table 10.2*b* shows a partial-width rectangle, $b < L$, which will cause the sides of the overfall to contract inward and reduce the flow rate. An adequate contraction correction [23, 24] is to reduce the effective weir width by $0.1H$, as shown in the table. It seems, however, that this type of weir is rather sensitive to small effects, such as plate thickness and sidewall boundary layer growth. Small heads ($H < 75$ mm) and small slot widths ($b < 30$ cm) are not recommended. See Refs. 23 and 24 for further details.

The V notch, in Table 10.2*c*, is intrinsically interesting in that its overfall has only one length scale, H—there is no separate "width." The discharge will thus be proportional to $H^{5/2}$, rather than a power of $\frac{3}{2}$. Application of Bernoulli's equation to the triangular opening, in the spirit of Eq. (10.52), leads to the following ideal flow rate for a V notch:

$$\text{V notch:} \qquad Q_{\text{ideal}} = \frac{8\sqrt{2}}{15}\tan\frac{\theta}{2}g^{1/2}H^{5/2} \tag{10.59}$$

where θ is the total included angle of the notch. The actual measured flow is about 40 percent less than this, due to contraction similar to a thin-plate orifice. In terms of an experimental discharge coefficient, the recommended formula is

$$Q_{\text{V notch}} \approx C_d \tan\frac{\theta}{2}g^{1/2}H^{5/2} \qquad C_d \approx 0.44 \quad \text{for} \quad 20° < \theta < 100° \tag{10.60}$$

Table 10.2 Thin-Plate Weirs for Flow Measurement

Thin-plate weir	Flow-rate correlation
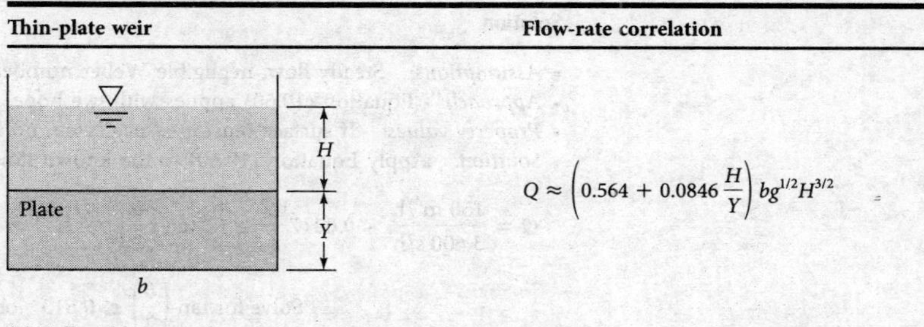 (*a*) Full-width rectangle.	$Q \approx \left(0.564 + 0.0846\dfrac{H}{Y}\right) bg^{1/2}H^{3/2}$
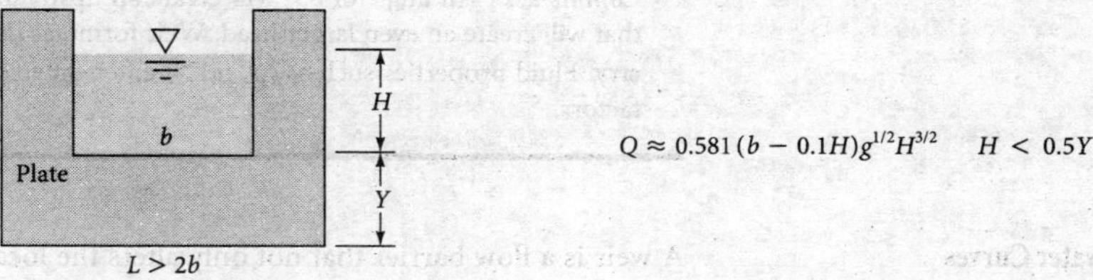 (*b*) Rectangle with side contractions.	$Q \approx 0.581\,(b - 0.1H)g^{1/2}H^{3/2} \quad H < 0.5Y$
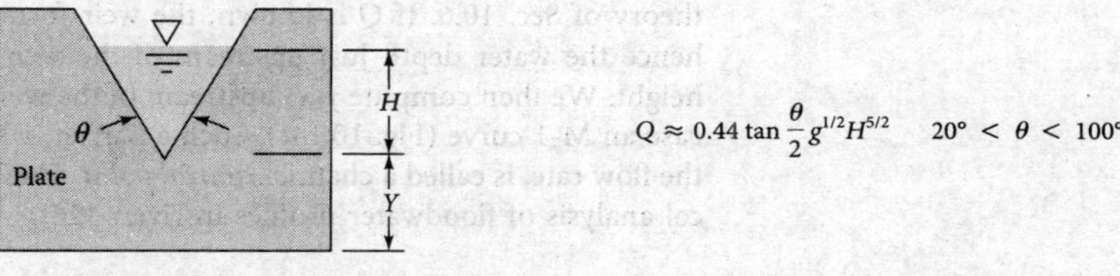 (*c*) V notch.	$Q \approx 0.44 \tan\dfrac{\theta}{2} g^{1/2}H^{5/2} \quad 20° < \theta < 100°$

for heads $H > 50$ mm. For smaller heads, both Reynolds number and Weber number effects may be important, and a recommended correction [23] is

$$\text{Low heads, } H < 50 \text{ mm:} \qquad C_{d,\text{V notch}} \approx 0.44 + \frac{0.9}{(\text{Re We})^{1/6}} \tag{10.61}$$

where $\text{Re} = \rho g^{1/2}H^{3/2}/\mu$ and $\text{We} = \rho g H^2/\Upsilon$, with Υ being the coefficient of surface tension. Liquids other than water may be used with this formula, as long as $\text{Re} > 300/\tan(\theta/2)^{3/4}$ and $\text{We} > 300$.

A number of other thin-plate weir designs—trapezoidal, parabolic, circular arc, and U-shaped—are discussed in Ref. 25, which also contains considerable data on broad-crested weirs. See also Refs. 29 and 30.

EXAMPLE 10.12

A V notch weir is to be designed to meter an irrigation channel flow. For ease in reading the upstream water-level gage, a reading $H \geq 30$ cm is desired for the design flow rate of 150 m^3/h. What is the appropriate angle θ for the V notch?

Solution

- *Assumptions:* Steady flow, negligible Weber number effect because $H > 50$ mm.
- *Approach:* Equation (10.60) applies with, we hope, a notch angle $20° < \theta < 100°$.
- *Property values:* If surface tension is neglected, no fluid properties are needed. Why?
- *Solution:* Apply Equation (10.60) to the known flow rate and solve for θ:

$$Q = \frac{150 \text{ m}^3/\text{h}}{3{,}600 \text{ s/h}} = 0.0417\frac{\text{m}^3}{\text{s}} \geq C_d \tan\left(\frac{\theta}{2}\right) g^{1/2} H^{5/2} = 0.44 \tan\left(\frac{\theta}{2}\right)\left(9.81\frac{\text{m}}{\text{s}^2}\right)^{1/2} (0.3 \text{ m})^{5/2}$$

$$\text{Solve for } \tan\left(\frac{\theta}{2}\right) \leq 0.613 \quad \text{or} \quad \theta \leq 63° \qquad \textit{Ans.}$$

- *Comments:* An angle of 63° will create an upstream head of 30 cm. Any angle less than that will create an even larger head. Weir formulas depend primarily on gravity and geometry. Fluid properties such as (ρ, μ, Υ) enter only as slight modifications or as correction factors.

Backwater Curves

A weir is a flow barrier that not only alters the local flow over the weir but also modifies the flow depth distribution far upstream. Any strong barrier in an open-channel flow creates a *backwater curve,* which can be computed by the gradually varied flow theory of Sec. 10.6. If Q is known, the weir formula, Eq. (10.55), determines H and hence the water depth just upstream of the weir, $y = H + Y$, where Y is the weir height. We then compute $y(x)$ upstream of the weir from Eq. (10.51), following in this case an M-1 curve (Fig. 10.14*c*). Such a barrier, where the water depth correlates with the flow rate, is called a channel *control point.* These are the starting points for numerical analysis of floodwater profiles in rivers [26].

EXAMPLE 10.13

A rectangular channel 8 m wide, with a flow rate of 30 m³/s, encounters a 4-m-high sharp-edged dam, as shown in Fig. E10.13*a*. Determine the water depth 2 km upstream if the channel slope is $S_0 = 0.0004$ and $n = 0.025$.

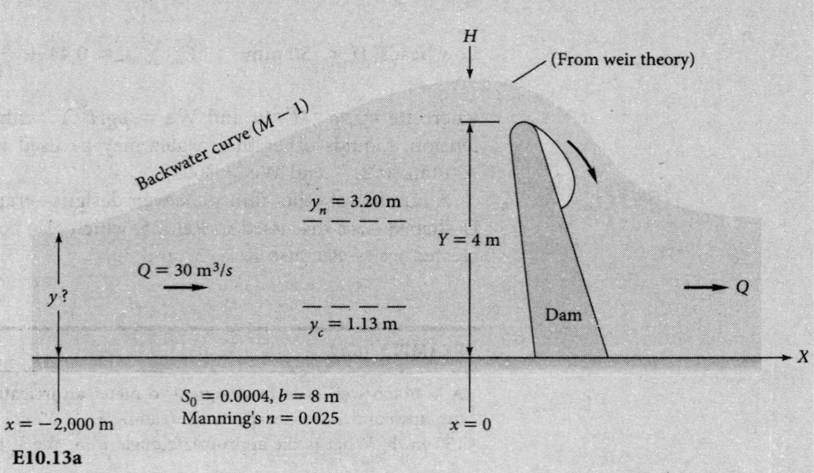

E10.13a

Solution

First, determine the head H produced by the dam, using sharp-crested full-width weir theory, Eq. (10.56):

$$Q = 30\ \text{m}^3/\text{s} = C_d b g^{1/2} H^{3/2} = \left(0.564 + 0.0846\frac{H}{4\ \text{m}}\right)(8\ \text{m})(9.81\ \text{m/s}^2)^{1/2} H^{3/2}$$

Since the term $0.0846H/4$ in parentheses is small, we may proceed by iteration to the solution $H \approx 1.59$ m. Then our initial condition at $x = 0$, just upstream of the dam, is $y(0) = Y + H = 4 + 1.59 = 5.59$ m. Compare this to the critical depth from Eq. (10.30):

$$y_c = \left(\frac{Q^2}{b^2 g}\right)^{1/3} = \left[\frac{(30\ \text{m}^3/\text{s})^2}{(8\ \text{m})^2(9.81\ \text{m/s}^2)}\right]^{1/3} = 1.13\ \text{m}$$

Since $y(0)$ is greater than y_c, the flow upstream is subcritical. Finally, for reference purposes, estimate the normal depth from the Chézy equation (10.19):

$$Q = 30\ \text{m}^3/\text{s} = \frac{\alpha}{n} b y R_h^{2/3} S_0^{1/2} = \frac{1.0}{0.025}(8\ \text{m}) y_n \left(\frac{8 y_n}{8 + 2 y_n}\right)^{2/3} (0.0004)^{1/2}$$

By trial and error, solve for $y_n \approx 3.20$ m. If there are no changes in channel width or slope, the water depth far upstream of the dam will approach this value. All these reference values $y(0)$, y_c, and y_n are shown in Fig. E10.13*b*.

Since $y(0) > y_n > y_c$, the solution will be an M-1 curve as computed from gradually varied theory, Eq. (10.51), for a rectangular channel with the given input data:

$$\frac{dy}{dx} \approx \frac{S_0 - n^2 Q^2/(\alpha^2 A^2 R_h^{4/3})}{1 - Q^2 b_0/(gA^3)} \qquad \alpha = 1.0 \quad A = 8y \quad n = 0.025 \quad R_h = \frac{8y}{8 + 2y} \quad b_0 = 8$$

Beginning with $y = 5.59$ m at $x = 0$, we integrate backward to $x = -2{,}000$ m. For the Runge-Kutta method, four-figure accuracy is achieved for $\Delta x = -100$ m. The complete solution curve is shown in Fig. E10.13*b*. The desired solution value is

At $x = -2{,}000$ m: $\qquad y \approx 5.00$ m $\qquad$ *Ans.*

E10.13b

Thus, even 2 km upstream, the dam has produced a "backwater" that is 1.8 m above the normal depth that would occur without a dam. For this example, a near-normal depth of, say, 10 cm greater than y_n, or $y \approx 3.3$ m, would not be achieved until $x = -13{,}400$ m. Backwater curves are quite far-reaching upstream, especially in flood stages.

Summary

This chapter has introduced open-channel flow analysis, limited to steady, one-dimensional flow conditions. The basic analysis combines the continuity equation with the extended Bernoulli equation including friction losses.

Open-channel flows are classified either by depth variation or by Froude number, the latter being analogous to the Mach number in compressible duct flow (Chap. 9). Flow at constant slope and depth is called uniform flow and satisfies the classical Chézy equation (10.19). Straight prismatic channels can be optimized to find the cross section that gives maximum flow rate with minimum friction losses. As the slope and flow velocity increase, the channel reaches a *critical* condition of Froude number unity, where velocity equals the speed of a small-amplitude surface wave in the channel. Every channel has a critical slope that varies with the flow rate and roughness. If the flow becomes supercritical (Fr > 1), it may undergo a hydraulic jump to a greater depth and lower (subcritical) velocity, analogous to a normal shock wave.

The analysis of gradually varied flow leads to a differential equation (10.51) that can be solved by numerical methods. The chapter ends with a discussion of the flow over a dam or weir, where the total flow rate can be correlated with upstream water depth.

Problems

Most of the problems herein are fairly straightforward. More difficult or open-ended assignments are labeled with an asterisk. Problems labeled with a computer icon may require the use of a computer. The standard end-of-chapter problems P10.1 to P10.128 (categorized in the problem list here) are followed by word problems W10.1 to W10.13, fundamentals of engineering exam problems FE10.1 to FE10.7, comprehensive problems C10.1 to C10.7, and design projects D10.1 and D10.2.

Problem Distribution

Section	Topic	Problems
10.1	Introduction: Froude number, wave speed	P10.1–P10.10
10.2	Uniform flow: the Chézy formula	P10.11–P10.36
10.3	Efficient uniform-flow channels	P10.37–P10.47
10.4	Specific energy: critical depth	P10.48–P10.58
10.4	Flow over a bump	P10.59–P10.68
10.4	Sluice gate flow	P10.69–P10.78
10.5	The hydraulic jump	P10.79–P10.96
10.6	Gradually varied flow	P10.97–P10.112
10.7	Weirs and flumes	P10.113–P10.123
10.7	Backwater curves	P10.124–P10.128

Introduction: Froude number, wave speed

P10.1 The formula for shallow-water wave propagation speed, Eq. (10.9) or (10.10), is independent of the physical properties of the liquid, like density, viscosity, or surface tension. Does this mean that waves propagate at the same speed in water, mercury, gasoline, and glycerin? Explain.

P10.2 Water at 20°C flows in a 30-cm-wide rectangular channel at a depth of 10 cm and a flow rate of 80,000 cm^3/s. Estimate (*a*) the Froude number and (*b*) the Reynolds number.

P10.3 Narragansett Bay is approximately 34 km long and has an average depth of 13 m. Tidal charts for the area indicate a time delay of 30 min between high tide at the mouth of the bay (Newport, Rhode Island) and its head (Providence, Rhode Island). Is this delay correlated with the propagation of a shallow-water tidal crest wave through the bay? Explain.

P10.4 The water flow in Fig. P10.4 has a free surface in three places. Does it qualify as an open-channel flow? Explain. What does the dashed line represent?

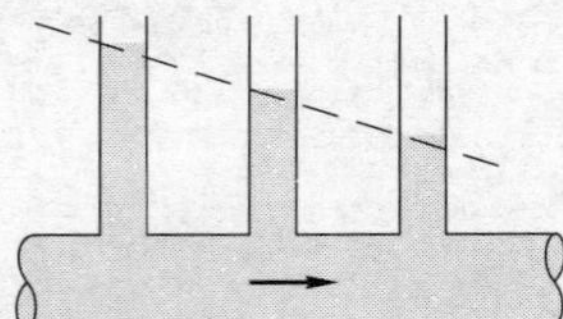

P10.4

P10.5 Water flows down a rectangular channel that is 1.2 m wide and 0.6 m deep. The flow rate is 1.25 m^3/s. Estimate the Froude number of the flow.

P10.6 Pebbles dropped successively at the same point, into a water channel flow of depth 42 cm, create two circular ripples, as in Fig. P10.6. From this information estimate (*a*) the Froude number and (*b*) the stream velocity.

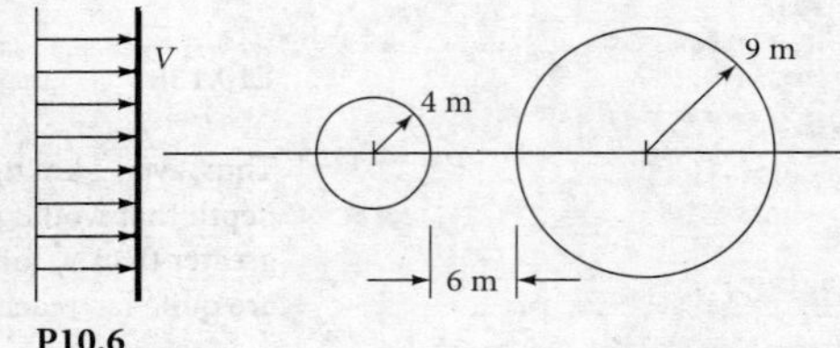

P10.6

P10.7 Pebbles dropped successively at the same point, into a water channel flow of depth 65 cm, create two circular ripples, as in Fig. P10.7. From this information estimate (*a*) the Froude number and (*b*) the stream velocity.

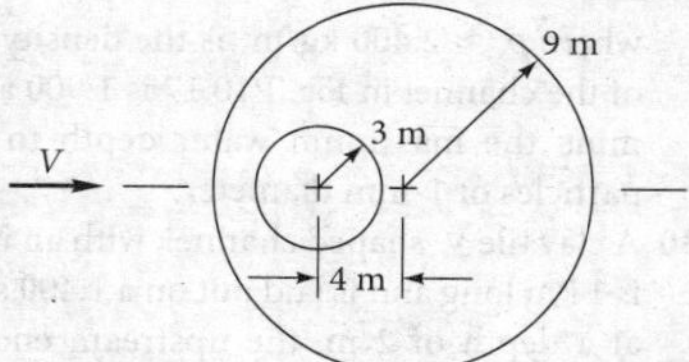

P10.7

P10.8 An earthquake near the Kenai Peninsula, Alaska, creates a single "tidal" wave (called a *tsunami*) that propagates southward across the Pacific Ocean. If the average ocean depth is 4 km and seawater density is 1,025 kg/m^3, estimate the time of arrival of this tsunami in Hilo, Hawaii.

P10.9 Equation (10.10) is for a single disturbance wave. For *periodic* small-amplitude surface waves of wavelength λ and period T, inviscid theory [8 to 10] predicts a wave propagation speed

$$c_0^2 = \frac{g\lambda}{2\pi}\tanh\frac{2\pi y}{\lambda}$$

where y is the water depth and surface tension is neglected. (*a*) Determine if this expression is affected by the Reynolds number, Froude number, or Weber number. Derive the limiting values of this expression for (*b*) $y \ll \lambda$ and (*c*) $y \gg \lambda$. (*d*) For what ratio y/λ is the wave speed within 1 percent of limit (*c*)?

P10.10 If surface tension Υ is included in the analysis of Prob. P10.9, the resulting wave speed is [8 to 10]

$$c_0^2 = \left(\frac{g\lambda}{2\pi} + \frac{2\pi\Upsilon}{\rho\lambda}\right)\tanh\frac{2\pi y}{\lambda}$$

(*a*) Determine if this expression is affected by the Reynolds number, Froude number, or Weber number. Derive the limiting values of this expression for (*b*) $y \ll \lambda$ and (*c*) $y \gg \lambda$. (*d*) Finally, determine the wavelength λ_{crit} for a minimum value of c_0, assuming that $y \gg \lambda$.

Uniform flow: the Chézy formula

P10.11 A rectangular channel is 2 m wide and contains water 3 m deep. If the slope is 0.85° and the lining is corrugated metal, estimate the discharge for uniform flow.

P10.12 (*a*) For laminar draining of a wide, thin sheet of water on pavement sloped at angle θ, as in Fig. P4.36, show that the flow rate is given by

$$Q = \frac{\rho g b h^3 \sin\theta}{3\mu}$$

where b is the sheet width and h its depth. (*b*) By (somewhat laborious) comparison with Eq. (10.13), show that this expression is compatible with a friction factor $f = 24/\text{Re}$, where $\text{Re} = V_{av}h/\nu$.

P10.13 A large pond drains down an asphalt rectangular channel that is 0.6 m wide. The channel slope is 0.8 degrees. If the flow is uniform, at a depth of 0.5 m, estimate the time to drain 1,000 m^3 of water.

P10.14 The Chézy formula (10.18) is independent of fluid density and viscosity. Does this mean that water, mercury, alcohol, and SAE 30 oil will all flow down a given open channel at the same rate? Explain.

P10.15 The painted-steel channel of Fig. P10.15 is designed, without the barrier, for a flow rate of 6 m^3/s at a normal depth of 1 m. Determine (*a*) the design slope of the channel and (*b*) the reduction in total flow rate if the proposed painted-steel central barrier is installed.

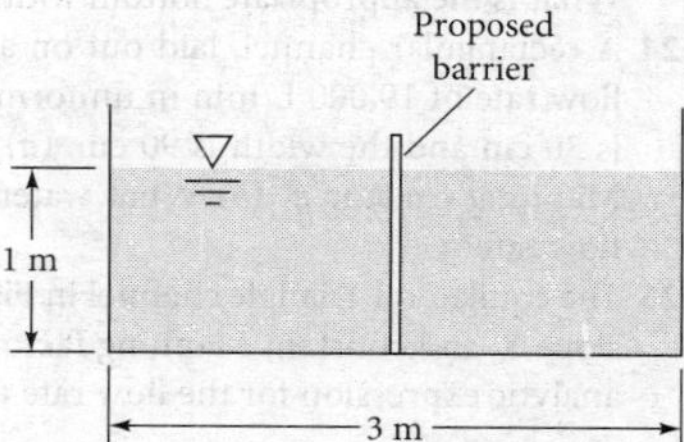

P10.15

P10.16 Water flows in a brickwork rectangular channel 2 m wide, on a slope of 5 m/km. (*a*) Find the flow rate when the normal depth is 50 cm. (*b*) If the normal depth remains 50 cm, find the channel width which will triple the flow rate. Comment on this result.

P10.17 The trapezoidal channel of Fig. P10.17 is made of brickwork and slopes at 1:500. Determine the flow rate if the normal depth is 80 cm.

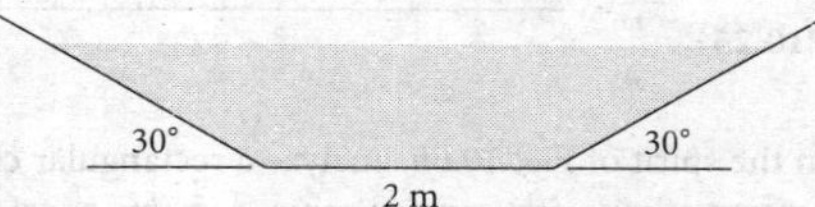

P10.17

P10.18 A V-shaped painted steel channel, similar to Fig. E10.6, has an included angle of 90°. If the slope, in uniform flow, is 3 m per km, estimate (*a*) the flow rate, in m^3/s and (*b*) the average wall shear stress. Take $y = 2$ m.

P10.19 Modify Prob. P10.18, the 90° V channel, to let the surface be clean earth, which erodes if the average velocity exceeds 2 m/s. Find the maximum depth that avoids erosion. The slope is still 3 m per km.

P10.20 An unfinished concrete sewer pipe, of diameter 1.2 m, is flowing half-full at 150 m^3 per minute. If this is the normal depth, what is the pipe slope, in degrees?

P10.21 An engineer makes careful measurements with a weir (see Sec. 10.7) that monitors a rectangular unfinished concrete channel laid on a slope of 1°. She finds, perhaps with surprise, that when the water depth doubles from 66 cm to 132 cm, the normal flow rate more than doubles, from

5.66 to 14.16 m^3/s. (*a*) Is this plausible? (*b*) If so, estimate the channel width.

P10.22 For more than a century, woodsmen harvested trees in Skowhegan, ME, elevation 52 m, and floated the logs down the Kennebec River to Bath, ME, elevation 19 m, a distance of 116 km. The river has an average depth of 4.25 m and an average width of 120 m. Assuming uniform flow and a stony bottom, estimate the travel time required for this trip.

P10.23 It is desired to excavate a clean-earth channel as a trapezoidal cross section with $\theta = 60°$ (see Fig. 10.7). The expected flow rate is 15 m^3/s, and the slope is 1.5 m per kilometer. The uniform flow depth is planned, for efficient performance, such that the flow cross section is half a hexagon. What is the appropriate bottom width of the channel?

P10.24 A rectangular channel, laid out on a 0.5° slope, delivers a flow rate of 19,000 L/min in uniform flow when the depth is 30 cm and the width is 90 cm. (*a*) Estimate the value of Manning's factor *n*. (*b*) What water depth will triple the flow rate?

P10.25 The equilateral-triangle channel in Fig. P10.25 has constant slope S_o and constant Manning factor *n*. If $y = a/2$, find an analytic expression for the flow rate *Q*.

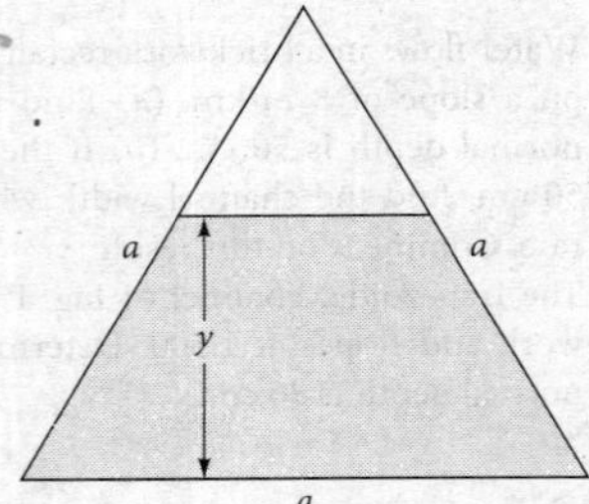

P10.25

P10.26 In the spirit of Fig. 10.6*b*, analyze a rectangular channel in uniform flow with constant area $A = by$, constant slope, but varying width *b* and depth *y*. Plot the resulting flow rate *Q*, normalized by its maximum value Q_{max}, in the range $0.2 < b/y < 4.0$, and comment on whether it is crucial for discharge efficiency to have the channel flow at a depth exactly equal to half the channel width.

P10.27 A circular corrugated-metal water channel has a slope of 1:800 and a diameter of 1.8 m. (*a*) Estimate the normal discharge, in L/min, when the water depth is 1.2 m. (*b*) For this condition, calculate the average wall shear stress.

P10.28 A new, finished-concrete trapezoidal channel, similar to Fig. 10.7, has $b = 2.4$ m, $y_n = 1.5$ m, and $\theta = 50°$. For this depth, the discharge is 15 m^3/s. (*a*) What is the slope of the channel? (*b*) As years pass, the channel corrodes and *n* doubles. What will be the new normal depth for the same flow rate?

P10.29 Suppose that the trapezoidal channel of Fig. P10.17 contains sand and silt that we wish not to erode. According to an empirical correlation by A. Shields in 1936, the average wall shear stress τ_{crit} required to erode sand particles of diameter d_p is approximated by

$$\frac{\tau_{crit}}{(\rho_s - \rho) g\, d_p} \approx 0.5$$

where $\rho_s \approx 2{,}400$ kg/m^3 is the density of sand. If the slope of the channel in Fig. P10.17 is 1:900 and $n \approx 0.014$, determine the maximum water depth to keep from eroding particles of 1-mm diameter.

P10.30 A clay tile V-shaped channel, with an included angle of 90°, is 1 km long and is laid out on a 1:400 slope. When running at a depth of 2 m, the upstream end is suddenly closed while the lower end continues to drain. Assuming quasi-steady normal discharge, find the time for the channel depth to drop to 20 cm.

P10.31 An unfinished-concrete 2 m-diameter sewer pipe flows half full. What is the appropriate slope to deliver 200 m^3/min of water in uniform flow?

P10.32 Does half a V-shaped channel perform as well as a full V-shaped channel? The answer to Prob. 10.18 is $Q = 12.4$ m^3/s. (Do not reveal this to your friends still working on P10.18.) For the painted-steel half-V in Fig. P10.32, at the same slope of 3:1,000, find the flow area that gives the same *Q* and compare with P10.18.

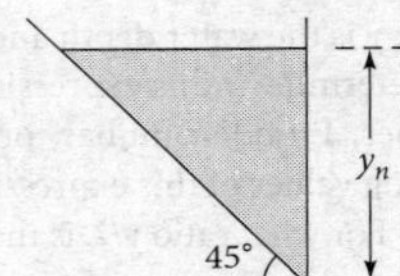

P10.32

P10.33 Five sewer pipes, each a 2-m-diameter clay tile pipe running half full on a slope of 0.25°, empty into a single asphalt pipe, also laid out at 0.25°. If the large pipe is also to run half full, what should be its diameter?

P10.34 A brick rectangular channel with $S_0 = 0.002$ is designed to carry 6.5 m^3/s of water in uniform flow. There is an argument over whether the channel width should be 120 or 240 cm. Which design needs fewer bricks? By what percentage?

P10.35 In flood stage a natural channel often consists of a deep main channel plus two floodplains, as in Fig. P10.35. The floodplains are often shallow and rough. If the channel has the same slope everywhere, how would you analyze this situation for the discharge? Suppose that $y_1 = 6$ m, $y_2 = 1.5$ m, $b_1 = 12$ m, $b_2 = 30$ m, $n_1 = 0.020$, and $n_2 = 0.040$, with a slope of 0.0002. Estimate the discharge in m^3/s.

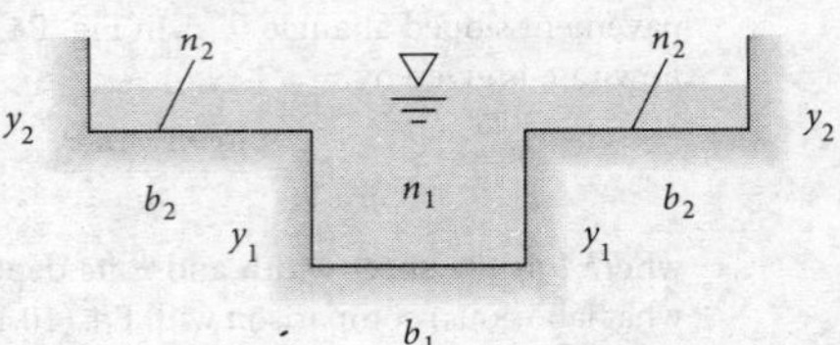

P10.35

P10.36 The Blackstone River in northern Rhode Island normally flows at about 25 m^3/s and resembles Fig. P10.35 with a clean-earth center channel, $b_1 \approx 20$ m and $y_1 \approx 3$ m. The bed slope is about 0.4 m per kilometer. The sides are heavy brush with $b_2 \approx 150$ m. During Hurricane Carol in 1954, a record flow rate of 1,000 m^3/s was estimated. Use this information to estimate the maximum flood depth y_2 during this event.

Efficient uniform-flow channels

P10.37 A triangular channel (see Fig. E10.6) is to be constructed of corrugated metal and will carry 8 m^3/s on a slope of 0.005. The supply of sheet metal is limited, so the engineers want to minimize the channel surface. What are (*a*) the best included angle θ for the channel, (*b*) the normal depth for part (*a*), and (*c*) the wetted perimeter for part (*b*)?

P10.38 For the half-Vee channel in Fig. P10.32, let the interior angle of the Vee be θ. For a given value of area, slope, and n, find the value of θ for which the flow rate is a maximum. To avoid cumbersome algebra, simply plot Q versus θ for constant A.

P10.39 A trapezoidal channel has $n = 0.022$ and $S_0 = 0.0003$ and is made in the shape of a half-hexagon for maximum efficiency. What should the length of the side of the hexagon be if the channel is to carry 6.4 m^3/s of water? What is the discharge of a semicircular channel of the same cross-sectional area and the same S_0 and n?

P10.40 Using the geometry of Fig. 10.6*a*, prove that the most efficient circular open channel (maximum hydraulic radius for a given flow area) is a semicircle.

P10.41 Determine the most efficient value of θ for the V-shaped channel of Fig. P10.41.

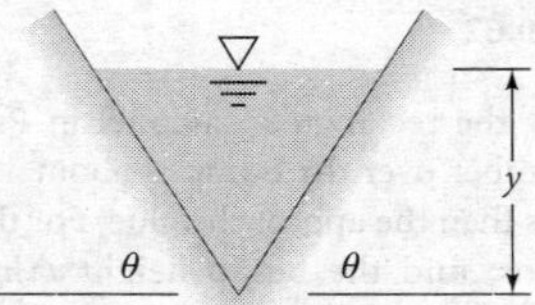

P10.41

P10.42 It is desired to deliver 113,500 L/min of water in a brickwork channel laid on a slope of 1:100. Which would require fewer bricks, in uniform flow: (*a*) a V channel with $\theta = 45°$, as in Fig. P10.41, or (*b*) an efficient rectangular channel with $b = 2y$?

P10.43 Determine the most efficient dimensions for a clay tile rectangular channel to carry 416,000 L/min on a slope of 0.002.

P10.44 What are the most efficient dimensions for a half-hexagon cast iron channel to carry 57,000 L/min on a slope of 0.16°?

P10.45 Calculus tells us that the most efficient wall angle for a V-shaped channel (Fig. P10.41) is $\theta = 45°$. It yields the highest normal flow rate for a given area. But is this a sharp or a flat maximum? For a flow area of 1 m^2 and an unfinished-concrete channel with a slope of 0.004, plot the normal flow rate Q, in m^3/s, versus angle for the range $30° \le \theta \le 60°$ and comment.

P10.46 It is suggested that a channel that reduces erosion has a parabolic shape, as in Fig. P10.46. Formulas for area and perimeter of the parabolic cross section are as follows [7, p. 36]:

$$A = \frac{2}{3}bh_0;\ P = \frac{b}{2}\left[\sqrt{1+\alpha^2} + \frac{1}{\alpha}\ln(\alpha + \sqrt{1+\alpha^2})\right]$$

$$\text{where} \quad \alpha = \frac{4\,h_0}{b}$$

For uniform flow conditions, determine the most efficient ratio h_0/b for this channel (minimum perimeter for a given constant area).

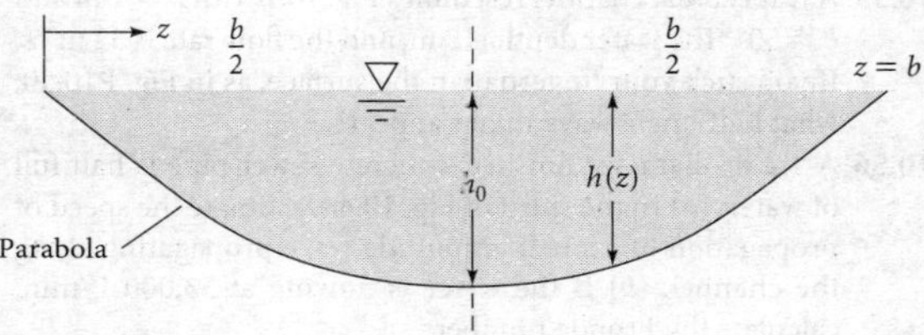

P10.46

P10.47 Calculus tells us that the most efficient water depth for a rectangular channel (such as Fig. E10.1) is $y/b = 1/2$. It yields the highest normal flow rate for a given area. But is this a sharp or a flat maximum? For a flow area of 1 m^2 and a clay tile channel with a slope of 0.006, plot the normal flow rate Q, in m^3/s, versus y/b for the range $0.3 \le y/b \le 0.7$ and comment

Specific energy: critical depth

P10.48 A wide, clean-earth river has a flow rate $q = 14$ $m^3/(m \cdot s)$. What is the critical depth? If the actual depth is 3.65 m, what is the Froude number of the river? Compute the critical slope by (*a*) Manning's formula and (*b*) the Moody chart.

P10.49 Find the critical depth of the brick channel in Prob. P10.34 for both the 120- and 245-cm widths. Are the normal flows subcritical or supercritical?

P10.50 A pencil point piercing the surface of a rectangular channel flow creates a wedgelike 25° half-angle wave, as in Fig. P10.50. If the channel surface is painted steel and the depth is 35 cm, determine (*a*) the Froude number, (*b*) the critical depth, and (*c*) the critical slope for uniform flow.

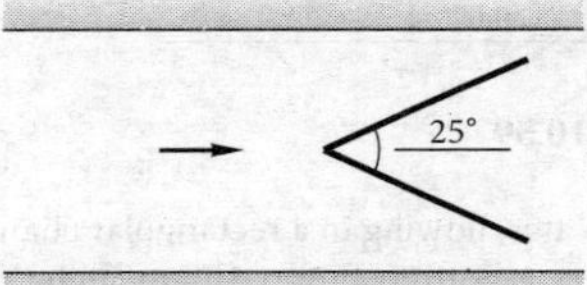

P10.50

P10.51 An unfinished concrete duct, of diameter 1.5 m, is flowing half-full at 8.0 m^3/s. (*a*) Is this a critical flow? If not, what is (*b*) the critical flow rate, (*c*) the critical slope, and (*d*) the Froude number? (*e*) If the flow is uniform, what is the slope of the duct?

P10.52 Water flows full in an asphalt half-hexagon channel of bottom width *W*. The flow rate is 12 m^3/s. Estimate *W* if the Froude number is exactly 0.60.

P10.53 For the river flow of Prob. P10.48, find the depth y_2 that has the same specific energy as the given depth y_1 = 365 cm. These are called *conjugate depths*. What is Fr_2?

P10.54 A clay tile V-shaped channel has an included angle of 70° and carries 8.5 m^3/s. Compute (*a*) the critical depth, (*b*) the critical velocity, and (*c*) the critical slope for uniform flow.

P10.55 A trapezoidal channel resembles Fig. 10.7 with $b = 1$ m and $\theta = 50°$. The water depth is 2 m, and the flow rate is 32 m^3/s. If you stick your fingernail in the surface, as in Fig. P10.50, what half-angle wave might appear?

P10.56 A 1.2 m-diameter finished-concrete sewer pipe is half full of water. (*a*) In the spirit of Fig. 10.4*a*, estimate the speed of propagation of a small-amplitude wave propagating along the channel. (*b*) If the water is flowing at 53,000 L/min, calculate the Froude number.

P10.57 Consider the V-shaped channel of arbitrary angle in Fig. P10.41. If the depth is *y*, (*a*) find an analytic expression for the propagation speed c_0 of a small-disturbance wave along this channel. [*Hint:* Eliminate flow rate from the analyses in Sec. 10.4.] If θ = 45° and the depth is 1 m, determine (*b*) the propagation speed and (*c*) the flow rate if the channel is running at a Froude number of 1/3.

P10.58 For a half-hexagon channel running full, find an analytic expression for the propagation speed of a small-disturbance wave travelling along this channel. Denote the bottom width as *b* and use Fig. 10.7 as a guide.

Flow over a bump

P10.59 Uniform water flow in a wide brick channel of slope 0.02° moves over a 10-cm bump as in Fig. P10.59. A slight depression in the water surface results. If the minimum water depth over the bump is 50 cm, compute (*a*) the velocity over the bump and (*b*) the flow rate per meter of width.

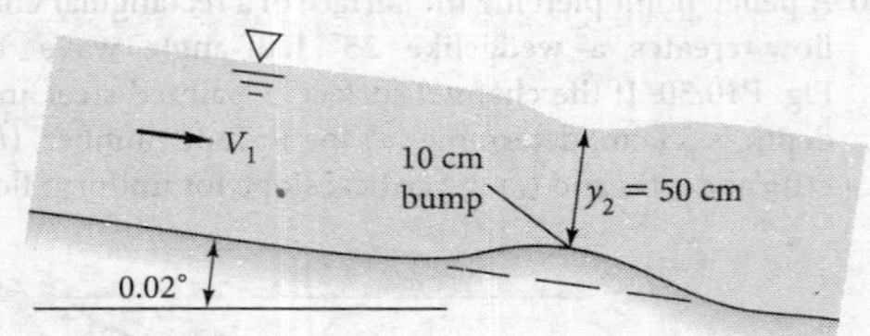

P10.59

P10.60 Water, flowing in a rectangular channel 2 m wide, encounters a bottom bump 10 cm high. The approach depth is 60 cm, and the flow rate 4.8 m^3/s. Determine (*a*) the water depth, (*b*) velocity, and (*c*) Froude number above the bump. *Hint:* The change in water depth is rather slight, only about 8 cm.

P10.61 Modify Prob. P10.59 as follows: Again assuming uniform subcritical approach flow (V_1, y_1), find (*a*) the flow rate and (*b*) y_2 for which the flow at the crest of the bump is exactly critical ($Fr_2 = 1.0$).

P10.62 Consider the flow in a wide channel over a bump, as in Fig. P10.62. One can estimate the water depth change or *transition* with frictionless flow. Use continuity and the Bernoulli equation to show that

$$\frac{dy}{dx} = -\frac{dh/dx}{1 - V^2/(gy)}$$

Is the drawdown of the water surface realistic in Fig. P10.62? Explain under what conditions the surface might rise above its upstream position y_0.

P10.63 In Fig. P10.62 let $V_0 = 1$ m/s and $y_0 = 1$ m. If the maximum bump height is 15 cm, estimate (*a*) the Froude number over the top of the bump and (*b*) the maximum depression in the water surface.

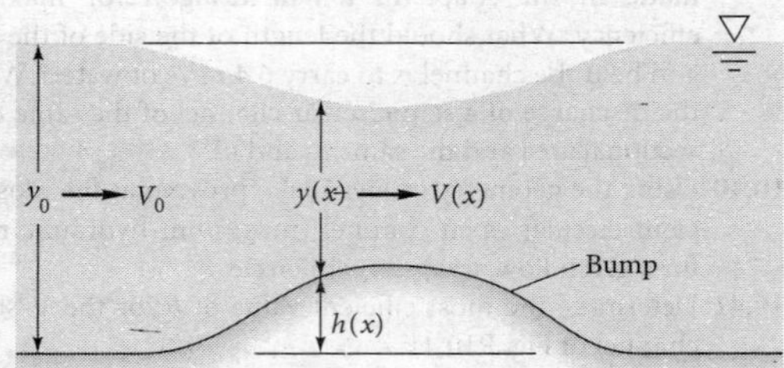

P10.62

P10.64 For the rectangular channel in Prob. P10.60, the Froude number over the bump is about 1.37, which is 17 percent less than the approach value. For the same entrance conditions, find the bump height Δh that causes the bump Froude number to be 1.00.

P10.65 Program and solve the differential equation of "frictionless flow over a bump," from Prob. P10.62, for entrance conditions $V_0 = 1$ m/s and $y_0 = 1$ m. Let the bump have the convenient shape $h = 0.5h_{max}[1 - \cos(2\pi x/L)]$, which simulates Fig. P10.62. Let $L = 3$ m, and generate a numerical solution for $y(x)$ in the bump region $0 < x < L$. If you have time for only one case, use $h_{max} = 15$ cm (Prob. P10.63), for which the maximum Froude number is 0.425. If more time is available, it is instructive to examine a complete family of surface profiles for $h_{max} \approx 1$ cm up to 35 cm (which is the solution of Prob. P10.64).

***P10.66** In Fig. P10.62, let $V_o = 5.5$ m/s and $y_o = 90$ cm. (*a*) Will the water rise or fall over the bump? (*b*) For a bump height of 30 cm, determine the Froude number over the bump. (*c*) Find the bump height that will cause critical flow over the bump.

P10.67 Modify Prob. P10.63 so that the 15-cm change in bottom level is a *depression*, not a bump. Estimate (*a*) the Froude number above the depression and (*b*) the maximum change in water depth.

P10.68 Modify Prob. P10.65 to have a supercritical approach condition $V_0 = 6$ m/s and $y_0 = 1$ m. If you have time for only one case, use $h_{max} = 35$ cm (Prob. P10.66), for which the maximum Froude number is 1.47. If more time is available, it is instructive to examine a complete family of surface profiles for 1 cm $< h_{max} <$ 52 cm (which is the solution to Prob. P10.67).

Sluice gate flow

***P10.69** Given is the flow of a channel of large width *b* under a sluice gate, as in Fig. P10.69. Assuming frictionless steady flow with negligible upstream kinetic energy, derive a formula for the dimensionless flow ratio $Q^2/(y_1^3 b^2 g)$ as a function of the ratio y_2/y_1. Show by differentiation that the maximum flow rate occurs at $y_2 = 2y_1/3$.

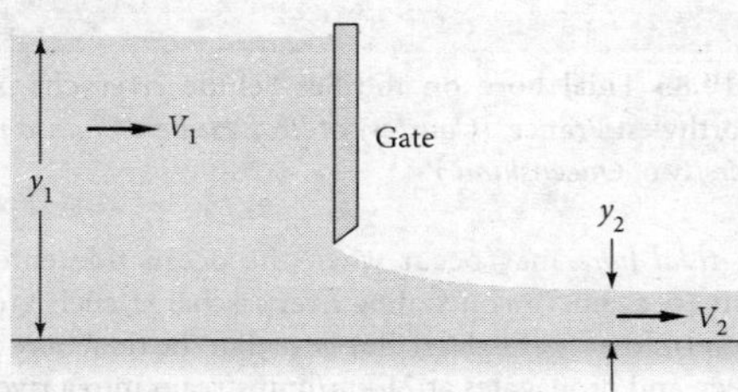

P10.69

P10.70 A periodic and spectacular water release, in China's Henan province, flows through a giant sluice gate. Assume that the gate is 23 m wide, and its opening is 8 m high. The water depth far upstream is 32 m. Assuming free discharge, estimate the volume flow rate through the gate.

P10.71 In Fig. P10.69 let $y_1 = 95$ cm and $y_2 = 50$ cm. Estimate the flow rate per unit width if the upstream kinetic energy is (*a*) neglected and (*b*) included.

***P10.72** Water approaches the wide sluice gate of Fig. P10.72 at $V_1 = 0.2$ m/s and $y_1 = 1$ m. Accounting for upstream kinetic energy, estimate at the outlet, section 2, the (*a*) depth, (*b*) velocity, and (*c*) Froude number.

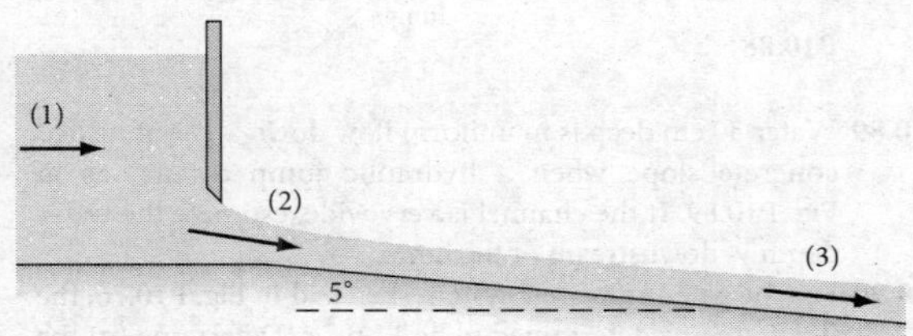

P10.72

P10.73 In Fig. P10.69, let $y_1 = 1.8$ m and the gate width $b = 2.4$ m. Find the gate opening *H* that would allow a free-discharge flow of 114,000 L/min under the gate.

P10.74 With respect to Fig. P10.69, show that, for frictionless flow, the upstream velocity may be related to the water levels by

$$V_1 = \sqrt{\frac{2g(y_1 - y_2)}{K^2 - 1}}$$

where $K = y_1/y_2$.

P10.75 A tank of water 1 m deep, 3 m long, and 4 m wide into the paper has a closed sluice gate on the right side, as in Fig. P10.75. At $t = 0$ the gate is opened to a gap of 10 cm. Assuming quasi-steady sluice gate theory, estimate the time required for the water level to drop to 50 cm. Assume free outflow.

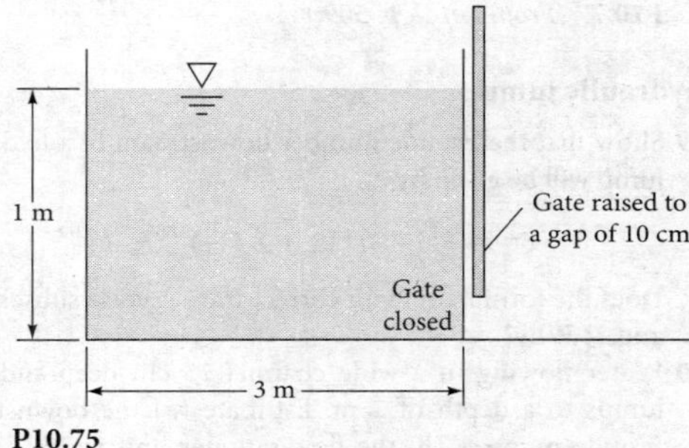

P10.75

P10.76 Figure P10.76 shows a horizontal flow of water through a sluice gate, a hydraulic jump, and over a 1.8 m sharp-crested weir. Channel, gate, jump, and weir are all 2.4 m wide unfinished concrete. Determine (*a*) the flow rate in m³/s and (*b*) the normal depth.

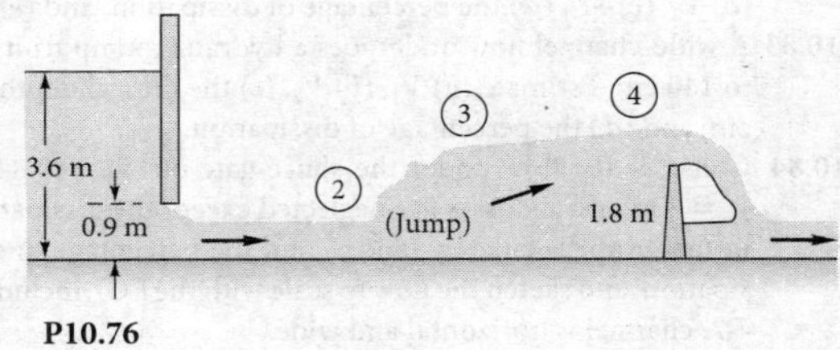

P10.76

P10.77 Equation (10.41) for sluice gate discharge is for free outflow. If the outflow is *drowned*, as in Fig. 10.10*c*, there is dissipation, and C_d drops sharply, as shown in Fig. P10.77, taken from Ref. 2. Use this data to restudy Prob. 10.73, with $H = 23$ cm. Plot the estimated flow rate, in L/min, versus y_2 in the range 0.15 m $< y_2 <$ 1.5 m.

P10.78 In Fig. P10.69, free discharge, a gate opening of 22 cm will allow a flow rate of 114,000 L/min. Recall $y_1 = 1.8$ m and the gate width $b = 2.4$ m. Suppose that the gate is drowned (Fig. P10.77), with $y_2 = 1.2$ m. What gate opening would then be required?

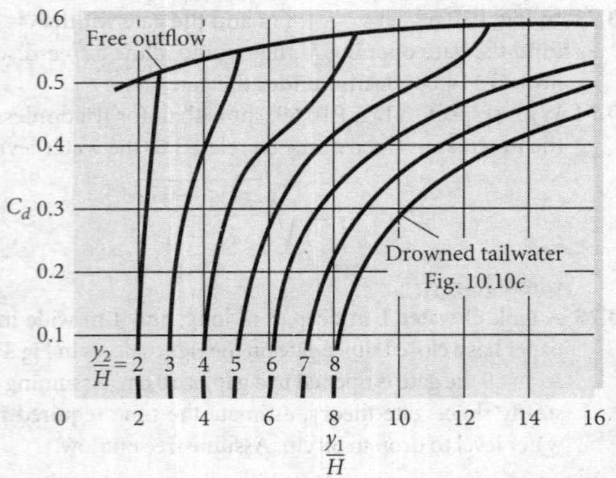

P10.77 *(From Ref. 2, p. 509.)*

The hydraulic jump

P10.79 Show that the Froude number downstream of a hydraulic jump will be given by

$$\text{Fr}_2 = 8^{1/2}\,\text{Fr}_1/[(1 + 8\,\text{Fr}_1^2)^{1/2} - 1]^{3/2}$$

Does the formula remain correct if we reverse subscripts 1 and 2? Why?

P10.80 Water flowing in a wide channel 25 cm deep suddenly jumps to a depth of 1 m. Estimate (*a*) the downstream Froude number; (*b*) the flow rate per unit width; (*c*) the critical depth; and (*d*) the percentage of dissipation.

P10.81 Water flows in a wide channel at $q = 2\ \text{m}^3/(\text{s} \cdot \text{m})$, $y_1 = 0.3$ m, and then undergoes a hydraulic jump. Compute y_2, V_2, Fr_2, h_f, the percentage of dissipation, and the horsepower dissipated per unit width. What is the critical depth?

P10.82 Downstream of a wide hydraulic jump the flow is 1.2 m deep and has a Froude number of 0.5. Estimate (*a*) y_1, (*b*) V_1, (*c*) Fr_1, (*d*) the percentage of dissipation, and (*e*) y_c.

P10.83 A wide-channel flow undergoes a hydraulic jump from 40 to 140 cm. Estimate (*a*) V_1, (*b*) V_2, (*c*) the critical depth, in cm, and (*d*) the percentage of dissipation.

***P10.84** Consider the flow under the sluice gate of Fig. P10.84. If $y_1 = 3$ m and all losses are neglected except the dissipation in the jump, calculate y_2 and y_3 and the percentage of dissipation, and sketch the flow to scale with the EGL included. The channel is horizontal and wide.

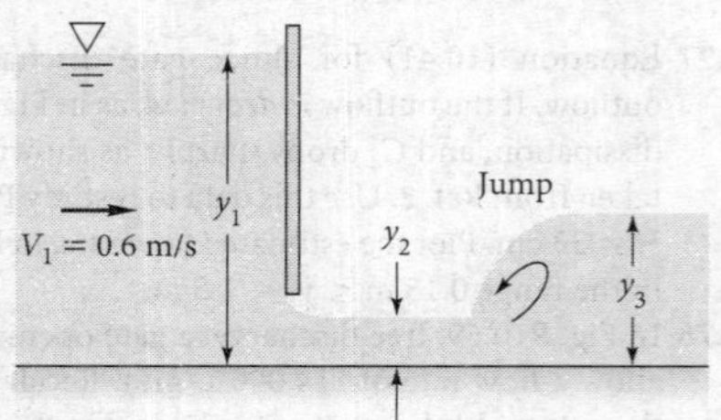

P10.84

P10.85 The analogy between a hydraulic jump and a normal shock equates Mach number and Froude number, air density and water depth, air pressure and the square of the water depth. Test this analogy for $\text{Ma}_1 = \text{Fr}_1 = 4.0$ and comment on the results.

P10.86 A *bore* is a hydraulic jump that propagates upstream into a still or slower-moving fluid, as in Fig. P10.86, on the Sée-Sélune channel, near Mont Saint Michel in northwest France. The bore is moving at about 3 m/s and is about 30 cm high. Estimate (*a*) the depth of the water in this area and (*b*) the velocity induced by the wave.

P10.86 Tidal bore on the Sée-Sélune river channel in northwest France. *(Courtesy of Prof. Hubert Chanson, University of Queensland.)*

P10.87 A *tidal bore* may occur when the ocean tide enters an estuary against an oncoming river discharge, such as on the Severn River in England. Suppose that the tidal bore is 3 m deep and propagates at 21 km/h upstream into a river that is 2 m deep. Estimate the river current in m/s.

***P10.88** Consider supercritical flow, $\text{Fr}_1 > 1$, down a shallow flat water channel toward a wedge of included angle 2θ, as in Fig. P10.88. By the compressible flow analogy, hydraulic jumps should form, similar to the shock waves in Fig. P9.132*a*. Using an approach similar to Fig. 9.20, develop and explain the equations that could be used to find the wave angle β and Fr_2.

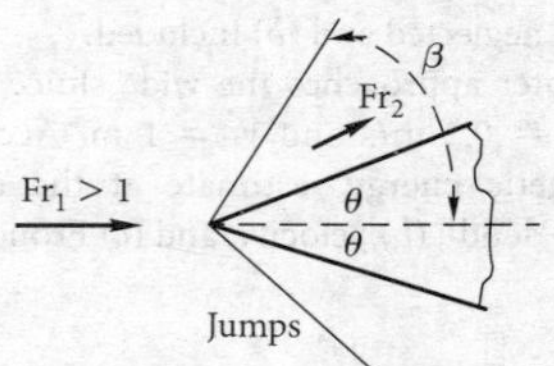

P10.88

P10.89 Water 30 cm deep is in uniform flow down a 1° unfinished concrete slope when a hydraulic jump occurs, as in Fig. P10.89. If the channel is very wide, estimate the water depth y_2 downstream of the jump.

P10.90 For the gate/jump/weir system sketched in Fig. P10.76, the flow rate was determined to be 11 m³/s. Determine (*a*) the water depths y_2 and y_3, and (*b*) the Froude numbers Fr_2 and Fr_3 before and after the hydraulic jump.

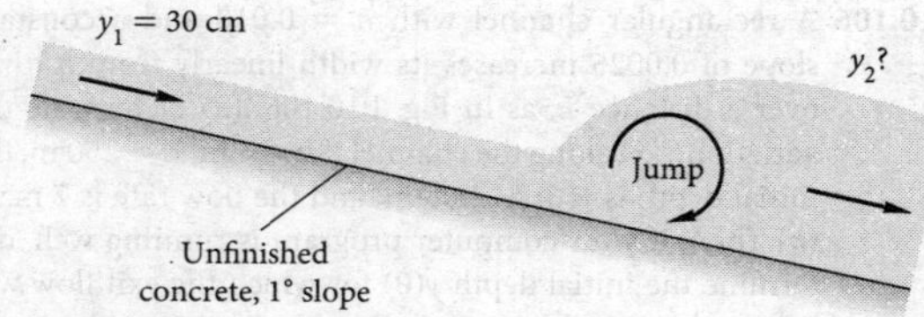

P10.89

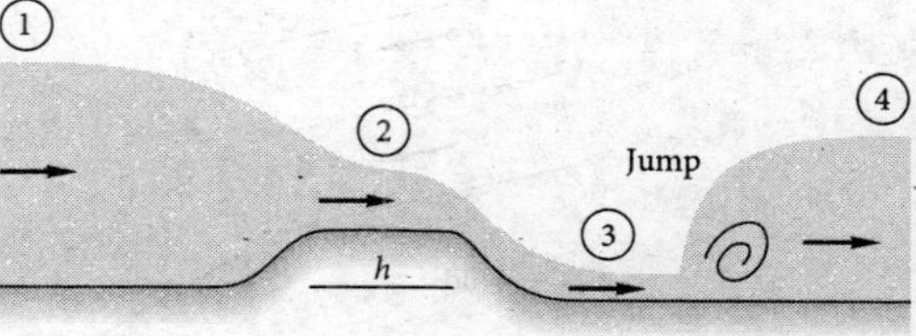

P10.93

*P10.91 Follow up Prob. P10.88 numerically with flow down a shallow, flat water channel 1 cm deep at an average velocity of 0.94 m/s. The wedge half-angle θ is 20°. Calculate (*a*) β; (*b*) Fr_2; and (*c*) y_2.

P10.92 A familiar sight is the circular hydraulic jump formed by a faucet jet falling onto a flat sink surface, as in Fig. P10.92. Because of the shallow depths, this jump is strongly dependent on bottom friction, viscosity, and surface tension [35]. It is also unstable and can form remarkable noncircular shapes, as shown in the website <http://web.mit.edu/jeffa/www/jump.htm>.

P10.92 A circular hydraulic jump in a kitchen sink. *(Courtesy of Prof. Hubert Chanson, University of Queensland.)*

For this problem, assume that two-dimensional jump theory is valid. If the water depth outside the jump is 4 mm, the radius at which the jump appears is $R = 3$ cm, and the faucet flow rate is 100 cm^3/s, find the conditions just upstream of the jump.

P10.93 Water in a horizontal channel accelerates smoothly over a bump and then undergoes a hydraulic jump, as in Fig. P10.93. If $y_1 = 1$ m and $y_3 = 40$ cm, estimate (*a*) V_1, (*b*) V_3, (*c*) y_4, and (*d*) the bump height *h*.

P10.94 In Fig. 10.11, the upstream flow is only 2.65 cm deep. The channel is 50 cm wide, and the flow rate is 0.0359 m^3/s. Determine (*a*) the upstream Froude number, (*b*) the downstream velocity, (*c*) the downstream depth, and (*d*) the percent dissipation.

P10.95 A 10-cm-high bump in a wide horizontal water channel creates a hydraulic jump just upstream and the flow pattern in Fig. P10.95. Neglecting losses except in the jump, for the case $y_3 = 30$ cm, estimate (*a*) V_4, (*b*) y_4, (*c*) V_1, and (*d*) y_1.

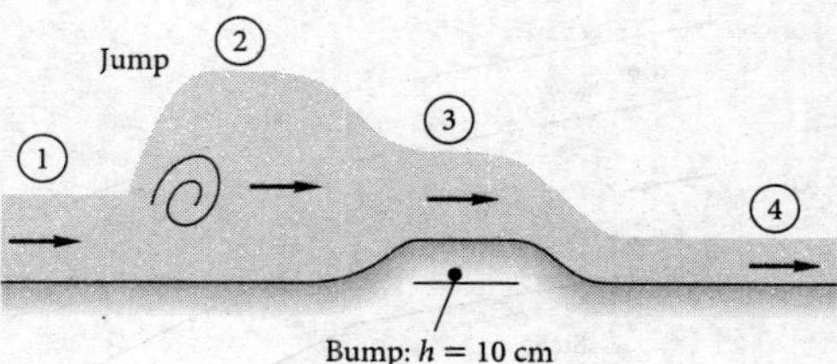

P10.95

P10.96 For the circular hydraulic jump in Fig. P10.92, the water depths before and after the jump are 2 mm and 4 mm, respectively. Assume that two-dimensional jump theory is valid. If the faucet flow rate is 150 cm^3/s, estimate the radius *R* at which the jump will appear.

Gradually varied flow

P10.97 A brickwork rectangular channel 4 m wide is flowing at 8.0 m^3/s on a slope of 0.1°. Is this a mild, critical, or steep slope? What type of gradually varied solution curve are we on if the local water depth is (*a*) 1 m, (*b*) 1.5 m, and (*c*) 2 m?

P10.98 A gravelly earth wide channel is flowing at 10 m^3/s per meter of width on a slope of 0.75°. Is this a mild, critical, or steep slope? What type of gradually varied solution curve are we on if the local water depth is (*a*) 1 m, (*b*) 2 m, or (*c*) 3 m?

P10.99 A clay tile V-shaped channel of included angle 60° is flowing at 1.98 m^3/s on a slope of 0.33°. Is this a mild, critical, or steep slope? What type of gradually varied solution curve are we on if the local water depth is (*a*) 1 m, (*b*) 2 m, or (*c*) 3 m?

P10.100 If bottom friction is included in the sluice gate flow of Prob. P10.84, the depths (y_1, y_2, y_3) will vary with *x*. Sketch the type and shape of gradually varied solution curve in each region (1, 2, 3), and show the regions of rapidly varied flow.

P10.101 Consider the gradual change from the profile beginning at point *a* in Fig. P10.101 on a mild slope S_{01} to a mild but steeper slope S_{02} downstream. Sketch and label the curve $y(x)$ expected.

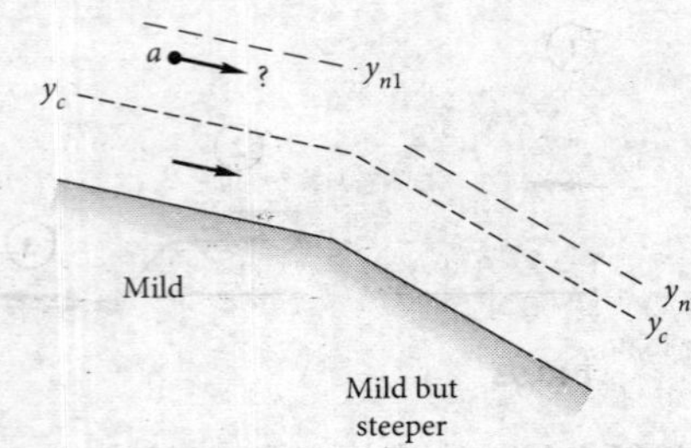

P10.101

***P10.102** The wide-channel flow in Fig. P10.102 changes from a steep slope to one even steeper. Beginning at points *a* and *b*, sketch and label the water surface profiles expected for gradually varied flow.

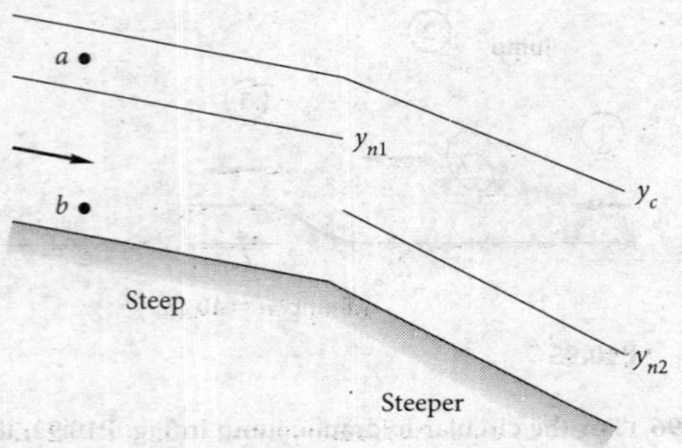

P10.102

P10.103 A gravelly rectangular channel, 7 m wide and 2 m deep, is flowing at 75 m^3/s on a slope of 0.013. (*a*) Is this on a mild, critical, or steep curve? (*b*) Approximately how many meters downstream will the gradually varied solution reach the normal depth?

P10.104 The rectangular-channel flow in Fig. P10.104 expands to a cross section 50 percent wider. Beginning at points *a* and *b*, sketch and label the water surface profiles expected for gradually varied flow.

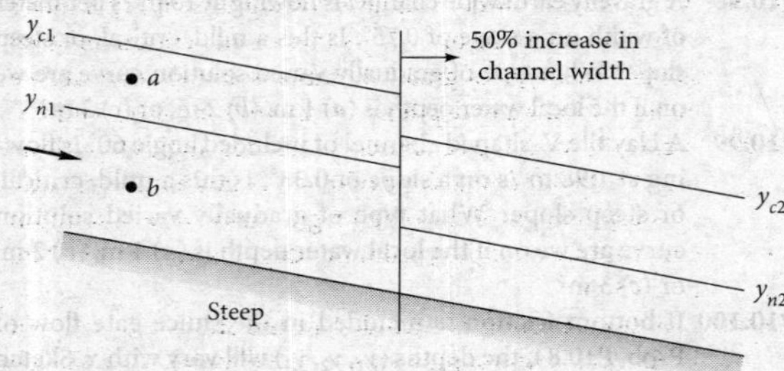

P10.104

P10.105 In Prob. P10.84 the frictionless solution is $y_2 = 0.25$ m, which we denote as $x = 0$ just downstream of the gate. If the channel is horizontal with $n = 0.018$ and there is no hydraulic jump, compute from gradually varied theory the downstream distance where $y = 0.6$ m.

P10.106 A rectangular channel with $n = 0.018$ and a constant slope of 0.0025 increases its width linearly from b to $2b$ over a distance L, as in Fig. P10.106. (*a*) Determine the variation $y(x)$ along the channel if $b = 4$ m, $L = 250$ m, the initial depth is $y(0) = 1.05$ m, and the flow rate is 7 m^3/s. (*b*) Then, if your computer program is running well, determine the initial depth $y(0)$ for which the exit flow will be exactly critical.

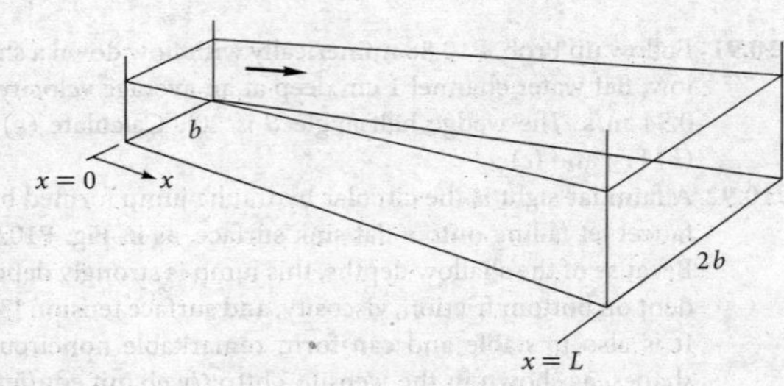

P10.106

P10.107 A clean-earth wide-channel flow is climbing an adverse slope with $S_0 = -0.002$. If the flow rate is $q = 4.5$ m^3/(s · m), use gradually varied theory to compute the distance for the depth to drop from 3.0 to 2.0 m.

P10.108 Water flows at 1.5 m^3/s along a straight, riveted-steel 90° V-shaped channel (see Fig. P10.41, $\theta = 45°$). At section 1, the water depth is 1.0 m. (*a*) As we proceed downstream, will the water depth rise or fall? Explain. (*b*) Depending upon your answer to part (*a*), calculate, in one numerical swoop, from gradually varied theory, the distance downstream for which the depth rises (or falls) 0.1 m.

P10.109 Figure P10.109 illustrates a free overfall or *dropdown* flow pattern, where a channel flow accelerates down a slope and falls freely over an abrupt edge. As shown, the flow reaches critical just before the overfall. Between y_c and the edge the flow is rapidly varied and does not satisfy gradually varied theory. Suppose that the flow rate is $q = 1.3$ m^3/(s · m) and the surface is unfinished concrete. Use Eq. (10.51) to estimate the water depth 300 m upstream as shown.

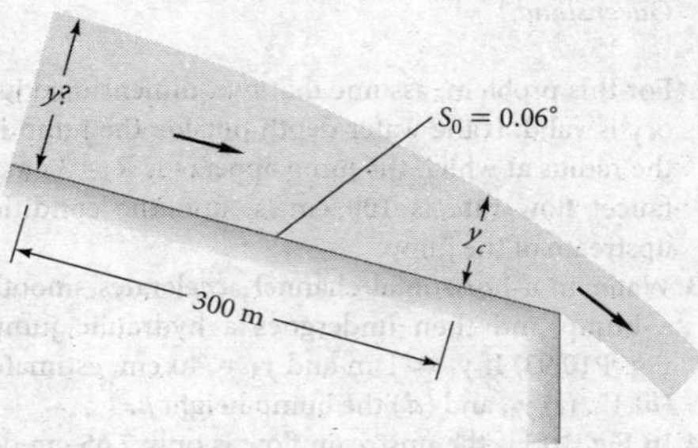

P10.109

P10.110 We assumed frictionless flow in solving the bump case, Prob. P10.65, for which $V_2 = 1.21$ m/s and $y_2 = 0.826$ m over the crest when $h_{max} = 15$ cm, $V_1 = 1$ m/s, and

$y_1 = 1$ m. However, if the bump is long and rough, friction may be important. Repeat Prob. P10.65 for the same bump shape, $h = 0.5h_{max}[1 - \cos(2\pi x/L)]$, to compute conditions (*a*) at the crest and (*b*) at the end of the bump, $x = L$. Let $h_{max} = 15$ cm and $L = 100$ m, and assume a clean-earth surface.

***P10.111** The Rolling Dam on the Blackstone River has a weedy bottom and an average flow rate of 25 m^3/s. Assume the river upstream is 45 m wide and slopes at 1.875 m per km. The water depth just upstream of the dam is 2.35 m. Calculate the water depth 1.5 km upstream (*a*) for the given initial depth, 2.35 m; and (*b*) if flashboards on the dam raise this depth to 3.25 m.

P10.112 The clean-earth channel in Fig. P10.112 is 6 m wide and slopes at 0.3°. Water flows at 30 m^3/s in the channel and enters a reservoir so that the channel depth is 3 m just before the entry. Assuming gradually varied flow, how far is the distance L to a point in the channel where $y = 2$ m? What type of curve is the water surface?

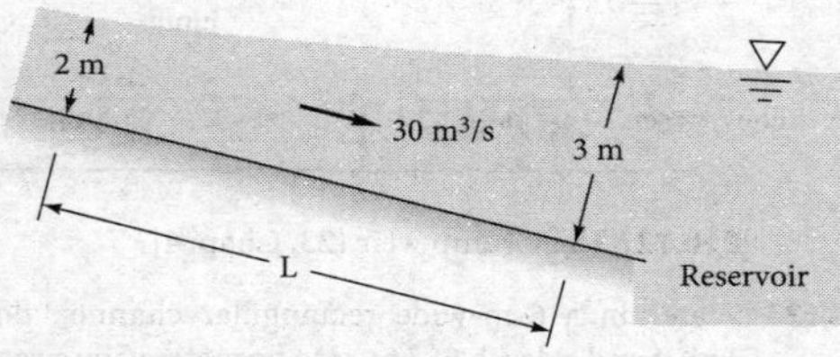

P10.112

Weirs and flumes

P10.113 Figure P10.113 shows a channel contraction section often called a *venturi flume* [23, p. 167] because measurements of y_1 and y_2 can be used to meter the flow rate. Show that if losses are neglected and the flow is one-dimensional and subcritical, the flow rate is given by

$$Q = \left[\frac{2g(y_1 - y_2)}{1/(b_2^2 y_2^2) - 1/(b_1^2 y_1^2)}\right]^{1/2}$$

Apply this to the special case $b_1 = 3$ m, $b_2 = 2$ m, and $y_1 = 1.9$ m. (*a*) Find the flow rate if $y_2 = 1.5$ m. (*b*) Also find the depth y_2 for which the flow becomes critical in the throat.

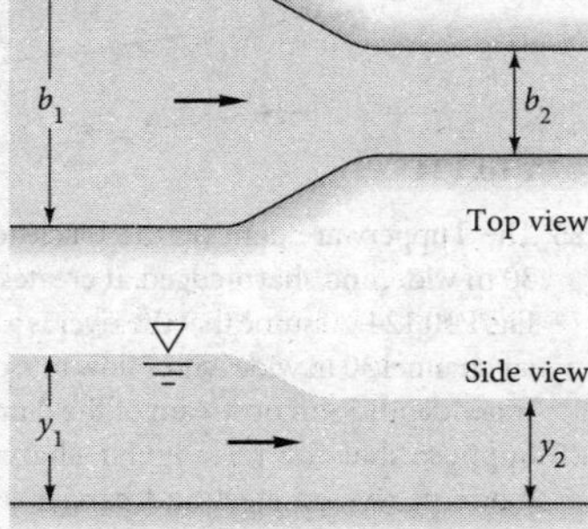

P10.113

P10.114 For the gate/jump/weir system sketched in Fig. P10.76, the flow rate was determined to be 11 m^3/s. Determine the water depth y_4 just upstream of the weir.

P10.115 Gradually varied theory, Eq. (10.49), neglects the effect of *width* changes, db/dx, assuming that they are small. But they are not small for a short, sharp contraction such as the venturi flume in Fig. P10.113. Show that, for a rectangular section with $b = b(x)$, Eq. (10.49) should be modified as follows:

$$\frac{dy}{dx} \approx \frac{S_0 - S + [V^2/(gb)](db/dx)}{1 - \mathrm{Fr}^2}$$

Investigate a criterion for reducing this relation to Eq. (10.49).

P10.116 A Cipolletti weir, popular in irrigation systems, is trapezoidal, with sides sloped at 1:4 horizontal to vertical, as in Fig. P10.116. The following are flow-rate values, from the U.S. Dept. of Agriculture, for a few different system parameters:

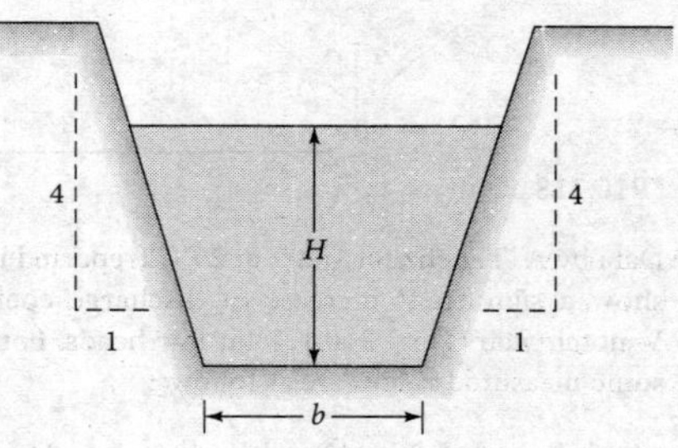

P10.116

H, m	0.24	0.3	0.4	0.45
b, m	0.45	0.6	0.75	10.5
Q, m^3/min	6.13	11.5	22.4	36.9

Source: U.S. Dept of Agriculture.

Use this data to correlate a Cipolletti weir formula with a reasonably constant weir coefficient.

P10.117 A popular flow-measurement device in agriculture is the *Parshall flume* [33], Fig. P10.117, named after its inventor, Ralph L. Parshall, who developed it in 1922 for the U.S. Bureau of Reclamation. The subcritical approach flow is driven, by a steep constriction, to go critical ($y = y_c$) and then supercritical. It gives a constant reading H for a wide range of tailwaters. Derive a formula for estimating Q from measurement of H and knowledge of constriction width b. Neglect the entrance velocity head.

***P10.118** Using a Bernoulli-type analysis similar to Fig. 10.16*a*, show that the theoretical discharge of the V-shaped weir in Fig. P10.118 is given by

$$Q = 0.7542 g^{1/2} \tan\alpha\, H^{5/2}$$

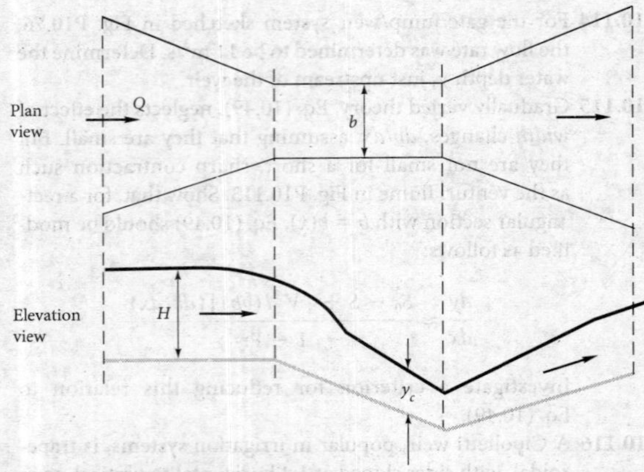

P10.117 The Parshall flume

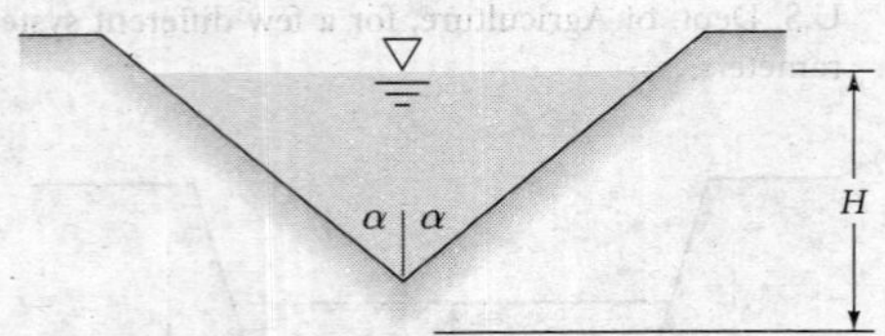

*P10.118

P10.119 Data by A. T. Lenz for water at 20°C (reported in Ref. 23) show a significant increase of discharge coefficient of V-notch weirs (Fig. P10.118) at low heads. For $\alpha = 20°$, some measured values are as follows:

H, cm	6	12	18	24	30
C_d	0.499	0.470	0.461	0.456	0.452

Determine if these data can be correlated with the Reynolds and Weber numbers vis-ã-vis Eq. (10.61). If not, suggest another correlation.

P10.120 The rectangular channel in Fig. P10.120 contains a V-notch weir as shown. The intent is to meter flow rates between 2.0 and 6.0 m³/s with an upstream hook gage set to measure water depths between 2.0 and 2.75 m. What are the most appropriate values for the notch height Y and the notch half-angle α?

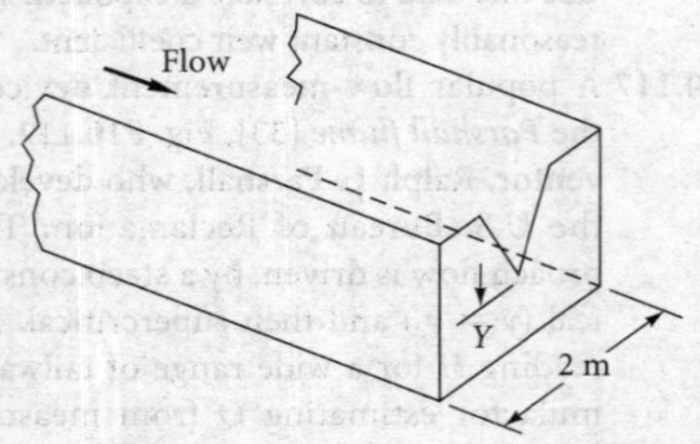

P10.120

P10.121 Water flow in a rectangular channel is to be metered by a thin-plate weir with side contractions, as in Table 10.2*b*, with $L = 1.8$ m and $Y = 0.3$ m. It is desired to measure flow rates between 6,000 and 12,000 L/min with only a 15-cm change in upstream water depth. What is the most appropriate length for the weir width b?

P10.122 In 1952 E. S. Crump developed the triangular weir shape shown in Fig. P10.122 [23, Chap. 4]. The front slope is 1:2 to avoid sediment deposition, and the rear slope is 1:5 to maintain a stable tailwater flow. The beauty of the design is that it has a unique discharge correlation up to near-drowning conditions, $H_2/H_1 \le 0.75$:

$$Q = C_d b g^{1/2}\left(H_1 + \frac{V_1^2}{2g} - k_h\right)^{3/2}$$

where $\quad C_d \approx 0.63 \quad$ and $\quad k_h \approx 0.3$ mm

The term k_h is a low-head loss factor. Suppose that the weir is 3 m wide and has a crest height $Y = 50$ cm. If the water depth upstream is 65 cm, estimate the flow rate in m³/min.

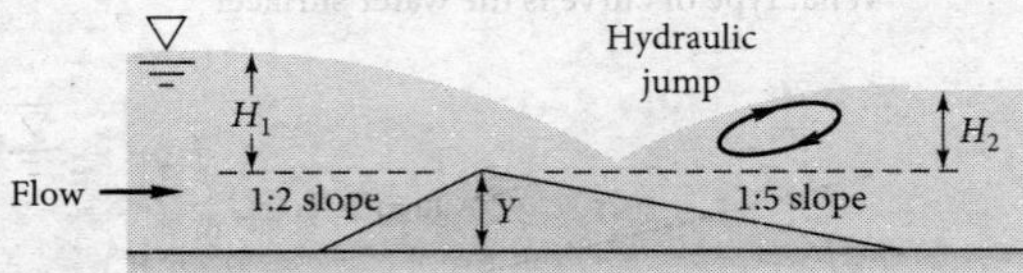

P10.122 The Crump weir [23, Chap. 4]

***P10.123** Water in a 6-m-wide rectangular channel, flowing at 3 m³/s and a depth of 3 m, is to be metered by a rectangular weir with side contractions, as in Table 10.2*b*. Suggest some appropriate design values of b, Y, and H to match the table conditions for this weir.

Backwater curves

P10.124 Water flows at 17 m³/s in a rectangular channel 7 m wide with $n \approx 0.024$ and a slope of 0.1°. A dam increases the depth to 5 m, as in Fig. P10.124. Using gradually varied theory, estimate the distance L upstream at which the water depth will be 3 m. What type of solution curve are we on? What should be the water depth asymptotically far upstream?

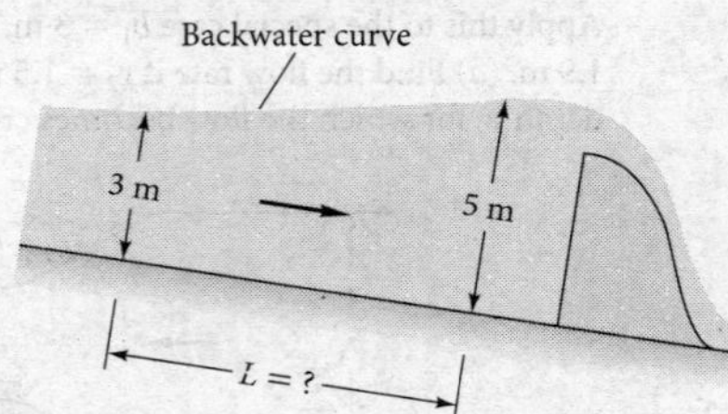

P10.124

P10.125 The Tupperware dam on the Blackstone River is 4 m high, 30 m wide, and sharp-edged. It creates a backwater similar to Fig. P10.124. Assume that the river is a weedy-earth rectangular channel 30 m wide with a flow rate of 23 m³/s. Estimate the water depth 3 km upstream of the dam if $S_0 = 0.001$.

P10.126 Suppose that the rectangular channel of Fig. P10.120 is made of riveted steel and carries a flow of 8 m³/s on a slope of 0.15°. If the V-notch weir has $\alpha = 30°$ and $Y =$ 50 cm, estimate, from gradually varied theory, the water depth 100 m upstream.

P10.127 A clean-earth river is 15 m wide and averages 17 m^3/s. It contains a dam that increases the water depth to 2.5 m, to provide head for a hydropower plant. The bed slope is 0.0025. (*a*) What is the normal depth of this river? (*b*) Engineers propose putting flashboards on the dam to raise the water level to 3 m. Residents a half kilometer upstream are worried about flooding above their present water depth of about 0.7 m. Using Eq. (10.52) in one big half-kilometer step, estimate the new water depth upstream.

P10.128 A rectangular channel 4 m wide is blocked by a broad-crested weir 2 m high, as in Fig. P10.128. The channel is horizontal for 200 m upstream and then slopes at 0.7° as shown. The flow rate is 12 m^3/s, and $n = 0.03$. Compute the water depth y at 300 m upstream from gradually varied theory.

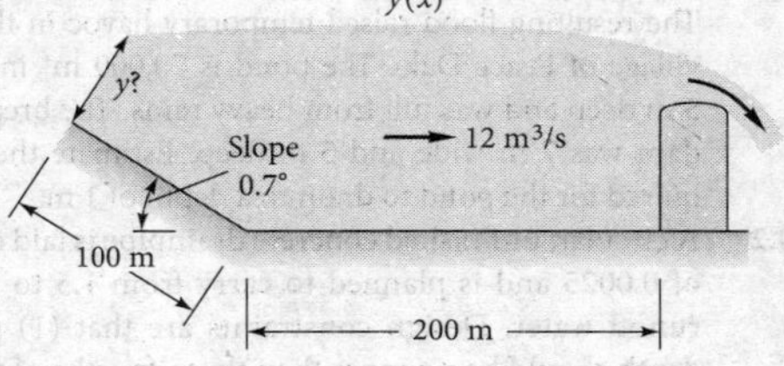

P10.128

Word Problems

W10.1 Free-surface problems are driven by gravity. Why do so many of the formulas in this chapter contain the *square root* of the acceleration of gravity?

W10.2 Explain why the flow under a sluice gate, Fig. 10.10, either is or is not analogous to compressible gas flow through a converging–diverging nozzle, Fig. 9.12.

W10.3 In uniform open-channel flow, what is the balance of forces? Can you use such a force balance to derive the Chézy equation (10.13)?

W10.4 A shallow-water wave propagates at the speed $c_0 \approx (gy)^{1/2}$. What makes it propagate? That is, what is the balance of forces in such wave motion? In which direction does such a wave propagate?

W10.5 Why is the Manning friction correlation, Eq. (10.16), used almost universally by hydraulics engineers, instead of the Moody friction factor?

W10.6 During horizontal channel flow over a bump, is the specific energy constant? Explain.

W10.7 Cite some similarities, and perhaps some dissimilarities, between a hydraulic jump and a gas dynamic normal shock wave.

W10.8 Give three examples of rapidly varied flow. For each case, cite reasons why it does not satisfy one or more of the five basic assumptions of gradually varied flow theory.

W10.9 Is a free overfall, Fig. 10.15*e*, similar to a weir? Could it be calibrated versus flow rate in the same manner as a weir? Explain.

W10.10 Cite some similarities, and perhaps some dissimilarities, between a weir and a Bernoulli obstruction flowmeter from Sec. 6.12.

W10.11 Is a bump, Fig. 10.9*a*, similar to a weir? If not, when does a bump become large enough, or sharp enough, to be a weir?

W10.12 After doing some reading and/or thinking, explain the design and operation of a *long-throated flume*.

W10.13 Describe the design and operation of a *critical-depth flume*. What are its advantages compared to the venturi flume of Prob. P10.113?

Fundamentals of Engineering Exam Problems

The FE Exam is fairly light on open-channel problems in the general (morning) session, but this subject plays a big part in the specialized civil engineering (afternoon) exam.

FE10.1 Consider a rectangular channel 3 m wide laid on a 1° slope. If the water depth is 2 m, the hydraulic radius is (*a*) 0.43 m, (*b*) 0.6 m, (*c*) 0.86 m, (*d*) 1.0 m, (*e*) 1.2 m

FE10.2 For the channel of Prob. FE10.1, the most efficient water depth (best flow for a given slope and resistance) is (*a*) 1 m, (*b*) 1.5 m, (*c*) 2 m, (*d*) 2.5 m, (*e*) 3 m

FE10.3 If the channel of Prob. FE10.1 is built of rubble cement (Manning's $n \approx 0.020$), what is the uniform flow rate when the water depth is 2 m?
(*a*) 6 m^3/s, (*b*) 18 m^3/s, (*c*) 36 m^3/s, (*d*) 40 m^3/s, (*e*) 53 m^3/s

FE10.4 For the channel of Prob. FE10.1, if the water depth is 2 m and the uniform flow rate is 24 m^3/s, what is the approximate value of Manning's roughness factor n?
(*a*) 0.015, (*b*) 0.020, (*c*) 0.025, (*d*) 0.030, (*e*) 0.035

FE10.5 For the channel of Prob. FE10.1, if Manning's roughness factor $n \approx 0.020$ and $Q \approx 29$ m^3/s, what is the normal depth y_n?
(*a*) 1 m, (*b*) 1.5 m, (*c*) 2 m, (*d*) 2.5 m, (*e*) 3 m

FE10.6 For the channel of Prob. FE10.1, if $Q \approx 24$ m^3/s, what is the critical depth y_c?
(*a*) 1.0 m, (*b*) 1.26 m, (*c*) 1.5 m, (*d*) 1.87 m, (*e*) 2.0 m

FE10.7 For the channel of Prob. FE10.1, if $Q \approx 24$ m^3/s and the depth is 2 m, what is the Froude number of the flow?
(*a*) 0.50, (*b*) 0.77, (*c*) 0.90, (*d*) 1.00, (*e*) 1.11

Comprehensive Problems

C10.1 February 1998 saw the failure of the earthen dam impounding California Jim's Pond in southern Rhode Island. The resulting flood raised temporary havoc in the nearby village of Peace Dale. The pond is 70,000 m^2 in area and 5 m deep and was full from heavy rains. The breach in the dam was 7 m wide and 5 m deep. Estimate the time required for the pond to drain to a depth of 1 m.

C10.2 A circular, unfinished concrete drainpipe is laid on a slope of 0.0025 and is planned to carry from 1.5 to 9 m^3/s of runoff water. Design constraints are that (1) the water depth should be no more than three-fourths of the diameter and (2) the flow should always be subcritical. What is the appropriate pipe diameter to satisfy these requirements? If no commercial pipe is exactly this calculated size, should you buy the next smallest or the next largest pipe?

C10.3 Extend Prob. P10.72, whose solution was $V_2 \approx 4.33$ m/s. (*a*) Use gradually varied theory to estimate the water depth 10 m downstream at section (3) for the 5° unfinished concrete slope shown in Fig. P10.72. (*b*) Repeat your calculation for an *upward* (adverse) slope of 5°. (*c*) When you find that part (*b*) is impossible with gradually varied theory, explain why and repeat for an adverse slope of 1°.

C10.4 It is desired to meter an asphalt rectangular channel of width 1.5 m, which is designed for uniform flow at a depth of 70 cm and a slope of 0.0036. The vertical sides of the channel are 1.2 m high. Consider using a thin-plate rectangular weir, either full or partial width (Table 10.2*a*,*b*) for this purpose. Sturm [7, p. 51] recommends, for accurate correlation, that such a weir have $Y \geq 9$ cm and $H/Y \leq 2.0$. Determine the feasibility of installing such a weir that will be accurate and yet not cause the water to overflow the sides of the channel.

C10.5 Figure C10.5 shows a hydraulic model of a *compound weir*, one that combines two different shapes. (*a*) Other than measurement, for which it might be poor, what could be the engineering reason for such a weir? (*b*) For the prototype river, assume that both sections have sides at a 70° angle to the vertical, with the bottom section having a base width of 2 m and the upper section having a base width of 4.5 m, including the cut-out portion. The heights of lower and upper horizontal sections are 1 m and 2 m, respectively. Use engineering estimates and make a plot of upstream water depth versus Petaluma River flow rate in the range 0 to 4 m^3/s. (*c*) For what river flow rate will the water overflow the top of the dam?

C10.6 Figure C10.6 shows a horizontal flow of water through a sluice gate, a hydraulic jump, and over a 2-m sharp-crested weir. Channel, gate, jump, and weir are all 2.5 m wide unfinished concrete. Determine (*a*) the flow rate, (*b*) the normal depth, (*c*) y_2, (*d*) y_3, and (*e*) y_4.

C10.5 *(Courtesy of the U.S. Army Corps of Engineers Waterways Experiment Station.)*

C10.6

C10.7 Consider the V-shaped channel in Fig. C10.7, with an arbitrary angle θ. Make a continuity and momentum analysis of a small disturbance $\delta y \ll y$, as in Fig. 10.4. Show that the wave propagation speed in this channel is independent of θ and does *not* equal the wide-channel result $c_0 = (gy)^{1/2}$.

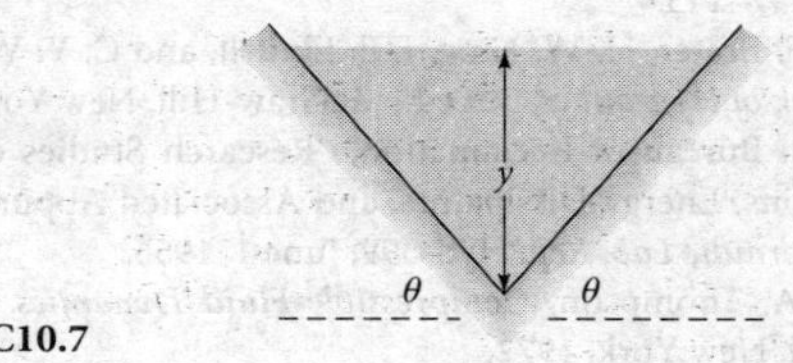

C10.7

Design Projects

D10.1 A straight weedy-earth channel has the trapezoidal shape of Fig. 10.7, with $b = 4$ m and $\theta = 35°$. The channel has a constant bottom slope of 0.001. The flow rate varies seasonally from 5 up to 10 m^3/s. It is desired to place a sharp-edged weir across the channel so that the water depth 1 km upstream remains at 2.0 m ± 10 percent throughout the year. Investigate the possibility of accomplishing this with a full-width weir; if successful, determine the proper weir height Y. If unsuccessful, try other alternatives, such as (*a*) a full-width broad-crested weir or (*b*) a weir with side contractions or (*c*) a V-notch weir. Whatever your final design, cite the seasonal variation of normal depths and critical depths for comparison with the desired year-round depth of 2 m.

D10.2 The Caroselli Dam on the Pawcatuck River is 3 m high, 30 m wide, and sharp edged. The Coakley Company uses this head to generate hydropower electricity and wants *more* head. They ask the town for permission to raise the dam higher. The river above the dam may be approximated as rectangular, 30 m wide, sloping upstream at 2.5 m per kilometer, and with a stony, cobbled bed. The average flow rate is 11 m^3/s, with a 30-year predicted flood rate of 34 m^3/s. The river sides are steep until 1 km upstream, where there are low-lying residences. The town council agrees the dam may be heightened if the new river level near these houses, during the 30-year flood, is no more than 1 m higher than the present level during average flow conditions. You, as project engineer, have to predict how high the dam crest can be raised and still meet this requirement.

References

1. B. A. Bakhmeteff, *Hydraulics of Open Channels,* McGraw-Hill, New York, 1932.
2. V. T. Chow, *Open Channel Hydraulics,* Blackburn Press, Caldwell, NJ, 2009.
3. M. H. Chaudhry, *Open Channel Flow,* 2d ed., Springer, New York, 2007.
4. R. Srivastava, *Flow Through Open Channels,* Oxford University Press, New York, 2008.
5. H. Chanson, *The Hydraulics of Open Channel Flow,* 2d ed., Elsevier, New York, 2004.
6. J. O. Akan, *Open Channel Hydraulics,* Butterworth-Heinemann, Woburn, MA, 2006.
7. T. W. Sturm, *Open Channel Hydraulics,* McGraw-Hill, New York, 2001.
8. J. Pedlosky, *Waves in the Ocean and Atmosphere: Introduction to Wave Dynamics,* Springer, New York, 2003.
9. L. H. Holthuijsen, *Waves in Oceanic and Coastal Waters,* Cambridge University Press, New York, 2007.
10. M. K. Ochi, *Ocean Waves: The Stochastic Approach,* Cambridge University Press, London, 2008.
11. G. Masselink, M. Hughes, and J. Knight, *Introduction to Coastal Processes and Geomorphology,* 2d ed., Routledge, New York, 2011.
12. M. B. Abbott and W. A. Price, *Coastal, Estuarial, and Harbor Engineers Reference Book,* Taylor & Francis, New York, 1994.
13. P. D. Komar, *Beach Processes and Sedimentation,* 2d ed., Pearson Education, Upper Saddle River, NJ, 1998.
14. W. Yue, C.-L. Lin, and V. C. Patel, "Large Eddy Simulation of Turbulent Open Channel Flow with Free Surface Simulated by Level Set Method," *Physics of Fluids,* vol. 17, no. 2, Feb. 2005, pp. 1–12.
15. J. M. Robertson and H. Rouse, "The Four Regimes of Open Channel Flow," *Civ. Eng.,* vol. 11, no. 3, March 1941, pp. 169–171.
16. R. W. Powell, "Resistance to Flow in Rough Channels," *Trans. Am. Geophys. Union,* vol. 31, no. 4, August 1950, pp. 575–582.
17. R. Manning, "On the Flow of Water in Open Channels and Pipes," *Trans. I.C.E. Ireland,* vol. 20, 1891, pp. 161–207.

18. "Friction Factors in Open Channels, Report of the Committee on Hydromechanics," *ASCE J. Hydraul. Div.*, March 1963, pp. 97–143.
19. E. F. Brater, H. W. King, J. E. Lindell, and C. Y. Wei, *Handbook of Hydraulics*, 7th ed., McGraw-Hill, New York, 1996.
20. U.S. Bureau of Reclamation, "Research Studies on Stilling Basins, Energy Dissipators, and Associated Appurtenances," *Hydraulic Lab. Rep.* Hyd-399, June 1, 1955.
21. P. A. Thompson, *Compressible-Fluid Dynamics*, McGraw-Hill, New York, 1972.
22. R. M. Olson and S. J. Wright, *Essentials of Engineering Fluid Mechanics*, 5th ed., Harper & Row, New York, 1990.
23. P. Ackers et al., *Weirs and Flumes for Flow Measurement*, Wiley, New York, 1978.
24. M. G. Bos, J. A. Replogle, and A. J. Clemmens, *Flow Measuring Flumes for Open Channel Systems*, American Soc. Agricultural and Biological Engineers, St. Joseph, MI, 1991.
25. M. G. Bos, *Long-Throated Plumes and Broad-Crested Weirs*, Springer-Verlag, New York, 1984.
26. D. H. Hoggan, *Computer-Assisted Floodplain Hydrology and Hydraulics*, 2d ed., McGraw-Hill, New York, 1996.
27. R. Jeppson, *Open Channel Flow: Numerical Methods and Computer Applications*, CRC Press, Boca Raton, FL, 2010.
28. R. Szymkiewicz, *Numerical Modeling in Open Channel Hydraulics*, Springer, New York, 2010.
29. R. Baban, *Design of Diversion Weirs: Small Scale Irrigation in Hot Climates*, Wiley, New York, 1995.
30. H. Chanson, *Hydraulic Design of Stepped Cascades, Channels, Weirs, and Spillways*, Pergamon Press, New York, 1994.
31. D. Kampion and A. Brewer, *The Book of Waves: Form and Beauty on the Ocean*, 3d ed., Rowman and Littlefield, Lanham, MD, 1997.
32. L. Mays, *Water Resources Engineering*, Wiley, New York, 2005.
33. D. K. Walkowiak (ed.), *Isco Open Channel Flow Measurement Handbook*, 5th ed., Teledyne Isco, Inc., Lincoln, NE, 2006.
34. H. Chanson, "Photographic Observations of Tidal Bores (Mascarets) in France," Hydraulic Model Report CH71/08, The University of Queensland, 2008, 104 pages.
35. E. J. Watson, "The Spread of a Liquid Jet over a Horizontal Plane," *J. Fluid Mechanics*, vol. 20, 1964, pp. 481–499.
36. Z. Arendze and B. W. Skews, "Experimental and Numerical Study of the Hydraulic Analogy to Supersonic Flow," *South African Institution of Mechanical Engineering R&D Journal*, vol. 24, 2008, pp. 9–15.
37. T. J. Mueller and W. L. Oberkampf, "Hydraulic Analog for the Expansion Deflection Nozzle," *AIAA Journal*, vol. 5, 1967, pp. 1200–1202.

Wind turbines will play a large role in our energy future. The photo shows a 100 kW HAWT, installed in 2011 at the Fisherman's Memorial State Camp Ground in Narragansett, Rhode Island. It is programmed to generate 100 kW in winds from 13 to 25 m/s and supplies half the electricity needed for the camp's 18,000 annual visitors. Wind energy is good, but expensive. This turbine cost more, just to install, than it will recover in power savings over its 20-year life span. [*Photo courtesy of F. M. White*]

Chapter 11
Turbomachinery

Motivation. The most common practical engineering application for fluid mechanics is the design of fluid machinery. The most numerous types are machines that *add* energy to the fluid (the pump family), but also important are those that *extract* energy (turbines). Both types are usually connected to a rotating shaft, hence the name *turbomachinery.*

The purpose of this chapter is to make elementary engineering estimates of the performance of fluid machines. The emphasis will be on nearly incompressible flow: liquids or low-velocity gases. Basic flow principles are discussed, but not the detailed construction of the machines.

11.1 Introduction and Classification

Turbomachines divide naturally into those that add energy (pumps) and those that extract energy (turbines). The prefix *turbo-* is a Latin word meaning "spin" or "whirl," appropriate for rotating devices.

The pump is the oldest fluid energy transfer device known. At least two designs date before Christ: (1) the undershot-bucket waterwheels, or *norias,* used in Asia and Africa (1000 B.C.) and (2) Archimedes' screw pump (250 B.C.), still being manufactured today to handle solid–liquid mixtures. Paddlewheel turbines were used by the Romans in 70 B.C., and Babylonian windmills date back to 700 B.C. [1].

Machines that deliver liquids are simply called *pumps,* but if gases are involved, three different terms are in use, depending on the pressure rise achieved. If the pressure rise is very small (a few centimeters of water), a gas pump is called a *fan;* up to 1 atm, it is usually called a *blower;* and above 1 atm it is commonly termed a *compressor.*

Classification of Pumps

There are two basic types of pumps: positive-displacement and dynamic or momentum-change pumps. There are several billion of each type in use in the world today.

Positive-displacement pumps (PDPs) force the fluid along by volume changes. A cavity opens, and the fluid is admitted through an inlet. The cavity then closes, and the fluid is squeezed through an outlet. The mammalian heart is a good example, and many mechanical designs are in wide use. References 35–38 give a summary of PDPs. A brief classification of PDP designs is as follows:

A. Reciprocating
   1. Piston or plunger
   2. Diaphragm

B. Rotary
   1. Single rotor
      a. Sliding vane
      b. Flexible tube or lining
      c. Screw
      d. Peristaltic (wave contraction)
   2. Multiple rotors
      a. Gear
      b. Lobe
      c. Screw
      d. Circumferential piston

All PDPs deliver a pulsating or periodic flow as the cavity volume opens, traps, and squeezes the fluid. Their great advantage is the delivery of any fluid regardless of its viscosity.

Figure 11.1 shows schematics of the operating principles of seven of these PDPs. It is rare for such devices to be run backward, so to speak, as turbines or energy extractors, the steam engine (reciprocating piston) being a classic exception.

Since PDPs compress mechanically against a cavity filled with liquid, a common feature is that they develop immense pressures if the outlet is shut down for any reason. Sturdy construction is required, and complete shutoff would cause damage if pressure relief valves were not used.

Dynamic pumps simply add momentum to the fluid by means of fast-moving blades or vanes or certain special designs. There is no closed volume: The fluid increases momentum while moving through open passages and then converts its high velocity to a pressure increase by exiting into a diffuser section. Dynamic pumps can be classified as follows:

A. Rotary
   1. Centrifugal or radial exit flow
   2. Axial flow
   3. Mixed flow (between radial and axial)

B. Special designs
   1. Jet pump or ejector (see Fig. P3.36)
   2. Electromagnetic pumps for liquid metals
   3. Fluid-actuated: gas lift or hydraulic ram

We shall concentrate in this chapter on the rotary designs, sometimes called *rotodynamic pumps.* Other designs of both PDP and dynamic pumps are discussed in specialized texts [for example, 3, 31].

Dynamic pumps generally provide a higher flow rate than PDPs and a much steadier discharge but are ineffective in handling high-viscosity liquids. Dynamic pumps also generally need *priming;* if they are filled with gas, they cannot suck up a liquid from below into their inlet. The PDP, on the other hand, is self-priming for most applications. A dynamic pump can provide very high flow rates (up to 1,200 m^3/min), but usually with moderate pressure rises (a few atmospheres). In contrast, a PDP can operate up to very high pressures (300 atm), but typically produces low flow rates (0.50 m^3/min).

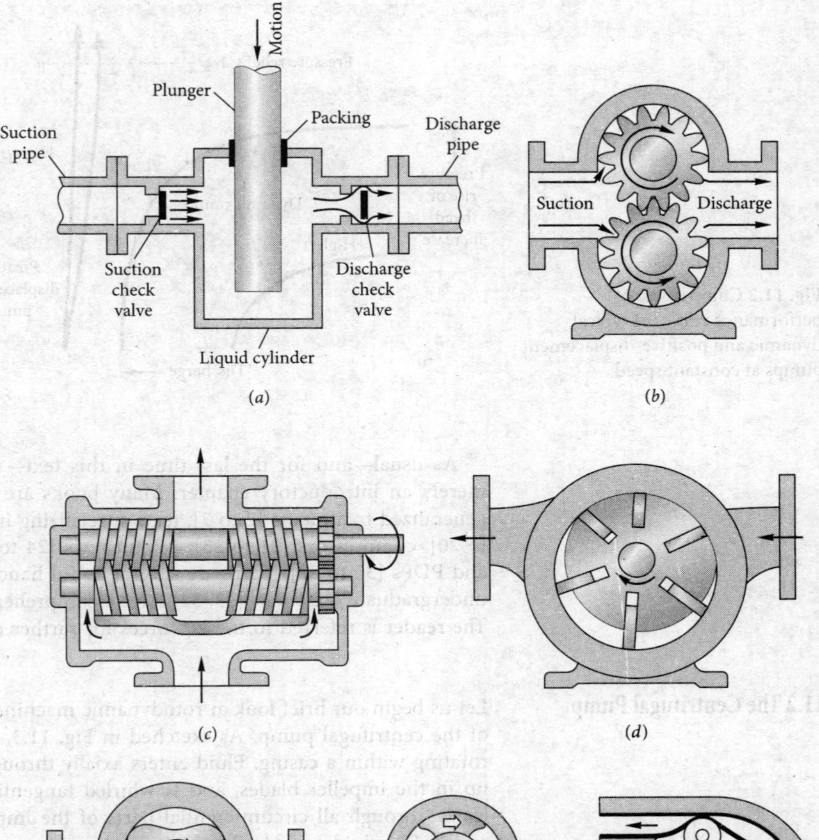

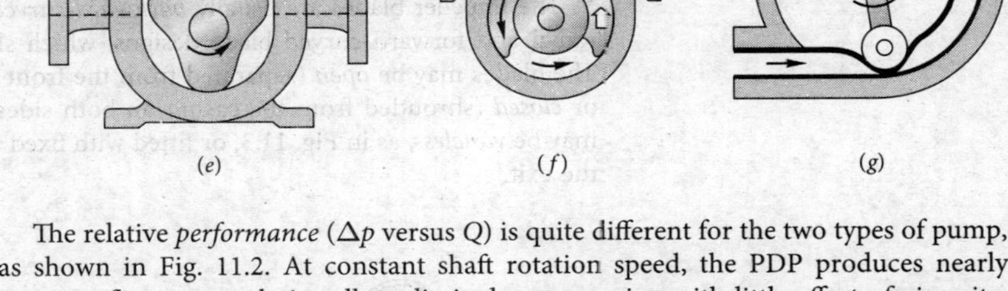

Fig. 11.1 Schematic design of positive-displacement pumps: (*a*) reciprocating piston or plunger, (*b*) external gear pump, (*c*) double-screw pump, (*d*) sliding vane, (*e*) three-lobe pump, (*f*) double circumferential piston, (*g*) flexible-tube squeegee.

The relative *performance* (Δp versus Q) is quite different for the two types of pump, as shown in Fig. 11.2. At constant shaft rotation speed, the PDP produces nearly constant flow rate and virtually unlimited pressure rise, with little effect of viscosity. The flow rate of a PDP cannot be varied except by changing the displacement or the speed. The reliable constant-speed discharge from PDPs has led to their wide use in metering flows [35].

The dynamic pump, by contrast in Fig. 11.2, provides a continuous constant-speed variation of performance, from near-maximum Δp at zero flow (shutoff conditions) to zero Δp at maximum flow rate. High-viscosity fluids sharply degrade the performance of a dynamic pump.

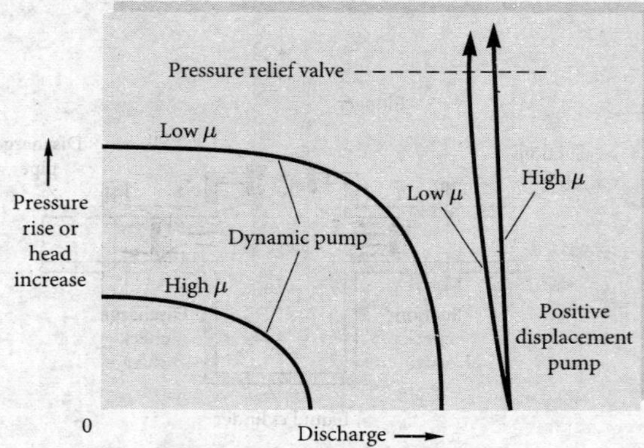

Fig. 11.2 Comparison of performance curves of typical dynamic and positive-displacement pumps at constant speed.

As usual—and for the last time in this text—we remind the reader that this is merely an introductory chapter. Many books are devoted solely to turbomachines: generalized treatments [2 to 7], texts specializing in pumps [8 to 16, 30, 31], fans [17 to 20], compressors [21 to 23], gas turbines [24 to 26], hydropower [27, 28, 29, 32], and PDPs [35 to 38]. There are several useful handbooks [30 to 32], and at least two undergraduate textbooks [33, 34] have a comprehensive discussion of turbomachines. The reader is referred to these sources for further details.

11.2 The Centrifugal Pump

Let us begin our brief look at rotodynamic machines by examining the characteristics of the centrifugal pump. As sketched in Fig. 11.3, this pump consists of an impeller rotating within a casing. Fluid enters axially through the *eye* of the casing, is caught up in the impeller blades, and is whirled tangentially and radially outward until it leaves through all circumferential parts of the impeller into the diffuser part of the casing. The fluid gains both velocity and pressure while passing through the impeller. The doughnut-shaped diffuser, or *scroll,* section of the casing decelerates the flow and further increases the pressure.

The impeller blades are usually *backward-curved,* as in Fig. 11.3, but there are also radial and forward-curved blade designs, which slightly change the output pressure. The blades may be *open* (separated from the front casing only by a narrow clearance) or *closed* (shrouded from the casing on both sides by an impeller wall). The diffuser may be *vaneless,* as in Fig. 11.3, or fitted with fixed vanes to help guide the flow toward the exit.

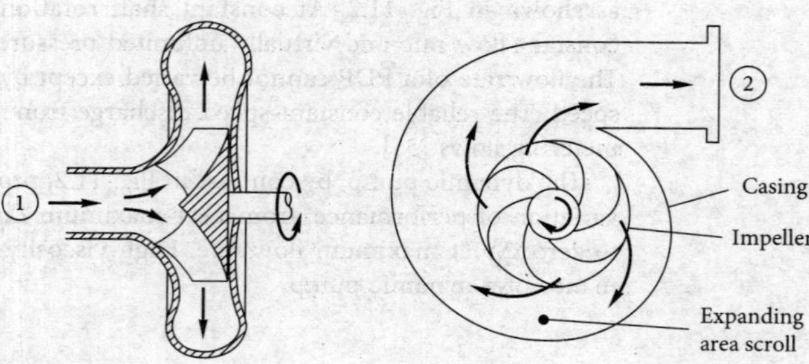

Fig. 11.3 Cutaway schematic of a typical centrifugal pump.

Basic Output Parameters

Assuming steady flow, the pump basically increases the Bernoulli head of the flow between point 1, the eye, and point 2, the exit. From Eq. (3.73), neglecting viscous work and heat transfer, this change is denoted by H:

$$H = \left(\frac{p}{\rho g} + \frac{V^2}{2g} + z\right)_2 - \left(\frac{p}{\rho g} + \frac{V^2}{2g} + z\right)_1 = h_s - h_f \tag{11.1}$$

where h_s is the pump head supplied and h_f the losses. The net head H is a primary output parameter for any turbomachine. Since Eq. (11.1) is for incompressible flow, it must be modified for gas compressors with large density changes.

Usually V_2 and V_1 are about the same, $z_2 - z_1$ is no more than a meter or so, and the net pump head is essentially equal to the change in pressure head:

$$H \approx \frac{p_2 - p_1}{\rho g} = \frac{\Delta p}{\rho g} \tag{11.2}$$

The power delivered to the fluid simply equals the specific weight times the discharge times the net head change:

$$P_w = \rho g Q H \tag{11.3}$$

This is traditionally called the *water horsepower*. The power required to drive the pump is the *brake horsepower*[1]

$$\text{bhp} = \omega T \tag{11.4}$$

where ω is the shaft angular velocity and T the shaft torque. If there were no losses, P_w and brake horsepower would be equal, but of course P_w is actually less, and the *efficiency* η of the pump is defined as

$$\eta = \frac{P_w}{\text{bhp}} = \frac{\rho g Q H}{\omega T} \tag{11.5}$$

The chief aim of the pump designer is to make η as high as possible over as broad a range of discharge Q as possible.

The efficiency is basically composed of three parts: volumetric, hydraulic, and mechanical. The *volumetric efficiency* is

$$\eta_v = \frac{Q}{Q + Q_L} \tag{11.6}$$

where Q_L is the loss of fluid due to leakage in the impeller casing clearances. The *hydraulic efficiency* is

$$\eta_h = 1 - \frac{h_f}{h_s} \tag{11.7}$$

where h_f has three parts: (1) *shock* loss at the eye due to imperfect match between inlet flow and the blade entrances, (2) *friction* losses in the blade passages, and (3) *circulation* loss due to imperfect match at the exit side of the blades.

Finally, the *mechanical efficiency* is

$$\eta_m = 1 - \frac{P_f}{\text{bhp}} \tag{11.8}$$

where P_f is the power loss due to mechanical friction in the bearings, packing glands, and other contact points in the machine.

[1]Conversion factors may be needed: 1 hp = 746 Nm/s = 746 W.

By definition, the total efficiency is simply the product of its three parts:

$$\eta \equiv \eta_v \eta_h \eta_m \tag{11.9}$$

The designer has to work in all three areas to improve the pump.

Elementary Pump Theory

You may have thought that Eqs. (11.1) to (11.9) were formulas from pump *theory*. Not so; they are merely definitions of performance parameters and cannot be used in any predictive mode. To actually *predict* the head, power, efficiency, and flow rate of a pump, two theoretical approaches are possible: (1) simple one-dimensional flow formulas and (2) complex computer models that account for viscosity and three-dimensionality. Many of the best design improvements still come from testing and experience, and pump research remains a very active field [39]. The last 10 years have seen considerable advances in *computational fluid dynamics* (CFD) modeling of flow in turbomachines [42], and at least eight commercial turbulent flow three-dimensional CFD codes are now available.

To construct an elementary theory of pump performance, we assume one-dimensional flow and combine idealized fluid velocity vectors through the impeller with the angular momentum theorem for a control volume, Eq. (3.59).

The idealized velocity diagrams are shown in Fig. 11.4. The fluid is assumed to enter the impeller at $r = r_1$ with velocity component w_1 tangent to the blade angle β_1 plus circumferential speed $u_1 = \omega r_1$ matching the tip speed of the impeller. Its absolute entrance velocity is thus the vector sum of w_1 and u_1, shown as V_1. Similarly, the flow exits at $r = r_2$ with component w_2 parallel to the blade angle β_2 plus tip speed $u_2 = \omega r_2$, with resultant velocity V_2.

We applied the angular momentum theorem to a turbomachine in Example 3.18 (Fig. 3.15) and arrived at a result for the applied torque T:

$$T = \rho Q(r_2 V_{t2} - r_1 V_{t1}) \tag{11.10}$$

where V_{t1} and V_{t2} are the absolute circumferential velocity components of the flow. The power delivered to the fluid is thus

$$P_w = \omega T = \rho Q(u_2 V_{t2} - u_1 V_{t1})$$

or

$$H = \frac{P_w}{\rho g Q} = \frac{1}{g}(u_2 V_{t2} - u_1 V_{t1}) \tag{11.11}$$

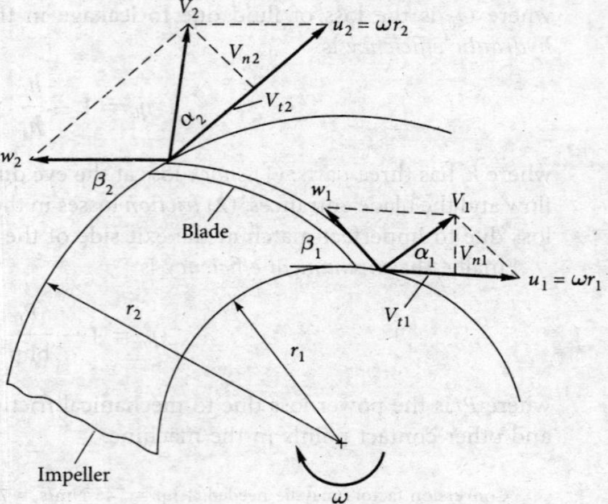

Fig. 11.4 Inlet and exit velocity diagrams for an idealized pump impeller.

These are the *Euler turbomachine equations,* showing that the torque, power, and ideal head are functions only of the rotor-tip velocities $u_{1,2}$ and the absolute fluid tangential velocities $V_{t1,2}$, independent of the axial velocities (if any) through the machine.

Additional insight is gained by rewriting these relations in another form. From the geometry of Fig. 11.4

$$V^2 = u^2 + w^2 - 2uw \cos\beta \qquad w\cos\beta = u - V_t$$

or

$$uV_t = \tfrac{1}{2}(V^2 + u^2 - w^2) \tag{11.12}$$

Substituting this into Eq. (11.11) gives

$$H = \frac{1}{2g}[(V_2^2 - V_1^2) + (u_2^2 - u_1^2) - (w_2^2 - w_1^2)] \tag{11.13}$$

Thus, the ideal head relates to the absolute plus the relative kinetic energy change of the fluid minus the rotor-tip kinetic energy change. Finally, substituting for H from its definition in Eq. (11.1) and rearranging, we obtain the classic relation

$$\frac{p}{\rho g} + z + \frac{w^2}{2g} - \frac{r^2\omega^2}{2g} = \text{const} \tag{11.14}$$

This is the *Bernoulli equation in rotating coordinates* and applies to either two- or three-dimensional ideal incompressible flow.

For a centrifugal pump, the power can be related to the radial velocity $V_n = V_t \tan\alpha$ and the continuity relation

$$P_w = \rho Q(u_2 V_{n2} \cot\alpha_2 - u_1 V_{n1} \cot\alpha_1) \tag{11.15}$$

where

$$V_{n2} = \frac{Q}{2\pi r_2 b_2} \qquad \text{and} \qquad V_{n1} = \frac{Q}{2\pi r_1 b_1}$$

and where b_1 and b_2 are the blade widths at inlet and exit. With the pump parameters r_1, r_2, β_1, β_2, and ω known, Eq. (11.11) or Eq. (11.15) is used to compute idealized power and head versus discharge. The "design" flow rate Q^* is commonly estimated by assuming that the flow enters exactly normal to the impeller:

$$\alpha_1 = 90° \qquad V_{n1} = V_1 \tag{11.16}$$

We can expect this simple analysis to yield estimates within ±25 percent for the head, water horsepower, and discharge of a pump. Let us illustrate with an example.

EXAMPLE 11.1

Given are the following data for a commercial centrifugal water pump: r_1 = 10 cm, r_2 = 18 cm, $\beta_1 = 30°$, $\beta_2 = 20°$, speed = 1,440 r/min. Estimate (*a*) the design point discharge, (*b*) the water horsepower, and (*c*) the head if $b_1 = b_2$ = 4 cm.

Solution

Part (a) The angular velocity is $\omega = 2\pi$ r/s $= 2\pi(1{,}440/60) = 150.8$ rad/s. Thus the tip speeds are $u_1 = \omega r_1 = 150.8(0.1) = 15.08$ m/s and $u_2 = \omega r_2 = 150.8(0.18) = 27.14$ m/s. From the inlet velocity diagram, Fig. E11.1*a*, with $\alpha_1 = 90°$ for design point, we compute

$$V_{n1} = u_1 \tan 30° = 8.71 \text{ m/s}$$

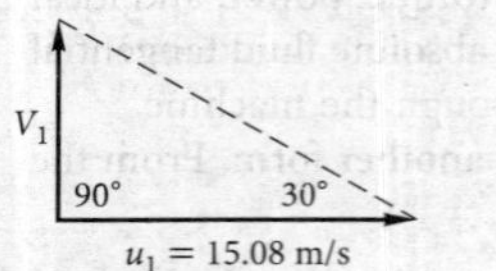

E11.1a

whence the discharge is

$$Q = 2\pi r_1 b_1 V_{n1} = (2\pi)(0.1\ \text{m})(0.04\ \text{m})(8.71\ \text{m/s})$$
$$= (0.219\ \text{m}^3/\text{s})(60\ \text{s/min})$$
$$= 13.12\ \text{m}^3/\text{min} \qquad \textit{Ans. (a)}$$

(The actual pump produces about 13.25 m³/min.)

Part (b) The outlet radial velocity follows from Q:

$$V_{n2} = \frac{Q}{2\pi r_2 b_2} = \frac{0.219\ \text{m}^3/\text{min}}{2\pi(0.18\,\text{m})(0.04\,\text{m})} = 4.84\ \text{m/s}$$

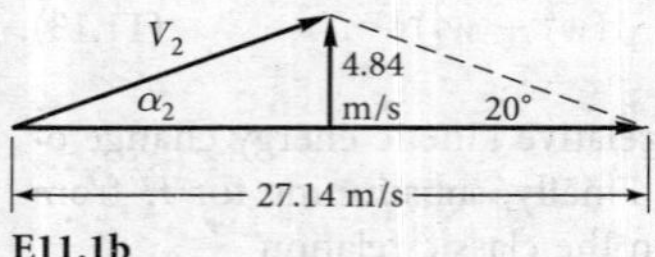

E11.1b

This enables us to construct the outlet velocity diagram as in Fig. E11.1*b*, given $\beta_2 = 20°$. The tangential component is

$$V_{t2} = u_2 - V_{n2}\cot\beta_2 = 27.14 - 4.84\cot 20° = 13.84\ \text{m/s}$$

$$\alpha_2 = \tan^{-1}\frac{4.84\ \text{m/s}}{13.84\ \text{m/s}} = 19.3°$$

The power is then computed from Eq. (11.11) with $V_{t1} = 0$ at the design point:

$$P_w = \rho Q u_2 V_{t2} = (998\ \text{kg/m}^3)(0.219\ \text{m}^3/\text{s})(27.14\ \text{m/s})(13.84\ \text{m/s})$$

$$= \frac{82{,}100\ \text{Nm/s}}{746\ \text{Nm/(s-hp)}} = 110\ \text{hp} \qquad \textit{Ans. (b)}$$

(The actual pump delivers about 125 water horsepower, requiring 147 bhp at 85 percent efficiency.)

Part (c) Finally, the head is estimated from Eq. (11.11):

$$H \approx \frac{P_w}{\rho g Q} = \frac{82{,}100\ \text{Nm/s}}{(9{,}790\ \text{N/m}^3)(0.219\ \text{m}^3/\text{s})} = 38.3\ \text{m} \qquad \textit{Ans. (c)}$$

(The actual pump develops about 42 m head.) Improved methods for obtaining closer estimates are given in advanced references [for example, 7, 8, and 31].

Effect of Blade Angle on Pump Head

The simple theory just discussed can be used to predict an important blade-angle effect. If we neglect inlet angular momentum, the theoretical water horsepower is

$$P_w = \rho Q u_2 V_{t2} \tag{11.17}$$

where $\quad V_{t2} = u_2 - V_{n2}\cot\beta_2 \qquad V_{n2} = \dfrac{Q}{2\pi r_2 b_2}$

Then, the theoretical head from Eq. (11.11) becomes

$$H \approx \frac{u_2^2}{g} - \frac{u_2\cot\beta_2}{2\pi r_2 b_2 g}Q \tag{11.18}$$

The head varies linearly with discharge Q, having a shutoff value u_2^2/g, where u_2 is the exit blade-tip speed. The slope is negative if $\beta_2 < 90°$ (backward-curved blades) and positive for $\beta_2 > 90°$ (forward-curved blades). This effect is shown in Fig. 11.5 and is accurate only at low flow rates.

The measured shutoff head of centrifugal pumps is only about 60 percent of the theoretical value $H_0 = \omega^2 r_2^2/g$. With the advent of the laser-doppler anemometer,

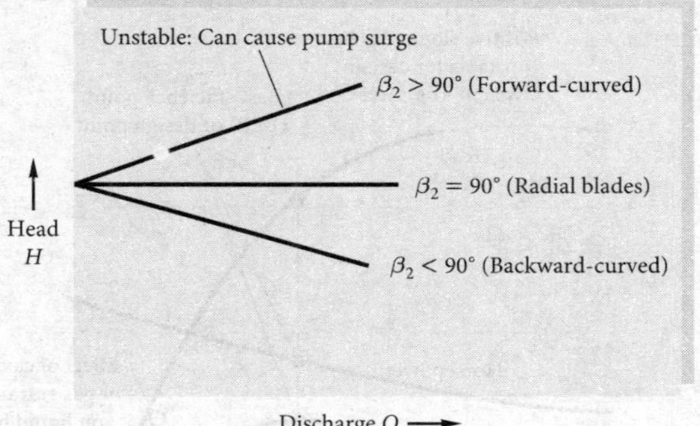

Fig. 11.5 Theoretical effect of blade exit angle on pump head versus discharge.

researchers can now make detailed three-dimensional flow measurements inside pumps and can even animate the data into a movie [40].

The positive slope condition in Fig. 11.5 can be unstable and can cause pump *surge*, an oscillatory condition where the pump "hunts" for the proper operating point. Surge may cause only rough operation in a liquid pump, but it can be a major problem in gas compressor operation. For this reason a backward-curved or radial blade design is generally preferred. A survey of the problem of pump stability is given by Greitzer [41].

11.3 Pump Performance Curves and Similarity Rules

Since the theory of the previous section is rather qualitative, the only solid indicator of a pump's performance lies in extensive testing. For the moment let us discuss the centrifugal pump in particular. The general principles and the presentation of data are exactly the same for mixed flow and axial flow pumps and compressors.

Performance charts are almost always plotted for constant shaft rotation speed n (in r/min usually). The basic independent variable is taken to be discharge Q (in L/min usually for liquids and m^3/min for gases). The dependent variables, or "output," are taken to be head H (pressure rise Δp for gases), brake horsepower (bhp), and efficiency η.

Figure 11.6 shows typical performance curves for a centrifugal pump. The head is approximately constant at low discharge and then drops to zero at $Q = Q_{max}$. At this speed and impeller size, the pump cannot deliver any more fluid than Q_{max}. The positive slope part of the head is shown dashed; as mentioned earlier, this region can be unstable and can cause hunting for the operating point.

The efficiency η is always zero at no flow and at Q_{max}, and it reaches a maximum, perhaps 80 to 90 percent, at about $0.6Q_{max}$. This is the *design flow rate* Q^* or *best efficiency point* (BEP), $\eta = \eta_{max}$. The head and horsepower at BEP will be termed H^* and P^* (or bhp*), respectively. It is desirable that the efficiency curve be flat near η_{max}, so that a wide range of efficient operation is achieved. However, some designs simply do not achieve flat efficiency curves. Note that η is not independent of H and P but rather is calculated from the relation in Eq. (11.5), $\eta = \rho gQH/P$.

As shown in Fig. 11.6, the horsepower required to drive the pump typically rises monotonically with the flow rate. Sometimes there is a large power rise beyond the BEP, especially for radial-tipped and forward-curved blades. This is considered undesirable because a much larger motor is then needed to provide high flow rates. Backward-curved blades typically have their horsepower level off above BEP ("nonoverloading" type of curve).

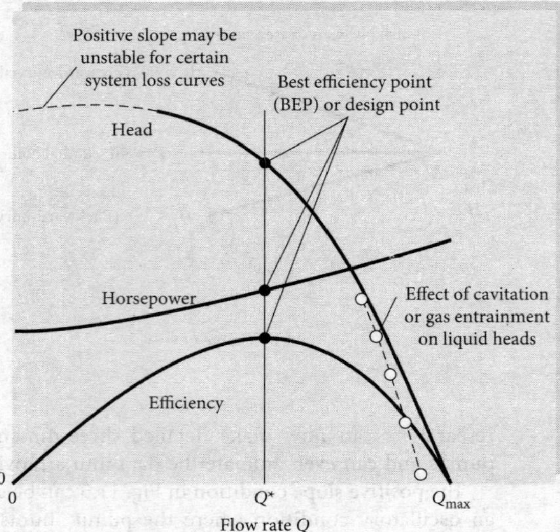

Fig. 11.6 Typical centrifugal pump performance curves at constant impeller rotation speed. The units are arbitrary.

Measured Performance Curves

Figure 11.7 shows actual performance data for a commercial centrifugal pump. Figure 11.7*a* is for a basic casing size with three different impeller diameters. The head curves $H(Q)$ are shown, but the horsepower and efficiency curves have to be inferred from the contour plots. Maximum discharges are not shown, being far outside the normal operating range near the BEP. Everything is plotted raw, of course [meters, horsepower, cubic meters per minute] since it is to be used directly by designers. Figure 11.7*b* is the same pump design with a 20 percent larger casing, a lower speed, and three larger impeller diameters. Comparing the two pumps may be a little confusing: The larger pump produces exactly the same discharge but only half the horsepower and half the head. This will be readily understood from the scaling or similarity laws we are about to formulate.

A point often overlooked is that raw curves like Fig. 11.7 are strictly applicable to a fluid of a certain density and viscosity, in this case water. If the pump were used to deliver, say, mercury, the brake horsepower would be about 13 times higher while Q, H, and η would be about the same. But in that case H should be interpreted as meters of *mercury,* not meters of water. If the pump were used for SAE 30 oil, *all* data would change (brake horsepower, Q, H, and η) due to the large change in viscosity (Reynolds number). Again, this should become clear with the similarity rules.

Net Positive-Suction Head

In the top of Fig. 11.7 is plotted the *net positive-suction head* (NPSH), which is the head required at the pump inlet to keep the liquid from cavitating or boiling. The pump inlet or suction side is the low-pressure point where cavitation will first occur. The NPSH is defined as

$$\text{NPSH} = \frac{p_i}{\rho g} + \frac{V_i^2}{2g} - \frac{p_v}{\rho g} \tag{11.19}$$

where p_i and V_i are the pressure and velocity at the pump inlet and p_v is the vapor pressure of the liquid. Given the left-hand side, NPSH, from the pump performance curve, we must ensure that the right-hand side is equal or greater in the actual system to avoid cavitation.

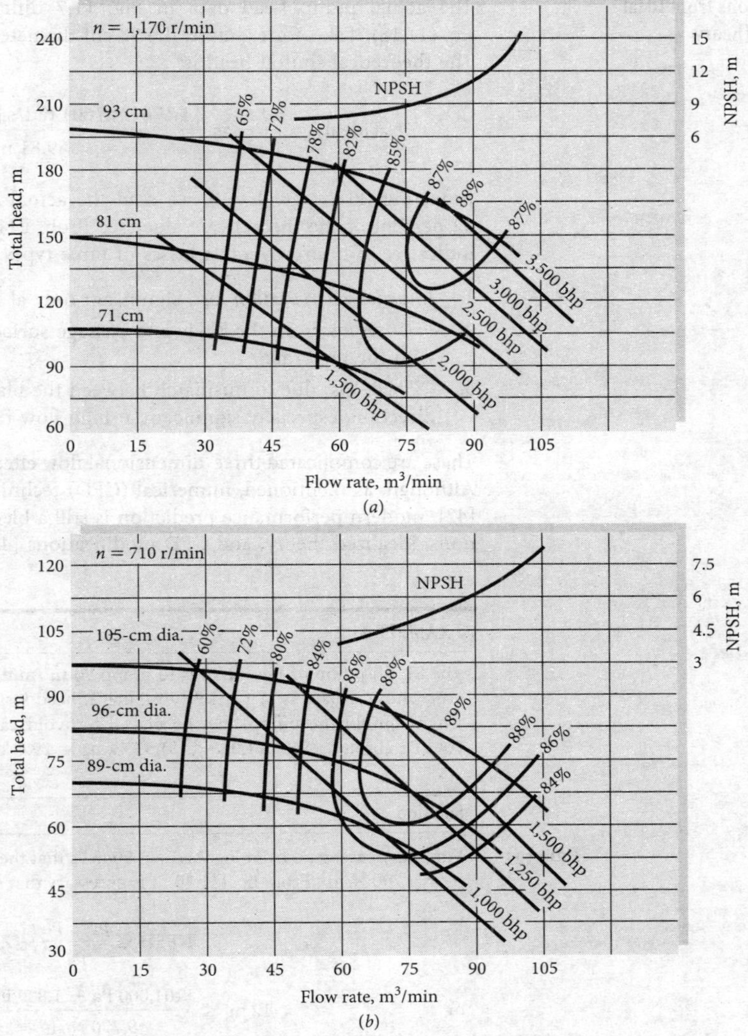

Fig. 11.7 Measured-performance curves for two models of a centrifugal water pump: (*a*) basic casing with three impeller sizes; (*b*) 20 percent larger casing with three larger impellers at slower speed. *(Courtesy of Ingersoll-Rand Corporation, Cameron Pump Division.)*

If the pump inlet is placed at a height Z_i above a reservoir whose free surface is at pressure p_a, we can use Bernoulli's equation to rewrite NPSH as

$$\text{NPSH} = \frac{p_a}{\rho g} - Z_i - h_{fi} - \frac{p_v}{\rho g} \tag{11.20}$$

where h_{fi} is the friction head loss between the reservoir and the pump inlet. Knowing p_a and h_{fi}, we can set the pump at a height Z_i that will keep the right-hand side greater than the "required" NPSH plotted in Fig. 11.7.

If cavitation does occur, there will be pump noise and vibration, pitting damage to the impeller, and a sharp dropoff in pump head and discharge. In some liquids, this deterioration starts before actual boiling, as dissolved gases and light hydrocarbons are liberated.

Deviations from Ideal Pump Theory

The actual pump head data in Fig. 11.7 differ considerably from ideal theory, Eq. (11.18). Take, for example, the 93 cm-diameter pump at 1,170 r/min in Fig. 11.7*a*. The theoretical shutoff head is

$$H_0(\text{ideal}) = \frac{\omega^2 r_2^2}{g} = \frac{[1{,}170(2\pi/60)\ \text{rad/s}]^2[(93/2)/100\ \text{m}]^2}{9.81\ \text{m/s}^2} = 331\ \text{m}$$

From Fig. 11.7*a*, at $Q = 0$, we read the actual shutoff head to be only 204 m, or 62 percent of the theoretical value (see Prob. P11.24). This is a sharp dropoff and is indicative of nonrecoverable losses of three types:

1. *Impeller recirculation loss*, significant only at low flow rates.
2. *Friction losses* on the blade and passage surfaces, which increase monotonically with the flow rate.
3. *"Shock" loss* due to mismatch between the blade angles and the inlet flow direction, especially significant at high flow rates.

These are complicated three-dimensional flow effects and hence are difficult to predict. Although, as mentioned, numerical (CFD) techniques are becoming more important [42], modern performance prediction is still a blend of experience, empirical correlations, idealized theory, and CFD modifications [45].

EXAMPLE 11.2

The 81 cm pump of Fig. 11.7*a* is to pump 90 m^3/min of water at 1,170 r/min from a reservoir whose surface is at 101 kPa absolute. If head loss from reservoir to pump inlet is 2 m, where should the pump inlet be placed to avoid cavitation for water at (*a*) 15°C, p_v = 1.8 kPa absolute, SG = 1.0 and (*b*) 93°C, p_v = 79.4 kPa absolute, SG = 0.9635?

Solution

Part (a)

For either case, read from Fig. 11.7*a* at 90 m^3/s that the required NPSH is 12 m. For this case, ρg = 9,790 N/m^3. From Eq. (11.20) it is necessary that

$$\text{NPSH} \le \frac{p_a - p_v}{\rho g} - Z_i - h_{fi}$$

or

$$12\ \text{m} \le \frac{(101{,}000\ \text{Pa} - 1{,}800\ \text{Pa})}{9{,}790\ \text{N/m}^3} - Z_i - 2\ \text{m}$$

or

$$Z_i \le 8.13 - 12 = -3.87\ \text{m} \qquad \textit{Ans. (a)}$$

The pump must be placed at least −3.87 m below the reservoir surface to avoid cavitation.

Part (b)

For this case, ρg = 9,790(0.9635) = 9,433 N/m^3. Equation (11.20) applies again with the higher p_v:

$$12\ \text{m} \le \frac{(101{,}000\ \text{Pa} - 79{,}400\ \text{Pa})}{9{,}433\ \text{N/m}^3} - Z_i - 2\ \text{m}$$

or

$$Z_i \le 0.29 - 12 = -11.71\ \text{m} \qquad \textit{Ans. (b)}$$

The pump must now be placed at least 11.71 m below the reservoir surface. These are unusually stringent conditions because a large, high-discharge pump requires a large NPSH.

Dimensionless Pump Performance

For a given pump design, the output variables H and brake horsepower should be dependent on discharge Q, impeller diameter D, and shaft speed n, at least. Other possible parameters are the fluid density ρ, viscosity μ, and surface roughness ε. Thus the performance curves in Fig. 11.7 are equivalent to the following assumed functional relations:[2]

$$gH = f_1(Q, D, n, \rho, \mu, \varepsilon) \qquad \text{bhp} = f_2(Q, D, n, \rho, \mu, \varepsilon) \tag{11.21}$$

This is a straightforward application of dimensional analysis principles from Chap. 5. As a matter of fact, it was given as an exercise (Example 5.3). For each function in Eq. (11.21) there are seven variables and three primary dimensions (M, L, and T); hence we expect $7 - 3 = 4$ dimensionless pi groups, and that is what we get. You can verify as an exercise that appropriate dimensionless forms for Eqs. (11.21) are

$$\begin{aligned} \frac{gH}{n^2D^2} &= g_1\left(\frac{Q}{nD^3}, \frac{\rho n D^2}{\mu}, \frac{\varepsilon}{D}\right) \\ \frac{\text{bhp}}{\rho n^3 D^5} &= g_2\left(\frac{Q}{nD^3}, \frac{\rho n D^2}{\mu}, \frac{\varepsilon}{D}\right) \end{aligned} \tag{11.22}$$

The quantities $\rho nD^2/\mu$ and ε/D are recognized as the Reynolds number and roughness ratio, respectively. Three new pump parameters have arisen:

$$\begin{aligned} \text{Capacity coefficient } C_Q &= \frac{Q}{nD^3} \\ \text{Head coefficient } C_H &= \frac{gH}{n^2D^2} \\ \text{Power coefficient } C_P &= \frac{\text{bhp}}{\rho n^3 D^5} \end{aligned} \tag{11.23}$$

Note that only the power coefficient contains fluid density, the parameters C_Q and C_H being kinematic types.

Figure 11.7 gives no warning of viscous or roughness effects. The Reynolds numbers are from 0.8 to 1.5×10^7, or fully turbulent flow in all passages probably. The roughness is not given and varies greatly among commercial pumps. But, at such high Reynolds numbers we expect more or less the same percentage effect on all these pumps. Therefore, it is common to assume that the Reynolds number and the roughness ratio have a constant effect, so that Eqs. (11.23) reduce to, approximately,

$$C_H \approx C_H(C_Q) \qquad C_P \approx C_P(C_Q) \tag{11.24}$$

For geometrically similar pumps, we expect head and power coefficients to be (nearly) unique functions of the capacity coefficient. We have to watch out that the pumps are geometrically similar or nearly so because (1) manufacturers put different-sized impellers in the same casing, thus violating geometric similarity, and (2) large pumps have smaller ratios of roughness and clearances to impeller diameter than small pumps. In addition, the more viscous liquids will have significant Reynolds number effects; for example, a factor-of-3 or more viscosity increase causes a clearly visible effect on C_H and C_P.

[2] We adopt gH as a variable instead of H for dimensional reasons.

The efficiency η is already dimensionless and is uniquely related to the other three. It varies with C_Q also:

$$\eta \equiv \frac{C_H C_Q}{C_P} = \eta(C_Q) \tag{11.25}$$

We can test Eqs. (11.24) and (11.25) from the data of Fig. 11.7. The impeller diameters of 81 and 96 cm are approximately 20 percent different in size, and so their ratio of impeller to casing size is the same. The parameters C_Q, C_H, and C_P are computed with n in r/s, Q in m^3/s, H and D in m, $g = 9.81$ m/s^2, and brake horsepower in horsepower times 746 Nm/(s–hp). The nondimensional data are then plotted in Fig. 11.8. A dimensionless suction head coefficient is also defined:

$$C_{HS} = \frac{g(\text{NPSH})}{n^2 D^2} = C_{HS}(C_Q) \tag{11.26}$$

The coefficients C_P and C_{HS} are seen to correlate almost perfectly into a single function of C_Q, while η and C_H data deviate by a few percent. The last two parameters are more sensitive to slight discrepancies in model similarity; since the larger pump has smaller roughness and clearance ratios and a 40 percent larger Reynolds number, it develops slightly more head and is more efficient. The overall effect is a resounding victory for dimensional analysis.

The best efficiency point in Fig. 11.8 is approximately

$$\begin{aligned} C_{Q^*} &\approx 0.115 \qquad C_{P^*} \approx 0.65 \\ \eta_{\max} &\approx 0.88 \\ C_{H^*} &\approx 5.0 \qquad C_{HS^*} \approx 0.37 \end{aligned} \tag{11.27}$$

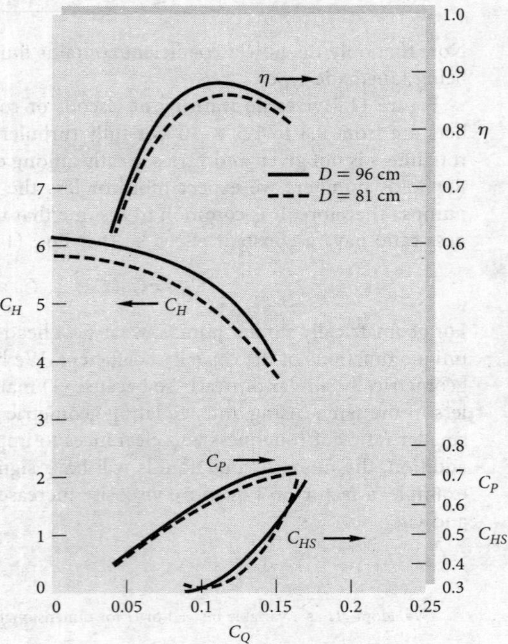

Fig. 11.8 Nondimensional plot of the pump performance data from Fig. 11.7. These numbers are not representative of other pump designs.

These values can be used to estimate the BEP performance of any size pump in this geometrically similar family. In like manner, the shutoff head is $C_H(0) \approx 6.0$, and by extrapolation the shutoff power is $C_P(0) \approx 0.25$ and the maximum discharge is $C_{Q,max} \approx 0.23$. Note, however, that Fig. 11.8 gives no reliable information about, say, the 71- or 89 cm impellers in Fig. 11.7, which have a different impeller-to-casing-size ratio and thus must be correlated separately.

By comparing values of n^2D^2, nD^3, and n^3D^5 for two pumps in Fig. 11.7, we can see readily why the large pump had the same discharge, but less power and head:

	D, m	***n*, r/s**	**Discharge nD^3, m³/s**	**Head n^2D^2/g, m**	**Power $\rho n^3D^5/746$ hp**
Fig. 11.7*a*	0.81	1,170/60	10.36	25.4	3,459
Fig. 11.7*b*	0.96	710/60	10.47	13.2	1,807
Ratio	—	—	1.011	0.520	0.522

Discharge goes as nD^3, which is about the same for both pumps. Head goes as n^2D^2 and power as n^3D^5 for the same ρ (water), and these are about half as much for the larger pump. The NPSH goes as n^2D^2 and is also half as much for the 96-cm pump.

EXAMPLE 11.3

A pump from the family of Fig. 11.8 has $D = 53$ cm and $n = 1{,}500$ r/min. Estimate (*a*) discharge, (*b*) head, (*c*) pressure rise, and (*d*) brake horsepower of this pump for water at 15°C and best efficiency.

Solution

Part (a) In SI units take $D = 53/100 = 0.53$ m and $n = 1{,}500/60 = 25$ r/s. At 15°C, ρ of water is 998 kg/m³. The BEP parameters are known from Fig. 11.8 or Eqs. (11.27). The BEP discharge is thus

$$Q^* = C_{Q^*}nD^3 = 0.115\,(25\ \text{r/s})(0.53\ \text{m})^3 = 0.428\ \text{m}^3/\text{s} = 25.7\ \text{m}^3/\text{min} \qquad \textit{Ans. (a)}$$

Part (b) Similarly, the BEP head is

$$H^* = \frac{C_{H^*}n^2D^2}{g} = \frac{5.0(25)^2(0.53)^2}{9.81} = 89.5\ \text{m water} \qquad \textit{Ans. (b)}$$

Part (c) Since we are not given elevation or velocity head changes across the pump, we neglect them and estimate

$$\Delta p \approx \rho gH = 998\ \text{kg/m}^3(9.81\ \text{m/s}^2)(89.5\ \text{m}) = 876{,}239\ \text{N/m}^2 = 876\ \text{kPa} \qquad \textit{Ans. (c)}$$

Part (d) Finally, the BEP power is

$$P^* = C_{P^*}\rho n^3D^5 = 0.65(998\ \text{kg/m}^3)(25)^3(0.53\ \text{m})^5$$

$$= \frac{423{,}880\ \text{Nm/s}}{746\ \text{Nm/(s} - \text{hp)}} = 570\ \text{hp} \qquad \textit{Ans. (d)}$$

EXAMPLE 11.4

We want to build a pump from the family of Fig. 11.8, which delivers 11.35 m³/min water at 1,200 r/min at best efficiency. Estimate (*a*) the impeller diameter, (*b*) the maximum discharge, (*c*) the shutoff head, and (*d*) the NPSH at best efficiency.

Solution

Part (a) $Q^* = 11.35\ \text{m}^3/\text{min} = 0.19\ \text{m}^3/\text{s}$ and n = 1,200 r/min = 20 r/s. At BEP we have

$$Q^* = C_{Q^*} n D^3 = 0.19\ \text{m}^3/\text{s} = (0.115)(20)D^3$$

$$D = \left[\frac{0.19\ \text{m}^3/\text{s}}{0.115(20)}\right]^{1/3} = 0.44\ \text{m} = 44\ \text{cm} \qquad \textit{Ans. (a)}$$

Part (b) The maximum Q is related to Q^* by a ratio of capacity coefficients:

$$Q_{\max} = \frac{Q^* C_{Q,\max}}{C_{Q^*}} \approx \frac{11.35\ \text{m}^3/\text{min}(0.23)}{0.115} = 22.7\ \text{m}^3/\text{min} \qquad \textit{Ans. (b)}$$

Part (c) From Fig. 11.8, we estimated the shutoff head coefficient to be 6.0. Thus

$$H(0) \approx \frac{C_H(0)n^2D^2}{g} = \frac{6.0(20)^2(0.44)^2}{9.81} = 47.4\ \text{m} \qquad \textit{Ans. (c)}$$

Part (d) Finally, from Eq. (11.27), the NPSH at BEP is approximately

$$\text{NPSH}^* = \frac{C_{HS^*} n^2 D^2}{g} = \frac{0.37(20)^2(0.44)^2}{9.81} = 2.9\ \text{m} \qquad \textit{Ans. (d)}$$

Since this is a small pump, it will be less efficient than the pumps in Fig. 11.8, probably about 85 percent maximum.

Similarity Rules

The success of Fig. 11.8 in correlating pump data leads to simple rules for comparing pump performance. If pump 1 and pump 2 are from the same geometric family and are operating at homologous points (the same dimensionless position on a chart such as Fig. 11.8), their flow rates, heads, and powers will be related as follows:

$$\frac{Q_2}{Q_1} = \frac{n_2}{n_1}\left(\frac{D_2}{D_1}\right)^3 \qquad \frac{H_2}{H_1} = \left(\frac{n_2}{n_1}\right)^2\left(\frac{D_2}{D_1}\right)^2$$
$$\frac{P_2}{P_1} = \frac{\rho_2}{\rho_1}\left(\frac{n_2}{n_1}\right)^3\left(\frac{D_2}{D_1}\right)^5 \qquad (11.28)$$

These are the *similarity rules,* which can be used to estimate the effect of changing the fluid, speed, or size on any dynamic turbomachine—pump or turbine—within a geometrically similar family. A graphic display of these rules is given in Fig. 11.9, showing the effect of speed and diameter changes on pump performance. In Fig. 11.9*a* the size is held constant and the speed is varied 20 percent, while Fig. 11.9*b* shows a 20 percent size change at constant speed. The curves are plotted to scale but with arbitrary units. The speed effect (Fig. 11.9*a*) is substantial, but the size effect (Fig. 11.9*b*) is even more dramatic, especially for power, which varies as D^5. Generally, we see that a given pump family can be adjusted in size and speed to fit a variety of system characteristics.

Strictly speaking, we would expect for perfect similarity that $\eta_1 = \eta_2$, but we have seen that larger pumps are more efficient, having a higher Reynolds number and lower roughness and clearance ratios. Two empirical correlations are recommended for maximum efficiency. One, developed by Moody [43] for turbines but also used for

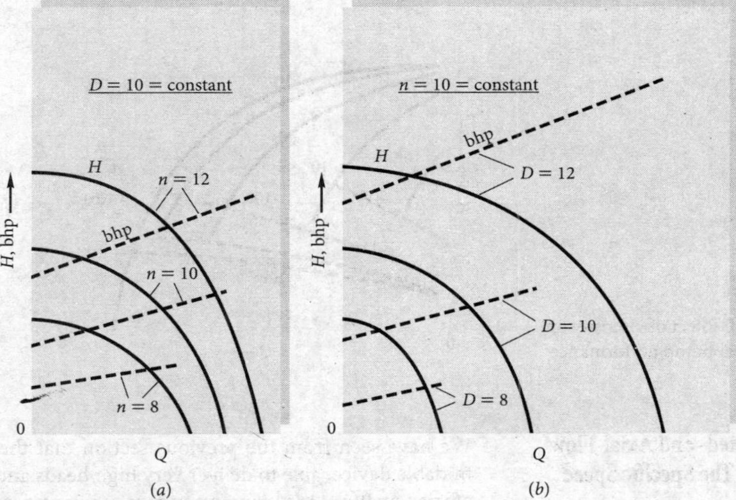

Fig. 11.9 Effect of changes in size and speed on homologous pump performance: (*a*) 20 percent change in speed at constant size; (*b*) 20 percent change in size at constant speed.
Source: Courtesy of Vickers Inc., PDN/PACE Division.

pumps, is a size effect. The other, suggested by Anderson [44] from thousands of pump tests, is a flow rate effect:

$$\text{Size changes [43]:} \qquad \frac{1-\eta_2}{1-\eta_1} \approx \left(\frac{D_1}{D_2}\right)^{1/4} \qquad (11.29a)$$

$$\text{Flow rate changes [44]:} \qquad \frac{0.94-\eta_2}{0.94-\eta_1} \approx \left(\frac{Q_1}{Q_2}\right)^{0.32} \qquad (11.29b)$$

Anderson's formula (11.29*b*) makes the practical observation that even an infinitely large pump will have losses. He thus proposes a maximum possible efficiency of 94 percent, rather than 100 percent. Anderson recommends that the same formula be used for turbines if the constant 0.94 is replaced by 0.95. The formulas in Eq. (11.29) assume the same value of surface roughness for both machines—one could micropolish a small pump and achieve the efficiency of a larger machine.

Effect of Viscosity

Centrifugal pumps are often used to pump oils and other viscous liquids up to 1,000 times the viscosity of water. But the Reynolds numbers become low turbulent or even laminar, with a strong effect on performance. Figure 11.10 shows typical test curves of head and brake horsepower versus discharge. High viscosity causes a dramatic drop in head and discharge and increases in power requirements. The efficiency also drops substantially according to the following typical results:

μ/μ_{water}	1.0	10.0	100	1,000
η_{max}, %	85	76	52	11

Beyond about $300\mu_{water}$ the deterioration in performance is so great that a positive-displacement pump is recommended.

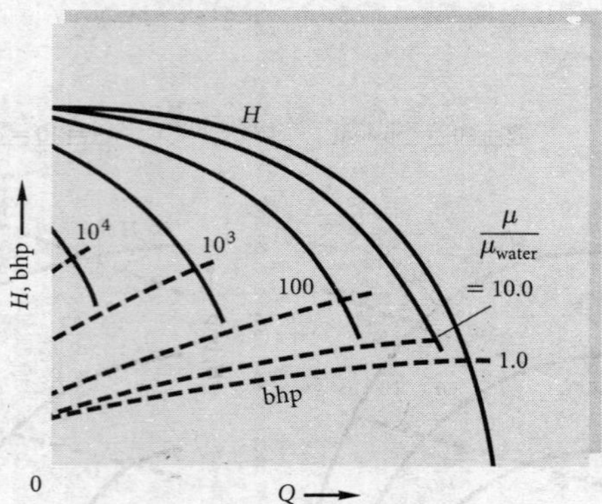

Fig. 11.10 Effect of viscosity on centrifugal pump performance.

11.4 Mixed- and Axial-Flow Pumps: The Specific Speed

We have seen from the previous section that the modern centrifugal pump is a formidable device, able to deliver very high heads and reasonable flow rates with excellent efficiency. It can match many system requirements. But, basically the centrifugal pump is a high-head, low-flow machine, whereas there are many applications requiring low head and high discharge. To see that the centrifugal design is not convenient for such systems, consider the following example.

EXAMPLE 11.5

We want to use a centrifugal pump from the family of Fig. 11.8 to deliver 360 m^3/min of water at 15°C with a head of 7.5 m. What should be (*a*) the pump size and speed and (*b*) brake horsepower, assuming operation at best efficiency?

Solution

Part (a) Enter the known head and discharge into the BEP parameters from Eq. (11.27):

$$H^* = 7.5\text{ m} = \frac{C_{H^*}n^2D^2}{g} = \frac{5.0n^2D^2}{9.81}$$

$$Q^* = 360\text{ m}^3/\text{min} = 6\text{ m}^3/\text{s} = C_{Q^*}nD^3 = 0.115nD^3$$

The two unknowns are *n* and *D*. The algebra is quite simple, so we don't really need Excel. Solve for *n* in the Q^* equation and substitute into the H^* equation:

$$n = \frac{6}{0.115\,D^3} = \frac{52.17}{D^3};\quad 7.5 = \frac{5.0D^2}{9.81}\left(\frac{52.17}{D^3}\right)^2 = \frac{1{,}387}{D^4},\quad \text{or: } D^4 = 185$$

Solve for $D = 3.69$ m $\quad n = 1.04$ r/s $= 62$ r/min *Ans. (a)*

Part (b) The most efficient horsepower is then, from Eq. (11.27),

$$\text{bhp}^* \approx C_{P^*}\rho n^3D^5 = \frac{0.65(998\text{ kg/m}^3)(1.04\text{ r/s})^3(3.69\text{ m})^5}{746\text{ Nm/(s} - \text{hp)}} = 670\text{ hp}$$ *Ans. (b)*

The solution to Example 11.5 is mathematically correct but results in a grotesque pump: an impeller more than 3.6 m in diameter, rotating so slowly one can visualize oxen walking in a circle turning the shaft.

Other dynamic pump designs provide low head and high discharge. For example, there is a type of 96 cm, 710 r/min pump, with the same input parameters as Fig. 11.7*b*, which will deliver the 7.5 m head and 6 m^3/s flow rate called for in Example 11.5. This is done by allowing the flow to pass through the impeller with an axial-flow component and less centrifugal component. The passages can be opened up to the increased flow rate with very little size increase, but the drop in radial outlet velocity decreases the head produced. These are the mixed-flow (part radial, part axial) and axial-flow (propeller-type) families of dynamic pump. Some vane designs are sketched in Fig. 11.11, which introduces an interesting new "design" parameter, the specific speed N_s or N_s'.

The Specific Speed

Most pump applications involve a known head and discharge for the particular system, plus a speed range dictated by electric motor speeds or cavitation requirements. The designer then selects the best size and shape (centrifugal, mixed, axial) for the pump. To help this selection, we need a dimensionless parameter involving speed, discharge, and head but not size. This is accomplished by eliminating the diameter between C_Q and C_H, applying the result only to the BEP. This ratio is called the *specific speed* and has both a dimensionless form and a somewhat lazy, practical form:

Rigorous form:
$$N_s' = \frac{C_{Q^*}^{1/2}}{C_{H^*}^{3/4}} = \frac{n(Q^*)^{1/2}}{(gH^*)^{3/4}} \tag{11.30a}$$

Lazy but common:
$$N_s = \frac{(\text{r/min})\,(\text{m}^3/\text{min})^{1/2}}{[H(\text{m})]^{3/4}} \tag{11.30b}$$

η_{max} 1.0 0.9 0.8 0.7 0.6

Centrifugal pump

Mixed flow

Axial flow

75 150 300 600 750 1,500 2,250

N_s

r/min (m^3/min)$^{1/2}$/(H, m)$^{3/4}$

(*a*)

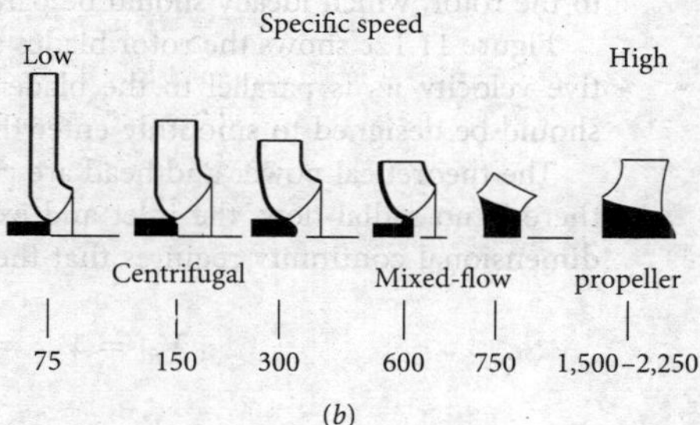

(*b*)

Fig. 11.11 (*a*) Optimum efficiency and (*b*) vane design of dynamic pump families as a function of specific speed.

In other words, practicing engineers do not bother to change n to revolutions per second or Q^* to cubic meters per second or to include gravity with head, although the latter would be necessary for, say, a pump on the moon. The conversion factor is

$$N_s = 2{,}576 N_s'$$

Note that N_s is applied only to BEP; thus a single number characterizes an entire family of pumps. For example, the family of Fig. 11.8 has $N_s' \approx (0.115)^{1/2}/(5.0)^{3/4} = 0.1014$, $N_s = 261$, regardless of size or speed.

It turns out that the specific speed is directly related to the most efficient pump design, as shown in Fig. 11.11. Low N_s means low Q and high H, hence a centrifugal pump, and large N_s implies an axial pump. The centrifugal pump is best for N_s between 75 and 600, the mixed-flow pump for N_s between 600 and 1,500, and the axial-flow pump for N_s above 1,500. Note the changes in impeller shape as N_s increases.

Suction Specific Speed

If we use NPSH rather than H in Eq. (11.30), the result is called *suction-specific speed:*

Rigorous:
$$N_{ss}' = \frac{nQ^{1/2}}{(g\,\text{NPSH})^{3/4}} \tag{11.31a}$$

Lazy:
$$N_{ss} = \frac{(\text{r/min})(\text{m}^3/\text{min})^{1/2}}{[\text{NPSH (m)}]^{3/4}} \tag{11.31b}$$

where NPSH denotes the available suction head of the system. Data from Wislicenus [4] show that a given pump is in danger of inlet cavitation if

$$N_{ss}' \geq 0.47 \qquad N_{ss} \geq 1{,}200$$

In the absence of test data, this relation can be used, given n and Q, to estimate the minimum required NPSH.

Axial-Flow Pump Theory

A multistage axial-flow geometry is shown in Fig. 11.12*a*. The fluid essentially passes almost axially through alternate rows of fixed *stator* blades and moving *rotor* blades. The incompressible flow assumption is frequently used even for gases because the pressure rise per stage is usually small.

The simplified vector diagram analysis assumes that the flow is one-dimensional and leaves each blade row at a relative velocity exactly parallel to the exit blade angle. Figure 11.12*b* shows the stator blades and their exit velocity diagram. Since the stator is fixed, ideally the absolute velocity V_1 is parallel to the trailing edge of the blade. After vectorially subtracting the rotor tangential velocity u from V_1, we obtain the velocity w_1 relative to the rotor, which ideally should be parallel to the rotor leading edge.

Figure 11.12*c* shows the rotor blades and their exit velocity diagram. Here the relative velocity w_2 is parallel to the blade trailing edge, while the absolute velocity V_2 should be designed to smoothly enter the next row of stator blades.

The theoretical power and head are given by Euler's turbine relation (11.11). Since there is no radial flow, the inlet and exit rotor speeds are equal, $u_1 = u_2$, and one-dimensional continuity requires that the axial-velocity component remain constant:

$$V_{n1} = V_{n2} = V_n = \frac{Q}{A} = \text{const}$$

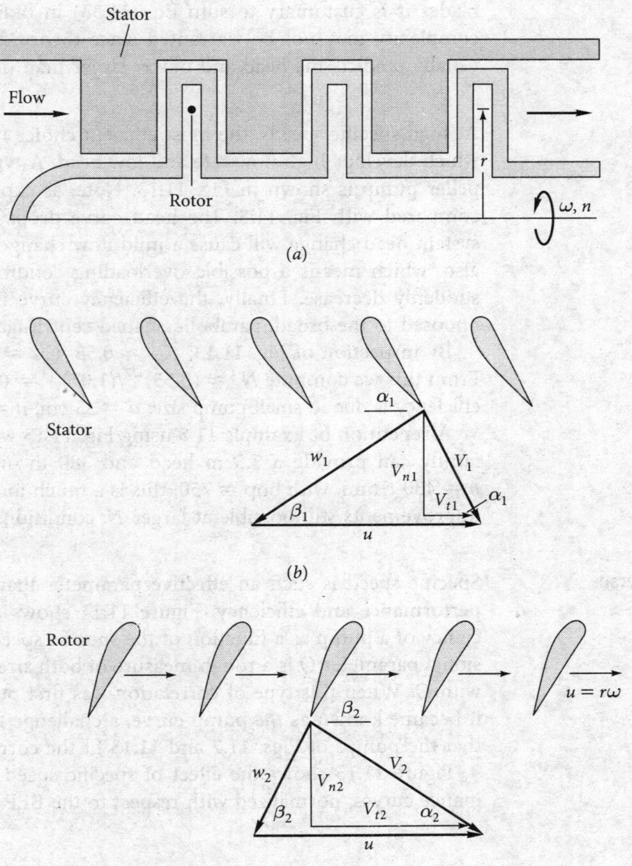

Fig. 11.12 Analysis of an axial-flow pump: (*a*) basic geometry; (*b*) stator blades and exit velocity diagram; (*c*) rotor blades and exit velocity diagram.

From the geometry of the velocity diagrams, the normal velocity (or volume flow) can be directly related to the blade rotational speed u:

$$u = \omega r_{\text{av}} = V_{n1}(\cot \alpha_1 + \cot \beta_1) = V_{n2}(\cot \alpha_2 + \cot \beta_2) \tag{11.32}$$

Thus, the flow rate can be predicted from the rotational speed and the blade angles. Meanwhile, since $V_{t1} = V_{n1} \cot \alpha_1$ and $V_{t2} = u - V_{n2} \cot \beta_2$, Euler's relation (11.11) for the pump head becomes

$$\begin{aligned} gH &= uV_n(\cot \alpha_2 - \cot \alpha_1) \\ &= u^2 - uV_n(\cot \alpha_1 + \cot \beta_2) \end{aligned} \tag{11.33}$$

the preferred form because it relates to the blade angles α_1 and β_2. The shutoff or no-flow head is seen to be $H_0 = u^2/g$, just as in Eq. (11.18) for a centrifugal pump. The blade-angle parameter $\cot \alpha_1 + \cot \beta_2$ can be designed to be negative, zero, or positive, corresponding to a rising, flat, or falling head curve, as in Fig. 11.5.

Strictly speaking, Eq. (11.33) applies only to a single streamtube of radius r, but it is a good approximation for very short blades if r denotes the average radius. For long blades it is customary to sum Eq. (11.33) in radial strips over the blade area. Such complexity may not be warranted since theory, being idealized, neglects losses and usually predicts the head and power larger than those in actual pump performance.

Performance of an Axial-Flow Pump

At high specific speeds, the most efficient choice is an axial-flow, or propeller, pump, which develops high flow rate and low head. A typical dimensionless chart for a propeller pump is shown in Fig. 11.13. Note, as expected, the higher C_Q and lower C_H compared with Fig. 11.8. The head curve drops sharply with discharge, so a large system head change will cause a mild flow change. The power curve drops with head also, which means a possible overloading condition if the system discharge should suddenly decrease. Finally, the efficiency curve is rather narrow and triangular, as opposed to the broad, parabolic-shaped centrifugal pump efficiency (Fig. 11.8).

By inspection of Fig. 11.13, $C_{Q^*} \approx 0.55$, $C_{H^*} \approx 1.07$, $C_{P^*} \approx 0.70$, and $\eta_{max} \approx 0.84$. From this we compute $N_s' \approx (0.55)^{1/2}/(1.07)^{3/4} = 0.705$, $N_s = 1{,}800$. The relatively low efficiency is due to small pump size: $d = 35$ cm, $n = 690$ r/min, $Q^* = 16.7$ m^3/min.

A repetition of Example 11.5 using Fig. 11.13 would show that this propeller pump family can provide a 7.5 m head and 380 m^3/min discharge if $D = 117$ cm and $n = 430$ r/min, with bhp $= 750$; this is a much more reasonable design solution, with improvements still possible at larger-N_s conditions.

Pump Performance versus Specific Speed

Specific speed is such an effective parameter that it is used as an indicator of both performance and efficiency. Figure 11.14 shows a correlation of the optimum efficiency of a pump as a function of the specific speed and capacity. Because the dimensional parameter Q is a rough measure of both size and Reynolds number, η increases with Q. When this type of correlation was first published by Wislicenus [4] in 1947, it became known as *the* pump curve, a challenge to all manufacturers. We can check that the pumps of Figs. 11.7 and 11.13 fit the correlation very well.

Figure 11.15 shows the effect of specific speed on the shape of the pump performance curves, normalized with respect to the BEP point. The numerical values shown

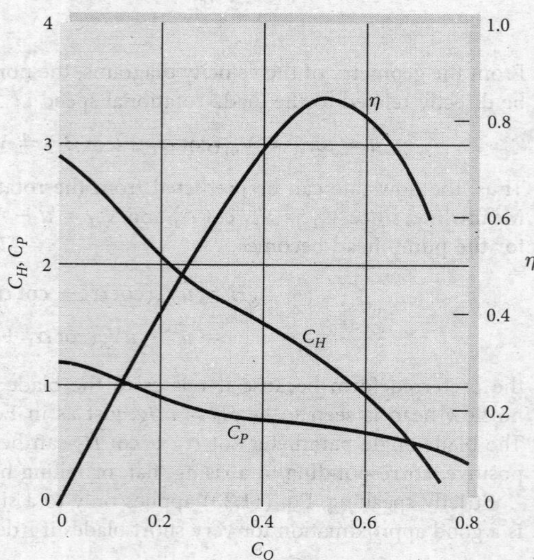

Fig. 11.13 Dimensionless performance curves for a typical axial-flow pump, $N_s = 1{,}800$. Constructed from data given by Stepanoff [8] for a 35 cm pump at 690 r/min.

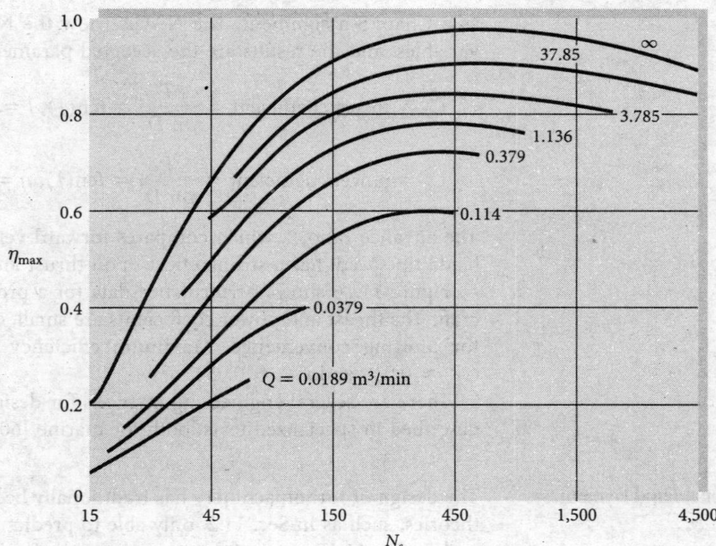

Fig. 11.14 Optimum efficiency of pumps versus capacity and specific speed. *(Adapted from Refs. 4 and 31.)*

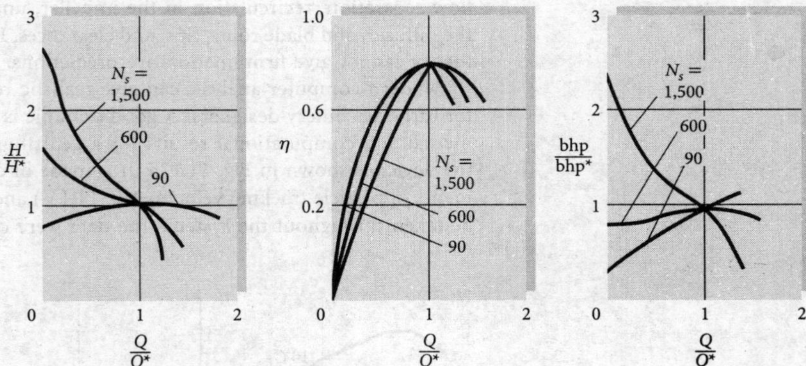

Fig. 11.15 Effect of specific speed on pump performance curves.

are representative but somewhat qualitative. The high-specific-speed pumps ($N_s \approx 1{,}500$) have head and power curves that drop sharply with discharge, implying overload or start-up problems at low flow. Their efficiency curve is very narrow.

A low-specific-speed pump ($N_s = 90$) has a broad efficiency curve, a rising power curve, and a head curve that "droops" at shutoff, implying possible surge or hunting problems.

The Free Propeller

The propeller-style pump of Fig. 11.12 is enclosed in a duct and captures all the approach flow. In contrast, the *free propeller*, for either aircraft or marine applications, acts in an unbounded fluid and thus is much less effective. The analog of propeller-pump pressure rise is the free propeller *thrust* per unit area ($\pi D^2/4$) swept out by the blades. In a customary dimensional analysis, thrust T and power required P are functions of fluid density ρ, rotation rate n (rev/s), forward velocity V, and propeller diameter D. Viscosity effects are small and neglected. You might enjoy analyzing this as a Chap. 5 assignment. The NACA (now the NASA) chose (ρ, n, D) as repeating variables, and the results are the accepted parameters:

$$C_T = \text{thrust coefficient} = \frac{T}{\rho n^2 D^4} = \text{fcn}(J), J = \text{advance ratio} = \frac{V}{nD}$$

$$C_P = \text{power coefficient} = \frac{P}{\rho n^3 D^5} = \text{fcn}(J), \eta = \text{efficiency} = \frac{VT}{P} = \frac{JC_T}{C_P} \qquad (11.34)$$

The advance ratio, J, which compares forward velocity to a measure proportional to blade tip speed, has a strong effect upon thrust and power.

Figure 11.16 shows performance data for a propeller used on the Cessna 172 aircraft. The thrust and power coefficients are small, of $\mathcal{O}(0.05)$, and are multiplied by 10 for plotting convenience. Maximum efficiency is 83 percent at $J = 0.7$, where $C_T^* \approx 0.040$ and $C_P^* \approx 0.034$.

There are several engineering methods for designing propellers. These theories are described in specialized texts, both for marine [60] and aircraft [61] propellers.

Computational Fluid Dynamics

The design of turbomachinery has traditionally been highly experimental, with simple theories, such as in Sec. 11.2, only able to predict trends. Dimensionless correlations, such as Fig. 11.15, are useful but require extensive experimentation. Consider that flow in a pump is three-dimensional; unsteady (both periodic and turbulent); and involves flow separation, recirculation in the impeller, unsteady blade wakes passing through the diffuser, and blade roots, tips, and clearances. It is no wonder that one-dimensional theory cannot give firm quantitative predictions.

Modern computer analysis can give realistic results and is becoming a useful tool for turbomachinery designers. A good example is Ref. 56, reporting combined experimental and computational results for a centrifugal pump diffuser. A photograph of the device is shown in Fig. 11.17*a*. It is made of clear Perspex, so that laser measurements of particle tracking velocimetry (LPTV) and doppler anemometry (LDA) could be taken throughout the system. The data were compared with a CFD simulation of

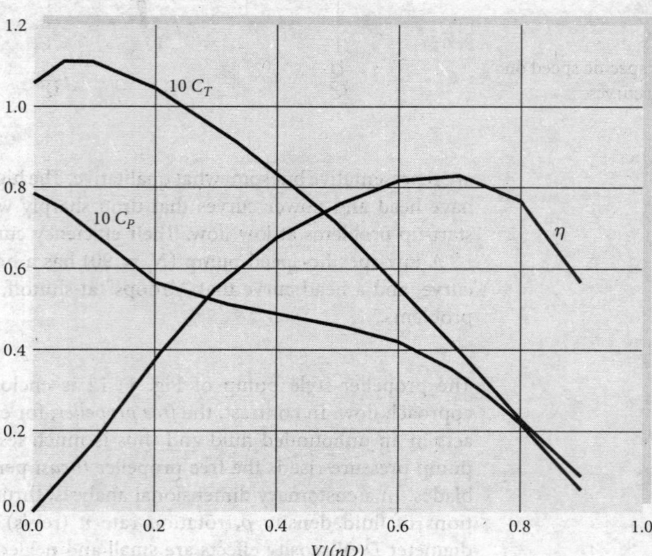

Fig. 11.16 Performance data for a free propeller used on the Cessna 172 aircraft. Compare to Fig. 11.13 for a (ducted) propeller pump. The thrust and power coefficients are much smaller for the free propeller.

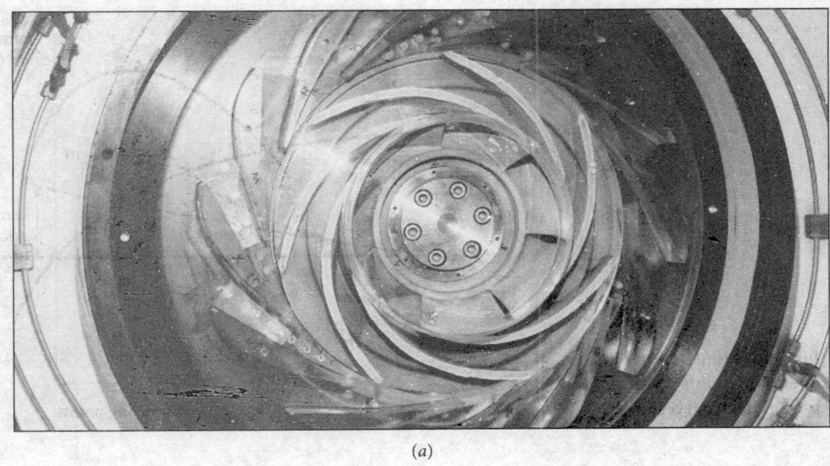

(*a*)

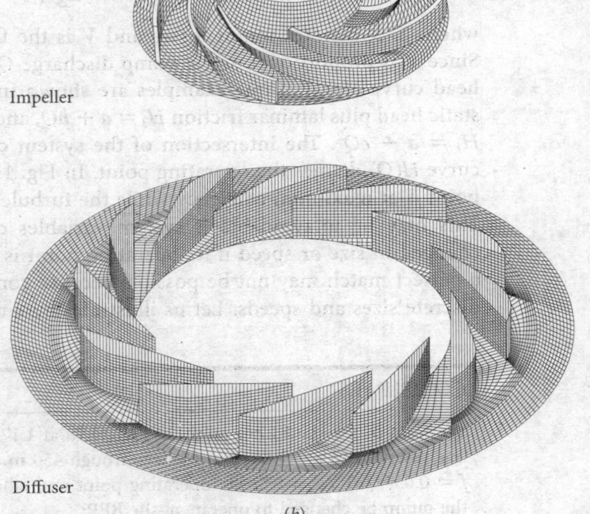

(*b*)

Fig. 11.17 Turbomachinery design now involves both experimentation and computational fluid dynamics (CFD): (*a*) a centrifugal impeller and diffuser; (*b*) a three-dimensional CFD model grid for this system.
Sources: (a) courtesy of K. Eisele et al., "Flow Analysis in a Pump Diffuser: Part 1, Measurements: Part 2, CFD," Journal of Fluids Eng. Vol. 119, December 1997, pp. 967–984/American Society of Mechanical Engineers (b) From K. Eisele et al ., "Row Analysis in a Pump Diffuser: Part I Measurements; Part 2, CFD," J. Fluids Eng., vol. 119, December 1997, pp. 967–984. by permission of the American Society of Mechanical Engineers.

the impeller and diffuser, using the grids shown in Fig. 11.17*b*. The computations used a turbulence formulation called the *k*-ε model, popular in commercial CFD codes (see Sec. 8.9). Results were good but not excellent. The CFD model predicted velocity and pressure data adequately up until flow separation, after which it was only qualitative. Clearly, CFD is developing a significant role in turbomachinery design [42, 45].

11.5 Matching Pumps to System Characteristics

The ultimate test of a pump is its match with the operating system characteristics. Physically, the system head must match the head produced by the pump, and this intersection should occur in the region of best efficiency.

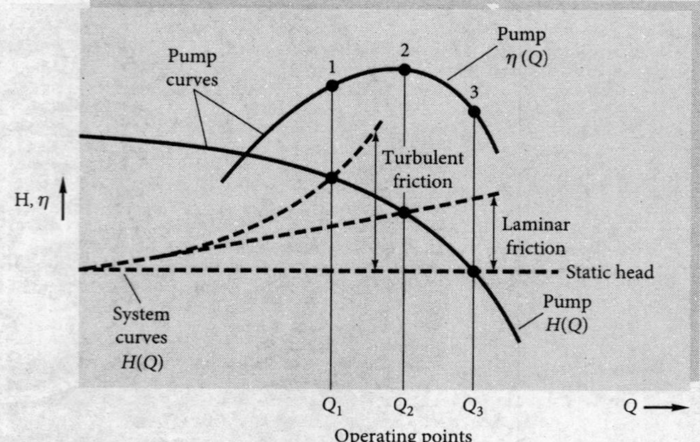

Fig. 11.18 Illustration of pump operating points for three types of system head curves.

The system head will probably contain a static elevation change $z_2 - z_1$ plus friction losses in pipes and fittings:

$$H_{sys} = (z_2 - z_1) + \frac{V^2}{2g}\left(\sum \frac{fL}{D} + \sum K\right)$$

where ΣK denotes minor losses and V is the flow velocity in the principal pipe. Since V is proportional to the pump discharge Q, the equation represents a system head curve $H_s(Q)$. Three examples are shown in Fig. 11.18: a static head $H_s = a$, static head plus laminar friction $H_s = a + bQ$, and static head plus turbulent friction $H_s = a + cQ^2$. The intersection of the system curve with the pump performance curve $H(Q)$ defines the operating point. In Fig. 11.18 the laminar friction operating point is at maximum efficiency while the turbulent and static curves are off design. This may be unavoidable if system variables change, but the pump should be changed in size or speed if its operating point is consistently off design. Of course, a perfect match may not be possible because commercial pumps have only certain discrete sizes and speeds. Let us illustrate these concepts with an example.

EXAMPLE 11.6

We want to use the 81 cm pump of Fig. 11.7*a* at 1,170 r/min to pump water at 15°C from one reservoir to another 36 m higher through 450 m of 40-cm-ID pipe with friction factor $f = 0.030$. (*a*) What will the operating point and efficiency be? (*b*) To what speed should the pump be changed to operate at the BEP?

Solution

Part (a)

For reservoirs the initial and final velocities are zero; thus the system head is

$$H_s = z_2 - z_1 + \frac{V^2}{2g}\frac{fL}{D} = 36 \text{ m} + \frac{V^2}{2g}\frac{0.030(450 \text{ m})}{0.4 \text{ m}}$$

From continuity in the pipe, $V = Q/A = Q/\left[\frac{1}{4}\pi(0.4 \text{ m})^2\right]$, and so we substitute for V to get

$$H_s = 36 + 109Q^2 \qquad Q \text{ in m}^3\text{/s} \tag{1}$$

Since Fig. 11.7*a* uses m³/min for the abscissa, we convert Q in Eq. (1) to this unit:

$$H_s = 36 + 0.03Q^2 \qquad Q \text{ in m}^3/\text{min} \tag{2}$$

We can plot Eq. (2) on Fig. 11.7*a* and see where it intersects the 81-cm pump head curve, as in Fig. E11.6. A graphical solution gives approximately

$$H \approx 130 \text{ m} \qquad Q \approx 56 \text{ m}^3/\text{min}$$

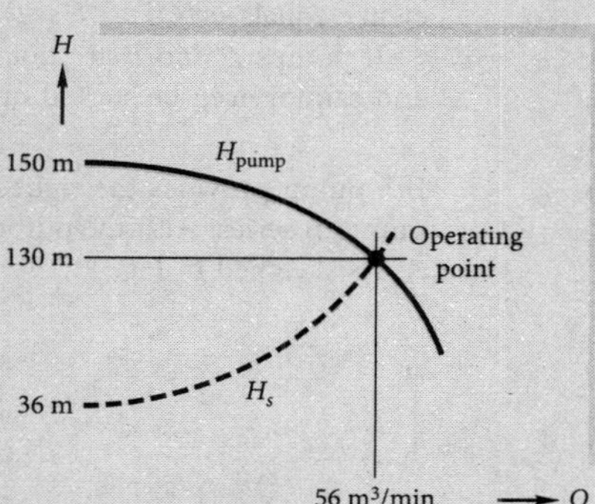

E11.6

The efficiency is about 82 percent, slightly off design.

An anaiytic solution is possible if we fit the pump head curve to a parabola, which is very accurate:

$$H_{\text{pump}} \approx 150 - 0.0064Q^2 \qquad Q \text{ in m}^3/\text{min} \tag{3}$$

Equations (2) and (3) must match at the operating point:

$$150 - 0.0064Q^2 = 36 + 0.03Q^2$$

or

$$Q^2 = \frac{150 - 36}{0.0064 + 0.03} = 3{,}130$$

$$Q = 56 \text{ m}^3/\text{min} = 0.933 \text{ m}^3/\text{s} \qquad \textit{Ans. (a)}$$

$$H = 150 - 0.0064(56)^2 = 130 \text{ m} \qquad \textit{Ans. (a)}$$

Part (b) To move the operating point to BEP, we change n, which changes both $Q \propto n$ and $H \propto n^2$. From Fig. 11.7*a*, at BEP, $H^* \approx 118$ m; thus for any n, $H^* = 118(n/1{,}170)^2$. Also read $Q^* \approx 75$ m³/min; thus for any n, $Q^* = 75(n/1{,}170)$. Match H^* to the system characteristics, Eq. (2):

$$H^* = 118\left(\frac{n}{1{,}170}\right)^2 \approx 36 + 0.03\left(75\frac{n}{1{,}170}\right)^2 \qquad \textit{Ans. (b)}$$

which gives $n^2 < 0$. Thus it is impossible to operate at maximum efficiency with this particular system and pump.

Pumps Combined in Parallel

If a pump provides the right head but too little discharge, a possible remedy is to combine two similar pumps in parallel, sharing the same suction and inlet conditions. A parallel arrangement is also used if delivery demand varies, so that one pump is used at low flow and the second pump is started up for higher discharges. Both pumps should have check valves to avoid backflow when one is shut down.

The two pumps in parallel need not be identical. Physically, their flow rates will sum for the same head, as illustrated in Fig. 11.19*a*. If pump *A* has more head than pump *B*, pump *B* cannot be added in until the operating head is below the shutoff head of pump *B*. Since the system curve rises with *Q*, the combined delivery Q_{A+B} will be less than the separate operating discharges $Q_A + Q_B$ but certainly greater than either one. For a very flat (static) curve two similar pumps in parallel will deliver nearly twice the flow. The combined brake horsepower is found by adding brake horsepower for each of pumps *A* and *B* at the same head as the operating point. The combined efficiency equals $\rho g(Q_{A+B})(H_{A+B})/(746\ \text{bhp}_{A+B})$.

If pumps *A* and *B* are not identical, as in Fig. 11.19*a*, pump *B* should not be run and cannot even be started up if the operating point is above its shutoff head.

Pumps Combined in Series

If a pump provides the right discharge but too little head, consider adding a similar pump in series, with the output of pump *B* fed directly into the suction side of pump *A*. As sketched in Fig. 11.19*b*, the physical principle for summing in series is that the

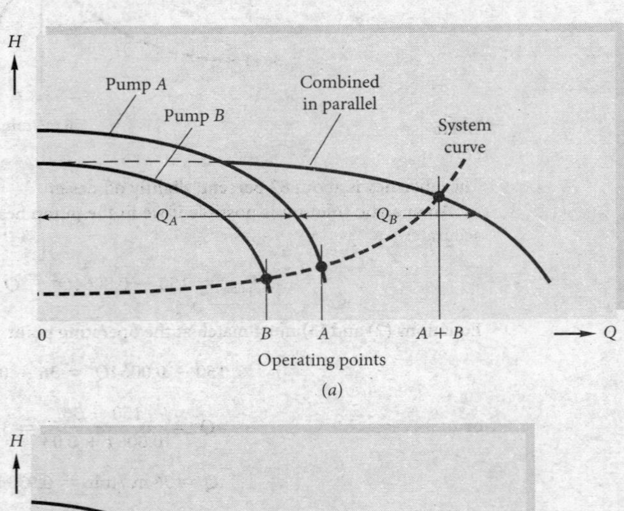

(*a*)

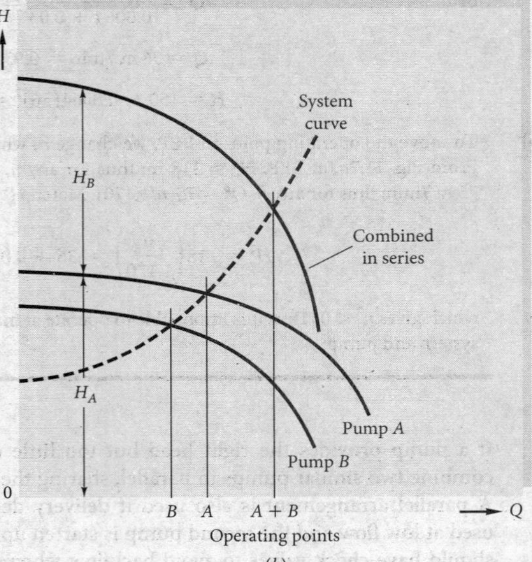

(*b*)

Fig. 11.19 Performance and operating points of two pumps operating singly and (*a*) in parallel or (*b*) in series.
Source: Copyright United Technologies Corporation 2008. Used with permission.

two heads add at the same flow rate to give the combined performance curve. The two need not be identical at all, since they merely handle the same discharge; they may even have different speeds, although normally both are driven by the same shaft.

The need for a series arrangement implies that the system curve is steep, that is, it requires higher head than either pump A or B can provide. The combined operating point head will be more than either A or B separately but not as great as their sum. The combined power is the sum of brake horsepower for A and B at the operating point flow rate. The combined efficiency is

$$\frac{\rho g(Q_{A+B})(H_{A+B})}{746\ \text{bhp}_{A+B}}$$

similar to parallel pumps.

Whether pumps are used in series or in parallel, the arrangement will be uneconomical unless both pumps are operating near their best efficiency.

Multistage Pumps

For very high heads in continuous operation, the solution is a multistage pump, with the exit of one impeller feeding directly into the eye of the next. Centrifugal, mixed-flow, and axial-flow pumps have all been grouped in as many as 50 stages, with heads up to 2,500 m of water and pressure rises up to 35,000 kPa absolute. Figure 11.20 shows a

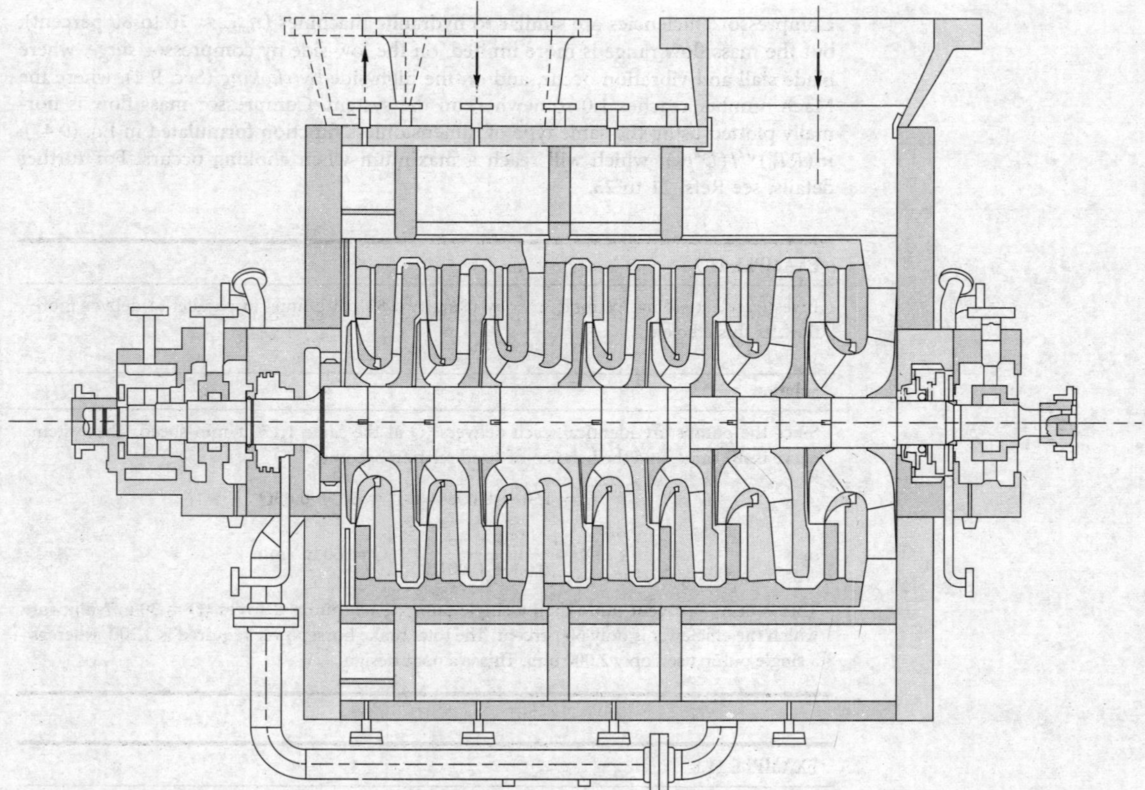

Fig. 11.20 Cross section of a seven-stage centrifugal propane compressor that delivers 1,150 m^3/min at 35,000 bhp and a pressure rise of 2,000 kPa. Note the second inlet at stage 5 and the varying impeller designs.
Source: Courtesy of DeLavai-Stork V.OF., Centrifugal Compressor Division.

section of a seven-stage centrifugal propane compressor that develops 2,000 kPa rise at 1,150 m^3/min and 35,000 bhp.

Compressors

Most of the discussion in this chapter concerns incompressible flow—that is, negligible change in fluid density. Even the pump of Fig. 11.7, which can produce 180 m of head at 1,170 r/min, will increase standard air pressure only by 2,200 Pa, about a 2 percent change in density. The picture changes at higher speeds, $\Delta p \propto n^2$, and multiple stages, where very large changes in pressure and density are achieved. Such devices are called *compressors,* as in Fig. 11.20. The concept of static head, $H = \Delta p/\rho g$, becomes inappropriate, since ρ varies. Compressor performance is measured by (1) the pressure ratio across the stage p_2/p_1 and (2) the change in stagnation enthalpy $(h_{02} - h_{01})$, where $h_0 = h + \frac{1}{2}V^2$ (see Sec. 9.3). Combining m stages in series results in $p_{final}/p_{initial} \approx (p_2/p_1)^m$. As density increases, less area is needed: note the decrease in impeller size from right to left in Fig. 11.20. Compressors may be either of the centrifugal or axial-flow type [21 to 23].

Compressor efficiency, from inlet condition 1 to final outlet f, is defined by the change in gas enthalpy, assuming an adiabatic process:

$$\eta_{comp} = \frac{h_f - h_{01}}{h_{0f} - h_{01}} \approx \frac{T_f - T_{01}}{T_{0f} - T_{01}}$$

Compressor efficiencies are similar to hydraulic machines ($\eta_{max} \approx$ 70 to 80 percent), but the mass flow range is more limited: on the low side by compressor *surge,* where blade stall and vibration occur, and on the high side by *choking* (Sec. 9.4), where the Mach number reaches 1.0 somewhere in the system. Compressor mass flow is normally plotted using the same type of dimensionless function formulated in Eq. (9.47): $\dot{m}(RT_0)^{1/2}/(D^2 p_0)$, which will reach a maximum when choking occurs. For further details, see Refs. 21 to 23.

EXAMPLE 11.7

Investigate extending Example 11.6 by using two 81 cm pumps in parallel to deliver more flow. Is this efficient?

Solution

Since the pumps are identical, each delivers $\frac{1}{2}Q$ at the same 1,170 r/min speed. The system curve is the same, and the balance-of-head relation becomes

$$H = 150 - 0.0064(\tfrac{1}{2}Q)^2 = 36 + 0.03Q^2$$

or

$$Q^2 = \frac{150 - 36}{0.03 + 0.0016} \qquad Q = 60 \text{ m}^3/\text{min} \qquad \textit{Ans.}$$

This is only 7 percent more than a single pump. Each pump delivers $\frac{1}{2}Q = 30$ m^3/min, for which the efficiency is only 60 percent. The total brake horsepower required is 3,200, whereas a single pump used only 2,000 bhp. This is a poor design.

EXAMPLE 11.8

Suppose the elevation change in Example 11.6 is raised from 36 to 150 m, greater than a single 81 cm pump can supply. Investigate using two 81 cm pumps in series at 1,170 r/min.

Solution

Since the pumps are identical, the total head is twice as much and the constant 36 m in the system head curve is replaced by 150 m. The balance of heads becomes

$$H = 2(150 - 0.0064Q^2) = 150 + 0.03Q^2$$

or
$$Q^2 = \frac{300 - 150}{0.03 + 0.0128} \qquad Q = 59.2\,\text{m}^3/\text{min} \qquad \textit{Ans.}$$

The operating head is $150 + 0.03(59.2)^2 = 255$ m, or 96 percent more than that for a single pump in Example 11.6. Each pump is operating at 59.2 m³/min, which from Fig. 11.7*a* is 83 percent efficient, a pretty good match to the system. To pump at this operating point requires 4,100 bhp, or about 2,050 bhp for each pump.

Gas Turbines

Some modern devices contain both pumps and turbines. A classic case is the *gas turbine,* which combines a compressor, a combustion chamber, a turbine, and, often, a fan. Gas turbines are used to drive aircraft, helicopters, Army tanks, and small electric power plants. They have a higher power-to-weight ratio than reciprocating engines, but they spin at very high speeds and require high-temperature materials and thus are costly. The compressor raises the inlet air to pressures as much as 30 to 40 times higher, before entering the combustion chamber. The heated air then passes through a turbine, which drives the compressor. The airflow then exits to provide the thrust and is generally a supersonic flow.

The example illustrated in Fig. 11.21 is a Pratt & Whitney 6000 turbofan aircraft engine. The large entrance fan greatly increases the airflow into the engine, some of which bypasses the compressors. The central flow enters a low-pressure (LPC) and a high-pressure (HPC) compressor and thence into the combustor. After combustion, the hot, high-velocity gases pass through a high-pressure turbine (HPT), which drives the HPC, and a low-pressure turbine (LPT), which separately drives both the LPC and the fan. The exhaust gases then create the thrust in the usual momentum–exchange manner. The engine shown, designed for shorter airline flights, has a maximum thrust of 107 kN.

11.6 Turbines

A turbine extracts energy from a fluid that possesses high head, but it is fatuous to say a turbine is a pump run backward. Basically there are two types, reaction and impulse, the difference lying in the manner of head conversion. In the *reaction turbine,*

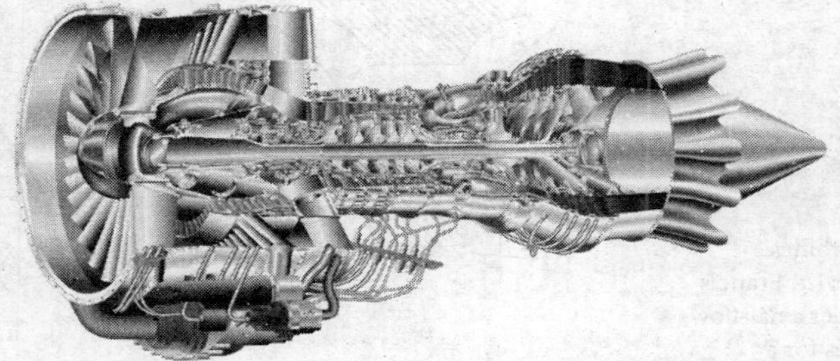

Fig. 11.21 Cutaway view of a Pratt & Whitney 6000 turbofan aircraft engine. *(Copyright United Technologies Corporation 2008. Used with permission.)*

the fluid fills the blade passages, and the head change or pressure drop occurs within the impeller. Reaction designs are of the radial-flow, mixed-flow, and axial-flow types and are essentially dynamic devices designed to admit the high-energy fluid and extract its momentum. An *impulse turbine* first converts the high head through a nozzle into a high-velocity jet, which then strikes the blades at one position as they pass by. The impeller passages are not fluid-filled, and the jet flow past the blades is essentially at constant pressure. Reaction turbines are smaller because fluid fills all the blades at one time.

Reaction Turbines

Reaction turbines are low-head, high-flow devices. The flow is opposite that in a pump, entering at the larger-diameter section and discharging through the eye after giving up most of its energy to the impeller. Early designs were very inefficient because they lacked stationary guide vanes at the entrance to direct the flow smoothly into the impeller passages. The first efficient inward-flow turbine was built in 1849 by James B. Francis, a U.S. engineer, and all radial- or mixed-flow designs are now called *Francis turbines*. At still lower heads, a turbine can be designed more compactly with purely axial flow and is termed a *propeller turbine* [52]. The propeller may be either fixed-blade or adjustable (Kaplan type), the latter being complicated mechanically, but much more efficient at low-power settings. Figure 11.22 shows sketches of runner designs for Francis radial, Francis mixed-flow, and propeller-type turbines.

Idealized Radial Turbine Theory

The Euler turbomachine formulas (11.11) also apply to energy-extracting machines if we reverse the flow direction and reshape the blades. Figure 11.23 shows a radial turbine runner. Again, assume one-dimensional frictionless flow through the blades. Adjustable inlet guide vanes are absolutely necessary for good efficiency. They bring the inlet flow to the blades at angle α_2 and absolute velocity V_2 for minimum "shock" or directional-mismatch loss. After vectorially adding in the runner tip speed $u_2 = \omega r_2$, the outer blade angle should be set at angle β_2 to accommodate the relative velocity w_2, as shown in the figure. (See Fig. 11.4 for the analogous radial pump velocity diagrams.)

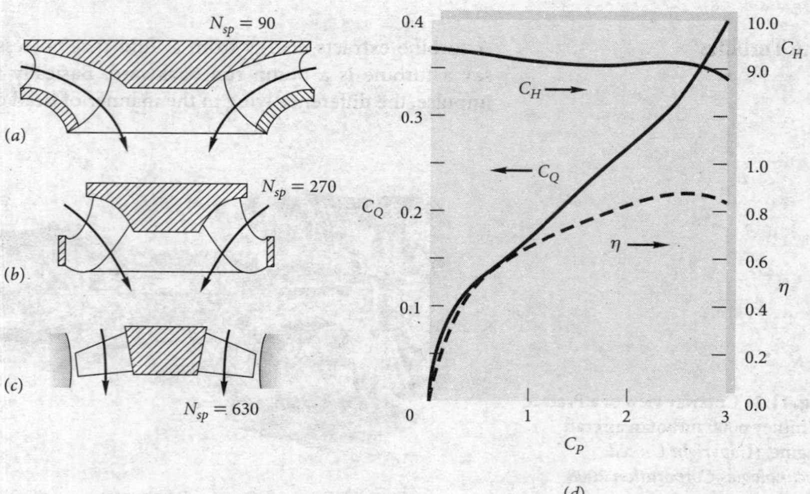

Fig. 11.22 Reaction turbines: (*a*) Francis, radial type; (*b*) Francis mixed-flow; (*c*) propeller axial-flow; (*d*) performance curves for a Francis turbine, $n = 600$ r/min, $D = 69$ cm, $N_{sp} = 126$.

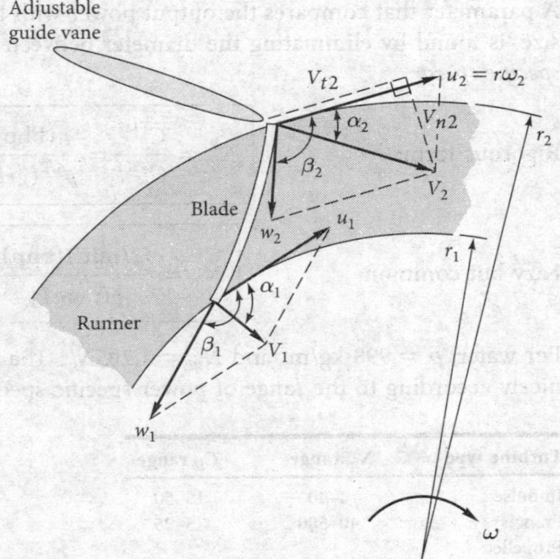

Fig. 11.23 Inlet and outlet velocity diagrams for an idealized radial-flow reaction turbine runner.

Application of the angular momentum control volume theorem, Eq. (3.59), to Fig. 11.23 (see Example 3.18 for a similar case) yields an idealized formula for the power P extracted by the runner:

$$P = \omega T = \rho\omega Q(r_2 V_{t2} - r_1 V_{t1}) = \rho Q(u_2 V_2 \cos\alpha_2 - u_1 V_1 \cos\alpha_1) \qquad (11.35)$$

where V_{t2} and V_{t1} are the absolute inlet and outlet circumferential velocity components of the flow. Note that Eq. (11.35) is identical to Eq. (11.11) for a radial pump, except that the blade shapes are different.

The absolute inlet normal velocity $V_{n2} = V_2 \sin\alpha_2$ is proportional to the flow rate Q. If the flow rate changes and the runner speed u_2 is constant, the vanes must be adjusted to a new angle α_2 so that w_2 still follows the blade surface. Thus, adjustable inlet vanes are very important to avoid shock loss.

Power Specific Speed

Turbine parameters are similar to those of a pump, but the dependent variable is the output brake horsepower, which depends on the inlet flow rate Q, available head H, impeller speed n, and diameter D. The efficiency is the output brake horsepower divided by the available water horsepower ρgQH. The dimensionless forms are C_Q, C_H, and C_P, defined just as for a pump, Eqs. (11.23). If we neglect Reynolds number and roughness effects, the functional relationships are written with C_P as the independent variable:

$$C_H = \frac{gH}{n^2D^2} = C_H(C_P) \qquad C_Q = \frac{Q}{nD^3} = C_Q(C_P) \qquad \eta = \frac{\text{bhp}}{\rho gQH} = \eta(C_P) \qquad (11.36)$$

where $$C_P = \frac{\text{bhp}}{\rho n^3 D^5}$$

Figure 11.22*d* shows typical performance curves for a small Francis radial turbine. The maximum efficiency point is called the *normal power,* and the values for this particular turbine are

$$\eta_{\max} = 0.89 \qquad C_{P^*} = 2.70 \qquad C_{Q^*} = 0.34 \qquad C_{H^*} = 9.03$$

A parameter that compares the output power with the available head, independent of size, is found by eliminating the diameter between C_H and C_P. It is called the *power specific speed:*

Rigorous form:
$$N'_{sp} = \frac{C_P^{*1/2}}{C_H^{*5/4}} = \frac{n(\text{bhp})^{1/2}}{\rho^{1/2}(gH)^{5/4}} \tag{11.37a}$$

Lazy but common:
$$N_{sp} = \frac{(\text{r/min})(\text{bhp})^{1/2}}{[H\ (\text{m})]^{5/4}} \tag{11.37b}$$

For water, $\rho = 998\ \text{kg/m}^3$ and $N_{sp} = 1{,}205N'_{sp}$. The various turbine designs divide up nicely according to the range of power specific speed, as follows:

Turbine type	N_{sp} range	C_H range
Impulse	4–40	15–50
Francis	40–500	5–25
Propeller:		
Water	400–1,100	1–4
Gas, steam	100–1,300	10–80

Note that N_{sp}, like N_s for pumps, is defined only with respect to the BEP and has a single value for a given turbine family. In Fig. 11.22*d*, $N_{sp} = 1{,}205(2.70)^{1/2}/(9.03)^{5/4} = 126$, regardless of size.

Like pumps, turbines of large size are generally more efficient, and Eqs. (11.29) can be used as an estimate when data are lacking.

The design of a complete large-scale power-generating turbine system is a major engineering project, involving inlet and outlet ducts, trash racks, guide vanes, wicket gates, spiral cases, generator with cooling coils, bearings and transmission gears, runner blades, draft tubes, and automatic controls. Some typical large-scale reaction turbine designs are shown in Fig. 11.24. The reversible pump-and-turbine design of Fig. 11.24*d* requires special care for adjustable guide vanes to be efficient for flow in either direction.

The largest (1,000-MW) hydropower designs are awesome when viewed on a human scale, as shown in Fig. 11.25. The economic advantages of small-scale model testing are evident from this photograph of the Francis turbine units at Grand Coulee Dam.

Impulse Turbines

For high head and relatively low power (that is, low N_{sp}), not only would a reaction turbine require too high a speed but also the high pressure in the runner would require a massive casing thickness. The impulse turbine of Fig. 11.26 is ideal for this situation. Since N_{sp} is low, n will be low and the high pressure is confined to the small nozzle, which converts the head to an atmospheric pressure jet of high velocity V_j. The jet strikes the buckets and imparts a momentum change similar to that in our control volume analysis for a moving vane in Example 3.9 or Prob. P3.51. The buckets have an elliptical split-cup shape, as in Fig. 11.26*b*. They are named *Pelton wheels,* after Lester A. Pelton (1829–1908), who produced the first efficient design.

In Example 3.10, we found that the force per unit mass flow on a single moving vane, or in this case a single Pelton bucket, was $(V_j - u)(1 - \cos\beta)$, where u is the vane velocity and β is the exit angle of the jet. For a single vane, as in Example 3.10, the mass flow would be $\rho A_j(V_j - u)$, but for a Pelton wheel, where buckets keep entering the jet and capture all the flow, the mass flow would be $\rho Q = \rho A_j V_j$. An alternative

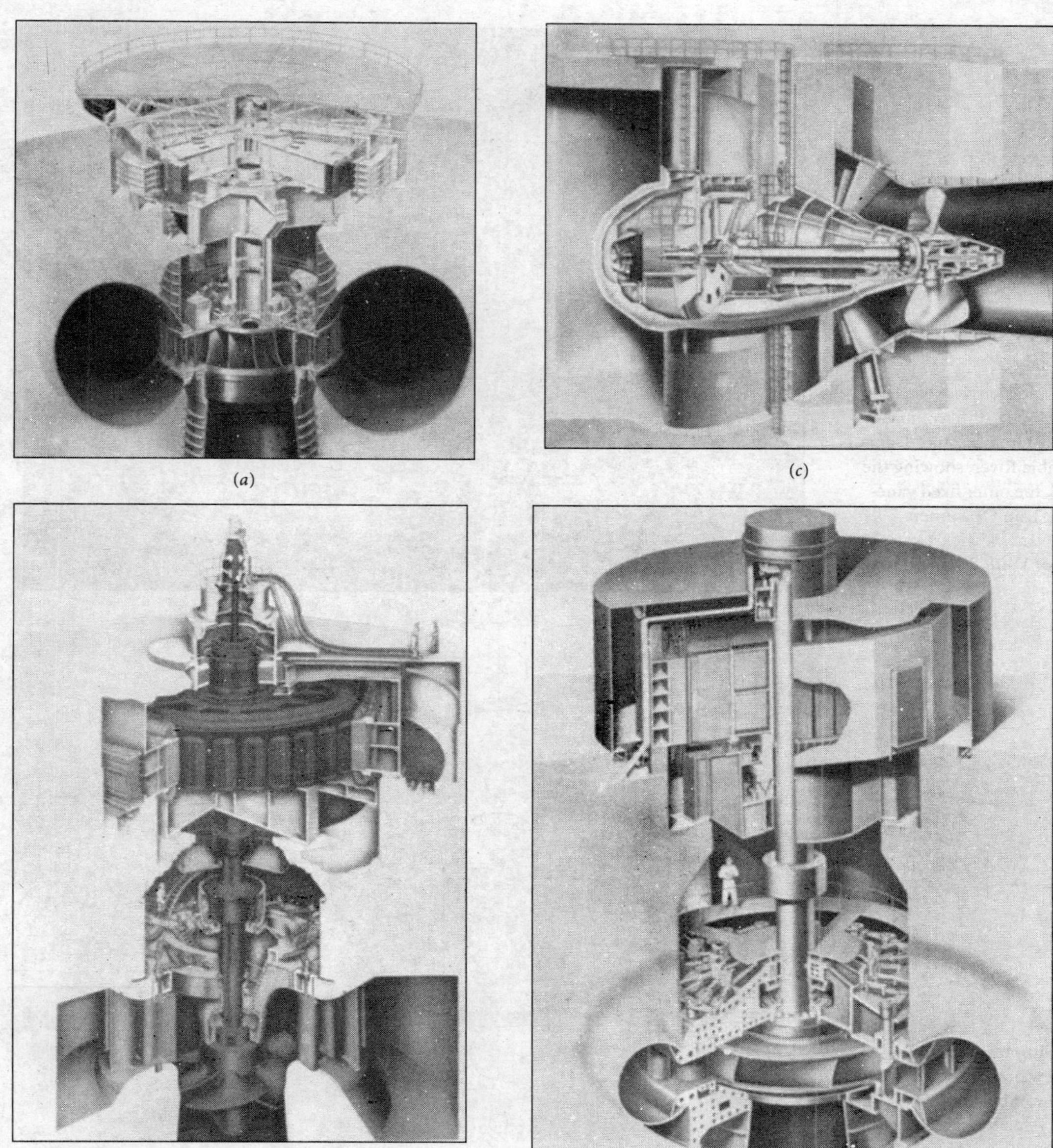

Fig. 11.24 Large-scale turbine designs depend on available head and flow rate and operating conditions: (*a*) Francis (radial); (*b*) Kaplan (propeller); (*c*) bulb mounting with propeller runner; (*d*) reversible pump turbine with radial runner. *(Courtesy of Voith Siemens Hydro Power.)*

Fig. 11.25 Interior view of the 1.1-million hp (820-MW) turbine units on the Grand Coulee Dam of the Columbia River, showing the spiral case, the outer fixed vanes ("stay ring"), and the inner adjustable vanes ("wicket gates"). *(Courtesy of Voith Siemens Hydro Power.)*

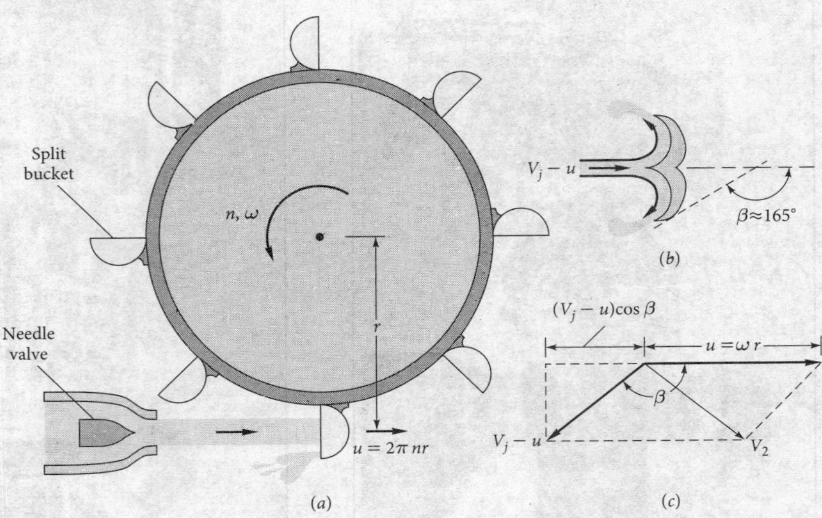

Fig. 11.26 Impulse turbine: (*a*) side view of wheel and jet; (*b*) top view of bucket; (*c*) typical velocity diagram.

analysis uses the Euler turbomachine equation (11.11) and the velocity diagram of Fig. 11.26*c*. Noting that $u_1 = u_2 = u$, we substitute the absolute exit and inlet tangential velocities into the turbine power relation:

$$P = \rho Q(u_1 V_{t1} - u_2 V_{t2}) = \rho Q\{uV_j - u[u + (V_j - u)\cos\beta]\}$$

or

$$P = \rho Q u(V_j - u)(1 - \cos\beta) \tag{11.38}$$

where $u = 2\pi nr$ is the bucket linear velocity and r is the *pitch radius,* or distance to the jet centerline. A bucket angle $\beta = 180°$ gives maximum power but is physically impractical. In practice, $\beta \approx 165°$, or $1 - \cos\beta \approx 1.966$ or only 2 percent less than maximum power.

From Eq. (11.38), the theoretical power of an impulse turbine is parabolic in bucket speed u and is maximum when $dP/du = 0$, or

$$u^* = 2\pi n^* r = \tfrac{1}{2}V_j \tag{11.39}$$

For a perfect nozzle, the entire available head would be converted to jet velocity $V_j = (2gH)^{1/2}$. Actually, since there are 2 to 8 percent nozzle losses, a velocity coefficient C_v is used:

$$V_j = C_v(2gH)^{1/2} \qquad 0.92 \le C_v \le 0.98 \tag{11.40}$$

By combining Eqs. (11.36) and (11.40), the theoretical impulse turbine efficiency becomes

$$\eta = 2(1 - \cos\beta)\phi(C_v - \phi) \tag{11.41}$$

where
$$\phi = \frac{u}{(2gH)^{1/2}} = \text{peripheral velocity factor}$$

Maximum efficiency occurs at $\phi = \frac{1}{2}C_v \approx 0.47$.

Figure 11.27 shows Eq. (11.41) plotted for an ideal turbine ($\beta = 180°$, $C_v = 1.0$) and for typical working conditions ($\beta = 160°$, $C_v = 0.94$). The latter case predicts $\eta_{max} = 85$ percent at $\phi = 0.47$, but the actual data for a 24-in Pelton wheel test are somewhat less efficient due to windage, mechanical friction, backsplashing, and nonuniform bucket flow. For this test $\eta_{max} = 80$ percent, and, generally speaking, an impulse turbine is not quite as efficient as the Francis or propeller turbines at their BEPs.

Figure 11.28 shows the optimum efficiency of the three turbine types, and the importance of the power specific speed N_{sp} as a selection tool for the designer. These efficiencies are optimum and are obtained in careful design of large machines.

The water power available to a turbine may vary due to either net head or flow rate changes, both of which are common in field installations such as hydroelectric plants.

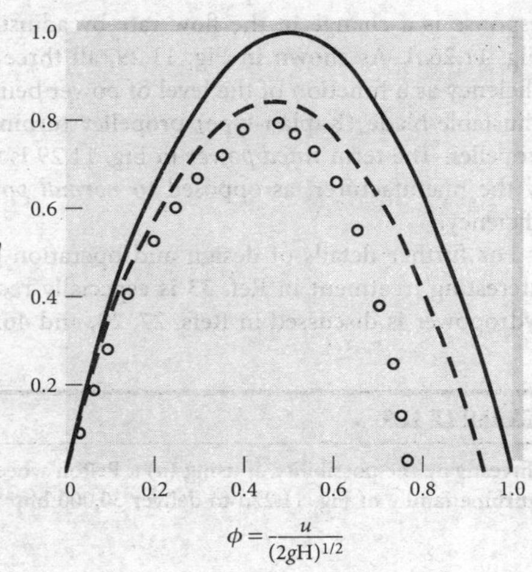

Fig. 11.27 Efficiency of an impulse turbine calculated from Eq. (11.41): solid curve = ideal, $\beta = 180°$, $C_v = 1.0$; dashed curve = actual, $\beta = 160°$, $C_v = 0.94$; open circles = data, Pelton wheel, diameter = 0.6 m.

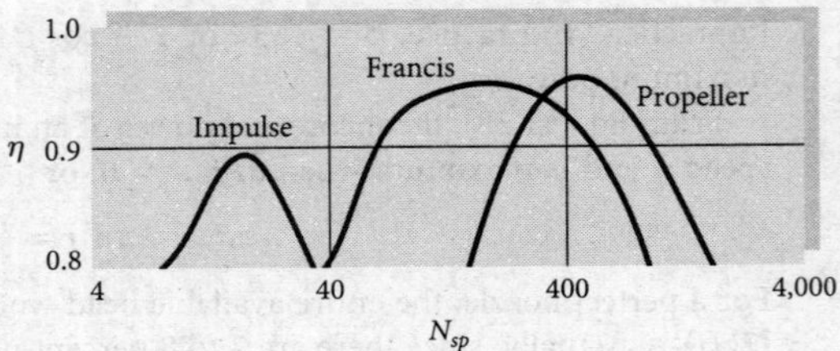

Fig. 11.28 Optimum efficiency of turbine designs.

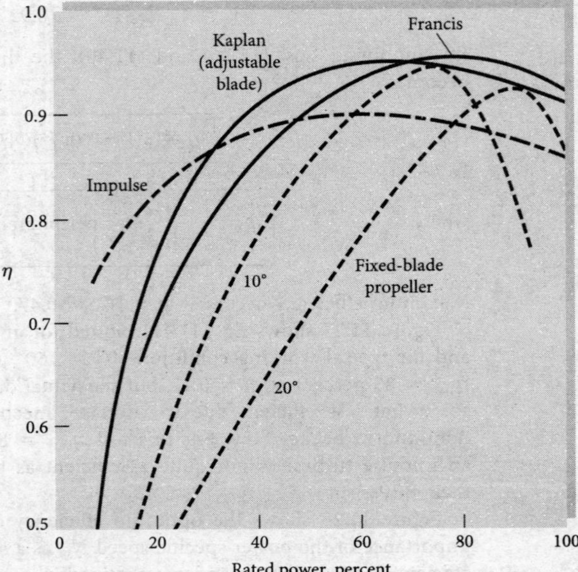

Fig. 11.29 Efficiency versus power level for various turbine designs at constant speed and head.

The demand for turbine power also varies from light to heavy, and the operating response is a change in the flow rate by adjustment of a gate valve or needle valve (Fig. 11.26*a*). As shown in Fig. 11.29, all three turbine types achieve fairly uniform efficiency as a function of the level of power being extracted. Especially effective is the adjustable-blade (Kaplan-type) propeller turbine, while the poorest is a fixed-blade propeller. The term *rated power* in Fig. 11.29 is the largest power delivery guaranteed by the manufacturer, as opposed to *normal power,* which is delivered at maximum efficiency.

For further details of design and operation of turbomachinery, the readable and interesting treatment in Ref. 33 is especially recommended. The feasibility of micro-hydropower is discussed in Refs. 27, 28, and 46.

EXAMPLE 11.9

Investigate the possibility of using (*a*) a Pelton wheel similar to Fig. 11.27 or (*b*) the Francis turbine family of Fig. 11.22*d* to deliver 30,000 bhp from a net head of 365 m.

Solution

Part (a) From Fig. 11.28, the most efficient Pelton wheel occurs at about

$$N_{sp} \approx 20 = \frac{(\text{r/min})(30{,}000 \text{ bhp})^{1/2}}{(365 \text{ m})^{1.25}}$$

or $$n = 184 \text{ r/min} = 3.07 \text{ r/s}$$

From Fig. 11.27, the best operating point is

$$\phi \approx 0.47 = \frac{\pi D(3.07 \text{ r/s})}{[2(9.81)(365)]^{1/2}}$$

or $$D = 4.12 \text{ m}$$ *Ans. (a)*

This Pelton wheel is perhaps a little slow and a trifle large. You could reduce D and increase n by increasing N_{sp} to, say, 25 or 30 and accepting the slight reduction in efficiency. Or you could use a double-hung, two-wheel configuration, each delivering 15,000 bhp, which changes D and n by the factor $2^{1/2}$:

Double wheel: $$n = (184)2^{1/2} = 260 \text{ r/min} \qquad D = \frac{4.12}{2^{1/2}} = 2.91 \text{ m}$$ *Ans. (a)*

Part (b) The Francis wheel of Fig. 11.22*d* must have

$$N_{sp} = 126 = \frac{(\text{r/min})(30{,}000 \text{ bhp})^{1/2}}{(365 \text{ m})^{1.25}}$$

or $$n = 1{,}161 \text{ r/min} = 19.35 \text{ r/s}$$

Then the optimum power coefficient is

$$C_{P^*} = 2.70 = \frac{P}{\rho n^3 D^5} = \frac{30{,}000(746)}{(998)(19.35)^3 D^5}$$

or $$D^5 = 1.146 \qquad D = 1.03 \text{ m}$$ *Ans. (b)*

This is a faster speed than normal practice, and the casing would have to withstand 365 m of water or about 3,600 kPa internal pressure, but the 1.03 m size is extremely attractive. Francis turbines are now being operated at heads up to 450 m.

Wind Turbines

Wind energy has long been used as a source of mechanical power. The familiar four-bladed windmills of Holland, England, and the Greek islands have been used for centuries to pump water, grind grain, and saw wood. Modern research concentrates on the ability of wind turbines to generate electric power. Koeppl [47] stresses the potential for propeller-type machines. See also Refs. 47 to 51.

Some examples of wind turbine designs are shown in Fig. 11.30. The familiar American multiblade farm windmill (Fig. 11.30*a*) is of low efficiency, but thousands are in use as a rugged, reliable, and inexpensive way to pump water. A more efficient design is the propeller mill in Fig. 11.30*b*, similar to the pioneering Smith-Putnam 1,250-kW two-bladed system that operated on Grampa's Knob, 20 km west of Rutland, Vermont, from 1941 to 1945. The Smith-Putnam design broke because of inadequate blade strength, but it withstood winds up to 185 km/h and its efficiency was amply demonstrated [47].

The Dutch, American multiblade, and propeller mills are examples of *horizontal-axis wind turbines* (HAWTs), which are efficient but somewhat awkward in that they require extensive bracing and gear systems when combined with an electric generator.

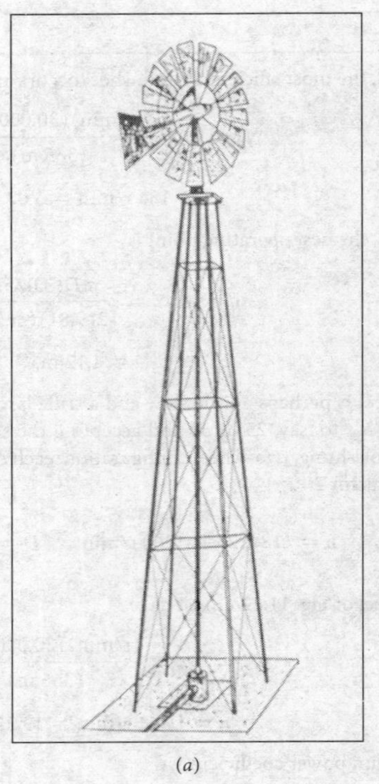
(a)

(c)

(b)

(d)

Fig. 11.30 Wind turbine designs: (*a*) the American multiblade farm HAWT; (*b*) a modern three-blade, 750 kW HAWT on a wind farm in Plouarzel, France *(courtesy of Hubert Chanson)*; (*c*) The Darrieus VAWT *(courtesy of National Research Council Canada)*; (*d*) a four-blade Dutch-type windmill, built in 1787 in Jamestown, Rhode Island *(courtesy of F. M. White)*.

Thus, a competing family of *vertical-axis wind turbines* (VAWTs) has been proposed to simplify gearing and strength requirements. Figure 11.30*c* shows the "eggbeater" VAWT invented by G. J. M. Darrieus in 1925. To minimize centrifugal stresses, the twisted blades of the Darrieus turbine follow a *troposkien* curve formed by a chain anchored at two points on a spinning vertical rod. The Darrieus design has the advantage that the generator and gearbox may be placed on the ground for easy access. It is not as efficient, though, as a HAWT, and, furthermore, it is not self-starting. The largest Darrieus device known to the writer is a 4.2 MW turbine, 100 m in diameter, at Cap Chat, Québec, Canada.

The four-arm Dutch-type windmill in Fig. 11.30*d* is found throughout Europe and the Middle East and dates back to the 9th century. In the Netherlands, they are primarily used to drain lowlands. The mill in the photo was built in 1787 to grind corn. It is now a Rhode Island tourist attraction.

Idealized Wind Turbine Theory

The ideal, frictionless efficiency of a propeller windmill was predicted by A. Betz in 1920, using the simulation shown in Fig. 11.31. The propeller is represented by an *actuator disk*, which creates across the propeller plane a pressure discontinuity of area A and local velocity V. The wind is represented by a streamtube of approach velocity V_1 and a slower downstream wake velocity V_2. The pressure rises to p_b just before the disk and drops to p_a just after, returning to free-stream pressure in the far wake. To hold the propeller rigid when it is extracting energy from the wind, there must be a leftward force F on its support, as shown.

A control-volume–horizontal-momentum relation applied between sections 1 and 2 gives

$$\sum F_x = -F = \dot{m}(V_2 - V_1)$$

A similar relation for a control volume just before and after the disk gives

$$\sum F_x = -F + (p_b - p_a)A = \dot{m}(V_a - V_b) = 0$$

Equating these two yields the propeller force:

$$F = (p_b - p_a)A = \dot{m}(V_1 - V_2) \tag{11.42}$$

Assuming ideal flow, the pressures can be found by applying the incompressible Bernoulli relation up to the disk:

From 1 to b: $$p_\infty + \tfrac{1}{2}\rho V_1^2 = p_b + \tfrac{1}{2}\rho V^2$$

From a to 2: $$p_a + \tfrac{1}{2}\rho V^2 = p_\infty + \tfrac{1}{2}\rho V_2^2$$

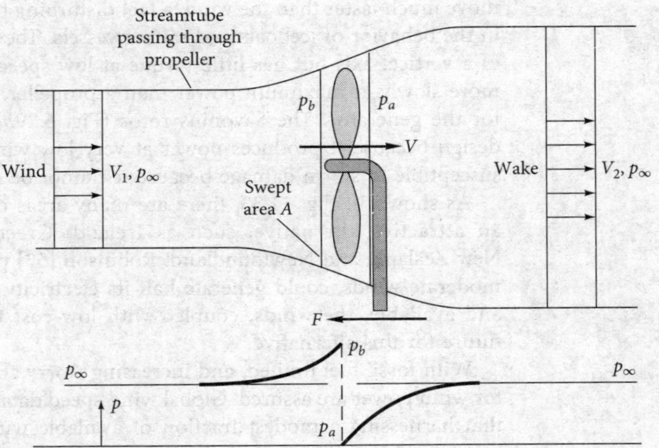

Fig. 11.31 Idealized actuator disk and streamtube analysis of flow through a windmill.

Subtracting these and noting that $\dot{m} = \rho AV$ through the propeller, we can substitute for $p_b - p_a$ in Eq. (11.42) to obtain

$$p_b - p_a = \tfrac{1}{2}\rho(V_1^2 - V_2^2) = \rho V(V_1 - V_2)$$

or

$$V = \tfrac{1}{2}(V_1 + V_2) \tag{11.43}$$

Continuity and momentum thus require that the velocity V through the disk equal the average of the wind and far-wake speeds.

Finally, the power extracted by the disk can be written in terms of V_1 and V_2 by combining Eqs. (11.42) and (11.43):

$$P = FV = \rho AV^2(V_1 - V_2) = \tfrac{1}{4}\rho A(V_1^2 - V_2^2)(V_1 + V_2) \tag{11.44}$$

For a given wind speed V_1, we can find the maximum possible power by differentiating P with respect to V_2 and setting equal to zero. The result is

$$P = P_{\max} = \tfrac{8}{27}\rho AV_1^3 \quad \text{at } V_2 = \tfrac{1}{3}V_1 \tag{11.45}$$

which corresponds to $V = 2V_1/3$ through the disk.

The maximum available power to the propeller is the mass flow through the propeller times the total kinetic energy of the wind:

$$P_{\text{avail}} = \tfrac{1}{2}\dot{m}V_1^2 = \tfrac{1}{2}\rho AV_1^3$$

Thus the maximum possible efficiency of an ideal frictionless wind turbine is usually stated in terms of the *power coefficient:*

$$C_P = \frac{P}{\tfrac{1}{2}\rho AV_1^3} \tag{11.46}$$

Equation (11.45) states that the total power coefficient is

$$C_{p,\max} = \tfrac{16}{27} = 0.593 \tag{11.47}$$

This is called the *Betz number* and serves as an ideal with which to compare the actual performance of real windmills.

Figure 11.32 shows the measured power coefficients of various wind turbine designs. The independent variable is not V_2/V_1 (which is artificial and convenient only in the ideal theory) but the ratio of blade-tip speed ωr to wind speed. Note that the tip can move much faster than the wind, a fact disturbing to the laity but familiar to engineers in the behavior of iceboats and sailing vessels. The Darrieus has the many advantages of a vertical axis but has little torque at low speeds (see Fig. 11.32) and also rotates more slowly at maximum power than a propeller, thus requiring a higher gear ratio for the generator. The Savonius rotor (Fig. 6.29*b*) has been suggested as a VAWT design because it produces power at very low wind speeds, but it is inefficient and susceptible to storm damage because it cannot be feathered in high winds.

As shown in Fig. 11.33, there are many areas of the world where wind energy is an attractive alternative, such as Ireland, Greenland, Iceland, Argentina, Chile, New Zealand, and Newfoundland. Robinson [53] points out that Australia, with only moderate winds, could generate half its electricity with wind turbines. Inexhaustible and available, the winds, coupled with low-cost turbine designs, promise a bright future for this alternative.

With fossil fuel limited, and increasing worry about global warming, the prospects for wind power are assured. Global wind speed data by Archer and Jacobsen [62] show that harnessing a modest fraction of available wind energy could supply *all* of the

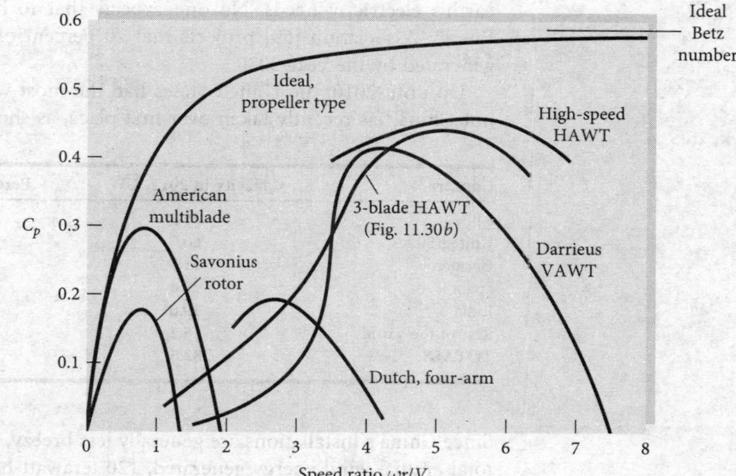

Fig. 11.32 Estimated performance of various wind turbine designs as a function of blade-tip speed ratio. *(From Ref. 53.)*

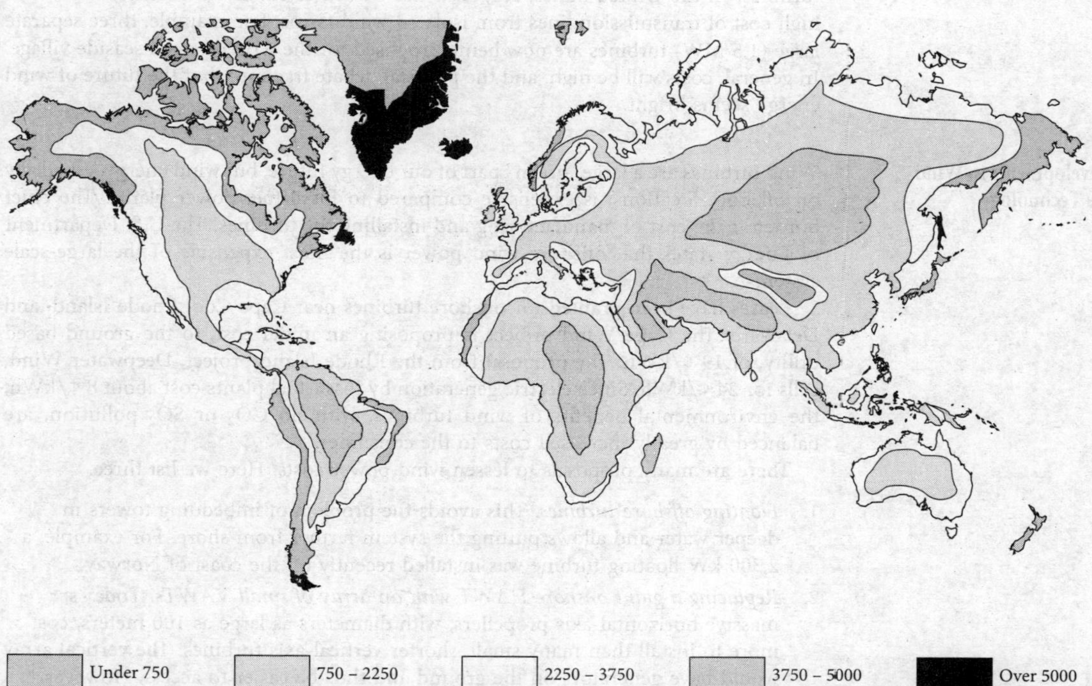

Fig. 11.33 World availability of land-based wind energy: estimated annual electric output in kWh/kW of a wind turbine rated at 11.2 m/s (25 mi/h).

Source: D. F. Warne and P. G. Calnan, "Generation of Electricity from lhe Wmd," IEE Rev., vol. 124, no. 11R, November 1977, pp. 963–985.

earth's electricity needs. No one expects that to happen, but the American Wind Energy Association [59] projects that 20 percent of America's power could be wind generated by the year 2030.

Up until 2010, the United States had the most wind-power capacity in the world, but China has recently taken over first place, as shown in the following table.

Country	Capacity in 2012, GW	Percent of world total
China	75.6	26.8
United States	60.0	21.2
Germany	31.3	11.1
Spain	22.8	8.1
India	19.6	6.9
Rest of the world	73.2	25.9
TOTALS:	**282.5**	**100.0**

Since China's installations are generally less breezy, the United States actually leads in total annual wind energy generated, 120 terawatt-hours to 73 for China. The growth in wind energy generation is excellent, but 283 GW is still only 2.5 percent of total world energy needs. Most of the future growth will probably be land-based, because of the high costs of construction, transmission, and maintenance of offshore wind turbines. In the United States, emphasis has shifted to localized turbines, due to the high cost of transmission lines from isolated wind farms. For example, three separate large (1.5 MW) turbines are now being proposed for the writer's small seaside village. In general, costs will be high, and the political debate fractious, but the future of wind energy seems bright.

New Developments in Wind Turbine Technology

Wind turbines are a large "green" part of our energy future, but wind energy, especially on offshore locations, is expensive compared to fossil-fuel power plants. The chief burden is the cost of manufacturing and installing the turbines. The U.S. Department of Energy states that offshore wind power is the most expensive of the large-scale systems.

Leases have been granted for offshore turbines near Cape Cod, Rhode Island, and Delaware. The Cape Wind project is proposing an initial cost, to the ground-based utility, of 19 ¢/kWh. The proposal from the Rhode Island project, Deepwater Wind, calls for 24 ¢/kWh. Since electric generation by fossil-fuel plants cost about 8 ¢/kWh, the environmental benefits of wind turbines, with no CO_2 or SO_2 pollution, are balanced by greatly increased costs to the consumer.

There are many proposals to lessen wind power costs. Here we list three.

1. *Floating offshore turbines.* This avoids the problem of imbedding towers in deeper water and allows putting the system farther from shore. For example, a 2,300 kW floating turbine was installed recently off the coast of Norway.
2. *Replacing a giant onshore HAWT with an array of small VAWTs.* Today's massive horizontal-axis propellers, with diameters as large as 100 meters, cost more to install than many small, shorter vertical-axis turbines. The vertical array would have generators on the ground and thus be easier to access. However, many, perhaps hundreds, of the smaller turbines would be required. This idea is best suited for remote locations.
3. *Tethered shrouded-balloon turbines.* A buoyant, ring-shaped shroud, with a turbine inside, can be deployed at high altitudes, where winds are two to three times higher.

Fig. 11.34 An experimental 10.5 m diameter buoyant-ring high-altitude wind turbine. Tested in Limestone, Maine, at an altitude of 100 m. [*Image courtesy of Altaeros Energies*]

Figure 11.34 shows an experimental design by Altaeros Energies, Inc. The shroud also helps draw in more wind than a free propeller. Installation costs are less. For the test in Fig. 11.34, the turbine at 100 m produced twice as much power compared to ground level. In 2014 a large Altaeros balloon turbine was purchased by the city of Fairbanks, Alaska.

Summary

Turbomachinery design is perhaps the most practical and most active application of the principles of fluid mechanics. There are billions of pumps and turbines in use in the world, and thousands of companies are seeking improvements. This chapter has discussed both positive-displacement devices and, more extensively, rotodynamic machines. With the centrifugal pump as an example, the basic concepts of torque, power, head, flow rate, and efficiency were developed for a turbomachine. Nondimensionalization leads to the pump similarity rules and some typical dimensionless performance curves for axial and centrifugal machines. The single most useful pump parameter is the specific speed, which delineates the type of design needed. An interesting design application is the theory of pumps combined in series and in parallel.

Turbines extract energy from flowing fluids and are of two types: impulse turbines, which convert the momentum of a high-speed stream, and reaction turbines, where the pressure drop occurs within the blade passages in an internal flow. By analogy with pumps, the power specific speed is important for turbines and is used to classify them into impulse, Francis, and propeller-type designs. A special case of reaction turbine with unconfined flow is the wind turbine. Several types of windmills were discussed and their relative performances compared.

Problems

Most of the problems herein are fairly straightforward. More difficult or open-ended assignments are labeled with an asterisk. Problems labeled with a computer icon may require the use of a computer. The standard end-of-chapter problems P11.1 to P11.108 (categorized in the problem list here) are followed by word problems W11.1 to W11.10, comprehensive problems C11.1 to C11.8, and design project D11.1.

Problem Distribution

Section	Topic	Problems
11.1	Introduction and classification	P11.1–P11.14
11.2	Centrifugal pump theory	P11.15–P11.21
11.3	Pump performance and similarity rules	P11.22–P11.41
11.3	Net positive-suction head	P11.42–P11.44
11.4	Specific speed: mixed- and axial-flow pumps	P11.45–P11.62
11.5	Matching pumps to system characteristics	P11.63–P11.73
11.5	Pumps in parallel or series	P11.74–P11.81
11.5	Pump instability	P11.82–P11.83
11.6	Reaction and impulse turbines	P11.84–P11.99
11.6	Wind turbines	P11.100–P11.108

Introduction and classification

P11.1 Describe the geometry and operation of a human peristaltic PDP that is cherished by every romantic person on earth. How do the two ventricles differ?

P11.2 What would be the technical classification of the following turbomachines: (*a*) a household fan, (*b*) a windmill, (*c*) an aircraft propeller, (*d*) a fuel pump in a car, (*e*) an eductor, (*f*) a fluid-coupling transmission, and (*g*) a power plant steam turbine?

P11.3 A PDP can deliver almost any fluid, but there is always a limiting very high viscosity for which performance will deteriorate. Can you explain the probable reason?

P11.4 Figure P11.4 shows the impeller on a common device which, when operating, turns at up to 300,000 r/min. Can you guess what it is and offer a description?

P11.4 [*Image provided by Sigma-Aldrich Corporation.*]

P11.5 What type of pump is shown in Fig. P11.5? How does it operate?

P11.6 Figure P11.6 shows two points a half-period apart in the operation of a pump. What type of pump is this [13]? How does it work? Sketch your best guess of flow rate versus time for a few cycles.

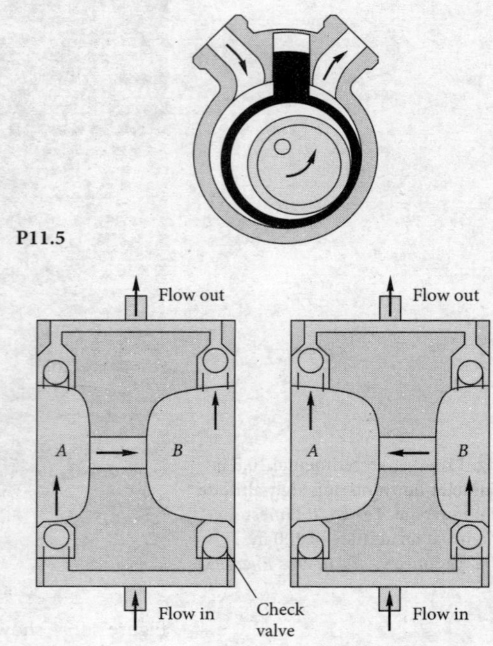

P11.5

P11.6

P11.7 A piston PDP has a 12.5 cm diameter and a 5 cm stroke and operates at 750 r/min with 92 percent volumetric efficiency. (*a*) What is its delivery, in m^3/min? (*b*) If the pump delivers SAE 10W oil at 20°C against a head of 15 m, what horsepower is required when the overall efficiency is 84 percent?

P11.8 A Bell and Gossett pump at best efficiency, running at 1,750 r/min and a brake horsepower of 32.4, delivers 4 m^3/min against a head of 32 m. (*a*) What is its efficiency? (*b*) What type of pump is this?

P11.9 Figure P11.9 shows the measured performance of the Vickers model PVQ40 piston pump when delivering SAE 10W oil at 82°C ($\rho \approx 910$ kg/m^3). Make some general observations about these data vis-à-vis Fig. 11.2 and your intuition about the behavior of piston pumps.

P11.10 Suppose that the pump of Fig. P11.9 is run at 1,100 r/min against a pressure rise of 213 kPa. (*a*) Using the measured displacement, estimate the theoretical delivery in m^3/min. From the chart, estimate (*b*) the actual delivery and (*c*) the overall efficiency.

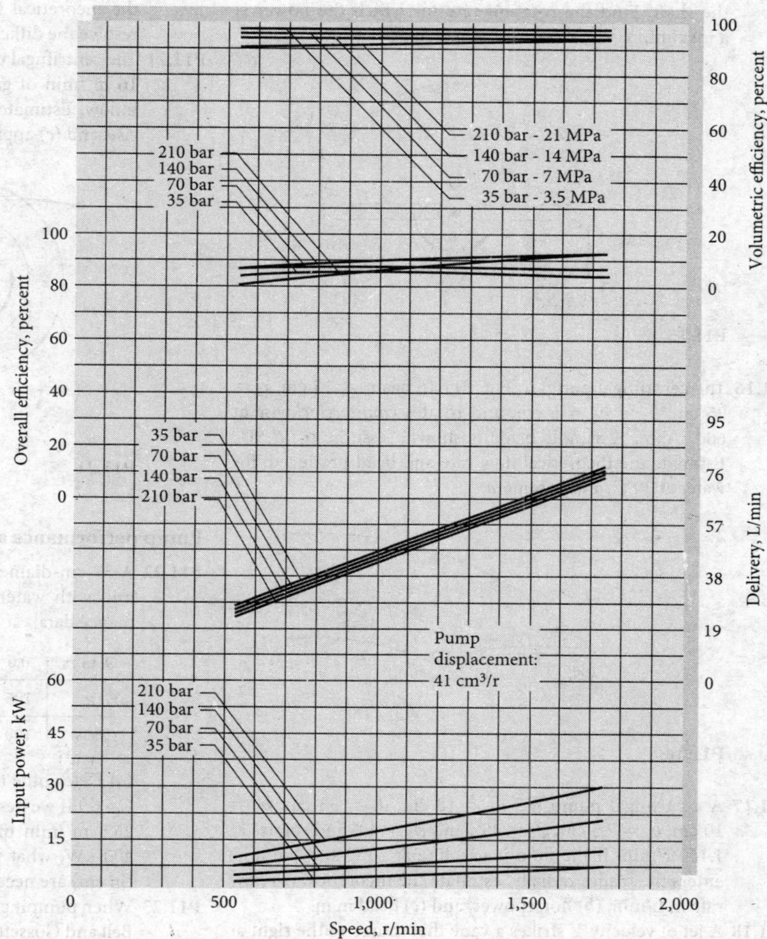

P11.9 Performance of the model PVQ40 piston pump delivering SAE 10W oil at 82°C. (*Courtesy of Vickers Inc., PDN/PACE Division.*)

P11.11 A pump delivers 1500 L/min of water at 20°C against a pressure rise of 270 kPa. Kinetic and potential energy changes are negligible. If the driving motor supplies 9 kW, what is the overall efficiency?

P11.12 In a test of the centrifugal pump shown in Fig. P11.12, the following data are taken: p_1 = 100 mmHg (vacuum) and p_2 = 500 mmHg (gage). The pipe diameters are D_1 = 12 cm and D_2 = 5 cm. The flow rate is 680 L/min of light oil (SG = 0.91). Estimate (*a*) the head developed, in meters, and (*b*) the input power required at 75 percent efficiency.

P11.13 A 3.5 hp pump delivers 5,000 N of ethylene glycol at 20°C in 12 seconds, against a head of 5 m. Calculate the efficiency of the pump.

P11.14 A pump delivers gasoline at 20°C and 12 m^3/h. At the inlet p_1 = 100 kPa, z_1 = 1 m, and V_1 = 2 m/s. At the exit p_2 = 500 kPa, z_2 = 4 m, and V_2 = 3 m/s. How much power is required if the motor efficiency is 75 percent?

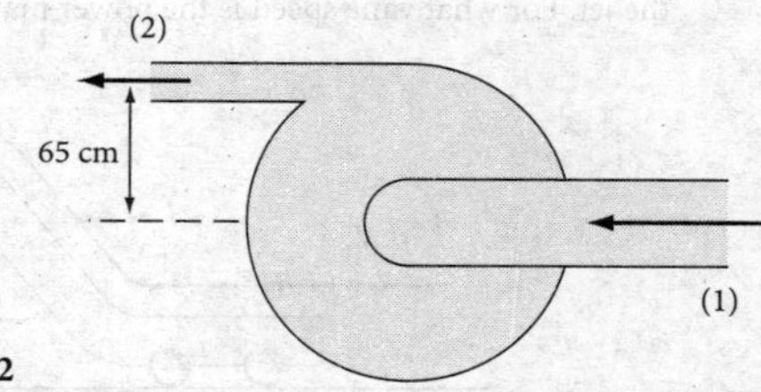

P11.12

Centrifugal pump theory

P11.15 A lawn sprinkler can be used as a simple turbine. As shown in Fig. P11.15, flow enters normal to the paper in the center and splits evenly into $Q/2$ and V_{rel} leaving each nozzle. The arms rotate at angular velocity ω and do work on a shaft. Draw the velocity diagram for this turbine. Neglecting friction, find an expression for the power delivered to the shaft. Find the rotation rate for which the power is a maximum.

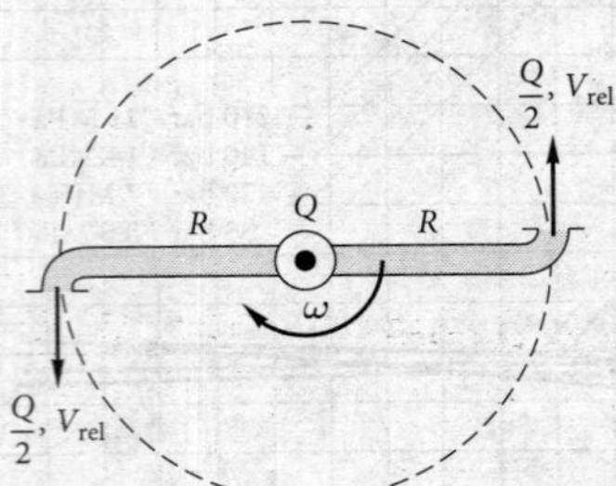

P11.15

P11.16 The centrifugal pump in Fig. P11.16 has $r_1 = 15$ cm, $r_2 = 25$ cm, $b_1 = b_2 = 6$ cm, and rotates *counterclockwise* at 600 r/min. A sample blade is shown. Assume $\alpha_1 = 90°$. Estimate the theoretical flow rate and head produced, for water at 20°C, and comment.

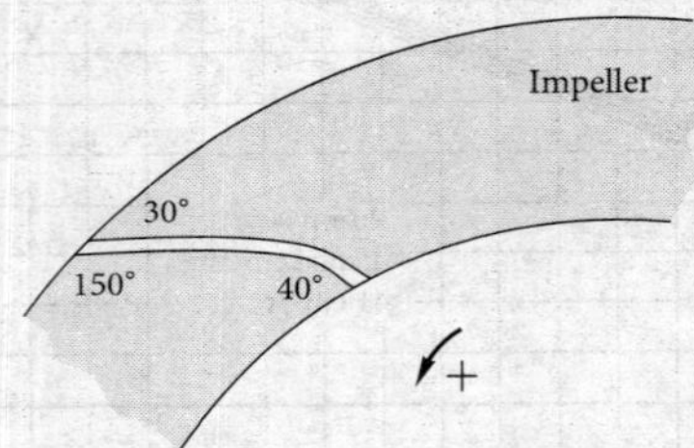

P11.16

P11.17 A centrifugal pump has $d_1 = 18$ cm, $d_2 = 33$ cm, $b_1 = 10$ cm, $b_2 = 7.5$ cm, $\beta_1 = 25°$, and $\beta_2 = 40°$ and rotates at 1,160 r/min. If the fluid is gasoline at 20°C and the flow enters the blades radially, estimate the theoretical (*a*) flow rate in L/min, (*b*) horsepower, and (*c*) head in m.

P11.18 A jet of velocity V strikes a vane that moves to the right at speed V_c, as in Fig. P11.18. The vane has a turning angle θ. Derive an expression for the power delivered to the vane by the jet. For what vane speed is the power maximum?

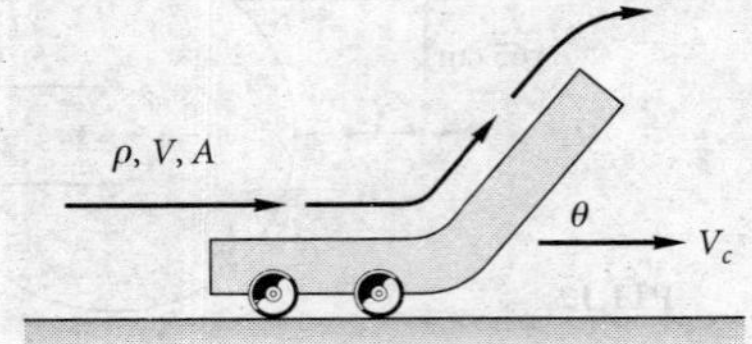

P11.18

P11.19 A centrifugal pump has $r_2 = 23$ cm, $b_2 = 5$ cm, and $\beta_2 = 35°$ and rotates at 1,060 r/min. If it generates a head of 55 m, determine the theoretical (*a*) flow rate in L/min and (*b*) horsepower. Assume near-radial entry flow.

P11.20 Suppose that Prob. P11.19 is reversed into a statement of the theoretical power $P_w \approx 153$ hp. Can you then compute the theoretical (*a*) flow rate and (*b*) head? Explain and resolve the difficulty that arises.

P11.21 The centrifugal pump of Fig. P11.21 develops a flow rate of 16 m³/min of gasoline at 20°C with near-radial absolute inflow. Estimate the theoretical (*a*) horsepower, (*b*) head rise, and (*c*) appropriate blade angle at the inner radius.

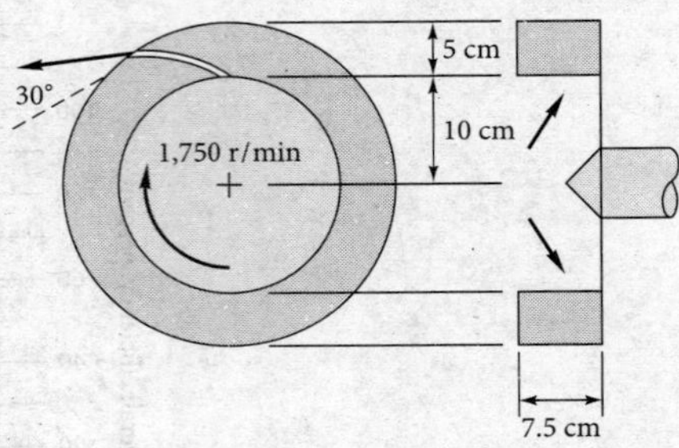

P11.21

Pump performance and similarity rules

P11.22 A 37-cm-diameter centrifugal pump, running at 2,140 r/min with water at 20°C, produces the following performance data:

Q, m³/s	0.0	0.05	0.10	0.15	0.20	0.25	0.30
H, m	105	104	102	100	95	85	67
P, kW	100	115	135	171	202	228	249

(*a*) Determine the best efficiency point. (*b*) Plot C_H versus C_Q. (*c*) If we desire to use this same pump family to deliver 26.5 m³/min of kerosene at 20°C at an input power of 400 kW, what pump speed (in r/min) and impeller size (in cm) are needed? What head will be developed?

P11.23 When pumping water, (*a*) at what speed should the 27.5 cm Bell and Gossett centrifugal pump of Prob. P11.8 be run, at best efficiency, to deliver 3 m³/min? Estimate the resulting (*b*) head, and (*c*) brake horsepower.

P11.24 Figure P11.24 shows performance data for the Taco, Inc., model 4013 pump. Compute the ratios of measured shutoff head to the ideal value U^2/g for all seven impeller sizes. Determine the average and standard deviation of this ratio and compare it to the average for the six impellers in Fig. 11.7.

P11.25 At what speed in r/min should the 89 cm-diameter pump of Fig. 11.7*b* be run to produce a head of 120 m at a discharge of 75 m³/min? What brake horsepower will be required? *Hint:* Fit $H(Q)$ to a formula.

P11.26 Would the smallest, or the largest, of the seven Taco, Inc. pumps in Fig. P11.24 be better (*a*) for producing, near best efficiency, a water flow rate of 2.3 m³/min and a head of 29 m? (*b*) At what speed, in r/min, should this pump be run? (*c*) What input power is required?

P11.27 The 27.5 cm Bell and Gossett pump of Prob. P11.8 is to be scaled up to provide, at best efficiency, a head of 75 m

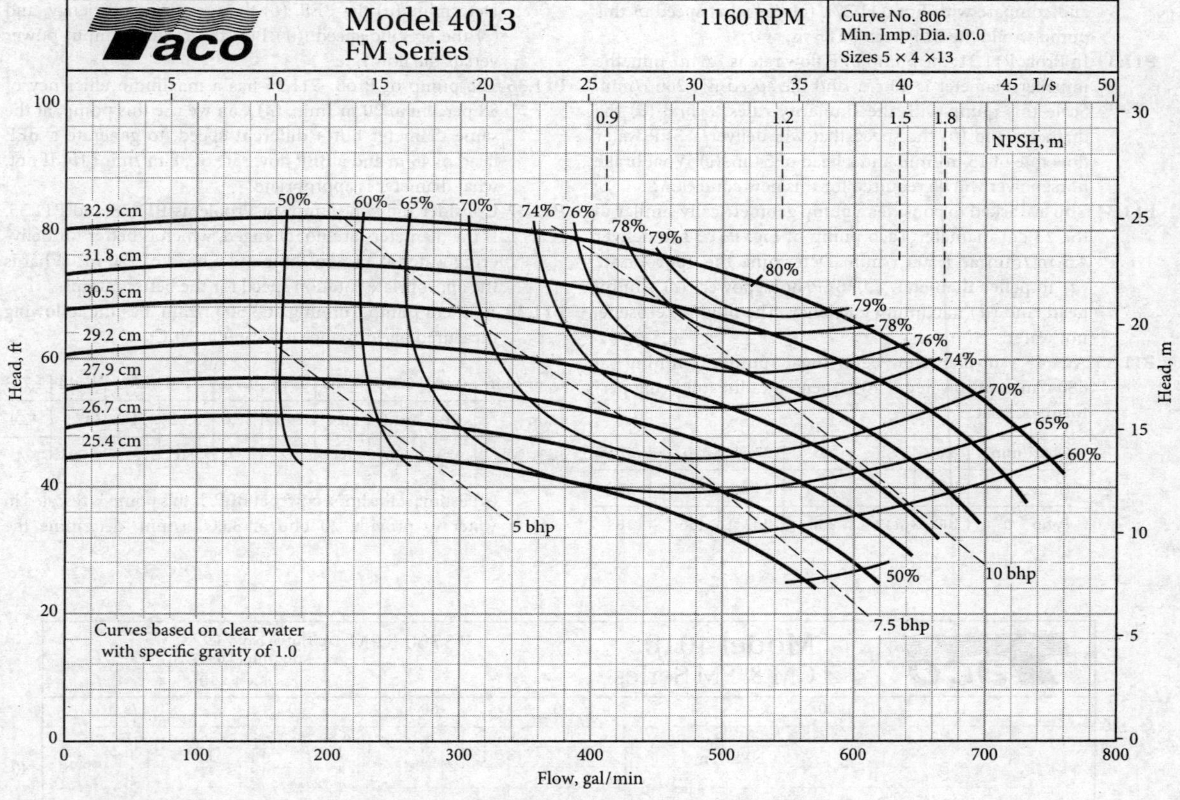

P11.24 Performance data for a centrifugal pump. *(Courtesy of Taco, Inc., Cranston, Rhode Island.)*

and a flow rate of 11.5 m^3/min. Find the appropriate (*a*) impeller diameter; (*b*) speed in r/min; and (*c*) horsepower required.

P11.28 Tests by the Byron Jackson Co. of a 37 cm-diameter centrifugal water pump at 2,134 r/min yield the following data:

Q, L/s	0	57	113	170	227	283
H, m	104	104	104	100	91	67
bhp	135	160	205	255	330	330

What is the BEP? What is the specific speed? Estimate the maximum discharge possible.

P11.29 If the scaling laws are applied to the pump of Prob. P11.28 for the same impeller diameter, determine (*a*) the speed for which the shutoff head will be 84 m, (*b*) the speed for which the BEP flow rate will be 225 L/s, and (*c*) the speed for which the BEP conditions will require 80 hp.

P11.30 A pump, geometrically similar to the 33 cm model in Fig. P11.24, has a diameter of 60 cm and is to develop 30 hp at BEP when pumping *gasoline* (not water). Determine (*a*) the appropriate speed, in r/min; (*b*) the BEP head, in m; and (*c*) the BEP flow rate, in m^3/min.

P11.31 A centrifugal pump with backward-curved blades has the following measured performance when tested with water at 20°C:

Q, m^3/min	0	1.5	3	4.5	6	7.5	9
H, m	37.5	35	33	30.8	28.3	24.7	18.9
P, hp	30	36	40	44	47	48	46

(*a*) Estimate the best efficiency point and the maximum efficiency. (*b*) Estimate the most efficient flow rate, and the resulting head and brake horsepower, if the diameter is doubled and the rotation speed increased by 50 percent.

P11.32 The data of Prob. P11.31 correspond to a pump speed of 1200 r/min. (Were you able to solve Prob. P11.31 without this knowledge?) (*a*) Estimate the diameter of the impeller. [*Hint:* See Prob. P11.24 for a clue.] (*b*) Using your estimate from part (*a*), calculate the BEP parameters C_Q^*, C_H^*, and C_P^*

and compare with Eqs. (11.27). (*c*) For what speed of this pump would the BEP head be 85 m?

P11.33 In Prob. P11.31, the pump BEP flow rate is 7.6 m^3/min, the impeller diameter is 40 cm, and the speed is 1,200 r/min. Scale this pump with the similarity rules to find (*a*) the diameter and (*b*) the speed that will deliver a BEP water flow rate of 15 m^3/min and a head of 54 m. (*c*) What brake horsepower will be required for this new condition?

P11.34 You are asked to consider a pump geometrically similar to the 23 cm-diameter Taco pump of Fig. P11.34 to deliver 4.5 m^3/min at 1,500 r/min. Determine the appropriate (*a*) impeller diameter, (*b*) BEP horsepower, (*c*) shutoff head, and (*d*) maximum efficiency. The fluid is kerosene, not water.

P11.35 An 45 cm-diameter centrifugal pump, running at 880 r/min with water at 20°C, generates the following performance data:

Q, m^3/min	0.0	7.5	15	22.5	30	37.5
H, m	28.0	27.1	25.6	23.8	20.7	15.2
P, hp	100	112	130	143	156	163

Determine (*a*) the BEP, (*b*) the maximum efficiency, and (*c*) the specific speed. (*d*) Plot the required input power versus the flow rate.

P11.36 The pump of Prob. P11.35 has a maximum efficiency of 88 percent at 30 m^3/min. (*a*) Can we use this pump, at the same diameter but a different speed, to generate a BEP head of 45 m and a BEP flow rate of 38 m^3/min? (*b*) If not, what diameter is appropriate?

P11.37 Consider the two pumps of Problems P11.28 and P11.35. If the diameters are not changed, which is better for delivering water at 11.5 m^3/min and a head of 120 m? What is the appropriate rotation speed for the better pump?

P11.38 A 17 cm pump, running at 3,500 r/min, has the following measured performance for water at 20°C:

Q, m^3/min	0.189	0.379	0.568	0.757	0.946	1.136	1.325	1.514	1.703
H, m	61.3	61.0	60.4	59.1	57.6	55.2	51.5	47.5	42.4
η, %	29	50	64	72	77	80	81	79	74

(*a*) Estimate the horsepower at BEP. If this pump is rescaled in water to provide 20 bhp at 3,000 r/min, determine the

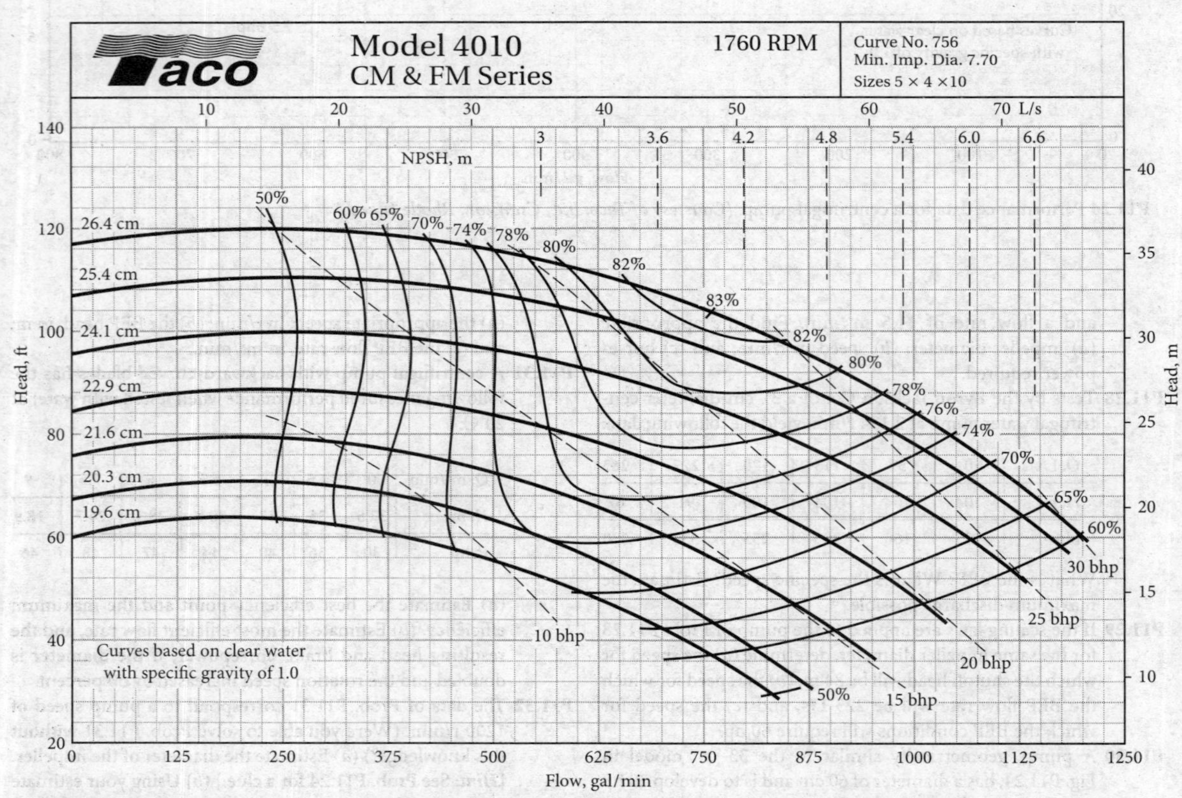

P11.34 Performance data for a family of centrifugal pump impellers. *(Courtesy of Taco, Inc., Cranston, Rhode Island.)*

appropriate (*b*) impeller diameter, (*c*) flow rate, and (*d*) efficiency for this new condition.

P11.39 The Allis-Chalmers D30LR centrifugal compressor delivers 950 m^3/min of SO_2 with a pressure change from 96.5 to 124 kPa absolute using an 800-hp motor at 3,550 r/min. What is the overall efficiency? What will the flow rate and Δp be at 3,000 r/min? Estimate the diameter of the impeller.

P11.40 The specific speed N_s, as defined by Eqs. (11.30), does not contain the impeller diameter. How then should we size the pump for a given N_s? An alternate parameter is the *specific diameter*, D_s, which is a dimensionless combination of Q, gH, and D. (*a*) If D_s is proportional to D, determine its form. (*b*) What is the relationship, if any, of D_s to C_{Q^*}, C_{H^*}, and C_{P^*}? (*c*) Estimate D_s for the two pumps of Figs. 11.8 and 11.13.

P11.41 It is desired to build a centrifugal pump geometrically similar to that of Prob. P11.28 to deliver 25 m^3/min of gasoline at 20°C at 1,060 r/min. Estimate the resulting (*a*) impeller diameter, (*b*) head, (*c*) brake horsepower, and (*d*) maximum efficiency.

Net positive-suction head

P11.42 An 20 cm model pump delivering 80°C water at 3 m^3/min and 2,400 r/min begins to cavitate when the inlet pressure and velocity are 82.7 kPa absolute and 6 m/s, respectively. Find the required NPSH of a prototype that is 4 times larger and runs at 1,000 r/min.

P11.43 The 71 cm-diameter pump in Fig. 11.7*a* at 1,170 r/min is used to pump water at 20°C through a piping system at 53 m^3/min. (*a*) Determine the required brake horsepower. The average friction factor is 0.018. (*b*) If there is 20 m of 30 cm-diameter pipe upstream of the pump, how far below the surface should the pump inlet be placed to avoid cavitation?

P11.44 The pump of Prob. P11.28 is scaled up to an 46 cm diameter, operating in water at best efficiency at 1,760 r/min. The measured NPSH is 4.8 m, and the friction loss between the inlet and the pump is 6.6 m. Will it be sufficient to avoid cavitation if the pump inlet is placed 2.7 m below the surface of a sea-level reservoir?

Specific speed: mixed- and axial-flow pumps

P11.45 Determine the specific speeds of the seven Taco, Inc., pump impellers in Fig. P11.24. Are they appropriate for centrifugal designs? Are they approximately equal within experimental uncertainty? If not, why not?

P11.46 The answer to Prob. P11.40 is that the dimensionless "specific diameter" takes the form $D_s = D(gH^*)^{1/4}/Q^{*1/2}$, evaluated at the BEP. Data collected by the author for 30 different pumps indicate, in Fig. P11.46, that D_s correlates well with specific speed N_s. Use this figure to estimate the appropriate impeller diameter for a pump that delivers 76 m^3/min of water and a head of 120 m when running at 1200 r/min. Suggest a curve-fitted formula to the data. *Hint:* Use a hyperbolic formula.

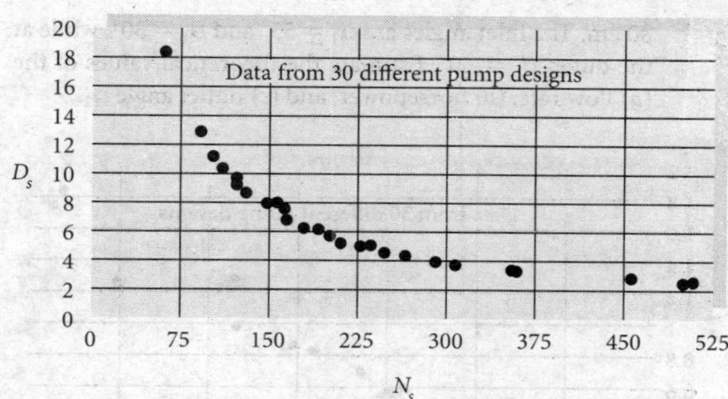

P11.46 Specific diameter at BEP for 30 commercial pumps.

P11.47 A pump must be designed to deliver 6 m^3/s of water against a head of 28 m. The specified shaft speed is 20 r/s. What type of pump do you recommend?

P11.48 Using the data for the pump in Prob. P11.8, (*a*) determine its type: PDP, centrifugal, mixed-flow, or axial-flow. (*b*) Estimate the shutoff head at 1,750 r/min. (*c*) Does this data fit on Fig. 11.14? (*d*) What speed and flow rate would result if the head were increased to 48 m?

P11.49 Data collected by the author for flow coefficient at BEP for 30 different pumps are plotted versus specific speed in Fig. P11.49. Determine if the values of C_Q^* for the three pumps in Probs. P11.28, P11.35, and P11.38 also fit on this correlation. If so, suggest a curve-fitted formula for the data.

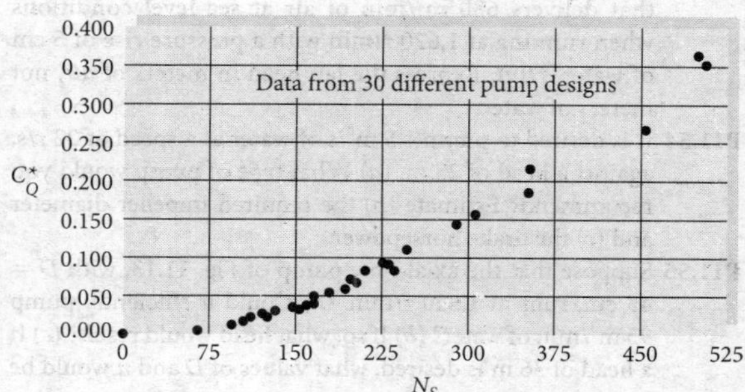

P11.49 Flow coefficient at BEP for 30 commercial pumps.

P11.50 Data collected by the author for power coefficient at BEP for 30 different pumps are plotted versus specific speed in Fig. P11.50. Determine if the values of C_P^* for the three pumps in Prob. P11.49 also fit on this correlation. If so, suggest a curve-fitted formula for the data.

P11.51 An axial-flow blower delivers 1 m^3/s of air that enters at 20°C and 1 atm. The flow passage has a 25 cm outer radius and an 20 cm inner radius. Blade angles are $\alpha_1 = 60°$ and $\beta_2 = 70°$, and the rotor runs at 1,800 r/min. For the first stage compute (*a*) the head rise and (*b*) the power required.

P11.52 An axial-flow fan operates in sea-level air at 1,200 r/min and has a blade-tip diameter of 1 m and a root diameter of

80 cm. The inlet angles are $\alpha_1 = 55°$ and $\beta_1 = 30°$, while at the outlet $\beta_2 = 60°$. Estimate the theoretical values of the (a) flow rate, (b) horsepower, and (c) outlet angle α_2.

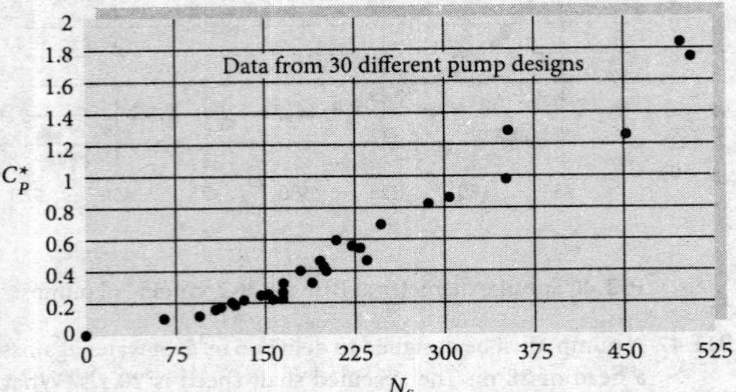

P11.50 Power coefficient at BEP for 30 commercial pumps.

P11.53 Figure P11.46 is an example of a centrifugal pump correlation, where D_s is defined in the problem. From data in the literature, we can suggest the following correlation for axial-flow pumps and fans:

$$D_s \approx \frac{873}{N_s^{0.946}} \quad \text{for } N_s > 100$$

where N_s is the *dimensional* specific speed, Eq. (11.30b). Use this correlation to find the appropriate size for a fan that delivers 680 m^3/min of air at sea-level conditions when running at 1,620 r/min with a pressure rise of 5 cm of water. *Hint:* Express the fan head in meters of *air*, not meters of water.

P11.54 It is desired to pump 1.5 m^3/s of water at a speed of 22 r/s, against a head of 25 m. (a) What type of pump would you recommend? Estimate (b) the required impeller diameter and (c) the brake horsepower.

P11.55 Suppose that the axial-flow pump of Fig. 11.13, with $D =$ 45 cm, runs at 1,800 r/min. (a) Could it efficiently pump 95 m^3/min of water? (b) If so, what head would result? (c) If a head of 36 m is desired, what values of D and n would be better?

P11.56 Determine if the Bell and Gossett pump of Prob. P11.8 (a) fits the three correlations in Figs. P11.46, P11.49, and P11.50. (b) If so, use these correlations to find the flow rate and horsepower that would result if the pump is scaled up to D = 60 cm but still runs at 1,750 r/min.

P11.57 Performance data for a 53 cm-diameter air blower running at 3550 r/min are as follows:

Δp, cm H_2O	74	76	71	53	25
Q, m^3/min	14	28	56.5	85	113
bhp	6	8	12	18	25

Note the fictitious expression of pressure rise in terms of water rather than air. What is the specific speed? How does the performance compare with Fig. 11.8? What are C_Q^*, C_H^*, and C_P^*?

P11.58 Aircraft propeller specialists claim that dimensionless propeller data, when plotted as (C_T/J^2) versus (C_P/J^2), form a nearly straight line, $y = mx + b$. (a) Test this hypothesis for the data of Fig. 11.16, in the high efficiency range $J = V/(nD)$ equal to 0.6, 0.7, and 0.8. (b) If successful, try this straight line to predict the rotation rate, in r/min, for a propeller with D = 1.5 m, P = 30 hp, T = 420 N, and V = 153 km/h, for sea level standard conditions. Comment.

P11.59 Suppose it is desired to deliver 20 m^3/min of propane gas (molecular weight = 44.06) at 1 atm and 20°C with a single-stage pressure rise of 20 cm H_2O. Determine the appropriate size and speed for using the pump families of (a) Prob. P11.57 and (b) Fig. 11.13. Which is the better design?

P11.60 Performance curves for a certain free propeller, comparable to Fig. 11.16, can be plotted as shown in Fig. P11.60, for thrust T versus speed V for constant power P. (a) What is striking, at least to the writer, about these curves? (b) Can you deduce this behavior by rearranging, or replotting, the data of Fig. 11.16?

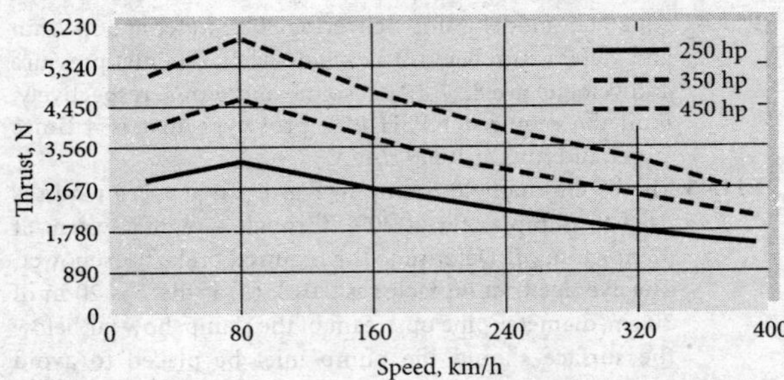

P11.60

P11.61 A mine ventilation fan, running at 295 r/min, delivers 500 m^3/s of sea-level air with a pressure rise of 1,100 Pa. Is this fan axial, centrifugal, or mixed? Estimate its diameter in m. If the flow rate is increased 50 percent for the same diameter, by what percentage will the pressure rise change?

P11.62 The actual mine ventilation fan discussed in Prob. P11.61 had a diameter of 6 m [Ref. 20, p. 339]. What would be the proper diameter for the pump family of Fig. 11.14 to provide 500 m^3/s at 295 r/min and BEP? What would be the resulting pressure rise in Pa?

Matching pumps to system characteristics

P11.63 A good curve-fit to the head vs. flow for the 81 cm pump in Fig. 11.7a is

$$H \text{ (in m)} \approx 150 - (0.00617)\, Q^2 \qquad Q \text{ in m}^3/\text{min}$$

Assume the same rotation rate, 1,170 r/min, and estimate the flow rate this pump will provide to deliver water from a reservoir, through 270 m of 30 cm pipe, to a point 45 m above the reservoir surface. Assume a friction factor $f = 0.019$.

P11.64 A leaf blower is essentially a centrifugal impeller exiting to a tube. Suppose that the tube is smooth PVC pipe, 1.2 m long, with a diameter of 6.35 cm. The desired exit velocity is 117 km/h in sea-level standard air. If we use the pump family of Eqs. (11.27) to drive the blower, what approximate (*a*) diameter and (*b*) rotation speed are appropriate? (*c*) Is this a good design?

***P11.65** An 29 cm-diameter centrifugal pump, running at 1,750 r/min, delivers 3.2 m^3/min and a head of 32 m at best efficiency (82 percent). (*a*) Can this pump operate efficiently when delivering water at 20°C through 200 m of 10-cm-diameter smooth pipe? Neglect minor losses. (*b*) If your answer to (*a*) is negative, can the speed *n* be changed to operate efficiently? (*c*) If your answer to (*b*) is also negative, can the impeller diameter be changed to operate efficiently and still run at 1,750 rev/min?

P11.66 It is proposed to run the pump of Prob. P11.35 at 880 r/min to pump water at 20°C through the system in Fig. P11.66. The pipe is 20-cm-diameter commercial steel. What flow rate in m^3/min will result? Is this an efficient application?

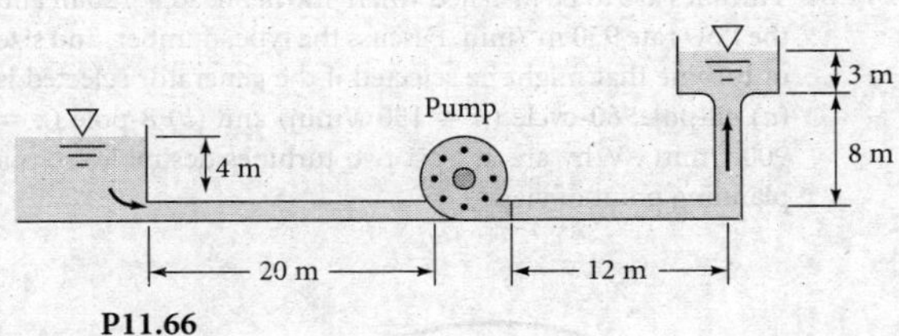

P11.66

P11.67 The pump of Prob. P11.35, running at 880 r/min, is to pump water at 20°C through 75 m of horizontal galvanized iron pipe. All other system losses are neglected. Determine the flow rate and input power for (*a*) pipe diameter = 20 cm and (*b*) the pipe diameter found to yield maximum pump efficiency.

P11.68 A popular small aircraft cruises at 230 km/h at 2,500 m altitude. It weighs 10 kN, has a 180-hp engine, a 193 cm-diameter propeller, and a drag-area $C_DA \approx 0.52$ m^2. The propeller data in Fig. P11.68 is proposed to drive this aircraft. Estimate the required rotation rate, in r/min, and power delivered, in hp. [*NOTE*: Simply use the coefficient pairs. The actual advance ratio is too high.]

P11.69 The pump of Prob. P11.38, running at 3,500 r/min, is used to deliver water at 20°C through 180 m of cast iron pipe to an elevation 30 m higher. Determine (*a*) the proper pipe diameter for BEP operation and (*b*) the flow rate that results if the pipe diameter is 7.5 cm.

P11.70 The pump of Prob. P11.28, operating at 2134 r/min, is used with 20°C water in the system of Fig. P11.70. (*a*) If it is operating at BEP, what is the proper elevation z_2? (*b*) If z_2 = 68 m, what is the flow rate if d = 20 cm.?

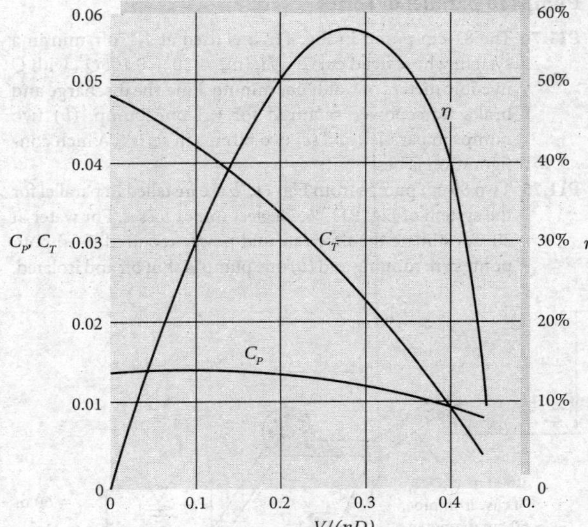

P11.68

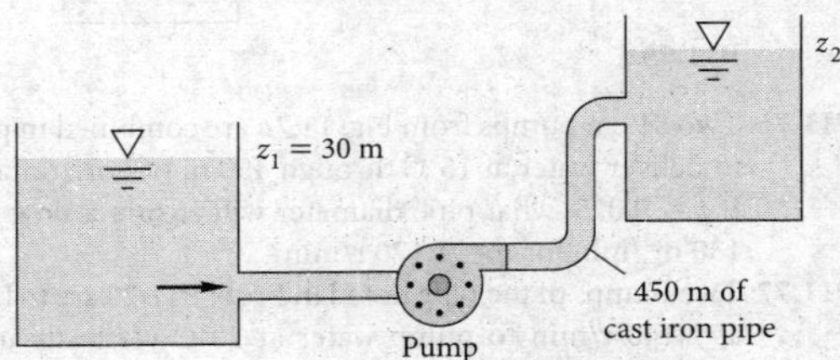

P11.70

P11.71 The pump of Prob. P11.38, running at 3,500 r/min, delivers water at 20°C through 2,200 m of horizontal 12.5 cm-diameter commercial steel pipe. There are a sharp entrance, sharp exit, four 90° elbows, and a gate valve. Estimate (*a*) the flow rate if the valve is wide open and (*b*) the valve closing percentage that causes the pump to operate at BEP. (*c*) If the latter condition holds continuously for 1 year, estimate the energy cost at 10 ¢/kWh.

P11.72 Performance data for a small commercial pump are as follows:

Q, L/min	0	38	76	114	151	189	227	265
H, m	22.9	22.9	22.6	21.9	20.7	18.9	14.3	7.3

This pump supplies 20°C water to a horizontal 1.6 cm-diameter garden hose ($\varepsilon \approx 0.025$ cm) that is 15 m long. Estimate (*a*) the flow rate and (*b*) the hose diameter that would cause the pump to operate at BEP.

P11.73 The Bell and Gossett pump of Prob. P11.8, running under the same conditions, delivers water at 20°C through a long, smooth, 20 cm-diameter pipe. Neglect minor losses. How long is the pipe?

Pumps in parallel or series

P11.74 The 81 cm pump in Fig. 11.7*a* is used at 1,170 r/min in a system whose head curve is H_s (m) $= 30 + 0.105Q^2$, with Q in cubic meters of water per minute. Find the discharge and brake horsepower required for (*a*) one pump, (*b*) two pumps in parallel, and (*c*) two pumps in series. Which configuration is best?

P11.75 Two 89 cm pumps from Fig. 11.7*b* are installed in parallel for the system of Fig. P11.75. Neglect minor losses. For water at 20°C, estimate the flow rate and power required if (*a*) both pumps are running and (*b*) one pump is shut off and isolated.

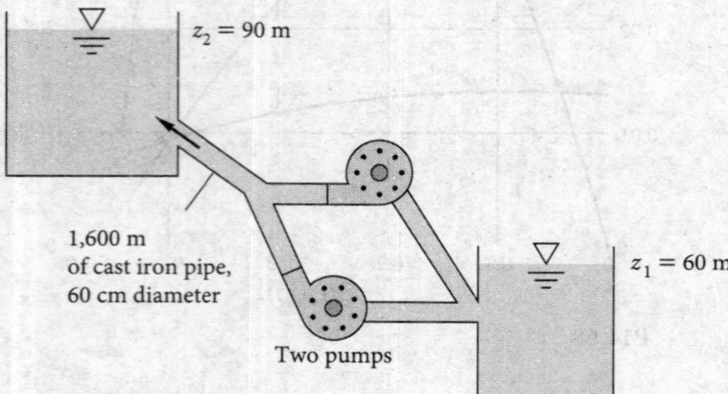

P11.75

P11.76 Two 81 cm pumps from Fig. 11.7*a* are combined in parallel to deliver water at 15°C through 450 m of horizontal pipe. If $f = 0.025$, what pipe diameter will ensure a flow rate of 130 m^3/min for $n = 1{,}170$ r/min?

P11.77 Two pumps of the type tested in Prob. P11.22 are to be used at 2,140 r/min to pump water at 20°C vertically upward through 100 m of commercial steel pipe. Should they be in series or in parallel? What is the proper pipe diameter for most efficient operation?

P11.78 Consider the axial-flow pump of Fig. 11.13, running at 4,200 r/min, with an impeller diameter of 90 cm. The fluid is propane gas (molecular weight 44.06). (*a*) How many pumps in series are needed to increase the gas pressure from 1 atm to 2 atm? (*b*) Estimate the mass flow of gas.

P11.79 Two 81 cm pumps from Fig. 11.7*a* are to be used in series at 1,170 r/min to lift water through 150 m of vertical cast iron pipe. What should the pipe diameter be for most efficient operation? Neglect minor losses.

P11.80 Determine if either (*a*) the smallest or (*b*) the largest of the seven Taco pumps in Fig. P11.24, running in series at 1,160 r/min, can efficiently pump water at 20°C through 1 km of horizontal 12-cm-diameter commercial steel pipe.

P11.81 Reconsider the system of Fig. P6.62. Use the Byron Jackson pump of Prob. P11.28 running at 2,134 r/min, no scaling, to drive the flow. Determine the resulting flow rate between the reservoirs. What is the pump efficiency?

Pump instability

P11.82 The S-shaped head-versus-flow curve in Fig. P11.82 occurs in some axial-flow pumps. Explain how a fairly flat system loss curve might cause instabilities in the operation of the pump. How might we avoid instability?

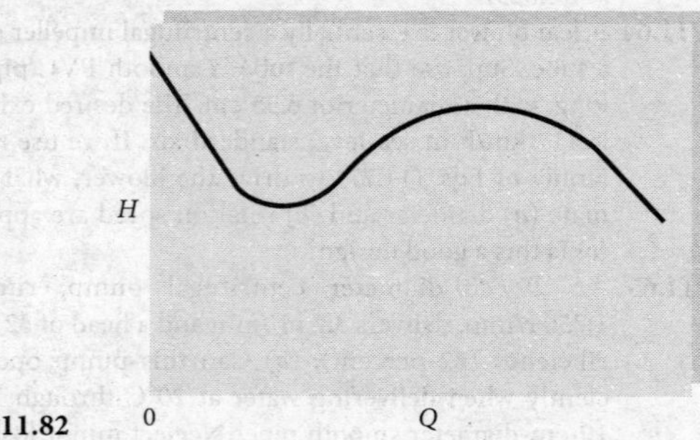

P11.82

P11.83 The low-shutoff head-versus-flow curve in Fig. P11.83 occurs in some centrifugal pumps. Explain how a fairly flat system loss curve might cause instabilities in the operation of the pump. What additional vexation occurs when two of these pumps are in parallel? How might we avoid instability?

Reaction and impulse turbines

P11.84 Turbines are to be installed where the net head is 120 m and the flow rate 950 m^3/min. Discuss the type, number, and size of turbine that might be selected if the generator selected is (*a*) 48-pole, 60-cycle ($n = 150$ r/min) and (*b*) 8-pole ($n = 900$ r/min). Why are at least two turbines desirable from a planning point of view?

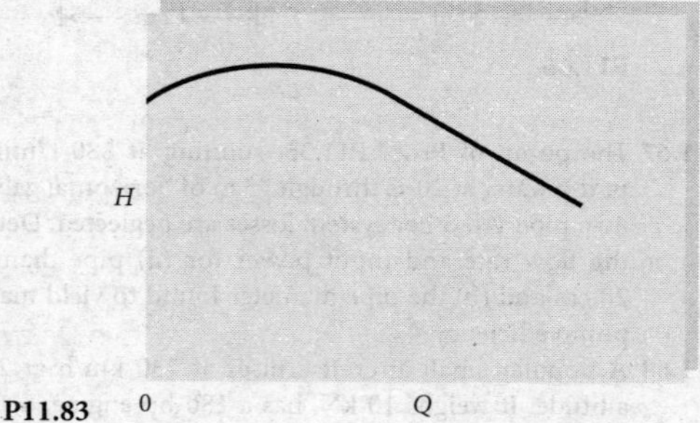

P11.83

P11.85 For a high-flow site with a head of 13 m, it is desired to design a single 2 m-diameter turbine that develops 4,000 bhp at a speed of 360 r/min and 88-percent efficiency. It is decided first to test a geometrically similar model of diameter 30 cm, running at 1,180 r/min. (*a*) What likely type of turbine is in the prototype? What are the appropriate (*b*) head and (*c*) flow rate for the model test? (*d*) Estimate the power expected to be delivered by the model turbine.

P11.86 The Tupperware hydroelectric plant on the Blackstone River has four 90 cm-diameter turbines, each providing 447 kW at 200 r/min and 5.8 m^3/s for a head of 9 m. What type of turbine are these? How does their performance compare with Fig. 11.22?

P11.87 An idealized radial turbine is shown in Fig. P11.87. The absolute flow enters at 30° and leaves radially inward. The flow rate is 3.5 m^3/s of water at 20°C. The blade thickness is constant at 10 cm. Compute the theoretical power developed.

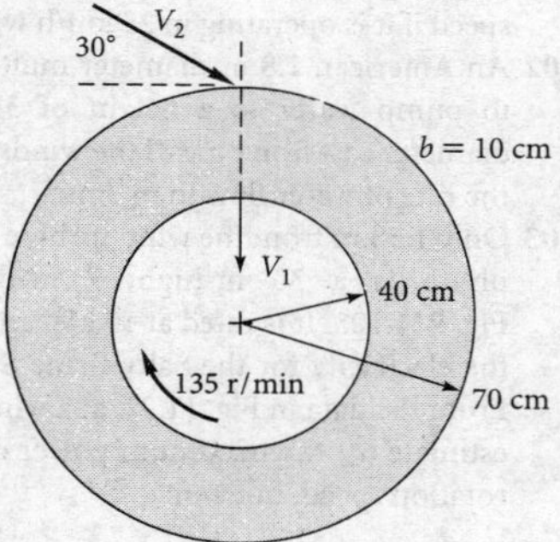

P11.87

P11.88 Performance data for a very small (D = 8.25 cm) model water turbine, operating with an available head of 15 m, are as follows:

Q, m^3/h	18.7	18.7	18.5	18.3	17.6	16.7	15.1	11.5
RPM	0	500	1,000	1,500	2,000	2,500	3,000	3,500
η	0	14%	27%	38%	50%	65%	61%	11%

(*a*) What type of turbine is this likely to be? (*b*) What is so different about these data compared to the dimensionless performance plot in Fig. 11.22*d*? Suppose it is desired to use a geometrically similar turbine to serve where the available head and flow are 45 m and 0.2 m^3/s, respectively. Estimate the most efficient (*c*) turbine diameter, (*d*) rotation speed, and (*e*) horsepower.

P11.89 A Pelton wheel of 3.6 m pitch diameter operates under a net head of 600 m. Estimate the speed, power output, and flow rate for best efficiency if the nozzle exit diameter is 10 cm.

P11.90 An idealized radial turbine is shown in Fig. P11.90. The absolute flow enters at 25° with the blade angles as shown. The flow rate is 8 m^3/s of water at 20°C. The blade thickness is constant at 20 cm. Compute the theoretical power developed.

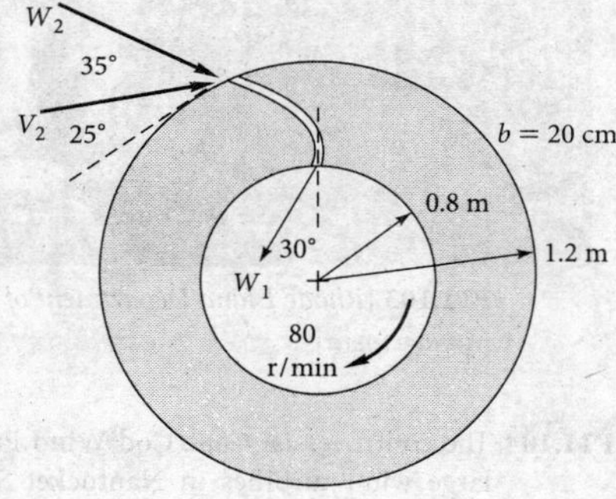

P11.90

P11.91 The flow through an axial-flow *turbine* can be idealized by modifying the stator–rotor diagrams of Fig. 11.12 for energy absorption. Sketch a suitable blade and flow arrangement and the associated velocity vector diagrams.

P11.92 A dam on a river is being sited for a hydraulic turbine. The flow rate is 1,500 m^3/h, the available head is 24 m, and the turbine speed is to be 480 r/min. Discuss the estimated turbine size and feasibility for (*a*) a Francis turbine and (*b*) a Pelton wheel.

P11.93 Figure P11.93 shows a cutaway of a *cross-flow* or "Banki" turbine [55], which resembles a squirrel cage with slotted curved blades. The flow enters at about 2 o'clock and passes through the center and then again through the blades, leaving at about 8 o'clock. Report to the class on the operation and advantages of this design, including idealized velocity vector diagrams.

P11.94 A simple cross-flow turbine, Fig. P11.93, was constructed and tested at the University of Rhode Island. The blades were made of PVC pipe cut lengthwise into three 120°-arc pieces. When it was tested in water at a head of 165 cm and a flow rate of 2.4 m^3/min, the measured power output was 0.6 hp. Estimate (*a*) the efficiency and (*b*) the power specific speed if n = 200 r/min.

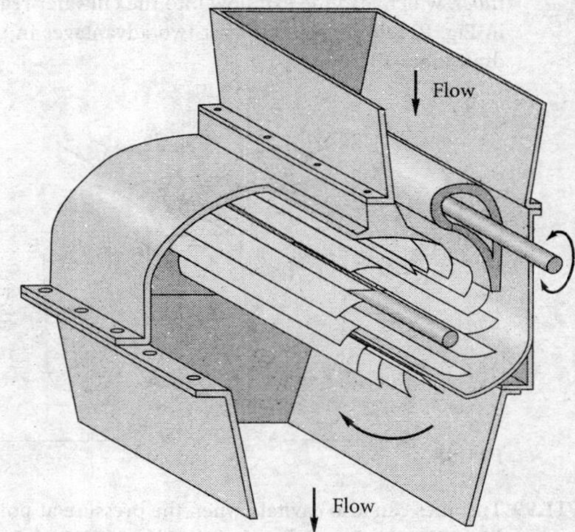

P11.93

***P11.95** One can make a theoretical estimate of the proper diameter for a penstock in an impulse turbine installation, as in Fig. P11.95. Let L and H be known, and let the turbine performance be idealized by Eqs. (11.38) and (11.39). Account for friction loss h_f in the penstock, but neglect minor losses. Show that (*a*) the maximum power is generated when $h_f = H/3$, (*b*) the optimum jet velocity is $(4gH/3)^{1/2}$, and (*c*) the best nozzle diameter is $D_j = [D^5/(2fL)]^{1/4}$, where f is the pipe friction factor.

P11.96 Apply the results of Prob. P11.95 to determine the optimum (*a*) penstock diameter and (*b*) nozzle diameter for a head of 330 m and a flow rate of 5,400 m^3/h with a cast iron penstock of length 600 m.

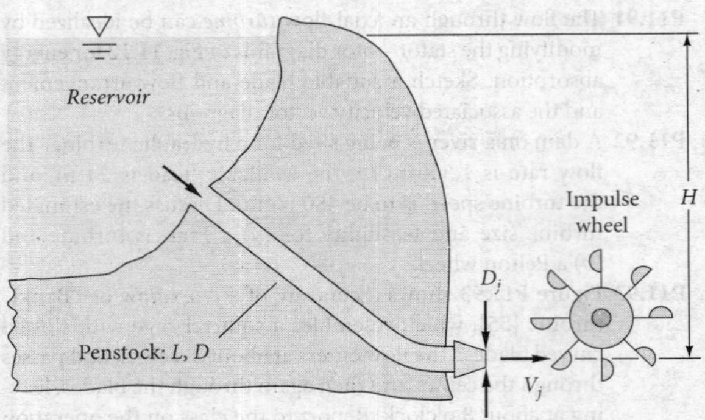

P11.95

P11.97 Consider the following nonoptimum version of Prob. P11.95: $H = 450$ m, $L = 5$ km, $D = 1.2$ m, $D_j = 20$ cm. The penstock is concrete, $\varepsilon = 1$ mm. The impulse wheel diameter is 3.2 m. Estimate (*a*) the power generated by the wheel at 80 percent efficiency and (*b*) the best speed of the wheel in r/min. Neglect minor losses.

P11.98 Francis and Kaplan turbines are often provided with *draft tubes,* which lead the exit flow into the tailwater region, as in Fig. P11.98. Explain at least two advantages in using a draft tube.

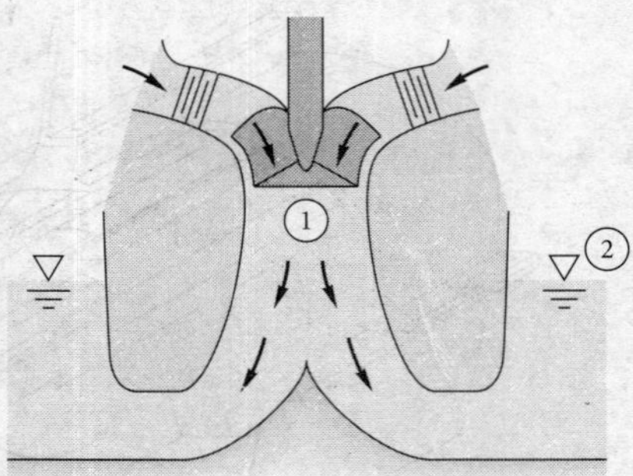

P11.98

P11.99 Turbines can also cavitate when the pressure at point 1 in Fig. P11.98 drops too low. With NPSH defined by Eq. (11.20), the empirical criterion given by Wislicenus [4] for cavitation is

$$N_{ss} = \frac{(\text{r/min})(\text{m}^3/\text{min})^{1/2}}{[\text{NPSH (m)}]^{3/4}} \geq 1{,}650$$

Use this criterion to compute how high $z_1 - z_2$, the impeller eye in Fig. P11.98, can be placed for a Francis turbine with a head of 90 m, $N_{sp} = 180$, and $p_a = 96.5$ kPa absolute before cavitation occurs in 15°C water.

Wind turbines

P11.100 The manufacturer of the wind turbine in the chapter-opener photo claims that it develops exactly 100 kW at a wind speed of 15 m/s. Compare this with an estimate from the correlations in Fig. 11.32.

P11.101 A Darrieus VAWT in operation in Lumsden, Saskatchewan, that is 10 m high and 6 m in diameter sweeps out an area of 40 m^2. Estimate (*a*) the maximum power and (*b*) the rotor speed if it is operating in 25 km/h winds.

P11.102 An American 1.8 m-diameter multiblade HAWT is used to pump water to a height of 3 m through 7.5 cm-diameter cast iron pipe. If the winds are 20 km/h, estimate the rate of water flow in m^3/min.

P11.103 Only 1.5 km from the wind turbine in the chapter-opener photo is a 30 m-high, 7 m-diameter HAWT, in Fig. P11.103. It is rated at 10 kW and provides one-half of the electricity for the Salty Brine State Beach bathhouse. From the data in Fig. 11.32, at a wind velocity of 32 km/h, estimate (*a*) the maximum power developed, and (*b*) the rotation speed, in r/min.

P11.103 [*Rhode Island Department of Environmental Management*]

P11.104 The controversial Cape Cod Wind Project proposes 130 large wind turbines in Nantucket Sound, intended to provide 75 percent of the electric power needs of Cape Cod and the Islands. The turbine diameter is 100 m. For an average wind velocity of 22 km/h, what are the best rotation rate and total power output estimates for (*a*) a HAWT and (*b*) a VAWT?

P11.105 In 2007, a wind-powered-vehicle contest, held in North Holland [64], was won with a design by students at the University of Stuttgart. A schematic of the winning three-wheeler is shown in Fig. P11.105. It is powered by a shrouded wind turbine, not a propeller, and, unlike a sail-boat, can move directly into the wind. (*a*) How does it work? (*b*) What if the wind is off to the side? (*c*) Cite some design questions you might have.

P11.106 Analyze the wind-powered-vehicle of Fig. P11.105 with the following data: turbine diameter $D = 1.8$ m, power coefficient (Fig. 11.32) $= 0.3$, vehicle $C_DA = 0.4$ m^2, and turbine rotation 240 r/min. The vehicle moves directly into a head wind, $W = 40$ km/h. The wind backward thrust on the turbine is approximately $T \approx C_T(\rho/2)\,V_{rel}^{\,2}A_{turbine}$, where V_{rel} is the air velocity relative to the turbine, and $C_T \approx 0.7$. Eighty percent of the turbine power is delivered by gears to the wheels, to propel the vehicle. Estimate the sea-level vehicle velocity V, in km/h.

P11.107 Figure 11.32 showed the typical *power* performance of a wind turbine. The wind also causes a *thrust* force that must be resisted by the structure. The thrust coefficient C_T of a wind turbine may be defined as follows:

$$C_T = \frac{\textit{Thrust force}}{(\rho/2)\,AV^2} = \frac{T}{(\rho/2)\,[(\pi/4)D^2]\,V^2}$$

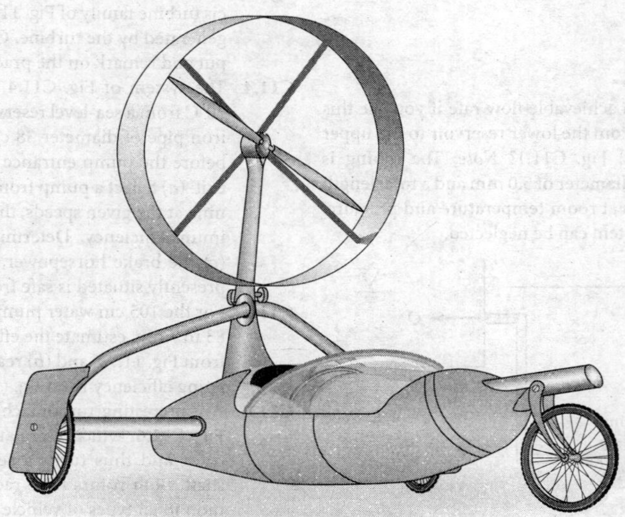

P11.105

Values of C_T for a typical horizontal-axis wind turbine are shown in Fig. P11.107. The abscissa is the same as in Fig. 11.32. Consider the turbine of Prob. P11.103. If the wind is 32 km/h and the rotation rate 115 r/min, estimate the bending moment about the tower base.

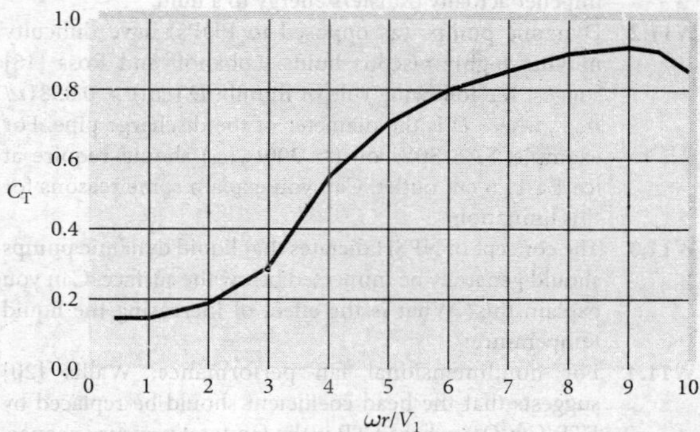

P11.107 Thrust coefficient for a typical HAWT.

P11.108 To avoid the bulky tower and impeller and generator in the HAWT of the chapter-opener photo, we could instead build a number of Darrieus turbines of height 4 m and diameter 3 m. (*a*) How many of these would we need to match the HAWT's 100 kW output for 15 m/s wind speed and maximum power? (*b*) How fast would they rotate? Assume the area swept out by a Darrieus turbine is two-thirds the height times the diameter.

Word Problems

W11.1 We know that an enclosed rotating bladed impeller will impart energy to a fluid, usually in the form of a pressure rise, but how does it actually happen? Discuss, with sketches, the physical mechanisms through which an impeller actually transfers energy to a fluid.

W11.2 Dynamic pumps (as opposed to PDPs) have difficulty moving highly viscous fluids. Lobanoff and Ross [15] suggest the following rule of thumb: D (cm) $> 0.0381\nu/\nu_{water}$, where D is the diameter of the discharge pipe. For example, SAE 30W oil ($\approx 300\nu_{water}$) should require at least a 11.5 cm outlet. Can you explain some reasons for this limitation?

W11.3 The concept of NPSH dictates that liquid dynamic pumps should generally be immersed below the surface. Can you explain this? What is the effect of increasing the liquid temperature?

W11.4 For nondimensional fan performance, Wallis [20] suggests that the head coefficient should be replaced by FTP/$(\rho n^2 D^2)$, where FTP is the fan total pressure change. Explain the usefulness of this modification.

W11.5 Performance data for centrifugal pumps, even if well scaled geometrically, show a decrease in efficiency with decreasing impeller size. Discuss some physical reasons why this is so.

W11.6 Consider a dimensionless pump performance chart such as Fig. 11.8. What additional dimensionless parameters might modify or even destroy the similarity indicated in such data?

W11.7 One parameter not discussed in this text is the *number of blades* on an impeller. Do some reading on this subject, and report to the class about its effect on pump performance.

W11.8 Explain why some pump performance curves may lead to unstable operating conditions.

W11.9 Why are Francis and Kaplan turbines generally considered unsuitable for hydropower sites where the available head exceeds 300 m?

W11.10 Do some reading on the performance of the *free propeller* that is used on small, low-speed aircraft. What dimensionless parameters are typically reported for the data? How do the performance and efficiency compare with those for the axial-flow pump?

Comprehensive Problems

C11.1 The net head of a little aquarium pump is given by the manufacturer as a function of volume flow rate as listed below:

Q, m³/s	H, mH₂O
0	1.10
1.0 E-6	1.00
2.0 E-6	0.80
3.0 E-6	0.60
4.0 E-6	0.35
5.0 E-6	0.0

What is the maximum achievable flow rate if you use this pump to move water from the lower reservoir to the upper reservoir as shown in Fig. C11.1? *Note:* The tubing is smooth with an inner diameter of 5.0 mm and a total length of 29.8 m. The water is at room temperature and pressure. Minor losses in the system can be neglected.

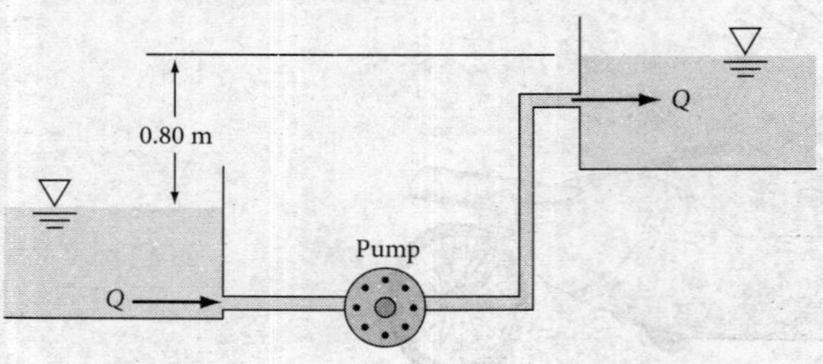

C11.1

C11.2 Reconsider Prob. P6.62 as an exercise in pump selection. Select an impeller size and rotational speed from the Byron Jackson pump family of Prob. P11.28 to deliver a flow rate of 0.085 m³/s to the system of Fig. P6.62 at minimum input power. Calculate the horsepower required.

C11.3 Reconsider Prob. P6.77 as an exercise in turbine selection. Select an impeller size and rotational speed from the Francis turbine family of Fig. 11.22*d* to deliver maximum power generated by the turbine. Calculate the turbine power output and remark on the practicality of your design.

C11.4 The system of Fig. C11.4 is designed to deliver water at 20°C from a sea-level reservoir to another through new cast iron pipe of diameter 38 cm. Minor losses are $\Sigma K_1 = 0.5$ before the pump entrance and $\Sigma K_2 = 7.2$ after the pump exit. (*a*) Select a pump from either Fig. 11.7*a* or 11.7*b*, running at the given speeds, that can perform this task at maximum efficiency. Determine (*b*) the resulting flow rate, (*c*) the brake horsepower, and (*d*) whether the pump as presently situated is safe from cavitation.

C11.5 For the 105 cm water pump of Fig. 11.7*b*, at 710 r/min and 83 m³/min, estimate the efficiency by (*a*) reading it directly from Fig. 11.7*b*; and (*b*) reading *H* and *bhp* and then calculating efficiency from Eq. (11.5). Compare your results.

C11.6 An interesting turbomachine [58] is the *fluid coupling* of Fig. C11.6, which circulates fluid from a primary pump rotor and thus turns a secondary turbine on a separate shaft. Both rotors have radial blades. Couplings are common in all types of vehicle and machine transmissions and

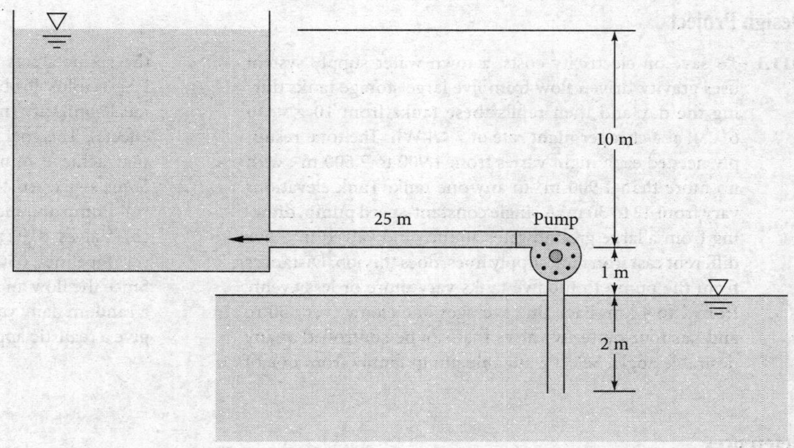

C11.4

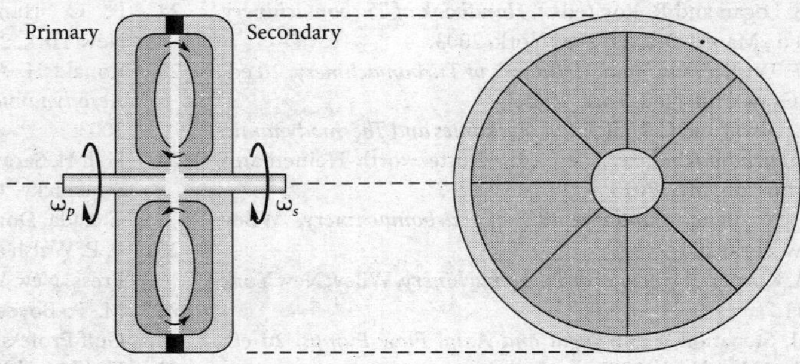

C11.6

drives. The *slip* of the coupling is defined as the dimensionless difference between shaft rotation rates, $s = 1 - \omega_s/\omega_p$. For a given volume of fluid, the torque T transmitted is a function of s, ρ, ω_p, and impeller diameter D. (*a*) Nondimensionalize this function into two pi groups, with one pi proportional to T. Tests on a 30 cm-diameter coupling at 2,500 r/min, filled with hydraulic fluid of density 896 kg/m^3, yield the following torque versus slip data:

Slip, s	0%	5%	10%	15%	20%	25%
Torque T, N-m	0	122	373	597	786	922

(*b*) If this coupling is run at 3,600 r/min, at what slip value will it transmit a torque of 1,220 N-m? (*c*) What is the proper diameter for a geometrically similar coupling to run at 3,000 r/min and 5 percent slip and transmit 800 N-m of torque?

C11.7 Report to the class on the *Cordier method* [63] for optimal design of turbomachinery. The method is related to, and greatly expanded from, Prob. P11.46 and uses both software and charts to develop an efficient design for any given pump or compressor application.

C11.8 A *pump-turbine* is a reversible device that uses a reservoir to generate power in the daytime and then pumps water back up to the reservoir at night. Let us reset Prob. P6.62 as a pump-turbine. Recall that $\Delta z = 36$ m, and the water flows through 600 m of 15 cm-diameter cast iron pipe. For simplicity, assume that the pump operates at BEP (92%) with $H^*_p = 60$ m and the turbine operates at BEP (89%) with $H^*_t = 30$ m. Neglect minor losses. Estimate (*a*) the input power, in watts, required by the pump; and (*b*) the power, in watts, generated by the turbine. For further technical reading, consult the URL www.usbr.gov/pmts/hydraulics_lab/pubs/EM/EM39.pdf.

Design Project

D11.1 To save on electricity costs, a town water supply system uses gravity-driven flow from five large storage tanks during the day and then refills these tanks from 10 P.M. to 6 A.M. at a cheaper night rate of 7 ¢/kWh. The total resupply needed each night varies from 1,900 to 7,600 m^3, with no more than 1,900 m^3 to any one tank. Tank elevations vary from 12 to 30 m. A single constant-speed pump, drawing from a large groundwater aquifer and valved into five different cast iron tank supply lines, does this job. Distances from the pump to the five tanks vary more or less evenly from 2 to 4 km. Each line averages one elbow every 50 m and has four butterfly valves that can be controlled at any desirable angle. Select a suitable pump family from one of the six data sets in this chapter: Figs. 11.8, P11.24, and P11.34 plus Probs. P11.28, P11.35, and P11.38. Assume ideal similarity (no Reynolds number or pump roughness effects). The goal is to determine pump and pipeline sizes that achieve minimum total cost over a 5-year period. Some suggested cost data are

(*a*) Pump and motor: $2,500 plus $600 per cm of pipe size
(*b*) Valves: $100 plus $40 per cm of pipe size
(*c*) Pipelines: 65¢ per cm of diameter per meter of length

Since the flow and elevation parameters vary considerably, a random daily variation within the specified ranges might give a realistic approach.

References

1. D. G. Wilson, "Turbomachinery—From Paddle Wheels to Turbojets," *Mech. Eng.*, vol. 104, Oct. 1982, pp. 28–40.
2. D. Japikse and N. C. Baines, *Introduction to Turbomachinery*, Concepts ETI Inc., Hanover, NH, 1997.
3. E. S. Logan and R. Roy (eds.), *Handbook of Turbomachinery*, 2d ed., Marcel Dekker, New York, 2003.
4. G. F. Wislicenus, *Fluid Mechanics of Turbomachinery*, 2d ed., McGraw-Hill, New York, 1965.
5. S. L. Dixon and C. Hall, *Fluid Mechanics and Thermodynamics of Turbomachinery*, 7th ed., Butterworth-Heinemann, Burlington, MA, 2013.
6. W. W. Peng, *Fundamentals of Turbomachinery*, Wiley, New York, 2007.
7. S. A. Korpela, *Principles of Turbomachinery*, Wiley, New York, 2011.
8. A. J. Stepanoff, *Centrifugal and Axial Flow Pumps*, 2d ed., Wiley, New York, 1957.
9. J. Tuzson, *Centrifugal Pump Design*, Wiley, New York, 2000.
10. P. Girdhar and O. Moniz, *Practical Centrifugal Pumps*, Elsevier, New York, 2004.
11. L. Bachus and A. Custodio, *Know and Understand Centrifugal Pumps*, Elsevier, New York, 2003.
12. J. F. Gülich, *Centrifugal Pumps*, Springer, New York, 2010.
13. R. K. Turton, *Rotodynamic Pump Design*, Cambridge University Press, Cambridge, UK, 2005.
14. I. J. Karassik and T. McGuire, *Centrifugal Pumps*, 2d ed., Springer-Verlag, New York, 1996.
15. V. L. Lobanoff and R. R. Ross, *Centrifugal Pumps: Design and Application*, 2d ed., Elsevier, New York, 1992.
16. H. L. Stewart, *Pumps*, 5th ed. Macmillan, New York, 1991.
17. A. B. McKenzie, *Axial Flow Fans and Compressors: Aerodynamic Design and Performance*, Ashgate Publishing, Brookfield, VT, 1997.
18. A. J. Wennerstrom, *Design of Highly Loaded Axial-Flow Fans and Compressors*, Concepts ETI Inc., Hanover, NH, 2001.
19. F. P. Bleier, *Fan Handbook: Selection, Application, and Design*, McGraw-Hill, New York, 1997.
20. R. A. Wallis, *Axial Flow Fans and Ducts*, Wiley, New York, 1983.
21. H. P. Bloch, *A Practical Guide to Compressor Technology*, 2d ed., McGraw-Hill, New York, 2006.
22. P. C. Hanlon, *Compressor Handbook*, McGraw-Hill, New York, 2001.
23. Ronald H. Aungier, *Axial-Flow Compressors: A Strategy for Aerodynamic Design and Analysis*, ASME Press, New York, 2003.
24. H. I. H. Saravanamuttoo, G. F. C. Rogers, H. Cohen, and Paul Straznicky, *Gas Turbine Theory*, 6th ed., Pearson Education Canada, Don Mills, Ontario, 2008.
25. P. P. Walsh and P. Fletcher, *Gas Turbine Performance*, ASME Press, New York, 2004.
26. M. P. Boyce, *Gas Turbine Engineering Handbook*, 4th ed., Gulf Professional Publishing, Burlington, MA, 2011.
27. Fluid Machinery Group, Institution of Mechanical Engineers, *Hydropower*, Wiley, New York, 2005.
28. Jeremy Thake, *The Micro-Hydro Pelton Turbine Manual*, Intermediate Technology Pub., Colchester, Essex, UK, 2000.
29. L. Rodriguez and T. Sanchez, *Designing and Building Mini and Micro Hydro Power Schemes*, Practical Action Publishing, Warwickshire, UK, 2011.
30. Hydraulic Institute, *Hydraulic Institute Pump Standards Complete*, 4th ed., New York, 1994.
31. P. Cooper, J. Messina, C. Heald, and I. J. Karassik (eds.), *Pump Handbook*, 4th ed., McGraw-Hill, New York, 2008.
32. J. S. Gulliver and R. E. A. Arndt, *Hydropower Engineering Handbook*, McGraw-Hill, New York, 1990.
33. R. L. Daugherty, J. B. Franzini, and E. J. Finnemore, *Fluid Mechanics and Engineering Applications*, 9th ed., McGraw-Hill, New York, 1997.
34. R. H. Sabersky, E. M. Gates, A. J. Acosta, and E. G. Hauptmann, *Fluid Flow: A First Course in Fluid Mechanics*, 4th ed., Pearson Education, Upper Saddle River, NJ, 1994.
35. J. P. Poynton, *Metering Pumps*, Marcel Dekker, New York, 1983.

36. Hydraulic Institute, *Reciprocating Pump Test Standard,* New York, 1994.
37. T. L. Henshaw, *Reciprocating Pumps,* Wiley, New York, 1987.
38. J. E. Miller, *The Reciprocating Pump: Theory, Design and Use,* Wiley, NewYork, 1987.
39. D. G. Wilson and T. Korakianitis, *The Design of High Efficiency Turbomachinery and Gas Turbines,* 2d ed., Pearson Education, Upper Saddle River, NJ, 1998.
40. S. O. Kraus et al., "Periodic Velocity Measurements in a Wide and Large Radius Ratio Automotive Torque Converter at the Pump/Turbine Interface," *J. Fluids Engineering,* vol. 127, no. 2, 2005, pp. 308–316.
41. E. M. Greitzer, "The Stability of Pumping Systems: The 1980 Freeman Scholar Lecture," *J. Fluids Eng.,* vol. 103, June 1981, pp. 193–242.
42. R. Elder et al. (eds.), *Advances of CFD in Fluid Machinery Design,* Wiley, New York, 2003.
43. L. F. Moody, "The Propeller Type Turbine," *ASCE Trans.,* vol. 89, 1926, p. 628.
44. H. H. Anderson, "Prediction of Head, Quantity, and Efficiency in Pumps—The Area-Ratio Principle," in *Performance Prediction of Centrifugal Pumps and Compressors,* vol. 100127, ASME Symp., New York, 1980, pp. 201–211.
45. M. Schobeiri, *Turbomachinery Flow Physics and Dynamic Performance,* Springer, New York, 2004.
46. D. J. Mahoney (ed.), *Proceedings of the 1997 International Conference on Hydropower,* ASCE, Reston, VA, 1997.
47. G. W. Koeppl, *Putnam's Power from the Wind,* 2d ed., Van Nostrand Reinhold, New York, 1982.
48. P. Jain, *Wind Energy Engineering,* McGraw-Hill, New York, 2010.
49. D. Wood, *Small Wind Turbines: Analysis, Design, and Application,* Springer, New York, 2011.
50. E. Hau, *Wind Turbines: Fundamentals, Technologies, Application, Economics,* 2d ed., Springer-Verlag, New York, 2005.
51. R. Harrison, E. Hau, and H. Snel, *Large Wind Turbines,* Wiley, New York, 2000.
52. R. H. Aungier, *Turbine Aerodynamics: Axial-Flow and Radial-Flow Turbine Design and Analysis,* ASME Press, New York, 2006.
53. M. L. Robinson, "The Darrieus Wind Turbine for Electrical Power Generation," *Aeronaut. J.,* June 1981, pp. 244–255.
54. D. F. Warne and P. G. Calnan, "Generation of Electricity from the Wind," *IEE Rev.,* vol. 124, no. 11R, November 1977, pp. 963–985.
55. L. A. Haimerl, "The Crossflow Turbine," *Waterpower,* January 1960, pp. 5–13; see also *ASME Symp. Small Hydropower Fluid Mach.,* vol. 1, 1980, and vol. 2, 1982.
56. K. Eisele et al., "Flow Analysis in a Pump Diffuser: Part 1, Measurements; Part 2, CFD," *J. Fluids Eng.,* vol. 119, December 1997, pp. 968–984.
57. D. Japikse and N. C. Baines, *Turbomachinery Diffuser Design Technology,* Concepts ETI Inc., Hanover, NH, 1998.
58. B. Massey and J. Ward-Smith, *Mechanics of Fluids,* 7th ed., Nelson Thornes Publishing, Cheltenham, UK, 1998.
59. American Wind Energy Association, "Global Wind Energy Market Report," URL: <http://www.awea.org/pubs/documents/globalmarket2004.pdf>.
60. J. Carlton, *Marine Propellers and Propulsion,* 3d ed., Butterworth-Heinemann, New York, 2012.
61. M. Hollmann, *Modern Propeller and Duct Design,* Aircraft Designs, Inc., Monterey, CA, 1993.
62. C. L. Archer and M. Z. Jacobson, "Evaluation of Global Wind Power," *J. Geophys. Res.-Atm.,* vol. 110, 2005, doi:10.1029/2004JD005462.
63. M. Farinas and A. Garon, "Application of DOE for Optimal Turbomachinery Design," Paper AIAA-2004-2139, AIAA Fluid Dynamics Conference, Portland, OR, June 2004.
64. C. Crain, "Running Against the Wind," *Popular Science,* March 2009, pp. 69–70.

Appendix A Physical Properties of Fluids

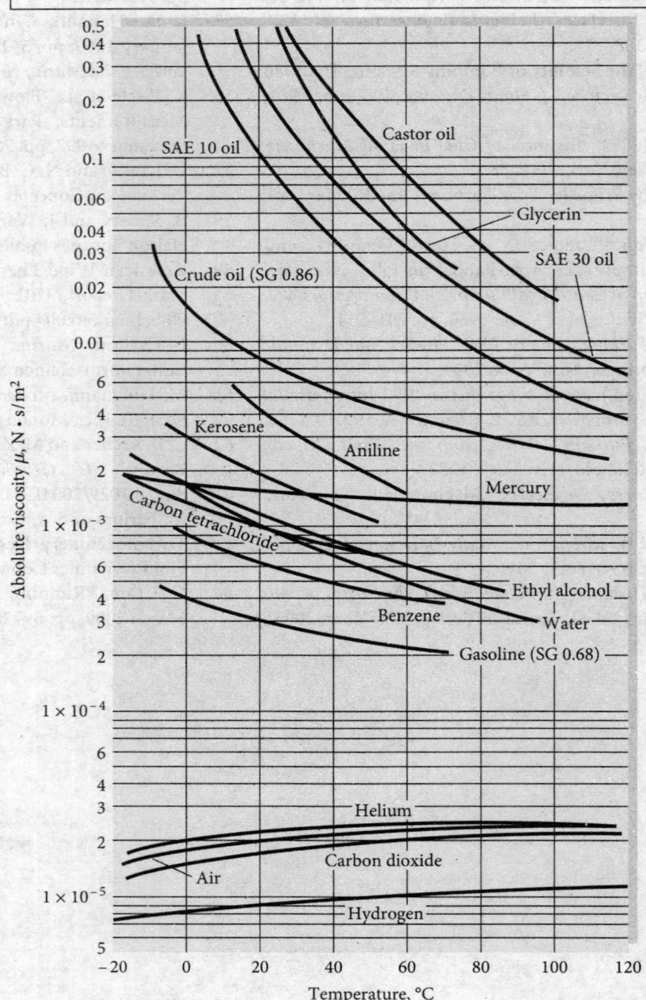

Fig. A.1 Absolute viscosity of common fluids at 1 atm.

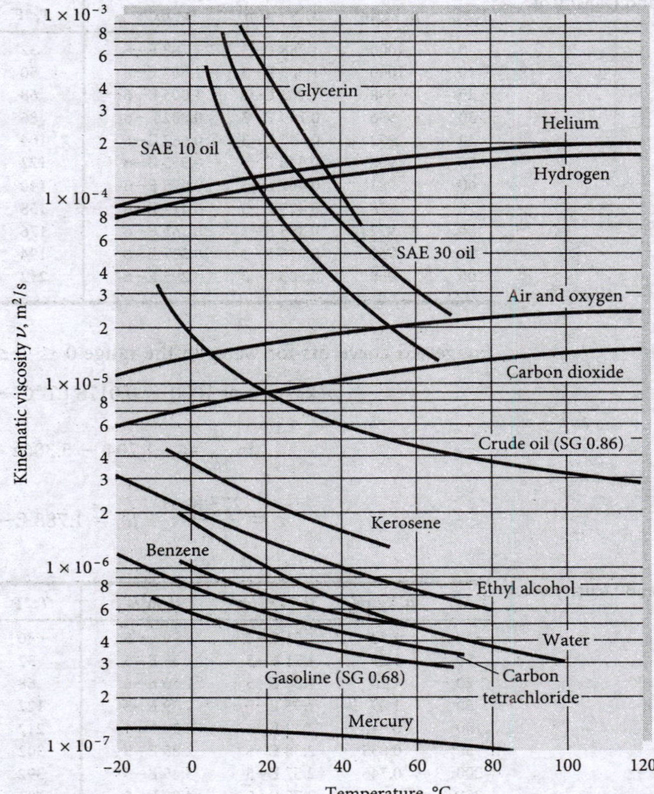

Fig. A.2 Kinematic viscosity of common fluids at 1 atm.

Table A.1 Viscosity and Density of Water at 1 atm

T, °C	ρ, kg/m³	μ, N · s/m²	ν, m²/s	T, °F	ρ, slug/ft³	μ, lb · s/ft²	ν, ft²/s
0	1000	1.788 E−3	1.788 E−6	32	1.940	3.73 E−5	1.925 E−5
10	1000	1.307 E−3	1.307 E−6	50	1.940	2.73 E−5	1.407 E−5
20	998	1.003 E−3	1.005 E−6	68	1.937	2.09 E−5	1.082 E−5
30	996	0.799 E−3	0.802 E−6	86	1.932	1.67 E−5	0.864 E−5
40	992	0.657 E−3	0.662 E−6	104	1.925	1.37 E−5	0.713 E−5
50	988	0.548 E−3	0.555 E−6	122	1.917	1.14 E−5	0.597 E−5
60	983	0.467 E−3	0.475 E−6	140	1.908	0.975 E−5	0.511 E−5
70	978	0.405 E−3	0.414 E−6	158	1.897	0.846 E−5	0.446 E−5
80	972	0.355 E−3	0.365 E−6	176	1.886	0.741 E−5	0.393 E−5
90	965	0.316 E−3	0.327 E−6	194	1.873	0.660 E−5	0.352 E−5
100	958	0.283 E−3	0.295 E−6	212	1.859	0.591 E−5	0.318 E−5

Suggested curve fits for water in the range $0 \le T \le 100°C$:

$$\rho(\text{kg/m}^3) \approx 1000 - 0.0178\,|\,T°C - 4°C\,|^{1.7} \pm 0.2\%$$

$$\ln\frac{\mu}{\mu_0} \approx -1.704 - 5.306z + 7.003z^2$$

$$z = \frac{273\text{ K}}{T\text{ K}} \qquad \mu_0 = 1.788\text{ E}{-3}\text{ kg/(m}\cdot\text{s)}$$

Table A.2 Viscosity and Density of Air at 1 atm

T, °C	ρ, kg/m³	μ, N · s/m²	ν, m²/s	T, °F	ρ, slug/ft³	μ, lb · s/ft²	ν, ft²/s
−40	1.52	1.51 E−5	0.99 E−5	−40	2.94 E−3	3.16 E−7	1.07 E−4
0	1.29	1.71 E−5	1.33 E−5	32	2.51 E−3	3.58 E−7	1.43 E−4
20	1.20	1.80 E−5	1.50 E−5	68	2.34 E−3	3.76 E−7	1.61 E−4
50	1.09	1.95 E−5	1.79 E−5	122	2.12 E−3	4.08 E−7	1.93 E−4
100	0.946	2.17 E−5	2.30 E−5	212	1.84 E−3	4.54 E−7	2.47 E−4
150	0.835	2.38 E−5	2.85 E−5	302	1.62 E−3	4.97 E−7	3.07 E−4
200	0.746	2.57 E−5	3.45 E−5	392	1.45 E−3	5.37 E−7	3.71 E−4
250	0.675	2.75 E−5	4.08 E−5	482	1.31 E−3	5.75 E−7	4.39 E−4
300	0.616	2.93 E−5	4.75 E−5	572	1.20 E−3	6.11 E−7	5.12 E−4
400	0.525	3.25 E−5	6.20 E−5	752	1.02 E−3	6.79 E−7	6.67 E−4
500	0.457	3.55 E−5	7.77 E−5	932	0.89 E−3	7.41 E−7	8.37 E−4

Suggested curve fits for air:

$$\rho = \frac{p}{RT} \qquad R_{air} \approx 287\text{ J/(kg}\cdot\text{K)}$$

Power law: $$\frac{\mu}{\mu_0} \approx \left(\frac{T}{T_0}\right)^{0.7}$$

Sutherland law: $$\frac{\mu}{\mu_0} \approx \left(\frac{T}{T_0}\right)^{3/2}\left(\frac{T_0 + S}{T + S}\right) \qquad S_{air} \approx 110.4\text{ K}$$

with $T_0 = 273$ K, $\mu_0 = 1.71$ E−5 kg/(m · s), and T in kelvins.

Table A.3 Properties of Common Liquids at 1 atm and 20°C (68°F)

Liquid	ρ, kg/m³	μ, kg/(m • s)	Υ, N/m*	p_v, N/m²	Bulk modulus K, N/m²	Viscosity parameter C†
Ammonia	608	2.20 E−4	2.13 E−2	9.10 E+5	1.82 E+9	1.05
Benzene	881	6.51 E−4	2.88 E−2	1.01 E+4	1.47 E+9	4.34
Carbon tetrachloride	1590	9.67 E−4	2.70 E−2	1.20 E+4	1.32 E+9	4.45
Ethanol	789	1.20 E−3	2.28 E−2	5.7 E+3	1.09 E+9	5.72
Ethylene glycol	1117	2.14 E−2	4.84 E−2	1.2 E+1	3.05 E+9	11.7
Freon 12	1327	2.62 E−4	—	—	7.95 E+8	1.76
Gasoline	680	2.92 E−4	2.16 E−2	5.51 E+4	1.3 E+9	3.68
Glycerin	1260	1.49	6.33 E−2	1.4 E−2	4.35 E+9	28.0
Kerosene	804	1.92 E−3	2.8 E−2	3.11 E+3	1.41 E+9	5.56
Mercury	13,550	1.56 E−3	4.84 E−1	1.1 E−3	2.85 E+10	1.07
Methanol	791	5.98 E−4	2.25 E−2	1.34 E+4	1.03 E+9	4.63
SAE 10W oil	870	1.04 E−1‡	3.6 E−2	—	1.31 E+9	15.7
SAE 10W30 oil	876	1.7 E−1‡	—	—	—	14.0
SAE 30W oil	891	2.9 E−1‡	3.5 E−2	—	1.38 E+9	18.3
SAE 50W oil	902	8.6 E−1‡	—	—	—	20.2
Water	998	1.00 E−3	7.28 E−2	2.34 E+3	2.19 E+9	Table A.1
Seawater (30‰)	1025	1.07 E−3	7.28 E−2	2.34 E+3	2.33 E+9	7.28

*In contact with air.

†The viscosity–temperature variation of these liquids may be fitted to the empirical expression

$$\frac{\mu}{\mu_{20^\circ\mathrm{C}}} \approx \exp\left[C\left(\frac{293\ \mathrm{K}}{T\ \mathrm{K}} - 1\right)\right]$$

with accuracy of ±6 percent in the range $0 \leq T \leq 100$°C.

‡Representative values. The SAE oil classifications allow a viscosity variation of up to ±50 percent, especially at lower temperatures.

Table A.4 Properties of Common Gases at 1 atm and 20°C (68°F)

Gas	Molecular weight	R, m²/(s² • K)	ρg, N/m³	μ, N • s/m²	ν, m²/s	Specific-heat ratio	Power-law exponent n*
H_2	2.016	4124	0.822	9.05 E−6	1.08 E−04	1.41	0.68
He	4.003	2077	1.63	1.97 E−5	1.18 E−04	1.66	0.67
H_2O	18.02	461	7.35	1.02 E−5	1.36 E−05	1.33	1.15
Ar	39.944	208	16.3	2.24 E−5	1.35 E−05	1.67	0.72
Dry air	28.96	287	11.8	1.80 E−5	1.49 E−05	1.40	0.67
CO_2	44.01	189	17.9	1.48 E−5	8.09 E−06	1.30	0.79
CO	28.01	297	11.4	1.82 E−5	1.56 E−05	1.40	0.71
N_2	28.02	297	11.4	1.76 E−5	1.51 E−05	1.40	0.67
O_2	32.00	260	13.1	2.00 E−5	1.50 E−05	1.40	0.69
NO	30.01	277	12.1	1.90 E−5	1.52 E−05	1.40	0.78
N_2O	44.02	189	17.9	1.45 E−5	7.93 E−06	1.31	0.89
Cl_2	70.91	117	28.9	1.03 E−5	3.49 E−06	1.34	1.00
CH_4	16.04	518	6.54	1.34 E−5	2.01 E−05	1.32	0.87

*The power-law curve fit, Eq. (1.27), $\mu/\mu_{293K} \approx (T/293)^n$, fits these gases to within ±4 percent in the range $250 \leq T \leq 1000$ K. The temperature must be in kelvins.

Table A.5 Surface Tension, Vapor Pressure, and Sound Speed of Water

T, °C	Υ, N/m	p_v, kPa	a, m/s
0	0.0756	0.611	1402
10	0.0742	1.227	1447
20	0.0728	2.337	1482
30	0.0712	4.242	1509
40	0.0696	7.375	1529
50	0.0679	12.34	1542
60	0.0662	19.92	1551
70	0.0644	31.16	1553
80	0.0626	47.35	1554
90	0.0608	70.11	1550
100	0.0589	101.3	1543
120	0.0550	198.5	1518
140	0.0509	361.3	1483
160	0.0466	617.8	1440
180	0.0422	1002	1389
200	0.0377	1554	1334
220	0.0331	2318	1268
240	0.0284	3344	1192
260	0.0237	4688	1110
280	0.0190	6412	1022
300	0.0144	8581	920
320	0.0099	11,274	800
340	0.0056	14,586	630
360	0.0019	18,651	370
374*	0.0*	22,090*	0*

*Critical point.

Table A.6 Properties of the Standard Atmosphere

z, m	T, K	p, Pa	ρ, kg/m³	a, m/s
−500	291.41	107,508	1.2854	342.2
0	288.16	101,350	1.2255	340.3
500	284.91	95,480	1.1677	338.4
1000	281.66	89,889	1.1120	336.5
1500	278.41	84,565	1.0583	334.5
2000	275.16	79,500	1.0067	332.6
2500	271.91	74,684	0.9570	330.6
3000	268.66	70,107	0.9092	328.6
3500	265.41	65,759	0.8633	326.6
4000	262.16	61,633	0.8191	324.6
4500	258.91	57,718	0.7768	322.6
5000	255.66	54,008	0.7361	320.6
5500	252.41	50,493	0.6970	318.5
6000	249.16	47,166	0.6596	316.5
6500	245.91	44,018	0.6237	314.4
7000	242.66	41,043	0.5893	312.3
7500	239.41	38,233	0.5564	310.2
8000	236.16	35,581	0.5250	308.1
8500	232.91	33,080	0.4949	306.0
9000	229.66	30,723	0.4661	303.8
9500	226.41	28,504	0.4387	301.7
10,000	223.16	26,416	0.4125	299.5
10,500	219.91	24,455	0.3875	297.3
11,000	216.66	22,612	0.3637	295.1
11,500	216.66	20,897	0.3361	295.1
12,000	216.66	19,312	0.3106	295.1
12,500	216.66	17,847	0.2870	295.1
13,000	216.66	16,494	0.2652	295.1
13,500	216.66	15,243	0.2451	295.1
14,000	216.66	14,087	0.2265	295.1
14,500	216.66	13,018	0.2094	295.1
15,000	216.66	12,031	0.1935	295.1
15,500	216.66	11,118	0.1788	295.1
16,000	216.66	10,275	0.1652	295.1
16,500	216.66	9496	0.1527	295.1
17,000	216.66	8775	0.1411	295.1
17,500	216.66	8110	0.1304	295.1
18,000	216.66	7495	0.1205	295.1
18,500	216.66	6926	0.1114	295.1
19,000	216.66	6401	0.1029	295.1
19,500	216.66	5915	0.0951	295.1
20,000	216.66	5467	0.0879	295.1
22,000	218.60	4048	0.0645	296.4
24,000	220.60	2972	0.0469	297.8
26,000	222.50	2189	0.0343	299.1
28,000	224.50	1616	0.0251	300.4
30,000	226.50	1197	0.0184	301.7
40,000	250.40	287	0.0040	317.2
50,000	270.70	80	0.0010	329.9
60,000	255.70	22	0.0003	320.6
70,000	219.70	6	0.0001	297.2

Appendix B
Compressible Flow Tables

Table B.1
Isentropic Flow of a Perfect Gas, $k = 1.4$

Ma	p/p_0	ρ/ρ_0	T/T_0	A/A^*
0.00	1.0000	1.0000	1.0000	∞
0.10	0.9930	0.9950	0.9980	5.8218
0.20	0.9725	0.9803	0.9921	2.9635
0.30	0.9395	0.9564	0.9823	2.0351
0.40	0.8956	0.9243	0.9690	1.5901
0.50	0.8430	0.8852	0.9524	1.3398
0.60	0.7840	0.8405	0.9328	1.1882
0.70	0.7209	0.7916	0.9107	1.0944
0.80	0.6560	0.7400	0.8865	1.0382
0.90	0.5913	0.6870	0.8606	1.0089
1.00	0.5283	0.6339	0.8333	1.0000
1.10	0.4684	0.5817	0.8052	1.0079
1.20	0.4124	0.5311	0.7764	1.0304
1.30	0.3609	0.4829	0.7474	1.0663
1.40	0.3142	0.4374	0.7184	1.1149
1.50	0.2724	0.3950	0.6897	1.1762
1.60	0.2353	0.3557	0.6614	1.2502
1.70	0.2026	0.3197	0.6337	1.3376
1.80	0.1740	0.2868	0.6068	1.4390
1.90	0.1492	0.2570	0.5807	1.5553
2.00	0.1278	0.2300	0.5556	1.6875
2.10	0.1094	0.2058	0.5313	1.8369
2.20	0.0935	0.1841	0.5081	2.0050
2.30	0.0800	0.1646	0.4859	2.1931
2.40	0.0684	0.1472	0.4647	2.4031
2.50	0.0585	0.1317	0.4444	2.6367
2.60	0.0501	0.1179	0.4252	2.8960
2.70	0.0430	0.1056	0.4068	3.1830
2.80	0.0368	0.0946	0.3894	3.5001
2.90	0.0317	0.0849	0.3729	3.8498
3.00	0.0272	0.0762	0.3571	4.2346
3.10	0.0234	0.0685	0.3422	4.6573
3.20	0.0202	0.0617	0.3281	5.1210
3.30	0.0175	0.0555	0.3147	5.6286
3.40	0.0151	0.0501	0.3019	6.1837
3.50	0.0131	0.0452	0.2899	6.7896
3.60	0.0114	0.0409	0.2784	7.4501
3.70	0.0099	0.0370	0.2675	8.1691
3.80	0.0086	0.0335	0.2572	8.9506
3.90	0.0075	0.0304	0.2474	9.7990
4.00	0.0066	0.0277	0.2381	10.7188

Table B.2 Normal Shock Relations for a Perfect Gas, $k = 1.4$

Ma_{n1}	Ma_{n2}	p_2/p_1	$V_1/V_2 = \rho_2/\rho_1$	T_2/T_1	p_{02}/p_{01}	A_2^*/A_1^*
1.00	1.0000	1.0000	1.0000	1.0000	1.0000	1.0000
1.10	0.9118	1.2450	1.1691	1.0649	0.9989	1.0011
1.20	0.8422	1.5133	1.3416	1.1280	0.9928	1.0073
1.30	0.7860	1.8050	1.5157	1.1909	0.9794	1.0211
1.40	0.7397	2.1200	1.6897	1.2547	0.9582	1.0436
1.50	0.7011	2.4583	1.8621	1.3202	0.9298	1.0755
1.60	0.6684	2.8200	2.0317	1.3880	0.8952	1.1171
1.70	0.6405	3.2050	2.1977	1.4583	0.8557	1.1686
1.80	0.6165	3.6133	2.3592	1.5316	0.8127	1.2305
1.90	0.5956	4.0450	2.5157	1.6079	0.7674	1.3032
2.00	0.5774	4.5000	2.6667	1.6875	0.7209	1.3872
2.10	0.5613	4.9783	2.8119	1.7705	0.6742	1.4832
2.20	0.5471	5.4800	2.9512	1.8569	0.6281	1.5920
2.30	0.5344	6.0050	3.0845	1.9468	0.5833	1.7144
2.40	0.5231	6.5533	3.2119	2.0403	0.5401	1.8514
2.50	0.5130	7.1250	3.3333	2.1375	0.4990	2.0039
2.60	0.5039	7.7200	3.4490	2.2383	0.4601	2.1733
2.70	0.4956	8.3383	3.5590	2.3429	0.4236	2.3608
2.80	0.4882	8.9800	3.6636	2.4512	0.3895	2.5676
2.90	0.4814	9.6450	3.7629	2.5632	0.3577	2.7954
3.00	0.4752	10.3333	3.8571	2.6790	0.3283	3.0456
3.10	0.4695	11.0450	3.9466	2.7986	0.3012	3.3199
3.20	0.4643	11.7800	4.0315	2.9220	0.2762	3.6202
3.30	0.4596	12.5383	4.1120	3.0492	0.2533	3.9483
3.40	0.4552	13.3200	4.1884	3.1802	0.2322	4.3062
3.50	0.4512	14.1250	4.2609	3.3151	0.2129	4.6960
3.60	0.4474	14.9533	4.3296	3.4537	0.1953	5.1200
3.70	0.4439	15.8050	4.3949	3.5962	0.1792	5.5806
3.80	0.4407	16.6800	4.4568	3.7426	0.1645	6.0801
3.90	0.4377	17.5783	4.5156	3.8928	0.1510	6.6213
4.00	0.4350	18.5000	4.5714	4.0469	0.1388	7.2069
4.10	0.4324	19.4450	4.6245	4.2048	0.1276	7.8397
4.20	0.4299	20.4133	4.6749	4.3666	0.1173	8.5227
4.30	0.4277	21.4050	4.7229	4.5322	0.1080	9.2591
4.40	0.4255	22.4200	4.7685	4.7017	0.0995	10.0522
4.50	0.4236	23.4583	4.8119	4.8751	0.0917	10.9054
4.60	0.4217	24.5200	4.8532	5.0523	0.0846	11.8222
4.70	0.4199	25.6050	4.8926	5.2334	0.0781	12.8065
4.80	0.4183	26.7133	4.9301	5.4184	0.0721	13.8620
4.90	0.4167	27.8450	4.9659	5.6073	0.0667	14.9928
5.00	0.4152	29.0000	5.0000	5.8000	0.0617	16.2032

Table B.3 Adiabatic Frictional Flow in a Constant-Area Duct for $k = 1.4$

Ma	$\bar{f}L/D$	p/p^*	T/T^*	$\rho^*/\rho = V/V^*$	p_0/p_0^*
0.00	∞	∞	1.2000	0.0000	∞
0.10	66.9216	10.9435	1.1976	0.1094	5.8218
0.20	14.5333	5.4554	1.1905	0.2182	2.9635
0.30	5.2993	3.6191	1.1788	0.3257	2.0351
0.40	2.3085	2.6958	1.1628	0.4313	1.5901
0.50	1.0691	2.1381	1.1429	0.5345	1.3398
0.60	0.4908	1.7634	1.1194	0.6348	1.1882
0.70	0.2081	1.4935	1.0929	0.7318	1.0944
0.80	0.0723	1.2893	1.0638	0.8251	1.0382
0.90	0.0145	1.1291	1.0327	0.9146	1.0089
1.00	0.0000	1.0000	1.0000	1.0000	1.0000
1.10	0.0099	0.8936	0.9662	1.0812	1.0079
1.20	0.0336	0.8044	0.9317	1.1583	1.0304
1.30	0.0648	0.7285	0.8969	1.2311	1.0663
1.40	0.0997	0.6632	0.8621	1.2999	1.1149
1.50	0.1361	0.6065	0.8276	1.3646	1.1762
1.60	0.1724	0.5568	0.7937	1.4254	1.2502
1.70	0.2078	0.5130	0.7605	1.4825	1.3376
1.80	0.2419	0.4741	0.7282	1.5360	1.4390
1.90	0.2743	0.4394	0.6969	1.5861	1.5553
2.00	0.3050	0.4082	0.6667	1.6330	1.6875
2.10	0.3339	0.3802	0.6376	1.6769	1.8369
2.20	0.3609	0.3549	0.6098	1.7179	2.0050
2.30	0.3862	0.3320	0.5831	1.7563	2.1931
2.40	0.4099	0.3111	0.5576	1.7922	2.4031
2.50	0.4320	0.2921	0.5333	1.8257	2.6367
2.60	0.4526	0.2747	0.5102	1.8571	2.8960
2.70	0.4718	0.2588	0.4882	1.8865	3.1830
2.80	0.4898	0.2441	0.4673	1.9140	3.5001
2.90	0.5065	0.2307	0.4474	1.9398	3.8498
3.00	0.5222	0.2182	0.4286	1.9640	4.2346
3.10	0.5368	0.2067	0.4107	1.9866	4.6573
3.20	0.5504	0.1961	0.3937	2.0079	5.1210
3.30	0.5632	0.1862	0.3776	2.0278	5.6286
3.40	0.5752	0.1770	0.3623	2.0466	6.1837
3.50	0.5864	0.1685	0.3478	2.0642	6.7896
3.60	0.5970	0.1606	0.3341	2.0808	7.4501
3.70	0.6068	0.1531	0.3210	2.0964	8.1691
3.80	0.6161	0.1462	0.3086	2.1111	8.9506
3.90	0.6248	0.1397	0.2969	2.1250	9.7990
4.00	0.6331	0.1336	0.2857	2.1381	10.7187

Table B.4 Frictionless Duct Flow with Heat Transfer for $k = 1.4$

Ma	T_0/T_0^*	p/p^*	T/T^*	$\rho^*/\rho = V/V^*$	p_0/p_0^*
0.00	0.0000	2.4000	0.0000	0.0000	1.2679
0.10	0.0468	2.3669	0.0560	0.0237	1.2591
0.20	0.1736	2.2727	0.2066	0.0909	1.2346
0.30	0.3469	2.1314	0.4089	0.1918	1.1985
0.40	0.5290	1.9608	0.6151	0.3137	1.1566
0.50	0.6914	1.7778	0.7901	0.4444	1.1141
0.60	0.8189	1.5957	0.9167	0.5745	1.0753
0.70	0.9085	1.4235	0.9929	0.6975	1.0431
0.80	0.9639	1.2658	1.0255	0.8101	1.0193
0.90	0.9921	1.1246	1.0245	0.9110	1.0049
1.00	1.0000	1.0000	1.0000	1.0000	1.0000
1.10	0.9939	0.8909	0.9603	1.0780	1.0049
1.20	0.9787	0.7958	0.9118	1.1459	1.0194
1.30	0.9580	0.7130	0.8592	1.2050	1.0437
1.40	0.9343	0.6410	0.8054	1.2564	1.0777
1.50	0.9093	0.5783	0.7525	1.3012	1.1215
1.60	0.8842	0.5236	0.7017	1.3403	1.1756
1.70	0.8597	0.4756	0.6538	1.3746	1.2402
1.80	0.8363	0.4335	0.6089	1.4046	1.3159
1.90	0.8141	0.3964	0.5673	1.4311	1.4033
2.00	0.7934	0.3636	0.5289	1.4545	1.5031
2.10	0.7741	0.3345	0.4936	1.4753	1.6162
2.20	0.7561	0.3086	0.4611	1.4938	1.7434
2.30	0.7395	0.2855	0.4312	1.5103	1.8860
2.40	0.7242	0.2648	0.4038	1.5252	2.0451
2.50	0.7101	0.2462	0.3787	1.5385	2.2218
2.60	0.6970	0.2294	0.3556	1.5505	2.4177
2.70	0.6849	0.2142	0.3344	1.5613	2.6343
2.80	0.6738	0.2004	0.3149	1.5711	2.8731
2.90	0.6635	0.1879	0.2969	1.5801	3.1359
3.00	0.6540	0.1765	0.2803	1.5882	3.4245
3.10	0.6452	0.1660	0.2650	1.5957	3.7408
3.20	0.6370	0.1565	0.2508	1.6025	4.0871
3.30	0.6294	0.1477	0.2377	1.6088	4.4655
3.40	0.6224	0.1397	0.2255	1.6145	4.8783
3.50	0.6158	0.1322	0.2142	1.6198	5.3280
3.60	0.6097	0.1254	0.2037	1.6247	5.8173
3.70	0.6040	0.1190	0.1939	1.6293	6.3488
3.80	0.5987	0.1131	0.1848	1.6335	6.9256
3.90	0.5937	0.1077	0.1763	1.6374	7.5505
4.00	0.5891	0.1026	0.1683	1.6410	8.2268

Table B.5 Prandtl-Meyer Supersonic Expansion Function for $k = 1.4$

Ma	ω, deg	Ma	ω, deg	Ma	ω, deg	Ma	ω, deg
1.00	0.00	3.10	51.65	5.10	77.84	7.10	91.49
1.10	1.34	3.20	53.47	5.20	78.73	7.20	92.00
1.20	3.56	3.30	55.22	5.30	79.60	7.30	92.49
1.30	6.17	3.40	56.91	5.40	80.43	7.40	92.97
1.40	8.99	3.50	58.53	5.50	81.24	7.50	93.44
1.50	11.91	3.60	60.09	5.60	82.03	7.60	93.90
1.60	14.86	3.70	61.60	5.70	82.80	7.70	94.34
1.70	17.81	3.80	63.04	5.80	83.54	7.80	94.78
1.80	20.73	3.90	64.44	5.90	84.26	7.90	95.21
1.90	23.59	4.00	65.78	6.00	84.96	8.00	95.62
2.00	26.38	4.10	67.08	6.10	85.63	8.10	96.03
2.10	29.10	4.20	68.33	6.20	86.29	8.20	96.43
2.20	31.73	4.30	69.54	6.30	86.94	8.30	96.82
2.30	34.28	4.40	70.71	6.40	87.56	8.40	97.20
2.40	36.75	4.50	71.83	6.50	88.17	8.50	97.57
2.50	39.12	4.60	72.92	6.60	88.76	8.60	97.94
2.60	41.41	4.70	73.97	6.70	89.33	8.70	98.29
2.70	43.62	4.80	74.99	6.80	89.89	8.80	98.64
2.80	45.75	4.90	75.97	6.90	90.44	8.90	98.98
2.90	47.79	5.00	76.92	7.00	90.97	9.00	99.32
3.00	49.76						

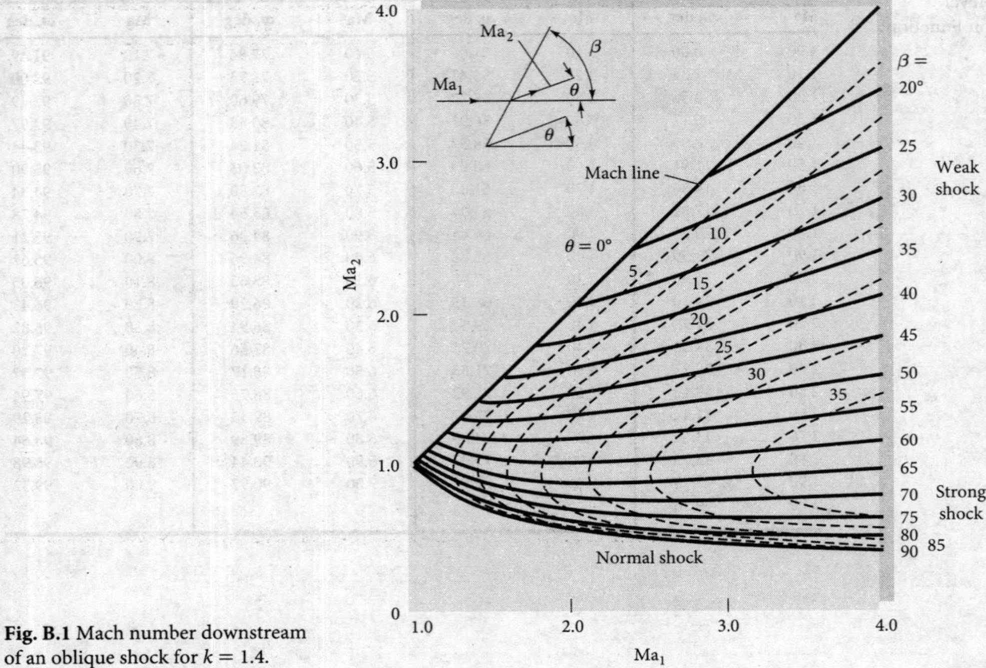

Fig. B.1 Mach number downstream of an oblique shock for $k = 1.4$.

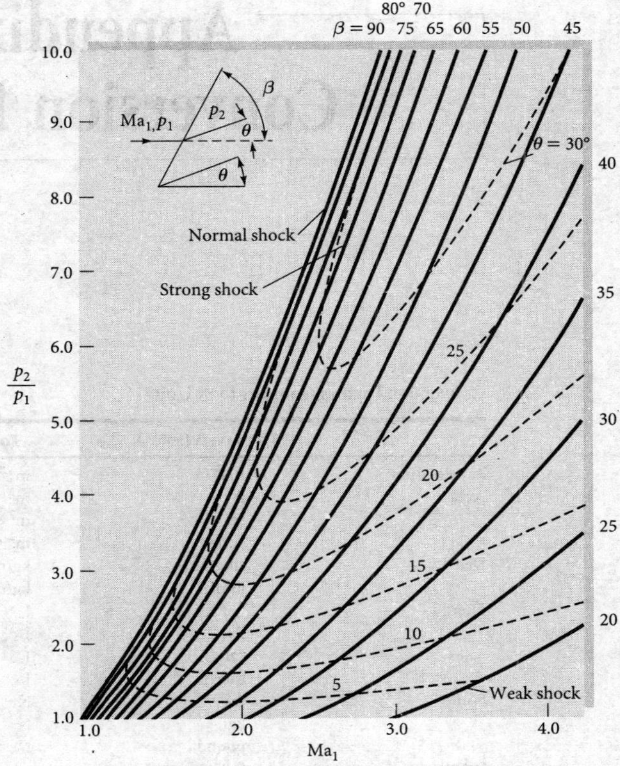

Fig. B.2 Pressure ratio downstream of an oblique shock for $k = 1.4$.

Appendix C
Conversion Factors

Conversion Factors from BG to SI Units

	To convert from	To	Multiply by
Acceleration	ft/s^2	m/s^2	0.3048
Area	ft^2	m^2	9.2903 E − 2
	mi^2	m^2	2.5900 E + 6
	acres	m^2	4.0469 E + 3
Density	$slug/ft^3$	kg/m^3	3 5.1538 E + 2
	lbm/ft^3	kg/m^3	1.6019 E + 1
Energy	ft-lbf	J	1.3558
	Btu	J	1.0551 E + 3
	cal	J	4.1868
	therm	J	1.0551 E8
Force	lbf	N	4.4482
	kgf	N	9.8067
	pounda	N	0.13826
Length	ft	m	0.3048
	in	m	2.5400 E − 2
	mi (statute)	m	1.6093 E + 3
	nmi (nautical)	m	1.8520 E + 3
Mass	slug	kg	1.4594 E + 1
	lbm	kg	4.5359 E − 1
	U.S. ounce	kg	2.835 E − 2
Mass flow	slug/s	kg/s	1.4594 E + 1
	lbm/s	kg/s	4.5359 E − 1
Power	ft • lbf/s	W	1.3558
	hp	W	7.4570 E + 2
Pressure	lbf/ft^2	Pa	4.7880 E + 1
	lbf/in^2	Pa	6.8948 E + 3
	atm	Pa	1.0133 E + 5
	mm Hg	Pa	1.3332 E + 2
	torr	Pa	1.3332 E2
Specific weight	lbf/ft^3	N/m^3	1.5709 E + 2
Specific heat	$ft^2/(s^2 \cdot °R)$	$m^2/(s2 \cdot K)$	1.6723 E − 1
Surface tension	lbf/ft	N/m	1.4594 E + 1
Temperature	°F	°C	$t_C = \frac{5}{9}(t_F - 32°)$
	°R	K	0.5556
Velocity	ft/s	m/s	0.3048
	mi/h	m/s	4.4704 E − 1
	knot	m/s	5.1444 E − 1
	knot	mi/h	1.1508

Conversion Factors from BG to SI Units (Continued)

	To convert from	To	Multiply by
Viscosity	lbf • s/ft^2	N • s/m^2	4.7880 E + 1
	g/(cm • s)	N • s/m^2	0.1
Volume	ft^3	m^3	2.8317 E − 2
	L	m^3	0.001
	gal (U.S.)	m^3	3.7854 E − 3
	fluid ounce (U.S.)	m^3	2.9574 E − 5
	barrel	U.S. gal	42
Volume flow	ft^3/s	m^3/s	2.8317 E − 2
	gal/min	m^3/s	6.3090 E − 5
	ft^3/s	m^3/s	2.8317 E − 2

Additional Conversions

Length	Volume
1 ft = 12 in = 0.3048 m 1 mi = 5280 ft = 1609.344 m 1 nautical mile (nmi) = 6076 ft = 1852 m 1 yd = 3 ft = 0.9144 m 1 angstrom (Å) = 1.0 E−10 m	1 ft^3 = 0.028317 m^3 1 U.S. gal = 231 in^3 = 0.0037854 m^3 1 L = 0.001 m^3 = 0.035315 ft^3 1 U.S. fluid ounce = 2.9574 E−5 m^3 1 U.S. quart (qt) = 9.4635 E−4 m^3 1 barrel = 42 U.S. gal = 0.15899 m^3
Mass	**Area**
1 slug = 32.174 lbm = 14.594 kg 1 lbm = 0.4536 kg 1 short ton = 2000 lbm = 907.185 kg 1 tonne = 1000 kg 1 U.S. ounce = 0.02835 kg	1 ft^2 = 0.092903 m^2 1 mi^2 = 2.78784 E7 ft^2 = 2.59 E6 m^2 1 acre = 43,560 ft^2 = 4046.9 m^2 1 hectare (ha) = 10,000 m^2
Velocity	**Acceleration**
1 ft/s = 0.3048 m/s 1 mi/h = 1.466666 ft/s = 0.44704 m/s 1 kn = 1 nmi/h = 1.6878 ft/s = 0.5144 m/s 1 kn = 1.1508 mi/h	1 ft/s^2 = 0.3048 m/s^2
Mass flow	**Volume flow**
1 slug/s = 14.594 kg/s 1 lbm/s = 0.4536 kg/s	1 gal/min = 0.002228 ft^3/s = 0.06309 L/s 1 × 10^6 gal/day = 1.5472 ft^3/s = 0.04381 m^3/s 1 ft^3/s = 0.028317 m^3/s
Pressure	**Force**
1 lbf/ft^2 = 47.88 Pa 1 lbf/in^2 = 144 lbf/ft^2 = 6895 Pa 1 atm = 2116.2 lbf/ft^2 = 14.696 lbf/in^2 = 101,325 Pa 1 inHg (at 20°C) = 3375 Pa 1 bar = 1.0 E5 Pa 1 torr = (1/760)atm = 133.32 Pa = 2.7845 lbf/ft^2	1 lbf = 4.448222 N = 16 oz 1 kgf = 2.2046 lbf = 9.80665 N 1 U.S. (short) ton = 2000 lbf 1 dyne = 1.0 E−5 N 1 ounce (avoirdupois) (oz) = 0.27801 N 1 poundal = 0.13826 N
Energy	**Power**
1 ft · lbf = 1.35582 J 1 Btu = 252 cal = 1055.056 J = 778.17 ft · lbf 1 kilowatt hour (kWh) = 3.6 E6 J 1 calorie = 4.1868 J 1 therm = 1 E5 Btu = 1.0551 E8 J	1 hp = 550 ft · lbf/s = 745.7 W 1 ft · lbf/s = 1.3558 W
Specific weight	**Density**
1 lbf/ft^3 = 157.09 N/m^3	1 $slug/ft^3$ = 515.38 kg/m^3 1 lbm/ft^3 = 16.0185 kg/m^3 1 g/cm^3 = 1000 kg/m^3
Viscosity	**Kinematic viscosity**
1 slug/(ft · s) = 47.88 kg/(m · s) 1 poise (P) = 1 g/(cm · s) = 0.1 kg/(m · s)	1 ft^2/h = 0.000025806 m^2/s 1 stokes (St) = 1 cm^2/s = 0.0001 m^2/s

Additional Conversions (Continued)

Temperature scale readings
$T_F = \frac{9}{5}T_C + 32$ $T_C = \frac{5}{9}(T_F - 32)$ $T_R = T_F + 459.69$ $T_K = T_C + 273.16$ where subscripts F, C, K, and R refer to readings on the Fahrenheit, Celsius, Kelvin, and Rankine scales, respectively.

Specific heat or gas constant*	**Thermal conductivity***
1 ft · lbf/(slug · °R) = 0.16723 N · m/(kg · K) 1 Btu/(lbm · °R) = 4186.8 J/(kg · K)	1 Btu/(h · ft · °R) = 1.7307 W/(m · K)

*Although the absolute (Kelvin) and Celsius temperature scales have different starting points, the intervals are the same size: 1 kelvin = 1 Celsius degree. The same holds true for the nonmetric absolute (Rankine) and Fahrenheit scales: 1 Rankine degree = 1 Fahrenheit degree. It is customary to express temperature differences in absolute temperature units.

Appendix D Equations of Motion in Cylindrical Coordinates

The equations of motion of an incompressible newtonian fluid with constant μ, k, and c_p are given here in cylindrical coordinates (r, θ, z), which are related to cartesian coordinates (x, y, z) as in Fig. 4.2:

$$x = r\cos\theta \qquad y = r\sin\theta \qquad z = z \tag{D.1}$$

The velocity components are υ_r, υ_θ, and υ_z. Here are the equations:
Continuity:

$$\frac{1}{r}\frac{\partial}{\partial r}(r\upsilon_r) + \frac{1}{r}\frac{\partial}{\partial \theta}(\upsilon_\theta) + \frac{\partial}{\partial z}(\upsilon_z) = 0 \tag{D.2}$$

Convective time derivative:

$$\mathbf{V}\cdot\nabla = \upsilon_r\frac{\partial}{\partial r} + \frac{1}{r}\upsilon_\theta\frac{\partial}{\partial \theta} + \upsilon_z\frac{\partial}{\partial z} \tag{D.3}$$

Laplacian operator:

$$\nabla^2 = \frac{1}{r}\frac{\partial}{\partial r}\left(r\frac{\partial}{\partial r}\right) + \frac{1}{r^2}\frac{\partial^2}{\partial \theta^2} + \frac{\partial^2}{\partial z^2} \tag{D.4}$$

The r-momentum equation:

$$\frac{\partial \upsilon_r}{\partial t} + (\mathbf{V}\cdot\nabla)\upsilon_r - \frac{1}{r}\upsilon_\theta^2 = -\frac{1}{\rho}\frac{\partial p}{\partial r} + g_r + \nu\left(\nabla^2\upsilon_r - \frac{\upsilon_r}{r^2} - \frac{2}{r^2}\frac{\partial \upsilon_\theta}{\partial \theta}\right) \tag{D.5}$$

The θ-momentum equation:

$$\frac{\partial \upsilon_\theta}{\partial t} + (\mathbf{V}\cdot\nabla)\,\upsilon_\theta + \frac{1}{r}\upsilon_r\upsilon_\theta = -\frac{1}{\rho r}\frac{\partial p}{\partial \theta} + g_\theta + \nu\left(\nabla^2\upsilon_\theta - \frac{\upsilon_\theta}{r^2} + \frac{2}{r^2}\frac{\partial \upsilon_r}{\partial \theta}\right) \tag{D.6}$$

The z-momentum equation:

$$\frac{\partial \upsilon_z}{\partial t} + (\mathbf{V}\cdot\nabla)\,\upsilon_z = -\frac{1}{\rho}\frac{\partial p}{\partial z} + g_z + \nu\nabla^2\upsilon_z \tag{D.7}$$

The energy equation:

$$\rho c_p \left[\frac{\partial T}{\partial t} + (\mathbf{V} \cdot \nabla)T\right] = k\nabla^2 T + \mu[2(\varepsilon_{rr}^2 + \varepsilon_{\theta\theta}^2 + \varepsilon_{zz}^2) + \varepsilon_{\theta z}^2 + \varepsilon_{rz}^2 + \varepsilon_{r\theta}^2] \quad \text{(D.8)}$$

where

$$\varepsilon_{rr} = \frac{\partial v_r}{\partial r} \qquad \varepsilon_{\theta\theta} = \frac{1}{r}\left(\frac{\partial v_\theta}{\partial \theta} + v_r\right)$$

$$\varepsilon_{zz} = \frac{\partial v_z}{\partial z} \qquad \varepsilon_{\theta z} = \frac{1}{r}\frac{\partial v_z}{\partial \theta} + \frac{\partial v_\theta}{\partial z} \quad \text{(D.9)}$$

$$\varepsilon_{rz} = \frac{\partial v_r}{\partial z} + \frac{\partial v_z}{\partial r} \qquad \varepsilon_{r\theta} = \frac{1}{r}\left(\frac{\partial v_r}{\partial \theta} - v_\theta\right) + \frac{\partial v_\theta}{\partial r}$$

Viscous stress components:

$$\tau_{rr} = 2\mu\varepsilon_{rr} \qquad \tau_{\theta\theta} = 2\mu\varepsilon_{\theta\theta} \qquad \tau_{zz} = 2\mu\varepsilon_{zz}$$

$$\tau_{r\theta} = \mu\varepsilon_{r\theta} \qquad \tau_{\theta z} = \mu\varepsilon_{\theta z} \qquad \tau_{rz} = \mu\varepsilon_{rz} \quad \text{(D.10)}$$

Angular velocity components:

$$2\omega_r = \frac{1}{r}\frac{\partial v_z}{\partial \theta} - \frac{\partial v_\theta}{\partial z}$$

$$2\omega_\theta = \frac{\partial v_r}{\partial z} - \frac{\partial v_z}{\partial r} \quad \text{(D.11)}$$

$$2\omega_z = \frac{1}{r}\frac{\partial}{\partial r}(rv_\theta) - \frac{1}{r}\frac{\partial v_r}{\partial \theta}$$

Appendix E
Estimating Uncertainty in Experimental Data

Uncertainty is a fact of life in engineering. We rarely know any engineering properties or variables to an extreme degree of accuracy. The *uncertainty* of data is normally defined as the band within which one is 95 percent confident that the true value lies. Recall from Fig. 1.7 that the uncertainty of the ratio μ/μ_c was estimated as ± 20 percent. There are whole monographs devoted to the subject of experimental uncertainty, so we give only a brief summary here.

All experimental data have uncertainty, separated into two causes: (1) a *systematic* error due to the instrument or its environment and (2) a *random* error due to scatter in repeated readings. We minimize the systematic error by careful calibration and then estimate the random error statistically. The judgment of the experimenter is of crucial importance.

Here is the accepted mathematical estimate. Suppose a desired result P depends upon a single experimental variable x. If x has an uncertainty δx, then the uncertainty δP is estimated from the calculus:

$$\delta P \approx \frac{\partial P}{\partial x}\delta x$$

If there are multiple variables, $P = P(x_1, x_2, x_3, \ldots x_N)$, the overall uncertainty δP is calculated as a root-mean-square estimate [4]:

$$\delta P = \left[\left(\frac{\partial P}{\partial x_1}\delta x_1\right)^2 + \left(\frac{\partial P}{\partial x_2}\delta x_2\right)^2 + \cdots + \left(\frac{\partial P}{\partial x_N}\delta x_N\right)^2\right]^{1/2} \tag{E.1}$$

This calculation is statistically much more probable than simply adding linearly the various uncertainties δx_i, thereby making the unlikely assumption that all variables simultaneously attain maximum error. Note that it is the responsibility of the experimenter to establish and report accurate estimates of all the relevant uncertainties δx_i.

If the quantity P is a simple power-law expression of the other variables, for example, $P = \text{Const}\, x_1^{n_1} x_2^{n_2} x_3^{n_3} \ldots$, then each derivative in Eq. (E.1) is proportional to P and the relevant power-law exponent and is inversely proportional to that variable.

If $P = \text{Const}\, x_1^{n_1} x_2^{n_2} x_3^{n_3} \ldots$, then

$$\frac{\partial P}{\partial x_1} = \frac{n_1 P}{x_1}, \quad \frac{\partial P}{\partial x_2} = \frac{n_2 P}{x_2}, \quad \frac{\partial P}{\partial x_3} = \frac{n_3 P}{x_3}, \ldots$$

Thus. from Eq. (E.1),

$$\frac{\delta P}{P} = \left[\left(n_1 \frac{\delta x_1}{x_1}\right)^2 + \left(n_2 \frac{\delta x_2}{x_2}\right)^2 + \left(n_3 \frac{\delta x_3}{x_3}\right)^2 + \cdots\right]^{1/2} \tag{E.2}$$

Evaluation of δP is then a straightforward procedure, as in the following example.

EXAMPLE

The so-called dimensionless Moody pipe friction factor *f*, plotted in Fig. 6.13, is calculated in experiments from the following formula involving pipe diameter *D*, pressure drop Δp, density ρ, volume flow rate *Q*, and pipe length *L*:

$$f = \frac{\pi^2}{8} \frac{D^5 \Delta p}{\rho Q^2 L}$$

Measurement uncertainties are given for a certain experiment: D = 0.5 percent, Δp = 2.0 percent, ρ = 1.0 percent. Q = 3.5 percent, and L = 0.4 percent. Estimate the overall uncertainty of the friction factor *f*.

Solution

The coefficient $\pi^2/8$ is assumed to be a pure theoretical number, with no uncertainty. The other variables may be collected using Eqs. (E.1) and (E.2):

$$U = \frac{\delta f}{f} = \left[\left(5\frac{\delta D}{D}\right)^2 + \left(1\frac{\delta \Delta p}{\Delta p}\right)^2 + \left(1\frac{\delta \rho}{\rho}\right)^2 + \left(2\frac{\delta Q}{Q}\right)^2 + \left(1\frac{\delta L}{L}\right)^2\right]^{1/2}$$

$$= [\{5(0.5\%)\}^2 + (2.0\%)^2 + (1.0\%)^2 + \{2(3.5\%)\}^2 + (0.4\%)^2]^{1/2} \approx 7.8\% \quad \textit{Ans.}$$

By far the dominant effect in this particular calculation is the 3.5 percent error in *Q*, which is amplified by doubling, due to the power of 2 on flow rate. The diameter uncertainty, which is quintupled, would have contributed more had δD been larger than 0.5 percent.

References

1. I. Hughes and J. Hase, *Measurements and their Uncertainties*, Oxford University Press, New York, 2010.
2. H. W. Coleman and W. G. Steele, *Experimentation and Uncertainty Analysis for Engineers*, 3d ed., Wiley, New York, 2009.
3. S. E. Serrano, *Engineering Uncertainty and Risk Analysis*, Hydroscience Inc., Toms River, NJ, 2011.
4. S. J. Kline and F. A. McClintock, "Describing Uncertainties in Single-Sample Experiments," *Mechanical Engineering*, January, 1953, pp. 3–9.

Answers to Selected Problems

Chapter 1

P1.2	5.7 E18 kg; 1.2 E44 molecules
P1.6	$\{\alpha\} = \{L^{-1}\}$
P1.8	$\sigma \approx 1.00\, My/I$
P1.10	Yes, all terms are $\{ML/T^2\}$
P1.12	$\{B\} = \{L^{-1}\}$
P1.14	$Q = \text{Const}\, B\, g^{1/2} H^{3/2}$
P1.16	All terms are $\{ML^{-2}T^{-2}\}$
P1.18	$V = V_0 e^{-mt/K}$
P1.20	(*b*) 2080
P1.24	(*a*) 41 kPa; (*b*) 0.65 kg/m^3
P1.26	$W_{air} = 0.71$ lbf
P1.28	$\rho_{wet} = 1.10$ kg/m^3, $\rho_{dry} = 1.13$ kg/m^3
P1.30	$W_{1\text{-}2} = 21$ ft • lbf
P1.32	(*a*) 76 kN; (*b*) 501 kN
P1.34	(*a*) $\rho_1 = 5.05$ kg/m^3; (*b*) $\rho_2 = 2.12$ kg/m^3 (ideal gas)
P1.36	(*b*) $\rho \approx 628$ kg/m^3
P1.38	$\tau = 1380$ Pa, $Re_L = 28$
P1.40	Approximately 25 N • m per meter
P1.42	$T \approx 539$°C
P1.44	$\mu \approx 0.040$ kg/m · s
P1.46	(*d*) 3.0 m/s; (*e*) 0.79 m/s; (*f*) 22 m/s
P1.48	$F \approx (\mu_1/h_1 + \mu_2/h_2)AV$
P1.50	(*a*) Yes; (*b*) $\mu \approx 0.40$ kg/(m • s)
P1.52	$P \approx 73$ W
P1.54	$M \approx \pi\mu\Omega R^4/h$
P1.56	$\mu = 3M \sin\theta/(2\pi\Omega R^3)$
P1.58	$\mu = 0.040$ kg/(m • s), last 2 points are *turbulent* flow
P1.60	39,500 Pa
P1.62	28,500 Pa
P1.64	$D > 5$ mm
P1.66	$F = 0.014$ N
P1.68	$h = (\Upsilon/\rho g)^{1/2} \cot\theta$
P1.70	$h = 2\Upsilon \cos\theta/(\rho g W)$
P1.72	$z \approx 4800$ m
P1.74	8.6 km
P1.76	(*b*) $\beta_{steam} \approx 262$ kPa
P1.78	(*a*) 25°C; (*b*) 4°C
P1.80	Ma = 1.20
P1.82	$y = x \tan\theta + \text{constant}$
P1.86	Approximately 5.0 percent

Chapter 2

P2.2	$\sigma_{xy} = -13{,}800$ Pa, $\tau_{AA} = -27{,}600$ Pa
P2.4	Approximately 100 degrees
P2.6	(*a*) 8.9 m; (*b*) 760 mm; (*c*) 10.35 m; (*d*) 13,100 mm
P2.8	Approximately 2.36 E6 Pa
P2.10	10,500 Pa
P2.12	8.0 cm
P2.14	$h_1 = 6.0$ cm, $h_2 = 52$ cm
P2.16	(*a*) 90,260 Pa; (*b*) 103,650 Pa
P2.18	1.56
P2.20	61.2 N
P2.22	0.94 cm
P2.24	$p_{sealevel} \approx 115$ kPa, $m_{exact} = 5.3$ E18 kg
P2.28	136.08 m
P2.30	(*a*) 29.6 kPa; (*b*) $K = 0.98$
P2.32	22.6 cm
P2.34	$\Delta p = \Delta h[\gamma_{water}(1 + d^2/D^2) - \gamma_{oil}(1 - d^2/D^2)]$
P2.36	25°
P2.38	$p_A = 219$ kPa
P2.40	$p_B = 121.36$ kN/m^2
P2.42	$p_A - p_B = (\rho_2 - \rho_1)gh$
P2.44	(*a*) 8100 N/m^2; (*b*) 18,800 N/m^2; manometer reads friction loss
P2.46	1.45
P2.48	(*a*) 132 kPa; (*b*) 1.38 m

P2.50 (a) 61 m; (b) 10.050 kN/m^3
P2.52 (b) 339,225 N; 2.78 m from A
P2.56 0.73 m
P2.58 0.40 m
P2.60 (a) Approximately 336,500 N
P2.62 3.18 m
P2.64 1.35 m
P2.66 $F = 1.18$ E9 N, $M_C = 3.13$ E9 N · m counterclockwise, no tipping
P2.68 18,040 N
P2.70 0.79 m
P2.72 $M_A = 32{,}700$ N · m
P2.74 $H = R[\pi/4 + \{(\pi/4)^2 + 2/3\}^{1/2}]$
P2.76 (a) 239 kN; (c) 388 kN · m
P2.78 (b) $F_{AB} = 4390$ N, $F_{CD} = 4220$ N
P2.80 $\theta > 77.4°$
P2.82 $F_H = 97.9$ MN, $F_V = 153.8$ MN
P2.84 (a) $F_V = 2940$ N; $F_H = 6880$ N
P2.86 $P = 59$ kN
P2.88 $F_H = 176$ kN, $F_V = 31.9$ kN, yes
P2.90 $F_V = 22{,}600$ N; $F_H = 16{,}500$ N
P2.92 $F_{\text{one bolt}} \approx 11{,}300$ N
P2.94 Forces on each panel are equal.
P2.96 $F_H = 110$ kN , $F_V = 279$ kN
P2.98 $F_H = 245$ kN, $F_V = 51$ kN
P2.100 $F_H = 0$, $F_V = 297$ kN
P2.102 (a) 238 kN; (b) 125 kN
P2.104 5.0 N
P2.106 $D \approx 2.0$ m
P2.108 (a) 0.0427 m; (b) 1592 kg/m^3
P2.110 (a) 14.95 N, SG = 0.50
P2.112 (a) 39 N; (b) 0.64
P2.114 0.636
P2.116 (a) Yes; (b) Yes; (c) 3.51 in
P2.118 1.85 m
P2.120 34.3°
P2.122 $a/b \approx 0.834$
P2.124 6850 m
P2.126 3130 Pa (vacuum)
P2.128 Yes, stable if $S > 0.789$
P2.130 Slightly unstable, MG = −0.007 m
P2.132 Stable if $R/h > 3.31$
P2.134 (a) unstable; (b) stable
P2.136 $MG = L^2/(3\pi R) - 4R/(3\pi) > 0$ if $L > 2R$
P2.138 2.77 in deep; volume = 10.8 fluid ounces
P2.140 $a_x = -4.9$ m^2/s (decelerating)
P2.142 (a) 16.3 cm; (b) 15.7 N
P2.144 (a) $a_x \approx 319$ m/s^2; (b) no effect, $p_A = p_B$
P2.146 Leans to the right at $\theta = 27°$
P2.148 (a) backward; (b) forward
P2.150 5.5 cm; linear scale OK
P2.152 (a) 224 r/min; (b) 275 r/min
P2.154 552 r/min
P2.156 420 r/min
P2.158 10.57 r/min

Chapter 3

P3.2 **r** = position vector from point O
P3.4 $V = 4.38$ m/s
P3.6 $Q = (2b/3)(2g)^{1/2}[(h + L)^{3/2} - (h - L)^{3/2}]$
P3.8 (a) 5.45 m/s; (b) 5.89 m/s; (c) 5.24 m/s
P3.10 (a) 3 m/s; (b) 18 m/s; (c) 5 cm/s out
P3.12 $\Delta t = 46$ s
P3.14 $dh/dt = (Q_1 + Q_3 - Q_2)/(\pi d^2/4)$
P3.16 $Q_{top} = 3U_0 b\delta/8$
P3.18 $V_3 = 14.7$ m/s (out)
P3.20 (a) 7.8 mL/s; (b) 1.24 cm/s
P3.22 (a) 0.06 kg/s; (b) 1060 m/s; (c) 3.4
P3.24 $h = [3Kt^2d^2/(8\tan^2\theta)]^{1/3}$
P3.26 $Q = 2U_0 bh/3$
P3.28 (a) 0.131 kg/s; (b) 115 kPa
P3.30 (a) $Ma_1 = 1.00$; (b) $T_2 = 216$ K
P3.32 $V_{hole} = 6.1$ m/s
P3.34 $V_2 = 1470$ m/s
P3.36 $U_3 = 6.33$ m/s
P3.38 $V = V_0 r/(2h)$
P3.40 500 N to the left
P3.42 $F = (p_1 - p_a)A_1 - \rho_1 A_1 V_1^2[(D_1/D_2)^2 - 1]$
P3.44 $F = \rho U^2 Lb/3$
P3.46 $\alpha = (1 + \cos\theta)/2$
P3.48 $V_0 \approx 2.27$ m/s
P3.50 102 kN
P3.52 35.36 N
P3.54 163 N
P3.56 (a) 18.5 N to left; (b) 7.1 N up
P3.58 40 N
P3.60 2100 N
P3.62 3100 N
P3.64 980 N
P3.66 8800 N
P3.70 3,464 N
P3.72 Drag ≈ 4260 N
P3.74 $F_x = 0$, $F_y = -17$ N, $F_z = 126$ N
P3.76 (a) 1670 N/m; (b) 3.0 cm; (c) 9.4 cm
P3.80 $F = (\rho/2)gb(h_1^2 - h_2^2) - \rho h_1 b V_1^2(h_1/h_2 - 1)$
P3.82 25 m/s
P3.84 23 N
P3.86 274 kPa
P3.88 $V = \zeta + [\zeta^2 + 2\zeta V_j]^{1/2}$, $\zeta = \rho Q/2k$
P3.90 $dV/dt = g$
P3.92 $dV/dt = gh/(L + h)$
P3.94 $V_{jet} = 6.84$ m/s
P3.96 $d^2Z/dt^2 + 2gZ/L = 0$
P3.100 (a) 507 m/s and 1393 m; (b) 14.5 km
P3.102 $h_2/h_1 = -\frac{1}{2} + \frac{1}{2}[1 + 8V_1^2/(gh_1)]^{1/2}$
P3.104 $\Omega = (-V_e/R)\ln(1 - \dot{m}t/M_0)$
P3.106 $F = (\pi/8)\,\rho\, D^2\, V^2$
P3.108 (a) $V = V_0/(1 + CV_0 t/M)$, $C = \rho bh(1 - \cos\theta)$
P3.110 $F_{bolts} = 5{,}720$ N
P3.112 (a) 10.3 kg/s; (b) 760 kPa
P3.114 (a) $V_2 = 9.96$ m/s; (b) $h = 40$ cm
P3.116 $X = 2[h(H - h)]^{1/2}$

P3.120 (*a*) 0.017 m^3/s; (*b*) 0.42 m^3/s
P3.122 $h_{max} = 25.4$ cm
P3.126 (*a*) 0.007 m^3/s; (*b*) 0.003 m^3/s
P3.130 104 kPa (gage)
P3.132 $Q = 127\ cm^3/s$
P3.134 (*a*) 15 m; (*b*) 25 mm
P3.136 0.037 kg/s
P3.140 5.93 m (subcritical); 0.432 m (supercritical)
P3.142 $t_0 = [2(\alpha - 1)h_0/g]^{1/2}, \alpha = (D/d)^4$
P3.144 Approximately 294 gal/min
P3.150 $T_0 = -305\ \mathbf{k}$ N · m
P3.154 $T_B = 58$ Nm
P3.156 2.44 kg/s
P3.158 $P = \rho Q\omega R V_n(\cot \alpha_1 + \cot \alpha_2)$
P3.160 $\mathbf{T}_B = -2400\ \mathbf{k}$ N · m
P3.162 197 hp; max power at 179 r/min
P3.164 $\mathbf{T}_0 = \rho Q V_w (R + L/2)\ \mathbf{k}$
P3.166 $Q_0 = 2.5\ m^3/s$; $T_0 = 23.15°C$
P3.168 $\Delta T \approx 0.7°F$
P3.170 (*a*) 699 kJ/kg; (*b*) 7.0 MW
P3.172 8.7 m
P3.174 (*a*) 410 hp; (*b*) 540 hp
P3.176 97 hp
P3.178 8.4 kW
P3.180 58,740 Nm/s
P3.182 3.2°C/s
P3.184 76.5 m^3/s and 138 m^3/s

Chapter 4

P4.2 (*a*) $du/dt = (2V_0^2/L)(1 + 2x/L)$
P4.4 (*b*) $a_x = (U_0^2/L)(1 + x/L)$; $a_y = (U_0^2/L)(y/L)$
P4.6 (*b*) $a_x = 16\,x$; $a_y = 16\,y$
P4.8 (*a*) 0.0196 U^2/L; (*b*) at $t = 1.05\ L/U$
P4.10 (*b*) $\mathbf{a} = 8\,\mathbf{r}$
P4.12 If $\upsilon_\theta = \upsilon_\phi = 0$, $\upsilon_r = r^{-2}$ fcn (θ, ϕ)
P4.14 $\upsilon_\theta =$ fcn(*r*) only
P4.16 (*a*) Yes, continuity is satisfied.
P4.18 $\rho = \rho_0 L_0/(L_0 - Vt)$
P4.20 $\upsilon = \upsilon_0 =$ const, $\{K\} = \{L/T\}$, $\{a\} = \{L^{-1}\}$
P4.22 $v_r = -B\,r/2 + f(z)/r$
P4.24 (*b*) $B = 3v_w/(2h^3)$
P4.28 (*a*) Yes; (*b*) Yes
P4.30 (*a, b*) Yes, continuity and Navier-Stokes are satisfied.
P4.32 $f_1 = C_1 r$; $f_2 = C_2/r$
P4.36 $C = \rho g \sin\theta/(2\mu)$
P4.40 $T = T_w + (\mu U^2/3k)(1 - y^4/h^4)$
P4.48 $\psi = U_0\,r \sin\theta - V_0\,r\cos\theta +$ const
P4.50 (*a*) Yes, ψ exists.
P4.52 $\psi = -4Q\theta/(\pi b)$
P4.54 $Q = ULb$
P4.60 Irrotational, $z_C = H - \omega^2 R^2/(2g)$
P4.62 $\psi = Vy^2/(2h) +$ const
P4.66 $\psi = -K \sin\theta/r$
P4.68 (*a*) Yes, a velocity potential exists.
P4.70 $V \approx 2.33$ m/s, $\alpha \approx -16.5°$
P4.72 (*a*) $\psi = -0.0008\ \theta$; (*b*) $\phi = -0.0008 \ln(r)$
P4.74 $\psi = B\,r \sin\theta + B\,L \ln r +$ const
P4.76 Yes, ψ exists.
P4.78 $y = A\,r^n \cos(n\theta) +$ const
P4.80 (*a*) $w = (\rho g/2\mu)(2\delta x - x^2)$
P4.82 Obsessive result: $\upsilon_\theta = \Omega R^2/r$
P4.84 $\upsilon_z = (\rho g b^2/2\mu) \ln(r/a) - (\rho g/4\mu)(r^2 - a^2)$
P4.86 $Q = 0.0031\ m^3/(s \cdot m)$
P4.88 $\upsilon_z = U \ln(r/b)/[\ln(a/b)]$
P4.90 (*a*) 54 kg/h; (*b*) 7 mm
P4.92 $h = h_0 \exp[-\pi D^4 \rho g t/(128\mu L A_0)]$
P4.94 $v_\theta = \Omega R^2/r$
P4.96 (*a*) 1130 Pa

Chapter 5

P5.2 Prototype $V = 36.5$ km/h
P5.4 $V = 1.55$ m/s, $F = 1.3$ N
P5.6 (*a*) 1.39; (*b*) 0.45
P5.8 $Ar = gH^3(\Delta\rho)^2/\mu^2$
P5.10 (*a*) $\{ML^{-2}T^{-2}\}$; (*b*) $\{MLT^{-2}\}$
P5.12 $St = \mu U/(\rho g D^2)$
P5.14 One possible group is hL/k.
P5.16 Stanton number $= h/(\rho V c_p)$
P5.18 $Q\mu/[(\Delta p/L)b^4] =$ const
P5.20 One possible group is $\Omega D/U$.
P5.22 $\Omega D/V =$ fcn$(N, H/L)$
P5.24 $F/(\rho V^2 L^2) =$ fcn$(\alpha, \rho VL/\mu, L/D, V/a)$
P5.26 (*a*) Indeterminate; (*b*) $T = 2.75$ s
P5.28 $\delta/L =$ fcn$[L/D, \rho VD/\mu, E/(\rho V^2)]$
P5.30 $\dot{m}(RT_o)^{1/2}/(p_o D^2) =$ fcn(c_p/R)
P5.32 $Q/(bg^{1/2}H^{3/2}) =$ const
P5.34 $k_{hydrogen} \approx 0.182$ W/(m · K)
P5.36 (*a*) $Q_{loss}R/(A\Delta T) =$ constant
P5.38 $d/D =$ fcn$(\rho UD/\mu, \rho U^2 D/Y)$
P5.42 Halving *m* increases *f* by about 41 percent.
P5.44 (*a*) $\{\sigma\} = \{L^2\}$
P5.48 $F \approx 0.17$ N; (doubling *U* quadruples *F*)
P5.50 Approximately 8880 N (on earth)
P5.52 (*a*) 0.44 s; (*b*) 768,000
P5.54 Power ≈ 7 hp
P5.56 $F_{air} \approx 25$ N/m
P5.58 $V \approx 2.8$ m/s
P5.60 (*b*) 4300 N
P5.62 (*a*) $\omega = 14.4$ r/min
P5.64 $\omega_{aluminum} = 0.77$ Hz
P5.66 (*a*) $V = 27$ m/s; (*b*) $z = 27$ m
P5.68 (*b*) Approximately 1800 N
P5.70 $F = 385$ N (extrapolated)
P5.72 About 44 kN (extrapolated)
P5.74 Prototype moment = 88 kN · m
P5.76 Drag = 463,200 N
P5.78 Weber no. ≈ 100 if $L_m/L_p = 0.0090$
P5.80 (*a*) 1.86 m/s; (*b*) 42,900; (*c*) 254,000
P5.82 561 kN
P5.84 $V_m = 39$ cm/s; $T_m = 3.1$ s; $H_m = 0.20$ m
P5.88 At 340 W, $D = 0.109$ m

Chapter 6

P6.2 (*a*) Yes
P6.4 (*a*) 106 m^3/h; (*b*) 3.6 m^3/h

P6.6 (*a*) hydrogen, $x = 43$ m
P6.8 (*a*) -3600 Pa/m; (*b*) $-13{,}400$ Pa/m
P6.10 (*a*) From *A* to *B*; (*b*) $h_f = 7.8$ m
P6.12 $\mu = 0.29$ kg/m • s
P6.14 $Q = 0.0067$ m^3/h if $H = 50$ cm
P6.16 19 mm
P6.18 4.3 m^3/h
P6.20 4500 cc/h
P6.22 $F = 4.0$ N
P6.24 (*a*) 0.019 m^3/h, laminar; (*b*) $d = 2.67$ mm
P6.26 (*a*) $D_2 = 5.95$ cm
P6.28 $\Delta p = 65$ Pa
P6.30 (*a*) 19.3 m^3/h; (*b*) flow is *up*
P6.32 (*a*) flow is *up*; (*b*) 1.86 m^3/h
P6.36 (*a*) 1.41 kg/m^2; (*b*) 34 m/s
P6.38 5.72 m/s
P6.42 16.7 mm
P6.44 $h_f = 10.4$ m, $\Delta p = 1.4$ MPa
P6.46 20,800 Watts
P6.48 238,000 barrels/day
P6.50 (*a*) -4000 Pa/m; (*b*) 50 Pa; (*c*) 46 percent
P6.52 $p_1 = 2.38$ MPa
P6.54 $t_{drain} = [4WY/(\pi D^2)][2h_0(1 + f_{av}L/D)/g]^{1/2}$
P6.56 (*a*) 2.0E7 Pa; (*b*) 5,760 hp
P6.58 80 m^3/h
P6.60 (*a*) Not identical to Haaland
P6.62 216 hP
P6.64 $Q = 19.6$ m^3/h (laminar, Re = 1450)
P6.66 (*a*) 56 kPa; (*b*) 85 m^3/h; (*c*) $u = 3.3$ m/s at $r = 1$ cm
P6.70 $Q = 31$ m^3/h
P6.72 $D \approx 9.2$ cm
P6.74 $L = 205$ m
P6.76 $Q = 15$ m^3/h or 9.0 m^3/h
P6.78 $Q = 25$ m^3/h (to the left)
P6.80 $Q = 0.905$ m^3/s
P6.82 (*a*) 10.9 m^3/h; (*b*) 100 m^3/h
P6.84 $D \approx 0.104$ m
P6.86 (*a*) 3.0 m/s; (*b*) 0.325 m/m; (*c*) 2770 Pa/m
P6.88 About 17 passages
P6.90 $H = 1$ m
P6.92 (*a*) 1530 m^3/h; (*b*) 6.5 Pa (vacuum)
P6.94 $a = 18.3$ cm
P6.96 (*b*) 12,800 Pa
P6.98 Approximately 128 squares
P6.100 4.85 m^3/h
P6.102 (*a*) 4.25 hp; (*b*) 4.07 hp with 6° cone
P6.104 Approximately 34 kPa
P6.106 $Q = 8.2\times10^{-4}$ m^3/s
P6.108 (*a*) $K \approx 9.09$; (*b*) $Q \approx 0.013$ m^3/s
P6.110 840 W
P6.112 $Q = 4.08 \times 10^{-4}$ m^3/s
P6.114 Short duct: $Q = 0.15$ m^3/s
P6.116 $Q = 0.027$ m^3/s
P6.118 $\Delta p = 768200$ N/m^2
P6.120 $Q_1 = 0.0281$ m^3/s, $Q_2 = 0.0111$ m^3/s, $Q_3 = 0.0164$ m^3/s
P6.122 Increased ε/d and L/d are the causes
P6.124 $Q_1 = -0.0592$ m^3/s, $Q_2 = 0.0456$ m^3/s, $Q_3 = 0.044$ m^3/s
P6.126 $\theta_{opening} = 35°$
P6.128 $Q_{AB} = 0.098$, $Q_{BC} = 0.082$, $Q_{BD} = 0.016$, $Q_{CD} = 0.150$, $Q_{AC} = 0.067$ m^3/s (all)
P6.130 $Q_{AB} = 0.027$, $Q_{BC} = 0.007$, $Q_{BD} = 0.005$, $Q_{CD} = 0.009$, $Q_{AC} = 0.030$ m^3/s (all)
P6.132 $2\theta = 6°$; $D_e = 2.0$ m, $p_e = 224$ kPa
P6.134 $2\theta = 10°$, $W_e = 2.57$ m, $p_e = 104{,}200$ Pa
P6.136 (*a*) 25.5 m/s, (*b*) 0.109 m^3/s, (*c*) 1.23 Pa
P6.138 46.7 m/s
P6.140 333 Pa
P6.142 $Q = 18.6$ gal/min, $d_{reducer} = 0.84$ cm
P6.144 (*a*) $h = 58$ cm
P6.146 (*a*) 0.00653 m^3/s; (*b*) 100 kPa
P6.148 (*a*) 1.58 m; (*b*) 1.7 m
P6.150 $\Delta p = 27$ kPa
P6.152 $D = 4.12$ cm
P6.154 106 gal/min
P6.156 $Q = 0.025$ m^3/s
P6.158 (*a*) 49 m^3/h; (*b*) 6200 Pa

Chapter 7

P7.2 This is probably helium.
P7.4 (*a*) 4 μm; (*b*) 1 m
P7.6 $H = 2.5$ (versus 2.59 for Blasius)
P7.8 Approximately 0.073 N per meter of width
P7.12 Does not satisfy $\partial^2 u/\partial y^2 = 0$ at $y = 0$
P7.14 $C = \rho v_0/\mu = \text{const} < 0$ (wall suction)
P7.16 (*a*) F = 181 N; (*b*) 256 N
P7.18 (*a*) 3.41 m/s; (*b*) 0.0223 Pa
P7.20 $x \approx 0.91$ m
P7.22 (*a*) $y = 3.2$ mm
P7.24 $h_1 = 9.2$ mm; $h_2 = 5.5$ mm
P7.26 $F_a = 2.83\ F_1$, $F_b = 2.0\ F_1$
P7.28 (*a*) $F_{drag} = 2.66\ N^2(\rho\mu L)^{1/2}U^{3/2}a$
P7.30 Predicted thickness is about 10 percent higher
P7.32 $F = 0.0245\ \rho\nu^{1/7}\ L^{6/7}\ U_0^{13/7}\ \delta$
P7.34 45 percent
P7.36 7.2 m/s = 14 kn
P7.38 (*a*) 7.6 m/s; (*b*) 6.2 m/s
P7.40 $L = 3.51$ m, $b = 1.14$ m
P7.42 (*a*) 5.2 N/m
P7.44 Accurate to about ±6 percent
P7.46 $\varepsilon \approx 9$ mm, $U = 11.1$ m/s = 22 kn
P7.48 Separation at $x/L = 0.158$ (1 percent error)
P7.50 Separation at $\theta \approx 2.3$ degrees
P7.52 (*a*) $Re_b = 0.84 < 1$; (*b*) $2a = 30$ mm
P7.54 $z^* = T_0/[B(n+1)]$, $n = g/(RB) - 1$
P7.56 (*a*) 14 N; (*b*) crosswind creates a very large side force
P7.60 Tow power = 140 hp
P7.62 Square side length ≈ 0.83 m
P7.64 $\Delta t_{1000-2000m} = 202$ s
P7.68 69 m/s
P7.70 40 ft
P7.72 (*a*) $L = 6.3$ m; (*b*) 120 m
P7.74 About 212 km/hr
P7.76 (*a*) 296 hp
P7.78 28,400 hp

P7.80 $\theta = 72°$

P7.82 (*a*) 46 s

P7.84 $V = 9$ m/s

P7.86 Approximately 2.9 m by 5.8 m

P7.88 (*a*) 62 hp; (*b*) 86 hp

P7.90 $V_{\text{overturn}} \approx 158$ km/h

P7.94 (*a*) 45 m/s; (*b*) 40 m/s

P7.96 $\Omega_{\text{avg}} \approx 0.21\, U/D$

P7.98 (*b*) $h \approx 0.18$ m

P7.100 (*a*) 118 km/h; (*b*) 128 km/h

P7.104 29.5 knots

P7.106 1130 m^2

P7.108 $\Delta x_{\text{ball}} \approx 13$ m

P7.110 $\Delta y \approx 0.6$ m

P7.114 $V_{\text{down}} \approx 25$ m/min; $V_{\text{up}} \approx 30$ m/min

P7.116 (*a*) 140 km/h; (*b*) 680 hp

P7.118 (*a*) 27 m/s; (*b*) 360 m

P7.120 $(L/D)_{\max} = 21$; $\alpha = 4.8°$

P7.122 (*a*) 6.7 m/s; (*b*) 13.5 m/s = 26 kn

P7.124 $\Omega_{\text{crude theory}} \approx 340$ r/s

P7.126 Approximately 520 m

Chapter 8

P8.2 $\Gamma = \pi\Omega(R_2^2 - R_1^2)$

P8.4 No, $1/r$ is not a proper two-dimensional potential

P8.6 $\psi = B r^2 \sin(2\theta)$

P8.8 $\Gamma = 4B$

P8.10 (*a*) 1.27 cm

P8.12 $\Gamma = 0$

P8.14 Irrotational outer, rotational inner; minimum $p = p_\infty - \rho\omega^2 R^2$ at $r = 0$

P8.16 (*a*) 0.106 m to the left of *A*

P8.18 From afar: a single source 4*m*

P8.20 Vortex near a wall (see Fig. 8.17*b*)

P8.22 Stagnation flow toward a bump

P8.24 $C_p = -\{2(x/a)/[1 + (x/a)^2]\}^2$, $C_{p,\min} = -1.0$ at $x = a$

P8.26 (*a*) 8.75 m; (*b*) 27.5 m on each side

P8.28 Creates a source in a square corner

P8.30 $r = 25$ m

P8.32 $m_2 = 40\ m^2/s$

P8.34 Two stagnation points, at $x = \pm a/\sqrt{3}$

P8.36 $U_\infty = 12.9$ m/s, $2L = 53$ cm, $V_{\max} = 22.5$ m/s

P8.40 1.47 m

P8.42 111 kPa

P8.44 $K = 3.44\ m^2/s$; (*a*) 218 kPa; (*b*) 205 kPa *upper*, 40 kPa *lower*

P8.46 $F_{\text{1-bolt}} = 5060$ N

P8.50 $h = 3a/2$, $U_{\max} = 5U/4$

P8.52 $V_{\text{boat}} = 3.2$ m/s with wind at 47°

P8.54 $F_{\text{parallel}} = 35{,}400$ N, $F_{\text{normal}} = 12{,}000$ N, power ≈ 770 hp (very approximate)

P8.60 This is Fig. 8.18*a*, flow in a 60° corner

P8.62 Stagnation flow near a "bump"

P8.66 $\lambda = 0.45m/(5m + 1)$ if $U = Cx^m$

P8.68 Flow past a Rankine oval

P8.70 Applied to wind tunnel "blockage"

P8.72 Adverse gradient for $x > a$

P8.74 $V_{B,\text{total}} = (8K\mathbf{i} + 4K\mathbf{j})/(15a)$

P8.78 Need an infinite array of images

P8.82 (*a*) 4.5 m/s; (*b*) 1.13; (*c*) 1.26 hp

P8.84 (*a*) 0.21; (*b*) 1.9°

P8.86 (*a*) 26 m; (*b*) 8.7; (*c*) 1600 N

P8.88 (*a*) 11.1; (*b*) 0.56

P8.92 (*a*) 0.77 m; (*b*) $V = 4.5$ m/s at $(r, \theta) = (1.81, 51°)$ and $(1.11, 88°)$

P8.94 Yes, they are orthogonal

P8.96 (*a*) $0.61\, U_\infty^2/a$

P8.98 Yes, a closed teardrop shape appears

P8.100 $V = 14.1$ m/s, $p_A = 115$ kPa

P8.102 (*a*) 370 m; (*b*) 445 m (crudely)

Chapter 9

P9.2 (*a*) $V_2 = 450$ m/s, $\Delta s = 515$ J/(kg · K); (*b*) $V_2 = 453$ m/s, $\Delta s = 512$ J/(kg · K)

P9.4 (*a*) +372 J/(kg · K)

P9.6 (*a*) 381 K

P9.8 410 K

P9.10 (*a*) 0.80

P9.12 (*a*) 2.13 E9 Pa and 1460 m/s; (*b*) 2.91 E9 Pa and 1670 m/s; (*c*) 2645 m/s

P9.14 Approximately 1300 m/s

P9.18 Ma ≈ 0.24

P9.20 (*a*) 41 kPa; (*b*) 421 m/s; (*c*) 1.27

P9.22 (*a*) 267 m/s; (*b*) 286 m/s

P9.24 (*b*) at Ma ≈ 0.576

P9.28 (*b*) 232 K

P9.30 Deviation less than 1 percent at Ma = 0.3

P9.32 (*a*) 141 kPa; (*b*) 101 kPa; (*c*) 0.706

P9.34 (*a*) 3.74 cm

P9.36 (*a*) 0.142 kg/s

P9.40 (*a*) 0.192 kg/s

P9.42 (*a*) Ma = 0.90, $T = 260$ K, $V = 291$ m/s

P9.44 $V_e = 1730$ m/s, $p_e = 110$ kPa, $T_e = 884$ K, thrust = 21,310 N

P9.46 (*a*) 0.0020 m^2

P9.48 (*a*) 313 m/s; (*b*) 0.124 m/s; (*c*) 0.00331 kg/s

P9.50 (*a*) 0.0970 kg/s

P9.52 (*a*) 5.9 cm^2; (*b*) 773 kPa

P9.54 (*a*) $Ma_2 = 0.513$

P9.56 At about $A_1 \approx 24.7\ cm^2$

P9.58 (*a*) 3.50

P9.60 Upstream: Ma = 1.92, $V = 585$ m/s

P9.62 $C = 5831$ m/s, $V_{\text{inside}} = 4843$ m/s

P9.64 (*a*) 4.0 cm^2; (*b*) 325 kPa

P9.66 $h = 1.09$ m

P9.68 $p_{\text{atm}} = 92.6$ kPa; max flow = 0.140 kg/s

P9.70 119 kPa

P9.72 $D \approx 9.3$ mm

P9.74 0.191 kg/s

P9.76 $\Delta t_{\text{shocks}} \approx 23$ s; $\Delta t_{\text{choking-stops}} \approx 39$ s

P9.78 Case *A*: 0.071 kg/s; *B*: 0.068 kg/s

P9.80 $A^* = 2.2$ E-6 m^2 or $D_{\text{hole}} = 0.0005$ m

P9.82 $V_e = 110$ m/s, $Ma_e = 0.67$ (yes)

P9.84 (*a*) 0.96 kg/s; (*b*) 0.27; (*c*) 435 kPa

P9.86 $V_2 = 107$ m/s, $p_2 = 371$ kPa, $T_2 = 330$ K, $p_{02} = 394$ kPa

P9.88 (*a*) 12.7 m
P9.90 (*a*) 0.030; (*b*) 114 kPa
P9.92 (*a*) 14.46 m
P9.96 (*a*) 128 m; (*b*) 80 m; (*c*) 105 m
P9.98 (*a*) 430; (*b*) 0.12; (*c*) 0.00243 kg/h
P9.100 0.345 kg/s
P9.102 Flow is choked at 0.56 kg/s
P9.104 $p_{tank} = 190$ kPa
P9.106 about 91 s
P9.108 Mass flow drops by about 32 percent
P9.112 (*b*) 129 kPa
P9.114 (*a*) 2.21; (*b*) 779 kPa; (*c*) 1146 K
P9.116 $V_{plane} < 810$ m/s
P9.118 $V = 204$ m/s, Ma = 0.6
P9.120 *P* is 3 m ahead of the small circle, Ma = 2.0, $T_{stag} = 518$ K
P9.122 $\beta = 23.13°$, $Ma_2 = 2.75$, $p_2 = 145$ kPa
P9.124 (*a*) 1.87; (*b*) 293 kPa; (*c*) 404 K; (*d*) 415 m/s
P9.126 (*a*) 2.11
P9.128 $\delta_{wedge} \approx 15.5°$
P9.132 (*a*) $p_A = 124$ kN/m^2; (*b*) $p_B = 833$ kPa
P9.134 $Ma_3 = 1.02$, $p_3 = 727$ kPa, $\phi = 42.8°$
P9.136 (*a*) $h = 0.40$ m; (*b*) $Ma_3 = 2.43$
P9.138 $p_r = 21.7$ kPa
P9.140 $Ma_2 = 2.75$, $p_2 = 145$ kPa
P9.142 (*a*) $Ma_2 = 2.641$, $p_2 = 60.3$ kPa; (*b*) $Ma_2 = 2.299$, $p_2 = 24.1$ kPa
P9.144 (*a*) 10.34 degrees
P9.146 (*a*) 2.385; (*b*) 47 kPa
P9.148 (*a*) 4.44; (*b*) 9.6 kPa
P9.150 (*a*) $\alpha = 4.10°$; (*b*) drag = 2150 N/m
P9.152 Approximately 234 N
P9.156 (*a*) $C_L = 0.139$; $C_D = 0.0146$

Chapter 10

P10.2 (*a*) Fr = 2.69
P10.4 These are piezometer tubes (no flow)
P10.6 (*a*) Fr = 3.8; (*b*) $V_{current} = 7.7$ m/s
P10.8 $\Delta t_{travel} = 6.3$ h
P10.10 $\lambda_{crit} = 2\pi(\Upsilon/\rho g)^{1/2}$
P10.14 Flow must be fully rough turbulent (high Re) for Chézy to be valid
P10.16 (*a*) 2.27 m^3/s
P10.18 (*a*) 12.4 m^3/s; (*b*) about 22 Pa
P10.20 0.0174 or 1.0°
P10.22 $S_0 = 0.00038$ (or 0.38 m/km)
P10.24 (*a*) $n \approx 0.027$; (*b*) 2.28 ft
P10.28 (*a*) 0.00106
P10.30 $\Delta t \approx 32$ min
P10.32 $A = 4.39$ m^2, 10 percent larger
P10.34 If $b = 1.2$ m, $y = 2.89$ m, $P = 6.98$ m; if $b = 2.45$ m, $y = 1.23$ m, $P = 4.91$ m
P10.36 $y_2 = 3.6$ m
P10.38 Maximum flow at $\theta = 60°$
P10.42 The two are equally efficient.
P10.44 Hexagon side length $b = 0.65$ m
P10.46 $h_0/b \approx 0.49$
P10.48 (*a*) 0.00634; (*b*) 0.00637
P10.50 (*a*) 2.37; (*b*) 0.62 m; (*c*) 0.0023
P10.52 $W = 2.06$ m
P10.54 (*a*) 1.98 m; (*b*) 3.11 m/s; (*c*) 0.00405
P10.56 (*a*) 2.15 m/s; (*b*) 0.72
P10.58 (*a*) 0.0033; (*b*) 0.0016
P10.60 $y_2 = 0.679$ m; $V_2 = 3.53$ m/s
P10.64 $\Delta h = 15.94$ cm
P10.66 (*b*) 1.39
P10.70 2600 m^3/s
P10.72 (*a*) 0.046 m; (*b*) 4.33 m/s; (*c*) 6.43
P10.76 (*a*) 11 m^3/s
P10.78 $H \approx 0.315$ m
P10.80 (*a*) 0.395
P10.82 (*a*) 0.44 m; (*b*) 4.69 m/s; (*c*) 2.26; (*d*) 13 percent; (*e*) 0.77 m
P10.84 $y_2 = 0.82$ ft; $y_3 = 5.11$ ft; 47 percent
P10.86 (*a*) 1.18 ft; (*b*) 4.58 ft/s
P10.88 (*a*) 2.22 m^3/s/m; (*b*) 0.79 m; (*c*) 5.17 m; (*d*) 60 percent; (*e*) 0.37 m
P10.90 (*a*) $y_2 = 0.56$ m; $y_3 = 2.4$ m
P10.92 $y_1 = 1.71$ mm; $V_1 = 0.310$ m/s
P10.94 (*a*) 5.32; (*b*) 0.385 m/s; (*c*) 18.7 cm
P10.96 $R \approx 4.92$ cm
P10.98 (*a*) steep S-3; (*b*) S-2; (*c*) S-1
P10.106 No entry depth leads to critical flow
P10.108 Approximately 6.6 m
P10.110 (*a*) $y_{crest} \approx 0.782$ m; (*b*) $y(L) \approx 0.909$ m
P10.112 M-1 curve, with $y = 2$ m at $L \approx 214$ m
P10.114 3.51 m
P10.120 $Y = 0.64$ m, $\alpha = 34°$
P10.122 20940 L/min
P10.124 M-1 curve, $y = 3$ m at $x = -926.5$ m
P10.126 At $x = -100$ m, $y = 2.81$ m
P10.128 At 300 m upstream, $y = 2.37$ m

Chapter 11

P11.6 This is a diaphragm pump.
P11.8 (*a*) 86 percent
P11.10 (*a*) 0.045 m^3/min; (*b*) 0.047 m^3/min; (*c*) 87 percent
P11.12 (*a*) 11.3 m; (*b*) 1520 W
P11.14 1870 W
P11.16 $Q \approx 0.447$ m^3/s; $H \approx 38$ m
P11.18 $V_{vane} = (1/3)V_{jet}$ for max power
P11.20 (*a*) 2 roots: $Q = 0.17$ and 1.07 m^3/s; (*b*) 2 roots; $H = 55$ m and 9 m
P11.22 (*a*) BEP = 92 percent at $Q = 0.20$ m^3/s
P11.26 (*a*) Both are fine, the largest is more efficient.
P11.28 BEP at about 0.17 m^3/s; $N_s \approx 28$, $Q_{max} \approx 12.4$ m^3/s
P11.30 (*a*) 660 r/min; (*b*) 24 m
P11.32 (*a*) $D \approx 0.39$ m; (*c*) $n \approx 2220$ r/min
P11.34 (*a*) 0.29 m; (*b*) 20.8 kW; (*c*) 28 m; (*d*) 78 percent
P11.36 (*a*) No; (*b*) 24.5 in at 960 r/min
P11.38 (*a*) 13.8 kW; (*b*) 0.194 m; (*c*) 1.53 m^3/min; (*d*) 81 percent
P11.40 (*a*) $D_s = D(gH^*)^{1/4}/Q^{*1/2}$
P11.42 $NPSH_{proto} \approx 7$ m
P11.44 No cavitation, required depth is only 2.7 m
P11.46 $D_s \approx C/N_s$, $C = 0.78 \pm 0.06$ m
P11.48 (*b*) Approximately 39 m

P11.52 (*a*) 6.56 m^3/s; (*b*) 12.0 kW; (*c*) 28.3°
P11.53 (*b*) 383 kW
P11.55 (*a*) $C_P^* \approx 0.68$
(*b*) $P_2 = 1264$ kW
P11.57 (*b*) Approximately 2500 r/min
P11.59 (*b*) No.
P11.61 $D = 5.71$ m, $\Delta p = 1160$ Pa
P11.62 (*a*) 0.37 m; (*b*) 960 r/min
P11.64 20.45 m^3/min, non-BEP efficiency 78 percent
P11.66 (*a*) 3,800 r/min; (*b*) 75 kW
P11.68 (*a*) 58 m; (*b*) 0.16 m^3/s
P11.70 (*a*) 0.037 m^3/min; (*b*) 2.9 cm
P11.72 (*a*) 14.9; (*b*) 15.9; (*c*) 20.7 kgal/min (all)
P11.75 Approximately 10 stages
P11.77 Both pumps work with three each in series, the largest being more efficient.
P11.81 Two turbines: (*a*) $D \approx 2.9$ m; (*b*) $D \approx 1.0$ m
P11.83 $N_{sp} \approx 270$, hence Francis turbines
P11.85 (*a*) Francis; (*c*) 0.40 m; (*d*) 948 r/min; (*e*) 56.4 kW
P11.87 $P \approx 800$ kW
P11.91 (*a*) 71 percent; (*b*) $N_{sp} \approx 75$
P11.93 (*a*) 0.45 m; (*b*) 0.17 m
P11.96 About 5.7 MW
P11.98 $Q \approx 29$ gal/min
P11.100 (*a*) 6.8 r/min and 489 kW
P11.102 Approximately 23.5 km/h
P11.104 (*a*) About 15 Darrieus turbines

Index

A

B

C

D

E

F

G

H

I

J

K

L

M

N

O

P

Q

R

S

T

U

V

W

Y

Z

EQUATION SHEET

Ideal-gas law: $p = \rho RT$, $R_{\text{air}} = 287$ J/kg-K	Surface tension: $\Delta p = Y(R_1^{-1} + R_2^{-1})$
Hydrostatics, constant density: $p_2 - p_1 = -\gamma(z_2 - z_1)$, $\gamma = \rho g$	Hydrostatic panel force: $F = \gamma h_{\text{CG}} A$, $y_{\text{CP}} = -I_{xx}\sin\theta/(h_{\text{CG}}A)$, $x_{\text{CP}} = -I_{xy}\sin\theta/(h_{\text{CG}}A)$
Buoyant force: $F_{\text{B}} = \gamma_{\text{fluid}}(\text{displaced volume})$	CV mass: $d/dt(\int_{\text{CV}} \rho dv) + \sum(\rho AV)_{\text{out}}$ $- \sum(\rho AV)_{\text{in}} = 0$
CV momentum: $d/dt(\int_{\text{CV}} \rho \mathbf{V} dv)$ $+ \sum[(\rho AV)\mathbf{V}]_{\text{out}} - \sum[(\rho AV)\mathbf{V}]_{\text{in}} = \sum \mathbf{F}$	CV angular momentum: $d/dt(\int_{\text{CV}} \rho(\mathbf{r_0} \times \mathbf{V})dv)$ $+ \sum \rho AV(\mathbf{r_0} \times \mathbf{V})_{\text{out}} - \sum \rho AV(\mathbf{r_0} \times \mathbf{V})_{\text{in}} = \sum \mathbf{M_0}$
Steady flow energy: $(p/\gamma + \alpha V^2/2g + z)_{\text{in}} =$ $(p/\gamma + \alpha V^2/2g + z)_{\text{out}} + h_{\text{friction}} - h_{\text{pump}} + h_{\text{turbine}}$	Acceleration: $d\mathbf{V}/dt = \partial \mathbf{V}/\partial t$ $+ u(\partial \mathbf{V}/\partial x) + v(\partial \mathbf{V}/\partial y) + w(\partial \mathbf{V}/\partial z)$
Incompressible continuity: $\nabla \cdot \mathbf{V} = 0$	Navier-Stokes: $\rho(d\mathbf{V}/dt) = \rho \mathbf{g} - \nabla p + \mu \nabla^2 \mathbf{V}$
Incompressible stream function $\psi(x,y)$: $u = \partial\psi/\partial y$; $v = -\partial\psi/\partial x$	Velocity potential $\phi(x,y,z)$: $u = \partial\phi/\partial x$; $v = \partial\phi/\partial y$; $w = \partial\phi/\partial z$
Bernoulli unsteady irrotational flow: $\partial\phi/\partial t + \int dp/\rho + V^2/2 + gz = \text{Const}$	Turbulent friction factor: $1/\sqrt{f} =$ $-2.0 \log_{10}[\varepsilon/(3.7d) + 2.51/(\text{Re}_d \sqrt{f})]$
Pipe head loss: $h_f = f(L/d)V^2/(2g)$ where f = Moody chart friction factor	Orifice, nozzle, venturi flow: $Q = C_d A_{\text{throat}}[2\Delta p/\{\rho(1-\beta^4)\}]^{1/2}$, $\beta = d/D$
Laminar flat plate flow: $\delta/x = 5.0/\text{Re}_x^{1/2}$, $c_f = 0.664/\text{Re}_x^{1/2}$, $C_D = 1.328/\text{Re}_L^{1/2}$	Turbulent flat plate flow: $\delta/x = 0.16/\text{Re}_x^{1/7}$, $c_f = 0.027/\text{Re}_x^{1/7}$, $C_D = 0.031/\text{Re}_L^{1/7}$
$C_D = \text{Drag}/(\frac{1}{2}\rho V^2 A)$; $C_L = \text{Lift}/(\frac{1}{2}\rho V^2 A)$	2-D potential flow: $\nabla^2\phi = \nabla^2\psi = 0$
Isentropic flow: $T_0/T = 1 + \{(k-1)/2\}\text{Ma}^2$, $\rho_0/\rho = (T_0/T)^{1/(k-1)}$, $p_0/p = (T_0/T)^{k(k-1)}$	One-dimensional isentropic area change: $A/A^* = (1/\text{Ma})[1 + \{(k-1)/2\}\text{Ma}^2]^{(1/2)(k+1)/(k-1)}$
Prandtl-Meyer expansion: $K = (k+1)/(k-1)$, $\omega = K^{1/2}\tan^{-1}[(\text{Ma}^2-1)/K]^{1/2} - \tan^{-1}(\text{Ma}^2-1)^{1/2}$	Uniform flow, Manning's n, SI units: $V_0(\text{m/s}) = (1.0/n)[R_h(\text{m})]^{2/3} S_0^{1/2}$
Gradually varied channel flow: $dy/dx = (S_0 - S)/(1 - \text{Fr}^2)$, $\text{Fr} = V/V_{\text{crit}}$	Euler turbine formula: $\text{Power} = \rho Q(u_2 V_{t2} - u_1 V_{t1})$, $u = r\omega$